$196.95 per copy (in United States)
Price is subject to change without prior notice.

0034

The 2014 *Electrical Cost Data,* 37th annual edition, and all of the RSMeans annual cost data books for 2014, are dedicated to the memory of our respected colleague and friend, Paula Reale-Camelio.

RSMeans® Electrical Cost Data

37TH ANNUAL EDITION

2014

RSMeans
A division of Reed Construction Data, LLC
Construction Publishers & Consultants
700 Longwater Drive
Norwell, MA 02061
USA
1-800-334-3509
www.rsmeans.com

Copyright 2013 by RSMeans
A division of Reed Construction Data, LLC
All rights reserved.

Printed in the United States of America
ISSN 1540-6318
ISBN 978-1-940238-04-3

Senior Editor
Adrian C. Charest, PE *(26, 27, 28, 48)**

Contributing Editors
Christopher Babbitt
Gary W. Christensen
David G. Drain, PE *(33)*
Cheryl Elsmore
David Fiske
Robert Fortier, PE *(2, 31, 32, 34, 35, 46)*
Robert J. Kuchta *(8)*
Thomas Lane
Robert C. McNichols
Genevieve Medeiros
Melville J. Mossman, PE *(21, 22, 23)*
Jeannene D. Murphy
Marilyn Phelan, AIA *(9, 10, 11, 12)*
Stephen C. Plotner *(3, 5)*
Siobhan Soucie
Phillip R. Waier, PE *(1,13)*

Cover Design
Wendy McBay
Paul Robertson
Wayne Engebretson

Chief Operating Officer
Richard Remington

Vice President, Business Development
Steven Ritchie

Vice President, Finance & Operations
Renée Pinczes

Senior Director, Inside Sales
Derek F. Dean

Sales Manager
Reid Clanton

Director, Engineering
Bob Mewis, CCC *(4, 6, 7, 14, 41, 44)*

Senior Director, Product Management
Jennifer Johnson

Senior Product Marketing Manager
Andrea Sillah

Senior Director, Operations
Rodney L. Sharples

Production Manager
Michael Kokernak

Production
Jill Goodman
Jonathan Forgit
Debbie Panarelli
Sara Rutan
Mary Lou Geary

Technical Support
Christopher Anderson
Gary L. Hoitt
Kathryn S. Rodriguez

* Numbers in italics are the divisional responsibilities for each editor. Please contact the designated editor directly with any questions.

This book is printed on 90% recycled paper using soy-based ink and is entirely recyclable.

Related RSMeans Products and Services

The engineers at RSMeans suggest the following products and services as companion information resources to *RSMeans Electrical Cost Data:*

Construction Cost Data Books

Electrical Change Order Cost Data 2014
Building Construction Cost Data 2014
Mechanical Cost Data 2014

Reference Books

Electrical Estimating Methods
Estimating Building Costs
RSMeans Estimating Handbook
Green Building: Project Planning & Estimating
How to Estimate with RSMeans Data
Plan Reading & Material Takeoff
Project Scheduling & Management for Construction

Seminars and In-House Training

Unit Price Estimating
RSMeans Online® Training
RSMeans Data for Job Order Contracting (JOC)
Practical Project Management for Construction Professionals
Scheduling with MSProject for Construction Professionals
Mechanical & Electrical Estimating

RSMeans Online Bookstore

Visit RSMeans at **www.rsmeans.com** for a one-stop portal for the most reliable and current resources available on the market. Here you can learn about our more than 20 annual cost data books, also available on CD and online versions, as well as our libary of reference books and RSMeans seminars.

RSMeans Electronic Data

Get the information found in RSMeans cost books electronically at **RSMeansOnline.com**.

RSMeans Business Solutions

Cost engineers and analysts conduct facility life cycle and benchmark research studies and predictive cost modeling, as well as offer consultation services for conceptual estimating, real property management, and job order contracting. Research studies are designed with the application of proprietary cost and project data from Reed Construction Data and RSMeans' extensive North American databases. Analysts offer building product manufacturers qualitative and quantitative market research, as well as time/motion studies for new products, and Web-based dashboards for market opportunity analysis. Clients are from both the public and private sectors, including federal, state, and municipal agencies; corporations; institutions; construction management firms; hospitals; and associations.

RSMeans for Job Order Contracting (JOC)

Best practice job order contracting (JOC) services support the owner in the development of their program. RSMeans develops all required contracting documents, works with stakeholders, and develops qualified and experienced contractor lists. RSMeans JOCWorks® software is the best of class in the functional tools that meet JOC specific requirements. RSMeans data is organized to meet JOC cost estimating requirements.

Construction Costs for Software Applications

More than 25 unit price and assemblies cost databases are available through a number of leading estimating and facilities management software providers (listed below). For more information see the "Other RSMeans Products" pages at the back of this publication.

RSMeansData™ is also available to federal, state, and local government agencies as multi-year, multi-seat licenses.

- 4Clicks-Solutions, LLC
- Assetworks
- Beck Technology
- BSD – Building Systems Design, Inc.
- CMS – Construction Management Software
- Corecon Technologies, Inc.
- CorVet Systems
- Hard Dollar Corporation
- Maxwell Systems
- MC^2
- MOCA Systems
- Parsons Corporation
- ProEst
- Refined Data
- Sage Timberline Office
- UDA Technologies, Inc.
- US Cost, Inc.
- VFA – Vanderweil Facility Advisers
- WinEstimator, Inc.
- R & K Solutions
- Tririga

Table of Contents

Foreword   iv

How the Book Is Built: An Overview   v

Absolute Essentials for a Quick Start   vi

Estimating with RSMeans Unit Price Cost Data   vii

How to Use the Book: The Details   ix

Unit Price Section   1

Assemblies Section   357

Reference Section   441

     Construction Equipment Rental Costs   443

     Crew Listings   455

     Historical Cost Indexes   491

     City Cost Indexes   492

     Location Factors   535

     Reference Tables   541

     Change Orders   614

     Square Foot Costs   619

     Abbreviations   629

Index   633

Other RSMeans Products and Services   645

Labor Trade Rates including
     Overhead & Profit   Inside Back Cover

Foreword

Our Mission

Since 1942, RSMeans has been actively engaged in construction cost publishing and consulting throughout North America.

Today, more than 70 years after RSMeans began, our primary objective remains the same: to provide you, the construction and facilities professional, with the most current and comprehensive construction cost data possible.

Whether you are a contractor, owner, architect, engineer, facilities manager, or anyone else who needs a reliable construction cost estimate, you'll find this publication to be a highly useful and necessary tool.

With the constant flow of new construction methods and materials today, it's difficult to find the time to look at and evaluate all the different construction cost possibilities. In addition, because labor and material costs keep changing, last year's cost information is not a reliable basis for today's estimate or budget.

That's why so many construction professionals turn to RSMeans. We keep track of the costs for you, along with a wide range of other key information, from city cost indexes . . . to productivity rates . . . to crew composition . . . to contractor's overhead and profit rates.

RSMeans performs these functions by collecting data from all facets of the industry and organizing it in a format that is instantly accessible to you. From the preliminary budget to the detailed unit price estimate, you'll find the data in this book useful for all phases of construction cost determination.

The Staff, the Organization, and Our Services

When you purchase one of RSMeans' publications, you are, in effect, hiring the services of a full-time staff of construction and engineering professionals.

Our thoroughly experienced and highly qualified staff works daily at collecting, analyzing, and disseminating comprehensive cost information for your needs. These staff members have years of practical construction experience and engineering training prior to joining the firm. As a result, you can count on them not only for accurate cost figures, but also for additional background reference information that will help you create a realistic estimate.

The RSMeans organization is always prepared to help you solve construction problems through its variety of data solutions, including online, CD, and print book formats, as well as cost estimating expertise available via our business solutions, training, and seminars.

Besides a full array of construction cost estimating books, RSMeans also publishes a number of other reference works for the construction industry. Subjects include construction estimating and project and business management, special topics such as green building and job order contracting, and a library of facility management references.

In addition, you can access all of our construction cost data electronically in convenient CD format or on the web. Visit **RSMeansOnline.com** for more information on our 24/7 online cost data.

What's more, you can increase your knowledge and improve your construction estimating and management performance with an RSMeans construction seminar or in-house training program. These two-day seminar programs offer unparalleled opportunities for everyone in your organization to become updated on a wide variety of construction-related issues.

RSMeans is also a worldwide provider of construction cost management and analysis services for commercial and government owners.

In short, RSMeans can provide you with the tools and expertise for constructing accurate and dependable construction estimates and budgets in a variety of ways.

Robert Snow Means Established a Tradition of Quality That Continues Today

Robert Snow Means spent years building RSMeans, making certain he always delivered a quality product.

Today, at RSMeans, we do more than talk about the quality of our data and the usefulness of our books. We stand behind all of our data, from historical cost indexes to construction materials and techniques to current costs.

If you have any questions about our products or services, please call us toll-free at 1-800-334-3509. Our customer service representatives will be happy to assist you. You can also visit our website at **www.rsmeans.com**

How the Book is Built: An Overview

The Construction Specifications Institute (CSI) and Construction Specifications Canada (CSC) have produced the 2012 edition of MasterFormat®, a system of titles and numbers used extensively to organize construction information.

All unit price data in the RSMeans cost data books is now arranged in the 50-division MasterFormat® 2012 system.

A Powerful Construction Tool

You have in your hands one of the most powerful construction tools available today. A successful project is built on the foundation of an accurate and dependable estimate. This book will enable you to construct just such an estimate.

For the casual user the book is designed to be:

- quickly and easily understood so you can get right to your estimate.
- filled with valuable information so you can understand the necessary factors that go into the cost estimate.

For the regular user, the book is designed to be:

- a handy desk reference that can be quickly referred to for key costs.
- a comprehensive, fully reliable source of current construction costs and productivity rates so you'll be prepared to estimate any project.
- a source book for preliminary project cost, product selections, and alternate materials and methods.

To meet all of these requirements, we have organized the book into the following clearly defined sections.

Quick Start

See our "Quick Start" instructions on the following page to get started right away.

Estimating with RSMeans Unit Price Cost Data

Please refer to these steps for guidance on completing an estimate using RSMeans unit price cost data.

How to Use the Book: The Details

This section contains an in-depth explanation of how the book is arranged . . . and how you can use it to determine a reliable construction cost estimate. It includes information about how we develop our cost figures and how to completely prepare your estimate.

Unit Price Section

All cost data has been divided into the 50 divisions according to the MasterFormat system of classification and numbering. For a listing of these divisions and an outline of their subdivisions, see the Unit Price Section Table of Contents.

Estimating tips are included at the beginning of each division.

Assemblies Section

The cost data in this section has been organized in an "Assemblies" format. These assemblies are the functional elements of a building and are arranged according to the 7 elements of the UNIFORMAT II classification system. For a complete explanation of a typical "Assemblies" page, see "How RSMeans Assemblies Data Works."

Reference Section

This section includes information on Equipment Rental Costs, Crew Listings, Historical Cost Indexes, City Cost Indexes, Location Factors, Reference Tables, Change Orders, Square Foot Costs, and a listing of Abbreviations.

Equipment Rental Costs: This section contains the average costs to rent and operate hundreds of pieces of construction equipment.

Crew Listings: This section lists all of the crews referenced in the book. For the purposes of this book, a crew is composed of more than one trade classification and/or the addition of power equipment to any trade classification. Power equipment is included in the cost of the crew. Costs are shown both with bare labor rates and with the installing contractor's overhead and profit added. For each, the total crew cost per eight-hour day and the composite cost per labor-hour are listed.

Historical Cost Indexes: These indexes provide you with data to adjust construction costs over time.

City Cost Indexes: All costs in this book are U.S. national averages. Costs vary because of the regional economy. You can adjust costs by CSI Division to over 700 locations throughout the U.S. and Canada by using the data in this section.

Location Factors: You can adjust total project costs to over 900 locations throughout the U.S. and Canada by using the data in this section.

Reference Tables: At the beginning of selected major classifications in the Unit Price section are reference numbers shown in a shaded box. These numbers refer you to related information in the Reference Section. In this section, you'll find reference tables, explanations, estimating information that support how we develop the unit price data, technical data, and estimating procedures.

Change Orders: This section includes information on the factors that influence the pricing of change orders.

See Electrical Change Order Cost Data for an expanded reference as well as Pre- and Post-Installation cost data.

Square Foot Costs: This section contains costs for 59 different building types that allow you to make a rough estimate for the overall cost of a project or its major components.

Abbreviations: A listing of abbreviations used throughout this book, along with the terms they represent, is included in this section.

Index

A comprehensive listing of all terms and subjects in this book will help you quickly find what you need when you are not sure where it occurs in MasterFormat.

The Scope of This Book

This book is designed to be as comprehensive and as easy to use as possible. To that end we have made certain assumptions and limited its scope in two key ways:

1. We have established material prices based on a national average.
2. We have computed labor costs based on a 30-city national average of union wage rates.

For a more detailed explanation of how the cost data is developed, see "How To Use the Book: The Details."

Project Size/Type

The material prices in RSMeans cost data books are "contractor's prices." They are the prices that contractors can expect to pay at the lumberyards, suppliers/distributers warehouses, etc. Small orders of speciality items would be higher than the costs shown, while very large orders, such as truckload lots, would be less. The variation would depend on the size, timing, and negotiating power of the contractor. The labor costs are primarily for new construction or major renovation rather than repairs or minor alterations.

With reasonable exercise of judgment, the figures can be used for any building work.

Absolute Essentials for a Quick Start

If you feel you are ready to use this book and don't think you will need the detailed instructions that begin on the following page, this Absolute Essentials for a Quick Start page is for you. These steps will allow you to get started estimating in a matter of minutes.

1 Scope

Think through the project that you will be estimating, and identify the many individual work tasks that will need to be covered in your estimate.

2 Quantify

Determine the number of units that will be required for each work task that you identified.

3 Pricing

Locate individual Unit Price line items that match the work tasks you identified. The Unit Price Section Table of Contents that begins on page 1 and the Index in the back of the book will help you find these line items.

4 Multiply

Multiply the Total Incl O&P cost for a Unit Price line item in the book by your quantity for that item. The price you calculate will be an estimate for a completed item of work performed by a subcontractor. Keep adding line items in this manner to build your estimate.

5 Project Overhead

Include project overhead items in your estimate. These items are needed to make the job run and are typically, but not always, provided by the General Contractor. They can be found in Division 1. An alternate method of estimating project overhead costs is to apply a percentage of the total project cost.

Include rented tools not included in crews, waste, rubbish handling, and cleanup.

6 Estimate Summary

Include General Contractor's markup on subcontractors, General Contractor's office overhead and profit, and sales tax on materials and equipment.

Adjust your estimate to the project's location by using the City Cost Indexes or Location Factors found in the Reference Section.

Editors' Note: We urge you to spend time reading and understanding the supporting material in the front of this book. An accurate estimate requires experience, knowledge, and careful calculation. The more you know about how we at RSMeans developed the data, the more accurate your estimate will be. In addition, it is important to take into consideration the reference material in the back of the book such as Equipment Listings, Crew Listings, City Cost Indexes, Location Factors, and Reference Tables.

Estimating with RSMeans Unit Price Cost Data

Following these steps will allow you to complete an accurate estimate using RSMeans Unit Price cost data.

1 Scope Out the Project

- Identify the individual work tasks that will need to be covered in your estimate.
- The Unit Price data in this book has been divided into 50 Divisions according to the CSI MasterFormat 2012—their titles are listed on the back cover of your book.
- Think through the project and identify those CSI Divisions needed in your estimate.
- The Unit Price Section Table of Contents on page 1 may also be helpful when scoping out your project.
- Experienced estimators find it helpful to begin with Division 2 and continue through completion. Division 1 can be estimated after the full project scope is known.

2 Quantify

- Determine the number of units required for each work task that you identified.
- Experienced estimators include an allowance for waste in their quantities.
  (Waste is not included in RSMeans Unit Price line items unless so stated.)

3 Price the Quantities

- Use the Unit Price Table of Contents, and the Index, to locate individual Unit Price line items for your estimate.
- Reference Numbers indicated within a Unit Price section refer to additional information that you may find useful.
- The crew indicates who is performing the work for that task. Crew codes are expanded in the Crew Listings in the Reference Section to include all trades and equipment that comprise the crew.

- The Daily Output is the amount of work the crew is expected to do in one day.
- The Labor-Hours value is the amount of time it will take for the crew to install one unit of work.
- The abbreviated Unit designation indicates the unit of measure upon which the crew, productivity, and prices are based.
- Bare Costs are shown for materials, labor, and equipment needed to complete the Unit Price line item. Bare costs do not include waste, project overhead, payroll insurance, payroll taxes, main office overhead, or profit.
- The Total Incl O&P cost is the billing rate or invoice amount of the installing contractor or subcontractor who performs the work for the Unit Price line item.

4 Multiply

- Multiply the total number of units needed for your project by the Total Incl O&P cost for each Unit Price line item.
- Be careful that your take off unit of measure matches the unit of measure in the Unit column.
- The price you calculate is an estimate for a completed item of work.
- Keep scoping individual tasks, determining the number of units required for those tasks, matching each task with individual Unit Price line items in the book, and multiply quantities by Total Incl O&P costs.
- An estimate completed in this manner is priced as if a subcontractor, or set of subcontractors, is performing the work. The estimate does not yet include Project Overhead or Estimate Summary components such as General Contractor markups on subcontracted work, General Contractor office overhead and profit, contingency, and location factor.

5 Project Overhead

- Include project overhead items from Division 1 – General Requirements.
- These items are needed to make the job run. They are typically, but not always, provided by the General Contractor. Items include, but are not limited to, field personnel, insurance, performance bond, permits, testing, temporary utilities, field office and storage facilities, temporary scaffolding and platforms, equipment mobilization and demobilization, temporary roads and sidewalks, winter protection, temporary barricades and fencing, temporary security, temporary signs, field engineering and layout, final cleaning and commissioning.
- Each item should be quantified, and matched to individual Unit Price line items in Division 1, and then priced and added to your estimate.
- An alternate method of estimating project overhead costs is to apply a percentage of the total project cost, usually 5% to 15% with an average of 10% (see General Conditions, page ix).
- Include other project related expenses in your estimate such as:
  - Rented equipment not itemized in the Crew Listings
  - Rubbish handling throughout the project (see 02 41 19.19)

6 Estimate Summary

- Include sales tax as required by laws of your state or county.
- Include the General Contractor's markup on self-performed work, usually 5% to 15% with an average of 10%.
- Include the General Contractor's markup on subcontracted work, usually 5% to 15% with an average of 10%.
- Include General Contractor's main office overhead and profit:
  - RSMeans gives general guidelines on the General Contractor's main office overhead (see section 01 31 13.60 and Reference Number R013113-50).
  - RSMeans gives no guidance on the General Contractor's profit.
  - Markups will depend on the size of the General Contractor's operations, his projected annual revenue, the level of risk he is taking on, and on the level of competition in the local area and for this project in particular.
- Include a contingency, usually 3% to 5%, if appropriate.
- Adjust your estimate to the project's location by using the City Cost Indexes or the Location Factors in the Reference Section:
  - Look at the rules on the pages for How to Use the City Cost Indexes to see how to apply the Indexes for your location.
  - When the proper Index or Factor has been identified for the project's location, convert it to a multiplier by dividing it by 100, and then multiply that multiplier by your estimated total cost. The original estimated total cost will now be adjusted up or down from the national average to a total that is appropriate for your location.

Editors' Notes:

1) We urge you to spend time reading and understanding the supporting material in the front of this book. An accurate estimate requires experience, knowledge, and careful calculation. The more you know about how we at RSMeans developed the data, the more accurate your estimate will be. In addition, it is important to take into consideration the reference material in the back of the book such as Equipment Listings, Crew Listings, City Cost Indexes, Location Factors, and Reference Tables.

2) Contractors who are bidding or are involved in JOC, DOC, SABER, or IDIQ type contracts are cautioned that workers' compensation insurance, federal and state payroll taxes, waste, project supervision, project overhead, main office overhead, and profit are not included in bare costs. The coefficient or multiplier must cover these costs.

How to Use the Book: The Details

What's Behind the Numbers? The Development of Cost Data

The staff at RSMeans continually monitors developments in the construction industry in order to ensure reliable, thorough, and up-to-date cost information. While overall construction costs may vary relative to general economic conditions, price fluctuations within the industry are dependent upon many factors. Individual price variations may, in fact, be opposite to overall economic trends. Therefore, costs are constantly tracked and complete updates are published yearly. Also, new items are frequently added in response to changes in materials and methods.

Costs—$ (U.S.)

All costs represent U.S. national averages and are given in U.S. dollars. The RSMeans City Cost Indexes can be used to adjust costs to a particular location. The City Cost Indexes for Canada can be used to adjust U.S. national averages to local costs in Canadian dollars. No exchange rate conversion is necessary.

G̲ The processes or products identified by the green symbol in our publications have been determined to be environmentally responsible and/or resource-efficient solely by the RSMeans engineering staff. The inclusion of the green symbol does not represent compliance with any specific industry association or standard.

Material Costs

The RSMeans staff contacts manufacturers, dealers, distributors, and contractors all across the U.S. and Canada to determine national average material costs. If you have access to current material costs for your specific location, you may wish to make adjustments to reflect differences from the national average. Included within material costs are fasteners for a normal installation. RSMeans engineers use manufacturers' recommendations, written specifications, and/or standard construction practice for size and spacing of fasteners. Adjustments to material costs may be required for your specific application or location. The manufacturer's warranty is assumed. Extended warranties are not included in the material costs. Material costs do not include sales tax.

Labor Costs

Labor costs are based on the average of wage rates from 30 major U.S. cities. Rates are determined from labor union agreements or prevailing wages for construction trades for the current year. Rates, along with overhead and profit markups, are listed on the inside back cover of this book.

- If wage rates in your area vary from those used in this book, or if rate increases are expected within a given year, labor costs should be adjusted accordingly.

Labor costs reflect productivity based on actual working conditions. In addition to actual installation, these figures include time spent during a normal weekday on tasks such as, material receiving and handling, mobilization at site, site movement, breaks, and cleanup.

Productivity data is developed over an extended period so as not to be influenced by abnormal variations and reflects a typical average.

Equipment Costs

Equipment costs include not only rental, but also operating costs for equipment under normal use. The operating costs include parts and labor for routine servicing such as repair and replacement of pumps, filters, and worn lines. Normal operating expendables, such as fuel, lubricants, tires, and electricity (where applicable), are also included. Extraordinary operating expendables with highly variable wear patterns, such as diamond bits and blades, are excluded. These costs are included under materials. Equipment rental rates are obtained from industry sources throughout North America—contractors, suppliers, dealers, manufacturers, and distributors.

Rental rates can also be treated as reimbursement costs for contractor-owned equipment. Owned equipment costs include depreciation, loan payments, interest, taxes, insurance, storage and major repairs.

Equipment costs do not include operators' wages; nor do they include the cost to move equipment to a job site (mobilization) or from a job site (demobilization).

Equipment Cost/Day——The cost of power equipment required for each crew is included in the Crew Listings in the Reference Section (small tools that are considered as essential everyday tools are not listed out separately). The Crew Listings itemize specialized tools and heavy equipment along with labor trades. The daily cost of itemized equipment included in a crew is based on dividing the weekly bare rental rate by 5 (number of working days per week) and then adding the hourly operating cost times 8 (the number of hours per day). This Equipment Cost/Day is shown in the last column of the Equipment Rental Cost pages in the Reference Section.

Mobilization/Demobilization——The cost to move construction equipment from an equipment yard or rental company to the job site and back again is not included in equipment costs. Mobilization (to the site) and demobilization (from the site) costs can be found in the Unit Price Section. If a piece of equipment is already at the job site, it is not appropriate to utilize mob./demob. costs again in an estimate.

Overhead and Profit

Total Cost including O&P for the *Installing Contractor* is shown in the last column on the Unit Price and/or the Assemblies pages of this book. This figure is the sum of the bare material cost plus 10% for profit, the bare labor cost plus total overhead and profit, and the bare equipment cost plus 10% for profit. Details for the calculation of Overhead and Profit on labor are shown on the inside back cover and in the Reference Section of this book. (See "How RSMeans Data Works" for an example of this calculation.)

General Conditions

Cost data in this book is presented in two ways: Bare Costs and Total Cost including O&P (Overhead and Profit). General Conditions, or General Requirements, of the contract should also be added to the Total Cost including O&P when applicable. Costs for General Conditions are listed in Division 1 of the Unit Price Section and the Reference

Section of this book. General Conditions for the *Installing Contractor* may range from 0% to 10% of the Total Cost including O&P. For the *General* or *Prime Contractor*, costs for General Conditions may range from 5% to 15% of the Total Cost including O&P, with a figure of 10% as the most typical allowance. If applicable, the Assemblies and Models sections of this book use costs that include the installing contractor's overhead and profit (O&P).

Factors Affecting Costs

Costs can vary depending upon a number of variables. Here's how we have handled the main factors affecting costs.

Quality——The prices for materials and the workmanship upon which productivity is based represent sound construction work. They are also in line with U.S. government specifications.

Overtime——We have made no allowance for overtime. If you anticipate premium time or work beyond normal working hours, be sure to make an appropriate adjustment to your labor costs.

Productivity——The productivity, daily output, and labor-hour figures for each line item are based on working an eight-hour day in daylight hours in moderate temperatures. For work that extends beyond normal work hours or is performed under adverse conditions, productivity may decrease. (See "How RSMeans Data Works" for more on productivity.)

Size of Project——The size, scope of work, and type of construction project will have a significant impact on cost. Economies of scale can reduce costs for large projects. Unit costs can often run higher for small projects.

Location——Material prices in this book are for metropolitan areas. However, in dense urban areas, traffic and site storage limitations may increase costs. Beyond a 20-mile radius of large cities, extra trucking or transportation charges may also increase the material costs slightly. On the other hand, lower wage rates may be in effect. Be sure to consider both of these factors when preparing an estimate, particularly if the job site is located in a central city or remote rural location. In addition, highly specialized subcontract items may require travel and per-diem expenses for mechanics.

Other Factors——
- season of year
- contractor management
- weather conditions
- local union restrictions
- building code requirements
- availability of:
  - adequate energy
  - skilled labor
  - building materials
- owner's special requirements/restrictions
- safety requirements
- environmental considerations

Unpredictable Factors——General business conditions influence "in-place" costs of all items. Substitute materials and construction methods may have to be employed. These may affect the installed cost and/or life cycle costs. Such factors may be difficult to evaluate and cannot necessarily be predicted on the basis of the job's location in a particular section of the country. Thus, where these factors apply, you may find significant but unavoidable cost variations for which you will have to apply a measure of judgment to your estimate.

Rounding of Costs

In general, all unit prices in excess of $5.00 have been rounded to make them easier to use and still maintain adequate precision of the results. The rounding rules we have chosen are in the following table.

Prices from . . .	Rounded to the nearest . . .
$.01 to $5.00	$.01
$5.01 to $20.00	$.05
$20.01 to $100.00	$.50
$100.01 to $300.00	$1.00
$300.01 to $1,000.00	$5.00
$1,000.01 to $10,000.00	$25.00
$10,000.01 to $50,000.00	$100.00
$50,000.01 and above	$500.00

How Subcontracted Items Affect Costs

A considerable portion of all large construction jobs is usually subcontracted. In fact, the percentage done by subcontractors is constantly increasing and may run over 90%. Since the workers employed by these companies do nothing else but install their particular product, they soon become expert in that line. The result is, installation by these firms is accomplished so efficiently that the total in-place cost, even adding the general contractor's overhead and profit, is no more, and often less, than if the principal contractor had handled the installation himself/herself. Companies that deal with construction specialties are anxious to have their product perform well and, consequently, the installation will be the best possible.

Contingencies

The allowance for contingencies generally provides for unforeseen construction difficulties. On alterations or repair jobs, 20% is not too much. If drawings are final and only field contingencies are being considered, 2% or 3% is probably sufficient, and often nothing need be added. Contractually, changes in plans will be covered by extras. The contractor should consider inflationary price trends and possible material shortages during the course of the job. These escalation factors are dependent upon both economic conditions and the anticipated time between the estimate and actual construction. If drawings are not complete or approved, or a budget cost is wanted, it is wise to add 5% to 10%. Contingencies, then, are a matter of judgment. Additional allowances for contingencies are shown in Division 1.

Important Estimating Considerations

The "productivity," or daily output, of each craftsman includes mobilization and cleanup time, break time and plan layout time, as well as an allowance to carry stock from the storage trailer or location on the job site up to 200' into the building and to the first or second floor. If material has to be transported over greater distances or to higher floors, an additional allowance should be considered by the estimator. An allowance has also been included in the piping and fittings installation time for leak checking and minor tightening. Equipment installation time includes the following applicable items: positioning, leveling and securing the unit in place, connecting all associated piping, ducts, vents, etc., which shall have been estimated separately, connecting to an adjacent power source, filling/bleeding, startup, adjusting the controls up and down to ensure proper response, setting the integral controls/valves/ regulators/thermostats for proper operation (does not include external building type control systems, DDC systems, etc.), explaining/training owner's operator, and warranty. A reasonable breakdown of the labor costs is as follows:

1. Movement into building, installation/ setting of equipment — 35%
2. Connection to piping/duct/power, etc. — 25%
3. Filling/flushing/cleaning/touchup, etc. — 15%
4. Startup/running adjustments — 5%
5. Training owner's representative — 5%
6. Warranty/call back/service — 15%

Note that cranes or other lifting equipment are not included on any lines in the Mechanical divisions. For example, if a crane is required to lift a heavy piece of pipe into place high above a gym floor, or to put a rooftop unit on the roof of a four-story building, etc., it must be added. Due to the potential for extreme variation— from nothing additional required to a major crane or helicopter—we feel that including a nominal amount for "lifting contingency" would be useless and detract from the accuracy of the estimate. When using equipment rental from RSMeans, remember to include the cost of the operator(s).

Estimating Labor-Hours

The labor-hours expressed in this publication are based on Average Installation time, using an efficiency level of approximately 60%–65% (see item 7), which has been found reasonable and acceptable by many contractors. The book uses this national efficiency average to establish a consistent benchmark. For bid situations, adjustments to this efficiency level should be the responsibility of the contractor bidding the project. The unit labor-hour is divided in the following manner. A typical day for a journeyman might be:

1. Study Plans	3%	14.4 min.
2. Material Procurement	3%	14.4 min.
3. Receiving and Storing	3%	14.4 min.
4. Mobilization	5%	24.0 min.
5. Site Movement	5%	24.0 min.
6. Layout and Marking	8%	38.4 min.
7. Actual Installation	64%	307.2 min.
8. Cleanup	3%	14.4 min.
9. Breaks, Nonproductive	6%	28.8 min.
	100%	480.0 min.

If any of the percentages expressed in this breakdown do not apply to the particular work or project situation, then that percentage or a portion of it may be deducted from or added to labor-hours.

Final Checklist

Estimating can be a straightforward process provided you remember the basics. Here's a checklist of some of the steps you should remember to complete before finalizing your estimate.

Did you remember to . . .

- factor in the City Cost Index for your locale?
- take into consideration which items have been marked up and by how much?
- mark up the entire estimate sufficiently for your purposes?
- read the background information on techniques and technical matters that could impact your project time span and cost?
- include all components of your project in the final estimate?
- double check your figures for accuracy?
- call RSMeans if you have any questions about your estimate or the data you've found in our publications? Remember, RSMeans stands behind its publications. If you have any questions about your estimate . . . about the costs you've used from our books . . . or even about the technical aspects of the job that may affect your estimate, feel free to call the RSMeans editors at 1-800-334-3509.

Free Quarterly Updates

Stay up-to-date throughout 2014 with RSMeans' free cost data updates four times a year. Sign up online to make sure you have access to the newest data. Every quarter we provide the city cost adjustment factors for hundreds of cities and key materials. Register at: **www.rsmeans2014updates.com**

Unit Price Section

Table of Contents

Sect. No.		Page
General Requirements		
01 11	Summary of Work	8
01 21	Allowances	8
01 31	Project Management and Coordination	10
01 32	Construction Progress Documentation	11
01 41	Regulatory Requirements	11
01 51	Temporary Utilities	11
01 52	Construction Facilities	13
01 54	Construction Aids	14
01 55	Vehicular Access and Parking	16
01 56	Temporary Barriers and Enclosures	17
01 58	Project Identification	18
01 74	Cleaning and Waste Management	18
01 91	Commissioning	18
01 93	Facility Maintenance	18
Existing Conditions		
02 41	Demolition	22
02 82	Asbestos Remediation	24
Concrete		
03 01	Maintenance of Concrete	26
03 11	Concrete Forming	26
03 15	Concrete Accessories	26
03 21	Reinforcement Bars	27
03 22	Fabric and Grid Reinforcing	28
03 30	Cast-In-Place Concrete	28
03 31	Structural Concrete	29
03 35	Concrete Finishing	30
03 54	Cast Underlayment	30
03 63	Epoxy Grouting	31
03 81	Concrete Cutting	31
03 82	Concrete Boring	31
Metals		
05 05	Common Work Results for Metals	34
05 12	Structural Steel Framing	37
05 35	Raceway Decking Assemblies	37
Wood, Plastics & Composites		
06 16	Sheathing	40
Thermal & Moisture Protection		
07 84	Firestopping	42
07 91	Preformed Joint Seals	43
07 92	Joint Sealants	44
Openings		
08 31	Access Doors and Panels	48
08 74	Access Control Hardware	48
Finishes		
09 22	Supports for Plaster and Gypsum Board	52
09 69	Access Flooring	52
Specialties		
10 28	Toilet, Bath, and Laundry Accessories	54
Equipment		
11 12	Parking Control Equipment	56
11 23	Commercial Laundry and Dry Cleaning Equipment	56
11 26	Unit Kitchens	57
11 31	Residential Appliances	57
11 41	Foodservice Storage Equipment	58
11 42	Food Preparation Equipment	59
11 44	Food Cooking Equipment	59
11 46	Food Dispensing Equipment	59

Sect. No.		Page
11 48	Cleaning and Disposal Equipment	59
11 52	Audio-Visual Equipment	60
11 53	Laboratory Equipment	60
11 61	Broadcast, Theater, and Stage Equipment	61
11 66	Athletic Equipment	61
11 71	Medical Sterilizing Equipment	61
11 72	Examination and Treatment Equipment	61
11 76	Operating Room Equipment	61
11 78	Mortuary Equipment	62
11 82	Solid Waste Handling Equipment	62
Furnishings		
12 46	Furnishing Accessories	64
Special Construction		
13 11	Swimming Pools	66
13 21	Controlled Environment Rooms	66
13 24	Special Activity Rooms	67
13 49	Radiation Protection	68
Fire Suppression		
21 21	Carbon-Dioxide Fire-Extinguishing Systems	72
21 22	Clean-Agent Fire-Extinguishing Systems	72
Plumbing		
22 01	Operation and Maintenance of Plumbing	74
22 05	Common Work Results for Plumbing	75
22 33	Electric Domestic Water Heaters	77
22 34	Fuel-Fired Domestic Water Heaters	78
Heating Ventilation Air Conditioning		
23 05	Common Work Results for HVAC	80
23 09	Instrumentation and Control for HVAC	80
23 34	HVAC Fans	83
23 52	Heating Boilers	87
23 54	Furnaces	88
23 82	Convection Heating and Cooling Units	88
23 83	Radiant Heating Units	90
Electrical		
26 01	Operation and Maintenance of Electrical Systems	96
26 05	Common Work Results for Electrical	96
26 09	Instrumentation and Control for Electrical Systems	203
26 12	Medium-Voltage Transformers	205
26 13	Medium-Voltage Switchgear	205
26 22	Low-Voltage Transformers	207
26 24	Switchboards and Panelboards	210
26 25	Enclosed Bus Assemblies	234
26 27	Low-Voltage Distribution Equipment	255
26 28	Low-Voltage Circuit Protective Devices	271
26 29	Low-Voltage Controllers	276
26 31	Photovoltaic Collectors	278
26 32	Packaged Generator Assemblies	279
26 33	Battery Equipment	280
26 35	Power Filters and Conditioners	281
26 36	Transfer Switches	284
26 41	Facility Lightning Protection	285
26 42	Cathodic Protection	286

Sect. No.		Page
26 51	Interior Lighting	287
26 52	Emergency Lighting	295
26 53	Exit Signs	295
26 54	Classified Location Lighting	295
26 55	Special Purpose Lighting	296
26 56	Exterior Lighting	297
26 61	Lighting Systems and Accessories	300
26 71	Electrical Machines	302
Communications		
27 01	Operation and Maintenance of Communications Systems	306
27 05	Common Work Results for Communications	306
27 11	Communications Equipment Room Fittings	307
27 13	Communications Backbone Cabling	307
27 15	Communications Horizontal Cabling	308
27 21	Data Communications Network Equipment	310
27 32	Voice Communications Terminal Equipment	311
27 41	Audio-Video Systems	311
27 51	Distributed Audio-Video Communications Systems	312
27 52	Healthcare Communications and Monitoring Systems	313
27 53	Distributed Systems	313
Electronic Safety & Security		
28 01	Operation and Maint. of Electronic Safety and Security	316
28 05	Common Work Results for Electronic Safety and Security	316
28 13	Access Control	316
28 16	Intrusion Detection	317
28 23	Video Surveillance	318
28 31	Fire Detection and Alarm	319
28 39	Mass Notification Systems	320
Earthwork		
31 23	Excavation and Fill	322
Exterior Improvements		
32 31	Fences and Gates	336
Utilities		
33 05	Common Work Results for Utilities	338
33 44	Storm Utility Water Drains	339
33 49	Storm Drainage Structures	340
33 71	Electrical Utility Transmission and Distribution	340
33 72	Utility Substations	349
33 73	Utility Transformers	349
33 75	High-Voltage Switchgear and Protection Devices	350
33 79	Site Grounding	351
33 81	Communications Structures	351
Transportation		
34 43	Airfield Signaling and Control Equipment	354
Electrical Power Generation		
48 15	Wind Energy Electrical Power Generation Equipment	356
48 18	Fuel Cell Electrical Power Generation Equipment	356

1

How RSMeans Unit Price Data Works

All RSMeans unit price data is organized in the same way. It is important to understand the structure, so that you can find information easily and use it correctly.

RSMeans **Line Numbers** consist of 12 characters, which identify a unique location in the database for each task. The first 6 or 8 digits conform to the Construction Specifications Institute MasterFormat® 2012. The remainder of the digits are a further breakdown by RSMeans in order to arrange items in understandable groups of similar tasks. Line numbers are consistent across all RSMeans publications, so a line number in any RSMeans product will always refer to the same unit of work.

RSMeans engineers have created **reference** information to assist you in your estimate. If there is information that applies to a section, it will be indicated at the start of the section. In this case, R033105-10 provides information on the proportionate quantities of formwork, reinforcing, and concrete used in cast-in-place concrete items such as footings, slabs, beams, and columns. The Reference Section is located in the back of the book on the pages with a gray edge.

RSMeans **Descriptions** are shown in a hierarchical structure to make them readable. In order to read a complete description, read up through the indents to the top of the section. Include everything that is above and to the left that is not contradicted by information below. For instance, the complete description for line 03 30 53.40 3550 is "Concrete in place, including forms (4 uses), Grade 60 rebar, concrete (Portland cement Type 1), placement and finishing unless otherwise indicated; Equipment pad (3000 psi), 4' x 4' x 6" thick."

03 30 Cast-In-Place Concrete

03 30 53 – Miscellaneous Cast-In-Place Concrete

03 30 53.40 Concrete In Place

0010	**CONCRETE IN PLACE**	R033105-10
0020	Including forms (4 uses), Grade 60 rebar, concrete (Portland cement	R033105-20
0050	Type I), placement and finishing unless otherwise indicated	R033105-50
0300	Beams (3500 psi), 5 kip per L.F., 10' span	R033105-65
0350	25' span	R033105-70
0500	Chimney foundations (5000 psi), over 5 C.Y.	R033105-85
0510	(3500 psi), under 5 C.Y.	
0700	Columns, square (4000 psi), 12" x 12", less than 2% reinforcing	
3450	Over 10,000 S.F.	
3500	Add per floor for 3 to 6 stories high	
3520	For 7 to 20 stories high	
3540	Equipment pad (3000 psi), 3' x 3' x 6" thick	
3550	4' x 4' x 6" thick	
3560	5' x 5' x 8" thick	
3570	6' x 6' x 8" thick	

When using **RSMeans data**, it is important to read through an entire section to ensure that you use the data that most closely matches your work. Note that sometimes there is additional information shown in the section that may improve your price. There are frequently lines that further describe, add to, or adjust data for specific situations.

The data published in RSMeans print books represents a "national average" cost. This data should be modified to the project location using the **City Cost Indexes** or **Location Factors** tables found in the Reference Section (see pages 492–540). Use the location factors to adjust estimate totals if the project covers multiple trades. Use the city cost indexes (CCI) for single trade projects or projects where a more detailed analysis is required. All figures in the two tables are derived from the same research. The last row of data in the CCI, the weighted average, is the same as the numbers reported for each location in the location factor table.

Crews include labor or labor and equipment necessary to accomplish each task. In this case, Crew C-14H is used. RSMeans selects a crew to represent the workers and equipment that are typically used for that task. In this case, Crew C-14H consists of one carpenter foreman (outside), two carpenters, one rodman, one laborer, one cement finisher, and one gas engine vibrator. Details of all crews can be found in the reference section.

Crews

Crew No.	Bare Costs		Incl. Subs O&P		Cost Per Labor-Hour	
Crew C-14H	Hr.	Daily	Hr.	Daily	Bare Costs	Incl. O&P
1 Carpenter Foreman (outside)	$47.85	$382.80	$73.85	$590.80	$45.15	$69.42
2 Carpenters	45.85	733.60	70.75	1132.00		
1 Rodman (reinf.)	50.65	405.20	79.55	636.40		
1 Laborer	36.65	293.20	56.55	452.40		
1 Cement Finisher	44.05	352.40	65.10	520.80		
1 Gas Engine Vibrator		33.00		36.30	.69	.76
48 L.H., Daily Totals		$2200.20		$3368.70	$45.84	$70.18

The **Daily Output** is the amount of work that the crew can do in a normal 8-hour workday, including mobilization, layout, movement of materials, and cleanup. In this case, crew C-14H can install thirty 4' x 4' x 6" thick concrete pads in a day. Daily output is variable, based on many factors, including the size of the job, location, and environmental conditions. RSMeans data represents work done in daylight (or adequate lighting) and temperate conditions.

Bare Costs are the costs of materials, labor, and equipment that the installing contractor pays. They represent the cost, in U.S. dollars, for one unit of work. They do not include any markups for profit or labor burden.

Crew	Daily Output	Labor-Hours	Unit	Material	2014 Bare Costs Labor	Equipment	Total	Total Incl O&P
C-14A	15.62	12.804	C.Y.	315	590	47.50	952.50	1,325
"	18.55	10.782		335	495	40	870	1,175
C-14C	32.22	3.476		147	153	1.01	301.01	400
"	23.71	4.724		173	207	1.37	381.37	515
C-14A	11.96	16.722		355	770	62	1,187	1,675
	2200	.025		.80	1.04	.32	2.16	2.83
	31800	.002			.07	.02	.09	.13
	21200	.003			.11	.03	.14	.20
C-14H	45	1.067	Ea.	44.50	48	.74	93.24	124
	30	1.600		66.50	72	1.10	139.60	185
	18	2.667		119	120	1.84	240.84	320
	14	3.429		160	155	2.37	317.37	415

The **Total Incl O&P column** is the total cost, including overhead and profit, that the installing contractor will charge the customer. This represents the cost of materials plus 10% profit, the cost of labor plus labor burden and 10% profit, and the cost of equipment plus 10% profit. It does not include the general contractor's overhead and profit. Note: See the inside back cover for details of how RSMeans calculates labor burden.

The **Total column** represents the total bare cost for the installing contractor, in U.S. dollars. In this case, the sum of $66.50 for material + $72.00 for labor + $1.10 for equipment is $139.60.

The figure in the **Labor Hours** column is the amount of labor required to perform one unit of work—in this case the amount of labor required to construct one 4' x 4' equipment pad. This figure is calculated by dividing the number of hours of labor in the crew by the daily output (48 labor hours divided by 30 pads = 1.6 hours of labor per pad). Multiply 1.600 times 60 to see the value in minutes: 60 x 1.6 = 96 minutes. Note: the labor hour figure is not dependent on the crew size. A change in crew size will result in a corresponding change in daily output, but the labor hours per unit of work will not change.

All RSMeans unit cost data includes the typical **Unit of Measure** used for estimating that item. For concrete-in-place the typical unit is cubic yards (C.Y.) or each (Ea.). For installing broadloom carpet it is square yard, and for gypsum board it is square foot. The estimator needs to take special care that the unit in the data matches the unit in the take-off. Unit conversions may be found in the Reference Section.

Sample Estimate

This sample demonstrates the elements of an estimate, including a tally of the RSMeans data lines, and a summary of the markups on a contractor's work to arrive at a total cost to the owner. The RSMeans Location Factor is added at the bottom of the estimate to adjust the cost of the work to a specific location.

Work Performed: The body of the estimate shows the RSMeans data selected, including line number, a brief description of each item, its take-off unit and quantity, and the bare costs of materials, labor, and equipment. This estimate also includes a column titled "SubContract." This data is taken from the RSMeans column "Total Incl O&P," and represents the total that a subcontractor would charge a general contractor for the work, including the sub's markup for overhead and profit.

Division 1, General Requirements: This is the first division numerically, but the last division estimated. Division 1 includes project-wide needs provided by the general contractor. These requirements vary by project, but may include temporary facilities and utilities, security, testing, project cleanup, etc. For small projects a percentage can be used, typically between 5% and 15% of project cost. For large projects the costs may be itemized and priced individually.

Bonds: Bond costs should be added to the estimate. The figures here represent a typical performance bond, ensuring the owner that if the general contractor does not complete the obligations in the construction contract the bonding company will pay the cost for completion of the work.

Location Adjustment: RSMeans published data is based on national average costs. If necessary, adjust the total cost of the project using a location factor from the "Location Factor" table or the "City Cost Index" table. Use location factors if the work is general, covering multiple trades. If the work is by a single trade (e.g., masonry) use the more specific data found in the "City Cost Indexes."

This estimate is based on an interactive spreadsheet. A copy of this spreadsheet is located on the RSMeans website at **http://www.reedconstructiondata.com/rsmeans/extras/546011**. You are free to download it and adjust it to your methodology.

Line Number	Description	Qty	Unit	Material
Project Name: Pre-Engineered Steel Building			**Architect: As Shown**	
Location:	**Anywhere, USA**			
03 30 53.40 3940	Strip footing, 12" x 24", reinforced	34	C.Y.	$4,624.00
03 30 53.40 3950	Strip footing, 12" x 36", reinforced	15	C.Y.	$1,950.00
03 11 13.65 3000	Concrete slab edge forms	500	L.F.	$155.00
03 22 11.10 0200	Welded wire fabric reinforcing	150	C.S.F.	$2,580.00
03 31 13.35 0300	Ready mix concrete, 4000 psi for slab on grade	278	C.Y.	$28,912.00
03 31 13.70 4300	Place, strike off & consolidate concrete slab	278	C.Y.	$0.00
03 35 13.30 0250	Machine float & trowel concrete slab	15,000	S.F.	$0.00
03 15 16.20 0140	Cut control joints in concrete slab	950	L.F.	$57.00
03 39 23.13 0300	Sprayed concrete curing membrane	150	C.S.F.	$1,642.50
Division 03	**Subtotal**			**$39,920.50**
08 36 13.10 2650	Manual 10' x 10' steel sectional overhead door	8	Ea.	$8,800.00
08 36 13.10 2860	Insulation and steel back panel for OH door	800	S.F.	$3,760.00
Division 08	**Subtotal**			**$12,560.00**
13 34 19.50 1100	Pre-Engineered Steel Building, 100' x 150' x 24'	15,000	SF Flr.	$0.00
13 34 19.50 6050	Framing for PESB door opening, 3' x 7'	4	Opng.	$0.00
13 34 19.50 6100	Framing for PESB door opening, 10' x 10'	8	Opng.	$0.00
13 34 19.50 6200	Framing for PESB window opening, 4' x 3'	6	Opng.	$0.00
13 34 19.50 5750	PESB door, 3' x 7', single leaf	4	Opng.	$2,340.00
13 34 19.50 7750	PESB sliding window, 4' x 3' with screen	6	Opng.	$2,250.00
13 34 19.50 6550	PESB gutter, eave type, 26 ga., painted	300	L.F.	$2,010.00
13 34 19.50 8650	PESB roof vent, 12" wide x 10' long	15	Ea.	$540.00
13 34 19.50 6900	PESB insulation, vinyl faced, 4" thick	27,400	S.F.	$10,960.00
Division 13	**Subtotal**			**$18,100.00**
			Subtotal	$70,580.50
Division 01	**General Requirements @ 7%**			4,940.64
			Estimate Subtotal	$75,521.14
			Sales Tax @ 5%	3,776.06
			Subtotal A	79,297.19
			GC O & P	7,929.72
			Subtotal B	87,226.91
			Contingency @ 5%	
			Subtotal C	
			Bond @ $12/1000 +10% O&P	
			Subtotal D	
			Location Adjustment Factor	
			Grand Total	

This example shows the cost to construct a pre-engineered steel building. The foundation, doors, windows, and insulation will be installed by the general contractor. A subcontractor will install the structural steel, roofing, and siding.

Labor	Equipment	SubContract	01/01/14 STD Estimate Total	
$3,468.00	$23.12	$0.00		
$1,230.00	$8.10	$0.00		
$1,160.00	$0.00	$0.00		
$3,900.00	$0.00	$0.00		
$0.00	$0.00	$0.00		
$4,642.60	$166.80	$0.00		
$8,700.00	$450.00	$0.00		
$370.50	$95.00	$0.00		
$922.50	$0.00	$0.00		
$24,393.60	**$743.02**	**$0.00**	**$65,057.12**	**Division 03**
$3,280.00	$0.00	$0.00		
$0.00	$0.00	$0.00		
$3,280.00	**$0.00**	**$0.00**	**$15,840.00**	**Division 08**
$0.00	$0.00	$345,000.00		
$0.00	$0.00	$2,260.00		
$0.00	$0.00	$9,200.00		
$0.00	$0.00	$3,300.00		
$656.00	$0.00	$0.00		
$558.00	$65.70	$0.00		
$768.00	$0.00	$0.00		
$3,060.00	$0.00	$0.00		
$8,768.00	$0.00	$0.00		
$13,810.00	**$65.70**	**$359,760.00**	**$391,735.70**	**Division 13**
$41,483.60	**$808.72**	**$359,760.00**	**$472,632.82**	**Subtotal**
2,903.85	**56.61**	**25,183.20**		**Gen. Requirements**
$44,387.45	$865.33	$384,943.20	$472,632.82	Estimate Subtotal
	43.27	9,623.58		Sales tax
44,387.45	908.60	394,566.78		Subtotal
24,368.71	90.86	39,456.68		GC O & P
68,756.16	999.46	434,023.46	$591,005.99	Subtotal
			29,550.30	Contingency
			$620,556.29	Subtotal
			8,191.34	Bond
			$628,747.63	Subtotal
102.30			14,461.20	Location Adjustment
			$643,208.83	**Grand Total**

Sales Tax: If the work is subject to state or local sales taxes, the amount must be added to the estimate. Sales tax may be added to material costs, equipment costs, and subcontracted work. In this case, sales tax was added in all three categories. It was assumed that approximately half the subcontracted work would be material cost, so the tax was applied to 50% of the subcontract total.

GC O&P: This entry represents the general contractor's markup on material, labor, equipment, and subcontractor costs. RSMeans' standard markup on materials, equipment, and subcontracted work is 10%. In this estimate, the markup on the labor performed by the GC's workers uses "Skilled Workers Average" shown in Column F on the table "Installing Contractor's Overhead & Profit," which can be found on the inside-back cover of the book.

Contingency: A factor for contingency may be added to any estimate to represent the cost of unknowns that may occur between the time that the estimate is performed and the time the project is constructed. The amount of the allowance will depend on the stage of design at which the estimate is done, and the contractor's assessment of the risk involved. Refer to section 01 21 16.50 for contingency allowances.

Estimating Tips

01 20 00 Price and Payment Procedures

- Allowances that should be added to estimates to cover contingencies and job conditions that are not included in the national average material and labor costs are shown in section 01 21.

- When estimating historic preservation projects (depending on the condition of the existing structure and the owner's requirements), a 15%–20% contingency or allowance is recommended, regardless of the stage of the drawings.

01 30 00 Administrative Requirements

- Before determining a final cost estimate, it is a good practice to review all the items listed in Subdivisions 01 31 and 01 32 to make final adjustments for items that may need customizing to specific job conditions.

- Requirements for initial and periodic submittals can represent a significant cost to the General Requirements of a job. Thoroughly check the submittal specifications when estimating a project to determine any costs that should be included.

01 40 00 Quality Requirements

- All projects will require some degree of Quality Control. This cost is not included in the unit cost of construction listed in each division. Depending upon the terms of the contract, the various costs of inspection and testing can be the responsibility of either the owner or the contractor. Be sure to include the required costs in your estimate.

01 50 00 Temporary Facilities and Controls

- Barricades, access roads, safety nets, scaffolding, security, and many more requirements for the execution of a safe project are elements of direct cost. These costs can easily be overlooked when preparing an estimate. When looking through the major classifications of this subdivision, determine which items apply to each division in your estimate.

- Construction Equipment Rental Costs can be found in the Reference Section in section 01 54 33. Operators' wages are not included in equipment rental costs.

- Equipment mobilization and demobilization costs are not included in equipment rental costs and must be considered separately in section 01 54 36.50.

- The cost of small tools provided by the installing contractor for his workers is covered in the "Overhead" column on the "Installing Contractor's Overhead and Profit" table that lists labor trades, base rates and markups and, therefore, is included in the "Total Incl. O&P" cost of any Unit Price line item. For those users who are constrained by contract terms to use only bare costs, there are two line items in section 01 54 39.70 for small tools as a percentage of bare labor cost. If some of those users are further constrained by contract terms to refrain from using line items from Division 1, you are advised to cover the cost of small tools within your coefficient or multiplier.

01 70 00 Execution and Closeout Requirements

- When preparing an estimate, thoroughly read the specifications to determine the requirements for Contract Closeout. Final cleaning, record documentation, operation and maintenance data, warranties and bonds, and spare parts and maintenance materials can all be elements of cost for the completion of a contract. Do not overlook these in your estimate.

Reference Numbers

Reference numbers are shown in shaded boxes at the beginning of some major classifications. These numbers refer to related items in the Reference Section. The reference information may be an estimating procedure, an alternate pricing method, or technical information.

Note: Not all subdivisions listed here necessarily appear in this publication.

01 11 Summary of Work

01 11 31 – Professional Consultants

01 11 31.20 Construction Management Fees	Crew	Daily Output	Labor-Hours	Unit	Material	2014 Bare Costs Labor	Equipment	Total	Total Incl O&P
0010 **CONSTRUCTION MANAGEMENT FEES**									
0020 $1,000,000 job, minimum				Project				4.50%	4.50%
0050 Maximum								7.50%	7.50%
0060 For work to $100,000								10%	10%
0300 $50,000,000 job, minimum								2.50%	2.50%
0350 Maximum				↓				4%	4%

01 11 31.30 Engineering Fees

	Crew	Daily Output	Labor-Hours	Unit	Material	2014 Bare Costs Labor	Equipment	Total	Total Incl O&P
0010 **ENGINEERING FEES** R011110-30									
0020 Educational planning consultant, minimum				Project				.50%	.50%
0100 Maximum				"				2.50%	2.50%
0200 Electrical, minimum				Contrct				4.10%	4.10%
0300 Maximum								10.10%	10.10%
0400 Elevator & conveying systems, minimum								2.50%	2.50%
0500 Maximum								5%	5%
0600 Food service & kitchen equipment, minimum								8%	8%
0700 Maximum								12%	12%
1000 Mechanical (plumbing & HVAC), minimum								4.10%	4.10%
1100 Maximum				↓				10.10%	10.10%

01 21 Allowances

01 21 16 – Contingency Allowances

01 21 16.50 Contingencies

	Crew	Daily Output	Labor-Hours	Unit	Material	2014 Bare Costs Labor	Equipment	Total	Total Incl O&P
0010 **CONTINGENCIES**, Add to estimate									
0020 Conceptual stage				Project				20%	20%
0050 Schematic stage								15%	15%
0100 Preliminary working drawing stage (Design Dev.)								10%	10%
0150 Final working drawing stage				↓				3%	3%

01 21 53 – Factors Allowance

01 21 53.50 Factors

	Crew	Daily Output	Labor-Hours	Unit	Material	2014 Bare Costs Labor	Equipment	Total	Total Incl O&P
0010 **FACTORS** Cost adjustments									
0100 Add to construction costs for particular job requirements									
0500 Cut & patch to match existing construction, add, minimum				Costs	2%	3%			
0550 Maximum					5%	9%			
0800 Dust protection, add, minimum					1%	2%			
0850 Maximum					4%	11%			
1100 Equipment usage curtailment, add, minimum					1%	1%			
1150 Maximum					3%	10%			
1400 Material handling & storage limitation, add, minimum					1%	1%			
1450 Maximum					6%	7%			
1700 Protection of existing work, add, minimum					2%	2%			
1750 Maximum					5%	7%			
2000 Shift work requirements, add, minimum						5%			
2050 Maximum						30%			
2300 Temporary shoring and bracing, add, minimum					2%	5%			
2350 Maximum					5%	12%			
2400 Work inside prisons and high security areas, add, minimum						30%			
2450 Maximum				↓		50%			

01 21 Allowances

01 21 55 – Job Conditions Allowance

01 21 55.50 Job Conditions	Crew	Daily Output	Labor-Hours	Unit	Material	2014 Bare Costs Labor	Equipment	Total	Total Incl O&P
0010 **JOB CONDITIONS** Modifications to applicable									
0020 cost summaries									
0100 Economic conditions, favorable, deduct				Project				2%	2%
0200 Unfavorable, add								5%	5%
0300 Hoisting conditions, favorable, deduct								2%	2%
0400 Unfavorable, add								5%	5%
0700 Labor availability, surplus, deduct								1%	1%
0800 Shortage, add								10%	10%
0900 Material storage area, available, deduct								1%	1%
1000 Not available, add								2%	2%
1100 Subcontractor availability, surplus, deduct								5%	5%
1200 Shortage, add								12%	12%
1300 Work space, available, deduct								2%	2%
1400 Not available, add								5%	5%

01 21 57 – Overtime Allowance

01 21 57.50 Overtime

	Crew	Daily Output	Labor-Hours	Unit	Material	2014 Bare Costs Labor	Equipment	Total	Total Incl O&P
0010 **OVERTIME** for early completion of projects or where	R012909-90								
0020 labor shortages exist, add to usual labor, up to				Costs		100%			

01 21 61 – Cost Indexes

01 21 61.10 Construction Cost Index

	Crew	Daily Output	Labor-Hours	Unit	Material	2014 Bare Costs Labor	Equipment	Total	Total Incl O&P
0010 **CONSTRUCTION COST INDEX** (Reference) over 930 zip code locations in									
0020 the U.S. and Canada, total bldg. cost, min. (Longview, TX)				%					73.50%
0050 Average									100%
0100 Maximum (New York, NY)									131.10%

01 21 61.20 Historical Cost Indexes

	Crew	Daily Output	Labor-Hours	Unit	Material	2014 Bare Costs Labor	Equipment	Total	Total Incl O&P
0010 **HISTORICAL COST INDEXES** (See Reference Section)									

01 21 61.30 Labor Index

	Crew	Daily Output	Labor-Hours	Unit	Material	2014 Bare Costs Labor	Equipment	Total	Total Incl O&P
0010 **LABOR INDEX** (Reference) For over 930 zip code locations in									
0020 the U.S. and Canada, minimum (Longview, TX)				%		40.50%			
0050 Average						100%			
0100 Maximum (New York, NY)						166.40%			

01 21 61.50 Material Index

	Crew	Daily Output	Labor-Hours	Unit	Material	2014 Bare Costs Labor	Equipment	Total	Total Incl O&P
0010 **MATERIAL INDEX** (Reference) For over 930 zip code locations in									
0020 the U.S. and Canada, minimum (Vale, OR)				%	91.30%				
0040 Average					100%				
0060 Maximum (Ketchikan, AK)					132%				

01 21 63 – Taxes

01 21 63.10 Taxes

	Crew	Daily Output	Labor-Hours	Unit	Material	2014 Bare Costs Labor	Equipment	Total	Total Incl O&P
0010 **TAXES**	R012909-80								
0020 Sales tax, State, average				%	5.06%				
0050 Maximum	R012909-85				7.25%				
0200 Social Security, on first $115,800 of wages						7.65%			
0300 Unemployment, combined Federal and State, minimum						.60%			
0350 Average						7.80%			
0400 Maximum						13.50%			

01 31 13 – Project Coordination

01 31 13.20 Field Personnel

		Crew	Daily Output	Labor-Hours	Unit	Material	2014 Bare Costs Labor	2014 Bare Costs Equipment	Total	Total Incl O&P
0010	**FIELD PERSONNEL**									
0020	Clerk, average				Week		435		435	675
0100	Field engineer, minimum						1,025		1,025	1,600
0120	Average						1,350		1,350	2,100
0140	Maximum						1,525		1,525	2,375
0160	General purpose laborer, average						1,475		1,475	2,250
0180	Project manager, minimum						1,925		1,925	2,975
0200	Average						2,225		2,225	3,425
0220	Maximum						2,525		2,525	3,925
0240	Superintendent, minimum						1,875		1,875	2,900
0260	Average						2,050		2,050	3,200
0280	Maximum						2,350		2,350	3,625
0290	Timekeeper, average				↓		1,200		1,200	1,850

01 31 13.30 Insurance

		Crew	Daily Output	Labor-Hours	Unit	Material	2014 Bare Costs Labor	2014 Bare Costs Equipment	Total	Total Incl O&P
0010	**INSURANCE** R013113-40									
0020	Builders risk, standard, minimum				Job				.24%	.24%
0050	Maximum R013113-60								.64%	.64%
0200	All-risk type, minimum								.25%	.25%
0250	Maximum				↓				.62%	.62%
0400	Contractor's equipment floater, minimum				Value				.50%	.50%
0450	Maximum				"				1.50%	1.50%
0600	Public liability, average				Job				2.02%	2.02%
0800	Workers' compensation & employer's liability, average									
0850	by trade, carpentry, general				Payroll		15.38%			
1000	Electrical						5.83%			
1150	Insulation						11.67%			
1450	Plumbing						7.08%			
1550	Sheet metal work (HVAC)				↓		8.79%			

01 31 13.50 General Contractor's Mark-Up

		Crew	Daily Output	Labor-Hours	Unit	Material	2014 Bare Costs Labor	2014 Bare Costs Equipment	Total	Total Incl O&P
0010	**GENERAL CONTRACTOR'S MARK-UP** on Change Orders									
0200	Extra work, by subcontractors, add				%				10%	10%
0250	By General Contractor, add								15%	15%
0400	Omitted work, by subcontractors, deduct all but								5%	5%
0450	By General Contractor, deduct all but								7.50%	7.50%
0600	Overtime work, by subcontractors, add								15%	15%
0650	By General Contractor, add				↓				10%	10%

01 31 13.60 Installing Contractor's Main Office Overhead

		Crew	Daily Output	Labor-Hours	Unit	Material	2014 Bare Costs Labor	2014 Bare Costs Equipment	Total	Total Incl O&P
0010	**INSTALLING CONTRACTOR'S MAIN OFFICE OVERHEAD**									
0020	As percent of direct costs, minimum				%				5%	
0050	Average								13%	
0100	Maximum				↓				30%	

01 31 13.80 Overhead and Profit

		Crew	Daily Output	Labor-Hours	Unit	Material	2014 Bare Costs Labor	2014 Bare Costs Equipment	Total	Total Incl O&P
0010	**OVERHEAD & PROFIT** Allowance to add to items in this									
0020	book that do not include Subs O&P, average				%				25%	
0100	Allowance to add to items in this book that									
0110	do include Subs O&P, minimum				%				5%	5%
0150	Average								10%	10%
0200	Maximum								15%	15%
0300	Typical, by size of project, under $100,000								30%	
0350	$500,000 project								25%	
0400	$2,000,000 project								20%	
0450	Over $10,000,000 project				↓				15%	

01 31 Project Management and Coordination

01 31 13 – Project Coordination

01 31 13.90 Performance Bond		Crew	Daily Output	Labor-Hours	Unit	Material	2014 Bare Costs Labor	Equipment	Total	Total Incl O&P
0010	**PERFORMANCE BOND** R013113-80									
0020	For buildings, minimum				Job				.60%	.60%
0100	Maximum				"				2.50%	2.50%

01 32 Construction Progress Documentation

01 32 33 – Photographic Documentation

01 32 33.50 Photographs

		Crew	Daily Output	Labor-Hours	Unit	Material	2014 Bare Costs Labor	Equipment	Total	Total Incl O&P
0010	**PHOTOGRAPHS**									
0020	8" x 10", 4 shots, 2 prints ea., std. mounting				Set	475			475	525
0100	Hinged linen mounts					530			530	580
0200	8" x 10", 4 shots, 2 prints each, in color					425			425	470
0300	For I.D. slugs, add to all above				↓	5.30			5.30	5.85
1500	Time lapse equipment, camera and projector, buy				Ea.	2,650			2,650	2,925
1550	Rent per month				"	1,600			1,600	1,750
1700	Cameraman and film, including processing, B.&W.				Day	1,225			1,225	1,350
1720	Color				"	1,425			1,425	1,550

01 41 Regulatory Requirements

01 41 26 – Permit Requirements

01 41 26.50 Permits

		Crew	Daily Output	Labor-Hours	Unit	Material	2014 Bare Costs Labor	Equipment	Total	Total Incl O&P
0010	**PERMITS**									
0020	Rule of thumb, most cities, minimum				Job				.50%	.50%
0100	Maximum				"				2%	2%

01 51 Temporary Utilities

01 51 13 – Temporary Electricity

01 51 13.50 Temporary Power Equip (Pro-Rated Per Job)

		Crew	Daily Output	Labor-Hours	Unit	Material	2014 Bare Costs Labor	Equipment	Total	Total Incl O&P
0010	**TEMPORARY POWER EQUIP (PRO-RATED PER JOB)** R015113-65									
0020	Service, overhead feed, 3 use									
0030	100 Amp	1 Elec	1.25	6.400	Ea.	665	340		1,005	1,250
0040	200 Amp		1	8		860	425		1,285	1,600
0050	400 Amp		.75	10.667		1,575	570		2,145	2,600
0060	600 Amp	↓	.50	16	↓	2,300	855		3,155	3,800
0100	Underground feed, 3 use									
0110	100 Amp	1 Elec	2	4	Ea.	630	213		843	1,025
0120	200 Amp		1.15	6.957		855	370		1,225	1,500
0130	400 Amp		1	8		1,575	425		2,000	2,375
0140	600 Amp		.75	10.667		1,975	570		2,545	3,025
0150	800 Amp		.50	16		2,975	855		3,830	4,550
0160	1000 Amp		.35	22.857		3,325	1,225		4,550	5,475
0170	1200 Amp		.25	32		3,725	1,700		5,425	6,650
0180	2000 Amp	↓	.20	40	↓	4,525	2,125		6,650	8,175
0200	Transformers, 3 use									
0210	30 kVA	1 Elec	1	8	Ea.	1,650	425		2,075	2,475
0220	45 kVA		.75	10.667		1,975	570		2,545	3,025
0230	75 kVA		.50	16		3,325	855		4,180	4,925
0240	112.5 kVA	↓	.40	20		3,600	1,075		4,675	5,575

01 51 13.50 Temporary Power Equip (Pro-Rated Per Job)	Crew	Daily Output	Labor-Hours	Unit	Material	2014 Bare Costs Labor	Equipment	Total	Total Incl O&P
0250	Feeder, PVC, CU wire in trench									
0260	60 Amp	1 Elec	96	.083	L.F.	4.09	4.45		8.54	11.15
0270	100 Amp		85	.094		7.80	5		12.80	16.10
0280	200 Amp		59	.136		18.75	7.25		26	31.50
0290	400 Amp		42	.190		43.50	10.15		53.65	62.50
0300	Feeder, PVC, aluminum wire in trench									
0310	60 Amp	1 Elec	96	.083	L.F.	4.49	4.45		8.94	11.60
0320	100 Amp		85	.094		5	5		10	13
0330	200 Amp		59	.136		10.75	7.25		18	22.50
0340	400 Amp		42	.190		22.50	10.15		32.65	39.50
0350	Feeder, EMT, CU wire									
0360	60 Amp	1 Elec	90	.089	L.F.	3.43	4.74		8.17	10.85
0370	100 Amp		80	.100		8	5.35		13.35	16.80
0380	200 Amp		60	.133		17.15	7.10		24.25	29.50
0390	400 Amp		35	.229		50.50	12.20		62.70	74
0400	Feeder, EMT, alum. wire									
0410	60 Amp	1 Elec	90	.089	L.F.	4.71	4.74		9.45	12.30
0420	100 Amp		80	.100		5.90	5.35		11.25	14.50
0430	200 Amp		60	.133		16	7.10		23.10	28.50
0440	400 Amp		35	.229		28	12.20		40.20	49.50
0500	Equipment, 3 use									
0510	Spider box 50 Amp	1 Elec	8	1	Ea.	895	53.50		948.50	1,075
0520	Lighting cord 100'		8	1		116	53.50		169.50	208
0530	Light stanchion		8	1		70.50	53.50		124	158
0540	Temporary cords, 100', 3 use									
0550	Feeder cord, 50 Amp	1 Elec	16	.500	Ea.	390	26.50		416.50	470
0560	Feeder cord, 100 Amp		12	.667		535	35.50		570.50	645
0570	Tap cord, 50 Amp		12	.667		221	35.50		256.50	296
0580	Tap cord, 100 Amp		6	1.333		535	71		606	695
0590	Temporary cords, 50', 3 use									
0600	Feeder cord, 50 Amp	1 Elec	16	.500	Ea.	198	26.50		224.50	258
0610	Feeder cord, 100 Amp		12	.667		370	35.50		405.50	460
0620	Tap cord, 50 Amp		12	.667		109	35.50		144.50	173
0630	Tap cord, 100 Amp		6	1.333		370	71		441	510
0700	Connections									
0710	Compressor or pump									
0720	30 Amp	1 Elec	7	1.143	Ea.	23.50	61		84.50	117
0730	60 Amp		5.30	1.509		43.50	80.50		124	169
0740	100 Amp		4	2		97	107		204	267
0750	Tower crane									
0760	60 Amp	1 Elec	4.50	1.778	Ea.	43.50	95		138.50	190
0770	100 Amp	"	3	2.667	"	97	142		239	320
0780	Manlift									
0790	Single	1 Elec	3	2.667	Ea.	42	142		184	260
0800	Double	"	2	4	"	76.50	213		289.50	405
0810	Welder with disconnect									
0820	50 Amp	1 Elec	5	1.600	Ea.	271	85.50		356.50	425
0830	100 Amp		3.80	2.105		455	112		567	670
0840	200 Amp		2.50	3.200		820	171		991	1,150
0850	400 Amp		1	8		1,975	425		2,400	2,825
0860	Office trailer									
0870	60 Amp	1 Elec	4.50	1.778	Ea.	101	95		196	253
0880	100 Amp		3	2.667		194	142		336	425

01 51 Temporary Utilities

01 51 13 – Temporary Electricity

01 51 13.50 Temporary Power Equip (Pro-Rated Per Job)	Crew	Daily Output	Labor-Hours	Unit	Material	2014 Bare Costs Labor	Equipment	Total	Total Incl O&P
0890	200 Amp	1 Elec	2	4	Ea.	655	213		868	1,050
0900	Lamping, add per floor				Total				625	690
0910	Maintenance, total temp. power cost				Job				5%	5%

01 51 13.80 Temporary Utilities

		Crew	Daily Output	Labor-Hours	Unit	Material	2014 Bare Costs Labor	Equipment	Total	Total Incl O&P
0010	**TEMPORARY UTILITIES**									
0100	Heat, incl. fuel and operation, per week, 12 hrs. per day	1 Skwk	100	.080	CSF Flr	28	3.78		31.78	36.50
0200	24 hrs. per day	"	60	.133		53.50	6.30		59.80	68.50
0350	Lighting, incl. service lamps, wiring & outlets, minimum	1 Elec	34	.235		2.80	12.55		15.35	22
0360	Maximum	"	17	.471		5.85	25		30.85	44
0400	Power for temp lighting only, 6.6 KWH, per month								.92	1.01
0430	11.8 KWH, per month								1.65	1.82
0450	23.6 KWH, per month								3.30	3.63
0600	Power for job duration incl. elevator, etc., minimum								47	51.50
0650	Maximum								110	121
1000	Toilet, portable, see Equip. Rental 01 54 33 in Reference Section									

01 52 Construction Facilities

01 52 13 – Field Offices and Sheds

01 52 13.20 Office and Storage Space

		Crew	Daily Output	Labor-Hours	Unit	Material	2014 Bare Costs Labor	Equipment	Total	Total Incl O&P
0010	**OFFICE AND STORAGE SPACE**									
0020	Office trailer, furnished, no hookups, 20' x 8', buy	2 Skwk	1	16	Ea.	9,150	755		9,905	11,300
0250	Rent per month					169			169	185
0300	32' x 8', buy	2 Skwk	.70	22.857		14,000	1,075		15,075	17,100
0350	Rent per month					203			203	223
0400	50' x 10', buy	2 Skwk	.60	26.667		22,400	1,250		23,650	26,600
0450	Rent per month					305			305	335
0500	50' x 12', buy	2 Skwk	.50	32		27,000	1,525		28,525	32,100
0550	Rent per month					350			350	385
0700	For air conditioning, rent per month, add					48.50			48.50	53.50
0800	For delivery, add per mile				Mile	11			11	12.10
0890	Delivery each way				Ea.	200			200	220
0900	Bunk house trailer, 8' x 40' duplex dorm with kitchen, no hookups, buy	2 Carp	1	16		37,300	735		38,035	42,100
0910	9 man with kitchen and bath, no hookups, buy		1	16		38,000	735		38,735	42,900
0920	18 man sleeper with bath, no hookups, buy		1	16		49,000	735		49,735	55,000
1000	Portable buildings, prefab, on skids, economy, 8' x 8'		265	.060	S.F.	29.50	2.77		32.27	37
1100	Deluxe, 8' x 12'		150	.107	"	24	4.89		28.89	34
1200	Storage boxes, 20' x 8', buy	2 Skwk	1.80	8.889	Ea.	2,775	420		3,195	3,700
1250	Rent per month					78.50			78.50	86
1300	40' x 8', buy	2 Skwk	1.40	11.429		3,700	540		4,240	4,900
1350	Rent per month					99			99	109

01 52 13.40 Field Office Expense

		Crew	Daily Output	Labor-Hours	Unit	Material	2014 Bare Costs Labor	Equipment	Total	Total Incl O&P
0010	**FIELD OFFICE EXPENSE**									
0100	Office equipment rental average				Month	200			200	220
0120	Office supplies, average				"	75			75	82.50
0125	Office trailer rental, see Section 01 52 13.20									
0140	Telephone bill; avg. bill/month incl. long dist.				Month	81			81	89
0160	Lights & HVAC				"	152			152	167

01 54 Construction Aids

01 54 16 – Temporary Hoists

01 54 16.50 Weekly Forklift Crew	Crew	Daily Output	Labor-Hours	Unit	Material	2014 Bare Costs Labor	Equipment	Total	Total Incl O&P
0010 **WEEKLY FORKLIFT CREW**									
0100 All-terrain forklift, 45' lift, 35' reach, 9000 lb. capacity	A-3P	.20	40	Week		1,875	2,650	4,525	5,775

01 54 19 – Temporary Cranes

01 54 19.50 Daily Crane Crews	Crew	Daily Output	Labor-Hours	Unit	Material	2014 Bare Costs Labor	Equipment	Total	Total Incl O&P
0010 **DAILY CRANE CREWS** for small jobs, portal to portal									
0100 12-ton truck-mounted hydraulic crane	A-3H	1	8	Day		400	875	1,275	1,575
0200 25-ton	A-3I	1	8			400	1,000	1,400	1,700
0300 40-ton	A-3J	1	8			400	1,250	1,650	1,975
0400 55-ton	A-3K	1	16			750	1,650	2,400	2,975
0500 80-ton	A-3L	1	16			750	2,400	3,150	3,775
0600 100-ton	A-3M	1	16	↓		750	2,375	3,125	3,775
0900 If crane is needed on a Saturday, Sunday or Holiday									
0910 At time-and-a-half, add				Day		50%			
0920 At double time, add				"		100%			

01 54 19.60 Monthly Tower Crane Crew	Crew	Daily Output	Labor-Hours	Unit	Material	2014 Bare Costs Labor	Equipment	Total	Total Incl O&P
0010 **MONTHLY TOWER CRANE CREW**, excludes concrete footing									
0100 Static tower crane, 130' high, 106' jib, 6200 lb. capacity	A-3N	.05	176	Month		8,850	24,200	33,050	40,000

01 54 23 – Temporary Scaffolding and Platforms

01 54 23.70 Scaffolding	Crew	Daily Output	Labor-Hours	Unit	Material	2014 Bare Costs Labor	Equipment	Total	Total Incl O&P
0010 **SCAFFOLDING** R015423-10									
0015 Steel tube, regular, no plank, labor only to erect & dismantle									
0090 Building exterior, wall face, 1 to 5 stories, 6'-4" x 5' frames	3 Carp	8	3	C.S.F.		138		138	212
0200 6 to 12 stories	4 Carp	8	4			183		183	283
0301 13 to 20 stories	5 Clab	8	5			183		183	283
0460 Building interior, wall face area, up to 16' high	3 Carp	12	2			91.50		91.50	142
0560 16' to 40' high		10	2.400	↓		110		110	170
0800 Building interior floor area, up to 30' high	↓	150	.160	C.C.F.		7.35		7.35	11.30
0900 Over 30' high	4 Carp	160	.200	"		9.15		9.15	14.15
0906 Complete system for face of walls, no plank, material only rent/mo				C.S.F.	35			35	38.50
0908 Interior spaces, no plank, material only rent/mo				C.C.F.	3.35			3.35	3.68
0910 Steel tubular, heavy duty shoring, buy									
0920 Frames 5' high 2' wide				Ea.	81.50			81.50	90
0925 5' high 4' wide					93			93	102
0930 6' high 2' wide					93.50			93.50	103
0935 6' high 4' wide				↓	109			109	120
0940 Accessories									
0945 Cross braces				Ea.	15.50			15.50	17.05
0950 U-head, 8" x 8"					19.10			19.10	21
0955 J-head, 4" x 8"					13.90			13.90	15.30
0960 Base plate, 8" x 8"					15.50			15.50	17.05
0965 Leveling jack				↓	33.50			33.50	36.50
1000 Steel tubular, regular, buy									
1100 Frames 3' high 5' wide				Ea.	72.50			72.50	80
1150 5' high 5' wide					86.50			86.50	95
1200 6'-4" high 5' wide					118			118	129
1350 7'-6" high 6' wide					158			158	173
1500 Accessories cross braces					15.50			15.50	17.05
1550 Guardrail post					16.40			16.40	18.05
1600 Guardrail 7' section					6.30			6.30	6.95
1650 Screw jacks & plates					21.50			21.50	24
1700 Sidearm brackets				↓	25.50			25.50	28.50

01 54 23 – Temporary Scaffolding and Platforms

01 54 23.70 Scaffolding	Crew	Daily Output	Labor-Hours	Unit	Material	2014 Bare Costs Labor	Equipment	Total	Total Incl O&P
1750	8" casters				Ea.	31			31	34
1800	Plank 2" x 10" x 16'-0"					52			52	57.50
1900	Stairway section					276			276	305
1910	Stairway starter bar					32			32	35
1920	Stairway inside handrail					53			53	58
1930	Stairway outside handrail					81.50			81.50	90
1940	Walk-thru frame guardrail					41.50			41.50	46
2000	Steel tubular, regular, rent/mo.									
2100	Frames 3' high 5' wide				Ea.	5			5	5.50
2150	5' high 5' wide					5			5	5.50
2200	6'-4" high 5' wide					5.15			5.15	5.65
2250	7'-6" high 6' wide					7			7	7.70
2500	Accessories, cross braces					1			1	1.10
2550	Guardrail post					1			1	1.10
2600	Guardrail 7' section					1			1	1.10
2650	Screw jacks & plates					2			2	2.20
2700	Sidearm brackets					2			2	2.20
2750	8" casters					8			8	8.80
2800	Outrigger for rolling tower					3			3	3.30
2850	Plank 2" x 10" x 16'-0"					6			6	6.60
2900	Stairway section					35.50			35.50	39
2940	Walk-thru frame guardrail					2.50			2.50	2.75
3000	Steel tubular, heavy duty shoring, rent/mo.									
3250	5' high 2' & 4' wide				Ea.	5			5	5.50
3300	6' high 2' & 4' wide					5			5	5.50
3500	Accessories, cross braces					1			1	1.10
3600	U - head, 8" x 8"					1			1	1.10
3650	J - head, 4" x 8"					1			1	1.10
3700	Base plate, 8" x 8"					1			1	1.10
3750	Leveling jack					2			2	2.20
5700	Planks, 2" x 10" x 16'-0", labor only to erect & remove to 50' H	3 Carp	72	.333			15.30		15.30	23.50
5800	Over 50' high	4 Carp	80	.400			18.35		18.35	28.50

01 54 23.75 Scaffolding Specialties

		Crew	Daily Output	Labor-Hours	Unit	Material	2014 Bare Costs Labor	Equipment	Total	Total Incl O&P
0010	**SCAFFOLDING SPECIALTIES**									
1500	Sidewalk bridge using tubular steel scaffold frames including									
1510	planking (material cost is rent/month)	3 Carp	45	.533	L.F.	5.60	24.50		30.10	43.50
1600	For 2 uses per month, deduct from all above					50%				
1700	For 1 use every 2 months, add to all above					100%				
1900	Catwalks, 20" wide, no guardrails, 7' span, buy				Ea.	133			133	146
2000	10' span, buy					186			186	205
3720	Putlog, standard, 8' span, with hangers, buy					66.50			66.50	73
3750	12' span, buy					99			99	109
3760	Trussed type, 16' span, buy					229			229	252
3790	22' span, buy					274			274	300
3795	Rent per month					30			30	33
3800	Rolling ladders with handrails, 30" wide, buy, 2 step					231			231	254
4000	7 step					760			760	835
4050	10 step					1,050			1,050	1,150
4100	Rolling towers, buy, 5' wide, 7' long, 10' high					1,150			1,150	1,275
4200	For 5' high added sections, to buy, add					204			204	224
4300	Complete incl. wheels, railings, outriggers,									
4350	21' high, to buy				Ea.	1,975			1,975	2,175

01 54 Construction Aids

01 54 23 – Temporary Scaffolding and Platforms

01 54 23.75 Scaffolding Specialties	Crew	Daily Output	Labor-Hours	Unit	Material	2014 Bare Costs Labor	Equipment	Total	Total Incl O&P
4400	Rent/month = 5% of purchase cost				Ea.	170			170	187

01 54 26 – Temporary Swing Staging

01 54 26.50 Swing Staging

		Crew	Daily Output	Labor-Hours	Unit	Material	Labor	Equipment	Total	Total Incl O&P
0010	**SWING STAGING**, 500 lb. cap., 2' wide to 24' long, hand operated									
0020	steel cable type, with 60' cables, buy				Ea.	4,900			4,900	5,375
0030	Rent per month				"	490			490	540
2200	Move swing staging (setup and remove)	E-4	2	16	Move		825	71	896	1,525

01 54 36 – Equipment Mobilization

01 54 36.50 Mobilization

		Crew	Daily Output	Labor-Hours	Unit	Material	Labor	Equipment	Total	Total Incl O&P
0010	**MOBILIZATION** (Use line item again for demobilization) R015433-10									
0015	Up to 25 mi. haul dist. (50 mi. RT for mob/demob crew)									
0020	Dozer, loader, backhoe, excav., grader, paver, roller, 70 to 150 H.P.	B-34N	4	2	Ea.		75	142	217	272
0100	Above 150 HP	B-34K	3	2.667			100	320	420	510
0900	Shovel or dragline, 3/4 C.Y.	"	3.60	2.222			83.50	267	350.50	420
1100	Small equipment, placed in rear of, or towed by pickup truck	A-3A	8	1			36.50	21	57.50	78.50
1150	Equip up to 70 HP, on flatbed trailer behind pickup truck	A-3D	4	2			73	70.50	143.50	189
2000	Crane, truck-mounted, up to 75 ton, (driver only, one-way)	1 Eqhv	7.20	1.111			56		56	84.50
2100	Crane, truck-mounted, over 75 ton	A-3E	2.50	6.400			281	67.50	348.50	500
2200	Crawler-mounted, up to 75 ton	A-3F	2	8			350	500	850	1,075
2300	Over 75 ton	A-3G	1.50	10.667			470	755	1,225	1,550
2500	For each additional 5 miles haul distance, add						10%	10%		
3000	For large pieces of equipment, allow for assembly/knockdown									
3100	For mob/demob of micro-tunneling equip, see Section 33 05 23.19									

01 54 39 – Construction Equipment

01 54 39.70 Small Tools

		Crew	Daily Output	Labor-Hours	Unit	Material	Labor	Equipment	Total	Total Incl O&P
0010	**SMALL TOOLS**									
0020	As % of contractor's bare labor cost for project, minimum				Total		.50%			
0100	Maximum				"		2%			

01 55 Vehicular Access and Parking

01 55 23 – Temporary Roads

01 55 23.50 Roads and Sidewalks

		Crew	Daily Output	Labor-Hours	Unit	Material	Labor	Equipment	Total	Total Incl O&P
0010	**ROADS AND SIDEWALKS** Temporary									
0050	Roads, gravel fill, no surfacing, 4" gravel depth	B-14	715	.067	S.Y.	3.98	2.60	.51	7.09	8.95
0100	8" gravel depth	"	615	.078	"	7.95	3.02	.59	11.56	14.05
1000	Ramp, 3/4" plywood on 2" x 6" joists, 16" O.C.	2 Carp	300	.053	S.F.	1.44	2.45		3.89	5.35
1100	On 2" x 10" joists, 16" O.C.	"	275	.058	"	1.99	2.67		4.66	6.30

01 56 Temporary Barriers and Enclosures

01 56 13 – Temporary Air Barriers

01 56 13.90 Winter Protection

		Crew	Daily Output	Labor-Hours	Unit	Material	2014 Bare Costs Labor	Equipment	Total	Total Incl O&P
0010	**WINTER PROTECTION**									
0100	Framing to close openings	2 Clab	500	.032	S.F.	.41	1.17		1.58	2.26
0200	Tarpaulins hung over scaffolding, 8 uses, not incl. scaffolding		1500	.011		.25	.39		.64	.88
0300	Prefab fiberglass panels, steel frame, 8 uses		1200	.013		2.20	.49		2.69	3.17

01 56 16 – Temporary Dust Barriers

01 56 16.10 Dust Barriers, Temporary

		Crew	Daily Output	Labor-Hours	Unit	Material	2014 Bare Costs Labor	Equipment	Total	Total Incl O&P
0010	**DUST BARRIERS, TEMPORARY**									
0020	Spring loaded telescoping pole & head, to 12', erect and dismantle	1 Clab	240	.033	Ea.		1.22		1.22	1.88
0025	Cost per day (based upon 250 days)				Day	.24			.24	.26
0030	To 21', erect and dismantle	1 Clab	240	.033	Ea.		1.22		1.22	1.88
0035	Cost per day (based upon 250 days)				Day	.38			.38	.42
0040	Accessories, caution tape reel, erect and dismantle	1 Clab	480	.017	Ea.		.61		.61	.94
0045	Cost per day (based upon 250 days)				Day	.36			.36	.40
0060	Foam rail and connector, erect and dismantle	1 Clab	240	.033	Ea.		1.22		1.22	1.88
0065	Cost per day (based upon 250 days)				Day	.10			.10	.11
0070	Caution tape	1 Clab	384	.021	C.L.F.	2.70	.76		3.46	4.15
0080	Zipper, standard duty		60	.133	Ea.	15.95	4.89		20.84	25
0090	Heavy duty		48	.167	"	18.90	6.10		25	30.50
0100	Polyethylene sheet, 4 mil		37	.216	Sq.	2.91	7.90		10.81	15.45
0110	6 mil		37	.216	"	3.98	7.90		11.88	16.65
1000	Dust partition, 6 mil polyethylene, 1" x 3" frame	2 Carp	2000	.008	S.F.	.29	.37		.66	.89
1080	2" x 4" frame	"	2000	.008	"	.32	.37		.69	.92

01 56 23 – Temporary Barricades

01 56 23.10 Barricades

		Crew	Daily Output	Labor-Hours	Unit	Material	2014 Bare Costs Labor	Equipment	Total	Total Incl O&P
0010	**BARRICADES**									
0020	5' high, 3 rail @ 2" x 8", fixed	2 Carp	20	.800	L.F.	5.75	36.50		42.25	63
0150	Movable		30	.533		4.77	24.50		29.27	43
1000	Guardrail, wooden, 3' high, 1" x 6", on 2" x 4" posts		200	.080		1.29	3.67		4.96	7.05
1100	2" x 6", on 4" x 4" posts		165	.097		2.35	4.45		6.80	9.45
1200	Portable metal with base pads, buy					15.55			15.55	17.10
1250	Typical installation, assume 10 reuses	2 Carp	600	.027		2.55	1.22		3.77	4.70
1300	Barricade tape, polyethylene, 7 mil, 3" wide x 500' long roll				Ea.	25			25	27.50
3000	Detour signs, set up and remove									
3010	Reflective aluminum, MUTCD, 24" x 24", post mounted	1 Clab	20	.400	Ea.	2.35	14.65		17	25

01 56 26 – Temporary Fencing

01 56 26.50 Temporary Fencing

		Crew	Daily Output	Labor-Hours	Unit	Material	2014 Bare Costs Labor	Equipment	Total	Total Incl O&P
0010	**TEMPORARY FENCING**									
0020	Chain link, 11 ga., 4' high	2 Clab	400	.040	L.F.	2.95	1.47		4.42	5.50
0100	6' high		300	.053		2.95	1.95		4.90	6.25
0200	Rented chain link, 6' high, to 1000' (up to 12 mo.)		400	.040		4.39	1.47		5.86	7.10
0250	Over 1000' (up to 12 mo.)		300	.053		4.29	1.95		6.24	7.75
0350	Plywood, painted, 2" x 4" frame, 4' high	A-4	135	.178		6.25	7.80		14.05	18.75
0400	4" x 4" frame, 8' high	"	110	.218		11.45	9.55		21	27
0500	Wire mesh on 4" x 4" posts, 4' high	2 Carp	100	.160		9.75	7.35		17.10	22
0550	8' high	"	80	.200		14.70	9.15		23.85	30.50

01 56 29 – Temporary Protective Walkways

01 56 29.50 Protection

		Crew	Daily Output	Labor-Hours	Unit	Material	2014 Bare Costs Labor	Equipment	Total	Total Incl O&P
0010	**PROTECTION**									
0020	Stair tread, 2" x 12" planks, 1 use	1 Carp	75	.107	Tread	4.57	4.89		9.46	12.60
0100	Exterior plywood, 1/2" thick, 1 use		65	.123		1.79	5.65		7.44	10.65

01 56 Temporary Barriers and Enclosures

01 56 29 – Temporary Protective Walkways

01 56 29.50 Protection		Crew	Daily Output	Labor-Hours	Unit	Material	2014 Bare Costs Labor	Equipment	Total	Total Incl O&P
0200	3/4" thick, 1 use	1 Carp	60	.133	Tread	2.64	6.10		8.74	12.35
2200	Sidewalks, 2" x 12" planks, 2 uses		350	.023	S.F.	.76	1.05		1.81	2.46
2300	Exterior plywood, 2 uses, 1/2" thick		750	.011		.30	.49		.79	1.08
2400	5/8" thick		650	.012		.37	.56		.93	1.28
2500	3/4" thick		600	.013		.44	.61		1.05	1.42

01 58 Project Identification

01 58 13 – Temporary Project Signage

01 58 13.50 Signs

		Crew	Daily Output	Labor-Hours	Unit	Material	2014 Bare Costs Labor	Equipment	Total	Total Incl O&P
0010	**SIGNS**									
0020	High intensity reflectorized, no posts, buy				S.F.	31.50			31.50	34.50

01 74 Cleaning and Waste Management

01 74 13 – Progress Cleaning

01 74 13.20 Cleaning Up

		Crew	Daily Output	Labor-Hours	Unit	Material	2014 Bare Costs Labor	Equipment	Total	Total Incl O&P
0010	**CLEANING UP**									
0020	After job completion, allow, minimum				Job				.30%	.30%
0040	Maximum				"				1%	1%
0050	Cleanup of floor area, continuous, per day, during const.	A-5	24	.750	M.S.F.	.81	27.50	2.81	31.12	46.50
0100	Final by GC at end of job	"	11.50	1.565	"	.86	57.50	5.85	64.21	96

01 91 Commissioning

01 91 13 – General Commissioning Requirements

01 91 13.50 Building Commissioning

		Crew	Daily Output	Labor-Hours	Unit	Material	2014 Bare Costs Labor	Equipment	Total	Total Incl O&P
0010	**BUILDING COMMISSIONING**									
0100	Basic building commissioning, minimum				%				.25%	.25%
0150	Maximum								.50%	.50%
0200	Enhanced building commissioning, minimum								.50%	.50%
0250	Maximum								1%	1%

01 93 Facility Maintenance

01 93 13 – Facility Maintenance Procedures

01 93 13.16 Electrical Facilities Maintenance

		Crew	Daily Output	Labor-Hours	Unit	Material	2014 Bare Costs Labor	Equipment	Total	Total Incl O&P
0010	**ELECTRICAL FACILITIES MAINTENANCE**									
0700	Cathodic protection systems									
0720	Check and adjust reading on rectifier	1 Elec	20	.400	Ea.		21.50		21.50	32
0730	Check pipe to soil potential		20	.400			21.50		21.50	32
0740	Replace lead connection		4	2			107		107	160
0800	Control device, install		5.70	1.404			75		75	112
0810	Disassemble, clean and reinstall		7	1.143			61		61	91.50
0820	Replace		10.70	.748			40		40	59.50
0830	Trouble shoot		10	.800			42.50		42.50	64
0900	Demolition, for electrical demolition see Section 26 05 05.10									
1000	Distribution systems and equipment install or repair a breaker									
1010	In power panels up to 200 amps	1 Elec	7	1.143	Ea.		61		61	91.50
1020	Over 200 amps		2	4			213		213	320

01 93 13 – Facility Maintenance Procedures

01 93 13.16 Electrical Facilities Maintenance	Crew	Daily Output	Labor-Hours	Unit	Material	2014 Bare Costs Labor	Equipment	Total	Total Incl O&P
1030 Reset breaker or replace fuse	1 Elec	20	.400	Ea.		21.50		21.50	32
1100 Megger test MCC (each stack)		4	2			107		107	160
1110 MCC vacuum and clean (each stack)		5.30	1.509			80.50		80.50	121
2500 Remove and replace or maint., road fixture & lamp		3	2.667		915	142		1,057	1,225
2510 Fluorescent fixture		7	1.143		58	61		119	156
2515 Relamp (fluor.) facility area each tube, spot	1 Clab	24	.333		3.99	12.20		16.19	23
2516 Group		100	.080		3.99	2.93		6.92	8.90
2518 Fluorescent fixture, clean (area)		44	.182		.11	6.65		6.76	10.40
2520 Incandescent fixture	1 Elec	11	.727		61	39		100	125
2530 Lamp (incandescent or fluorescent)	1 Clab	60	.133		3.99	4.89		8.88	11.95
2535 Replace cord in socket lamp	1 Elec	13	.615		2.45	33		35.45	51.50
2540 Ballast electronic type for two tubes		8	1		37	53.50		90.50	121
2541 Starter		30	.267		1.10	14.25		15.35	22.50
2545 Replace other lighting parts		11	.727		15.90	39		54.90	75.50
2550 Switch		11	.727		8.80	39		47.80	67.50
2555 Receptacle		11	.727		8.35	39		47.35	67
2560 Floodlight		4	2		270	107		377	455
2570 Christmas lighting, indoor, per string	1 Clab	16	.500			18.35		18.35	28.50
2580 Outdoor		13	.615			22.50		22.50	35
2590 Test battery operated emergency lights		40	.200			7.35		7.35	11.30
2600 Repair and replace component in communication system	1 Elec	6	1.333		54	71		125	166
2700 Repair misc. appliances (incl. clocks, vent fan, blower, etc.)	"	6	1.333			71		71	106
2710 Reset clocks & timers	1 Clab	50	.160			5.85		5.85	9.05
2720 Adjust time delay relays	1 Elec	16	.500			26.50		26.50	40
2730 Test specific gravity of lead-acid batteries	1 Clab	80	.100			3.67		3.67	5.65
3000 Motors and generators									
3020 Disassemble, clean and reinstall motor, up to 1/4 HP	1 Elec	4	2	Ea.		107		107	160
3030 Up to 3/4 HP		3	2.667			142		142	213
3040 Up to 10 HP		2	4			213		213	320
3050 Replace part, up to 1/4 HP		6	1.333			71		71	106
3060 Up to 3/4 HP		4	2			107		107	160
3070 Up to 10 HP		3	2.667			142		142	213
3080 Megger test motor windings		5.33	1.501			80		80	120
3082 Motor vibration check		16	.500			26.50		26.50	40
3084 Oil motor bearings		25	.320			17.05		17.05	25.50
3086 Run test emergency generator for 30 minutes		11	.727			39		39	58
3090 Rewind motor, up to 1/4 HP		3	2.667			142		142	213
3100 Up to 3/4 HP		2	4			213		213	320
3110 Up to 10 HP		1.50	5.333			285		285	425
3150 Generator, repair or replace part		4	2			107		107	160
3160 Repair DC generator		2	4			213		213	320
4000 Stub pole, install or remove		3	2.667			142		142	213
4500 Transformer maintenance up to 15 kVA		2.70	2.963			158		158	237

Division Notes

	CREW	DAILY OUTPUT	LABOR-HOURS	UNIT	BARE COSTS				TOTAL INCL O&P
					MAT.	LABOR	EQUIP.	TOTAL	

Estimating Tips

02 30 00 Subsurface Investigation

In preparing estimates on structures involving earthwork or foundations, all information concerning soil characteristics should be obtained. Look particularly for hazardous waste, evidence of prior dumping of debris, and previous stream beds.

02 40 00 Demolition and Structure Moving

The costs shown for selective demolition do not include rubbish handling or disposal. These items should be estimated separately using RSMeans data or other sources.

- Historic preservation often requires that the contractor remove materials from the existing structure, rehab them, and replace them. The estimator must be aware of any related measures and precautions that must be taken when doing selective demolition and cutting and patching. Requirements may include special handling and storage, as well as security.

- In addition to Subdivision 02 41 00, you can find selective demolition items in each division. Example: Roofing demolition is in Division 7.

02 40 00 Building Deconstruction

This section provides costs for the careful dismantling and recycling of most of low-rise building materials.

02 50 00 Containment of Hazardous Waste

This section addresses on-site hazardous waste disposal costs.

02 80 00 Hazardous Material Disposal/ Remediation

This subdivision includes information on hazardous waste handling, asbestos remediation, lead remediation, and mold remediation. See reference R028213-20 and R028319-60 for further guidance in using these unit price lines.

02 90 00 Monitoring Chemical Sampling, Testing Analysis

This section provides costs for on-site sampling and testing hazardous waste.

Reference Numbers

Reference numbers are shown in shaded boxes at the beginning of some major classifications. These numbers refer to related items in the Reference Section. The reference information may be an estimating procedure, an alternate pricing method, or technical information.

Note: Not all subdivisions listed here necessarily appear in this publication.

02 41 Demolition

02 41 13 – Selective Site Demolition

02 41 13.17 Demolish, Remove Pavement and Curb

		Crew	Daily Output	Labor-Hours	Unit	Material	2014 Bare Costs Labor	Equipment	Total	Total Incl O&P
0010	**DEMOLISH, REMOVE PAVEMENT AND CURB** R024119-10									
5010	Pavement removal, bituminous roads, up to 3" thick	B-38	690	.058	S.Y.		2.41	1.86	4.27	5.75
5050	4" to 6" thick		420	.095			3.96	3.05	7.01	9.40
5100	Bituminous driveways		640	.063			2.60	2	4.60	6.20
5200	Concrete to 6" thick, hydraulic hammer, mesh reinforced		255	.157			6.50	5	11.50	15.50
5300	Rod reinforced		200	.200			8.30	6.40	14.70	19.80
5400	Concrete, 7" to 24" thick, plain		33	1.212	C.Y.		50.50	39	89.50	120
5500	Reinforced		24	1.667	"		69.50	53.50	123	165
5600	With hand held air equipment, bituminous, to 6" thick	B-39	1900	.025	S.F.		.98	.12	1.10	1.64
5700	Concrete to 6" thick, no reinforcing		1600	.030			1.16	.15	1.31	1.95
5800	Mesh reinforced		1400	.034			1.33	.17	1.50	2.22
5900	Rod reinforced		765	.063			2.43	.31	2.74	4.08

02 41 13.33 Minor Site Demolition

		Crew	Daily Output	Labor-Hours	Unit	Material	2014 Bare Costs Labor	Equipment	Total	Total Incl O&P
0010	**MINOR SITE DEMOLITION** R024119-10									
0015	No hauling, abandon catch basin or manhole	B-6	7	3.429	Ea.		138	52	190	269
0020	Remove existing catch basin or manhole, masonry		4	6			241	91.50	332.50	470
0030	Catch basin or manhole frames and covers, stored		13	1.846			74	28	102	145
0040	Remove and reset		7	3.429			138	52	190	269
4000	Sidewalk removal, bituminous, 2" thick		350	.069	S.Y.		2.75	1.04	3.79	5.35
4010	2-1/2" thick		325	.074			2.96	1.12	4.08	5.80
4050	Brick, set in mortar		185	.130			5.20	1.97	7.17	10.15
4100	Concrete, plain, 4"		160	.150			6	2.28	8.28	11.70
4110	Plain, 5"		140	.171			6.90	2.61	9.51	13.40
4120	Plain, 6"		120	.200			8	3.04	11.04	15.65
4200	Mesh reinforced, concrete, 4"		150	.160			6.40	2.44	8.84	12.55
4210	5" thick		131	.183			7.35	2.79	10.14	14.30
4220	6" thick		112	.214			8.60	3.26	11.86	16.75

02 41 19 – Selective Demolition

02 41 19.13 Selective Building Demolition

		Crew	Daily Output	Labor-Hours	Unit	Material	2014 Bare Costs Labor	Equipment	Total	Total Incl O&P
0010	**SELECTIVE BUILDING DEMOLITION**									
0020	Costs related to selective demolition of specific building components									
0025	are included under Common Work Results (XX 05)									
0030	in the component's appropriate division.									

02 41 19.16 Selective Demolition, Cutout

		Crew	Daily Output	Labor-Hours	Unit	Material	2014 Bare Costs Labor	Equipment	Total	Total Incl O&P
0010	**SELECTIVE DEMOLITION, CUTOUT** R024119-10									
0020	Concrete, elev. slab, light reinforcement, under 6 C.F.	B-9	65	.615	C.F.		23	3.59	26.59	39
0050	Light reinforcing, over 6 C.F.	"	75	.533	"		19.75	3.11	22.86	34
6000	Walls, interior, not including re-framing,									
6010	openings to 5 S.F.									
6100	Drywall to 5/8" thick	1 Clab	24	.333	Ea.		12.20		12.20	18.85
6200	Paneling to 3/4" thick		20	.400			14.65		14.65	22.50
6300	Plaster, on gypsum lath		20	.400			14.65		14.65	22.50
6340	On wire lath		14	.571			21		21	32.50

02 41 19.19 Selective Facility Services Demolition

		Crew	Daily Output	Labor-Hours	Unit	Material	2014 Bare Costs Labor	Equipment	Total	Total Incl O&P
0010	**SELECTIVE FACILITY SERVICES DEMOLITION**, Rubbish Handling R024119-10									
0020	The following are to be added to the demolition prices									
0600	Dumpster, weekly rental, 1 dump/week, 6 C.Y. capacity (2 Tons)				Week	415			415	455
0700	10 C.Y. capacity (3 Tons)					480			480	530
0725	20 C.Y. capacity (5 Tons) R024119-20					565			565	625
0800	30 C.Y. capacity (7 Tons)					730			730	800
0840	40 C.Y. capacity (10 Tons)					775			775	850

02 41 19 – Selective Demolition

02 41 19.19 Selective Facility Services Demolition

		Crew	Daily Output	Labor-Hours	Unit	Material	2014 Bare Costs Labor	Equipment	Total	Total Incl O&P
2000	Load, haul, dump and return, 50' haul, hand carried	2 Clab	24	.667	C.Y.		24.50		24.50	37.50
2005	Wheeled		37	.432			15.85		15.85	24.50
2040	51' to 100' haul, hand carried		16.50	.970			35.50		35.50	55
2045	Wheeled		25	.640			23.50		23.50	36
2080	Over 100' haul, add per 100 L.F., hand carried		35.50	.451			16.50		16.50	25.50
2085	Wheeled		54	.296			10.85		10.85	16.75
2120	In elevators, per 10 floors, add		140	.114			4.19		4.19	6.45
2130	Load, haul, dump and return, up to 50' haul, incl. up to 5 rsr stair, hand		23	.696			25.50		25.50	39.50
2135	Wheeled		35	.457			16.75		16.75	26
2140	6 – 10 riser stairs, hand carried		22	.727			26.50		26.50	41
2145	Wheeled		34	.471			17.25		17.25	26.50
2150	11 – 20 riser stairs, hand carried		20	.800			29.50		29.50	45
2155	Wheeled		31	.516			18.90		18.90	29
2160	21 – 40 riser stairs, hand carried		16	1			36.50		36.50	56.50
2165	Wheeled		24	.667			24.50		24.50	37.50
2170	100' haul, incl. 5 riser stair, hand carried		15	1.067			39		39	60.50
2175	Wheeled		23	.696			25.50		25.50	39.50
2180	6 – 10 riser stair, hand carried		14	1.143			42		42	64.50
2185	Wheeled		21	.762			28		28	43
2190	11 – 20 riser stair, hand carried		12	1.333			49		49	75.50
2195	Wheeled		18	.889			32.50		32.50	50.50
2200	21 – 40 riser stair, hand carried		8	2			73.50		73.50	113
2205	Wheeled		12	1.333			49		49	75.50
2210	Over 100' haul, add per 100 L.F., hand carried		35.50	.451			16.50		16.50	25.50
2215	Wheeled		54	.296			10.85		10.85	16.75
2220	For each additional flight of stairs, up to 5 risers, add		550	.029	Flight		1.07		1.07	1.65
2225	6 – 10 risers, add		275	.058			2.13		2.13	3.29
2230	11 – 20 risers, add		138	.116			4.25		4.25	6.55
2235	21 – 40 risers, add		69	.232			8.50		8.50	13.10
3000	Loading & trucking, including 2 mile haul, chute loaded	B-16	45	.711	C.Y.		26.50	15.40	41.90	58
3040	Hand loading truck, 50' haul	"	48	.667			25	14.40	39.40	54.50
3080	Machine loading truck	B-17	120	.267			10.55	6.50	17.05	23.50
5000	Haul, per mile, up to 8 C.Y. truck	B-34B	1165	.007			.26	.59	.85	1.04
5100	Over 8 C.Y. truck	"	1550	.005			.19	.45	.64	.79

02 41 19.20 Selective Demolition, Dump Charges

		Crew	Daily Output	Labor-Hours	Unit	Material	2014 Bare Costs Labor	Equipment	Total	Total Incl O&P
0010	**SELECTIVE DEMOLITION, DUMP CHARGES** R024119-10									
0020	Dump charges, typical urban city, tipping fees only									
0100	Building construction materials				Ton	74			74	81
0200	Trees, brush, lumber					63			63	69.50
0300	Rubbish only					63			63	69.50
0500	Reclamation station, usual charge					74			74	81

02 41 19.27 Selective Demolition, Torch Cutting

		Crew	Daily Output	Labor-Hours	Unit	Material	2014 Bare Costs Labor	Equipment	Total	Total Incl O&P
0010	**SELECTIVE DEMOLITION, TORCH CUTTING** R024119-10									
0020	Steel, 1" thick plate	E-25	333	.024	L.F.	.76	1.28	.03	2.07	3.12
0040	1" diameter bar	"	600	.013	Ea.	.13	.71	.02	.86	1.41
1000	Oxygen lance cutting, reinforced concrete walls									
1040	12" to 16" thick walls	1 Clab	10	.800	L.F.		29.50		29.50	45
1080	24" thick walls	"	6	1.333	"		49		49	75.50

02 82 Asbestos Remediation

02 82 13 – Asbestos Abatement

02 82 13.47 Asbestos Waste Pkg., Handling, and Disp.	Crew	Daily Output	Labor-Hours	Unit	Material	2014 Bare Costs Labor	Equipment	Total	Total Incl O&P
0010 **ASBESTOS WASTE PACKAGING, HANDLING, AND DISPOSAL**									
0100 Collect and bag bulk material, 3 C.F. bags, by hand	A-9	400	.160	Ea.	1.65	8.20		9.85	14.55
0200 Large production vacuum loader	A-12	880	.073		.84	3.72	.88	5.44	7.70
1000 Double bag and decontaminate	A-9	960	.067		1.65	3.41		5.06	7.10
2000 Containerize bagged material in drums, per 3 C.F. drum	"	800	.080		17.85	4.10		21.95	26
3000 Cart bags 50' to dumpster	2 Asbe	400	.040			2.05		2.05	3.18
5000 Disposal charges, not including haul, minimum				C.Y.				61	67
5020 Maximum				"				355	395
9000 For type B (supplied air) respirator equipment, add				%				10%	10%

Estimating Tips

General

- Carefully check all the plans and specifications. Concrete often appears on drawings other than structural drawings, including mechanical and electrical drawings for equipment pads. The cost of cutting and patching is often difficult to estimate. See Subdivision 03 81 for Concrete Cutting, Subdivision 02 41 19.16 for Cutout Demolition, Subdivision 03 05 05.10 for Concrete Demolition, and Subdivision 02 41 19.19 for Rubbish Handling (handling, loading and hauling of debris).

- Always obtain concrete prices from suppliers near the job site. A volume discount can often be negotiated, depending upon competition in the area. Remember to add for waste, particularly for slabs and footings on grade.

03 10 00 Concrete Forming and Accessories

- A primary cost for concrete construction is forming. Most jobs today are constructed with prefabricated forms. The selection of the forms best suited for the job and the total square feet of forms required for efficient concrete forming and placing are key elements in estimating concrete construction. Enough forms must be available for erection to make efficient use of the concrete placing equipment and crew.

- Concrete accessories for forming and placing depend upon the systems used. Study the plans and specifications to ensure that all special accessory requirements have been included in the cost estimate, such as anchor bolts, inserts, and hangers.

- Included within costs for forms-in-place are all necessary bracing and shoring.

03 20 00 Concrete Reinforcing

- Ascertain that the reinforcing steel supplier has included all accessories, cutting, bending, and an allowance for lapping, splicing, and waste. A good rule of thumb is 10% for lapping, splicing, and waste. Also, 10% waste should be allowed for welded wire fabric.

- The unit price items in the subdivisions for Reinforcing In Place, Glass Fiber Reinforcing, and Welded Wire Fabric include the labor to install accessories such as beam and slab bolsters, high chairs, and bar ties and tie wire. The material cost for these accessories is not included; they may be obtained from the Accessories Division.

03 30 00 Cast-In-Place Concrete

- When estimating structural concrete, pay particular attention to requirements for concrete additives, curing methods, and surface treatments. Special consideration for climate, hot or cold, must be included in your estimate. Be sure to include requirements for concrete placing equipment, and concrete finishing.

- For accurate concrete estimating, the estimator must consider each of the following major components individually: forms, reinforcing steel, ready-mix concrete, placement of the concrete, and finishing of the top surface. For faster estimating, Subdivision 03 30 53.40 for Concrete-In-Place can be used; here, various items of concrete work are presented that include the costs of all five major components (unless specifically stated otherwise).

03 40 00 Precast Concrete
03 50 00 Cast Decks and Underlayment

- The cost of hauling precast concrete structural members is often an important factor. For this reason, it is important to get a quote from the nearest supplier. It may become economically feasible to set up precasting beds on the site if the hauling costs are prohibitive.

Reference Numbers

Reference numbers are shown in shaded boxes at the beginning of some major classifications. These numbers refer to related items in the Reference Section. The reference information may be an estimating procedure, an alternate pricing method, or technical information.

Note: Not all subdivisions listed here necessarily appear in this publication.

Division 3 - Concrete

03 01 Maintenance of Concrete

03 01 30 – Maintenance of Cast-In-Place Concrete

03 01 30.62 Concrete Patching		Crew	Daily Output	Labor-Hours	Unit	Material	2014 Bare Costs Labor	Equipment	Total	Total Incl O&P
0010	**CONCRETE PATCHING**									
0100	Floors, 1/4" thick, small areas, regular grout	1 Cefi	170	.047	S.F.	1.42	2.07		3.49	4.62
0150	Epoxy grout	"	100	.080	"	7.80	3.52		11.32	13.75
2000	Walls, including chipping, cleaning and epoxy grout									
2100	1/4" deep	1 Cefi	65	.123	S.F.	7.20	5.40		12.60	15.90
2150	1/2" deep	↓	50	.160		14.40	7.05		21.45	26.50
2200	3/4" deep	↓	40	.200	↓	21.50	8.80		30.30	37

03 11 Concrete Forming

03 11 13 – Structural Cast-In-Place Concrete Forming

03 11 13.40 Forms In Place, Equipment Foundations

		Crew	Daily Output	Labor-Hours	Unit	Material	Labor	Equipment	Total	Total Incl O&P
0010	**FORMS IN PLACE, EQUIPMENT FOUNDATIONS**									
0020	1 use	C-2	160	.300	SFCA	3.09	13.40		16.49	24
0050	2 use		190	.253		1.70	11.30		13	19.25
0100	3 use		200	.240		1.24	10.70		11.94	17.90
0150	4 use	↓	205	.234	↓	1.01	10.45		11.46	17.25

03 11 13.45 Forms In Place, Footings

		Crew	Daily Output	Labor-Hours	Unit	Material	Labor	Equipment	Total	Total Incl O&P
0010	**FORMS IN PLACE, FOOTINGS**									
0020	Continuous wall, plywood, 1 use	C-1	375	.085	SFCA	6.25	3.72		9.97	12.60
0050	2 use		440	.073		3.42	3.17		6.59	8.65
0100	3 use		470	.068		2.49	2.97		5.46	7.30
0150	4 use		485	.066		2.03	2.87		4.90	6.65
5000	Spread footings, job-built lumber, 1 use		305	.105		2	4.57		6.57	9.25
5050	2 use		371	.086		1.11	3.76		4.87	7
5100	3 use		401	.080		.80	3.48		4.28	6.25
5150	4 use	↓	414	.077	↓	.65	3.37		4.02	5.90

03 11 13.65 Forms In Place, Slab On Grade

		Crew	Daily Output	Labor-Hours	Unit	Material	Labor	Equipment	Total	Total Incl O&P
0010	**FORMS IN PLACE, SLAB ON GRADE**									
3000	Edge forms, wood, 4 use, on grade, to 6" high	C-1	600	.053	L.F.	.31	2.32		2.63	3.92
6000	Trench forms in floor, wood, 1 use		160	.200	SFCA	1.68	8.70		10.38	15.30
6050	2 use		175	.183		.92	7.95		8.87	13.30
6100	3 use		180	.178		.67	7.75		8.42	12.70
6150	4 use	↓	185	.173	↓	.55	7.55		8.10	12.20

03 15 Concrete Accessories

03 15 05 – Concrete Forming Accessories

03 15 05.75 Sleeves and Chases

		Crew	Daily Output	Labor-Hours	Unit	Material	Labor	Equipment	Total	Total Incl O&P
0010	**SLEEVES AND CHASES**									
0100	Plastic, 1 use, 12" long, 2" diameter	1 Carp	100	.080	Ea.	2.44	3.67		6.11	8.35
0150	4" diameter		90	.089		6.65	4.08		10.73	13.60
0200	6" diameter		75	.107		14.25	4.89		19.14	23.50
0250	12" diameter		60	.133		46.50	6.10		52.60	60.50
5000	Sheet metal, 2" diameter	G	100	.080		1.36	3.67		5.03	7.15
5100	4" diameter	G	90	.089		1.70	4.08		5.78	8.15
5150	6" diameter	G	75	.107		1.90	4.89		6.79	9.65
5200	12" diameter	G	60	.133		3.63	6.10		9.73	13.45
6000	Steel pipe, 2" diameter	G	100	.080		4.65	3.67		8.32	10.75
6100	4" diameter	G	90	.089		16.70	4.08		20.78	24.50
6150	6" diameter	G	75	.107	↓	36.50	4.89		41.39	47.50

03 15 Concrete Accessories

03 15 05 – Concrete Forming Accessories

03 15 05.75 Sleeves and Chases		Crew	Daily Output	Labor-Hours	Unit	Material	2014 Bare Costs Labor	Equipment	Total	Total Incl O&P
6200	12" diameter	G	1 Carp	60	.133	Ea.	86.50	6.10		92.60	104

03 15 16 – Concrete Construction Joints

03 15 16.20 Control Joints, Saw Cut

		Crew	Daily Output	Labor-Hours	Unit	Material	Labor	Equipment	Total	Total Incl O&P
0010	**CONTROL JOINTS, SAW CUT**									
0100	Sawcut control joints in green concrete									
0120	1" depth	C-27	2000	.008	L.F.	.04	.35	.09	.48	.66
0140	1-1/2" depth		1800	.009		.06	.39	.10	.55	.75
0160	2" depth		1600	.010		.07	.44	.11	.62	.85
0180	Sawcut joint reservoir in cured concrete									
0182	3/8" wide x 3/4" deep, with single saw blade	C-27	1000	.016	L.F.	.06	.70	.18	.94	1.29
0184	1/2" wide x 1" deep, with double saw blades		900	.018		.11	.78	.19	1.08	1.49
0186	3/4" wide x 1-1/2" deep, with double saw blades		800	.020		.22	.88	.22	1.32	1.79
0190	Water blast joint to wash away laitance, 2 passes	C-29	2500	.003			.12	.03	.15	.21
0200	Air blast joint to blow out debris and air dry, 2 passes	C-28	2000	.004			.18	.01	.19	.27
0300	For backer rod, see Section 07 91 23.10									
0342	For joint sealant, see Section 07 92 13.20									

03 15 19 – Cast-In Concrete Anchors

03 15 19.10 Anchor Bolts

			Crew	Daily Output	Labor-Hours	Unit	Material	Labor	Equipment	Total	Total Incl O&P
0010	**ANCHOR BOLTS**										
0015	Made from recycled materials										
0025	Single bolts installed in fresh concrete, no templates										
0030	Hooked w/nut and washer, 1/2" diameter, 8" long	G	1 Carp	132	.061	Ea.	1.49	2.78		4.27	5.95
0040	12" long	G		131	.061		1.66	2.80		4.46	6.15
0050	5/8" diameter, 8" long	G		129	.062		3.25	2.84		6.09	7.95
0060	12" long	G		127	.063		4	2.89		6.89	8.85
0070	3/4" diameter, 8" long	G		127	.063		4	2.89		6.89	8.85
0080	12" long	G		125	.064		5	2.93		7.93	10.05

03 15 19.45 Machinery Anchors

			Crew	Daily Output	Labor-Hours	Unit	Material	Labor	Equipment	Total	Total Incl O&P
0010	**MACHINERY ANCHORS**, heavy duty, incl. sleeve, floating base nut,										
0020	lower stud & coupling nut, fiber plug, connecting stud, washer & nut.										
0030	For flush mounted embedment in poured concrete heavy equip. pads.										
0200	Stud & bolt, 1/2" diameter	G	E-16	40	.400	Ea.	72.50	21	3.55	97.05	121
0300	5/8" diameter	G		35	.457		80.50	24	4.06	108.56	135
0500	3/4" diameter	G		30	.533		93	28	4.74	125.74	156
0600	7/8" diameter	G		25	.640		101	33.50	5.70	140.20	176
0800	1" diameter	G		20	.800		117	41.50	7.10	165.60	210
0900	1-1/4" diameter	G		15	1.067		141	55.50	9.45	205.95	263

03 21 Reinforcement Bars

03 21 11 – Plain Steel Reinforcement Bars

03 21 11.60 Reinforcing In Place

			Crew	Daily Output	Labor-Hours	Unit	Material	Labor	Equipment	Total	Total Incl O&P
0010	**REINFORCING IN PLACE**, 50-60 ton lots, A615 Grade 60										
0020	Includes labor, but not material cost, to install accessories										
0030	Made from recycled materials										
0502	Footings, #4 to #7	G	4 Rodm	4200	.008	Lb.	.50	.39		.89	1.16
0552	#8 to #18	G		7200	.004		.50	.22		.72	.90
0602	Slab on grade, #3 to #7	G		4200	.008		.50	.39		.89	1.16
0900	For other than 50 – 60 ton lots										
1000	Under 10 ton job, #3 to #7, add						25%	10%			
1010	#8 to #18, add						20%	10%			

03 21 Reinforcement Bars

03 21 11 – Plain Steel Reinforcement Bars

03 21 11.60 Reinforcing In Place	Crew	Daily Output	Labor-Hours	Unit	Material	2014 Bare Costs Labor	Equipment	Total	Total Incl O&P
1050	10 – 50 ton job, #3 to #7, add					10%				
1060	#8 to #18, add					5%				
1100	60 – 100 ton job, #3 to #7, deduct					5%				
1110	#8 to #18, deduct					10%				
1150	Over 100 ton job, #3 to #7, deduct					10%				
1160	#8 to #18, deduct					15%				

03 22 Fabric and Grid Reinforcing

03 22 11 – Plain Welded Wire Fabric Reinforcing

03 22 11.10 Plain Welded Wire Fabric

		Crew	Daily Output	Labor-Hours	Unit	Material	Labor	Equipment	Total	Total Incl O&P
0010	**PLAIN WELDED WIRE FABRIC** ASTM A185									
0020	Includes labor, but not material cost, to install accessories									
0050	Sheets									
0100	6 x 6 - W1.4 x W1.4 (10 x 10) 21 lb. per C.S.F. **G**	2 Rodm	35	.457	C.S.F.	14.50	23		37.50	52.50

03 22 13 – Galvanized Welded Wire Fabric Reinforcing

03 22 13.10 Galvanized Welded Wire Fabric

		Crew	Daily Output	Labor-Hours	Unit	Material	Labor	Equipment	Total	Total Incl O&P
0010	**GALVANIZED WELDED WIRE FABRIC**									
0100	Add to plain welded wire pricing for galvanized welded wire				Lb.	.23			.23	.25

03 22 16 – Epoxy-Coated Welded Wire Fabric Reinforcing

03 22 16.10 Epoxy-Coated Welded Wire Fabric

		Crew	Daily Output	Labor-Hours	Unit	Material	Labor	Equipment	Total	Total Incl O&P
0010	**EPOXY-COATED WELDED WIRE FABRIC**									
0100	Add to plain welded wire pricing for epoxy-coated welded wire				Lb.	.21			.21	.23

03 30 Cast-In-Place Concrete

03 30 53 – Miscellaneous Cast-In-Place Concrete

03 30 53.40 Concrete In Place

		Crew	Daily Output	Labor-Hours	Unit	Material	2014 Bare Costs Labor	Equipment	Total	Total Incl O&P
0010	**CONCRETE IN PLACE**									
0020	Including forms (4 uses), Grade 60 rebar, concrete (Portland cement									
0050	Type I), placement and finishing unless otherwise indicated									
3540	Equipment pad (3000 psi), 3' x 3' x 6" thick	C-14H	45	1.067	Ea.	44.50	48	.74	93.24	124
3550	4' x 4' x 6" thick		30	1.600		66.50	72	1.10	139.60	185
3560	5' x 5' x 8" thick		18	2.667		119	120	1.84	240.84	320
3570	6' x 6' x 8" thick		14	3.429		160	155	2.37	317.37	415
3580	8' x 8' x 10" thick		8	6		335	271	4.14	610.14	790
3590	10' x 10' x 12" thick		5	9.600		575	435	6.60	1,016.60	1,300
3800	Footings (3000 psi), spread under 1 C.Y.	C-14C	28	4	C.Y.	164	176	1.16	341.16	455
3825	1 C.Y. to 5 C.Y.		43	2.605		198	114	.76	312.76	395
3850	Over 5 C.Y.		75	1.493		180	65.50	.43	245.93	299
3900	Footings, strip (3000 psi), 18" x 9", unreinforced	C-14L	40	2.400		120	103	.82	223.82	291
3920	18" x 9", reinforced	C-14C	35	3.200		144	141	.93	285.93	375
3925	20" x 10", unreinforced	C-14L	45	2.133		117	91.50	.73	209.23	269
3930	20" x 10", reinforced	C-14C	40	2.800		136	123	.81	259.81	340
3935	24" x 12", unreinforced	C-14L	55	1.745		116	74.50	.59	191.09	243
3940	24" x 12", reinforced	C-14C	48	2.333		136	102	.68	238.68	310
3945	36" x 12", unreinforced	C-14L	70	1.371		112	58.50	.47	170.97	214
3950	36" x 12", reinforced	C-14C	60	1.867		130	82	.54	212.54	271
4000	Foundation mat (3000 psi), under 10 C.Y.		38.67	2.896		204	127	.84	331.84	420
4050	Over 20 C.Y.		56.40	1.986		178	87	.58	265.58	330

03 30 Cast-In-Place Concrete

03 30 53 – Miscellaneous Cast-In-Place Concrete

03 30 53.40 Concrete In Place

		Crew	Daily Output	Labor-Hours	Unit	Material	2014 Bare Costs Labor	2014 Bare Costs Equipment	Total	Total Incl O&P
4650	Slab on grade (3500 psi), not including finish, 4" thick	C-14E	60.75	1.449	C.Y.	120	65.50	.55	186.05	234
4700	6" thick	"	92	.957	↓	115	43	.36	158.36	194
4701	Thickened slab edge (3500 psi), for slab on grade poured									
4702	monolithically with slab; depth is in addition to slab thickness;									
4703	formed vertical outside edge, earthen bottom and inside slope									
4705	8" deep x 8" wide bottom, unreinforced	C-14L	2190	.044	L.F.	3.33	1.88	.01	5.22	6.55
4710	8" x 8", reinforced	C-14C	1670	.067		5.70	2.95	.02	8.67	10.80
4715	12" deep x 12" wide bottom, unreinforced	C-14L	1800	.053		6.75	2.28	.02	9.05	11
4720	12" x 12", reinforced	C-14C	1310	.086		11.05	3.76	.02	14.83	18
4725	16" deep x 16" wide bottom, unreinforced	C-14L	1440	.067		11.40	2.85	.02	14.27	16.95
4730	16" x 16", reinforced	C-14C	1120	.100		16.50	4.39	.03	20.92	25
4735	20" deep x 20" wide bottom, unreinforced	C-14L	1150	.083		17.25	3.57	.03	20.85	24.50
4740	20" x 20", reinforced	C-14C	920	.122		24	5.35	.04	29.39	34.50
4745	24" deep x 24" wide bottom, unreinforced	C-14L	930	.103		24.50	4.42	.04	28.96	34
4750	24" x 24", reinforced	C-14C	740	.151	↓	33	6.65	.04	39.69	46.50

03 31 Structural Concrete

03 31 13 – Heavyweight Structural Concrete

03 31 13.35 Heavyweight Concrete, Ready Mix

		Crew	Daily Output	Labor-Hours	Unit	Material	2014 Bare Costs Labor	2014 Bare Costs Equipment	Total	Total Incl O&P
0010	**HEAVYWEIGHT CONCRETE, READY MIX,** delivered									
0012	Includes local aggregate, sand, Portland cement (Type I) and water									
0015	Excludes all additives and treatments									
0020	2000 psi				C.Y.	93			93	102
0100	2500 psi					95.50			95.50	105
0150	3000 psi					99			99	109
0200	3500 psi					101			101	111
0300	4000 psi					104			104	114
1000	For high early strength (Portland cement Type III), add					10%				
1300	For winter concrete (hot water), add					4.50			4.50	4.95
1400	For hot weather concrete (ice), add					7.65			7.65	8.40
1410	For mid-range water reducer, add					3.69			3.69	4.06
1420	For high-range water reducer/superplasticizer, add					5.70			5.70	6.30
1430	For retarder, add					3.06			3.06	3.37
1440	For non-Chloride accelerator, add					5.65			5.65	6.20
1450	For Chloride accelerator, per 1%, add					3.32			3.32	3.65
1460	For fiber reinforcing, synthetic (1 lb./C.Y.), add					6.95			6.95	7.60
1500	For Saturday delivery, add				↓	4.93			4.93	5.40
1510	For truck holding/waiting time past 1st hour per load, add				Hr.	97			97	107
1520	For short load (less than 4 C.Y.), add per load				Ea.	77.50			77.50	85
2000	For all lightweight aggregate, add				C.Y.	45%				

03 31 13.70 Placing Concrete

		Crew	Daily Output	Labor-Hours	Unit	Material	2014 Bare Costs Labor	2014 Bare Costs Equipment	Total	Total Incl O&P
0010	**PLACING CONCRETE**									
0020	Includes labor and equipment to place, level (strike off) and consolidate									
1900	Footings, continuous, shallow, direct chute	C-6	120	.400	C.Y.		15.30	.55	15.85	24
1950	Pumped	C-20	150	.427			16.80	5.20	22	31
2000	With crane and bucket	C-7	90	.800			32	13.60	45.60	63.50
2100	Footings, continuous, deep, direct chute	C-6	140	.343			13.10	.47	13.57	20.50
2150	Pumped	C-20	160	.400			15.75	4.85	20.60	29.50
2200	With crane and bucket	C-7	110	.655			26	11.15	37.15	52.50
2400	Footings, spread, under 1 C.Y., direct chute	C-6	55	.873			33.50	1.20	34.70	52.50
2450	Pumped	C-20	65	.985	↓		39	11.95	50.95	72.50

03 31 Structural Concrete

03 31 13 – Heavyweight Structural Concrete

03 31 13.70 Placing Concrete	Crew	Daily Output	Labor-Hours	Unit	Material	2014 Bare Costs Labor	Equipment	Total	Total Incl O&P
2500 With crane and bucket	C-7	45	1.600	C.Y.		63.50	27.50	91	128
2600 Over 5 C.Y., direct chute	C-6	120	.400			15.30	.55	15.85	24
2650 Pumped	C-20	150	.427			16.80	5.20	22	31
2700 With crane and bucket	C-7	100	.720			28.50	12.25	40.75	57.50
2900 Foundation mats, over 20 C.Y., direct chute	C-6	350	.137			5.25	.19	5.44	8.20
2950 Pumped	C-20	400	.160			6.30	1.94	8.24	11.80
3000 With crane and bucket	C-7	300	.240			9.55	4.09	13.64	19.10

03 35 Concrete Finishing

03 35 13 – High-Tolerance Concrete Floor Finishing

03 35 13.30 Finishing Floors, High Tolerance

		Crew	Daily Output	Labor-Hours	Unit	Material	2014 Bare Costs Labor	Equipment	Total	Total Incl O&P
0010	**FINISHING FLOORS, HIGH TOLERANCE**									
0012	Finishing requires that concrete first be placed, struck off & consolidated									
0015	Basic finishing for various unspecified flatwork									
0100	Bull float only	C-10	4000	.006	S.F.		.25		.25	.37
0125	Bull float & manual float		2000	.012			.50		.50	.75
0150	Bull float, manual float, & broom finish, w/edging & joints		1850	.013			.54		.54	.81
0200	Bull float, manual float & manual steel trowel		1265	.019			.79		.79	1.18
0210	For specified Random Access Floors in ACI Classes 1, 2, 3 and 4 to achieve									
0215	Composite Overall Floor Flatness and Levelness values up to FF35/FL25									
0250	Bull float, machine float & machine trowel (walk-behind)	C-10C	1715	.014	S.F.		.58	.03	.61	.90
0300	Power screed, bull float, machine float & trowel (walk-behind)	C-10D	2400	.010			.42	.05	.47	.67
0350	Power screed, bull float, machine float & trowel (ride-on)	C-10E	4000	.006			.25	.07	.32	.44
0352	For specified Random Access Floors in ACI Classes 5, 6, 7 and 8 to achieve									
0354	Composite Overall Floor Flatness and Levelness values up to FF50/FL50									
0356	Add for two-dimensional restraightening after power float	C-10	6000	.004	S.F.		.17		.17	.25
0358	For specified Random or Defined Access Floors in ACI Class 9 to achieve									
0360	Composite Overall Floor Flatness and Levelness values up to FF100/FL100									
0362	Add for two-dimensional restraightening after bull float & power float	C-10	3000	.008	S.F.		.33		.33	.50
0364	For specified Superflat Defined Access Floors in ACI Class 9 to achieve									
0366	Minimum Floor Flatness and Levelness values of FF100/FL100									
0368	Add for 2-dim'l restraightening after bull float, power float, power trowel	C-10	2000	.012	S.F.		.50		.50	.75

03 54 Cast Underlayment

03 54 16 – Hydraulic Cement Underlayment

03 54 16.50 Cement Underlayment

		Crew	Daily Output	Labor-Hours	Unit	Material	2014 Bare Costs Labor	Equipment	Total	Total Incl O&P
0010	**CEMENT UNDERLAYMENT**									
2510	Underlayment, P.C based, self-leveling, 4100 psi, pumped, 1/4" thick	C-8	20000	.003	S.F.	1.49	.11	.04	1.64	1.85
2520	1/2" thick		19000	.003		2.99	.12	.04	3.15	3.51
2530	3/4" thick		18000	.003		4.48	.13	.04	4.65	5.15
2540	1" thick		17000	.003		6	.13	.04	6.17	6.85
2550	1-1/2" thick		15000	.004		8.95	.15	.05	9.15	10.15
2560	Hand placed, 1/2" thick	C-18	450	.020		2.99	.74	.13	3.86	4.58
2610	Topping, P.C based, self-leveling, 6100 psi, pumped, 1/4" thick	C-8	20000	.003		2.18	.11	.04	2.33	2.61
2620	1/2" thick		19000	.003		4.37	.12	.04	4.53	5
2630	3/4" thick		18000	.003		6.55	.13	.04	6.72	7.45
2660	1" thick		17000	.003		8.75	.13	.04	8.92	9.85
2670	1-1/2" thick		15000	.004		13.10	.15	.05	13.30	14.70
2680	Hand placed, 1/2" thick	C-18	450	.020		4.37	.74	.13	5.24	6.10

03 63 Epoxy Grouting

03 63 05 – Grouting of Dowels and Fasteners

03 63 05.10 Epoxy Only		Crew	Daily Output	Labor-Hours	Unit	Material	2014 Bare Costs Labor	Equipment	Total	Total Incl O&P
0010	**EPOXY ONLY**									
1500	Chemical anchoring, epoxy cartridge, excludes layout, drilling, fastener									
1530	For fastener 3/4" diam. x 6" embedment	2 Skwk	72	.222	Ea.	5.55	10.50		16.05	22.50
1535	1" diam. x 8" embedment		66	.242		8.30	11.45		19.75	27
1540	1-1/4" diam. x 10" embedment		60	.267		16.60	12.60		29.20	38
1545	1-3/4" diam. x 12" embedment		54	.296		27.50	14		41.50	52
1550	14" embedment		48	.333		33	15.75		48.75	61
1555	2" diam. x 12" embedment		42	.381		44.50	18		62.50	76.50
1560	18" embedment		32	.500		55.50	23.50		79	97.50

03 81 Concrete Cutting

03 81 13 – Flat Concrete Sawing

03 81 13.50 Concrete Floor/Slab Cutting

		Crew	Daily Output	Labor-Hours	Unit	Material	2014 Bare Costs Labor	Equipment	Total	Total Incl O&P
0010	**CONCRETE FLOOR/SLAB CUTTING**									
0050	Includes blade cost, layout and set-up time									
0300	Saw cut concrete slabs, plain, up to 3" deep	B-89	1060	.015	L.F.	.13	.63	.50	1.26	1.66
0320	Each additional inch of depth		3180	.005		.04	.21	.17	.42	.55
0400	Mesh reinforced, up to 3" deep		980	.016		.15	.68	.54	1.37	1.79
0420	Each additional inch of depth		2940	.005		.05	.23	.18	.46	.60
0500	Rod reinforced, up to 3" deep		800	.020		.19	.84	.66	1.69	2.19
0520	Each additional inch of depth		2400	.007		.06	.28	.22	.56	.73

03 81 13.75 Concrete Saw Blades

		Crew	Daily Output	Labor-Hours	Unit	Material	2014 Bare Costs Labor	Equipment	Total	Total Incl O&P
0010	**CONCRETE SAW BLADES**									
3000	Blades for saw cutting, included in cutting line items									
3020	Diamond, 12" diameter				Ea.	244			244	269
3040	18" diameter					445			445	490
3080	24" diameter					755			755	830
3120	30" diameter					1,100			1,100	1,200
3160	36" diameter					1,425			1,425	1,575
3200	42" diameter					2,700			2,700	2,975

03 82 Concrete Boring

03 82 13 – Concrete Core Drilling

03 82 13.10 Core Drilling

		Crew	Daily Output	Labor-Hours	Unit	Material	2014 Bare Costs Labor	Equipment	Total	Total Incl O&P
0010	**CORE DRILLING**									
0015	Includes bit cost, layout and set-up time									
0020	Reinforced concrete slab, up to 6" thick									
0100	1" diameter core	B-89A	17	.941	Ea.	.18	39.50	6.85	46.53	68.50
0150	For each additional inch of slab thickness in same hole, add		1440	.011		.03	.47	.08	.58	.84
0200	2" diameter core		16.50	.970		.29	40.50	7.05	47.84	71
0250	For each additional inch of slab thickness in same hole, add		1080	.015		.05	.62	.11	.78	1.13
0300	3" diameter core		16	1		.39	42	7.25	49.64	73.50
0350	For each additional inch of slab thickness in same hole, add		720	.022		.07	.93	.16	1.16	1.69
0500	4" diameter core		15	1.067		.50	45	7.75	53.25	78
0550	For each additional inch of slab thickness in same hole, add		480	.033		.08	1.40	.24	1.72	2.52
0700	6" diameter core		14	1.143		.78	48	8.30	57.08	84
0750	For each additional inch of slab thickness in same hole, add		360	.044		.13	1.87	.32	2.32	3.38
0900	8" diameter core		13	1.231		1.08	51.50	8.95	61.53	91
0950	For each additional inch of slab thickness in same hole, add		288	.056		.18	2.33	.40	2.91	4.25

03 82 Concrete Boring

03 82 13 – Concrete Core Drilling

03 82 13.10 Core Drilling

		Crew	Daily Output	Labor-Hours	Unit	Material	2014 Bare Costs Labor	2014 Bare Costs Equipment	Total	Total Incl O&P
1100	10" diameter core	B-89A	12	1.333	Ea.	1.51	56	9.70	67.21	99
1150	For each additional inch of slab thickness in same hole, add		240	.067		.25	2.80	.48	3.53	5.15
1300	12" diameter core		11	1.455		1.79	61	10.55	73.34	108
1350	For each additional inch of slab thickness in same hole, add		206	.078		.30	3.26	.56	4.12	6
1500	14" diameter core		10	1.600		2.11	67	11.60	80.71	119
1550	For each additional inch of slab thickness in same hole, add		180	.089		.35	3.73	.65	4.73	6.85
1700	18" diameter core		9	1.778		2.84	74.50	12.90	90.24	132
1750	For each additional inch of slab thickness in same hole, add		144	.111		.47	4.66	.81	5.94	8.60
1754	24" diameter core		8	2		3.99	84	14.50	102.49	150
1756	For each additional inch of slab thickness in same hole, add	▼	120	.133	▼	.67	5.60	.97	7.24	10.45
1760	For horizontal holes, add to above						20%	20%		
1770	Prestressed hollow core plank, 8" thick									
1780	1" diameter core	B-89A	17.50	.914	Ea.	.24	38.50	6.65	45.39	67
1790	For each additional inch of plank thickness in same hole, add		3840	.004		.03	.18	.03	.24	.33
1794	2" diameter core		17.25	.928		.38	39	6.75	46.13	68
1796	For each additional inch of plank thickness in same hole, add		2880	.006		.05	.23	.04	.32	.45
1800	3" diameter core		17	.941		.52	39.50	6.85	46.87	69
1810	For each additional inch of plank thickness in same hole, add		1920	.008		.07	.35	.06	.48	.68
1820	4" diameter core		16.50	.970		.66	40.50	7.05	48.21	71.50
1830	For each additional inch of plank thickness in same hole, add		1280	.013		.08	.52	.09	.69	1
1840	6" diameter core		15.50	1.032		1.04	43.50	7.50	52.04	76.50
1850	For each additional inch of plank thickness in same hole, add		960	.017		.13	.70	.12	.95	1.35
1860	8" diameter core		15	1.067		1.44	45	7.75	54.19	79
1870	For each additional inch of plank thickness in same hole, add		768	.021		.18	.87	.15	1.20	1.72
1880	10" diameter core		14	1.143		2.01	48	8.30	58.31	85.50
1890	For each additional inch of plank thickness in same hole, add		640	.025		.25	1.05	.18	1.48	2.10
1900	12" diameter core		13.50	1.185		2.39	50	8.60	60.99	89
1910	For each additional inch of plank thickness in same hole, add	▼	548	.029	▼	.30	1.23	.21	1.74	2.46

03 82 16 – Concrete Drilling

03 82 16.10 Concrete Impact Drilling

		Crew	Daily Output	Labor-Hours	Unit	Material	2014 Bare Costs Labor	2014 Bare Costs Equipment	Total	Total Incl O&P
0010	**CONCRETE IMPACT DRILLING**									
0020	Includes bit cost, layout and set-up time, no anchors									
0050	Up to 4" deep in concrete/brick floors/walls									
0100	Holes, 1/4" diameter	1 Carp	75	.107	Ea.	.06	4.89		4.95	7.60
0150	For each additional inch of depth in same hole, add		430	.019		.02	.85		.87	1.34
0200	3/8" diameter		63	.127		.05	5.80		5.85	9.05
0250	For each additional inch of depth in same hole, add		340	.024		.01	1.08		1.09	1.67
0300	1/2" diameter		50	.160		.05	7.35		7.40	11.35
0350	For each additional inch of depth in same hole, add		250	.032		.01	1.47		1.48	2.27
0400	5/8" diameter		48	.167		.08	7.65		7.73	11.90
0450	For each additional inch of depth in same hole, add		240	.033		.02	1.53		1.55	2.38
0500	3/4" diameter		45	.178		.11	8.15		8.26	12.70
0550	For each additional inch of depth in same hole, add		220	.036		.03	1.67		1.70	2.60
0600	7/8" diameter		43	.186		.15	8.55		8.70	13.30
0650	For each additional inch of depth in same hole, add		210	.038		.04	1.75		1.79	2.74
0700	1" diameter		40	.200		.16	9.15		9.31	14.35
0750	For each additional inch of depth in same hole, add		190	.042		.04	1.93		1.97	3.02
0800	1-1/4" diameter		38	.211		.23	9.65		9.88	15.15
0850	For each additional inch of depth in same hole, add		180	.044		.06	2.04		2.10	3.20
0900	1-1/2" diameter		35	.229		.32	10.50		10.82	16.50
0950	For each additional inch of depth in same hole, add	▼	165	.048	▼	.08	2.22		2.30	3.52
1000	For ceiling installations, add						40%			

Estimating Tips

05 05 00 Common Work Results for Metals

- Nuts, bolts, washers, connection angles, and plates can add a significant amount to both the tonnage of a structural steel job and the estimated cost. As a rule of thumb, add 10% to the total weight to account for these accessories.

- Type 2 steel construction, commonly referred to as "simple construction," consists generally of field-bolted connections with lateral bracing supplied by other elements of the building, such as masonry walls or x-bracing. The estimator should be aware, however, that shop connections may be accomplished by welding or bolting. The method may be particular to the fabrication shop and may have an impact on the estimated cost.

05 10 00 Structural Steel

- Steel items can be obtained from two sources: a fabrication shop or a metals service center. Fabrication shops can fabricate items under more controlled conditions than can crews in the field. They are also more efficient and can produce items more economically. Metal service centers serve as a source of long mill shapes to both fabrication shops and contractors.

- Most line items in this structural steel subdivision, and most items in 05 50 00 Metal Fabrications, are indicated as being shop fabricated. The bare material cost for these shop fabricated items is the "Invoice Cost" from the shop and includes the mill base price of steel plus mill extras, transportation to the shop, shop drawings and detailing where warranted, shop fabrication and handling, sandblasting and a shop coat of primer paint, all necessary structural bolts, and delivery to the job site. The bare labor cost and bare equipment cost for these shop fabricated items is for field installation or erection.

- Line items in Subdivision 05 12 23.40 Lightweight Framing, and other items scattered in Division 5, are indicated as being field fabricated. The bare material cost for these field fabricated items is the "Invoice Cost" from the metals service center and includes the mill base price of steel plus mill extras, transportation to the metals service center, material handling, and delivery of long lengths of mill shapes to the job site. Material costs for structural bolts and welding rods should be added to the estimate. The bare labor cost and bare equipment cost for these items is for both field fabrication and field installation or erection, and include time for cutting, welding and drilling in the fabricated metal items. Drilling into concrete and fasteners to fasten field fabricated items to other work are not included and should be added to the estimate.

05 20 00 Steel Joist Framing

- In any given project the total weight of open web steel joists is determined by the loads to be supported and the design. However, economies can be realized in minimizing the amount of labor used to place the joists. This is done by maximizing the joist spacing, and therefore minimizing the number of joists required to be installed on the job. Certain spacings and locations may be required by the design, but in other cases maximizing the spacing and keeping it as uniform as possible will keep the costs down.

05 30 00 Steel Decking

- The takeoff and estimating of metal deck involves more than simply the area of the floor or roof and the type of deck specified or shown on the drawings. Many different sizes and types of openings may exist. Small openings for individual pipes or conduits may be drilled after the floor/roof is installed, but larger openings may require special deck lengths as well as reinforcing or structural support. The estimator should determine who will be supplying this reinforcing. Additionally, some deck terminations are part of the deck package, such as screed angles and pour stops, and others will be part of the steel contract, such as angles attached to structural members and cast-in-place angles and plates. The estimator must ensure that all pieces are accounted for in the complete estimate.

05 50 00 Metal Fabrications

- The most economical steel stairs are those that use common materials, standard details, and most importantly, a uniform and relatively simple method of field assembly. Commonly available A36 channels and plates are very good choices for the main stringers of the stairs, as are angles and tees for the carrier members. Risers and treads are usually made by specialty shops, and it is most economical to use a typical detail in as many places as possible. The stairs should be pre-assembled and shipped directly to the site. The field connections should be simple and straightforward to be accomplished efficiently, and with minimum equipment and labor.

Reference Numbers

Reference numbers are shown in shaded boxes at the beginning of some major classifications. These numbers refer to related items in the Reference Section. The reference information may be an estimating procedure, an alternate pricing method, or technical information. *Note:* Not all subdivisions listed here necessarily appear in this publication.

05 05 Common Work Results for Metals

05 05 19 – Post-Installed Concrete Anchors

05 05 19.10 Chemical Anchors

		Crew	Daily Output	Labor-Hours	Unit	Material	2014 Bare Costs Labor	Equipment	Total	Total Incl O&P
0010	**CHEMICAL ANCHORS**									
0020	Includes layout & drilling									
1430	Chemical anchor, w/rod & epoxy cartridge, 3/4" diam. x 9-1/2" long	B-89A	27	.593	Ea.	9.75	25	4.30	39.05	54
1435	1" diameter x 11-3/4" long		24	.667		16.70	28	4.84	49.54	67
1440	1-1/4" diameter x 14" long		21	.762		36	32	5.55	73.55	95.50
1445	1-3/4" diameter x 15" long		20	.800		64.50	33.50	5.80	103.80	129
1450	18" long		17	.941		77.50	39.50	6.85	123.85	154
1455	2" diameter x 18" long		16	1		102	42	7.25	151.25	185
1460	24" long		15	1.067		132	45	7.75	184.75	224

05 05 19.20 Expansion Anchors

			Crew	Daily Output	Labor-Hours	Unit	Material	2014 Bare Costs Labor	Equipment	Total	Total Incl O&P
0010	**EXPANSION ANCHORS**										
0100	Anchors for concrete, brick or stone, no layout and drilling										
0200	Expansion shields, zinc, 1/4" diameter, 1-5/16" long, single	G	1 Carp	90	.089	Ea.	.48	4.08		4.56	6.85
0300	1-3/8" long, double	G		85	.094		.59	4.32		4.91	7.30
0400	3/8" diameter, 1-1/2" long, single	G		85	.094		.81	4.32		5.13	7.55
0500	2" long, double	G		80	.100		1.32	4.59		5.91	8.55
0600	1/2" diameter, 2-1/16" long, single	G		80	.100		1.62	4.59		6.21	8.90
0700	2-1/2" long, double	G		75	.107		2.10	4.89		6.99	9.85
0800	5/8" diameter, 2-5/8" long, single	G		75	.107		2.85	4.89		7.74	10.70
0900	2-3/4" long, double	G		70	.114		2.99	5.25		8.24	11.40
1000	3/4" diameter, 2-3/4" long, single	G		70	.114		3.25	5.25		8.50	11.70
1100	3-15/16" long, double	G		65	.123		5.80	5.65		11.45	15.10
2100	Hollow wall anchors for gypsum wall board, plaster or tile										
2300	1/8" diameter, short	G	1 Carp	160	.050	Ea.	.23	2.29		2.52	3.79
2400	Long	G		150	.053		.23	2.45		2.68	4.02
2500	3/16" diameter, short	G		150	.053		.40	2.45		2.85	4.21
2600	Long	G		140	.057		.58	2.62		3.20	4.68
2700	1/4" diameter, short	G		140	.057		.62	2.62		3.24	4.72
2800	Long	G		130	.062		.61	2.82		3.43	5
3000	Toggle bolts, bright steel, 1/8" diameter, 2" long	G		85	.094		.19	4.32		4.51	6.85
3100	4" long	G		80	.100		.25	4.59		4.84	7.40
3400	1/4" diameter, 3" long	G		75	.107		.34	4.89		5.23	7.90
3500	6" long	G		70	.114		.50	5.25		5.75	8.65
3600	3/8" diameter, 3" long	G		70	.114		.82	5.25		6.07	9
3700	6" long	G		60	.133		1.24	6.10		7.34	10.80
3800	1/2" diameter, 4" long	G		60	.133		1.84	6.10		7.94	11.45
3900	6" long	G		50	.160		2.40	7.35		9.75	13.95
4000	Nailing anchors										
4100	Nylon nailing anchor, 1/4" diameter, 1" long		1 Carp	3.20	2.500	C	16.35	115		131.35	195
4200	1-1/2" long			2.80	2.857		21	131		152	226
4300	2" long			2.40	3.333		25.50	153		178.50	264
4400	Metal nailing anchor, 1/4" diameter, 1" long	G		3.20	2.500		17.60	115		132.60	196
4500	1-1/2" long	G		2.80	2.857		22.50	131		153.50	227
4600	2" long	G		2.40	3.333		28	153		181	267
5000	Screw anchors for concrete, masonry,										
5100	stone & tile, no layout or drilling included										
5700	Lag screw shields, 1/4" diameter, short	G	1 Carp	90	.089	Ea.	.37	4.08		4.45	6.70
5900	3/8" diameter, short	G		85	.094		.65	4.32		4.97	7.35
6100	1/2" diameter, short	G		80	.100		1.05	4.59		5.64	8.25
6300	5/8" diameter, short	G		70	.114		1.81	5.25		7.06	10.10
6600	Lead, #6 & #8, 3/4" long	G		260	.031		.18	1.41		1.59	2.38
6700	#10 - #14, 1-1/2" long	G		200	.040		.29	1.83		2.12	3.15

05 05 Common Work Results for Metals

05 05 19 - Post-Installed Concrete Anchors

05 05 19.20 Expansion Anchors

			Crew	Daily Output	Labor-Hours	Unit	Material	2014 Bare Costs Labor	Equipment	Total	Total Incl O&P
6800	#16 & #18, 1-1/2" long	G	1 Carp	160	.050	Ea.	.39	2.29		2.68	3.97
6900	Plastic, #6 & #8, 3/4" long			260	.031		.04	1.41		1.45	2.22
7100	#10 & #12, 1" long		↓	220	.036	↓	.05	1.67		1.72	2.63
8000	Wedge anchors, not including layout or drilling										
8050	Carbon steel, 1/4" diameter, 1-3/4" long	G	1 Carp	150	.053	Ea.	.40	2.45		2.85	4.21
8100	3-1/4" long	G		140	.057		.53	2.62		3.15	4.62
8150	3/8" diameter, 2-1/4" long	G		145	.055		.49	2.53		3.02	4.44
8200	5" long	G		140	.057		.86	2.62		3.48	4.99
8250	1/2" diameter, 2-3/4" long	G		140	.057		.97	2.62		3.59	5.10
8300	7" long	G		125	.064		1.66	2.93		4.59	6.35
8350	5/8" diameter, 3-1/2" long	G		130	.062		1.69	2.82		4.51	6.20
8400	8-1/2" long	G		115	.070		3.60	3.19		6.79	8.90
8450	3/4" diameter, 4-1/4" long	G		115	.070		2.81	3.19		6	8
8500	10" long	G		95	.084		6.40	3.86		10.26	13
8550	1" diameter, 6" long	G		100	.080		9.05	3.67		12.72	15.60
8575	9" long	G		85	.094		11.75	4.32		16.07	19.60
8600	12" long	G		75	.107		12.70	4.89		17.59	21.50
8650	1-1/4" diameter, 9" long	G		70	.114		24	5.25		29.25	34
8700	12" long	G	↓	60	.133	↓	30.50	6.10		36.60	43
8750	For type 303 stainless steel, add						350%				
8800	For type 316 stainless steel, add						450%				
8950	Self-drilling concrete screw, hex washer head, 3/16" diam. x 1-3/4" long	G	1 Carp	300	.027	Ea.	.20	1.22		1.42	2.11
8960	2-1/4" long	G		250	.032		.23	1.47		1.70	2.51
8970	Phillips flat head, 3/16" diam. x 1-3/4" long	G		300	.027		.20	1.22		1.42	2.11
8980	2-1/4" long	G	↓	250	.032	↓	.23	1.47		1.70	2.51

05 05 21 - Fastening Methods for Metal

05 05 21.15 Drilling Steel

			Crew	Daily Output	Labor-Hours	Unit	Material	2014 Bare Costs Labor	Equipment	Total	Total Incl O&P
0010	**DRILLING STEEL**										
1910	Drilling & layout for steel, up to 1/4" deep, no anchor										
1920	Holes, 1/4" diameter		1 Sswk	112	.071	Ea.	.08	3.65		3.73	6.55
1925	For each additional 1/4" depth, add			336	.024		.08	1.22		1.30	2.24
1930	3/8" diameter			104	.077		.09	3.93		4.02	7.05
1935	For each additional 1/4" depth, add			312	.026		.09	1.31		1.40	2.41
1940	1/2" diameter			96	.083		.09	4.26		4.35	7.60
1945	For each additional 1/4" depth, add			288	.028		.09	1.42		1.51	2.61
1950	5/8" diameter			88	.091		.13	4.65		4.78	8.35
1955	For each additional 1/4" depth, add			264	.030		.13	1.55		1.68	2.88
1960	3/4" diameter			80	.100		.17	5.10		5.27	9.20
1965	For each additional 1/4" depth, add			240	.033		.17	1.70		1.87	3.19
1970	7/8" diameter			72	.111		.22	5.70		5.92	10.25
1975	For each additional 1/4" depth, add			216	.037		.22	1.89		2.11	3.58
1980	1" diameter			64	.125		.23	6.40		6.63	11.55
1985	For each additional 1/4" depth, add		↓	192	.042		.23	2.13		2.36	4.01
1990	For drilling up, add					↓		40%			

05 05 23 - Metal Fastenings

05 05 23.10 Bolts and Hex Nuts

			Crew	Daily Output	Labor-Hours	Unit	Material	2014 Bare Costs Labor	Equipment	Total	Total Incl O&P
0010	**BOLTS & HEX NUTS**, Steel, A307										
0100	1/4" diameter, 1/2" long	G	1 Sswk	140	.057	Ea.	.06	2.92		2.98	5.20
0200	1" long	G		140	.057		.07	2.92		2.99	5.25
0300	2" long	G		130	.062		.10	3.14		3.24	5.65
0400	3" long	G		130	.062		.15	3.14		3.29	5.70
0500	4" long	G	↓	120	.067	↓	.17	3.41		3.58	6.20

05 05 Common Work Results for Metals

05 05 23 – Metal Fastenings

05 05 23.10 Bolts and Hex Nuts

			Crew	Daily Output	Labor-Hours	Unit	Material	2014 Bare Costs Labor	Equipment	Total	Total Incl O&P
0600	3/8" diameter, 1" long	G	1 Sswk	130	.062	Ea.	.14	3.14		3.28	5.70
0700	2" long	G		130	.062		.18	3.14		3.32	5.75
0800	3" long	G		120	.067		.24	3.41		3.65	6.25
0900	4" long	G		120	.067		.30	3.41		3.71	6.35
1000	5" long	G		115	.070		.38	3.56		3.94	6.70
1100	1/2" diameter, 1-1/2" long	G		120	.067		.40	3.41		3.81	6.45
1200	2" long	G		120	.067		.46	3.41		3.87	6.50
1300	4" long	G		115	.070		.75	3.56		4.31	7.15
1400	6" long	G		110	.073		1.05	3.72		4.77	7.70
1500	8" long	G		105	.076		1.38	3.89		5.27	8.35
1600	5/8" diameter, 1-1/2" long	G		120	.067		.98	3.41		4.39	7.10
1700	2" long	G		120	.067		1.09	3.41		4.50	7.20
1800	4" long	G		115	.070		1.59	3.56		5.15	8.05
1900	6" long	G		110	.073		2.05	3.72		5.77	8.80
2000	8" long	G		105	.076		3.06	3.89		6.95	10.20
2100	10" long	G		100	.080		3.87	4.09		7.96	11.45
2200	3/4" diameter, 2" long	G		120	.067		1.15	3.41		4.56	7.25
2300	4" long	G		110	.073		1.65	3.72		5.37	8.35
2400	6" long	G		105	.076		2.12	3.89		6.01	9.20
2500	8" long	G		95	.084		3.20	4.30		7.50	11.10
2600	10" long	G		85	.094		4.20	4.81		9.01	13.10
2700	12" long	G		80	.100		4.92	5.10		10.02	14.40
2800	1" diameter, 3" long	G		105	.076		2.69	3.89		6.58	9.80
2900	6" long	G		90	.089		3.94	4.54		8.48	12.35
3000	12" long	G		75	.107		7.10	5.45		12.55	17.40
3100	For galvanized, add						75%				
3200	For stainless, add						350%				

05 05 23.30 Lag Screws

			Crew	Daily Output	Labor-Hours	Unit	Material	Labor	Equipment	Total	Total Incl O&P
0010	**LAG SCREWS**										
0020	Steel, 1/4" diameter, 2" long	G	1 Carp	200	.040	Ea.	.09	1.83		1.92	2.93
0100	3/8" diameter, 3" long	G		150	.053		.29	2.45		2.74	4.09
0200	1/2" diameter, 3" long	G		130	.062		.63	2.82		3.45	5.05
0300	5/8" diameter, 3" long	G		120	.067		1.13	3.06		4.19	5.95

05 05 23.35 Machine Screws

			Crew	Daily Output	Labor-Hours	Unit	Material	Labor	Equipment	Total	Total Incl O&P
0010	**MACHINE SCREWS**										
0020	Steel, round head, #8 x 1" long	G	1 Carp	4.80	1.667	C	3.79	76.50		80.29	122
0110	#8 x 2" long	G		2.40	3.333		6.10	153		159.10	243
0200	#10 x 1" long	G		4	2		6	91.50		97.50	149
0300	#10 x 2" long	G		2	4		8.50	183		191.50	292

05 05 23.50 Powder Actuated Tools and Fasteners

			Crew	Daily Output	Labor-Hours	Unit	Material	Labor	Equipment	Total	Total Incl O&P
0010	**POWDER ACTUATED TOOLS & FASTENERS**										
0020	Stud driver, .22 caliber, single shot					Ea.	147			147	161
0100	.27 caliber, semi automatic, strip					"	435			435	475
0300	Powder load, single shot, .22 cal, power level 2, brown					C	5.30			5.30	5.80
0400	Strip, .27 cal, power level 4, red						7.60			7.60	8.35
0600	Drive pin, .300 x 3/4" long	G	1 Carp	4.80	1.667		4.86	76.50		81.36	123
0700	.300 x 3" long with washer	G	"	4	2		12.50	91.50		104	156

05 05 23.70 Structural Blind Bolts

			Crew	Daily Output	Labor-Hours	Unit	Material	Labor	Equipment	Total	Total Incl O&P
0010	**STRUCTURAL BLIND BOLTS**										
0100	1/4" diameter x 1/4" grip	G	1 Sswk	240	.033	Ea.	1.24	1.70		2.94	4.37
0150	1/2" grip	G		216	.037		1.33	1.89		3.22	4.80
0200	3/8" diameter x 1/2" grip	G		232	.034		1.75	1.76		3.51	5.05

05 05 Common Work Results for Metals

05 05 23 – Metal Fastenings

05 05 23.70 Structural Blind Bolts		Crew	Daily Output	Labor-Hours	Unit	Material	2014 Bare Costs Labor	Equipment	Total	Total Incl O&P
0250	3/4" grip	G	1 Sswk	208	.038	Ea.	1.84	1.97		3.81	5.50
0300	1/2" diameter x 1/2" grip	G		224	.036		3.99	1.82		5.81	7.60
0350	3/4" grip	G		200	.040		5.60	2.04		7.64	9.75
0400	5/8" diameter x 3/4" grip	G		216	.037		8.25	1.89		10.14	12.45
0450	1" grip	G		192	.042		9.50	2.13		11.63	14.20

05 12 Structural Steel Framing

05 12 23 – Structural Steel for Buildings

05 12 23.60 Pipe Support Framing

		Crew	Daily Output	Labor-Hours	Unit	Material	2014 Bare Costs Labor	Equipment	Total	Total Incl O&P
0010	**PIPE SUPPORT FRAMING**										
0020	Under 10#/L.F., shop fabricated	G	E-4	3900	.008	Lb.	1.78	.42	.04	2.24	2.74
0200	10.1 to 15#/L.F.	G		4300	.007		1.75	.38	.03	2.16	2.64
0400	15.1 to 20#/L.F.	G		4800	.007		1.72	.34	.03	2.09	2.53
0600	Over 20#/L.F.	G		5400	.006		1.70	.31	.03	2.04	2.44

05 35 Raceway Decking Assemblies

05 35 13 – Steel Cellular Decking

05 35 13.50 Cellular Decking

		Crew	Daily Output	Labor-Hours	Unit	Material	2014 Bare Costs Labor	Equipment	Total	Total Incl O&P
0010	**CELLULAR DECKING**										
0015	Made from recycled materials										
0200	Cellular units, galv, 1-1/2" deep, Type BC, 20-20 ga., over 15 squares	G	E-4	1460	.022	S.F.	7.75	1.13	.10	8.98	10.65
0250	18-20 ga.	G		1420	.023		8.85	1.16	.10	10.11	11.85
0300	18-18 ga.	G		1390	.023		9.05	1.19	.10	10.34	12.15
0320	16-18 ga.	G		1360	.024		10.80	1.21	.10	12.11	14.10
0340	16-16 ga.	G		1330	.024		12	1.24	.11	13.35	15.50
0400	3" deep, Type NC, galvanized, 20-20 ga.	G		1375	.023		8.55	1.20	.10	9.85	11.65
0500	18-20 ga.	G		1350	.024		10.35	1.22	.11	11.68	13.65
0600	18-18 ga.	G		1290	.025		10.30	1.28	.11	11.69	13.70
0700	16-18 ga.	G		1230	.026		11.60	1.34	.12	13.06	15.25
0800	16-16 ga.	G		1150	.028		12.65	1.44	.12	14.21	16.55
1000	4-1/2" deep, Type JC, galvanized, 18-20 ga.	G		1100	.029		11.90	1.50	.13	13.53	15.90
1100	18-18 ga.	G		1040	.031		11.85	1.59	.14	13.58	15.95
1200	16-18 ga.	G		980	.033		13.35	1.68	.15	15.18	17.80
1300	16-16 ga.	G		935	.034		14.55	1.77	.15	16.47	19.30
1900	For multi-story or congested site, add							50%			

37

Division Notes

	CREW	DAILY OUTPUT	LABOR-HOURS	UNIT	BARE COSTS				TOTAL INCL O&P
					MAT.	LABOR	EQUIP.	TOTAL	

Estimating Tips

06 05 00 Common Work Results for Wood, Plastics, and Composites

- Common to any wood-framed structure are the accessory connector items such as screws, nails, adhesives, hangers, connector plates, straps, angles, and hold-downs. For typical wood-framed buildings, such as residential projects, the aggregate total for these items can be significant, especially in areas where seismic loading is a concern. For floor and wall framing, the material cost is based on 10 to 25 lbs. per MBF. Hold-downs, hangers, and other connectors should be taken off by the piece.

  Included with material costs are fasteners for a normal installation. RSMeans engineers use manufacturer's recommendations, written specifications, and/or standard construction practice for size and spacing of fasteners. Prices for various fasteners are shown for informational purposes only. Adjustments should be made if unusual fastening conditions exist.

06 10 00 Carpentry

- Lumber is a traded commodity and therefore sensitive to supply and demand in the marketplace. Even in "budgetary" estimating of wood-framed projects, it is advisable to call local suppliers for the latest market pricing.

- Common quantity units for wood-framed projects are "thousand board feet" (MBF). A board foot is a volume of wood, 1" x 1' x 1', or 144 cubic inches. Board-foot quantities are generally calculated using nominal material dimensions—dressed sizes are ignored. Board foot per lineal foot of any stick of lumber can be calculated by dividing the nominal cross-sectional area by 12. As an example, 2,000 lineal feet of 2 x 12 equates to 4 MBF by dividing the nominal area, 2 x 12, by 12, which equals 2, and multiplying by 2,000 to give 4,000 board feet. This simple rule applies to all nominal dimensioned lumber.

- Waste is an issue of concern at the quantity takeoff for any area of construction. Framing lumber is sold in even foot lengths, i.e., 10', 12', 14', 16', and depending on spans, wall heights, and the grade of lumber, waste is inevitable. A rule of thumb for lumber waste is 5%–10% depending on material quality and the complexity of the framing.

- Wood in various forms and shapes is used in many projects, even where the main structural framing is steel, concrete, or masonry. Plywood as a back-up partition material and 2x boards used as blocking and cant strips around roof edges are two common examples. The estimator should ensure that the costs of all wood materials are included in the final estimate.

06 20 00 Finish Carpentry

- It is necessary to consider the grade of workmanship when estimating labor costs for erecting millwork and interior finish. In practice, there are three grades: premium, custom, and economy. The RSMeans daily output for base and case moldings is in the range of 200 to 250 L.F. per carpenter per day. This is appropriate for most average custom-grade projects. For premium projects, an adjustment to productivity of 25%–50% should be made, depending on the complexity of the job.

Reference Numbers

Reference numbers are shown in shaded boxes at the beginning of some major classifications. These numbers refer to related items in the Reference Section. The reference information may be an estimating procedure, an alternate pricing method, or technical information.

Note: Not all subdivisions listed here necessarily appear in this publication.

06 16 Sheathing

06 16 36 – Wood Panel Product Sheathing

06 16 36.10 Sheathing	Crew	Daily Output	Labor-Hours	Unit	Material	2014 Bare Costs Labor	Equipment	Total	Total Incl O&P
0010 **SHEATHING**									
0012 Plywood on roofs, CDX									
0030 5/16" thick	2 Carp	1600	.010	S.F.	.55	.46		1.01	1.32
0035 Pneumatic nailed		1952	.008		.55	.38		.93	1.19
0050 3/8" thick		1525	.010		.50	.48		.98	1.29
0055 Pneumatic nailed		1860	.009		.50	.39		.89	1.16
0100 1/2" thick		1400	.011		.60	.52		1.12	1.46
0105 Pneumatic nailed		1708	.009		.60	.43		1.03	1.31
0200 5/8" thick		1300	.012		.74	.56		1.30	1.69
0205 Pneumatic nailed		1586	.010		.74	.46		1.20	1.53
0300 3/4" thick		1200	.013		.88	.61		1.49	1.91
0305 Pneumatic nailed		1464	.011		.88	.50		1.38	1.74
0500 Plywood on walls, with exterior CDX, 3/8" thick		1200	.013		.50	.61		1.11	1.49
0505 Pneumatic nailed		1488	.011		.50	.49		.99	1.31
0600 1/2" thick		1125	.014		.60	.65		1.25	1.66
0605 Pneumatic nailed		1395	.011		.60	.53		1.13	1.46
0700 5/8" thick		1050	.015		.74	.70		1.44	1.90
0705 Pneumatic nailed		1302	.012		.74	.56		1.30	1.69
0800 3/4" thick		975	.016		.88	.75		1.63	2.13
0805 Pneumatic nailed		1209	.013		.88	.61		1.49	1.91

Estimating Tips

07 10 00 Dampproofing and Waterproofing

- Be sure of the job specifications before pricing this subdivision. The difference in cost between waterproofing and dampproofing can be great. Waterproofing will hold back standing water. Dampproofing prevents the transmission of water vapor. Also included in this section are vapor retarding membranes.

07 20 00 Thermal Protection

- Insulation and fireproofing products are measured by area, thickness, volume or R-value. Specifications may give only what the specific R-value should be in a certain situation. The estimator may need to choose the type of insulation to meet that R-value.

07 30 00 Steep Slope Roofing
07 40 00 Roofing and Siding Panels

- Many roofing and siding products are bought and sold by the square. One square is equal to an area that measures 100 square feet.

This simple change in unit of measure could create a large error if the estimator is not observant. Accessories necessary for a complete installation must be figured into any calculations for both material and labor.

07 50 00 Membrane Roofing
07 60 00 Flashing and Sheet Metal
07 70 00 Roofing and Wall Specialties and Accessories

- The items in these subdivisions compose a roofing system. No one component completes the installation, and all must be estimated. Built-up or single-ply membrane roofing systems are made up of many products and installation trades. Wood blocking at roof perimeters or penetrations, parapet coverings, reglets, roof drains, gutters, downspouts, sheet metal flashing, skylights, smoke vents, and roof hatches all need to be considered along with the roofing material. Several different installation trades will need to work together on the roofing system. Inherent difficulties in the scheduling and coordination of various trades must be accounted for when estimating labor costs.

07 90 00 Joint Protection

- To complete the weather-tight shell, the sealants and caulkings must be estimated. Where different materials meet—at expansion joints, at flashing penetrations, and at hundreds of other locations throughout a construction project—they provide another line of defense against water penetration. Often, an entire system is based on the proper location and placement of caulking or sealants. The detailed drawings that are included as part of a set of architectural plans show typical locations for these materials. When caulking or sealants are shown at typical locations, this means the estimator must include them for all the locations where this detail is applicable. Be careful to keep different types of sealants separate, and remember to consider backer rods and primers if necessary.

Reference Numbers

Reference numbers are shown in shaded boxes at the beginning of some major classifications. These numbers refer to related items in the Reference Section. The reference information may be an estimating procedure, an alternate pricing method, or technical information.

Note: Not all subdivisions listed here necessarily appear in this publication.

Division 7 - Thermal and Moisture Protection

07 84 13 – Penetration Firestopping

07 84 13.10 Firestopping	Crew	Daily Output	Labor-Hours	Unit	Material	2014 Bare Costs Labor	Equipment	Total	Total Incl O&P
0010 **FIRESTOPPING** R078413-30									
0100 Metallic piping, non insulated									
0110 Through walls, 2" diameter	1 Carp	16	.500	Ea.	15.90	23		38.90	53
0120 4" diameter		14	.571		24.50	26		50.50	67
0130 6" diameter		12	.667		32.50	30.50		63	83
0140 12" diameter		10	.800		57.50	36.50		94	120
0150 Through floors, 2" diameter		32	.250		9.65	11.45		21.10	28.50
0160 4" diameter		28	.286		13.85	13.10		26.95	35.50
0170 6" diameter		24	.333		18.20	15.30		33.50	43.50
0180 12" diameter		20	.400		31	18.35		49.35	62.50
0190 Metallic piping, insulated									
0200 Through walls, 2" diameter	1 Carp	16	.500	Ea.	22.50	23		45.50	60.50
0210 4" diameter		14	.571		31	26		57	74.50
0220 6" diameter		12	.667		39	30.50		69.50	90
0230 12" diameter		10	.800		64	36.50		100.50	127
0240 Through floors, 2" diameter		32	.250		16.30	11.45		27.75	35.50
0250 4" diameter		28	.286		20.50	13.10		33.60	42.50
0260 6" diameter		24	.333		25	15.30		40.30	51
0270 12" diameter		20	.400		31	18.35		49.35	62.50
0280 Non metallic piping, non insulated									
0290 Through walls, 2" diameter	1 Carp	12	.667	Ea.	65.50	30.50		96	119
0300 4" diameter		10	.800		82	36.50		118.50	147
0310 6" diameter		8	1		114	46		160	197
0330 Through floors, 2" diameter		16	.500		51.50	23		74.50	92
0340 4" diameter		6	1.333		63.50	61		124.50	165
0350 6" diameter		6	1.333		76.50	61		137.50	179
0370 Ductwork, insulated & non insulated, round									
0380 Through walls, 6" diameter	1 Carp	12	.667	Ea.	33.50	30.50		64	83.50
0390 12" diameter		10	.800		66	36.50		102.50	129
0400 18" diameter		8	1		108	46		154	190
0410 Through floors, 6" diameter		16	.500		18.20	23		41.20	55.50
0420 12" diameter		14	.571		33.50	26		59.50	77
0430 18" diameter		12	.667		57.50	30.50		88	111
0440 Ductwork, insulated & non insulated, rectangular									
0450 With stiffener/closure angle, through walls, 6" x 12"	1 Carp	8	1	Ea.	27.50	46		73.50	102
0460 12" x 24"		6	1.333		36.50	61		97.50	135
0470 24" x 48"		4	2		105	91.50		196.50	257
0480 With stiffener/closure angle, through floors, 6" x 12"		10	.800		14.95	36.50		51.45	73
0490 12" x 24"		8	1		27	46		73	101
0500 24" x 48"		6	1.333		52.50	61		113.50	153
0510 Multi trade openings									
0520 Through walls, 6" x 12"	1 Carp	2	4	Ea.	57.50	183		240.50	345
0530 12" x 24"	"	1	8		234	365		599	820
0540 24" x 48"	2 Carp	1	16		935	735		1,670	2,150
0550 48" x 96"	"	.75	21.333		3,750	980		4,730	5,625
0560 Through floors, 6" x 12"	1 Carp	2	4		57.50	183		240.50	345
0570 12" x 24"	"	1	8		234	365		599	820
0580 24" x 48"	2 Carp	.75	21.333		935	980		1,915	2,525
0590 48" x 96"	"	.50	32		3,750	1,475		5,225	6,400
0600 Structural penetrations, through walls									
0610 Steel beams, W8 x 10	1 Carp	8	1	Ea.	36.50	46		82.50	111
0620 W12 x 14		6	1.333		57.50	61		118.50	158
0630 W21 x 44		5	1.600		116	73.50		189.50	241

07 84 Firestopping

07 84 13 – Penetration Firestopping

07 84 13.10 Firestopping		Crew	Daily Output	Labor-Hours	Unit	Material	2014 Bare Costs Labor	Equipment	Total	Total Incl O&P
0640	W36 x 135	1 Carp	3	2.667	Ea.	282	122		404	500
0650	Bar joists, 18" deep		6	1.333		53.50	61		114.50	153
0660	24" deep		6	1.333		66	61		127	167
0670	36" deep		5	1.600		99.50	73.50		173	222
0680	48" deep	↓	4	2	↓	116	91.50		207.50	270
0690	Construction joints, floor slab at exterior wall									
0700	Precast, brick, block or drywall exterior									
0710	2" wide joint	1 Carp	125	.064	L.F.	8.30	2.93		11.23	13.65
0720	4" wide joint	"	75	.107	"	16.55	4.89		21.44	26
0730	Metal panel, glass or curtain wall exterior									
0740	2" wide joint	1 Carp	40	.200	L.F.	19.60	9.15		28.75	35.50
0750	4" wide joint	"	25	.320	"	27	14.65		41.65	52
0760	Floor slab to drywall partition									
0770	Flat joint	1 Carp	100	.080	L.F.	8.15	3.67		11.82	14.60
0780	Fluted joint		50	.160	↓	16.55	7.35		23.90	29.50
0790	Etched fluted joint	↓	75	.107	↓	10.75	4.89		15.64	19.40
0800	Floor slab to concrete/masonry partition									
0810	Flat joint	1 Carp	75	.107	L.F.	18.20	4.89		23.09	27.50
0820	Fluted joint	"	50	.160	"	21.50	7.35		28.85	35
0830	Concrete/CMU wall joints									
0840	1" wide	1 Carp	100	.080	L.F.	9.95	3.67		13.62	16.55
0850	2" wide		75	.107		18.20	4.89		23.09	27.50
0860	4" wide	↓	50	.160	↓	34.50	7.35		41.85	49.50
0870	Concrete/CMU floor joints									
0880	1" wide	1 Carp	200	.040	L.F.	4.99	1.83		6.82	8.35
0890	2" wide		150	.053		9.10	2.45		11.55	13.75
0900	4" wide	↓	100	.080	↓	17.40	3.67		21.07	25

07 91 Preformed Joint Seals

07 91 13 – Compression Seals

07 91 13.10 Compression Seals		Crew	Daily Output	Labor-Hours	Unit	Material	2014 Bare Costs Labor	Equipment	Total	Total Incl O&P
0010	**COMPRESSION SEALS**									
4900	O-ring type cord, 1/4"	1 Bric	472	.017	L.F.	.48	.77		1.25	1.71
4910	1/2"		440	.018		1.79	.83		2.62	3.23
4920	3/4"		424	.019		3.62	.86		4.48	5.30
4930	1"		408	.020		6.95	.89		7.84	9
4940	1-1/4"		384	.021		13.75	.95		14.70	16.55
4950	1-1/2"		368	.022		17.40	.99		18.39	20.50
4960	1-3/4"		352	.023		36.50	1.04		37.54	41.50
4970	2"	↓	344	.023	↓	64.50	1.06		65.56	72

07 91 23 – Backer Rods

07 91 23.10 Backer Rods		Crew	Daily Output	Labor-Hours	Unit	Material	2014 Bare Costs Labor	Equipment	Total	Total Incl O&P
0010	**BACKER RODS**									
0030	Backer rod, polyethylene, 1/4" diameter	1 Bric	4.60	1.739	C.L.F.	2.43	79.50		81.93	124
0050	1/2" diameter		4.60	1.739		3.66	79.50		83.16	125
0070	3/4" diameter		4.60	1.739		5.55	79.50		85.05	127
0090	1" diameter	↓	4.60	1.739	↓	8.80	79.50		88.30	131

07 91 26 – Joint Fillers

07 91 26.10 Joint Fillers	Crew	Daily Output	Labor-Hours	Unit	Material	2014 Bare Costs Labor	Equipment	Total	Total Incl O&P
0010	**JOINT FILLERS**								
4360	Butyl rubber filler, 1/4" x 1/4"	1 Bric	290	.028	L.F.	.23	1.26		1.49	2.17
4365	1/2" x 1/2"		250	.032		.92	1.46		2.38	3.23
4370	1/2" x 3/4"		210	.038		1.38	1.74		3.12	4.17
4375	3/4" x 3/4"		230	.035		2.08	1.59		3.67	4.70
4380	1" x 1"		180	.044		2.77	2.03		4.80	6.15
4390	For coloring, add					12%			
4980	Polyethylene joint backing, 1/4" x 2"	1 Bric	2.08	3.846	C.L.F.	12	175		187	280
4990	1/4" x 6"		1.28	6.250	"	28	285		313	465
5600	Silicone, room temp vulcanizing foam seal, 1/4" x 1/2"		1312	.006	L.F.	.33	.28		.61	.79
5610	1/2" x 1/2"		656	.012		.67	.56		1.23	1.59
5620	1/2" x 3/4"		442	.018		1	.83		1.83	2.36
5630	3/4" x 3/4"		328	.024		1.50	1.11		2.61	3.35
5640	1/8" x 1"		1312	.006		.33	.28		.61	.79
5650	1/8" x 3"		442	.018		1	.83		1.83	2.36
5670	1/4" x 3"		295	.027		2.01	1.24		3.25	4.09
5680	1/4" x 6"		148	.054		4.01	2.46		6.47	8.15
5690	1/2" x 6"		82	.098		8	4.45		12.45	15.60
5700	1/2" x 9"		52.50	.152		12.05	6.95		19	24
5710	1/2" x 12"		33	.242		16.05	11.05		27.10	34.50

07 92 Joint Sealants

07 92 13 – Elastomeric Joint Sealants

07 92 13.20 Caulking and Sealant Options

		Crew	Daily Output	Labor-Hours	Unit	Material	2014 Bare Costs Labor	Equipment	Total	Total Incl O&P
0010	**CAULKING AND SEALANT OPTIONS**									
0050	Latex acrylic based, bulk				Gal.	29.50			29.50	32.50
0055	Bulk in place 1/4" x 1/4" bead	1 Bric	300	.027	L.F.	.09	1.22		1.31	1.95
0060	1/4" x 3/8"		294	.027		.16	1.24		1.40	2.06
0065	1/4" x 1/2"		288	.028		.21	1.27		1.48	2.16
0075	3/8" x 3/8"		284	.028		.23	1.28		1.51	2.21
0080	3/8" x 1/2"		280	.029		.31	1.30		1.61	2.33
0085	3/8" x 5/8"		276	.029		.39	1.32		1.71	2.44
0095	3/8" x 3/4"		272	.029		.46	1.34		1.80	2.55
0100	1/2" x 1/2"		275	.029		.41	1.33		1.74	2.48
0105	1/2" x 5/8"		269	.030		.52	1.36		1.88	2.64
0110	1/2" x 3/4"		263	.030		.62	1.39		2.01	2.79
0115	1/2" x 7/8"		256	.031		.72	1.43		2.15	2.97
0120	1/2" x 1"		250	.032		.83	1.46		2.29	3.13
0125	3/4" x 3/4"		244	.033		.93	1.50		2.43	3.31
0130	3/4" x 1"		225	.036		1.24	1.62		2.86	3.84
0135	1" x 1"		200	.040		1.66	1.82		3.48	4.60
0190	Cartridges				Gal.	26.50			26.50	29
0200	11 fl. oz. cartridge				Ea.	2.28			2.28	2.51
0600	1/2" x 1/2"	1 Bric	275	.029	L.F.	.37	1.33		1.70	2.43
3200	Polyurethane, 1 or 2 component				Gal.	51			51	56
3300	Cartridges				"	64			64	70.50
3500	Bulk, in place, 1/4" x 1/4"	1 Bric	300	.027	L.F.	.17	1.22		1.39	2.03
3655	1/2" x 1/4"		288	.028		.33	1.27		1.60	2.29
3800	3/4" x 3/8"		272	.029		.75	1.34		2.09	2.87
3900	1" x 1/2"		250	.032		1.33	1.46		2.79	3.68

07 92 Joint Sealants

07 92 19 – Acoustical Joint Sealants

07 92 19.10 Acoustical Sealant	Crew	Daily Output	Labor-Hours	Unit	Material	2014 Bare Costs Labor	Equipment	Total	Total Incl O&P
0010 **ACOUSTICAL SEALANT**									
0020 Acoustical sealant, elastomeric, cartridges				Ea.	8.50			8.50	9.35
0025 In place, 1/4" x 1/4"	1 Bric	300	.027	L.F.	.35	1.22		1.57	2.23
0030 1/4" x 1/2"		288	.028		.69	1.27		1.96	2.69
0035 1/2" x 1/2"		275	.029		1.39	1.33		2.72	3.54
0040 1/2" x 3/4"		263	.030		2.08	1.39		3.47	4.40
0045 3/4" x 3/4"		244	.033		3.12	1.50		4.62	5.70
0050 1" x 1"		200	.040		5.55	1.82		7.37	8.90

Division Notes

	CREW	DAILY OUTPUT	LABOR-HOURS	UNIT	BARE COSTS				TOTAL INCL O&P
					MAT.	LABOR	EQUIP.	TOTAL	

Estimating Tips

08 10 00 Doors and Frames

All exterior doors should be addressed for their energy conservation (insulation and seals).

- Most metal doors and frames look alike, but there may be significant differences among them. When estimating these items, be sure to choose the line item that most closely compares to the specification or door schedule requirements regarding:
  - type of metal
  - metal gauge
  - door core material
  - fire rating
  - finish
- Wood and plastic doors vary considerably in price. The primary determinant is the veneer material. Lauan, birch, and oak are the most common veneers. Other variables include the following:
  - hollow or solid core
  - fire rating
  - flush or raised panel
  - finish
- Door pricing includes bore for cylindrical lockset and mortise for hinges.

08 30 00 Specialty Doors and Frames

- There are many varieties of special doors, and they are usually priced per each. Add frames, hardware, or operators required for a complete installation.

08 40 00 Entrances, Storefronts, and Curtain Walls

- Glazed curtain walls consist of the metal tube framing and the glazing material. The cost data in this subdivision is presented for the metal tube framing alone or the composite wall. If your estimate requires a detailed takeoff of the framing, be sure to add the glazing cost.

08 50 00 Windows

- Most metal windows are delivered preglazed. However, some metal windows are priced without glass. Refer to 08 80 00 Glazing for glass pricing. The grade C indicates commercial grade windows, usually ASTM C-35.
- All wood windows and vinyl are priced preglazed. The glazing is insulating glass. Some wood windows may have single pane float glass. Add the cost of screens and grills if required, and not already included.

08 70 00 Hardware

- Hardware costs add considerably to the cost of a door. The most efficient method to determine the hardware requirements for a project is to review the door and hardware schedule together.

- Door hinges are priced by the pair, with most doors requiring 1-1/2 pairs per door. The hinge prices do not include installation labor because it is included in door installation. Hinges are classified according to the frequency of use, base material, and finish.

08 80 00 Glazing

- Different openings require different types of glass. The most common types are:
  - float
  - tempered
  - insulating
  - impact-resistant
  - ballistic-resistant
- Most exterior windows are glazed with insulating glass. Entrance doors and window walls, where the glass is less than 18" from the floor, are generally glazed with tempered glass. Interior windows and some residential windows are glazed with float glass.
- Coastal communities require the use of impact-resistant glass, dependant on wind speed.
- The insulation or 'u' value is a strong consideration, along with solar heat gain, to determine total energy efficiency.

Reference Numbers

Reference numbers are shown in shaded boxes at the beginning of some major classifications. These numbers refer to related items in the Reference Section. The reference information may be an estimating procedure, an alternate pricing method, or technical information.

Note: Not all subdivisions listed here necessarily appear in this publication.

08 31 Access Doors and Panels

08 31 13 – Access Doors and Frames

08 31 13.10 Types of Framed Access Doors	Crew	Daily Output	Labor-Hours	Unit	Material	2014 Bare Costs Labor	Equipment	Total	Total Incl O&P
0010 **TYPES OF FRAMED ACCESS DOORS**									
4000 Recessed door for drywall									
4100 Metal 12" x 12"	1 Carp	6	1.333	Ea.	80	61		141	183
4150 12" x 24"		5.50	1.455		114	66.50		180.50	228
4200 24" x 36"		5	1.600		182	73.50		255.50	315

08 74 Access Control Hardware

08 74 13 – Card Key Access Control Hardware

08 74 13.50 Card Key Access

	Crew	Daily Output	Labor-Hours	Unit	Material	2014 Bare Costs Labor	Equipment	Total	Total Incl O&P
0010 **CARD KEY ACCESS**									
0020 Computerized system , processor, proximity reader and cards									
0030 Does not inculde door hardware, lockset or wiring									
0040 Card key system for 1 door				Ea.	1,175			1,175	1,300
0060 Card key system for 2 doors					2,125			2,125	2,350
0080 Card key system for 4 doors					2,650			2,650	2,900
0100 Processor for card key access system					850			850	935
0160 Magnetic lock for electric access, 600 Pound holding force					190			190	209
0170 Magnetic lock for electric access, 1200 Pound holding force					190			190	209
0200 Proximity card reader					125			125	138

08 74 13.60 Entrance Card Systems

	Crew	Daily Output	Labor-Hours	Unit	Material	2014 Bare Costs Labor	Equipment	Total	Total Incl O&P
0010 **ENTRANCE CARD SYSTEMS**									
0100 Entrance card, barium ferrite				Ea.	4.50			4.50	4.95
0120 Credential					5.50			5.50	6.05
0140 Proximity					8.50			8.50	9.35
0160 Weigand					8.75			8.75	9.65
0500 Entrance card reader, barium ferrite	R-19	4	5		305	267		572	735
0520 Credential		4	5		360	267		627	795
0540 Proximity		4	5		315	267		582	745
0560 Weigand		4	5		415	267		682	855
0600 Local processor for card system		4	5		1,900	267		2,167	2,500
0650 Scanner, eye retina		4	5		6,200	267		6,467	7,225
0700 Gate opener, cantilever	R-18	3	8.667		2,350	365		2,715	3,125
0710 Average		3	8.667		1,400	365		1,765	2,100
0712 Custom		3	8.667		1,900	365		2,265	2,650
0750 Switch, tamper	R-19	6	3.333		22	178		200	291
0900 Accessories, electric door strike/bolt		5	4		77.50	214		291.50	405
0920 Electromagnetic lock		5	4		169	214		383	505
0940 Keypad for card reader		4	5		775	267		1,042	1,250

08 74 16 – Keypad Access Control Hardware

08 74 16.50 Keypad Access

	Crew	Daily Output	Labor-Hours	Unit	Material	2014 Bare Costs Labor	Equipment	Total	Total Incl O&P
0010 **KEYPAD ACCESS**									
0340 Digital keypad, int/ext, basic, excl. striker/power/wiring	1 Elec	3	2.667	Ea.	70	142		212	290
0350 Lockset, mechanical push-button type, complete, incl hardware	1 Carp	4	2	"	335	91.50		426.50	510

08 74 19 – Biometric Identity Access Control Hardware

08 74 19.50 Biometric Identity Access

	Crew	Daily Output	Labor-Hours	Unit	Material	2014 Bare Costs Labor	Equipment	Total	Total Incl O&P
0010 **BIOMETRIC IDENTITY ACCESS**									
0200 Fingerprint scanner unit, excl striker/power supply	1 Elec	4	2	Ea.	1,400	107		1,507	1,700
0210 Fingerprint scanner unit, for computer keyboard access		8	1		1,100	53.50		1,153.50	1,275
0220 Hand geometry scanner, mem of 512 users, excl striker/power		3	2.667		2,100	142		2,242	2,525
0230 Memory upgrade for, adds 9,700 user profiles		8	1		300	53.50		353.50	410

08 74 Access Control Hardware

08 74 19 – Biometric Identity Access Control Hardware

08 74 19.50 Biometric Identity Access	Crew	Daily Output	Labor- Hours	Unit	Material	2014 Bare Costs Labor	Equipment	Total	Total Incl O&P
0240	Adds 32,500 user profiles	1 Elec	8	1	Ea.	600	53.50		653.50	740
0250	Prison type, memory of 256 users, excl striker, power		3	2.667		2,600	142		2,742	3,075
0260	Memory upgrade for, adds 3,300 user profiles		8	1		250	53.50		303.50	355
0270	Adds 9,700 user profiles		8	1		460	53.50		513.50	585
0280	Adds 27,900 user profiles		8	1		610	53.50		663.50	750
0290	All weather, mem of 512 users, excl striker/power		3	2.667		3,900	142		4,042	4,525
0300	Facial & fingerprint scanner, combination unit, excl striker/power		3	2.667		4,300	142		4,442	4,950
0310	Access for, for initial setup, excl striker/power		3	2.667		1,100	142		1,242	1,425

08 74 23 – Access Control Accessories

08 74 23.50 Security Access Control Accessories

		Crew	Daily Output	Labor- Hours	Unit	Material	Labor	Equipment	Total	Total Incl O&P
0010	**SECURITY ACCESS CONTROL ACCESSORIES**									
0360	Scanner/reader access, power supply/transf, 110 V to 12/24 V	1 Elec	4	2	Ea.	330	107		437	525
0370	Elec/mag strikers, 12/24V		4	2		85	107		192	254
0380	1 hour battery backup power supply		4	2		50	107		157	215
0390	Deadbolt, digital, batt-operated, indoor/outdoor, complete, incl hardware	1 Carp	4	2		240	91.50		331.50	405

Division Notes

	CREW	DAILY OUTPUT	LABOR-HOURS	UNIT	BARE COSTS				TOTAL INCL O&P
					MAT.	LABOR	EQUIP.	TOTAL	

Estimating Tips

General

- Room Finish Schedule: A complete set of plans should contain a room finish schedule. If one is not available, it would be well worth the time and effort to obtain one.

09 20 00 Plaster and Gypsum Board

- Lath is estimated by the square yard plus a 5% allowance for waste. Furring, channels, and accessories are measured by the linear foot. An extra foot should be allowed for each accessory miter or stop.

- Plaster is also estimated by the square yard. Deductions for openings vary by preference, from zero deduction to 50% of all openings over 2 feet in width. The estimator should allow one extra square foot for each linear foot of horizontal interior or exterior angle located below the ceiling level. Also, double the areas of small radius work.

- Drywall accessories, studs, track, and acoustical caulking are all measured by the linear foot. Drywall taping is figured by the square foot. Gypsum wallboard is estimated by the square foot. No material deductions should be made for door or window openings under 32 S.F.

09 60 00 Flooring

- Tile and terrazzo areas are taken off on a square foot basis. Trim and base materials are measured by the linear foot. Accent tiles are listed per each. Two basic methods of installation are used. Mud set is approximately 30% more expensive than thin set. In terrazzo work, be sure to include the linear footage of embedded decorative strips, grounds, machine rubbing, and power cleanup.

- Wood flooring is available in strip, parquet, or block configuration. The latter two types are set in adhesives with quantities estimated by the square foot. The laying pattern will influence labor costs and material waste. In addition to the material and labor for laying wood floors, the estimator must make allowances for sanding and finishing these areas unless the flooring is prefinished.

- Sheet flooring is measured by the square yard. Roll widths vary, so consideration should be given to use the most economical width, as waste must be figured into the total quantity. Consider also the installation methods available, direct glue down or stretched.

09 70 00 Wall Finishes

- Wall coverings are estimated by the square foot. The area to be covered is measured, length by height of wall above baseboards, to calculate the square footage of each wall. This figure is divided by the number of square feet in the single roll which is being used. Deduct, in full, the areas of openings such as doors and windows. Where a pattern match is required allow 25%–30% waste.

09 80 00 Acoustic Treatment

- Acoustical systems fall into several categories. The takeoff of these materials should be by the square foot of area with a 5% allowance for waste. Do not forget about scaffolding, if applicable, when estimating these systems.

09 90 00 Painting and Coating

- A major portion of the work in painting involves surface preparation. Be sure to include cleaning, sanding, filling, and masking costs in the estimate.

- Protection of adjacent surfaces is not included in painting costs. When considering the method of paint application, an important factor is the amount of protection and masking required. These must be estimated separately and may be the determining factor in choosing the method of application.

Reference Numbers

Reference numbers are shown in shaded boxes at the beginning of some major classifications. These numbers refer to related items in the Reference Section. The reference information may be an estimating procedure, an alternate pricing method, or technical information.

Note: Not all subdivisions listed here necessarily appear in this publication.

09 22 Supports for Plaster and Gypsum Board

09 22 03 – Fastening Methods for Finishes

09 22 03.20 Drilling Plaster/Drywall	Crew	Daily Output	Labor-Hours	Unit	Material	2014 Bare Costs Labor	Equipment	Total	Total Incl O&P
0010 **DRILLING PLASTER/DRYWALL**									
1100 Drilling & layout for drywall/plaster walls, up to 1" deep, no anchor									
1200 Holes, 1/4" diameter	1 Carp	150	.053	Ea.	.01	2.45		2.46	3.78
1300 3/8" diameter		140	.057		.01	2.62		2.63	4.05
1400 1/2" diameter		130	.062		.01	2.82		2.83	4.36
1500 3/4" diameter		120	.067		.01	3.06		3.07	4.74
1600 1" diameter		110	.073		.02	3.33		3.35	5.15
1700 1-1/4" diameter		100	.080		.03	3.67		3.70	5.70
1800 1-1/2" diameter		90	.089		.04	4.08		4.12	6.35
1900 For ceiling installations, add						40%			

09 69 Access Flooring

09 69 13 – Rigid-Grid Access Flooring

09 69 13.10 Access Floors	Crew	Daily Output	Labor-Hours	Unit	Material	2014 Bare Costs Labor	Equipment	Total	Total Incl O&P
0010 **ACCESS FLOORS**									
0015 Access floor package including panel, pedestal, stringers & laminate cover									
0100 Computer room, greater than 6,000 S.F.	4 Carp	750	.043	S.F.	7.50	1.96		9.46	11.25
0110 Less than 6,000 S.F.	2 Carp	375	.043		8.15	1.96		10.11	12
0120 Office, greater than 6,000 S.F.	4 Carp	1050	.030		5.35	1.40		6.75	8.05
0250 Panels, particle board or steel, 1250# load, no covering, under 6,000 S.F.	2 Carp	600	.027		3.75	1.22		4.97	6
0300 Over 6,000 S.F.		640	.025		3.21	1.15		4.36	5.30
0400 Aluminum, 24" panels		500	.032		32.50	1.47		33.97	38.50
0600 For carpet covering, add					8.75			8.75	9.65
0700 For vinyl floor covering, add					9.05			9.05	9.95
0900 For high pressure laminate covering, add					7.50			7.50	8.25
0910 For snap on stringer system, add	2 Carp	1000	.016		1.56	.73		2.29	2.85
0950 Office applications, steel or concrete panels,									
0960 no covering, over 6,000 S.F.	2 Carp	960	.017	S.F.	10.55	.76		11.31	12.80
1000 Machine cutouts after initial installation	1 Carp	50	.160	Ea.	20	7.35		27.35	33.50
1050 Pedestals, 6" to 12"	2 Carp	85	.188		8.30	8.65		16.95	22.50
1100 Air conditioning grilles, 4" x 12"	1 Carp	17	.471		68	21.50		89.50	109
1150 4" x 18"	"	14	.571		93.50	26		119.50	144
1200 Approach ramps, minimum	2 Carp	60	.267	S.F.	24.50	12.25		36.75	46
1300 Maximum	"	40	.400	"	34	18.35		52.35	66
1500 Handrail, 2 rail, aluminum	1 Carp	15	.533	L.F.	100	24.50		124.50	148

Estimating Tips

General

- The items in this division are usually priced per square foot or each.
- Many items in Division 10 require some type of support system or special anchors that are not usually furnished with the item. The required anchors must be added to the estimate in the appropriate division.
- Some items in Division 10, such as lockers, may require assembly before installation. Verify the amount of assembly required. Assembly can often exceed installation time.

10 20 00 Interior Specialties

- Support angles and blocking are not included in the installation of toilet compartments, shower/dressing compartments, or cubicles. Appropriate line items from Divisions 5 or 6 may need to be added to support the installations.
- Toilet partitions are priced by the stall. A stall consists of a side wall, pilaster, and door with hardware. Toilet tissue holders and grab bars are extra.
- The required acoustical rating of a folding partition can have a significant impact on costs. Verify the sound transmission coefficient rating of the panel priced to the specification requirements.

- Grab bar installation does not include supplemental blocking or backing to support the required load. When grab bars are installed at an existing facility, provisions must be made to attach the grab bars to solid structure.

Reference Numbers

Reference numbers are shown in shaded boxes at the beginning of some major classifications. These numbers refer to related items in the Reference Section. The reference information may be an estimating procedure, an alternate pricing method, or technical information.

Note: Not all subdivisions listed here necessarily appear in this publication.

Division 10 - Specialties

10 28 16 – Bath Accessories

10 28 16.20 Medicine Cabinets	Crew	Daily Output	Labor-Hours	Unit	Material	2014 Bare Costs Labor	Equipment	Total	Total Incl O&P
0010 **MEDICINE CABINETS**									
0020 With mirror, sst frame, 16" x 22", unlighted	1 Carp	14	.571	Ea.	95	26		121	146
0100 Wood frame		14	.571		136	26		162	191
0300 Sliding mirror doors, 20" x 16" x 4-3/4", unlighted		7	1.143		125	52.50		177.50	218
0400 24" x 19" x 8-1/2", lighted		5	1.600		179	73.50		252.50	310
0600 Triple door, 30" x 32", unlighted, plywood body		7	1.143		320	52.50		372.50	430
0700 Steel body		7	1.143		350	52.50		402.50	465
0900 Oak door, wood body, beveled mirror, single door		7	1.143		195	52.50		247.50	295
1000 Double door		6	1.333		390	61		451	525
1200 Hotel cabinets, stainless, with lower shelf, unlighted		10	.800		200	36.50		236.50	276
1300 Lighted		5	1.600		305	73.50		378.50	450

Estimating Tips

General

- The items in this division are usually priced per square foot or each. Many of these items are purchased by the owner for installation by the contractor. Check the specifications for responsibilities and include time for receiving, storage, installation, and mechanical and electrical hookups in the appropriate divisions.

- Many items in Division 11 require some type of support system that is not usually furnished with the item. Examples of these systems include blocking for the attachment of casework and support angles for ceiling-hung projection screens. The required blocking or supports must be added to the estimate in the appropriate division.

- Some items in Division 11 may require assembly or electrical hookups. Verify the amount of assembly required or the need for a hard electrical connection and add the appropriate costs.

Reference Numbers

Reference numbers are shown in shaded boxes at the beginning of some major classifications. These numbers refer to related items in the Reference Section. The reference information may be an estimating procedure, an alternate pricing method, or technical information.

Note: Not all subdivisions listed here necessarily appear in this publication.

11 12 Parking Control Equipment

11 12 13 – Parking Key and Card Control Units

11 12 13.10 Parking Control Units

		Crew	Daily Output	Labor-Hours	Unit	Material	2014 Bare Costs Labor	Equipment	Total	Total Incl O&P
0010	**PARKING CONTROL UNITS**									
5100	Card reader	1 Elec	2	4	Ea.	1,925	213		2,138	2,450
5120	Proximity with customer display	2 Elec	1	16		5,500	855		6,355	7,325
6000	Parking control software, minimum	1 Elec	.50	16		23,600	855		24,455	27,200
6020	Maximum	"	.20	40		103,000	2,125		105,125	116,500

11 12 16 – Parking Ticket Dispensers

11 12 16.10 Ticket Dispensers

		Crew	Daily Output	Labor-Hours	Unit	Material	Labor	Equipment	Total	Total Incl O&P
0010	**TICKET DISPENSERS**									
5900	Ticket spitter with time/date stamp, standard	2 Elec	2	8	Ea.	6,400	425		6,825	7,675
5920	Mag stripe encoding	"	2	8	"	18,600	425		19,025	21,000

11 12 26 – Parking Fee Collection Equipment

11 12 26.13 Parking Fee Coin Collection Equipment

		Crew	Daily Output	Labor-Hours	Unit	Material	Labor	Equipment	Total	Total Incl O&P
0010	**PARKING FEE COIN COLLECTION EQUIPMENT**									
5200	Cashier booth, average	B-22	1	30	Ea.	10,300	1,275	210	11,785	13,500
5300	Collector station, pay on foot	2 Elec	.20	80		109,500	4,275		113,775	127,000
5320	Credit card only	"	.50	32		20,200	1,700		21,900	24,900

11 12 26.23 Fee Equipment

		Crew	Daily Output	Labor-Hours	Unit	Material	Labor	Equipment	Total	Total Incl O&P
0010	**FEE EQUIPMENT**									
5600	Fee computer	1 Elec	1.50	5.333	Ea.	14,000	285		14,285	15,800

11 12 33 – Parking Gates

11 12 33.13 Lift Arm Parking Gates

		Crew	Daily Output	Labor-Hours	Unit	Material	Labor	Equipment	Total	Total Incl O&P
0010	**LIFT ARM PARKING GATES**									
5000	Barrier gate with programmable controller	2 Elec	3	5.333	Ea.	3,350	285		3,635	4,100
5020	Industrial		3	5.333		5,000	285		5,285	5,925
5500	Exit verifier		1	16		17,400	855		18,255	20,500
5700	Full sign, 4" letters	1 Elec	2	4		1,225	213		1,438	1,650
5800	Inductive loop	2 Elec	4	4		169	213		382	505
5950	Vehicle detector, microprocessor based	1 Elec	3	2.667		420	142		562	680

11 23 Commercial Laundry and Dry Cleaning Equipment

11 23 13 – Dry Cleaning Equipment

11 23 13.13 Dry Cleaners

		Crew	Daily Output	Labor-Hours	Unit	Material	Labor	Equipment	Total	Total Incl O&P
0010	**DRY CLEANERS** Not incl. rough-in									
2000	Dry cleaners, electric, 20 lb. capacity	L-1	.20	80	Ea.	33,300	4,425		37,725	43,300
2050	25 lb. capacity		.17	94.118		48,000	5,225		53,225	61,000
2100	30 lb. capacity		.15	106		50,500	5,925		56,425	64,500
2150	60 lb. capacity		.09	177		78,000	9,850		87,850	101,000

11 23 19 – Finishing Equipment

11 23 19.13 Folders and Spreaders

		Crew	Daily Output	Labor-Hours	Unit	Material	Labor	Equipment	Total	Total Incl O&P
0010	**FOLDERS AND SPREADERS**									
3500	Folders, blankets & sheets, minimum	1 Elec	.17	47.059	Ea.	32,700	2,500		35,200	39,800
3700	King size with automatic stacker		.10	80		59,000	4,275		63,275	71,500
3800	For conveyor delivery, add		.45	17.778		14,300	950		15,250	17,200

11 23 23 – Commercial Ironing Equipment

11 23 23.13 Irons and Pressers

		Crew	Daily Output	Labor-Hours	Unit	Material	Labor	Equipment	Total	Total Incl O&P
0010	**IRONS AND PRESSERS**									
4500	Ironers, institutional, 110", single roll	1 Elec	.20	40	Ea.	31,700	2,125		33,825	38,100

11 23 Commercial Laundry and Dry Cleaning Equipment

11 23 26 – Commercial Washers and Extractors

11 23 26.13 Washers and Extractors	Crew	Daily Output	Labor-Hours	Unit	Material	2014 Bare Costs Labor	Equipment	Total	Total Incl O&P
0010 **WASHERS AND EXTRACTORS**, not including rough-in									
6000 Combination washer/extractor, 20 lb. capacity	L-6	1.50	8	Ea.	5,875	450		6,325	7,125
6100 30 lb. capacity		.80	15		9,150	840		9,990	11,400
6200 50 lb. capacity		.68	17.647		10,700	990		11,690	13,300
6300 75 lb. capacity		.30	40		20,200	2,250		22,450	25,600
6350 125 lb. capacity	↓	.16	75	↓	27,300	4,200		31,500	36,500

11 23 33 – Coin-Operated Laundry Equipment

11 23 33.13 Coin Operated Washers and Dryers

	Crew	Daily Output	Labor-Hours	Unit	Material	Labor	Equipment	Total	Total Incl O&P
0010 **COIN OPERATED WASHERS AND DRYERS**									
5290 Clothes washer									
5300 Commercial, coin operated, average	1 Plum	3	2.667	Ea.	1,250	153		1,403	1,600

11 26 Unit Kitchens

11 26 13 – Metal Unit Kitchens

11 26 13.10 Commercial Unit Kitchens

	Crew	Daily Output	Labor-Hours	Unit	Material	Labor	Equipment	Total	Total Incl O&P
0010 **COMMERCIAL UNIT KITCHENS**									
1500 Combination range, refrigerator and sink, 30" wide, minimum	L-1	2	8	Ea.	1,100	445		1,545	1,875
1550 Maximum		1	16		1,450	885		2,335	2,925
1570 60" wide, average		1.40	11.429		1,525	635		2,160	2,625
1590 72" wide, average		1.20	13.333		1,600	740		2,340	2,850
1600 Office model, 48" wide		2	8		2,000	445		2,445	2,875
1620 Refrigerator and sink only	↓	2.40	6.667	↓	2,500	370		2,870	3,300
1640 Combination range, refrigerator, sink, microwave									
1660 Oven and ice maker	L-1	.80	20	Ea.	4,500	1,100		5,600	6,625

11 31 Residential Appliances

11 31 13 – Residential Kitchen Appliances

11 31 13.13 Cooking Equipment

	Crew	Daily Output	Labor-Hours	Unit	Material	Labor	Equipment	Total	Total Incl O&P
0010 **COOKING EQUIPMENT**									
0020 Cooking range, 30" free standing, 1 oven, minimum	2 Clab	10	1.600	Ea.	435	58.50		493.50	570
0050 Maximum		4	4		2,000	147		2,147	2,425
0700 Free-standing, 1 oven, 21" wide range, minimum		10	1.600		450	58.50		508.50	585
0750 21" wide, maximum	↓	4	4		445	147		592	715
0900 Countertop cooktops, 4 burner, standard, minimum	1 Elec	6	1.333		293	71		364	425
0950 Maximum		3	2.667		1,300	142		1,442	1,650
1050 As above, but with grill and griddle attachment, minimum		6	1.333		1,225	71		1,296	1,450
1100 Maximum		3	2.667		3,650	142		3,792	4,250
1200 Induction cooktop, 30" wide		3	2.667		1,175	142		1,317	1,525
1250 Microwave oven, minimum		4	2		119	107		226	291
1300 Maximum	↓	2	4	↓	460	213		673	825

11 31 13.33 Kitchen Cleaning Equipment

	Crew	Daily Output	Labor-Hours	Unit	Material	Labor	Equipment	Total	Total Incl O&P
0010 **KITCHEN CLEANING EQUIPMENT**									
2750 Dishwasher, built-in, 2 cycles, minimum	L-1	4	4	Ea.	235	222		457	595
2800 Maximum		2	8		430	445		875	1,125
2950 4 or more cycles, minimum		4	4		370	222		592	745
2960 Average		4	4		495	222		717	880
3000 Maximum	↓	2	8	↓	1,075	445		1,520	1,875

11 31 Residential Appliances

11 31 13 – Residential Kitchen Appliances

11 31 13.43 Waste Disposal Equipment

		Crew	Daily Output	Labor-Hours	Unit	Material	2014 Bare Costs Labor	2014 Bare Costs Equipment	Total	Total Incl O&P
0010	**WASTE DISPOSAL EQUIPMENT**									
3300	Garbage disposal, sink type, minimum	L-1	10	1.600	Ea.	86.50	88.50		175	228
3350	Maximum	"	10	1.600	"	217	88.50		305.50	370

11 31 13.53 Kitchen Ventilation Equipment

		Crew	Daily Output	Labor-Hours	Unit	Material	2014 Bare Costs Labor	2014 Bare Costs Equipment	Total	Total Incl O&P
0010	**KITCHEN VENTILATION EQUIPMENT**									
4150	Hood for range, 2 speed, vented, 30" wide, minimum	L-3	5	3.200	Ea.	69.50	160		229.50	320
4200	Maximum		3	5.333		770	266		1,036	1,250
4300	42" wide, minimum		5	3.200		166	160		326	425
4330	Custom		5	3.200		1,650	160		1,810	2,050
4350	Maximum		3	5.333		2,000	266		2,266	2,600
4500	For ventless hood, 2 speed, add					18.45			18.45	20.50
4650	For vented 1 speed, deduct from maximum					49.50			49.50	54.50

11 31 23 – Residential Laundry Appliances

11 31 23.13 Washers

		Crew	Daily Output	Labor-Hours	Unit	Material	2014 Bare Costs Labor	2014 Bare Costs Equipment	Total	Total Incl O&P
0010	**WASHERS**									
5000	Residential, 4 cycle, average	1 Plum	3	2.667	Ea.	865	153		1,018	1,175

11 31 33 – Miscellaneous Residential Appliances

11 31 33.23 Water Heaters

		Crew	Daily Output	Labor-Hours	Unit	Material	2014 Bare Costs Labor	2014 Bare Costs Equipment	Total	Total Incl O&P
0010	**WATER HEATERS**									
6900	Electric, glass lined, 30 gallon, minimum	L-1	5	3.200	Ea.	430	177		607	735
6950	Maximum		3	5.333		595	296		891	1,100
7100	80 gallon, minimum		2	8		1,225	445		1,670	2,025
7150	Maximum		1	16		1,700	885		2,585	3,175

11 31 33.43 Air Quality

		Crew	Daily Output	Labor-Hours	Unit	Material	2014 Bare Costs Labor	2014 Bare Costs Equipment	Total	Total Incl O&P
0010	**AIR QUALITY**									
2450	Dehumidifier, portable, automatic, 15 pint	1 Elec	4	2	Ea.	150	107		257	325
2550	40 pint					207			207	228
3550	Heater, electric, built-in, 1250 watt, ceiling type, minimum	1 Elec	4	2		106	107		213	276
3600	Maximum		3	2.667		172	142		314	405
3700	Wall type, minimum		4	2		170	107		277	345
3750	Maximum		3	2.667		183	142		325	415
3900	1500 watt wall type, with blower		4	2		170	107		277	345
3950	3000 watt		3	2.667		350	142		492	600
4850	Humidifier, portable, 8 gallons per day					131			131	144
5000	15 gallons per day					208			208	229

11 41 Foodservice Storage Equipment

11 41 13 – Refrigerated Food Storage Cases

11 41 13.20 Refrigerated Food Storage Equipment

		Crew	Daily Output	Labor-Hours	Unit	Material	2014 Bare Costs Labor	2014 Bare Costs Equipment	Total	Total Incl O&P
0010	**REFRIGERATED FOOD STORAGE EQUIPMENT**									
2350	Cooler, reach-in, beverage, 6' long	Q-1	6	2.667	Ea.	3,550	138		3,688	4,100

11 42 Food Preparation Equipment

11 42 10 – Commercial Food Preparation Equipment

11 42 10.10 Choppers, Mixers and Misc. Equipment	Crew	Daily Output	Labor-Hours	Unit	Material	2014 Bare Costs Labor	Equipment	Total	Total Incl O&P
0010 **CHOPPERS, MIXERS AND MISC. EQUIPMENT**									
1850 Coffee urn, twin 6 gallon urns	1 Plum	2	4	Ea.	2,375	230		2,605	2,975
3800 Food mixers, bench type, 20 quarts	L-7	7	4		2,925	177		3,102	3,500
3900 60 quarts	"	5	5.600	↓	11,200	248		11,448	12,700

11 44 Food Cooking Equipment

11 44 13 – Commercial Ranges

11 44 13.10 Cooking Equipment

	Crew	Daily Output	Labor-Hours	Unit	Material	2014 Bare Costs Labor	Equipment	Total	Total Incl O&P
0010 **COOKING EQUIPMENT**									
0020 Bake oven, gas, one section	Q-1	8	2	Ea.	5,550	104		5,654	6,250
1300 Broiler, without oven, standard	"	8	2		3,500	104		3,604	4,000
1550 Infrared	L-7	4	7		7,325	310		7,635	8,550
6350 Kettle, w/steam jacket, tilting, w/positive lock, SS, 20 gallons		7	4		8,125	177		8,302	9,225
6600 60 gallons		6	4.667		10,800	207		11,007	12,200
8850 Steamer, electric 27 KW		7	4		10,600	177		10,777	12,000
9100 Electric, 10 KW or gas 100,000 BTU	↓	5	5.600		6,250	248		6,498	7,250
9150 Toaster, conveyor type, 16-22 slices per minute				↓	1,075			1,075	1,175
9200 For deluxe models of above equipment, add					75%				
9400 Rule of thumb: Equipment cost based									
9410 on kitchen work area									
9420 Office buildings, minimum	L-7	77	.364	S.F.	89.50	16.10		105.60	123
9450 Maximum		58	.483		151	21.50		172.50	199
9550 Public eating facilities, minimum		77	.364		117	16.10		133.10	154
9600 Maximum		46	.609		191	27		218	252
9750 Hospitals, minimum		58	.483		121	21.50		142.50	166
9800 Maximum	↓	39	.718	↓	222	32		254	293

11 46 Food Dispensing Equipment

11 46 16 – Service Line Equipment

11 46 16.10 Commercial Food Dispensing Equipment

	Crew	Daily Output	Labor-Hours	Unit	Material	2014 Bare Costs Labor	Equipment	Total	Total Incl O&P
0010 **COMMERCIAL FOOD DISPENSING EQUIPMENT**									
3300 Food warmer, counter, 1.2 KW				Ea.	660			660	725
3550 1.6 KW					2,100			2,100	2,325
5750 Ice dispenser 567 pound	Q-1	6	2.667	↓	5,150	138		5,288	5,875

11 48 Cleaning and Disposal Equipment

11 48 13 – Commercial Dishwashers

11 48 13.10 Dishwashers

		Crew	Daily Output	Labor-Hours	Unit	Material	2014 Bare Costs Labor	Equipment	Total	Total Incl O&P
0010 **DISHWASHERS**										
2700 Dishwasher, commercial, rack type										
2720 10 to 12 racks per hour		Q-1	3.20	5	Ea.	3,475	259		3,734	4,225
2730 Energy star rated, 35 to 40 racks/hour	G		1.30	12.308		4,650	640		5,290	6,075
2740 50 to 60 racks/hour	G	↓	1.30	12.308	↓	10,100	640		10,740	12,100

11 52 Audio-Visual Equipment

11 52 16 – Projectors

11 52 16.10 Movie Equipment

11 52 16.10 Movie Equipment	Crew	Daily Output	Labor-Hours	Unit	Material	2014 Bare Costs Labor	Equipment	Total	Total Incl O&P
0010 **MOVIE EQUIPMENT**									
0020 Changeover, minimum				Ea.	465			465	510
0100 Maximum					905			905	995
0800 Lamphouses, incl. rectifiers, xenon, 1,000 watt	1 Elec	2	4		6,650	213		6,863	7,625
0900 1,600 watt		2	4		7,100	213		7,313	8,125
1000 2,000 watt		1.50	5.333		7,625	285		7,910	8,800
1100 4,000 watt		1.50	5.333		9,425	285		9,710	10,800
3700 Sound systems, incl. amplifier, mono, minimum		.90	8.889		3,300	475		3,775	4,325
3800 Dolby/Super Sound, maximum		.40	20		18,100	1,075		19,175	21,500
4100 Dual system, 2 channel, front surround, minimum		.70	11.429		4,625	610		5,235	6,025
4200 Dolby/Super Sound, 4 channel, maximum		.40	20		16,500	1,075		17,575	19,800
5300 Speakers, recessed behind screen, minimum		2	4		1,050	213		1,263	1,500
5400 Maximum		1	8		3,100	425		3,525	4,050
7000 For automation, varying sophistication, minimum		1	8	System	2,375	425		2,800	3,275
7100 Maximum	2 Elec	.30	53.333	"	5,575	2,850		8,425	10,400

11 52 16.20 Movie Equipment- Digital

	Crew	Daily Output	Labor-Hours	Unit	Material	2014 Bare Costs Labor	Equipment	Total	Total Incl O&P
0010 **MOVIE EQUIPMENT- DIGITAL**									
1000 Digital 2K projection system, 98" DMD	1 Elec	2	4	Ea.	44,000	213		44,213	48,700
1100 OEM lens		2	4		5,375	213		5,588	6,225
2000 Pedestal with power distribution		2	4		2,050	213		2,263	2,575
3000 Software		2	4		1,725	213		1,938	2,225

11 53 Laboratory Equipment

11 53 19 – Laboratory Sterilizers

11 53 19.13 Sterilizers

	Crew	Daily Output	Labor-Hours	Unit	Material	2014 Bare Costs Labor	Equipment	Total	Total Incl O&P
0010 **STERILIZERS**									
0700 Glassware washer, undercounter, minimum	L-1	1.80	8.889	Ea.	6,250	495		6,745	7,625
0710 Maximum	"	1	16	"	13,100	885		13,985	15,700

11 53 33 – Emergency Safety Appliances

11 53 33.13 Emergency Equipment

	Crew	Daily Output	Labor-Hours	Unit	Material	2014 Bare Costs Labor	Equipment	Total	Total Incl O&P
0010 **EMERGENCY EQUIPMENT**									
1400 Safety equipment, eye wash, hand held				Ea.	405			405	445
1450 Deluge shower				"	760			760	840

11 53 43 – Service Fittings and Accessories

11 53 43.13 Fittings

	Crew	Daily Output	Labor-Hours	Unit	Material	2014 Bare Costs Labor	Equipment	Total	Total Incl O&P
0010 **FITTINGS**									
8000 Alternate pricing method: as percent of lab furniture									
8050 Installation, not incl. plumbing & duct work				% Furn.				22%	22%
8100 Plumbing, final connections, simple system								10%	10%
8110 Moderately complex system								15%	15%
8120 Complex system								20%	20%
8150 Electrical, simple system								10%	10%
8160 Moderately complex system								20%	20%
8170 Complex system								35%	35%

11 61 Broadcast, Theater, and Stage Equipment

11 61 33 – Rigging Systems and Controls

11 61 33.10 Controls	Crew	Daily Output	Labor-Hours	Unit	Material	2014 Bare Costs Labor	Equipment	Total	Total Incl O&P
0010 **CONTROLS**									
0050　Control boards with dimmers and breakers, minimum	1 Elec	1	8	Ea.	12,500	425		12,925	14,300
0100　　Average		.50	16		39,200	855		40,055	44,400
0150　　Maximum		.20	40		127,500	2,125		129,625	143,500

11 66 Athletic Equipment

11 66 43 – Interior Scoreboards

11 66 43.10 Scoreboards	Crew	Daily Output	Labor-Hours	Unit	Material	2014 Bare Costs Labor	Equipment	Total	Total Incl O&P
0010 **SCOREBOARDS**									
7000　Baseball, minimum	R-3	1.30	15.385	Ea.	4,075	815	108	4,998	5,825
7200　　Maximum		.05	400		17,900	21,200	2,800	41,900	54,500
7300　Football, minimum		.86	23.256		5,275	1,225	163	6,663	7,825
7400　　Maximum		.20	100		15,700	5,300	700	21,700	26,000
7500　Basketball (one side), minimum		2.07	9.662		2,375	510	67.50	2,952.50	3,475
7600　　Maximum		.30	66.667		3,475	3,525	465	7,465	9,650
7700　Hockey-basketball (four sides), minimum		.25	80		5,600	4,225	560	10,385	13,100
7800　　Maximum		.15	133		5,675	7,050	935	13,660	17,900

11 71 Medical Sterilizing Equipment

11 71 10 – Medical Sterilizers & Distillers

11 71 10.10 Sterilizers and Distillers	Crew	Daily Output	Labor-Hours	Unit	Material	2014 Bare Costs Labor	Equipment	Total	Total Incl O&P
0010 **STERILIZERS AND DISTILLERS**									
6200　　Steam generators, electric 10 kW to 180 kW, freestanding									
6250　　Minimum	1 Elec	3	2.667	Ea.	8,675	142		8,817	9,775
6300　　Maximum	"	.70	11.429	"	28,800	610		29,410	32,600

11 72 Examination and Treatment Equipment

11 72 53 – Treatment Equipment

11 72 53.13 Medical Treatment Equipment	Crew	Daily Output	Labor-Hours	Unit	Material	2014 Bare Costs Labor	Equipment	Total	Total Incl O&P
0010 **MEDICAL TREATMENT EQUIPMENT**									
6700　Surgical lights, doctor's office, single arm	2 Elec	2	8	Ea.	2,325	425		2,750	3,200
6750　　Dual arm	"	1	16	"	4,525	855		5,380	6,250

11 76 Operating Room Equipment

11 76 10 – Operating Room Equipment

11 76 10.10 Surgical Equipment	Crew	Daily Output	Labor-Hours	Unit	Material	2014 Bare Costs Labor	Equipment	Total	Total Incl O&P
0010 **SURGICAL EQUIPMENT**									
6800　Surgical lights, major operating room, dual head, minimum	2 Elec	1	16	Ea.	4,450	855		5,305	6,150
6850　　Maximum	"	1	16	"	29,700	855		30,555	34,000

11 78 Mortuary Equipment

11 78 13 – Mortuary Refrigerators

11 78 13.10 Mortuary and Autopsy Equipment	Crew	Daily Output	Labor-Hours	Unit	Material	2014 Bare Costs Labor	Equipment	Total	Total Incl O&P
0010 **MORTUARY AND AUTOPSY EQUIPMENT**									
0015 Autopsy table, standard	1 Plum	1	8	Ea.	9,625	460		10,085	11,300

11 78 16 – Crematorium Equipment

11 78 16.10 Crematory

0010 **CREMATORY**									
1500 Crematory, not including building, 1 place	Q-3	.20	160	Ea.	71,500	8,775		80,275	92,000
1750 2 place	"	.10	320	"	102,500	17,500		120,000	139,000

11 82 Solid Waste Handling Equipment

11 82 26 – Facility Waste Compactors

11 82 26.10 Compactors

	Crew	Daily Output	Labor-Hours	Unit	Material	Labor	Equipment	Total	Total Incl O&P
0010 **COMPACTORS**									
0020 Compactors, 115 volt, 250#/hr., chute fed	L-4	1	24	Ea.	11,900	1,025		12,925	14,700
1000 Heavy duty industrial compactor, 0.5 C.Y. capacity		1	24		10,000	1,025		11,025	12,600
1050 1.0 C.Y. capacity		1	24		15,000	1,025		16,025	18,100
1100 3.0 C.Y. capacity		.50	48		25,600	2,075		27,675	31,400
1150 5.0 C.Y. capacity		.50	48		32,200	2,075		34,275	38,600
1200 Combination shredder/compactor (5,000 lb./hr.)		.50	48		62,500	2,075		64,575	72,000
1400 For handling hazardous waste materials, 55 gallon drum packer, std.					19,500			19,500	21,500
1410 55 gallon drum packer w/HEPA filter					24,300			24,300	26,700
1420 55 gallon drum packer w/charcoal & HEPA filter					32,400			32,400	35,600
1430 All of the above made explosion proof, add					1,425			1,425	1,575
5800 Shredder, industrial, minimum					23,700			23,700	26,100
5850 Maximum					127,000			127,000	139,500
5900 Baler, industrial, minimum					9,500			9,500	10,500
5950 Maximum					554,000			554,000	609,500

Estimating Tips

General

- The items in this division are usually priced per square foot or each. Most of these items are purchased by the owner and installed by the contractor. Do not assume the items in Division 12 will be purchased and installed by the contractor. Check the specifications for responsibilities and include receiving, storage, installation, and mechanical and electrical hookups in the appropriate divisions.

- Some items in this division require some type of support system that is not usually furnished with the item. Examples of these systems include blocking for the attachment of casework and heavy drapery rods. The required blocking must be added to the estimate in the appropriate division.

Reference Numbers

Reference numbers are shown in shaded boxes at the beginning of some major classifications. These numbers refer to related items in the Reference Section. The reference information may be an estimating procedure, an alternate pricing method, or technical information.

Note: Not all subdivisions listed here necessarily appear in this publication.

Division 12 - Furnishings

12 46 Furnishing Accessories

12 46 19 – Clocks

12 46 19.50 Wall Clocks		Crew	Daily Output	Labor-Hours	Unit	Material	2014 Bare Costs Labor	2014 Bare Costs Equipment	Total	Total Incl O&P
0010	**WALL CLOCKS**									
0080	12" diameter, single face	1 Elec	8	1	Ea.	128	53.50		181.50	220
0100	Double face	"	6.20	1.290	"	294	69		363	430

Estimating Tips

General

- The items and systems in this division are usually estimated, purchased, supplied, and installed as a unit by one or more subcontractors. The estimator must ensure that all parties are operating from the same set of specifications and assumptions, and that all necessary items are estimated and will be provided. Many times the complex items and systems are covered, but the more common ones, such as excavation or a crane, are overlooked for the very reason that everyone assumes nobody could miss them. The estimator should be the central focus and be able to ensure that all systems are complete.

- Another area where problems can develop in this division is at the interface between systems. The estimator must ensure, for instance, that anchor bolts, nuts, and washers are estimated and included for the air-supported structures and pre-engineered buildings to be bolted to their foundations. Utility supply is a common area where essential items or pieces of equipment can be missed or overlooked due to the fact that each subcontractor may feel it is another's responsibility. The estimator should also be aware of certain items which may be supplied as part of a package but installed by others, and ensure that the installing contractor's estimate includes the cost of installation. Conversely, the estimator must also ensure that items are not costed by two different subcontractors, resulting in an inflated overall estimate.

13 30 00 Special Structures

- The foundations and floor slab, as well as rough mechanical and electrical, should be estimated, as this work is required for the assembly and erection of the structure. Generally, as noted in the book, the pre-engineered building comes as a shell. Pricing is based on the size and structural design parameters stated in the reference section. Additional features, such as windows and doors with their related structural framing, must also be included by the estimator. Here again, the estimator must have a clear understanding of the scope of each portion of the work and all the necessary interfaces.

Reference Numbers

Reference numbers are shown in shaded boxes at the beginning of some major classifications. These numbers refer to related items in the Reference Section. The reference information may be an estimating procedure, an alternate pricing method, or technical information.

Note: Not all subdivisions listed here necessarily appear in this publication.

13 11 Swimming Pools

13 11 13 – Below-Grade Swimming Pools

13 11 13.50 Swimming Pools

		Crew	Daily Output	Labor-Hours	Unit	Material	2014 Bare Costs Labor	Equipment	Total	Total Incl O&P
0010	**SWIMMING POOLS** Residential in-ground, vinyl lined, concrete									
0020	Sides including equipment, sand bottom	B-52	300	.187	SF Surf	21.50	7.95	1.98	31.43	38
0100	Metal or polystyrene sides	B-14	410	.117		17.75	4.53	.89	23.17	27.50
0200	Add for vermiculite bottom					1.35			1.35	1.49
0500	Gunite bottom and sides, white plaster finish									
0600	12' x 30' pool	B-52	145	.386	SF Surf	39.50	16.40	4.11	60.01	73
0720	16' x 32' pool		155	.361		35.50	15.35	3.84	54.69	66.50
0750	20' x 40' pool		250	.224		32	9.50	2.38	43.88	52
0810	Concrete bottom and sides, tile finish									
0820	12' x 30' pool	B-52	80	.700	SF Surf	40	30	7.45	77.45	97.50
0830	16' x 32' pool		95	.589		33	25	6.25	64.25	82
0840	20' x 40' pool		130	.431		26.50	18.30	4.58	49.38	62
1100	Motel, gunite with plaster finish, incl. medium									
1150	capacity filtration & chlorination	B-52	115	.487	SF Surf	48.50	20.50	5.20	74.20	90.50
1200	Municipal, gunite with plaster finish, incl. high									
1250	capacity filtration & chlorination	B-52	100	.560	SF Surf	63	24	5.95	92.95	113
1350	Add for formed gutters				L.F.	92.50			92.50	102
1360	Add for stainless steel gutters				"	274			274	300
1700	Filtration and deck equipment only, as % of total				Total				20%	20%
1800	Deck equipment, rule of thumb, 20' x 40' pool				SF Pool				1.18	1.30
1900	5000 S.F. pool				"				1.73	1.90

13 11 46 – Swimming Pool Accessories

13 11 46.50 Swimming Pool Equipment

		Crew	Daily Output	Labor-Hours	Unit	Material	2014 Bare Costs Labor	Equipment	Total	Total Incl O&P
0010	**SWIMMING POOL EQUIPMENT**									
0020	Diving stand, stainless steel, 3 meter	2 Carp	.40	40	Ea.	14,500	1,825		16,325	18,700
2100	Lights, underwater, 12 volt, with transformer, 300 watt	1 Elec	1	8		335	425		760	1,000
2200	110 volt, 500 watt, standard		1	8		293	425		718	960
2400	Low water cutoff type		1	8		300	425		725	970
2800	Heaters, see Section 23 52 28.10									

13 21 Controlled Environment Rooms

13 21 13 – Clean Rooms

13 21 13.50 Clean Room Components

		Crew	Daily Output	Labor-Hours	Unit	Material	2014 Bare Costs Labor	Equipment	Total	Total Incl O&P
0010	**CLEAN ROOM COMPONENTS**									
1100	Clean room, soft wall, 12' x 12', Class 100	1 Carp	.18	44.444	Ea.	17,200	2,050		19,250	22,200
1110	Class 1,000		.18	44.444		14,000	2,050		16,050	18,600
1120	Class 10,000		.21	38.095		12,400	1,750		14,150	16,300
1130	Class 100,000		.21	38.095		11,600	1,750		13,350	15,500
2800	Ceiling grid support, slotted channel struts 4'-0" O.C., ea. way				S.F.				5.90	6.50
3000	Ceiling panel, vinyl coated foil on mineral substrate									
3020	Sealed, non-perforated				S.F.				1.27	1.40
4000	Ceiling panel seal, silicone sealant, 150 L.F./gal.	1 Carp	150	.053	L.F.	.34	2.45		2.79	4.14
4100	Two sided adhesive tape	"	240	.033	"	.12	1.53		1.65	2.49
4200	Clips, one per panel				Ea.	.99			.99	1.09
6000	HEPA filter, 2' x 4', 99.97% eff., 3" dp beveled frame (silicone seal)					480			480	530
6040	6" deep skirted frame (channel seal)					420			420	460
6100	99.99% efficient, 3" deep beveled frame (silicone seal)					555			555	610
6140	6" deep skirted frame (channel seal)					435			435	475
6200	99.999% efficient, 3" deep beveled frame (silicone seal)					625			625	690
6240	6" deep skirted frame (channel seal)					450			450	495

13 21 Controlled Environment Rooms

13 21 13 – Clean Rooms

13 21 13.50 Clean Room Components		Crew	Daily Output	Labor-Hours	Unit	Material	2014 Bare Costs Labor	Equipment	Total	Total Incl O&P
7000	Wall panel systems, including channel strut framing									
7020	Polyester coated aluminum, particle board				S.F.				18.20	20
7100	Porcelain coated aluminum, particle board								32	35
7400	Wall panel support, slotted channel struts, to 12' high				▼				16.35	18

13 24 Special Activity Rooms

13 24 16 – Saunas

13 24 16.50 Saunas and Heaters

		Crew	Daily Output	Labor-Hours	Unit	Material	2014 Bare Costs Labor	Equipment	Total	Total Incl O&P
0010	**SAUNAS AND HEATERS**									
0020	Prefabricated, incl. heater & controls, 7' high, 6' x 4', C/C	L-7	2.20	12.727	Ea.	5,375	565		5,940	6,800
0050	6' x 4', C/P		2	14		4,875	620		5,495	6,325
0400	6' x 5', C/C		2	14		5,675	620		6,295	7,200
0450	6' x 5', C/P		2	14		5,150	620		5,770	6,600
0600	6' x 6', C/C		1.80	15.556		6,275	690		6,965	7,950
0650	6' x 6', C/P		1.80	15.556		5,725	690		6,415	7,350
0800	6' x 9', C/C		1.60	17.500		7,750	775		8,525	9,725
0850	6' x 9', C/P		1.60	17.500		7,000	775		7,775	8,900
1000	8' x 12', C/C		1.10	25.455		11,600	1,125		12,725	14,500
1050	8' x 12', C/P		1.10	25.455		10,500	1,125		11,625	13,200
1200	8' x 8', C/C		1.40	20		9,150	885		10,035	11,500
1250	8' x 8', C/P		1.40	20		8,450	885		9,335	10,600
1400	8' x 10', C/C		1.20	23.333		10,100	1,025		11,125	12,800
1450	8' x 10', C/P		1.20	23.333		9,225	1,025		10,250	11,800
1600	10' x 12', C/C		1	28		12,300	1,250		13,550	15,400
1650	10' x 12', C/P	▼	1	28		10,900	1,250		12,150	13,900
2500	Heaters only (incl. above), wall mounted, to 200 C.F.					675			675	745
2750	To 300 C.F.					910			910	1,000
3000	Floor standing, to 720 C.F., 10,000 watts, w/controls	1 Elec	3	2.667		2,900	142		3,042	3,425
3250	To 1,000 C.F., 16,000 watts	"	3	2.667	▼	3,700	142		3,842	4,300

13 24 26 – Steam Baths

13 24 26.50 Steam Baths and Components

		Crew	Daily Output	Labor-Hours	Unit	Material	2014 Bare Costs Labor	Equipment	Total	Total Incl O&P
0010	**STEAM BATHS AND COMPONENTS**									
0020	Heater, timer & head, single, to 140 C.F.	1 Plum	1.20	6.667	Ea.	1,450	385		1,835	2,175
0500	To 300 C.F.		1.10	7.273		1,550	420		1,970	2,325
1000	Commercial size, with blow-down assembly, to 800 C.F.		.90	8.889		7,350	510		7,860	8,850
1500	To 2500 C.F.	▼	.80	10		9,125	575		9,700	10,900
2000	Multiple, motels, apts., 2 baths, w/blow-down assm., 500 C.F.	Q-1	1.30	12.308		7,950	640		8,590	9,725
2500	4 baths	"	.70	22.857	▼	10,700	1,175		11,875	13,600

13 49 Radiation Protection

13 49 13 – Integrated X-Ray Shielding Assemblies

13 49 13.50 Lead Sheets

		Crew	Daily Output	Labor-Hours	Unit	Material	2014 Bare Costs Labor	Equipment	Total	Total Incl O&P
0010	**LEAD SHEETS**									
0300	Lead sheets, 1/16" thick	2 Lath	135	.119	S.F.	10.05	4.81		14.86	18.15
0400	1/8" thick		120	.133		20	5.40		25.40	30
0500	Lead shielding, 1/4" thick		135	.119		49	4.81		53.81	61
0550	1/2" thick		120	.133		83.50	5.40		88.90	99.50
0950	Lead headed nails (average 1 lb. per sheet)				Lb.	8.05			8.05	8.85
1000	Butt joints in 1/8" lead or thicker, 2" batten strip x 7' long	2 Lath	240	.067	Ea.	28	2.71		30.71	35
1200	X-ray protection, average radiography or fluoroscopy									
1210	room, up to 300 S.F. floor, 1/16" lead, minimum	2 Lath	.25	64	Total	10,100	2,600		12,700	14,900
1500	Maximum, 7'-0" walls	"	.15	106	"	12,200	4,325		16,525	19,800
1600	Deep therapy X-ray room, 250 kV capacity,									
1800	up to 300 S.F. floor, 1/4" lead, minimum	2 Lath	.08	200	Total	28,300	8,125		36,425	43,000
1900	Maximum, 7'-0" walls	"	.06	266	"	34,900	10,800		45,700	54,500

13 49 19 – Lead-Lined Materials

13 49 19.50 Shielding Lead

		Crew	Daily Output	Labor-Hours	Unit	Material	2014 Bare Costs Labor	Equipment	Total	Total Incl O&P
0010	**SHIELDING LEAD**									
0100	Laminated lead in wood doors, 1/16" thick, no hardware				S.F.	45			45	49.50
0200	Lead lined door frame, not incl. hardware,									
0210	1/16" thick lead, butt prepared for hardware	1 Lath	2.40	3.333	Ea.	770	135		905	1,050
0850	Window frame with 1/16" lead and voice passage, 36" x 60"	2 Glaz	2	8		5,550	350		5,900	6,625
0870	24" x 36" frame		4	4		2,700	176		2,876	3,250
0900	Lead gypsum board, 5/8" thick with 1/16" lead		160	.100	S.F.	10.15	4.41		14.56	17.85
0910	1/8" lead		140	.114		22	5.05		27.05	31.50
0930	1/32" lead	2 Lath	200	.080		7.60	3.25		10.85	13.15

13 49 21 – Lead Glazing

13 49 21.50 Lead Glazing

		Crew	Daily Output	Labor-Hours	Unit	Material	2014 Bare Costs Labor	Equipment	Total	Total Incl O&P
0010	**LEAD GLAZING**									
0600	Lead glass, 1/4" thick, 2.0 mm LE, 12" x 16"	2 Glaz	13	1.231	Ea.	360	54		414	485
0700	24" x 36"		8	2		1,275	88		1,363	1,525
0800	36" x 60"		2	8		3,500	350		3,850	4,375
2000	X-ray viewing panels, clear lead plastic									
2010	7 mm thick, 0.3 mm LE, 2.3 lb./S.F.	H-3	139	.115	S.F.	234	4.53		238.53	264
2020	12 mm thick, 0.5 mm LE, 3.9 lb./S.F.		82	.195		345	7.70		352.70	390
2030	18 mm thick, 0.8 mm LE, 5.9 lb./S.F.		54	.296		395	11.65		406.65	455
2040	22 mm thick, 1.0 mm LE, 7.2 lb./S.F.		44	.364		515	14.30		529.30	585
2050	35 mm thick, 1.5 mm LE, 11.5 lb./S.F.		28	.571		790	22.50		812.50	905
2060	46 mm thick, 2.0 mm LE, 15.0 lb./S.F.		21	.762		1,025	30		1,055	1,175
2090	For panels 12 S.F. to 48 S.F., add crating charge				Ea.					50

13 49 23 – Integrated RFI/EMI Shielding Assemblies

13 49 23.50 Modular Shielding Partitions

		Crew	Daily Output	Labor-Hours	Unit	Material	2014 Bare Costs Labor	Equipment	Total	Total Incl O&P
0010	**MODULAR SHIELDING PARTITIONS**									
4000	X-ray barriers, modular, panels mounted within framework for									
4002	attaching to floor, wall or ceiling, upper portion is clear lead									
4005	plastic window panels 48"H, lower portion is opaque leaded									
4008	steel panels 36"H, structural supports not incl.									
4010	1-section barrier, 36"W x 84"H overall									
4020	0.5 mm LE panels	H-3	6.40	2.500	Ea.	5,725	98.50		5,823.50	6,450
4030	0.8 mm LE panels		6.40	2.500		6,350	98.50		6,448.50	7,125
4040	1.0 mm LE panels		5.33	3.002		7,775	118		7,893	8,725
4050	1.5 mm LE panels		5.33	3.002		11,100	118		11,218	12,400
4060	2-section barrier, 72"W x 84"H overall									

13 49 Radiation Protection

13 49 23 – Integrated RFI/EMI Shielding Assemblies

13 49 23.50 Modular Shielding Partitions		Crew	Daily Output	Labor-Hours	Unit	Material	2014 Bare Costs Labor	Equipment	Total	Total Incl O&P
4070	0.5 mm LE panels	H-3	4	4	Ea.	11,400	157		11,557	12,800
4080	0.8 mm LE panels		4	4		12,700	157		12,857	14,100
4090	1.0 mm LE panels		3.56	4.494		15,600	177		15,777	17,400
5000	1.5 mm LE panels		3.20	5		22,200	197		22,397	24,700
5010	3-section barrier, 108"W x 84"H overall									
5020	0.5 mm LE panels	H-3	3.20	5	Ea.	17,200	197		17,397	19,200
5030	0.8 mm LE panels		3.20	5		19,000	197		19,197	21,200
5040	1.0 mm LE panels		2.67	5.993		23,300	236		23,536	26,100
5050	1.5 mm LE panels		2.46	6.504		33,200	256		33,456	36,900
7000	X-ray barriers, mobile, mounted within framework w/casters on									
7005	bottom, clear lead plastic window panels on upper portion,									
7010	opaque on lower, 30"W x 75"H overall, incl. framework									
7020	24"H upper w/0.5 mm LE, 48"H lower w/0.8 mm LE	1 Carp	16	.500	Ea.	3,700	23		3,723	4,075
7030	48"W x 75"H overall, incl. framework									
7040	36"H upper w/0.5 mm LE, 36"H lower w/0.8 mm LE	1 Carp	16	.500	Ea.	6,000	23		6,023	6,625
7050	36"H upper w/1.0 mm LE, 36"H lower w/1.5 mm LE	"	16	.500	"	7,100	23		7,123	7,825
7060	72"W x 75"H overall, incl. framework									
7070	36"H upper w/0.5 mm LE, 36"H lower w/0.8 mm LE	1 Carp	16	.500	Ea.	7,100	23		7,123	7,825
7080	36"H upper w/1.0 mm LE, 36"H lower w/1.5 mm LE	"	16	.500	"	8,900	23		8,923	9,800

13 49 33 – Radio Frequency Shielding

13 49 33.50 Shielding, Radio Frequency

		Crew	Daily Output	Labor-Hours	Unit	Material	2014 Bare Costs Labor	Equipment	Total	Total Incl O&P
0010	**SHIELDING, RADIO FREQUENCY**									
0020	Prefabricated, galvanized steel	2 Carp	375	.043	SF Surf	4.46	1.96		6.42	7.95
0040	5 oz., copper floor panel		480	.033		3.68	1.53		5.21	6.40
0050	5 oz., copper wall/ceiling panel		155	.103		3.68	4.73		8.41	11.35
0100	12 oz., copper floor panel		470	.034		7.85	1.56		9.41	11.05
0110	12 oz., copper wall/ceiling panel		140	.114		7.85	5.25		13.10	16.75
0150	Door, copper/wood laminate, 4' x 7'		1.50	10.667	Ea.	7,400	490		7,890	8,875

Division Notes

	CREW	DAILY OUTPUT	LABOR-HOURS	UNIT	BARE COSTS				TOTAL INCL O&P
					MAT.	LABOR	EQUIP.	TOTAL	

Estimating Tips

Pipe for fire protection and all uses is located in Subdivisions 21 11 13 and 22 11 13.

The labor adjustment factors listed in Subdivision 22 01 02.20 also apply to Division 21.

Many, but not all, areas in the U.S. require backflow protection in the fire system. It is advisable to check local building codes for specific requirements.

For your reference, the following is a list of the most applicable Fire Codes and Standards which may be purchased from the NFPA, 1 Batterymarch Park, Quincy, MA 02169-7471.

NFPA 1: Uniform Fire Code
NFPA 10: Portable Fire Extinguishers
NFPA 11: Low-, Medium-, and High-Expansion Foam
NFPA 12: Carbon Dioxide Extinguishing Systems (Also companion 12A)
NFPA 13: Installation of Sprinkler Systems (Also companion 13D, 13E, and 13R)
NFPA 14: Installation of Standpipe and Hose Systems
NFPA 15: Water Spray Fixed Systems for Fire Protection
NFPA 16: Installation of Foam-Water Sprinkler and Foam-Water Spray Systems
NFPA 17: Dry Chemical Extinguishing Systems (Also companion 17A)
NFPA 18: Wetting Agents

NFPA 20: Installation of Stationary Pumps for Fire Protection
NFPA 22: Water Tanks for Private Fire Protection
NFPA 24: Installation of Private Fire Service Mains and their Appurtenances
NFPA 25: Inspection, Testing and Maintenance of Water-Based Fire Protection

Reference Numbers

Reference numbers are shown in shaded boxes at the beginning of some major classifications. These numbers refer to related items in the Reference Section. The reference information may be an estimating procedure, an alternate pricing method, or technical information.

Note: Not all subdivisions listed here necessarily appear in this publication.

Note: **Trade Service,** *in part, has been used as a reference source for some of the material prices used in Division 21.*

21 21 Carbon-Dioxide Fire-Extinguishing Systems

21 21 16 – Carbon-Dioxide Fire-Extinguishing Equipment

21 21 16.50 CO2 Fire Extinguishing System	Crew	Daily Output	Labor-Hours	Unit	Material	2014 Bare Costs Labor	Equipment	Total	Total Incl O&P
0010 **CO₂ FIRE EXTINGUISHING SYSTEM**									
0100 Control panel, single zone with batteries (2 zones det., 1 suppr.)	1 Elec	1	8	Ea.	1,675	425		2,100	2,500
0150 Multizone (4) with batteries (8 zones det., 4 suppr.)	"	.50	16		3,200	855		4,055	4,775
1000 Dispersion nozzle, CO₂, 3" x 5"	1 Plum	18	.444		65	25.50		90.50	110
2000 Extinguisher, CO₂ system, high pressure, 75 lb. cylinder	Q-1	6	2.667		1,250	138		1,388	1,550
2100 100 lb. cylinder	"	5	3.200		1,275	166		1,441	1,650
3000 Electro/mechanical release	L-1	4	4		163	222		385	515
3400 Manual pull station	1 Plum	6	1.333		58.50	76.50		135	181
4000 Pneumatic damper release	"	8	1	▼	217	57.50		274.50	325

21 22 Clean-Agent Fire-Extinguishing Systems

21 22 16 – Clean-Agent Fire-Extinguishing Equipment

21 22 16.50 FM200 Fire Extinguishing System

	Crew	Daily Output	Labor-Hours	Unit	Material	2014 Bare Costs Labor	Equipment	Total	Total Incl O&P
0010 **FM200 FIRE EXTINGUISHING SYSTEM**									
1100 Dispersion nozzle FM200, 1-1/2"	1 Plum	14	.571	Ea.	65	33		98	121
2400 Extinguisher, FM200 system, filled, with mounting bracket									
2460 26 lb. container	Q-1	8	2	Ea.	2,225	104		2,329	2,600
2480 44 lb. container		7	2.286		2,975	118		3,093	3,450
2500 63 lb. container		6	2.667		3,475	138		3,613	4,025
2520 101 lb. container		5	3.200		4,650	166		4,816	5,350
2540 196 lb. container	▼	4	4	▼	7,550	207		7,757	8,625
6000 Average FM200 system, minimum				C.F.	1.71			1.71	1.88
6020 Maximum				"	3.40			3.40	3.74

Estimating Tips

22 10 00 Plumbing Piping and Pumps

This subdivision is primarily basic pipe and related materials. The pipe may be used by any of the mechanical disciplines, i.e., plumbing, fire protection, heating, and air conditioning.

Note: CPVC plastic piping approved for fire protection is located in 21 11 13.

- The labor adjustment factors listed in Subdivision 22 01 02.20 apply throughout Divisions 21, 22, and 23. CAUTION: the correct percentage may vary for the same items. For example, the percentage add for the basic pipe installation should be based on the maximum height that the craftsman must install for that particular section. If the pipe is to be located 14' above the floor but it is suspended on threaded rod from beams, the bottom flange of which is 18' high (4' rods), then the height is actually 18' and the add is 20%. The pipe coverer, however, does not have to go above the 14', and so his or her add should be 10%.

- Most pipe is priced first as straight pipe with a joint (coupling, weld, etc.) every 10' and a hanger usually every 10'.

There are exceptions with hanger spacing such as for cast iron pipe (5') and plastic pipe (3 per 10'). Following each type of pipe there are several lines listing sizes and the amount to be subtracted to delete couplings and hangers. This is for pipe that is to be buried or supported together on trapeze hangers. The reason that the couplings are deleted is that these runs are usually long, and frequently longer lengths of pipe are used. By deleting the couplings, the estimator is expected to look up and add back the correct reduced number of couplings.

- When preparing an estimate, it may be necessary to approximate the fittings. Fittings usually run between 25% and 50% of the cost of the pipe. The lower percentage is for simpler runs, and the higher number is for complex areas, such as mechanical rooms.

- For historic restoration projects, the systems must be as invisible as possible, and pathways must be sought for pipes, conduit, and ductwork. While installations in accessible spaces (such as basements and attics) are relatively straightforward to estimate, labor costs may be more difficult to determine when delivery systems must be concealed.

22 40 00 Plumbing Fixtures

- Plumbing fixture costs usually require two lines: the fixture itself and its "rough-in, supply, and waste."

- In the Assemblies Section (Plumbing D2010) for the desired fixture, the System Components Group at the center of the page shows the fixture on the first line. The rest of the list (fittings, pipe, tubing, etc.) will total up to what we refer to in the Unit Price section as "Rough-in, supply, waste, and vent." Note that for most fixtures we allow a nominal 5' of tubing to reach from the fixture to a main or riser.

- Remember that gas- and oil-fired units need venting.

Reference Numbers

Reference numbers are shown in shaded boxes at the beginning of some major classifications. These numbers refer to related items in the Reference Section. The reference information may be an estimating procedure, an alternate pricing method, or technical information.

Note: Not all subdivisions listed here necessarily appear in this publication.

Note: **Trade Service,** *in part, has been used as a reference source for some of the material prices used in Division 22.*

22 01 Operation and Maintenance of Plumbing

22 01 02 – Labor Adjustments

22 01 02.10 Boilers, General	Crew	Daily Output	Labor- Hours	Unit	Material	2014 Bare Costs Labor	Equipment	Total	Total Incl O&P
0010 **BOILERS, GENERAL**, Prices do not include flue piping, elec. wiring,									
0020 gas or oil piping, boiler base, pad, or tankless unless noted									
0100 Boiler H.P.: 10 KW = 34 lb./steam/hr. = 33,475 BTU/hr.									
0150 To convert SFR to BTU rating: Hot water, 150 x SFR;									
0160 Forced hot water, 180 x SFR; steam, 240 x SFR									

22 01 02.20 Labor Adjustment Factors	Crew	Daily Output	Labor- Hours	Unit	Material	2014 Bare Costs Labor	Equipment	Total	Total Incl O&P
0010 **LABOR ADJUSTMENT FACTORS**, (For Div. 21, 22 and 23) R220102-20									
0100 Labor factors, The below are reasonable suggestions, however									
0110 each project must be evaluated for its own peculiarities, and									
0120 the adjustments be increased or decreased depending on the									
0130 severity of the special conditions.									
1000 Add to labor for elevated installation (Above floor level)									
1080 10' to 14.5' high						10%			
1100 15' to 19.5' high						20%			
1120 20' to 24.5' high						25%			
1140 25' to 29.5' high						35%			
1160 30' to 34.5' high						40%			
1180 35' to 39.5' high						50%			
1200 40' and higher						55%			
2000 Add to labor for crawl space									
2100 3' high						40%			
2140 4' high						30%			
3000 Add to labor for multi-story building									
3100 Add per floor for floors 3 thru 19						2%			
3140 Add per floor for floors 20 and up						4%			
4000 Add to labor for working in existing occupied buildings									
4100 Hospital						35%			
4140 Office building						25%			
4180 School						20%			
4220 Factory or warehouse						15%			
4260 Multi dwelling						15%			
5000 Add to labor, miscellaneous									
5100 Cramped shaft						35%			
5140 Congested area						15%			
5180 Excessive heat or cold						30%			
9000 Labor factors, The above are reasonable suggestions, however									
9010 each project should be evaluated for its own peculiarities.									
9100 Other factors to be considered are:									
9140 Movement of material and equipment through finished areas									
9180 Equipment room									
9220 Attic space									
9260 No service road									
9300 Poor unloading/storage area									
9340 Congested site area/heavy traffic									

22 05 29.10 Hangers & Supp. for Plumb'g/HVAC Pipe/Equip.	Crew	Daily Output	Labor-Hours	Unit	Material	2014 Bare Costs Labor	Equipment	Total	Total Incl O&P
0010	**HANGERS AND SUPPORTS FOR PLUMB'G/HVAC PIPE/EQUIP.**									
0011	TYPE numbers per MSS-SP58									
0050	Brackets									
0060	Beam side or wall, malleable iron, TYPE 34									
0070	3/8" threaded rod size	1 Plum	48	.167	Ea.	2.96	9.60		12.56	17.75
0080	1/2" threaded rod size		48	.167		4.18	9.60		13.78	19.10
0090	5/8" threaded rod size		48	.167		7.85	9.60		17.45	23
0100	3/4" threaded rod size		48	.167		10.75	9.60		20.35	26.50
0110	7/8" threaded rod size		48	.167		11.10	9.60		20.70	27
0120	For concrete installation, add						30%			
0150	Wall, welded steel, medium, TYPE 32									
0160	0 size, 12" wide, 18" deep	1 Plum	34	.235	Ea.	164	13.55		177.55	201
0170	1 size, 18" wide, 24" deep		34	.235		201	13.55		214.55	242
0180	2 size, 24" wide, 30" deep		34	.235		276	13.55		289.55	325
0300	Clamps									
0310	C-clamp, for mounting on steel beam flange, w/locknut, TYPE 23									
0320	3/8" threaded rod size	1 Plum	160	.050	Ea.	1.95	2.88		4.83	6.50
0330	1/2" threaded rod size		160	.050		2.09	2.88		4.97	6.65
0340	5/8" threaded rod size		160	.050		3.52	2.88		6.40	8.20
0350	3/4" threaded rod size		160	.050		4.34	2.88		7.22	9.10
0750	Riser or extension pipe, carbon steel, TYPE 8									
0760	3/4" pipe size	1 Plum	48	.167	Ea.	2.40	9.60		12	17.15
0770	1" pipe size		47	.170		2.52	9.80		12.32	17.55
0780	1-1/4" pipe size		46	.174		3.06	10		13.06	18.45
0790	1-1/2" pipe size		45	.178		3.26	10.25		13.51	19.05
0800	2" pipe size		43	.186		3.38	10.70		14.08	19.85
0810	2-1/2" pipe size		41	.195		3.51	11.25		14.76	21
0820	3" pipe size		40	.200		3.81	11.50		15.31	21.50
0830	3-1/2" pipe size		39	.205		4.28	11.80		16.08	22.50
0840	4" pipe size		38	.211		5.20	12.10		17.30	24
0850	5" pipe size		37	.216		6.50	12.45		18.95	26
0860	6" pipe size		36	.222		7.95	12.80		20.75	28
1150	Insert, concrete									
1160	Wedge type, carbon steel body, malleable iron nut, galvanized									
1170	1/4" threaded rod size	1 Plum	96	.083	Ea.	2.53	4.80		7.33	10.05
1180	3/8" threaded rod size		96	.083		2.53	4.80		7.33	10.05
1190	1/2" threaded rod size		96	.083		2.85	4.80		7.65	10.40
1200	5/8" threaded rod size		96	.083		3.38	4.80		8.18	10.95
1210	3/4" threaded rod size		96	.083		7.05	4.80		11.85	15.05
1220	7/8" threaded rod size		96	.083		7.85	4.80		12.65	15.90
2650	Rods, carbon steel									
2660	Continuous thread									
2670	1/4" thread size	1 Plum	144	.056	L.F.	1.62	3.20		4.82	6.60
2680	3/8" thread size		144	.056		1.72	3.20		4.92	6.70
2690	1/2" thread size		144	.056		2.72	3.20		5.92	7.80
2700	5/8" thread size		144	.056		3.86	3.20		7.06	9.10
2710	3/4" thread size		144	.056		6.80	3.20		10	12.35
2720	7/8" thread size		144	.056		8.55	3.20		11.75	14.25
2725	1/4" thread size, bright finish		144	.056		2.13	3.20		5.33	7.15
2726	1/2" thread size, bright finish		144	.056		3.83	3.20		7.03	9.05
2730	For galvanized, add						40%			
2750	Both ends machine threaded 18" length									
2760	3/8" thread size	1 Plum	240	.033	Ea.	4.83	1.92		6.75	8.20

22 05 29 – Hangers and Supports for Plumbing Piping and Equipment

22 05 29.10 Hangers & Supp. for Plumb'g/HVAC Pipe/Equip.	Crew	Daily Output	Labor-Hours	Unit	Material	2014 Bare Costs Labor	Equipment	Total	Total Incl O&P
2770	1/2" thread size	1 Plum	240	.033	Ea.	7.15	1.92		9.07	10.80
2780	5/8" thread size		240	.033		6.40	1.92		8.32	9.95
2790	3/4" thread size		240	.033		11.45	1.92		13.37	15.50
2800	7/8" thread size		240	.033		11.85	1.92		13.77	15.95
2810	1" thread size		240	.033		15.10	1.92		17.02	19.50
4400	U-bolt, carbon steel									
4410	Standard, with nuts, TYPE 42									
4420	1/2" pipe size	1 Plum	160	.050	Ea.	.93	2.88		3.81	5.35
4430	3/4" pipe size		158	.051		.98	2.91		3.89	5.50
4450	1" pipe size		152	.053		1.02	3.03		4.05	5.70
4460	1-1/4" pipe size		148	.054		1.29	3.11		4.40	6.10
4470	1-1/2" pipe size		143	.056		1.37	3.22		4.59	6.35
4480	2" pipe size		139	.058		1.48	3.31		4.79	6.65
4490	2-1/2" pipe size		134	.060		2.41	3.44		5.85	7.85
4500	3" pipe size		128	.063		2.72	3.60		6.32	8.45
4510	3-1/2" pipe size		122	.066		2.93	3.77		6.70	8.90
4520	4" pipe size		117	.068		2.97	3.94		6.91	9.20
4530	5" pipe size		114	.070		3.49	4.04		7.53	9.95
4540	6" pipe size		111	.072		6.65	4.15		10.80	13.55
4580	For plastic coating on 1/2" thru 6" size, add					150%				
8800	Wire cable support system									
8810	Cable with hook terminal and locking device									
8830	2 mm, (.079") dia cable, (100 lb. cap.)									
8840	1 m, (3.3') length, with hook	1 Shee	96	.083	Ea.	3.03	4.56		7.59	10.30
8850	2 m, (6.6') length, with hook		84	.095		3.50	5.20		8.70	11.80
8860	3 m, (9.9') length, with hook		72	.111		3.97	6.10		10.07	13.65
8870	5 m, (16.4') length, with hook	Q-9	60	.267		4.97	13.15		18.12	25.50
8880	10 m, (32.8') length, with hook	"	30	.533		7.20	26.50		33.70	48
8900	3mm, (.118") dia cable, (200 lb. cap.)									
8910	1 m, (3.3') length, with hook	1 Shee	96	.083	Ea.	3.88	4.56		8.44	11.20
8920	2 m, (6.6') length, with hook		84	.095		4.36	5.20		9.56	12.75
8930	3 m, (9.9') length, with hook		72	.111		4.81	6.10		10.91	14.60
8940	5 m, (16.4') length, with hook	Q-9	60	.267		5.95	13.15		19.10	26.50
8950	10 m, (32.8') length, with hook	"	30	.533		8.55	26.50		35.05	49.50
9000	Cable system accessories									
9010	Anchor bolt, 3/8", with nut	1 Shee	140	.057	Ea.	1.16	3.13		4.29	6.05
9020	Air duct corner protector		160	.050		.52	2.74		3.26	4.75
9030	Air duct support attachment		140	.057		.99	3.13		4.12	5.85
9040	Flange clip, hammer-on style									
9044	For flange thickness 3/32" - 9/64", 160 lb. cap.	1 Shee	180	.044	Ea.	.29	2.43		2.72	4.03
9048	For flange thickness 1/8" - 1/4", 200 lb. cap.		160	.050		.29	2.74		3.03	4.50
9052	For flange thickness 5/16" - 1/2", 200 lb. cap.		150	.053		.56	2.92		3.48	5.10
9056	For flange thickness 9/16" - 3/4", 200 lb. cap.		140	.057		.75	3.13		3.88	5.60
9060	Wire insulation protection tube		180	.044	L.F.	.34	2.43		2.77	4.08
9070	Wire cutter				Ea.	41.50			41.50	45.50

22 33 Electric Domestic Water Heaters

22 33 13 – Instantaneous Electric Domestic Water Heaters

22 33 13.10 Hot Water Dispensers

		Crew	Daily Output	Labor-Hours	Unit	Material	2014 Bare Costs Labor	2014 Bare Costs Equipment	Total	Total Incl O&P
0010	**HOT WATER DISPENSERS**									
0160	Commercial, 100 cup, 11.3 amp	1 Plum	14	.571	Ea.	510	33		543	615
3180	Household, 60 cup	"	14	.571	"	269	33		302	345

22 33 30 – Residential, Electric Domestic Water Heaters

22 33 30.13 Residential, Small-Capacity Elec. Water Heaters

				Crew	Daily Output	Labor-Hours	Unit	Material	Labor	Equipment	Total	Total Incl O&P
0010	**RESIDENTIAL, SMALL-CAPACITY ELECTRIC DOMESTIC WATER HEATERS**											
1000	Residential, electric, glass lined tank, 5 yr., 10 gal., single element	R224000-10		1 Plum	2.30	3.478	Ea.	330	200		530	660
1040	20 gallon, single element	R224000-20			2.20	3.636		410	209		619	765
1060	30 gallon, double element				2.20	3.636		475	209		684	840
1080	40 gallon, double element				2	4		800	230		1,030	1,225
1100	52 gallon, double element				2	4		895	230		1,125	1,325
1120	66 gallon, double element				1.80	4.444		1,200	256		1,456	1,700
1140	80 gallon, double element				1.60	5		1,350	288		1,638	1,900
1180	120 gallon, double element				1.40	5.714		1,900	330		2,230	2,575

22 33 33 – Light-Commercial Electric Domestic Water Heaters

22 33 33.10 Commercial Electric Water Heaters

		Crew	Daily Output	Labor-Hours	Unit	Material	Labor	Equipment	Total	Total Incl O&P
0010	**COMMERCIAL ELECTRIC WATER HEATERS**									
4000	Commercial, 100° rise. NOTE: for each size tank, a range of									
4010	heaters between the ones shown are available									
4020	Electric									
4100	5 gal., 3 kW, 12 GPH, 208 volt	1 Plum	2	4	Ea.	2,750	230		2,980	3,375
4120	10 gal., 6 kW, 25 GPH, 208 volt		2	4		3,050	230		3,280	3,700
4130	30 gal., 24 kW, 98 GPH, 208 volt		1.92	4.167		4,725	240		4,965	5,550
4136	40 gal., 36 kW, 148 GPH, 208 volt		1.88	4.255		5,650	245		5,895	6,600
4140	50 gal., 9 kW, 37 GPH, 208 volt		1.80	4.444		4,175	256		4,431	4,975
4160	50 gal., 36 kW, 148 GPH, 208 volt		1.80	4.444		6,400	256		6,656	7,400
4180	80 gal., 12 kW, 49 GPH, 208 volt		1.50	5.333		5,150	305		5,455	6,150
4200	80 gal., 36 kW, 148 GPH, 208 volt		1.50	5.333		6,925	305		7,230	8,100
4220	100 gal., 36 kW, 148 GPH, 208 volt		1.20	6.667		7,425	385		7,810	8,750
4240	120 gal., 36 kW, 148 GPH, 208 volt		1.20	6.667		7,750	385		8,135	9,100
4260	150 gal., 15 kW, 61 GPH, 480 volt		1	8		18,800	460		19,260	21,400
4280	150 gal., 120 kW, 490 GPH, 480 volt		1	8		25,500	460		25,960	28,800
4300	200 gal., 15 kW, 61 GPH, 480 volt	Q-1	1.70	9.412		19,700	490		20,190	22,400
4320	200 gal., 120 kW , 490 GPH, 480 volt		1.70	9.412		26,900	490		27,390	30,300
4340	250 gal., 15 kW, 61 GPH, 480 volt		1.50	10.667		20,700	555		21,255	23,600
4360	250 gal., 150 kW, 615 GPH, 480 volt		1.50	10.667		32,500	555		33,055	36,500
4380	300 gal., 30 kW, 123 GPH, 480 volt		1.30	12.308		22,400	640		23,040	25,600
4400	300 gal., 180 kW, 738 GPH, 480 volt		1.30	12.308		37,500	640		38,140	42,300
4420	350 gal., 30 kW, 123 GPH, 480 volt		1.10	14.545		23,600	755		24,355	27,200
4440	350 gal., 180 kW, 738 GPH, 480 volt		1.10	14.545		33,300	755		34,055	37,900
4460	400 gal., 30 kW, 123 GPH, 480 volt		1	16		26,800	830		27,630	30,800
4480	400 gal., 210 kW, 860 GPH, 480 volt		1	16		38,500	830		39,330	43,600
4500	500 gal., 30 kW, 123 GPH, 480 volt		.80	20		31,200	1,025		32,225	36,000
4520	500 gal., 240 kW, 984 GPH, 480 volt		.80	20		46,000	1,025		47,025	52,000
4540	600 gal., 30 kW, 123 GPH, 480 volt	Q-2	1.20	20		23,800	1,075		24,875	27,800
4560	600 gal., 300 kW, 1230 GPH, 480 volt		1.20	20		36,100	1,075		37,175	41,300
4580	700 gal., 30 kW, 123 GPH, 480 volt		1	24		25,000	1,300		26,300	29,500
4600	700 gal., 300 kW, 1230 GPH, 480 volt		1	24		37,600	1,300		38,900	43,400
4620	800 gal., 60 kW, 245 GPH, 480 volt		.90	26.667		27,000	1,425		28,425	32,000
4640	800 gal., 300 kW, 1230 DPH, 480 volt		.90	26.667		38,400	1,425		39,825	44,500
4660	1000 gal., 60 kW, 245 GPH, 480 volt		.70	34.286		29,600	1,850		31,450	35,400
4680	1000 gal., 480 kW, 1970 GPH, 480 volt		.70	34.286		49,200	1,850		51,050	57,000

22 33 Electric Domestic Water Heaters

22 33 33 – Light-Commercial Electric Domestic Water Heaters

22 33 33.10 Commercial Electric Water Heaters	Crew	Daily Output	Labor-Hours	Unit	Material	2014 Bare Costs Labor	Equipment	Total	Total Incl O&P
4700	1200 gal., 60 kW, 245 GPH, 480 volt	Q-2	.60	40	Ea.	50,500	2,150		52,650	59,000
4720	1200 gal., 480 kW, 1970 GPH, 480 volt		.60	40		76,500	2,150		78,650	87,500
4740	1500 gal., 60 kW, 245 GPH, 480 volt		.50	48		66,500	2,575		69,075	77,000
4760	1500 gal., 480 kW, 1970 GPH, 480 volt		.50	48		92,000	2,575		94,575	105,500
5400	Modulating step control for under 90 kW, 2-5 steps	1 Elec	5.30	1.509		810	80.50		890.50	1,000
5440	1 through 5 steps beyond standard		3.20	2.500		221	133		354	445
5460	6 through 10 steps beyond standard		2.70	2.963		455	158		613	735
5480	11 through 18 steps beyond standard		1.60	5		680	267		947	1,150

22 34 Fuel-Fired Domestic Water Heaters

22 34 30 – Residential Gas Domestic Water Heaters

22 34 30.13 Residential, Atmospheric, Gas Domestic Water Heaters

		Crew	Daily Output	Labor-Hours	Unit	Material	2014 Bare Costs Labor	Equipment	Total	Total Incl O&P
0010	**RESIDENTIAL, ATMOSPHERIC, GAS DOMESTIC WATER HEATERS**									
2000	Gas fired, foam lined tank, 10 yr., vent not incl.									
2040	30 gallon	1 Plum	2	4	Ea.	895	230		1,125	1,325
2060	40 gallon		1.90	4.211		895	242		1,137	1,350
2080	50 gallon		1.80	4.444		935	256		1,191	1,400
2090	60 gallon		1.70	4.706		1,325	271		1,596	1,850
2100	75 gallon		1.50	5.333		1,350	305		1,655	1,975
2120	100 gallon		1.30	6.154		1,525	355		1,880	2,200
2900	Water heater, safety-drain pan, 26" round		20	.400		37	23		60	76

22 34 46 – Oil-Fired Domestic Water Heaters

22 34 46.10 Residential Oil-Fired Water Heaters

		Crew	Daily Output	Labor-Hours	Unit	Material	2014 Bare Costs Labor	Equipment	Total	Total Incl O&P
0010	**RESIDENTIAL OIL-FIRED WATER HEATERS**									
3000	Oil fired, glass lined tank, 5 yr., vent not included, 30 gallon	1 Plum	2	4	Ea.	1,175	230		1,405	1,625
3040	50 gallon		1.80	4.444		1,375	256		1,631	1,900
3060	70 gallon		1.50	5.333		1,975	305		2,280	2,650

Estimating Tips

The labor adjustment factors listed in Subdivision 22 01 02.20 also apply to Division 23.

23 10 00 Facility Fuel Systems

- The prices in this subdivision for above- and below-ground storage tanks do not include foundations or hold-down slabs, unless noted. The estimator should refer to Divisions 3 and 31 for foundation system pricing. In addition to the foundations, required tank accessories, such as tank gauges, leak detection devices, and additional manholes and piping, must be added to the tank prices.

23 50 00 Central Heating Equipment

- When estimating the cost of an HVAC system, check to see who is responsible for providing and installing the temperature control system. It is possible to overlook controls, assuming that they would be included in the electrical estimate.
- When looking up a boiler, be careful on specified capacity. Some manufacturers rate their products on output while others use input.
- Include HVAC insulation for pipe, boiler, and duct (wrap and liner).
- Be careful when looking up mechanical items to get the correct pressure rating and connection type (thread, weld, flange).

23 70 00 Central HVAC Equipment

- Combination heating and cooling units are sized by the air conditioning requirements. (See Reference No. R236000-20 for preliminary sizing guide.)
- A ton of air conditioning is nominally 400 CFM.
- Rectangular duct is taken off by the linear foot for each size, but its cost is usually estimated by the pound. Remember that SMACNA standards now base duct on internal pressure.
- Prefabricated duct is estimated and purchased like pipe: straight sections and fittings.
- Note that cranes or other lifting equipment are not included on any lines in Division 23. For example, if a crane is required to lift a heavy piece of pipe into place high above a gym floor, or to put a rooftop unit on the roof of a four-story building, etc., it must be added. Due to the potential for extreme variation—from nothing additional required to a major crane or helicopter—we feel that including a nominal amount for "lifting contingency" would be useless and detract from the accuracy of the estimate. When using equipment rental cost data from RSMeans, do not forget to include the cost of the operator(s).

Reference Numbers

Reference numbers are shown in shaded boxes at the beginning of some major classifications. These numbers refer to related items in the Reference Section. The reference information may be an estimating procedure, an alternate pricing method, or technical information.

Note: Not all subdivisions listed here necessarily appear in this publication.

Note: **Trade Service,** *in part, has been used as a reference source for some of the material prices used in Division 23.*

23 05 Common Work Results for HVAC

23 05 02 – HVAC General

23 05 02.10 Air Conditioning, General	Crew	Daily Output	Labor-Hours	Unit	Material	2014 Bare Costs Labor	Equipment	Total	Total Incl O&P
0010 **AIR CONDITIONING, GENERAL** Prices are for standard efficiencies (SEER 13)									
0020 for upgrade to SEER 14 add					10%				

23 09 Instrumentation and Control for HVAC

23 09 23 – Direct-Digital Control System for HVAC

23 09 23.10 Control Components/DDC Systems

	Crew	Daily Output	Labor-Hours	Unit	Material	2014 Bare Costs Labor	Equipment	Total	Total Incl O&P
0010 **CONTROL COMPONENTS/DDC SYSTEMS** (Sub's quote incl. M & L)									
0100 Analog inputs									
0110 Sensors (avg. 50' run in 1/2" EMT)									
0120 Duct temperature				Ea.				360	395
0130 Space temperature								575	635
0140 Duct humidity, +/- 3%								605	665
0150 Space humidity, +/- 2%								925	1,025
0160 Duct static pressure								490	540
0170 CFM/Transducer								660	730
0180 KW/Transducer								1,175	1,300
0182 KWH totalization (not incl. elec. meter pulse xmtr.)								545	600
0190 Space static pressure								925	1,025
1000 Analog outputs (avg. 50' run in 1/2" EMT)									
1010 P/I Transducer				Ea.				550	605
1020 Analog output, matl. in MUX								265	291
1030 Pneumatic (not incl. control device)								560	615
1040 Electric (not incl. control device)								330	365
2000 Status (Alarms)									
2100 Digital inputs (avg. 50' run in 1/2" EMT)									
2110 Freeze				Ea.				380	415
2120 Fire								340	375
2130 Differential pressure, (air)								520	570
2140 Differential pressure, (water)								850	935
2150 Current sensor								380	415
2160 Duct high temperature thermostat								495	545
2170 Duct smoke detector								615	675
2200 Digital output (avg. 50' run in 1/2" EMT)									
2210 Start/stop				Ea.				297	325
2220 On/off (maintained contact)				"				510	560
3000 Controller MUX panel, incl. function boards									
3100 48 point				Ea.				4,575	5,050
3110 128 point				"				6,300	6,925
3200 DDC controller (avg. 50' run in conduit)									
3210 Mechanical room									
3214 16 point controller (incl. 120 volt/1 phase power supply)				Ea.				1,925	3,125
3229 32 point controller (incl. 120 volt/1 phase power supply)				"				4,725	5,200
3230 Includes software programming and checkout									
3260 Space									
3266 VAV terminal box (incl. space temp. sensor)				Ea.				735	805
3280 Host computer (avg. 50' run in conduit)									
3281 Package complete with PC, keyboard,									
3282 printer, monitor, basic software				Ea.				8,500	9,375
4000 Front end costs									
4100 Computer (P.C.)/software program				Ea.				5,675	6,250
4200 Color graphics software								3,400	3,750

23 09 23 – Direct-Digital Control System for HVAC

23 09 23.10 Control Components/DDC Systems		Crew	Daily Output	Labor-Hours	Unit	Material	2014 Bare Costs Labor	Equipment	Total	Total Incl O&P
4300	Color graphics slides				Ea.				425	470
4350	Additional printer				↓				850	935
4400	Communications trunk cable				L.F.				3.31	3.64
4500	Engineering labor, (not incl. dftg.)				Point				82	90
4600	Calibration labor								105	115
4700	Start-up, checkout labor				↓				105	115
4800	Programming labor, as req'd									
5000	Communications bus (data transmission cable)									
5010	#18 twisted shielded pair in 1/2" EMT conduit				C.L.F.				330	365
8000	Applications software									
8050	Basic maintenance manager software (not incl. data base entry)				Ea.				1,700	1,875
8100	Time program				Point				5.95	6.55
8120	Duty cycle								11.85	13.05
8140	Optimum start/stop								36	39.50
8160	Demand limiting								17.80	19.55
8180	Enthalpy program				↓				36	39.50
8200	Boiler optimization				Ea.				1,075	1,175
8220	Chiller optimization				"				1,425	1,550
8240	Custom applications									
8260	Cost varies with complexity									

23 09 43 – Pneumatic Control System for HVAC

23 09 43.10 Pneumatic Control Systems

		Crew	Daily Output	Labor-Hours	Unit	Material	2014 Bare Costs Labor	Equipment	Total	Total Incl O&P
0010	**PNEUMATIC CONTROL SYSTEMS**									
0011	Including a nominal 50 ft. of tubing. Add control panelboard if req'd.									
0100	Heating and ventilating, split system									
0200	Mixed air control, economizer cycle, panel readout, tubing									
0220	Up to 10 tons ⬛G	Q-19	.68	35.294	Ea.	4,075	1,875		5,950	7,275
0240	For 10 to 20 tons ⬛G		.63	37.915		4,350	2,000		6,350	7,825
0260	For over 20 tons ⬛G		.58	41.096		4,725	2,175		6,900	8,475
0270	Enthalpy cycle, up to 10 tons		.50	48.387		4,500	2,550		7,050	8,800
0280	For 10 to 20 tons		.46	52.174		4,850	2,750		7,600	9,500
0290	For over 20 tons	↓	.42	56.604	↓	5,250	3,000		8,250	10,300
0300	Heating coil, hot water, 3 way valve,									
0320	Freezestat, limit control on discharge, readout	Q-5	.69	23.088	Ea.	3,025	1,225		4,250	5,150
0500	Cooling coil, chilled water, room									
0520	Thermostat, 3 way valve	Q-5	2	8	Ea.	1,350	420		1,770	2,100
0600	Cooling tower, fan cycle, damper control,									
0620	Control system including water readout in/out at panel	Q-19	.67	35.821	Ea.	5,350	1,900		7,250	8,750
1000	Unit ventilator, day/night operation,									
1100	freezestat, ASHRAE, cycle 2	Q-19	.91	26.374	Ea.	2,975	1,400		4,375	5,350
2000	Compensated hot water from boiler, valve control,									
2100	readout and reset at panel, up to 60 GPM	Q-19	.55	43.956	Ea.	5,550	2,325		7,875	9,600
2120	For 120 GPM		.51	47.059		5,925	2,500		8,425	10,300
2140	For 240 GPM		.49	49.180		6,200	2,600		8,800	10,800
3000	Boiler room combustion air, damper to 5 S.F., controls		1.37	17.582		2,675	930		3,605	4,350
3500	Fan coil, heating and cooling valves, 4 pipe control system		3	8		1,200	425		1,625	1,950
3600	Heat exchanger system controls	↓	.86	27.907	↓	2,600	1,475		4,075	5,075
3900	Multizone control (one per zone), includes thermostat, damper									
3910	motor and reset of discharge temperature	Q-5	.51	31.373	Ea.	2,675	1,650		4,325	5,450
4000	Pneumatic thermostat, including controlling room radiator valve	"	2.43	6.593		805	345		1,150	1,400
4040	Program energy saving optimizer ⬛G	Q-19	1.21	19.786		6,700	1,050		7,750	8,925
4060	Pump control system	"	3	8	↓	1,225	425		1,650	1,975

23 09 43 – Pneumatic Control System for HVAC

23 09 43.10 Pneumatic Control Systems	Crew	Daily Output	Labor-Hours	Unit	Material	2014 Bare Costs Labor	Equipment	Total	Total Incl O&P
4080	Reheat coil control system, not incl. coil	Q-5	2.43	6.593	Ea.	1,050	345		1,395	1,675
4500	Air supply for pneumatic control system									
4600	Tank mounted duplex compressor, starter, alternator,									
4620	piping, dryer, PRV station and filter									
4630	1/2 HP	Q-19	.68	35.139	Ea.	9,950	1,850		11,800	13,700
4640	3/4 HP		.64	37.383		10,400	1,975		12,375	14,500
4650	1 HP		.61	39.539		11,400	2,100		13,500	15,700
4660	1-1/2 HP		.58	41.739		12,100	2,200		14,300	16,700
4680	3 HP		.55	43.956		16,500	2,325		18,825	21,700
4690	5 HP		.42	57.143		28,900	3,025		31,925	36,400
4800	Main air supply, includes 3/8" copper main and labor	Q-5	1.82	8.791	C.L.F.	335	465		800	1,075
4810	If poly tubing used, deduct								30%	30%
7000	Static pressure control for air handling unit, includes pressure									
7010	sensor, receiver controller, readout and damper motors	Q-19	.64	37.383	Ea.	7,925	1,975		9,900	11,700
7020	If return air fan requires control, add								70%	70%
8600	VAV boxes, incl. thermostat, damper motor, reheat coil & tubing	Q-5	1.46	10.989		1,250	580		1,830	2,250
8610	If no reheat coil, deduct									204

23 09 53 – Pneumatic and Electric Control System for HVAC

23 09 53.10 Control Components	Crew	Daily Output	Labor-Hours	Unit	Material	2014 Bare Costs Labor	Equipment	Total	Total Incl O&P
0010	**CONTROL COMPONENTS**									
0700	Controller, receiver									
0850	Electric, single snap switch	1 Elec	4	2	Ea.	465	107		572	670
0860	Dual snap switches	"	3	2.667	"	625	142		767	905
3590	Sensor, electric operated									
3620	Humidity	1 Elec	8	1	Ea.	124	53.50		177.50	217
3650	Pressure		8	1		171	53.50		224.50	268
3680	Temperature		10	.800		107	42.50		149.50	181
5000	Thermostats									
5200	24 hour, automatic, clock [G]	1 Shee	8	1	Ea.	158	54.50		212.50	258
5220	Electric, low voltage, 2 wire	1 Elec	13	.615		39.50	33		72.50	92.50
5230	3 wire	"	10	.800		36	42.50		78.50	104
7090	Valves, motor controlled, including actuator									
7100	Electric motor actuated									
7200	Brass, two way, screwed									
7210	1/2" pipe size	L-6	36	.333	Ea.	283	18.70		301.70	340
7220	3/4" pipe size		30	.400		495	22.50		517.50	580
7230	1" pipe size		28	.429		595	24		619	690
7240	1-1/2" pipe size		19	.632		640	35.50		675.50	760
7250	2" pipe size		16	.750		940	42		982	1,100
7350	Brass, three way, screwed									
7360	1/2" pipe size	L-6	33	.364	Ea.	280	20.50		300.50	335
7370	3/4" pipe size		27	.444		325	25		350	400
7380	1" pipe size		25.50	.471		425	26.50		451.50	505
7384	1-1/4" pipe size		21	.571		540	32		572	640
7390	1-1/2" pipe size		17	.706		615	39.50		654.50	735
7400	2" pipe size		14	.857		820	48		868	975
7550	Iron body, two way, flanged									
7560	2-1/2" pipe size	L-6	4	3	Ea.	1,225	168		1,393	1,600
7570	3" pipe size		3	4		1,275	225		1,500	1,775
7580	4" pipe size		2	6		2,225	335		2,560	2,950
7850	Iron body, three way, flanged									
7860	2-1/2" pipe size	L-6	3	4	Ea.	1,225	225		1,450	1,700

23 09 Instrumentation and Control for HVAC

23 09 53 – Pneumatic and Electric Control System for HVAC

23 09 53.10 Control Components	Crew	Daily Output	Labor-Hours	Unit	Material	2014 Bare Costs Labor	Equipment	Total	Total Incl O&P
7870	3" pipe size	L-6	2.50	4.800	Ea.	1,425	270		1,695	1,975
7880	4" pipe size	↓	2	6	↓	1,775	335		2,110	2,450

23 34 HVAC Fans

23 34 13 – Axial HVAC Fans

23 34 13.10 Axial Flow HVAC Fans

23 34 13.10 Axial Flow HVAC Fans	Crew	Daily Output	Labor-Hours	Unit	Material	2014 Bare Costs Labor	Equipment	Total	Total Incl O&P
0010	**AXIAL FLOW HVAC FANS**									
0020	Air conditioning and process air handling									
0030	Axial flow, compact, low sound, 2.5" S.P.									
0050	3,800 CFM, 5 HP	Q-20	3.40	5.882	Ea.	5,025	294		5,319	6,000
0080	6,400 CFM, 5 HP		2.80	7.143		5,625	360		5,985	6,750
0100	10,500 CFM, 7-1/2 HP		2.40	8.333		7,000	415		7,415	8,325
0120	15,600 CFM, 10 HP		1.60	12.500		8,825	625		9,450	10,700
0140	23,000 CFM, 15 HP		.70	28.571		13,500	1,425		14,925	17,100
0160	28,000 CFM, 20 HP		.40	50		15,100	2,500		17,600	20,400
1500	Vaneaxial, low pressure, 2000 CFM, 1/2 HP		3.60	5.556		2,150	278		2,428	2,800
1520	4,000 CFM, 1 HP		3.20	6.250		2,525	315		2,840	3,250
1540	8,000 CFM, 2 HP		2.80	7.143		3,200	360		3,560	4,050
1560	16,000 CFM, 5 HP	↓	2.40	8.333	↓	4,475	415		4,890	5,550

23 34 14 – Blower HVAC Fans

23 34 14.10 Blower Type HVAC Fans

23 34 14.10 Blower Type HVAC Fans	Crew	Daily Output	Labor-Hours	Unit	Material	2014 Bare Costs Labor	Equipment	Total	Total Incl O&P
0010	**BLOWER TYPE HVAC FANS**									
2000	Blowers, direct drive with motor, complete									
2020	1045 CFM @ .5" S.P., 1/5 HP	Q-20	18	1.111	Ea.	330	55.50		385.50	445
2040	1385 CFM @ .5" S.P., 1/4 HP		18	1.111		320	55.50		375.50	435
2060	1640 CFM @ .5" S.P., 1/3 HP		18	1.111		289	55.50		344.50	405
2080	1760 CFM @ .5" S.P., 1/2 HP	↓	18	1.111	↓	300	55.50		355.50	415
2090	4 speed									
2100	1164 to 1739 CFM @ .5" S.P., 1/3 HP	Q-20	16	1.250	Ea.	310	62.50		372.50	435
2120	1467 to 2218 CFM @ 1.0" S.P., 3/4 HP	"	14	1.429	"	355	71.50		426.50	505
2500	Ceiling fan, right angle, extra quiet, 0.10" S.P.									
2520	95 CFM	Q-20	20	1	Ea.	287	50		337	390
2540	210 CFM		19	1.053		340	52.50		392.50	455
2560	385 CFM		18	1.111		430	55.50		485.50	560
2580	885 CFM		16	1.250		850	62.50		912.50	1,025
2600	1,650 CFM		13	1.538		1,175	77		1,252	1,425
2620	2,960 CFM	↓	11	1.818		1,575	91		1,666	1,875
2680	For speed control switch, add	1 Elec	16	.500	↓	157	26.50		183.50	212
7500	Utility set, steel construction, pedestal, 1/4" S.P.									
7520	Direct drive, 150 CFM, 1/8 HP	Q-20	6.40	3.125	Ea.	845	156		1,001	1,175
7540	485 CFM, 1/6 HP		5.80	3.448		1,075	173		1,248	1,425
7560	1950 CFM, 1/2 HP		4.80	4.167		1,250	209		1,459	1,700
7580	2410 CFM, 3/4 HP		4.40	4.545		2,300	228		2,528	2,875
7600	3328 CFM, 1-1/2 HP	↓	3	6.667	↓	2,575	335		2,910	3,325
7680	V-belt drive, drive cover, 3 phase									
7700	800 CFM, 1/4 HP	Q-20	6	3.333	Ea.	955	167		1,122	1,300
7720	1,300 CFM, 1/3 HP		5	4		1,000	200		1,200	1,400
7740	2,000 CFM, 1 HP		4.60	4.348		1,175	218		1,393	1,625
7760	2,900 CFM, 3/4 HP		4.20	4.762		1,600	238		1,838	2,100
7780	3,600 CFM, 3/4 HP	↓	4	5		1,975	250		2,225	2,525

23 34 HVAC Fans

23 34 14 – Blower HVAC Fans

23 34 14.10 Blower Type HVAC Fans

		Crew	Daily Output	Labor-Hours	Unit	Material	2014 Bare Costs Labor	Equipment	Total	Total Incl O&P
7800	4,800 CFM, 1 HP	Q-20	3.50	5.714	Ea.	2,325	286		2,611	2,975
7820	6,700 CFM, 1-1/2 HP		3	6.667		2,850	335		3,185	3,650
7830	7,500 CFM, 2 HP		2.50	8		3,875	400		4,275	4,875
7840	11,000 CFM, 3 HP		2	10		5,200	500		5,700	6,450
7860	13,000 CFM, 3 HP		1.60	12.500		5,275	625		5,900	6,750
7880	15,000 CFM, 5 HP		1	20		5,450	1,000		6,450	7,525
7900	17,000 CFM, 7-1/2 HP		.80	25		5,825	1,250		7,075	8,325
7920	20,000 CFM, 7-1/2 HP		.80	25		6,950	1,250		8,200	9,550

23 34 16 – Centrifugal HVAC Fans

23 34 16.10 Centrifugal Type HVAC Fans

		Crew	Daily Output	Labor-Hours	Unit	Material	2014 Bare Costs Labor	Equipment	Total	Total Incl O&P
0010	**CENTRIFUGAL TYPE HVAC FANS**									
0200	In-line centrifugal, supply/exhaust booster									
0220	aluminum wheel/hub, disconnect switch, 1/4" S.P.									
0240	500 CFM, 10" diameter connection	Q-20	3	6.667	Ea.	1,250	335		1,585	1,875
0260	1,380 CFM, 12" diameter connection		2	10		1,325	500		1,825	2,200
0280	1,520 CFM, 16" diameter connection		2	10		1,450	500		1,950	2,350
0300	2,560 CFM, 18" diameter connection		1	20		1,575	1,000		2,575	3,250
0320	3,480 CFM, 20" diameter connection		.80	25		1,875	1,250		3,125	3,950
0326	5,080 CFM, 20" diameter connection		.75	26.667		2,025	1,325		3,350	4,250
3500	Centrifugal, airfoil, motor and drive, complete									
3520	1000 CFM, 1/2 HP	Q-20	2.50	8	Ea.	1,850	400		2,250	2,625
3540	2,000 CFM, 1 HP		2	10		2,075	500		2,575	3,050
3560	4,000 CFM, 3 HP		1.80	11.111		2,650	555		3,205	3,750
3580	8,000 CFM, 7-1/2 HP		1.40	14.286		3,975	715		4,690	5,450
3600	12,000 CFM, 10 HP		1	20		5,300	1,000		6,300	7,350
5000	Utility set, centrifugal, V belt drive, motor									
5020	1/4" S.P., 1200 CFM, 1/4 HP	Q-20	6	3.333	Ea.	1,650	167		1,817	2,075
5040	1520 CFM, 1/3 HP		5	4		2,125	200		2,325	2,650
5060	1850 CFM, 1/2 HP		4	5		2,125	250		2,375	2,700
5080	2180 CFM, 3/4 HP		3	6.667		2,525	335		2,860	3,275
5100	1/2" S.P., 3600 CFM, 1 HP		2	10		2,975	500		3,475	4,025
5120	4250 CFM, 1-1/2 HP		1.60	12.500		3,150	625		3,775	4,425
5140	4800 CFM, 2 HP		1.40	14.286		3,825	715		4,540	5,275
5160	6920 CFM, 5 HP		1.30	15.385		4,650	770		5,420	6,275
5180	7700 CFM, 7-1/2 HP		1.20	16.667		5,000	835		5,835	6,775
5200	For explosion proof motor, add					15%				
5500	Fans, industrial exhauster, for air which may contain granular matl.									
5520	1000 CFM, 1-1/2 HP	Q-20	2.50	8	Ea.	2,800	400		3,200	3,675
5540	2000 CFM, 3 HP		2	10		3,400	500		3,900	4,500
5560	4000 CFM, 7-1/2 HP		1.80	11.111		4,725	555		5,280	6,050
5580	8000 CFM, 15 HP		1.40	14.286		6,100	715		6,815	7,775
5600	12,000 CFM, 30 HP		1	20		9,925	1,000		10,925	12,400
7000	Roof exhauster, centrifugal, aluminum housing, 12" galvanized									
7020	curb, bird screen, back draft damper, 1/4" S.P.									
7100	Direct drive, 320 CFM, 11" sq. damper	Q-20	7	2.857	Ea.	665	143		808	950
7120	600 CFM, 11" sq. damper		6	3.333		865	167		1,032	1,200
7140	815 CFM, 13" sq. damper		5	4		865	200		1,065	1,250
7160	1450 CFM, 13" sq. damper		4.20	4.762		1,400	238		1,638	1,900
7180	2050 CFM, 16" sq. damper		4	5		1,675	250		1,925	2,225
7200	V-belt drive, 1650 CFM, 12" sq. damper		6	3.333		1,250	167		1,417	1,650
7220	2750 CFM, 21" sq. damper		5	4		1,500	200		1,700	1,950
7230	3500 CFM, 21" sq. damper		4.50	4.444		1,675	222		1,897	2,200

23 34 HVAC Fans

23 34 16 – Centrifugal HVAC Fans

23 34 16.10 Centrifugal Type HVAC Fans	Crew	Daily Output	Labor-Hours	Unit	Material	2014 Bare Costs Labor	Equipment	Total	Total Incl O&P
7240	4910 CFM, 23" sq. damper	Q-20	4	5	Ea.	2,050	250		2,300	2,650
7260	8525 CFM, 28" sq. damper		3	6.667		2,725	335		3,060	3,500
7280	13,760 CFM, 35" sq. damper		2	10		3,800	500		4,300	4,950
7300	20,558 CFM, 43" sq. damper		1	20		7,625	1,000		8,625	9,900
7320	For 2 speed winding, add					15%				
7340	For explosionproof motor, add					580			580	635
7360	For belt driven, top discharge, add					15%				
8500	Wall exhausters, centrifugal, auto damper, 1/8" S.P.									
8520	Direct drive, 610 CFM, 1/20 HP	Q-20	14	1.429	Ea.	405	71.50		476.50	555
8540	796 CFM, 1/12 HP		13	1.538		845	77		922	1,050
8560	822 CFM, 1/6 HP		12	1.667		1,025	83.50		1,108.50	1,250
8580	1,320 CFM, 1/4 HP		12	1.667		2,125	83.50		2,208.50	2,450
8600	1756 CFM, 1/4 HP		11	1.818		1,200	91		1,291	1,475
8620	1983 CFM, 1/4 HP		10	2		1,225	100		1,325	1,500
8640	2900 CFM, 1/2 HP		9	2.222		1,300	111		1,411	1,600
8660	3307 CFM, 3/4 HP		8	2.500		1,425	125		1,550	1,750
9500	V-belt drive, 3 phase									
9520	2,800 CFM, 1/4 HP	Q-20	9	2.222	Ea.	1,825	111		1,936	2,175
9540	3,740 CFM, 1/2 HP		8	2.500		1,900	125		2,025	2,300
9560	4400 CFM, 3/4 HP		7	2.857		1,925	143		2,068	2,350
9580	5700 CFM, 1-1/2 HP		6	3.333		2,000	167		2,167	2,450

23 34 23 – HVAC Power Ventilators

23 34 23.10 HVAC Power Circulators and Ventilators	Crew	Daily Output	Labor-Hours	Unit	Material	2014 Bare Costs Labor	Equipment	Total	Total Incl O&P
0010	**HVAC POWER CIRCULATORS AND VENTILATORS**									
3000	Paddle blade air circulator, 3 speed switch									
3020	42", 5,000 CFM high, 3000 CFM low [G]	1 Elec	2.40	3.333	Ea.	163	178		341	445
3040	52", 6,500 CFM high, 4000 CFM low [G]	"	2.20	3.636	"	170	194		364	475
3100	For antique white motor, same cost									
3200	For brass plated motor, same cost									
3300	For light adaptor kit, add [G]				Ea.	41			41	45
6000	Propeller exhaust, wall shutter									
6020	Direct drive, one speed, .075" S.P.									
6100	653 CFM, 1/30 HP	Q-20	10	2	Ea.	183	100		283	355
6120	1033 CFM, 1/20 HP		9	2.222		273	111		384	470
6140	1323 CFM, 1/15 HP		8	2.500		305	125		430	525
6160	2444 CFM, 1/4 HP		7	2.857		380	143		523	635
6300	V-belt drive, 3 phase									
6320	6175 CFM, 3/4 HP	Q-20	5	4	Ea.	2,850	200		3,050	3,425
6340	7500 CFM, 3/4 HP		5	4		2,925	200		3,125	3,500
6360	10,100 CFM, 1 HP		4.50	4.444		3,075	222		3,297	3,725
6380	14,300 CFM, 1-1/2 HP		4	5		3,325	250		3,575	4,025
6400	19,800 CFM, 2 HP		3	6.667		1,875	335		2,210	2,550
6420	26,250 CFM, 3 HP		2.60	7.692		2,050	385		2,435	2,825
6440	38,500 CFM, 5 HP		2.20	9.091		2,425	455		2,880	3,375
6460	46,000 CFM, 7-1/2 HP		2	10		2,600	500		3,100	3,600
6480	51,500 CFM, 10 HP		1.80	11.111		2,700	555		3,255	3,825
6650	Residential, bath exhaust, grille, back draft damper									
6660	50 CFM	Q-20	24	.833	Ea.	61	41.50		102.50	131
6670	110 CFM		22	.909		94.50	45.50		140	173
6680	Light combination, squirrel cage, 100 watt, 70 CFM		24	.833		108	41.50		149.50	182
6700	Light/heater combination, ceiling mounted									
6710	70 CFM, 1450 watt	Q-20	24	.833	Ea.	155	41.50		196.50	235

23 34 23 – HVAC Power Ventilators

23 34 23.10 HVAC Power Circulators and Ventilators		Crew	Daily Output	Labor-Hours	Unit	Material	2014 Bare Costs Labor	Equipment	Total	Total Incl O&P
6800	Heater combination, recessed, 70 CFM	Q-20	24	.833	Ea.	64.50	41.50		106	135
6820	With 2 infrared bulbs		23	.870		101	43.50		144.50	177
6900	Kitchen exhaust, grille, complete, 160 CFM		22	.909		104	45.50		149.50	183
6910	180 CFM		20	1		87	50		137	172
6920	270 CFM		18	1.111		164	55.50		219.50	266
6930	350 CFM		16	1.250		124	62.50		186.50	231
6940	Residential roof jacks and wall caps									
6944	Wall cap with back draft damper									
6946	3" & 4" diam. round duct	1 Shee	11	.727	Ea.	24.50	40		64.50	88
6948	6" diam. round duct	"	11	.727	"	61.50	40		101.50	129
6958	Roof jack with bird screen and back draft damper									
6960	3" & 4" diam. round duct	1 Shee	11	.727	Ea.	25	40		65	88.50
6962	3-1/4" x 10" rectangular duct	"	10	.800	"	46.50	44		90.50	118
6980	Transition									
6982	3-1/4" x 10" to 6" diam. round	1 Shee	20	.400	Ea.	31	22		53	67.50
8020	Attic, roof type									
8030	Aluminum dome, damper & curb									
8080	12" diameter, 1000 CFM (gravity)	1 Elec	10	.800	Ea.	545	42.50		587.50	665
8090	16" diameter, 1500 CFM (gravity)		9	.889		660	47.50		707.50	795
8100	20" diameter, 2500 CFM (gravity)		8	1		805	53.50		858.50	970
8110	26" diameter, 4000 CFM (gravity)		7	1.143		980	61		1,041	1,175
8120	32" diameter, 6500 CFM (gravity)		6	1.333		1,350	71		1,421	1,575
8130	38" diameter, 8000 CFM (gravity)		5	1.600		2,000	85.50		2,085.50	2,325
8140	50" diameter, 13,000 CFM (gravity)		4	2		2,875	107		2,982	3,325
8160	Plastic, ABS dome									
8180	1050 CFM	1 Elec	14	.571	Ea.	160	30.50		190.50	222
8200	1600 CFM	"	12	.667	"	240	35.50		275.50	315
8240	Attic, wall type, with shutter, one speed									
8250	12" diameter, 1000 CFM	1 Elec	14	.571	Ea.	385	30.50		415.50	470
8260	14" diameter, 1500 CFM		12	.667		420	35.50		455.50	515
8270	16" diameter, 2000 CFM		9	.889		475	47.50		522.50	590
8290	Whole house, wall type, with shutter, one speed									
8300	30" diameter, 4800 CFM	1 Elec	7	1.143	Ea.	1,025	61		1,086	1,225
8310	36" diameter, 7000 CFM		6	1.333		1,100	71		1,171	1,325
8320	42" diameter, 10,000 CFM		5	1.600		1,225	85.50		1,310.50	1,475
8330	48" diameter, 16,000 CFM		4	2		1,525	107		1,632	1,850
8340	For two speed, add					92.50			92.50	102
8350	Whole house, lay-down type, with shutter, one speed									
8360	30" diameter, 4500 CFM	1 Elec	8	1	Ea.	1,075	53.50		1,128.50	1,275
8370	36" diameter, 6500 CFM		7	1.143		1,150	61		1,211	1,375
8380	42" diameter, 9000 CFM		6	1.333		1,275	71		1,346	1,500
8390	48" diameter, 12,000 CFM		5	1.600		1,450	85.50		1,535.50	1,725
8440	For two speed, add					69.50			69.50	76.50
8450	For 12 hour timer switch, add	1 Elec	32	.250		69.50	13.35		82.85	96.50

23 52 Heating Boilers

23 52 13 – Electric Boilers

23 52 13.10 Electric Boilers, ASME	Crew	Daily Output	Labor-Hours	Unit	Material	2014 Bare Costs Labor	Equipment	Total	Total Incl O&P
0010 **ELECTRIC BOILERS, ASME**, Standard controls and trim									
1000 Steam, 6 KW, 20.5 MBH	Q-19	1.20	20	Ea.	3,950	1,050		5,000	5,950
1040 9 KW, 30.7 MBH		1.20	20		4,025	1,050		5,075	6,025
1060 18 KW, 61.4 MBH		1.20	20		4,150	1,050		5,200	6,175
1080 24 KW, 81.8 MBH		1.10	21.818		4,800	1,150		5,950	7,025
1120 36 KW, 123 MBH		1.10	21.818		5,325	1,150		6,475	7,575
1160 60 KW, 205 MBH		1	24		6,650	1,275		7,925	9,225
1220 112 KW, 382 MBH		.75	32		9,375	1,700		11,075	12,900
1240 148 KW, 505 MBH		.65	36.923		9,600	1,950		11,550	13,600
1260 168 KW, 573 MBH		.60	40		20,400	2,125		22,525	25,700
1280 222 KW, 758 MBH		.55	43.636		23,800	2,300		26,100	29,700
1300 296 KW, 1010 MBH		.45	53.333		25,000	2,825		27,825	31,900
1320 300 KW, 1023 MBH		.40	60		26,400	3,175		29,575	33,800
1340 370 KW, 1263 MBH		.35	68.571		29,200	3,625		32,825	37,600
1360 444 KW, 1515 MBH		.30	80		31,400	4,225		35,625	40,900
1380 518 KW, 1768 MBH	Q-21	.36	88.889		32,600	4,825		37,425	43,200
1400 592 KW, 2020 MBH		.34	94.118		35,400	5,100		40,500	46,600
1420 666 KW, 2273 MBH		.32	100		36,800	5,425		42,225	48,700
1460 740 KW, 2526 MBH		.28	114		38,300	6,200		44,500	51,500
1480 814 KW, 2778 MBH		.25	128		40,600	6,950		47,550	55,000
1500 962 KW, 3283 MBH		.22	145		44,900	7,900		52,800	61,500
1520 1036 KW, 3536 MBH		.20	160		46,700	8,675		55,375	64,500
1540 1110 KW, 3788 MBH		.19	168		49,600	9,150		58,750	68,500
1560 2070 KW, 7063 MBH		.18	177		68,000	9,650		77,650	89,500
1580 2250 KW, 7677 MBH		.17	188		80,000	10,200		90,200	103,500
1600 2,340 KW, 7984 MBH		.16	200		86,500	10,900		97,400	111,500
2000 Hot water, 7.5 KW, 25.6 MBH	Q-19	1.30	18.462		4,975	975		5,950	6,950
2020 15 KW, 51.2 MBH		1.30	18.462		5,000	975		5,975	6,975
2040 30 KW, 102 MBH		1.20	20		5,325	1,050		6,375	7,450
2060 45 KW, 164 MBH		1.20	20		5,425	1,050		6,475	7,550
2070 60 KW, 205 MBH		1.20	20		5,525	1,050		6,575	7,675
2080 75 KW, 256 MBH		1.10	21.818		5,900	1,150		7,050	8,225
2100 90 KW, 307 MBH		1.10	21.818		6,000	1,150		7,150	8,325
2120 105 KW, 358 MBH		1	24		6,275	1,275		7,550	8,800
2140 120 KW, 410 MBH		.90	26.667		6,375	1,400		7,775	9,150
2160 135 KW, 461 MBH		.75	32		7,400	1,700		9,100	10,700
2180 150 KW, 512 MBH		.65	36.923		7,500	1,950		9,450	11,200
2200 165 KW, 563 MBH		.60	40		7,600	2,125		9,725	11,600
2220 296 KW, 1010 MBH		.55	43.636		16,100	2,300		18,400	21,200
2280 370 KW, 1263 MBH		.40	60		17,900	3,175		21,075	24,500
2300 444 KW, 1515 MBH		.35	68.571		20,800	3,625		24,425	28,400
2340 518 KW, 1768 MBH	Q-21	.44	72.727		23,100	3,950		27,050	31,400
2360 592 KW, 2020 MBH		.43	74.419		26,100	4,050		30,150	34,900
2400 666 KW, 2273 MBH		.40	80		27,300	4,350		31,650	36,600
2420 740 KW, 2526 MBH		.39	82.051		28,500	4,450		32,950	38,000
2440 814 KW, 2778 MBH		.38	84.211		30,500	4,575		35,075	40,400
2460 888 KW, 3031 MBH		.37	86.486		31,100	4,700		35,800	41,300
2480 962 KW, 3283 MBH		.36	88.889		34,000	4,825		38,825	44,700
2500 1036 KW, 3536 MBH		.34	94.118		35,500	5,100		40,600	46,800
2520 1110 KW, 3788 MBH		.33	96.970		38,800	5,275		44,075	50,500
2540 1440 KW, 4915 MBH		.32	100		43,800	5,425		49,225	56,500
2560 1560 KW, 5323 MBH		.31	103		46,800	5,600		52,400	60,000
2580 1680 KW, 5733 MBH		.30	106		50,000	5,800		55,800	63,500

23 52 Heating Boilers

23 52 13 – Electric Boilers

23 52 13.10 Electric Boilers, ASME

		Crew	Daily Output	Labor-Hours	Unit	Material	2014 Bare Costs Labor	Equipment	Total	Total Incl O&P
2600	1800 KW, 6143 MBH	Q-21	.29	110	Ea.	53,500	6,000		59,500	67,500
2620	1980 KW, 6757 MBH		.28	114		58,000	6,200		64,200	73,500
2640	2100 KW, 7167 MBH		.27	118		59,500	6,425		65,925	75,000
2660	2220 KW, 7576 MBH		.26	123		65,000	6,675		71,675	81,500
2680	2400 KW, 8191 MBH		.25	128		68,000	6,950		74,950	85,000
2700	2610 KW, 8905 MBH		.24	133		71,000	7,250		78,250	89,000
2720	2790 KW, 9519 MBH		.23	139		75,500	7,550		83,050	94,500
2740	2970 KW, 10133 MBH		.21	152		77,000	8,275		85,275	97,500
2760	3150 KW, 10748 MBH		.19	168		82,500	9,150		91,650	105,000
2780	3240 KW, 11055 MBH		.18	177		84,500	9,650		94,150	107,500
2800	3420 KW, 11669 MBH		.17	188		90,000	10,200		100,200	114,500
2820	3600 KW, 12,283 MBH		.16	200		94,000	10,900		104,900	120,000

23 52 28 – Swimming Pool Boilers

23 52 28.10 Swimming Pool Heaters

		Crew	Daily Output	Labor-Hours	Unit	Material	2014 Bare Costs Labor	Equipment	Total	Total Incl O&P
0010	**SWIMMING POOL HEATERS**, Not including wiring, external									
0020	piping, base or pad,									
2000	Electric, 12 KW, 4,800 gallon pool	Q-19	3	8	Ea.	2,075	425		2,500	2,900
2020	15 KW, 7,200 gallon pool		2.80	8.571		2,100	455		2,555	3,000
2040	24 KW, 9,600 gallon pool		2.40	10		2,425	530		2,955	3,475
2060	30 KW, 12,000 gallon pool		2	12		2,475	635		3,110	3,675
2080	36 KW, 14,400 gallon pool		1.60	15		2,850	795		3,645	4,325
2100	57 KW, 24,000 gallon pool		1.20	20		3,575	1,050		4,625	5,525
9000	To select pool heater: 12 BTUH x S.F. pool area									
9010	X temperature differential = required output									
9050	For electric, KW = gallons x 2.5 divided by 1000									
9100	For family home type pool, double the									
9110	Rated gallon capacity = 1/2°F rise per hour									

23 54 Furnaces

23 54 13 – Electric-Resistance Furnaces

23 54 13.10 Electric Furnaces

		Crew	Daily Output	Labor-Hours	Unit	Material	2014 Bare Costs Labor	Equipment	Total	Total Incl O&P
0010	**ELECTRIC FURNACES**, Hot air, blowers, std. controls									
1000	Electric, UL listed									
1100	34.1 MBH	Q-20	4.40	4.545	Ea.	560	228		788	960
1120	51.6 MBH		4.20	4.762		630	238		868	1,050
1140	68.3 MBH		4	5		650	250		900	1,100
1160	85.3 MBH		3.80	5.263		660	263		923	1,125

23 82 Convection Heating and Cooling Units

23 82 16 – Air Coils

23 82 16.20 Duct Heaters

		Crew	Daily Output	Labor-Hours	Unit	Material	2014 Bare Costs Labor	Equipment	Total	Total Incl O&P
0010	**DUCT HEATERS**, Electric, 480 V, 3 Ph.									
0020	Finned tubular insert, 500°F									
0100	8" wide x 6" high, 4.0 kW	Q-20	16	1.250	Ea.	795	62.50		857.50	970
0120	12" high, 8.0 kW		15	1.333		1,325	66.50		1,391.50	1,550
0140	18" high, 12.0 kW		14	1.429		1,850	71.50		1,921.50	2,125
0160	24" high, 16.0 kW		13	1.538		2,375	77		2,452	2,725
0180	30" high, 20.0 kW		12	1.667		2,900	83.50		2,983.50	3,325
0300	12" wide x 6" high, 6.7 kW		15	1.333		845	66.50		911.50	1,025

23 82 16 – Air Coils

23 82 16.20 Duct Heaters		Crew	Daily Output	Labor-Hours	Unit	Material	2014 Bare Costs Labor	Equipment	Total	Total Incl O&P
0320	12" high, 13.3 kW	Q-20	14	1.429	Ea.	1,350	71.50		1,421.50	1,600
0340	18" high, 20.0 kW		13	1.538		1,900	77		1,977	2,225
0360	24" high, 26.7 kW		12	1.667		2,450	83.50		2,533.50	2,825
0380	30" high, 33.3 kW		11	1.818		3,000	91		3,091	3,450
0500	18" wide x 6" high, 13.3 kW		14	1.429		910	71.50		981.50	1,100
0520	12" high, 26.7 kW		13	1.538		1,550	77		1,627	1,850
0540	18" high, 40.0 kW		12	1.667		2,075	83.50		2,158.50	2,400
0560	24" high, 53.3 kW		11	1.818		2,750	91		2,841	3,175
0580	30" high, 66.7 kW		10	2		3,450	100		3,550	3,950
0700	24" wide x 6" high, 17.8 kW		13	1.538		1,000	77		1,077	1,225
0720	12" high, 35.6 kW		12	1.667		1,700	83.50		1,783.50	1,975
0740	18" high, 53.3 kW		11	1.818		2,375	91		2,466	2,750
0760	24" high, 71.1 kW		10	2		3,075	100		3,175	3,525
0780	30" high, 88.9 kW		9	2.222		3,750	111		3,861	4,300
0900	30" wide x 6" high, 22.2 kW		12	1.667		1,050	83.50		1,133.50	1,300
0920	12" high, 44.4 kW		11	1.818		1,775	91		1,866	2,100
0940	18" high, 66.7 kW		10	2		2,525	100		2,625	2,925
0960	24" high, 88.9 kW		9	2.222		3,100	111		3,211	3,575
0980	30" high, 111.0 kW		8	2.500		3,975	125		4,100	4,575
1400	Note decreased kW available for									
1410	each duct size at same cost									
1420	See line 5000 for modifications and accessories									
2000	Finned tubular flange with insulated									
2020	terminal box, 500°F									
2100	12" wide x 36" high, 54 kW	Q-20	10	2	Ea.	3,900	100		4,000	4,450
2120	40" high, 60 kW		9	2.222		4,500	111		4,611	5,125
2200	24" wide x 36" high, 118.8 kW		9	2.222		5,000	111		5,111	5,675
2220	40" high, 132 kW		8	2.500		5,225	125		5,350	5,950
2400	36" wide x 8" high, 40 kW		11	1.818		2,000	91		2,091	2,350
2420	16" high, 80 kW		10	2		2,650	100		2,750	3,075
2440	24" high, 120 kW		9	2.222		3,475	111		3,586	4,000
2460	32" high, 160 kW		8	2.500		4,450	125		4,575	5,100
2480	36" high, 180 kW		7	2.857		5,500	143		5,643	6,250
2500	40" high, 200 kW		6	3.333		6,075	167		6,242	6,925
2600	40" wide x 8" high, 45 kW		11	1.818		2,075	91		2,166	2,425
2620	16" high, 90 kW		10	2		2,850	100		2,950	3,300
2640	24" high, 135 kW		9	2.222		3,625	111		3,736	4,175
2660	32" high, 180 kW		8	2.500		4,850	125		4,975	5,525
2680	36" high, 202.5 kW		7	2.857		5,575	143		5,718	6,350
2700	40" high, 225 kW		6	3.333		6,350	167		6,517	7,225
2800	48" wide x 8" high, 54.8 kW		10	2		2,175	100		2,275	2,550
2820	16" high, 109.8 kW		9	2.222		3,025	111		3,136	3,500
2840	24" high, 164.4 kW		8	2.500		3,850	125		3,975	4,425
2860	32" high, 219.2 kW		7	2.857		5,125	143		5,268	5,875
2880	36" high, 246.6 kW		6	3.333		5,900	167		6,067	6,750
2900	40" high, 274 kW		5	4		6,675	200		6,875	7,625
3000	56" wide x 8" high, 64 kW		9	2.222		2,475	111		2,586	2,900
3020	16" high, 128 kW		8	2.500		3,450	125		3,575	3,975
3040	24" high, 192 kW		7	2.857		4,175	143		4,318	4,800
3060	32" high, 256 kW		6	3.333		5,825	167		5,992	6,650
3080	36" high, 288kW		5	4		6,675	200		6,875	7,650
3100	40" high, 320kW		4	5		7,375	250		7,625	8,500
3200	64" wide x 8" high, 74kW		8	2.500		2,550	125		2,675	3,000

23 82 Convection Heating and Cooling Units

23 82 16 – Air Coils

23 82 16.20 Duct Heaters	Crew	Daily Output	Labor-Hours	Unit	Material	2014 Bare Costs Labor	Equipment	Total	Total Incl O&P
3220	16" high, 148kW	Q-20	7	2.857	Ea.	3,525	143		3,668	4,125
3240	24" high, 222kW		6	3.333		4,500	167		4,667	5,200
3260	32" high, 296kW		5	4		6,025	200		6,225	6,925
3280	36" high, 333kW		4	5		7,200	250		7,450	8,275
3300	40" high, 370kW		3	6.667		7,925	335		8,260	9,225
3800	Note decreased kW available for									
3820	each duct size at same cost									
5000	Duct heater modifications and accessories									
5120	T.C.O. limit auto or manual reset	Q-20	42	.476	Ea.	130	24		154	179
5140	Thermostat		28	.714		555	36		591	665
5160	Overheat thermocouple (removable)		7	2.857		785	143		928	1,075
5180	Fan interlock relay		18	1.111		190	55.50		245.50	294
5200	Air flow switch		20	1		164	50		214	256
5220	Split terminal box cover		100	.200		54	10		64	74.50
8000	To obtain BTU multiply kW by 3413									

23 82 27 – Infrared Units

23 82 27.10 Infrared Type Heating Units

23 82 27.10		Crew	Daily Output	Labor-Hours	Unit	Material	Labor	Equipment	Total	Total Incl O&P
0010	**INFRARED TYPE HEATING UNITS**									
2000	Electric, single or three phase									
2050	6 kW, 20,478 BTU	1 Elec	2.30	3.478	Ea.	675	186		861	1,025
2100	13.5 KW, 40,956 BTU		2.20	3.636		835	194		1,029	1,200
2150	24 KW, 81,912 BTU		2	4		1,675	213		1,888	2,175

23 83 Radiant Heating Units

23 83 33 – Electric Radiant Heaters

23 83 33.10 Electric Heating

23 83 33.10		Crew	Daily Output	Labor-Hours	Unit	Material	Labor	Equipment	Total	Total Incl O&P
0010	**ELECTRIC HEATING**, not incl. conduit or feed wiring									
1100	Rule of thumb: Baseboard units, including control	1 Elec	4.40	1.818	kW	101	97		198	256
1300	Baseboard heaters, 2' long, 350 watt		8	1	Ea.	30	53.50		83.50	113
1400	3' long, 750 watt		8	1		34.50	53.50		88	118
1600	4' long, 1000 watt		6.70	1.194		41	63.50		104.50	141
1800	5' long, 935 watt		5.70	1.404		51	75		126	168
2000	6' long, 1500 watt		5	1.600		56	85.50		141.50	190
2200	7' long, 1310 watt		4.40	1.818		64	97		161	216
2400	8' long, 2000 watt		4	2		69.50	107		176.50	236
2600	9' long, 1680 watt		3.60	2.222		104	119		223	291
2800	10' long, 1875 watt		3.30	2.424		193	129		322	405
2950	Wall heaters with fan, 120 to 277 volt									
3160	Recessed, residential, 750 watt	1 Elec	6	1.333	Ea.	134	71		205	253
3170	1000 watt		6	1.333		134	71		205	253
3180	1250 watt		5	1.600		134	85.50		219.50	275
3190	1500 watt		4	2		134	107		241	305
3210	2000 watt		4	2		134	107		241	305
3230	2500 watt		3.50	2.286		335	122		457	555
3240	3000 watt		3	2.667		495	142		637	755
3250	4000 watt		2.70	2.963		495	158		653	780
3260	Commercial, 750 watt		6	1.333		189	71		260	315
3270	1000 watt		6	1.333		189	71		260	315
3280	1250 watt		5	1.600		189	85.50		274.50	335
3290	1500 watt		4	2		189	107		296	365

23 83 33 – Electric Radiant Heaters

23 83 33.10 Electric Heating		Crew	Daily Output	Labor-Hours	Unit	Material	2014 Bare Costs Labor	2014 Bare Costs Equipment	Total	Total Incl O&P
3300	2000 watt	1 Elec	4	2	Ea.	189	107		296	365
3310	2500 watt		3.50	2.286		189	122		311	390
3320	3000 watt		3	2.667		340	142		482	590
3330	4000 watt		2.70	2.963		340	158		498	610
3600	Thermostats, integral		16	.500		30	26.50		56.50	73
3800	Line voltage, 1 pole		8	1		20.50	53.50		74	103
3810	2 pole	▼	8	1	▼	29.50	53.50		83	113
4000	Heat trace system, 400 degree R238313-10									
4020	115 V, 2.5 watts per L.F.	1 Elec	530	.015	L.F.	7.90	.81		8.71	9.85
4030	5 watts per L.F. R238313-20		530	.015		7.90	.81		8.71	9.85
4050	10 watts per L.F.		530	.015		7.90	.81		8.71	9.85
4060	208 V, 5 watts per L.F.		530	.015		7.90	.81		8.71	9.85
4080	480 V, 8 watts per L.F.	▼	530	.015	▼	7.90	.81		8.71	9.85
4200	Heater raceway									
4260	Heat transfer cement									
4280	1 gallon				Ea.	57			57	62.50
4300	5 gallon				"	220			220	242
4320	Snap band, clamp									
4340	3/4" pipe size	1 Elec	470	.017	Ea.		.91		.91	1.36
4360	1" pipe size		444	.018			.96		.96	1.44
4380	1-1/4" pipe size		400	.020			1.07		1.07	1.60
4400	1-1/2" pipe size		355	.023			1.20		1.20	1.80
4420	2" pipe size		320	.025			1.33		1.33	2
4440	3" pipe size		160	.050			2.67		2.67	3.99
4460	4" pipe size		100	.080			4.27		4.27	6.40
4480	Thermostat NEMA 3R, 22 amp, 0-150 Deg, 10' cap.		8	1		241	53.50		294.50	345
4500	Thermostat NEMA 4X, 25 amp, 40 Deg, 5-1/2' cap.		7	1.143		241	61		302	355
4520	Thermostat NEMA 4X, 22 amp, 25-325 Deg, 10' cap.		7	1.143		650	61		711	805
4540	Thermostat NEMA 4X, 22 amp, 15-140 Deg,		6	1.333		545	71		616	705
4580	Thermostat NEMA 4,7,9, 22 amp, 25-325 Deg, 10' cap.		3.60	2.222		800	119		919	1,050
4600	Thermostat NEMA 4,7,9, 22 amp, 15-140 Deg,		3	2.667		795	142		937	1,075
4720	Fiberglass application tape, 36 yard roll		11	.727		77.50	39		116.50	144
5000	Radiant heating ceiling panels, 2' x 4', 500 watt		16	.500		291	26.50		317.50	360
5050	750 watt		16	.500		310	26.50		336.50	385
5200	For recessed plaster frame, add		32	.250		110	13.35		123.35	141
5300	Infrared quartz heaters, 120 volts, 1000 watts		6.70	1.194		230	63.50		293.50	350
5350	1500 watt		5	1.600		230	85.50		315.50	380
5400	240 volts, 1500 watt		5	1.600		230	85.50		315.50	380
5450	2000 watt		4	2		230	107		337	415
5500	3000 watt		3	2.667		255	142		397	495
5550	4000 watt		2.60	3.077		255	164		419	525
5570	Modulating control	▼	.80	10	▼	275	535		810	1,100
5600	Unit heaters, heavy duty, with fan & mounting bracket									
5650	Single phase, 208-240-277 volt, 3 kW	1 Elec	6	1.333	Ea.	435	71		506	585
5750	5 kW		5.50	1.455		460	77.50		537.50	620
5800	7 kW		5	1.600		710	85.50		795.50	910
5850	10 kW		4	2		810	107		917	1,050
5950	15 kW		3.80	2.105		1,300	112		1,412	1,600
6000	480 volt, 3 kW		6	1.333		480	71		551	635
6020	4 kW		5.80	1.379		490	73.50		563.50	650
6040	5 kW		5.50	1.455		510	77.50		587.50	675
6060	7 kW		5	1.600		755	85.50		840.50	965
6080	10 kW	▼	4	2	▼	845	107		952	1,100

23 83 Radiant Heating Units

23 83 33 – Electric Radiant Heaters

23 83 33.10 Electric Heating		Crew	Daily Output	Labor-Hours	Unit	Material	2014 Bare Costs Labor	Equipment	Total	Total Incl O&P
6100	13 kW	1 Elec	3.80	2.105	Ea.	1,300	112		1,412	1,600
6120	15 kW		3.70	2.162		1,300	115		1,415	1,600
6140	20 kW		3.50	2.286		1,700	122		1,822	2,025
6300	3 phase, 208-240 volt, 5 kW		5.50	1.455		430	77.50		507.50	590
6320	7 kW		5	1.600		670	85.50		755.50	865
6340	10 kW		4	2		750	107		857	985
6360	15 kW		3.70	2.162		1,375	115		1,490	1,675
6380	20 kW		3.50	2.286		1,725	122		1,847	2,075
6400	25 kW		3.30	2.424		2,025	129		2,154	2,425
6500	480 volt, 5 kW		5.50	1.455		605	77.50		682.50	780
6520	7 kW		5	1.600		765	85.50		850.50	970
6540	10 kW		4	2		810	107		917	1,050
6560	13 kW		3.80	2.105		1,475	112		1,587	1,800
6580	15 kW		3.70	2.162		1,475	115		1,590	1,800
6600	20 kW		3.50	2.286		1,925	122		2,047	2,275
6620	25 kW		3.30	2.424		2,025	129		2,154	2,425
6630	30 kW		3	2.667		2,350	142		2,492	2,800
6640	40 kW		2	4		3,000	213		3,213	3,625
6650	50 kW		1.60	5		3,625	267		3,892	4,375
6800	Vertical discharge heaters, with fan									
6820	Single phase, 208-240-277 volt, 10 kW	1 Elec	4	2	Ea.	750	107		857	985
6840	15 kW		3.70	2.162		1,250	115		1,365	1,550
6900	3 phase, 208-240 volt, 10 kW		4	2		750	107		857	985
6920	15 kW		3.70	2.162		1,250	115		1,365	1,550
6940	20 kW		3.50	2.286		1,775	122		1,897	2,150
6960	25 kW		3.30	2.424		2,025	129		2,154	2,425
6980	30 kW		3	2.667		2,425	142		2,567	2,900
7000	40 kW		2	4		3,000	213		3,213	3,625
7020	50 kW		1.60	5		3,750	267		4,017	4,525
7100	480 volt, 10 kW		4	2		810	107		917	1,050
7120	15 kW		3.70	2.162		1,300	115		1,415	1,600
7140	20 kW		3.50	2.286		1,700	122		1,822	2,025
7160	25 kW		3.30	2.424		2,025	129		2,154	2,425
7180	30 kW		3	2.667		2,350	142		2,492	2,800
7200	40 kW		2	4		3,000	213		3,213	3,625
7220	50 kW		1.60	5		3,625	267		3,892	4,375
7410	Sill height convector heaters, 5" high x 2' long, 500 watt		6.70	1.194		286	63.50		349.50	410
7420	3' long, 750 watt		6.50	1.231		340	65.50		405.50	470
7430	4' long, 1000 watt		6.20	1.290		390	69		459	535
7440	5' long, 1250 watt		5.50	1.455		445	77.50		522.50	605
7450	6' long, 1500 watt		4.80	1.667		505	89		594	690
7460	8' long, 2000 watt		3.60	2.222		685	119		804	925
7470	10' long, 2500 watt		3	2.667		850	142		992	1,150
7900	Cabinet convector heaters, 240 volt									
7920	3' long, 2000 watt	1 Elec	5.30	1.509	Ea.	1,975	80.50		2,055.50	2,300
7940	3000 watt		5.30	1.509		2,075	80.50		2,155.50	2,400
7960	4000 watt		5.30	1.509		2,125	80.50		2,205.50	2,450
7980	6000 watt		4.60	1.739		2,200	93		2,293	2,575
8000	8000 watt		4.60	1.739		2,300	93		2,393	2,675
8020	4' long, 4000 watt		4.60	1.739		2,150	93		2,243	2,525
8040	6000 watt		4	2		2,225	107		2,332	2,600
8060	8000 watt		4	2		2,325	107		2,432	2,700
8080	10,000 watt		4	2		2,325	107		2,432	2,725

23 83 33 – Electric Radiant Heaters

23 83 33.10 Electric Heating	Crew	Daily Output	Labor-Hours	Unit	Material	2014 Bare Costs Labor	Equipment	Total	Total Incl O&P
8100	Available also in 208 or 277 volt									
8200	Cabinet unit heaters, 120 to 277 volt, 1 pole,									
8220	wall mounted, 2 kW	1 Elec	4.60	1.739	Ea.	1,925	93		2,018	2,250
8230	3 kW		4.60	1.739		1,950	93		2,043	2,300
8240	4 kW		4.40	1.818		2,000	97		2,097	2,350
8250	5 kW		4.40	1.818		2,075	97		2,172	2,425
8260	6 kW		4.20	1.905		2,100	102		2,202	2,450
8270	8 kW		4	2		2,175	107		2,282	2,550
8280	10 kW		3.80	2.105		2,225	112		2,337	2,600
8290	12 kW		3.50	2.286		2,400	122		2,522	2,825
8300	13.5 kW		2.90	2.759		2,450	147		2,597	2,900
8310	16 kW		2.70	2.963		2,500	158		2,658	2,975
8320	20 kW		2.30	3.478		3,625	186		3,811	4,250
8330	24 kW		1.90	4.211		3,675	225		3,900	4,375
8350	Recessed, 2 kW		4.40	1.818		1,875	97		1,972	2,200
8370	3 kW		4.40	1.818		1,950	97		2,047	2,275
8380	4 kW		4.20	1.905		1,950	102		2,052	2,300
8390	5 kW		4.20	1.905		2,175	102		2,277	2,550
8400	6 kW		4	2		2,275	107		2,382	2,675
8410	8 kW		3.80	2.105		2,275	112		2,387	2,700
8420	10 kW		3.50	2.286		2,800	122		2,922	3,250
8430	12 kW		2.90	2.759		2,800	147		2,947	3,325
8440	13.5 kW		2.70	2.963		2,975	158		3,133	3,500
8450	16 kW		2.30	3.478		3,000	186		3,186	3,575
8460	20 kW		1.90	4.211		3,025	225		3,250	3,650
8470	24 kW		1.60	5		3,600	267		3,867	4,350
8490	Ceiling mounted, 2 kW		3.20	2.500		1,950	133		2,083	2,325
8510	3 kW		3.20	2.500		2,025	133		2,158	2,425
8520	4 kW		3	2.667		2,050	142		2,192	2,475
8530	5 kW		3	2.667		2,150	142		2,292	2,600
8540	6 kW		2.80	2.857		2,175	152		2,327	2,600
8550	8 kW		2.40	3.333		2,250	178		2,428	2,750
8560	10 kW		2.20	3.636		2,225	194		2,419	2,725
8570	12 kW		2	4		2,325	213		2,538	2,900
8580	13.5 kW		1.50	5.333		2,375	285		2,660	3,025
8590	16 kW		1.30	6.154		2,500	330		2,830	3,250
8600	20 kW		.90	8.889		3,500	475		3,975	4,550
8610	24 kW		.60	13.333		3,600	710		4,310	5,025
8630	208 to 480 V, 3 pole									
8650	Wall mounted, 2 kW	1 Elec	4.60	1.739	Ea.	2,075	93		2,168	2,450
8670	3 kW		4.60	1.739		2,175	93		2,268	2,525
8680	4 kW		4.40	1.818		2,225	97		2,322	2,600
8690	5 kW		4.40	1.818		2,300	97		2,397	2,675
8700	6 kW		4.20	1.905		2,300	102		2,402	2,675
8710	8 kW		4	2		2,400	107		2,507	2,775
8720	10 kW		3.80	2.105		2,425	112		2,537	2,850
8730	12 kW		3.50	2.286		2,475	122		2,597	2,900
8740	13.5 kW		2.90	2.759		2,475	147		2,622	2,950
8750	16 kW		2.70	2.963		2,500	158		2,658	2,975
8760	20 kW		2.30	3.478		3,200	186		3,386	3,775
8770	24 kW		1.90	4.211		3,750	225		3,975	4,450
8790	Recessed, 2 kW		4.40	1.818		2,075	97		2,172	2,450
8810	3 kW		4.40	1.818		2,175	97		2,272	2,525

23 83 Radiant Heating Units

23 83 33 – Electric Radiant Heaters

23 83 33.10 Electric Heating		Crew	Daily Output	Labor-Hours	Unit	Material	2014 Bare Costs Labor	Equipment	Total	Total Incl O&P
8820	4 kW	1 Elec	4.20	1.905	Ea.	2,250	102		2,352	2,625
8830	5 kW		4.20	1.905		2,300	102		2,402	2,675
8840	6 kW		4	2		2,300	107		2,407	2,675
8850	8 kW		3.80	2.105		2,400	112		2,512	2,800
8860	10 kW		3.50	2.286		2,425	122		2,547	2,850
8870	12 kW		2.90	2.759		2,475	147		2,622	2,950
8880	13.5 kW		2.70	2.963		2,525	158		2,683	3,000
8890	16 kW		2.30	3.478		3,650	186		3,836	4,275
8900	20 kW		1.90	4.211		3,750	225		3,975	4,450
8920	24 kW		1.60	5		3,800	267		4,067	4,575
8940	Ceiling mount, 2 kW		3.20	2.500		2,075	133		2,208	2,500
8950	3 kW		3.20	2.500		2,175	133		2,308	2,575
8960	4 kW		3	2.667		2,225	142		2,367	2,675
8970	5 kW		3	2.667		2,300	142		2,442	2,750
8980	6 kW		2.80	2.857		2,300	152		2,452	2,750
8990	8 kW		2.40	3.333		2,400	178		2,578	2,900
9000	10 kW		2.20	3.636		2,425	194		2,619	2,975
9020	13.5 kW		1.50	5.333		2,525	285		2,810	3,200
9030	16 kW		1.30	6.154		3,700	330		4,030	4,575
9040	20 kW		.90	8.889		3,700	475		4,175	4,775
9060	24 kW		.60	13.333		3,725	710		4,435	5,175

Estimating Tips

26 05 00 Common Work Results for Electrical

- Conduit should be taken off in three main categories—power distribution, branch power, and branch lighting—so the estimator can concentrate on systems and components, therefore making it easier to ensure all items have been accounted for.
- For cost modifications for elevated conduit installation, add the percentages to labor according to the height of installation, and only to the quantities exceeding the different height levels, not to the total conduit quantities.
- Remember that aluminum wiring of equal ampacity is larger in diameter than copper and may require larger conduit.
- If more than three wires at a time are being pulled, deduct percentages from the labor hours of that grouping of wires.
- When taking off grounding systems, identify separately the type and size of wire, and list each unique type of ground connection.

- The estimator should take the weights of materials into consideration when completing a takeoff. Topics to consider include: How will the materials be supported? What methods of support are available? How high will the support structure have to reach? Will the final support structure be able to withstand the total burden? Is the support material included or separate from the fixture, equipment, and material specified?
- Do not overlook the costs for equipment used in the installation. If scaffolding or highlifts are available in the field, contractors may use them in lieu of the proposed ladders and rolling staging.

26 20 00 Low-Voltage Electrical Transmission

- Supports and concrete pads may be shown on drawings for the larger equipment, or the support system may be only a piece of plywood for the back of a panelboard. In either case, it must be included in the costs.

26 40 00 Electrical and Cathodic Protection

- When taking off cathodic protections systems, identify the type and size of cable, and list each unique type of anode connection.

26 50 00 Lighting

- Fixtures should be taken off room by room, using the fixture schedule, specifications, and the ceiling plan. For large concentrations of lighting fixtures in the same area, deduct the percentages from labor hours.

Reference Numbers

Reference numbers are shown in shaded boxes at the beginning of some major classifications. These numbers refer to related items in the Reference Section. The reference information may be an estimating procedure, an alternate pricing method, or technical information.

Note: Not all subdivisions listed here necessarily appear in this publication.

Note: **Trade Service,** *in part, has been used as a reference source for some of the material prices used in Division 26.*

26 01 Operation and Maintenance of Electrical Systems

26 01 40 – Operation and Maintenance of Electrical and Cathodic Protection Systems

26 01 40.51 Electrical Systems Repair and Replacement	Crew	Daily Output	Labor-Hours	Unit	Material	2014 Bare Costs Labor	Equipment	Total	Total Incl O&P
0010 **ELECTRICAL SYSTEMS REPAIR AND REPLACEMENT**									
3000 Remove and replace (reinstall), switch cover	1 Elec	60	.133	Ea.		7.10		7.10	10.65
3020 Outlet cover	"	60	.133	"		7.10		7.10	10.65

26 01 50 – Operation and Maintenance of Lighting

26 01 50.81 Luminaire Replacement

	Crew	Daily Output	Labor-Hours	Unit	Material	Labor	Equipment	Total	Total Incl O&P
0010 **LUMINAIRE REPLACEMENT**									
0962 Remove and install new replacement lens, 2' x 2' troffer	1 Elec	16	.500	Ea.	15.75	26.50		42.25	57.50
0964 2' x 4' troffer		16	.500		14.75	26.50		41.25	56
3200 Remove and replace (reinstall), lighting fixture	↓	4	2	↓		107		107	160

26 05 Common Work Results for Electrical

26 05 05 – Selective Electrical Demolition

26 05 05.10 Electrical Demolition

	Crew	Daily Output	Labor-Hours	Unit	Material	Labor	Equipment	Total	Total Incl O&P
0010 **ELECTRICAL DEMOLITION** R260105-30									
0020 Conduit to 15' high, including fittings & hangers									
0100 Rigid galvanized steel, 1/2" to 1" diameter	1 Elec	242	.033	L.F.		1.76		1.76	2.64
0120 1-1/4" to 2"	"	200	.040			2.13		2.13	3.19
0140 2-1/2" to 3-1/2"	2 Elec	302	.053			2.83		2.83	4.23
0160 4" to 6"	"	160	.100			5.35		5.35	8
0170 PVC #40, 1/2" to 1"	1 Elec	410	.020			1.04		1.04	1.56
0172 1-1/4" to 2"		350	.023			1.22		1.22	1.83
0174 2-1/2"	↓	250	.032			1.71		1.71	2.56
0176 3" to 3-1/2"	2 Elec	340	.047			2.51		2.51	3.76
0178 4" to 6"	"	230	.070			3.71		3.71	5.55
0200 Electric metallic tubing (EMT), 1/2" to 1"	1 Elec	394	.020			1.08		1.08	1.62
0220 1-1/4" to 1-1/2"		326	.025			1.31		1.31	1.96
0240 2" to 3"	↓	236	.034			1.81		1.81	2.71
0260 3-1/2" to 4"	2 Elec	310	.052	↓		2.75		2.75	4.12
0270 Armored cable, (BX) avg. 50' runs									
0280 #14, 2 wire	1 Elec	690	.012	L.F.		.62		.62	.93
0290 #14, 3 wire		571	.014			.75		.75	1.12
0300 #12, 2 wire		605	.013			.71		.71	1.06
0310 #12, 3 wire		514	.016			.83		.83	1.24
0320 #10, 2 wire		514	.016			.83		.83	1.24
0330 #10, 3 wire		425	.019			1		1	1.50
0340 #8, 3 wire	↓	342	.023	↓		1.25		1.25	1.87
0350 Non metallic sheathed cable (Romex)									
0360 #14, 2 wire	1 Elec	720	.011	L.F.		.59		.59	.89
0370 #14, 3 wire		657	.012			.65		.65	.97
0380 #12, 2 wire		629	.013			.68		.68	1.02
0390 #10, 3 wire	↓	450	.018	↓		.95		.95	1.42
0400 Wiremold raceway, including fittings & hangers									
0420 No. 3000	1 Elec	250	.032	L.F.		1.71		1.71	2.56
0440 No. 4000		217	.037			1.97		1.97	2.94
0460 No. 6000		166	.048			2.57		2.57	3.85
0462 Plugmold with receptacle		114	.070	↓		3.74		3.74	5.60
0465 Telephone/power pole		12	.667	Ea.		35.50		35.50	53
0470 Non-metallic, straight section	↓	480	.017	L.F.		.89		.89	1.33
0500 Channels, steel, including fittings & hangers									
0520 3/4" x 1-1/2"	1 Elec	308	.026	L.F.		1.39		1.39	2.07

26 05 05 – Selective Electrical Demolition

26 05 05.10 Electrical Demolition	Crew	Daily Output	Labor-Hours	Unit	Material	2014 Bare Costs Labor	Equipment	Total	Total Incl O&P
0540	1-1/2" x 1-1/2"	1 Elec	269	.030	L.F.		1.59		1.59	2.37
0560	1-1/2" x 1-7/8"	↓	229	.035	↓		1.86		1.86	2.79
0600	Copper bus duct, indoor, 3 phase									
0610	Including hangers & supports									
0620	225 amp	2 Elec	135	.119	L.F.		6.30		6.30	9.45
0640	400 amp		106	.151			8.05		8.05	12.05
0660	600 amp		86	.186			9.95		9.95	14.85
0680	1000 amp		60	.267			14.25		14.25	21.50
0700	1600 amp		40	.400			21.50		21.50	32
0720	3000 amp	↓	10	1.600	↓		85.50		85.50	128
0800	Plug-in switches, 600V 3 ph., incl. disconnecting									
0820	wire, conduit terminations, 30 amp	1 Elec	15.50	.516	Ea.		27.50		27.50	41
0840	60 amp		13.90	.576			30.50		30.50	46
0850	100 amp		10.40	.769			41		41	61.50
0860	200 amp	↓	6.20	1.290			69		69	103
0880	400 amp	2 Elec	5.40	2.963			158		158	237
0900	600 amp		3.40	4.706			251		251	375
0920	800 amp		2.60	6.154			330		330	490
0940	1200 amp		2	8			425		425	640
0960	1600 amp	↓	1.70	9.412	↓		500		500	750
1010	Safety switches, 250 or 600V, incl. disconnection									
1050	of wire & conduit terminations									
1100	30 amp	1 Elec	12.30	.650	Ea.		34.50		34.50	52
1120	60 amp		8.80	.909			48.50		48.50	72.50
1140	100 amp		7.30	1.096			58.50		58.50	87.50
1160	200 amp	↓	5	1.600			85.50		85.50	128
1180	400 amp	2 Elec	6.80	2.353			126		126	188
1200	600 amp	"	4.60	3.478			186		186	278
1202	Explosion proof, 60 amp	1 Elec	8.80	.909			48.50		48.50	72.50
1204	100 amp	↓	5	1.600			85.50		85.50	128
1206	200 amp	↓	3.33	2.402	↓		128		128	192
1210	Panel boards, incl. removal of all breakers,									
1220	conduit terminations & wire connections									
1230	3 wire, 120/240 V, 100A, to 20 circuits	1 Elec	2.60	3.077	Ea.		164		164	246
1240	200 amps, to 42 circuits	2 Elec	2.60	6.154			330		330	490
1250	400 amps, to 42 circuits	"	2.20	7.273			390		390	580
1260	4 wire, 120/208 V, 125A, to 20 circuits	1 Elec	2.40	3.333			178		178	266
1270	200 amps, to 42 circuits	2 Elec	2.40	6.667			355		355	530
1280	400 amps, to 42 circuits		1.92	8.333			445		445	665
1285	600 amps, to 42 circuits	↓	1.60	10	↓		535		535	800
1300	Transformer, dry type, 1 phase, incl. removal of									
1320	supports, wire & conduit terminations									
1340	1 kVA	1 Elec	7.70	1.039	Ea.		55.50		55.50	83
1360	5 kVA		4.70	1.702			91		91	136
1380	10 kVA	↓	3.60	2.222			119		119	177
1400	37.5 kVA	2 Elec	3	5.333			285		285	425
1420	75 kVA	"	2.50	6.400	↓		340		340	510
1440	3 phase to 600V, primary									
1460	3 kVA	1 Elec	3.85	2.078	Ea.		111		111	166
1480	15 kVA	2 Elec	4.20	3.810			203		203	305
1500	30 kVA		3.50	4.571			244		244	365
1510	45 kVA		3.10	5.161			275		275	410
1520	75 kVA	↓	2.70	5.926			315		315	475

26 05 Common Work Results for Electrical

26 05 05 – Selective Electrical Demolition

26 05 05.10 Electrical Demolition		Crew	Daily Output	Labor-Hours	Unit	Material	2014 Bare Costs		Total	Total Incl O&P
							Labor	Equipment		
1530	112.5 kVA	R-3	2.90	6.897	Ea.		365	48.50	413.50	605
1540	150 kVA		2.70	7.407			390	52	442	645
1550	300 kVA		1.80	11.111			590	78	668	965
1560	500 kVA		1.40	14.286			755	100	855	1,225
1570	750 kVA	↓	1.10	18.182	↓		960	127	1,087	1,600
1600	Pull boxes & cabinets, sheet metal, incl. removal									
1620	of supports and conduit terminations									
1640	6" x 6" x 4"	1 Elec	31.10	.257	Ea.		13.70		13.70	20.50
1660	12" x 12" x 4"		23.30	.343			18.30		18.30	27.50
1680	24" x 24" x 6"		12.30	.650			34.50		34.50	52
1700	36" x 36" x 8"		7.70	1.039			55.50		55.50	83
1720	Junction boxes, 4" sq. & oct.		80	.100			5.35		5.35	8
1740	Handy box		107	.075			3.99		3.99	5.95
1760	Switch box		107	.075			3.99		3.99	5.95
1780	Receptacle & switch plates		257	.031			1.66		1.66	2.49
1790	Receptacles & switches, 15 to 30 amp	↓	135	.059	↓		3.16		3.16	4.73
1800	Wire, THW-THWN-THHN, removed from									
1810	in place conduit, to 15' high									
1830	#14	1 Elec	65	.123	C.L.F.		6.55		6.55	9.85
1840	#12		55	.145			7.75		7.75	11.60
1850	#10		45.50	.176			9.40		9.40	14.05
1860	#8		40.40	.198			10.55		10.55	15.80
1870	#6	↓	32.60	.245			13.10		13.10	19.60
1880	#4	2 Elec	53	.302			16.10		16.10	24
1890	#3		50	.320			17.05		17.05	25.50
1900	#2		44.60	.359			19.15		19.15	28.50
1910	1/0		33.20	.482			25.50		25.50	38.50
1920	2/0		29.20	.548			29		29	44
1930	3/0		25	.640			34		34	51
1940	4/0		22	.727			39		39	58
1950	250 kcmil		20	.800			42.50		42.50	64
1960	300 kcmil		19	.842			45		45	67
1970	350 kcmil		18	.889			47.50		47.50	71
1980	400 kcmil		17	.941			50		50	75
1990	500 kcmil	↓	16.20	.988	↓		52.50		52.50	79
2000	Interior fluorescent fixtures, incl. supports									
2010	& whips, to 15' high									
2100	Recessed drop-in 2' x 2', 2 lamp	2 Elec	35	.457	Ea.		24.50		24.50	36.50
2120	2' x 4', 2 lamp		33	.485			26		26	38.50
2140	2' x 4', 4 lamp		30	.533			28.50		28.50	42.50
2160	4' x 4', 4 lamp	↓	20	.800	↓		42.50		42.50	64
2180	Surface mount, acrylic lens & hinged frame									
2200	1' x 4', 2 lamp	2 Elec	44	.364	Ea.		19.40		19.40	29
2220	2' x 2', 2 lamp		44	.364			19.40		19.40	29
2260	2' x 4', 4 lamp		33	.485			26		26	38.50
2280	4' x 4', 4 lamp	↓	23	.696	↓		37		37	55.50
2300	Strip fixtures, surface mount									
2320	4' long, 1 lamp	2 Elec	53	.302	Ea.		16.10		16.10	24
2340	4' long, 2 lamp		50	.320			17.05		17.05	25.50
2360	8' long, 1 lamp		42	.381			20.50		20.50	30.50
2380	8' long, 2 lamp	↓	40	.400	↓		21.50		21.50	32
2400	Pendant mount, industrial, incl. removal									
2410	of chain or rod hangers, to 15' high									

26 05 05 – Selective Electrical Demolition

26 05 05.10 Electrical Demolition

		Crew	Daily Output	Labor-Hours	Unit	Material	2014 Bare Costs Labor	Equipment	Total	Total Incl O&P
2420	4' long, 2 lamp	2 Elec	35	.457	Ea.		24.50		24.50	36.50
2440	8' long, 2 lamp	"	27	.593	"		31.50		31.50	47.50
2460	Interior incandescent, surface, ceiling									
2470	or wall mount, to 12' high									
2480	Metal cylinder type, 75 Watt	2 Elec	62	.258	Ea.		13.75		13.75	20.50
2500	150 Watt		62	.258			13.75		13.75	20.50
2502	300 Watt	↓	53	.302	↓		16.10		16.10	24
2520	Metal halide, high bay									
2540	400 Watt	2 Elec	15	1.067	Ea.		57		57	85
2560	1000 Watt		12	1.333			71		71	106
2580	150 Watt, low bay		20	.800			42.50		42.50	64
2585	Globe type, ceiling mount		62	.258			13.75		13.75	20.50
2586	Can type, recessed mount	↓	62	.258	↓		13.75		13.75	20.50
2600	Exterior fixtures, incandescent, wall mount									
2620	100 Watt	2 Elec	50	.320	Ea.		17.05		17.05	25.50
2640	Quartz, 500 Watt		33	.485			26		26	38.50
2660	1500 Watt	↓	27	.593	↓		31.50		31.50	47.50
2680	Wall pack, mercury vapor									
2700	175 Watt	2 Elec	25	.640	Ea.		34		34	51
2720	250 Watt	"	25	.640			34		34	51
7000	Weatherhead/mast, 2"	1 Elec	16	.500			26.50		26.50	40
7002	3"		10	.800			42.50		42.50	64
7004	3-1/2"		8.50	.941			50		50	75
7006	4"		8	1	↓		53.50		53.50	80
7100	Service entry cable, #6, +#6 neutral		420	.019	L.F.		1.02		1.02	1.52
7102	#4, +#4 neutral		360	.022			1.19		1.19	1.77
7104	#2, +#4 neutral		345	.023			1.24		1.24	1.85
7106	#2, +#2 neutral	↓	330	.024	↓		1.29		1.29	1.94
9900	Add to labor for higher elevated installation									
9910	15' to 20' high, add						10%			
9920	20' to 25' high, add						20%			
9930	25' to 30' high, add						25%			
9940	30' to 35' high, add						30%			
9950	35' to 40' high, add						35%			
9960	Over 40' high, add						40%			

26 05 05.15 Electrical Demolition, Grounding

		Crew	Daily Output	Labor-Hours	Unit	Material	2014 Bare Costs Labor	Equipment	Total	Total Incl O&P
0010	**ELECTRICAL DEMOLITION, GROUNDING** Addition									
0100	Ground clamp, bronze	1 Elec	64	.125	Ea.		6.65		6.65	10
0140	Water pipe ground clamp, bronze, heavy duty		24	.333			17.80		17.80	26.50
0150	Ground rod, 8' to 10'	↓	13	.615	↓		33		33	49
0200	Ground wire, bare armored	2 Elec	900	.018	L.F.		.95		.95	1.42
0240	Bare copper or aluminum	"	2800	.006	"		.30		.30	.46

26 05 05.20 Electrical Demolition, Wiring Methods

			Crew	Daily Output	Labor-Hours	Unit	Material	2014 Bare Costs Labor	Equipment	Total	Total Incl O&P
0010	**ELECTRICAL DEMOLITION, WIRING METHODS** Addition										
0100	Armored cable, w/PVC jacket, in cable tray										
0110	#6	R024119-10	1 Elec	9.30	.860	C.L.F.		46		46	68.50
0120	#4		2 Elec	16.20	.988			52.50		52.50	79
0130	#2			13.80	1.159			62		62	92.50
0140	#1			12	1.333			71		71	106
0150	1/0			10.80	1.481			79		79	118
0160	2/0			10.20	1.569			83.50		83.50	125
0170	3/0		↓	9.60	1.667			89		89	133

26 05 05 – Selective Electrical Demolition

26 05 05.20 Electrical Demolition, Wiring Methods	Crew	Daily Output	Labor-Hours	Unit	Material	2014 Bare Costs Labor	Equipment	Total	Total Incl O&P
0180	4/0	2 Elec	9	1.778	C.L.F.		95		95	142
0190	250 kcmil	3 Elec	10.80	2.222			119		119	177
0210	350 kcmil		9.90	2.424			129		129	194
0230	500 kcmil		9	2.667			142		142	213
0240	750 kcmil		8.10	2.963			158		158	237
1100	Control cable, 600 V or less									
1110	3 wires	1 Elec	24	.333	C.L.F.		17.80		17.80	26.50
1120	5 wires		20	.400			21.50		21.50	32
1130	7 wires		16.50	.485			26		26	38.50
1140	9 wires		15	.533			28.50		28.50	42.50
1150	12 wires	2 Elec	26	.615			33		33	49
1160	15 wires		22	.727			39		39	58
1170	19 wires		19	.842			45		45	67
1180	25 wires		16	1			53.50		53.50	80
1500	Mineral insulated (MI) cable, 600 V									
1510	#10	1 Elec	4.80	1.667	C.L.F.		89		89	133
1520	#8		4.50	1.778			95		95	142
1530	#6		4.20	1.905			102		102	152
1540	#4	2 Elec	7.20	2.222			119		119	177
1550	#2		6.60	2.424			129		129	194
1560	#1		6.30	2.540			135		135	203
1570	1/0		6	2.667			142		142	213
1580	2/0		5.70	2.807			150		150	224
1590	3/0		5.40	2.963			158		158	237
1600	4/0		4.80	3.333			178		178	266
1610	250 kcmil	3 Elec	7.20	3.333			178		178	266
1620	500 kcmil	"	5.90	4.068			217		217	325
2000	Shielded cable, XLP shielding, to 35 kV									
2010	#4	2 Elec	13.20	1.212	C.L.F.		64.50		64.50	97
2020	#1		12	1.333			71		71	106
2030	1/0		11.40	1.404			75		75	112
2040	2/0		10.80	1.481			79		79	118
2050	4/0		9.60	1.667			89		89	133
2060	250 kcmil	3 Elec	13.50	1.778			95		95	142
2070	350 kcmil		11.70	2.051			109		109	164
2080	500 kcmil		11.20	2.143			114		114	171
2090	750 kcmil		10.80	2.222			119		119	177
3000	Modular flexible wiring									
3010	Cable set	1 Elec	120	.067	Ea.		3.56		3.56	5.30
3020	Conversion module		48	.167			8.90		8.90	13.30
3030	Switching assembly		96	.083			4.45		4.45	6.65
3200	Undercarpet									
3210	Power or telephone, flat cable	1 Elec	1320	.006	L.F.		.32		.32	.48
3220	Transition block assemblies w/fitting		75	.107	Ea.		5.70		5.70	8.50
3230	Floor box with fitting		60	.133			7.10		7.10	10.65
3300	Data system, cable with connection		50	.160			8.55		8.55	12.80
4000	Cable tray, including fitting & support									
4010	Galvanized steel, 6" wide	2 Elec	310	.052	L.F.		2.75		2.75	4.12
4020	9" wide		295	.054			2.89		2.89	4.33
4030	12" wide		285	.056			3		3	4.48
4040	18" wide		270	.059			3.16		3.16	4.73
4050	24" wide		260	.062			3.28		3.28	4.91
4060	30" wide		240	.067			3.56		3.56	5.30

26 05 05 – Selective Electrical Demolition

26 05 05.20 Electrical Demolition, Wiring Methods	Crew	Daily Output	Labor-Hours	Unit	Material	2014 Bare Costs Labor	Equipment	Total	Total Incl O&P
4070	36" wide	2 Elec	220	.073	L.F.		3.88		3.88	5.80
4110	Aluminum, 6" wide		420	.038			2.03		2.03	3.04
4120	9" wide		415	.039			2.06		2.06	3.08
4130	12" wide		390	.041			2.19		2.19	3.28
4140	18" wide		370	.043			2.31		2.31	3.45
4150	24" wide		350	.046			2.44		2.44	3.65
4160	30" wide		325	.049			2.63		2.63	3.93
4170	36" wide		300	.053			2.85		2.85	4.26
4310	Cable channel, aluminum, 4" wide straight	1 Elec	240	.033			1.78		1.78	2.66
4320	Cable tray fittings, 6" to 9" wide	2 Elec	27	.593	Ea.		31.50		31.50	47.50
4330	12" to 18" wide	"	19	.842	"		45		45	67
5000	Conduit nipples, with locknuts and bushings									
5020	1/2"	1 Elec	108	.074	Ea.		3.95		3.95	5.90
5040	3/4"		96	.083			4.45		4.45	6.65
5060	1"		81	.099			5.25		5.25	7.90
5080	1-1/4"		69	.116			6.20		6.20	9.25
5100	1-1/2"		60	.133			7.10		7.10	10.65
5120	2"		54	.148			7.90		7.90	11.85
5140	2-1/2"		45	.178			9.50		9.50	14.20
5160	3"		36	.222			11.85		11.85	17.75
5180	3-1/2"		33	.242			12.95		12.95	19.35
5200	4"		27	.296			15.80		15.80	23.50
5220	5"		21	.381			20.50		20.50	30.50
5240	6"		18	.444			23.50		23.50	35.50
5500	Electric nonmetallic tubing (ENT), flexible, 1/2" to 1" diameter		690	.012	L.F.		.62		.62	.93
5510	1-1/4" to 2" diameter		300	.027			1.42		1.42	2.13
5600	Flexible metallic tubing, steel, 3/8" to 3/4" diameter		600	.013			.71		.71	1.06
5610	1" to 1-1/4" diameter		260	.031			1.64		1.64	2.46
5620	1-1/2" to 2" diameter		140	.057			3.05		3.05	4.56
5630	2-1/2" diameter		100	.080			4.27		4.27	6.40
5640	3" to 3-1/2" diameter	2 Elec	150	.107			5.70		5.70	8.50
5650	4" diameter	"	100	.160			8.55		8.55	12.80
5700	Sealtite flexible conduit, 3/8" to 3/4" diameter	1 Elec	420	.019			1.02		1.02	1.52
5710	1" to 1-1/4" diameter		180	.044			2.37		2.37	3.55
5720	1-1/2" to 2" diameter		120	.067			3.56		3.56	5.30
5730	2-1/2" diameter		80	.100			5.35		5.35	8
5740	3" diameter	2 Elec	150	.107			5.70		5.70	8.50
5750	4" diameter	"	100	.160			8.55		8.55	12.80
5800	Wiring duct, plastic, 1-1/2" to 2-1/2" wide	1 Elec	360	.022			1.19		1.19	1.77
5810	3" wide		330	.024			1.29		1.29	1.94
5820	4" wide		300	.027			1.42		1.42	2.13
6000	Floor box and carpet flange		9	.889	Ea.		47.50		47.50	71
6300	Wireway, with fittings and supports, to 15' high									
6310	2-1/2" x 2-1/2"	1 Elec	180	.044	L.F.		2.37		2.37	3.55
6320	4" x 4"	"	160	.050			2.67		2.67	3.99
6330	6" x 6"	2 Elec	240	.067			3.56		3.56	5.30
6340	8" x 8"		160	.100			5.35		5.35	8
6350	10" x 10"		120	.133			7.10		7.10	10.65
6360	12" x 12"		80	.200			10.65		10.65	15.95
6400	Cable reel with receptacle, 120 V/208 V		13	1.231	Ea.		65.50		65.50	98.50
6500	Equipment connection, to 10 HP	1 Elec	23	.348			18.55		18.55	28
6510	To 15 to 30 HP		18	.444			23.50		23.50	35.50
6520	To 40 to 60 HP		15	.533			28.50		28.50	42.50

26 05 Common Work Results for Electrical

26 05 05 – Selective Electrical Demolition

26 05 05.20 Electrical Demolition, Wiring Methods		Crew	Daily Output	Labor-Hours	Unit	Material	2014 Bare Costs Labor	Equipment	Total	Total Incl O&P
6530	To 75 to 100 HP	1 Elec	10	.800	Ea.		42.50		42.50	64
7008	Receptacle, explosionproof, 120 V, to 30 A.		16	.500			26.50		26.50	40
7020	Dimmer switch, 2000 W or less	↓	62	.129	↓		6.90		6.90	10.30

26 05 05.25 Electrical Demolition, Electrical Power

26 05 05.25		Crew	Daily Output	Labor-Hours	Unit	Material	2014 Bare Costs Labor	Equipment	Total	Total Incl O&P
0010	**ELECTRICAL DEMOLITION, ELECTRICAL POWER** R024119-10									
0100	Meter centers and sockets									
0120	Meter socket, 4 terminal R260105-30	1 Elec	8.40	.952	Ea.		51		51	76
0140	Trans-socket, 13 terminal, 400 A	"	3.60	2.222			119		119	177
0160	800 A	2 Elec	4.40	3.636			194		194	290
0200	Meter center, 400 A		5.80	2.759			147		147	220
0210	600 A		4	4			213		213	320
0220	800 A		3.30	4.848			259		259	385
0230	1200 A		2.80	5.714			305		305	455
0240	1600 A		2.50	6.400			340		340	510
0300	Base meter devices, 3 meter		3.60	4.444			237		237	355
0310	4 meter		3.30	4.848			259		259	385
0320	5 meter		2.90	5.517			294		294	440
0330	6 meter		2.20	7.273			390		390	580
0340	7 meter		2	8			425		425	640
0350	8 meter	↓	1.90	8.421	↓		450		450	670
0400	Branch meter devices									
0410	Socket w/circuit breaker 200 A, 2 meter	2 Elec	3.30	4.848	Ea.		259		259	385
0420	3 meter		2.90	5.517			294		294	440
0430	4 meter		2.60	6.154			330		330	490
0450	Main circuit breaker, 400 A		5.80	2.759			147		147	220
0460	600 A		4	4			213		213	320
0470	800 A		3.30	4.848			259		259	385
0480	1200 A		2.80	5.714			305		305	455
0490	1600 A		2.50	6.400			340		340	510
0500	Main lug terminal box, 800 A		3.40	4.706			251		251	375
0510	1200 A	↓	2.60	6.154			330		330	490
1000	Motors, 230/460 V, 60 Hz, 3/4 HP	1 Elec	10.70	.748			40		40	59.50
1010	5 HP		9	.889			47.50		47.50	71
1020	10 HP		8	1			53.50		53.50	80
1030	15 HP	↓	6.40	1.250			66.50		66.50	100
1040	20 HP	2 Elec	10.40	1.538			82		82	123
1050	50 HP		9.60	1.667			89		89	133
1060	75 HP	↓	5.60	2.857			152		152	228
1070	100 HP	3 Elec	5.40	4.444			237		237	355
1080	150 HP		3.60	6.667			355		355	530
1090	200 HP	↓	3	8			425		425	640
1200	Variable frequency drive, 460 V, for 5 HP motor size	1 Elec	3.20	2.500			133		133	200
1210	10 HP motor size	"	2.70	2.963			158		158	237
1220	20 HP motor size	2 Elec	3.60	4.444			237		237	355
1230	50 HP motor size	"	2.10	7.619			405		405	610
1240	75 HP motor size	R-3	2.20	9.091			480	63.50	543.50	790
1250	100 HP motor size		2	10			530	70	600	870
1260	150 HP motor size		2	10			530	70	600	870
1270	200 HP motor size	↓	1.70	11.765	↓		625	82.50	707.50	1,025
2000	Generator set w/accessories, 3 phase 4 wire, 277/480 V									
2010	7.5 kW	3 Elec	2	12	Ea.		640		640	960
2020	20 kW	↓	1.70	14.118	↓		755		755	1,125

26 05 05.25 Electrical Demolition, Electrical Power	Crew	Daily Output	Labor-Hours	Unit	Material	2014 Bare Costs Labor	2014 Bare Costs Equipment	Total	Total Incl O&P
2040	30 kW	3 Elec	1.33	18.045	Ea.		965		965	1,450
2060	50 kW		1	24			1,275		1,275	1,925
2080	100 kW		.75	32			1,700		1,700	2,550
2100	150 kW		.63	38.095			2,025		2,025	3,050
2120	250 kW		.59	40.678			2,175		2,175	3,250
2140	400 kW		.50	48			2,550		2,550	3,825
2160	500 kW		.44	54.545			2,900		2,900	4,350
2180	750 kW	4 Elec	.56	57.143			3,050		3,050	4,575
2200	1000 kW	"	.46	69.565			3,700		3,700	5,550
3000	Uninterruptible power supply system (UPS)									
3010	Single phase, 120 V, 1 kVA	1 Elec	3.20	2.500	Ea.		133		133	200
3020	2 kVA	2 Elec	3.60	4.444			237		237	355
3030	5 kVA	3 Elec	2.50	9.600			510		510	765
3040	10 kVA		2.30	10.435			555		555	835
3050	15 kVA		1.80	13.333			710		710	1,075
4000	Transformer, incl support, wire & conduit termination									
4010	Buck-boost, single phase, 120/240 V, 0.1 kVA	1 Elec	25	.320	Ea.		17.05		17.05	25.50
4020	0.5 kVA		12.50	.640			34		34	51
4030	1 kVA		6.30	1.270			68		68	101
4040	5 kVA		3.80	2.105			112		112	168
4800	5 kV or 15 kV primary, 277/480 V second, 112.5 kVA	R-3	3.50	5.714			300	40	340	500
4810	150 kVA		2.70	7.407			390	52	442	645
4820	225 kVA		2.30	8.696			460	61	521	755
4830	500 kVA		1.45	13.793			730	96.50	826.50	1,200
4840	750 kVA		1.33	15.038			795	105	900	1,325
4850	1000 kVA		1.25	16			845	112	957	1,400
4860	2000 kVA		1.05	19.048			1,000	133	1,133	1,675
4870	3000 kVA		.75	26.667			1,400	187	1,587	2,325
6010	Isolation panel, 3 kVA	1 Elec	2.30	3.478			186		186	278
6020	5 kVA		2.20	3.636			194		194	290
6030	7.5 kVA		2.10	3.810			203		203	305
6040	10 kVA		1.80	4.444			237		237	355
6050	15 kVA		1.40	5.714			305		305	455
7000	Power filters & conditioners									
7100	Automatic voltage regulator	2 Elec	6	2.667	Ea.		142		142	213
7200	Capacitor, 1 kVAR	1 Elec	8.60	.930			49.50		49.50	74.50
7210	5 kVAR		5.80	1.379			73.50		73.50	110
7220	10 kVAR		4.80	1.667			89		89	133
7230	15 kVAR		4.20	1.905			102		102	152
7240	20 kVAR		3.50	2.286			122		122	183
7250	30 kVAR		3.40	2.353			126		126	188
7260	50 kVAR		3.20	2.500			133		133	200
7400	Computer isolator transformer									
7410	Single phase, 120/240 V, 0.5 kVAR	1 Elec	12.80	.625	Ea.		33.50		33.50	50
7420	1 kVAR		8.50	.941			50		50	75
7430	2.5 kVAR		6.40	1.250			66.50		66.50	100
7440	5 kVAR		3.70	2.162			115		115	173
7500	Computer regulator transformer									
7510	Single phase, 240 V, 0.5 kVAR	1 Elec	8.50	.941	Ea.		50		50	75
7520	1 kVAR		6.40	1.250			66.50		66.50	100
7530	2 kVAR		3.20	2.500			133		133	200
7540	Single phase, plug-in unit 120 V, 0.5 kVAR		26	.308			16.40		16.40	24.50
7550	1 kVAR		17	.471			25		25	37.50

26 05 05 – Selective Electrical Demolition

26 05 05.25 Electrical Demolition, Electrical Power

		Crew	Daily Output	Labor-Hours	Unit	Material	2014 Bare Costs Labor	2014 Bare Costs Equipment	Total	Total Incl O&P
7600	Power conditioner transformer									
7610	Single phase 115 V - 240 V, 3 kVA	2 Elec	5.10	3.137	Ea.		167		167	251
7620	5 kVA	"	3.70	4.324			231		231	345
7630	7.5 kVA	3 Elec	4.80	5			267		267	400
7640	10 kVA	"	4.30	5.581			298		298	445
7700	Transient suppressor/voltage regulator									
7710	Single phase 115 or 220 V, 1 kVA	1 Elec	8.50	.941	Ea.		50		50	75
7720	2 kVA		7.30	1.096			58.50		58.50	87.50
7730	4 kVA		6.80	1.176			63		63	94
7800	Transient voltage suppressor transformer									
7810	Single phase, 115 or 220 V, 3.6 kVA	1 Elec	12.80	.625	Ea.		33.50		33.50	50
7820	7.2 kVA		11.50	.696			37		37	55.50
7830	14.4 kVA		10.20	.784			42		42	62.50
7840	Single phase, plug-in, 120 V, 1.8 kVA		26	.308			16.40		16.40	24.50
8000	Power measurement & control									
8010	Switchboard instruments, 3 phase 4 wire, indicating unit	1 Elec	25	.320	Ea.		17.05		17.05	25.50
8020	Recording unit		12	.667			35.50		35.50	53
8100	3 current transformers, 3 phase 4 wire, 5 to 800 A		6.40	1.250			66.50		66.50	100
8110	1000 to 1500 A		4.20	1.905			102		102	152
8120	2000 to 4000 A		3.20	2.500			133		133	200

26 05 05.30 Electrical Demolition, Transmission and Distribution

		Crew	Daily Output	Labor-Hours	Unit	Material	2014 Bare Costs Labor	2014 Bare Costs Equipment	Total	Total Incl O&P
0010	**ELECTRICAL DEMOLITION, TRANSMISSION & DISTRIBUTION**									
0100	Load interrupter switch, 600 A, NEMA 1, 4.8 kV	R-3	1.33	15.038	Ea.		795	105	900	1,325
0120	13.8 kV	R024119-10	"	1.27	15.748			835	110	945	1,375
0200	Lightning arrester, 4.8 kV	1 Elec	9	.889			47.50		47.50	71
0220	13.8 kV	R260105-30		6.67	1.199			64		64	96
0300	Alarm or option items		3.33	2.402			128		128	192

26 05 05.35 Electrical Demolition, L.V. Distribution

		Crew	Daily Output	Labor-Hours	Unit	Material	2014 Bare Costs Labor	2014 Bare Costs Equipment	Total	Total Incl O&P
0010	**ELECTRICAL DEMOLITION, L.V. DISTRIBUTION** Addition R024119-10									
0100	Circuit breakers in enclosure									
0120	Enclosed (NEMA 1), 600 V, 3 pole, 30 A R260105-30	1 Elec	12.30	.650	Ea.		34.50		34.50	52
0140	60 A		8.80	.909			48.50		48.50	72.50
0160	100 A		7.30	1.096			58.50		58.50	87.50
0180	225 A		5	1.600			85.50		85.50	128
0200	400 A	2 Elec	6.80	2.353			126		126	188
0220	600 A		4.60	3.478			186		186	278
0240	800 A		3.60	4.444			237		237	355
0260	1200 A		3.10	5.161			275		275	410
0280	1600 A		2.80	5.714			305		305	455
0300	2000 A		2.50	6.400			340		340	510
0400	Enclosed (NEMA 7), 600 V, 3 pole, 50 A	1 Elec	8.80	.909			48.50		48.50	72.50
0410	100 A		5.80	1.379			73.50		73.50	110
0420	150 A		3.80	2.105			112		112	168
0430	250 A	2 Elec	6.20	2.581			138		138	206
0440	400 A	"	4.60	3.478			186		186	278
0460	Manual motor starter, NEMA 1	1 Elec	24	.333			17.80		17.80	26.50
0480	NEMA 4 or NEMA 7		15	.533			28.50		28.50	42.50
0500	Time switches, single pole single throw		15.40	.519			27.50		27.50	41.50
0520	Photo cell		30	.267			14.25		14.25	21.50
0540	Load management device, 4 loads		7.70	1.039			55.50		55.50	83
0550	8 loads		3.80	2.105			112		112	168
0560	Master light control panel	2 Elec	2	8			425		425	640

26 05 05 – Selective Electrical Demolition

26 05 05.35 Electrical Demolition, L.V. Distribution	Crew	Daily Output	Labor-Hours	Unit	Material	2014 Bare Costs Labor	2014 Bare Costs Equipment	Total	Total Incl O&P
0600	Transfer switches, enclosed, 30 A	1 Elec	12	.667	Ea.		35.50		35.50	53
0610	60 A		9.50	.842			45		45	67
0620	100 A		6.50	1.231			65.50		65.50	98.50
0630	150 A	2 Elec	12	1.333			71		71	106
0640	260 A		10	1.600			85.50		85.50	128
0660	400 A		8	2			107		107	160
0670	600 A		5	3.200			171		171	256
0680	800 A		4	4			213		213	320
0690	1200 A		3.50	4.571			244		244	365
0700	1600 A		3	5.333			285		285	425
0710	2000 A		2.50	6.400			340		340	510
0800	Air terminal with base and connector	1 Elec	24	.333			17.80		17.80	26.50
1000	Enclosed controller, NEMA 1, 30 A		13.80	.580			31		31	46.50
1020	60 A		11.50	.696			37		37	55.50
1040	100 A		9.60	.833			44.50		44.50	66.50
1060	150 A		7.70	1.039			55.50		55.50	83
1080	200 A		5.40	1.481			79		79	118
1100	400 A	2 Elec	6.90	2.319			124		124	185
1120	600 A		4.60	3.478			186		186	278
1140	800 A		3.80	4.211			225		225	335
1160	1200 A		3.10	5.161			275		275	410
1200	Control station, NEMA 1	1 Elec	30	.267			14.25		14.25	21.50
1220	NEMA 7		23	.348			18.55		18.55	28
1230	Control switches, push button		69	.116			6.20		6.20	9.25
1240	Indicating light unit		120	.067			3.56		3.56	5.30
1250	Relay		15	.533			28.50		28.50	42.50
2000	Motor control center components									
2020	Starter, NEMA 1, size 1	1 Elec	9	.889	Ea.		47.50		47.50	71
2040	Size 2	2 Elec	13.30	1.203			64		64	96
2060	Size 3		6.70	2.388			127		127	191
2080	Size 4		5.30	3.019			161		161	241
2100	Size 5		3.30	4.848			259		259	385
2120	NEMA 7, size 1	1 Elec	8.70	.920			49		49	73.50
2140	Size 2	2 Elec	12.70	1.260			67		67	101
2160	Size 3		6.30	2.540			135		135	203
2180	Size 4		5	3.200			171		171	256
2200	Size 5		3.20	5			267		267	400
2300	Fuse, light contactor, NEMA 1, 30 A	1 Elec	9	.889			47.50		47.50	71
2310	60 A		6.70	1.194			63.50		63.50	95.50
2320	100 A		3.30	2.424			129		129	194
2330	200 A		2.70	2.963			158		158	237
2350	Motor control center, incoming section	2 Elec	4	4			213		213	320
2352	Structure per section	"	4.20	3.810			203		203	305
2400	Starter & structure, 10 HP	1 Elec	9	.889			47.50		47.50	71
2410	25 HP	2 Elec	13.30	1.203			64		64	96
2420	50 HP		6.70	2.388			127		127	191
2430	75 HP		5.30	3.019			161		161	241
2440	100 HP		4.70	3.404			182		182	272
2450	200 HP		3.30	4.848			259		259	385
2460	400 HP		2.70	5.926			315		315	475
2500	Motor starter & control									
2510	Motor starter, NEMA 1, 5 HP	1 Elec	8.90	.899	Ea.		48		48	72
2520	10 HP	"	6.20	1.290			69		69	103

26 05 05.35 Electrical Demolition, L.V. Distribution	Crew	Daily Output	Labor-Hours	Unit	Material	2014 Bare Costs Labor	Equipment	Total	Total Incl O&P
2530	25 HP	2 Elec	8.40	1.905	Ea.		102		102	152
2540	50 HP		6.90	2.319			124		124	185
2550	100 HP		4.60	3.478			186		186	278
2560	200 HP		3.50	4.571			244		244	365
2570	400 HP		3.10	5.161			275		275	410
2610	Motor starter, NEMA 7, 5 HP	1 Elec	6.20	1.290			69		69	103
2620	10 HP	"	4.20	1.905			102		102	152
2630	25 HP	2 Elec	6.90	2.319			124		124	185
2640	50 HP		4.60	3.478			186		186	278
2650	100 HP		3.50	4.571			244		244	365
2660	200 HP		1.90	8.421			450		450	670
2710	Combination control unit, NEMA 1, 5 HP	1 Elec	6.90	1.159			62		62	92.50
2720	10 HP	"	5	1.600			85.50		85.50	128
2730	25 HP	2 Elec	7.70	2.078			111		111	166
2740	50 HP		5.10	3.137			167		167	251
2750	100 HP		3.10	5.161			275		275	410
2810	NEMA 7, 5 HP	1 Elec	5	1.600			85.50		85.50	128
2820	10 HP	"	3.90	2.051			109		109	164
2830	25 HP	2 Elec	5.10	3.137			167		167	251
2840	50 HP		3.10	5.161			275		275	410
2850	100 HP		2.30	6.957			370		370	555
2860	200 HP		1.50	10.667			570		570	850
3000	Panelboard or load center circuit breaker									
3010	Bolt-on or plug in, 15 A to 50 A	1 Elec	20	.400	Ea.		21.50		21.50	32
3020	60 A to 70 A		16	.500			26.50		26.50	40
3030	Bolt-on, 80 A to 100 A		14	.571			30.50		30.50	45.50
3040	Up to 250 A		7.30	1.096			58.50		58.50	87.50
3050	Motor operated, 30 A		13	.615			33		33	49
3060	60 A		10	.800			42.50		42.50	64
3070	100 A		8	1			53.50		53.50	80
3200	Switchboard circuit breaker									
3210	15 A - 60 A	1 Elec	19	.421	Ea.		22.50		22.50	33.50
3220	70 A - 100 A		14	.571			30.50		30.50	45.50
3230	125 A - 400 A		10	.800			42.50		42.50	64
3240	450 A - 600 A		5.30	1.509			80.50		80.50	121
3250	700 A - 800 A		4.30	1.860			99.50		99.50	149
3260	1000 A		3.30	2.424			129		129	194
3270	1200 A		2.70	2.963			158		158	237
3500	Switchboard, incoming section, 400 A	2 Elec	3.70	4.324			231		231	345
3510	600 A		3.30	4.848			259		259	385
3520	800 A		2.90	5.517			294		294	440
3530	1200 A		2.40	6.667			355		355	530
3540	1600 A		2.20	7.273			390		390	580
3550	2000 A		2.10	7.619			405		405	610
3560	3000 A		1.90	8.421			450		450	670
3570	4000 A		1.70	9.412			500		500	750
3610	Distribution section, 600 A		4	4			213		213	320
3620	800 A		3.60	4.444			237		237	355
3630	1200 A		3.10	5.161			275		275	410
3640	1600 A		2.90	5.517			294		294	440
3650	2000 A		2.70	5.926			315		315	475
3710	Transition section, 600 A		3.80	4.211			225		225	335
3720	800 A		3.30	4.848			259		259	385

26 05 05 – Selective Electrical Demolition

26 05 05.35 Electrical Demolition, L.V. Distribution

		Crew	Daily Output	Labor-Hours	Unit	Material	2014 Bare Costs Labor	2014 Bare Costs Equipment	Total	Total Incl O&P
3730	1200 A	2 Elec	2.70	5.926	Ea.		315		315	475
3740	1600 A		2.40	6.667			355		355	530
3750	2000 A		2.20	7.273			390		390	580
3760	2500 A		2.10	7.619			405		405	610
3770	3000 A		1.90	8.421			450		450	670
4000	Bus duct, aluminum or copper, 30 A	1 Elec	200	.040	L.F.		2.13		2.13	3.19
4020	60 A		160	.050			2.67		2.67	3.99
4040	100 A		140	.057			3.05		3.05	4.56
5000	Feedrail, trolley busway, up to 60 A		160	.050			2.67		2.67	3.99
5020	100 A		120	.067			3.56		3.56	5.30
5040	Busway, 50 A		200	.040			2.13		2.13	3.19
6000	Fuse, 30 A		133	.060	Ea.		3.21		3.21	4.80
6010	60 A		133	.060			3.21		3.21	4.80
6020	100 A		105	.076			4.06		4.06	6.10
6030	200 A		95	.084			4.49		4.49	6.70
6040	400 A		80	.100			5.35		5.35	8
6050	600 A		53	.151			8.05		8.05	12.05
6060	601 A - 1200 A		42	.190			10.15		10.15	15.20
6070	1500 A - 1600 A		34	.235			12.55		12.55	18.80
6080	1800 A - 2500 A		26	.308			16.40		16.40	24.50
6090	4000 A		21	.381			20.50		20.50	30.50
6100	4500 A - 5000 A		18	.444			23.50		23.50	35.50
6120	6000 A		15	.533			28.50		28.50	42.50
6200	Fuse plug or fustat		100	.080			4.27		4.27	6.40

26 05 05.50 Electrical Demolition, Lighting

		Crew	Daily Output	Labor-Hours	Unit	Material	2014 Bare Costs Labor	2014 Bare Costs Equipment	Total	Total Incl O&P
0010	**ELECTRICAL DEMOLITION, LIGHTING** Addition R024119-10									
0100	Fixture hanger, flexible, 1/2" diameter	1 Elec	36	.222	Ea.		11.85		11.85	17.75
0120	3/4" diameter R260105-30	"	30	.267	"		14.25		14.25	21.50
3000	Light pole, anchor base, excl concrete bases									
3010	Metal light pole, 10'	2 Elec	24	.667	Ea.		35.50		35.50	53
3020	16'	"	18	.889			47.50		47.50	71
3030	20'	R-3	8.70	2.299			122	16.10	138.10	201
3040	40'	"	6	3.333			176	23.50	199.50	291
3100	Wood light pole, 10'	2 Elec	36	.444			23.50		23.50	35.50
3120	20'	"	24	.667			35.50		35.50	53
3140	Bollard light, 42"	1 Elec	9	.889			47.50		47.50	71
3160	Walkway luminaire	"	8.10	.988			52.50		52.50	79
4000	Explosionproof									
4010	Metal halide, 175 W	1 Elec	8.70	.920	Ea.		49		49	73.50
4020	250 W	"	8.10	.988			52.50		52.50	79
4030	400 W	2 Elec	14.40	1.111			59.50		59.50	88.50
4050	High pressure sodium, 70 W	1 Elec	9	.889			47.50		47.50	71
4060	100 W		9	.889			47.50		47.50	71
4070	150 W		8.10	.988			52.50		52.50	79
4100	Incandescent		8.70	.920			49		49	73.50
4200	Fluorescent		8.10	.988			52.50		52.50	79
5000	Ballast, fluorescent fixture		24	.333			17.80		17.80	26.50
5040	High intensity discharge fixture		24	.333			17.80		17.80	26.50
5300	Exit and emergency lighting									
5310	Exit light	1 Elec	24	.333	Ea.		17.80		17.80	26.50
5320	Emergency battery pack lighting unit		12	.667			35.50		35.50	53
5330	Remote lamp only		80	.100			5.35		5.35	8

26 05 05 – Selective Electrical Demolition

26 05 05.50 Electrical Demolition, Lighting	Crew	Daily Output	Labor-Hours	Unit	Material	2014 Bare Costs Labor	2014 Bare Costs Equipment	Total	Total Incl O&P
5340	Self-contained fluorescent lamp pack	1 Elec	30	.267	Ea.		14.25		14.25	21.50
5500	Track lighting, 8' section	2 Elec	40	.400			21.50		21.50	32
5510	Track lighting fixture	1 Elec	64	.125	↓		6.65		6.65	10
5800	Energy saving devices									
5810	Occupancy sensor	1 Elec	21	.381	Ea.		20.50		20.50	30.50
5820	Automatic wall switch		72	.111			5.95		5.95	8.85
5830	Remote power pack		30	.267			14.25		14.25	21.50
5840	Photoelectric control		24	.333			17.80		17.80	26.50
5850	Fixture whip	↓	100	.080	↓		4.27		4.27	6.40
6000	Lamps									
6010	Fluorescent	1 Elec	200	.040	Ea.		2.13		2.13	3.19
6030	High intensity discharge lamp, up to 400 W		68	.118			6.30		6.30	9.40
6040	Up to 1000 W		45	.178			9.50		9.50	14.20
6050	Quartz		90	.089			4.74		4.74	7.10
6070	Incandescent		360	.022			1.19		1.19	1.77
6080	Exterior, PAR		290	.028			1.47		1.47	2.20
6090	Guards for fluorescent lamp	↓	24	.333	↓		17.80		17.80	26.50

26 05 13 – Medium-Voltage Cables

26 05 13.10 Cable Terminations

		Crew	Daily Output	Labor-Hours	Unit	Material	2014 Bare Costs Labor	2014 Bare Costs Equipment	Total	Total Incl O&P
0010	**CABLE TERMINATIONS**, 5 kV to 35 kV									
0100	Indoor, insulation diameter range .64" to 1.08"									
0300	Padmount, 5 kV	1 Elec	8	1	Ea.	85.50	53.50		139	175
0400	10 kV		6.40	1.250		97	66.50		163.50	206
0500	15 kV		6	1.333		127	71		198	246
0600	25 kV	↓	5.60	1.429	↓	156	76		232	286
0700	insulation diameter range 1.05" to 1.8"									
0800	Padmount, 5 kV	1 Elec	8	1	Ea.	99	53.50		152.50	189
0900	10 kV		6	1.333		131	71		202	250
1000	15 kV		5.60	1.429		151	76		227	280
1100	25 kV	↓	5.30	1.509	↓	193	80.50		273.50	335
1200	insulation diameter range 1.53" to 2.32"									
1300	Padmount, 5 KV	1 Elec	7.40	1.081	Ea.	117	57.50		174.50	216
1400	10 kV		5.60	1.429		164	76		240	295
1500	15 kV		5.30	1.509		195	80.50		275.50	335
1600	25 kV	↓	5	1.600	↓	231	85.50		316.50	380
1700	Outdoor systems, #4 stranded to 1/0 stranded									
1800	5 kV	1 Elec	7.40	1.081	Ea.	107	57.50		164.50	205
1900	15 kV		5.30	1.509		133	80.50		213.50	267
2000	25 kV		5	1.600		178	85.50		263.50	325
2100	35 kV		4.80	1.667		240	89		329	395
2200	#1 solid to 4/0 stranded, 5 kV		6.90	1.159		143	62		205	250
2300	15 kV		5	1.600		170	85.50		255.50	315
2400	25 kV		4.80	1.667		230	89		319	385
2500	35 kV		4.60	1.739		254	93		347	420
2600	2/0 solid to 350 kcmil stranded, 5 kV		6.40	1.250		192	66.50		258.50	310
2700	15 kV		4.80	1.667		192	89		281	345
2800	25 kV		4.60	1.739		297	93		390	465
2900	35 kV		4.40	1.818		315	97		412	490
3000	400 kcmil compact to 750 kcmil stranded, 5 kV		6	1.333		194	71		265	320
3100	15 kV		4.60	1.739		231	93		324	395
3200	25 kV		4.40	1.818		340	97		437	520
3300	35 kV	↓	4.20	1.905	↓	345	102		447	530

26 05 Common Work Results for Electrical

26 05 13 – Medium-Voltage Cables

26 05 13.10 Cable Terminations

		Crew	Daily Output	Labor-Hours	Unit	Material	2014 Bare Costs Labor	Equipment	Total	Total Incl O&P
3400	1000 kcmil, 5 kV	1 Elec	5.60	1.429	Ea.	315	76		391	460
3500	15 kV		4.40	1.818		315	97		412	490
3600	25 kV		4.20	1.905		345	102		447	530
3700	35 kV		4	2		345	107		452	540

26 05 13.16 Medium-Voltage, Single Cable

		Crew	Daily Output	Labor-Hours	Unit	Material	2014 Bare Costs Labor	Equipment	Total	Total Incl O&P
0010	**MEDIUM-VOLTAGE, SINGLE CABLE** Splicing & terminations not included									
0040	Copper, XLP shielding, 5 kV, #6	2 Elec	4.40	3.636	C.L.F.	158	194		352	465
0050	#4		4.40	3.636		205	194		399	515
0100	#2		4	4		242	213		455	585
0200	#1		4	4		278	213		491	625
0400	1/0		3.80	4.211		305	225		530	670
0600	2/0		3.60	4.444		385	237		622	780
0800	4/0		3.20	5		500	267		767	950
1000	250 kcmil	3 Elec	4.50	5.333		620	285		905	1,100
1200	350 kcmil		3.90	6.154		800	330		1,130	1,375
1400	500 kcmil		3.60	6.667		975	355		1,330	1,600
1600	15 kV, ungrounded neutral, #1	2 Elec	4	4		340	213		553	690
1800	1/0		3.80	4.211		405	225		630	780
2000	2/0		3.60	4.444		460	237		697	860
2200	4/0		3.20	5		610	267		877	1,075
2400	250 kcmil	3 Elec	4.50	5.333		675	285		960	1,175
2600	350 kcmil		3.90	6.154		855	330		1,185	1,425
2800	500 kcmil		3.60	6.667		1,050	355		1,405	1,675
3000	25 kV, grounded neutral, #1/0	2 Elec	3.60	4.444		560	237		797	970
3200	2/0		3.40	4.706		615	251		866	1,050
3400	4/0		3	5.333		770	285		1,055	1,275
3600	250 kcmil	3 Elec	4.20	5.714		960	305		1,265	1,500
3800	350 kcmil		3.60	6.667		1,125	355		1,480	1,750
3900	500 kcmil		3.30	7.273		1,325	390		1,715	2,025
4000	35 kV, grounded neutral, #1/0	2 Elec	3.40	4.706		590	251		841	1,025
4200	2/0		3.20	5		695	267		962	1,175
4400	4/0		2.80	5.714		875	305		1,180	1,425
4600	250 kcmil	3 Elec	3.90	6.154		1,025	330		1,355	1,625
4800	350 kcmil		3.30	7.273		1,225	390		1,615	1,925
5000	500 kcmil		3	8		1,450	425		1,875	2,250
5050	Aluminum, XLP shielding, 5 kV, #2	2 Elec	5	3.200		168	171		339	440
5070	#1		4.40	3.636		174	194		368	480
5090	1/0		4	4		202	213		415	540
5100	2/0		3.80	4.211		226	225		451	585
5150	4/0		3.60	4.444		268	237		505	650
5200	250 kcmil	3 Elec	4.80	5		325	267		592	755
5220	350 kcmil		4.50	5.333		380	285		665	845
5240	500 kcmil		3.90	6.154		490	330		820	1,025
5260	750 kcmil		3.60	6.667		650	355		1,005	1,250
5300	15 kV aluminum, XLP, #1	2 Elec	4.40	3.636		216	194		410	530
5320	1/0		4	4		223	213		436	565
5340	2/0		3.80	4.211		266	225		491	630
5360	4/0		3.60	4.444		295	237		532	680
5380	250 kcmil	3 Elec	4.80	5		355	267		622	790
5400	350 kcmil		4.50	5.333		395	285		680	860
5420	500 kcmil		3.90	6.154		545	330		875	1,100
5440	750 kcmil		3.60	6.667		725	355		1,080	1,325

26 05 19 – Low-Voltage Electrical Power Conductors and Cables

26 05 19.13 Undercarpet Electrical Power Cables	Crew	Daily Output	Labor-Hours	Unit	Material	2014 Bare Costs Labor	Equipment	Total	Total Incl O&P
0010 **UNDERCARPET ELECTRICAL POWER CABLES** R260519-80									
0020 Power System									
0100 Cable flat, 3 conductor, #12, w/attached bottom shield	1 Elec	982	.008	L.F.	5.10	.43		5.53	6.25
0200 Shield, top, steel		1768	.005	"	5.45	.24		5.69	6.35
0250 Splice, 3 conductor		48	.167	Ea.	16.60	8.90		25.50	31.50
0300 Top shield		96	.083		1.48	4.45		5.93	8.30
0350 Tap		40	.200		21.50	10.65		32.15	39.50
0400 Insulating patch, splice, tap, & end		48	.167		52.50	8.90		61.40	71
0450 Fold		230	.035			1.86		1.86	2.78
0500 Top shield, tap & fold		96	.083		1.48	4.45		5.93	8.30
0700 Transition, block assembly		77	.104		76.50	5.55		82.05	92.50
0750 Receptacle frame & base		32	.250		42	13.35		55.35	66
0800 Cover receptacle		120	.067		3.57	3.56		7.13	9.25
0850 Cover blank		160	.050		4.20	2.67		6.87	8.60
0860 Receptacle, direct connected, single		25	.320		89.50	17.05		106.55	124
0870 Dual		16	.500		147	26.50		173.50	202
0880 Combination high & low, tension		21	.381		108	20.50		128.50	150
0900 Box, floor with cover		20	.400		91	21.50		112.50	132
0920 Floor service w/barrier		4	2		257	107		364	440
1000 Wall, surface, with cover		20	.400		60	21.50		81.50	98
1100 Wall, flush, with cover		20	.400	▼	42	21.50		63.50	78
1450 Cable flat, 5 conductor #12, w/attached bottom shield		800	.010	L.F.	8.35	.53		8.88	9.95
1550 Shield, top, steel		1768	.005	"	8.30	.24		8.54	9.45
1600 Splice, 5 conductor		48	.167	Ea.	27	8.90		35.90	43
1650 Top shield		96	.083		1.48	4.45		5.93	8.30
1700 Tap		48	.167		35	8.90		43.90	52
1750 Insulating patch, splice tap, & end		83	.096		52.50	5.15		57.65	65
1800 Transition, block assembly		77	.104		55	5.55		60.55	69
1850 Box, wall, flush with cover		20	.400	▼	56	21.50		77.50	93.50
1900 Cable flat, 4 conductor, #12		933	.009	L.F.	6.75	.46		7.21	8.10
1950 3 conductor #10		982	.008		5.85	.43		6.28	7.05
1960 4 conductor #10		933	.009		7.65	.46		8.11	9.10
1970 5 conductor #10	▼	884	.009	▼	9.35	.48		9.83	10.95
2500 Telephone System									
2510 Transition fitting wall box, surface	1 Elec	24	.333	Ea.	45	17.80		62.80	76
2520 Flush		24	.333		45	17.80		62.80	76
2530 Flush, for PC board		24	.333		45	17.80		62.80	76
2540 Floor service box	▼	4	2		233	107		340	415
2550 Cover, surface					14.30			14.30	15.75
2560 Flush					14.30			14.30	15.75
2570 Flush for PC board					14.30			14.30	15.75
2700 Floor fitting w/duplex jack & cover	1 Elec	21	.381		44	20.50		64.50	79
2720 Low profile		53	.151		15.40	8.05		23.45	29
2740 Miniature w/duplex jack		53	.151		24	8.05		32.05	38.50
2760 25 pair kit		21	.381		46.50	20.50		67	81.50
2780 Low profile		53	.151		15.75	8.05		23.80	29.50
2800 Call director kit for 5 cable		19	.421		74	22.50		96.50	115
2820 4 pair kit		19	.421		88	22.50		110.50	131
2840 3 pair kit		19	.421		93	22.50		115.50	137
2860 Comb. 25 pair & 3 conductor power		21	.381		74	20.50		94.50	112
2880 5 conductor power		21	.381		84.50	20.50		105	124
2900 PC board, 8 per 3 pair	▼	161	.050	▼	64	2.65		66.65	74.50

26 05 19 – Low-Voltage Electrical Power Conductors and Cables

26 05 19.13 Undercarpet Electrical Power Cables		Crew	Daily Output	Labor-Hours	Unit	Material	2014 Bare Costs Labor	Equipment	Total	Total Incl O&P
2920	6 per 4 pair	1 Elec	161	.050	Ea.	64	2.65		66.65	74.50
2940	3 pair adapter		161	.050		58.50	2.65		61.15	68
2950	Plug		77	.104		2.72	5.55		8.27	11.30
2960	Couplers		321	.025		7.65	1.33		8.98	10.40
3000	Bottom shield for 25 pair cable		4420	.002	L.F.	.77	.10		.87	.99
3020	4 pair		4420	.002		.37	.10		.47	.55
3040	Top shield for 25 pair cable		4420	.002		.77	.10		.87	.99
3100	Cable assembly, double-end, 50', 25 pair		11.80	.678	Ea.	225	36		261	300
3110	3 pair		23.60	.339		66.50	18.10		84.60	100
3120	4 pair		23.60	.339		74	18.10		92.10	109
3140	Bulk 3 pair		1473	.005	L.F.	1.14	.29		1.43	1.68
3160	4 pair		1473	.005	"	1.42	.29		1.71	1.99
3500	Data System									
3520	Cable 25 conductor w/connection 40', 75 ohm	1 Elec	14.50	.552	Ea.	67	29.50		96.50	118
3530	Single lead		22	.364		182	19.40		201.40	229
3540	Dual lead		22	.364		226	19.40		245.40	278
3560	Shields same for 25 conductor as 25 pair telephone									
3570	Single & dual, none required									
3590	BNC coax connectors, Plug	1 Elec	40	.200	Ea.	9.95	10.65		20.60	27
3600	TNC coax connectors, Plug	"	40	.200	"	12.80	10.65		23.45	30
3700	Cable-bulk									
3710	Single lead	1 Elec	1473	.005	L.F.	2.43	.29		2.72	3.10
3720	Dual lead	"	1473	.005	"	3.52	.29		3.81	4.30
3730	Hand tool crimp				Ea.	505			505	555
3740	Hand tool notch				"	17.35			17.35	19.10
3750	Boxes & floor fitting same as telephone									
3790	Data cable notching, 90°	1 Elec	97	.082	Ea.		4.40		4.40	6.60
3800	180°		60	.133			7.10		7.10	10.65
8100	Drill floor		160	.050		1.94	2.67		4.61	6.10
8200	Marking floor		1600	.005	L.F.		.27		.27	.40
8300	Tape, hold down		6400	.001	"	.16	.07		.23	.28
8350	Tape primer, 500 ft. per can		96	.083	Ea.	29.50	4.45		33.95	39
8400	Tool, splicing				"	234			234	258

26 05 19.20 Armored Cable

0010	**ARMORED CABLE**	R260519-20									
0050	600 volt, copper (BX), #14, 2 conductor, solid		1 Elec	2.40	3.333	C.L.F.	54	178		232	325
0100	3 conductor, solid			2.20	3.636		83.50	194		277.50	380
0120	4 conductor, solid			2	4		116	213		329	445
0150	#12, 2 conductor, solid			2.30	3.478		54.50	186		240.50	340
0200	3 conductor, solid			2	4		88.50	213		301.50	420
0220	4 conductor, solid			1.80	4.444		121	237		358	490
0250	#10, 2 conductor, solid			2	4		100	213		313	430
0300	3 conductor, solid			1.60	5		139	267		406	555
0320	4 conductor, solid			1.40	5.714		215	305		520	690
0340	#8, 2 conductor, stranded			1.50	5.333		262	285		547	715
0350	3 conductor, stranded			1.30	6.154		262	330		592	780
0370	4 conductor, stranded			1.10	7.273		375	390		765	990
0380	#6, 2 conductor, stranded			1.30	6.154		274	330		604	790
0400	3 conductor with PVC jacket, in cable tray, #6			3.10	2.581		455	138		593	705
0450	#4		2 Elec	5.40	2.963		565	158		723	855
0500	#2			4.60	3.478		710	186		896	1,075
0550	#1			4	4		940	213		1,153	1,350

26 05 19.20 Armored Cable		Crew	Daily Output	Labor-Hours	Unit	Material	2014 Bare Costs Labor	Equipment	Total	Total Incl O&P
0600	1/0	2 Elec	3.60	4.444	C.L.F.	1,025	237		1,262	1,475
0650	2/0		3.40	4.706		1,250	251		1,501	1,750
0700	3/0		3.20	5		1,575	267		1,842	2,125
0750	4/0		3	5.333		1,825	285		2,110	2,450
0800	250 kcmil	3 Elec	3.60	6.667		2,100	355		2,455	2,825
0850	350 kcmil		3.30	7.273		2,825	390		3,215	3,675
0900	500 kcmil		3	8		3,800	425		4,225	4,825
0910	4 conductor with PVC jacket, in cable tray, #6	1 Elec	2.70	2.963		575	158		733	870
0920	#4	2 Elec	4.60	3.478		720	186		906	1,075
0930	#2		4	4		880	213		1,093	1,275
0940	#1		3.60	4.444		1,200	237		1,437	1,675
0950	1/0		3.40	4.706		1,300	251		1,551	1,800
0960	2/0		3.20	5		1,550	267		1,817	2,100
0970	3/0		3	5.333		1,950	285		2,235	2,575
0980	4/0		2.40	6.667		2,350	355		2,705	3,125
0990	250 kcmil	3 Elec	3.30	7.273		2,825	390		3,215	3,675
1000	350 kcmil		3	8		3,600	425		4,025	4,600
1010	500 kcmil		2.70	8.889		4,950	475		5,425	6,150
1050	5 kV, copper, 3 conductor with PVC jacket,									
1060	non-shielded, in cable tray, #4	2 Elec	380	.042	L.F.	7.35	2.25		9.60	11.40
1100	#2		360	.044		9.55	2.37		11.92	14.05
1200	#1		300	.053		12.15	2.85		15	17.60
1400	1/0		290	.055		14	2.94		16.94	19.80
1600	2/0		260	.062		16.20	3.28		19.48	23
2000	4/0		240	.067		21.50	3.56		25.06	29.50
2100	250 kcmil	3 Elec	330	.073		29.50	3.88		33.38	38.50
2150	350 kcmil		315	.076		36.50	4.06		40.56	46
2200	500 kcmil		270	.089		51	4.74		55.74	63
2400	15 kV, copper, 3 conductor with PVC jacket galv., steel armored									
2500	grounded neutral, in cable tray, #2	2 Elec	300	.053	L.F.	15.40	2.85		18.25	21
2600	#1		280	.057		16.40	3.05		19.45	22.50
2800	1/0		260	.062		18.75	3.28		22.03	25.50
2900	2/0		220	.073		25	3.88		28.88	33.50
3000	4/0		190	.084		28.50	4.49		32.99	37.50
3100	250 kcmil	3 Elec	270	.089		31.50	4.74		36.24	41.50
3150	350 kcmil		240	.100		37	5.35		42.35	49
3200	500 kcmil		210	.114		50	6.10		56.10	64
3400	15 kV, copper, 3 conductor with PVC jacket,									
3450	ungrounded neutral, in cable tray, #2	2 Elec	260	.062	L.F.	16.60	3.28		19.88	23
3500	#1		230	.070		18.40	3.71		22.11	26
3600	1/0		200	.080		21	4.27		25.27	29.50
3700	2/0		190	.084		26	4.49		30.49	35
3800	4/0		160	.100		31	5.35		36.35	42.50
4000	250 kcmil	3 Elec	210	.114		36.50	6.10		42.60	49
4050	350 kcmil		195	.123		48	6.55		54.55	62.50
4100	500 kcmil		180	.133		58.50	7.10		65.60	74.50
4200	600 volt, aluminum, 3 conductor in cable tray with PVC jacket									
4300	#2	2 Elec	540	.030	L.F.	3.92	1.58		5.50	6.70
4400	#1		460	.035		4.34	1.86		6.20	7.55
4500	#1/0		400	.040		5.40	2.13		7.53	9.15
4600	#2/0		360	.044		5.50	2.37		7.87	9.60
4700	#3/0		340	.047		6.40	2.51		8.91	10.80
4800	#4/0		320	.050		7.75	2.67		10.42	12.55

26 05 19 – Low-Voltage Electrical Power Conductors and Cables

26 05 19.20 Armored Cable	Crew	Daily Output	Labor-Hours	Unit	Material	2014 Bare Costs Labor	Equipment	Total	Total Incl O&P
4900	250 kcmil	3 Elec	450	.053	L.F.	9.35	2.85		12.20	14.50
5000	350 kcmil		360	.067		11.10	3.56		14.66	17.50
5200	500 kcmil		330	.073		13.80	3.88		17.68	21
5300	750 kcmil		285	.084		17.80	4.49		22.29	26.50
5400	600 volt, aluminum, 4 conductor in cable tray with PVC jacket									
5410	#2	2 Elec	520	.031	L.F.	4.49	1.64		6.13	7.40
5430	#1		440	.036		5.50	1.94		7.44	8.95
5450	1/0		380	.042		6.40	2.25		8.65	10.40
5470	2/0		340	.047		6.55	2.51		9.06	10.95
5480	3/0		320	.050		7.70	2.67		10.37	12.45
5500	4/0		300	.053		9.05	2.85		11.90	14.20
5520	250 kcmil	3 Elec	420	.057		9.70	3.05		12.75	15.20
5540	350 kcmil		330	.073		12.55	3.88		16.43	19.60
5560	500 kcmil		300	.080		15.65	4.27		19.92	23.50
5580	750 kcmil		270	.089		21.50	4.74		26.24	31
5600	5 kV, aluminum, unshielded in cable tray, #2 with PVC jacket	2 Elec	380	.042		5.50	2.25		7.75	9.40
5700	#1 with PVC jacket		360	.044		6.10	2.37		8.47	10.30
5800	1/0 with PVC jacket		300	.053		6.25	2.85		9.10	11.15
6000	2/0 with PVC jacket		290	.055		6.40	2.94		9.34	11.45
6200	3/0 with PVC jacket		260	.062		7.70	3.28		10.98	13.35
6300	4/0 with PVC jacket		240	.067		9.05	3.56		12.61	15.25
6400	250 kcmil with PVC jacket	3 Elec	330	.073		9.90	3.88		13.78	16.70
6500	350 kcmil with PVC jacket		315	.076		11.60	4.06		15.66	18.90
6600	500 kcmil with PVC jacket		300	.080		13.75	4.27		18.02	21.50
6800	750 kcmil with PVC jacket		270	.089		16.95	4.74		21.69	26
6900	15 kV, aluminum, shielded-grounded, #2 with PVC jacket	2 Elec	320	.050		12.45	2.67		15.12	17.70
7000	#1 with PVC jacket		300	.053		12.80	2.85		15.65	18.35
7200	1/0 with PVC jacket		280	.057		13.80	3.05		16.85	19.75
7300	2/0 with PVC jacket		260	.062		14.05	3.28		17.33	20.50
7400	3/0 with PVC jacket		240	.067		15.75	3.56		19.31	22.50
7500	4/0 with PVC jacket		220	.073		16.15	3.88		20.03	23.50
7600	250 kcmil with PVC jacket	3 Elec	300	.080		17.75	4.27		22.02	26
7700	350 kcmil with PVC jacket		270	.089		21	4.74		25.74	30
7800	500 kcmil with PVC jacket		240	.100		25	5.35		30.35	35.50
8000	750 kcmil with PVC jacket		204	.118		29.50	6.30		35.80	42
8200	15 kV, aluminum, shielded-ungrounded, #1 with PVC jacket	2 Elec	250	.064		15.60	3.41		19.01	22.50
8300	1/0 with PVC jacket		230	.070		16.15	3.71		19.86	23.50
8400	2/0 with PVC jacket		210	.076		17.75	4.06		21.81	25.50
8500	3/0 with PVC jacket		200	.080		18.05	4.27		22.32	26.50
8600	4/0 with PVC jacket		190	.084		19.60	4.49		24.09	28
8700	250 kcmil with PVC jacket	3 Elec	270	.089		21	4.74		25.74	30
8800	350 kcmil with PVC jacket		240	.100		24	5.35		29.35	34.50
8900	500 kcmil with PVC jacket		210	.114		29	6.10		35.10	41
8950	750 kcmil with PVC jacket		174	.138		36	7.35		43.35	50.50
9010	600 volt, copper (MC) steel clad, #14, 2 wire	1 Elec	2.40	3.333	C.L.F.	55	178		233	325
9020	3 wire		2.20	3.636		85	194		279	385
9030	4 wire		2	4		118	213		331	450
9040	#12, 2 wire		2.30	3.478		55.50	186		241.50	340
9050	3 wire		2	4		94	213		307	425
9060	4 wire		1.80	4.444		127	237		364	495
9070	#10, 2 wire		2	4		116	213		329	450
9080	3 wire		1.60	5		162	267		429	580
9090	4 wire		1.40	5.714		254	305		559	735

26 05 Common Work Results for Electrical

26 05 19 – Low-Voltage Electrical Power Conductors and Cables

26 05 19.20 Armored Cable		Crew	Daily Output	Labor-Hours	Unit	Material	2014 Bare Costs Labor	Equipment	Total	Total Incl O&P
9091	For health care facilities cable, add, minimum					5%				
9092	Maximum					20%				
9100	#8, 2 wire, stranded	1 Elec	1.80	4.444	C.L.F.	223	237		460	600
9110	3 wire, stranded		1.30	6.154		315	330		645	835
9120	4 wire, stranded		1.10	7.273		425	390		815	1,050
9130	#6, 2 wire, stranded		1.30	6.154		330	330		660	855
9200	600 volt, copper (MC) aluminum clad, #14, 2 wire		2.65	3.019		53.50	161		214.50	300
9210	3 wire		2.45	3.265		83.50	174		257.50	355
9220	4 wire		2.20	3.636		117	194		311	420
9230	#12, 2 wire		2.55	3.137		55.50	167		222.50	310
9240	3 wire		2.20	3.636		92.50	194		286.50	390
9250	4 wire		2	4		125	213		338	460
9260	#10, 2 wire		2.20	3.636		115	194		309	415
9270	3 wire		1.80	4.444		161	237		398	530
9280	4 wire		1.55	5.161		253	275		528	690
9600	Alum (MC) aluminum clad, #6, 3 conductor w/#6 grnd		1.67	4.790		315	256		571	730
9610	4 conductor w/#6 grnd		1.64	4.878		340	260		600	765
9620	#4, 3 conductor w/#6 grnd	2 Elec	2.86	5.594		325	298		623	805
9630	4 conductor w/#6 grnd		2.82	5.674		355	305		660	845
9640	#2, 3 conductor w/#4 grnd		2.50	6.400		375	340		715	925
9650	4 conductor w/#4 grnd		2.47	6.478		425	345		770	980
9660	#1, 3 conductor w/#4 grnd		2	8		485	425		910	1,175
9670	4 conductor w/#4 grnd		1.98	8.081		575	430		1,005	1,275
9680	1/0, 3 conductor w/#4 grnd		1.82	8.791		555	470		1,025	1,300
9690	4 conductor w/#4 grnd		1.79	8.939		675	475		1,150	1,450
9700	2/0, 3 conductor w/#4 grnd		1.75	9.143		585	490		1,075	1,375
9710	4 conductor w/#4 grnd		1.71	9.357		755	500		1,255	1,575
9720	3/0, 3 conductor w/#4 grnd		1.71	9.357		750	500		1,250	1,575
9730	4 conductor w/#4 grnd		1.67	9.581		935	510		1,445	1,800
9740	4/0, 3 conductor w/#2 grnd		1.67	9.581		840	510		1,350	1,700
9750	4 conductor w/#2 grnd		1.61	9.938		1,050	530		1,580	1,950
9760	250 kcmil, 3 conductor w/#1 grnd	3 Elec	2.42	9.917		1,075	530		1,605	1,975
9770	4 conductor w/#1 grnd		2.36	10.169		1,450	545		1,995	2,400
9775	300 kcmil, 4 conductor w/#1 grnd		2.22	10.811		1,550	575		2,125	2,600
9780	350 kcmil, 3 conductor w/1/0 grnd		2.19	10.959		1,225	585		1,810	2,225
9790	4 conductor w/1/0 grnd		2.10	11.429		1,775	610		2,385	2,875
9800	500 kcmil, 3 conductor w/#1 grnd		2.10	11.429		1,775	610		2,385	2,875
9810	4 conductor w/2/0 grnd		2	12		2,175	640		2,815	3,325
9840	750 kcmil, 3 conductor w/1/0 grnd		1.95	12.308		2,550	655		3,205	3,800
9850	4 conductor w/3/0 grnd		1.89	12.698		3,025	675		3,700	4,350

26 05 19.25 Cable Connectors

0010	CABLE CONNECTORS									
0100	600 volt, nonmetallic, #14-2 wire	1 Elec	160	.050	Ea.	1.41	2.67		4.08	5.55
0200	#14-3 wire to #12-2 wire		133	.060		1.41	3.21		4.62	6.35
0300	#12-3 wire to #10-2 wire		114	.070		1.41	3.74		5.15	7.15
0400	#10-3 wire to #14-4 and #12-4 wire		100	.080		1.41	4.27		5.68	7.95
0500	#8-3 wire to #10-4 wire		80	.100		2.66	5.35		8.01	10.95
0600	#6-3 wire		40	.200		3.57	10.65		14.22	19.90
0800	SER, 3 #8 insulated + 1 #8 ground		32	.250		2.73	13.35		16.08	23
0900	3 #6 + 1 #6 ground		24	.333		3.41	17.80		21.21	30.50
1000	3 #4 + 1 #6 ground		22	.364		4.26	19.40		23.66	33.50
1100	3 #2 + 1 #4 ground		20	.400		7.80	21.50		29.30	40.50

26 05 19 – Low-Voltage Electrical Power Conductors and Cables

26 05 19.25 Cable Connectors		Crew	Daily Output	Labor-Hours	Unit	Material	2014 Bare Costs Labor	Equipment	Total	Total Incl O&P
1200	3 1/0 + 1 #2 ground	1 Elec	18	.444	Ea.	20	23.50		43.50	57.50
1400	3 2/0 + 1 #1 ground		16	.500		24	26.50		50.50	66.50
1600	3 4/0 + 1 #2/0 ground		14	.571		26.50	30.50		57	75
1800	600 volt, armored, #14-2 wire		80	.100		.88	5.35		6.23	8.95
2200	#14-4, #12-3 and #10-2 wire		40	.200		.90	10.65		11.55	16.95
2400	#12-4, #10-3 and #8-2 wire		32	.250		2.07	13.35		15.42	22
2600	#8-3 and #10-4 wire		26	.308		3.55	16.40		19.95	28.50
2650	#8-4 wire		22	.364		5	19.40		24.40	34.50
2700	PVC jacket connector, #6-3 wire, #6-4 wire		16	.500		9.40	26.50		35.90	50.50
2800	#4-3 wire, #4-4 wire		16	.500		9.40	26.50		35.90	50.50
2900	#2-3 wire		12	.667		9.40	35.50		44.90	63.50
3000	#1-3 wire, #2-4 wire		12	.667		14.50	35.50		50	69
3200	1/0-3 wire		11	.727		14.50	39		53.50	74
3400	2/0-3 wire, 1/0-4 wire		10	.800		14.50	42.50		57	80
3500	3/0-3 wire, 2/0-4 wire		9	.889		24	47.50		71.50	97.50
3600	4/0-3 wire, 3/0-4 wire		7	1.143		24	61		85	118
3800	250 kcmil-3 wire, 4/0-4 wire		6	1.333		44	71		115	155
4000	350 kcmil-3 wire, 250 kcmil-4 wire		5	1.600		44	85.50		129.50	177
4100	350 kcmil-4 wire		4	2		185	107		292	365
4200	500 kcmil-3 wire		4	2		185	107		292	365
4250	500 kcmil-4 wire, 750 kcmil-3 wire		3.50	2.286		251	122		373	460
4300	750 kcmil-4 wire		3	2.667		251	142		393	490
4400	5 kV, armored, #4		8	1		67	53.50		120.50	154
4600	#2		8	1		67	53.50		120.50	154
4800	#1		8	1		86.50	53.50		140	175
5000	1/0		6.40	1.250		107	66.50		173.50	218
5200	2/0		5.30	1.509		107	80.50		187.50	239
5500	4/0		4	2		143	107		250	315
5600	250 kcmil		3.60	2.222		143	119		262	335
5650	350 kcmil		3.20	2.500		172	133		305	390
5700	500 kcmil		2.50	3.200		216	171		387	495
5720	750 kcmil		2.20	3.636		263	194		457	580
5750	1000 kcmil		2	4		325	213		538	675
5800	15 kV, armored, #1		4	2		126	107		233	299
5900	1/0		4	2		158	107		265	335
6000	3/0		3.60	2.222		187	119		306	380
6100	4/0		3.40	2.353		186	126		312	395
6200	250 kcmil		3.20	2.500		237	133		370	460
6300	350 kcmil		2.70	2.963		282	158		440	545
6400	500 kcmil		2	4		282	213		495	630

26 05 19.30 Cable Splicing		Crew	Daily Output	Labor-Hours	Unit	Material	2014 Bare Costs Labor	Equipment	Total	Total Incl O&P
0010	**CABLE SPLICING** URD or similar, ideal conditions									
0100	#6 stranded to #1 stranded, 5 kV	1 Elec	4	2	Ea.	118	107		225	290
0120	15 kV		3.60	2.222		159	119		278	350
0140	25 kV		3.20	2.500		160	133		293	375
0200	#1 stranded to 4/0 stranded, 5 kV		3.60	2.222		131	119		250	320
0210	15 kV		3.20	2.500		172	133		305	390
0220	25 kV		2.80	2.857		160	152		312	405
0300	4/0 stranded to 500 kcmil stranded, 5 kV		3.30	2.424		131	129		260	340
0310	15 kV		2.90	2.759		228	147		375	470
0320	25 kV		2.50	3.200		284	171		455	570
0400	500 kcmil, 5 kV		3.20	2.500		213	133		346	435

26 05 19 – Low-Voltage Electrical Power Conductors and Cables

26 05 19.30 Cable Splicing		Crew	Daily Output	Labor-Hours	Unit	Material	2014 Bare Costs Labor	Equipment	Total	Total Incl O&P
0410	15 kV	1 Elec	2.80	2.857	Ea.	288	152		440	545
0420	25 kV		2.30	3.478		284	186		470	595
0500	600 kcmil, 5 kV		2.90	2.759		213	147		360	455
0510	15 kV		2.40	3.333		290	178		468	585
0520	25 kV		2	4		288	213		501	635
0600	750 kcmil, 5 kV		2.60	3.077		213	164		377	480
0610	15 kV		2.20	3.636		290	194		484	610
0620	25 kV		1.90	4.211		288	225		513	650
0700	1000 kcmil, 5 kV		2.30	3.478		213	186		399	515
0710	15 kV		1.90	4.211		315	225		540	680
0720	25 kV		1.60	5		375	267		642	810

26 05 19.35 Cable Terminations

		Crew	Daily Output	Labor-Hours	Unit	Material	Labor	Equipment	Total	Total Incl O&P
0010	**CABLE TERMINATIONS**									
0015	Wire connectors, screw type, #22 to #14	1 Elec	260	.031	Ea.	.06	1.64		1.70	2.53
0020	#18 to #12		240	.033		.08	1.78		1.86	2.75
0025	#18 to #10		240	.033		.14	1.78		1.92	2.81
0030	Screw-on connectors, insulated, #18 to #12		240	.033		.22	1.78		2	2.90
0035	#16 to #10		230	.035		.25	1.86		2.11	3.06
0040	#14 to #8		210	.038		.30	2.03		2.33	3.37
0045	#12 to #6		180	.044		.62	2.37		2.99	4.23
0050	Terminal lugs, solderless, #16 to #10		50	.160		.37	8.55		8.92	13.20
0100	#8 to #4		30	.267		.73	14.25		14.98	22.50
0150	#2 to #1		22	.364		1.03	19.40		20.43	30
0200	1/0 to 2/0		16	.500		1.64	26.50		28.14	42
0250	3/0		12	.667		2.87	35.50		38.37	56
0300	4/0		11	.727		3.51	39		42.51	62
0350	250 kcmil		9	.889		3.41	47.50		50.91	75
0400	350 kcmil		7	1.143		4.44	61		65.44	96.50
0450	500 kcmil		6	1.333		8.20	71		79.20	115
0500	600 kcmil		5.80	1.379		9.30	73.50		82.80	120
0550	750 kcmil		5.20	1.538		10.95	82		92.95	135
0600	Split bolt connectors, tapped, #6		16	.500		3.78	26.50		30.28	44
0650	#4		14	.571		4.41	30.50		34.91	50.50
0700	#2		12	.667		6.75	35.50		42.25	60.50
0750	#1		11	.727		8.45	39		47.45	67.50
0800	1/0		10	.800		8.45	42.50		50.95	73.50
0850	2/0		9	.889		13.70	47.50		61.20	86
0900	3/0		7.20	1.111		19.05	59.50		78.55	110
1000	4/0		6.40	1.250		23.50	66.50		90	126
1100	250 kcmil		5.70	1.404		23.50	75		98.50	138
1200	300 kcmil		5.30	1.509		42.50	80.50		123	168
1400	350 kcmil		4.60	1.739		42.50	93		135.50	186
1500	500 kcmil		4	2		62.50	107		169.50	229
1600	Crimp 1 hole lugs, copper or aluminum, 600 volt									
1620	#14	1 Elec	60	.133	Ea.	.60	7.10		7.70	11.30
1630	#12		50	.160		.92	8.55		9.47	13.80
1640	#10		45	.178		.92	9.50		10.42	15.20
1780	#8		36	.222		1.91	11.85		13.76	19.85
1800	#6		30	.267		2.48	14.25		16.73	24
2000	#4		27	.296		3.38	15.80		19.18	27
2200	#2		24	.333		5.30	17.80		23.10	32.50
2400	#1		20	.400		5.50	21.50		27	38

26 05 19 – Low-Voltage Electrical Power Conductors and Cables

26 05 19.35 Cable Terminations		Crew	Daily Output	Labor-Hours	Unit	Material	2014 Bare Costs Labor	Equipment	Total	Total Incl O&P
2500	1/0	1 Elec	17.50	.457	Ea.	5.85	24.50		30.35	43
2600	2/0		15	.533		6.80	28.50		35.30	50
2800	3/0		12	.667		8	35.50		43.50	62
3000	4/0		11	.727		8.90	39		47.90	68
3200	250 kcmil		9	.889		10.50	47.50		58	82.50
3400	300 kcmil		8	1		12	53.50		65.50	93
3500	350 kcmil		7	1.143		13.60	61		74.60	106
3600	400 kcmil		6.50	1.231		15.10	65.50		80.60	115
3800	500 kcmil		6	1.333		17.70	71		88.70	125
4000	600 kcmil		5.80	1.379		31.50	73.50		105	145
4200	700 kcmil		5.50	1.455		37.50	77.50		115	158
4400	750 kcmil		5.20	1.538		37.50	82		119.50	165
4500	Crimp 2-way connectors, copper or alum., 600 volt,									
4510	#14	1 Elec	60	.133	Ea.	2.15	7.10		9.25	13
4520	#12		50	.160		2.28	8.55		10.83	15.30
4530	#10		45	.178		2.40	9.50		11.90	16.85
4540	#8		27	.296		2.43	15.80		18.23	26
4600	#6		25	.320		4.21	17.05		21.26	30
4800	#4		23	.348		4.80	18.55		23.35	33.50
5000	#2		20	.400		6.10	21.50		27.60	38.50
5200	#1		16	.500		6.95	26.50		33.45	47.50
5400	1/0		13	.615		7.65	33		40.65	57.50
5420	2/0		12	.667		8	35.50		43.50	62
5440	3/0		11	.727		9.55	39		48.55	68.50
5460	4/0		10	.800		10.15	42.50		52.65	75
5480	250 kcmil		9	.889		13.55	47.50		61.05	86
5500	300 kcmil		8.50	.941		14.85	50		64.85	91.50
5520	350 kcmil		8	1		15.80	53.50		69.30	97.50
5540	400 kcmil		7.30	1.096		22.50	58.50		81	113
5560	500 kcmil		6.20	1.290		23	69		92	129
5580	600 kcmil		5.50	1.455		33	77.50		110.50	153
5600	700 kcmil		4.50	1.778		43.50	95		138.50	190
5620	750 kcmil		4	2		43	107		150	208
7000	Compression equipment adapter, aluminum wire, #6		30	.267		10.30	14.25		24.55	33
7020	#4		27	.296		9.95	15.80		25.75	34.50
7040	#2		24	.333		10.40	17.80		28.20	38
7060	#1		20	.400		12.30	21.50		33.80	45.50
7080	1/0		18	.444		13.30	23.50		36.80	50
7100	2/0		15	.533		23	28.50		51.50	67.50
7140	4/0		11	.727		23	39		62	83.50
7160	250 kcmil		9	.889		26.50	47.50		74	100
7180	300 kcmil		8	1		27	53.50		80.50	110
7200	350 kcmil		7	1.143		30.50	61		91.50	125
7220	400 kcmil		6.50	1.231		39.50	65.50		105	142
7240	500 kcmil		6	1.333		41	71		112	152
7260	600 kcmil		5.80	1.379		48.50	73.50		122	163
7280	750 kcmil		5.20	1.538		50.50	82		132.50	179
8000	Compression tool, hand					1,175			1,175	1,300
8100	Hydraulic					1,700			1,700	1,875
8500	Hydraulic dies					320			320	355

26 05 19 – Low-Voltage Electrical Power Conductors and Cables

26 05 19.50 Mineral Insulated Cable	Crew	Daily Output	Labor-Hours	Unit	Material	2014 Bare Costs Labor	Equipment	Total	Total Incl O&P
0010 **MINERAL INSULATED CABLE** 600 volt									
0100 1 conductor, #12	1 Elec	1.60	5	C.L.F.	360	267		627	795
0200 #10		1.60	5		465	267		732	910
0400 #8		1.50	5.333		515	285		800	995
0500 #6		1.40	5.714		615	305		920	1,125
0600 #4	2 Elec	2.40	6.667		825	355		1,180	1,450
0800 #2		2.20	7.273		1,175	390		1,565	1,850
0900 #1		2.10	7.619		1,350	405		1,755	2,075
1000 1/0		2	8		1,575	425		2,000	2,375
1100 2/0		1.90	8.421		1,875	450		2,325	2,750
1200 3/0		1.80	8.889		2,250	475		2,725	3,175
1400 4/0		1.60	10		2,600	535		3,135	3,675
1410 250 kcmil	3 Elec	2.40	10		2,950	535		3,485	4,025
1420 350 kcmil		1.95	12.308		3,350	655		4,005	4,675
1430 500 kcmil		1.95	12.308		4,325	655		4,980	5,725
1500 2 conductor, #12	1 Elec	1.40	5.714		750	305		1,055	1,275
1600 #10		1.20	6.667		925	355		1,280	1,550
1800 #8		1.10	7.273		1,075	390		1,465	1,750
2000 #6		1.05	7.619		1,450	405		1,855	2,200
2100 #4	2 Elec	2	8		1,975	425		2,400	2,825
2200 3 conductor, #12	1 Elec	1.20	6.667		940	355		1,295	1,550
2400 #10		1.10	7.273		1,075	390		1,465	1,750
2600 #8		1.05	7.619		1,325	405		1,730	2,075
2800 #6		1	8		1,750	425		2,175	2,575
3000 #4	2 Elec	1.80	8.889		2,200	475		2,675	3,125
3100 4 conductor, #12	1 Elec	1.20	6.667		990	355		1,345	1,625
3200 #10		1.10	7.273		1,200	390		1,590	1,875
3400 #8		1	8		1,550	425		1,975	2,375
3600 #6		.90	8.889		2,025	475		2,500	2,925
3620 7 conductor, #12		1.10	7.273		1,275	390		1,665	1,975
3640 #10		1	8		1,675	425		2,100	2,475
3800 Terminations, 600 volt, 1 conductor, #12		8	1	Ea.	15.90	53.50		69.40	97.50
4000 #10		7.60	1.053		15.90	56		71.90	102
4100 #8		7.30	1.096		15.90	58.50		74.40	105
4200 #6		6.70	1.194		15.90	63.50		79.40	113
4400 #4		6.20	1.290		15.90	69		84.90	121
4600 #2		5.70	1.404		24	75		99	139
4800 #1		5.30	1.509		24	80.50		104.50	148
5000 1/0		5	1.600		24	85.50		109.50	155
5100 2/0		4.70	1.702		24	91		115	163
5200 3/0		4.30	1.860		24	99.50		123.50	176
5400 4/0		4	2		53	107		160	219
5410 250 kcmil		4	2		53	107		160	219
5420 350 kcmil		4	2		83.50	107		190.50	252
5430 500 kcmil		4	2		83.50	107		190.50	252
5500 2 conductor, #12		6.70	1.194		15.90	63.50		79.40	113
5600 #10		6.40	1.250		24	66.50		90.50	127
5800 #8		6.20	1.290		24	69		93	130
6000 #6		5.70	1.404		24	75		99	139
6200 #4		5.30	1.509		53	80.50		133.50	180
6400 3 conductor, #12		5.70	1.404		24	75		99	139
6500 #10		5.50	1.455		24	77.50		101.50	143

26 05 Common Work Results for Electrical

26 05 19 – Low-Voltage Electrical Power Conductors and Cables

26 05 19.50 Mineral Insulated Cable

		Crew	Daily Output	Labor-Hours	Unit	Material	2014 Bare Costs Labor	Equipment	Total	Total Incl O&P
6600	#8	1 Elec	5.20	1.538	Ea.	24	82		106	150
6800	#6		4.80	1.667		24	89		113	160
7200	#4		4.60	1.739		53	93		146	198
7400	4 conductor, #12		4.60	1.739		26.50	93		119.50	168
7500	#10		4.40	1.818		26.50	97		123.50	174
7600	#8		4.20	1.905		26.50	102		128.50	181
8400	#6		4	2		56	107		163	222
8500	7 conductor, #12		3.50	2.286		26.50	122		148.50	212
8600	#10	↓	3	2.667		57.50	142		199.50	277
8800	Crimping tool, plier type					59.50			59.50	65.50
9000	Stripping tool					224			224	247
9200	Hand vise				↓	67			67	73.50

26 05 19.55 Non-Metallic Sheathed Cable

		Crew	Daily Output	Labor-Hours	Unit	Material	2014 Bare Costs Labor	Equipment	Total	Total Incl O&P
0010	**NON-METALLIC SHEATHED CABLE** 600 volt									
0100	Copper with ground wire, (Romex)									
0150	#14, 2 conductor	1 Elec	2.70	2.963	C.L.F.	25.50	158		183.50	265
0200	3 conductor		2.40	3.333		35.50	178		213.50	305
0220	4 conductor		2.20	3.636		51	194		245	345
0250	#12, 2 conductor		2.50	3.200		38.50	171		209.50	299
0300	3 conductor		2.20	3.636		54	194		248	350
0320	4 conductor		2	4		79.50	213		292.50	410
0350	#10, 2 conductor		2.20	3.636		60	194		254	355
0400	3 conductor		1.80	4.444		85.50	237		322.50	450
0420	4 conductor		1.60	5		123	267		390	535
0430	#8, 2 conductor		1.60	5		94.50	267		361.50	505
0450	3 conductor		1.50	5.333		138	285		423	575
0500	#6, 3 conductor	↓	1.40	5.714		224	305		529	700
0520	#4, 3 conductor	2 Elec	2.40	6.667		510	355		865	1,100
0540	#2, 3 conductor	"	2.20	7.273	↓	760	390		1,150	1,425
0550	SE type SER aluminum cable, 3 RHW and									
0600	1 bare neutral, 3 #8 & 1 #8	1 Elec	1.60	5	C.L.F.	137	267		404	550
0650	3 #6 & 1 #6	"	1.40	5.714		156	305		461	625
0700	3 #4 & 1 #6	2 Elec	2.40	6.667		144	355		499	690
0750	3 #2 & 1 #4		2.20	7.273		257	390		647	860
0800	3 #1/0 & 1 #2		2	8		390	425		815	1,075
0850	3 #2/0 & 1 #1		1.80	8.889		460	475		935	1,225
0900	3 #4/0 & 1 #2/0		1.60	10		655	535		1,190	1,525
1000	URD - triplex underground distribution cable, alum. 2 #4 + #4 neutral		2.80	5.714		101	305		406	565
1010	2 #2 + #4 neutral		2.65	6.038		123	320		443	615
1020	2 #2 + #2 neutral		2.55	6.275		125	335		460	640
1030	2 1/0 + #2 neutral		2.40	6.667		155	355		510	700
1040	2 1/0 + 1/0 neutral		2.30	6.957		170	370		540	740
1050	2 2/0 + #1 neutral		2.20	7.273		183	390		573	780
1060	2 2/0 + 2/0 neutral		2.10	7.619		198	405		603	825
1070	2 3/0 + 1/0 neutral		1.95	8.205		207	440		647	885
1080	2 3/0 + 3/0 neutral		1.95	8.205		237	440		677	915
1090	2 4/0 + 2/0 neutral		1.85	8.649		252	460		712	970
1100	2 4/0 + 4/0 neutral	↓	1.85	8.649		280	460		740	1,000
1450	UF underground feeder cable, copper with ground, #14, 2 conductor	1 Elec	4	2		29.50	107		136.50	193
1500	#12, 2 conductor		3.50	2.286		44.50	122		166.50	232
1550	#10, 2 conductor		3	2.667		69.50	142		211.50	290
1600	#14, 3 conductor	↓	3.50	2.286	↓	41	122		163	228

26 05 19 – Low-Voltage Electrical Power Conductors and Cables

26 05 19.55 Non-Metallic Sheathed Cable

		Crew	Daily Output	Labor-Hours	Unit	Material	2014 Bare Costs Labor	Equipment	Total	Total Incl O&P
1650	#12, 3 conductor	1 Elec	3	2.667	C.L.F.	62	142		204	281
1700	#10, 3 conductor		2.50	3.200		97	171		268	365
1710	#8, 3 conductor		2	4		160	213		373	495
1720	#6, 3 conductor		1.80	4.444		258	237		495	640
2400	SEU service entrance cable, copper 2 conductors, #8 + #8 neutral		1.50	5.333		119	285		404	555
2600	#6 + #8 neutral		1.30	6.154		176	330		506	685
2800	#6 + #6 neutral		1.30	6.154		200	330		530	710
3000	#4 + #6 neutral	2 Elec	2.20	7.273		277	390		667	885
3200	#4 + #4 neutral		2.20	7.273		310	390		700	925
3400	#3 + #5 neutral		2.10	7.619		365	405		770	1,000
3600	#3 + #3 neutral		2.10	7.619		390	405		795	1,050
3800	#2 + #4 neutral		2	8		430	425		855	1,100
4000	#1 + #1 neutral		1.90	8.421		485	450		935	1,200
4200	1/0 + 1/0 neutral		1.80	8.889		780	475		1,255	1,575
4400	2/0 + 2/0 neutral		1.70	9.412		975	500		1,475	1,825
4600	3/0 + 3/0 neutral		1.60	10		1,225	535		1,760	2,150
4620	4/0 + 4/0 neutral		1.45	11.034		1,375	590		1,965	2,375
4800	Aluminum 2 conductors, #8 + #8 neutral	1 Elec	1.60	5		113	267		380	525
5000	#6 + #6 neutral	"	1.40	5.714		114	305		419	580
5100	#4 + #6 neutral	2 Elec	2.50	6.400		134	340		474	655
5200	#4 + #4 neutral		2.40	6.667		146	355		501	690
5300	#2 + #4 neutral		2.30	6.957		179	370		549	750
5400	#2 + #2 neutral		2.20	7.273		196	390		586	795
5450	1/0 + #2 neutral		2.10	7.619		283	405		688	920
5500	1/0 + 1/0 neutral		2	8		298	425		723	965
5550	2/0 + #1 neutral		1.90	8.421		320	450		770	1,025
5600	2/0 + 2/0 neutral		1.80	8.889		340	475		815	1,075
5800	3/0 + 1/0 neutral		1.70	9.412		375	500		875	1,150
6000	3/0 + 3/0 neutral		1.70	9.412		410	500		910	1,200
6200	4/0 + 2/0 neutral		1.60	10		440	535		975	1,275
6400	4/0 + 4/0 neutral		1.60	10		480	535		1,015	1,325
6500	Service entrance cap for copper SEU									
6600	100 amp	1 Elec	12	.667	Ea.	10.35	35.50		45.85	64.50
6700	150 amp		10	.800		18.90	42.50		61.40	85
6800	200 amp		8	1		29	53.50		82.50	112

26 05 19.70 Portable Cord

		Crew	Daily Output	Labor-Hours	Unit	Material	2014 Bare Costs Labor	Equipment	Total	Total Incl O&P
0010	**PORTABLE CORD** 600 volt									
0100	Type SO, #18, 2 conductor	1 Elec	980	.008	L.F.	.48	.44		.92	1.18
0110	3 conductor		980	.008		.59	.44		1.03	1.30
0120	#16, 2 conductor		840	.010		.56	.51		1.07	1.38
0130	3 conductor		840	.010		.81	.51		1.32	1.65
0140	4 conductor		840	.010		1.13	.51		1.64	2
0240	#14, 2 conductor		840	.010		.93	.51		1.44	1.78
0250	3 conductor		840	.010		1.15	.51		1.66	2.03
0260	4 conductor		840	.010		1.38	.51		1.89	2.28
0280	#12, 2 conductor		840	.010		1.31	.51		1.82	2.20
0290	3 conductor		840	.010		1.69	.51		2.20	2.62
0300	4 conductor		840	.010		2.01	.51		2.52	2.97
0320	#10, 2 conductor		765	.010		1.91	.56		2.47	2.94
0330	3 conductor		765	.010		2.31	.56		2.87	3.38
0340	4 conductor		765	.010		2.61	.56		3.17	3.71
0360	#8, 2 conductor		555	.014		3.87	.77		4.64	5.40

26 05 19 – Low-Voltage Electrical Power Conductors and Cables

26 05 19.70 Portable Cord		Crew	Daily Output	Labor-Hours	Unit	Material	2014 Bare Costs Labor	Equipment	Total	Total Incl O&P
0370	3 conductor	1 Elec	540	.015	L.F.	3.39	.79		4.18	4.91
0380	4 conductor		525	.015		3.98	.81		4.79	5.60
0400	#6, 2 conductor		525	.015		3.17	.81		3.98	4.71
0410	3 conductor		490	.016		5.15	.87		6.02	7
0420	4 conductor		415	.019		5.65	1.03		6.68	7.80
0440	#4, 2 conductor	2 Elec	830	.019		5.35	1.03		6.38	7.45
0450	3 conductor		700	.023		7.15	1.22		8.37	9.75
0460	4 conductor		660	.024		10.35	1.29		11.64	13.35
0480	#2, 2 conductor		450	.036		8.40	1.90		10.30	12.10
0490	3 conductor		350	.046		8.40	2.44		10.84	12.90
0500	4 conductor		280	.057		14	3.05		17.05	19.95

26 05 19.75 Modular Flexible Wiring System		Crew	Daily Output	Labor-Hours	Unit	Material	2014 Bare Costs Labor	Equipment	Total	Total Incl O&P
0010	**MODULAR FLEXIBLE WIRING SYSTEM**									
0020	Commercial system grid ceiling									
0100	Conversion Module	1 Elec	32	.250	Ea.	8.30	13.35		21.65	29
0120	Fixture cable, for fixture to fixture, 3 conductor, 15' long		40	.200		33	10.65		43.65	52.50
0150	Extender cable, 3 conductor, 15' long		48	.167		32.50	8.90		41.40	49
0200	Switch drop, 1 level, 9' long		32	.250		33	13.35		46.35	56
0220	2 level, 9' long		32	.250		30	13.35		43.35	53.50
0250	Power tee, 9' long		32	.250		37	13.35		50.35	60.50
1020	Industrial system open ceiling									
1100	Converter, interface between hardwiring and modular wiring	1 Elec	32	.250	Ea.	15.65	13.35		29	37
1120	Fixture cable, for fixture to fixture, 3 conductor, 21' long		32	.250		100	13.35		113.35	130
1125	3 conductor, 25' long		24	.333		116	17.80		133.80	154
1130	3 conductor, 31' long		19	.421		103	22.50		125.50	147
1150	Fixture cord drop, 10' long		40	.200		27.50	10.65		38.15	46.50

26 05 19.90 Wire			Crew	Daily Output	Labor-Hours	Unit	Material	2014 Bare Costs Labor	Equipment	Total	Total Incl O&P
0010	**WIRE**	R260519-90									
0020	600 volt, copper type THW, solid, #14		1 Elec	13	.615	C.L.F.	8.30	33		41.30	58
0030	#12			11	.727		12.50	39		51.50	72
0040	#10			10	.800		19.80	42.50		62.30	86
0050	Stranded, #14	R260519-91		13	.615		9.40	33		42.40	59.50
0100	#12			11	.727		14.45	39		53.45	74
0120	#10	R260519-92		10	.800		23	42.50		65.50	89
0140	#8			8	1		37.50	53.50		91	122
0160	#6	R260519-93		6.50	1.231		64	65.50		129.50	169
0180	#4		2 Elec	10.60	1.509		101	80.50		181.50	232
0200	#3	R260519-94		10	1.600		126	85.50		211.50	267
0220	#2			9	1.778		160	95		255	320
0240	#1	R260533-22		8	2		202	107		309	380
0260	1/0			6.60	2.424		251	129		380	470
0280	2/0			5.80	2.759		315	147		462	565
0300	3/0			5	3.200		395	171		566	690
0350	4/0			4.40	3.636		505	194		699	845
0400	250 kcmil		3 Elec	6	4		590	213		803	970
0420	300 kcmil			5.70	4.211		705	225		930	1,100
0450	350 kcmil			5.40	4.444		860	237		1,097	1,300
0480	400 kcmil			5.10	4.706		990	251		1,241	1,475
0490	500 kcmil			4.80	5		1,175	267		1,442	1,675
0500	600 kcmil			3.90	6.154		1,400	330		1,730	2,050
0510	750 kcmil			3.30	7.273		1,975	390		2,365	2,750
0520	1000 kcmil			2.70	8.889		2,825	475		3,300	3,825

26 05 19 – Low-Voltage Electrical Power Conductors and Cables

26 05 19.90 Wire	Crew	Daily Output	Labor-Hours	Unit	Material	2014 Bare Costs Labor	Equipment	Total	Total Incl O&P
0540	600 volt, aluminum type THHN, stranded, #6	1 Elec	8	1	C.L.F.	37.50	53.50		91	121
0560	#4	2 Elec	13	1.231		46.50	65.50		112	150
0580	#2		10.60	1.509		63	80.50		143.50	190
0600	#1		9	1.778		92	95		187	243
0620	1/0		8	2		110	107		217	281
0640	2/0		7.20	2.222		130	119		249	320
0680	3/0		6.60	2.424		162	129		291	370
0700	4/0		6.20	2.581		180	138		318	405
0720	250 kcmil	3 Elec	8.70	2.759		219	147		366	460
0740	300 kcmil		8.10	2.963		305	158		463	570
0760	350 kcmil		7.50	3.200		310	171		481	595
0780	400 kcmil		6.90	3.478		360	186		546	675
0800	500 kcmil		6	4		395	213		608	755
0850	600 kcmil		5.70	4.211		505	225		730	890
0880	700 kcmil		5.10	4.706		580	251		831	1,025
0900	750 kcmil		4.80	5		600	267		867	1,050
0910	1000 kcmil		3.78	6.349		890	340		1,230	1,475
0920	600 volt, copper type THWN-THHN, solid, #14	1 Elec	13	.615		8.95	33		41.95	59
0940	#12		11	.727		13.55	39		52.55	73
0960	#10		10	.800		21.50	42.50		64	87.50
1000	Stranded, #14		13	.615		10.25	33		43.25	60.50
1200	#12		11	.727		15.70	39		54.70	75.50
1250	#10		10	.800		24.50	42.50		67	90.50
1300	#8		8	1		38	53.50		91.50	122
1350	#6		6.50	1.231		65	65.50		130.50	170
1400	#4	2 Elec	10.60	1.509		103	80.50		183.50	234
1450	#3		10	1.600		129	85.50		214.50	270
1500	#2		9	1.778		161	95		256	320
1550	#1		8	2		212	107		319	395
1600	1/0		6.60	2.424		254	129		383	475
1650	2/0		5.80	2.759		320	147		467	570
1700	3/0		5	3.200		400	171		571	695
2000	4/0		4.40	3.636		500	194		694	840
2200	250 kcmil	3 Elec	6	4		605	213		818	985
2400	300 kcmil		5.70	4.211		730	225		955	1,150
2600	350 kcmil		5.40	4.444		855	237		1,092	1,300
2700	400 kcmil		5.10	4.706		960	251		1,211	1,425
2800	500 kcmil		4.80	5		1,175	267		1,442	1,700
2802	600 kcmil		3.90	6.154		1,525	330		1,855	2,175
2804	750 kcmil		3.30	7.273		2,525	390		2,915	3,350
2805	1000 kcmil		1.94	12.371		3,350	660		4,010	4,700
2900	600 volt, copper type XHHW, solid, #14	1 Elec	13	.615		13.30	33		46.30	63.50
2920	#12		11	.727		21	39		60	81
2940	#10		10	.800		32.50	42.50		75	100
3000	Stranded, #14		13	.615		12.75	33		45.75	63
3020	#12		11	.727		19.10	39		58.10	79
3040	#10		10	.800		29	42.50		71.50	96
3060	#8		8	1		44.50	53.50		98	129
3080	#6		6.50	1.231		73.50	65.50		139	180
3100	#4	2 Elec	10.60	1.509		113	80.50		193.50	246
3120	#2		9	1.778		177	95		272	335
3140	#1		8	2		242	107		349	425
3160	1/0		6.60	2.424		300	129		429	525

26 05 19 – Low-Voltage Electrical Power Conductors and Cables

26 05 19.90 Wire		Crew	Daily Output	Labor-Hours	Unit	Material	2014 Bare Costs Labor	Equipment	Total	Total Incl O&P
3180	2/0	2 Elec	5.80	2.759	C.L.F.	375	147		522	635
3200	3/0		5	3.200		470	171		641	775
3220	4/0		4.40	3.636		590	194		784	940
3240	250 kcmil	3 Elec	6	4		610	213		823	990
3260	300 kcmil		5.70	4.211		700	225		925	1,100
3280	350 kcmil		5.40	4.444		800	237		1,037	1,225
3300	400 kcmil		5.10	4.706		915	251		1,166	1,375
3320	500 kcmil		4.80	5		1,100	267		1,367	1,600
3340	600 kcmil		3.90	6.154		1,400	330		1,730	2,050
3360	750 kcmil		3.30	7.273		2,375	390		2,765	3,200
3380	1000 kcmil		2.40	10		3,075	535		3,610	4,200
5020	600 volt, aluminum type XHHW, stranded, #6	1 Elec	8	1		35.50	53.50		89	119
5040	#4	2 Elec	13	1.231		44	65.50		109.50	147
5060	#2		10.60	1.509		59.50	80.50		140	187
5080	#1		9	1.778		87	95		182	238
5100	1/0		8	2		104	107		211	275
5120	2/0		7.20	2.222		123	119		242	315
5140	3/0		6.60	2.424		153	129		282	365
5160	4/0		6.20	2.581		170	138		308	395
5180	250 kcmil	3 Elec	8.70	2.759		208	147		355	450
5200	300 kcmil		8.10	2.963		287	158		445	550
5220	350 kcmil		7.50	3.200		292	171		463	575
5240	400 kcmil		6.90	3.478		340	186		526	655
5260	500 kcmil		6	4		375	213		588	735
5280	600 kcmil		5.70	4.211		475	225		700	860
5300	700 kcmil		5.40	4.444		550	237		787	960
5320	750 kcmil		5.10	4.706		555	251		806	985
5340	1000 kcmil		3.60	6.667		830	355		1,185	1,450
5390	600 volt, copper type XLPE-USE (RHW), solid, #14	1 Elec	12	.667		15.20	35.50		50.70	69.50
5400	#12		11	.727		21	39		60	81
5420	#10		10	.800		31	42.50		73.50	98
5440	Stranded, #14		13	.615		20.50	33		53.50	72
5460	#12		11	.727		25	39		64	85
5480	#10		10	.800		37	42.50		79.50	105
5500	#8		8	1		51	53.50		104.50	136
5520	#6		6.50	1.231		85.50	65.50		151	193
5540	#4	2 Elec	10.60	1.509		129	80.50		209.50	263
5560	#2		9	1.778		210	95		305	375
5580	#1		8	2		283	107		390	470
5600	1/0		6.60	2.424		355	129		484	585
5620	2/0		5.80	2.759		445	147		592	710
5640	3/0		5	3.200		555	171		726	865
5660	4/0		4.40	3.636		635	194		829	990
5680	250 kcmil	3 Elec	6	4		670	213		883	1,050
5700	300 kcmil		5.70	4.211		825	225		1,050	1,250
5720	350 kcmil		5.40	4.444		935	237		1,172	1,375
5740	400 kcmil		5.10	4.706		1,075	251		1,326	1,550
5760	500 kcmil		4.80	5		1,250	267		1,517	1,775
5780	600 kcmil		3.90	6.154		1,500	330		1,830	2,150
5800	750 kcmil		3.30	7.273		2,675	390		3,065	3,525
5820	1000 kcmil		2.70	8.889		3,550	475		4,025	4,600
5840	600 volt, aluminum type XLPE-USE (RHW), stranded, #6	1 Elec	8	1		43	53.50		96.50	128
5860	#4	2 Elec	13	1.231		49.50	65.50		115	153

26 05 19 – Low-Voltage Electrical Power Conductors and Cables

26 05 19.90 Wire		Crew	Daily Output	Labor-Hours	Unit	Material	2014 Bare Costs Labor	Equipment	Total	Total Incl O&P
5880	#2	2 Elec	10.60	1.509	C.L.F.	68.50	80.50		149	197
5900	#1		9	1.778		95	95		190	247
5920	1/0		8	2		118	107		225	289
5940	2/0		7.20	2.222		138	119		257	330
5960	3/0		6.60	2.424		162	129		291	375
5980	4/0		6.20	2.581		180	138		318	405
6000	250 kcmil	3 Elec	8.70	2.759		245	147		392	490
6020	300 kcmil		8.10	2.963		320	158		478	585
6040	350 kcmil		7.50	3.200		325	171		496	615
6060	400 kcmil		6.90	3.478		395	186		581	715
6080	500 kcmil		6	4		435	213		648	800
6100	600 kcmil		5.70	4.211		560	225		785	950
6110	700 kcmil		5.40	4.444		645	237		882	1,075
6120	750 kcmil		5.10	4.706		650	251		901	1,100

26 05 23 – Control-Voltage Electrical Power Cables

26 05 23.10 Control Cable

		Crew	Daily Output	Labor-Hours	Unit	Material	2014 Bare Costs Labor	Equipment	Total	Total Incl O&P
0010	**CONTROL CABLE**									
0020	600 volt, copper, #14 THWN wire with PVC jacket, 2 wires	1 Elec	9	.889	C.L.F.	26.50	47.50		74	100
0030	3 wires		8	1		37	53.50		90.50	121
0100	4 wires		7	1.143		45.50	61		106.50	142
0150	5 wires		6.50	1.231		56	65.50		121.50	160
0200	6 wires		6	1.333		85.50	71		156.50	200
0300	8 wires		5.30	1.509		98	80.50		178.50	229
0400	10 wires		4.80	1.667		123	89		212	269
0500	12 wires		4.30	1.860		144	99.50		243.50	305
0600	14 wires		3.80	2.105		172	112		284	355
0700	16 wires		3.50	2.286		177	122		299	380
0800	18 wires		3.30	2.424		193	129		322	405
0810	19 wires		3.10	2.581		218	138		356	445
0900	20 wires		3	2.667		236	142		378	475
1000	22 wires		2.80	2.857		243	152		395	495

26 05 23.20 Special Wires and Fittings

		Crew	Daily Output	Labor-Hours	Unit	Material	2014 Bare Costs Labor	Equipment	Total	Total Incl O&P
0010	**SPECIAL WIRES & FITTINGS**									
0100	Fixture TFFN 600 volt 90°C stranded, #18	1 Elec	13	.615	C.L.F.	10.75	33		43.75	61
0150	#16		13	.615		14.55	33		47.55	65
0500	Thermostat, jacket non-plenum, twisted, #18-2 conductor		8	1		15	53.50		68.50	96.50
0550	#18-3 conductor		7	1.143		22	61		83	116
0600	#18-4 conductor		6.50	1.231		31	65.50		96.50	133
0650	#18-5 conductor		6	1.333		34	71		105	144
0700	#18-6 conductor		5.50	1.455		40.50	77.50		118	161
0750	#18-7 conductor		5	1.600		41	85.50		126.50	173
0800	#18-8 conductor		4.80	1.667		53	89		142	191
2460	Tray cable, type TC, copper, #16-2 conductor		9.40	.851		21	45.50		66.50	91
2464	#16-3 conductor		8.40	.952		27	51		78	106
2468	#16-4 conductor		7.30	1.096		32.50	58.50		91	123
2472	#16-5 conductor		6.70	1.194		40.50	63.50		104	141
2476	#16-7 conductor		5.50	1.455		55.50	77.50		133	177
2480	#16-9 conductor		5	1.600		69	85.50		154.50	204
2484	#16-12 conductor	2 Elec	8.80	1.818		90	97		187	244
2488	#16-15 conductor		7.20	2.222		121	119		240	310
2492	#16-19 conductor		6.60	2.424		149	129		278	360
2496	#16-25 conductor		6.40	2.500		211	133		344	430

26 05 Common Work Results for Electrical

26 05 23 – Control-Voltage Electrical Power Cables

26 05 23.20 Special Wires and Fittings		Crew	Daily Output	Labor-Hours	Unit	Material	2014 Bare Costs Labor	Equipment	Total	Total Incl O&P
2500	#14-2 conductor	1 Elec	9	.889	C.L.F.	25.50	47.50		73	99
2520	#14-3 conductor		8	1		35.50	53.50		89	119
2540	#14-4 conductor		7	1.143		44	61		105	140
2560	#14-5 conductor		6.50	1.231		54	65.50		119.50	158
2564	#14-7 conductor		5.30	1.509		82.50	80.50		163	212
2568	#14-9 conductor	2 Elec	9.60	1.667		101	89		190	245
2572	#14-12 conductor		8.60	1.860		139	99.50		238.50	300
2576	#14-15 conductor		7.20	2.222		166	119		285	360
2578	#14-19 conductor		6.20	2.581		210	138		348	435
2582	#14-25 conductor		5.30	3.019		274	161		435	540
2590	#12-2 conductor	1 Elec	8.40	.952		36.50	51		87.50	116
2592	#12-3 conductor		7.60	1.053		50.50	56		106.50	140
2594	#12-4 conductor		6.60	1.212		67	64.50		131.50	171
2596	#12-5 conductor		6.20	1.290		82.50	69		151.50	194
2598	#12-7 conductor	2 Elec	10.40	1.538		122	82		204	257
2602	#12-9 conductor		8.80	1.818		157	97		254	320
2604	#12-12 conductor		7.80	2.051		204	109		313	390
2606	#12-15 conductor		6.80	2.353		250	126		376	465
2608	#12-19 conductor		6	2.667		305	142		447	550
2610	#12-25 conductor	3 Elec	7.70	3.117		415	166		581	705
2618	#10-2 conductor	1 Elec	8	1		54.50	53.50		108	140
2622	#10-3 conductor	"	7.30	1.096		80.50	58.50		139	176
2624	#10-4 conductor	2 Elec	12.80	1.250		105	66.50		171.50	215
2626	#10-5 conductor		11.80	1.356		130	72.50		202.50	250
2628	#10-7 conductor		9.40	1.702		187	91		278	340
2630	#10-9 conductor		8.40	1.905		231	102		333	405
2632	#10-12 conductor		7.20	2.222		269	119		388	475
2640	300 V, copper braided shield, PVC jacket									
2650	2 conductor #18 stranded	1 Elec	7	1.143	C.L.F.	56	61		117	153
2660	3 conductor #18 stranded	"	6	1.333	"	79	71		150	193
3000	Strain relief grip for cable									
3050	Cord, top, #12-3	1 Elec	40	.200	Ea.	12.45	10.65		23.10	29.50
3060	#12-4		40	.200		12.45	10.65		23.10	29.50
3070	#12-5		39	.205		13.20	10.95		24.15	31
3100	#10-3		39	.205		13.20	10.95		24.15	31
3110	#10-4		38	.211		13.20	11.25		24.45	31.50
3120	#10-5		38	.211		14.35	11.25		25.60	32.50
3200	Bottom, #12-3		40	.200		31	10.65		41.65	50
3210	#12-4		40	.200		31	10.65		41.65	50
3220	#12-5		39	.205		31.50	10.95		42.45	51
3230	#10-3		39	.205		31.50	10.95		42.45	51
3300	#10-4		38	.211		31.50	11.25		42.75	51.50
3310	#10-5		38	.211		39.50	11.25		50.75	60
3400	Cable ties, standard, 4" length		190	.042		.17	2.25		2.42	3.55
3410	7" length		160	.050		.19	2.67		2.86	4.20
3420	14.5" length		90	.089		.46	4.74		5.20	7.60
3430	Heavy, 14.5" length		80	.100		.64	5.35		5.99	8.70

26 05 26 – Grounding and Bonding for Electrical Systems

26 05 26.80 Grounding

			Crew	Daily Output	Labor-Hours	Unit	Material	2014 Bare Costs Labor	Equipment	Total	Total Incl O&P
0010	**GROUNDING**	R260526-80									
0030	Rod, copper clad, 8' long, 1/2" diameter		1 Elec	5.50	1.455	Ea.	20.50	77.50		98	139
0040	5/8" diameter			5.50	1.455		19.50	77.50		97	138

125

26 05 26 – Grounding and Bonding for Electrical Systems

26 05 26.80 Grounding		Crew	Daily Output	Labor-Hours	Unit	Material	2014 Bare Costs Labor	Equipment	Total	Total Incl O&P
0050	3/4" diameter	1 Elec	5.30	1.509	Ea.	34	80.50		114.50	159
0080	10' long, 1/2" diameter		4.80	1.667		22.50	89		111.50	158
0090	5/8" diameter		4.60	1.739		24.50	93		117.50	166
0100	3/4" diameter		4.40	1.818		39	97		136	188
0130	15' long, 3/4" diameter		4	2		63.50	107		170.50	230
0150	Coupling, bronze, 1/2" diameter					4.06			4.06	4.47
0160	5/8" diameter					5.35			5.35	5.90
0170	3/4" diameter					13.25			13.25	14.60
0190	Drive studs, 1/2" diameter					10.65			10.65	11.70
0210	5/8" diameter					15.25			15.25	16.75
0220	3/4" diameter					17.80			17.80	19.60
0230	Clamp, bronze, 1/2" diameter	1 Elec	32	.250		4.48	13.35		17.83	25
0240	5/8" diameter		32	.250		5.35	13.35		18.70	26
0250	3/4" diameter		32	.250		6.40	13.35		19.75	27
0260	Wire ground bare armored, #8-1 conductor		2	4	C.L.F.	89.50	213		302.50	420
0270	#6-1 conductor		1.80	4.444		107	237		344	475
0280	#4-1 conductor		1.60	5		146	267		413	560
0320	Bare copper wire, #14 solid		14	.571		8.50	30.50		39	55
0330	#12		13	.615		13.15	33		46.15	63.50
0340	#10		12	.667		21	35.50		56.50	76
0350	#8		11	.727		33.50	39		72.50	95
0360	#6		10	.800		59	42.50		101.50	129
0370	#4		8	1		102	53.50		155.50	192
0380	#2		5	1.600		153	85.50		238.50	296
0390	Bare copper wire, stranded, #8		11	.727		35	39		74	96.50
0400	#6		10	.800		61	42.50		103.50	131
0450	#4	2 Elec	16	1		103	53.50		156.50	193
0600	#2		10	1.600		162	85.50		247.50	305
0650	#1		9	1.778		209	95		304	370
0700	1/0		8	2		231	107		338	415
0750	2/0		7.20	2.222		291	119		410	495
0800	3/0		6.60	2.424		365	129		494	600
1000	4/0		5.70	2.807		460	150		610	730
1200	250 kcmil	3 Elec	7.20	3.333		545	178		723	865
1210	300 kcmil		6.60	3.636		615	194		809	970
1220	350 kcmil		6	4		765	213		978	1,150
1230	400 kcmil		5.70	4.211		900	225		1,125	1,325
1240	500 kcmil		5.10	4.706		1,100	251		1,351	1,575
1260	750 kcmil		3.60	6.667		1,850	355		2,205	2,575
1270	1000 kcmil		3	8		2,475	425		2,900	3,350
1360	Bare aluminum, stranded, #6	1 Elec	9	.889		14.60	47.50		62.10	87
1370	#4	2 Elec	16	1		22.50	53.50		76	105
1380	#2		13	1.231		32	65.50		97.50	134
1390	#1		10.60	1.509		35.50	80.50		116	160
1400	1/0		9	1.778		40.50	95		135.50	187
1410	2/0		8	2		51	107		158	217
1420	3/0		7.20	2.222		69	119		188	253
1430	4/0		6.60	2.424		79	129		208	281
1440	250 kcmil	3 Elec	9.30	2.581		119	138		257	335
1450	300 kcmil		8.70	2.759		140	147		287	375
1460	400 kcmil		7.50	3.200		180	171		351	455
1470	500 kcmil		6.90	3.478		214	186		400	515
1480	600 kcmil		6	4		243	213		456	590

26 05 26.80 Grounding		Crew	Daily Output	Labor-Hours	Unit	Material	2014 Bare Costs Labor	Equipment	Total	Total Incl O&P
1490	700 kcmil	3 Elec	5.70	4.211	C.L.F.	267	225		492	630
1500	750 kcmil		5.10	4.706		277	251		528	680
1510	1000 kcmil		4.80	5		350	267		617	785
1800	Water pipe ground clamps, heavy duty									
2000	Bronze, 1/2" to 1" diameter	1 Elec	8	1	Ea.	23.50	53.50		77	106
2100	1-1/4" to 2" diameter		8	1		32.50	53.50		86	116
2200	2-1/2" to 3" diameter		6	1.333		49	71		120	160
2730	Exothermic weld, 4/0 wire to 1" ground rod		7	1.143		11.55	61		72.55	104
2740	4/0 wire to building steel		7	1.143		9.25	61		70.25	102
2750	4/0 wire to motor frame		7	1.143		9.25	61		70.25	102
2760	4/0 wire to 4/0 wire		7	1.143		8	61		69	100
2770	4/0 wire to #4 wire		7	1.143		8	61		69	100
2780	4/0 wire to #8 wire		7	1.143		8	61		69	100
2790	Mold, reusable, for above					138			138	152
2800	Brazed connections, #6 wire	1 Elec	12	.667		16	35.50		51.50	70.50
3000	#2 wire		10	.800		21.50	42.50		64	87.50
3100	3/0 wire		8	1		32	53.50		85.50	116
3200	4/0 wire		7	1.143		37	61		98	132
3400	250 kcmil wire		5	1.600		43	85.50		128.50	175
3600	500 kcmil wire		4	2		53	107		160	218
3700	Insulated ground wire, copper #14		13	.615	C.L.F.	10.25	33		43.25	60.50
3710	#12		11	.727		15.70	39		54.70	75.50
3720	#10		10	.800		24.50	42.50		67	90.50
3730	#8		8	1		38	53.50		91.50	122
3740	#6		6.50	1.231		65	65.50		130.50	170
3750	#4	2 Elec	10.60	1.509		103	80.50		183.50	234
3770	#2		9	1.778		161	95		256	320
3780	#1		8	2		212	107		319	395
3790	1/0		6.60	2.424		254	129		383	475
3800	2/0		5.80	2.759		320	147		467	570
3810	3/0		5	3.200		400	171		571	695
3820	4/0		4.40	3.636		500	194		694	840
3830	250 kcmil	3 Elec	6	4		605	213		818	985
3840	300 kcmil		5.70	4.211		730	225		955	1,150
3850	350 kcmil		5.40	4.444		855	237		1,092	1,300
3860	400 kcmil		5.10	4.706		960	251		1,211	1,425
3870	500 kcmil		4.80	5		1,175	267		1,442	1,700
3880	600 kcmil		3.90	6.154		1,400	330		1,730	2,050
3890	750 kcmil		3.30	7.273		1,975	390		2,365	2,750
3900	1000 kcmil		2.70	8.889		2,825	475		3,300	3,825
3960	Insulated ground wire, aluminum, #6	1 Elec	8	1		37.50	53.50		91	121
3970	#4	2 Elec	13	1.231		46.50	65.50		112	150
3980	#2		10.60	1.509		63	80.50		143.50	190
3990	#1		9	1.778		92	95		187	243
4000	1/0		8	2		110	107		217	281
4010	2/0		7.20	2.222		130	119		249	320
4020	3/0		6.60	2.424		162	129		291	370
4030	4/0		6.20	2.581		180	138		318	405
4040	250 kcmil	3 Elec	8.70	2.759		219	147		366	460
4050	300 kcmil		8.10	2.963		305	158		463	570
4060	350 kcmil		7.50	3.200		310	171		481	595
4070	400 kcmil		6.90	3.478		360	186		546	675
4080	500 kcmil		6	4		395	213		608	755

26 05 26 – Grounding and Bonding for Electrical Systems

26 05 26.80 Grounding		Crew	Daily Output	Labor-Hours	Unit	Material	2014 Bare Costs Labor	Equipment	Total	Total Incl O&P
4090	600 kcmil	3 Elec	5.70	4.211	C.L.F.	505	225		730	890
4100	700 kcmil		5.10	4.706		580	251		831	1,025
4110	750 kcmil	↓	4.80	5	↓	600	267		867	1,050
5000	Copper Electrolytic ground rod system									
5010	Includes augering hole, mixing bentonite clay,									
5020	Installing rod, and terminating ground wire									
5100	Straight Vertical type, 2" diam.									
5120	8.5' long, clamp connection	1 Elec	2.67	2.996	Ea.	760	160		920	1,075
5130	With exothermic weld connection		1.95	4.103		760	219		979	1,175
5140	10' long		2.35	3.404		890	182		1,072	1,250
5150	With exothermic weld connection		1.78	4.494		890	240		1,130	1,350
5160	12' long		2.16	3.704		1,000	198		1,198	1,400
5170	With exothermic weld connection		1.67	4.790		1,000	256		1,256	1,475
5180	20' long		1.74	4.598		1,450	245		1,695	1,975
5190	With exothermic weld connection	↓	1.40	5.714		1,750	305		2,055	2,375
5195	40' long with exothermic weld connection	2 Elec	2	8	↓	3,025	425		3,450	4,000
5200	L-Shaped, 2" diam.									
5220	4' Vert. x 10' Horz., clamp connection	1 Elec	5.33	1.501	Ea.	1,275	80		1,355	1,525
5230	With exothermic weld connection	"	3.08	2.597	"	1,275	139		1,414	1,600
5300	Protective Box at grade level, with breather slots									
5320	Round 12" long, fiberlyte	1 Elec	32	.250	Ea.	64.50	13.35		77.85	91
5330	Concrete	"	16	.500		99.50	26.50		126	150
5400	Bentonite Clay, 50# bag, 1 per 10' of rod					41			41	45
5500	Equipotential earthing bar	1 Elec	2	4	↓	175	213		388	510

26 05 29 – Hangers and Supports for Electrical Systems

26 05 29.20 Hangers

0010	HANGERS	R260533-20	Crew	Daily Output	Labor-Hours	Unit	Material	Labor	Equipment	Total	Total Incl O&P
0015	See section 22 05 29.10 for additional items										
0030	Conduit supports										
0050	Strap w/2 holes, rigid steel conduit										
0100	1/2" diameter		1 Elec	470	.017	Ea.	.15	.91		1.06	1.53
0150	3/4" diameter			440	.018		.17	.97		1.14	1.64
0200	1" diameter			400	.020		.26	1.07		1.33	1.89
0300	1-1/4" diameter			355	.023		.41	1.20		1.61	2.25
0350	1-1/2" diameter			320	.025		.47	1.33		1.80	2.52
0400	2" diameter			266	.030		.50	1.60		2.10	2.95
0500	2-1/2" diameter			160	.050		.98	2.67		3.65	5.05
0550	3" diameter			133	.060		1.27	3.21		4.48	6.20
0600	3-1/2" diameter			100	.080		2.76	4.27		7.03	9.45
0650	4" diameter			80	.100		1.72	5.35		7.07	9.90
0700	EMT, 1/2" diameter			470	.017		.16	.91		1.07	1.54
0800	3/4" diameter			440	.018		.21	.97		1.18	1.68
0850	1" diameter			400	.020		.37	1.07		1.44	2.01
0900	1-1/4" diameter			355	.023		.61	1.20		1.81	2.47
0950	1-1/2" diameter			320	.025		.65	1.33		1.98	2.72
1000	2" diameter			266	.030		.94	1.60		2.54	3.43
1100	2-1/2" diameter			160	.050		1.58	2.67		4.25	5.75
1150	3" diameter			133	.060		2.20	3.21		5.41	7.20
1200	3-1/2" diameter			100	.080		2.47	4.27		6.74	9.10
1250	4" diameter			80	.100		2.86	5.35		8.21	11.15
1400	Hanger, with bolt, 1/2" diameter			200	.040		.62	2.13		2.75	3.87
1450	3/4" diameter		↓	190	.042	↓	.64	2.25		2.89	4.06

26 05 29 – Hangers and Supports for Electrical Systems

26 05 29.20 Hangers	Crew	Daily Output	Labor-Hours	Unit	Material	2014 Bare Costs Labor	Equipment	Total	Total Incl O&P
1500	1" diameter	1 Elec	176	.045	Ea.	1.10	2.42		3.52	4.84
1550	1-1/4" diameter		160	.050		1.48	2.67		4.15	5.60
1600	1-1/2" diameter		140	.057		1.84	3.05		4.89	6.60
1650	2" diameter		130	.062		2.39	3.28		5.67	7.55
1700	2-1/2" diameter		100	.080		2.49	4.27		6.76	9.15
1750	3" diameter		64	.125		3.42	6.65		10.07	13.75
1800	3-1/2" diameter		50	.160		4.32	8.55		12.87	17.55
1850	4" diameter		40	.200		9.85	10.65		20.50	27
1900	Riser clamps, conduit, 1/2" diameter		40	.200		9.85	10.65		20.50	27
1950	3/4" diameter		36	.222		10.70	11.85		22.55	29.50
2000	1" diameter		30	.267		11.95	14.25		26.20	34.50
2100	1-1/4" diameter		27	.296		13.15	15.80		28.95	38
2150	1-1/2" diameter		27	.296		13.80	15.80		29.60	38.50
2200	2" diameter		20	.400		14.45	21.50		35.95	48
2250	2-1/2" diameter		20	.400		15.30	21.50		36.80	49
2300	3" diameter		18	.444		16.60	23.50		40.10	54
2350	3-1/2" diameter		18	.444		18.90	23.50		42.40	56.50
2400	4" diameter		14	.571	▼	22.50	30.50		53	70.50
2500	Threaded rod, painted, 1/4" diameter		260	.031	L.F.	2.09	1.64		3.73	4.76
2600	3/8" diameter		200	.040		3.40	2.13		5.53	6.95
2700	1/2" diameter		140	.057		4.43	3.05		7.48	9.45
2800	5/8" diameter		100	.080		6	4.27		10.27	13
2900	3/4" diameter	▼	60	.133	▼	8.90	7.10		16	20.50
2940	Couplings painted, 1/4" diameter				C	490			490	540
2960	3/8" diameter					565			565	625
2970	1/2" diameter					1,100			1,100	1,200
2980	5/8" diameter					1,675			1,675	1,850
2990	3/4" diameter					2,000			2,000	2,200
3000	Nuts, galvanized, 1/4" diameter					15.15			15.15	16.65
3050	3/8" diameter					22.50			22.50	25
3100	1/2" diameter					49			49	53.50
3150	5/8" diameter					108			108	119
3200	3/4" diameter					161			161	177
3250	Washers, galvanized, 1/4" diameter					15.15			15.15	16.65
3300	3/8" diameter					20.50			20.50	23
3350	1/2" diameter					33			33	36.50
3400	5/8" diameter					89			89	98
3450	3/4" diameter					131			131	144
3500	Lock washers, galvanized, 1/4" diameter					15.55			15.55	17.10
3550	3/8" diameter					17.75			17.75	19.50
3600	1/2" diameter					20.50			20.50	22.50
3650	5/8" diameter					43.50			43.50	47.50
3700	3/4" diameter				▼	82.50			82.50	91
3800	Channels, steel, 3/4" x 1-1/2", 14 Ga	1 Elec	80	.100	L.F.	4.83	5.35		10.18	13.30
3900	1-1/2" x 1-1/2", 12 Ga		70	.114		6.35	6.10		12.45	16.15
4000	1-7/8" x 1-1/2"		60	.133		22.50	7.10		29.60	35
4100	3" x 1-1/2"		50	.160		38	8.55		46.55	54.50
4110	1-5/8" x 13/16", 16 Ga.		80	.100	▼	5.45	5.35		10.80	14
4114	Trapeze channel support, 12" wide, steel, 12 Ga		8.80	.909	Ea.	34	48.50		82.50	110
4116	18" wide, steel		8.30	.964	"	37	51.50		88.50	118
4120	1-5/8" x 1-5/8", 14 Ga.		70	.114	L.F.	6.20	6.10		12.30	16
4130	1-5/8" x 7/8", 12 Ga.		70	.114		5.80	6.10		11.90	15.55
4140	1-5/8" x 1-3/8", 12 Ga.	▼	60	.133	▼	7.85	7.10		14.95	19.30

26 05 29.20 Hangers	Crew	Daily Output	Labor-Hours	Unit	Material	2014 Bare Costs Labor	Equipment	Total	Total Incl O&P
4150	1-5/8" x 1-5/8", 12 Ga.	1 Elec	50	.160	L.F.	7.30	8.55		15.85	21
4160	Flat plate fitting, 2 hole, 3-1/2", material only				Ea.	3.35			3.35	3.69
4170	3 hole, 1-7/16" x 4-1/8"					4.23			4.23	4.65
4200	Spring nuts, long, 1/4"	1 Elec	120	.067		1.37	3.56		4.93	6.80
4250	3/8"		100	.080		1.44	4.27		5.71	8
4300	1/2"		80	.100		1.56	5.35		6.91	9.70
4350	Spring nuts, short, 1/4"		120	.067		1.36	3.56		4.92	6.80
4400	3/8"		100	.080		1.40	4.27		5.67	7.95
4450	1/2"		80	.100		1.67	5.35		7.02	9.85
4500	Closure strip		200	.040	L.F.	4.49	2.13		6.62	8.15
4550	End cap		60	.133	Ea.	1.62	7.10		8.72	12.45
4600	End connector 3/4" conduit		40	.200		6.45	10.65		17.10	23
4650	Junction box, 1 channel		16	.500		51.50	26.50		78	96.50
4700	2 channel		14	.571		62	30.50		92.50	114
4750	3 channel		12	.667		72	35.50		107.50	132
4800	4 channel		10	.800		78.50	42.50		121	150
4850	Splice plate		40	.200		10.70	10.65		21.35	27.50
4900	Continuous concrete insert, 1-1/2" deep, 1' long		16	.500		19.85	26.50		46.35	62
4950	2' long		14	.571		24.50	30.50		55	72.50
5000	3' long		12	.667		22.50	35.50		58	78
5050	4' long		10	.800		34	42.50		76.50	102
5100	6' long		8	1		51	53.50		104.50	136
5150	3/4" deep, 1' long		16	.500		14.15	26.50		40.65	55.50
5200	2' long		14	.571		17.60	30.50		48.10	65
5250	3' long		12	.667		18.30	35.50		53.80	73
5300	4' long		10	.800		24.50	42.50		67	90.50
5350	6' long		8	1		36.50	53.50		90	120
5400	90° angle fitting 2-1/8" x 2-1/8"		60	.133		3.86	7.10		10.96	14.90
5450	Supports, suspension rod type, small		60	.133		21	7.10		28.10	33.50
5500	Large		40	.200		23	10.65		33.65	41
5550	Beam clamp, small		60	.133		8.55	7.10		15.65	20
5600	Large		40	.200		10.10	10.65		20.75	27
5650	U-support, small		60	.133		5.30	7.10		12.40	16.50
5700	Large		40	.200		12.30	10.65		22.95	29.50
5750	Concrete insert, cast, for up to 1/2" threaded rod		16	.500		5.20	26.50		31.70	45.50
5800	Beam clamp, 1/4" clamp, for 1/4" threaded drop rod		32	.250		2.63	13.35		15.98	23
5900	3/8" clamp, for 3/8" threaded drop rod		32	.250		6.85	13.35		20.20	27.50
6000	Strap, rigid conduit, 1/2" diameter		540	.015		1.68	.79		2.47	3.03
6050	3/4" diameter		440	.018		1.48	.97		2.45	3.08
6100	1" diameter		420	.019		2.07	1.02		3.09	3.80
6150	1-1/4" diameter		400	.020		1.86	1.07		2.93	3.65
6200	1-1/2" diameter		400	.020		2.23	1.07		3.30	4.05
6250	2" diameter		267	.030		2.40	1.60		4	5.05
6300	2-1/2" diameter		267	.030		3.23	1.60		4.83	5.95
6350	3" diameter		160	.050		3.04	2.67		5.71	7.35
6400	3-1/2" diameter		133	.060		4.12	3.21		7.33	9.35
6450	4" diameter		100	.080		4.24	4.27		8.51	11.05
6500	5" diameter		80	.100		9.55	5.35		14.90	18.50
6550	6" diameter		60	.133		18.30	7.10		25.40	30.50
6600	EMT, 1/2" diameter		540	.015		1.33	.79		2.12	2.64
6650	3/4" diameter		440	.018		1.42	.97		2.39	3.01
6700	1" diameter		420	.019		1.58	1.02		2.60	3.26
6750	1-1/4" diameter		400	.020		1.78	1.07		2.85	3.56

26 05 29.20 Hangers		Crew	Daily Output	Labor-Hours	Unit	Material	2014 Bare Costs Labor	Equipment	Total	Total Incl O&P
6800	1-1/2" diameter	1 Elec	400	.020	Ea.	2.18	1.07		3.25	4
6850	2" diameter		267	.030		2.48	1.60		4.08	5.10
6900	2-1/2" diameter		267	.030		2.64	1.60		4.24	5.30
6950	3" diameter		160	.050		2.85	2.67		5.52	7.15
6970	3-1/2" diameter		133	.060		3.20	3.21		6.41	8.30
6990	4" diameter		100	.080		3.68	4.27		7.95	10.45
7000	Clip, 1 hole for rigid conduit, 1/2" diameter		500	.016		.56	.85		1.41	1.90
7050	3/4" diameter		470	.017		.82	.91		1.73	2.26
7100	1" diameter		440	.018		1.02	.97		1.99	2.57
7150	1-1/4" diameter		400	.020		2.19	1.07		3.26	4.01
7200	1-1/2" diameter		355	.023		2.53	1.20		3.73	4.58
7250	2" diameter		320	.025		4.92	1.33		6.25	7.40
7300	2-1/2" diameter		266	.030		10.95	1.60		12.55	14.40
7350	3" diameter		160	.050		15.70	2.67		18.37	21.50
7400	3-1/2" diameter		133	.060		22.50	3.21		25.71	29.50
7450	4" diameter		100	.080		50	4.27		54.27	61.50
7500	5" diameter		80	.100		157	5.35		162.35	181
7550	6" diameter		60	.133		168	7.10		175.10	196
7820	Conduit hangers, with bolt & 12" rod, 1/2" diameter		150	.053		4.02	2.85		6.87	8.70
7830	3/4" diameter		145	.055		4.04	2.94		6.98	8.85
7840	1" diameter		135	.059		4.50	3.16		7.66	9.70
7850	1-1/4" diameter		120	.067		4.88	3.56		8.44	10.65
7860	1-1/2" diameter		110	.073		5.25	3.88		9.13	11.55
7870	2" diameter		100	.080		6.80	4.27		11.07	13.90
7880	2-1/2" diameter		80	.100		6.90	5.35		12.25	15.60
7890	3" diameter		60	.133		7.85	7.10		14.95	19.30
7900	3-1/2" diameter		45	.178		8.75	9.50		18.25	24
7910	4" diameter		35	.229		14.25	12.20		26.45	34
7920	5" diameter		30	.267		15.80	14.25		30.05	39
7930	6" diameter		25	.320		30.50	17.05		47.55	59
7950	Jay clamp, 1/2" diameter		32	.250		4.42	13.35		17.77	25
7960	3/4" diameter		32	.250		4.71	13.35		18.06	25
7970	1" diameter		32	.250		6.65	13.35		20	27.50
7980	1-1/4" diameter		30	.267		5.90	14.25		20.15	28
7990	1-1/2" diameter		30	.267		9	14.25		23.25	31.50
8000	2" diameter		30	.267		13.30	14.25		27.55	36
8010	2-1/2" diameter		28	.286		17.25	15.25		32.50	42
8020	3" diameter		28	.286		17.25	15.25		32.50	42
8030	3-1/2" diameter		25	.320		23	17.05		40.05	51
8040	4" diameter		25	.320		23	17.05		40.05	51
8050	5" diameter		20	.400		98.50	21.50		120	140
8060	6" diameter		16	.500		185	26.50		211.50	244
8070	Channels, 3/4" x 1-1/2" w/12" rods for 1/2" to 1" conduit		30	.267		10.65	14.25		24.90	33
8080	1-1/2" x 1-1/2" w/12" rods for 1-1/4" to 2" conduit		28	.286		11.35	15.25		26.60	35.50
8090	1-1/2" x 1-1/2" w/12" rods for 2-1/2" to 4" conduit		26	.308		13.95	16.40		30.35	40
8100	1-1/2" x 1-7/8" w/12" rods for 5" to 6" conduit		24	.333		53.50	17.80		71.30	85.50
8110	Beam clamp, conduit, plastic coated steel, 1/2" diam.		30	.267		21.50	14.25		35.75	45
8120	3/4" diameter		30	.267		22.50	14.25		36.75	46.50
8130	1" diameter		30	.267		25.50	14.25		39.75	49.50
8140	1-1/4" diameter		28	.286		31	15.25		46.25	57
8150	1-1/2" diameter		28	.286		38.50	15.25		53.75	65
8160	2" diameter		28	.286		53	15.25		68.25	81.50
8170	2-1/2" diameter		26	.308		54.50	16.40		70.90	84.50

26 05 29 – Hangers and Supports for Electrical Systems

26 05 29.20 Hangers		Crew	Daily Output	Labor-Hours	Unit	Material	2014 Bare Costs Labor	Equipment	Total	Total Incl O&P
8180	3" diameter	1 Elec	26	.308	Ea.	58.50	16.40		74.90	89
8190	3-1/2" diameter		23	.348		63	18.55		81.55	97.50
8200	4" diameter		23	.348		69	18.55		87.55	104
8210	5" diameter	↓	18	.444	↓	206	23.50		229.50	263
8220	Channels, plastic coated									
8250	3/4" x 1-1/2", w/12" rods for 1/2" to 1" conduit	1 Elec	28	.286	Ea.	34.50	15.25		49.75	61
8260	1-1/2" x 1-1/2", w/12" rods for 1-1/4" to 2" conduit		26	.308		38.50	16.40		54.90	67
8270	1-1/2" x 1-1/2", w/12" rods for 2-1/2" to 3-1/2" conduit		24	.333		42	17.80		59.80	73
8280	1-1/2" x 1-7/8", w/12" rods for 4" to 5" conduit		22	.364		70	19.40		89.40	106
8290	1-1/2" x 1-7/8", w/12" rods for 6" conduit		20	.400		76.50	21.50		98	117
8320	Conduit hangers, plastic coated steel, with bolt & 12" rod, 1/2" diam.		140	.057		19.80	3.05		22.85	26.50
8330	3/4" diameter		135	.059		20.50	3.16		23.66	27
8340	1" diameter		125	.064		20.50	3.41		23.91	28
8350	1-1/4" diameter		110	.073		22	3.88		25.88	30
8360	1-1/2" diameter		100	.080		24.50	4.27		28.77	33.50
8370	2" diameter		90	.089		27	4.74		31.74	37
8380	2-1/2" diameter		70	.114		32.50	6.10		38.60	44.50
8390	3" diameter		50	.160		39	8.55		47.55	56
8400	3-1/2" diameter		35	.229		39.50	12.20		51.70	62
8410	4" diameter		25	.320		58	17.05		75.05	89.50
8420	5" diameter		20	.400		63.50	21.50		85	102
9000	Parallel type, conduit beam clamp, 1/2"		32	.250		3.70	13.35		17.05	24
9010	3/4"		32	.250		3.98	13.35		17.33	24.50
9020	1"		32	.250		4.25	13.35		17.60	24.50
9030	1-1/4"		30	.267		5.75	14.25		20	28
9040	1-1/2"		30	.267		6.60	14.25		20.85	29
9050	2"		30	.267		8.80	14.25		23.05	31
9060	2-1/2"		28	.286		11.40	15.25		26.65	35.50
9070	3"		28	.286		14.75	15.25		30	39.50
9090	4"		25	.320		18.40	17.05		35.45	46
9110	Right angle, conduit beam clamp, 1/2"		32	.250		2.37	13.35		15.72	22.50
9120	3/4"		32	.250		2.44	13.35		15.79	22.50
9130	1"		32	.250		2.71	13.35		16.06	23
9140	1-1/4"		30	.267		3.36	14.25		17.61	25
9150	1-1/2"		30	.267		3.80	14.25		18.05	25.50
9160	2"		30	.267		5.50	14.25		19.75	27.50
9170	2-1/2"		28	.286		6.80	15.25		22.05	30.50
9180	3"		28	.286		7.80	15.25		23.05	31.50
9190	3-1/2"		25	.320		9.85	17.05		26.90	36.50
9200	4"		25	.320		10.60	17.05		27.65	37
9230	Adjustable, conduit hanger, 1/2"		32	.250		7.20	13.35		20.55	28
9240	3/4"		32	.250		7.35	13.35		20.70	28
9250	1"		32	.250		7.45	13.35		20.80	28
9260	1-1/4"		30	.267		7.55	14.25		21.80	30
9270	1-1/2"		30	.267		7.95	14.25		22.20	30.50
9280	2"		30	.267		5.55	14.25		19.80	27.50
9290	2-1/2"		28	.286		15.45	15.25		30.70	40
9300	3"		28	.286		16.40	15.25		31.65	41
9310	3-1/2"		25	.320		17.40	17.05		34.45	44.50
9320	4"		25	.320		27	17.05		44.05	55.50
9330	5"		20	.400		31	21.50		52.50	66
9340	6"		16	.500		46	26.50		72.50	90.50
9350	Combination conduit hanger, 3/8"	↓	32	.250	↓	11.15	13.35		24.50	32

26 05 Common Work Results for Electrical

26 05 29 – Hangers and Supports for Electrical Systems

26 05 29.20 Hangers		Crew	Daily Output	Labor-Hours	Unit	Material	2014 Bare Costs Labor	Equipment	Total	Total Incl O&P
9360	Adjustable flange 3/8"	1 Elec	32	.250	Ea.	13.05	13.35		26.40	34.50

26 05 33 – Raceway and Boxes for Electrical Systems

26 05 33.13 Conduit

		Crew	Daily Output	Labor-Hours	Unit	Material	2014 Bare Costs Labor	Equipment	Total	Total Incl O&P
0010	**CONDUIT** To 15' high, includes 2 terminations, 2 elbows, R260533-20									
0020	11 beam clamps, and 11 couplings per 100 L.F.									
0300	Aluminum, 1/2" diameter	1 Elec	100	.080	L.F.	1.83	4.27		6.10	8.40
0500	3/4" diameter R260533-21		90	.089		2.46	4.74		7.20	9.80
0700	1" diameter		80	.100		3.53	5.35		8.88	11.90
1000	1-1/4" diameter		70	.114		4.35	6.10		10.45	13.95
1030	1-1/2" diameter		65	.123		5.50	6.55		12.05	15.90
1050	2" diameter		60	.133		7.80	7.10		14.90	19.20
1070	2-1/2" diameter		50	.160		11.40	8.55		19.95	25.50
1100	3" diameter	2 Elec	90	.178		15.95	9.50		25.45	32
1130	3-1/2" diameter		80	.200		20.50	10.65		31.15	38.50
1140	4" diameter		70	.229		25	12.20		37.20	46
1150	5" diameter		50	.320		56.50	17.05		73.55	87.50
1160	6" diameter		40	.400		84.50	21.50		106	125
1161	Field bends, 45° to 90°, 1/2" diameter	1 Elec	53	.151	Ea.		8.05		8.05	12.05
1162	3/4" diameter		47	.170			9.10		9.10	13.60
1163	1" diameter		44	.182			9.70		9.70	14.50
1164	1-1/4" diameter		23	.348			18.55		18.55	28
1165	1-1/2" diameter		21	.381			20.50		20.50	30.50
1166	2" diameter		16	.500			26.50		26.50	40
1170	Elbows, 1/2" diameter		40	.200		8.80	10.65		19.45	25.50
1200	3/4" diameter		32	.250		12	13.35		25.35	33
1230	1" diameter		28	.286		16.65	15.25		31.90	41.50
1250	1-1/4" diameter		24	.333		26.50	17.80		44.30	55.50
1270	1-1/2" diameter		20	.400		35	21.50		56.50	70.50
1300	2" diameter		16	.500		52.50	26.50		79	97.50
1330	2-1/2" diameter		12	.667		88	35.50		123.50	150
1350	3" diameter		8	1		134	53.50		187.50	228
1370	3-1/2" diameter		6	1.333		211	71		282	340
1400	4" diameter		5	1.600		250	85.50		335.50	405
1410	5" diameter		4	2		745	107		852	975
1420	6" diameter		2.50	3.200		1,125	171		1,296	1,475
1430	Couplings, 1/2" diameter		320	.025		2.89	1.33		4.22	5.20
1450	3/4" diameter		192	.042		4.41	2.22		6.63	8.20
1470	1" diameter		160	.050		5.75	2.67		8.42	10.35
1500	1-1/4" diameter		120	.067		7.10	3.56		10.66	13.10
1530	1-1/2" diameter		96	.083		8.15	4.45		12.60	15.65
1550	2" diameter		80	.100		11.60	5.35		16.95	21
1570	2-1/2" diameter		69	.116		26	6.20		32.20	38
1600	3" diameter		64	.125		34	6.65		40.65	47.50
1630	3-1/2" diameter		56	.143		47	7.60		54.60	63.50
1650	4" diameter		53	.151		56	8.05		64.05	74
1670	5" diameter		48	.167		190	8.90		198.90	222
1690	6" diameter		46	.174		305	9.30		314.30	355
1691	See note on line 26 05 33.13 9995 R260533-30									
1750	Rigid galvanized steel, 1/2" diameter	1 Elec	90	.089	L.F.	2.40	4.74		7.14	9.75
1770	3/4" diameter		80	.100		2.56	5.35		7.91	10.80
1800	1" diameter		65	.123		3.69	6.55		10.24	13.90
1830	1-1/4" diameter		60	.133		4.84	7.10		11.94	16

133

26 05 33.13 Conduit		Crew	Daily Output	Labor-Hours	Unit	Material	2014 Bare Costs Labor	Equipment	Total	Total Incl O&P
1850	1-1/2" diameter	1 Elec	55	.145	L.F.	5.75	7.75		13.50	17.90
1870	2" diameter		45	.178		7.45	9.50		16.95	22.50
1900	2-1/2" diameter		35	.229		13.30	12.20		25.50	33
1930	3" diameter	2 Elec	50	.320		15.50	17.05		32.55	42.50
1950	3-1/2" diameter		44	.364		19.45	19.40		38.85	50.50
1970	4" diameter		40	.400		23	21.50		44.50	57
1980	5" diameter		30	.533		45	28.50		73.50	92
1990	6" diameter		20	.800		68.50	42.50		111	140
1991	Field bends, 45° to 90°, 1/2" diameter	1 Elec	44	.182	Ea.		9.70		9.70	14.50
1992	3/4" diameter		40	.200			10.65		10.65	15.95
1993	1" diameter		36	.222			11.85		11.85	17.75
1994	1-1/4" diameter		19	.421			22.50		22.50	33.50
1995	1-1/2" diameter		18	.444			23.50		23.50	35.50
1996	2" diameter		13	.615			33		33	49
2000	Elbows, 1/2" diameter		32	.250		7.85	13.35		21.20	28.50
2030	3/4" diameter		28	.286		8.20	15.25		23.45	32
2050	1" diameter		24	.333		12.60	17.80		30.40	40.50
2070	1-1/4" diameter		18	.444		17.35	23.50		40.85	54.50
2100	1-1/2" diameter		16	.500		21.50	26.50		48	63.50
2130	2" diameter		12	.667		31	35.50		66.50	87
2150	2-1/2" diameter		8	1		57.50	53.50		111	143
2170	3" diameter		6	1.333		80	71		151	194
2200	3-1/2" diameter		4.20	1.905		126	102		228	291
2220	4" diameter		4	2		143	107		250	320
2230	5" diameter		3.50	2.286		395	122		517	620
2240	6" diameter		2	4		600	213		813	975
2250	Couplings, 1/2" diameter		267	.030		1.93	1.60		3.53	4.51
2270	3/4" diameter		160	.050		2.37	2.67		5.04	6.60
2300	1" diameter		133	.060		3.50	3.21		6.71	8.65
2330	1-1/4" diameter		100	.080		4.39	4.27		8.66	11.25
2350	1-1/2" diameter		80	.100		5.55	5.35		10.90	14.10
2370	2" diameter		67	.119		7.35	6.35		13.70	17.60
2400	2-1/2" diameter		57	.140		18.05	7.50		25.55	31
2430	3" diameter		53	.151		23.50	8.05		31.55	37.50
2450	3-1/2" diameter		47	.170		31.50	9.10		40.60	48
2470	4" diameter		44	.182		31.50	9.70		41.20	49
2480	5" diameter		40	.200		73	10.65		83.65	96.50
2490	6" diameter		38	.211		102	11.25		113.25	129
2491	See note on line 26 05 33.13 9995	R260533-30								
2500	Steel, intermediate conduit (IMC), 1/2" diameter	1 Elec	100	.080	L.F.	1.72	4.27		5.99	8.30
2530	3/4" diameter		90	.089		2.06	4.74		6.80	9.35
2550	1" diameter		70	.114		3.04	6.10		9.14	12.50
2570	1-1/4" diameter		65	.123		3.73	6.55		10.28	13.95
2600	1-1/2" diameter		60	.133		4.92	7.10		12.02	16.05
2630	2" diameter		50	.160		5.95	8.55		14.50	19.30
2650	2-1/2" diameter		40	.200		10.60	10.65		21.25	27.50
2670	3" diameter	2 Elec	60	.267		14.50	14.25		28.75	37.50
2700	3-1/2" diameter		54	.296		19.15	15.80		34.95	44.50
2730	4" diameter		50	.320		21	17.05		38.05	48.50
2731	Field bends, 45° to 90°, 1/2" diameter	1 Elec	44	.182	Ea.		9.70		9.70	14.50
2732	3/4" diameter		40	.200			10.65		10.65	15.95
2733	1" diameter		36	.222			11.85		11.85	17.75
2734	1-1/4" diameter		19	.421			22.50		22.50	33.50

26 05 33.13 Conduit		Crew	Daily Output	Labor-Hours	Unit	Material	2014 Bare Costs Labor	Equipment	Total	Total Incl O&P
2735	1-1/2" diameter	1 Elec	18	.444	Ea.		23.50		23.50	35.50
2736	2" diameter		13	.615			33		33	49
2750	Elbows, 1/2" diameter		32	.250		10.55	13.35		23.90	31.50
2770	3/4" diameter		28	.286		11.25	15.25		26.50	35.50
2800	1" diameter		24	.333		16.45	17.80		34.25	44.50
2830	1-1/4" diameter		18	.444		25	23.50		48.50	63
2850	1-1/2" diameter		16	.500		31	26.50		57.50	74
2870	2" diameter		12	.667		33	35.50		68.50	89.50
2900	2-1/2" diameter		8	1		60	53.50		113.50	146
2930	3" diameter		6	1.333		86.50	71		157.50	202
2950	3-1/2" diameter		4.20	1.905		184	102		286	355
2970	4" diameter		4	2		177	107		284	355
3000	Couplings, 1/2" diameter		293	.027		1.93	1.46		3.39	4.30
3030	3/4" diameter		176	.045		2.37	2.42		4.79	6.25
3050	1" diameter		147	.054		3.50	2.90		6.40	8.20
3070	1-1/4" diameter		110	.073		4.39	3.88		8.27	10.65
3100	1-1/2" diameter		88	.091		5.55	4.85		10.40	13.35
3130	2" diameter		73	.110		7.35	5.85		13.20	16.80
3150	2-1/2" diameter		63	.127		18.05	6.75		24.80	30
3170	3" diameter		59	.136		23.50	7.25		30.75	36.50
3200	3-1/2" diameter		52	.154		31.50	8.20		39.70	47
3230	4" diameter		49	.163		31.50	8.70		40.20	47.50
3231	See note on line 26 05 33.13 9995	R260533-30								
4100	Rigid steel, plastic coated, 40 mil thick									
4130	1/2" diameter	1 Elec	80	.100	L.F.	7.10	5.35		12.45	15.80
4150	3/4" diameter		70	.114		7.55	6.10		13.65	17.45
4170	1" diameter		55	.145		9.35	7.75		17.10	22
4200	1-1/4" diameter		50	.160		13.50	8.55		22.05	27.50
4230	1-1/2" diameter		45	.178		14.15	9.50		23.65	30
4250	2" diameter		35	.229		19	12.20		31.20	39.50
4270	2-1/2" diameter		25	.320		31	17.05		48.05	59.50
4300	3" diameter	2 Elec	44	.364		35	19.40		54.40	67.50
4330	3-1/2" diameter		40	.400		47	21.50		68.50	83.50
4350	4" diameter		36	.444		49	23.50		72.50	89
4370	5" diameter		30	.533		121	28.50		149.50	176
4400	Elbows, 1/2" diameter	1 Elec	28	.286	Ea.	19.75	15.25		35	44.50
4430	3/4" diameter		24	.333		15.25	17.80		33.05	43.50
4450	1" diameter		18	.444		17.45	23.50		40.95	54.50
4470	1-1/4" diameter		16	.500		29	26.50		55.50	71.50
4500	1-1/2" diameter		12	.667		26.50	35.50		62	82
4530	2" diameter		8	1		37	53.50		90.50	121
4550	2-1/2" diameter		6	1.333		80	71		151	194
4570	3" diameter		4.20	1.905		132	102		234	297
4600	3-1/2" diameter		4	2		163	107		270	340
4630	4" diameter		3.80	2.105		178	112		290	365
4650	5" diameter		3.50	2.286		440	122		562	670
4680	Couplings, 1/2" diameter		213	.038		5.75	2		7.75	9.35
4700	3/4" diameter		128	.063		4.36	3.33		7.69	9.80
4730	1" diameter		107	.075		5.65	3.99		9.64	12.15
4750	1-1/4" diameter		80	.100		9	5.35		14.35	17.90
4770	1-1/2" diameter		64	.125		7.85	6.65		14.50	18.60
4800	2" diameter		53	.151		11.35	8.05		19.40	24.50
4830	2-1/2" diameter		46	.174		33	9.30		42.30	50

26 05 33 – Raceway and Boxes for Electrical Systems

26 05 33.13 Conduit		Crew	Daily Output	Labor-Hours	Unit	Material	2014 Bare Costs Labor	Equipment	Total	Total Incl O&P
4850	3" diameter	1 Elec	43	.186	Ea.	40.50	9.95		50.45	59.50
4870	3-1/2" diameter		38	.211		49.50	11.25		60.75	71.50
4900	4" diameter		36	.222		59	11.85		70.85	83
4950	5" diameter		32	.250		188	13.35		201.35	226
4951	See note on line 26 05 33.13 9995 R260533-30									
5000	Electric metallic tubing (EMT), 1/2" diameter	1 Elec	170	.047	L.F.	.56	2.51		3.07	4.38
5020	3/4" diameter		130	.062		.86	3.28		4.14	5.85
5040	1" diameter		115	.070		1.48	3.71		5.19	7.20
5060	1-1/4" diameter		100	.080		2.45	4.27		6.72	9.10
5080	1-1/2" diameter		90	.089		3.16	4.74		7.90	10.60
5100	2" diameter		80	.100		4.02	5.35		9.37	12.40
5120	2-1/2" diameter		60	.133		9.55	7.10		16.65	21
5140	3" diameter	2 Elec	100	.160		11.05	8.55		19.60	25
5160	3-1/2" diameter		90	.178		14	9.50		23.50	29.50
5180	4" diameter		80	.200		15.35	10.65		26	33
5200	Field bends, 45° to 90°, 1/2" diameter	1 Elec	89	.090	Ea.		4.80		4.80	7.20
5220	3/4" diameter		80	.100			5.35		5.35	8
5240	1" diameter		73	.110			5.85		5.85	8.75
5260	1-1/4" diameter		38	.211			11.25		11.25	16.80
5280	1-1/2" diameter		36	.222			11.85		11.85	17.75
5300	2" diameter		26	.308			16.40		16.40	24.50
5320	Offsets, 1/2" diameter		65	.123			6.55		6.55	9.85
5340	3/4" diameter		62	.129			6.90		6.90	10.30
5360	1" diameter		53	.151			8.05		8.05	12.05
5380	1-1/4" diameter		30	.267			14.25		14.25	21.50
5400	1-1/2" diameter		28	.286			15.25		15.25	23
5420	2" diameter		20	.400			21.50		21.50	32
5700	Elbows, 1" diameter		40	.200		6.30	10.65		16.95	23
5720	1-1/4" diameter		32	.250		7.85	13.35		21.20	28.50
5740	1-1/2" diameter		24	.333		9.10	17.80		26.90	36.50
5760	2" diameter		20	.400		13.40	21.50		34.90	47
5780	2-1/2" diameter		12	.667		36.50	35.50		72	93
5800	3" diameter		9	.889		48.50	47.50		96	125
5820	3-1/2" diameter		7	1.143		65	61		126	163
5840	4" diameter		6	1.333		76.50	71		147.50	190
6200	Couplings, set screw, steel, 1/2" diameter		235	.034		1.53	1.82		3.35	4.40
6220	3/4" diameter		188	.043		2.32	2.27		4.59	5.95
6240	1" diameter		157	.051		3.87	2.72		6.59	8.35
6260	1-1/4" diameter		118	.068		8.10	3.62		11.72	14.30
6280	1-1/2" diameter		94	.085		11.25	4.54		15.79	19.20
6300	2" diameter		78	.103		15.10	5.45		20.55	25
6320	2-1/2" diameter		67	.119		43.50	6.35		49.85	57.50
6340	3" diameter		59	.136		47.50	7.25		54.75	63
6360	3-1/2" diameter		47	.170		53.50	9.10		62.60	72.50
6380	4" diameter		43	.186		59	9.95		68.95	80
6381	See note on line 26 05 33.13 9995 R260533-30									
6500	Box connectors, set screw, steel, 1/2" diameter	1 Elec	120	.067	Ea.	1.17	3.56		4.73	6.60
6520	3/4" diameter		110	.073		1.90	3.88		5.78	7.90
6540	1" diameter		90	.089		3.59	4.74		8.33	11.05
6560	1-1/4" diameter		70	.114		7	6.10		13.10	16.85
6580	1-1/2" diameter		60	.133		10.10	7.10		17.20	22
6600	2" diameter		50	.160		13.90	8.55		22.45	28
6620	2-1/2" diameter		36	.222		49	11.85		60.85	72

26 05 33.13 Conduit		Crew	Daily Output	Labor-Hours	Unit	Material	2014 Bare Costs Labor	Equipment	Total	Total Incl O&P
6640	3" diameter	1 Elec	27	.296	Ea.	58	15.80		73.80	87.50
6680	3-1/2" diameter		21	.381		81.50	20.50		102	120
6700	4" diameter		16	.500		90.50	26.50		117	140
6740	Insulated box connectors, set screw, steel, 1/2" diameter		120	.067		1.60	3.56		5.16	7.05
6760	3/4" diameter		110	.073		2.52	3.88		6.40	8.55
6780	1" diameter		90	.089		4.64	4.74		9.38	12.20
6800	1-1/4" diameter		70	.114		8.45	6.10		14.55	18.40
6820	1-1/2" diameter		60	.133		11.95	7.10		19.05	24
6840	2" diameter		50	.160		17.35	8.55		25.90	32
6860	2-1/2" diameter		36	.222		82	11.85		93.85	108
6880	3" diameter		27	.296		97	15.80		112.80	131
6900	3-1/2" diameter		21	.381		135	20.50		155.50	179
6920	4" diameter		16	.500		147	26.50		173.50	202
7000	EMT to conduit adapters, 1/2" diameter (compression)		70	.114		3.74	6.10		9.84	13.25
7020	3/4" diameter		60	.133		5.30	7.10		12.40	16.45
7040	1" diameter		50	.160		8.15	8.55		16.70	22
7060	1-1/4" diameter		40	.200		15.20	10.65		25.85	32.50
7080	1-1/2" diameter		30	.267		18.65	14.25		32.90	42
7100	2" diameter		25	.320		27	17.05		44.05	55.50
7200	EMT to Greenfield adapters, 1/2" to 3/8" diameter (compression)		90	.089		2.58	4.74		7.32	9.95
7220	1/2" diameter		90	.089		4.73	4.74		9.47	12.30
7240	3/4" diameter		80	.100		6.10	5.35		11.45	14.75
7260	1" diameter		70	.114		17	6.10		23.10	28
7270	1-1/4" diameter		60	.133		20	7.10		27.10	32.50
7280	1-1/2" diameter		50	.160		23	8.55		31.55	38.50
7290	2" diameter		40	.200		34	10.65		44.65	53.50
7400	EMT,LB, LR or LL fittings with covers, 1/2" dia, set screw		24	.333		6.40	17.80		24.20	33.50
7420	3/4" diameter		20	.400		8.65	21.50		30.15	41.50
7440	1" diameter		16	.500		11.70	26.50		38.20	53
7450	1-1/4" diameter		13	.615		17.75	33		50.75	68.50
7460	1-1/2" diameter		11	.727		23	39		62	83.50
7470	2" diameter		9	.889		38	47.50		85.50	113
7600	EMT, "T" fittings with covers, 1/2" diameter, set screw		16	.500		9.25	26.50		35.75	50
7620	3/4" diameter		15	.533		11.90	28.50		40.40	55.50
7640	1" diameter		12	.667		16.35	35.50		51.85	71
7650	1-1/4" diameter		11	.727		22.50	39		61.50	83
7660	1-1/2" diameter		10	.800		28	42.50		70.50	95
7670	2" diameter		8	1		37	53.50		90.50	121
8000	EMT, expansion fittings, no jumper, 1/2" diameter		24	.333		61	17.80		78.80	93.50
8020	3/4" diameter		20	.400		72	21.50		93.50	111
8040	1" diameter		16	.500		85	26.50		111.50	134
8060	1-1/4" diameter		13	.615		115	33		148	175
8080	1-1/2" diameter		11	.727		156	39		195	230
8100	2" diameter		9	.889		231	47.50		278.50	325
8110	2-1/2" diameter		7	1.143		370	61		431	495
8120	3" diameter		6	1.333		450	71		521	605
8140	4" diameter		5	1.600		790	85.50		875.50	1,000
8200	Split adapter, 1/2" diameter		110	.073		2.68	3.88		6.56	8.75
8210	3/4" diameter		90	.089		2.38	4.74		7.12	9.70
8220	1" diameter		70	.114		3.27	6.10		9.37	12.75
8230	1-1/4" diameter		60	.133		5.25	7.10		12.35	16.45
8240	1-1/2" diameter		50	.160		8.10	8.55		16.65	22
8250	2" diameter		36	.222		24	11.85		35.85	44

26 05 33.13 Conduit		Crew	Daily Output	Labor-Hours	Unit	Material	2014 Bare Costs Labor	Equipment	Total	Total Incl O&P
8300	1 hole clips, 1/2" diameter	1 Elec	500	.016	Ea.	.46	.85		1.31	1.79
8320	3/4" diameter		470	.017		.60	.91		1.51	2.02
8340	1" diameter		444	.018		1.10	.96		2.06	2.65
8360	1-1/4" diameter		400	.020		1.47	1.07		2.54	3.22
8380	1-1/2" diameter		355	.023		2.25	1.20		3.45	4.28
8400	2" diameter		320	.025		3.54	1.33		4.87	5.90
8420	2-1/2" diameter		266	.030		6.55	1.60		8.15	9.60
8440	3" diameter		160	.050		8.25	2.67		10.92	13.10
8460	3-1/2" diameter		133	.060		13.05	3.21		16.26	19.15
8480	4" diameter		100	.080		16.85	4.27		21.12	25
8500	Clamp back spacers, 1/2" diameter		500	.016		1.37	.85		2.22	2.79
8510	3/4" diameter		470	.017		1.64	.91		2.55	3.16
8520	1" diameter		444	.018		2.87	.96		3.83	4.60
8530	1-1/4" diameter		400	.020		5.15	1.07		6.22	7.25
8540	1-1/2" diameter		355	.023		5.75	1.20		6.95	8.15
8550	2" diameter		320	.025		9.70	1.33		11.03	12.65
8560	2-1/2" diameter		266	.030		18.60	1.60		20.20	23
8570	3" diameter		160	.050		26.50	2.67		29.17	33.50
8580	3-1/2" diameter		133	.060		36	3.21		39.21	44.50
8590	4" diameter		100	.080		78	4.27		82.27	92.50
8600	Offset connectors, 1/2" diameter		40	.200		5.05	10.65		15.70	21.50
8610	3/4" diameter		32	.250		7.30	13.35		20.65	28
8620	1" diameter		24	.333		8.05	17.80		25.85	35.50
8650	90° pulling elbows, female, 1/2" diameter, with gasket		24	.333		6.65	17.80		24.45	34
8660	3/4" diameter		20	.400		11.30	21.50		32.80	44.50
8700	Couplings, compression, 1/2" diameter, steel		78	.103		3.64	5.45		9.09	12.20
8710	3/4" diameter		67	.119		5.10	6.35		11.45	15.15
8720	1" diameter		59	.136		8.75	7.25		16	20.50
8730	1-1/4" diameter		47	.170		16.55	9.10		25.65	32
8740	1-1/2" diameter		38	.211		24	11.25		35.25	43.50
8750	2" diameter		31	.258		33	13.75		46.75	56.50
8760	2-1/2" diameter		24	.333		147	17.80		164.80	189
8770	3" diameter		21	.381		184	20.50		204.50	233
8780	3-1/2" diameter		19	.421		299	22.50		321.50	365
8790	4" diameter		16	.500		305	26.50		331.50	375
8791	See note on line 26 05 33.13 9995 R260533-30									
8800	Box connectors, compression, 1/2" diam., steel	1 Elec	120	.067	Ea.	3.02	3.56		6.58	8.60
8810	3/4" diameter		110	.073		4.24	3.88		8.12	10.45
8820	1" diameter		90	.089		6.30	4.74		11.04	14
8830	1-1/4" diameter		70	.114		16.05	6.10		22.15	27
8840	1-1/2" diameter		60	.133		18.85	7.10		25.95	31
8850	2" diameter		50	.160		27	8.55		35.55	43
8860	2-1/2" diameter		36	.222		67	11.85		78.85	91.50
8870	3" diameter		27	.296		92.50	15.80		108.30	126
8880	3-1/2" diameter		21	.381		140	20.50		160.50	185
8890	4" diameter		16	.500		143	26.50		169.50	197
8900	Box connectors, insulated compression, 1/2" diam., steel		120	.067		2.92	3.56		6.48	8.50
8910	3/4" diameter		110	.073		4.05	3.88		7.93	10.25
8920	1" diameter		90	.089		6.80	4.74		11.54	14.60
8930	1-1/4" diameter		70	.114		13.70	6.10		19.80	24.50
8940	1-1/2" diameter		60	.133		20.50	7.10		27.60	33
8950	2" diameter		50	.160		30	8.55		38.55	46
8960	2-1/2" diameter		36	.222		78	11.85		89.85	104

26 05 33.13 Conduit		Crew	Daily Output	Labor-Hours	Unit	Material	2014 Bare Costs Labor	2014 Bare Costs Equipment	Total	Total Incl O&P
8970	3" diameter	1 Elec	27	.296	Ea.	102	15.80		117.80	136
8980	3-1/2" diameter		21	.381		149	20.50		169.50	195
8990	4" diameter		16	.500	▼	152	26.50		178.50	207
9100	PVC, schedule 40, 1/2" diameter		190	.042	L.F.	.97	2.25		3.22	4.43
9110	3/4" diameter		145	.055		1.07	2.94		4.01	5.60
9120	1" diameter		125	.064		2.09	3.41		5.50	7.40
9130	1-1/4" diameter		110	.073		2.76	3.88		6.64	8.85
9140	1-1/2" diameter		100	.080		3.14	4.27		7.41	9.85
9150	2" diameter		90	.089		4.09	4.74		8.83	11.60
9160	2-1/2" diameter		65	.123		6.15	6.55		12.70	16.60
9170	3" diameter	2 Elec	110	.145		7.05	7.75		14.80	19.35
9180	3-1/2" diameter		100	.160		9.25	8.55		17.80	23
9190	4" diameter		90	.178		10.70	9.50		20.20	26
9200	5" diameter		70	.229		13.95	12.20		26.15	33.50
9210	6" diameter	▼	60	.267	▼	18.45	14.25		32.70	42
9220	Elbows, 1/2" diameter	1 Elec	50	.160	Ea.	1.06	8.55		9.61	13.95
9225	3/4" diameter		42	.190		1.04	10.15		11.19	16.35
9230	1" diameter		35	.229		1.57	12.20		13.77	20
9235	1-1/4" diameter		28	.286		2.41	15.25		17.66	25.50
9240	1-1/2" diameter		20	.400		3.11	21.50		24.61	35.50
9245	2" diameter		16	.500		4.22	26.50		30.72	44.50
9250	2-1/2" diameter		11	.727		7.60	39		46.60	66.50
9255	3" diameter		9	.889		12.70	47.50		60.20	85
9260	3-1/2" diameter		7	1.143		16.45	61		77.45	110
9265	4" diameter		6	1.333		20.50	71		91.50	129
9270	5" diameter		4	2		33	107		140	197
9275	6" diameter		3	2.667		76	142		218	297
9312	Couplings, 1/2" diameter		50	.160		.25	8.55		8.80	13.10
9314	3/4" diameter		42	.190		.27	10.15		10.42	15.50
9316	1" diameter		35	.229		.43	12.20		12.63	18.70
9318	1-1/4" diameter		28	.286		.60	15.25		15.85	23.50
9320	1-1/2" diameter		20	.400		.57	21.50		22.07	32.50
9322	2" diameter		16	.500		1.26	26.50		27.76	41.50
9324	2-1/2" diameter		11	.727		1.80	39		40.80	60
9326	3" diameter		9	.889		2.31	47.50		49.81	73.50
9328	3-1/2" diameter		7	1.143		2.95	61		63.95	95
9330	4" diameter		6	1.333		3.62	71		74.62	110
9332	5" diameter		4	2		7.55	107		114.55	168
9334	6" diameter	▼	3	2.667	▼	10.60	142		152.60	225
9335	See note on line 26 05 33.13 9995 R260533-30									
9340	Field bends, 45° & 90°, 1/2" diameter	1 Elec	45	.178	Ea.		9.50		9.50	14.20
9350	3/4" diameter		40	.200			10.65		10.65	15.95
9360	1" diameter		35	.229			12.20		12.20	18.25
9370	1-1/4" diameter		32	.250			13.35		13.35	19.95
9380	1-1/2" diameter		27	.296			15.80		15.80	23.50
9390	2" diameter		20	.400			21.50		21.50	32
9400	2-1/2" diameter		16	.500			26.50		26.50	40
9410	3" diameter		13	.615			33		33	49
9420	3-1/2" diameter		12	.667			35.50		35.50	53
9430	4" diameter		10	.800			42.50		42.50	64
9440	5" diameter		9	.889			47.50		47.50	71
9450	6" diameter		8	1			53.50		53.50	80
9460	PVC adapters, 1/2" diameter		50	.160	▼	.30	8.55		8.85	13.15

26 05 33.13 Conduit		Crew	Daily Output	Labor-Hours	Unit	Material	2014 Bare Costs Labor	Equipment	Total	Total Incl O&P
9470	3/4" diameter	1 Elec	42	.190	Ea.	.49	10.15		10.64	15.75
9480	1" diameter		38	.211		.66	11.25		11.91	17.55
9490	1-1/4" diameter		35	.229		.87	12.20		13.07	19.20
9500	1-1/2" diameter		32	.250		1.14	13.35		14.49	21
9510	2" diameter		27	.296		1.44	15.80		17.24	25
9520	2-1/2" diameter		23	.348		2.52	18.55		21.07	31
9530	3" diameter		18	.444		3.50	23.50		27	39.50
9540	3-1/2" diameter		13	.615		5.40	33		38.40	55
9550	4" diameter		11	.727		6.20	39		45.20	65
9560	5" diameter		8	1		13.20	53.50		66.70	94.50
9570	6" diameter	▼	6	1.333	▼	22	71		93	130
9580	PVC-LB, LR or LL fittings & covers									
9590	1/2" diameter	1 Elec	20	.400	Ea.	3.58	21.50		25.08	36
9600	3/4" diameter		16	.500		5.45	26.50		31.95	46
9610	1" diameter		12	.667		5.90	35.50		41.40	59.50
9620	1-1/4" diameter		9	.889		9.15	47.50		56.65	81
9630	1-1/2" diameter		7	1.143		10.90	61		71.90	104
9640	2" diameter		6	1.333		17.30	71		88.30	125
9650	2-1/2" diameter		6	1.333		51	71		122	162
9660	3" diameter		5	1.600		49.50	85.50		135	183
9670	3-1/2" diameter		4	2		62.50	107		169.50	229
9680	4" diameter	▼	3	2.667	▼	83	142		225	305
9690	PVC-tee fitting & cover									
9700	1/2"	1 Elec	14	.571	Ea.	6.85	30.50		37.35	53
9710	3/4"		13	.615		6.85	33		39.85	56.50
9720	1"		10	.800		7.70	42.50		50.20	72.50
9730	1-1/4"		9	.889		13.50	47.50		61	86
9740	1-1/2"		8	1		15.85	53.50		69.35	97.50
9750	2"	▼	7	1.143		24	61		85	118
9760	PVC-reducers, 3/4" x 1/2" diameter					1.66			1.66	1.83
9770	1" x 1/2" diameter					3.64			3.64	4
9780	1" x 3/4" diameter					3.93			3.93	4.32
9790	1-1/4" x 3/4" diameter					4.83			4.83	5.30
9800	1-1/4" x 1" diameter					5.10			5.10	5.60
9810	1-1/2" x 1-1/4" diameter					5.30			5.30	5.85
9820	2" x 1-1/4" diameter					6.40			6.40	7.05
9830	2-1/2" x 2" diameter					19.85			19.85	22
9840	3" x 2" diameter					20.50			20.50	22.50
9850	4" x 3" diameter					23.50			23.50	26
9860	Cement, quart					19.55			19.55	21.50
9870	Gallon					102			102	112
9880	Heat bender, to 6" diameter				▼	1,850			1,850	2,025
9900	Add to labor for higher elevated installation									
9910	15' to 20' high, add						10%			
9920	20' to 25' high, add						20%			
9930	25' to 30' high, add						25%			
9940	30' to 35' high, add						30%			
9950	35' to 40' high, add						35%			
9960	Over 40' high, add						40%			
9980	Allowance for conduit fittings 5% min.-20% max.									
9995	Do not include labor when adding couplings to a fitting installation R260533-30									

26 05 33.14 Conduit		Crew	Daily Output	Labor-Hours	Unit	Material	2014 Bare Costs Labor	Equipment	Total	Total Incl O&P
0010	**CONDUIT** To 15' high, includes 11 couplings per 100' R260533-23									
0200	Electric metallic tubing, 1/2"diameter	1 Elec	435	.018	L.F.	.37	.98		1.35	1.88
0220	3/4" diameter		253	.032		.67	1.69		2.36	3.25
0240	1" diameter		207	.039		1.12	2.06		3.18	4.32
0260	1-1/4" diameter		173	.046		1.98	2.47		4.45	5.85
0280	1-1/2" diameter		153	.052		2.59	2.79		5.38	7.05
0300	2" diameter		130	.062		3.26	3.28		6.54	8.50
0320	2-1/2" diameter		92	.087		7.40	4.64		12.04	15.10
0340	3" diameter	2 Elec	148	.108		8.45	5.75		14.20	17.95
0360	3-1/2" diameter		134	.119		10.55	6.35		16.90	21
0380	4" diameter		114	.140		11.20	7.50		18.70	23.50
0500	Steel rigid galvanized, 1/2" diameter R260533-24	1 Elec	146	.055		1.74	2.92		4.66	6.30
0520	3/4" diameter		125	.064		1.86	3.41		5.27	7.15
0540	1" diameter		93	.086		2.68	4.59		7.27	9.80
0560	1-1/4" diameter		88	.091		3.80	4.85		8.65	11.45
0580	1-1/2" diameter		80	.100		4.24	5.35		9.59	12.65
0600	2" diameter		65	.123		5.25	6.55		11.80	15.65
0620	2-1/2" diameter		48	.167		10.35	8.90		19.25	24.50
0640	3" diameter	2 Elec	64	.250		11.95	13.35		25.30	33
0660	3-1/2" diameter		60	.267		15.30	14.25		29.55	38.50
0680	4" diameter		52	.308		16.95	16.40		33.35	43
0700	5" diameter		50	.320		33.50	17.05		50.55	62
0720	6" diameter		48	.333		47.50	17.80		65.30	78.50
1000	Steel intermediate conduit (IMC), 1/2 diameter R260533-25	1 Elec	155	.052		1.09	2.75		3.84	5.30
1010	3/4" diameter		130	.062		1.33	3.28		4.61	6.35
1020	1" diameter		100	.080		2	4.27		6.27	8.60
1030	1-1/4" diameter		93	.086		2.58	4.59		7.17	9.70
1040	1-1/2" diameter		85	.094		3.34	5		8.34	11.15
1050	2" diameter		70	.114		4.17	6.10		10.27	13.75
1060	2-1/2" diameter		53	.151		8.55	8.05		16.60	21.50
1070	3" diameter	2 Elec	80	.200		11.30	10.65		21.95	28.50
1080	3-1/2" diameter		70	.229		13.70	12.20		25.90	33.50
1090	4" diameter		60	.267		15	14.25		29.25	38

26 05 33.15 Conduit Nipples

26 05 33.15 Conduit Nipples		Crew	Daily Output	Labor-Hours	Unit	Material	2014 Bare Costs Labor	Equipment	Total	Total Incl O&P
0010	**CONDUIT NIPPLES** With locknuts and bushings									
0100	Aluminum, 1/2" diameter, close	1 Elec	36	.222	Ea.	7.60	11.85		19.45	26
0120	1-1/2" long		36	.222		7.75	11.85		19.60	26.50
0140	2" long		36	.222		8	11.85		19.85	26.50
0160	2-1/2" long		36	.222		8.75	11.85		20.60	27.50
0180	3" long		36	.222		8.85	11.85		20.70	27.50
0200	3-1/2" long		36	.222		9.55	11.85		21.40	28.50
0220	4" long		36	.222		9.70	11.85		21.55	28.50
0240	5" long		36	.222		11.05	11.85		22.90	30
0260	6" long		36	.222		11.25	11.85		23.10	30
0280	8" long		36	.222		12	11.85		23.85	31
0300	10" long		36	.222		15.20	11.85		27.05	34.50
0320	12" long		36	.222		15.15	11.85		27	34.50
0340	3/4" diameter, close		32	.250		9.95	13.35		23.30	31
0360	1-1/2" long		32	.250		10.10	13.35		23.45	31
0380	2" long		32	.250		10.20	13.35		23.55	31
0400	2-1/2" long		32	.250		11.05	13.35		24.40	32
0420	3" long		32	.250		11.20	13.35		24.55	32.50

26 05 33.15 Conduit Nipples		Crew	Daily Output	Labor-Hours	Unit	Material	2014 Bare Costs Labor	Equipment	Total	Total Incl O&P
0440	3-1/2" long	1 Elec	32	.250	Ea.	11.90	13.35		25.25	33
0460	4" long		32	.250		11.85	13.35		25.20	33
0480	5" long		32	.250		13.45	13.35		26.80	35
0500	6" long		32	.250		14	13.35		27.35	35.50
0520	8" long		32	.250		16.05	13.35		29.40	37.50
0540	10" long		32	.250		18.90	13.35		32.25	41
0560	12" long		32	.250		26.50	13.35		39.85	49
0580	1" diameter, close		27	.296		15.45	15.80		31.25	40.50
0600	2" long		27	.296		17.45	15.80		33.25	42.50
0620	2-1/2" long		27	.296		16.60	15.80		32.40	42
0640	3" long		27	.296		19.50	15.80		35.30	45
0660	3-1/2" long		27	.296		18.80	15.80		34.60	44
0680	4" long		27	.296		21.50	15.80		37.30	47
0700	5" long		27	.296		21	15.80		36.80	46.50
0720	6" long		27	.296		22.50	15.80		38.30	48.50
0740	8" long		27	.296		26	15.80		41.80	52
0760	10" long		27	.296		30.50	15.80		46.30	57
0780	12" long		27	.296		33	15.80		48.80	60
0800	1-1/4" diameter, close		23	.348		22.50	18.55		41.05	53
0820	2" long		23	.348		23	18.55		41.55	53.50
0840	2-1/2" long		23	.348		24	18.55		42.55	54
0860	3" long		23	.348		25	18.55		43.55	55.50
0880	3-1/2" long		23	.348		26.50	18.55		45.05	57.50
0900	4" long		23	.348		27	18.55		45.55	58
0920	5" long		23	.348		30	18.55		48.55	61
0940	6" long		23	.348		32	18.55		50.55	63.50
0960	8" long		23	.348		40.50	18.55		59.05	72.50
0980	10" long		23	.348		38.50	18.55		57.05	70.50
1000	12" long		23	.348		56	18.55		74.55	89.50
1020	1-1/2" diameter, close		20	.400		30	21.50		51.50	65
1040	2" long		20	.400		31	21.50		52.50	66
1060	2-1/2" long		20	.400		31.50	21.50		53	67
1080	3" long		20	.400		33	21.50		54.50	68.50
1100	3-1/2" long		20	.400		35.50	21.50		57	71
1120	4" long		20	.400		35.50	21.50		57	71
1140	5" long		20	.400		39	21.50		60.50	75
1160	6" long		20	.400		40	21.50		61.50	76
1180	8" long		20	.400		43.50	21.50		65	80
1200	10" long		20	.400		56.50	21.50		78	94
1220	12" long		20	.400		68	21.50		89.50	107
1240	2" diameter, close		18	.444		41.50	23.50		65	81.50
1260	2-1/2" long		18	.444		44	23.50		67.50	83.50
1280	3" long		18	.444		45.50	23.50		69	85.50
1300	3-1/2" long		18	.444		48	23.50		71.50	88.50
1320	4" long		18	.444		48.50	23.50		72	88.50
1340	5" long		18	.444		52.50	23.50		76	93.50
1360	6" long		18	.444		55	23.50		78.50	96
1380	8" long		18	.444		70.50	23.50		94	113
1400	10" long		18	.444		63.50	23.50		87	105
1420	12" long		18	.444		78.50	23.50		102	122
1440	2-1/2" diameter, close		15	.533		101	28.50		129.50	155
1460	3" long		15	.533		103	28.50		131.50	156
1480	3-1/2" long		15	.533		101	28.50		129.50	155

26 05 33 – Raceway and Boxes for Electrical Systems

26 05 33.15 Conduit Nipples	Crew	Daily Output	Labor-Hours	Unit	Material	2014 Bare Costs Labor	Equipment	Total	Total Incl O&P
1500	4" long	1 Elec	15	.533	Ea.	103	28.50		131.50	157
1520	5" long		15	.533		112	28.50		140.50	167
1540	6" long		15	.533		111	28.50		139.50	165
1560	8" long		15	.533		121	28.50		149.50	176
1580	10" long		15	.533		130	28.50		158.50	186
1600	12" long		15	.533		138	28.50		166.50	195
1620	3" diameter, close		12	.667		112	35.50		147.50	176
1640	3" long		12	.667		120	35.50		155.50	185
1660	3-1/2" long		12	.667		116	35.50		151.50	181
1680	4" long		12	.667		127	35.50		162.50	193
1700	5" long		12	.667		123	35.50		158.50	188
1720	6" long		12	.667		132	35.50		167.50	198
1740	8" long		12	.667		162	35.50		197.50	231
1760	10" long		12	.667		160	35.50		195.50	229
1780	12" long		12	.667		182	35.50		217.50	254
1800	3-1/2" diameter, close		11	.727		217	39		256	297
1820	4" long		11	.727		228	39		267	310
1840	5" long		11	.727		230	39		269	310
1860	6" long		11	.727		235	39		274	315
1880	8" long		11	.727		248	39		287	330
1900	10" long		11	.727		263	39		302	345
1920	12" long		11	.727		310	39		349	405
1940	4" diameter, close		9	.889		254	47.50		301.50	350
1960	4" long		9	.889		278	47.50		325.50	375
1980	5" long		9	.889		286	47.50		333.50	385
2000	6" long		9	.889		275	47.50		322.50	375
2020	8" long		9	.889		325	47.50		372.50	425
2040	10" long		9	.889		335	47.50		382.50	440
2060	12" long		9	.889		325	47.50		372.50	425
2080	5" diameter, close		7	1.143		450	61		511	590
2100	5" long		7	1.143		465	61		526	605
2120	6" long		7	1.143		480	61		541	615
2140	8" long		7	1.143		500	61		561	640
2160	10" long		7	1.143		535	61		596	680
2180	12" long		7	1.143		555	61		616	700
2200	6" diameter, close		6	1.333		750	71		821	930
2220	5" long		6	1.333		770	71		841	955
2240	6" long		6	1.333		780	71		851	965
2260	8" long		6	1.333		820	71		891	1,000
2280	10" long		6	1.333		865	71		936	1,050
2300	12" long		6	1.333		880	71		951	1,075
2320	Rigid galvanized steel, 1/2" diameter, close		32	.250		3.03	13.35		16.38	23.50
2340	1-1/2" long		32	.250		3.36	13.35		16.71	23.50
2360	2" long		32	.250		3.57	13.35		16.92	24
2380	2-1/2" long		32	.250		3.73	13.35		17.08	24
2400	3" long		32	.250		3.94	13.35		17.29	24.50
2420	3-1/2" long		32	.250		4.15	13.35		17.50	24.50
2440	4" long		32	.250		4.37	13.35		17.72	25
2460	5" long		32	.250		4.70	13.35		18.05	25
2480	6" long		32	.250		5.35	13.35		18.70	26
2500	8" long		32	.250		8.15	13.35		21.50	29
2520	10" long		32	.250		9.10	13.35		22.45	30
2540	12" long		32	.250		10.25	13.35		23.60	31.50

26 05 33.15 Conduit Nipples		Crew	Daily Output	Labor-Hours	Unit	Material	2014 Bare Costs Labor	Equipment	Total	Total Incl O&P
2560	3/4" diameter, close	1 Elec	27	.296	Ea.	3.99	15.80		19.79	28
2580	2" long		27	.296		4.36	15.80		20.16	28.50
2600	2-1/2" long		27	.296		4.63	15.80		20.43	28.50
2620	3" long		27	.296		4.83	15.80		20.63	29
2640	3-1/2" long		27	.296		4.95	15.80		20.75	29
2660	4" long		27	.296		5.35	15.80		21.15	29.50
2680	5" long		27	.296		5.85	15.80		21.65	30
2700	6" long		27	.296		6.50	15.80		22.30	30.50
2720	8" long		27	.296		9.40	15.80		25.20	34
2740	10" long		27	.296		10.80	15.80		26.60	35.50
2760	12" long		27	.296		11.95	15.80		27.75	36.50
2780	1" diameter, close		23	.348		6.60	18.55		25.15	35.50
2800	2" long		23	.348		6.85	18.55		25.40	35.50
2820	2-1/2" long		23	.348		7.10	18.55		25.65	36
2840	3" long		23	.348		7.55	18.55		26.10	36.50
2860	3-1/2" long		23	.348		8.10	18.55		26.65	37
2880	4" long		23	.348		8.40	18.55		26.95	37.50
2900	5" long		23	.348		9.05	18.55		27.60	38
2920	6" long		23	.348		9.55	18.55		28.10	38.50
2940	8" long		23	.348		12.90	18.55		31.45	42
2960	10" long		23	.348		15.60	18.55		34.15	45
2980	12" long		23	.348		17.25	18.55		35.80	47
3000	1-1/4" diameter, close		20	.400		9	21.50		30.50	42
3020	2" long		20	.400		9.25	21.50		30.75	42
3040	3" long		20	.400		10	21.50		31.50	43
3060	3-1/2" long		20	.400		10.65	21.50		32.15	44
3080	4" long		20	.400		11	21.50		32.50	44
3100	5" long		20	.400		11.90	21.50		33.40	45
3120	6" long		20	.400		12.75	21.50		34.25	46
3140	8" long		20	.400		17.60	21.50		39.10	51.50
3160	10" long		20	.400		21	21.50		42.50	55
3180	12" long		20	.400		23.50	21.50		45	58
3200	1-1/2" diameter, close		18	.444		12.55	23.50		36.05	49.50
3220	2" long		18	.444		12.80	23.50		36.30	49.50
3240	2-1/2" long		18	.444		13.40	23.50		36.90	50.50
3260	3" long		18	.444		13.85	23.50		37.35	50.50
3280	3-1/2" long		18	.444		14.70	23.50		38.20	51.50
3300	4" long		18	.444		15.35	23.50		38.85	52.50
3320	5" long		18	.444		16.15	23.50		39.65	53.50
3340	6" long		18	.444		18.15	23.50		41.65	55.50
3360	8" long		18	.444		23.50	23.50		47	61.50
3380	10" long		18	.444		26.50	23.50		50	65
3400	12" long		18	.444		28	23.50		51.50	66.50
3420	2" diameter, close		16	.500		15.65	26.50		42.15	57
3440	2-1/2" long		16	.500		16.65	26.50		43.15	58.50
3460	3" long		16	.500		17.65	26.50		44.15	59.50
3480	3-1/2" long		16	.500		18.75	26.50		45.25	60.50
3500	4" long		16	.500		19.65	26.50		46.15	61.50
3520	5" long		16	.500		21.50	26.50		48	63.50
3540	6" long		16	.500		23	26.50		49.50	65
3560	8" long		16	.500		28.50	26.50		55	71.50
3580	10" long		16	.500		32	26.50		58.50	75.50
3600	12" long		16	.500		35	26.50		61.50	78.50

26 05 33.15 Conduit Nipples		Crew	Daily Output	Labor-Hours	Unit	Material	2014 Bare Costs Labor	Equipment	Total	Total Incl O&P
3620	2-1/2" diameter, close	1 Elec	13	.615	Ea.	46	33		79	99.50
3640	3" long		13	.615		45.50	33		78.50	99
3660	3-1/2" long		13	.615		49	33		82	103
3680	4" long		13	.615		49.50	33		82.50	104
3700	5" long		13	.615		53.50	33		86.50	108
3720	6" long		13	.615		56.50	33		89.50	111
3740	8" long		13	.615		64.50	33		97.50	120
3760	10" long		13	.615		70	33		103	126
3780	12" long		13	.615		77	33		110	134
3800	3" diameter, close		12	.667		56	35.50		91.50	115
3820	3" long		12	.667		57	35.50		92.50	116
3900	3-1/2" long		12	.667		59.50	35.50		95	119
3920	4" long		12	.667		61.50	35.50		97	121
3940	5" long		12	.667		65.50	35.50		101	125
3960	6" long		12	.667		69	35.50		104.50	129
3980	8" long		12	.667		78.50	35.50		114	140
4000	10" long		12	.667		86	35.50		121.50	148
4020	12" long		12	.667		96.50	35.50		132	159
4040	3-1/2" diameter, close		10	.800		95.50	42.50		138	169
4060	4" long		10	.800		102	42.50		144.50	176
4080	5" long		10	.800		105	42.50		147.50	180
4100	6" long		10	.800		110	42.50		152.50	185
4120	8" long		10	.800		120	42.50		162.50	196
4140	10" long		10	.800		129	42.50		171.50	206
4160	12" long		10	.800		139	42.50		181.50	217
4180	4" diameter, close		8	1		107	53.50		160.50	198
4200	4" long		8	1		113	53.50		166.50	204
4220	5" long		8	1		119	53.50		172.50	210
4240	6" long		8	1		123	53.50		176.50	215
4260	8" long		8	1		134	53.50		187.50	227
4280	10" long		8	1		146	53.50		199.50	241
4300	12" long		8	1		158	53.50		211.50	254
4320	5" diameter, close		6	1.333		214	71		285	340
4340	5" long		6	1.333		237	71		308	365
4360	6" long		6	1.333		243	71		314	375
4380	8" long		6	1.333		258	71		329	390
4400	10" long		6	1.333		275	71		346	405
4420	12" long		6	1.333		299	71		370	435
4440	6" diameter, close		5	1.600		450	85.50		535.50	625
4460	5" long		5	1.600		475	85.50		560.50	655
4480	6" long		5	1.600		485	85.50		570.50	665
4500	8" long		5	1.600		505	85.50		590.50	685
4520	10" long		5	1.600		535	85.50		620.50	715
4540	12" long		5	1.600		550	85.50		635.50	735
4560	Plastic coated, 40 mil thick, 1/2" diameter, 2" long		32	.250		17.55	13.35		30.90	39.50
4580	2-1/2" long		32	.250		20.50	13.35		33.85	42.50
4600	3" long		32	.250		20.50	13.35		33.85	42.50
4680	3-1/2" long		32	.250		23.50	13.35		36.85	46
4700	4" long		32	.250		23	13.35		36.35	45.50
4720	5" long		32	.250		23	13.35		36.35	45.50
4740	6" long		32	.250		24.50	13.35		37.85	47
4760	8" long		32	.250		23	13.35		36.35	45
4780	10" long		32	.250		24	13.35		37.35	46.50

26 05 33.15 Conduit Nipples		Crew	Daily Output	Labor-Hours	Unit	Material	2014 Bare Costs Labor	Equipment	Total	Total Incl O&P
4800	12" long	1 Elec	32	.250	Ea.	25	13.35		38.35	47.50
4820	3/4" diameter, 2" long		26	.308		18.65	16.40		35.05	45
4840	2-1/2" long		26	.308		21	16.40		37.40	48
4860	3" long		26	.308		20.50	16.40		36.90	47
4880	3-1/2" long		26	.308		24	16.40		40.40	51
4900	4" long		26	.308		24	16.40		40.40	51
4920	5" long		26	.308		25	16.40		41.40	52
4940	6" long		26	.308		24	16.40		40.40	51
4960	8" long		26	.308		24	16.40		40.40	51
4980	10" long		26	.308		25	16.40		41.40	52
5000	12" long		26	.308		29	16.40		45.40	56.50
5020	1" diameter, 2" long		22	.364		20.50	19.40		39.90	51.50
5040	2-1/2" long		22	.364		24.50	19.40		43.90	56
5060	3" long		22	.364		24	19.40		43.40	55
5080	3-1/2" long		22	.364		25	19.40		44.40	56.50
5100	4" long		22	.364		27.50	19.40		46.90	59.50
5120	5" long		22	.364		28.50	19.40		47.90	60
5140	6" long		22	.364		26	19.40		45.40	58
5160	8" long		22	.364		27.50	19.40		46.90	59
5180	10" long		22	.364		28.50	19.40		47.90	60.50
5200	12" long		22	.364		35	19.40		54.40	67.50
5220	1-1/4" diameter, 2" long		18	.444		25.50	23.50		49	63.50
5240	2-1/2" long		18	.444		28	23.50		51.50	66
5260	3" long		18	.444		30	23.50		53.50	68.50
5280	3-1/2" long		18	.444		29.50	23.50		53	68
5300	4" long		18	.444		30.50	23.50		54	69
5320	5" long		18	.444		31	23.50		54.50	70
5340	6" long		18	.444		35.50	23.50		59	74.50
5360	8" long		18	.444		32.50	23.50		56	71.50
5380	10" long		18	.444		36.50	23.50		60	75.50
5400	12" long		18	.444		47.50	23.50		71	87.50
5420	1-1/2" diameter, 2" long		16	.500		29	26.50		55.50	72
5440	2-1/2" long		16	.500		31.50	26.50		58	74.50
5460	3" long		16	.500		31	26.50		57.50	74.50
5480	3-1/2" long		16	.500		33	26.50		59.50	76.50
5500	4" long		16	.500		37.50	26.50		64	81
5520	5" long		16	.500		38.50	26.50		65	82.50
5540	6" long		16	.500		40	26.50		66.50	84
5560	8" long		16	.500		40.50	26.50		67	84.50
5580	10" long		16	.500		48.50	26.50		75	93.50
5600	12" long		16	.500		61.50	26.50		88	108
5620	2" diameter, 2-1/2" long		14	.571		39	30.50		69.50	88.50
5640	3" long		14	.571		39	30.50		69.50	88.50
5660	3-1/2" long		14	.571		40	30.50		70.50	89.50
5680	4" long		14	.571		43	30.50		73.50	92.50
5700	5" long		14	.571		47	30.50		77.50	97
5720	6" long		14	.571		48.50	30.50		79	99
5740	8" long		14	.571		51.50	30.50		82	103
5760	10" long		14	.571		62.50	30.50		93	114
5780	12" long		14	.571		75.50	30.50		106	129
5800	2-1/2" diameter, 3-1/2" long		12	.667		77.50	35.50		113	138
5820	4" long		12	.667		79	35.50		114.50	140
5840	5" long		12	.667		91	35.50		126.50	153

26 05 33 – Raceway and Boxes for Electrical Systems

26 05 33.15 Conduit Nipples		Crew	Daily Output	Labor-Hours	Unit	Material	2014 Bare Costs Labor	Equipment	Total	Total Incl O&P
5860	6" long	1 Elec	12	.667	Ea.	100	35.50		135.50	163
5880	8" long		12	.667		105	35.50		140.50	169
5900	10" long		12	.667		114	35.50		149.50	179
5920	12" long		12	.667		129	35.50		164.50	195
5940	3" diameter, 3-1/2" long		11	.727		96.50	39		135.50	164
5960	4" long		11	.727		96	39		135	164
5980	5" long		11	.727		103	39		142	171
6000	6" long		11	.727		112	39		151	181
6020	8" long		11	.727		124	39		163	195
6040	10" long		11	.727		142	39		181	214
6060	12" long		11	.727		162	39		201	236
6080	3-1/2" diameter, 4" long		9	.889		146	47.50		193.50	231
6100	5" long		9	.889		147	47.50		194.50	233
6120	6" long		9	.889		165	47.50		212.50	252
6140	8" long		9	.889		175	47.50		222.50	264
6160	10" long		9	.889		192	47.50		239.50	283
6180	12" long		9	.889		210	47.50		257.50	300
6200	4" diameter, 4" long		7.50	1.067		159	57		216	259
6220	5" long		7.50	1.067		162	57		219	264
6240	6" long		7.50	1.067		176	57		233	279
6260	8" long		7.50	1.067		196	57		253	300
6280	10" long		7.50	1.067		232	57		289	340
6300	12" long		7.50	1.067		241	57		298	350
6320	5" diameter, 5" long		5.50	1.455		267	77.50		344.50	410
6340	6" long		5.50	1.455		283	77.50		360.50	425
6360	8" long		5.50	1.455		295	77.50		372.50	440
6380	10" long		5.50	1.455		315	77.50		392.50	465
6400	12" long		5.50	1.455		320	77.50		397.50	470
6420	6" diameter, 5" long		4.50	1.778		520	95		615	710
6440	6" long		4.50	1.778		535	95		630	725
6460	8" long		4.50	1.778		550	95		645	745
6480	10" long		4.50	1.778		570	95		665	770
6500	12" long		4.50	1.778		600	95		695	800

26 05 33.16 Outlet Boxes

		Crew	Daily Output	Labor-Hours	Unit	Material	2014 Bare Costs Labor	Equipment	Total	Total Incl O&P
0010	**OUTLET BOXES** R260533-65									
0020	Pressed steel, octagon, 4"	1 Elec	20	.400	Ea.	3.07	21.50		24.57	35.50
0040	For Romex or BX		20	.400		4.60	21.50		26.10	37
0050	For Romex or BX, with bracket		20	.400		7.25	21.50		28.75	40
0060	Covers, blank		64	.125		1.29	6.65		7.94	11.40
0100	Extension rings		40	.200		5.10	10.65		15.75	21.50
0150	Square, 4"		20	.400		2.10	21.50		23.60	34.50
0160	For Romex or BX		20	.400		8.20	21.50		29.70	41
0170	For Romex or BX, with bracket		20	.400		8.95	21.50		30.45	42
0200	Extension rings		40	.200		5.15	10.65		15.80	21.50
0220	2-1/8" deep, 1" KO		20	.400		6.80	21.50		28.30	39.50
0250	Covers, blank		64	.125		1.39	6.65		8.04	11.55
0260	Raised device		64	.125		2.61	6.65		9.26	12.85
0300	Plaster rings		64	.125		2.83	6.65		9.48	13.10
0350	Square, 4-11/16"		20	.400		7.20	21.50		28.70	40
0370	2-1/8" deep, 3/4" to 1-1/4" KO		20	.400		8.65	21.50		30.15	41.50
0400	Extension rings		40	.200		11.55	10.65		22.20	28.50
0450	Covers, blank		53	.151		2.60	8.05		10.65	14.90

26 05 33.16 Outlet Boxes		Crew	Daily Output	Labor-Hours	Unit	Material	2014 Bare Costs Labor	Equipment	Total	Total Incl O&P
0460	Raised device	1 Elec	53	.151	Ea.	7.75	8.05		15.80	20.50
0500	Plaster rings		53	.151		7.70	8.05		15.75	20.50
0550	Handy box		27	.296		3.39	15.80		19.19	27
0560	Covers, device		64	.125		1.19	6.65		7.84	11.30
0600	Extension rings		54	.148		4.05	7.90		11.95	16.30
0650	Switchbox		27	.296		4.91	15.80		20.71	29
0660	Romex or BX		27	.296		7.05	15.80		22.85	31.50
0670	with bracket		27	.296		8.15	15.80		23.95	32.50
0680	Partition, metal		27	.296		3.08	15.80		18.88	27
0700	Masonry, 1 gang, 2-1/2" deep		27	.296		9.40	15.80		25.20	34
0710	3-1/2" deep		27	.296		9.40	15.80		25.20	34
0750	2 gang, 2-1/2" deep		20	.400		17.60	21.50		39.10	51.50
0760	3-1/2" deep		20	.400		14.20	21.50		35.70	47.50
0800	3 gang, 2-1/2" deep		13	.615		21	33		54	72
0850	4 gang, 2-1/2" deep		10	.800		24.50	42.50		67	91
0860	5 gang, 2-1/2" deep		9	.889		26	47.50		73.50	99.50
0870	6 gang, 2-1/2" deep		8	1		53	53.50		106.50	139
0880	Masonry thru-the-wall, 1 gang, 4" block		16	.500		35	26.50		61.50	78.50
0890	6" block		16	.500		52.50	26.50		79	98
0900	8" block		16	.500		59	26.50		85.50	105
0920	2 gang, 6" block		16	.500		79	26.50		105.50	127
0940	Bar hanger with 3/8" stud, for wood and masonry boxes		53	.151		6.55	8.05		14.60	19.30
0950	Concrete, set flush, 4" deep		20	.400		14.40	21.50		35.90	48
1000	Plate with 3/8" stud		80	.100		9.55	5.35		14.90	18.50
1100	Concrete, floor, 1 gang		5.30	1.509		89	80.50		169.50	219
1150	2 gang		4	2		135	107		242	310
1200	3 gang		2.70	2.963		199	158		357	455
1250	For duplex receptacle, pedestal mounted, add		24	.333		95.50	17.80		113.30	132
1270	Flush mounted, add		27	.296		30	15.80		45.80	56.50
1300	For telephone, pedestal mounted, add		30	.267		124	14.25		138.25	159
1350	Carpet flange, 1 gang		53	.151		50	8.05		58.05	67
1400	Cast, 1 gang, FS (2" deep), 1/2" hub		12	.667		18.25	35.50		53.75	73
1410	3/4" hub		12	.667		19.50	35.50		55	74.50
1420	FD (2-11/16" deep), 1/2" hub		12	.667		17	35.50		52.50	71.50
1430	3/4" hub		12	.667		18.50	35.50		54	73.50
1450	2 gang, FS, 1/2" hub		10	.800		33	42.50		75.50	100
1460	3/4" hub		10	.800		35.50	42.50		78	103
1470	FD, 1/2" hub		10	.800		39.50	42.50		82	108
1480	3/4" hub		10	.800		39.50	42.50		82	108
1500	3 gang, FS, 3/4" hub		9	.889		56	47.50		103.50	133
1510	Switch cover, 1 gang, FS		64	.125		4.87	6.65		11.52	15.35
1520	2 gang		53	.151		7.40	8.05		15.45	20
1530	Duplex receptacle cover, 1 gang, FS		64	.125		4.97	6.65		11.62	15.45
1540	2 gang, FS		53	.151		7.55	8.05		15.60	20.50
1542	Weatherproof blank cover, 1 gang		64	.125		1.64	6.65		8.29	11.80
1544	2 gang		53	.151		3.50	8.05		11.55	15.90
1550	Weatherproof switch cover, 1 gang		64	.125		6.90	6.65		13.55	17.60
1554	2 gang		53	.151		14.65	8.05		22.70	28
1600	Weatherproof receptacle cover, 1 gang		64	.125		5.65	6.65		12.30	16.25
1604	2 gang		53	.151		14.65	8.05		22.70	28
1620	Weatherproof receptacle cover, tamper resistant, 1 gang		58	.138		13.70	7.35		21.05	26
1624	2 gang		48	.167		27	8.90		35.90	43.50
1750	FSC, 1 gang, 1/2" hub		11	.727		20	39		59	80

26 05 33 – Raceway and Boxes for Electrical Systems

26 05 33.16 Outlet Boxes	Crew	Daily Output	Labor-Hours	Unit	Material	2014 Bare Costs Labor	Equipment	Total	Total Incl O&P
1760	3/4" hub	1 Elec	11	.727	Ea.	22.50	39		61.50	83
1770	2 gang, 1/2" hub		9	.889		36.50	47.50		84	111
1780	3/4" hub		9	.889		39.50	47.50		87	115
1790	FDC, 1 gang, 1/2" hub		11	.727		23	39		62	83
1800	3/4" hub		11	.727		25.50	39		64.50	86.50
1810	2 gang, 1/2" hub		9	.889		47.50	47.50		95	123
1820	3/4" hub		9	.889		44.50	47.50		92	120
1850	Weatherproof in-use cover, 1 gang		64	.125		22.50	6.65		29.15	35
1870	2 gang		53	.151		31	8.05		39.05	46
2000	Poke-thru fitting, fire rated, for 3-3/4" floor		6.80	1.176		129	63		192	235
2040	For 7" floor		6.80	1.176		143	63		206	251
2100	Pedestal, 15 amp, duplex receptacle & blank plate		5.25	1.524		136	81.50		217.50	272
2120	Duplex receptacle and telephone plate		5.25	1.524		136	81.50		217.50	272
2140	Pedestal, 20 amp, duplex recept. & phone plate		5	1.600		137	85.50		222.50	279
2160	Telephone plate, both sides		5.25	1.524		130	81.50		211.50	265
2200	Abandonment plate		32	.250		38	13.35		51.35	62

26 05 33.17 Outlet Boxes, Plastic

		Crew	Daily Output	Labor-Hours	Unit	Material	2014 Bare Costs Labor	Equipment	Total	Total Incl O&P
0010	**OUTLET BOXES, PLASTIC**									
0050	4" diameter, round with 2 mounting nails	1 Elec	25	.320	Ea.	2.57	17.05		19.62	28.50
0100	Bar hanger mounted		25	.320		5.30	17.05		22.35	31.50
0200	4", square with 2 mounting nails		25	.320		4.54	17.05		21.59	30.50
0300	Plaster ring		64	.125		2.02	6.65		8.67	12.20
0400	Switch box with 2 mounting nails, 1 gang		30	.267		4.31	14.25		18.56	26
0500	2 gang		25	.320		3.68	17.05		20.73	29.50
0600	3 gang		20	.400		5.80	21.50		27.30	38.50
0700	Old work box		30	.267		6.70	14.25		20.95	29
1400	PVC, FSS, 1 gang, 1/2" hub		14	.571		19.25	30.50		49.75	66.50
1410	3/4" hub		14	.571		15.70	30.50		46.20	63
1420	FD, 1 gang for variable terminations		14	.571		12.95	30.50		43.45	60
1450	FS, 2 gang for variable terminations		12	.667		15.25	35.50		50.75	70
1480	Weatherproof blank cover, FS, 1 gang		64	.125		5.25	6.65		11.90	15.80
1500	2 gang		53	.151		5.75	8.05		13.80	18.40
1510	Weatherproof switch cover, FS, 1 gang		64	.125		9.65	6.65		16.30	20.50
1520	2 gang		53	.151		17.45	8.05		25.50	31.50
1530	Weatherproof duplex receptacle cover, FS, 1 gang		64	.125		14.60	6.65		21.25	26
1540	2 gang		53	.151		16.40	8.05		24.45	30
1750	FSC, 1 gang, 1/2" hub		13	.615		12.75	33		45.75	63
1760	3/4" hub		13	.615		13.70	33		46.70	64
1770	FSC, 2 gang, 1/2" hub		11	.727		19.60	39		58.60	79.50
1780	3/4" hub		11	.727		19.45	39		58.45	79.50
1790	FDC, 1 gang, 1/2" hub		13	.615		13.15	33		46.15	63.50
1800	3/4" hub		13	.615		14.35	33		47.35	65
1810	Weatherproof, T box w/3 holes		14	.571		11.30	30.50		41.80	58
1820	4" diameter round w/5 holes		14	.571		11.15	30.50		41.65	58
1850	In-use cover, 1 gang		64	.125		8.35	6.65		15	19.20
1870	2 gang		53	.151		10.50	8.05		18.55	23.50

26 05 33.18 Pull Boxes

		Crew	Daily Output	Labor-Hours	Unit	Material	2014 Bare Costs Labor	Equipment	Total	Total Incl O&P
0010	**PULL BOXES** R260533-70									
0100	Sheet metal, pull box, NEMA 1, type SC, 6" W x 6" H x 4" D	1 Elec	8	1	Ea.	10.75	53.50		64.25	92
0180	8" W x 6" H x 4" D		8	1		15.15	53.50		68.65	96.50
0200	8" W x 8" H x 4" D		8	1		15.30	53.50		68.80	97
0210	10" W x 10" H x 4" D		7	1.143		20	61		81	114

26 05 33.18 Pull Boxes		Crew	Daily Output	Labor-Hours	Unit	Material	2014 Bare Costs Labor	Equipment	Total	Total Incl O&P
0220	12" W x 12" H x 4" D	1 Elec	6.50	1.231	Ea.	26	65.50		91.50	127
0230	15" W x 15" H x 4" D		5.20	1.538		39.50	82		121.50	167
0240	18" W x 18" H x 4" D		4.40	1.818		40	97		137	189
0250	6" W x 6" H x 6" D		8	1		11.30	53.50		64.80	92.50
0260	8" W x 8" H x 6" D		7.50	1.067		15.25	57		72.25	102
0270	10" W x 10" H x 6" D		5.50	1.455		19.90	77.50		97.40	138
0300	10" W x 12" H x 6" D		5.30	1.509		29	80.50		109.50	153
0310	12" W x 12" H x 6" D		5.20	1.538		25	82		107	151
0320	15" W x 15" H x 6" D		4.60	1.739		43.50	93		136.50	187
0330	18" W x 18" H x 6" D		4.20	1.905		45	102		147	202
0340	24" W x 24" H x 6" D		3.20	2.500		90	133		223	299
0350	12" W x 12" H x 8" D		5	1.600		28.50	85.50		114	160
0360	15" W x 15" H x 8" D		4.50	1.778		44.50	95		139.50	191
0370	18" W x 18" H x 8" D		4	2		60.50	107		167.50	227
0380	24" W x 18" H x 6" D		3.70	2.162		82.50	115		197.50	264
0400	16" W x 20" H x 8" D		4	2		86	107		193	255
0500	20" W x 24" H x 8" D		3.20	2.500		97.50	133		230.50	305
0510	24" W x 24" H x 8" D		3	2.667		112	142		254	335
0600	24" W x 36" H x 8" D		2.70	2.963		140	158		298	390
0610	30" W x 30" H x 8" D		2.70	2.963		168	158		326	420
0620	36" W x 36" H x 8" D		2	4		210	213		423	550
0630	24" W x 24" H x 10" D		2.50	3.200		161	171		332	435
0650	Pull box, hinged , NEMA 1, 6" W x 6" H x 4" D		8	1		10.90	53.50		64.40	92
0660	8" W x 8" H x 4" D		8	1		15.45	53.50		68.95	97
0670	10" W x 10" H x 4" D		7	1.143		21	61		82	115
0680	12" W x 12" H x 4" D		6	1.333		26	71		97	135
0690	15" W x 15" H x 4" D		5.20	1.538		30	82		112	156
0700	18" W x 18" H x 4" D		4.40	1.818		34.50	97		131.50	183
0710	6" W x 6" H x 6" D		8	1		14.20	53.50		67.70	95.50
0720	8" W x 8" H x 6" D		7.50	1.067		18.85	57		75.85	106
0730	10" W x 10" H x 6" D		5.50	1.455		25.50	77.50		103	144
0740	12" W x 12" H x 6" D		5.20	1.538		30	82		112	157
0800	12" W x 16" H x 6" D		4.70	1.702		43	91		134	184
0810	15" W x 15" H x 6" D		4.60	1.739		32	93		125	175
0820	18" W x 18" H x 6" D		4.20	1.905		54.50	102		156.50	212
1000	20" W x 20" H x 6" D		3.60	2.222		74.50	119		193.50	259
1010	24" W x 24" H x 6" D		3.20	2.500		108	133		241	320
1020	12" W x 12" H x 8" D		5	1.600		65.50	85.50		151	200
1030	15" W x 15" H x 8" D		4.50	1.778		84.50	95		179.50	235
1040	18" W x 18" H x 8" D		4	2		119	107		226	291
1200	20" W x 20" H x 8" D		3.20	2.500		148	133		281	365
1210	24" W x 24" H x 8" D		3	2.667		163	142		305	390
1220	30" W x 30" H x 8" D		2.70	2.963		235	158		393	495
1400	24" W x 36" H x 8" D		2.70	2.963		228	158		386	490
1600	24" W x 42" H x 8" D		2	4		345	213		558	700
1610	36" W x 36" H x 8" D		2	4		320	213		533	670
2100	Pull box, NEMA 3R, type SC, raintight & weatherproof									
2150	6" L x 6" W x 6" D	1 Elec	10	.800	Ea.	18.85	42.50		61.35	84.50
2200	8" L x 6" W x 6" D		8	1		28	53.50		81.50	111
2250	10" L x 6" W x 6" D		7	1.143		37	61		98	132
2300	12" L x 12" W x 6" D		5	1.600		44	85.50		129.50	177
2350	16" L x 16" W x 6" D		4.50	1.778		115	95		210	268
2400	20" L x 20" W x 6" D		4	2		142	107		249	315

26 05 33 – Raceway and Boxes for Electrical Systems

26 05 33.18 Pull Boxes

		Crew	Daily Output	Labor-Hours	Unit	Material	2014 Bare Costs Labor	Equipment	Total	Total Incl O&P
2450	24" L x 18" W x 8" D	1 Elec	3	2.667	Ea.	125	142		267	350
2500	24" L x 24" W x 10" D		2.50	3.200		161	171		332	435
2550	30" L x 24" W x 12" D		2	4		380	213		593	735
2600	36" L x 36" W x 12" D		1.50	5.333		430	285		715	895
2800	Cast iron, pull boxes for surface mounting R260533-75									
3000	NEMA 4, watertight & dust tight									
3050	6" L x 6" W x 6" D	1 Elec	4	2	Ea.	248	107		355	430
3100	8" L x 6" W x 6" D		3.20	2.500		345	133		478	580
3150	10" L x 6" W x 6" D		2.50	3.200		385	171		556	680
3200	12" L x 12" W x 6" D		2.30	3.478		705	186		891	1,050
3250	16" L x 16" W x 6" D		1.30	6.154		1,125	330		1,455	1,750
3300	20" L x 20" W x 6" D		.80	10		2,100	535		2,635	3,125
3350	24" L x 18" W x 8" D		.70	11.429		2,600	610		3,210	3,800
3400	24" L x 24" W x 10" D		.50	16		4,925	855		5,780	6,700
3450	30" L x 24" W x 12" D		.40	20		9,025	1,075		10,100	11,500
3500	36" L x 36" W x 12" D		.20	40		7,025	2,125		9,150	10,900
3510	NEMA 4 clamp cover, 6" L x 6" W x 4" D		4	2		197	107		304	375
3520	8" L x 6" W x 4" D		4	2		230	107		337	415
4000	NEMA 7, explosionproof									
4050	6" L x 6" W x 6" D	1 Elec	2	4	Ea.	685	213		898	1,075
4100	8" L x 6" W x 6" D		1.80	4.444		920	237		1,157	1,350
4150	10" L x 6" W x 6" D		1.60	5		1,275	267		1,542	1,800
4200	12" L x 12" W x 6" D		1	8		2,150	425		2,575	3,000
4250	16" L x 14" W x 6" D		.60	13.333		3,025	710		3,735	4,400
4300	18" L x 18" W x 8" D		.50	16		5,575	855		6,430	7,425
4350	24" L x 18" W x 8" D		.40	20		7,100	1,075		8,175	9,400
4400	24" L x 24" W x 10" D		.30	26.667		9,675	1,425		11,100	12,700
4450	30" L x 24" W x 12" D		.20	40		14,100	2,125		16,225	18,700
5000	NEMA 9, dust tight 6" L x 6" W x 6" D		3.20	2.500		355	133		488	595
5050	8" L x 6" W x 6" D		2.70	2.963		425	158		583	705
5100	10" L x 6" W x 6" D		2	4		580	213		793	955
5150	12" L x 12" W x 6" D		1.60	5		1,125	267		1,392	1,625
5200	16" L x 16" W x 6" D		1	8		1,975	425		2,400	2,825
5250	18" L x 18" W x 8" D		.70	11.429		3,100	610		3,710	4,325
5300	24" L x 18" W x 8" D		.60	13.333		4,375	710		5,085	5,875
5350	24" L x 24" W x 10" D		.40	20		5,625	1,075		6,700	7,775
5400	30" L x 24" W x 12" D		.30	26.667		8,875	1,425		10,300	11,900
6000	J.I.C. wiring boxes, NEMA 12, dust tight & drip tight									
6050	6" L x 8" W x 4" D	1 Elec	10	.800	Ea.	51.50	42.50		94	121
6100	8" L x 10" W x 4" D		8	1		64	53.50		117.50	151
6150	12" L x 14" W x 6" D		5.30	1.509		109	80.50		189.50	240
6200	14" L x 16" W x 6" D		4.70	1.702		127	91		218	276
6250	16" L x 20" W x 6" D		4.40	1.818		224	97		321	390
6300	24" L x 30" W x 6" D		3.20	2.500		330	133		463	560
6350	24" L x 30" W x 8" D		2.90	2.759		355	147		502	610
6400	24" L x 36" W x 8" D		2.70	2.963		390	158		548	665
6450	24" L x 42" W x 8" D		2.30	3.478		425	186		611	750
6500	24" L x 48" W x 8" D		2	4		465	213		678	830

26 05 33.23 Wireway

		Crew	Daily Output	Labor-Hours	Unit	Material	2014 Bare Costs Labor	Equipment	Total	Total Incl O&P
0010	**WIREWAY** to 15' high R260533-60									
0020	For higher elevations, see Section 26 05 36.40									
0100	NEMA 1, Screw cover w/fittings and supports, 2-1/2" x 2-1/2"	1 Elec	45	.178	L.F.	10.35	9.50		19.85	25.50

26 05 33 – Raceway and Boxes for Electrical Systems

26 05 33.23 Wireway		Crew	Daily Output	Labor-Hours	Unit	Material	2014 Bare Costs Labor	Equipment	Total	Total Incl O&P
0200	4" x 4"	1 Elec	40	.200	L.F.	11.50	10.65		22.15	28.50
0400	6" x 6"	2 Elec	60	.267		19.50	14.25		33.75	43
0600	8" x 8"		40	.400		28	21.50		49.50	63
0620	10" x 10"		30	.533		47.50	28.50		76	95
0640	12" x 12"		20	.800		60.50	42.50		103	131
0800	Elbows, 90°, 2-1/2"	1 Elec	24	.333	Ea.	32.50	17.80		50.30	62
1000	4"		20	.400		36.50	21.50		58	72
1200	6"		18	.444		41	23.50		64.50	81
1400	8"		16	.500		66.50	26.50		93	113
1420	10"		12	.667		95	35.50		130.50	157
1440	12"		10	.800		121	42.50		163.50	197
1500	Elbows, 45°, 2-1/2"		24	.333		32.50	17.80		50.30	62
1510	4"		20	.400		39.50	21.50		61	75.50
1520	6"		18	.444		41	23.50		64.50	81
1530	8"		16	.500		66.50	26.50		93	113
1540	10"		12	.667		94.50	35.50		130	157
1550	12"		10	.800		168	42.50		210.50	249
1600	"T" box, 2-1/2"		18	.444		37	23.50		60.50	76.50
1800	4"		16	.500		45	26.50		71.50	89.50
2000	6"		14	.571		50	30.50		80.50	101
2200	8"		12	.667		92.50	35.50		128	155
2220	10"		10	.800		123	42.50		165.50	199
2240	12"		8	1		175	53.50		228.50	272
2300	Cross, 2-1/2"		16	.500		41	26.50		67.50	85.50
2310	4"		14	.571		50	30.50		80.50	101
2320	6"		12	.667		62	35.50		97.50	122
2400	Panel adapter, 2-1/2"		24	.333		11	17.80		28.80	38.50
2600	4"		20	.400		15.90	21.50		37.40	49.50
2800	6"		18	.444		17.05	23.50		40.55	54.50
3000	8"		16	.500		23.50	26.50		50	65.50
3020	10"		14	.571		43	30.50		73.50	93
3040	12"		12	.667		55	35.50		90.50	114
3200	Reducer, 4" to 2-1/2"		24	.333		17.65	17.80		35.45	46
3400	6" to 4"		20	.400		35.50	21.50		57	71
3600	8" to 6"		18	.444		38	23.50		61.50	77.50
3620	10" to 8"		16	.500		47.50	26.50		74	92
3640	12" to 10"		14	.571		54.50	30.50		85	106
3780	End cap, 2-1/2"		24	.333		4.60	17.80		22.40	31.50
3800	4"		20	.400		5.70	21.50		27.20	38.50
4000	6"		18	.444		6.80	23.50		30.30	43
4200	8"		16	.500		9.25	26.50		35.75	50
4220	10"		14	.571		15.05	30.50		45.55	62
4240	12"		12	.667		21	35.50		56.50	76
4300	U-connector, 2-1/2"		200	.040		4.60	2.13		6.73	8.25
4320	4"		200	.040		5.70	2.13		7.83	9.45
4340	6"		180	.044		6.80	2.37		9.17	11.05
4360	8"		170	.047		13.80	2.51		16.31	18.95
4380	10"		150	.053		20	2.85		22.85	26.50
4400	12"		130	.062		27.50	3.28		30.78	35.50
4420	Hanger, 2-1/2"		100	.080		11.75	4.27		16.02	19.30
4430	4"		100	.080		15	4.27		19.27	23
4440	6"		80	.100		19.25	5.35		24.60	29
4450	8"		65	.123		24.50	6.55		31.05	36.50

26 05 33.23 Wireway		Crew	Daily Output	Labor-Hours	Unit	Material	2014 Bare Costs Labor	Equipment	Total	Total Incl O&P
4460	10"	1 Elec	50	.160	Ea.	44.50	8.55		53.05	62
4470	12"		40	.200		64.50	10.65		75.15	87
4475	NEMA 3R, Screw cover w/fittings and supports, 4" x 4"		36	.222	L.F.	19.45	11.85		31.30	39.50
4480	6" x 6"	2 Elec	55	.291		26	15.50		41.50	51.50
4485	8" x 8"		36	.444		40	23.50		63.50	79.50
4490	12" x 12"		18	.889		64	47.50		111.50	142
4500	Hinged cover, with fittings and supports, 2-1/2" x 2-1/2"	1 Elec	60	.133		13.10	7.10		20.20	25
4520	4" x 4"	"	45	.178		16.70	9.50		26.20	32.50
4540	6" x 6"	2 Elec	80	.200		27	10.65		37.65	45.50
4560	8" x 8"		60	.267		43.50	14.25		57.75	69.50
4580	10" x 10"		50	.320		55.50	17.05		72.55	86.50
4600	12" x 12"		24	.667		79.50	35.50		115	141
4700	Elbows 90°, 2-1/2" x 2-1/2"	1 Elec	32	.250	Ea.	46	13.35		59.35	71
4720	4"		27	.296		54	15.80		69.80	82.50
4730	6"		23	.348		63	18.55		81.55	97.50
4740	8"		18	.444		97	23.50		120.50	142
4750	10"		14	.571		108	30.50		138.50	165
4760	12"		12	.667		173	35.50		208.50	244
4800	Tee box, hinged cover, 2-1/2" x 2-1/2"		23	.348		53.50	18.55		72.05	87
4810	4"		20	.400		65	21.50		86.50	104
4820	6"		18	.444		72	23.50		95.50	115
4830	8"		16	.500		135	26.50		161.50	189
4840	10"		12	.667		155	35.50		190.50	223
4860	12"		10	.800		194	42.50		236.50	277
4880	Cross box, hinged cover, 2-1/2" x 2-1/2"		18	.444		59.50	23.50		83	101
4900	4"		16	.500		73	26.50		99.50	120
4920	6"		13	.615		91.50	33		124.50	150
4940	8"		11	.727		135	39		174	206
4960	10"		10	.800		229	42.50		271.50	315
4980	12"		9	.889		251	47.50		298.50	345
5000	NEMA 12, Hinged cover, 2-1/2" x 2-1/2"		40	.200	L.F.	44	10.65		54.65	64.50
5020	4" x 4"		35	.229		51.50	12.20		63.70	75.50
5040	6" x 6"	2 Elec	60	.267		73.50	14.25		87.75	103
5060	8" x 8"	"	50	.320		100	17.05		117.05	136
5120	Elbows 90°, flanged, 2-1/2" x 2-1/2"	1 Elec	23	.348	Ea.	80	18.55		98.55	116
5140	4"		20	.400		101	21.50		122.50	143
5160	6"		18	.444		128	23.50		151.50	177
5180	8"		15	.533		187	28.50		215.50	248
5240	Tee box, flanged, 2-1/2" x 2-1/2"		18	.444		109	23.50		132.50	156
5260	4"		16	.500		128	26.50		154.50	181
5280	6"		15	.533		173	28.50		201.50	234
5300	8"		13	.615		252	33		285	325
5360	Cross box, flanged, 2-1/2" x 2-1/2"		15	.533		134	28.50		162.50	191
5380	4"		13	.615		175	33		208	241
5400	6"		12	.667		227	35.50		262.50	300
5420	8"		10	.800		267	42.50		309.50	355
5480	Flange gasket, 2-1/2"		160	.050		3.97	2.67		6.64	8.35
5500	4"		80	.100		5.50	5.35		10.85	14.05
5520	6"		53	.151		7.60	8.05		15.65	20.50
5530	8"		40	.200		9.85	10.65		20.50	27

26 05 33.25 Conduit Fittings for Rigid Galvanized Steel	Crew	Daily Output	Labor-Hours	Unit	Material	2014 Bare Costs Labor	Equipment	Total	Total Incl O&P
0010 **CONDUIT FITTINGS FOR RIGID GALVANIZED STEEL**									
0050 Standard, locknuts, 1/2" diameter				Ea.	.20			.20	.22
0100 3/4" diameter					.36			.36	.40
0300 1" diameter					.61			.61	.67
0500 1-1/4" diameter					.71			.71	.78
0700 1-1/2" diameter					1.19			1.19	1.31
1000 2" diameter					1.72			1.72	1.89
1030 2-1/2" diameter					4.81			4.81	5.30
1050 3" diameter					6.15			6.15	6.75
1070 3-1/2" diameter					10.35			10.35	11.40
1100 4" diameter					12.95			12.95	14.25
1110 5" diameter					28			28	30.50
1120 6" diameter					66.50			66.50	73.50
1130 Bushings, plastic, 1/2" diameter	1 Elec	40	.200		.19	10.65		10.84	16.15
1150 3/4" diameter		32	.250		.31	13.35		13.66	20.50
1170 1" diameter		28	.286		.51	15.25		15.76	23.50
1200 1-1/4" diameter		24	.333		.77	17.80		18.57	27.50
1230 1-1/2" diameter		18	.444		1.04	23.50		24.54	36.50
1250 2" diameter		15	.533		1.93	28.50		30.43	44.50
1270 2-1/2" diameter		13	.615		4.48	33		37.48	54
1300 3" diameter		12	.667		4.62	35.50		40.12	58
1330 3-1/2" diameter		11	.727		6.10	39		45.10	64.50
1350 4" diameter		9	.889		7.50	47.50		55	79.50
1360 5" diameter		7	1.143		16.70	61		77.70	110
1370 6" diameter		5	1.600		36	85.50		121.50	168
1390 Steel, 1/2" diameter		40	.200		.57	10.65		11.22	16.60
1400 3/4" diameter		32	.250		.67	13.35		14.02	20.50
1430 1" diameter		28	.286		1.22	15.25		16.47	24.50
1450 Steel insulated, 1-1/4" diameter		24	.333		5.40	17.80		23.20	32.50
1470 1-1/2" diameter		18	.444		6.85	23.50		30.35	43
1500 2" diameter		15	.533		9.75	28.50		38.25	53.50
1530 2-1/2" diameter		13	.615		18.15	33		51.15	69
1550 3" diameter		12	.667		24.50	35.50		60	80
1570 3-1/2" diameter		11	.727		31	39		70	92.50
1600 4" diameter		9	.889		39.50	47.50		87	115
1610 5" diameter		7	1.143		86.50	61		147.50	187
1620 6" diameter		5	1.600		162	85.50		247.50	305
1630 Sealing locknuts, 1/2" diameter		40	.200		1.28	10.65		11.93	17.35
1650 3/4" diameter		32	.250		1.37	13.35		14.72	21.50
1670 1" diameter		28	.286		2.01	15.25		17.26	25
1700 1-1/4" diameter		24	.333		3.25	17.80		21.05	30
1730 1-1/2" diameter		18	.444		4.05	23.50		27.55	40
1750 2" diameter		15	.533		5.25	28.50		33.75	48.50
1760 Grounding bushing, insulated, 1/2" diameter		32	.250		5.80	13.35		19.15	26.50
1770 3/4" diameter		28	.286		6.65	15.25		21.90	30.50
1780 1" diameter		20	.400		8.25	21.50		29.75	41
1800 1-1/4" diameter		18	.444		10.55	23.50		34.05	47
1830 1-1/2" diameter		16	.500		12.15	26.50		38.65	53.50
1850 2" diameter		13	.615		16.90	33		49.90	67.50
1870 2-1/2" diameter		12	.667		27.50	35.50		63	83.50
1900 3" diameter		11	.727		29	39		68	90
1930 3-1/2" diameter		9	.889		33	47.50		80.50	108

26 05 33 – Raceway and Boxes for Electrical Systems

26 05 33.25 Conduit Fittings for Rigid Galvanized Steel	Crew	Daily Output	Labor-Hours	Unit	Material	2014 Bare Costs Labor	Equipment	Total	Total Incl O&P
1950	4" diameter	1 Elec	8	1	Ea.	45	53.50		98.50	130
1960	5" diameter		6	1.333		88.50	71		159.50	204
1970	6" diameter		4	2		135	107		242	310
1990	Coupling, with set screw, 1/2" diameter		50	.160		4.13	8.55		12.68	17.35
2000	3/4" diameter		40	.200		5.25	10.65		15.90	22
2030	1" diameter		35	.229		8.65	12.20		20.85	28
2050	1-1/4" diameter		28	.286		14.75	15.25		30	39
2070	1-1/2" diameter		23	.348		18.95	18.55		37.50	49
2090	2" diameter		20	.400		42.50	21.50		64	78.50
2100	2-1/2" diameter		18	.444		90.50	23.50		114	135
2110	3" diameter		15	.533		107	28.50		135.50	161
2120	3-1/2" diameter		12	.667		154	35.50		189.50	222
2130	4" diameter		10	.800		201	42.50		243.50	285
2140	5" diameter		9	.889		325	47.50		372.50	430
2150	6" diameter		8	1		530	53.50		583.50	665
2160	Box connector with set screw, plain, 1/2" diameter		70	.114		2.70	6.10		8.80	12.10
2170	3/4" diameter		60	.133		3.86	7.10		10.96	14.90
2180	1" diameter		50	.160		6.10	8.55		14.65	19.50
2190	Insulated, 1-1/4" diameter		40	.200		12.80	10.65		23.45	30
2200	1-1/2" diameter		30	.267		20.50	14.25		34.75	44
2210	2" diameter		20	.400		38	21.50		59.50	74
2220	2-1/2" diameter		18	.444		101	23.50		124.50	147
2230	3" diameter		15	.533		147	28.50		175.50	205
2240	3-1/2" diameter		12	.667		207	35.50		242.50	281
2250	4" diameter		10	.800		267	42.50		309.50	355
2260	5" diameter		9	.889		258	47.50		305.50	355
2270	6" diameter		8	1		330	53.50		383.50	445
2280	LB, LR or LL fittings & covers, 1/2" diameter		16	.500		8.35	26.50		34.85	49
2290	3/4" diameter		13	.615		10.35	33		43.35	60.50
2300	1" diameter		11	.727		15.10	39		54.10	74.50
2330	1-1/4" diameter		8	1		23.50	53.50		77	106
2350	1-1/2" diameter		6	1.333		31	71		102	141
2370	2" diameter		5	1.600		51	85.50		136.50	184
2380	2-1/2" diameter		4	2		107	107		214	278
2390	3" diameter		3.50	2.286		138	122		260	335
2400	3-1/2" diameter		3	2.667		240	142		382	475
2410	4" diameter		2.50	3.200		262	171		433	545
2420	T fittings, with cover, 1/2" diameter		12	.667		10.70	35.50		46.20	65
2430	3/4" diameter		11	.727		12.35	39		51.35	71.50
2440	1" diameter		9	.889		21	47.50		68.50	94
2450	1-1/4" diameter		6	1.333		27	71		98	136
2470	1-1/2" diameter		5	1.600		37.50	85.50		123	169
2500	2" diameter		4	2		55	107		162	221
2510	2-1/2" diameter		3.50	2.286		121	122		243	315
2520	3" diameter		3	2.667		168	142		310	395
2530	3-1/2" diameter		2.50	3.200		293	171		464	580
2540	4" diameter		2	4		350	213		563	705
2550	Nipples chase, plain, 1/2" diameter		40	.200		.81	10.65		11.46	16.85
2560	3/4" diameter		32	.250		1.02	13.35		14.37	21
2570	1" diameter		28	.286		2.08	15.25		17.33	25.50
2600	Insulated, 1-1/4" diameter		24	.333		7.70	17.80		25.50	35
2630	1-1/2" diameter		18	.444		10.20	23.50		33.70	47
2650	2" diameter		15	.533		15.35	28.50		43.85	59.50

26 05 33.25 Conduit Fittings for Rigid Galvanized Steel	Crew	Daily Output	Labor-Hours	Unit	Material	2014 Bare Costs Labor	Equipment	Total	Total Incl O&P
2660	2-1/2" diameter	1 Elec	12	.667	Ea.	43	35.50		78.50	100
2670	3" diameter		10	.800		39.50	42.50		82	108
2680	3-1/2" diameter		9	.889		72	47.50		119.50	150
2690	4" diameter		8	1		85	53.50		138.50	174
2700	5" diameter		7	1.143		249	61		310	365
2710	6" diameter		6	1.333		385	71		456	530
2720	Nipples offset, plain, 1/2" diameter		40	.200		5.20	10.65		15.85	21.50
2730	3/4" diameter		32	.250		5.45	13.35		18.80	26
2740	1" diameter		24	.333		7.15	17.80		24.95	34.50
2750	Insulated, 1-1/4" diameter		20	.400		29	21.50		50.50	64
2760	1-1/2" diameter		18	.444		36.50	23.50		60	75.50
2770	2" diameter		16	.500		56	26.50		82.50	102
2780	3" diameter		14	.571		114	30.50		144.50	172
2850	Coupling, expansion, 1/2" diameter		12	.667		42.50	35.50		78	99.50
2880	3/4" diameter		10	.800		49.50	42.50		92	119
2900	1" diameter		8	1		57.50	53.50		111	144
2920	1-1/4" diameter		6.40	1.250		70.50	66.50		137	178
2940	1-1/2" diameter		5.30	1.509		109	80.50		189.50	241
2960	2" diameter		4.60	1.739		164	93		257	320
2980	2-1/2" diameter		3.60	2.222		229	119		348	430
3000	3" diameter		3	2.667		305	142		447	550
3020	3-1/2" diameter		2.80	2.857		400	152		552	670
3040	4" diameter		2.40	3.333		565	178		743	885
3060	5" diameter		2	4		1,000	213		1,213	1,425
3080	6" diameter		1.80	4.444		1,625	237		1,862	2,125
3100	Expansion deflection, 1/2" diameter		12	.667		247	35.50		282.50	325
3120	3/4" diameter		12	.667		277	35.50		312.50	360
3140	1" diameter		10	.800		288	42.50		330.50	380
3160	1-1/4" diameter		6.40	1.250		345	66.50		411.50	480
3180	1-1/2" diameter		5.30	1.509		400	80.50		480.50	560
3200	2" diameter		4.60	1.739		500	93		593	685
3220	2-1/2" diameter		3.60	2.222		675	119		794	920
3240	3" diameter		3	2.667		830	142		972	1,125
3260	3-1/2" diameter		2.80	2.857		1,075	152		1,227	1,400
3280	4" diameter		2.40	3.333		1,200	178		1,378	1,600
3300	5" diameter		2	4		1,850	213		2,063	2,375
3320	6" diameter		1.80	4.444		3,225	237		3,462	3,900
3340	Ericson, 1/2" diameter		16	.500		5.80	26.50		32.30	46.50
3360	3/4" diameter		14	.571		7.45	30.50		37.95	53.50
3380	1" diameter		11	.727		13.75	39		52.75	73
3400	1-1/4" diameter		8	1		29.50	53.50		83	112
3420	1-1/2" diameter		7	1.143		35	61		96	130
3440	2" diameter		5	1.600		70.50	85.50		156	206
3460	2-1/2" diameter		4	2		171	107		278	350
3480	3" diameter		3.50	2.286		260	122		382	470
3500	3-1/2" diameter		3	2.667		430	142		572	690
3520	4" diameter		2.70	2.963		510	158		668	795
3540	5" diameter		2.50	3.200		940	171		1,111	1,275
3560	6" diameter		2.30	3.478		1,250	186		1,436	1,650
3580	Split, 1/2" diameter		32	.250		3.87	13.35		17.22	24
3600	3/4" diameter		27	.296		4.85	15.80		20.65	29
3620	1" diameter		20	.400		8.95	21.50		30.45	42
3640	1-1/4" diameter		16	.500		12.95	26.50		39.45	54.50

26 05 33.25 Conduit Fittings for Rigid Galvanized Steel		Crew	Daily Output	Labor-Hours	Unit	Material	2014 Bare Costs Labor	Equipment	Total	Total Incl O&P
3660	1-1/2" diameter	1 Elec	14	.571	Ea.	16.20	30.50		46.70	63.50
3680	2" diameter		12	.667		32.50	35.50		68	88.50
3700	2-1/2" diameter		10	.800		75	42.50		117.50	147
3720	3" diameter		9	.889		113	47.50		160.50	195
3740	3-1/2" diameter		8	1		181	53.50		234.50	280
3760	4" diameter		7	1.143		213	61		274	325
3780	5" diameter		6	1.333		370	71		441	515
3800	6" diameter		5	1.600		500	85.50		585.50	680
4600	Reducing bushings, 3/4" to 1/2" diameter		54	.148		2.05	7.90		9.95	14.10
4620	1" to 3/4" diameter		46	.174		2.62	9.30		11.92	16.80
4640	1-1/4" to 1" diameter		40	.200		6.10	10.65		16.75	22.50
4660	1-1/2" to 1-1/4" diameter		36	.222		8.80	11.85		20.65	27.50
4680	2" to 1-1/2" diameter		32	.250		15.90	13.35		29.25	37.50
4740	2-1/2" to 2" diameter		30	.267		17.65	14.25		31.90	41
4760	3" to 2-1/2" diameter		28	.286		19.40	15.25		34.65	44.50
4800	Through-wall seal, 1/2" diameter		8	1		189	53.50		242.50	288
4820	3/4" diameter		7.50	1.067		229	57		286	335
4840	1" diameter		6.50	1.231		229	65.50		294.50	350
4860	1-1/4" diameter		5.50	1.455		281	77.50		358.50	425
4880	1-1/2" diameter		5	1.600		355	85.50		440.50	520
4900	2" diameter		4.20	1.905		320	102		422	500
4920	2-1/2" diameter		3.50	2.286		430	122		552	660
4940	3" diameter		3	2.667		460	142		602	720
4960	3-1/2" diameter		2.50	3.200		695	171		866	1,025
4980	4" diameter		2	4		705	213		918	1,100
5000	5" diameter		1.50	5.333		910	285		1,195	1,425
5020	6" diameter		1	8		910	425		1,335	1,650
5100	Cable supports, 2 or more wires									
5120	1-1/2" diameter	1 Elec	8	1	Ea.	132	53.50		185.50	225
5140	2" diameter		6	1.333		166	71		237	288
5160	2-1/2" diameter		4	2		181	107		288	360
5180	3" diameter		3.50	2.286		231	122		353	435
5200	3-1/2" diameter		2.60	3.077		310	164		474	590
5220	4" diameter		2	4		385	213		598	745
5240	5" diameter		1.50	5.333		575	285		860	1,050
5260	6" diameter		1	8		935	425		1,360	1,675
5280	Service entrance cap, 1/2" diameter		16	.500		6.65	26.50		33.15	47.50
5300	3/4" diameter		13	.615		7.60	33		40.60	57.50
5320	1" diameter		10	.800		9.25	42.50		51.75	74
5340	1-1/4" diameter		8	1		12.40	53.50		65.90	93.50
5360	1-1/2" diameter		6.50	1.231		37	65.50		102.50	140
5380	2" diameter		5.50	1.455		37	77.50		114.50	157
5400	2-1/2" diameter		4	2		99	107		206	269
5420	3" diameter		3.40	2.353		161	126		287	365
5440	3-1/2" diameter		3	2.667		231	142		373	465
5460	4" diameter		2.70	2.963		300	158		458	565
5600	Fire stop fittings, to 3/4" diameter		24	.333		5.15	17.80		22.95	32
5610	1" diameter		22	.364		5.75	19.40		25.15	35.50
5620	1-1/2" diameter		20	.400		8.95	21.50		30.45	42
5640	2" diameter		16	.500		8.50	26.50		35	49.50
5660	3" diameter		12	.667		11.45	35.50		46.95	65.50
5680	4" diameter		10	.800		16.70	42.50		59.20	82.50
5700	6" diameter		8	1		58	53.50		111.50	144

26 05 33.25 Conduit Fittings for Rigid Galvanized Steel		Crew	Daily Output	Labor-Hours	Unit	Material	2014 Bare Costs Labor	Equipment	Total	Total Incl O&P
5750	90° pull elbows steel, female, 1/2" diameter	1 Elec	16	.500	Ea.	7.70	26.50		34.20	48.50
5760	3/4" diameter		13	.615		9	33		42	59
5780	1" diameter		11	.727		15.25	39		54.25	75
5800	1-1/4" diameter		8	1		22.50	53.50		76	105
5820	1-1/2" diameter		6	1.333		32	71		103	142
5840	2" diameter	↓	5	1.600	↓	52	85.50		137.50	186
6000	Explosion proof, flexible coupling									
6010	1/2" diameter, 4" long	1 Elec	12	.667	Ea.	118	35.50		153.50	182
6020	6" long		12	.667		104	35.50		139.50	167
6050	12" long		12	.667		157	35.50		192.50	226
6070	18" long		12	.667		186	35.50		221.50	258
6090	24" long		12	.667		236	35.50		271.50	315
6110	30" long		12	.667		310	35.50		345.50	400
6130	36" long		12	.667		298	35.50		333.50	385
6140	3/4" diameter, 4" long		10	.800		144	42.50		186.50	223
6150	6" long		10	.800		126	42.50		168.50	202
6180	12" long		10	.800		188	42.50		230.50	271
6200	18" long		10	.800		236	42.50		278.50	325
6220	24" long		10	.800		295	42.50		337.50	390
6240	30" long		10	.800		410	42.50		452.50	515
6260	36" long		10	.800		350	42.50		392.50	450
6270	1" diameter, 6" long		8	1		335	53.50		388.50	445
6300	12" long		8	1		320	53.50		373.50	430
6320	18" long		8	1		395	53.50		448.50	510
6340	24" long		8	1		515	53.50		568.50	650
6360	30" long		8	1		690	53.50		743.50	840
6380	36" long		8	1		755	53.50		808.50	910
6390	1-1/4" diameter, 12" long		6.40	1.250		585	66.50		651.50	740
6410	18" long		6.40	1.250		685	66.50		751.50	855
6430	24" long		6.40	1.250		790	66.50		856.50	965
6450	30" long		6.40	1.250		830	66.50		896.50	1,025
6470	36" long		6.40	1.250		1,300	66.50		1,366.50	1,525
6480	1-1/2" diameter, 12" long		5.30	1.509		605	80.50		685.50	785
6500	18" long		5.30	1.509		730	80.50		810.50	925
6520	24" long		5.30	1.509		1,075	80.50		1,155.50	1,300
6540	30" long		5.30	1.509		1,200	80.50		1,280.50	1,450
6560	36" long		5.30	1.509		1,275	80.50		1,355.50	1,525
6570	2" diameter, 12" long		4.60	1.739		1,025	93		1,118	1,275
6590	18" long		4.60	1.739		1,325	93		1,418	1,600
6610	24" long		4.60	1.739		1,425	93		1,518	1,725
6630	30" long		4.60	1.739		1,375	93		1,468	1,650
6650	36" long		4.60	1.739		1,800	93		1,893	2,150
7000	Close up plug, 1/2" diameter, explosion proof		40	.200		2.20	10.65		12.85	18.35
7010	3/4" diameter		32	.250		2.59	13.35		15.94	23
7020	1" diameter		28	.286		2.98	15.25		18.23	26.50
7030	1-1/4" diameter		24	.333		3.29	17.80		21.09	30
7040	1-1/2" diameter		18	.444		4.54	23.50		28.04	40.50
7050	2" diameter		15	.533		7.85	28.50		36.35	51
7060	2-1/2" diameter		13	.615		12.90	33		45.90	63
7070	3" diameter		12	.667		19.15	35.50		54.65	74
7080	3-1/2" diameter		11	.727		24	39		63	84.50
7090	4" diameter		9	.889		30	47.50		77.50	104
7091	Elbow female, 45°, 1/2"		16	.500		11.30	26.50		37.80	52.50

26 05 33.25 Conduit Fittings for Rigid Galvanized Steel	Crew	Daily Output	Labor-Hours	Unit	Material	2014 Bare Costs Labor	Equipment	Total	Total Incl O&P
7092	3/4"	1 Elec	13	.615	Ea.	14.05	33		47.05	64.50
7093	1"		11	.727		16.85	39		55.85	76.50
7094	1-1/4"		8	1		25.50	53.50		79	108
7095	1-1/2"		6	1.333		27.50	71		98.50	136
7096	2"		5	1.600		32	85.50		117.50	163
7097	2-1/2"		4.50	1.778		93.50	95		188.50	245
7098	3"		4.20	1.905		94.50	102		196.50	256
7099	3-1/2"		4	2		146	107		253	320
7100	4"		3.80	2.105		158	112		270	340
7101	90°, 1/2"		16	.500		11.45	26.50		37.95	52.50
7102	3/4"		13	.615		12.35	33		45.35	62.50
7103	1"		11	.727		16.85	39		55.85	76.50
7104	1-1/4"		8	1		27	53.50		80.50	110
7105	1-1/2"		6	1.333		37.50	71		108.50	147
7106	2"		5	1.600		59	85.50		144.50	193
7107	2-1/2"		4.50	1.778		138	95		233	293
7110	Elbows 90°, long male & female, 1/2" diameter, explosion proof		16	.500		16.70	26.50		43.20	58.50
7120	3/4" diameter		13	.615		19.30	33		52.30	70
7130	1" diameter		11	.727		25	39		64	85.50
7140	1-1/4" diameter		8	1		29.50	53.50		83	113
7150	1-1/2" diameter		6	1.333		36	71		107	146
7160	2" diameter		5	1.600		53.50	85.50		139	187
7170	Capped elbow, 1/2" diameter, explosion proof		11	.727		18.25	39		57.25	78
7180	3/4" diameter		8	1		22	53.50		75.50	105
7190	1" diameter		6	1.333		24.50	71		95.50	133
7200	1-1/4" diameter		5	1.600		48.50	85.50		134	182
7210	Pulling elbow, 1/2" diameter, explosion proof		11	.727		72	39		111	138
7220	3/4" diameter		8	1		126	53.50		179.50	219
7230	1" diameter		6	1.333		181	71		252	305
7240	1-1/4" diameter		5	1.600		241	85.50		326.50	395
7250	1-1/2" diameter		5	1.600		300	85.50		385.50	460
7260	2" diameter		4	2		293	107		400	480
7270	2-1/2" diameter		3.50	2.286		725	122		847	985
7280	3" diameter		3	2.667		1,000	142		1,142	1,325
7290	3-1/2" diameter		2.50	3.200		1,275	171		1,446	1,650
7300	4" diameter		2.20	3.636		1,300	194		1,494	1,725
7310	LB conduit body, 1/2" diameter		11	.727		42.50	39		81.50	105
7320	3/4" diameter		8	1		51.50	53.50		105	137
7330	T conduit body, 1/2" diameter		9	.889		46	47.50		93.50	122
7340	3/4" diameter		6	1.333		54.50	71		125.50	166
7350	Explosion proof, round box w/cover, 3 threaded hubs, 1/2" diameter		8	1		44.50	53.50		98	129
7351	3/4" diameter		8	1		63	53.50		116.50	149
7352	1" diameter		7.50	1.067		68	57		125	160
7353	1-1/4" diameter		7	1.143		77.50	61		138.50	177
7354	1-1/2" diameter		7	1.143		177	61		238	286
7355	2" diameter		6	1.333		183	71		254	305
7356	Round box w/cover & mtng flange, 3 threaded hubs, 1/2" diameter		8	1		49.50	53.50		103	135
7357	3/4" diameter		8	1		56	53.50		109.50	142
7358	4 threaded hubs, 1" diameter		7	1.143		62	61		123	160
7400	Unions, 1/2" diameter		20	.400		11.40	21.50		32.90	44.50
7410	3/4" - 1/2" diameter		16	.500		16.45	26.50		42.95	58
7420	3/4" diameter		16	.500		15.65	26.50		42.15	57
7430	1" diameter		14	.571		28.50	30.50		59	76.50

26 05 33.25 Conduit Fittings for Rigid Galvanized Steel	Crew	Daily Output	Labor-Hours	Unit	Material	2014 Bare Costs Labor	Equipment	Total	Total Incl O&P
7440	1-1/4" diameter	1 Elec	12	.667	Ea.	41.50	35.50		77	99
7450	1-1/2" diameter		10	.800		53	42.50		95.50	123
7460	2" diameter		8.50	.941		68.50	50		118.50	150
7480	2-1/2" diameter		8	1		107	53.50		160.50	198
7490	3" diameter		7	1.143		137	61		198	243
7500	3-1/2" diameter		6	1.333		258	71		329	390
7510	4" diameter		5	1.600		260	85.50		345.50	415
7680	Reducer, 3/4" to 1/2"		54	.148		2.32	7.90		10.22	14.40
7690	1" to 1/2"		46	.174		4.16	9.30		13.46	18.50
7700	1" to 3/4"		46	.174		4.20	9.30		13.50	18.50
7710	1-1/4" to 3/4"		40	.200		5.25	10.65		15.90	21.50
7720	1-1/4" to 1"		40	.200		6.30	10.65		16.95	23
7730	1-1/2" to 1"		36	.222		9.80	11.85		21.65	28.50
7740	1-1/2" to 1-1/4"		36	.222		10.25	11.85		22.10	29
7750	2" to 3/4"		32	.250		13.30	13.35		26.65	34.50
7760	2" to 1-1/4"		32	.250		11.55	13.35		24.90	32.50
7770	2" to 1-1/2"		32	.250		12.25	13.35		25.60	33.50
7780	2-1/2" to 1-1/2"		30	.267		20.50	14.25		34.75	44
7790	3" to 2"		30	.267		21	14.25		35.25	44.50
7800	3-1/2" to 2-1/2"		28	.286		42.50	15.25		57.75	70
7810	4" to 3"		28	.286		51.50	15.25		66.75	79.50
7820	Sealing fitting, vertical/horizontal, 1/2" diameter		14.50	.552		14.65	29.50		44.15	60
7830	3/4" diameter		13.30	.602		17.10	32		49.10	67
7840	1" diameter		11.40	.702		21.50	37.50		59	79.50
7850	1-1/4" diameter		10	.800		27	42.50		69.50	94
7860	1-1/2" diameter		8.80	.909		41	48.50		89.50	118
7870	2" diameter		8	1		53.50	53.50		107	139
7880	2-1/2" diameter		6.70	1.194		74	63.50		137.50	177
7890	3" diameter		5.70	1.404		95.50	75		170.50	217
7900	3-1/2" diameter		4.70	1.702		252	91		343	415
7910	4" diameter		4	2		425	107		532	625
7920	Sealing hubs, 1" by 1-1/2"		12	.667		37	35.50		72.50	94
7930	1-1/4" by 2"		10	.800		47	42.50		89.50	116
7940	1-1/2" by 2"		9	.889		62	47.50		109.50	139
7950	2" by 2-1/2"		8	1		93.50	53.50		147	183
7960	3" by 4"		7	1.143		137	61		198	243
7970	4" by 5"		6	1.333		285	71		356	420
7980	Drain, 1/2"		32	.250		83	13.35		96.35	111
7990	Breather, 1/2"		32	.250		83	13.35		96.35	111
8000	Plastic coated 40 mil thick									
8010	LB, LR or LL conduit body w/cover, 1/2" diameter	1 Elec	13	.615	Ea.	49.50	33		82.50	104
8020	3/4" diameter		11	.727		44.50	39		83.50	107
8030	1" diameter		8	1		72	53.50		125.50	160
8040	1-1/4" diameter		6	1.333		107	71		178	224
8050	1-1/2" diameter		5	1.600		130	85.50		215.50	271
8060	2" diameter		4.50	1.778		190	95		285	350
8070	2-1/2" diameter		4	2		300	107		407	490
8080	3" diameter		3.50	2.286		440	122		562	670
8090	3-1/2" diameter		3	2.667		550	142		692	815
8100	4" diameter		2.50	3.200		615	171		786	930
8150	T conduit body with cover, 1/2" diameter		11	.727		63	39		102	127
8160	3/4" diameter		9	.889		58	47.50		105.50	135
8170	1" diameter		6	1.333		94	71		165	209

26 05 33 – Raceway and Boxes for Electrical Systems

26 05 33.25 Conduit Fittings for Rigid Galvanized Steel		Crew	Daily Output	Labor-Hours	Unit	Material	2014 Bare Costs Labor	Equipment	Total	Total Incl O&P
8180	1-1/4" diameter	1 Elec	5	1.600	Ea.	126	85.50		211.50	267
8190	1-1/2" diameter		4.50	1.778		149	95		244	305
8200	2" diameter		4	2		203	107		310	385
8210	2-1/2" diameter		3.50	2.286		405	122		527	630
8220	3" diameter		3	2.667		525	142		667	795
8230	3-1/2" diameter		2.50	3.200		830	171		1,001	1,175
8240	4" diameter		2	4		900	213		1,113	1,300
8300	FS conduit body, 1 gang, 3/4" diameter		11	.727		61	39		100	125
8310	1" diameter		10	.800		61.50	42.50		104	132
8350	2 gang, 3/4" diameter		9	.889		110	47.50		157.50	192
8360	1" diameter		8	1		126	53.50		179.50	219
8400	Duplex receptacle cover		64	.125		44	6.65		50.65	58.50
8410	Switch cover		64	.125		64.50	6.65		71.15	81
8420	Switch, vaportight cover		53	.151		183	8.05		191.05	213
8430	Blank, cover		64	.125		43	6.65		49.65	57
8520	FSC conduit body, 1 gang, 3/4" diameter		10	.800		66.50	42.50		109	137
8530	1" diameter		9	.889		74	47.50		121.50	152
8550	2 gang, 3/4" diameter		8	1		124	53.50		177.50	217
8560	1" diameter		7	1.143		116	61		177	219
8590	Conduit hubs, 1/2" diameter		18	.444		37	23.50		60.50	76
8600	3/4" diameter		16	.500		37.50	26.50		64	81.50
8610	1" diameter		14	.571		47	30.50		77.50	97.50
8620	1-1/4" diameter		12	.667		65	35.50		100.50	125
8630	1-1/2" diameter		10	.800		62	42.50		104.50	132
8640	2" diameter		8.80	.909		89	48.50		137.50	171
8650	2-1/2" diameter		8.50	.941		149	50		199	239
8660	3" diameter		8	1		191	53.50		244.50	290
8670	3-1/2" diameter		7.50	1.067		296	57		353	410
8680	4" diameter		7	1.143		325	61		386	450
8690	5" diameter		6	1.333		425	71		496	575
8700	Plastic coated 40 mil thick									
8710	Pipe strap, stamped 1 hole, 1/2" diameter	1 Elec	470	.017	Ea.	11.85	.91		12.76	14.35
8720	3/4" diameter		440	.018		10.25	.97		11.22	12.70
8730	1" diameter		400	.020		10.55	1.07		11.62	13.20
8740	1-1/4" diameter		355	.023		18.60	1.20		19.80	22.50
8750	1-1/2" diameter		320	.025		19.60	1.33		20.93	23.50
8760	2" diameter		266	.030		27.50	1.60		29.10	33
8770	2-1/2" diameter		200	.040		37.50	2.13		39.63	44.50
8780	3" diameter		133	.060		49.50	3.21		52.71	59.50
8790	3-1/2" diameter		110	.073		63.50	3.88		67.38	76
8800	4" diameter		90	.089		82	4.74		86.74	97
8810	5" diameter		70	.114		300	6.10		306.10	340
8840	Clamp back spacers, 3/4" diameter		440	.018		17.95	.97		18.92	21
8850	1" diameter		400	.020		20	1.07		21.07	23.50
8860	1-1/4" diameter		355	.023		29.50	1.20		30.70	34
8870	1-1/2" diameter		320	.025		40.50	1.33		41.83	46.50
8880	2" diameter		266	.030		61.50	1.60		63.10	70
8900	3" diameter		133	.060		143	3.21		146.21	162
8920	4" diameter		90	.089		121	4.74		125.74	141
8950	Touch-up plastic coating, spray, 12 oz.					53.50			53.50	58.50
8960	Sealing fittings, 1/2" diameter	1 Elec	11	.727		63.50	39		102.50	128
8970	3/4" diameter		9	.889		63.50	47.50		111	141
8980	1" diameter		7.50	1.067		74.50	57		131.50	167

26 05 33 – Raceway and Boxes for Electrical Systems

26 05 33.25 Conduit Fittings for Rigid Galvanized Steel		Crew	Daily Output	Labor-Hours	Unit	Material	2014 Bare Costs Labor	Equipment	Total	Total Incl O&P
8990	1-1/4" diameter	1 Elec	6.50	1.231	Ea.	99.50	65.50		165	208
9000	1-1/2" diameter		5.50	1.455		130	77.50		207.50	259
9010	2" diameter		4.80	1.667		151	89		240	299
9020	2-1/2" diameter		4	2		269	107		376	455
9030	3" diameter		3.50	2.286		287	122		409	500
9040	3-1/2" diameter		3	2.667		775	142		917	1,075
9050	4" diameter		2.50	3.200		1,025	171		1,196	1,375
9060	5" diameter		1.70	4.706		2,075	251		2,326	2,675
9070	Unions, 1/2" diameter		18	.444		56.50	23.50		80	97.50
9080	3/4" diameter		15	.533		48.50	28.50		77	96
9090	1" diameter		13	.615		65	33		98	121
9100	1-1/4" diameter		11	.727		128	39		167	198
9110	1-1/2" diameter		9.50	.842		154	45		199	237
9120	2" diameter		8	1		169	53.50		222.50	266
9130	2-1/2" diameter		7.50	1.067		305	57		362	420
9140	3" diameter		6.80	1.176		400	63		463	535
9150	3-1/2" diameter		5.80	1.379		525	73.50		598.50	690
9160	4" diameter		4.80	1.667		630	89		719	830
9170	5" diameter	▼	4	2	▼	935	107		1,042	1,175

26 05 33.30 Electrical Nonmetallic Tubing (ENT)

		Crew	Daily Output	Labor-Hours	Unit	Material	2014 Bare Costs Labor	Equipment	Total	Total Incl O&P
0010	**ELECTRICAL NONMETALLIC TUBING (ENT)**									
0050	Flexible, 1/2" diameter	1 Elec	270	.030	L.F.	.73	1.58		2.31	3.17
0100	3/4" diameter		230	.035		1.18	1.86		3.04	4.08
0200	1" diameter		145	.055		1.99	2.94		4.93	6.60
0210	1-1/4" diameter		125	.064		1.94	3.41		5.35	7.25
0220	1-1/2" diameter		100	.080		2.88	4.27		7.15	9.55
0230	2" diameter		75	.107	▼	3.81	5.70		9.51	12.70
0300	Connectors, to outlet box, 1/2" diameter		230	.035	Ea.	1.20	1.86		3.06	4.10
0310	3/4" diameter		210	.038		2.12	2.03		4.15	5.35
0320	1" diameter		200	.040		3.28	2.13		5.41	6.80
0400	Couplings, to conduit, 1/2" diameter		145	.055		1.27	2.94		4.21	5.80
0410	3/4" diameter		130	.062		1.50	3.28		4.78	6.55
0420	1" diameter	▼	125	.064	▼	3.14	3.41		6.55	8.55

26 05 33.35 Flexible Metallic Conduit

		Crew	Daily Output	Labor-Hours	Unit	Material	2014 Bare Costs Labor	Equipment	Total	Total Incl O&P
0010	**FLEXIBLE METALLIC CONDUIT**									
0050	Steel, 3/8" diameter	1 Elec	200	.040	L.F.	.51	2.13		2.64	3.75
0100	1/2" diameter		200	.040		.57	2.13		2.70	3.82
0200	3/4" diameter		160	.050		.78	2.67		3.45	4.85
0250	1" diameter		100	.080		1.42	4.27		5.69	7.95
0300	1-1/4" diameter		70	.114		1.84	6.10		7.94	11.15
0350	1-1/2" diameter		50	.160		3	8.55		11.55	16.10
0370	2" diameter		40	.200		3.66	10.65		14.31	20
0380	2-1/2" diameter	▼	30	.267		4.43	14.25		18.68	26.50
0390	3" diameter	2 Elec	50	.320		7.75	17.05		24.80	34
0400	3-1/2" diameter		40	.400		8.75	21.50		30.25	41.50
0410	4" diameter	▼	30	.533	▼	9.95	28.50		38.45	53.50
0420	Connectors, plain, 3/8" diameter	1 Elec	100	.080	Ea.	2.31	4.27		6.58	8.95
0430	1/2" diameter		80	.100		2.67	5.35		8.02	10.95
0440	3/4" diameter		70	.114		3.06	6.10		9.16	12.50
0450	1" diameter		50	.160		6.50	8.55		15.05	19.95
0452	1-1/4" diameter		45	.178		8	9.50		17.50	23
0454	1-1/2" diameter	▼	40	.200	▼	11.75	10.65		22.40	29

26 05 33 – Raceway and Boxes for Electrical Systems

26 05 33.35 Flexible Metallic Conduit		Crew	Daily Output	Labor-Hours	Unit	Material	2014 Bare Costs Labor	Equipment	Total	Total Incl O&P
0456	2" diameter	1 Elec	28	.286	Ea.	17.60	15.25		32.85	42.50
0458	2-1/2" diameter		25	.320		30	17.05		47.05	58.50
0460	3" diameter		20	.400		43	21.50		64.50	79.50
0462	3-1/2" diameter		16	.500		126	26.50		152.50	178
0464	4" diameter		13	.615		160	33		193	225
0490	Insulated, 1" diameter		40	.200		7.10	10.65		17.75	24
0500	1-1/4" diameter		40	.200		13.10	10.65		23.75	30.50
0550	1-1/2" diameter		32	.250		19.70	13.35		33.05	41.50
0600	2" diameter		23	.348		32.50	18.55		51.05	64
0610	2-1/2" diameter		20	.400		85.50	21.50		107	127
0620	3" diameter		17	.471		113	25		138	163
0630	3-1/2" diameter		13	.615		193	33		226	261
0640	4" diameter		10	.800		254	42.50		296.50	345
0650	Connectors 90°, plain, 3/8" diameter		80	.100		2.80	5.35		8.15	11.10
0660	1/2" diameter		60	.133		4.70	7.10		11.80	15.80
0700	3/4" diameter		50	.160		7.70	8.55		16.25	21.50
0750	1" diameter		40	.200		13.15	10.65		23.80	30.50
0790	Insulated, 1" diameter		40	.200		14.65	10.65		25.30	32
0800	1-1/4" diameter		30	.267		34	14.25		48.25	59
0850	1-1/2" diameter		23	.348		64	18.55		82.55	98
0900	2" diameter		18	.444		78	23.50		101.50	122
0910	2-1/2" diameter		16	.500		150	26.50		176.50	205
0920	3" diameter		14	.571		192	30.50		222.50	257
0930	3-1/2" diameter		11	.727		565	39		604	680
0940	4" diameter		8	1		770	53.50		823.50	930
0960	Couplings, to flexible conduit, 1/2" diameter		50	.160		1.75	8.55		10.30	14.75
0970	3/4" diameter		40	.200		2.98	10.65		13.63	19.25
0980	1" diameter		35	.229		5.20	12.20		17.40	24
0990	1-1/4" diameter		28	.286		11.40	15.25		26.65	35.50
1000	1-1/2" diameter		23	.348		14.75	18.55		33.30	44
1010	2" diameter		20	.400		30	21.50		51.50	65
1020	2-1/2" diameter		18	.444		34.50	23.50		58	73.50
1030	3" diameter		15	.533		87.50	28.50		116	139
1032	Aluminum, 3/8" diameter		210	.038	L.F.	.48	2.03		2.51	3.57
1034	1/2" diameter		210	.038		.55	2.03		2.58	3.65
1036	3/4" diameter		165	.048		.77	2.59		3.36	4.72
1038	1" diameter		105	.076		1.46	4.06		5.52	7.70
1040	1-1/4" diameter		75	.107		2.04	5.70		7.74	10.75
1042	1-1/2" diameter		53	.151		2.89	8.05		10.94	15.25
1044	2" diameter		42	.190		3.68	10.15		13.83	19.25
1046	2-1/2" diameter		32	.250		4.74	13.35		18.09	25
1048	3" diameter	2 Elec	53	.302		8.20	16.10		24.30	33
1050	3-1/2" diameter		42	.381		8.95	20.50		29.45	40.50
1052	4" diameter		32	.500		9.95	26.50		36.45	51
1070	Sealtite, 3/8" diameter	1 Elec	140	.057		1.24	3.05		4.29	5.90
1080	1/2" diameter		140	.057		1.34	3.05		4.39	6.05
1090	3/4" diameter		100	.080		1.88	4.27		6.15	8.45
1100	1" diameter		70	.114		2.83	6.10		8.93	12.25
1200	1-1/4" diameter		50	.160		3.89	8.55		12.44	17.10
1300	1-1/2" diameter		40	.200		4.50	10.65		15.15	21
1400	2" diameter		30	.267		5.65	14.25		19.90	28
1410	2-1/2" diameter		27	.296		9.65	15.80		25.45	34
1420	3" diameter	2 Elec	50	.320		13.35	17.05		30.40	40

26 05 33.35 Flexible Metallic Conduit		Crew	Daily Output	Labor-Hours	Unit	Material	2014 Bare Costs Labor	Equipment	Total	Total Incl O&P
1440	4" diameter	2 Elec	30	.533	L.F.	20	28.50		48.50	64.50
1490	Connectors, plain, 3/8" diameter	1 Elec	70	.114	Ea.	3.26	6.10		9.36	12.75
1500	1/2" diameter		70	.114		4.42	6.10		10.52	14
1700	3/4" diameter		50	.160		6.55	8.55		15.10	20
1900	1" diameter		40	.200		11.50	10.65		22.15	28.50
1910	Insulated, 1" diameter		40	.200		14.55	10.65		25.20	32
2000	1-1/4" diameter		32	.250		22.50	13.35		35.85	45
2100	1-1/2" diameter		27	.296		30	15.80		45.80	56
2200	2" diameter		20	.400		53	21.50		74.50	90.50
2210	2-1/2" diameter		15	.533		315	28.50		343.50	390
2220	3" diameter		12	.667		350	35.50		385.50	440
2240	4" diameter		8	1		435	53.50		488.50	555
2290	Connectors, 90°, 3/8" diameter		70	.114		4.15	6.10		10.25	13.70
2300	1/2" diameter		70	.114		4.90	6.10		11	14.55
2400	3/4" diameter		50	.160		7.60	8.55		16.15	21
2600	1" diameter		40	.200		15.40	10.65		26.05	33
2790	Insulated, 1" diameter		40	.200		26	10.65		36.65	44.50
2800	1-1/4" diameter		32	.250		42	13.35		55.35	66
3000	1-1/2" diameter		27	.296		51.50	15.80		67.30	80.50
3100	2" diameter		20	.400		72	21.50		93.50	112
3110	2-1/2" diameter		14	.571		340	30.50		370.50	415
3120	3" diameter		11	.727		330	39		369	425
3140	4" diameter		7	1.143		505	61		566	650
4300	Coupling sealtite to rigid, 1/2" diameter		20	.400		4.20	21.50		25.70	36.50
4500	3/4" diameter		18	.444		5.95	23.50		29.45	42
4800	1" diameter		14	.571		8.15	30.50		38.65	54.50
4900	1-1/4" diameter		12	.667		13.60	35.50		49.10	68
5000	1-1/2" diameter		11	.727		34	39		73	95.50
5100	2" diameter		10	.800		42	42.50		84.50	110
5110	2-1/2" diameter		9.50	.842		164	45		209	248
5120	3" diameter		9	.889		182	47.50		229.50	271
5130	3-1/2" diameter		9	.889		203	47.50		250.50	295
5140	4" diameter		8.50	.941		225	50		275	320

26 05 33.95 Cutting and Drilling

		Crew	Daily Output	Labor-Hours	Unit	Material	2014 Bare Costs Labor	Equipment	Total	Total Incl O&P
0010	**CUTTING AND DRILLING**									
0100	Hole drilling to 10' high, concrete wall									
0110	8" thick, 1/2" pipe size	R-31	12	.667	Ea.	.24	35.50	3.85	39.59	57.50
0120	3/4" pipe size		12	.667		.24	35.50	3.85	39.59	57.50
0130	1" pipe size		9.50	.842		.38	45	4.87	50.25	73
0140	1-1/4" pipe size		9.50	.842		.38	45	4.87	50.25	73
0150	1-1/2" pipe size		9.50	.842		.38	45	4.87	50.25	73
0160	2" pipe size		4.40	1.818		.52	97	10.50	108.02	157
0170	2-1/2" pipe size		4.40	1.818		.52	97	10.50	108.02	157
0180	3" pipe size		4.40	1.818		.52	97	10.50	108.02	157
0190	3-1/2" pipe size		3.30	2.424		.66	129	14	143.66	210
0200	4" pipe size		3.30	2.424		.66	129	14	143.66	210
0500	12" thick, 1/2" pipe size		9.40	.851		.36	45.50	4.92	50.78	74
0520	3/4" pipe size		9.40	.851		.36	45.50	4.92	50.78	74
0540	1" pipe size		7.30	1.096		.58	58.50	6.35	65.43	95
0560	1-1/4" pipe size		7.30	1.096		.58	58.50	6.35	65.43	95
0570	1-1/2" pipe size		7.30	1.096		.58	58.50	6.35	65.43	95
0580	2" pipe size		3.60	2.222		.78	119	12.85	132.63	192

26 05 33.95 Cutting and Drilling		Crew	Daily Output	Labor-Hours	Unit	Material	2014 Bare Costs Labor	Equipment	Total	Total Incl O&P
0590	2-1/2" pipe size	R-31	3.60	2.222	Ea.	.78	119	12.85	132.63	192
0600	3" pipe size		3.60	2.222		.78	119	12.85	132.63	192
0610	3-1/2" pipe size		2.80	2.857		.99	152	16.50	169.49	247
0630	4" pipe size		2.50	3.200		.99	171	18.50	190.49	278
0650	16" thick, 1/2" pipe size		7.60	1.053		.48	56	6.10	62.58	91
0670	3/4" pipe size		7	1.143		.48	61	6.60	68.08	99.50
0690	1" pipe size		6	1.333		.77	71	7.70	79.47	115
0710	1-1/4" pipe size		5.50	1.455		.77	77.50	8.40	86.67	126
0730	1-1/2" pipe size		5.50	1.455		.77	77.50	8.40	86.67	126
0750	2" pipe size		3	2.667		1.05	142	15.40	158.45	231
0770	2-1/2" pipe size		2.70	2.963		1.05	158	17.15	176.20	257
0790	3" pipe size		2.50	3.200		1.05	171	18.50	190.55	278
0810	3-1/2" pipe size		2.30	3.478		1.32	186	20	207.32	300
0830	4" pipe size		2	4		1.32	213	23	237.32	345
0850	20" thick, 1/2" pipe size		6.40	1.250		.60	66.50	7.25	74.35	109
0870	3/4" pipe size		6	1.333		.60	71	7.70	79.30	115
0890	1" pipe size		5	1.600		.96	85.50	9.25	95.71	139
0910	1-1/4" pipe size		4.80	1.667		.96	89	9.65	99.61	145
0930	1-1/2" pipe size		4.60	1.739		.96	93	10.05	104.01	151
0950	2" pipe size		2.70	2.963		1.31	158	17.15	176.46	257
0970	2-1/2" pipe size		2.40	3.333		1.31	178	19.25	198.56	288
0990	3" pipe size		2.20	3.636		1.31	194	21	216.31	315
1010	3-1/2" pipe size		2	4		1.66	213	23	237.66	345
1030	4" pipe size		1.70	4.706		1.66	251	27	279.66	405
1050	24" thick, 1/2" pipe size		5.50	1.455		.72	77.50	8.40	86.62	126
1070	3/4" pipe size		5.10	1.569		.72	83.50	9.05	93.27	136
1090	1" pipe size		4.30	1.860		1.15	99.50	10.75	111.40	162
1110	1-1/4" pipe size		4	2		1.15	107	11.55	119.70	174
1130	1-1/2" pipe size		4	2		1.15	107	11.55	119.70	174
1150	2" pipe size		2.40	3.333		1.57	178	19.25	198.82	289
1170	2-1/2" pipe size		2.20	3.636		1.57	194	21	216.57	315
1190	3" pipe size		2	4		1.57	213	23	237.57	345
1210	3-1/2" pipe size		1.80	4.444		1.99	237	25.50	264.49	385
1230	4" pipe size		1.50	5.333		1.99	285	31	317.99	460
1500	Brick wall, 8" thick, 1/2" pipe size		18	.444		.24	23.50	2.57	26.31	38.50
1520	3/4" pipe size		18	.444		.24	23.50	2.57	26.31	38.50
1540	1" pipe size		13.30	.602		.38	32	3.48	35.86	52
1560	1-1/4" pipe size		13.30	.602		.38	32	3.48	35.86	52
1580	1-1/2" pipe size		13.30	.602		.38	32	3.48	35.86	52
1600	2" pipe size		5.70	1.404		.52	75	8.10	83.62	121
1620	2-1/2" pipe size		5.70	1.404		.52	75	8.10	83.62	121
1640	3" pipe size		5.70	1.404		.52	75	8.10	83.62	121
1660	3-1/2" pipe size		4.40	1.818		.66	97	10.50	108.16	157
1680	4" pipe size		4	2		.66	107	11.55	119.21	173
1700	12" thick, 1/2" pipe size		14.50	.552		.36	29.50	3.19	33.05	48
1720	3/4" pipe size		14.50	.552		.36	29.50	3.19	33.05	48
1740	1" pipe size		11	.727		.58	39	4.20	43.78	63.50
1760	1-1/4" pipe size		11	.727		.58	39	4.20	43.78	63.50
1780	1-1/2" pipe size		11	.727		.58	39	4.20	43.78	63.50
1800	2" pipe size		5	1.600		.78	85.50	9.25	95.53	139
1820	2-1/2" pipe size		5	1.600		.78	85.50	9.25	95.53	139
1840	3" pipe size		5	1.600		.78	85.50	9.25	95.53	139
1860	3-1/2" pipe size		3.80	2.105		.99	112	12.15	125.14	182

26 05 33.95 Cutting and Drilling		Crew	Daily Output	Labor-Hours	Unit	Material	2014 Bare Costs			Total Incl O&P
							Labor	Equipment	Total	
1880	4" pipe size	R-31	3.30	2.424	Ea.	.99	129	14	143.99	210
1900	16" thick, 1/2" pipe size		12.30	.650		.48	34.50	3.76	38.74	56.50
1920	3/4" pipe size		12.30	.650		.48	34.50	3.76	38.74	56.50
1940	1" pipe size		9.30	.860		.77	46	4.97	51.74	75
1960	1-1/4" pipe size		9.30	.860		.77	46	4.97	51.74	75
1980	1-1/2" pipe size		9.30	.860		.77	46	4.97	51.74	75
2000	2" pipe size		4.40	1.818		1.05	97	10.50	108.55	158
2010	2-1/2" pipe size		4.40	1.818		1.05	97	10.50	108.55	158
2030	3" pipe size		4.40	1.818		1.05	97	10.50	108.55	158
2050	3-1/2" pipe size		3.30	2.424		1.32	129	14	144.32	211
2070	4" pipe size		3	2.667		1.32	142	15.40	158.72	231
2090	20" thick, 1/2" pipe size		10.70	.748		.60	40	4.32	44.92	65
2110	3/4" pipe size		10.70	.748		.60	40	4.32	44.92	65
2130	1" pipe size		8	1		.96	53.50	5.80	60.26	87.50
2150	1-1/4" pipe size		8	1		.96	53.50	5.80	60.26	87.50
2170	1-1/2" pipe size		8	1		.96	53.50	5.80	60.26	87.50
2190	2" pipe size		4	2		1.31	107	11.55	119.86	174
2210	2-1/2" pipe size		4	2		1.31	107	11.55	119.86	174
2230	3" pipe size		4	2		1.31	107	11.55	119.86	174
2250	3-1/2" pipe size		3	2.667		1.66	142	15.40	159.06	232
2270	4" pipe size		2.70	2.963		1.66	158	17.15	176.81	258
2290	24" thick, 1/2" pipe size		9.40	.851		.72	45.50	4.92	51.14	74
2310	3/4" pipe size		9.40	.851		.72	45.50	4.92	51.14	74
2330	1" pipe size		7.10	1.127		1.15	60	6.50	67.65	98.50
2350	1-1/4" pipe size		7.10	1.127		1.15	60	6.50	67.65	98.50
2370	1-1/2" pipe size		7.10	1.127		1.15	60	6.50	67.65	98.50
2390	2" pipe size		3.60	2.222		1.57	119	12.85	133.42	193
2410	2-1/2" pipe size		3.60	2.222		1.57	119	12.85	133.42	193
2430	3" pipe size		3.60	2.222		1.57	119	12.85	133.42	193
2450	3-1/2" pipe size		2.80	2.857		1.99	152	16.50	170.49	248
2470	4" pipe size		2.50	3.200		1.99	171	18.50	191.49	279
3000	Knockouts to 8' high, metal boxes & enclosures									
3020	With hole saw, 1/2" pipe size	1 Elec	53	.151	Ea.		8.05		8.05	12.05
3040	3/4" pipe size		47	.170			9.10		9.10	13.60
3050	1" pipe size		40	.200			10.65		10.65	15.95
3060	1-1/4" pipe size		36	.222			11.85		11.85	17.75
3070	1-1/2" pipe size		32	.250			13.35		13.35	19.95
3080	2" pipe size		27	.296			15.80		15.80	23.50
3090	2-1/2" pipe size		20	.400			21.50		21.50	32
4010	3" pipe size		16	.500			26.50		26.50	40
4030	3-1/2" pipe size		13	.615			33		33	49
4050	4" pipe size		11	.727			39		39	58
4070	With hand punch set, 1/2" pipe size		40	.200			10.65		10.65	15.95
4090	3/4" pipe size		32	.250			13.35		13.35	19.95
4110	1" pipe size		30	.267			14.25		14.25	21.50
4130	1-1/4" pipe size		28	.286			15.25		15.25	23
4150	1-1/2" pipe size		26	.308			16.40		16.40	24.50
4170	2" pipe size		20	.400			21.50		21.50	32
4190	2-1/2" pipe size		17	.471			25		25	37.50
4200	3" pipe size		15	.533			28.50		28.50	42.50
4220	3-1/2" pipe size		12	.667			35.50		35.50	53
4240	4" pipe size		10	.800			42.50		42.50	64
4260	With hydraulic punch, 1/2" pipe size		44	.182			9.70		9.70	14.50

26 05 33 – Raceway and Boxes for Electrical Systems

26 05 33.95 Cutting and Drilling		Crew	Daily Output	Labor-Hours	Unit	Material	2014 Bare Costs Labor	Equipment	Total	Total Incl O&P
4280	3/4" pipe size	1 Elec	38	.211	Ea.		11.25		11.25	16.80
4300	1" pipe size		38	.211			11.25		11.25	16.80
4320	1-1/4" pipe size		38	.211			11.25		11.25	16.80
4340	1-1/2" pipe size		38	.211			11.25		11.25	16.80
4360	2" pipe size		32	.250			13.35		13.35	19.95
4380	2-1/2" pipe size		27	.296			15.80		15.80	23.50
4400	3" pipe size		23	.348			18.55		18.55	28
4420	3-1/2" pipe size		20	.400			21.50		21.50	32
4440	4" pipe size		18	.444			23.50		23.50	35.50

26 05 36 – Cable Trays for Electrical Systems

26 05 36.10 Cable Tray Ladder Type

		Crew	Daily Output	Labor-Hours	Unit	Material	2014 Bare Costs Labor	Equipment	Total	Total Incl O&P
0010	**CABLE TRAY LADDER TYPE** w/ftngs. & supports, 4" dp., to 15' elev.									
0100	For higher elevations, see Section 26 05 36.40									
0160	Galvanized steel tray R260536-10									
0170	4" rung spacing, 6" wide	2 Elec	98	.163	L.F.	17.95	8.70		26.65	33
0180	9" wide R260536-11		92	.174		19.60	9.30		28.90	35.50
0200	12" wide		86	.186		21.50	9.95		31.45	38.50
0400	18" wide		82	.195		25	10.40		35.40	43
0600	24" wide		78	.205		29	10.95		39.95	48.50
0650	30" wide		68	.235		36.50	12.55		49.05	59
0700	36" wide		60	.267		40.50	14.25		54.75	66
0800	6" rung spacing, 6" wide		100	.160		16.35	8.55		24.90	31
0850	9" wide		94	.170		17.95	9.10		27.05	33.50
0860	12" wide		88	.182		18.85	9.70		28.55	35.50
0870	18" wide		84	.190		21.50	10.15		31.65	38.50
0880	24" wide		80	.200		24	10.65		34.65	42.50
0890	30" wide		70	.229		28.50	12.20		40.70	50
0900	36" wide		64	.250		31	13.35		44.35	54.50
0910	9" rung spacing, 6" wide		102	.157		15.40	8.35		23.75	29.50
0920	9" wide		98	.163		15.75	8.70		24.45	30.50
0930	12" wide		94	.170		18.10	9.10		27.20	33.50
0940	18" wide		90	.178		20.50	9.50		30	36.50
0950	24" wide		86	.186		23.50	9.95		33.45	40.50
0960	30" wide		80	.200		25.50	10.65		36.15	44
0970	36" wide		74	.216		27	11.55		38.55	47.50
0980	12" rung spacing, 6" wide		106	.151		15.15	8.05		23.20	28.50
0990	9" wide		104	.154		15.50	8.20		23.70	29.50
1000	12" wide		100	.160		16.35	8.55		24.90	31
1010	18" wide		96	.167		17.60	8.90		26.50	32.50
1020	24" wide		94	.170		18.85	9.10		27.95	34.50
1030	30" wide		88	.182		21.50	9.70		31.20	38
1040	36" wide		84	.190		22.50	10.15		32.65	40
1041	18" rung spacing, 6" wide		108	.148		15	7.90		22.90	28.50
1042	9" wide		106	.151		15.25	8.05		23.30	29
1043	12" wide		102	.157		16.25	8.35		24.60	30.50
1044	18" wide		98	.163		17.10	8.70		25.80	32
1045	24" wide		96	.167		17.60	8.90		26.50	32.50
1046	30" wide		90	.178		18.55	9.50		28.05	34.50
1047	36" wide		86	.186		19.05	9.95		29	36
1050	Elbows horiz. 9" rung spacing, 90°, 12" radius, 6" wide		9.60	1.667	Ea.	70	89		159	210
1060	9" wide		8.40	1.905		74	102		176	234
1070	12" wide		7.60	2.105		80.50	112		192.50	257

26 05 36 – Cable Trays for Electrical Systems

26 05 36.10 Cable Tray Ladder Type		Crew	Daily Output	Labor-Hours	Unit	Material	2014 Bare Costs Labor	Equipment	Total	Total Incl O&P
1080	18" wide	2 Elec	6.20	2.581	Ea.	106	138		244	325
1090	24" wide		5.40	2.963		119	158		277	370
1100	30" wide		4.80	3.333		157	178		335	440
1110	36" wide		4.20	3.810		182	203		385	505
1120	90° 24" radius, 6" wide		9.20	1.739		144	93		237	298
1130	9" wide		8	2		148	107		255	325
1140	12" wide		7.20	2.222		153	119		272	345
1150	18" wide		5.80	2.759		165	147		312	400
1160	24" wide		5	3.200		178	171		349	450
1170	30" wide		4.40	3.636		193	194		387	500
1180	36" wide		3.80	4.211		229	225		454	585
1190	90° 36" radius, 6" wide		8.80	1.818		159	97		256	320
1200	9" wide		7.60	2.105		167	112		279	350
1210	12" wide		6.80	2.353		176	126		302	380
1220	18" wide		5.40	2.963		201	158		359	460
1230	24" wide		4.60	3.478		220	186		406	520
1240	30" wide		4	4		265	213		478	610
1250	36" wide		3.40	4.706		288	251		539	690
1260	45° 12" radius, 6" wide		13.20	1.212		54	64.50		118.50	157
1270	9" wide		11	1.455		56	77.50		133.50	178
1280	12" wide		9.60	1.667		58.50	89		147.50	198
1290	18" wide		7.60	2.105		61.50	112		173.50	236
1300	24" wide		6.20	2.581		79.50	138		217.50	294
1310	30" wide		5.40	2.963		101	158		259	350
1320	36" wide		4.60	3.478		106	186		292	395
1330	45° 24" radius, 6" wide		12.80	1.250		68.50	66.50		135	175
1340	9" wide		10.60	1.509		72.50	80.50		153	201
1350	12" wide		9.20	1.739		81.50	93		174.50	229
1360	18" wide		7.20	2.222		83.50	119		202.50	269
1370	24" wide		5.80	2.759		101	147		248	330
1380	30" wide		5	3.200		121	171		292	390
1390	36" wide		4.20	3.810		139	203		342	460
1400	45° 36" radius, 6" wide		12.40	1.290		98	69		167	211
1410	9" wide		10.20	1.569		104	83.50		187.50	239
1420	12" wide		8.80	1.818		106	97		203	261
1430	18" wide		6.80	2.353		114	126		240	315
1440	24" wide		5.40	2.963		121	158		279	370
1450	30" wide		4.60	3.478		157	186		343	450
1460	36" wide		3.80	4.211		174	225		399	525
1470	Elbows horizontal, 4" rung spacing, use 9" rung x 1.50									
1480	6" rung spacing use 9" rung x 1.20									
1490	12" rung spacing use 9" rung x .93									
1500	Elbows vert. 9" rung spacing, 90°, 12" radius, 6" wide	2 Elec	9.60	1.667	Ea.	96	89		185	238
1510	9" wide		8.40	1.905		98	102		200	260
1520	12" wide		7.60	2.105		101	112		213	279
1530	18" wide		6.20	2.581		106	138		244	320
1540	24" wide		5.40	2.963		110	158		268	360
1550	30" wide		4.80	3.333		119	178		297	395
1560	36" wide		4.20	3.810		123	203		326	440
1570	24" radius, 6" wide		9.20	1.739		136	93		229	288
1580	9" wide		8	2		140	107		247	315
1590	12" wide		7.20	2.222		144	119		263	335
1600	18" wide		5.80	2.759		153	147		300	390

26 05 36 – Cable Trays for Electrical Systems

26 05 36.10 Cable Tray Ladder Type	Crew	Daily Output	Labor-Hours	Unit	Material	2014 Bare Costs Labor	Equipment	Total	Total Incl O&P
1610	24" wide	2 Elec	5	3.200	Ea.	161	171		332	435
1620	30" wide		4.40	3.636		167	194		361	475
1630	36" wide		3.80	4.211		189	225		414	540
1640	36" radius, 6" wide		8.80	1.818		182	97		279	345
1650	9" wide		7.60	2.105		189	112		301	375
1660	12" wide		6.80	2.353		191	126		317	400
1670	18" wide		5.40	2.963		208	158		366	465
1680	24" wide		4.60	3.478		210	186		396	510
1690	30" wide		4	4		237	213		450	580
1700	36" wide		3.40	4.706		254	251		505	655
1710	Elbows vertical, 4" rung spacing, use 9" rung x 1.25									
1720	6" rung spacing, use 9" rung x 1.15									
1730	12" rung spacing, use 9" rung x .90									
1740	Tee horizontal, 9" rung spacing, 12" radius, 6" wide	2 Elec	5	3.200	Ea.	148	171		319	420
1750	9" wide		4.60	3.478		159	186		345	455
1760	12" wide		4.40	3.636		167	194		361	475
1770	18" wide		4	4		191	213		404	530
1780	24" wide		3.60	4.444		225	237		462	600
1790	30" wide		3.40	4.706		269	251		520	670
1800	36" wide		3	5.333		305	285		590	760
1810	24" radius, 6" wide		4.60	3.478		231	186		417	530
1820	9" wide		4.20	3.810		242	203		445	570
1830	12" wide		4	4		250	213		463	595
1840	18" wide		3.60	4.444		271	237		508	655
1850	24" wide		3.20	5		295	267		562	725
1860	30" wide		3	5.333		320	285		605	775
1870	36" wide		2.60	6.154		410	330		740	940
1880	36" radius, 6" wide		4.20	3.810		345	203		548	685
1890	9" wide		3.80	4.211		360	225		585	730
1900	12" wide		3.60	4.444		385	237		622	775
1910	18" wide		3.20	5		410	267		677	850
1920	24" wide		2.80	5.714		465	305		770	965
1930	30" wide		2.60	6.154		505	330		835	1,050
1940	36" wide		2.20	7.273		545	390		935	1,175
1980	Tee vertical, 9" rung spacing, 12" radius, 6" wide		5.40	2.963		252	158		410	515
1990	9" wide		5.20	3.077		254	164		418	525
2000	12" wide		5	3.200		256	171		427	540
2010	18" wide		4.60	3.478		259	186		445	560
2020	24" wide		4.40	3.636		265	194		459	580
2030	30" wide		4	4		282	213		495	630
2040	36" wide		3.60	4.444		292	237		529	675
2050	24" radius, 6" wide		5	3.200		455	171		626	755
2060	9" wide		4.80	3.333		460	178		638	770
2070	12" wide		4.60	3.478		465	186		651	795
2080	18" wide		4.20	3.810		480	203		683	835
2090	24" wide		4	4		495	213		708	865
2100	30" wide		3.60	4.444		505	237		742	910
2110	36" wide		3.20	5		530	267		797	985
2120	36" radius, 6" wide		4.60	3.478		850	186		1,036	1,200
2130	9" wide		4.40	3.636		860	194		1,054	1,225
2140	12" wide		4.20	3.810		870	203		1,073	1,275
2150	18" wide		3.80	4.211		880	225		1,105	1,300
2160	24" wide		3.60	4.444		890	237		1,127	1,325

26 05 36.10 Cable Tray Ladder Type	Crew	Daily Output	Labor-Hours	Unit	Material	2014 Bare Costs Labor	2014 Bare Costs Equipment	Total	Total Incl O&P
2170	30" wide	2 Elec	3.20	5	Ea.	925	267		1,192	1,425
2180	36" wide	↓	2.80	5.714	↓	935	305		1,240	1,475
2190	Tee, 4" rung spacing, use 9" rung x 1.30									
2200	6" rung spacing, use 9" rung x 1.20									
2210	12" rung spacing, use 9" rung x .90									
2220	Cross horizontal, 9" rung spacing, 12" radius, 6" wide	2 Elec	4	4	Ea.	183	213		396	520
2230	9" wide		3.80	4.211		192	225		417	545
2240	12" wide		3.60	4.444		210	237		447	585
2250	18" wide		3.40	4.706		229	251		480	625
2260	24" wide		3	5.333		254	285		539	705
2270	30" wide		2.80	5.714		305	305		610	795
2280	36" wide		2.60	6.154		340	330		670	865
2290	24" radius, 6" wide		3.60	4.444		355	237		592	745
2300	9" wide		3.40	4.706		365	251		616	780
2310	12" wide		3.20	5		375	267		642	815
2320	18" wide		3	5.333		405	285		690	870
2330	24" wide		2.60	6.154		430	330		760	960
2340	30" wide		2.40	6.667		455	355		810	1,025
2350	36" wide		2.20	7.273		530	390		920	1,175
2360	36" radius, 6" wide		3.20	5		495	267		762	945
2370	9" wide		3	5.333		515	285		800	995
2380	12" wide		2.80	5.714		540	305		845	1,050
2390	18" wide		2.60	6.154		595	330		925	1,150
2400	24" wide		2.20	7.273		660	390		1,050	1,300
2410	30" wide		2	8		715	425		1,140	1,425
2420	36" wide	↓	1.80	8.889	↓	770	475		1,245	1,550
2430	Cross horizontal, 4" rung spacing, use 9" rung x 1.30									
2440	6" rung spacing, use 9" rung x 1.20									
2450	12" rung spacing, use 9" rung x .90									
2460	Reducer, 9" to 6" wide tray	2 Elec	13	1.231	Ea.	102	65.50		167.50	211
2470	12" to 9" wide tray		12	1.333		104	71		175	220
2480	18" to 12" wide tray		10.40	1.538		104	82		186	237
2490	24" to 18" wide tray		9	1.778		106	95		201	259
2500	30" to 24" wide tray		8	2		108	107		215	279
2510	36" to 30" wide tray		7	2.286		121	122		243	315
2511	Reducer, 18" to 6" wide tray		10.40	1.538		108	82		190	242
2512	24" to 12" wide tray		9	1.778		110	95		205	263
2513	30" to 18" wide tray		8	2		114	107		221	286
2514	30" to 12" wide tray		8	2		119	107		226	291
2515	36" to 24" wide tray		7	2.286		123	122		245	320
2516	36" to 18" wide tray		7	2.286		123	122		245	320
2517	36" to 12" wide tray		7	2.286		123	122		245	320
2520	Dropout or end plate, 6" wide		32	.500		11.35	26.50		37.85	52.50
2530	9" wide		28	.571		13.35	30.50		43.85	60
2540	12" wide		26	.615		14.60	33		47.60	65
2550	18" wide		22	.727		17.15	39		56.15	77
2560	24" wide		20	.800		21	42.50		63.50	87
2570	30" wide		18	.889		22.50	47.50		70	96
2580	36" wide		16	1		26	53.50		79.50	109
2590	Tray connector		48	.333	↓	20	17.80		37.80	48.50
3200	Aluminum tray, 4" deep, 6" rung spacing, 6" wide		134	.119	L.F.	17.80	6.35		24.15	29
3210	9" wide		128	.125		18.90	6.65		25.55	31
3220	12" wide	↓	124	.129	↓	19.95	6.90		26.85	32.50

26 05 36.10 Cable Tray Ladder Type	Crew	Daily Output	Labor-Hours	Unit	Material	2014 Bare Costs Labor	Equipment	Total	Total Incl O&P
3230	18" wide	2 Elec	114	.140	L.F.	22	7.50		29.50	35.50
3240	24" wide		106	.151		26	8.05		34.05	40.50
3250	30" wide		100	.160		28.50	8.55		37.05	44.50
3260	36" wide		94	.170		31	9.10		40.10	47.50
3270	9" rung spacing, 6" wide		140	.114		14.50	6.10		20.60	25
3280	9" wide		134	.119		15.15	6.35		21.50	26.50
3290	12" wide		130	.123		15.45	6.55		22	27
3300	18" wide		122	.131		18.05	7		25.05	30.50
3310	24" wide		116	.138		21	7.35		28.35	34
3320	30" wide		108	.148		23	7.90		30.90	37.50
3330	36" wide		100	.160		24.50	8.55		33.05	40
3340	12" rung spacing, 6" wide		146	.110		15.35	5.85		21.20	25.50
3350	9" wide		140	.114		15.55	6.10		21.65	26.50
3360	12" wide		134	.119		16.05	6.35		22.40	27
3370	18" wide		128	.125		17	6.65		23.65	28.50
3380	24" wide		124	.129		18.90	6.90		25.80	31.50
3390	30" wide		114	.140		20	7.50		27.50	33
3400	36" wide		106	.151		21.50	8.05		29.55	35.50
3401	18" rung spacing, 6" wide		150	.107		15.25	5.70		20.95	25.50
3402	9" wide tray		144	.111		15.45	5.95		21.40	26
3403	12" wide tray		140	.114		16.05	6.10		22.15	27
3404	18" wide tray		134	.119		16.90	6.35		23.25	28
3405	24" wide tray		130	.123		18.20	6.55		24.75	30
3406	30" wide tray		120	.133		18.85	7.10		25.95	31.50
3407	36" wide tray		110	.145		19.70	7.75		27.45	33
3410	Elbows horiz. 9" rung spacing, 90° 12" radius, 6" wide		9.60	1.667	Ea.	68.50	89		157.50	209
3420	9" wide		8.40	1.905		70.50	102		172.50	230
3430	12" wide		7.60	2.105		83.50	112		195.50	260
3440	18" wide		6.20	2.581		94	138		232	310
3450	24" wide		5.40	2.963		117	158		275	365
3460	30" wide		4.80	3.333		123	178		301	400
3470	36" wide		4.20	3.810		149	203		352	470
3480	24" radius, 6" wide		9.20	1.739		122	93		215	273
3490	9" wide		8	2		127	107		234	300
3500	12" wide		7.20	2.222		133	119		252	325
3510	18" wide		5.80	2.759		145	147		292	380
3520	24" wide		5	3.200		159	171		330	430
3530	30" wide		4.40	3.636		172	194		366	480
3540	36" wide		3.80	4.211		196	225		421	550
3550	90°, 36" radius, 6" wide		8.80	1.818		134	97		231	292
3560	9" wide		7.60	2.105		137	112		249	320
3570	12" wide		6.80	2.353		154	126		280	355
3580	18" wide		5.40	2.963		172	158		330	425
3590	24" wide		4.60	3.478		194	186		380	490
3600	30" wide		4	4		214	213		427	555
3610	36" wide		3.40	4.706		246	251		497	645
3620	45° 12" radius, 6" wide		13.20	1.212		45	64.50		109.50	147
3630	9" wide		11	1.455		47	77.50		124.50	168
3640	12" wide		9.60	1.667		48.50	89		137.50	187
3650	18" wide		7.60	2.105		57.50	112		169.50	231
3660	24" wide		6.20	2.581		71.50	138		209.50	285
3670	30" wide		5.40	2.963		82	158		240	325
3680	36" wide		4.60	3.478		85.50	186		271.50	370

26 05 36.10 Cable Tray Ladder Type	Crew	Daily Output	Labor-Hours	Unit	Material	2014 Bare Costs Labor	Equipment	Total	Total Incl O&P
3690	45° 24" radius, 6" wide	2 Elec	12.80	1.250	Ea.	60	66.50		126.50	167
3700	9" wide		10.60	1.509		68.50	80.50		149	197
3710	12" wide		9.20	1.739		70.50	93		163.50	217
3720	18" wide		7.20	2.222		75.50	119		194.50	260
3730	24" wide		5.80	2.759		87	147		234	315
3740	30" wide		5	3.200		104	171		275	370
3750	36" wide		4.20	3.810		110	203		313	425
3760	45° 36" radius, 6" wide		12.40	1.290		82	69		151	193
3770	9" wide		10.20	1.569		83.50	83.50		167	217
3780	12" wide		8.80	1.818		85.50	97		182.50	239
3790	18" wide		6.80	2.353		93.50	126		219.50	291
3800	24" wide		5.40	2.963		115	158		273	365
3810	30" wide		4.60	3.478		122	186		308	410
3820	36" wide		3.80	4.211		134	225		359	480
3830	Elbows horizontal, 4" rung spacing, use 9" rung x 1.50									
3840	6" rung spacing, use 9" rung x 1.20									
3850	12" rung spacing, use 9" rung x .93									
3860	Elbows vertical 9" rung spacing, 90° 12" radius, 6" wide	2 Elec	9.60	1.667	Ea.	95.50	89		184.50	238
3870	9" wide		8.40	1.905		100	102		202	262
3880	12" wide		7.60	2.105		102	112		214	280
3890	18" wide		6.20	2.581		105	138		243	320
3900	24" wide		5.40	2.963		112	158		270	360
3910	30" wide		4.80	3.333		117	178		295	395
3920	36" wide		4.20	3.810		120	203		323	440
3930	24" radius, 6" wide		9.20	1.739		104	93		197	253
3940	9" wide		8	2		107	107		214	278
3950	12" wide		7.20	2.222		110	119		229	298
3960	18" wide		5.80	2.759		117	147		264	350
3970	24" wide		5	3.200		124	171		295	390
3980	30" wide		4.40	3.636		131	194		325	435
3990	36" wide		3.80	4.211		139	225		364	490
4000	36" radius, 6" wide		8.80	1.818		161	97		258	320
4010	9" wide		7.60	2.105		164	112		276	350
4020	12" wide		6.80	2.353		174	126		300	380
4030	18" wide		5.40	2.963		181	158		339	435
4040	24" wide		4.60	3.478		187	186		373	485
4050	30" wide		4	4		202	213		415	545
4060	36" wide		3.40	4.706		211	251		462	605
4070	Elbows vertical, 4" rung spacing, use 9" rung x 1.25									
4080	6" rung spacing, use 9" rung x 1.15									
4090	12" rung spacing, use 9" rung x .90									
4100	Tee horizontal 9" rung spacing, 12" radius, 6" wide	2 Elec	5	3.200	Ea.	106	171		277	375
4110	9" wide		4.60	3.478		114	186		300	405
4120	12" wide		4.40	3.636		126	194		320	430
4130	18" wide		4.20	3.810		137	203		340	455
4140	24" wide		4	4		151	213		364	485
4150	30" wide		3.60	4.444		176	237		413	550
4160	36" wide		3.40	4.706		211	251		462	605
4170	24" radius, 6" wide		4.60	3.478		181	186		367	475
4180	9" wide		4.20	3.810		191	203		394	515
4190	12" wide		4	4		196	213		409	535
4200	18" wide		3.80	4.211		214	225		439	570
4210	24" wide		3.60	4.444		234	237		471	615

26 05 36 – Cable Trays for Electrical Systems

26 05 36.10 Cable Tray Ladder Type	Crew	Daily Output	Labor-Hours	Unit	Material	2014 Bare Costs Labor	Equipment	Total	Total Incl O&P
4220	30" wide	2 Elec	3.20	5	Ea.	254	267		521	680
4230	36" wide		3	5.333		390	285		675	855
4240	36" radius, 6" wide		4.20	3.810		256	203		459	585
4250	9" wide		3.80	4.211		263	225		488	625
4260	12" wide		3.60	4.444		288	237		525	670
4270	18" wide		3.40	4.706		305	251		556	710
4280	24" wide		3.20	5		335	267		602	770
4290	30" wide		2.80	5.714		425	305		730	920
4300	36" wide		2.60	6.154		470	330		800	1,000
4310	Tee vertical 9" rung spacing, 12" radius, 6" wide		5.40	2.963		238	158		396	500
4320	9" wide		5.20	3.077		246	164		410	515
4330	12" wide		5	3.200		249	171		420	530
4340	18" wide		4.60	3.478		253	186		439	555
4350	24" wide		4.40	3.636		273	194		467	590
4360	30" wide		4.20	3.810		278	203		481	610
4370	36" wide		4	4		288	213		501	635
4380	24" radius, 6" wide		5	3.200		405	171		576	700
4390	9" wide		4.80	3.333		410	178		588	715
4400	12" wide		4.60	3.478		415	186		601	735
4410	18" wide		4.20	3.810		425	203		628	775
4420	24" wide		4	4		435	213		648	800
4430	30" wide		3.80	4.211		450	225		675	830
4440	36" wide		3.60	4.444		485	237		722	890
4450	36" radius, 6" wide		4.60	3.478		885	186		1,071	1,250
4460	9" wide		4.40	3.636		905	194		1,099	1,275
4470	12" wide		4.20	3.810		940	203		1,143	1,325
4480	18" wide		3.80	4.211		950	225		1,175	1,375
4490	24" wide		3.60	4.444		975	237		1,212	1,425
4500	30" wide		3.40	4.706		985	251		1,236	1,450
4510	36" wide		3.20	5		1,000	267		1,267	1,500
4520	Tees, 4" rung spacing, use 9" rung x 1.30									
4530	6" rung spacing, use 9" rung x 1.20									
4540	12" rung spacing, use 9" rung x .90									
4550	Cross horizontal 9" rung spacing, 12" radius, 6" wide	2 Elec	4.40	3.636	Ea.	147	194		341	450
4560	9" wide		4.20	3.810		164	203		367	485
4570	12" wide		4	4		182	213		395	520
4580	18" wide		3.60	4.444		201	237		438	575
4590	24" wide		3.40	4.706		207	251		458	605
4600	30" wide		3	5.333		244	285		529	695
4610	36" wide		2.80	5.714		305	305		610	790
4620	24" radius, 6" wide		4	4		283	213		496	630
4630	9" wide		3.80	4.211		293	225		518	655
4640	12" wide		3.60	4.444		300	237		537	685
4650	18" wide		3.20	5		315	267		582	750
4660	24" wide		3	5.333		345	285		630	805
4670	30" wide		2.60	6.154		370	330		700	895
4680	36" wide		2.40	6.667		415	355		770	990
4690	36" radius, 6" wide		3.60	4.444		335	237		572	725
4700	9" wide		3.40	4.706		350	251		601	760
4710	12" wide		3.20	5		370	267		637	805
4720	18" wide		2.80	5.714		395	305		700	890
4730	24" wide		2.60	6.154		485	330		815	1,025
4740	30" wide		2.20	7.273		540	390		930	1,175

26 05 36 – Cable Trays for Electrical Systems

26 05 36.10 Cable Tray Ladder Type	Crew	Daily Output	Labor-Hours	Unit	Material	2014 Bare Costs Labor	Equipment	Total	Total Incl O&P
4750	36" wide	2 Elec	2	8	Ea.	600	425		1,025	1,300
4760	Cross horizontal, 4" rung spacing, use 9" rung x 1.30									
4770	6" rung spacing, use 9" rung x 1.20									
4780	12" rung spacing, use 9" rung x .90									
4790	Reducer, 9" to 6" wide tray	2 Elec	16	1	Ea.	78.50	53.50		132	167
4800	12" to 9" wide tray		14	1.143		82	61		143	182
4810	18" to 12" wide tray		12.40	1.290		82	69		151	193
4820	24" to 18" wide tray		10.60	1.509		83.50	80.50		164	213
4830	30" to 24" wide tray		9.20	1.739		85.50	93		178.50	233
4840	36" to 30" wide tray		8	2		90.50	107		197.50	260
4841	Reducer, 18" to 6" wide tray		12.40	1.290		82	69		151	193
4842	24" to 12" wide tray		10.60	1.509		83.50	80.50		164	213
4843	30" to 18" wide tray		9.20	1.739		85.50	93		178.50	233
4844	30" to 12" wide tray		9.20	1.739		85.50	93		178.50	233
4845	36" to 24" wide tray		8	2		90.50	107		197.50	260
4846	36" to 18" wide tray		8	2		90.50	107		197.50	260
4847	36" to 12" wide tray		8	2		90.50	107		197.50	260
4850	Dropout or end plate, 6" wide		32	.500		10.40	26.50		36.90	51.50
4860	9" wide tray		28	.571		11.05	30.50		41.55	57.50
4870	12" wide tray		26	.615		12.05	33		45.05	62.50
4880	18" wide tray		22	.727		16.05	39		55.05	75.50
4890	24" wide tray		20	.800		18.90	42.50		61.40	85
4900	30" wide tray		18	.889		22.50	47.50		70	95.50
4910	36" wide tray		16	1		25	53.50		78.50	108
4920	Tray connector		48	.333		15.05	17.80		32.85	43
8000	Elbow 36" radius horiz., 60°, 6" wide tray		10.60	1.509		114	80.50		194.50	246
8010	9" wide tray		9	1.778		124	95		219	278
8020	12" wide tray		7.80	2.051		127	109		236	305
8030	18" wide tray		6.20	2.581		136	138		274	355
8040	24" wide tray		5	3.200		159	171		330	430
8050	30" wide tray		4.40	3.636		162	194		356	470
8060	30°, 6" wide tray		14	1.143		87	61		148	187
8070	9" wide tray		11.40	1.404		92	75		167	213
8080	12" wide tray		9.80	1.633		93.50	87		180.50	233
8090	18" wide tray		7.40	2.162		98.50	115		213.50	282
8100	24" wide tray		5.80	2.759		120	147		267	355
8110	30" wide tray		4.80	3.333		124	178		302	400
8120	Adjustable, 6" wide tray		12.40	1.290		115	69		184	230
8130	9" wide tray		10.20	1.569		122	83.50		205.50	259
8140	12" wide tray		8.80	1.818		124	97		221	281
8150	18" wide tray		6.80	2.353		136	126		262	335
8160	24" wide tray		5.40	2.963		144	158		302	395
8170	30" wide tray		4.60	3.478		161	186		347	455
8180	Wye 36" radius horiz., 45°, 6" wide tray		4.60	3.478		126	186		312	415
8190	9" wide tray		4.40	3.636		132	194		326	435
8200	12" wide tray		4.20	3.810		137	203		340	455
8210	18" wide tray		3.80	4.211		156	225		381	505
8220	24" wide tray		3.60	4.444		174	237		411	545
8230	30" wide tray		3.40	4.706		194	251		445	590
8240	Elbow 36" radius vert. in/outside, 60°, 6" wide tray		10.60	1.509		126	80.50		206.50	259
8250	9" wide tray		9	1.778		131	95		226	286
8260	12" wide tray		7.80	2.051		132	109		241	310
8270	18" wide tray		6.20	2.581		137	138		275	355

26 05 36.10 Cable Tray Ladder Type		Crew	Daily Output	Labor-Hours	Unit	Material	2014 Bare Costs Labor	2014 Bare Costs Equipment	Total	Total Incl O&P
8280	24" wide tray	2 Elec	5	3.200	Ea.	146	171		317	415
8290	30" wide tray		4.40	3.636		156	194		350	460
8300	45°, 6" wide tray		12.40	1.290		113	69		182	227
8310	9" wide tray		10.20	1.569		120	83.50		203.50	257
8320	12" wide tray		8.80	1.818		122	97		219	279
8330	18" wide tray		6.80	2.353		126	126		252	325
8340	24" wide tray		5.40	2.963		132	158		290	380
8350	30" wide tray		4.60	3.478		136	186		322	425
8360	30°, 6" wide tray		14	1.143		99	61		160	201
8370	9" wide tray		11.40	1.404		101	75		176	223
8380	12" wide tray		9.80	1.633		108	87		195	249
8390	18" wide tray		7.40	2.162		110	115		225	294
8400	24" wide tray		5.80	2.759		112	147		259	345
8410	30" wide tray		4.80	3.333		118	178		296	395
8660	Adjustable, 6" wide tray		12.40	1.290		113	69		182	227
8670	9" wide tray		10.20	1.569		120	83.50		203.50	257
8680	12" wide tray		8.80	1.818		122	97		219	279
8690	18" wide tray		6.80	2.353		136	126		262	335
8700	24" wide tray		5.40	2.963		144	158		302	395
8710	30" wide tray		4.60	3.478		159	186		345	455
8720	Cross, vertical, 24" radius, 6" wide tray		3.60	4.444		775	237		1,012	1,200
8730	9" wide tray		3.40	4.706		785	251		1,036	1,250
8740	12" wide tray		3.20	5		810	267		1,077	1,300
8750	18" wide tray		2.80	5.714		845	305		1,150	1,375
8760	24" wide tray		2.60	6.154		875	330		1,205	1,450
8770	30" wide tray		2.20	7.273		925	390		1,315	1,600
9200	Splice plate	1 Elec	48	.167	Pr.	15.70	8.90		24.60	30.50
9210	Expansion joint		48	.167		18.90	8.90		27.80	34.50
9220	Horizontal hinged		48	.167		15.70	8.90		24.60	30.50
9230	Vertical hinged		48	.167		18.90	8.90		27.80	34.50
9240	Ladder hanger, vertical		28	.286	Ea.	4.68	15.25		19.93	28
9250	Ladder to channel connector		24	.333		55	17.80		72.80	87
9260	Ladder to box connector 30" wide		19	.421		55	22.50		77.50	94
9270	24" wide		20	.400		55	21.50		76.50	92.50
9280	18" wide		21	.381		53	20.50		73.50	89
9290	12" wide		22	.364		49.50	19.40		68.90	83.50
9300	9" wide		23	.348		44.50	18.55		63.05	77
9310	6" wide		24	.333		39.50	17.80		57.30	70
9320	Ladder floor flange		24	.333		35.50	17.80		53.30	65.50
9330	Cable roller for tray 30" wide		10	.800		292	42.50		334.50	385
9340	24" wide		11	.727		248	39		287	330
9350	18" wide		12	.667		237	35.50		272.50	315
9360	12" wide		13	.615		186	33		219	254
9370	9" wide		14	.571		165	30.50		195.50	227
9380	6" wide		15	.533		134	28.50		162.50	190
9390	Pulley, single wheel		12	.667		305	35.50		340.50	390
9400	Triple wheel		10	.800		610	42.50		652.50	735
9440	Nylon cable tie, 14" long		80	.100		.46	5.35		5.81	8.50
9450	Ladder, hold down clamp		60	.133		10.05	7.10		17.15	21.50
9460	Cable clamp		60	.133		10.20	7.10		17.30	22
9470	Wall bracket, 30" wide tray		19	.421		54	22.50		76.50	93
9480	24" wide tray		20	.400		42	21.50		63.50	78
9490	18" wide tray		21	.381		37.50	20.50		58	72

26 05 36 – Cable Trays for Electrical Systems

26 05 36.10 Cable Tray Ladder Type		Crew	Daily Output	Labor-Hours	Unit	Material	2014 Bare Costs Labor	Equipment	Total	Total Incl O&P
9500	12" wide tray	1 Elec	22	.364	Ea.	35	19.40		54.40	67.50
9510	9" wide tray		23	.348		31	18.55		49.55	62
9520	6" wide tray		24	.333		30	17.80		47.80	59.50

26 05 36.20 Cable Tray Solid Bottom

			Crew	Daily Output	Labor-Hours	Unit	Material	Labor	Equipment	Total	Total Incl O&P
0010	**CABLE TRAY SOLID BOTTOM** w/ftngs. & supports, 3" dp, to 15' high										
0200	For higher elevations, see Section 26 05 36.40										
0220	Galvanized steel, tray, 6" wide	R260536-10	2 Elec	120	.133	L.F.	14.50	7.10		21.60	26.50
0240	12" wide			100	.160		18.60	8.55		27.15	33.50
0260	18" wide			70	.229		22.50	12.20		34.70	43.50
0280	24" wide	R260536-11		60	.267		26.50	14.25		40.75	50.50
0300	30" wide			50	.320		31.50	17.05		48.55	60
0320	36" wide			44	.364		45	19.40		64.40	78.50
0340	Elbow horizontal 90°, 12" radius, 6" wide			9.60	1.667	Ea.	110	89		199	254
0360	12" wide			6.80	2.353		123	126		249	325
0370	18" wide			5.40	2.963		141	158		299	390
0380	24" wide			4.40	3.636		172	194		366	480
0390	30" wide			3.80	4.211		206	225		431	560
0400	36" wide			3.40	4.706		233	251		484	630
0420	24" radius, 6" wide			9.20	1.739		157	93		250	310
0440	12" wide			6.40	2.500		182	133		315	400
0450	18" wide			5	3.200		212	171		383	490
0460	24" wide			4	4		246	213		459	590
0470	30" wide			3.40	4.706		288	251		539	690
0480	36" wide			3	5.333		325	285		610	780
0500	36" radius, 6" wide			8.80	1.818		229	97		326	395
0520	12" wide			6	2.667		259	142		401	495
0530	18" wide			4.60	3.478		315	186		501	625
0540	24" wide			3.60	4.444		340	237		577	730
0550	30" wide			3	5.333		415	285		700	880
0560	36" wide			2.60	6.154		465	330		795	1,000
0580	Elbow vertical 90°, 12" radius, 6" wide			9.60	1.667		134	89		223	280
0600	12" wide			6.80	2.353		145	126		271	345
0610	18" wide			5.40	2.963		156	158		314	410
0620	24" wide			4.40	3.636		167	194		361	475
0630	30" wide			3.80	4.211		178	225		403	530
0640	36" wide			3.40	4.706		184	251		435	580
0670	24" radius, 6" wide			9.20	1.739		189	93		282	345
0690	12" wide			6.40	2.500		208	133		341	430
0700	18" wide			5	3.200		225	171		396	505
0710	24" wide			4	4		244	213		457	590
0720	30" wide			3.40	4.706		259	251		510	660
0730	36" wide			3	5.333		282	285		567	735
0750	36" radius, 6" wide			8.80	1.818		259	97		356	430
0770	12" wide			6.60	2.424		288	129		417	510
0780	18" wide			4.60	3.478		315	186		501	625
0790	24" wide			3.60	4.444		345	237		582	735
0800	30" wide			3	5.333		375	285		660	840
0810	36" wide			2.60	6.154		410	330		740	940
0840	Tee horizontal, 12" radius, 6" wide			5	3.200		155	171		326	425
0860	12" wide			4	4		167	213		380	505
0870	18" wide			3.40	4.706		199	251		450	595
0880	24" wide			2.80	5.714		223	305		528	700

26 05 36 – Cable Trays for Electrical Systems

26 05 36.20 Cable Tray Solid Bottom		Crew	Daily Output	Labor-Hours	Unit	Material	2014 Bare Costs Labor	Equipment	Total	Total Incl O&P
0890	30" wide	2 Elec	2.60	6.154	Ea.	254	330		584	770
0900	36" wide		2.20	7.273		290	390		680	900
0940	24" radius, 6" wide		4.60	3.478		239	186		425	540
0960	12" wide		3.60	4.444		280	237		517	665
0970	18" wide		3	5.333		310	285		595	770
0980	24" wide		2.40	6.667		420	355		775	995
0990	30" wide		2.20	7.273		460	390		850	1,075
1000	36" wide		1.80	8.889		495	475		970	1,250
1020	36" radius, 6" wide		4.20	3.810		375	203		578	720
1040	12" wide		3.20	5		420	267		687	865
1050	18" wide		2.60	6.154		460	330		790	995
1060	24" wide		2.20	7.273		590	390		980	1,225
1070	30" wide		2	8		640	425		1,065	1,350
1080	36" wide		1.60	10		660	535		1,195	1,525
1100	Tee vertical, 12" radius, 6" wide		5	3.200		254	171		425	535
1120	12" wide		4	4		265	213		478	610
1130	18" wide		3.60	4.444		269	237		506	650
1140	24" wide		3.40	4.706		295	251		546	700
1150	30" wide		3	5.333		315	285		600	770
1160	36" wide		2.60	6.154		325	330		655	850
1180	24" radius, 6" wide		4.60	3.478		375	186		561	695
1200	12" wide		3.60	4.444		390	237		627	785
1210	18" wide		3.20	5		410	267		677	850
1220	24" wide		3	5.333		430	285		715	900
1230	30" wide		2.60	6.154		475	330		805	1,000
1240	36" wide		2.20	7.273		495	390		885	1,125
1260	36" radius, 6" wide		4.20	3.810		585	203		788	950
1280	12" wide		3.20	5		595	267		862	1,050
1290	18" wide		2.80	5.714		640	305		945	1,150
1300	24" wide		2.60	6.154		660	330		990	1,225
1310	30" wide		2.20	7.273		770	390		1,160	1,425
1320	36" wide		2	8		815	425		1,240	1,550
1340	Cross horizontal, 12" radius, 6" wide		4	4		184	213		397	525
1360	12" wide		3.40	4.706		201	251		452	595
1370	18" wide		2.80	5.714		235	305		540	715
1380	24" wide		2.40	6.667		265	355		620	820
1390	30" wide		2	8		295	425		720	965
1400	36" wide		1.80	8.889		325	475		800	1,075
1420	24" radius, 6" wide		3.60	4.444		335	237		572	725
1440	12" wide		3	5.333		375	285		660	840
1450	18" wide		2.40	6.667		420	355		775	995
1460	24" wide		2	8		540	425		965	1,225
1470	30" wide		1.80	8.889		585	475		1,060	1,350
1480	36" wide		1.60	10		615	535		1,150	1,475
1500	36" radius, 6" wide		3.20	5		550	267		817	1,000
1520	12" wide		2.60	6.154		605	330		935	1,150
1530	18" wide		2	8		660	425		1,085	1,375
1540	24" wide		1.80	8.889		780	475		1,255	1,575
1550	30" wide		1.60	10		850	535		1,385	1,725
1560	36" wide		1.40	11.429		905	610		1,515	1,900
1580	Drop out or end plate, 6" wide		32	.500		22.50	26.50		49	64.50
1600	12" wide		26	.615		28	33		61	80
1610	18" wide		22	.727		30	39		69	91

26 05 36.20 Cable Tray Solid Bottom		Crew	Daily Output	Labor-Hours	Unit	Material	2014 Bare Costs Labor	Equipment	Total	Total Incl O&P
1620	24" wide	2 Elec	20	.800	Ea.	35.50	42.50		78	103
1630	30" wide		18	.889		39.50	47.50		87	115
1640	36" wide		16	1		42.50	53.50		96	127
1660	Reducer, 12" to 6" wide		12	1.333		108	71		179	225
1680	18" to 12" wide		10.60	1.509		110	80.50		190.50	242
1700	18" to 6" wide		10.60	1.509		110	80.50		190.50	242
1720	24" to 18" wide		9.20	1.739		114	93		207	265
1740	24" to 12" wide		9.20	1.739		114	93		207	265
1760	30" to 24" wide		8	2		123	107		230	295
1780	30" to 18" wide		8	2		123	107		230	295
1800	30" to 12" wide		8	2		123	107		230	295
1820	36" to 30" wide		7.20	2.222		127	119		246	315
1840	36" to 24" wide		7.20	2.222		127	119		246	315
1860	36" to 18" wide		7.20	2.222		127	119		246	315
1880	36" to 12" wide		7.20	2.222		127	119		246	315
2000	Aluminum tray, 6" wide		150	.107	L.F.	13.90	5.70		19.60	24
2020	12" wide		130	.123		18.20	6.55		24.75	30
2030	18" wide		100	.160		23	8.55		31.55	38
2040	24" wide		90	.178		29	9.50		38.50	45.50
2050	30" wide		70	.229		34	12.20		46.20	56
2060	36" wide		64	.250		42.50	13.35		55.85	67
2080	Elbow horizontal 90°, 12" radius, 6" wide		9.60	1.667	Ea.	111	89		200	256
2100	12" wide		7.60	2.105		125	112		237	305
2110	18" wide		6.80	2.353		154	126		280	355
2120	24" wide		5.80	2.759		176	147		323	415
2130	30" wide		5	3.200		214	171		385	490
2140	36" wide		4.40	3.636		241	194		435	555
2160	24" radius, 6" wide		9.20	1.739		159	93		252	315
2180	12" wide		7.20	2.222		187	119		306	385
2190	18" wide		6.40	2.500		214	133		347	435
2200	24" wide		5.40	2.963		264	158		422	530
2210	30" wide		4.60	3.478		305	186		491	615
2220	36" wide		4	4		350	213		563	705
2240	36" radius, 6" wide		8.80	1.818		234	97		331	405
2260	12" wide		6.80	2.353		281	126		407	500
2270	18" wide		6	2.667		335	142		477	585
2280	24" wide		5	3.200		360	171		531	650
2290	30" wide		4.20	3.810		405	203		608	750
2300	36" wide		3.60	4.444		475	237		712	880
2320	Elbow vertical 90°, 12" radius, 6" wide		9.60	1.667		132	89		221	278
2340	12" wide		7.60	2.105		137	112		249	320
2350	18" wide		6.80	2.353		156	126		282	360
2360	24" wide		5.80	2.759		164	147		311	400
2370	30" wide		5	3.200		174	171		345	445
2380	36" wide		4.40	3.636		179	194		373	485
2400	24" radius, 6" wide		9.20	1.739		187	93 ·		280	345
2420	12" wide		7.20	2.222		204	119		323	400
2430	18" wide		6.40	2.500		219	133		352	440
2440	24" wide		5.40	2.963		238	158		396	500
2450	30" wide		4.60	3.478		253	186		439	555
2460	36" wide		4	4		278	213		491	625
2480	36" radius, 6" wide		8.80	1.818		248	97		345	415
2500	12" wide		6.80	2.353		278	126		404	495

26 05 36 – Cable Trays for Electrical Systems

26 05 36.20 Cable Tray Solid Bottom		Crew	Daily Output	Labor-Hours	Unit	Material	2014 Bare Costs Labor	Equipment	Total	Total Incl O&P
2510	18" wide	2 Elec	6	2.667	Ea.	305	142		447	550
2520	24" wide		5	3.200		315	171		486	600
2530	30" wide		4.20	3.810		350	203		553	690
2540	36" wide		3.60	4.444		360	237		597	755
2560	Tee horizontal, 12" radius, 6" wide		5	3.200		172	171		343	445
2580	12" wide		4.40	3.636		209	194		403	520
2590	18" wide		4	4		234	213		447	580
2600	24" wide		3.60	4.444		263	237		500	645
2610	30" wide		3	5.333		305	285		590	760
2620	36" wide		2.40	6.667		350	355		705	915
2640	24" radius, 6" wide		4.60	3.478		284	186		470	595
2660	12" wide		4	4		320	213		533	670
2670	18" wide		3.60	4.444		360	237		597	755
2680	24" wide		3	5.333		445	285		730	915
2690	30" wide		2.40	6.667		505	355		860	1,075
2700	36" wide		2.20	7.273		565	390		955	1,200
2720	36" radius, 6" wide		4.20	3.810		460	203		663	815
2740	12" wide		3.60	4.444		505	237		742	910
2750	18" wide		3.20	5		580	267		847	1,050
2760	24" wide		2.60	6.154		680	330		1,010	1,225
2770	30" wide		2	8		750	425		1,175	1,475
2780	36" wide		1.80	8.889		835	475		1,310	1,625
2800	Tee vertical, 12" radius, 6" wide		5	3.200		246	171		417	525
2820	12" wide		4.40	3.636		251	194		445	565
2830	18" wide		4.20	3.810		261	203		464	590
2840	24" wide		4	4		278	213		491	625
2850	30" wide		3.60	4.444		305	237		542	690
2860	36" wide		3	5.333		320	285		605	775
2880	24" radius, 6" wide		4.60	3.478		360	186		546	680
2900	12" wide		4	4		385	213		598	740
2910	18" wide		3.80	4.211		410	225		635	785
2920	24" wide		3.60	4.444		445	237		682	845
2930	30" wide		3.20	5		485	267		752	935
2940	36" wide		2.60	6.154		515	330		845	1,050
2960	36" radius, 6" wide		4.20	3.810		565	203		768	925
2980	12" wide		3.40	4.706		590	251		841	1,025
2990	18" wide		3.40	4.706		625	251		876	1,075
3000	24" wide		3.20	5		645	267		912	1,100
3010	30" wide		2.80	5.714		695	305		1,000	1,225
3020	36" wide		2.20	7.273		770	390		1,160	1,425
3040	Cross horizontal, 12" radius, 6" wide		4.40	3.636		219	194		413	530
3060	12" wide		4	4		248	213		461	590
3070	18" wide		3.40	4.706		281	251		532	685
3080	24" wide		2.80	5.714		320	305		625	805
3090	30" wide		2.60	6.154		360	330		690	890
3100	36" wide		2.20	7.273		400	390		790	1,025
3120	24" radius, 6" wide		4	4		385	213		598	740
3140	12" wide		3.60	4.444		445	237		682	845
3150	18" wide		3	5.333		480	285		765	955
3160	24" wide		2.40	6.667		580	355		935	1,175
3170	30" wide		2.20	7.273		645	390		1,035	1,300
3180	36" wide		1.80	8.889		670	475		1,145	1,450
3200	36" radius, 6" wide		3.60	4.444		670	237		907	1,100

26 05 36.20 Cable Tray Solid Bottom

		Crew	Daily Output	Labor-Hours	Unit	Material	2014 Bare Costs Labor	Equipment	Total	Total Incl O&P
3220	12" wide	2 Elec	3.20	5	Ea.	705	267		972	1,175
3230	18" wide		2.60	6.154		785	330		1,115	1,350
3240	24" wide		2	8		855	425		1,280	1,575
3250	30" wide		1.80	8.889		920	475		1,395	1,725
3260	36" wide		1.60	10		1,075	535		1,610	1,975
3280	Dropout, or end plate, 6" wide		32	.500		22	26.50		48.50	64.50
3300	12" wide		26	.615		25	33		58	76.50
3310	18" wide		22	.727		30	39		69	91
3320	24" wide		20	.800		33	42.50		75.50	100
3330	30" wide		18	.889		38.50	47.50		86	113
3340	36" wide		16	1		44.50	53.50		98	129
3380	Reducer, 12" to 6" wide		14	1.143		108	61		169	211
3400	18" to 12" wide		12	1.333		111	71		182	229
3420	18" to 6" wide		12	1.333		111	71		182	229
3440	24" to 18" wide		10.60	1.509		118	80.50		198.50	251
3460	24" to 12" wide		10.60	1.509		118	80.50		198.50	251
3480	30" to 24" wide		9.20	1.739		129	93		222	281
3500	30" to 18" wide		9.20	1.739		129	93		222	281
3520	30" to 12" wide		9.20	1.739		129	93		222	281
3540	36" to 30" wide		8	2		136	107		243	310
3560	36" to 24" wide		8	2		137	107		244	310
3580	36" to 18" wide		8	2		137	107		244	310
3600	36" to 12" wide		8	2		141	107		248	315

26 05 36.30 Cable Tray Trough

		Crew	Daily Output	Labor-Hours	Unit	Material	2014 Bare Costs Labor	Equipment	Total	Total Incl O&P
0010	**CABLE TRAY TROUGH** vented, w/ftngs. & supports, 6" dp, to 15' high									
0020	For higher elevations, see Section 26 05 36.40									
0200	Galvanized steel, tray, 6" wide R260536-10	2 Elec	90	.178	L.F.	16.05	9.50		25.55	32
0240	12" wide		80	.200		19.60	10.65		30.25	37.50
0260	18" wide		70	.229		23.50	12.20		35.70	44.50
0280	24" wide R260536-11		60	.267		34	14.25		48.25	59
0300	30" wide		50	.320		47.50	17.05		64.55	77.50
0320	36" wide		40	.400		59	21.50		80.50	96.50
0340	Elbow horizontal 90°, 12" radius, 6" wide		7.60	2.105	Ea.	118	112		230	298
0360	12" wide		5.60	2.857		140	152		292	380
0370	18" wide		4.40	3.636		161	194		355	465
0380	24" wide		3.60	4.444		199	237		436	575
0390	30" wide		3.20	5		225	267		492	645
0400	36" wide		2.80	5.714		261	305		566	740
0420	24" radius, 6" wide		7.20	2.222		174	119		293	370
0440	12" wide		5.20	3.077		201	164		365	465
0450	18" wide		4	4		242	213		455	585
0460	24" wide		3.20	5		273	267		540	700
0470	30" wide		2.80	5.714		315	305		620	800
0480	36" wide		2.40	6.667		365	355		720	935
0500	36" radius, 6" wide		6.80	2.353		261	126		387	475
0520	12" wide		4.80	3.333		297	178		475	590
0530	18" wide		3.60	4.444		355	237		592	745
0540	24" wide		2.80	5.714		375	305		680	870
0550	30" wide		2.40	6.667		425	355		780	995
0560	36" wide		2	8		475	425		900	1,150
0580	Elbow vertical 90°, 12" radius, 6" wide		7.60	2.105		153	112		265	335
0600	12" wide		5.60	2.857		170	152		322	415

26 05 36 – Cable Trays for Electrical Systems

26 05 36.30 Cable Tray Trough		Crew	Daily Output	Labor-Hours	Unit	Material	2014 Bare Costs Labor	Equipment	Total	Total Incl O&P
0610	18" wide	2 Elec	4.40	3.636	Ea.	172	194		366	480
0620	24" wide		3.60	4.444		197	237		434	570
0630	30" wide		3.20	5		201	267		468	620
0640	36" wide		2.80	5.714		212	305		517	690
0660	24" radius, 6" wide		7.20	2.222		218	119		337	415
0680	12" wide		5.20	3.077		231	164		395	500
0690	18" wide		4	4		256	213		469	600
0700	24" wide		3.20	5		265	267		532	690
0710	30" wide		2.80	5.714		297	305		602	780
0720	36" wide		2.40	6.667		310	355		665	875
0740	36" radius, 6" wide		6.80	2.353		295	126		421	515
0760	12" wide		4.80	3.333		315	178		493	610
0770	18" wide		3.60	4.444		350	237		587	735
0780	24" wide		2.80	5.714		375	305		680	865
0790	30" wide		2.40	6.667		380	355		735	950
0800	36" wide		2	8		425	425		850	1,100
0820	Tee horizontal, 12" radius, 6" wide		4	4		178	213		391	515
0840	12" wide		3.20	5		199	267		466	620
0850	18" wide		2.80	5.714		225	305		530	700
0860	24" wide		2.40	6.667		261	355		616	815
0870	30" wide		2.20	7.273		297	390		687	905
0880	36" wide		2	8		325	425		750	1,000
0900	24" radius, 6" wide		3.60	4.444		295	237		532	680
0920	12" wide		2.80	5.714		320	305		625	810
0930	18" wide		2.40	6.667		360	355		715	925
0940	24" wide		2	8		465	425		890	1,150
0950	30" wide		1.80	8.889		505	475		980	1,275
0960	36" wide		1.60	10		560	535		1,095	1,425
0980	36" radius, 6" wide		3.20	5		425	267		692	865
1000	12" wide		2.40	6.667		485	355		840	1,050
1010	18" wide		2	8		530	425		955	1,225
1020	24" wide		1.60	10		635	535		1,170	1,500
1030	30" wide		1.40	11.429		675	610		1,285	1,650
1040	36" wide		1.20	13.333		795	710		1,505	1,950
1060	Tee vertical, 12" radius, 6" wide		4	4		297	213		510	645
1080	12" wide		3.20	5		300	267		567	730
1090	18" wide		3	5.333		310	285		595	765
1100	24" wide		2.80	5.714		325	305		630	815
1110	30" wide		2.60	6.154		345	330		675	870
1120	36" wide		2.20	7.273		350	390		740	965
1140	24" radius, 6" wide		3.60	4.444		400	237		637	795
1160	12" wide		2.80	5.714		420	305		725	915
1170	18" wide		2.60	6.154		440	330		770	975
1180	24" wide		2.40	6.667		475	355		830	1,050
1190	30" wide		2.20	7.273		495	390		885	1,125
1200	36" wide		1.80	8.889		540	475		1,015	1,300
1220	36" radius, 6" wide		3.20	5		635	267		902	1,100
1240	12" wide		2.40	6.667		650	355		1,005	1,250
1250	18" wide		2.20	7.273		675	390		1,065	1,325
1260	24" wide		2	8		710	425		1,135	1,425
1270	30" wide		1.80	8.889		830	475		1,305	1,625
1280	36" wide		1.40	11.429		835	610		1,445	1,825
1300	Cross horizontal, 12" radius, 6" wide		3.20	5		229	267		496	650

26 05 36 – Cable Trays for Electrical Systems

26 05 36.30 Cable Tray Trough		Crew	Daily Output	Labor-Hours	Unit	Material	2014 Bare Costs Labor	Equipment	Total	Total Incl O&P
1320	12" wide	2 Elec	2.80	5.714	Ea.	231	305		536	710
1330	18" wide		2.40	6.667		259	355		614	815
1340	24" wide		2	8		271	425		696	940
1350	30" wide		1.80	8.889		305	475		780	1,050
1360	36" wide		1.60	10		340	535		875	1,175
1380	24" radius, 6" wide		2.80	5.714		360	305		665	850
1400	12" wide		2.40	6.667		375	355		730	945
1410	18" wide		2	8		415	425		840	1,100
1420	24" wide		1.60	10		530	535		1,065	1,375
1430	30" wide		1.40	11.429		585	610		1,195	1,550
1440	36" wide		1.20	13.333		615	710		1,325	1,750
1460	36" radius, 6" wide		2.40	6.667		615	355		970	1,200
1480	12" wide		2	8		640	425		1,065	1,350
1490	18" wide		1.60	10		650	535		1,185	1,525
1500	24" wide		1.20	13.333		780	710		1,490	1,925
1510	30" wide		1	16		860	855		1,715	2,225
1520	36" wide		.80	20		925	1,075		2,000	2,625
1540	Dropout or end plate, 6" wide		26	.615		26.50	33		59.50	78
1560	12" wide		22	.727		32.50	39		71.50	93.50
1580	18" wide		20	.800		35.50	42.50		78	103
1600	24" wide		18	.889		41	47.50		88.50	116
1620	30" wide		16	1		44	53.50		97.50	129
1640	36" wide		13.40	1.194		48.50	63.50		112	149
1660	Reducer, 12" to 6" wide		9.40	1.702		111	91		202	258
1680	18" to 12" wide		8.40	1.905		116	102		218	279
1700	18" to 6" wide		8.40	1.905		116	102		218	279
1720	24" to 18" wide		7.20	2.222		123	119		242	315
1740	24" to 12" wide		7.20	2.222		123	119		242	315
1760	30" to 24" wide		6.40	2.500		128	133		261	340
1780	30" to 18" wide		6.40	2.500		128	133		261	340
1800	30" to 12" wide		6.40	2.500		128	133		261	340
1820	36" to 30" wide		5.80	2.759		137	147		284	370
1840	36" to 24" wide		5.80	2.759		137	147		284	370
1860	36" to 18" wide		5.80	2.759		139	147		286	375
1880	36" to 12" wide		5.80	2.759		139	147		286	375
2000	Aluminum, tray, vented, 6" wide		120	.133	L.F.	18.20	7.10		25.30	30.50
2010	9" wide		110	.145		20.50	7.75		28.25	34
2020	12" wide		100	.160		23	8.55		31.55	38
2030	18" wide		90	.178		27.50	9.50		37	44
2040	24" wide		80	.200		32.50	10.65		43.15	51.50
2050	30" wide		70	.229		43	12.20		55.20	66
2060	36" wide		60	.267		47.50	14.25		61.75	73.50
2080	Elbow horiz. 90°, 12" radius, 6" wide		7.60	2.105	Ea.	120	112		232	300
2090	9" wide		7	2.286		131	122		253	325
2100	12" wide		6.20	2.581		139	138		277	360
2110	18" wide		5.60	2.857		162	152		314	405
2120	24" wide		4.60	3.478		187	186		373	485
2130	30" wide		4	4		231	213		444	575
2140	36" wide		3.60	4.444		253	237		490	635
2160	24" radius, 6" wide		7.20	2.222		174	119		293	370
2180	12" wide		5.80	2.759		204	147		351	445
2190	18" wide		5.20	3.077		234	164		398	505
2200	24" wide		4.20	3.810		268	203		471	600

26 05 36 – Cable Trays for Electrical Systems

26 05 36.30 Cable Tray Trough	Crew	Daily Output	Labor-Hours	Unit	Material	2014 Bare Costs Labor	Equipment	Total	Total Incl O&P
2210	30" wide	2 Elec	3.60	4.444	Ea.	305	237		542	690
2220	36" wide		3.20	5		330	267		597	765
2240	36" radius, 6" wide		6.80	2.353		246	126		372	460
2260	12" wide		5.40	2.963		279	158		437	540
2270	18" wide		4.80	3.333		320	178		498	615
2280	24" wide		3.80	4.211		350	225		575	720
2290	30" wide		3.40	4.706		410	251		661	825
2300	36" wide		2.80	5.714		445	305		750	945
2320	Elbow vertical 90°, 12" radius, 6" wide		7.60	2.105		146	112		258	330
2330	9" wide		7	2.286		156	122		278	355
2340	12" wide		6.20	2.581		157	138		295	380
2350	18" wide		5.60	2.857		162	152		314	405
2360	24" wide		4.60	3.478		179	186		365	475
2370	30" wide		4	4		187	213		400	525
2380	36" wide		3.60	4.444		189	237		426	565
2400	24" radius, 6" wide		7.20	2.222		197	119		316	395
2420	12" wide		5.80	2.759		214	147		361	455
2430	18" wide		5.20	3.077		233	164		397	500
2440	24" wide		4.20	3.810		236	203		439	565
2450	30" wide		3.60	4.444		249	237		486	630
2460	36" wide		3.20	5		261	267		528	685
2480	36" radius, 6" wide		6.80	2.353		244	126		370	455
2500	12" wide		5.40	2.963		261	158		419	525
2510	18" wide		4.80	3.333		293	178		471	585
2520	24" wide		3.80	4.211		315	225		540	685
2530	30" wide		3.40	4.706		350	251		601	760
2540	36" wide		2.80	5.714		360	305		665	850
2560	Tee horizontal, 12" radius, 6" wide		4	4		182	213		395	520
2570	9" wide		3.80	4.211		187	225		412	540
2580	12" wide		3.60	4.444		204	237		441	580
2590	18" wide		3.20	5		236	267		503	660
2600	24" wide		2.80	5.714		271	305		576	755
2610	30" wide		2.40	6.667		291	355		646	850
2620	36" wide		2.20	7.273		330	390		720	945
2640	24" radius, 6" wide		3.60	4.444		278	237		515	660
2660	12" wide		3.20	5		315	267		582	750
2670	18" wide		2.80	5.714		350	305		655	840
2680	24" wide		2.40	6.667		435	355		790	1,000
2690	30" wide		2	8		470	425		895	1,150
2700	36" wide		1.80	8.889		550	475		1,025	1,325
2720	36" radius, 6" wide		3.20	5		445	267		712	890
2740	12" wide		2.80	5.714		515	305		820	1,025
2750	18" wide		2.40	6.667		580	355		935	1,175
2760	24" wide		2	8		680	425		1,105	1,375
2770	30" wide		1.60	10		725	535		1,260	1,600
2780	36" wide		1.40	11.429		835	610		1,445	1,825
2800	Tee vertical, 12" radius, 6" wide		4	4		251	213		464	595
2810	9" wide		3.80	4.211		253	225		478	615
2820	12" wide		3.60	4.444		271	237		508	655
2830	18" wide		3.40	4.706		273	251		524	675
2840	24" wide		3.20	5		279	267		546	705
2850	30" wide		3	5.333		293	285		578	745
2860	36" wide		2.60	6.154		305	330		635	825

26 05 36.30 Cable Tray Trough		Crew	Daily Output	Labor-Hours	Unit	Material	2014 Bare Costs Labor	2014 Bare Costs Equipment	Total	Total Incl O&P
2880	24" radius, 6" wide	2 Elec	3.60	4.444	Ea.	355	237		592	745
2900	12" wide		3.20	5		385	267		652	820
2910	18" wide		3	5.333		410	285		695	875
2920	24" wide		2.80	5.714		435	305		740	935
2930	30" wide		2.60	6.154		470	330		800	1,000
2940	36" wide		2.20	7.273		480	390		870	1,100
2960	36" radius, 6" wide		3.20	5		550	267		817	1,000
2980	12" wide		2.80	5.714		580	305		885	1,100
2990	18" wide		2.60	6.154		600	330		930	1,150
3000	24" wide		2.40	6.667		645	355		1,000	1,250
3010	30" wide		2.20	7.273		695	390		1,085	1,350
3020	36" wide		1.80	8.889		730	475		1,205	1,525
3040	Cross horizontal, 12" radius, 6" wide		3.60	4.444		228	237		465	605
3050	9" wide		3.40	4.706		244	251		495	645
3060	12" wide		3.20	5		251	267		518	675
3070	18" wide		2.80	5.714		263	305		568	745
3080	24" wide		2.40	6.667		293	355		648	850
3090	30" wide		2.20	7.273		355	390		745	970
3100	36" wide		1.80	8.889		430	475		905	1,175
3120	24" radius, 6" wide		3.20	5		410	267		677	850
3140	12" wide		2.80	5.714		445	305		750	945
3150	18" wide		2.40	6.667		480	355		835	1,050
3160	24" wide		2	8		555	425		980	1,250
3170	30" wide		1.80	8.889		600	475		1,075	1,375
3180	36" wide		1.40	11.429		680	610		1,290	1,650
3200	36" radius, 6" wide		2.80	5.714		680	305		985	1,200
3220	12" wide		2.40	6.667		715	355		1,070	1,325
3230	18" wide		2	8		775	425		1,200	1,500
3240	24" wide		1.60	10		895	535		1,430	1,775
3250	30" wide		1.40	11.429		975	610		1,585	2,000
3260	36" wide		1.20	13.333		1,125	710		1,835	2,325
3280	Dropout, or end plate, 6" wide		26	.615		23	33		56	74
3300	12" wide		22	.727		27	39		66	87.50
3310	18" wide		20	.800		32.50	42.50		75	99.50
3320	24" wide		18	.889		39	47.50		86.50	114
3330	30" wide		16	1		42	53.50		95.50	126
3340	36" wide		14	1.143		48	61		109	144
3370	Reducer, 9" to 6" wide		12	1.333		111	71		182	229
3380	12" to 6" wide		11.40	1.404		117	75		192	241
3390	12" to 9" wide		11.40	1.404		117	75		192	241
3400	18" to 12" wide		9.60	1.667		126	89		215	271
3420	18" to 6" wide		9.60	1.667		126	89		215	271
3430	18" to 9" wide		9.60	1.667		126	89		215	271
3440	24" to 18" wide		8.40	1.905		134	102		236	299
3460	24" to 12" wide		8.40	1.905		134	102		236	299
3470	24" to 9" wide		8.40	1.905		136	102		238	300
3475	24" to 6" wide		8.40	1.905		136	102		238	300
3480	30" to 24" wide		7.20	2.222		137	119		256	330
3500	30" to 18" wide		7.20	2.222		137	119		256	330
3520	30" to 12" wide		7.20	2.222		141	119		260	330
3540	36" to 30" wide		6.40	2.500		142	133		275	355
3560	36" to 24" wide		6.40	2.500		142	133		275	355
3580	36" to 18" wide		6.40	2.500		142	133		275	355

26 05 36 – Cable Trays for Electrical Systems

26 05 36.30 Cable Tray Trough		Crew	Daily Output	Labor-Hours	Unit	Material	2014 Bare Costs Labor	Equipment	Total	Total Incl O&P
3600	36" to 12" wide	2 Elec	6.40	2.500	Ea.	142	133		275	355
3610	Elbow horizontal 60°, 12" radius, 6" wide		7.80	2.051		97	109		206	271
3620	9" wide		7.20	2.222		106	119		225	294
3630	12" wide		6.40	2.500		117	133		250	330
3640	18" wide		5.80	2.759		126	147		273	360
3650	24" wide		4.80	3.333		154	178		332	435
3680	Elbow horizontal 45°, 12" radius, 6" wide		8	2		83.50	107		190.50	252
3690	9" wide		7.40	2.162		88.50	115		203.50	271
3700	12" wide		6.60	2.424		93.50	129		222.50	297
3710	18" wide		6	2.667		105	142		247	330
3720	24" wide		5	3.200		122	171		293	390
3750	Elbow horizontal, 30° 12" radius, 6" wide		8.20	1.951		73	104		177	237
3760	9" wide		7.60	2.105		76.50	112		188.50	253
3770	12" wide		6.80	2.353		82	126		208	278
3780	18" wide		6.20	2.581		88.50	138		226.50	305
3790	24" wide		5.20	3.077		97	164		261	355
3820	Elbow vertical 60° in/outside, 12" radius, 6" wide		7.80	2.051		121	109		230	297
3830	9" wide		7.20	2.222		122	119		241	310
3840	12" wide		6.40	2.500		124	133		257	335
3850	18" wide		5.80	2.759		129	147		276	360
3860	24" wide		4.80	3.333		134	178		312	415
3890	Elbow vertical 45° in/outside, 12" radius, 6" wide		8	2		97	107		204	267
3900	9" wide		7.40	2.162		103	115		218	286
3910	12" wide		6.60	2.424		106	129		235	310
3920	18" wide		6	2.667		110	142		252	335
3930	24" wide		5	3.200		120	171		291	390
3960	Elbow vertical 30° in/outside, 12" radius, 6" wide		8.20	1.951		83.50	104		187.50	248
3970	9" wide		7.60	2.105		88.50	112		200.50	266
3980	12" wide		6.80	2.353		90.50	126		216.50	288
3990	18" wide		6.20	2.581		94	138		232	310
4000	24" wide		5.20	3.077		95.50	164		259.50	350
4250	Reducer, left or right hand, 24" to 18" wide		8.40	1.905		127	102		229	292
4260	24" to 12" wide		8.40	1.905		127	102		229	292
4270	24" to 9" wide		8.40	1.905		127	102		229	292
4280	24" to 6" wide		8.40	1.905		129	102		231	294
4290	18" to 12" wide		9.60	1.667		119	89		208	264
4300	18" to 9" wide		9.60	1.667		119	89		208	264
4310	18" to 6" wide		9.60	1.667		119	89		208	264
4320	12" to 9" wide		11.40	1.404		114	75		189	237
4330	12" to 6" wide		11.40	1.404		114	75		189	237
4340	9" to 6" wide		12	1.333		110	71		181	228
4350	Splice plate	1 Elec	48	.167		10.55	8.90		19.45	25
4360	Splice plate, expansion joint		48	.167		10.80	8.90		19.70	25
4370	Splice plate, hinged, horizontal		48	.167		8.70	8.90		17.60	23
4380	Vertical		48	.167		12.40	8.90		21.30	27
4390	Trough, hanger, vertical		28	.286		35	15.25		50.25	61.50
4400	Box connector, 24" wide		20	.400		47	21.50		68.50	83.50
4410	18" wide		21	.381		39	20.50		59.50	73.50
4420	12" wide		22	.364		37.50	19.40		56.90	70.50
4430	9" wide		23	.348		35.50	18.55		54.05	67
4440	6" wide		24	.333		34	17.80		51.80	64
4450	Floor flange		24	.333		35.50	17.80		53.30	65.50
4460	Hold down clamp		60	.133		4.10	7.10		11.20	15.15

26 05 36.30 Cable Tray Trough

		Crew	Daily Output	Labor-Hours	Unit	Material	2014 Bare Costs Labor	Equipment	Total	Total Incl O&P
4520	Wall bracket, 24" wide tray	1 Elec	20	.400	Ea.	36	21.50		57.50	71.50
4530	18" wide tray		21	.381		35	20.50		55.50	69
4540	12" wide tray		22	.364		18.75	19.40		38.15	49.50
4550	9" wide tray		23	.348		16.75	18.55		35.30	46.50
4560	6" wide tray		24	.333		15.20	17.80		33	43.50
5000	Cable channel aluminum, vented, 1-1/4" deep, 4" wide, straight		80	.100	L.F.	13.25	5.35		18.60	22.50
5010	Elbow horizontal, 36" radius, 90°		5	1.600	Ea.	238	85.50		323.50	390
5020	60°		5.50	1.455		183	77.50		260.50	315
5030	45°		6	1.333		149	71		220	270
5040	30°		6.50	1.231		128	65.50		193.50	240
5050	Adjustable		6	1.333		122	71		193	240
5060	Elbow vertical, 36" radius, 90°		5	1.600		256	85.50		341.50	410
5070	60°		5.50	1.455		200	77.50		277.50	335
5080	45°		6	1.333		165	71		236	287
5090	30°		6.50	1.231		146	65.50		211.50	259
5100	Adjustable		6	1.333		122	71		193	240
5110	Splice plate, hinged, horizontal		48	.167		9.15	8.90		18.05	23.50
5120	Splice plate, hinged, vertical		48	.167		12.95	8.90		21.85	27.50
5130	Hanger, vertical		28	.286		14.30	15.25		29.55	38.50
5140	Single		28	.286		21.50	15.25		36.75	47
5150	Double		20	.400		22	21.50		43.50	56.50
5160	Channel to box connector		24	.333		29	17.80		46.80	58.50
5170	Hold down clip		80	.100		4.12	5.35		9.47	12.55
5180	Wall bracket, single		28	.286		11.85	15.25		27.10	36
5190	Double		20	.400		15.20	21.50		36.70	48.50
5200	Cable roller		16	.500		186	26.50		212.50	245
5210	Splice plate		48	.167		5.70	8.90		14.60	19.60

26 05 36.40 Cable Tray, Covers and Dividers

		Crew	Daily Output	Labor-Hours	Unit	Material	2014 Bare Costs Labor	Equipment	Total	Total Incl O&P
0010	**CABLE TRAY, COVERS AND DIVIDERS** To 15' high　R260536-10									
0011	For higher elevations, see lines 9900 – 9960									
0100	Covers, ventilated galv. steel, straight, 6" wide tray size	2 Elec	520	.031	L.F.	7	1.64		8.64	10.15
0200	9" wide tray size		460	.035		8.60	1.86		10.46	12.25
0300	12" wide tray size　R260536-11		400	.040		10.30	2.13		12.43	14.50
0400	18" wide tray size		300	.053		14	2.85		16.85	19.65
0500	24" wide tray size		220	.073		17.15	3.88		21.03	24.50
0600	30" wide tray size		180	.089		21	4.74		25.74	30
0700	36" wide tray size		160	.100		23.50	5.35		28.85	34
1000	Elbow horizontal 90°, 12" radius, 6" wide tray size		150	.107	Ea.	51.50	5.70		57.20	65.50
1020	9" wide tray size		128	.125		57.50	6.65		64.15	73
1040	12" wide tray size		108	.148		60.50	7.90		68.40	78.50
1060	18" wide tray size		84	.190		83.50	10.15		93.65	107
1080	24" wide tray size		66	.242		101	12.95		113.95	130
1100	30" wide tray size		60	.267		126	14.25		140.25	160
1120	36" wide tray size		50	.320		152	17.05		169.05	193
1160	24" radius, 6" wide tray size		136	.118		86	6.30		92.30	104
1180	9" wide tray size		116	.138		88	7.35		95.35	108
1200	12" wide tray size		96	.167		98	8.90		106.90	121
1220	18" wide tray size		76	.211		121	11.25		132.25	150
1240	24" wide tray size		60	.267		148	14.25		162.25	184
1260	30" wide tray size		52	.308		194	16.40		210.40	238
1280	36" wide tray size		44	.364		225	19.40		244.40	276
1320	36" radius, 6" wide tray size		120	.133		126	7.10		133.10	149

26 05 36 – Cable Trays for Electrical Systems

26 05 36.40 Cable Tray, Covers and Dividers		Crew	Daily Output	Labor-Hours	Unit	Material	2014 Bare Costs Labor	Equipment	Total	Total Incl O&P
1340	9" wide tray size	2 Elec	104	.154	Ea.	139	8.20		147.20	165
1360	12" wide tray size		84	.190		148	10.15		158.15	178
1380	18" wide tray size		72	.222		189	11.85		200.85	226
1400	24" wide tray size		52	.308		225	16.40		241.40	272
1420	30" wide tray size		46	.348		267	18.55		285.55	320
1440	36" wide tray size		40	.400		315	21.50		336.50	375
1480	Elbow horizontal 45°, 12" radius, 6" wide tray size		150	.107		37	5.70		42.70	49
1500	9" wide tray size		128	.125		44	6.65		50.65	58.50
1520	12" wide tray size		108	.148		48	7.90		55.90	64.50
1540	18" wide tray size		88	.182		58.50	9.70		68.20	79
1560	24" wide tray size		76	.211		68	11.25		79.25	92
1580	30" wide tray size		66	.242		80.50	12.95		93.45	108
1600	36" wide tray size		60	.267		90.50	14.25		104.75	121
1640	24" radius, 6" wide tray size		136	.118		54	6.30		60.30	69
1660	9" wide tray size		116	.138		60.50	7.35		67.85	77.50
1680	12" wide tray size		96	.167		68	8.90		76.90	88.50
1700	18" wide tray size		80	.200		78	10.65		88.65	102
1720	24" wide tray size		70	.229		90.50	12.20		102.70	118
1740	30" wide tray size		60	.267		115	14.25		129.25	148
1760	36" wide tray size		52	.308		126	16.40		142.40	163
1800	36" radius, 6" wide tray size		120	.133		78	7.10		85.10	96.50
1820	9" wide tray size		104	.154		88	8.20		96.20	109
1840	12" wide tray size		84	.190		97	10.15		107.15	122
1860	18" wide tray size		76	.211		111	11.25		122.25	139
1880	24" wide tray size		62	.258		136	13.75		149.75	170
1900	30" wide tray size		52	.308		148	16.40		164.40	187
1920	36" wide tray size		48	.333		178	17.80		195.80	223
1960	Elbow vertical 90°, 12" radius, 6" wide tray size		150	.107		43	5.70		48.70	56
1980	9" wide tray size		128	.125		44	6.65		50.65	58.50
2000	12" wide tray size		108	.148		47.50	7.90		55.40	64
2020	18" wide tray size		88	.182		54	9.70		63.70	74
2040	24" wide tray size		68	.235		55	12.55		67.55	79.50
2060	30" wide tray size		60	.267		60.50	14.25		74.75	88
2080	36" wide tray size		50	.320		78	17.05		95.05	112
2120	24" radius, 6" wide tray size		136	.118		53	6.30		59.30	68
2140	9" wide tray size		116	.138		58.50	7.35		65.85	75.50
2160	12" wide tray size		96	.167		60.50	8.90		69.40	80
2180	18" wide tray size		80	.200		80.50	10.65		91.15	104
2200	24" wide tray size		62	.258		90.50	13.75		104.25	120
2220	30" wide tray size		52	.308		103	16.40		119.40	138
2240	36" wide tray size		44	.364		117	19.40		136.40	157
2280	36" radius, 6" wide tray size		120	.133		61.50	7.10		68.60	78.50
2300	9" wide tray size		104	.154		78	8.20		86.20	98.50
2320	12" wide tray size		84	.190		88	10.15		98.15	112
2340	18" wide tray size		76	.211		111	11.25		122.25	139
2350	24" wide tray size		54	.296		123	15.80		138.80	160
2360	30" wide tray size		46	.348		148	18.55		166.55	190
2370	36" wide tray size		40	.400		174	21.50		195.50	224
2400	Tee horizontal, 12" radius, 6" wide tray size		92	.174		78.50	9.30		87.80	100
2410	9" wide tray size		80	.200		80.50	10.65		91.15	104
2420	12" wide tray size		68	.235		91	12.55		103.55	119
2430	18" wide tray size		60	.267		110	14.25		124.25	143
2440	24" wide tray size		52	.308		139	16.40		155.40	178

26 05 36 – Cable Trays for Electrical Systems

26 05 36.40 Cable Tray, Covers and Dividers	Crew	Daily Output	Labor-Hours	Unit	Material	2014 Bare Costs Labor	Equipment	Total	Total Incl O&P
2460	30" wide tray size	2 Elec	36	.444	Ea.	163	23.50		186.50	216
2470	36" wide tray size		30	.533		197	28.50		225.50	260
2500	24" radius, 6" wide tray size		88	.182		123	9.70		132.70	150
2510	9" wide tray size		76	.211		141	11.25		152.25	172
2520	12" wide tray size		64	.250		145	13.35		158.35	180
2530	18" wide tray size		56	.286		187	15.25		202.25	228
2540	24" wide tray size		48	.333		286	17.80		303.80	340
2560	30" wide tray size		32	.500		325	26.50		351.50	400
2570	36" wide tray size		26	.615		360	33		393	445
2600	36" radius, 6" wide tray size		84	.190		223	10.15		233.15	260
2610	9" wide tray size		72	.222		227	11.85		238.85	267
2620	12" wide tray size		60	.267		252	14.25		266.25	299
2630	18" wide tray size		52	.308		295	16.40		311.40	350
2640	24" wide tray size		44	.364		390	19.40		409.40	455
2660	30" wide tray size		28	.571		420	30.50		450.50	510
2670	36" wide tray size		22	.727		480	39		519	585
2700	Cross horizontal, 12" radius, 6" wide tray size		68	.235		117	12.55		129.55	147
2710	9" wide tray size		64	.250		123	13.35		136.35	156
2720	12" wide tray size		60	.267		138	14.25		152.25	174
2730	18" wide tray size		52	.308		163	16.40		179.40	205
2740	24" wide tray size		36	.444		199	23.50		222.50	255
2760	30" wide tray size		30	.533		227	28.50		255.50	292
2770	36" wide tray size		28	.571		265	30.50		295.50	335
2800	24" radius, 6" wide tray size		64	.250		223	13.35		236.35	265
2810	9" wide tray size		60	.267		242	14.25		256.25	288
2820	12" wide tray size		56	.286		261	15.25		276.25	310
2830	18" wide tray size		48	.333		310	17.80		327.80	370
2840	24" wide tray size		32	.500		380	26.50		406.50	455
2860	30" wide tray size		26	.615		430	33		463	520
2870	36" wide tray size		24	.667		480	35.50		515.50	585
2900	36" radius, 6" wide tray size		60	.267		375	14.25		389.25	430
2910	9" wide tray size		56	.286		385	15.25		400.25	450
2920	12" wide tray size		52	.308		405	16.40		421.40	470
2930	18" wide tray size		44	.364		465	19.40		484.40	545
2940	24" wide tray size		28	.571		595	30.50		625.50	700
2960	30" wide tray size		22	.727		640	39		679	765
2970	36" wide tray size		20	.800		680	42.50		722.50	815
3000	Reducer, 9" to 6" wide tray size		128	.125		49.50	6.65		56.15	64.50
3010	12" to 6" wide tray size		108	.148		51.50	7.90		59.40	69
3020	12" to 9" wide tray size		108	.148		51.50	7.90		59.40	69
3030	18" to 12" wide tray size		88	.182		55	9.70		64.70	75
3050	18" to 6" wide tray size		88	.182		55	9.70		64.70	75
3060	24" to 18" wide tray size		80	.200		77	10.65		87.65	101
3070	24" to 12" wide tray size		80	.200		70.50	10.65		81.15	93.50
3090	30" to 24" wide tray size		70	.229		81.50	12.20		93.70	108
3100	30" to 18" wide tray size		70	.229		81.50	12.20		93.70	108
3110	30" to 12" wide tray size		70	.229		70.50	12.20		82.70	96
3140	36" to 30" wide tray size		64	.250		89	13.35		102.35	118
3150	36" to 24" wide tray size		64	.250		89	13.35		102.35	118
3160	36" to 18" wide tray size		64	.250		89	13.35		102.35	118
3170	36" to 12" wide tray size		64	.250		89	13.35		102.35	118
3250	Covers, aluminum, straight, 6" wide tray size		520	.031	L.F.	5.35	1.64		6.99	8.35
3270	9" wide tray size		460	.035		6.45	1.86		8.31	9.90

26 05 36 – Cable Trays for Electrical Systems

26 05 36.40 Cable Tray, Covers and Dividers	Crew	Daily Output	Labor-Hours	Unit	Material	2014 Bare Costs Labor	Equipment	Total	Total Incl O&P
3290	12" wide tray size	2 Elec	400	.040	L.F.	7.70	2.13		9.83	11.65
3310	18" wide tray size		320	.050		10.20	2.67		12.87	15.25
3330	24" wide tray size		260	.062		12.70	3.28		15.98	18.90
3350	30" wide tray size		200	.080		14.05	4.27		18.32	22
3370	36" wide tray size		180	.089	↓	14.90	4.74		19.64	23.50
3400	Elbow horizontal 90°, 12" radius, 6" wide tray size		150	.107	Ea.	40	5.70		45.70	52.50
3410	9" wide tray size		128	.125		42	6.65		48.65	56
3420	12" wide tray size		108	.148		45	7.90		52.90	61.50
3430	18" wide tray size		88	.182		59.50	9.70		69.20	79.50
3440	24" wide tray size		70	.229		75	12.20		87.20	101
3460	30" wide tray size		64	.250		89.50	13.35		102.85	118
3470	36" wide tray size		54	.296		110	15.80		125.80	145
3500	24" radius, 6" wide tray size		136	.118		55	6.30		61.30	70
3510	9" wide tray size		116	.138		68	7.35		75.35	85.50
3520	12" wide tray size		96	.167		75	8.90		83.90	96
3530	18" wide tray size		80	.200		88.50	10.65		99.15	113
3540	24" wide tray size		64	.250		110	13.35		123.35	141
3560	30" wide tray size		56	.286		134	15.25		149.25	170
3570	36" wide tray size		48	.333		166	17.80		183.80	209
3600	36" radius, 6" wide tray size		120	.133		94	7.10		101.10	114
3610	9" wide tray size		104	.154		103	8.20		111.20	125
3620	12" wide tray size		84	.190		117	10.15		127.15	143
3630	18" wide tray size		76	.211		139	11.25		150.25	170
3640	24" wide tray size		56	.286		169	15.25		184.25	209
3660	30" wide tray size		50	.320		194	17.05		211.05	240
3670	36" wide tray size		44	.364		228	19.40		247.40	279
3700	Elbow horizontal 45°, 12" radius, 6" wide tray size		150	.107		29.50	5.70		35.20	41
3710	9" wide tray size		128	.125		30.50	6.65		37.15	43.50
3720	12" wide tray size		108	.148		34	7.90		41.90	49.50
3730	18" wide tray size		88	.182		39	9.70		48.70	57.50
3740	24" wide tray size		80	.200		43.50	10.65		54.15	64
3760	30" wide tray size		70	.229		55	12.20		67.20	79
3770	36" wide tray size		64	.250		66	13.35		79.35	92.50
3800	24" radius, 6" wide tray size		136	.118		34.50	6.30		40.80	47
3810	9" wide tray size		116	.138		43.50	7.35		50.85	59
3820	12" wide tray size		96	.167		45.50	8.90		54.40	63.50
3830	18" wide tray size		80	.200		55	10.65		65.65	76.50
3840	24" wide tray size		72	.222		68	11.85		79.85	92.50
3860	30" wide tray size		64	.250		78.50	13.35		91.85	106
3870	36" wide tray size		56	.286		90.50	15.25		105.75	123
3900	36" radius, 6" wide tray size		120	.133		59.50	7.10		66.60	75.50
3910	9" wide tray size		104	.154		64.50	8.20		72.70	83.50
3920	12" wide tray size		84	.190		68	10.15		78.15	89.50
3930	18" wide tray size		76	.211		82	11.25		93.25	107
3940	24" wide tray size		64	.250		99	13.35		112.35	129
3960	30" wide tray size		56	.286		113	15.25		128.25	147
3970	36" wide tray size		50	.320		129	17.05		146.05	168
4000	Elbow vertical 90°, 12" radius, 6" wide tray size		150	.107		34	5.70		39.70	46
4010	9" wide tray size		128	.125		34.50	6.65		41.15	47.50
4020	12" wide tray size		108	.148		37.50	7.90		45.40	53.50
4030	18" wide tray size		88	.182		43.50	9.70		53.20	62.50
4040	24" wide tray size		70	.229		45.50	12.20		57.70	68.50
4060	30" wide tray size		64	.250		46	13.35		59.35	71

26 05 36 – Cable Trays for Electrical Systems

26 05 36.40 Cable Tray, Covers and Dividers		Crew	Daily Output	Labor-Hours	Unit	Material	2014 Bare Costs Labor	Equipment	Total	Total Incl O&P
4070	36" wide tray size	2 Elec	54	.296	Ea.	55.50	15.80		71.30	85
4100	24" radius, 6" wide tray size		136	.118		39	6.30		45.30	52.50
4110	9" wide tray size		116	.138		42.50	7.35		49.85	58
4120	12" wide tray size		96	.167		46	8.90		54.90	64.50
4130	18" wide tray size		80	.200		55.50	10.65		66.15	77.50
4140	24" wide tray size		64	.250		61	13.35		74.35	87
4160	30" wide tray size		56	.286		78.50	15.25		93.75	109
4170	36" wide tray size		48	.333		85.50	17.80		103.30	121
4200	36" radius, 6" wide tray size		120	.133		45.50	7.10		52.60	60.50
4210	9" wide tray size		104	.154		55.50	8.20		63.70	74
4220	12" wide tray size		84	.190		64.50	10.15		74.65	86
4230	18" wide tray size		76	.211		78.50	11.25		89.75	103
4240	24" wide tray size		56	.286		92	15.25		107.25	124
4260	30" wide tray size		50	.320		117	17.05		134.05	154
4270	36" wide tray size		44	.364		125	19.40		144.40	167
4300	Tee horizontal, 12" radius, 6" wide tray size		108	.148		55.50	7.90		63.40	73.50
4310	9" wide tray size		88	.182		59.50	9.70		69.20	79.50
4320	12" wide tray size		80	.200		66	10.65		76.65	88.50
4330	18" wide tray size		68	.235		78.50	12.55		91.05	105
4340	24" wide tray size		56	.286		99	15.25		114.25	132
4360	30" wide tray size		44	.364		117	19.40		136.40	157
4370	36" wide tray size		36	.444		141	23.50		164.50	191
4400	24" radius, 6" wide tray size		96	.167		90.50	8.90		99.40	113
4410	9" wide tray size		80	.200		101	10.65		111.65	127
4420	12" wide tray size		72	.222		113	11.85		124.85	142
4430	18" wide tray size		60	.267		132	14.25		146.25	168
4440	24" wide tray size		48	.333		211	17.80		228.80	259
4460	30" wide tray size		40	.400		233	21.50		254.50	288
4470	36" wide tray size		32	.500		259	26.50		285.50	325
4500	36" radius, 6" wide tray size		88	.182		166	9.70		175.70	197
4510	9" wide tray size		72	.222		169	11.85		180.85	204
4520	12" wide tray size		64	.250		186	13.35		199.35	224
4530	18" wide tray size		56	.286		211	15.25		226.25	255
4540	24" wide tray size		44	.364		268	19.40		287.40	325
4560	30" wide tray size		36	.444		300	23.50		323.50	365
4570	36" wide tray size		28	.571		340	30.50		370.50	420
4600	Cross horizontal, 12" radius, 6" wide tray size		80	.200		85.50	10.65		96.15	110
4610	9" wide tray size		72	.222		90.50	11.85		102.35	117
4620	12" wide tray size		64	.250		100	13.35		113.35	130
4630	18" wide tray size		56	.286		122	15.25		137.25	157
4640	24" wide tray size		48	.333		141	17.80		158.80	182
4660	30" wide tray size		40	.400		167	21.50		188.50	216
4670	36" wide tray size		32	.500		192	26.50		218.50	252
4700	24" radius, 6" wide tray size		72	.222		167	11.85		178.85	202
4710	9" wide tray size		64	.250		179	13.35		192.35	217
4720	12" wide tray size		56	.286		192	15.25		207.25	235
4730	18" wide tray size		48	.333		224	17.80		241.80	274
4740	24" wide tray size		40	.400		268	21.50		289.50	325
4760	30" wide tray size		32	.500		315	26.50		341.50	385
4770	36" wide tray size		24	.667		340	35.50		375.50	430
4800	36" radius, 6" wide tray size		64	.250		268	13.35		281.35	315
4810	9" wide tray size		56	.286		279	15.25		294.25	330
4820	12" wide tray size		50	.320		300	17.05		317.05	355

26 05 36.40 Cable Tray, Covers and Dividers	Crew	Daily Output	Labor-Hours	Unit	Material	2014 Bare Costs Labor	Equipment	Total	Total Incl O&P
4830	18" wide tray size	2 Elec	44	.364	Ea.	330	19.40		349.40	395
4840	24" wide tray size		36	.444		410	23.50		433.50	485
4860	30" wide tray size		28	.571		470	30.50		500.50	560
4870	36" wide tray size		22	.727		515	39		554	625
4900	Reducer, 9" to 6" wide tray size		128	.125		42.50	6.65		49.15	57
4910	12" to 6" wide tray size		108	.148		44.50	7.90		52.40	61
4920	12" to 9" wide tray size		108	.148		44.50	7.90		52.40	61
4930	18" to 12" wide tray size		88	.182		48.50	9.70		58.20	68
4950	18" to 6" wide tray size		88	.182		48.50	9.70		58.20	68
4960	24" to 18" wide tray size		80	.200		62.50	10.65		73.15	85
4970	24" to 12" wide tray size		80	.200		54	10.65		64.65	75.50
4990	30" to 24" wide tray size		70	.229		65.50	12.20		77.70	90.50
5000	30" to 18" wide tray size		70	.229		65.50	12.20		77.70	90.50
5010	30" to 12" wide tray size		70	.229		65.50	12.20		77.70	90.50
5040	36" to 30" wide tray size		64	.250		72.50	13.35		85.85	99.50
5050	36" to 24" wide tray size		64	.250		72.50	13.35		85.85	99.50
5060	36" to 18" wide tray size		64	.250		72.50	13.35		85.85	99.50
5070	36" to 12" wide tray size		64	.250		72.50	13.35		85.85	99.50
5710	Tray cover hold down clamp	1 Elec	60	.133		11.55	7.10		18.65	23.50
8000	Divider strip, straight, galvanized, 3" deep		200	.040	L.F.	6.15	2.13		8.28	10
8020	4" deep		180	.044		7.55	2.37		9.92	11.85
8040	6" deep		160	.050		9.85	2.67		12.52	14.85
8060	Aluminum, straight, 3" deep		210	.038		6.15	2.03		8.18	9.85
8080	4" deep		190	.042		7.65	2.25		9.90	11.75
8100	6" deep		170	.047		9.75	2.51		12.26	14.45
8110	Divider strip vertical fitting 3" deep									
8120	12" radius, galvanized, 30°	1 Elec	28	.286	Ea.	30	15.25		45.25	56
8140	45°		27	.296		37.50	15.80		53.30	64.50
8160	60°		26	.308		39.50	16.40		55.90	68
8180	90°		25	.320		49.50	17.05		66.55	80
8200	Aluminum, 30°		29	.276		19.90	14.70		34.60	44
8220	45°		28	.286		23.50	15.25		38.75	49
8240	60°		27	.296		27.50	15.80		43.30	53.50
8260	90°		26	.308		35	16.40		51.40	63
8280	24" radius, galvanized, 30°		25	.320		45	17.05		62.05	75
8300	45°		24	.333		50.50	17.80		68.30	82
8320	60°		23	.348		64	18.55		82.55	98.50
8340	90°		22	.364		87	19.40		106.40	125
8360	Aluminum, 30°		26	.308		33	16.40		49.40	61
8380	45°		25	.320		38.50	17.05		55.55	67.50
8400	60°		24	.333		48	17.80		65.80	79
8420	90°		23	.348		65.50	18.55		84.05	100
8440	36" radius, galvanized, 30°		22	.364		60.50	19.40		79.90	95.50
8460	45°		21	.381		70.50	20.50		91	108
8480	60°		20	.400		82.50	21.50		104	123
8500	90°		19	.421		115	22.50		137.50	160
8520	Aluminum, 30°		23	.348		50.50	18.55		69.05	83.50
8540	45°		22	.364		65.50	19.40		84.90	101
8560	60°		21	.381		84.50	20.50		105	124
8570	90°		20	.400		111	21.50		132.50	154
8590	Divider strip vertical fitting 4" deep									
8600	12" radius, galvanized, 30°	1 Elec	27	.296	Ea.	38	15.80		53.80	65.50
8610	45°		26	.308		44	16.40		60.40	73

26 05 36.40 Cable Tray, Covers and Dividers	Crew	Daily Output	Labor-Hours	Unit	Material	2014 Bare Costs Labor	Equipment	Total	Total Incl O&P
8620	60°	1 Elec	25	.320	Ea.	49.50	17.05		66.55	80
8630	90°		24	.333		60.50	17.80		78.30	93
8640	Aluminum, 30°		28	.286		28	15.25		43.25	54
8650	45°		27	.296		32.50	15.80		48.30	59.50
8660	60°		26	.308		37.50	16.40		53.90	66
8670	90°		25	.320		44.50	17.05		61.55	74.50
8680	24" radius, galvanized, 30°		24	.333		60.50	17.80		78.30	93
8690	45°		23	.348		76	18.55		94.55	112
8700	60°		22	.364		86	19.40		105.40	124
8710	90°		21	.381		115	20.50		135.50	157
8720	Aluminum, 30°		25	.320		44.50	17.05		61.55	74.50
8730	45°		24	.333		54	17.80		71.80	86
8740	60°		23	.348		63.50	18.55		82.05	98
8750	90°		22	.364		87	19.40		106.40	125
8760	36" radius, galvanized, 30°		23	.348		70.50	18.55		89.05	106
8770	45°		22	.364		80.50	19.40		99.90	118
8780	60°		21	.381		101	20.50		121.50	142
8790	90°		20	.400		138	21.50		159.50	184
8800	Aluminum, 30°		24	.333		68	17.80		85.80	101
8810	45°		23	.348		87	18.55		105.55	124
8820	60°		22	.364		106	19.40		125.40	146
8830	90°		21	.381		134	20.50		154.50	178
8840	Divider strip vertical fitting 6" deep									
8850	12" radius, galvanized, 30°	1 Elec	24	.333	Ea.	41.50	17.80		59.30	72
8860	45°		23	.348		46	18.55		64.55	79
8870	60°		22	.364		53	19.40		72.40	87.50
8880	90°		21	.381		66	20.50		86.50	103
8890	Aluminum, 30°		25	.320		31	17.05		48.05	59.50
8900	45°		24	.333		36.50	17.80		54.30	66.50
8910	60°		23	.348		38.50	18.55		57.05	70
8920	90°		22	.364		45.50	19.40		64.90	79
8930	24" radius, galvanized, 30°		23	.348		60.50	18.55		79.05	94.50
8940	45°		22	.364		76	19.40		95.40	113
8950	60°		21	.381		87	20.50		107.50	127
8960	90°		20	.400		115	21.50		136.50	158
8970	Aluminum, 30°		24	.333		45.50	17.80		63.30	76.50
8980	45°		23	.348		62	18.55		80.55	96
8990	60°		22	.364		68	19.40		87.40	104
9000	90°		21	.381		92	20.50		112.50	132
9010	36" radius, galvanized, 30°		22	.364		70.50	19.40		89.90	107
9020	45°		21	.381		87	20.50		107.50	127
9030	60°		20	.400		115	21.50		136.50	158
9040	90°		19	.421		148	22.50		170.50	196
9050	Aluminum, 30°		23	.348		69	18.55		87.55	104
9060	45°		22	.364		92	19.40		111.40	130
9070	60°		21	.381		109	20.50		129.50	151
9080	90°		20	.400		131	21.50		152.50	176
9120	Divider strip, horizontal fitting, galvanized, 3" deep		33	.242		42.50	12.95		55.45	66
9130	4" deep		30	.267		47.50	14.25		61.75	73.50
9140	6" deep		27	.296		60.50	15.80		76.30	90
9150	Aluminum, 3" deep		35	.229		31.50	12.20		43.70	53.50
9160	4" deep		32	.250		35	13.35		48.35	58.50
9170	6" deep		29	.276		46	14.70		60.70	73

26 05 36 – Cable Trays for Electrical Systems

26 05 36.40 Cable Tray, Covers and Dividers	Crew	Daily Output	Labor-Hours	Unit	Material	2014 Bare Costs Labor	Equipment	Total	Total Incl O&P
9300	Divider strip protector	1 Elec	300	.027	L.F.	3.78	1.42		5.20	6.30
9310	Fastener, ladder tray				Ea.	.68			.68	.75
9320	Trough or solid bottom tray				"	.46			.46	.51
9900	Add to labor for higher elevated installation									
9910	15' to 20' high add						10%			
9920	20' to 25' high add						20%			
9930	25' to 30' high add						25%			
9940	30' to 35' high add						30%			
9960	Over 40' high add						40%			

26 05 39 – Underfloor Raceways for Electrical Systems

26 05 39.30 Conduit In Concrete Slab

		Crew	Daily Output	Labor-Hours	Unit	Material	2014 Bare Costs Labor	Equipment	Total	Total Incl O&P
0010	**CONDUIT IN CONCRETE SLAB** Including terminations, R260539-30									
0020	fittings and supports									
3230	PVC, schedule 40, 1/2" diameter	1 Elec	270	.030	L.F.	.64	1.58		2.22	3.07
3250	3/4" diameter		230	.035		.75	1.86		2.61	3.60
3270	1" diameter		200	.040		.94	2.13		3.07	4.23
3300	1-1/4" diameter		170	.047		1.36	2.51		3.87	5.25
3330	1-1/2" diameter		140	.057		1.63	3.05		4.68	6.35
3350	2" diameter		120	.067		2.07	3.56		5.63	7.55
3370	2-1/2" diameter		90	.089		3.53	4.74		8.27	11
3400	3" diameter	2 Elec	160	.100		4.47	5.35		9.82	12.90
3430	3-1/2" diameter		120	.133		5.70	7.10		12.80	16.95
3440	4" diameter		100	.160		6.25	8.55		14.80	19.65
3450	5" diameter		80	.200		9.55	10.65		20.20	26.50
3460	6" diameter		60	.267		13.65	14.25		27.90	36.50
3530	Sweeps, 1" diameter, 30" radius	1 Elec	32	.250	Ea.	37	13.35		50.35	60.50
3550	1-1/4" diameter		24	.333		44	17.80		61.80	74.50
3570	1-1/2" diameter		21	.381		46	20.50		66.50	81
3600	2" diameter		18	.444		50	23.50		73.50	90.50
3630	2-1/2" diameter		14	.571		66.50	30.50		97	119
3650	3" diameter		10	.800		84.50	42.50		127	157
3670	3-1/2" diameter		8	1		103	53.50		156.50	194
3700	4" diameter		7	1.143		141	61		202	248
3710	5" diameter		6	1.333		165	71		236	287
3730	Couplings, 1/2" diameter					.25			.25	.28
3750	3/4" diameter					.28			.28	.31
3770	1" diameter					.42			.42	.46
3800	1-1/4" diameter					.65			.65	.72
3830	1-1/2" diameter					.80			.80	.88
3850	2" diameter					1.04			1.04	1.14
3870	2-1/2" diameter					1.85			1.85	2.04
3900	3" diameter					2.99			2.99	3.29
3930	3-1/2" diameter					3.77			3.77	4.15
3950	4" diameter					4.21			4.21	4.63
3960	5" diameter					10.70			10.70	11.75
3970	6" diameter					14.55			14.55	16
4030	End bells 1" diameter, PVC	1 Elec	60	.133		4.59	7.10		11.69	15.70
4050	1-1/4" diameter		53	.151		5.65	8.05		13.70	18.30
4100	1-1/2" diameter		48	.167		5.65	8.90		14.55	19.50
4150	2" diameter		34	.235		8.40	12.55		20.95	28
4170	2-1/2" diameter		27	.296		9.30	15.80		25.10	34
4200	3" diameter		20	.400		9.80	21.50		31.30	43

26 05 39 – Underfloor Raceways for Electrical Systems

26 05 39.30 Conduit In Concrete Slab	Crew	Daily Output	Labor-Hours	Unit	Material	2014 Bare Costs Labor	Equipment	Total	Total Incl O&P
4250 3-1/2" diameter	1 Elec	16	.500	Ea.	11.65	26.50		38.15	53
4300 4" diameter		14	.571		12.70	30.50		43.20	59.50
4310 5" diameter		12	.667		13.40	35.50		48.90	68
4320 6" diameter		9	.889		15.90	47.50		63.40	88.50
4350 Rigid galvanized steel, 1/2" diameter		200	.040	L.F.	2.31	2.13		4.44	5.75
4400 3/4" diameter		170	.047		2.44	2.51		4.95	6.45
4450 1" diameter		130	.062		3.36	3.28		6.64	8.60
4500 1-1/4" diameter		110	.073		4.59	3.88		8.47	10.85
4600 1-1/2" diameter		100	.080		5.15	4.27		9.42	12.05
4800 2" diameter		90	.089		6.40	4.74		11.14	14.15

26 05 39.40 Conduit In Trench

	Crew	Daily Output	Labor-Hours	Unit	Material	2014 Bare Costs Labor	Equipment	Total	Total Incl O&P
0010 **CONDUIT IN TRENCH** Includes terminations and fittings R260539-40									
0020 Does not include excavation or backfill, see Section 31 23 16.00									
0200 Rigid galvanized steel, 2" diameter	1 Elec	150	.053	L.F.	6	2.85		8.85	10.85
0400 2-1/2" diameter	"	100	.080		11.75	4.27		16.02	19.35
0600 3" diameter	2 Elec	160	.100		13.90	5.35		19.25	23.50
0800 3-1/2" diameter		140	.114		18.35	6.10		24.45	29
1000 4" diameter		100	.160		20.50	8.55		29.05	35.50
1200 5" diameter		80	.200		42.50	10.65		53.15	63
1400 6" diameter		60	.267		62.50	14.25		76.75	90.50

26 05 43 – Underground Ducts and Raceways for Electrical Systems

26 05 43.10 Trench Duct

	Crew	Daily Output	Labor-Hours	Unit	Material	2014 Bare Costs Labor	Equipment	Total	Total Incl O&P
0010 **TRENCH DUCT** Steel with cover									
0020 Standard adjustable, depths to 4"									
0100 Straight, single compartment, 9" wide	2 Elec	40	.400	L.F.	116	21.50		137.50	160
0200 12" wide		32	.500		133	26.50		159.50	186
0400 18" wide		26	.615		163	33		196	228
0600 24" wide		22	.727		195	39		234	272
0700 27" wide		21	.762		211	40.50		251.50	293
0800 30" wide		20	.800		226	42.50		268.50	315
1000 36" wide		16	1		259	53.50		312.50	365
1020 Two compartment, 9" wide		38	.421		117	22.50		139.50	163
1030 12" wide		30	.533		134	28.50		162.50	190
1040 18" wide		24	.667		164	35.50		199.50	233
1050 24" wide		20	.800		214	42.50		256.50	300
1060 30" wide		18	.889		261	47.50		308.50	360
1070 36" wide		14	1.143		305	61		366	425
1090 Three compartment, 9" wide		36	.444		134	23.50		157.50	183
1100 12" wide		28	.571		147	30.50		177.50	208
1110 18" wide		22	.727		185	39		224	262
1120 24" wide		18	.889		233	47.50		280.50	325
1130 30" wide		16	1		281	53.50		334.50	390
1140 36" wide		12	1.333		330	71		401	465
1200 Horizontal elbow, 9" wide		5.40	2.963	Ea.	380	158		538	655
1400 12" wide		4.60	3.478		440	186		626	765
1600 18" wide		4	4		565	213		778	940
1800 24" wide		3.20	5		795	267		1,062	1,275
1900 27" wide		3	5.333		900	285		1,185	1,425
2000 30" wide		2.60	6.154		1,050	330		1,380	1,675
2200 36" wide		2.40	6.667		1,400	355		1,755	2,050
2220 Two compartment, 9" wide		3.80	4.211		620	225		845	1,025
2230 12" wide		3	5.333		685	285		970	1,175

26 05 43 – Underground Ducts and Raceways for Electrical Systems

26 05 43.10 Trench Duct		Crew	Daily Output	Labor-Hours	Unit	Material	2014 Bare Costs Labor	Equipment	Total	Total Incl O&P
2240	18" wide	2 Elec	2.40	6.667	Ea.	820	355		1,175	1,425
2250	24" wide		2	8		1,050	425		1,475	1,800
2260	30" wide		1.80	8.889		1,400	475		1,875	2,225
2270	36" wide		1.60	10		1,700	535		2,235	2,650
2290	Three compartment, 9" wide		3.60	4.444		645	237		882	1,075
2300	12" wide		2.80	5.714		725	305		1,030	1,250
2310	18" wide		2.20	7.273		880	390		1,270	1,550
2320	24" wide		1.80	8.889		1,125	475		1,600	1,925
2330	30" wide		1.60	10		1,450	535		1,985	2,400
2350	36" wide		1.40	11.429		1,775	610		2,385	2,875
2400	Vertical elbow, 9" wide		5.40	2.963		134	158		292	385
2600	12" wide		4.60	3.478		145	186		331	435
2800	18" wide		4	4		166	213		379	505
3000	24" wide		3.20	5		207	267		474	625
3100	27" wide		3	5.333		216	285		501	660
3200	30" wide		2.60	6.154		228	330		558	740
3400	36" wide		2.40	6.667		251	355		606	805
3600	Cross, 9" wide		4	4		625	213		838	1,000
3800	12" wide		3.20	5		665	267		932	1,125
4000	18" wide		2.60	6.154		795	330		1,125	1,375
4200	24" wide		2.20	7.273		1,000	390		1,390	1,675
4300	27" wide		2.20	7.273		1,175	390		1,565	1,875
4400	30" wide		2	8		1,300	425		1,725	2,100
4600	36" wide		1.80	8.889		1,625	475		2,100	2,500
4620	Two compartment, 9" wide		3.80	4.211		655	225		880	1,050
4630	12" wide		3	5.333		690	285		975	1,175
4640	18" wide		2.40	6.667		830	355		1,185	1,450
4650	24" wide		2	8		1,050	425		1,475	1,825
4660	30" wide		1.80	8.889		1,400	475		1,875	2,225
4670	36" wide		1.60	10		1,675	535		2,210	2,650
4690	Three compartment, 9" wide		3.60	4.444		665	237		902	1,075
4700	12" wide		2.80	5.714		760	305		1,065	1,300
4710	18" wide		2.20	7.273		900	390		1,290	1,575
4720	24" wide		1.80	8.889		1,125	475		1,600	1,925
4730	30" wide		1.60	10		1,475	535		2,010	2,400
4740	36" wide		1.40	11.429		1,800	610		2,410	2,925
4800	End closure, 9" wide		14.40	1.111		39	59.50		98.50	132
5000	12" wide		12	1.333		45	71		116	156
5200	18" wide		10	1.600		68.50	85.50		154	204
5400	24" wide		8	2		90.50	107		197.50	260
5500	27" wide		7	2.286		105	122		227	298
5600	30" wide		6.60	2.424		113	129		242	320
5800	36" wide		5.80	2.759		135	147		282	370
6000	Tees, 9" wide		4	4		380	213		593	740
6200	12" wide		3.60	4.444		440	237		677	840
6400	18" wide		3.20	5		565	267		832	1,025
6600	24" wide		3	5.333		815	285		1,100	1,325
6700	27" wide		2.80	5.714		900	305		1,205	1,450
6800	30" wide		2.60	6.154		1,050	330		1,380	1,675
7000	36" wide		2	8		1,400	425		1,825	2,175
7020	Two compartment, 9" wide		3.80	4.211		440	225		665	820
7030	12" wide		3.40	4.706		475	251		726	900
7040	18" wide		3	5.333		635	285		920	1,125

26 05 43.10 Trench Duct		Crew	Daily Output	Labor-Hours	Unit	Material	2014 Bare Costs Labor	Equipment	Total	Total Incl O&P
7050	24" wide	2 Elec	2.80	5.714	Ea.	860	305		1,165	1,400
7060	30" wide		2.40	6.667		1,175	355		1,530	1,800
7070	36" wide		1.90	8.421		1,450	450		1,900	2,275
7090	Three compartment, 9" wide		3.60	4.444		505	237		742	910
7100	12" wide		3.20	5		530	267		797	980
7110	18" wide		2.80	5.714		670	305		975	1,200
7120	24" wide		2.60	6.154		925	330		1,255	1,525
7130	30" wide		2.20	7.273		1,200	390		1,590	1,900
7140	36" wide		1.80	8.889		1,550	475		2,025	2,400
7200	Riser, and cabinet connector, 9" wide		5.40	2.963		166	158		324	420
7400	12" wide		4.60	3.478		194	186		380	490
7600	18" wide		4	4		238	213		451	580
7800	24" wide		3.20	5		289	267		556	715
7900	27" wide		3	5.333		297	285		582	750
8000	30" wide		2.60	6.154		330	330		660	855
8200	36" wide		2	8		385	425		810	1,075
8400	Insert assembly, cell to conduit adapter, 1-1/4"	1 Elec	16	.500		66	26.50		92.50	113
8500	Adjustable partition	"	320	.025	L.F.	23.50	1.33		24.83	28
8600	Depth of duct over 4", per 1", add					10.20			10.20	11.25
8700	Support post	1 Elec	240	.033		23.50	1.78		25.28	28
8800	Cover double tile trim, 2 sides					36			36	40
8900	4 sides					107			107	118
9160	Trench duct 3-1/2" x 4-1/2", add					9.70			9.70	10.65
9170	Trench duct 4" x 5", add					9.70			9.70	10.65
9200	For carpet trim, add					33			33	36
9210	For double carpet trim, add					100			100	110

26 05 43.20 Underfloor Duct		Crew	Daily Output	Labor-Hours	Unit	Material	2014 Bare Costs Labor	Equipment	Total	Total Incl O&P
0010	**UNDERFLOOR DUCT** R260543-50									
0100	Duct, 1-3/8" x 3-1/8" blank, standard	2 Elec	160	.100	L.F.	14.10	5.35		19.45	23.50
0200	1-3/8" x 7-1/4" blank, super duct		120	.133		28	7.10		35.10	41.50
0400	7/8" or 1-3/8" insert type, 24" O.C., 1-3/8" x 3-1/8", std.		140	.114		18.85	6.10		24.95	30
0600	1-3/8" x 7-1/4", super duct		100	.160		33	8.55		41.55	49.50
0800	Junction box, single duct, 1 level, 3-1/8"	1 Elec	4	2	Ea.	425	107		532	625
0820	3-1/8" x 7-1/4"		4	2		495	107		602	700
0840	2 level, 3-1/8" upper & lower		3.20	2.500		495	133		628	740
0860	3-1/8" upper, 7-1/4" lower		2.70	2.963		495	158		653	775
0880	Carpet pan for above		80	.100		350	5.35		355.35	395
0900	Terrazzo pan for above		67	.119		820	6.35		826.35	910
1000	Junction box, single duct, 1 level, 7-1/4"		2.70	2.963		495	158		653	775
1020	2 level, 7-1/4" upper & lower		2.70	2.963		565	158		723	855
1040	2 duct, two 3-1/8" upper & lower		3.20	2.500		750	133		883	1,025
1200	1 level, 2 duct, 3-1/8"		3.20	2.500		565	133		698	820
1220	Carpet pan for above boxes		80	.100		350	5.35		355.35	395
1240	Terrazzo pan for above boxes		67	.119		820	6.35		826.35	910
1260	Junction box, 1 level, two 3-1/8" x one 3-1/8" + one 7-1/4"		2.30	3.478		960	186		1,146	1,325
1280	2 level, two 3-1/8" upper, one 3-1/8" + one 7-1/4" lower		2	4		1,050	213		1,263	1,475
1300	Carpet pan for above boxes		80	.100		350	5.35		355.35	395
1320	Terrazzo pan for above boxes		67	.119		820	6.35		826.35	910
1400	Junction box, 1 level, 2 duct, 7-1/4"		2.30	3.478		1,450	186		1,636	1,875
1420	Two 3-1/8" + one 7-1/4"		2	4		1,450	213		1,663	1,925
1440	Carpet pan for above		80	.100		350	5.35		355.35	395
1460	Terrazzo pan for above		67	.119		820	6.35		826.35	910

26 05 43 – Underground Ducts and Raceways for Electrical Systems

26 05 43.20 Underfloor Duct		Crew	Daily Output	Labor-Hours	Unit	Material	2014 Bare Costs Labor	Equipment	Total	Total Incl O&P
1580	Junction box, 1 level, one 3-1/8" + one 7-1/4" x same	1 Elec	2.30	3.478	Ea.	960	186		1,146	1,325
1600	Triple duct, 3-1/8"		2.30	3.478		960	186		1,146	1,325
1700	Junction box, 1 level, one 3-1/8" + two 7-1/4"		2	4		1,625	213		1,838	2,100
1720	Carpet pan for above		80	.100		350	5.35		355.35	395
1740	Terrazzo pan for above		67	.119		820	6.35		826.35	910
1800	Insert to conduit adapter, 3/4" & 1"		32	.250		34.50	13.35		47.85	58
2000	Support, single cell		27	.296		52	15.80		67.80	81
2200	Super duct		16	.500		52	26.50		78.50	97.50
2400	Double cell		16	.500		52	26.50		78.50	97.50
2600	Triple cell		11	.727		52	39		91	116
2800	Vertical elbow, standard duct		10	.800		93.50	42.50		136	167
3000	Super duct		8	1		93.50	53.50		147	183
3200	Cabinet connector, standard duct		32	.250		70.50	13.35		83.85	97.50
3400	Super duct		27	.296		70.50	15.80		86.30	101
3600	Conduit adapter, 1" to 1-1/4"		32	.250		70.50	13.35		83.85	97.50
3800	2" to 1-1/4"		27	.296		84.50	15.80		100.30	117
4000	Outlet, low tension (tele, computer, etc.)		8	1		98.50	53.50		152	188
4200	High tension, receptacle (120 volt)		8	1		98.50	53.50		152	188
4300	End closure, standard duct		160	.050		3.94	2.67		6.61	8.30
4310	Super duct		160	.050		7.45	2.67		10.12	12.20
4350	Elbow, horiz., standard duct		26	.308		235	16.40		251.40	284
4360	Super duct		26	.308		235	16.40		251.40	284
4380	Elbow, offset, standard duct		26	.308		93.50	16.40		109.90	128
4390	Super duct		26	.308		93.50	16.40		109.90	128
4400	Marker screw assembly for inserts		50	.160		16.60	8.55		25.15	31
4410	Y take off, standard duct		26	.308		141	16.40		157.40	180
4420	Super duct		26	.308		141	16.40		157.40	180
4430	Box opening plug, standard duct		160	.050		13.20	2.67		15.87	18.50
4440	Super duct		160	.050		13.20	2.67		15.87	18.50
4450	Sleeve coupling, standard duct		160	.050		35	2.67		37.67	42.50
4460	Super duct		160	.050		43.50	2.67		46.17	51.50
4470	Conduit adapter, standard duct, 3/4"		32	.250		70.50	13.35		83.85	97.50
4480	1" or 1-1/4"		32	.250		70.50	13.35		83.85	97.50
4500	1-1/2"		32	.250		70.50	13.35		83.85	97.50

26 05 80 – Wiring Connections

26 05 80.10 Motor Connections

0010	MOTOR CONNECTIONS									
0020	Flexible conduit and fittings, 115 volt, 1 phase, up to 1 HP motor	1 Elec	8	1	Ea.	7.25	53.50		60.75	88
0050	2 HP motor	R260580-75	6.50	1.231		13.75	65.50		79.25	114
0100	3 HP motor		5.50	1.455		12.25	77.50		89.75	129
0110	230 volt, 3 phase, 3 HP motor		6.78	1.180		8.70	63		71.70	104
0112	5 HP motor		5.47	1.463		7.40	78		85.40	125
0114	7-1/2 HP motor		4.61	1.735		10.50	92.50		103	151
0120	10 HP motor		4.20	1.905		23.50	102		125.50	178
0150	15 HP motor		3.30	2.424		23.50	129		152.50	220
0200	25 HP motor		2.70	2.963		33.50	158		191.50	274
0400	50 HP motor		2.20	3.636		67	194		261	365
0600	100 HP motor		1.50	5.333		158	285		443	600
1500	460 volt, 5 HP motor, 3 phase		8	1		7.75	53.50		61.25	88.50
1520	10 HP motor		8	1		7.75	53.50		61.25	88.50
1530	25 HP motor		6	1.333		14.70	71		85.70	122
1540	30 HP motor		6	1.333		14.70	71		85.70	122

26 05 80 – Wiring Connections

26 05 80.10 Motor Connections		Crew	Daily Output	Labor-Hours	Unit	Material	2014 Bare Costs Labor	Equipment	Total	Total Incl O&P
1550	40 HP motor	1 Elec	5	1.600	Ea.	22.50	85.50		108	153
1560	50 HP motor		5	1.600		26	85.50		111.50	157
1570	60 HP motor		3.80	2.105		29	112		141	200
1580	75 HP motor		3.50	2.286		39	122		161	226
1590	100 HP motor		2.50	3.200		59	171		230	320
1600	125 HP motor		2	4		77	213		290	405
1610	150 HP motor		1.80	4.444		81.50	237		318.50	445
1620	200 HP motor		1.50	5.333		131	285		416	570
2005	460 Volt, 5 HP motor, 3 Phase, w/sealtite		8	1		12.40	53.50		65.90	93.50
2010	10 HP motor		8	1		12.40	53.50		65.90	93.50
2015	25 HP motor		6	1.333		23.50	71		94.50	132
2020	30 HP motor		6	1.333		23.50	71		94.50	132
2025	40 HP motor		5	1.600		43.50	85.50		129	176
2030	50 HP motor		5	1.600		44.50	85.50		130	177
2035	60 HP motor		3.80	2.105		68.50	112		180.50	244
2040	75 HP motor		3.50	2.286		72	122		194	262
2045	100 HP motor		2.50	3.200		97	171		268	365
2055	150 HP motor		1.80	4.444		103	237		340	470
2060	200 HP motor		1.50	5.333		710	285		995	1,200

26 05 90 – Residential Applications

26 05 90.10 Residential Wiring		Crew	Daily Output	Labor-Hours	Unit	Material	2014 Bare Costs Labor	Equipment	Total	Total Incl O&P
0010	**RESIDENTIAL WIRING**									
0020	20' avg. runs and #14/2 wiring incl. unless otherwise noted									
1000	Service & panel, includes 24' SE-AL cable, service eye, meter,									
1010	Socket, panel board, main bkr., ground rod, 15 or 20 amp									
1020	1-pole circuit breakers, and misc. hardware									
1100	100 amp, with 10 branch breakers	1 Elec	1.19	6.723	Ea.	540	360		900	1,125
1110	With PVC conduit and wire		.92	8.696		595	465		1,060	1,350
1120	With RGS conduit and wire		.73	10.959		740	585		1,325	1,700
1150	150 amp, with 14 branch breakers		1.03	7.767		840	415		1,255	1,550
1170	With PVC conduit and wire		.82	9.756		955	520		1,475	1,825
1180	With RGS conduit and wire		.67	11.940		1,225	635		1,860	2,300
1200	200 amp, with 18 branch breakers	2 Elec	1.80	8.889		1,125	475		1,600	1,925
1220	With PVC conduit and wire		1.46	10.959		1,225	585		1,810	2,225
1230	With RGS conduit and wire		1.24	12.903		1,600	690		2,290	2,775
1800	Lightning surge suppressor	1 Elec	32	.250		50.50	13.35		63.85	76
2000	Switch devices									
2100	Single pole, 15 amp, Ivory, with a 1-gang box, cover plate,									
2110	Type NM (Romex) cable	1 Elec	17.10	.468	Ea.	12.30	25		37.30	51
2120	Type MC (BX) cable		14.30	.559		23.50	30		53.50	70.50
2130	EMT & wire		5.71	1.401		29.50	75		104.50	145
2150	3-way, #14/3, type NM cable		14.55	.550		15.60	29.50		45.10	61
2170	Type MC cable		12.31	.650		31	34.50		65.50	86
2180	EMT & wire		5	1.600		32.50	85.50		118	164
2200	4-way, #14/3, type NM cable		14.55	.550		22.50	29.50		52	68.50
2220	Type MC cable		12.31	.650		37.50	34.50		72	93.50
2230	EMT & wire		5	1.600		39.50	85.50		125	172
2250	S.P., 20 amp, #12/2, type NM cable		13.33	.600		21	32		53	71.50
2270	Type MC cable		11.43	.700		30	37.50		67.50	89
2280	EMT & wire		4.85	1.649		40	88		128	176
2290	S.P. rotary dimmer, 600W, no wiring		17	.471		27	25		52	67
2300	S.P. rotary dimmer, 600W, type NM cable		14.55	.550		32	29.50		61.50	79

26 05 90.10 Residential Wiring	Crew	Daily Output	Labor-Hours	Unit	Material	2014 Bare Costs Labor	Equipment	Total	Total Incl O&P
2320	Type MC cable	1 Elec	12.31	.650	Ea.	43.50	34.50		78	99.50
2330	EMT & wire		5	1.600		51	85.50		136.50	184
2350	3-way rotary dimmer, type NM cable		13.33	.600		31.50	32		63.50	82.50
2370	Type MC cable		11.43	.700		42.50	37.50		80	103
2380	EMT & wire		4.85	1.649		50	88		138	187
2400	Interval timer wall switch, 20 amp, 1-30 min., #12/2									
2410	Type NM cable	1 Elec	14.55	.550	Ea.	54	29.50		83.50	104
2420	Type MC cable		12.31	.650		59	34.50		93.50	117
2430	EMT & wire		5	1.600		73	85.50		158.50	209
2500	Decorator style									
2510	S.P., 15 amp, type NM cable	1 Elec	17.10	.468	Ea.	16.15	25		41.15	55.50
2520	Type MC cable		14.30	.559		27.50	30		57.50	74.50
2530	EMT & wire		5.71	1.401		33.50	75		108.50	149
2550	3-way, #14/3, type NM cable		14.55	.550		19.40	29.50		48.90	65.50
2570	Type MC cable		12.31	.650		34.50	34.50		69	90
2580	EMT & wire		5	1.600		36.50	85.50		122	168
2600	4-way, #14/3, type NM cable		14.55	.550		26.50	29.50		56	73
2620	Type MC cable		12.31	.650		41.50	34.50		76	97.50
2630	EMT & wire		5	1.600		43	85.50		128.50	176
2650	S.P., 20 amp, #12/2, type NM cable		13.33	.600		25	32		57	75.50
2670	Type MC cable		11.43	.700		34	37.50		71.50	93
2680	EMT & wire		4.85	1.649		44	88		132	180
2700	S.P., slide dimmer, type NM cable		17.10	.468		32	25		57	73
2720	Type MC cable		14.30	.559		43.50	30		73.50	92.50
2730	EMT & wire		5.71	1.401		51	75		126	168
2750	S.P., touch dimmer, type NM cable		17.10	.468		29.50	25		54.50	69.50
2770	Type MC cable		14.30	.559		40.50	30		70.50	89
2780	EMT & wire		5.71	1.401		48	75		123	165
2800	3-way touch dimmer, type NM cable		13.33	.600		51.50	32		83.50	105
2820	Type MC cable		11.43	.700		63	37.50		100.50	125
2830	EMT & wire		4.85	1.649		70.50	88		158.50	210
3000	Combination devices									
3100	S.P. switch/15 amp recpt., Ivory, 1-gang box, plate									
3110	Type NM cable	1 Elec	11.43	.700	Ea.	24	37.50		61.50	82.50
3120	Type MC cable		10	.800		35.50	42.50		78	103
3130	EMT & wire		4.40	1.818		43	97		140	192
3150	S.P. switch/pilot light, type NM cable		11.43	.700		24	37.50		61.50	82.50
3170	Type MC cable		10	.800		35.50	42.50		78	103
3180	EMT & wire		4.43	1.806		43	96.50		139.50	192
3190	2-S.P. switches, 2-#14/2, no wiring		14	.571		6.95	30.50		37.45	53
3200	2-S.P. switches, 2-#14/2, type NM cables		10	.800		27.50	42.50		70	94.50
3220	Type MC cable		8.89	.900		44.50	48		92.50	121
3230	EMT & wire		4.10	1.951		46.50	104		150.50	207
3250	3-way switch/15 amp recpt., #14/3, type NM cable		10	.800		31.50	42.50		74	98.50
3270	Type MC cable		8.89	.900		46.50	48		94.50	123
3280	EMT & wire		4.10	1.951		48	104		152	209
3300	2-3 way switches, 2-#14/3, type NM cables		8.89	.900		41	48		89	117
3320	Type MC cable		8	1		66	53.50		119.50	153
3330	EMT & wire		4	2		56	107		163	222
3350	S.P. switch/20 amp recpt., #12/2, type NM cable		10	.800		32	42.50		74.50	99.50
3370	Type MC cable		8.89	.900		37	48		85	113
3380	EMT & wire		4.10	1.951		51	104		155	212
3400	Decorator style									

26 05 90.10 Residential Wiring	Crew	Daily Output	Labor-Hours	Unit	Material	2014 Bare Costs Labor	Equipment	Total	Total Incl O&P
3410	S.P. switch/15 amp recpt., type NM cable	1 Elec	11.43	.700	Ea.	28	37.50		65.50	86.50
3420	Type MC cable		10	.800		39	42.50		81.50	107
3430	EMT & wire		4.40	1.818		46.50	97		143.50	197
3450	S.P. switch/pilot light, type NM cable		11.43	.700		28	37.50		65.50	87
3470	Type MC cable		10	.800		39.50	42.50		82	108
3480	EMT & wire		4.40	1.818		47	97		144	197
3500	2-S.P. switches, 2-#14/2, type NM cables		10	.800		31.50	42.50		74	98.50
3520	Type MC cable		8.89	.900		48.50	48		96.50	125
3530	EMT & wire		4.10	1.951		50	104		154	211
3550	3-way/15 amp recpt., #14/3, type NM cable		10	.800		35	42.50		77.50	103
3570	Type MC cable		8.89	.900		50.50	48		98.50	128
3580	EMT & wire		4.10	1.951		52	104		156	213
3650	2-3 way switches, 2-#14/3, type NM cables		8.89	.900		45	48		93	122
3670	Type MC cable		8	1		69.50	53.50		123	157
3680	EMT & wire		4	2		59.50	107		166.50	226
3700	S.P. switch/20 amp recpt., #12/2, type NM cable		10	.800		36	42.50		78.50	104
3720	Type MC cable		8.89	.900		41	48		89	117
3730	EMT & wire		4.10	1.951		55	104		159	216
4000	Receptacle devices									
4010	Duplex outlet, 15 amp recpt., Ivory, 1-gang box, plate									
4015	Type NM cable	1 Elec	14.55	.550	Ea.	11	29.50		40.50	56
4020	Type MC cable		12.31	.650		22.50	34.50		57	76.50
4030	EMT & wire		5.33	1.501		28	80		108	151
4050	With #12/2, type NM cable		12.31	.650		13.65	34.50		48.15	67
4070	Type MC cable		10.67	.750		22.50	40		62.50	84.50
4080	EMT & wire		4.71	1.699		32.50	90.50		123	172
4100	20 amp recpt., #12/2, type NM cable		12.31	.650		21	34.50		55.50	75
4120	Type MC cable		10.67	.750		29.50	40		69.50	92.50
4130	EMT & wire		4.71	1.699		39.50	90.50		130	180
4140	For GFI see Section 26 05 90.10 line 4300 below									
4150	Decorator style, 15 amp recpt., type NM cable	1 Elec	14.55	.550	Ea.	14.80	29.50		44.30	60.50
4170	Type MC cable		12.31	.650		26	34.50		60.50	80.50
4180	EMT & wire		5.33	1.501		32	80		112	155
4200	With #12/2, type NM cable		12.31	.650		17.45	34.50		51.95	71
4220	Type MC cable		10.67	.750		26	40		66	89
4230	EMT & wire		4.71	1.699		36	90.50		126.50	176
4250	20 amp recpt. #12/2, type NM cable		12.31	.650		24.50	34.50		59	79
4270	Type MC cable		10.67	.750		33.50	40		73.50	96.50
4280	EMT & wire		4.71	1.699		43.50	90.50		134	184
4300	GFI, 15 amp recpt., type NM cable		12.31	.650		47.50	34.50		82	104
4320	Type MC cable		10.67	.750		58.50	40		98.50	125
4330	EMT & wire		4.71	1.699		64.50	90.50		155	207
4350	GFI with #12/2, type NM cable		10.67	.750		50	40		90	115
4370	Type MC cable		9.20	.870		58.50	46.50		105	134
4380	EMT & wire		4.21	1.900		68.50	101		169.50	228
4400	20 amp recpt., #12/2 type NM cable		10.67	.750		52	40		92	118
4420	Type MC cable		9.20	.870		61	46.50		107.50	137
4430	EMT & wire		4.21	1.900		71	101		172	230
4500	Weather-proof cover for above receptacles, add		32	.250		4.08	13.35		17.43	24.50
4550	Air conditioner outlet, 20 amp-240 volt recpt.									
4560	30' of #12/2, 2 pole circuit breaker									
4570	Type NM cable	1 Elec	10	.800	Ea.	63	42.50		105.50	133
4580	Type MC cable		9	.889		73	47.50		120.50	152

26 05 90.10 Residential Wiring	Crew	Daily Output	Labor-Hours	Unit	Material	2014 Bare Costs Labor	Equipment	Total	Total Incl O&P
4590	EMT & wire	1 Elec	4	2	Ea.	81.50	107		188.50	250
4600	Decorator style, type NM cable		10	.800		67.50	42.50		110	139
4620	Type MC cable		9	.889		78	47.50		125.50	157
4630	EMT & wire		4	2		86.50	107		193.50	255
4650	Dryer outlet, 30 amp-240 volt recpt., 20' of #10/3									
4660	2 pole circuit breaker									
4670	Type NM cable	1 Elec	6.41	1.248	Ea.	64	66.50		130.50	170
4680	Type MC cable		5.71	1.401		69	75		144	188
4690	EMT & wire		3.48	2.299		75.50	123		198.50	267
4700	Range outlet, 50 amp-240 volt recpt., 30' of #8/3									
4710	Type NM cable	1 Elec	4.21	1.900	Ea.	92	101		193	253
4720	Type MC cable		4	2		136	107		243	310
4730	EMT & wire		2.96	2.703		106	144		250	335
4750	Central vacuum outlet, Type NM cable		6.40	1.250		60	66.50		126.50	166
4770	Type MC cable		5.71	1.401		75.50	75		150.50	195
4780	EMT & wire		3.48	2.299		81.50	123		204.50	274
4800	30 amp-110 volt locking recpt., #10/2 circ. bkr.									
4810	Type NM cable	1 Elec	6.20	1.290	Ea.	69	69		138	179
4820	Type MC cable		5.40	1.481		86.50	79		165.50	213
4830	EMT & wire		3.20	2.500		92	133		225	300
4900	Low voltage outlets									
4910	Telephone recpt., 20' of 4/C phone wire	1 Elec	26	.308	Ea.	10.95	16.40		27.35	36.50
4920	TV recpt., 20' of RG59U coax wire, F type connector	"	16	.500	"	19.20	26.50		45.70	61
4950	Door bell chime, transformer, 2 buttons, 60' of bellwire									
4970	Economy model	1 Elec	11.50	.696	Ea.	58	37		95	120
4980	Custom model		11.50	.696		123	37		160	191
4990	Luxury model, 3 buttons		9.50	.842		305	45		350	400
6000	Lighting outlets									
6050	Wire only (for fixture), type NM cable	1 Elec	32	.250	Ea.	7.20	13.35		20.55	28
6070	Type MC cable		24	.333		13.85	17.80		31.65	41.50
6080	EMT & wire		10	.800		18.65	42.50		61.15	84.50
6100	Box (4"), and wire (for fixture), type NM cable		25	.320		16.10	17.05		33.15	43.50
6120	Type MC cable		20	.400		23	21.50		44.50	57
6130	EMT & wire		11	.727		27.50	39		66.50	88.50
6200	Fixtures (use with lines 6050 or 6100 above)									
6210	Canopy style, economy grade	1 Elec	40	.200	Ea.	37.50	10.65		48.15	57
6220	Custom grade		40	.200		57.50	10.65		68.15	79
6250	Dining room chandelier, economy grade		19	.421		93.50	22.50		116	137
6260	Custom grade		19	.421		274	22.50		296.50	335
6270	Luxury grade		15	.533		620	28.50		648.50	725
6310	Kitchen fixture (fluorescent), economy grade		30	.267		69.50	14.25		83.75	98
6320	Custom grade		25	.320		190	17.05		207.05	235
6350	Outdoor, wall mounted, economy grade		30	.267		35.50	14.25		49.75	60.50
6360	Custom grade		30	.267		121	14.25		135.25	156
6370	Luxury grade		25	.320		274	17.05		291.05	325
6410	Outdoor PAR floodlights, 1 lamp, 150 watt		20	.400		38	21.50		59.50	74
6420	2 lamp, 150 watt each		20	.400		63	21.50		84.50	101
6425	Motion sensing, 2 lamp, 150 watt each		20	.400		76	21.50		97.50	116
6430	For infrared security sensor, add		32	.250		127	13.35		140.35	160
6450	Outdoor, quartz-halogen, 300 watt flood		20	.400		46	21.50		67.50	82.50
6600	Recessed downlight, round, pre-wired, 50 or 75 watt trim		30	.267		47	14.25		61.25	73
6610	With shower light trim		30	.267		58	14.25		72.25	85.50
6620	With wall washer trim		28	.286		70	15.25		85.25	100

26 05 90.10 Residential Wiring	Crew	Daily Output	Labor-Hours	Unit	Material	2014 Bare Costs Labor	Equipment	Total	Total Incl O&P
6630	With eye-ball trim	1 Elec	28	.286	Ea.	70	15.25		85.25	100
6640	For direct contact with insulation, add					2.49			2.49	2.74
6700	Porcelain lamp holder	1 Elec	40	.200		3.50	10.65		14.15	19.80
6710	With pull switch		40	.200		5.70	10.65		16.35	22.50
6750	Fluorescent strip, 1-20 watt tube, wrap around diffuser, 24"		24	.333		58.50	17.80		76.30	91
6760	1-34 watt tube, 48"		24	.333		75.50	17.80		93.30	110
6770	2-34 watt tubes, 48"		20	.400		89.50	21.50		111	131
6780	With residential ballast		20	.400		100	21.50		121.50	142
6800	Bathroom heat lamp, 1-250 watt		28	.286		51.50	15.25		66.75	79.50
6810	2-250 watt lamps	↓	28	.286	↓	82	15.25		97.25	114
6820	For timer switch, see Section 26 05 90.10 line 2400									
6900	Outdoor post lamp, incl. post, fixture, 35' of #14/2									
6910	Type NMC cable	1 Elec	3.50	2.286	Ea.	226	122		348	430
6920	Photo-eye, add		27	.296		38	15.80		53.80	65.50
6950	Clock dial time switch, 24 hr., w/enclosure, type NM cable		11.43	.700		70.50	37.50		108	134
6970	Type MC cable		11	.727		82	39		121	148
6980	EMT & wire	↓	4.85	1.649	↓	88	88		176	229
7000	Alarm systems									
7050	Smoke detectors, box, #14/3, type NM cable	1 Elec	14.55	.550	Ea.	35	29.50		64.50	83
7070	Type MC cable		12.31	.650		46.50	34.50		81	103
7080	EMT & wire	↓	5	1.600		48.50	85.50		134	181
7090	For relay output to security system, add				↓	14.10			14.10	15.55
8000	Residential equipment									
8050	Disposal hook-up, incl. switch, outlet box, 3' of flex									
8060	20 amp-1 pole circ. bkr., and 25' of #12/2									
8070	Type NM cable	1 Elec	10	.800	Ea.	32.50	42.50		75	100
8080	Type MC cable	↓	8	1		42	53.50		95.50	127
8090	EMT & wire	↓	5	1.600	↓	53.50	85.50		139	187
8100	Trash compactor or dishwasher hook-up, incl. outlet box,									
8110	3' of flex, 15 amp-1 pole circ. bkr., and 25' of #14/2									
8120	Type NM cable	1 Elec	10	.800	Ea.	24	42.50		66.50	90.50
8130	Type MC cable	↓	8	1		37	53.50		90.50	121
8140	EMT & wire	↓	5	1.600	↓	45	85.50		130.50	178
8150	Hot water sink dispensor hook-up, use line 8100									
8200	Vent/exhaust fan hook-up, type NM cable	1 Elec	32	.250	Ea.	7.20	13.35		20.55	28
8220	Type MC cable		24	.333		13.85	17.80		31.65	41.50
8230	EMT & wire	↓	10	.800	↓	18.65	42.50		61.15	84.50
8250	Bathroom vent fan, 50 CFM (use with above hook-up)									
8260	Economy model	1 Elec	15	.533	Ea.	27	28.50		55.50	72.50
8270	Low noise model		15	.533		37	28.50		65.50	83
8280	Custom model	↓	12	.667	↓	127	35.50		162.50	193
8300	Bathroom or kitchen vent fan, 110 CFM									
8310	Economy model	1 Elec	15	.533	Ea.	75.50	28.50		104	126
8320	Low noise model	"	15	.533	"	93.50	28.50		122	146
8350	Paddle fan, variable speed (w/o lights)									
8360	Economy model (AC motor)	1 Elec	10	.800	Ea.	125	42.50		167.50	201
8362	With light kit		10	.800		166	42.50		208.50	246
8370	Custom model (AC motor)		10	.800		197	42.50		239.50	281
8372	With light kit		10	.800		238	42.50		280.50	325
8380	Luxury model (DC motor)		8	1		390	53.50		443.50	510
8382	With light kit		8	1		430	53.50		483.50	555
8390	Remote speed switch for above, add	↓	12	.667	↓	32.50	35.50		68	88.50
8500	Whole house exhaust fan, ceiling mount, 36", variable speed									

26 05 90 – Residential Applications

26 05 90.10 Residential Wiring	Crew	Daily Output	Labor-Hours	Unit	Material	2014 Bare Costs Labor	Equipment	Total	Total Incl O&P
8510 Remote switch, incl. shutters, 20 amp-1 pole circ. bkr.									
8520 30' of #12/2, type NM cable	1 Elec	4	2	Ea.	1,250	107		1,357	1,525
8530 Type MC cable	↓	3.50	2.286		1,275	122		1,397	1,575
8540 EMT & wire	↓	3	2.667	↓	1,275	142		1,417	1,625
8600 Whirlpool tub hook-up, incl. timer switch, outlet box									
8610 3' of flex, 20 amp-1 pole GFI circ. bkr.									
8620 30' of #12/2, type NM cable	1 Elec	5	1.600	Ea.	127	85.50		212.50	267
8630 Type MC cable		4.20	1.905		131	102		233	297
8640 EMT & wire	↓	3.40	2.353	↓	141	126		267	345
8650 Hot water heater hook-up, incl. 1-2 pole circ. bkr., box;									
8660 3' of flex, 20' of #10/2, type NM cable	1 Elec	5	1.600	Ea.	34	85.50		119.50	165
8670 Type MC cable		4.20	1.905		47.50	102		149.50	204
8680 EMT & wire	↓	3.40	2.353		48.50	126		174.50	241
9000 Heating/air conditioning									
9050 Furnace/boiler hook-up, incl. firestat, local on-off switch									
9060 Emergency switch, and 40' of type NM cable	1 Elec	4	2	Ea.	52.50	107		159.50	218
9070 Type MC cable		3.50	2.286		69.50	122		191.50	260
9080 EMT & wire	↓	1.50	5.333	↓	81	285		366	515
9100 Air conditioner hook-up, incl. local 60 amp disc. switch									
9110 3' sealtite, 40 amp, 2 pole circuit breaker									
9130 40' of #8/2, type NM cable	1 Elec	3.50	2.286	Ea.	182	122		304	385
9140 Type MC cable		3	2.667		249	142		391	485
9150 EMT & wire	↓	1.30	6.154		218	330		548	730
9200 Heat pump hook-up, 1-40 & 1-100 amp 2 pole circ. bkr.									
9210 Local disconnect switch, 3' sealtite									
9220 40' of #8/2 & 30' of #3/2									
9230 Type NM cable	1 Elec	1.30	6.154	Ea.	525	330		855	1,075
9240 Type MC cable		1.08	7.407		600	395		995	1,250
9250 EMT & wire	↓	.94	8.511	↓	590	455		1,045	1,325
9500 Thermostat hook-up, using low voltage wire									
9520 Heating only, 25' of #18-3	1 Elec	24	.333	Ea.	9.60	17.80		27.40	37
9530 Heating/cooling, 25' of #18-4	"	20	.400	"	11.75	21.50		33.25	45

26 09 13 – Electrical Power Monitoring

26 09 13.10 Switchboard Instruments

	Crew	Daily Output	Labor-Hours	Unit	Material	2014 Bare Costs Labor	Equipment	Total	Total Incl O&P
0010 **SWITCHBOARD INSTRUMENTS** 3 phase, 4 wire R260913-80									
0100 AC indicating, ammeter & switch	1 Elec	8	1	Ea.	2,175	53.50		2,228.50	2,450
0200 Voltmeter & switch		8	1		2,175	53.50		2,228.50	2,450
0300 Wattmeter		8	1		4,300	53.50		4,353.50	4,800
0400 AC recording, ammeter		4	2		7,650	107		7,757	8,550
0500 Voltmeter		4	2		7,650	107		7,757	8,550
0600 Ground fault protection, zero sequence		2.70	2.963		6,750	158		6,908	7,650
0700 Ground return path		2.70	2.963		6,750	158		6,908	7,650
0800 3 current transformers, 5 to 800 amp		2	4		3,150	213		3,363	3,775
0900 1000 to 1500 amp		1.30	6.154		4,525	330		4,855	5,475
1200 2000 to 4000 amp		1	8		5,325	425		5,750	6,525
1300 Fused potential transformer, maximum 600 volt	↓	8	1	↓	1,175	53.50		1,228.50	1,375

26 09 13 – Electrical Power Monitoring

26 09 13.20 Voltage Monitor Systems

		Crew	Daily Output	Labor-Hours	Unit	Material	2014 Bare Costs Labor	Equipment	Total	Total Incl O&P
0010	**VOLTAGE MONITOR SYSTEMS** (test equipment)									
0100	AC voltage monitor system, 120/240 V, one-channel				Ea.	2,725			2,725	3,000
0110	Modem adapter					340			340	375
0120	Add-on detector only					1,425			1,425	1,575
0150	AC voltage remote monitor sys., 3 channel, 120, 230, or 480 V					4,950			4,950	5,450
0160	With internal modem					5,225			5,225	5,750
0170	Combination temperature and humidity probe					770			770	845
0180	Add-on detector only					3,600			3,600	3,950
0190	With internal modem					3,925			3,925	4,300

26 09 13.30 Smart Metering

		Crew	Daily Output	Labor-Hours	Unit	Material	2014 Bare Costs Labor	Equipment	Total	Total Incl O&P
0010	**SMART METERING**, In panel									
0100	Single phase, 120/208 volt, 100 amp	G	1 Elec	8.78	.911	Ea.	410	48.50		458.50	530
0120	200 amp	G		8.78	.911		410	48.50		458.50	530
0200	277 volt, 100 amp	G		8.78	.911		430	48.50		478.50	550
0220	200 amp	G		8.78	.911		430	48.50		478.50	550
1100	Three phase, 120/208 volt, 100 amp	G		4.69	1.706		605	91		696	800
1120	200 amp	G		4.69	1.706		605	91		696	800
1130	400 amp	G		4.69	1.706		605	91		696	800
1140	800 amp	G		4.69	1.706		715	91		806	925
1150	1600 amp	G		4.69	1.706		715	91		806	925
1200	277/480 volt, 100 amp	G		4.69	1.706		670	91		761	875
1220	200 amp	G		4.69	1.706		670	91		761	875
1230	400 amp	G		4.69	1.706		670	91		761	875
1240	800 amp	G		4.69	1.706		670	91		761	875
1250	1600 amp	G		4.69	1.706		785	91		876	995
2000	Data recorder, 8 meters	G		10.97	.729		1,375	39		1,414	1,550
2100	16 meters	G		8.53	.938		1,925	50		1,975	2,200
3000	Software package, per meter, basic	G					234			234	257
3100	Premium	G					605			605	665

26 09 23 – Lighting Control Devices

26 09 23.10 Energy Saving Lighting Devices

		Crew	Daily Output	Labor-Hours	Unit	Material	2014 Bare Costs Labor	Equipment	Total	Total Incl O&P
0010	**ENERGY SAVING LIGHTING DEVICES**									
0100	Occupancy sensors, passive infrared ceiling mounted	G	1 Elec	7	1.143	Ea.	95	61		156	197
0110	Ultrasonic ceiling mounted	G		7	1.143		105	61		166	208
0120	Dual technology ceiling mounted	G		6.50	1.231		152	65.50		217.50	267
0150	Automatic wall switches	G		24	.333		64.50	17.80		82.30	97.50
0160	Daylighting sensor, manual control, ceiling mounted	G		7	1.143		111	61		172	214
0170	Remote and dimming control with remote controller	G		6.50	1.231		174	65.50		239.50	291
0200	Remote power pack	G		10	.800		33	42.50		75.50	100
0250	Photoelectric control, S.P.S.T. 120 V	G		8	1		16.90	53.50		70.40	98.50
0300	S.P.S.T. 208 V/277 V	G		8	1		27	53.50		80.50	110
0350	D.P.S.T. 120 V	G		6	1.333		196	71		267	320
0400	D.P.S.T. 208 V/277 V	G		6	1.333		154	71		225	275
0450	S.P.D.T. 208 V/277 V	G		6	1.333		183	71		254	305
0460	Daylight level sensor, wall mounted, on/off or dimming	G		8	1		135	53.50		188.50	229

26 09 36 – Modular Dimming Controls

26 09 36.13 Manual Modular Dimming Controls

		Crew	Daily Output	Labor-Hours	Unit	Material	2014 Bare Costs Labor	Equipment	Total	Total Incl O&P
0010	**MANUAL MODULAR DIMMING CONTROLS**									
2000	Lighting control module	G	1 Elec	2	4	Ea.	305	213		518	655

26 12 19 – Pad-Mounted, Liquid-Filled, Medium-Voltage Transformers

26 12 19.10 Transformer, Oil-Filled

		Crew	Daily Output	Labor-Hours	Unit	Material	2014 Bare Costs Labor	Equipment	Total	Total Incl O&P
0010	**TRANSFORMER, OIL-FILLED** primary delta or Y, R262213-60									
0050	Pad mounted 5 kV or 15 kV, with taps, 277/480 V secondary, 3 phase									
0100	150 kVA	R-3	.65	30.769	Ea.	8,875	1,625	215	10,715	12,500
0110	225 kVA		.55	36.364		10,100	1,925	255	12,280	14,300
0200	300 kVA		.45	44.444		12,600	2,350	310	15,260	17,800
0300	500 kVA		.40	50		18,000	2,650	350	21,000	24,200
0400	750 kVA		.38	52.632		22,700	2,775	370	25,845	29,600
0500	1000 kVA		.26	76.923		27,000	4,075	540	31,615	36,400
0600	1500 kVA		.23	86.957		32,000	4,600	610	37,210	42,800
0700	2000 kVA		.20	100		40,500	5,300	700	46,500	53,000
0710	2500 kVA		.19	105		48,900	5,575	735	55,210	63,000
0720	3000 kVA		.17	117		59,000	6,225	825	66,050	75,500
0800	3750 kVA	▼	.16	125	▼	76,000	6,625	875	83,500	94,500
1990	Pole mounted distribution type, single phase									
2000	13.8 kV primary, 120/240 V secondary, 10 kVA	R-15	7.45	6.443	Ea.	880	335	43.50	1,258.50	1,525
2010	50 kVA		3.70	12.973		1,750	680	88	2,518	3,050
2020	100 kVA		2.75	17.455		3,075	915	118	4,108	4,875
2030	167 kVA		2.15	22.326		5,450	1,175	151	6,776	7,925
2900	2400 V primary, 120/240 V secondary, 10 kVA		7.45	6.443		925	335	43.50	1,303.50	1,575
2910	15 kVA		6.70	7.164		1,175	375	48.50	1,598.50	1,925
2920	25 kVA		6	8		1,475	420	54	1,949	2,325
2930	37.5 kVA		4.30	11.163		1,800	585	75.50	2,460.50	2,925
2940	50 kVA		3.70	12.973		2,125	680	88	2,893	3,475
2950	75 kVA		3	16		3,100	840	108	4,048	4,800
2960	100 kVA	▼	2.75	17.455	▼	3,100	915	118	4,133	4,900

26 12 19.20 Transformer, Liquid-Filled

		Crew	Daily Output	Labor-Hours	Unit	Material	2014 Bare Costs Labor	Equipment	Total	Total Incl O&P
0010	**TRANSFORMER, LIQUID-FILLED** Pad mounted									
0020	5 kV or 15 kV primary, 277/480 volt secondary, 3 phase									
0050	225 kVA	R-3	.55	36.364	Ea.	13,500	1,925	255	15,680	18,000
0100	300 kVA		.45	44.444		16,000	2,350	310	18,660	21,500
0200	500 kVA		.40	50		20,200	2,650	350	23,200	26,600
0250	750 kVA		.38	52.632		26,100	2,775	370	29,245	33,300
0300	1000 kVA		.26	76.923		30,300	4,075	540	34,915	40,100
0350	1500 kVA		.23	86.957		35,400	4,600	610	40,610	46,500
0400	2000 kVA		.20	100		43,800	5,300	700	49,800	57,000
0450	2500 kVA	▼	.19	105	▼	52,000	5,575	735	58,310	66,500

26 13 Medium-Voltage Switchgear

26 13 16 – Medium-Voltage Fusible Interrupter Switchgear

26 13 16.10 Switchgear

		Crew	Daily Output	Labor-Hours	Unit	Material	2014 Bare Costs Labor	Equipment	Total	Total Incl O&P
0010	**SWITCHGEAR**, Incorporate switch with cable connections, transformer,									
0100	& Low Voltage section									
0200	Load interrupter switch, 600 amp, 2 position									
0300	NEMA 1, 4.8 kV, 300 kVA & below w/CLF fuses	R-3	.40	50	Ea.	21,200	2,650	350	24,200	27,800
0400	400 kVA & above w/CLF fuses		.38	52.632		23,700	2,775	370	26,845	30,700
0500	Non fusible		.41	48.780		17,200	2,575	340	20,115	23,200
0600	13.8 kV, 300 kVA & below w/CLF fuses		.38	52.632		26,700	2,775	370	29,845	33,900
0700	400 kVA & above w/CLF fuses		.36	55.556		26,700	2,950	390	30,040	34,100
0800	Non fusible	▼	.40	50		20,200	2,650	350	23,200	26,600
0900	Cable lugs for 2 feeders 4.8 kV or 13.8 kV	1 Elec	8	1		645	53.50		698.50	790
1000	Pothead, one 3 conductor or three 1 conductor	▼	4	2	▼	3,100	107		3,207	3,575

26 13 16 – Medium-Voltage Fusible Interrupter Switchgear

26 13 16.10 Switchgear	Crew	Daily Output	Labor-Hours	Unit	Material	2014 Bare Costs Labor	Equipment	Total	Total Incl O&P
1100	Two 3 conductor or six 1 conductor	1 Elec	2	4	Ea.	6,125	213		6,338	7,075
1200	Key interlocks	↓	8	1	↓	715	53.50		768.50	870
1300	Lightning arresters, Distribution class (no charge)									
1400	Intermediate class or line type 4.8 kV	1 Elec	2.70	2.963	Ea.	3,475	158		3,633	4,050
1500	13.8 kV		2	4		4,625	213		4,838	5,400
1600	Station class, 4.8 kV		2.70	2.963		5,975	158		6,133	6,800
1700	13.8 kV	↓	2	4		10,300	213		10,513	11,600
1800	Transformers, 4800 volts to 480/277 volts, 75 kVA	R-3	.68	29.412		18,200	1,550	206	19,956	22,600
1900	112.5 kVA		.65	30.769		22,200	1,625	215	24,040	27,100
2000	150 kVA		.57	35.088		25,200	1,850	246	27,296	30,800
2100	225 kVA		.48	41.667		29,000	2,200	292	31,492	35,500
2200	300 kVA		.41	48.780		32,400	2,575	340	35,315	39,900
2300	500 kVA		.36	55.556		42,600	2,950	390	45,940	51,500
2400	750 kVA		.29	68.966		48,400	3,650	485	52,535	59,000
2500	13,800 volts to 480/277 volts, 75 kVA		.61	32.787		25,600	1,725	230	27,555	31,000
2600	112.5 kVA		.55	36.364		33,900	1,925	255	36,080	40,500
2700	150 kVA		.49	40.816		34,300	2,150	286	36,736	41,300
2800	225 kVA		.41	48.780		39,600	2,575	340	42,515	47,800
2900	300 kVA		.37	54.054		40,400	2,850	380	43,630	49,100
3000	500 kVA		.31	64.516		44,700	3,425	450	48,575	55,000
3100	750 kVA	↓	.26	76.923		49,200	4,075	540	53,815	60,500
3200	Forced air cooling & temperature alarm	1 Elec	1	8	↓	3,950	425		4,375	5,000
3300	Low voltage components									
3400	Maximum panel height 49-1/2", single or twin row									
3500	Breaker heights, type FA or FH, 6"									
3600	type KA or KH, 8"									
3700	type LA, 11"									
3800	type MA, 14"									
3900	Breakers, 2 pole, 15 to 60 amp, type FA	1 Elec	5.60	1.429	Ea.	340	76		416	490
4000	70 to 100 amp, type FA		4.20	1.905		430	102		532	620
4100	15 to 60 amp, type FH		5.60	1.429		560	76		636	730
4200	70 to 100 amp, type FH		4.20	1.905		650	102		752	870
4300	125 to 225 amp, type KA		3.40	2.353		1,000	126		1,126	1,300
4400	125 to 225 amp, type KH		3.40	2.353		2,525	126		2,651	3,000
4500	125 to 400 amp, type LA		2.50	3.200		2,275	171		2,446	2,775
4600	125 to 600 amp, type MA		1.80	4.444		2,975	237		3,212	3,625
4700	700 & 800 amp, type MA		1.50	5.333		3,850	285		4,135	4,675
4800	3 pole, 15 to 60 amp, type FA		5.30	1.509		435	80.50		515.50	600
4900	70 to 100 amp, type FA		4	2		540	107		647	750
5000	15 to 60 amp, type FH		5.30	1.509		655	80.50		735.50	840
5100	70 to 100 amp, type FH		4	2		745	107		852	975
5200	125 to 225 amp, type KA		3.20	2.500		1,250	133		1,383	1,575
5300	125 to 225 amp, type KH		3.20	2.500		2,825	133		2,958	3,300
5400	125 to 400 amp, type LA		2.30	3.478		2,825	186		3,011	3,375
5500	125 to 600 amp, type MA		1.60	5		4,075	267		4,342	4,900
5600	700 & 800 amp, type MA	↓	1.30	6.154	↓	5,300	330		5,630	6,325

26 22 13 – Low-Voltage Distribution Transformers

26 22 13.10 Transformer, Dry-Type		Crew	Daily Output	Labor-Hours	Unit	Material	2014 Bare Costs Labor	Equipment	Total	Total Incl O&P
0010	**TRANSFORMER, DRY-TYPE** R262213-10									
0050	Single phase, 240/480 volt primary, 120/240 volt secondary									
0100	1 kVA R262213-60	1 Elec	2	4	Ea.	285	213		498	635
0300	2 kVA		1.60	5		430	267		697	870
0500	3 kVA		1.40	5.714		530	305		835	1,050
0700	5 kVA		1.20	6.667		730	355		1,085	1,325
0900	7.5 kVA	2 Elec	2.20	7.273		1,025	390		1,415	1,700
1100	10 kVA		1.60	10		1,300	535		1,835	2,225
1300	15 kVA R262213-65		1.20	13.333		1,475	710		2,185	2,700
1500	25 kVA		1	16		2,025	855		2,880	3,500
1700	37.5 kVA		.80	20		2,625	1,075		3,700	4,475
1900	50 kVA		.70	22.857		3,100	1,225		4,325	5,250
2100	75 kVA		.65	24.615		4,100	1,325		5,425	6,500
2110	100 kVA	R-3	.90	22.222		5,350	1,175	156	6,681	7,825
2120	167 kVA	"	.80	25		8,875	1,325	175	10,375	11,900
2190	480 V primary 120/240 V secondary, nonvent., 15 kVA	2 Elec	1.20	13.333		1,500	710		2,210	2,725
2200	25 kVA		.90	17.778		2,200	950		3,150	3,850
2210	37 kVA		.75	21.333		2,625	1,150		3,775	4,575
2220	50 kVA		.65	24.615		3,100	1,325		4,425	5,400
2230	75 kVA		.60	26.667		4,100	1,425		5,525	6,650
2240	100 kVA		.50	32		5,350	1,700		7,050	8,425
2250	Low operating temperature(80°C), 25 kVA		1	16		4,000	855		4,855	5,675
2260	37 kVA		.80	20		4,325	1,075		5,400	6,350
2270	50 kVA		.70	22.857		5,625	1,225		6,850	8,000
2280	75 kVA		.65	24.615		8,975	1,325		10,300	11,900
2290	100 kVA		.55	29.091		9,325	1,550		10,875	12,600
2300	3 phase, 480 volt primary 120/208 volt secondary									
2310	Ventilated, 3 kVA	1 Elec	1	8	Ea.	785	425		1,210	1,500
2700	6 kVA		.80	10		895	535		1,430	1,775
2900	9 kVA		.70	11.429		1,025	610		1,635	2,075
3100	15 kVA	2 Elec	1.10	14.545		1,075	775		1,850	2,325
3300	30 kVA		.90	17.778		1,175	950		2,125	2,725
3500	45 kVA		.80	20		1,425	1,075		2,500	3,150
3700	75 kVA		.70	22.857		2,125	1,225		3,350	4,175
3900	112.5 kVA	R-3	.90	22.222		2,825	1,175	156	4,156	5,075
4100	150 kVA		.85	23.529		3,700	1,250	165	5,115	6,125
4300	225 kVA		.65	30.769		5,000	1,625	215	6,840	8,200
4500	300 kVA		.55	36.364		6,325	1,925	255	8,505	10,200
4700	500 kVA		.45	44.444		10,500	2,350	310	13,160	15,400
4800	750 kVA		.35	57.143		18,400	3,025	400	21,825	25,200
4820	1000 kVA		.32	62.500		21,100	3,300	440	24,840	28,700
4850	K-4 rated, 15 kVA	2 Elec	1.10	14.545		3,325	775		4,100	4,800
4855	30 kVA		.90	17.778		5,000	950		5,950	6,900
4860	45 kVA		.80	20		6,000	1,075		7,075	8,200
4865	75 kVA		.70	22.857		9,050	1,225		10,275	11,800
4870	112.5 kVA	R-3	.90	22.222		12,000	1,175	156	13,331	15,200
4875	150 kVA		.85	23.529		15,700	1,250	165	17,115	19,400
4880	225 kVA		.65	30.769		21,900	1,625	215	23,740	26,800
4885	300 kVA		.55	36.364		28,700	1,925	255	30,880	34,700
4890	500 kVA		.45	44.444		40,100	2,350	310	42,760	48,000
4900	K-13 rated, 15 kVA	2 Elec	1.10	14.545		3,775	775		4,550	5,300
4905	30 kVA		.90	17.778		5,675	950		6,625	7,675
4910	45 kVA		.80	20		6,825	1,075		7,900	9,100

26 22 13.10 Transformer, Dry-Type		Crew	Daily Output	Labor-Hours	Unit	Material	2014 Bare Costs Labor	Equipment	Total	Total Incl O&P
4915	75 kVA	2 Elec	.70	22.857	Ea.	10,300	1,225		11,525	13,100
4920	112.5 kVA	R-3	.90	22.222		13,700	1,175	156	15,031	17,000
4925	150 kVA		.85	23.529		17,900	1,250	165	19,315	21,800
4930	225 kVA		.65	30.769		24,400	1,625	215	26,240	29,500
4935	300 kVA		.55	36.364		32,500	1,925	255	34,680	38,900
4940	500 kVA		.45	44.444		51,500	2,350	310	54,160	60,500
5020	480 volt primary 120/208 volt secondary									
5030	Nonventilated, 15 kVA	2 Elec	1.10	14.545	Ea.	3,800	775		4,575	5,325
5040	30 kVA		.80	20		4,250	1,075		5,325	6,275
5050	45 kVA		.70	22.857		5,800	1,225		7,025	8,200
5060	75 kVA		.65	24.615		8,450	1,325		9,775	11,300
5070	112.5 kVA	R-3	.85	23.529		11,500	1,250	165	12,915	14,700
5081	150 kVA		.85	23.529		13,900	1,250	165	15,315	17,400
5090	225 kVA		.60	33.333		17,000	1,775	233	19,008	21,600
5100	300 kVA		.50	40		20,300	2,125	280	22,705	25,900
5200	Low operating temperature (80°C), 30 kVA	2 Elec	.90	17.778		3,475	950		4,425	5,250
5210	45 kVA		.80	20		5,225	1,075		6,300	7,350
5220	75 kVA		.70	22.857		6,950	1,225		8,175	9,475
5230	112.5 kVA	R-3	.90	22.222		12,300	1,175	156	13,631	15,400
5240	150 kVA		.85	23.529		15,500	1,250	165	16,915	19,200
5250	225 kVA		.65	30.769		21,500	1,625	215	23,340	26,300
5260	300 kVA		.55	36.364		25,700	1,925	255	27,880	31,500
5270	500 kVA		.45	44.444		38,700	2,350	310	41,360	46,400
5380	3 phase, 5 kV primary 277/480 volt secondary									
5400	High voltage, 112.5 kVA	R-3	.85	23.529	Ea.	16,200	1,250	165	17,615	20,000
5410	150 kVA		.65	30.769		17,700	1,625	215	19,540	22,100
5420	225 kVA		.55	36.364		20,900	1,925	255	23,080	26,200
5430	300 kVA		.45	44.444		26,700	2,350	310	29,360	33,300
5440	500 kVA		.35	57.143		34,400	3,025	400	37,825	42,800
5450	750 kVA		.32	62.500		53,500	3,300	440	57,240	64,500
5460	1000 kVA		.30	66.667		63,000	3,525	465	66,990	75,000
5470	1500 kVA		.27	74.074		73,000	3,925	520	77,445	87,000
5480	2000 kVA		.25	80		86,000	4,225	560	90,785	101,500
5490	2500 kVA		.20	100		96,500	5,300	700	102,500	114,500
5500	3000 kVA		.18	111		126,500	5,875	780	133,155	149,000
5590	15 kV primary 277/480 volt secondary									
5600	High voltage, 112.5 kVA	R-3	.85	23.529	Ea.	24,800	1,250	165	26,215	29,400
5610	150 kVA		.65	30.769		28,900	1,625	215	30,740	34,500
5620	225 kVA		.55	36.364		32,000	1,925	255	34,180	38,400
5630	300 kVA		.45	44.444		37,600	2,350	310	40,260	45,300
5640	500 kVA		.35	57.143		47,700	3,025	400	51,125	57,500
5650	750 kVA		.32	62.500		62,000	3,300	440	65,740	74,000
5660	1000 kVA		.30	66.667		71,000	3,525	465	74,990	84,000
5670	1500 kVA		.27	74.074		81,500	3,925	520	85,945	96,500
5680	2000 kVA		.25	80		91,000	4,225	560	95,785	107,000
5690	2500 kVA		.20	100		105,000	5,300	700	111,000	124,000
5700	3000 kVA		.18	111		125,000	5,875	780	131,655	147,000
6000	2400 volt primary, 480 volt secondary, 300 kVA		.45	44.444		26,700	2,350	310	29,360	33,300
6010	500 kVA		.35	57.143		34,400	3,025	400	37,825	42,800
6020	750 kVA		.32	62.500		49,500	3,300	440	53,240	60,000

26 22 13 – Low-Voltage Distribution Transformers

26 22 13.20 Isolating Panels		Crew	Daily Output	Labor-Hours	Unit	Material	2014 Bare Costs Labor	2014 Bare Costs Equipment	Total	Total Incl O&P
0010	**ISOLATING PANELS** used with isolating transformers									
0020	For hospital applications									
0100	Critical care area, 8 circuit, 3 kVA	1 Elec	.58	13.793	Ea.	6,750	735		7,485	8,525
0200	5 kVA		.54	14.815		6,925	790		7,715	8,775
0400	7.5 kVA		.52	15.385		7,025	820		7,845	8,950
0600	10 kVA		.44	18.182		7,325	970		8,295	9,500
0800	Operating room power & lighting, 8 circuit, 3 kVA		.58	13.793		5,650	735		6,385	7,300
1000	5 kVA		.54	14.815		5,950	790		6,740	7,700
1200	7.5 kVA		.52	15.385		6,500	820		7,320	8,375
1400	10 kVA		.44	18.182		6,925	970		7,895	9,075
1600	X-ray systems, 15 kVA, 90 amp		.44	18.182		14,400	970		15,370	17,300
1800	25 kVA, 125 amp		.36	22.222		14,800	1,175		15,975	18,100

26 22 13.30 Isolating Transformer

26 22 13.30 Isolating Transformer			Crew	Daily Output	Labor-Hours	Unit	Material	2014 Bare Costs Labor	2014 Bare Costs Equipment	Total	Total Incl O&P
0010	**ISOLATING TRANSFORMER**										
0100	Single phase, 120/240 volt primary, 120/240 volt secondary										
0200	0.50 kVA	R262213-60	1 Elec	4	2	Ea.	375	107		482	570
0400	1 kVA			2	4		470	213		683	835
0600	2 kVA			1.60	5		700	267		967	1,175
0800	3 kVA			1.40	5.714		805	305		1,110	1,350
1000	5 kVA			1.20	6.667		1,050	355		1,405	1,700
1200	7.5 kVA			1.10	7.273		1,350	390		1,740	2,050
1400	10 kVA			.80	10		1,700	535		2,235	2,675
1600	15 kVA			.60	13.333		2,200	710		2,910	3,475
1800	25 kVA			.50	16		3,150	855		4,005	4,725
1810	37.5 kVA		2 Elec	.80	20		5,200	1,075		6,275	7,325
1820	75 kVA		"	.65	24.615		7,800	1,325		9,125	10,600
1830	3 phase, 120/240 V primary, 120/240 V secondary, 112.5 kVA		R-3	.90	22.222		9,850	1,175	156	11,181	12,700
1840	150 kVA			.85	23.529		12,500	1,250	165	13,915	15,900
1850	225 kVA			.65	30.769		17,400	1,625	215	19,240	21,900
1860	300 kVA			.55	36.364		23,200	1,925	255	25,380	28,700
1870	500 kVA			.45	44.444		38,700	2,350	310	41,360	46,400
1880	750 kVA			.35	57.143		39,000	3,025	400	42,425	47,900

26 22 13.90 Transformer Handling

26 22 13.90 Transformer Handling		Crew	Daily Output	Labor-Hours	Unit	Material	2014 Bare Costs Labor	2014 Bare Costs Equipment	Total	Total Incl O&P
0010	**TRANSFORMER HANDLING** Add to normal labor cost in restricted areas									
5000	Transformers									
5150	15 kVA, approximately 200 pounds	2 Elec	2.70	5.926	Ea.		315		315	475
5160	25 kVA, approximately 300 pounds		2.50	6.400			340		340	510
5170	37.5 kVA, approximately 400 pounds		2.30	6.957			370		370	555
5180	50 kVA, approximately 500 pounds		2	8			425		425	640
5190	75 kVA, approximately 600 pounds		1.80	8.889			475		475	710
5200	100 kVA, approximately 700 pounds		1.60	10			535		535	800
5210	112.5 kVA, approximately 800 pounds	3 Elec	2.20	10.909			580		580	870
5220	125 kVA, approximately 900 pounds		2	12			640		640	960
5230	150 kVA, approximately 1000 pounds		1.80	13.333			710		710	1,075
5240	167 kVA, approximately 1200 pounds		1.60	15			800		800	1,200
5250	200 kVA, approximately 1400 pounds		1.40	17.143			915		915	1,375
5260	225 kVA, approximately 1600 pounds		1.30	18.462			985		985	1,475
5270	250 kVA, approximately 1800 pounds		1.10	21.818			1,175		1,175	1,750
5280	300 kVA, approximately 2000 pounds		1	24			1,275		1,275	1,925
5290	500 kVA, approximately 3000 pounds		.75	32			1,700		1,700	2,550
5300	600 kVA, approximately 3500 pounds		.67	35.821			1,900		1,900	2,850
5310	750 kVA, approximately 4000 pounds		.60	40			2,125		2,125	3,200

26 22 Low-Voltage Transformers

26 22 13 – Low-Voltage Distribution Transformers

26 22 13.90 Transformer Handling	Crew	Daily Output	Labor-Hours	Unit	Material	2014 Bare Costs Labor	Equipment	Total	Total Incl O&P
5320 1000 kVA, approximately 5000 pounds	3 Elec	.50	48	Ea.		2,550		2,550	3,825

26 22 16 – Low-Voltage Buck-Boost Transformers

26 22 16.10 Buck-Boost Transformer

		Crew	Daily Output	Labor-Hours	Unit	Material	Labor	Equipment	Total	Total Incl O&P
0010	**BUCK-BOOST TRANSFORMER** R262213-60									
0100	Single phase, 120/240 V primary, 12/24 V secondary									
0200	0.10 kVA	1 Elec	8	1	Ea.	95.50	53.50		149	185
0400	0.25 kVA		5.70	1.404		141	75		216	267
0600	0.50 kVA		4	2		192	107		299	370
0800	0.75 kVA		3.10	2.581		247	138		385	480
1000	1.0 kVA		2	4		305	213		518	660
1200	1.5 kVA		1.80	4.444		375	237		612	770
1400	2.0 kVA		1.60	5		455	267		722	900
1600	3.0 kVA		1.40	5.714		615	305		920	1,125
1800	5.0 kVA		1.20	6.667		810	355		1,165	1,425
2000	3 phase, 240 V primary, 208/120 V secondary, 15 kVA	2 Elec	2.40	6.667		2,100	355		2,455	2,850
2200	30 kVA		1.60	10		2,350	535		2,885	3,400
2400	45 kVA		1.40	11.429		2,825	610		3,435	4,050
2600	75 kVA		1.20	13.333		3,425	710		4,135	4,850
2800	112.5 kVA	R-3	1.40	14.286		4,250	755	100	5,105	5,900
3000	150 kVA		1.10	18.182		5,675	960	127	6,762	7,825
3200	225 kVA		1	20		7,400	1,050	140	8,590	9,875
3400	300 kVA		.90	22.222		10,000	1,175	156	11,331	12,900

26 24 Switchboards and Panelboards

26 24 13 – Switchboards

26 24 13.10 Incoming Switchboards

		Crew	Daily Output	Labor-Hours	Unit	Material	Labor	Equipment	Total	Total Incl O&P
0010	**INCOMING SWITCHBOARDS** main service section R262419-84									
0100	Aluminum bus bars, not including CT's or PT's									
0200	No main disconnect, includes CT compartment									
0300	120/208 volt, 4 wire, 600 amp	2 Elec	1	16	Ea.	3,975	855		4,830	5,650
0400	800 amp		.88	18.182		3,975	970		4,945	5,825
0500	1000 amp		.80	20		4,775	1,075		5,850	6,850
0600	1200 amp		.72	22.222		4,775	1,175		5,950	7,025
0700	1600 amp		.66	24.242		4,775	1,300		6,075	7,175
0800	2000 amp		.62	25.806		5,150	1,375		6,525	7,700
1000	3000 amp		.56	28.571		6,800	1,525		8,325	9,750
1200	277/480 volt, 4 wire, 600 amp		1	16		3,975	855		4,830	5,650
1300	800 amp		.88	18.182		3,975	970		4,945	5,825
1400	1000 amp		.80	20		5,050	1,075		6,125	7,150
1500	1200 amp		.72	22.222		5,050	1,175		6,225	7,325
1600	1600 amp		.66	24.242		5,050	1,300		6,350	7,475
1700	2000 amp		.62	25.806		5,150	1,375		6,525	7,700
1800	3000 amp		.56	28.571		6,800	1,525		8,325	9,750
1900	4000 amp		.52	30.769		8,425	1,650		10,075	11,700
2000	Fused switch & CT compartment									
2100	120/208 volt, 4 wire, 400 amp	2 Elec	1.12	14.286	Ea.	2,675	760		3,435	4,075
2200	600 amp		.94	17.021		3,150	910		4,060	4,825
2300	800 amp		.84	19.048		10,800	1,025		11,825	13,300
2400	1200 amp		.68	23.529		14,000	1,250		15,250	17,300
2500	277/480 volt, 4 wire, 400 amp		1.14	14.035		2,800	750		3,550	4,200

26 24 Switchboards and Panelboards

26 24 13 – Switchboards

26 24 13.10 Incoming Switchboards	Crew	Daily Output	Labor-Hours	Unit	Material	2014 Bare Costs Labor	Equipment	Total	Total Incl O&P
2600	600 amp	2 Elec	.94	17.021	Ea.	3,200	910		4,110	4,875
2700	800 amp		.84	19.048		10,800	1,025		11,825	13,300
2800	1200 amp		.68	23.529		14,000	1,250		15,250	17,300
2900	Pressure switch & CT compartment									
3000	120/208 volt, 4 wire, 800 amp	2 Elec	.80	20	Ea.	9,650	1,075		10,725	12,200
3100	1200 amp		.66	24.242		18,700	1,300		20,000	22,500
3200	1600 amp		.62	25.806		19,900	1,375		21,275	24,000
3300	2000 amp		.56	28.571		21,100	1,525		22,625	25,500
3310	2500 amp		.50	32		26,000	1,700		27,700	31,200
3320	3000 amp		.44	36.364		35,100	1,950		37,050	41,500
3330	4000 amp		.40	40		45,000	2,125		47,125	52,500
3340	120/208 volt, 4 wire, 800 amp, with ground fault		.80	20		16,000	1,075		17,075	19,200
3350	1200 amp, with ground fault		.66	24.242		20,600	1,300		21,900	24,600
3360	1600 amp, with ground fault		.62	25.806		22,400	1,375		23,775	26,700
3370	2000 amp, with ground fault		.56	28.571		24,200	1,525		25,725	28,900
3400	277/480 volt, 4 wire, 800 amp, with ground fault		.80	20		16,000	1,075		17,075	19,200
3600	1200 amp, with ground fault		.66	24.242		20,600	1,300		21,900	24,600
4000	1600 amp, with ground fault		.62	25.806		22,400	1,375		23,775	26,700
4200	2000 amp, with ground fault		.56	28.571		24,200	1,525		25,725	28,900
4400	Circuit breaker, molded case & CT compartment									
4600	3 pole, 4 wire, 600 amp	2 Elec	.94	17.021	Ea.	8,325	910		9,235	10,500
4800	800 amp		.84	19.048		9,950	1,025		10,975	12,500
5000	1200 amp		.68	23.529		13,600	1,250		14,850	16,800
5100	Copper bus bars, not incl. CT's or PT's, add, minimum					15%				

26 24 13.20 In Plant Distribution Switchboards

		Crew	Daily Output	Labor-Hours	Unit	Material	2014 Bare Costs Labor	Equipment	Total	Total Incl O&P
0010	**IN PLANT DISTRIBUTION SWITCHBOARDS**									
0100	Main lugs only, to 600 volt, 3 pole, 3 wire, 200 amp	2 Elec	1.20	13.333	Ea.	1,275	710		1,985	2,475
0110	400 amp		1.20	13.333		1,275	710		1,985	2,475
0120	600 amp		1.20	13.333		1,325	710		2,035	2,525
0130	800 amp		1.08	14.815		1,425	790		2,215	2,750
0140	1200 amp		.92	17.391		1,725	930		2,655	3,300
0150	1600 amp		.86	18.605		2,275	995		3,270	3,975
0160	2000 amp		.82	19.512		2,525	1,050		3,575	4,325
0250	To 480 volt, 3 pole, 4 wire, 200 amp		1.20	13.333		1,100	710		1,810	2,300
0260	400 amp		1.20	13.333		1,275	710		1,985	2,475
0270	600 amp		1.20	13.333		1,400	710		2,110	2,600
0280	800 amp		1.08	14.815		1,525	790		2,315	2,850
0290	1200 amp		.92	17.391		1,875	930		2,805	3,475
0300	1600 amp		.86	18.605		2,150	995		3,145	3,825
0310	2000 amp		.82	19.512		2,500	1,050		3,550	4,300
0400	Main circuit breaker, to 600 volt, 3 pole, 3 wire, 200 amp		1.20	13.333		3,050	710		3,760	4,425
0410	400 amp		1.14	14.035		3,050	750		3,800	4,475
0420	600 amp		1.10	14.545		3,850	775		4,625	5,400
0430	800 amp		1.04	15.385		6,450	820		7,270	8,325
0440	1200 amp		.88	18.182		8,425	970		9,395	10,700
0450	1600 amp		.84	19.048		13,500	1,025		14,525	16,400
0460	2000 amp		.80	20		14,400	1,075		15,475	17,500
0550	277/480 volt, 3 pole, 4 wire, 200 amp		1.20	13.333		3,200	710		3,910	4,600
0560	400 amp		1.14	14.035		3,200	750		3,950	4,650
0570	600 amp		1.10	14.545		4,025	775		4,800	5,575
0580	800 amp		1.04	15.385		6,750	820		7,570	8,650
0590	1200 amp		.88	18.182		8,725	970		9,695	11,100

26 24 13 – Switchboards

26 24 13.20 In Plant Distribution Switchboards	Crew	Daily Output	Labor-Hours	Unit	Material	2014 Bare Costs Labor	Equipment	Total	Total Incl O&P
0600	1600 amp	2 Elec	.84	19.048	Ea.	13,500	1,025		14,525	16,400
0610	2000 amp		.80	20		14,400	1,075		15,475	17,500
0700	Main fusible switch w/fuse, 208/240 volt, 3 pole, 3 wire, 200 amp		1.20	13.333		3,350	710		4,060	4,775
0710	400 amp		1.14	14.035		3,350	750		4,100	4,825
0720	600 amp		1.10	14.545		4,075	775		4,850	5,650
0730	800 amp		1.04	15.385		8,350	820		9,170	10,400
0740	1200 amp		.88	18.182		9,725	970		10,695	12,200
0800	120/208, 120/240 volt, 3 pole, 4 wire, 200 amp		1.20	13.333		2,975	710		3,685	4,350
0810	400 amp		1.14	14.035		2,975	750		3,725	4,400
0820	600 amp		1.10	14.545		3,900	775		4,675	5,425
0830	800 amp		1.04	15.385		6,200	820		7,020	8,025
0840	1200 amp		.88	18.182		7,200	970		8,170	9,350
0900	480 or 600 volt, 3 pole, 3 wire, 200 amp		1.20	13.333		3,225	710		3,935	4,625
0910	400 amp		1.14	14.035		3,225	750		3,975	4,675
0920	600 amp		1.10	14.545		3,875	775		4,650	5,400
0930	800 amp		1.04	15.385		5,975	820		6,795	7,775
0940	1200 amp		.88	18.182		6,900	970		7,870	9,050
1000	277 or 480 volt, 3 pole, 4 wire, 200 amp		1.20	13.333		3,350	710		4,060	4,750
1010	400 amp		1.14	14.035		3,350	750		4,100	4,800
1020	600 amp		1.10	14.545		4,050	775		4,825	5,600
1030	800 amp		1.04	15.385		6,200	820		7,020	8,050
1040	1200 amp		.88	18.182		7,200	970		8,170	9,350
1120	1600 amp		.76	21.053		13,100	1,125		14,225	16,100
1130	2000 amp		.68	23.529		17,300	1,250		18,550	20,900
1150	Pressure switch, bolted, 3 pole, 208/240 volt, 3 wire, 800 amp		.96	16.667		9,725	890		10,615	12,000
1160	1200 amp		.80	20		12,400	1,075		13,475	15,300
1170	1600 amp		.76	21.053		14,200	1,125		15,325	17,300
1180	2000 amp		.68	23.529		16,200	1,250		17,450	19,700
1200	120/208 or 120/240 volt, 3 pole, 4 wire, 800 amp		.96	16.667		7,675	890		8,565	9,775
1210	1200 amp		.80	20		8,975	1,075		10,050	11,500
1220	1600 amp		.76	21.053		14,200	1,125		15,325	17,300
1230	2000 amp		.68	23.529		16,200	1,250		17,450	19,700
1300	480 or 600 volt, 3 wire, 800 amp		.96	16.667		9,725	890		10,615	12,000
1310	1200 amp		.80	20		13,600	1,075		14,675	16,600
1320	1600 amp		.76	21.053		15,300	1,125		16,425	18,600
1330	2000 amp		.68	23.529		17,300	1,250		18,550	20,900
1400	277-480 volt, 4 wire, 800 amp		.96	16.667		9,725	890		10,615	12,000
1410	1200 amp		.80	20		13,600	1,075		14,675	16,600
1420	1600 amp		.76	21.053		15,300	1,125		16,425	18,600
1430	2000 amp		.68	23.529		17,300	1,250		18,550	20,900
1500	Main ground fault protector, 1200-2000 amp		5.40	2.963		3,250	158		3,408	3,800
1600	Busway connection, 200 amp		5.40	2.963		485	158		643	770
1610	400 amp		4.60	3.478		485	186		671	815
1620	600 amp		4	4		485	213		698	855
1630	800 amp		3.20	5		485	267		752	935
1640	1200 amp		2.60	6.154		485	330		815	1,025
1650	1600 amp		2.40	6.667		1,000	355		1,355	1,625
1660	2000 amp		2	8		1,000	425		1,425	1,750
1700	Shunt trip for remote operation 200 amp		8	2		645	107		752	870
1710	400 amp		8	2		1,025	107		1,132	1,275
1720	600 amp		8	2		1,200	107		1,307	1,475
1730	800 amp		8	2		1,575	107		1,682	1,875
1740	1200-2000 amp		8	2		3,325	107		3,432	3,800

26 24 13 – Switchboards

26 24 13.20 In Plant Distribution Switchboards

		Crew	Daily Output	Labor-Hours	Unit	Material	2014 Bare Costs Labor	2014 Bare Costs Equipment	Total	Total Incl O&P
1800	Motor operated main breaker 200 amp	2 Elec	8	2	Ea.	3,200	107		3,307	3,675
1810	400 amp		8	2		3,200	107		3,307	3,675
1820	600 amp		8	2		3,275	107		3,382	3,775
1830	800 amp		8	2		3,275	107		3,382	3,775
1840	1200-2000 amp		8	2		3,375	107		3,482	3,850
1900	Current/potential transformer metering compartment 200-800 amp		5.40	2.963		2,775	158		2,933	3,300
1940	1200 amp		5.40	2.963		4,625	158		4,783	5,300
1950	1600-2000 amp		5.40	2.963		5,950	158		6,108	6,775
2000	With watt meter 200-800 amp		4	4		8,225	213		8,438	9,375
2040	1200 amp		4	4		10,200	213		10,413	11,500
2050	1600-2000 amp		4	4		10,200	213		10,413	11,500
2100	Split bus 60-200 amp	1 Elec	5.30	1.509		195	80.50		275.50	335
2130	400 amp	2 Elec	4.60	3.478		340	186		526	650
2140	600 amp		3.60	4.444		415	237		652	810
2150	800 amp		2.60	6.154		525	330		855	1,075
2170	1200 amp		2	8		600	425		1,025	1,300
2250	Contactor control 60 amp	1 Elec	2	4		1,275	213		1,488	1,725
2260	100 amp		1.50	5.333		1,425	285		1,710	2,000
2270	200 amp		1	8		2,175	425		2,600	3,025
2280	400 amp	2 Elec	1	16		6,575	855		7,430	8,500
2290	600 amp		.84	19.048		7,375	1,025		8,400	9,625
2300	800 amp		.72	22.222		8,725	1,175		9,900	11,400
2500	Modifier, two distribution sections, add		.80	20		2,750	1,075		3,825	4,600
2520	Three distribution sections, add		.40	40		6,150	2,125		8,275	9,950
2560	Auxiliary pull section, 20", add		2	8		1,100	425		1,525	1,850
2580	24", add		1.80	8.889		1,100	475		1,575	1,900
2600	30", add		1.60	10		1,100	535		1,635	2,000
2620	36", add		1.40	11.429		1,300	610		1,910	2,350
2640	Dog house, 12", add		2.40	6.667		225	355		580	780
2660	18", add		2	8		450	425		875	1,125
3000	Transition section between switchboard and transformer									
3050	or motor control center, 4 wire alum. bus, 600 amp	2 Elec	1.14	14.035	Ea.	2,225	750		2,975	3,575
3100	800 amp		1	16		2,475	855		3,330	4,000
3150	1000 amp		.88	18.182		2,800	970		3,770	4,525
3200	1200 amp		.80	20		3,075	1,075		4,150	4,975
3250	1600 amp		.72	22.222		3,625	1,175		4,800	5,775
3300	2000 amp		.66	24.242		4,175	1,300		5,475	6,525
3350	2500 amp		.62	25.806		4,850	1,375		6,225	7,400
3400	3000 amp		.56	28.571		5,600	1,525		7,125	8,425
4000	Weatherproof construction, per vertical section		1.76	9.091		2,550	485		3,035	3,525

26 24 13.30 Distribution Switchboards Section

		Crew	Daily Output	Labor-Hours	Unit	Material	2014 Bare Costs Labor	2014 Bare Costs Equipment	Total	Total Incl O&P
0010	**DISTRIBUTION SWITCHBOARDS SECTION** R262419-80									
0100	Aluminum bus bars, not including breakers									
0160	Subfeed lug-rated at 60 amp	2 Elec	1.30	12.308	Ea.	1,025	655		1,680	2,125
0170	100 amp		1.26	12.698		1,200	675		1,875	2,350
0180	200 amp		1.20	13.333		1,225	710		1,935	2,400
0190	400 amp		1.10	14.545		1,225	775		2,000	2,475
0195	120/208 or 277/480 volt, 4 wire, 400 amp		1.10	14.545		1,300	775		2,075	2,575
0200	600 amp		1	16		1,600	855		2,455	3,050
0300	800 amp		.88	18.182		2,100	970		3,070	3,750
0400	1000 amp		.80	20		2,600	1,075		3,675	4,450
0500	1200 amp		.72	22.222		3,100	1,175		4,275	5,175

26 24 Switchboards and Panelboards

26 24 13 – Switchboards

26 24 13.30 Distribution Switchboards Section	Crew	Daily Output	Labor-Hours	Unit	Material	2014 Bare Costs Labor	Equipment	Total	Total Incl O&P
0600	1600 amp	2 Elec	.66	24.242	Ea.	3,550	1,300		4,850	5,825
0700	2000 amp		.62	25.806		4,150	1,375		5,525	6,600
0800	2500 amp		.60	26.667		4,675	1,425		6,100	7,275
0900	3000 amp		.56	28.571		5,675	1,525		7,200	8,500
0950	4000 amp		.52	30.769		8,300	1,650		9,950	11,600

26 24 13.40 Switchboards Feeder Section

		Crew	Daily Output	Labor-Hours	Unit	Material	Labor	Equipment	Total	Total Incl O&P
0010	**SWITCHBOARDS FEEDER SECTION** group mounted devices R262419-82									
0030	Circuit breakers									
0160	FA frame, 15 to 60 amp, 240 volt, 1 pole	1 Elec	8	1	Ea.	119	53.50		172.50	211
0170	2 pole		7	1.143		241	61		302	355
0180	3 pole		5.30	1.509		360	80.50		440.50	515
0210	480 volt, 1 pole		8	1		151	53.50		204.50	246
0220	2 pole		7	1.143		440	61		501	575
0230	3 pole		5.30	1.509		565	80.50		645.50	745
0260	600 volt, 2 pole		7	1.143		510	61		571	650
0270	3 pole		5.30	1.509		655	80.50		735.50	840
0280	FA frame, 70 to 100 amp, 240 volt, 1 pole		7	1.143		189	61		250	300
0310	2 pole		5	1.600		395	85.50		480.50	565
0320	3 pole		4	2		510	107		617	720
0330	480 volt, 1 pole		7	1.143		226	61		287	340
0360	2 pole		5	1.600		570	85.50		655.50	760
0370	3 pole		4	2		670	107		777	895
0380	600 volt, 2 pole		5	1.600		645	85.50		730.50	840
0410	3 pole		4	2		810	107		917	1,050
0420	KA frame, 70 to 225 amp		3.20	2.500		1,225	133		1,358	1,550
0430	LA frame, 125 to 400 amp		2.30	3.478		2,775	186		2,961	3,325
0460	MA frame, 450 to 600 amp		1.60	5		4,950	267		5,217	5,850
0470	700 to 800 amp		1.30	6.154		6,425	330		6,755	7,575
0480	MAL frame, 1000 amp		1	8		6,675	425		7,100	7,975
0490	PA frame, 1200 amp		.80	10		13,600	535		14,135	15,700
0500	Branch circuit, fusible switch, 600 volt, double 30/30 amp		4	2		885	107		992	1,125
0550	60/60 amp		3.20	2.500		910	133		1,043	1,200
0600	100/100 amp		2.70	2.963		1,150	158		1,308	1,475
0650	Single, 30 amp		5.30	1.509		730	80.50		810.50	920
0700	60 amp		4.70	1.702		805	91		896	1,025
0750	100 amp		4	2		1,050	107		1,157	1,325
0800	200 amp		2.70	2.963		1,425	158		1,583	1,775
0850	400 amp		2.30	3.478		2,600	186		2,786	3,125
0900	600 amp		1.80	4.444		3,175	237		3,412	3,825
0950	800 amp		1.30	6.154		5,325	330		5,655	6,350
1000	1200 amp		.80	10		6,100	535		6,635	7,500
1080	Branch circuit, circuit breakers, high interrupting capacity									
1100	60 amp, 240, 480 or 600 volt, 1 pole	1 Elec	8	1	Ea.	257	53.50		310.50	360
1120	2 pole		7	1.143		660	61		721	815
1140	3 pole		5.30	1.509		665	80.50		745.50	855
1150	100 amp, 240, 480 or 600 volt, 1 pole		7	1.143		289	61		350	410
1160	2 pole		5	1.600		770	85.50		855.50	975
1180	3 pole		4	2		725	107		832	955
1200	225 amp, 240, 480 or 600 volt, 2 pole		3.50	2.286		2,125	122		2,247	2,525
1220	3 pole		3.20	2.500		2,250	133		2,383	2,675
1240	400 amp, 240, 480 or 600 volt, 2 pole		2.50	3.200		3,375	171		3,546	3,975
1260	3 pole		2.30	3.478		3,025	186		3,211	3,625

26 24 Switchboards and Panelboards

26 24 13 – Switchboards

26 24 13.40 Switchboards Feeder Section	Crew	Daily Output	Labor-Hours	Unit	Material	2014 Bare Costs Labor	Equipment	Total	Total Incl O&P
1280	600 amp, 240, 480 or 600 volt, 2 pole	1 Elec	1.80	4.444	Ea.	3,925	237		4,162	4,675
1300	3 pole		1.60	5		3,725	267		3,992	4,475
1320	800 amp, 240, 480 or 600 volt, 2 pole		1.50	5.333		5,050	285		5,335	5,975
1340	3 pole		1.30	6.154		5,550	330		5,880	6,625
1360	1000 amp, 240, 480 or 600 volt, 2 pole		1.10	7.273		6,600	390		6,990	7,825
1380	3 pole		1	8		7,225	425		7,650	8,600
1400	1200 amp, 240, 480 or 600 volt, 2 pole		.90	8.889		6,875	475		7,350	8,250
1420	3 pole		.80	10		7,850	535		8,385	9,425
1700	Fusible switch, 240 V, 60 amp, 2 pole		3.20	2.500		425	133		558	665
1720	3 pole		3	2.667		585	142		727	855
1740	100 amp, 2 pole		2.70	2.963		430	158		588	710
1760	3 pole		2.50	3.200		655	171		826	975
1780	200 amp, 2 pole		2	4		900	213		1,113	1,325
1800	3 pole		1.90	4.211		1,050	225		1,275	1,475
1820	400 amp, 2 pole		1.50	5.333		1,675	285		1,960	2,275
1840	3 pole		1.30	6.154		2,125	330		2,455	2,825
1860	600 amp, 2 pole		1	8		2,400	425		2,825	3,300
1880	3 pole		.90	8.889		2,950	475		3,425	3,950
1900	240-600 V, 800 amp, 2 pole		.70	11.429		5,350	610		5,960	6,800
1920	3 pole		.60	13.333		6,550	710		7,260	8,275
2000	600 V, 60 amp, 2 pole		3.20	2.500		640	133		773	905
2040	100 amp, 2 pole		2.70	2.963		660	158		818	965
2080	200 amp, 2 pole		2	4		1,150	213		1,363	1,600
2120	400 amp, 2 pole		1.50	5.333		2,275	285		2,560	2,925
2160	600 amp, 2 pole		1	8		2,775	425		3,200	3,700
2500	Branch circuit, circuit breakers, 60 amp, 600 volt, 3 pole		5.30	1.509		615	80.50		695.50	795
2520	240, 480 or 600 volt, 1 pole		8	1		99.50	53.50		153	189
2540	240 volt, 2 pole		7	1.143		170	61		231	279
2560	480 or 600 volt, 2 pole		7	1.143		360	61		421	490
2580	240 volt, 3 pole		5.30	1.509		277	80.50		357.50	425
2600	480 volt, 3 pole		5.30	1.509		480	80.50		560.50	650
2620	100 amp, 600 volt, 2 pole		5	1.600		460	85.50		545.50	640
2640	3 pole		4	2		585	107		692	800
2660	480 volt, 2 pole		5	1.600		400	85.50		485.50	565
2680	240 volt, 2 pole		5	1.600		227	85.50		312.50	380
2700	3 pole		4	2		365	107		472	560
2720	480 volt, 3 pole		4	2		535	107		642	750
2740	225 amp, 240, 480 or 600 volt, 2 pole		3.50	2.286		605	122		727	850
2760	3 pole		3.20	2.500		695	133		828	965
2780	400 amp, 240, 480 or 600 volt, 2 pole		2.50	3.200		1,400	171		1,571	1,775
2800	3 pole		2.30	3.478		1,625	186		1,811	2,050
2820	600 amp, 240 or 480 volt, 2 pole		1.80	4.444		2,275	237		2,512	2,850
2840	3 pole		1.60	5		2,800	267		3,067	3,475
2860	800 amp, 240, 480 volt or 600 volt, 2 pole		1.50	5.333		3,425	285		3,710	4,175
2880	3 pole		1.30	6.154		4,000	330		4,330	4,900
2900	1000 amp, 240, 480 or 600 volt, 2 pole		1.10	7.273		4,200	390		4,590	5,200
2920	480 volt, 600 volt, 3 pole		1	8		4,800	425		5,225	5,925
2940	1200 amp, 240, 480 or 600 volt, 2 pole		.90	8.889		5,900	475		6,375	7,200
2960	3 pole		.80	10		6,450	535		6,985	7,900
2980	600 volt, 3 pole		.80	10		6,450	535		6,985	7,900

26 24 16.10 Load Centers		Crew	Daily Output	Labor-Hours	Unit	Material	2014 Bare Costs Labor	Equipment	Total	Total Incl O&P
0010	**LOAD CENTERS** (residential type) R262416-50									
0100	3 wire, 120/240 V, 1 phase, including 1 pole plug-in breakers									
0200	100 amp main lugs, indoor, 8 circuits	1 Elec	1.40	5.714	Ea.	174	305		479	645
0300	12 circuits		1.20	6.667		243	355		598	795
0400	Rainproof, 8 circuits		1.40	5.714		209	305		514	685
0500	12 circuits		1.20	6.667		294	355		649	855
0600	200 amp main lugs, indoor, 16 circuits	R-1A	1.80	8.889		325	390		715	950
0700	20 circuits		1.50	10.667		405	470		875	1,150
0800	24 circuits		1.30	12.308		530	540		1,070	1,400
0900	30 circuits		1.20	13.333		610	585		1,195	1,575
1000	40 circuits		.80	20		850	880		1,730	2,250
1200	Rainproof, 16 circuits		1.80	8.889		385	390		775	1,025
1300	20 circuits		1.50	10.667		460	470		930	1,225
1400	24 circuits		1.30	12.308		690	540		1,230	1,575
1500	30 circuits		1.20	13.333		770	585		1,355	1,750
1600	40 circuits		.80	20		1,000	880		1,880	2,425
1800	400 amp main lugs, indoor, 42 circuits		.72	22.222		1,200	980		2,180	2,775
1900	Rainproof, 42 circuit		.72	22.222		1,400	980		2,380	3,025
2200	Plug in breakers, 20 amp, 1 pole, 4 wire, 120/208 volts									
2210	125 amp main lugs, indoor, 12 circuits	1 Elec	1.20	6.667	Ea.	305	355		660	865
2300	18 circuits		.80	10		430	535		965	1,275
2400	Rainproof, 12 circuits		1.20	6.667		350	355		705	915
2500	18 circuits		.80	10		470	535		1,005	1,325
2600	200 amp main lugs, indoor, 24 circuits	R-1A	1.30	12.308		580	540		1,120	1,450
2700	30 circuits		1.20	13.333		660	585		1,245	1,625
2800	36 circuits		1	16		800	705		1,505	1,950
2900	42 circuits		.80	20		875	880		1,755	2,300
3000	Rainproof, 24 circuits		1.30	12.308		645	540		1,185	1,525
3100	30 circuits		1.20	13.333		725	585		1,310	1,675
3200	36 circuits		1	16		1,000	705		1,705	2,175
3300	42 circuits		.80	20		1,075	880		1,955	2,525
3500	400 amp main lugs, indoor, 42 circuits		.72	22.222		1,325	980		2,305	2,925
3600	Rainproof, 42 circuits		.72	22.222		1,575	980		2,555	3,200
3700	Plug-in breakers, 20 amp, 1 pole, 3 wire, 120/240 volts									
3800	100 amp main breaker, indoor, 12 circuits	1 Elec	1.20	6.667	Ea.	325	355		680	890
3900	18 circuits	"	.80	10		440	535		975	1,275
4000	200 amp main breaker, indoor, 20 circuits	R-1A	1.50	10.667		590	470		1,060	1,350
4200	24 circuits		1.30	12.308		715	540		1,255	1,600
4300	30 circuits		1.20	13.333		795	585		1,380	1,775
4400	40 circuits		.90	17.778		990	780		1,770	2,275
4500	Rainproof, 20 circuits		1.50	10.667		645	470		1,115	1,425
4600	24 circuits		1.30	12.308		800	540		1,340	1,700
4700	30 circuits		1.20	13.333		880	585		1,465	1,850
4800	40 circuits		.90	17.778		1,075	780		1,855	2,375
5000	400 amp main breaker, indoor, 42 circuits		.72	22.222		2,850	980		3,830	4,600
5100	Rainproof, 42 circuits		.72	22.222		3,175	980		4,155	4,975
5300	Plug in breakers, 20 amp, 1 pole, 4 wire, 120/208 volts									
5400	200 amp main breaker, indoor, 30 circuits	R-1A	1.20	13.333	Ea.	1,200	585		1,785	2,225
5500	42 circuits		.80	20		1,475	880		2,355	2,950
5600	Rainproof, 30 circuits		1.20	13.333		1,275	585		1,860	2,300
5700	42 circuits		.80	20		1,550	880		2,430	3,050

26 24 16 – Panelboards

26 24 16.20 Panelboard and Load Center Circuit Breakers	Crew	Daily Output	Labor-Hours	Unit	Material	2014 Bare Costs Labor	Equipment	Total	Total Incl O&P
0010 **PANELBOARD AND LOAD CENTER CIRCUIT BREAKERS** R262419-82									
0050　Bolt-on, 10,000 amp I.C., 120 volt, 1 pole									
0100　　　　15 to 50 amp	1 Elec	10	.800	Ea.	14.40	42.50		56.90	80
0200　　　　60 amp		8	1		18.70	53.50		72.20	101
0300　　　　70 amp	↓	8	1	↓	27.50	53.50		81	110
0350　　　240 volt, 2 pole									
0400　　　　15 to 50 amp	1 Elec	8	1	Ea.	31.50	53.50		85	115
0500　　　　60 amp		7.50	1.067		42.50	57		99.50	132
0600　　　　80 to 100 amp		5	1.600		81.50	85.50		167	218
0700　　　　3 pole, 15 to 60 amp		6.20	1.290		100	69		169	213
0800　　　　70 amp		5	1.600		126	85.50		211.50	267
0900　　　　80 to 100 amp		3.60	2.222		143	119		262	335
1000　22,000 amp I.C., 240 volt, 2 pole, 70 – 225 amp		2.70	2.963		615	158		773	910
1100　　　3 pole, 70 – 225 amp		2.30	3.478		685	186		871	1,025
1200　14,000 amp I.C., 277 volts, 1 pole, 15 – 30 amp		8	1		38	53.50		91.50	122
1300　22,000 amp I.C., 480 volts, 2 pole, 70 – 225 amp		2.70	2.963		615	158		773	910
1400　　　3 pole, 70 – 225 amp		2.30	3.478		760	186		946	1,125
2000　Plug-in panel or load center, 120/240 volt, to 60 amp, 1 pole		12	.667		11.15	35.50		46.65	65.50
2004　Circuit breaker, 20 A, 1 pole with NM cable		6.50	1.231		18.90	65.50		84.40	120
2006　　　30 A, 1 pole with NM cable		6.50	1.231		18.90	65.50		84.40	120
2010　　　2 pole		9	.889		25	47.50		72.50	98.50
2014　　　50 A, 2 pole with NM cable		5.50	1.455		33	77.50		110.50	152
2020　　　3 pole		7.50	1.067		87	57		144	181
2030　　　100 amp, 2 pole		6	1.333		117	71		188	234
2040　　　3 pole		4.50	1.778		130	95		225	285
2050　　　150 to 200 amp, 2 pole		3	2.667		213	142		355	445
2060　Plug-in tandem, 120/240 V, 2-15 A, 1 pole		11	.727		24.50	39		63.50	85
2070　　　1-15 A & 1-20 A		11	.727		24.50	39		63.50	85
2080　　　2-20 A		11	.727		24.50	39		63.50	85
2082　Arc fault circuit interrupter, 120/240 V, 1-15 A & 1-20 A, 1 pole		11	.727		76.50	39		115.50	142
2100　High interrupting capacity, 120/240 volt, plug-in, 30 amp, 1 pole		12	.667		22	35.50		57.50	77
2110　　　60 amp, 2 pole		9	.889		51	47.50		98.50	127
2120　　　3 pole		7.50	1.067		132	57		189	231
2130　　　100 amp, 2 pole		6	1.333		147	71		218	268
2140　　　3 pole		4.50	1.778		190	95		285	350
2150　　　125 amp, 2 pole		3	2.667		390	142		532	645
2200　　　Bolt-on, 30 amp, 1 pole		10	.800		28	42.50		70.50	95
2210　　　60 amp, 2 pole		7.50	1.067		59.50	57		116.50	151
2220　　　3 pole		6.20	1.290		153	69		222	271
2230　　　100 amp, 2 pole		5	1.600		169	85.50		254.50	315
2240　　　3 pole		3.60	2.222		221	119		340	420
2300　Ground fault, 240 volt, 30 amp, 1 pole		7	1.143		86.50	61		147.50	187
2310　　　2 pole		6	1.333		155	71		226	276
2350　Key operated, 240 volt, 1 pole, 30 amp		7	1.143		78.50	61		139.50	178
2360　Switched neutral, 240 volt, 30 amp, 2 pole		6	1.333		36	71		107	146
2370　　　3 pole		5.50	1.455		54	77.50		131.50	176
2400　Shunt trip, for 240 volt breaker, 60 amp, 1 pole		4	2		70.50	107		177.50	238
2410　　　2 pole		3.50	2.286		70.50	122		192.50	261
2420　　　3 pole		3	2.667		70.50	142		212.50	291
2430　　　100 amp, 2 pole		3	2.667		70.50	142		212.50	291
2440　　　3 pole		2.50	3.200		70.50	171		241.50	335
2450　　　150 amp, 2 pole	↓	2	4	↓	168	213		381	505

26 24 16 – Panelboards

26 24 16.20 Panelboard and Load Center Circuit Breakers	Crew	Daily Output	Labor-Hours	Unit	Material	2014 Bare Costs Labor	Equipment	Total	Total Incl O&P
2500	Auxiliary switch, for 240 volt breaker, 60 amp, 1 pole	1 Elec	4	2	Ea.	64	107		171	231
2510	2 pole		3.50	2.286		64	122		186	254
2520	3 pole		3	2.667		64	142		206	284
2530	100 amp, 2 pole		3	2.667		64	142		206	284
2540	3 pole		2.50	3.200		64	171		235	325
2550	150 amp, 2 pole		2	4		107	213		320	435
2600	Panel or load center, 277/480 volt, plug-in, 30 amp, 1 pole		12	.667		52	35.50		87.50	110
2610	60 amp, 2 pole		9	.889		156	47.50		203.50	242
2620	3 pole		7.50	1.067		228	57		285	335
2650	Bolt-on, 60 amp, 2 pole		7.50	1.067		156	57		213	256
2660	3 pole		6.20	1.290		228	69		297	355
2700	I-line, 277/480 volt, 30 amp, 1 pole		8	1		59.50	53.50		113	146
2710	60 amp, 2 pole		7.50	1.067		209	57		266	315
2720	3 pole		6.20	1.290		271	69		340	400
2730	100 amp, 1 pole		7.50	1.067		114	57		171	211
2740	2 pole		5	1.600		271	85.50		356.50	425
2750	3 pole		3.50	2.286		320	122		442	535
2800	High interrupting capacity, 277/480 volt, plug-in, 30 amp, 1 pole		12	.667		300	35.50		335.50	385
2810	60 amp, 2 pole		9	.889		460	47.50		507.50	580
2820	3 pole		7	1.143		540	61		601	680
2830	Bolt-on, 30 amp, 1 pole		8	1		300	53.50		353.50	410
2840	60 amp, 2 pole		7.50	1.067		515	57		572	650
2850	3 pole		6.20	1.290		595	69		664	760
2900	I-line, 30 amp, 1 pole		8	1		300	53.50		353.50	410
2910	60 amp, 2 pole		7.50	1.067		425	57		482	555
2920	3 pole		6.20	1.290		475	69		544	625
2930	100 amp, 1 pole		7.50	1.067		485	57		542	615
2940	2 pole		5	1.600		485	85.50		570.50	660
2950	3 pole		3.60	2.222		535	119		654	765
2960	Shunt trip, 277/480 volt breaker, remote oper., 30 amp, 1 pole		4	2		320	107		427	515
2970	60 amp, 2 pole		3.50	2.286		470	122		592	700
2980	3 pole		3	2.667		535	142		677	800
2990	100 amp, 1 pole		3.50	2.286		375	122		497	600
3000	2 pole		3	2.667		535	142		677	800
3010	3 pole		2.50	3.200		580	171		751	895
3050	Under voltage trip, 277/480 volt breaker, 30 amp, 1 pole		4	2		320	107		427	515
3060	60 amp, 2 pole		3.50	2.286		470	122		592	700
3070	3 pole		3	2.667		535	142		677	800
3080	100 amp, 1 pole		3.50	2.286		375	122		497	600
3090	2 pole		3	2.667		535	142		677	800
3100	3 pole		2.50	3.200		580	171		751	895
3150	Motor operated, 277/480 volt breaker, 30 amp, 1 pole		4	2		570	107		677	785
3160	60 amp, 2 pole		3.50	2.286		720	122		842	975
3170	3 pole		3	2.667		785	142		927	1,075
3180	100 amp, 1 pole		3.50	2.286		625	122		747	875
3190	2 pole		3	2.667		785	142		927	1,075
3200	3 pole		2.50	3.200		830	171		1,001	1,175
3250	Panelboard spacers, per pole		40	.200		4.32	10.65		14.97	20.50

26 24 16.30 Panelboards Commercial Applications	Crew	Daily Output	Labor-Hours	Unit	Material	2014 Bare Costs Labor	Equipment	Total	Total Incl O&P
0010 **PANELBOARDS COMMERCIAL APPLICATIONS** R262416-50									
0050 NQOD, w/20 amp 1 pole bolt-on circuit breakers									
0100 3 wire, 120/240 volts, 100 amp main lugs									
0150 10 circuits	1 Elec	1	8	Ea.	535	425		960	1,225
0200 14 circuits		.88	9.091		655	485		1,140	1,450
0250 18 circuits		.75	10.667		710	570		1,280	1,625
0300 20 circuits		.65	12.308		795	655		1,450	1,850
0350 225 amp main lugs, 24 circuits	2 Elec	1.20	13.333		900	710		1,610	2,075
0400 30 circuits		.90	17.778		1,050	950		2,000	2,575
0450 36 circuits		.80	20		1,175	1,075		2,250	2,900
0500 38 circuits		.72	22.222		1,275	1,175		2,450	3,175
0550 42 circuits		.66	24.242		1,325	1,300		2,625	3,375
0600 4 wire, 120/208 volts, 100 amp main lugs, 12 circuits	1 Elec	1	8		630	425		1,055	1,325
0650 16 circuits		.75	10.667		725	570		1,295	1,650
0700 20 circuits		.65	12.308		835	655		1,490	1,900
0750 24 circuits		.60	13.333		905	710		1,615	2,075
0800 30 circuits		.53	15.094		1,025	805		1,830	2,350
0850 225 amp main lugs, 32 circuits	2 Elec	.90	17.778		1,150	950		2,100	2,700
0900 34 circuits		.84	19.048		1,200	1,025		2,225	2,825
0950 36 circuits		.80	20		1,225	1,075		2,300	2,950
1000 42 circuits		.68	23.529		1,350	1,250		2,600	3,375
1040 225 amp main lugs, NEMA 7, 12 circuits		1	16		3,675	855		4,530	5,300
1100 24 circuits		.40	40		4,400	2,125		6,525	8,025
1200 NEHB, w/20 amp, 1 pole bolt-on circuit breakers									
1250 4 wire, 277/480 volts, 100 amp main lugs, 12 circuits	1 Elec	.88	9.091	Ea.	1,225	485		1,710	2,050
1300 20 circuits	"	.60	13.333		1,800	710		2,510	3,050
1350 225 amp main lugs, 24 circuits	2 Elec	.90	17.778		2,050	950		3,000	3,675
1400 30 circuits		.80	20		2,450	1,075		3,525	4,300
1450 36 circuits		.72	22.222		2,850	1,175		4,025	4,900
1500 42 circuits		.60	26.667		3,250	1,425		4,675	5,700
1510 225 amp main lugs, NEMA 7, 12 circuits		.90	17.778		5,450	950		6,400	7,425
1590 24 circuits		.30	53.333		7,350	2,850		10,200	12,300
1600 NQOD panel, w/20 amp, 1 pole, circuit breakers									
1650 3 wire, 120/240 volt with main circuit breaker									
1700 100 amp main, 12 circuits	1 Elec	.80	10	Ea.	795	535		1,330	1,675
1750 20 circuits	"	.60	13.333		1,000	710		1,710	2,175
1800 225 amp main, 30 circuits	2 Elec	.68	23.529		1,900	1,250		3,150	3,950
1850 42 circuits		.52	30.769		2,175	1,650		3,825	4,850
1900 400 amp main, 30 circuits		.54	29.630		2,600	1,575		4,175	5,250
1950 42 circuits		.50	32		2,900	1,700		4,600	5,725
2000 4 wire, 120/208 volts with main circuit breaker									
2050 100 amp main, 24 circuits	1 Elec	.47	17.021	Ea.	1,175	910		2,085	2,625
2100 30 circuits	"	.40	20		1,325	1,075		2,400	3,050
2200 225 amp main, 32 circuits	2 Elec	.72	22.222		2,200	1,175		3,375	4,175
2250 42 circuits		.56	28.571		2,400	1,525		3,925	4,900
2300 400 amp main, 42 circuits		.48	33.333		3,225	1,775		5,000	6,200
2350 600 amp main, 42 circuits		.40	40		4,775	2,125		6,900	8,450
2400 NEHB, with 20 amp, 1 pole circuit breaker									
2450 4 wire, 277/480 volts with main circuit breaker									
2500 100 amp main, 24 circuits	1 Elec	.42	19.048	Ea.	2,350	1,025		3,375	4,125
2550 30 circuits	"	.38	21.053		2,750	1,125		3,875	4,700
2600 225 amp main, 30 circuits	2 Elec	.72	22.222		3,475	1,175		4,650	5,575

26 24 Switchboards and Panelboards

26 24 16 – Panelboards

26 24 16.30 Panelboards Commercial Applications	Crew	Daily Output	Labor-Hours	Unit	Material	2014 Bare Costs Labor	Equipment	Total	Total Incl O&P
2650	42 circuits	2 Elec	.56	28.571	Ea.	4,275	1,525		5,800	6,975
2700	400 amp main, 42 circuits		.46	34.783		5,125	1,850		6,975	8,400
2750	600 amp main, 42 circuits		.38	42.105		7,025	2,250		9,275	11,100
2900	Note: the following line items don't include branch circuit breakers									
2910	For branch circuit breakers information, see Section 26 24 16.20									
3010	Main lug, no main breaker, 240 volt, 1 pole, 3 wire, 100 amp	1 Elec	2.30	3.478	Ea.	530	186		716	865
3020	225 amp	2 Elec	2.40	6.667		635	355		990	1,225
3030	400 amp	"	1.80	8.889		930	475		1,405	1,725
3060	3 pole, 3 wire, 100 amp	1 Elec	2.30	3.478		570	186		756	910
3070	225 amp	2 Elec	2.40	6.667		670	355		1,025	1,275
3080	400 amp		1.80	8.889		1,000	475		1,475	1,800
3090	600 amp		1.60	10		1,175	535		1,710	2,075
3110	3 pole, 4 wire, 100 amp	1 Elec	2.30	3.478		630	186		816	975
3120	225 amp	2 Elec	2.40	6.667		790	355		1,145	1,400
3130	400 amp		1.80	8.889		1,000	475		1,475	1,800
3140	600 amp		1.60	10		1,175	535		1,710	2,075
3160	480 volt, 3 pole, 3 wire, 100 amp	1 Elec	2.30	3.478		750	186		936	1,100
3170	225 amp	2 Elec	2.40	6.667		910	355		1,265	1,525
3180	400 amp		1.80	8.889		1,275	475		1,750	2,100
3190	600 amp		1.60	10		1,400	535		1,935	2,350
3210	277/480 volt, 3 pole, 4 wire, 100 amp	1 Elec	2.30	3.478		735	186		921	1,100
3220	225 amp	2 Elec	2.40	6.667		890	355		1,245	1,500
3230	400 amp		1.80	8.889		1,250	475		1,725	2,075
3240	600 amp		1.60	10		1,375	535		1,910	2,325
3260	Main circuit breaker, 240 volt, 1 pole, 3 wire, 100 amp	1 Elec	2	4		710	213		923	1,100
3270	225 amp	2 Elec	2	8		1,450	425		1,875	2,250
3280	400 amp	"	1.60	10		2,275	535		2,810	3,300
3310	3 pole, 3 wire, 100 amp	1 Elec	2	4		815	213		1,028	1,225
3320	225 amp	2 Elec	2	8		1,650	425		2,075	2,475
3330	400 amp		1.60	10		2,600	535		3,135	3,675
3360	120/208 volt, 3 pole, 4 wire, 100 amp		4	4		815	213		1,028	1,225
3370	225 amp		2	8		1,650	425		2,075	2,475
3380	400 amp		1.60	10		2,600	535		3,135	3,675
3410	480 volt, 3 pole, 3 wire, 100 amp	1 Elec	2	4		1,125	213		1,338	1,575
3420	225 amp	2 Elec	2	8		1,950	425		2,375	2,800
3430	400 amp		1.60	10		2,925	535		3,460	4,025
3460	277/480 volt, 3 pole, 4 wire, 100 amp		4	4		1,100	213		1,313	1,550
3470	225 amp		2	8		1,900	425		2,325	2,750
3480	400 amp		1.60	10		2,950	535		3,485	4,025
3510	Main circuit breaker, HIC, 240 volt, 1 pole, 3 wire, 100 amp	1 Elec	2	4		1,075	213		1,288	1,525
3520	225 amp	2 Elec	2	8		2,850	425		3,275	3,775
3530	400 amp	"	1.60	10		3,800	535		4,335	4,975
3560	3 pole, 3 wire, 100 amp	1 Elec	2	4		1,200	213		1,413	1,650
3570	225 amp	2 Elec	2	8		3,200	425		3,625	4,175
3580	400 amp	"	1.60	10		4,225	535		4,760	5,450
3610	120/208 volt, 3 pole, 4 wire, 100 amp	1 Elec	2	4		1,200	213		1,413	1,650
3620	225 amp	2 Elec	2	8		3,200	425		3,625	4,175
3630	400 amp	"	1.60	10		4,225	535		4,760	5,450
3660	480 volt, 3 pole, 3 wire, 100 amp	1 Elec	2	4		1,800	213		2,013	2,300
3670	225 amp	2 Elec	2	8		3,550	425		3,975	4,550
3680	400 amp	"	1.60	10		4,525	535		5,060	5,775
3710	277/480 volt, 3 pole, 4 wire, 100 amp	1 Elec	2	4		1,725	213		1,938	2,200
3720	225 amp	2 Elec	2	8		3,400	425		3,825	4,375

26 24 Switchboards and Panelboards

26 24 16 – Panelboards

26 24 16.30 Panelboards Commercial Applications

		Crew	Daily Output	Labor-Hours	Unit	Material	2014 Bare Costs Labor	Equipment	Total	Total Incl O&P
3730	400 amp	2 Elec	1.60	10	Ea.	4,500	535		5,035	5,750
3760	Main circuit breaker, shunt trip, 100 amp	1 Elec	1.20	6.667		910	355	·	1,265	1,525
3770	225 amp	2 Elec	1.60	10		2,100	535		2,635	3,100
3780	400 amp	"	1.40	11.429		3,050	610		3,660	4,275

26 24 19 – Motor-Control Centers

26 24 19.20 Motor Control Center Components

		Crew	Daily Output	Labor-Hours	Unit	Material	2014 Bare Costs Labor	Equipment	Total	Total Incl O&P
0010	**MOTOR CONTROL CENTER COMPONENTS** R262419-60									
0100	Starter, size 1, FVNR, NEMA 1, type A, fusible	1 Elec	2.70	2.963	Ea.	1,575	158		1,733	1,950
0120	Circuit breaker		2.70	2.963		1,700	158		1,858	2,100
0140	Type B, fusible		2.70	2.963		1,725	158		1,883	2,125
0160	Circuit breaker		2.70	2.963		1,875	158		2,033	2,275
0180	NEMA 12, type A, fusible		2.60	3.077		1,600	164		1,764	2,000
0200	Circuit breaker		2.60	3.077		1,725	164		1,889	2,150
0220	Type B, fusible		2.60	3.077		1,750	164		1,914	2,175
0240	Circuit breaker		2.60	3.077		1,900	164		2,064	2,325
0300	Starter, size 1, FVR, NEMA 1, type A, fusible		2	4		2,250	213		2,463	2,800
0320	Circuit breaker		2	4		2,250	213		2,463	2,800
0340	Type B, fusible		2	4		2,475	213		2,688	3,050
0360	Circuit breaker		2	4		2,475	213		2,688	3,050
0380	NEMA 12, type A, fusible		1.90	4.211		2,275	225		2,500	2,825
0400	Circuit breaker		1.90	4.211		2,275	225		2,500	2,825
0420	Type B, fusible		1.90	4.211		2,500	225		2,725	3,075
0440	Circuit breaker		1.90	4.211		2,500	225		2,725	3,075
0490	Starter size 1, 2 speed, separate winding									
0500	NEMA 1, type A, fusible	1 Elec	2.60	3.077	Ea.	2,950	164		3,114	3,500
0520	Circuit breaker		2.60	3.077		2,950	164		3,114	3,500
0540	Type B, fusible		2.60	3.077		3,250	164		3,414	3,825
0560	Circuit breaker		2.60	3.077		3,250	164		3,414	3,825
0580	NEMA 12, type A, fusible		2.50	3.200		3,000	171		3,171	3,550
0600	Circuit breaker		2.50	3.200		3,000	171		3,171	3,550
0620	Type B, fusible		2.50	3.200		3,300	171		3,471	3,875
0640	Circuit breaker		2.50	3.200		3,300	171		3,471	3,875
0650	Starter size 1, 2 speed, consequent pole									
0660	NEMA 1, type A, fusible	1 Elec	2.60	3.077	Ea.	2,950	164		3,114	3,500
0680	Circuit breaker		2.60	3.077		2,950	164		3,114	3,500
0700	Type B, fusible		2.60	3.077		3,250	164		3,414	3,825
0720	Circuit breaker		2.60	3.077		3,250	164		3,414	3,825
0740	NEMA 12, type A, fusible		2.50	3.200		3,000	171		3,171	3,550
0760	Circuit breaker		2.50	3.200		3,000	171		3,171	3,550
0780	Type B, fusible		2.50	3.200		3,275	171		3,446	3,875
0800	Circuit breaker		2.50	3.200		3,300	171		3,471	3,875
0810	Starter size 1, 2 speed, space only									
0820	NEMA 1, type A, fusible	1 Elec	16	.500	Ea.	665	26.50		691.50	770
0840	Circuit breaker		16	.500		665	26.50		691.50	770
0860	Type B, fusible		16	.500		665	26.50		691.50	770
0880	Circuit breaker		16	.500		665	26.50		691.50	770
0900	NEMA 12, type A, fusible		15	.533		690	28.50		718.50	805
0920	Circuit breaker		15	.533		690	28.50		718.50	805
0940	Type B, fusible		15	.533		690	28.50		718.50	805
0960	Circuit breaker		15	.533		690	28.50		718.50	805
1100	Starter size 2, FVNR, NEMA 1, type A, fusible	2 Elec	4	4		1,775	213		1,988	2,275
1120	Circuit breaker		4	4		1,950	213		2,163	2,450

26 24 Switchboards and Panelboards

26 24 19 – Motor-Control Centers

26 24 19.20 Motor Control Center Components	Crew	Daily Output	Labor-Hours	Unit	Material	2014 Bare Costs Labor	Equipment	Total	Total Incl O&P
1140	Type B, fusible	2 Elec	4	4	Ea.	1,950	213		2,163	2,475
1160	Circuit breaker		4	4		2,125	213		2,338	2,650
1180	NEMA 12, type A, fusible		3.80	4.211		1,800	225		2,025	2,300
1200	Circuit breaker		3.80	4.211		1,975	225		2,200	2,475
1220	Type B, fusible		3.80	4.211		1,975	225		2,200	2,475
1240	Circuit breaker		3.80	4.211		2,150	225		2,375	2,675
1300	FVR, NEMA 1, type A, fusible		3.20	5		3,000	267		3,267	3,700
1320	Circuit breaker		3.20	5		3,000	267		3,267	3,700
1340	Type B, fusible		3.20	5		3,300	267		3,567	4,025
1360	Circuit breaker		3.20	5		3,300	267		3,567	4,025
1380	NEMA type 12, type A, fusible		3	5.333		3,050	285		3,335	3,775
1400	Circuit breaker		3	5.333		3,050	285		3,335	3,775
1420	Type B, fusible		3	5.333		3,350	285		3,635	4,100
1440	Circuit breaker	↓	3	5.333	↓	3,350	285		3,635	4,100
1490	Starter size 2, 2 speed, separate winding									
1500	NEMA 1, type A, fusible	2 Elec	3.80	4.211	Ea.	3,350	225		3,575	4,000
1520	Circuit breaker		3.80	4.211		3,350	225		3,575	4,000
1540	Type B, fusible		3.80	4.211		3,675	225		3,900	4,375
1560	Circuit breaker		3.80	4.211		3,675	225		3,900	4,375
1570	NEMA 12, type A, fusible		3.60	4.444		3,400	237		3,637	4,075
1580	Circuit breaker		3.60	4.444		3,400	237		3,637	4,075
1600	Type B, fusible		3.60	4.444		3,725	237		3,962	4,450
1620	Circuit breaker	↓	3.60	4.444	↓	3,725	237		3,962	4,450
1630	Starter size 2, 2 speed, consequent pole									
1640	NEMA 1, type A, fusible	2 Elec	3.80	4.211	Ea.	3,800	225		4,025	4,500
1660	Circuit breaker		3.80	4.211		3,800	225		4,025	4,500
1680	Type B, fusible		3.80	4.211		4,000	225		4,225	4,750
1700	Circuit breaker		3.80	4.211		4,000	225		4,225	4,750
1720	NEMA 12, type A, fusible		3.80	4.211		3,800	225		4,025	4,500
1740	Circuit breaker		3.60	4.444		3,850	237		4,087	4,575
1760	Type B, fusible		3.60	4.444		4,075	237		4,312	4,825
1780	Circuit breaker	↓	3.60	4.444	↓	4,075	237		4,312	4,825
1830	Starter size 2, autotransformer									
1840	NEMA 1, type A, fusible	2 Elec	3.40	4.706	Ea.	5,925	251		6,176	6,875
1860	Circuit breaker		3.40	4.706		6,050	251		6,301	7,025
1880	Type B, fusible		3.40	4.706		6,450	251		6,701	7,475
1900	Circuit breaker		3.40	4.706		6,450	251		6,701	7,475
1920	NEMA 12, type A, fusible		3.20	5		6,025	267		6,292	7,025
1940	Circuit breaker		3.20	5		6,025	267		6,292	7,025
1960	Type B, fusible		3.20	5		6,575	267		6,842	7,625
1980	Circuit breaker	↓	3.20	5	↓	6,575	267		6,842	7,625
2030	Starter size 2, space only									
2040	NEMA 1, type A, fusible	1 Elec	16	.500	Ea.	665	26.50		691.50	770
2060	Circuit breaker		16	.500		665	26.50		691.50	770
2080	Type B, fusible		16	.500		665	26.50		691.50	770
2100	Circuit breaker		16	.500		665	26.50		691.50	770
2120	NEMA 12, type A, fusible		15	.533		690	28.50		718.50	805
2140	Circuit breaker		15	.533		690	28.50		718.50	805
2160	Type B, fusible		15	.533		690	28.50		718.50	805
2180	Circuit breaker	↓	15	.533	↓	690	28.50		718.50	805
2300	Starter size 3, FVNR, NEMA 1, type A, fusible	2 Elec	2	8		3,375	425		3,800	4,350
2320	Circuit breaker		2	8		3,000	425		3,425	3,950
2340	Type B, fusible	↓	2	8	↓	3,700	425		4,125	4,725

26 24 19.20 Motor Control Center Components		Crew	Daily Output	Labor-Hours	Unit	Material	2014 Bare Costs Labor	Equipment	Total	Total Incl O&P
2360	Circuit breaker	2 Elec	2	8	Ea.	3,300	425		3,725	4,275
2380	NEMA 12, type A, fusible		1.90	8.421		3,425	450		3,875	4,450
2400	Circuit breaker		1.90	8.421		3,050	450		3,500	4,025
2420	Type B, fusible		1.90	8.421		3,825	450		4,275	4,875
2440	Circuit breaker		1.90	8.421		3,350	450		3,800	4,375
2500	Starter size 3, FVR, NEMA 1, type A, fusible		1.60	10		4,600	535		5,135	5,850
2520	Circuit breaker		1.60	10		4,375	535		4,910	5,625
2540	Type B, fusible		1.60	10		5,000	535		5,535	6,300
2560	Circuit breaker		1.60	10		4,800	535		5,335	6,075
2580	NEMA 12, type A, fusible		1.50	10.667		4,675	570		5,245	6,000
2600	Circuit breaker		1.50	10.667		4,475	570		5,045	5,775
2620	Type B, fusible		1.50	10.667		5,100	570		5,670	6,450
2640	Circuit breaker	▼	1.50	10.667	▼	5,275	570		5,845	6,650
2690	Starter size 3, 2 speed, separate winding									
2700	NEMA 1, type A, fusible	2 Elec	2	8	Ea.	5,275	425		5,700	6,450
2720	Circuit breaker		2	8		4,650	425		5,075	5,775
2740	Type B, fusible		2	8		5,775	425		6,200	7,000
2760	Circuit breaker		2	8		5,100	425		5,525	6,250
2780	NEMA 12, type A, fusible		1.90	8.421		5,400	450		5,850	6,600
2800	Circuit breaker		1.90	8.421		4,750	450		5,200	5,900
2820	Type B, fusible		1.90	8.421		5,900	450		6,350	7,150
2840	Circuit breaker	▼	1.90	8.421	▼	5,200	450		5,650	6,400
2850	Starter size 3, 2 speed, consequent pole									
2860	NEMA 1, type A, fusible	2 Elec	2	8	Ea.	5,900	425		6,325	7,125
2880	Circuit breaker		2	8		5,275	425		5,700	6,450
2900	Type B, fusible		2	8		6,400	425		6,825	7,675
2920	Circuit breaker		2	8		5,700	425		6,125	6,925
2940	NEMA 12, type A, fusible		1.90	8.421		6,000	450		6,450	7,275
2960	Circuit breaker		1.90	8.421		5,375	450		5,825	6,575
2980	Type B, fusible		1.90	8.421		6,500	450		6,950	7,825
3000	Circuit breaker		1.90	8.421		5,800	450		6,250	7,075
3100	Starter size 3, autotransformer, NEMA 1, type A, fusible		1.60	10		7,400	535		7,935	8,950
3120	Circuit breaker		1.60	10		7,325	535		7,860	8,875
3140	Type B, fusible		1.60	10		8,000	535		8,535	9,600
3160	Circuit breaker		1.60	10		8,000	535		8,535	9,600
3180	NEMA 12, type A, fusible		1.50	10.667		7,475	570		8,045	9,075
3200	Circuit breaker		1.50	10.667		7,475	570		8,045	9,075
3220	Type B, fusible		1.50	10.667		8,150	570		8,720	9,800
3240	Circuit breaker	▼	1.50	10.667		8,150	570		8,720	9,800
3260	Starter size 3, space only, NEMA 1, type A, fusible	1 Elec	15	.533		1,150	28.50		1,178.50	1,300
3280	Circuit breaker		15	.533		885	28.50		913.50	1,025
3300	Type B, fusible		15	.533		1,150	28.50		1,178.50	1,300
3320	Circuit breaker		15	.533		885	28.50		913.50	1,025
3340	NEMA 12, type A, fusible		14	.571		1,200	30.50		1,230.50	1,375
3360	Circuit breaker		14	.571		940	30.50		970.50	1,075
3380	Type B, fusible		14	.571		1,200	30.50		1,230.50	1,375
3400	Circuit breaker	▼	14	.571		940	30.50		970.50	1,075
3500	Starter size 4, FVNR, NEMA 1, type A, fusible	2 Elec	1.60	10		4,375	535		4,910	5,625
3520	Circuit breaker		1.60	10		3,975	535		4,510	5,175
3540	Type B, fusible		1.60	10		4,800	535		5,335	6,075
3560	Sircuit breaker		1.60	10		4,375	535		4,910	5,625
3580	NEMA 12, type A, fusible		1.50	10.667		4,500	570		5,070	5,800
3600	Circuit breaker	▼	1.50	10.667	▼	4,050	570		4,620	5,300

26 24 19 – Motor-Control Centers

26 24 19.20 Motor Control Center Components	Crew	Daily Output	Labor-Hours	Unit	Material	2014 Bare Costs Labor	Equipment	Total	Total Incl O&P
3620	Type B, fusible	2 Elec	1.50	10.667	Ea.	4,900	570		5,470	6,250
3640	Circuit breaker		1.50	10.667		4,450	570		5,020	5,750
3700	Starter size 4, FVR, NEMA 1, type A, fusible		1.20	13.333		6,000	710		6,710	7,675
3720	Circuit breaker		1.20	13.333		5,400	710		6,110	7,025
3740	Type B, fusible		1.20	13.333		6,550	710		7,260	8,275
3760	Circuit breaker		1.20	13.333		5,950	710		6,660	7,625
3780	NEMA 12, type A, fusible		1.16	13.793		6,125	735		6,860	7,850
3800	Circuit breaker		1.16	13.793		5,500	735		6,235	7,125
3820	Type B, fusible		1.16	13.793		5,550	735		6,285	7,200
3840	Circuit breaker		1.16	13.793		6,025	735		6,760	7,725
3890	Starter size 4, 2 speed, separate windings									
3900	NEMA 1, type A, fusible	2 Elec	1.60	10	Ea.	7,575	535		8,110	9,125
3920	Circuit breaker		1.60	10		5,675	535		6,210	7,050
3940	Type B, fusible		1.60	10		8,325	535		8,860	9,950
3960	Circuit breaker		1.60	10		6,225	535		6,760	7,650
3980	NEMA 12, type A, fusible		1.50	10.667		7,700	570		8,270	9,325
4000	Circuit breaker		1.50	10.667		5,775	570		6,345	7,200
4020	Type B, fusible		1.50	10.667		8,450	570		9,020	10,200
4040	Circuit breaker		1.50	10.667		6,325	570		6,895	7,800
4050	Starter size 4, 2 speed, consequent pole									
4060	NEMA 1, type A, fusible	2 Elec	1.60	10	Ea.	8,850	535		9,385	10,500
4080	Circuit breaker		1.60	10		6,575	535		7,110	8,025
4100	Type B, fusible		1.60	10		9,725	535		10,260	11,500
4120	Circuit breaker		1.60	10		7,200	535		7,735	8,725
4140	NEMA 12, type A, fusible		1.50	10.667		8,975	570		9,545	10,700
4160	Circuit breaker		1.50	10.667		6,675	570		7,245	8,200
4180	Type B, fusible		1.50	10.667		9,850	570		10,420	11,700
4200	Circuit breaker		1.50	10.667		7,300	570		7,870	8,875
4300	Starter size 4, autotransformer, NEMA 1, type A, fusible		1.30	12.308		8,700	655		9,355	10,600
4320	Circuit breaker		1.30	12.308		8,775	655		9,430	10,600
4340	Type B, fusible		1.30	12.308		9,600	655		10,255	11,500
4360	Circuit breaker		1.30	12.308		9,600	655		10,255	11,500
4380	NEMA 12, type A, fusible		1.24	12.903		8,850	690		9,540	10,800
4400	Circuit breaker		1.24	12.903		8,900	690		9,590	10,800
4420	Type B, fusible		1.24	12.903		9,750	690		10,440	11,700
4440	Circuit breaker		1.24	12.903		9,700	690		10,390	11,700
4500	Starter size 4, space only, NEMA 1, type A, fusible	1 Elec	14	.571		1,575	30.50		1,605.50	1,800
4520	Circuit breaker		14	.571		1,150	30.50		1,180.50	1,300
4540	Type B, fusible		14	.571		1,575	30.50		1,605.50	1,800
4560	Circuit breaker		14	.571		1,150	30.50		1,180.50	1,300
4580	NEMA 12, type A, fusible		13	.615		1,675	33		1,708	1,900
4600	Circuit breaker		13	.615		1,200	33		1,233	1,375
4620	Type B, fusible		13	.615		1,675	33		1,708	1,900
4640	Circuit breaker		13	.615		1,200	33		1,233	1,375
4800	Starter size 5, FVNR, NEMA 1, type A, fusible	2 Elec	1	16		9,050	855		9,905	11,200
4820	Circuit breaker		1	16		6,425	855		7,280	8,350
4840	Type B, fusible		1	16		9,950	855		10,805	12,200
4860	Circuit breaker		1	16		7,075	855		7,930	9,050
4880	NEMA 12, type A, fusible		.96	16.667		9,175	890		10,065	11,400
4900	Circuit breaker		.96	16.667		6,550	890		7,440	8,525
4920	Type B, fusible		.96	16.667		10,100	890		10,990	12,400
4940	Circuit breaker		.96	16.667		7,200	890		8,090	9,250
5000	Starter size 5, FVR, NEMA 1, type A, fusible		.80	20		13,300	1,075		14,375	16,300

26 24 19 – Motor-Control Centers

26 24 19.20 Motor Control Center Components		Crew	Daily Output	Labor-Hours	Unit	Material	2014 Bare Costs Labor	Equipment	Total	Total Incl O&P
5020	Circuit breaker	2 Elec	.80	20	Ea.	10,500	1,075		11,575	13,100
5040	Type B, fusible		.80	20		14,600	1,075		15,675	17,700
5060	Circuit breaker		.80	20		11,500	1,075		12,575	14,200
5080	NEMA 12, type A, fusible		.76	21.053		13,600	1,125		14,725	16,700
5100	Circuit breaker		.76	21.053		10,600	1,125		11,725	13,400
5120	Type B, fusible		.76	21.053		14,900	1,125		16,025	18,100
5140	Circuit breaker		.76	21.053		11,700	1,125		12,825	14,500
5190	Starter size 5, 2 speed, separate windings									
5200	NEMA 1, type A, fusible	2 Elec	1	16	Ea.	18,400	855		19,255	21,500
5220	Circuit breaker		1	16		13,700	855		14,555	16,400
5240	Type B, fusible		1	16		20,200	855		21,055	23,500
5260	Circuit breaker		1	16		15,100	855		15,955	17,900
5280	NEMA 12, type A, fusible		.96	16.667		18,600	890		19,490	21,800
5300	Circuit breaker		.96	16.667		13,900	890		14,790	16,600
5320	Type B, fusible		.96	16.667		20,500	890		21,390	23,800
5340	Circuit breaker		.96	16.667		15,200	890		16,090	18,100
5400	Starter size 5, autotransformer, NEMA 1, type A, fusible		.70	22.857		15,100	1,225		16,325	18,400
5420	Circuit breaker		.70	22.857		12,500	1,225		13,725	15,600
5440	Type B, fusible		.70	22.857		16,500	1,225		17,725	20,000
5460	Circuit breaker		.70	22.857		13,700	1,225		14,925	16,900
5480	NEMA 12, type A, fusible		.68	23.529		15,200	1,250		16,450	18,600
5500	Circuit breaker		.68	23.529		12,600	1,250		13,850	15,800
5520	Type B, fusible		.68	23.529		16,700	1,250		17,950	20,300
5540	Circuit breakers		.68	23.529		13,900	1,250		15,150	17,100
5600	Starter size 5, space only, NEMA 1, type A, fusible	1 Elec	12	.667		2,025	35.50		2,060.50	2,275
5620	Circuit breaker		12	.667		2,025	35.50		2,060.50	2,275
5640	Type B, fusible		12	.667		2,025	35.50		2,060.50	2,275
5660	Circuit breaker		12	.667		1,325	35.50		1,360.50	1,500
5680	NEMA 12, type A, fusible		11	.727		2,150	39		2,189	2,400
5700	Circuit breaker		11	.727		1,400	39		1,439	1,600
5720	Type B, fusible		11	.727		2,150	39		2,189	2,400
5740	Circuit breaker		11	.727		1,400	39		1,439	1,600
5800	Fuse, light contactor NEMA 1, type A, 30 amp		2.70	2.963		1,575	158		1,733	1,950
5820	60 amp		2	4		1,775	213		1,988	2,275
5840	100 amp		1	8		3,375	425		3,800	4,350
5860	200 amp		.80	10		7,825	535		8,360	9,400
5880	Type B, 30 amp		2.70	2.963		1,700	158		1,858	2,100
5900	60 amp		2	4		1,900	213		2,113	2,425
5920	100 amp		1	8		3,700	425		4,125	4,725
5940	200 amp	2 Elec	1.60	10		8,575	535		9,110	10,200
5960	NEMA 12, type A, 30 amp	1 Elec	2.60	3.077		1,600	164		1,764	2,000
5980	60 amp		1.90	4.211		1,800	225		2,025	2,300
6000	100 amp		.95	8.421		3,425	450		3,875	4,450
6020	200 amp	2 Elec	1.50	10.667		7,950	570		8,520	9,575
6040	Type B, 30 amp	1 Elec	2.60	3.077		1,725	164		1,889	2,150
6060	60 amp		1.90	4.211		1,925	225		2,150	2,450
6080	100 amp		.95	8.421		3,775	450		4,225	4,825
6100	200 amp	2 Elec	1.50	10.667		8,700	570		9,270	10,400
6200	Circuit breaker, light contactor NEMA 1, type A, 30 amp	1 Elec	2.70	2.963		1,700	158		1,858	2,100
6220	60 amp		2	4		1,950	213		2,163	2,450
6240	100 amp		1	8		3,000	425		3,425	3,950
6260	200 amp	2 Elec	1.60	10		6,475	535		7,010	7,925
6280	Type B, 30 amp	1 Elec	2.70	2.963		1,875	158		2,033	2,275

26 24 19 – Motor-Control Centers

26 24 19.20 Motor Control Center Components	Crew	Daily Output	Labor-Hours	Unit	2014 Bare Costs			Total	Total Incl O&P
					Material	Labor	Equipment	Total	
6300	60 amp	1 Elec	2	4	Ea.	2,150	213		2,363	2,675
6320	100 amp	↓	1	8		3,275	425		3,700	4,250
6340	200 amp	2 Elec	1.60	10		7,075	535		7,610	8,575
6360	NEMA 12, type A, 30 amp	1 Elec	2.60	3.077		1,725	164		1,889	2,150
6380	60 amp		1.90	4.211		1,975	225		2,200	2,475
6400	100 amp	↓	.95	8.421		3,050	450		3,500	4,025
6420	200 amp	2 Elec	1.50	10.667		6,500	570		7,070	8,000
6440	Type B, 30 amp	1 Elec	2.60	3.077		2,025	164		2,189	2,475
6460	60 amp		1.90	4.211		2,175	225		2,400	2,700
6480	100 amp	↓	.95	8.421		3,325	450		3,775	4,325
6500	200 amp	2 Elec	1.50	10.667		7,150	570		7,720	8,725
6600	Fusible switch, NEMA 1, type A, 30 amp	1 Elec	5.30	1.509		1,025	80.50		1,105.50	1,275
6620	60 amp		5	1.600		1,100	85.50		1,185.50	1,350
6640	100 amp		4	2		1,225	107		1,332	1,500
6660	200 amp	↓	3.20	2.500		2,100	133		2,233	2,525
6680	400 amp	2 Elec	4.60	3.478		4,975	186		5,161	5,725
6700	600 amp		3.20	5		5,375	267		5,642	6,300
6720	800 amp	↓	2.60	6.154		14,700	330		15,030	16,700
6740	NEMA 12, type A, 30 amp	1 Elec	5.20	1.538		1,075	82		1,157	1,300
6760	60 amp		4.90	1.633		1,125	87		1,212	1,375
6780	100 amp		3.90	2.051		1,250	109		1,359	1,550
6800	200 amp	↓	3.10	2.581		2,150	138		2,288	2,575
6820	400 amp	2 Elec	4.40	3.636		5,075	194		5,269	5,875
6840	600 amp		3	5.333		5,525	285		5,810	6,500
6860	800 amp	↓	2.40	6.667		14,900	355		15,255	16,900
6900	Circuit breaker, NEMA 1, type A, 30 amp	1 Elec	5.30	1.509		950	80.50		1,030.50	1,175
6920	60 amp		5	1.600		950	85.50		1,035.50	1,175
6940	100 amp		4	2		950	107		1,057	1,200
6960	225 amp	↓	3.20	2.500		1,700	133		1,833	2,075
6980	400 amp	2 Elec	4.60	3.478		3,225	186		3,411	3,825
7000	600 amp		3.20	5		3,775	267		4,042	4,550
7020	800 amp	↓	2.60	6.154		7,900	330		8,230	9,175
7040	NEMA 12, type A, 30 amp	1 Elec	5.20	1.538		980	82		1,062	1,200
7060	60 amp		4.90	1.633		980	87		1,067	1,200
7080	100 amp		3.90	2.051		980	109		1,089	1,250
7100	225 amp	↓	3.10	2.581		1,725	138		1,863	2,100
7120	400 amp	2 Elec	4.40	3.636		3,275	194		3,469	3,925
7140	600 amp		3	5.333		3,825	285		4,110	4,625
7160	800 amp		2.40	6.667		8,050	355		8,405	9,375
7300	Incoming line, main lug only, 600 amp, alum., NEMA 1		1.60	10		1,200	535		1,735	2,125
7320	NEMA 12		1.50	10.667		1,225	570		1,795	2,200
7340	Copper, NEMA 1		1.60	10		1,275	535		1,810	2,200
7360	800 amp, alum., NEMA 1		1.50	10.667		3,025	570		3,595	4,175
7380	NEMA 12		1.40	11.429		3,075	610		3,685	4,300
7400	Copper, NEMA 1		1.50	10.667		3,150	570		3,720	4,300
7420	1200 amp, copper, NEMA 1		1.40	11.429		3,250	610		3,860	4,500
7440	Incoming line, fusible switch, 400 amp, alum., NEMA 1		1.20	13.333		4,150	710		4,860	5,650
7460	NEMA 12		1.10	14.545		4,225	775		5,000	5,800
7480	Copper, NEMA 1		1.20	13.333		4,225	710		4,935	5,725
7500	600 amp, alum., NEMA 1		1.10	14.545		5,075	775		5,850	6,725
7520	NEMA 12		1	16		5,150	855		6,005	6,950
7540	Copper, NEMA 1		1.10	14.545		5,125	775		5,900	6,800
7560	Incoming line, circuit breaker, 225 amp, alum., NEMA 1	↓	1.20	13.333	↓	2,150	710		2,860	3,450

26 24 19 – Motor-Control Centers

26 24 19.20 Motor Control Center Components	Crew	Daily Output	Labor-Hours	Unit	Material	2014 Bare Costs Labor	Equipment	Total	Total Incl O&P
7580	NEMA 12	2 Elec	1.10	14.545	Ea.	2,200	775		2,975	3,575
7600	Copper, NEMA 1		1.20	13.333		2,225	710		2,935	3,525
7620	400 amp., alum., NEMA 1		1.20	13.333		3,225	710		3,935	4,625
7640	NEMA 12		1.10	14.545		3,275	775		4,050	4,775
7660	Copper, NEMA 1		1.20	13.333		3,300	710		4,010	4,700
7680	600 amp., alum., NEMA 1		1.10	14.545		3,775	775		4,550	5,300
7700	NEMA 12		1	16		3,825	855		4,680	5,475
7720	Copper, NEMA 1		1.10	14.545		3,850	775		4,625	5,375
7740	800 amp., copper, NEMA 1	↓	.90	17.778		7,900	950		8,850	10,100
7760	Incoming line, for copper bus, add					119			119	131
7780	For 65000 amp bus bracing, add					177			177	194
7800	For NEMA 3R enclosure, add					5,550			5,550	6,100
7820	For NEMA 12 enclosure, add					153			153	168
7840	For 1/4" x 1" ground bus, add	1 Elec	16	.500		98.50	26.50		125	148
7860	For 1/4" x 2" ground bus, add	"	12	.667		98.50	35.50		134	161
7900	Main rating basic section, alum., NEMA 1, 800 amp	2 Elec	1.40	11.429		325	610		935	1,275
7920	1200 amp	"	1.20	13.333		635	710		1,345	1,775
7940	For copper bus, add					475			475	525
7960	For 65000 amp bus bracing, add					320			320	350
7980	For NEMA 3R enclosure, add					5,550			5,550	6,100
8000	For NEMA 12, enclosure, add					153			153	168
8020	For 1/4" x 1" ground bus, add	1 Elec	16	.500		98.50	26.50		125	148
8040	For 1/4" x 2" ground bus, add		12	.667		98.50	35.50		134	161
8060	Unit devices, pilot light, standard		16	.500		85	26.50		111.50	134
8080	Pilot light, push to test		16	.500		119	26.50		145.50	171
8100	Pilot light, standard, and push button		12	.667		204	35.50		239.50	277
8120	Pilot light, push to test, and push button		12	.667		238	35.50		273.50	315
8140	Pilot light, standard, and select switch		12	.667		204	35.50		239.50	277
8160	Pilot light, push to test, and select switch	↓	12	.667	↓	238	35.50		273.50	315

26 24 19.30 Motor Control Center

26 24 19.30 Motor Control Center		Crew	Daily Output	Labor-Hours	Unit	Material	Labor	Equipment	Total	Total Incl O&P
0010	**MOTOR CONTROL CENTER** Consists of starters & structures	R262419-60								
0050	Starters, class 1, type B, comb. MCP, FVNR, with									
0100	control transformer, 10 HP, size 1, 12" high	1 Elec	2.70	2.963	Ea.	1,700	158		1,858	2,100
0200	25 HP, size 2, 18" high	2 Elec	4	4		1,950	213		2,163	2,450
0300	50 HP, size 3, 24" high		2	8		3,000	425		3,425	3,950
0350	75 HP, size 4, 24" high		1.60	10		3,975	535		4,510	5,175
0400	100 HP, size 4, 30" high		1.40	11.429		5,250	610		5,860	6,700
0500	200 HP, size 5, 48" high		1	16		7,825	855		8,680	9,900
0600	400 HP, size 6, 72" high	↓	.80	20	↓	16,100	1,075		17,175	19,300
0800	Structures, 600 amp, 22,000 rms, takes any									
0900	combination of starters up to 72" high	2 Elec	1.60	10	Ea.	1,900	535		2,435	2,900
1000	Back to back, 72" front & 66" back	"	1.20	13.333		2,575	710		3,285	3,900
1100	For copper bus add per structure					265			265	292
1200	For NEMA 12, add per structure					153			153	168
1300	For 42,000 rms, add per structure					199			199	219
1400	For 100,000 rms, size 1 & 2, add					680			680	750
1500	Size 3, add					1,100			1,100	1,200
1600	Size 4, add					885			885	970
1700	For pilot lights, add per starter	1 Elec	16	.500		119	26.50		145.50	171
1800	For push button, add per starter		16	.500		119	26.50		145.50	171
1900	For auxiliary contacts, add per starter	↓	16	.500	↓	177	26.50		203.50	234

26 24 19.40 Motor Starters and Controls		Crew	Daily Output	Labor-Hours	Unit	Material	2014 Bare Costs Labor	Equipment	Total	Total Incl O&P
0010	**MOTOR STARTERS AND CONTROLS** R262419-65									
0050	Magnetic, FVNR, with enclosure and heaters, 480 volt									
0080	2 HP, size 00	1 Elec	3.50	2.286	Ea.	195	122		317	395
0100	5 HP, size 0		2.30	3.478		236	186		422	535
0200	10 HP, size 1		1.60	5		265	267		532	690
0300	25 HP, size 2 R263413-33	2 Elec	2.20	7.273		500	390		890	1,125
0400	50 HP, size 3		1.80	8.889		810	475		1,285	1,600
0500	100 HP, size 4		1.20	13.333		1,800	710		2,510	3,050
0600	200 HP, size 5		.90	17.778		4,225	950		5,175	6,050
0610	400 HP, size 6		.80	20		18,200	1,075		19,275	21,700
0620	NEMA 7, 5 HP, size 0	1 Elec	1.60	5		1,350	267		1,617	1,900
0630	10 HP, size 1	"	1.10	7.273		1,425	390		1,815	2,125
0640	25 HP, size 2	2 Elec	1.80	8.889		2,300	475		2,775	3,225
0650	50 HP, size 3		1.20	13.333		3,425	710		4,135	4,850
0660	100 HP, size 4		.90	17.778		5,550	950		6,500	7,525
0670	200 HP, size 5		.50	32		13,200	1,700		14,900	17,200
0700	Combination, with motor circuit protectors, 5 HP, size 0	1 Elec	1.80	4.444		765	237		1,002	1,200
0800	10 HP, size 1	"	1.30	6.154		795	330		1,125	1,375
0900	25 HP, size 2	2 Elec	2	8		1,125	425		1,550	1,875
1000	50 HP, size 3		1.32	12.121		1,600	645		2,245	2,750
1200	100 HP, size 4		.80	20		3,500	1,075		4,575	5,450
1220	NEMA 7, 5 HP, size 0	1 Elec	1.30	6.154		2,725	330		3,055	3,475
1230	10 HP, size 1	"	1	8		2,775	425		3,200	3,700
1240	25 HP, size 2	2 Elec	1.32	12.121		3,725	645		4,370	5,050
1250	50 HP, size 3		.80	20		6,125	1,075		7,200	8,350
1260	100 HP, size 4		.60	26.667		9,550	1,425		10,975	12,600
1270	200 HP, size 5		.40	40		20,800	2,125		22,925	26,000
1400	Combination, with fused switch, 5 HP, size 0	1 Elec	1.80	4.444		585	237		822	1,000
1600	10 HP, size 1	"	1.30	6.154		625	330		955	1,175
1800	25 HP, size 2	2 Elec	2	8		1,025	425		1,450	1,775
2000	50 HP, size 3		1.32	12.121		1,725	645		2,370	2,875
2200	100 HP, size 4		.80	20		3,000	1,075		4,075	4,900
2610	NEMA 4, with start-stop pushbutton size 1	1 Elec	1.30	6.154		1,575	330		1,905	2,225
2620	Size 2	2 Elec	2	8		2,100	425		2,525	2,975
2630	Size 3		1.32	12.121		3,325	645		3,970	4,650
2640	Size 4		.80	20		5,125	1,075		6,200	7,225
2650	NEMA 4, FVNR, including control transformer									
2660	Size 1	2 Elec	2.60	6.154	Ea.	1,450	330		1,780	2,100
2670	Size 2		2	8		2,050	425		2,475	2,925
2680	Size 3		1.32	12.121		3,275	645		3,920	4,600
2690	Size 4		.80	20		5,075	1,075		6,150	7,175
2710	Magnetic, FVR, control circuit transformer, NEMA 1, size 1	1 Elec	1.30	6.154		855	330		1,185	1,425
2720	Size 2	2 Elec	2	8		1,375	425		1,800	2,150
2730	Size 3		1.32	12.121		2,075	645		2,720	3,250
2740	Size 4		.80	20		4,500	1,075		5,575	6,550
2760	NEMA 4, size 1	1 Elec	1.10	7.273		1,225	390		1,615	1,925
2770	Size 2	2 Elec	1.60	10		1,975	535		2,510	2,975
2780	Size 3		1.20	13.333		2,975	710		3,685	4,325
2790	Size 4		.70	22.857		6,100	1,225		7,325	8,525
2820	NEMA 12, size 1	1 Elec	1.10	7.273		1,000	390		1,390	1,675
2830	Size 2	2 Elec	1.60	10		1,575	535		2,110	2,550
2840	Size 3		1.20	13.333		2,500	710		3,210	3,825

26 24 19.40 Motor Starters and Controls		Crew	Daily Output	Labor-Hours	Unit	Material	2014 Bare Costs Labor	Equipment	Total	Total Incl O&P
2850	Size 4	2 Elec	.70	22.857	Ea.	5,150	1,225		6,375	7,500
2870	Combination FVR, fused, w/control XFMR & PB, NEMA 1, size 1	1 Elec	1	8		1,475	425		1,900	2,275
2880	Size 2	2 Elec	1.50	10.667		2,100	570		2,670	3,150
2890	Size 3		1.10	14.545		3,100	775		3,875	4,575
2900	Size 4		.70	22.857		6,300	1,225		7,525	8,750
2910	NEMA 4, size 1	1 Elec	.90	8.889		2,175	475		2,650	3,100
2920	Size 2	2 Elec	1.40	11.429		3,175	610		3,785	4,425
2930	Size 3		1	16		5,000	855		5,855	6,775
2940	Size 4		.60	26.667		8,325	1,425		9,750	11,300
2950	NEMA 12, size 1	1 Elec	1	8		1,675	425		2,100	2,475
2960	Size 2	2 Elec	1.40	11.429		2,350	610		2,960	3,500
2970	Size 3		1	16		3,450	855		4,305	5,075
2980	Size 4		.60	26.667		6,850	1,425		8,275	9,650
3010	Manual, single phase, w/pilot, 1 pole 120 V NEMA 1	1 Elec	6.40	1.250		61	66.50		127.50	167
3020	NEMA 4		4	2		345	107		452	540
3030	2 pole, 120/240 V, NEMA 1		6.40	1.250		81.50	66.50		148	190
3040	NEMA 4		4	2		310	107		417	500
3041	3 phase, 3 pole 600 V, NEMA 1		5.50	1.455		239	77.50		316.50	380
3042	NEMA 4		3.50	2.286		465	122		587	695
3043	NEMA 12		3.50	2.286		279	122		401	490
3070	Auxiliary contact, normally open					80.50			80.50	88.50
3500	Magnetic FVNR with NEMA 12, enclosure & heaters, 480 volt									
3600	5 HP, size 0	1 Elec	2.20	3.636	Ea.	222	194		416	535
3700	10 HP, size 1	"	1.50	5.333		335	285		620	795
3800	25 HP, size 2	2 Elec	2	8		625	425		1,050	1,325
3900	50 HP, size 3		1.60	10		965	535		1,500	1,850
4000	100 HP, size 4		1	16		2,300	855		3,155	3,800
4100	200 HP, size 5		.80	20		5,525	1,075		6,600	7,675
4200	Combination, with motor circuit protectors, 5 HP, size 0	1 Elec	1.70	4.706		735	251		986	1,175
4300	10 HP, size 1	"	1.20	6.667		765	355		1,120	1,375
4400	25 HP, size 2	2 Elec	1.80	8.889		1,150	475		1,625	1,950
4500	50 HP, size 3		1.20	13.333		1,875	710		2,585	3,125
4600	100 HP, size 4		.74	21.622		4,200	1,150		5,350	6,350
4700	Combination, with fused switch, 5 HP, size 0	1 Elec	1.70	4.706		715	251		966	1,150
4800	10 HP, size 1	"	1.20	6.667		740	355		1,095	1,350
4900	25 HP, size 2	2 Elec	1.80	8.889		1,125	475		1,600	1,950
5000	50 HP, size 3		1.20	13.333		1,825	710		2,535	3,075
5100	100 HP, size 4		.74	21.622		3,675	1,150		4,825	5,775
5200	Factory installed controls, adders to size 0 thru 5									
5300	Start-stop push button	1 Elec	32	.250	Ea.	46.50	13.35		59.85	71
5400	Hand-off-auto-selector switch		32	.250		46.50	13.35		59.85	71
5500	Pilot light		32	.250		87.50	13.35		100.85	116
5600	Start-stop-pilot		32	.250		134	13.35		147.35	167
5700	Auxiliary contact, NO or NC		32	.250		64	13.35		77.35	90.50
5800	NO-NC		32	.250		128	13.35		141.35	161
5810	Magnetic FVR, NEMA 7, w/heaters, size 1		.66	12.121		2,575	645		3,220	3,825
5830	Size 2	2 Elec	1.10	14.545		4,325	775		5,100	5,900
5840	Size 3		.70	22.857		6,950	1,225		8,175	9,450
5850	Size 4		.60	26.667		7,900	1,425		9,325	10,800
5860	Combination w/circuit breakers, heaters, control XFMR PB, size 1	1 Elec	.60	13.333		1,625	710		2,335	2,875
5870	Size 2	2 Elec	.80	20		2,100	1,075		3,175	3,900
5880	Size 3		.50	32		2,800	1,700		4,500	5,625
5890	Size 4		.40	40		5,525	2,125		7,650	9,275

26 24 19.40 Motor Starters and Controls		Crew	Daily Output	Labor-Hours	Unit	Material	2014 Bare Costs Labor	Equipment	Total	Total Incl O&P
5900	Manual, 240 volt, .75 HP motor	1 Elec	4	2	Ea.	49.50	107		156.50	215
5910	2 HP motor		4	2		134	107		241	305
6000	Magnetic, 240 volt, 1 or 2 pole, .75 HP motor		4	2		195	107		302	375
6020	2 HP motor		4	2		215	107		322	395
6040	5 HP motor		3	2.667		310	142		452	555
6060	10 HP motor		2.30	3.478		765	186		951	1,125
6100	3 pole, .75 HP motor		3	2.667		195	142		337	425
6120	5 HP motor		2.30	3.478		265	186		451	570
6140	10 HP motor		1.60	5		500	267		767	945
6160	15 HP motor		1.60	5		500	267		767	945
6180	20 HP motor	▼	1.10	7.273		810	390		1,200	1,475
6200	25 HP motor	2 Elec	2.20	7.273		810	390		1,200	1,475
6210	30 HP motor		1.80	8.889		810	475		1,285	1,600
6220	40 HP motor		1.80	8.889		1,800	475		2,275	2,675
6230	50 HP motor		1.80	8.889		1,800	475		2,275	2,675
6240	60 HP motor		1.20	13.333		4,225	710		4,935	5,700
6250	75 HP motor		1.20	13.333		4,225	710		4,935	5,700
6260	100 HP motor		1.20	13.333		4,225	710		4,935	5,700
6270	125 HP motor		.90	17.778		11,800	950		12,750	14,400
6280	150 HP motor		.90	17.778		11,800	950		12,750	14,400
6290	200 HP motor	▼	.90	17.778		11,800	950		12,750	14,400
6400	Starter & nonfused disconnect, 240 volt, 1-2 pole, .75 HP motor	1 Elec	2	4		264	213		477	610
6410	2 HP motor		2	4		285	213		498	635
6420	5 HP motor		1.80	4.444		380	237		617	770
6430	10 HP motor		1.40	5.714		855	305		1,160	1,400
6440	3 pole, .75 HP motor		1.60	5		264	267		531	690
6450	5 HP motor		1.40	5.714		335	305		640	825
6460	10 HP motor		1.10	7.273		590	390		980	1,225
6470	15 HP motor	▼	1	8		590	425		1,015	1,300
6480	20 HP motor	2 Elec	1.50	10.667		1,025	570		1,595	1,975
6490	25 HP motor		1.50	10.667		1,025	570		1,595	1,975
6500	30 HP motor		1.30	12.308		1,025	655		1,680	2,100
6510	40 HP motor		1.24	12.903		2,200	690		2,890	3,450
6520	50 HP motor		1.12	14.286		2,200	760		2,960	3,575
6530	60 HP motor		.90	17.778		4,600	950		5,550	6,500
6540	75 HP motor		.76	21.053		4,600	1,125		5,725	6,750
6550	100 HP motor		.70	22.857		4,600	1,225		5,825	6,900
6560	125 HP motor		.60	26.667		12,800	1,425		14,225	16,200
6570	150 HP motor		.52	30.769		12,800	1,650		14,450	16,600
6580	200 HP motor	▼	.50	32		12,800	1,700		14,500	16,700
6600	Starter & fused disconnect, 240 volt, 1-2 pole, .75 HP motor	1 Elec	2	4		281	213		494	630
6610	2 HP motor		2	4		300	213		513	650
6620	5 HP motor		1.80	4.444		395	237		632	790
6630	10 HP motor		1.40	5.714		910	305		1,215	1,450
6640	3 pole, .75 HP motor		1.60	5		281	267		548	710
6650	5 HP motor		1.40	5.714		350	305		655	840
6660	10 HP motor		1.10	7.273		645	390		1,035	1,300
6690	15 HP motor	▼	1	8		645	425		1,070	1,350
6700	20 HP motor	2 Elec	1.60	10		1,050	535		1,585	1,975
6710	25 HP motor		1.60	10		1,050	535		1,585	1,975
6720	30 HP motor		1.40	11.429		1,050	610		1,660	2,100
6730	40 HP motor		1.20	13.333		2,325	710		3,035	3,650
6740	50 HP motor	▼	1.20	13.333	▼	2,325	710		3,035	3,650

26 24 19 – Motor-Control Centers

26 24 19.40 Motor Starters and Controls		Crew	Daily Output	Labor-Hours	Unit	Material	2014 Bare Costs Labor	Equipment	Total	Total Incl O&P
6750	60 HP motor	2 Elec	.90	17.778	Ea.	4,750	950		5,700	6,650
6760	75 HP motor		.90	17.778		4,750	950		5,700	6,650
6770	100 HP motor		.70	22.857		4,750	1,225		5,975	7,050
6780	125 HP motor		.54	29.630		13,200	1,575		14,775	16,900
6790	Combination starter & nonfusible disconnect									
6800	240 volt, 1-2 pole, .75 HP motor	1 Elec	2	4	Ea.	555	213		768	930
6810	2 HP motor		2	4		555	213		768	930
6820	5 HP motor		1.50	5.333		585	285		870	1,075
6830	10 HP motor		1.20	6.667		910	355		1,265	1,525
6840	3 pole, .75 HP motor		1.80	4.444		555	237		792	965
6850	5 HP motor		1.30	6.154		585	330		915	1,125
6860	10 HP motor		1	8		910	425		1,335	1,650
6870	15 HP motor		1	8		910	425		1,335	1,650
6880	20 HP motor	2 Elec	1.32	12.121		1,500	645		2,145	2,625
6890	25 HP motor		1.32	12.121		1,500	645		2,145	2,625
6900	30 HP motor		1.32	12.121		1,500	645		2,145	2,625
6910	40 HP motor		.80	20		2,875	1,075		3,950	4,750
6920	50 HP motor		.80	20		2,875	1,075		3,950	4,750
6930	60 HP motor		.70	22.857		6,400	1,225		7,625	8,875
6940	75 HP motor		.70	22.857		6,400	1,225		7,625	8,875
6950	100 HP motor		.70	22.857		6,400	1,225		7,625	8,875
6960	125 HP motor		.60	26.667		16,800	1,425		18,225	20,600
6970	150 HP motor		.60	26.667		16,800	1,425		18,225	20,600
6980	200 HP motor		.60	26.667		16,800	1,425		18,225	20,600
6990	Combination starter and fused disconnect									
7000	240 volt, 1-2 pole, .75 HP motor	1 Elec	2	4	Ea.	575	213		788	950
7010	2 HP motor		2	4		575	213		788	950
7020	5 HP motor		1.50	5.333		600	285		885	1,100
7030	10 HP motor		1.20	6.667		935	355		1,290	1,550
7040	3 pole, .75 HP motor		1.80	4.444		575	237		812	985
7050	5 HP motor		1.30	6.154		600	330		930	1,150
7060	10 HP motor		1	8		935	425		1,360	1,675
7070	15 HP motor		1	8		935	425		1,360	1,675
7080	20 HP motor	2 Elec	1.32	12.121		1,550	645		2,195	2,700
7090	25 HP motor		1.32	12.121		1,550	645		2,195	2,700
7100	30 HP motor		1.32	12.121		1,550	645		2,195	2,700
7110	40 HP motor		.80	20		2,975	1,075		4,050	4,850
7120	50 HP motor		.80	20		2,975	1,075		4,050	4,850
7130	60 HP motor		.80	20		6,600	1,075		7,675	8,875
7140	75 HP motor		.70	22.857		6,600	1,225		7,825	9,100
7150	100 HP motor		.70	22.857		6,600	1,225		7,825	9,100
7160	125 HP motor		.70	22.857		17,500	1,225		18,725	21,000
7170	150 HP motor		.60	26.667		17,500	1,425		18,925	21,300
7180	200 HP motor		.60	26.667		17,500	1,425		18,925	21,300
7190	Combination starter & circuit breaker disconnect									
7200	240 volt, 1-2 pole, .75 HP motor	1 Elec	2	4	Ea.	580	213		793	955
7210	2 HP motor		2	4		580	213		793	955
7220	5 HP motor		1.50	5.333		610	285		895	1,100
7230	10 HP motor		1.20	6.667		930	355		1,285	1,550
7240	3 pole, .75 HP motor		1.80	4.444		595	237		832	1,000
7250	5 HP motor		1.30	6.154		625	330		955	1,175
7260	10 HP motor		1	8		945	425		1,370	1,700
7270	15 HP motor		1	8		945	425		1,370	1,700

26 24 19.40 Motor Starters and Controls		Crew	Daily Output	Labor-Hours	Unit	Material	2014 Bare Costs Labor	Equipment	Total	Total Incl O&P
7280	20 HP motor	2 Elec	1.32	12.121	Ea.	1,600	645		2,245	2,750
7290	25 HP motor		1.32	12.121		1,600	645		2,245	2,750
7300	30 HP motor		1.32	12.121		1,600	645		2,245	2,750
7310	40 HP motor		.80	20		3,500	1,075		4,575	5,450
7320	50 HP motor		.80	20		3,500	1,075		4,575	5,450
7330	60 HP motor		.80	20		8,075	1,075		9,150	10,500
7340	75 HP motor		.70	22.857		8,075	1,225		9,300	10,700
7350	100 HP motor		.70	22.857		8,075	1,225		9,300	10,700
7360	125 HP motor		.70	22.857		17,500	1,225		18,725	21,100
7370	150 HP motor		.60	26.667		17,500	1,425		18,925	21,400
7380	200 HP motor	▼	.60	26.667	▼	17,500	1,425		18,925	21,400
7400	Magnetic FVNR with enclosure & heaters, 2 pole,									
7410	230 volt, 1 HP size 00	1 Elec	4	2	Ea.	178	107		285	355
7420	2 HP, size 0		4	2		198	107		305	380
7430	3 HP, size 1		3	2.667		227	142		369	465
7440	5 HP, size 1p		3	2.667		291	142		433	535
7450	115 volt, 1/3 HP, size 00		4	2		178	107		285	355
7460	1 HP, size 0		4	2		198	107		305	380
7470	2 HP, size 1		3	2.667		227	142		369	465
7480	3 HP, size 1P	▼	3	2.667		282	142		424	525
7500	3 pole, 480 volt, 600 HP, size 7	2 Elec	.70	22.857	▼	15,700	1,225		16,925	19,100
7590	Magnetic FVNR with heater, NEMA 1									
7600	600 volt, 3 pole, 5 HP motor	1 Elec	2.30	3.478	Ea.	236	186		422	535
7610	10 HP motor	"	1.60	5		265	267		532	690
7620	25 HP motor	2 Elec	2.20	7.273		500	390		890	1,125
7630	30 HP motor		1.80	8.889		810	475		1,285	1,600
7640	40 HP motor		1.80	8.889		810	475		1,285	1,600
7650	50 HP motor		1.80	8.889		810	475		1,285	1,600
7660	60 HP motor		1.20	13.333		1,800	710		2,510	3,050
7670	75 HP motor		1.20	13.333		2,050	710		2,760	3,325
7680	100 HP motor		1.20	13.333		2,375	710		3,085	3,700
7690	125 HP motor		.90	17.778		4,125	950		5,075	5,950
7700	150 HP motor		.90	17.778		4,325	950		5,275	6,175
7710	200 HP motor	▼	.90	17.778		4,500	950		5,450	6,375
7750	Starter & nonfused disconnect, 600 volt, 3 pole, 5 HP motor	1 Elec	1.40	5.714		370	305		675	860
7760	10 HP motor	"	1.10	7.273		400	390		790	1,025
7770	25 HP motor	2 Elec	1.50	10.667		630	570		1,200	1,550
7780	30 HP motor		1.30	12.308		1,025	655		1,680	2,100
7790	40 HP motor		1.30	12.308		1,025	655		1,680	2,100
7800	50 HP motor		1.30	12.308		1,025	655		1,680	2,100
7810	60 HP motor		.92	17.391		2,150	930		3,080	3,775
7820	75 HP motor		.92	17.391		2,400	930		3,330	4,050
7830	100 HP motor		.84	19.048		2,725	1,025		3,750	4,525
7840	125 HP motor		.70	22.857		4,650	1,225		5,875	6,950
7850	150 HP motor		.70	22.857		4,850	1,225		6,075	7,150
7860	200 HP motor	▼	.60	26.667		5,050	1,425		6,475	7,675
7870	Starter & fused disconnect, 600 volt, 3 pole, 5 HP motor	1 Elec	1.40	5.714		470	305		775	975
7880	10 HP motor	"	1.10	7.273		500	390		890	1,125
7890	25 HP motor	2 Elec	1.50	10.667		735	570		1,305	1,650
7900	30 HP motor		1.30	12.308		1,100	655		1,755	2,175
7910	40 HP motor		1.30	12.308		1,100	655		1,755	2,175
7920	50 HP motor		1.30	12.308		1,100	655		1,755	2,175
7930	60 HP motor	▼	.92	17.391		2,325	930		3,255	3,950

26 24 19.40 Motor Starters and Controls		Crew	Daily Output	Labor-Hours	Unit	Material	2014 Bare Costs Labor	Equipment	Total	Total Incl O&P
7940	75 HP motor	2 Elec	.92	17.391	Ea.	2,575	930		3,505	4,225
7950	100 HP motor		.84	19.048		2,900	1,025		3,925	4,725
7960	125 HP motor		.70	22.857		4,875	1,225		6,100	7,200
7970	150 HP motor		.70	22.857		5,075	1,225		6,300	7,400
7980	200 HP motor		.60	26.667		5,275	1,425		6,700	7,925
7990	Combination starter and nonfusible disconnect									
8000	600 volt, 3 pole, 5 HP motor	1 Elec	1.80	4.444	Ea.	555	237		792	965
8010	10 HP motor	"	1.30	6.154		690	330		1,020	1,250
8020	25 HP motor	2 Elec	2	8		910	425		1,335	1,650
8030	30 HP motor		1.32	12.121		1,550	645		2,195	2,700
8040	40 HP motor		1.32	12.121		1,550	645		2,195	2,700
8050	50 HP motor		1.32	12.121		1,850	645		2,495	3,000
8060	60 HP motor		.80	20		2,975	1,075		4,050	4,850
8070	75 HP motor		.80	20		2,975	1,075		4,050	4,850
8080	100 HP motor		.80	20		3,675	1,075		4,750	5,650
8090	125 HP motor		.70	22.857		6,425	1,225		7,650	8,875
8100	150 HP motor		.70	22.857		6,700	1,225		7,925	9,200
8110	200 HP motor		.70	22.857		7,000	1,225		8,225	9,525
8140	Combination starter and fused disconnect									
8150	600 volt, 3 pole, 5 HP motor	1 Elec	1.80	4.444	Ea.	575	237		812	985
8160	10 HP motor	"	1.30	6.154		625	330		955	1,175
8170	25 HP motor	2 Elec	2	8		1,025	425		1,450	1,775
8180	30 HP motor		1.32	12.121		1,575	645		2,220	2,725
8190	40 HP motor		1.32	12.121		1,700	645		2,345	2,850
8200	50 HP motor		1.32	12.121		1,700	645		2,345	2,850
8210	60 HP motor		.80	20		3,000	1,075		4,075	4,875
8220	75 HP motor		.80	20		3,000	1,075		4,075	4,875
8230	100 HP motor		.80	20		3,000	1,075		4,075	4,875
8240	125 HP motor		.70	22.857		6,600	1,225		7,825	9,100
8250	150 HP motor		.70	22.857		6,600	1,225		7,825	9,100
8260	200 HP motor		.70	22.857		6,600	1,225		7,825	9,100
8290	Combination starter & circuit breaker disconnect									
8300	600 volt, 3 pole, 5 HP motor	1 Elec	1.80	4.444	Ea.	765	237		1,002	1,200
8310	10 HP motor	"	1.30	6.154		795	330		1,125	1,375
8320	25 HP motor	2 Elec	2	8		1,125	425		1,550	1,875
8330	30 HP motor		1.32	12.121		1,600	645		2,245	2,750
8340	40 HP motor		1.32	12.121		1,600	645		2,245	2,750
8350	50 HP motor		1.32	12.121		1,600	645		2,245	2,750
8360	60 HP motor		.80	20		3,500	1,075		4,575	5,450
8370	75 HP motor		.80	20		3,500	1,075		4,575	5,450
8380	100 HP motor		.80	20		3,500	1,075		4,575	5,450
8390	125 HP motor		.70	22.857		8,075	1,225		9,300	10,700
8400	150 HP motor		.70	22.857		8,075	1,225		9,300	10,700
8410	200 HP motor		.70	22.857		8,075	1,225		9,300	10,700
8430	Starter & circuit breaker disconnect									
8440	600 volt, 3 pole, 5 HP motor	1 Elec	1.40	5.714	Ea.	735	305		1,040	1,250
8450	10 HP motor	"	1.10	7.273		760	390		1,150	1,425
8460	25 HP motor	2 Elec	1.50	10.667		995	570		1,565	1,950
8470	30 HP motor		1.30	12.308		1,425	655		2,080	2,550
8480	40 HP motor		1.30	12.308		1,425	655		2,080	2,550
8490	50 HP motor		1.30	12.308		1,425	655		2,080	2,550
8500	60 HP motor		.92	17.391		3,425	930		4,355	5,150
8510	75 HP motor		.92	17.391		3,675	930		4,605	5,425

26 24 19.40 Motor Starters and Controls	Crew	Daily Output	Labor-Hours	Unit	Material	2014 Bare Costs Labor	2014 Bare Costs Equipment	Total	Total Incl O&P
8520	100 HP motor	2 Elec	.84	19.048	Ea.	4,000	1,025		5,025	5,925
8530	125 HP motor		.70	22.857		5,750	1,225		6,975	8,150
8540	150 HP motor		.70	22.857		5,925	1,225		7,150	8,350
8550	200 HP motor		.60	26.667		7,275	1,425		8,700	10,100
8900	240 volt, 1-2 pole, .75 HP motor	1 Elec	2	4		690	213		903	1,075
8910	2 HP motor		2	4		715	213		928	1,100
8920	5 HP motor		1.80	4.444		805	237		1,042	1,250
8930	10 HP motor		1.40	5.714		1,375	305		1,680	1,975
8950	3 pole, .75 HP motor		1.60	5		690	267		957	1,150
8970	5 HP motor		1.40	5.714		760	305		1,065	1,300
8980	10 HP motor		1.10	7.273		1,100	390		1,490	1,800
8990	15 HP motor		1	8		1,100	425		1,525	1,875
9100	20 HP motor	2 Elec	1.50	10.667		1,500	570		2,070	2,500
9110	25 HP motor		1.50	10.667		1,500	570		2,070	2,500
9120	30 HP motor		1.30	12.308		1,500	655		2,155	2,625
9130	40 HP motor		1.24	12.903		3,425	690		4,115	4,775
9140	50 HP motor		1.12	14.286		3,425	760		4,185	4,900
9150	60 HP motor		.90	17.778		6,975	950		7,925	9,100
9160	75 HP motor		.76	21.053		6,975	1,125		8,100	9,350
9170	100 HP motor		.70	22.857		6,975	1,225		8,200	9,500
9180	125 HP motor		.60	26.667		15,800	1,425		17,225	19,500
9190	150 HP motor		.52	30.769		15,800	1,650		17,450	19,900
9200	200 HP motor		.50	32		15,800	1,700		17,500	20,000

26 25 Enclosed Bus Assemblies
26 25 13 – Bus Duct/Busway and Fittings
26 25 13.10 Aluminum Bus Duct

		Crew	Daily Output	Labor-Hours	Unit	Material	2014 Bare Costs Labor	2014 Bare Costs Equipment	Total	Total Incl O&P
0010	**ALUMINUM BUS DUCT** 10 ft. long	R262513-10								
0050	Indoor 3 pole 4 wire, plug-in, straight section, 225 amp	2 Elec	44	.364	L.F.	113	19.40		132.40	153
0100	400 amp		36	.444		133	23.50		156.50	182
0150	600 amp		32	.500		150	26.50		176.50	205
0200	800 amp	R262513-15	26	.615		169	33		202	235
0250	1000 amp		24	.667		194	35.50		229.50	266
0300	1350 amp		22	.727		249	39		288	330
0310	1600 amp		18	.889		264	47.50		311.50	360
0320	2000 amp		16	1		345	53.50		398.50	460
0330	2500 amp		14	1.143		395	61		456	525
0340	3000 amp		12	1.333		480	71		551	635
0350	Feeder, 600 amp		34	.471		109	25		134	158
0400	800 amp		28	.571		128	30.50		158.50	186
0450	1000 amp		26	.615		146	33		179	209
0455	1200 amp		25	.640		146	34		180	211
0500	1350 amp		24	.667		228	35.50		263.50	305
0550	1600 amp		20	.800		264	42.50		306.50	355
0600	2000 amp		18	.889		310	47.50		357.50	410
0620	2500 amp		14	1.143		400	61		461	530
0630	3000 amp		12	1.333		455	71		526	605
0640	4000 amp		10	1.600		655	85.50		740.50	850
0650	Elbow, 225 amp		4.40	3.636	Ea.	645	194		839	1,000
0700	400 amp		3.80	4.211		645	225		870	1,050
0750	600 amp		3.40	4.706		645	251		896	1,075

26 25 13 – Bus Duct/Busway and Fittings

26 25 13.10 Aluminum Bus Duct		Crew	Daily Output	Labor-Hours	Unit	Material	2014 Bare Costs Labor	Equipment	Total	Total Incl O&P
0800	800 amp	2 Elec	3	5.333	Ea.	675	285		960	1,175
0850	1000 amp		2.80	5.714		870	305		1,175	1,425
0870	1200 amp		2.70	5.926		1,075	315		1,390	1,650
0900	1350 amp		2.60	6.154		1,150	330		1,480	1,775
0950	1600 amp		2.40	6.667		1,375	355		1,730	2,025
1000	2000 amp		2	8		1,500	425		1,925	2,325
1020	2500 amp		1.80	8.889		1,775	475		2,250	2,675
1030	3000 amp		1.60	10		2,050	535		2,585	3,075
1040	4000 amp		1.40	11.429		3,325	610		3,935	4,575
1100	Cable tap box end, 225 amp		3.60	4.444		905	237		1,142	1,350
1150	400 amp		3.20	5		905	267		1,172	1,400
1200	600 amp		2.60	6.154		905	330		1,235	1,475
1250	800 amp		2.20	7.273		985	390		1,375	1,650
1300	1000 amp		2	8		1,075	425		1,500	1,825
1320	1200 amp		2	8		1,275	425		1,700	2,050
1350	1350 amp		1.60	10		1,400	535		1,935	2,350
1400	1600 amp		1.40	11.429		1,575	610		2,185	2,650
1450	2000 amp		1.20	13.333		1,800	710		2,510	3,075
1460	2500 amp		1	16		2,200	855		3,055	3,675
1470	3000 amp		.80	20		2,425	1,075		3,500	4,275
1480	4000 amp		.60	26.667		3,225	1,425		4,650	5,675
1500	Switchboard stub, 225 amp		5.80	2.759		795	147		942	1,100
1550	400 amp		5.40	2.963		795	158		953	1,100
1600	600 amp		4.60	3.478		795	186		981	1,150
1650	800 amp		4	4		900	213		1,113	1,300
1700	1000 amp		3.20	5		1,000	267		1,267	1,500
1720	1200 amp		3.10	5.161		1,125	275		1,400	1,625
1750	1350 amp		3	5.333		1,250	285		1,535	1,800
1800	1600 amp		2.60	6.154		1,375	330		1,705	2,025
1850	2000 amp		2.40	6.667		1,575	355		1,930	2,275
1860	2500 amp		2.20	7.273		1,900	390		2,290	2,650
1870	3000 amp		2	8		2,200	425		2,625	3,075
1880	4000 amp		1.80	8.889		2,750	475		3,225	3,725
1890	Tee fittings, 225 amp		3.20	5		865	267		1,132	1,350
1900	400 amp		2.80	5.714		865	305		1,170	1,400
1950	600 amp		2.60	6.154		865	330		1,195	1,450
2000	800 amp		2.40	6.667		920	355		1,275	1,525
2050	1000 amp		2.20	7.273		975	390		1,365	1,650
2070	1200 amp		2.10	7.619		1,125	405		1,530	1,850
2100	1350 amp		2	8		1,575	425		2,000	2,375
2150	1600 amp		1.60	10		1,900	535		2,435	2,875
2200	2000 amp		1.20	13.333		2,100	710		2,810	3,375
2220	2500 amp		1	16		2,500	855		3,355	4,050
2230	3000 amp		.80	20		2,875	1,075		3,950	4,750
2240	4000 amp		.60	26.667		4,775	1,425		6,200	7,375
2300	Wall flange, 600 amp		20	.800		194	42.50		236.50	277
2310	800 amp		16	1		194	53.50		247.50	293
2320	1000 amp		13	1.231		194	65.50		259.50	310
2325	1200 amp		12	1.333		194	71		265	320
2330	1350 amp		10.80	1.481		194	79		273	330
2340	1600 amp		9	1.778		194	95		289	355
2350	2000 amp		8	2		194	107		301	375
2360	2500 amp		6.60	2.424		194	129		323	405

26 25 13.10 Aluminum Bus Duct		Crew	Daily Output	Labor-Hours	Unit	Material	2014 Bare Costs Labor	Equipment	Total	Total Incl O&P
2370	3000 amp	2 Elec	5.40	2.963	Ea.	194	158		352	450
2380	4000 amp		4	4		281	213		494	630
2390	5000 amp		3	5.333		281	285		566	735
2400	Vapor barrier		8	2		355	107		462	550
2420	Roof flange kit		4	4		630	213		843	1,025
2600	Expansion fitting, 225 amp		10	1.600		1,050	85.50		1,135.50	1,300
2610	400 amp		8	2		1,050	107		1,157	1,325
2620	600 amp		6	2.667		1,050	142		1,192	1,400
2630	800 amp		4.60	3.478		1,225	186		1,411	1,625
2640	1000 amp		4	4		1,375	213		1,588	1,825
2650	1350 amp		3.60	4.444		2,000	237		2,237	2,550
2660	1600 amp		3.20	5		2,400	267		2,667	3,050
2670	2000 amp		2.80	5.714		2,675	305		2,980	3,400
2680	2500 amp		2.40	6.667		3,250	355		3,605	4,100
2690	3000 amp		2	8		3,725	425		4,150	4,750
2700	4000 amp		1.60	10		4,900	535		5,435	6,200
2800	Reducer nonfused, 400 amp		8	2		870	107		977	1,125
2810	600 amp		6	2.667		870	142		1,012	1,175
2820	800 amp		4.60	3.478		1,050	186		1,236	1,425
2830	1000 amp		4	4		1,225	213		1,438	1,650
2840	1350 amp		3.60	4.444		1,600	237		1,837	2,100
2850	1600 amp		3.20	5		2,150	267		2,417	2,775
2860	2000 amp		2.80	5.714		2,475	305		2,780	3,175
2870	2500 amp		2.40	6.667		3,100	355		3,455	3,950
2880	3000 amp		2	8		3,600	425		4,025	4,600
2890	4000 amp		1.60	10		4,800	535		5,335	6,100
2950	Reducer fuse included, 225 amp		4.40	3.636		2,550	194		2,744	3,100
2960	400 amp		4.20	3.810		2,575	203		2,778	3,150
2970	600 amp		3.60	4.444		3,050	237		3,287	3,700
2980	800 amp		3.20	5		4,850	267		5,117	5,750
2990	1000 amp		3	5.333		5,550	285		5,835	6,525
3000	1200 amp		2.80	5.714		5,550	305		5,855	6,550
3010	1600 amp		2.20	7.273		12,700	390		13,090	14,500
3020	2000 amp		1.80	8.889		14,100	475		14,575	16,200
3100	Reducer circuit breaker, 225 amp		4.40	3.636		2,500	194		2,694	3,050
3110	400 amp		4.20	3.810		3,050	203		3,253	3,650
3120	600 amp		3.60	4.444		4,325	237		4,562	5,125
3130	800 amp		3.20	5		5,075	267		5,342	5,975
3140	1000 amp		3	5.333		5,750	285		6,035	6,750
3150	1200 amp		2.80	5.714		6,925	305		7,230	8,075
3160	1600 amp		2.20	7.273		10,200	390		10,590	11,900
3170	2000 amp		1.80	8.889		11,200	475		11,675	13,000
3250	Reducer circuit breaker, 75,000 AIC, 225 amp		4.40	3.636		3,925	194		4,119	4,600
3260	400 amp		4.20	3.810		3,925	203		4,128	4,600
3270	600 amp		3.60	4.444		5,250	237		5,487	6,100
3280	800 amp		3.20	5		5,750	267		6,017	6,725
3290	1000 amp		3	5.333		9,250	285		9,535	10,600
3300	1200 amp		2.80	5.714		9,250	305		9,555	10,700
3310	1600 amp		2.20	7.273		10,200	390		10,590	11,900
3320	2000 amp		1.80	8.889		11,200	475		11,675	13,000
3400	Reducer circuit breaker CLF 225 amp		4.40	3.636		4,025	194		4,219	4,725
3410	400 amp		4.20	3.810		4,775	203		4,978	5,575
3420	600 amp		3.60	4.444		7,150	237		7,387	8,225

26 25 13.10 Aluminum Bus Duct		Crew	Daily Output	Labor-Hours	Unit	Material	2014 Bare Costs Labor	Equipment	Total	Total Incl O&P
3430	800 amp	2 Elec	3.20	5	Ea.	7,475	267		7,742	8,625
3440	1000 amp		3	5.333		7,800	285		8,085	9,000
3450	1200 amp		2.80	5.714		10,200	305		10,505	11,700
3460	1600 amp		2.20	7.273		10,200	390		10,590	11,900
3470	2000 amp		1.80	8.889		11,200	475		11,675	13,000
3550	Ground bus added to bus duct, 225 amp		320	.050	L.F.	32.50	2.67		35.17	39.50
3560	400 amp		320	.050		32.50	2.67		35.17	39.50
3570	600 amp		280	.057		32.50	3.05		35.55	40
3580	800 amp		240	.067		32.50	3.56		36.06	41
3590	1000 amp		200	.080		32.50	4.27		36.77	42
3600	1350 amp		180	.089		32.50	4.74		37.24	42.50
3610	1600 amp		160	.100		32.50	5.35		37.85	43.50
3620	2000 amp		160	.100		32.50	5.35		37.85	43.50
3630	2500 amp		140	.114		32.50	6.10		38.60	44.50
3640	3000 amp		120	.133		32.50	7.10		39.60	46
3650	4000 amp		100	.160		32.50	8.55		41.05	48.50
3810	High short circuit, 400 amp		36	.444		109	23.50		132.50	156
3820	600 amp		32	.500		109	26.50		135.50	160
3830	800 amp		26	.615		128	33		161	189
3840	1000 amp		24	.667		137	35.50		172.50	203
3850	1350 amp		22	.727		182	39		221	258
3860	1600 amp		18	.889		210	47.50		257.50	300
3870	2000 amp		16	1		246	53.50		299.50	350
3880	2500 amp		14	1.143		410	61		471	540
3890	3000 amp		12	1.333		465	71		536	615
3920	Cross, 225 amp		5.60	2.857	Ea.	1,300	152		1,452	1,650
3930	400 amp		4.60	3.478		1,300	186		1,486	1,700
3940	600 amp		4	4		1,300	213		1,513	1,750
3950	800 amp		3.40	4.706		1,375	251		1,626	1,875
3960	1000 amp		3	5.333		1,425	285		1,710	2,000
3970	1350 amp		2.80	5.714		2,325	305		2,630	3,000
3980	1600 amp		2.20	7.273		2,750	390		3,140	3,600
3990	2000 amp		1.80	8.889		3,025	475		3,500	4,025
4000	2500 amp		1.60	10		3,575	535		4,110	4,750
4010	3000 amp		1.20	13.333		4,125	710		4,835	5,600
4020	4000 amp		1	16		6,300	855		7,155	8,200
4040	Cable tap box center, 225 amp		3.60	4.444		905	237		1,142	1,350
4050	400 amp		3.20	5		905	267		1,172	1,400
4060	600 amp		2.60	6.154		905	330		1,235	1,475
4070	800 amp		2.20	7.273		985	390		1,375	1,650
4080	1000 amp		2	8		1,075	425		1,500	1,825
4090	1350 amp		1.60	10		1,400	535		1,935	2,350
4100	1600 amp		1.40	11.429		1,575	610		2,185	2,650
4110	2000 amp		1.20	13.333		1,800	710		2,510	3,075
4120	2500 amp		1	16		2,200	855		3,055	3,675
4130	3000 amp		.80	20		2,425	1,075		3,500	4,275
4140	4000 amp		.60	26.667		3,225	1,425		4,650	5,675
4500	Weatherproof 3 pole 4 wire, feeder, 600 amp		30	.533	L.F.	131	28.50		159.50	187
4520	800 amp		24	.667		153	35.50		188.50	222
4540	1000 amp		22	.727		175	39		214	250
4550	1200 amp		21	.762		242	40.50		282.50	325
4560	1350 amp		20	.800		273	42.50		315.50	365
4580	1600 amp		17	.941		315	50		365	425

26 25 13.10 Aluminum Bus Duct	Crew	Daily Output	Labor-Hours	Unit	Material	2014 Bare Costs Labor	Equipment	Total	Total Incl O&P
4600	2000 amp	2 Elec	16	1	L.F.	370	53.50		423.50	490
4620	2500 amp		12	1.333		480	71		551	635
4640	3000 amp		10	1.600		545	85.50		630.50	730
4660	4000 amp		8	2		785	107		892	1,025
5000	Indoor 3 pole, 3 wire, feeder, 600 amp		40	.400		100	21.50		121.50	142
5010	800 amp		32	.500		118	26.50		144.50	170
5020	1000 amp		30	.533		128	28.50		156.50	183
5025	1200 amp		29	.552		128	29.50		157.50	184
5030	1350 amp		28	.571		173	30.50		203.50	236
5040	1600 amp		24	.667		200	35.50		235.50	274
5050	2000 amp		20	.800		237	42.50		279.50	325
5060	2500 amp		16	1		330	53.50		383.50	440
5070	3000 amp		14	1.143		390	61		451	520
5080	4000 amp		12	1.333		475	71		546	625
5200	Plug-in type, 225 amp		50	.320		109	17.05		126.05	146
5210	400 amp		42	.381		109	20.50		129.50	151
5220	600 amp		36	.444		109	23.50		132.50	156
5230	800 amp		30	.533		128	28.50		156.50	183
5240	1000 amp		28	.571		137	30.50		167.50	196
5245	1200 amp		27	.593		155	31.50		186.50	218
5250	1350 amp		26	.615		182	33		215	249
5260	1600 amp		20	.800		210	42.50		252.50	295
5270	2000 amp		18	.889		246	47.50		293.50	340
5280	2500 amp		16	1		335	53.50		388.50	450
5290	3000 amp		14	1.143		400	61		461	530
5300	4000 amp		12	1.333		485	71		556	635
5330	High short circuit, 400 amp		42	.381		109	20.50		129.50	151
5340	600 amp		36	.444		109	23.50		132.50	156
5350	800 amp		30	.533		128	28.50		156.50	183
5360	1000 amp		28	.571		137	30.50		167.50	196
5370	1350 amp		26	.615		182	33		215	249
5380	1600 amp		20	.800		210	42.50		252.50	295
5390	2000 amp		18	.889		246	47.50		293.50	340
5400	2500 amp		16	1		385	53.50		438.50	505
5410	3000 amp		14	1.143		400	61		461	530
5440	Elbow, 225 amp		5	3.200	Ea.	555	171		726	865
5450	400 amp		4.40	3.636		555	194		749	900
5460	600 amp		4	4		555	213		768	930
5470	800 amp		3.40	4.706		595	251		846	1,025
5480	1000 amp		3.20	5		610	267		877	1,075
5485	1200 amp		3.10	5.161		650	275		925	1,125
5490	1350 amp		3	5.333		705	285		990	1,200
5500	1600 amp		2.80	5.714		1,075	305		1,380	1,625
5510	2000 amp		2.40	6.667		1,175	355		1,530	1,825
5520	2500 amp		2	8		1,450	425		1,875	2,250
5530	3000 amp		1.80	8.889		1,725	475		2,200	2,600
5540	4000 amp		1.60	10		2,300	535		2,835	3,350
5560	Tee fittings, 225 amp		3.60	4.444		775	237		1,012	1,200
5570	400 amp		3.20	5		775	267		1,042	1,250
5580	600 amp		3	5.333		775	285		1,060	1,275
5590	800 amp		2.80	5.714		830	305		1,135	1,375
5600	1000 amp		2.60	6.154		855	330		1,185	1,425
5605	1200 amp		2.50	6.400		915	340		1,255	1,500

26 25 13.10 Aluminum Bus Duct		Crew	Daily Output	Labor-Hours	Unit	Material	2014 Bare Costs Labor	Equipment	Total	Total Incl O&P
5610	1350 amp	2 Elec	2.40	6.667	Ea.	1,275	355		1,630	1,925
5620	1600 amp		1.80	8.889		1,500	475		1,975	2,350
5630	2000 amp		1.40	11.429		1,675	610		2,285	2,750
5640	2500 amp		1.20	13.333		2,075	710		2,785	3,350
5650	3000 amp		1	16		2,425	855		3,280	3,950
5660	4000 amp		.70	22.857		3,550	1,225		4,775	5,750
5680	Cross, 225 amp		6.40	2.500		1,250	133		1,383	1,575
5690	400 amp		5.40	2.963		1,250	158		1,408	1,600
5700	600 amp		4.60	3.478		1,250	186		1,436	1,650
5710	800 amp		4	4		1,325	213		1,538	1,775
5720	1000 amp		3.60	4.444		1,375	237		1,612	1,850
5730	1350 amp		3.20	5		2,025	267		2,292	2,625
5740	1600 amp		2.60	6.154		2,375	330		2,705	3,100
5750	2000 amp		2.20	7.273		2,600	390		2,990	3,425
5760	2500 amp		1.80	8.889		3,125	475		3,600	4,150
5770	3000 amp		1.40	11.429		3,725	610		4,335	5,025
5780	4000 amp		1.20	13.333		5,200	710		5,910	6,800
5800	Expansion fitting, 225 amp		11.60	1.379		910	73.50		983.50	1,100
5810	400 amp		9.20	1.739		910	93		1,003	1,150
5820	600 amp		7	2.286		910	122		1,032	1,175
5830	800 amp		5.20	3.077		1,075	164		1,239	1,425
5840	1000 amp		4.60	3.478		1,175	186		1,361	1,575
5850	1350 amp		4.20	3.810		1,425	203		1,628	1,875
5860	1600 amp		3.60	4.444		1,775	237		2,012	2,300
5870	2000 amp		3.20	5		2,100	267		2,367	2,700
5880	2500 amp		2.80	5.714		2,375	305		2,680	3,050
5890	3000 amp		2.40	6.667		2,775	355		3,130	3,575
5900	4000 amp		1.80	8.889		3,800	475		4,275	4,875
5940	Reducer, nonfused, 400 amp		9.20	1.739		745	93		838	960
5950	600 amp		7	2.286		745	122		867	1,000
5960	800 amp		5.20	3.077		780	164		944	1,100
5970	1000 amp		4.60	3.478		960	186		1,146	1,325
5980	1350 amp		4.20	3.810		1,425	203		1,628	1,875
5990	1600 amp		3.60	4.444		1,625	237		1,862	2,125
6000	2000 amp		3.20	5		1,900	267		2,167	2,475
6010	2500 amp		2.80	5.714		2,425	305		2,730	3,125
6020	3000 amp		2.20	7.273		2,825	390		3,215	3,700
6030	4000 amp		1.80	8.889		3,900	475		4,375	4,975
6050	Reducer, fuse included, 225 amp		5	3.200		1,950	171		2,121	2,400
6060	400 amp		4.80	3.333		2,575	178		2,753	3,125
6070	600 amp		4.20	3.810		3,325	203		3,528	3,950
6080	800 amp		3.60	4.444		5,125	237		5,362	6,000
6090	1000 amp		3.40	4.706		5,600	251		5,851	6,525
6100	1350 amp		3.20	5		10,200	267		10,467	11,700
6110	1600 amp		2.60	6.154		12,100	330		12,430	13,800
6120	2000 amp		2	8		14,100	425		14,525	16,100
6160	Reducer, circuit breaker, 225 amp		5	3.200		2,400	171		2,571	2,875
6170	400 amp		4.80	3.333		2,925	178		3,103	3,500
6180	600 amp		4.20	3.810		4,225	203		4,428	4,925
6190	800 amp		3.60	4.444		4,925	237		5,162	5,775
6200	1000 amp		3.40	4.706		5,600	251		5,851	6,525
6210	1350 amp		3.20	5		6,775	267		7,042	7,850
6220	1600 amp		2.60	6.154		10,100	330		10,430	11,600

26 25 13.10 Aluminum Bus Duct

		Crew	Daily Output	Labor-Hours	Unit	Material	2014 Bare Costs Labor	Equipment	Total	Total Incl O&P
6230	2000 amp	2 Elec	2	8	Ea.	11,000	425		11,425	12,700
6270	Cable tap box center, 225 amp		4.20	3.810		875	203		1,078	1,275
6280	400 amp		3.60	4.444		875	237		1,112	1,325
6290	600 amp		3	5.333		875	285		1,160	1,375
6300	800 amp		2.60	6.154		955	330		1,285	1,550
6310	1000 amp		2.40	6.667		1,000	355		1,355	1,625
6320	1350 amp		1.80	8.889		1,225	475		1,700	2,050
6330	1600 amp		1.60	10		1,375	535		1,910	2,300
6340	2000 amp		1.40	11.429		1,575	610		2,185	2,650
6350	2500 amp		1.20	13.333		1,950	710		2,660	3,225
6360	3000 amp		1	16		2,200	855		3,055	3,700
6370	4000 amp		.70	22.857		2,625	1,225		3,850	4,700
6390	Cable tap box end, 225 amp		4.20	3.810		535	203		738	890
6400	400 amp		3.60	4.444		535	237		772	940
6410	600 amp		3	5.333		535	285		820	1,000
6420	800 amp		2.60	6.154		585	330		915	1,125
6430	1000 amp		2.40	6.667		630	355		985	1,225
6435	1200 amp		2.10	7.619		685	405		1,090	1,350
6440	1350 amp		1.80	8.889		760	475		1,235	1,550
6450	1600 amp		1.60	10		860	535		1,395	1,750
6460	2000 amp		1.40	11.429		970	610		1,580	2,000
6470	2500 amp		1.20	13.333		1,125	710		1,835	2,300
6480	3000 amp		1	16		1,325	855		2,180	2,750
6490	4000 amp		.70	22.857	▼	1,600	1,225		2,825	3,575
7000	Weatherproof 3 pole 3 wire, feeder, 600 amp		34	.471	L.F.	129	25		154	180
7020	800 amp		28	.571		142	30.50		172.50	203
7040	1000 amp		26	.615		153	33		186	218
7050	1200 amp		25	.640		179	34		213	248
7060	1350 amp		24	.667		208	35.50		243.50	282
7080	1600 amp		20	.800		240	42.50		282.50	330
7100	2000 amp		18	.889		284	47.50		331.50	385
7120	2500 amp		14	1.143		395	61		456	525
7140	3000 amp		12	1.333		470	71		541	625
7160	4000 amp	▼	10	1.600	▼	570	85.50		655.50	755

26 25 13.20 Bus Duct

		Crew	Daily Output	Labor-Hours	Unit	Material	2014 Bare Costs Labor	Equipment	Total	Total Incl O&P
0010	**BUS DUCT** 100 amp and less, aluminum or copper, plug-in									
0080	Bus duct, 3 pole 3 wire, 100 amp	1 Elec	42	.190	L.F.	22	10.15		32.15	39
0110	Elbow		4	2	Ea.	57	107		164	223
0120	Tee		2	4		83	213		296	410
0130	Wall flange		8	1		7.25	53.50		60.75	88
0140	Ground kit		16	.500	▼	15.30	26.50		41.80	57
0180	3 pole 4 wire, 100 amp		40	.200	L.F.	25	10.65		35.65	43.50
0200	Cable tap box		3.10	2.581	Ea.	68	138		206	281
0300	End closure		16	.500		9.10	26.50		35.60	50
0400	Elbow		4	2		57	107		164	223
0500	Tee		2	4		82.50	213		295.50	410
0600	Hangers		10	.800		5.45	42.50		47.95	70
0700	Circuit breakers, 15 to 50 amp, 1 pole		8	1		150	53.50		203.50	245
0800	15 to 60 amp, 2 pole		6.70	1.194		430	63.50		493.50	570
0900	3 pole		5.30	1.509		430	80.50		510.50	595
1000	60 to 100 amp, 1 pole		6.70	1.194		585	63.50		648.50	740
1100	70 to 100 amp, 2 pole	▼	5.30	1.509	▼	585	80.50		665.50	765

26 25 13 – Bus Duct/Busway and Fittings

26 25 13.20 Bus Duct		Crew	Daily Output	Labor-Hours	Unit	Material	2014 Bare Costs Labor	Equipment	Total	Total Incl O&P
1200	3 pole	1 Elec	4.50	1.778	Ea.	580	95		675	780
1220	Switch, nonfused, 3 pole, 4 wire		8	1		69.50	53.50		123	157
1240	Fused, 3 fuses, 4 wire, 30 amp		8	1		252	53.50		305.50	355
1260	60 amp		5.30	1.509		265	80.50		345.50	410
1280	100 amp		4.50	1.778		440	95		535	625
1300	Plug, fusible, 3 pole 250 volt, 30 amp		5.30	1.509		252	80.50		332.50	400
1310	60 amp		5.30	1.509		299	80.50		379.50	450
1320	100 amp		4.50	1.778		440	95		535	625
1330	3 pole 480 volt, 30 amp		5.30	1.509		288	80.50		368.50	435
1340	60 amp		5.30	1.509		310	80.50		390.50	460
1350	100 amp		4.50	1.778		450	95		545	635
1360	Circuit breaker, 3 pole 250 volt, 60 amp		5.30	1.509		510	80.50		590.50	685
1370	3 pole 480 volt, 100 amp		4.50	1.778	▼	510	95		605	705
2000	Bus duct, 2 wire, 250 volt, 30 amp		60	.133	L.F.	2.79	7.10		9.89	13.70
2100	60 amp		50	.160		2.79	8.55		11.34	15.85
2200	300 volt, 30 amp		60	.133		2.79	7.10		9.89	13.70
2300	60 amp		50	.160		2.79	8.55		11.34	15.85
2400	3 wire, 250 volt, 30 amp		60	.133		2.79	7.10		9.89	13.70
2500	60 amp		50	.160		2.79	8.55		11.34	15.85
2600	480/277 volt, 30 amp		60	.133		2.79	7.10		9.89	13.70
2700	60 amp		50	.160	▼	2.79	8.55		11.34	15.85
2750	End feed, 300 volt 2 wire max. 30 amp		6	1.333	Ea.	41.50	71		112.50	152
2800	60 amp		5.50	1.455		41.50	77.50		119	162
2850	30 amp miniature		6	1.333		41.50	71		112.50	152
2900	3 wire, 30 amp		6	1.333		52.50	71		123.50	164
2950	60 amp		5.50	1.455		52.50	77.50		130	174
3000	30 amp miniature		6	1.333		52.50	71		123.50	164
3050	Center feed, 300 volt 2 wire, 30 amp		6	1.333		57.50	71		128.50	170
3100	60 amp		5.50	1.455		57.50	77.50		135	180
3150	3 wire, 30 amp		6	1.333		65.50	71		136.50	178
3200	60 amp		5.50	1.455		65.50	77.50		143	188
3220	Elbow, 30 amp		6	1.333		40	71		111	150
3240	60 amp		5.50	1.455		40	77.50		117.50	160
3260	End cap		40	.200		8	10.65		18.65	25
3280	Strength beam, 10 ft.		15	.533		22.50	28.50		51	67
3300	Hanger		24	.333		4.80	17.80		22.60	32
3320	Tap box, nonfusible		6.30	1.270		68	68		136	176
3340	Fusible switch 30 amp, 1 fuse		6	1.333		296	71		367	430
3360	2 fuse		6	1.333		300	71		371	435
3380	3 fuse		6	1.333		305	71		376	440
3400	Circuit breaker handle on cover, 1 pole		6	1.333		47	71		118	158
3420	2 pole		6	1.333		57.50	71		128.50	170
3440	3 pole		6	1.333		73.50	71		144.50	187
3460	Circuit breaker external operhandle, 1 pole		6	1.333		48.50	71		119.50	160
3480	2 pole		6	1.333		58.50	71		129.50	170
3500	3 pole		6	1.333		85	71		156	200
3520	Terminal plug only		16	.500		8.80	26.50		35.30	49.50
3540	Terminal with receptacle		16	.500		11.40	26.50		37.90	52.50
3560	Fixture plug		16	.500	▼	7.85	26.50		34.35	48.50
4000	Copper bus duct, lighting, 2 wire 300 volt, 20 amp		70	.114	L.F.	5.15	6.10		11.25	14.80
4020	35 amp		60	.133		5.15	7.10		12.25	16.30
4040	50 amp		55	.145		5.15	7.75		12.90	17.25
4060	60 amp		50	.160	▼	5.15	8.55		13.70	18.45

26 25 Enclosed Bus Assemblies

26 25 13 – Bus Duct/Busway and Fittings

26 25 13.20 Bus Duct

		Crew	Daily Output	Labor-Hours	Unit	Material	2014 Bare Costs Labor	Equipment	Total	Total Incl O&P
4080	3 wire 300 volt, 20 amp	1 Elec	70	.114	L.F.	5.60	6.10		11.70	15.30
4100	35 amp		60	.133		5.60	7.10		12.70	16.80
4120	50 amp		55	.145		5.60	7.75		13.35	17.75
4140	60 amp		50	.160		5.60	8.55		14.15	18.95
4160	Feeder in box, end, 1 circuit		6	1.333	Ea.	77	71		148	191
4180	2 circuit		5.50	1.455		79.50	77.50		157	204
4200	Center, 1 circuit		6	1.333		105	71		176	222
4220	2 circuit		5.50	1.455		109	77.50		186.50	236
4240	End cap		40	.200		12.85	10.65		23.50	30
4260	Hanger, surface mount		24	.333		7.70	17.80		25.50	35
4280	Coupling		40	.200		9.75	10.65		20.40	26.50

26 25 13.30 Copper Bus Duct

		Crew	Daily Output	Labor-Hours	Unit	Material	2014 Bare Costs Labor	Equipment	Total	Total Incl O&P
0010	**COPPER BUS DUCT** R262513-15									
0100	Weatherproof 3 pole 4 wire, feeder duct, 600 amp	2 Elec	24	.667	L.F.	249	35.50		284.50	325
0110	800 amp		18	.889		340	47.50		387.50	445
0120	1000 amp		17	.941		380	50		430	495
0125	1200 amp		16.50	.970		445	51.50		496.50	570
0130	1350 amp		16	1		465	53.50		518.50	590
0140	1600 amp		12	1.333		500	71		571	650
0150	2000 amp		10	1.600		555	85.50		640.50	740
0160	2500 amp		7	2.286		605	122		727	850
0170	3000 amp		5	3.200		835	171		1,006	1,175
0180	4000 amp		3.60	4.444		1,300	237		1,537	1,800
0200	Indoor 3 pole 4 wire, plug-in, bus duct high short circuit, 400 amp		32	.500		239	26.50		265.50	305
0210	600 amp		26	.615		239	33		272	310
0220	800 amp		20	.800		284	42.50		326.50	375
0230	1000 amp		18	.889		315	47.50		362.50	415
0240	1350 amp		16	1		425	53.50		478.50	550
0250	1600 amp		12	1.333		480	71		551	635
0260	2000 amp		10	1.600		610	85.50		695.50	800
0270	2500 amp		8	2		750	107		857	985
0280	3000 amp		6	2.667		860	142		1,002	1,150
0310	Cross, 225 amp		3	5.333	Ea.	2,825	285		3,110	3,550
0320	400 amp		2.80	5.714		2,825	305		3,130	3,575
0330	600 amp		2.60	6.154		2,825	330		3,155	3,625
0340	800 amp		2.20	7.273		3,075	390		3,465	3,950
0350	1000 amp		2	8		3,425	425		3,850	4,425
0360	1350 amp		1.80	8.889		3,675	475		4,150	4,750
0370	1600 amp		1.70	9.412		4,100	500		4,600	5,250
0380	2000 amp		1.60	10		6,725	535		7,260	8,200
0390	2500 amp		1.40	11.429		8,125	610		8,735	9,875
0400	3000 amp		1.20	13.333		8,900	710		9,610	10,900
0410	4000 amp		1	16		11,600	855		12,455	14,100
0430	Expansion fitting, 225 amp		5.40	2.963		1,475	158		1,633	1,850
0440	400 amp		4.60	3.478		1,675	186		1,861	2,125
0450	600 amp		4	4		2,075	213		2,288	2,625
0460	800 amp		3.40	4.706		2,450	251		2,701	3,050
0470	1000 amp		3	5.333		2,825	285		3,110	3,525
0480	1350 amp		2.80	5.714		3,350	305		3,655	4,125
0490	1600 amp		2.60	6.154		4,675	330		5,005	5,650
0500	2000 amp		2.20	7.273		5,375	390		5,765	6,500
0510	2500 amp		1.80	8.889		6,550	475		7,025	7,900

26 25 13 – Bus Duct/Busway and Fittings

26 25 13.30 Copper Bus Duct	Crew	Daily Output	Labor-Hours	Unit	Material	2014 Bare Costs Labor	Equipment	Total	Total Incl O&P
0520	3000 amp	2 Elec	1.60	10	Ea.	7,325	535		7,860	8,875
0530	4000 amp		1.20	13.333		9,450	710		10,160	11,500
0550	Reducer nonfused, 225 amp		5.40	2.963		1,750	158		1,908	2,125
0560	400 amp		4.60	3.478		1,750	186		1,936	2,175
0570	600 amp		4	4		1,750	213		1,963	2,225
0580	800 amp		3.40	4.706		2,100	251		2,351	2,700
0590	1000 amp		3	5.333		2,525	285		2,810	3,200
0600	1350 amp		2.80	5.714		3,650	305		3,955	4,450
0610	1600 amp		2.60	6.154		4,175	330		4,505	5,100
0620	2000 amp		2.20	7.273		5,025	390		5,415	6,125
0630	2500 amp		1.80	8.889		6,350	475		6,825	7,675
0640	3000 amp		1.60	10		7,150	535		7,685	8,675
0650	4000 amp		1.20	13.333		9,325	710		10,035	11,400
0670	Reducer fuse included, 225 amp		4.40	3.636		3,725	194		3,919	4,400
0680	400 amp		4.20	3.810		4,650	203		4,853	5,425
0690	600 amp		3.60	4.444		5,750	237		5,987	6,675
0700	800 amp		3.20	5		8,100	267		8,367	9,325
0710	1000 amp		3	5.333		10,200	285		10,485	11,600
0720	1350 amp		2.80	5.714		15,800	305		16,105	17,900
0730	1600 amp		2.20	7.273		21,400	390		21,790	24,100
0740	2000 amp		1.80	8.889		24,100	475		24,575	27,200
0790	Reducer, circuit breaker, 225 amp		4.40	3.636		4,700	194		4,894	5,475
0800	400 amp		4.20	3.810		5,500	203		5,703	6,350
0810	600 amp		3.60	4.444		7,800	237		8,037	8,925
0820	800 amp		3.20	5		9,150	267		9,417	10,500
0830	1000 amp		3	5.333		10,400	285		10,685	11,800
0840	1350 amp		2.80	5.714		11,900	305		12,205	13,500
0850	1600 amp		2.20	7.273		17,500	390		17,890	19,900
0860	2000 amp		1.80	8.889		19,200	475		19,675	21,800
0910	Cable tap box, center, 225 amp		3.20	5		1,700	267		1,967	2,275
0920	400 amp		2.60	6.154		1,700	330		2,030	2,375
0930	600 amp		2.20	7.273		1,700	390		2,090	2,450
0940	800 amp		2	8		1,900	425		2,325	2,725
0950	1000 amp		1.60	10		2,050	535		2,585	3,050
0960	1350 amp		1.40	11.429		2,475	610		3,085	3,625
0970	1600 amp		1.20	13.333		2,750	710		3,460	4,100
0980	2000 amp		1	16		3,300	855		4,155	4,900
1040	2500 amp		.80	20		3,925	1,075		5,000	5,900
1060	3000 amp		.60	26.667		4,275	1,425		5,700	6,850
1080	4000 amp		.40	40		5,400	2,125		7,525	9,150
1800	Weatherproof 3 pole 3 wire, feeder duct, 600 amp		28	.571	L.F.	251	30.50		281.50	320
1820	800 amp		22	.727		305	39		344	395
1840	1000 amp		20	.800		340	42.50		382.50	440
1850	1200 amp		19	.842		350	45		395	450
1860	1350 amp		18	.889		475	47.50		522.50	595
1880	1600 amp		14	1.143		545	61		606	690
1900	2000 amp		12	1.333		700	71		771	870
1920	2500 amp		8	2		870	107		977	1,125
1940	3000 amp		6	2.667		1,000	142		1,142	1,325
1960	4000 amp		4	4		1,325	213		1,538	1,775
2000	Indoor 3 pole 3 wire, feeder duct, 600 amp		32	.500		209	26.50		235.50	270
2010	800 amp		26	.615		254	33		287	330
2020	1000 amp		24	.667		284	35.50		319.50	365

26 25 13.30 Copper Bus Duct		Crew	Daily Output	Labor-Hours	Unit	Material	2014 Bare Costs Labor	Equipment	Total	Total Incl O&P
2025	1200 amp	2 Elec	22	.727	L.F.	375	39		414	470
2030	1350 amp		20	.800		395	42.50		437.50	500
2040	1600 amp		16	1		455	53.50		508.50	580
2050	2000 amp		14	1.143		580	61		641	730
2060	2500 amp		10	1.600		725	85.50		810.50	925
2070	3000 amp		8	2		835	107		942	1,075
2080	4000 amp		6	2.667		1,100	142		1,242	1,425
2090	5000 amp		5	3.200		1,350	171		1,521	1,725
2200	Indoor 3 pole 3 wire, bus duct plug-in, 225 amp		46	.348		239	18.55		257.55	291
2210	400 amp		36	.444		239	23.50		262.50	299
2220	600 amp		30	.533		239	28.50		267.50	305
2230	800 amp		24	.667		284	35.50		319.50	365
2240	1000 amp		20	.800		315	42.50		357.50	410
2250	1350 amp		18	.889		425	47.50		472.50	540
2260	1600 amp		14	1.143		480	61		541	620
2270	2000 amp		12	1.333		610	71		681	775
2280	2500 amp		10	1.600		750	85.50		835.50	955
2290	3000 amp		8	2		860	107		967	1,100
2330	High short circuit, 400 amp		36	.444		239	23.50		262.50	299
2340	600 amp		30	.533		239	28.50		267.50	305
2350	800 amp		24	.667		284	35.50		319.50	365
2360	1000 amp		20	.800		315	42.50		357.50	410
2370	1350 amp		18	.889		425	47.50		472.50	540
2380	1600 amp		14	1.143		480	61		541	620
2390	2000 amp		12	1.333		610	71		681	775
2400	2500 amp		10	1.600		750	85.50		835.50	955
2410	3000 amp		8	2		860	107		967	1,100
2440	Elbows, 225 amp		4.60	3.478	Ea.	1,200	186		1,386	1,600
2450	400 amp		4.20	3.810		1,200	203		1,403	1,625
2460	600 amp		3.60	4.444		1,200	237		1,437	1,675
2470	800 amp		3.20	5		1,275	267		1,542	1,800
2480	1000 amp		3	5.333		1,350	285		1,635	1,900
2485	1200 amp		2.90	5.517		1,450	294		1,744	2,050
2490	1350 amp		2.80	5.714		1,525	305		1,830	2,125
2500	1600 amp		2.60	6.154		1,650	330		1,980	2,300
2510	2000 amp		2	8		2,000	425		2,425	2,850
2520	2500 amp		1.80	8.889		3,025	475		3,500	4,025
2530	3000 amp		1.60	10		3,325	535		3,860	4,450
2540	4000 amp		1.40	11.429		4,250	610		4,860	5,600
2560	Tee fittings, 225 amp		2.80	5.714		1,400	305		1,705	2,000
2570	400 amp		2.40	6.667		1,400	355		1,755	2,075
2580	600 amp		2	8		1,400	425		1,825	2,200
2590	800 amp		1.80	8.889		1,525	475		2,000	2,400
2600	1000 amp		1.60	10		1,725	535		2,260	2,675
2605	1200 amp		1.50	10.667		1,825	570		2,395	2,875
2610	1350 amp		1.40	11.429		1,900	610		2,510	3,025
2620	1600 amp		1.20	13.333		2,225	710		2,935	3,525
2630	2000 amp		1	16		3,700	855		4,555	5,350
2640	2500 amp		.70	22.857		4,325	1,225		5,550	6,600
2650	3000 amp		.60	26.667		4,775	1,425		6,200	7,375
2660	4000 amp		.50	32		6,150	1,700		7,850	9,325
2680	Cross, 225 amp		3.60	4.444		2,150	237		2,387	2,725
2690	400 amp		3.20	5		2,150	267		2,417	2,775

26 25 13 – Bus Duct/Busway and Fittings

26 25 13.30 Copper Bus Duct		Crew	Daily Output	Labor-Hours	Unit	Material	2014 Bare Costs Labor	Equipment	Total	Total Incl O&P
2700	600 amp	2 Elec	3	5.333	Ea.	2,150	285		2,435	2,800
2710	800 amp		2.60	6.154		2,575	330		2,905	3,325
2720	1000 amp		2.40	6.667		2,675	355		3,030	3,475
2730	1350 amp		2.20	7.273		3,075	390		3,465	3,950
2740	1600 amp		2	8		3,300	425		3,725	4,275
2750	2000 amp		1.80	8.889		5,200	475		5,675	6,425
2760	2500 amp		1.60	10		6,075	535		6,610	7,475
2770	3000 amp		1.40	11.429		6,625	610		7,235	8,225
2780	4000 amp		1	16		8,475	855		9,330	10,600
2800	Expansion fitting, 225 amp		6.40	2.500		1,725	133		1,858	2,100
2810	400 amp		5.40	2.963		1,725	158		1,883	2,125
2820	600 amp		4.60	3.478		1,725	186		1,911	2,175
2830	800 amp		4	4		2,050	213		2,263	2,575
2840	1000 amp		3.60	4.444		2,250	237		2,487	2,825
2850	1350 amp		3.20	5		2,675	267		2,942	3,350
2860	1600 amp		3	5.333		2,925	285		3,210	3,650
2870	2000 amp		2.60	6.154		3,550	330		3,880	4,400
2880	2500 amp		2.20	7.273		4,950	390		5,340	6,000
2890	3000 amp		1.80	8.889		5,500	475		5,975	6,775
2900	4000 amp		1.40	11.429		7,050	610		7,660	8,675
2920	Reducer nonfused, 225 amp		6.40	2.500		1,425	133		1,558	1,775
2930	400 amp		5.40	2.963		1,425	158		1,583	1,800
2940	600 amp		4.60	3.478		1,425	186		1,611	1,850
2950	800 amp		4	4		1,675	213		1,888	2,150
2960	1000 amp		3.60	4.444		1,875	237		2,112	2,400
2970	1350 amp		3.20	5		2,350	267		2,617	3,000
2980	1600 amp		3	5.333		2,675	285		2,960	3,350
2990	2000 amp		2.60	6.154		3,200	330		3,530	4,025
3000	2500 amp		2.20	7.273		4,675	390		5,065	5,725
3010	3000 amp		1.80	8.889		5,250	475		5,725	6,475
3020	4000 amp		1.40	11.429		6,825	610		7,435	8,425
3040	Reducer fuse included, 225 amp		5	3.200		3,375	171		3,546	3,975
3050	400 amp		4.80	3.333		4,500	178		4,678	5,225
3060	600 amp		4.20	3.810		5,350	203		5,553	6,200
3070	800 amp		3.60	4.444		7,625	237		7,862	8,750
3080	1000 amp		3.40	4.706		9,050	251		9,301	10,300
3090	1350 amp		3.20	5		8,750	267		9,017	10,000
3100	1600 amp		2.60	6.154		20,400	330		20,730	22,900
3110	2000 amp		2	8		23,100	425		23,525	26,000
3160	Reducer circuit breaker, 225 amp		5	3.200		4,325	171		4,496	5,000
3170	400 amp		4.80	3.333		5,300	178		5,478	6,100
3180	600 amp		4.20	3.810		7,600	203		7,803	8,650
3190	800 amp		3.60	4.444		8,875	237		9,112	10,100
3200	1000 amp		3.40	4.706		10,100	251		10,351	11,500
3210	1350 amp		3.20	5		11,600	267		11,867	13,100
3220	1600 amp		2.60	6.154		17,300	330		17,630	19,500
3230	2000 amp		2	8		18,900	425		19,325	21,400
3280	3 pole, 3 wire, cable tap box center, 225 amp		3.60	4.444		1,950	237		2,187	2,500
3290	400 amp		3	5.333		1,950	285		2,235	2,575
3300	600 amp		2.60	6.154		1,950	330		2,280	2,650
3310	800 amp		2.40	6.667		2,175	355		2,530	2,925
3320	1000 amp		1.80	8.889		2,350	475		2,825	3,300
3330	1350 amp		1.60	10		2,900	535		3,435	3,975

26 25 13.30 Copper Bus Duct		Crew	Daily Output	Labor-Hours	Unit	Material	2014 Bare Costs Labor	Equipment	Total	Total Incl O&P
3340	1600 amp	2 Elec	1.40	11.429	Ea.	3,225	610		3,835	4,475
3350	2000 amp		1.20	13.333		3,925	710		4,635	5,375
3360	2500 amp		1	16		4,675	855		5,530	6,400
3370	3000 amp		.70	22.857		5,150	1,225		6,375	7,475
3380	4000 amp		.50	32		6,525	1,700		8,225	9,725
3400	Cable tap box end, 225 amp		3.60	4.444		1,050	237		1,287	1,500
3410	400 amp		3	5.333		1,050	285		1,335	1,575
3420	600 amp		2.60	6.154		1,175	330		1,505	1,800
3430	800 amp		2.40	6.667		1,175	355		1,530	1,825
3440	1000 amp		1.80	8.889		1,300	475		1,775	2,125
3445	1200 amp		1.70	9.412		1,475	500		1,975	2,350
3450	1350 amp		1.60	10		1,550	535		2,085	2,500
3460	1600 amp		1.40	11.429		1,775	610		2,385	2,875
3470	2000 amp		1.20	13.333		2,025	710		2,735	3,300
3480	2500 amp		1	16		2,425	855		3,280	3,950
3490	3000 amp		.70	22.857		2,650	1,225		3,875	4,750
3500	4000 amp	▼	.50	32		3,375	1,700		5,075	6,250
4600	Plug-in, fusible switch w/3 fuses, 3 pole, 250 volt, 30 amp	1 Elec	4	2		425	107		532	625
4610	60 amp		3.60	2.222		550	119		669	780
4620	100 amp	▼	2.70	2.963		790	158		948	1,100
4630	200 amp	2 Elec	3.20	5		1,325	267		1,592	1,850
4640	400 amp		1.40	11.429		3,450	610		4,060	4,725
4650	600 amp	▼	.90	17.778		4,800	950		5,750	6,700
4700	4 pole, 120/208 volt, 30 amp	1 Elec	3.90	2.051		580	109		689	800
4710	60 amp		3.50	2.286		625	122		747	870
4720	100 amp	▼	2.60	3.077		870	164		1,034	1,200
4730	200 amp	2 Elec	3	5.333		1,450	285		1,735	2,025
4740	400 amp		1.30	12.308		3,425	655		4,080	4,725
4750	600 amp	▼	.80	20		4,800	1,075		5,875	6,875
4800	3 pole, 480 volt, 30 amp	1 Elec	4	2		435	107		542	635
4810	60 amp		3.60	2.222		460	119		579	680
4820	100 amp	▼	2.70	2.963		775	158		933	1,075
4830	200 amp	2 Elec	3.20	5		1,325	267		1,592	1,850
4840	400 amp		1.40	11.429		3,100	610		3,710	4,325
4850	600 amp		.90	17.778		4,400	950		5,350	6,275
4860	800 amp		.66	24.242		15,100	1,300		16,400	18,600
4870	1000 amp		.60	26.667		16,100	1,425		17,525	19,800
4880	1200 amp		.50	32		16,900	1,700		18,600	21,200
4890	1600 amp	▼	.44	36.364		18,600	1,950		20,550	23,400
4900	4 pole, 277/480 volt, 30 amp	1 Elec	3.90	2.051		625	109		734	855
4910	60 amp		3.50	2.286		670	122		792	920
4920	100 amp	▼	2.60	3.077		975	164		1,139	1,325
4930	200 amp	2 Elec	3	5.333		1,950	285		2,235	2,550
4940	400 amp		1.30	12.308		3,650	655		4,305	5,000
4950	600 amp		.80	20		5,000	1,075		6,075	7,100
5050	800 amp		.60	26.667		16,500	1,425		17,925	20,200
5060	1000 amp		.56	28.571		19,000	1,525		20,525	23,200
5070	1200 amp		.48	33.333		19,100	1,775		20,875	23,700
5080	1600 amp	▼	.42	38.095		21,100	2,025		23,125	26,300
5150	Fusible with starter, 3 pole 250 volt, 30 amp	1 Elec	3.50	2.286		2,325	122		2,447	2,750
5160	60 amp		3.20	2.500		2,475	133		2,608	2,925
5170	100 amp		2.50	3.200		2,800	171		2,971	3,325
5180	200 amp	2 Elec	2.80	5.714		4,600	305		4,905	5,525

26 25 13.30 Copper Bus Duct		Crew	Daily Output	Labor-Hours	Unit	Material	2014 Bare Costs Labor	Equipment	Total	Total Incl O&P
5200	3 pole 480 volt, 30 amp	1 Elec	3.50	2.286	Ea.	2,325	122		2,447	2,750
5210	60 amp		3.20	2.500		2,475	133		2,608	2,925
5220	100 amp		2.50	3.200		2,800	171		2,971	3,325
5230	200 amp	2 Elec	2.80	5.714		4,600	305		4,905	5,525
5300	Fusible with contactor, 3 pole 250 volt, 30 amp	1 Elec	3.50	2.286		2,275	122		2,397	2,675
5310	60 amp		3.20	2.500		2,900	133		3,033	3,375
5320	100 amp		2.50	3.200		4,050	171		4,221	4,700
5330	200 amp	2 Elec	2.80	5.714		4,625	305		4,930	5,550
5400	3 pole 480 volt, 30 amp	1 Elec	3.50	2.286		2,450	122		2,572	2,875
5410	60 amp		3.20	2.500		3,450	133		3,583	4,000
5420	100 amp		2.50	3.200		4,750	171		4,921	5,475
5430	200 amp	2 Elec	2.80	5.714		4,850	305		5,155	5,800
5450	Fusible with capacitor, 3 pole 250 volt, 30 amp	1 Elec	3	2.667		5,900	142		6,042	6,725
5460	60 amp		2	4		6,875	213		7,088	7,900
5500	3 pole 480 volt, 30 amp		3	2.667		4,950	142		5,092	5,675
5510	60 amp		2	4		6,175	213		6,388	7,125
5600	Circuit breaker, 3 pole, 250 volt, 60 amp		4.50	1.778		595	95		690	795
5610	100 amp		3.20	2.500		730	133		863	1,000
5650	4 pole, 120/208 volt, 60 amp		4.40	1.818		670	97		767	885
5660	100 amp		3.10	2.581		795	138		933	1,075
5700	3 pole, 4 wire 277/480 volt, 60 amp		4.30	1.860		905	99.50		1,004.50	1,150
5710	100 amp		3	2.667		1,000	142		1,142	1,325
5720	225 amp	2 Elec	3.20	5		2,225	267		2,492	2,850
5730	400 amp		1.20	13.333		4,625	710		5,335	6,175
5740	600 amp		.96	16.667		6,225	890		7,115	8,175
5750	700 amp		.60	26.667		7,875	1,425		9,300	10,800
5760	800 amp		.60	26.667		7,875	1,425		9,300	10,800
5770	900 amp		.54	29.630		10,400	1,575		11,975	13,800
5780	1000 amp		.54	29.630		10,400	1,575		11,975	13,800
5790	1200 amp		.42	38.095		12,500	2,025		14,525	16,800
5810	Circuit breaker w/HIC fuses, 3 pole 480 volt, 60 amp	1 Elec	4.40	1.818		1,150	97		1,247	1,400
5820	100 amp	"	3.10	2.581		1,250	138		1,388	1,575
5830	225 amp	2 Elec	3.40	4.706		3,975	251		4,226	4,750
5840	400 amp		1.40	11.429		6,325	610		6,935	7,875
5850	600 amp		1	16		6,400	855		7,255	8,325
5860	700 amp		.64	25		8,475	1,325		9,800	11,300
5870	800 amp		.64	25		8,475	1,325		9,800	11,300
5880	900 amp		.56	28.571		18,400	1,525		19,925	22,500
5890	1000 amp		.56	28.571		18,400	1,525		19,925	22,500
5950	3 pole 4 wire 277/480 volt, 60 amp	1 Elec	4.30	1.860		1,150	99.50		1,249.50	1,400
5960	100 amp	"	3	2.667		1,250	142		1,392	1,600
5970	225 amp	2 Elec	3	5.333		3,975	285		4,260	4,800
5980	400 amp		1.10	14.545		6,325	775		7,100	8,100
5990	600 amp		.94	17.021		6,400	910		7,310	8,400
6000	700 amp		.58	27.586		8,475	1,475		9,950	11,500
6010	800 amp		.58	27.586		8,475	1,475		9,950	11,500
6020	900 amp		.52	30.769		18,400	1,650		20,050	22,700
6030	1000 amp		.52	30.769		18,400	1,650		20,050	22,700
6040	1200 amp		.40	40		18,400	2,125		20,525	23,400
6100	Circuit breaker with starter, 3 pole 250 volt, 60 amp	1 Elec	3.20	2.500		1,625	133		1,758	1,975
6110	100 amp	"	2.50	3.200		2,250	171		2,421	2,725
6120	225 amp	2 Elec	3	5.333		2,875	285		3,160	3,600
6130	3 pole 480 volt, 60 amp	1 Elec	3.20	2.500		1,625	133		1,758	1,975

26 25 13 – Bus Duct/Busway and Fittings

26 25 13.30 Copper Bus Duct		Crew	Daily Output	Labor-Hours	Unit	Material	2014 Bare Costs Labor	Equipment	Total	Total Incl O&P
6140	100 amp	1 Elec	2.50	3.200	Ea.	2,250	171		2,421	2,725
6150	225 amp	2 Elec	3	5.333		2,750	285		3,035	3,450
6200	Circuit breaker with contactor, 3 pole 250 volt, 60 amp	1 Elec	3.20	2.500		1,525	133		1,658	1,875
6210	100 amp	"	2.50	3.200		2,075	171		2,246	2,525
6220	225 amp	2 Elec	3	5.333		2,675	285		2,960	3,375
6250	3 pole 480 volt, 60 amp	1 Elec	3.20	2.500		1,525	133		1,658	1,875
6260	100 amp	"	2.50	3.200		2,075	171		2,246	2,525
6270	225 amp	2 Elec	3	5.333		2,375	285		2,660	3,025
6300	Circuit breaker with capacitor, 3 pole 250 volt, 60 amp	1 Elec	2	4		7,675	213		7,888	8,775
6310	3 pole 480 volt, 60 amp		2	4		7,700	213		7,913	8,800
6400	Add control transformer with pilot light to above starter		16	.500		455	26.50		481.50	540
6410	Switch, fusible, mechanically held contactor optional		16	.500		1,200	26.50		1,226.50	1,375
6430	Circuit breaker, mechanically held contactor optional		16	.500		1,200	26.50		1,226.50	1,375
6450	Ground neutralizer, 3 pole		16	.500		55.50	26.50		82	101

26 25 13.40 Copper Bus Duct		Crew	Daily Output	Labor-Hours	Unit	Material	2014 Bare Costs Labor	Equipment	Total	Total Incl O&P
0010	**COPPER BUS DUCT** 10 ft. long									
0050	Indoor 3 pole 4 wire, plug-in, straight section, 225 amp	2 Elec	40	.400	L.F.	230	21.50		251.50	285
1000	400 amp		32	.500		230	26.50		256.50	293
1500	600 amp		26	.615		230	33		263	300
2400	800 amp		20	.800		273	42.50		315.50	365
2450	1000 amp		18	.889		300	47.50		347.50	400
2470	1200 amp		17	.941		390	50		440	500
2500	1350 amp		16	1		410	53.50		463.50	530
2510	1600 amp		12	1.333		465	71		536	615
2520	2000 amp		10	1.600		590	85.50		675.50	775
2530	2500 amp		8	2		725	107		832	955
2540	3000 amp		6	2.667		830	142		972	1,125
2550	Feeder, 600 amp		28	.571		202	30.50		232.50	268
2600	800 amp		22	.727		245	39		284	325
2700	1000 amp		20	.800		273	42.50		315.50	365
2750	1200 amp		19	.842		360	45		405	460
2800	1350 amp		18	.889		385	47.50		432.50	490
2900	1600 amp		14	1.143		440	61		501	570
3000	2000 amp		12	1.333		560	71		631	720
3010	2500 amp		8	2		695	107		802	925
3020	3000 amp		6	2.667		805	142		947	1,100
3030	4000 amp		4	4		1,050	213		1,263	1,500
3040	5000 amp		2	8		1,300	425		1,725	2,075
3100	Elbows, 225 amp		4	4	Ea.	1,375	213		1,588	1,825
3200	400 amp		3.60	4.444		1,375	237		1,612	1,850
3300	600 amp		3.20	5		1,375	267		1,642	1,900
3400	800 amp		2.80	5.714		1,475	305		1,780	2,075
3500	1000 amp		2.60	6.154		1,650	330		1,980	2,325
3550	1200 amp		2.50	6.400		1,825	340		2,165	2,500
3600	1350 amp		2.40	6.667		1,850	355		2,205	2,550
3700	1600 amp		2.20	7.273		2,000	390		2,390	2,775
3800	2000 amp		1.80	8.889		2,475	475		2,950	3,425
3810	2500 amp		1.60	10		3,900	535		4,435	5,100
3820	3000 amp		1.40	11.429		4,300	610		4,910	5,650
3830	4000 amp		1.20	13.333		5,550	710		6,260	7,175
3840	5000 amp		1	16		8,975	855		9,830	11,200
4000	End box, 225 amp		34	.471		184	25		209	241

26 25 13 – Bus Duct/Busway and Fittings

26 25 13.40 Copper Bus Duct	Crew	Daily Output	Labor-Hours	Unit	Material	2014 Bare Costs Labor	Equipment	Total	Total Incl O&P
4100	400 amp	2 Elec	32	.500	Ea.	184	26.50		210.50	243
4200	600 amp		28	.571		184	30.50		214.50	249
4300	800 amp		26	.615		184	33		217	252
4400	1000 amp		24	.667		184	35.50		219.50	256
4410	1200 amp		23	.696		184	37		221	259
4500	1350 amp		22	.727		175	39		214	251
4600	1600 amp		20	.800		175	42.50		217.50	257
4700	2000 amp		18	.889		215	47.50		262.50	305
4710	2500 amp		16	1		215	53.50		268.50	315
4720	3000 amp		14	1.143		203	61		264	315
4730	4000 amp		12	1.333		246	71		317	375
4740	5000 amp		10	1.600		246	85.50		331.50	400
4800	Cable tap box end, 225 amp		3.20	5		1,500	267		1,767	2,050
5000	400 amp		2.60	6.154		1,500	330		1,830	2,150
5100	600 amp		2.20	7.273		1,500	390		1,890	2,225
5200	800 amp		2	8		1,725	425		2,150	2,550
5300	1000 amp		1.60	10		1,850	535		2,385	2,825
5350	1200 amp		1.50	10.667		1,950	570		2,520	3,000
5400	1350 amp		1.40	11.429		2,175	610		2,785	3,325
5500	1600 amp		1.20	13.333		2,450	710		3,160	3,775
5600	2000 amp		1	16		2,725	855		3,580	4,275
5610	2500 amp		.80	20		3,025	1,075		4,100	4,925
5620	3000 amp		.60	26.667		3,525	1,425		4,950	6,000
5630	4000 amp		.40	40		4,075	2,125		6,200	7,675
5640	5000 amp		.20	80		4,800	4,275		9,075	11,700
5700	Switchboard stub, 225 amp		5.40	2.963		1,375	158		1,533	1,750
5800	400 amp		4.60	3.478		1,375	186		1,561	1,800
5900	600 amp		4	4		1,375	213		1,588	1,850
6000	800 amp		3.20	5		1,675	267		1,942	2,225
6100	1000 amp		3	5.333		1,925	285		2,210	2,550
6150	1200 amp		2.80	5.714		2,250	305		2,555	2,925
6200	1350 amp		2.60	6.154		2,375	330		2,705	3,100
6300	1600 amp		2.40	6.667		2,675	355		3,030	3,475
6400	2000 amp		2	8		3,250	425		3,675	4,200
6410	2500 amp		1.80	8.889		3,925	475		4,400	5,025
6420	3000 amp		1.60	10		4,375	535		4,910	5,600
6430	4000 amp		1.40	11.429		5,675	610		6,285	7,175
6440	5000 amp		1.20	13.333		6,975	710		7,685	8,750
6490	Tee fittings, 225 amp		2.40	6.667		1,875	355		2,230	2,600
6500	400 amp		2	8		1,875	425		2,300	2,725
6600	600 amp		1.80	8.889		1,875	475		2,350	2,775
6700	800 amp		1.60	10		2,175	535		2,710	3,175
6750	1000 amp		1.40	11.429		2,500	610		3,110	3,675
6770	1200 amp		1.30	12.308		2,800	655		3,455	4,050
6800	1350 amp		1.20	13.333		2,950	710		3,660	4,325
7000	1600 amp		1	16		3,375	855		4,230	4,975
7100	2000 amp		.80	20		4,000	1,075		5,075	6,000
7110	2500 amp		.60	26.667		4,900	1,425		6,325	7,525
7120	3000 amp		.50	32		5,425	1,700		7,125	8,500
7130	4000 amp		.40	40		7,000	2,125		9,125	10,900
7140	5000 amp		.20	80		8,275	4,275		12,550	15,500
7200	Plug-in fusible switches w/3 fuses, 600 volt, 3 pole, 30 amp	1 Elec	4	2		840	107		947	1,075
7300	60 amp		3.60	2.222		945	119		1,064	1,225

26 25 13.40 Copper Bus Duct		Crew	Daily Output	Labor-Hours	Unit	Material	2014 Bare Costs Labor	Equipment	Total	Total Incl O&P
7400	100 amp	1 Elec	2.70	2.963	Ea.	1,450	158		1,608	1,800
7500	200 amp	2 Elec	3.20	5		2,575	267		2,842	3,225
7600	400 amp		1.40	11.429		7,550	610		8,160	9,225
7700	600 amp		.90	17.778		8,550	950		9,500	10,900
7800	800 amp		.66	24.242		12,000	1,300		13,300	15,100
7900	1200 amp		.50	32		22,500	1,700		24,200	27,400
7910	1600 amp		.44	36.364		21,400	1,950		23,350	26,400
8000	Plug-in circuit breakers, molded case, 15 to 50 amp	1 Elec	4.40	1.818		795	97		892	1,025
8100	70 to 100 amp	"	3.10	2.581		885	138		1,023	1,175
8200	150 to 225 amp	2 Elec	3.40	4.706		2,400	251		2,651	3,025
8300	250 to 400 amp		1.40	11.429		4,200	610		4,810	5,550
8400	500 to 600 amp		1	16		5,675	855		6,530	7,525
8500	700 to 800 amp		.64	25		7,000	1,325		8,325	9,700
8600	900 to 1000 amp		.56	28.571		10,000	1,525		11,525	13,300
8700	1200 amp		.44	36.364		12,000	1,950		13,950	16,200
8720	1400 amp		.40	40		16,900	2,125		19,025	21,800
8730	1600 amp		.40	40		18,500	2,125		20,625	23,600
8750	Circuit breakers, with current limiting fuse, 15 to 50 amp	1 Elec	4.40	1.818		1,600	97		1,697	1,900
8760	70 to 100 amp	"	3.10	2.581		1,900	138		2,038	2,275
8770	150 to 225 amp	2 Elec	3.40	4.706		4,050	251		4,301	4,850
8780	250 to 400 amp		1.40	11.429		6,275	610		6,885	7,825
8790	500 to 600 amp		1	16		7,225	855		8,080	9,225
8800	700 to 800 amp		.64	25		11,900	1,325		13,225	15,100
8810	900 to 1000 amp		.56	28.571		13,600	1,525		15,125	17,200
8850	Combination starter FVNR, fusible switch, NEMA size 0, 30 amp	1 Elec	2	4		2,150	213		2,363	2,700
8860	NEMA size 1, 60 amp		1.80	4.444		2,275	237		2,512	2,850
8870	NEMA size 2, 100 amp		1.30	6.154		2,875	330		3,205	3,650
8880	NEMA size 3, 200 amp	2 Elec	2	8		4,575	425		5,000	5,675
8900	Circuit breaker, NEMA size 0, 30 amp	1 Elec	2	4		2,200	213		2,413	2,750
8910	NEMA size 1, 60 amp		1.80	4.444		2,300	237		2,537	2,875
8920	NEMA size 2, 100 amp		1.30	6.154		3,300	330		3,630	4,125
8930	NEMA size 3, 200 amp	2 Elec	2	8		4,175	425		4,600	5,250
8950	Combination contactor, fusible switch, NEMA size 0, 30 amp	1 Elec	2	4		1,250	213		1,463	1,700
8960	NEMA size 1, 60 amp		1.80	4.444		1,300	237		1,537	1,775
8970	NEMA size 2, 100 amp		1.30	6.154		1,925	330		2,255	2,600
8980	NEMA size 3, 200 amp	2 Elec	2	8		2,225	425		2,650	3,075
9000	Circuit breaker, NEMA size 0, 30 amp	1 Elec	2	4		1,450	213		1,663	1,900
9010	NEMA size 1, 60 amp		1.80	4.444		1,475	237		1,712	1,975
9020	NEMA size 2, 100 amp		1.30	6.154		2,275	330		2,605	3,000
9030	NEMA size 3, 200 amp	2 Elec	2	8		2,775	425		3,200	3,700
9050	Control transformer for above, NEMA size 0, 30 amp	1 Elec	8	1		243	53.50		296.50	350
9060	NEMA size 1, 60 amp		8	1		243	53.50		296.50	350
9070	NEMA size 2, 100 amp		7	1.143		340	61		401	465
9080	NEMA size 3, 200 amp	2 Elec	14	1.143		470	61		531	610
9100	Comb. fusible switch & lighting control, electrically held, 30 amp	1 Elec	2	4		995	213		1,208	1,425
9110	60 amp		1.80	4.444		1,425	237		1,662	1,925
9120	100 amp		1.30	6.154		1,850	330		2,180	2,525
9130	200 amp	2 Elec	2	8		4,550	425		4,975	5,650
9150	Mechanically held, 30 amp	1 Elec	2	4		1,250	213		1,463	1,700
9160	60 amp		1.80	4.444		1,850	237		2,087	2,400
9170	100 amp		1.30	6.154		2,400	330		2,730	3,125
9180	200 amp	2 Elec	2	8		4,875	425		5,300	6,000
9200	Ground bus added to bus duct, 225 amp		320	.050	L.F.	43	2.67		45.67	51.50

26 25 13 – Bus Duct/Busway and Fittings

26 25 13.40 Copper Bus Duct

		Crew	Daily Output	Labor-Hours	Unit	Material	2014 Bare Costs Labor	Equipment	Total	Total Incl O&P
9210	400 amp	2 Elec	240	.067	L.F.	43	3.56		46.56	53
9220	600 amp		240	.067		43	3.56		46.56	53
9230	800 amp		160	.100		50.50	5.35		55.85	63.50
9240	1000 amp		160	.100		57.50	5.35		62.85	71.50
9250	1350 amp		140	.114		82	6.10		88.10	99.50
9260	1600 amp		120	.133		89	7.10		96.10	109
9270	2000 amp		110	.145		116	7.75		123.75	140
9280	2500 amp		100	.160		144	8.55		152.55	171
9290	3000 amp		90	.178		162	9.50		171.50	192
9300	4000 amp		80	.200		214	10.65		224.65	251
9310	5000 amp		70	.229		259	12.20		271.20	305
9320	High short circuit bracing, add					18.10			18.10	19.90

26 25 13.60 Copper or Aluminum Bus Duct Fittings

		Crew	Daily Output	Labor-Hours	Unit	Material	2014 Bare Costs Labor	Equipment	Total	Total Incl O&P
0010	**COPPER OR ALUMINUM BUS DUCT FITTINGS**									
0100	Flange, wall, with vapor barrier, 225 amp	2 Elec	6.20	2.581	Ea.	785	138		923	1,075
0110	400 amp		6	2.667		785	142		927	1,075
0120	600 amp		5.80	2.759		785	147		932	1,075
0130	800 amp		5.40	2.963		785	158		943	1,100
0140	1000 amp		5	3.200		785	171		956	1,125
0145	1200 amp		4.80	3.333		785	178		963	1,125
0150	1350 amp		4.60	3.478		785	186		971	1,150
0160	1600 amp		4.20	3.810		785	203		988	1,175
0170	2000 amp		4	4		785	213		998	1,175
0180	2500 amp		3.60	4.444		785	237		1,022	1,225
0190	3000 amp		3.20	5		785	267		1,052	1,275
0200	4000 amp		2.60	6.154		785	330		1,115	1,350
0300	Roof, 225 amp		6.20	2.581		905	138		1,043	1,200
0310	400 amp		6	2.667		905	142		1,047	1,225
0320	600 amp		5.80	2.759		905	147		1,052	1,225
0330	800 amp		5.40	2.963		905	158		1,063	1,225
0340	1000 amp		5	3.200		905	171		1,076	1,250
0345	1200 amp		4.80	3.333		905	178		1,083	1,275
0350	1350 amp		4.60	3.478		905	186		1,091	1,275
0360	1600 amp		4.20	3.810		905	203		1,108	1,300
0370	2000 amp		4	4		905	213		1,118	1,325
0380	2500 amp		3.60	4.444		905	237		1,142	1,350
0390	3000 amp		3.20	5		905	267		1,172	1,400
0400	4000 amp		2.60	6.154		905	330		1,235	1,500
0420	Support, floor mounted, 225 amp		20	.800		165	42.50		207.50	245
0430	400 amp		20	.800		165	42.50		207.50	245
0440	600 amp		18	.889		165	47.50		212.50	252
0450	800 amp		16	1		165	53.50		218.50	261
0460	1000 amp		13	1.231		165	65.50		230.50	280
0465	1200 amp		11.80	1.356		165	72.50		237.50	289
0470	1350 amp		10.60	1.509		165	80.50		245.50	300
0480	1600 amp		9.20	1.739		165	93		258	320
0490	2000 amp		8	2		165	107		272	340
0500	2500 amp		6.40	2.500		165	133		298	380
0510	3000 amp		5.40	2.963		165	158		323	420
0520	4000 amp		4	4		165	213		378	500
0540	Weather stop, 225 amp		12	1.333		510	71		581	665
0550	400 amp		10	1.600		510	85.50		595.50	690

26 25 13.60 Copper or Aluminum Bus Duct Fittings		Crew	Daily Output	Labor-Hours	Unit	Material	2014 Bare Costs Labor	Equipment	Total	Total Incl O&P
0560	600 amp	2 Elec	9	1.778	Ea.	510	95		605	700
0570	800 amp		8	2		510	107		617	720
0580	1000 amp		6.40	2.500		510	133		643	760
0585	1200 amp		5.90	2.712		510	145		655	775
0590	1350 amp		5.40	2.963		510	158		668	795
0600	1600 amp		4.60	3.478		510	186		696	840
0610	2000 amp		4	4		510	213		723	880
0620	2500 amp		3.20	5		510	267		777	960
0630	3000 amp		2.60	6.154		510	330		840	1,050
0640	4000 amp		2	8		510	425		935	1,200
0660	End closure, 225 amp		34	.471		164	25		189	218
0670	400 amp		32	.500		164	26.50		190.50	220
0680	600 amp		28	.571		164	30.50		194.50	226
0690	800 amp		26	.615		164	33		197	229
0700	1000 amp		24	.667		164	35.50		199.50	233
0705	1200 amp		23	.696		164	37		201	236
0710	1350 amp		22	.727		164	39		203	238
0720	1600 amp		20	.800		164	42.50		206.50	244
0730	2000 amp		18	.889		201	47.50		248.50	292
0740	2500 amp		16	1		201	53.50		254.50	300
0750	3000 amp		14	1.143		201	61		262	315
0760	4000 amp		12	1.333		243	71		314	375
0780	Switchboard stub, 3 pole 3 wire, 225 amp		6	2.667		985	142		1,127	1,300
0790	400 amp		5.20	3.077		985	164		1,149	1,325
0800	600 amp		4.60	3.478		985	186		1,171	1,350
0810	800 amp		3.60	4.444		1,200	237		1,437	1,650
0820	1000 amp		3.40	4.706		1,400	251		1,651	1,900
0825	1200 amp		3.20	5		1,400	267		1,667	1,925
0830	1350 amp		3	5.333		1,675	285		1,960	2,275
0840	1600 amp		2.80	5.714		1,975	305		2,280	2,600
0850	2000 amp		2.40	6.667		2,300	355		2,655	3,050
0860	2500 amp		2	8		2,800	425		3,225	3,725
0870	3000 amp		1.80	8.889		3,200	475		3,675	4,200
0880	4000 amp		1.60	10		4,125	535		4,660	5,350
0890	5000 amp		1.40	11.429		5,075	610		5,685	6,500
0900	3 pole 4 wire, 225 amp		5.40	2.963		1,225	158		1,383	1,575
0910	400 amp		4.60	3.478		1,225	186		1,411	1,625
0920	600 amp		4	4		1,225	213		1,438	1,675
0930	800 amp		3.20	5		1,475	267		1,742	2,025
0940	1000 amp		3	5.333		1,725	285		2,010	2,325
0950	1350 amp		2.60	6.154		2,225	330		2,555	2,925
0960	1600 amp		2.40	6.667		2,500	355		2,855	3,275
0970	2000 amp		2	8		3,025	425		3,450	3,975
0980	2500 amp		1.80	8.889		3,700	475		4,175	4,750
0990	3000 amp		1.60	10		4,325	535		4,860	5,550
1000	4000 amp		1.40	11.429		5,600	610		6,210	7,075
1050	Service head, weatherproof, 3 pole 3 wire, 225 amp		3	5.333		1,675	285		1,960	2,275
1060	400 amp		2.80	5.714		1,675	305		1,980	2,300
1070	600 amp		2.60	6.154		1,675	330		2,005	2,350
1080	800 amp		2.40	6.667		1,875	355		2,230	2,575
1090	1000 amp		2	8		2,025	425		2,450	2,875
1100	1350 amp		1.80	8.889		2,625	475		3,100	3,600
1110	1600 amp		1.60	10		2,950	535		3,485	4,050

26 25 13 – Bus Duct/Busway and Fittings

26 25 13.60 Copper or Aluminum Bus Duct Fittings		Crew	Daily Output	Labor-Hours	Unit	Material	2014 Bare Costs Labor	Equipment	Total	Total Incl O&P
1120	2000 amp	2 Elec	1.40	11.429	Ea.	3,575	610		4,185	4,875
1130	2500 amp		1.20	13.333		4,275	710		4,985	5,775
1140	3000 amp		.90	17.778		5,000	950		5,950	6,925
1150	4000 amp		.70	22.857		6,350	1,225		7,575	8,800
1200	3 pole 4 wire, 225 amp		2.60	6.154		1,875	330		2,205	2,550
1210	400 amp		2.40	6.667		1,875	355		2,230	2,575
1220	600 amp		2.20	7.273		1,875	390		2,265	2,625
1230	800 amp		2	8		2,175	425		2,600	3,050
1240	1000 amp		1.70	9.412		2,500	500		3,000	3,500
1250	1350 amp		1.50	10.667		2,975	570		3,545	4,125
1260	1600 amp		1.40	11.429		3,250	610		3,860	4,500
1270	2000 amp		1.20	13.333		4,350	710		5,060	5,850
1280	2500 amp		1	16		5,350	855		6,205	7,150
1290	3000 amp		.80	20		6,250	1,075		7,325	8,475
1300	4000 amp		.60	26.667		8,125	1,425		9,550	11,100
1350	Flanged end, 3 pole 3 wire, 225 amp		6	2.667		910	142		1,052	1,225
1360	400 amp		5.20	3.077		910	164		1,074	1,250
1370	600 amp		4.60	3.478		910	186		1,096	1,275
1380	800 amp		3.60	4.444		1,000	237		1,237	1,450
1390	1000 amp		3.40	4.706		1,125	251		1,376	1,625
1395	1200 amp		3.20	5		1,275	267		1,542	1,800
1400	1350 amp		3	5.333		1,350	285		1,635	1,925
1410	1600 amp		2.80	5.714		1,550	305		1,855	2,150
1420	2000 amp		2.40	6.667		1,825	355		2,180	2,525
1430	2500 amp		2	8		2,125	425		2,550	2,975
1440	3000 amp		1.80	8.889		2,425	475		2,900	3,375
1450	4000 amp		1.60	10		3,000	535		3,535	4,100
1500	3 pole 4 wire, 225 amp		5.40	2.963		1,050	158		1,208	1,375
1510	400 amp		4.60	3.478		1,050	186		1,236	1,425
1520	600 amp		4	4		1,050	213		1,263	1,475
1530	800 amp		3.20	5		1,225	267		1,492	1,750
1540	1000 amp		3	5.333		1,375	285		1,660	1,950
1545	1200 amp		2.80	5.714		1,550	305		1,855	2,175
1550	1350 amp		2.60	6.154		1,675	330		2,005	2,350
1560	1600 amp		2.40	6.667		1,925	355		2,280	2,650
1570	2000 amp		2	8		2,275	425		2,700	3,150
1580	2500 amp		1.80	8.889		2,700	475		3,175	3,675
1590	3000 amp		1.60	10		3,075	535		3,610	4,200
1600	4000 amp		1.40	11.429		4,000	610		4,610	5,325
1650	Hanger, standard, 225 amp		64	.250		19.20	13.35		32.55	41
1660	400 amp		48	.333		19.20	17.80		37	47.50
1670	600 amp		40	.400		19.20	21.50		40.70	53
1680	800 amp		32	.500		19.20	26.50		45.70	61
1690	1000 amp		24	.667		19.20	35.50		54.70	74
1695	1200 amp		22	.727		19.20	39		58.20	79
1700	1350 amp		20	.800		19.20	42.50		61.70	85
1710	1600 amp		20	.800		19.20	42.50		61.70	85
1720	2000 amp		18	.889		19.20	47.50		66.70	92
1730	2500 amp		16	1		19.20	53.50		72.70	101
1740	3000 amp		16	1		19.20	53.50		72.70	101
1750	4000 amp		16	1		19.20	53.50		72.70	101
1800	Spring type, 225 amp		16	1		85.50	53.50		139	174
1810	400 amp		14	1.143		85.50	61		146.50	186

26 25 13.60 Copper or Aluminum Bus Duct Fittings	Crew	Daily Output	Labor-Hours	Unit	Material	2014 Bare Costs Labor	Equipment	Total	Total Incl O&P
1820	600 amp	2 Elec	14	1.143	Ea.	85.50	61		146.50	186
1830	800 amp		14	1.143		85.50	61		146.50	186
1840	1000 amp		14	1.143		85.50	61		146.50	186
1845	1200 amp		14	1.143		85.50	61		146.50	186
1850	1350 amp		14	1.143		85.50	61		146.50	186
1860	1600 amp		12	1.333		85.50	71		156.50	200
1870	2000 amp		12	1.333		85.50	71		156.50	200
1880	2500 amp		12	1.333		85.50	71		156.50	200
1890	3000 amp		10	1.600		85.50	85.50		171	222
1900	4000 amp		10	1.600		85.50	85.50		171	222

26 25 13.70 Feedrail

		Crew	Daily Output	Labor-Hours	Unit	Material	Labor	Equipment	Total	Total Incl O&P
0010	**FEEDRAIL**, 12 foot mounting									
0050	Trolley busway, 3 pole									
0100	300 volt 60 amp, plain, 10 ft. lengths	1 Elec	50	.160	L.F.	26.50	8.55		35.05	42.50
0300	Door track		50	.160		33.50	8.55		42.05	50
0500	Curved track		30	.267		390	14.25		404.25	450
0700	Coupling				Ea.	17.60			17.60	19.35
0900	Center feed	1 Elec	5.30	1.509		66	80.50		146.50	194
1100	End feed		5.30	1.509		58	80.50		138.50	185
1300	Hanger set		24	.333		5.25	17.80		23.05	32.50
3000	600 volt 100 amp, plain, 10 ft. lengths		35	.229	L.F.	61.50	12.20		73.70	86
3300	Door track		35	.229	"	75	12.20		87.20	101
3700	Coupling				Ea.	53			53	58.50
4000	End cap	1 Elec	40	.200		42.50	10.65		53.15	62.50
4200	End feed		4	2		256	107		363	440
4500	Trolley, 600 volt, 20 amp		5.30	1.509		345	80.50		425.50	500
4700	30 amp		5.30	1.509		345	80.50		425.50	500
4900	Duplex, 40 amp		4	2		840	107		947	1,075
5000	60 amp		4	2		840	107		947	1,075
5300	Fusible, 20 amp		4	2		955	107		1,062	1,200
5500	30 amp		4	2		955	107		1,062	1,200
5900	300 volt, 20 amp		5.30	1.509		296	80.50		376.50	445
6000	30 amp		5.30	1.509		330	80.50		410.50	485
6300	Fusible, 20 amp		4.70	1.702		390	91		481	565
6500	30 amp		4.70	1.702		560	91		651	750
7300	Busway, 250 volt 50 amp, 2 wire		70	.114	L.F.	19.55	6.10		25.65	30.50
7330	Coupling				Ea.	36			36	39.50
7340	Center feed	1 Elec	6	1.333		420	71		491	565
7350	End feed		6	1.333		89	71		160	204
7360	End cap		40	.200		27	10.65		37.65	45.50
7370	Hanger set		24	.333		3.07	17.80		20.87	30
7400	125/250 volt 50 amp, 3 wire		60	.133	L.F.	19.80	7.10		26.90	32.50
7430	Coupling		6	1.333	Ea.	50	71		121	161
7440	Center feed		6	1.333		445	71		516	595
7450	End feed		6	1.333		89	71		160	204
7460	End cap		40	.200		34	10.65		44.65	53
7470	Hanger set		24	.333		3.07	17.80		20.87	30
7480	Trolley, 250 volt, 2 pole, 20 amp		6	1.333		46	71		117	157
7490	30 amp		6	1.333		46	71		117	157
7500	125/250 volt, 3 pole, 20 amp		6	1.333		46	71		117	157
7510	30 amp		6	1.333		46	71		117	157
8000	Cleaning tools, 300 volt, dust remover					93			93	102

26 25 Enclosed Bus Assemblies

26 25 13 – Bus Duct/Busway and Fittings

26 25 13.70 Feedrail	Crew	Daily Output	Labor-Hours	Unit	Material	2014 Bare Costs Labor	Equipment	Total	Total Incl O&P
8100	Bus bar cleaner				Ea.	165			165	182
8300	600 volt, dust remover, 60 amp					350			350	385
8400	100 amp					480			480	530
8600	Bus bar cleaner, 60 amp					500			500	550
8700	100 amp					590			590	645

26 27 Low-Voltage Distribution Equipment

26 27 13 – Electricity Metering

26 27 13.10 Meter Centers and Sockets

		Crew	Daily Output	Labor-Hours	Unit	Material	2014 Bare Costs Labor	Equipment	Total	Total Incl O&P
0010	**METER CENTERS AND SOCKETS**									
0100	Sockets, single position, 4 terminal, 100 amp	1 Elec	3.20	2.500	Ea.	42	133		175	246
0200	150 amp		2.30	3.478		54.50	186		240.50	340
0300	200 amp		1.90	4.211		89.50	225		314.50	435
0400	Transformer rated, 20 amp		3.20	2.500		147	133		280	360
0500	Double position, 4 terminal, 100 amp		2.80	2.857		198	152		350	445
0600	150 amp		2.10	3.810		233	203		436	560
0700	200 amp		1.70	4.706		440	251		691	855
0800	Trans-socket, 13 terminal, 3 CT mounts, 400 amp		1	8		1,300	425		1,725	2,075
0900	800 amp	2 Elec	1.20	13.333		1,525	710		2,235	2,750
1100	Meter centers and sockets, three phase, single pos, 7 terminal, 100 amp	1 Elec	2.80	2.857		144	152		296	385
1200	200 amp		2.10	3.810		278	203		481	610
1400	400 amp		1.70	4.706		715	251		966	1,150
2000	Meter center, main fusible switch, 1P 3W 120/240 volt									
2030	400 amp	2 Elec	1.60	10	Ea.	1,600	535		2,135	2,550
2040	600 amp		1.10	14.545		2,775	775		3,550	4,225
2050	800 amp		.90	17.778		4,350	950		5,300	6,225
2060	Rainproof 1P 3W 120/240 volt, 400 amp		1.60	10		1,600	535		2,135	2,550
2070	600 amp		1.10	14.545		2,775	775		3,550	4,225
2080	800 amp		.90	17.778		4,350	950		5,300	6,225
2100	3P 4W 120/208 V, 400 amp		1.60	10		1,825	535		2,360	2,825
2110	600 amp		1.10	14.545		3,425	775		4,200	4,900
2120	800 amp		.90	17.778		6,350	950		7,300	8,400
2130	Rainproof 3P 4W 120/208 V, 400 amp		1.60	10		1,825	535		2,360	2,825
2140	600 amp		1.10	14.545		3,425	775		4,200	4,900
2150	800 amp		.90	17.778		6,350	950		7,300	8,400
2170	Main circuit breaker, 1P 3W 120/240 V									
2180	400 amp	2 Elec	1.60	10	Ea.	2,875	535		3,410	3,950
2190	600 amp		1.10	14.545		3,850	775		4,625	5,400
2200	800 amp		.90	17.778		4,500	950		5,450	6,375
2210	1000 amp		.80	20		6,200	1,075		7,275	8,425
2220	1200 amp		.76	21.053		8,350	1,125		9,475	10,900
2230	1600 amp		.68	23.529		18,700	1,250		19,950	22,500
2240	Rainproof 1P 3W 120/240 V, 400 amp		1.60	10		2,875	535		3,410	3,950
2250	600 amp		1.10	14.545		3,850	775		4,625	5,400
2260	800 amp		.90	17.778		4,500	950		5,450	6,375
2270	1000 amp		.80	20		6,200	1,075		7,275	8,425
2280	1200 amp		.76	21.053		8,350	1,125		9,475	10,900
2300	3P 4W 120/208 V, 400 amp		1.60	10		3,250	535		3,785	4,375
2310	600 amp		1.10	14.545		4,600	775		5,375	6,200
2320	800 amp		.90	17.778		5,450	950		6,400	7,425
2330	1000 amp		.80	20		7,175	1,075		8,250	9,475

26 27 13.10 Meter Centers and Sockets		Crew	Daily Output	Labor-Hours	Unit	Material	2014 Bare Costs Labor	Equipment	Total	Total Incl O&P
2340	1200 amp	2 Elec	.76	21.053	Ea.	9,175	1,125		10,300	11,800
2350	1600 amp		.68	23.529		18,700	1,250		19,950	22,500
2360	Rainproof 3P 4W 120/208 V, 400 amp		1.60	10		3,250	535		3,785	4,375
2370	600 amp		1.10	14.545		4,600	775		5,375	6,200
2380	800 amp		.90	17.778		5,450	950		6,400	7,425
2390	1000 amp		.76	21.053		7,175	1,125		8,300	9,550
2400	1200 amp	↓	.68	23.529	↓	9,175	1,250		10,425	12,000
2420	Main lugs terminal box, 1P 3W 120/240 V									
2430	800 amp	2 Elec	.94	17.021	Ea.	535	910		1,445	1,950
2440	1200 amp		.72	22.222		1,075	1,175		2,250	2,950
2450	Rainproof 1P 3W 120/240 V, 225 amp		2.40	6.667		365	355		720	935
2460	800 amp		.94	17.021		535	910		1,445	1,950
2470	1200 amp		.72	22.222		1,075	1,175		2,250	2,950
2500	3P 4W 120/208 V, 800 amp		.94	17.021		595	910		1,505	2,000
2510	1200 amp		.72	22.222		1,200	1,175		2,375	3,100
2520	Rainproof 3P 4W 120/208 V, 225 amp		2.40	6.667		365	355		720	935
2530	800 amp		.94	17.021		595	910		1,505	2,000
2540	1200 amp	↓	.72	22.222	↓	1,200	1,175		2,375	3,100
2590	Basic meter device									
2600	1P 3W 120/240 V 4 jaw 125A sockets, 3 meter	2 Elec	1	16	Ea.	615	855		1,470	1,950
2610	4 meter		.90	17.778		735	950		1,685	2,225
2620	5 meter		.80	20		925	1,075		2,000	2,625
2630	6 meter		.60	26.667		1,050	1,425		2,475	3,300
2640	7 meter		.56	28.571		1,350	1,525		2,875	3,775
2650	8 meter		.52	30.769		1,475	1,650		3,125	4,075
2660	10 meter	↓	.48	33.333	↓	1,850	1,775		3,625	4,675
2680	Rainproof 1P 3W 120/240 V 4 jaw 125A sockets									
2690	3 meter	2 Elec	1	16	Ea.	615	855		1,470	1,950
2700	4 meter		.90	17.778		735	950		1,685	2,225
2710	6 meter		.60	26.667		1,050	1,425		2,475	3,300
2720	7 meter		.56	28.571		1,350	1,525		2,875	3,775
2730	8 meter	↓	.52	30.769	↓	1,475	1,650		3,125	4,075
2750	1P 3W 120/240 V 4 jaw sockets									
2760	with 125A circuit breaker, 3 meter	2 Elec	1	16	Ea.	1,150	855		2,005	2,550
2770	4 meter		.90	17.778		1,450	950		2,400	3,025
2780	5 meter		.80	20		1,825	1,075		2,900	3,600
2790	6 meter		.60	26.667		2,125	1,425		3,550	4,475
2800	7 meter		.56	28.571		2,600	1,525		4,125	5,150
2810	8 meter		.52	30.769		2,900	1,650		4,550	5,650
2820	10 meter	↓	.48	33.333	↓	3,625	1,775		5,400	6,650
2830	Rainproof 1P 3W 120/240 V 4 jaw sockets									
2840	with 125A circuit breaker, 3 meter	2 Elec	1	16	Ea.	1,150	855		2,005	2,550
2850	4 meter		.90	17.778		1,450	950		2,400	3,025
2870	6 meter		.60	26.667		2,125	1,425		3,550	4,475
2880	7 meter		.56	28.571		2,600	1,525		4,125	5,150
2890	8 meter	↓	.52	30.769	↓	2,900	1,650		4,550	5,650
2920	1P 3W on 3P 4W 120/208 V system 5 jaw									
2930	125A sockets, 3 meter	2 Elec	1	16	Ea.	615	855		1,470	1,950
2940	4 meter		.90	17.778		735	950		1,685	2,225
2950	5 meter		.80	20		925	1,075		2,000	2,625
2960	6 meter		.60	26.667		1,050	1,425		2,475	3,300
2970	7 meter		.56	28.571		1,350	1,525		2,875	3,775
2980	8 meter	↓	.52	30.769		1,475	1,650		3,125	4,075

26 27 13.10 Meter Centers and Sockets	Crew	Daily Output	Labor-Hours	Unit	Material	2014 Bare Costs Labor	Equipment	Total	Total Incl O&P
2990	10 meter	2 Elec	.48	33.333	Ea.	1,850	1,775		3,625	4,675
3000	Rainproof 1P 3W on 3P 4W 120/208 V system									
3020	5 jaw 125A sockets, 3 meter	2 Elec	1	16	Ea.	615	855		1,470	1,950
3030	4 meter		.90	17.778		735	950		1,685	2,225
3050	6 meter		.60	26.667		1,050	1,425		2,475	3,300
3060	7 meter		.56	28.571		1,350	1,525		2,875	3,775
3070	8 meter		.52	30.769		1,475	1,650		3,125	4,075
3090	1P 3W on 3P 4W 120/208 V system 5 jaw sockets									
3100	With 125A circuit breaker, 3 meter	2 Elec	1	16	Ea.	1,150	855		2,005	2,550
3110	4 meter		.90	17.778		1,450	950		2,400	3,025
3120	5 meter		.80	20		1,825	1,075		2,900	3,600
3130	6 meter		.60	26.667		2,125	1,425		3,550	4,475
3140	7 meter		.56	28.571		2,600	1,525		4,125	5,150
3150	8 meter		.52	30.769		2,900	1,650		4,550	5,650
3160	10 meter		.48	33.333		3,625	1,775		5,400	6,650
3170	Rainproof 1P 3W on 3P 4W 120/208 V system									
3180	5 jaw sockets w/125A circuit breaker, 3 meter	2 Elec	1	16	Ea.	1,150	855		2,005	2,550
3190	4 meter		.90	17.778		1,450	950		2,400	3,025
3210	6 meter		.60	26.667		2,125	1,425		3,550	4,475
3220	7 meter		.56	28.571		2,600	1,525		4,125	5,150
3230	8 meter		.52	30.769		2,900	1,650		4,550	5,650
3250	1P 3W 120/240 V 4 jaw sockets									
3260	with 200A circuit breaker, 3 meter	2 Elec	1	16	Ea.	1,725	855		2,580	3,175
3270	4 meter		.90	17.778		2,325	950		3,275	4,000
3290	6 meter		.60	26.667		3,450	1,425		4,875	5,925
3300	7 meter		.56	28.571		4,050	1,525		5,575	6,750
3310	8 meter		.56	28.571		4,675	1,525		6,200	7,400
3330	Rainproof 1P 3W 120/240 V 4 jaw sockets									
3350	with 200A circuit breaker, 3 meter	2 Elec	1	16	Ea.	1,725	855		2,580	3,175
3360	4 meter		.90	17.778		2,325	950		3,275	4,000
3380	6 meter		.60	26.667		3,450	1,425		4,875	5,925
3390	7 meter		.56	28.571		4,050	1,525		5,575	6,750
3400	8 meter		.52	30.769		4,675	1,650		6,325	7,575
3420	1P 3W on 3P 4W 120/208 V 5 jaw sockets									
3430	with 200A circuit breaker, 3 meter	2 Elec	1	16	Ea.	1,725	855		2,580	3,175
3440	4 meter		.90	17.778		2,325	950		3,275	4,000
3460	6 meter		.60	26.667		3,450	1,425		4,875	5,925
3470	7 meter		.56	28.571		4,050	1,525		5,575	6,750
3480	8 meter		.52	30.769		4,675	1,650		6,325	7,575
3500	Rainproof 1P 3W on 3P 4W 120/208 V 5 jaw socket									
3510	with 200A circuit breaker, 3 meter	2 Elec	1	16	Ea.	1,725	855		2,580	3,175
3520	4 meter		.90	17.778		2,325	950		3,275	4,000
3540	6 meter		.60	26.667		3,450	1,425		4,875	5,925
3550	7 meter		.56	28.571		4,050	1,525		5,575	6,750
3560	8 meter		.52	30.769		4,675	1,650		6,325	7,575
3600	Automatic circuit closing, add					67			67	73.50
3610	Manual circuit closing, add					76			76	84
3650	Branch meter device									
3660	3P 4W 208/120 or 240/120 V 7 jaw sockets									
3670	with 200A circuit breaker, 2 meter	2 Elec	.90	17.778	Ea.	2,875	950		3,825	4,600
3680	3 meter		.80	20		4,325	1,075		5,400	6,350
3690	4 meter		.70	22.857		5,775	1,225		7,000	8,175
3700	Main circuit breaker 42,000 rms, 400 amp		1.60	10		2,525	535		3,060	3,575

26 27 13 – Electricity Metering

26 27 13.10 Meter Centers and Sockets		Crew	Daily Output	Labor-Hours	Unit	Material	2014 Bare Costs Labor	Equipment	Total	Total Incl O&P
3710	600 amp	2 Elec	1.10	14.545	Ea.	4,075	775		4,850	5,625
3720	800 amp		.90	17.778		5,450	950		6,400	7,425
3730	Rainproof main circ. breaker 42,000 rms, 400 amp		1.60	10		2,525	535		3,060	3,575
3740	600 amp		1.10	14.545		4,075	775		4,850	5,625
3750	800 amp		.90	17.778		5,450	950		6,400	7,425
3760	Main circuit breaker 65,000 rms, 400 amp		1.60	10		3,425	535		3,960	4,575
3770	600 amp		1.10	14.545		4,825	775		5,600	6,450
3780	800 amp		.90	17.778		5,450	950		6,400	7,425
3790	1000 amp		.80	20		7,175	1,075		8,250	9,475
3800	1200 amp		.76	21.053		9,175	1,125		10,300	11,800
3810	1600 amp		.68	23.529		18,700	1,250		19,950	22,500
3820	Rainproof main circ. breaker 65,000 rms, 400 amp		1.60	10		3,425	535		3,960	4,575
3830	600 amp		1.10	14.545		4,825	775		5,600	6,450
3840	800 amp		.90	17.778		5,450	950		6,400	7,425
3850	1000 amp		.80	20		7,175	1,075		8,250	9,475
3860	1200 amp		.76	21.053		9,175	1,125		10,300	11,800
3880	Main circuit breaker 100,000 rms, 400 amp		1.60	10		3,425	535		3,960	4,575
3890	600 amp		1.10	14.545		4,825	775		5,600	6,450
3900	800 amp		.90	17.778		5,650	950		6,600	7,650
3910	Rainproof main circ. breaker 100,000 rms, 400 amp		1.60	10		3,425	535		3,960	4,575
3920	600 amp		1.10	14.545		4,825	775		5,600	6,450
3930	800 amp		.90	17.778		5,650	950		6,600	7,650
3940	Main lugs terminal box, 800 amp		.94	17.021		595	910		1,505	2,000
3950	1600 amp		.72	22.222		2,025	1,175		3,200	4,000
3960	Rainproof, 800 amp		.94	17.021		595	910		1,505	2,000
3970	1600 amp		.72	22.222		2,025	1,175		3,200	4,000

26 27 16 – Electrical Cabinets and Enclosures

26 27 16.10 Cabinets

			Crew	Daily Output	Labor-Hours	Unit	Material	Labor	Equipment	Total	Total Incl O&P
0010	**CABINETS**	R260533-70									
7000	Cabinets, current transformer										
7050	Single door, 24" H x 24" W x 10" D	R262716-40	1 Elec	1.60	5	Ea.	178	267		445	595
7100	30" H x 24" W x 10" D			1.30	6.154		166	330		496	675
7150	36" H x 24" W x 10" D			1.10	7.273		186	390		576	785
7200	30" H x 30" W x 10" D			1	8		194	425		619	855
7250	36" H x 30" W x 10" D			.90	8.889		227	475		702	960
7300	36" H x 36" W x 10" D			.80	10		235	535		770	1,050
7500	Double door, 48" H x 36" W x 10" D			.60	13.333		555	710		1,265	1,675
7550	24" H x 24" W x 12" D			1	8		165	425		590	820
8000	NEMA 12, double door, floor mounted										
8020	54" H x 42" W x 8" D		2 Elec	6	2.667	Ea.	1,300	142		1,442	1,650
8040	60" H x 48" W x 8" D			5.40	2.963		1,725	158		1,883	2,125
8060	60" H x 48" W x 10" D			5.40	2.963		1,775	158		1,933	2,200
8080	60" H x 60" W x 10" D			5	3.200		2,000	171		2,171	2,450
8100	72" H x 60" W x 10" D			4	4		2,350	213		2,563	2,900
8120	72" H x 72" W x 10" D			3.40	4.706		2,550	251		2,801	3,200
8140	60" H x 48" W x 12" D			3.40	4.706		1,825	251		2,076	2,375
8160	60" H x 60" W x 12" D			3.20	5		2,050	267		2,317	2,675
8180	72" H x 60" W x 12" D			3	5.333		2,325	285		2,610	3,000
8200	72" H x 72" W x 12" D			3	5.333		2,600	285		2,885	3,275
8220	60" H x 48" W x 16" D			3.20	5		1,900	267		2,167	2,475
8240	72" H x 72" W x 16" D			2.60	6.154		2,750	330		3,080	3,525
8260	60" H x 48" W x 20" D			3	5.333		2,075	285		2,360	2,700

26 27 Low-Voltage Distribution Equipment

26 27 16 – Electrical Cabinets and Enclosures

26 27 16.10 Cabinets		Crew	Daily Output	Labor-Hours	Unit	Material	2014 Bare Costs Labor	Equipment	Total	Total Incl O&P
8280	72" H x 72" W x 20" D	2 Elec	2.20	7.273	Ea.	2,950	390		3,340	3,825
8300	60" H x 48" W x 24" D		2.60	6.154		2,200	330		2,530	2,925
8320	72" H x 72" W x 24" D	↓	2	8	↓	3,125	425		3,550	4,075
8340	Pushbutton enclosure, oiltight									
8360	3-1/2" H x 3-1/4" W x 2-3/4" D, for 1 P.B.	1 Elec	12	.667	Ea.	43	35.50		78.50	100
8380	5-3/4" H x 3-1/4" W x 2-3/4" D, for 2 P.B.		11	.727		72	39		111	137
8400	8" H x 3-1/4" W x 2-3/4" D, for 3 P.B.		10.50	.762		80	40.50		120.50	149
8420	10-1/4" H x 3-1/4" W x 2-3/4" D, for 4 P.B.		10.50	.762		65.50	40.50		106	134
8460	12-1/2" H x 3-1/4" W x 3" D, for 5 P.B.		9	.889		73.50	47.50		121	152
8480	9-1/2" H x 6-1/4" W x 3" D, for 6 P.B.		8.50	.941		108	50		158	194
8500	9-1/2" H x 8-1/2" W x 3" D, for 9 P.B.		8	1		90.50	53.50		144	180
8510	11-3/4" H x 8-1/2" W x 3" D, for 12 P.B.		7	1.143		100	61		161	202
8520	11-3/4" H x 10-3/4" W x 3" D, for 16 P.B.		6.50	1.231		129	65.50		194.50	241
8540	14" H x 10-3/4" W x 3" D, for 20 P.B.		5	1.600		146	85.50		231.50	289
8560	14" H x 13" W x 3" D, for 25 P.B.	↓	4.50	1.778	↓	168	95		263	325
8580	Sloping front pushbutton enclosures									
8600	3-1/2" H x 7-3/4" W x 4-7/8" D, for 3 P.B.	1 Elec	10	.800	Ea.	77	42.50		119.50	149
8620	7-1/4" H x 8-1/2" W x 6-3/4" D, for 6 P.B.		8	1		112	53.50		165.50	203
8640	9-1/2" H x 8-1/2" W x 7-7/8" D, for 9 P.B.		7	1.143		133	61		194	238
8660	11-1/4" H x 8-1/2" W x 9" D, for 12 P.B.		5	1.600		155	85.50		240.50	298
8680	11-3/4" H x 10" W x 9" D, for 16 P.B.		5	1.600		165	85.50		250.50	310
8700	11-3/4" H x 13" W x 9" D, for 20 P.B.		5	1.600		192	85.50		277.50	340
8720	14" H x 13" W x 10-1/8" D, for 25 P.B.	↓	4.50	1.778	↓	217	95		312	380
8740	Pedestals, not including P.B. enclosure or base									
8760	Straight column 4" x 4"	1 Elec	4.50	1.778	Ea.	269	95		364	435
8780	6" x 6"		4	2		420	107		527	620
8800	Angled column 4" x 4"		4.50	1.778		310	95		405	480
8820	6" x 6"		4	2		465	107		572	670
8840	Pedestal, base 18" x 18"		10	.800		159	42.50		201.50	238
8860	24" x 24"	↓	9	.889	↓	365	47.50		412.50	470
8900	Electronic rack enclosures									
8920	72" H x 19" W x 24" D	1 Elec	1.50	5.333	Ea.	1,850	285		2,135	2,475
8940	72" H x 23" W x 24" D		1.50	5.333		2,100	285		2,385	2,725
8960	72" H x 19" W x 30" D		1.30	6.154		2,225	330		2,555	2,950
8980	72" H x 19" W x 36" D		1.20	6.667		2,550	355		2,905	3,325
9000	72" H x 23" W x 36" D	↓	1.20	6.667	↓	2,825	355		3,180	3,625
9020	NEMA 12 & 4 enclosure panels									
9040	12" x 24"	1 Elec	20	.400	Ea.	34.50	21.50		56	70
9060	16" x 12"		20	.400		25.50	21.50		47	60
9080	20" x 16"		20	.400		36.50	21.50		58	72.50
9100	20" x 20"		19	.421		44	22.50		66.50	82
9120	24" x 20"		18	.444		57	23.50		80.50	98
9140	24" x 24"		17	.471		65.50	25		90.50	110
9160	30" x 20"		16	.500		65.50	26.50		92	112
9180	30" x 24"		16	.500		79.50	26.50		106	128
9200	36" x 24"		15	.533		94	28.50		122.50	147
9220	36" x 30"		15	.533		126	28.50		154.50	182
9240	42" x 24"		15	.533		107	28.50		135.50	161
9260	42" x 30"		14	.571		142	30.50		172.50	203
9280	42" x 36"		14	.571		167	30.50		197.50	230
9300	48" x 24"		14	.571		128	30.50		158.50	187
9320	48" x 30"		14	.571		166	30.50		196.50	229
9340	48" x 36"	↓	13	.615	↓	193	33		226	261

26 27 16.10 Cabinets

		Crew	Daily Output	Labor-Hours	Unit	Material	2014 Bare Costs Labor	Equipment	Total	Total Incl O&P
9360	60" x 36"	1 Elec	12	.667	Ea.	234	35.50		269.50	310
9400	Wiring trough steel JIC, clamp cover									
9490	4" x 4", 12" long	1 Elec	12	.667	Ea.	90	35.50		125.50	152
9510	24" long		10	.800		114	42.50		156.50	189
9530	36" long		8	1		142	53.50		195.50	236
9540	48" long		7	1.143		165	61		226	273
9550	60" long		6	1.333		194	71		265	320
9560	6" x 6", 12" long		11	.727		120	39		159	189
9580	24" long		9	.889		158	47.50		205.50	245
9600	36" long		7	1.143		194	61		255	305
9610	48" long		6	1.333		236	71		307	365
9620	60" long		5	1.600		278	85.50		363.50	435

26 27 16.20 Cabinets and Enclosures

		Crew	Daily Output	Labor-Hours	Unit	Material	2014 Bare Costs Labor	Equipment	Total	Total Incl O&P
0010	**CABINETS AND ENCLOSURES** Nonmetallic									
0080	Enclosures fiberglass NEMA 4X									
0100	Wall mount, quick release latch door, 20"H x 16"W x 6"D	1 Elec	4.80	1.667	Ea.	600	89		689	795
0110	20"H x 20"W x 6"D		4.50	1.778		720	95		815	935
0120	24"H x 20"W x 6"D		4.20	1.905		770	102		872	1,000
0130	20"H x 16"W x 8"D		4.50	1.778		685	95		780	895
0140	20"H x 20"W x 8"D		4.20	1.905		770	102		872	995
0150	24"H x 24"W x 8"D		3.80	2.105		870	112		982	1,125
0160	30"H x 24"W x 8"D		3.20	2.500		940	133		1,073	1,225
0170	36"H x 30"W x 8"D		3	2.667		1,325	142		1,467	1,675
0180	20"H x 16"W x 10"D		3.50	2.286		800	122		922	1,075
0190	20"H x 20"W x 10"D		3.20	2.500		850	133		983	1,125
0200	24"H x 20"W x 10"D		3	2.667		905	142		1,047	1,200
0210	30"H x 24"W x 10"D		2.80	2.857		1,025	152		1,177	1,350
0220	20"H x 16"W x 12"D		3	2.667		850	142		992	1,150
0230	20"H x 20"W x 12"D		2.80	2.857		910	152		1,062	1,225
0240	24"H x 24"W x 12"D		2.60	3.077		1,000	164		1,164	1,350
0250	30"H x 24"W x 12"D		2.40	3.333		1,125	178		1,303	1,500
0260	36"H x 30"W x 12"D		2.20	3.636		1,550	194		1,744	2,000
0270	36"H x 36"W x 12"D		2.10	3.810		1,725	203		1,928	2,200
0280	48"H x 36"W x 12"D		2	4		1,975	213		2,188	2,500
0290	60"H x 36"W x 12"D		1.80	4.444		2,250	237		2,487	2,800
0300	30"H x 24"W x 16"D		1.40	5.714		1,300	305		1,605	1,875
0310	48"H x 36"W x 16"D		1.20	6.667		2,175	355		2,530	2,925
0320	60"H x 36"W x 16"D		1	8		2,475	425		2,900	3,375
0480	Freestanding, one door, 72"H x 25"W x 25"D		.80	10		4,075	535		4,610	5,300
0490	Two doors with two panels, 72"H x 49"W x 24"D		.50	16		10,200	855		11,055	12,500
0500	Floor stand kits, for NEMA 4 & 12, 20"W or more, 6"H x 8"D		24	.333		149	17.80		166.80	191
0510	6"H x 10"D		24	.333		163	17.80		180.80	206
0520	6"H x 12"D		24	.333		182	17.80		199.80	227
0530	6"H x 18"D		24	.333		211	17.80		228.80	259
0540	12"H x 8"D		22	.364		188	19.40		207.40	236
0550	12"H x 10"D		22	.364		199	19.40		218.40	248
0560	12"H x 12"D		22	.364		213	19.40		232.40	263
0570	12"H x 16"D		22	.364		247	19.40		266.40	300
0580	12"H x 18"D		22	.364		253	19.40		272.40	310
0590	12"H x 20"D		22	.364		270	19.40		289.40	325
0600	18"H x 8"D		20	.400		224	21.50		245.50	279
0610	18"H x 10"D		20	.400		241	21.50		262.50	297

26 27 Low-Voltage Distribution Equipment

26 27 16 – Electrical Cabinets and Enclosures

26 27 16.20 Cabinets and Enclosures	Crew	Daily Output	Labor-Hours	Unit	Material	2014 Bare Costs Labor	Equipment	Total	Total Incl O&P
0620	18"H x 12"D	1 Elec	20	.400	Ea.	258	21.50		279.50	315
0630	18"H x 16"D		20	.400		286	21.50		307.50	345
0640	24"H x 8"D		16	.500		277	26.50		303.50	345
0650	24"H x 10"D		16	.500		286	26.50		312.50	355
0660	24"H x 12"D		16	.500		310	26.50		336.50	380
0670	24"H x 16"D		16	.500		325	26.50		351.50	395
0680	Small, screw cover, 5-1/2"H x 4"W x 4-15/16"D		12	.667		74	35.50		109.50	135
0690	7-1/2"H x 4"W x 4-15/16"D		12	.667		79	35.50		114.50	140
0700	7-1/2"H x 6"W x 5-3/16"D		10	.800		91.50	42.50		134	165
0710	9-1/2"H x 6"W x 5-11/16"D		10	.800		94	42.50		136.50	167
0720	11-1/2"H x 8"W x 6-11/16"D		8	1		139	53.50		192.50	233
0730	13-1/2"H x 10"W x 7-3/16"D		7	1.143		168	61		229	277
0740	15-1/2"H x 12"W x 8-3/16"D		6	1.333		220	71		291	350
0750	17-1/2"H x 14"W x 8-11/16"D		5	1.600		262	85.50		347.50	415
0760	Screw cover with window, 6"H x 4"W x 5"D		12	.667		158	35.50		193.50	226
0770	8"H x 4"W x 5"D		11	.727		165	39		204	240
0780	8"H x 6"W x 5"D		11	.727		189	39		228	266
0790	10"H x 6"W x 6"D		10	.800		204	42.50		246.50	288
0800	12"H x 8"W x 7"D		8	1		274	53.50		327.50	380
0810	14"H x 10"W x 7"D		7	1.143		298	61		359	420
0820	16"H x 12"W x 8"D		6	1.333		375	71		446	515
0830	18"H x 14"W x 9"D		5	1.600		450	85.50		535.50	625
0840	Quick-release latch cover, 5-1/2"H x 4"W x 5"D		12	.667		101	35.50		136.50	164
0850	7-1/2"H x 4"W x 5"D		12	.667		108	35.50		143.50	172
0860	7-1/2"H x 6"W x 5-1/4"D		10	.800		116	42.50		158.50	192
0870	9-1/2"H x 6"W x 5-3/4"D		10	.800		128	42.50		170.50	205
0880	11-1/2"H x 8"W x 6-3/4"D		8	1		182	53.50		235.50	281
0890	13-1/2"H x 10"W x 7-1/4"D		7	1.143		229	61		290	345
0900	15-1/2"H x 12"W x 8-1/4"D		6	1.333		281	71		352	415
0910	17-1/2"H x 14"W x 8-3/4"D		5	1.600		340	85.50		425.50	505
0920	Pushbutton, 1 hole 5-1/2"H x 4"W x 4-15/16"D		12	.667		80.50	35.50		116	142
0930	2 hole 7-1/2"H x 4"W x 4-15/16"D		11	.727		91.50	39		130.50	159
0940	4 hole 7-1/2"H x 6"W x 5-3/16"D		10.50	.762		108	40.50		148.50	179
0950	6 hole 9-1/2"H x 6"W x 5-11/16"D		9	.889		132	47.50		179.50	217
0960	8 hole 11-1/2"H x 8"W x 6-11/16"D		8.50	.941		168	50		218	260
0970	12 hole 13-1/2"H x 10"W x 7-3/16"D		8	1		216	53.50		269.50	320
0980	20 hole 15-1/2"H x 12"W x 8-3/16"D		5	1.600		296	85.50		381.50	455
0990	30 hole 17-1/2"H x 14"W x 8-11/16"D	▼	4.50	1.778	▼	330	95		425	500
1450	Enclosures polyester NEMA 4X									
1460	Small, screw cover,									
1500	3-15/16"H x 3-15/16"W x 3-1/16"D	1 Elec	12	.667	Ea.	68	35.50		103.50	128
1510	5-3/16"H x 3-5/16"W x 3-1/16"D		12	.667		67	35.50		102.50	127
1520	5-7/8"H x 3-7/8"W x 4-3/16"D		12	.667		70.50	35.50		106	131
1530	5-7/8"H x 5-7/8"W x 4-3/16"D		12	.667		79	35.50		114.50	140
1540	7-5/8"H x 3-5/16"W x 3-1/16"D		12	.667		73	35.50		108.50	134
1550	10-3/16"H x 3-5/16"W x 3-1/16"D		10	.800		86.50	42.50		129	160
1560	Clear cover, 3-15/16"H x 3-15/16"W x 2-7/8"D		12	.667		76.50	35.50		112	138
1570	5-3/16"H x 3-5/16"W x 2-7/8"D		12	.667		81.50	35.50		117	143
1580	5-7/8"H x 3-7/8"W x 4"D		12	.667		101	35.50		136.50	165
1590	5-7/8"H x 5-7/8"W x 4"D		12	.667		125	35.50		160.50	190
1600	7-5/8"H x 3-5/16"W x 2-7/8"D		12	.667		91.50	35.50		127	154
1610	10-3/16"H x 3-5/16"W x 2-7/8"D		10	.800		115	42.50		157.50	191
1620	Pushbutton, 1 hole, 5-5/16"H x 3-5/16"W x 3-1/16"D		12	.667		59.50	35.50		95	119

26 27 16 – Electrical Cabinets and Enclosures

26 27 16.20 Cabinets and Enclosures		Crew	Daily Output	Labor-Hours	Unit	Material	2014 Bare Costs Labor	Equipment	Total	Total Incl O&P
1630	2 hole, 7-5/8"H x 3-5/16"W x 3-1/8"D	1 Elec	11	.727	Ea.	67	39		106	132
1640	3 hole, 10-3/16"H x 3-5/16"W x 3-1/16"D		10.50	.762		79	40.50		119.50	148
8000	Wireway fiberglass, straight sect. screwcover, 12" L, 4" W x 4" D		40	.200		187	10.65		197.65	222
8010	6" W x 6" D		30	.267		249	14.25		263.25	295
8020	24" L, 4" W x 4" D		20	.400		227	21.50		248.50	281
8030	6" W x 6" D		15	.533		375	28.50		403.50	455
8040	36" L, 4" W x 4" D		13.30	.602		284	32		316	365
8050	6" W x 6" D		10	.800		445	42.50		487.50	555
8060	48" L, 4" W x 4" D		10	.800		340	42.50		382.50	440
8070	6" W x 6" D		7.50	1.067		560	57		617	700
8080	60" L, 4" W x 4" D		8	1		535	53.50		588.50	665
8090	6" W x 6" D		6	1.333		650	71		721	820
8100	Elbow, 90°, 4" W x 4" D		20	.400		192	21.50		213.50	243
8110	6" W x 6" D		18	.444		385	23.50		408.50	455
8120	Elbow, 45°, 4" W x 4" D		20	.400		182	21.50		203.50	232
8130	6" W x 6" D		18	.444		360	23.50		383.50	435
8140	Tee, 4" W x 4" D		16	.500		243	26.50		269.50	305
8150	6" W x 6" D		14	.571		440	30.50		470.50	530
8160	Cross, 4" W x 4" D		14	.571		340	30.50		370.50	415
8170	6" W x 6" D		12	.667		730	35.50		765.50	860
8180	Cut-off fitting, w/flange & adhesive, 4" W x 4" D		18	.444		136	23.50		159.50	185
8190	6" W x 6" D		16	.500		295	26.50		321.50	365
8200	Flexible ftng., hvy. neoprene coated nylon, 4" W x 4" D		20	.400		264	21.50		285.50	320
8210	6" W x 6" D		18	.444		390	23.50		413.50	465
8220	Closure plate, fiberglass, 4" W x 4" D		20	.400		52.50	21.50		74	90
8230	6" W x 6" D		18	.444		57.50	23.50		81	98.50
8240	Box connector, stainless steel type 304, 4" W x 4" D		20	.400		85.50	21.50		107	126
8250	6" W x 6" D		18	.444		100	23.50		123.50	146
8260	Hanger, 4" W x 4" D		100	.080		22	4.27		26.27	31
8270	6" W x 6" D		80	.100		28	5.35		33.35	38.50
8280	Straight tube section fiberglass, 4"W x 4"D, 12" long		40	.200		161	10.65		171.65	194
8290	24" long		20	.400		195	21.50		216.50	246
8300	36" long		13.30	.602		243	32		275	315
8310	48" long		10	.800		235	42.50		277.50	325
8320	60" long		8	1		290	53.50		343.50	400
8330	120" long		4	2		445	107		552	650

26 27 19 – Multi-Outlet Assemblies

26 27 19.10 Wiring Duct

		Crew	Daily Output	Labor-Hours	Unit	Material	2014 Bare Costs Labor	Equipment	Total	Total Incl O&P
0010	**WIRING DUCT** Plastic									
1250	PVC, snap-in slots, adhesive backed									
1270	1-1/2"W x 2"H	2 Elec	120	.133	L.F.	4.86	7.10		11.96	16
1280	1-1/2"W x 3"H		120	.133		5.85	7.10		12.95	17.05
1290	1-1/2"W x 4"H		120	.133		6.90	7.10		14	18.25
1300	2"W x 1"H		120	.133		4.36	7.10		11.46	15.45
1310	2"W x 1-1/2"H		120	.133		4.73	7.10		11.83	15.85
1320	2"W x 2"H		120	.133		4.75	7.10		11.85	15.90
1340	2"W x 3"H		120	.133		6.20	7.10		13.30	17.45
1350	2"W x 4"H		120	.133		7.50	7.10		14.60	18.90
1360	2-1/2"W x 3"H		120	.133		6.65	7.10		13.75	18
1370	3"W x 1"H		110	.145		4.95	7.75		12.70	17.05
1390	3"W x 2"H		110	.145		5.90	7.75		13.65	18.10
1400	3"W x 3"H		110	.145		7.15	7.75		14.90	19.45

26 27 19 – Multi-Outlet Assemblies

26 27 19.10 Wiring Duct		Crew	Daily Output	Labor-Hours	Unit	Material	2014 Bare Costs Labor	Equipment	Total	Total Incl O&P
1410	3"W x 4"H	2 Elec	110	.145	L.F.	8.95	7.75		16.70	21.50
1420	3"W x 5"H		110	.145		11.75	7.75		19.50	24.50
1430	4"W x 1-1/2"H		100	.160		5.90	8.55		14.45	19.30
1440	4"W x 2"H		100	.160		6.90	8.55		15.45	20.50
1450	4"W x 3"H		100	.160		7.80	8.55		16.35	21.50
1460	4"W x 4"H		100	.160		9.45	8.55		18	23
1470	4"W x 5"H		100	.160		13.30	8.55		21.85	27.50
1550	Cover, 1-1/2"W		200	.080		.99	4.27		5.26	7.50
1560	2"W		200	.080		1.22	4.27		5.49	7.75
1570	2-1/2"W		200	.080		1.53	4.27		5.80	8.10
1580	3"W		200	.080		1.86	4.27		6.13	8.45
1590	4"W		200	.080		2.24	4.27		6.51	8.85

26 27 23 – Indoor Service Poles

26 27 23.40 Surface Raceway		Crew	Daily Output	Labor-Hours	Unit	Material	2014 Bare Costs Labor	Equipment	Total	Total Incl O&P
0010	**SURFACE RACEWAY**									
0090	Metal, straight section									
0100	No. 500	1 Elec	100	.080	L.F.	1.04	4.27		5.31	7.55
0110	No. 700		100	.080		1.17	4.27		5.44	7.70
0400	No. 1500, small pancake		90	.089		2.15	4.74		6.89	9.45
0600	No. 2000, base & cover, blank		90	.089		2.19	4.74		6.93	9.50
0610	Receptacle, 6" O.C.		40	.200		15.40	10.65		26.05	33
0620	12" O.C.		44	.182		9.85	9.70		19.55	25.50
0630	18" O.C.		46	.174		8.25	9.30		17.55	23
0650	30" O.C.		50	.160		4.89	8.55		13.44	18.20
0670	No. 2400, base & cover, blank		80	.100		2.12	5.35		7.47	10.35
0680	Receptacle, 6" O.C.		42	.190		34.50	10.15		44.65	53
0690	12" O.C.		53	.151		20.50	8.05		28.55	34.50
0700	18" O.C.		55	.145		14.85	7.75		22.60	28
0710	24" O.C.		57	.140		9.40	7.50		16.90	21.50
0720	30" O.C.		59	.136		6.85	7.25		14.10	18.40
0730	60" O.C.		61	.131		5.20	7		12.20	16.15
0800	No. 3000, base & cover, blank		75	.107		4.40	5.70		10.10	13.35
0810	Receptacle, 6" O.C.		45	.178		37	9.50		46.50	55
0820	12" O.C.		62	.129		21	6.90		27.90	33.50
0830	18" O.C.		64	.125		16.95	6.65		23.60	28.50
0840	24" O.C.		66	.121		12.75	6.45		19.20	23.50
0850	30" O.C.		68	.118		11.45	6.30		17.75	22
0860	60" O.C.		70	.114		8.40	6.10		14.50	18.35
1000	No. 4000, base & cover, blank		65	.123		7.30	6.55		13.85	17.90
1010	Receptacle, 6" O.C.		41	.195		50	10.40		60.40	70.50
1020	12" O.C.		52	.154		31	8.20		39.20	46.50
1030	18" O.C.		54	.148		25.50	7.90		33.40	40
1040	24" O.C.		56	.143		21	7.60		28.60	34.50
1050	30" O.C.		58	.138		18.80	7.35		26.15	31.50
1060	60" O.C.		60	.133		14.45	7.10		21.55	26.50
1200	No. 6000, base & cover, blank		50	.160		11.40	8.55		19.95	25.50
1210	Receptacle, 6" O.C.		30	.267		61	14.25		75.25	88.50
1220	12" O.C.		37	.216		40.50	11.55		52.05	62
1230	18" O.C.		39	.205		34	10.95		44.95	54
1240	24" O.C.		41	.195		28	10.40		38.40	46.50
1250	30" O.C.		43	.186		26.50	9.95		36.45	44.50
1260	60" O.C.		45	.178		20.50	9.50		30	36.50

26 27 23.40 Surface Raceway		Crew	Daily Output	Labor-Hours	Unit	Material	2014 Bare Costs Labor	Equipment	Total	Total Incl O&P
2400	Fittings, elbows, No. 500	1 Elec	40	.200	Ea.	1.90	10.65		12.55	18.05
2800	Elbow cover, No. 2000		40	.200		3.57	10.65		14.22	19.90
2880	Tee, No. 500		42	.190		3.66	10.15		13.81	19.25
2900	No. 2000		27	.296		11.85	15.80		27.65	36.50
3000	Switch box, No. 500		16	.500		11	26.50		37.50	52
3400	Telephone outlet, No. 1500		16	.500		14.10	26.50		40.60	55.50
3600	Junction box, No. 1500	▼	16	.500	▼	9.65	26.50		36.15	50.50
3800	Plugmold wired sections, No. 2000									
4000	1 circuit, 6 outlets, 3 ft. long	1 Elec	8	1	Ea.	36	53.50		89.50	120
4100	2 circuits, 8 outlets, 6 ft. long		5.30	1.509		53	80.50		133.50	180
4110	Tele-power pole, alum, w/2 recept, 10'		4	2		205	107		312	385
4120	12'		3.85	2.078		196	111		307	380
4130	15'		3.70	2.162		310	115		425	515
4140	Steel, w/2 recept, 10'		4	2		130	107		237	300
4150	One phone fitting, 10'		4	2		146	107		253	320
4160	Alum, 4 outlets, 10'	▼	3.70	2.162	▼	262	115		377	460
4300	Overhead distribution systems, 125 volt									
4800	No. 2000, entrance end fitting	1 Elec	20	.400	Ea.	5.50	21.50		27	38
5000	Blank end fitting		40	.200		2.23	10.65		12.88	18.40
5200	Supporting clip		40	.200		1.32	10.65		11.97	17.40
5800	No. 3000, entrance end fitting		20	.400		9.15	21.50		30.65	42
6000	Blank end fitting		40	.200		2.59	10.65		13.24	18.80
6020	Internal elbow		20	.400		11.95	21.50		33.45	45
6030	External elbow		20	.400		16.95	21.50		38.45	50.50
6040	Device bracket		53	.151		4.06	8.05		12.11	16.50
6400	Hanger clamp		32	.250	▼	6.20	13.35		19.55	27
7000	No. 4000 Base		90	.089	L.F.	4.36	4.74		9.10	11.90
7200	Divider		100	.080	"	.84	4.27		5.11	7.30
7400	Entrance end fitting		16	.500	Ea.	22	26.50		48.50	64.50
7600	Blank end fitting		40	.200		6.15	10.65		16.80	22.50
7610	Recpt. & tele. cover		53	.151		10.45	8.05		18.50	23.50
7620	External elbow		16	.500		34	26.50		60.50	77.50
7630	Coupling		53	.151		5.20	8.05		13.25	17.80
7640	Divider clip & coupling		80	.100		1.03	5.35		6.38	9.15
7650	Panel connector		16	.500		21.50	26.50		48	64
7800	Take off connector		16	.500		72	26.50		98.50	119
8000	No. 6000, take off connector		16	.500		85.50	26.50		112	134
8100	Take off fitting		16	.500		63.50	26.50		90	110
8200	Hanger clamp	▼	32	.250		15.40	13.35		28.75	37
8230	Coupling					8.40			8.40	9.25
8240	One gang device plate	1 Elec	53	.151		8.20	8.05		16.25	21
8250	Two gang device plate		40	.200		10.05	10.65		20.70	27
8260	Blank end fitting		40	.200		8.70	10.65		19.35	25.50
8270	Combination elbow		14	.571		35.50	30.50		66	84.50
8300	Panel connector	▼	16	.500	▼	15.60	26.50		42.10	57
8500	Chan-L-Wire system installed in 1-5/8" x 1-5/8" strut. Strut									
8600	not incl., 30 amp, 4 wire, 3 phase	1 Elec	200	.040	L.F.	4.90	2.13		7.03	8.60
8700	Junction box		8	1	Ea.	32.50	53.50		86	116
8800	Insulating end cap		40	.200		8.70	10.65		19.35	25.50
8900	Strut splice plate		40	.200		11.50	10.65		22.15	28.50
9000	Tap		40	.200		23	10.65		33.65	41
9100	Fixture hanger	▼	60	.133		9.90	7.10		17	21.50
9200	Pulling tool				▼	88			88	96.50

26 27 23 – Indoor Service Poles

26 27 23.40 Surface Raceway	Crew	Daily Output	Labor-Hours	Unit	Material	2014 Bare Costs Labor	2014 Bare Costs Equipment	Total	Total Incl O&P
9300	Non-metallic, straight section								
9310	7/16" x 7/8", base & cover, blank	1 Elec	160	.050	L.F.	1.70	2.67		4.37	5.85
9320	Base & cover w/adhesive		160	.050		1.42	2.67		4.09	5.55
9340	7/16" x 1-5/16", base & cover, blank		145	.055		1.86	2.94		4.80	6.45
9350	Base & cover w/adhesive		145	.055		2.17	2.94		5.11	6.80
9370	11/16" x 2-1/4", base & cover, blank		130	.062		2.63	3.28		5.91	7.80
9380	Base & cover w/adhesive		130	.062		3.01	3.28		6.29	8.20
9385	1-11/16" x 5-1/4", two compartment base & cover w/screws		80	.100	▼	7.45	5.35		12.80	16.20
9400	Fittings, elbows, 7/16" x 7/8"		50	.160	Ea.	1.88	8.55		10.43	14.85
9410	7/16" x 1-5/16"		45	.178		1.95	9.50		11.45	16.35
9420	11/16" x 2-1/4"		40	.200		2.11	10.65		12.76	18.25
9425	1-11/16" x 5-1/4"		28	.286		11.05	15.25		26.30	35
9430	Tees, 7/16" x 7/8"		35	.229		2.42	12.20		14.62	21
9440	7/16" x 1-5/16"		32	.250		2.49	13.35		15.84	22.50
9450	11/16" x 2-1/4"		30	.267		2.56	14.25		16.81	24.50
9455	1-11/16" x 5-1/4"		24	.333		17.75	17.80		35.55	46
9460	Cover clip, 7/16" x 7/8"		80	.100		.49	5.35		5.84	8.55
9470	7/16" x 1-5/16"		72	.111		.44	5.95		6.39	9.35
9480	11/16" x 2-1/4"		64	.125		.75	6.65		7.40	10.85
9484	1-11/16" x 5-1/4"		42	.190		2.61	10.15		12.76	18.05
9486	Wire clip, 1-11/16" x 5-1/4"		68	.118		.45	6.30		6.75	9.90
9490	Blank end, 7/16" x 7/8"		50	.160		.70	8.55		9.25	13.55
9500	7/16" x 1-5/16"		45	.178		.78	9.50		10.28	15.05
9510	11/16" x 2-1/4"		40	.200		1.18	10.65		11.83	17.25
9515	1-11/16" x 5-1/4"		38	.211		5.45	11.25		16.70	23
9520	Round fixture box, 5.5" dia x 1"		25	.320		11.10	17.05		28.15	37.50
9530	Device box, 1 gang		30	.267		5	14.25		19.25	27
9540	2 gang	▼	25	.320	▼	7.30	17.05		24.35	33.50

26 27 26 – Wiring Devices

26 27 26.10 Low Voltage Switching

		Crew	Daily Output	Labor-Hours	Unit	Material	2014 Bare Costs Labor	2014 Bare Costs Equipment	Total	Total Incl O&P
0010	**LOW VOLTAGE SWITCHING**									
3600	Relays, 120 V or 277 V standard	1 Elec	12	.667	Ea.	41	35.50		76.50	98.50
3800	Flush switch, standard		40	.200		11.40	10.65		22.05	28.50
4000	Interchangeable		40	.200		14.90	10.65		25.55	32.50
4100	Surface switch, standard		40	.200		8.05	10.65		18.70	25
4200	Transformer 115 V to 25 V		12	.667		129	35.50		164.50	195
4400	Master control, 12 circuit, manual		4	2		125	107		232	297
4500	25 circuit, motorized		4	2		139	107		246	310
4600	Rectifier, silicon		12	.667		45	35.50		80.50	103
4800	Switchplates, 1 gang, 1, 2 or 3 switch, plastic		80	.100		4.95	5.35		10.30	13.45
5000	Stainless steel		80	.100		11.25	5.35		16.60	20.50
5400	2 gang, 3 switch, stainless steel		53	.151		22.50	8.05		30.55	37
5500	4 switch, plastic		53	.151		10.25	8.05		18.30	23.50
5600	2 gang, 4 switch, stainless steel		53	.151		21.50	8.05		29.55	35.50
5700	6 switch, stainless steel		53	.151		42.50	8.05		50.55	59
5800	3 gang, 9 switch, stainless steel		32	.250		64	13.35		77.35	90.50
5900	Receptacle, triple, 1 return, 1 feed		26	.308		42.50	16.40		58.90	71
6000	2 feed		20	.400		42.50	21.50		64	78.50
6100	Relay gang boxes, flush or surface, 6 gang		5.30	1.509		89	80.50		169.50	219
6200	12 gang		4.70	1.702		100	91		191	246
6400	18 gang		4	2		111	107		218	282
6500	Frame, to hold up to 6 relays	▼	12	.667	▼	82.50	35.50		118	144

26 27 Low-Voltage Distribution Equipment

26 27 26 – Wiring Devices

26 27 26.10 Low Voltage Switching

26 27 26.10 Low Voltage Switching	Crew	Daily Output	Labor-Hours	Unit	Material	2014 Bare Costs Labor	Equipment	Total	Total Incl O&P
7200	Control wire, 2 conductor	1 Elec	6.30	1.270	C.L.F.	25.50	68		93.50	129
7400	3 conductor		5	1.600		35.50	85.50		121	167
7600	19 conductor		2.50	3.200		305	171		476	590
7800	26 conductor		2	4		415	213		628	775
8000	Weatherproof, 3 conductor		5	1.600		75	85.50		160.50	211

26 27 26.20 Wiring Devices Elements

	26 27 26.20 Wiring Devices Elements	Crew	Daily Output	Labor-Hours	Unit	Material	2014 Bare Costs Labor	Equipment	Total	Total Incl O&P
0010	**WIRING DEVICES ELEMENTS** R262726-90									
0200	Toggle switch, quiet type, single pole, 15 amp	1 Elec	40	.200	Ea.	5.70	10.65		16.35	22.50
0500	20 amp		27	.296		8.80	15.80		24.60	33
0510	30 amp		23	.348		19.50	18.55		38.05	49.50
0530	Lock handle, 20 amp		27	.296		23	15.80		38.80	48.50
0540	Security key, 20 amp		26	.308		102	16.40		118.40	137
0550	Rocker, 15 amp		40	.200		4.84	10.65		15.49	21.50
0560	20 amp		27	.296		11.20	15.80		27	36
0600	3 way, 15 amp		23	.348		5.95	18.55		24.50	34.50
0800	20 amp		18	.444		8.30	23.50		31.80	44.50
0810	30 amp		9	.889		24.50	47.50		72	98
0830	Lock handle, 20 amp		18	.444		25.50	23.50		49	63.50
0840	Security key, 20 amp		17	.471		105	25		130	153
0850	Rocker, 15 amp		23	.348		6.80	18.55		25.35	35.50
0860	20 amp		18	.444		16.10	23.50		39.60	53.50
0900	4 way, 15 amp		15	.533		10.75	28.50		39.25	54.50
1000	20 amp		11	.727		42	39		81	105
1020	Lock handle, 20 amp		11	.727		47	39		86	110
1030	Rocker, 15 amp		15	.533		37	28.50		65.50	83.50
1040	20 amp		11	.727		39.50	39		78.50	102
1100	Toggle switch, quiet type, double pole, 15 amp		15	.533		12.35	28.50		40.85	56
1200	20 amp		11	.727		16.90	39		55.90	76.50
1210	30 amp		9	.889		24.50	47.50		72	98
1230	Lock handle, 20 amp		11	.727		21.50	39		60.50	82
1250	Security key, 20 amp		10	.800		109	42.50		151.50	184
1420	Toggle switch quiet type, 1 pole, 2 throw center off, 15 amp		23	.348		57.50	18.55		76.05	91
1440	20 amp		18	.444		49.50	23.50		73	90
1460	2 pole, 2 throw center off, lock handle, 20 amp		11	.727		61	39		100	125
1480	1 pole, momentary contact, 15 amp		23	.348		17.95	18.55		36.50	48
1500	20 amp		18	.444		29.50	23.50		53	68
1520	Momentary contact, lock handle, 20 amp		18	.444		30.50	23.50		54	69
1650	Dimmer switch, 120 volt, incandescent, 600 watt, 1 pole G		16	.500		18.40	26.50		44.90	60
1700	600 watt, 3 way G		12	.667		17.75	35.50		53.25	72.50
1750	1000 watt, 1 pole G		16	.500		44.50	26.50		71	89
1800	1000 watt, 3 way G		12	.667		72.50	35.50		108	133
2000	1500 watt, 1 pole G		11	.727		83	39		122	150
2100	2000 watt, 1 pole G		8	1		127	53.50		180.50	220
2110	Fluorescent, 600 watt G		15	.533		108	28.50		136.50	162
2120	1000 watt G		15	.533		131	28.50		159.50	187
2130	1500 watt G		10	.800		244	42.50		286.50	330
2160	Explosionproof, toggle switch, wall, single pole 20 amp		5.30	1.509		186	80.50		266.50	325
2180	Receptacle, single outlet, 20 amp		5.30	1.509		315	80.50		395.50	465
2190	30 amp		4	2		680	107		787	910
2290	60 amp		2.50	3.200		765	171		936	1,100
2360	Plug, 20 amp		16	.500		172	26.50		198.50	230
2370	30 amp		12	.667		293	35.50		328.50	375

26 27 26.20 Wiring Devices Elements		Crew	Daily Output	Labor-Hours	Unit	Material	2014 Bare Costs Labor	Equipment	Total	Total Incl O&P
2380	60 amp	1 Elec	8	1	Ea.	375	53.50		428.50	495
2410	Furnace, thermal cutoff switch with plate		26	.308		15.60	16.40		32	41.50
2460	Receptacle, duplex, 120 volt, grounded, 15 amp		40	.200		1.22	10.65		11.87	17.30
2470	20 amp		27	.296		8.35	15.80		24.15	32.50
2480	Ground fault interrupting, 15 amp		27	.296		37.50	15.80		53.30	65
2482	20 amp		27	.296		39.50	15.80		55.30	67
2486	Clock receptacle, 15 amp		40	.200		27	10.65		37.65	46
2490	Dryer, 30 amp		15	.533		5.20	28.50		33.70	48
2500	Range, 50 amp		11	.727		11.85	39		50.85	71
2530	Surge suppresser receptacle, duplex, 20 amp		27	.296		40	15.80		55.80	67.50
2532	Quad, 20 amp		20	.400		79	21.50		100.50	119
2540	Isolated ground receptacle, duplex, 20 amp		27	.296		20.50	15.80		36.30	46
2542	Quad, 20 amp		20	.400		37	21.50		58.50	72.50
2550	Simplex, 20 amp		27	.296		22	15.80		37.80	48
2560	Simplex, 30 amp		15	.533		27	28.50		55.50	72
2570	Cable reel w/receptacle 50' w/3#12, 120 V, 20 A	2 Elec	2.67	5.999		895	320		1,215	1,475
2600	Wall plates, stainless steel, 1 gang	1 Elec	80	.100		2.53	5.35		7.88	10.80
2800	2 gang		53	.151		4.29	8.05		12.34	16.75
3000	3 gang		32	.250		8.40	13.35		21.75	29
3100	4 gang		27	.296		10.90	15.80		26.70	35.50
3110	Brown plastic, 1 gang		80	.100		.38	5.35		5.73	8.40
3120	2 gang		53	.151		.75	8.05		8.80	12.90
3130	3 gang		32	.250		1.11	13.35		14.46	21
3140	4 gang		27	.296		2.85	15.80		18.65	26.50
3150	Brushed brass, 1 gang		80	.100		4.19	5.35		9.54	12.60
3160	Anodized aluminum, 1 gang		80	.100		2.98	5.35		8.33	11.30
3170	Switch cover, weatherproof, 1 gang		60	.133		8.25	7.10		15.35	19.75
3180	Vandal proof lock, 1 gang		60	.133		13.90	7.10		21	26
3200	Lampholder, keyless		26	.308		11	16.40		27.40	36.50
3400	Pullchain with receptacle		22	.364		20.50	19.40		39.90	51.50
3500	Pilot light, neon with jewel		27	.296		11.25	15.80		27.05	36
3600	Receptacle, 20 amp, 250 volt, NEMA 6		27	.296		21	15.80		36.80	47
3620	277 volt NEMA 7		27	.296		15.60	15.80		31.40	40.50
3640	125/250 volt NEMA 10		27	.296		17.50	15.80		33.30	43
3680	125/250 volt NEMA 14		25	.320		22.50	17.05		39.55	50
3700	3 pole, 250 volt NEMA 15		25	.320		22.50	17.05		39.55	50
3720	120/208 volt NEMA 18		25	.320		23	17.05		40.05	51
3740	30 amp, 125 volt NEMA 5		15	.533		17.30	28.50		45.80	61.50
3760	250 volt NEMA 6		15	.533		19.75	28.50		48.25	64
3780	277 volt NEMA 7		15	.533		26	28.50		54.50	71
3820	125/250 volt NEMA 14		14	.571		45.50	30.50		76	95.50
3840	3 pole, 250 volt NEMA 15		14	.571		44.50	30.50		75	94.50
3880	50 amp, 125 volt NEMA 5		11	.727		22	39		61	82.50
3900	250 volt NEMA 6		11	.727		22.50	39		61.50	83
3920	277 volt NEMA 7		11	.727		26	39		65	86.50
3960	125/250 volt NEMA 14		10	.800		59	42.50		101.50	129
3980	3 pole 250 volt NEMA 15		10	.800		60.50	42.50		103	131
4020	60 amp, 125/250 volt, NEMA 14		8	1		69	53.50		122.50	156
4040	3 pole, 250 volt NEMA 15		8	1		72.50	53.50		126	160
4060	120/208 volt NEMA 18		8	1		102	53.50		155.50	192
4100	Receptacle locking, 20 amp, 125 volt NEMA L5		27	.296		16.35	15.80		32.15	41.50
4120	250 volt NEMA L6		27	.296		16.25	15.80		32.05	41.50
4140	277 volt NEMA L7		27	.296		16.85	15.80		32.65	42

26 27 26 – Wiring Devices

26 27 26.20 Wiring Devices Elements	Crew	Daily Output	Labor-Hours	Unit	Material	2014 Bare Costs Labor	Equipment	Total	Total Incl O&P
4150	3 pole, 250 volt, NEMA L11	1 Elec	27	.296	Ea.	23	15.80		38.80	49
4160	20 amp, 480 volt NEMA L8		27	.296		19.10	15.80		34.90	44.50
4180	600 volt NEMA L9		27	.296		28.50	15.80		44.30	55
4200	125/250 volt NEMA L10		27	.296		23	15.80		38.80	49
4230	125/250 volt NEMA L14		25	.320		26.50	17.05		43.55	54.50
4280	250 volt NEMA L15		25	.320		28	17.05		45.05	56.50
4300	480 volt NEMA L16		25	.320		24.50	17.05		41.55	52
4320	3 phase, 120/208 volt NEMA L18		25	.320		31.50	17.05		48.55	60
4340	277/480 volt NEMA L19		25	.320		32	17.05		49.05	60.50
4360	347/600 volt NEMA L20		25	.320		31.50	17.05		48.55	60.50
4380	120/208 volt NEMA L21		23	.348		28.50	18.55		47.05	59.50
4400	277/480 volt NEMA L22		23	.348		30	18.55		48.55	61
4420	347/600 volt NEMA L23		23	.348		31.50	18.55		50.05	62.50
4440	30 amp, 125 volt NEMA L5		15	.533		23	28.50		51.50	67.50
4460	250 volt NEMA L6		15	.533		25	28.50		53.50	70
4480	277 volt NEMA L7		15	.533		25.50	28.50		54	70.50
4500	480 volt NEMA L8		15	.533		27	28.50		55.50	72
4520	600 volt NEMA L9		15	.533		27	28.50		55.50	72
4540	125/250 volt NEMA L10		15	.533		29.50	28.50		58	75
4560	3 phase, 250 volt NEMA L11		15	.533		27	28.50		55.50	72.50
4620	125/250 volt NEMA L14		14	.571		35.50	30.50		66	85
4640	250 volt NEMA L15		14	.571		36.50	30.50		67	85.50
4660	480 volt NEMA L16		14	.571		38	30.50		68.50	87.50
4680	600 volt NEMA L17		14	.571		37	30.50		67.50	86.50
4700	120/208 volt NEMA L18		14	.571		46.50	30.50		77	96.50
4720	277/480 volt NEMA L19		14	.571		46.50	30.50		77	96.50
4740	347/600 volt NEMA L20		14	.571		46.50	30.50		77	96.50
4760	120/208 volt NEMA L21		13	.615		39.50	33		72.50	92.50
4780	277/480 volt NEMA L22		13	.615		41	33		74	94.50
4800	347/600 volt NEMA L23		13	.615		52	33		85	107
4840	Receptacle, corrosion resistant, 15 or 20 amp, 125 volt NEMA L5		27	.296		25	15.80		40.80	51
4860	250 volt NEMA L6		27	.296		21.50	15.80		37.30	47
4900	Receptacle, cover plate, phenolic plastic, NEMA 5 & 6		80	.100		.61	5.35		5.96	8.65
4910	NEMA 7-23		80	.100		.74	5.35		6.09	8.80
4920	Stainless steel, NEMA 5 & 6		80	.100		2.68	5.35		8.03	10.95
4930	NEMA 7-23		80	.100		2.49	5.35		7.84	10.75
4940	Brushed brass NEMA 5 & 6		80	.100		5.35	5.35		10.70	13.90
4950	NEMA 7-23		80	.100		5.80	5.35		11.15	14.40
4960	Anodized aluminum, NEMA 5 & 6		80	.100		3.48	5.35		8.83	11.85
4970	NEMA 7-23		80	.100		5.75	5.35		11.10	14.35
4980	Weatherproof NEMA 7-23		60	.133		32	7.10		39.10	45.50
5100	Plug, 20 amp, 250 volt, NEMA 6		30	.267		15.20	14.25		29.45	38.50
5110	277 volt NEMA 7		30	.267		19	14.25		33.25	42.50
5120	3 pole, 120/250 volt, NEMA 10		26	.308		17.70	16.40		34.10	44
5130	125/250 volt NEMA 14		26	.308		36	16.40		52.40	64
5140	250 volt NEMA 15		26	.308		39.50	16.40		55.90	68
5150	120/208 volt NEMA 8		26	.308		42.50	16.40		58.90	71.50
5160	30 amp, 125 volt NEMA 5		13	.615		54.50	33		87.50	109
5170	250 volt NEMA 6		13	.615		52	33		85	106
5180	277 volt NEMA 7		13	.615		60.50	33		93.50	116
5190	125/250 volt NEMA 14		13	.615		41	33		74	94.50
5200	3 pole, 250 volt NEMA 15		12	.667		43	35.50		78.50	100
5210	50 amp, 125 volt NEMA 5		9	.889		61	47.50		108.50	138

26 27 26 – Wiring Devices

26 27 26.20 Wiring Devices Elements	Crew	Daily Output	Labor-Hours	Unit	Material	2014 Bare Costs Labor	Equipment	Total	Total Incl O&P
5220	250 volt NEMA 6	1 Elec	9	.889	Ea.	64	47.50		111.50	142
5230	277 volt NEMA 7		9	.889		68	47.50		115.50	146
5240	125/250 volt NEMA 14		9	.889		52.50	47.50		100	129
5250	3 pole, 250 volt NEMA 15		8	1		53.50	53.50		107	139
5260	60 amp, 125/250 volt NEMA 14		7	1.143		60	61		121	158
5270	3 pole, 250 volt NEMA 15		7	1.143		63.50	61		124.50	162
5280	120/208 volt NEMA 18		7	1.143		82	61		143	182
5300	Plug angle, 20 amp, 250 volt NEMA 6		30	.267		23.50	14.25		37.75	47.50
5310	30 amp, 125 volt NEMA 5		13	.615		42	33		75	95
5320	250 volt NEMA 6		13	.615		43.50	33		76.50	96.50
5330	277 volt NEMA 7		13	.615		49.50	33		82.50	104
5340	125/250 volt NEMA 14		13	.615		45.50	33		78.50	99.50
5350	3 pole, 250 volt NEMA 15		12	.667		47	35.50		82.50	105
5360	50 amp, 125 volt NEMA 5		9	.889		44	47.50		91.50	120
5370	250 volt NEMA 6		9	.889		45	47.50		92.50	121
5380	277 volt NEMA 7		9	.889		50.50	47.50		98	127
5390	125/250 volt NEMA 14		9	.889		57	47.50		104.50	134
5400	3 pole, 250 volt NEMA 15		8	1		59	53.50		112.50	145
5410	60 amp, 125/250 volt NEMA 14		7	1.143		66.50	61		127.50	165
5420	3 pole, 250 volt NEMA 15		7	1.143		69.50	61		130.50	168
5430	120/208 volt NEMA 18		7	1.143		70	61		131	169
5500	Plug, locking, 20 amp, 125 volt NEMA L5		30	.267		12.95	14.25		27.20	36
5510	250 volt NEMA L6		30	.267		12.90	14.25		27.15	35.50
5520	277 volt NEMA L7		30	.267		13.25	14.25		27.50	36
5530	480 volt NEMA L8		30	.267		14.75	14.25		29	38
5540	600 volt NEMA L9		30	.267		17.15	14.25		31.40	40.50
5550	3 pole, 125/250 volt NEMA L10		26	.308		21.50	16.40		37.90	48.50
5560	250 volt NEMA L11		26	.308		21.50	16.40		37.90	48.50
5570	480 volt NEMA L12		26	.308		25.50	16.40		41.90	52.50
5580	125/250 volt NEMA L14		26	.308		19.55	16.40		35.95	46
5590	250 volt NEMA L15		26	.308		20.50	16.40		36.90	47.50
5600	480 volt NEMA L16		26	.308		22	16.40		38.40	48.50
5610	4 pole, 120/208 volt NEMA L18		24	.333		28	17.80		45.80	57.50
5620	277/480 volt NEMA L19		24	.333		28	17.80		45.80	57.50
5630	347/600 volt NEMA L20		24	.333		28.50	17.80		46.30	57.50
5640	120/208 volt NEMA L21		24	.333		25	17.80		42.80	54.50
5650	277/480 volt NEMA L22		24	.333		26.50	17.80		44.30	56
5660	347/600 volt NEMA L23		24	.333		27.50	17.80		45.30	56.50
5670	30 amp, 125 volt NEMA L5		13	.615		19.85	33		52.85	71
5680	250 volt NEMA L6		13	.615		20.50	33		53.50	71.50
5690	277 volt NEMA L7		13	.615		19.85	33		52.85	71
5700	480 volt NEMA L8		13	.615		21.50	33		54.50	72.50
5710	600 volt NEMA L9		13	.615		20.50	33		53.50	72
5720	3 pole, 125/250 volt NEMA L10		11	.727		23.50	39		62.50	84
5730	250 volt NEMA L11		11	.727		23.50	39		62.50	84
5760	125/250 volt NEMA L14		11	.727		26.50	39		65.50	87.50
5770	250 volt NEMA L15		11	.727		26.50	39		65.50	87.50
5780	480 volt NEMA L16		11	.727		28	39		67	88.50
5790	600 volt NEMA L17		11	.727		29.50	39		68.50	90
5800	4 pole, 120/208 volt NEMA L18		10	.800		35	42.50		77.50	103
5810	120/208 volt NEMA L19		10	.800		35	42.50		77.50	103
5820	347/600 volt NEMA L20		10	.800		35.50	42.50		78	103
5830	120/208 volt NEMA L21		10	.800		30	42.50		72.50	97

26 27 26.20 Wiring Devices Elements	Crew	Daily Output	Labor-Hours	Unit	Material	2014 Bare Costs Labor	Equipment	Total	Total Incl O&P
5840	277/480 volt NEMA L22	1 Elec	10	.800	Ea.	32	42.50		74.50	99
5850	347/600 volt NEMA L23		10	.800		39.50	42.50		82	108
6000	Connector, 20 amp, 250 volt NEMA 6		30	.267		23.50	14.25		37.75	47.50
6010	277 volt NEMA 7		30	.267		30.50	14.25		44.75	55
6020	3 pole, 120/250 volt NEMA 10		26	.308		34.50	16.40		50.90	62.50
6030	125/250 volt NEMA 14		26	.308		34.50	16.40		50.90	62.50
6040	250 volt NEMA 15		26	.308		34.50	16.40		50.90	62.50
6050	120/208 volt NEMA 18		26	.308		37.50	16.40		53.90	66
6060	30 amp, 125 volt NEMA 5		13	.615		59.50	33		92.50	115
6070	250 volt NEMA 6		13	.615		59.50	33		92.50	115
6080	277 volt NEMA 7		13	.615		59.50	33		92.50	115
6110	50 amp, 125 volt NEMA 5		9	.889		82	47.50		129.50	161
6120	250 volt NEMA 6		9	.889		82	47.50		129.50	161
6130	277 volt NEMA 7		9	.889		82	47.50		129.50	161
6200	Connector, locking, 20 amp, 125 volt NEMA L5		30	.267		19.90	14.25		34.15	43.50
6210	250 volt NEMA L6		30	.267		20	14.25		34.25	43.50
6220	277 volt NEMA L7		30	.267		20.50	14.25		34.75	44
6230	480 volt NEMA L8		30	.267		26	14.25		40.25	50
6240	600 volt NEMA L9		30	.267		26	14.25		40.25	50
6250	3 pole, 125/250 volt NEMA L10		26	.308		33	16.40		49.40	61
6260	250 volt NEMA L11		26	.308		33	16.40		49.40	61
6280	125/250 volt NEMA L14		26	.308		27.50	16.40		43.90	54.50
6290	250 volt NEMA L15		26	.308		27.50	16.40		43.90	54.50
6300	480 volt NEMA L16		26	.308		29	16.40		45.40	56.50
6310	4 pole, 120/208 volt NEMA L18		24	.333		37.50	17.80		55.30	68
6320	277/480 volt NEMA L19		24	.333		40	17.80		57.80	70.50
6330	347/600 volt NEMA L20		24	.333		39.50	17.80		57.30	70
6340	120/208 volt NEMA L21		24	.333		42.50	17.80		60.30	73
6350	277/480 volt NEMA L22		24	.333		45	17.80		62.80	76
6360	347/600 volt NEMA L23		24	.333		52.50	17.80		70.30	84
6370	30 amp, 125 volt NEMA L5		13	.615		40.50	33		73.50	93.50
6380	250 volt NEMA L6		13	.615		40.50	33		73.50	93.50
6390	277 volt NEMA L7		13	.615		41	33		74	94
6400	480 volt NEMA L8		13	.615		43	33		76	96.50
6410	600 volt NEMA L9		13	.615		42	33		75	95.50
6420	3 pole, 125/250 volt NEMA L10		11	.727		64	39		103	129
6430	250 volt NEMA L11		11	.727		64	39		103	129
6460	125/250 volt NEMA L14		11	.727		55	39		94	119
6470	250 volt NEMA L15		11	.727		55.50	39		94.50	119
6480	480 volt NEMA L16		11	.727		58.50	39		97.50	123
6490	600 volt NEMA L17		11	.727		60.50	39		99.50	125
6500	4 pole, 120/208 volt NEMA L18		10	.800		72	42.50		114.50	143
6510	120/208 volt NEMA L19		10	.800		71.50	42.50		114	143
6520	347/600 volt NEMA L20		10	.800		72	42.50		114.50	143
6530	120/208 volt NEMA L21		10	.800		56	42.50		98.50	126
6540	277/480 volt NEMA L22		10	.800		57	42.50		99.50	127
6550	347/600 volt NEMA L23		10	.800		70.50	42.50		113	142
7000	Receptacle computer, 250 volt, 15 amp, 3 pole 4 wire		8	1		82	53.50		135.50	170
7010	20 amp, 2 pole 3 wire		8	1		82	53.50		135.50	170
7020	30 amp, 2 pole 3 wire		6.50	1.231		133	65.50		198.50	245
7030	30 amp, 3 pole 4 wire		6.50	1.231		144	65.50		209.50	257
7040	60 amp, 3 pole 4 wire		4.50	1.778		242	95		337	410
7050	100 amp, 3 pole 4 wire		3	2.667		300	142		442	545

26 27 Low-Voltage Distribution Equipment

26 27 26 – Wiring Devices

26 27 26.20 Wiring Devices Elements

		Crew	Daily Output	Labor-Hours	Unit	Material	2014 Bare Costs Labor	Equipment	Total	Total Incl O&P
7100	Connector computer, 250 volt, 15 amp, 3 pole 4 wire	1 Elec	27	.296	Ea.	128	15.80		143.80	165
7110	20 amp, 2 pole 3 wire		27	.296		114	15.80		129.80	149
7120	30 amp, 2 pole 3 wire		15	.533		175	28.50		203.50	235
7130	30 amp, 3 pole 4 wire		15	.533		181	28.50		209.50	242
7140	60 amp, 3 pole 4 wire		8	1		300	53.50		353.50	410
7150	100 amp, 3 pole 4 wire		4	2		415	107		522	615
7200	Plug, computer, 250 volt, 15 amp, 3 pole 4 wire		27	.296		115	15.80		130.80	151
7210	20 amp, 2 pole, 3 wire		27	.296		106	15.80		121.80	141
7220	30 amp, 2 pole, 3 wire		15	.533		188	28.50		216.50	250
7230	30 amp, 3 pole, 4 wire		15	.533		182	28.50		210.50	243
7240	60 amp, 3 pole, 4 wire		8	1		274	53.50		327.50	380
7250	100 amp, 3 pole, 4 wire		4	2		340	107		447	535
7300	Connector adapter to flexible conduit, 1/2"		60	.133		3.25	7.10		10.35	14.25
7310	3/4"		50	.160		4.61	8.55		13.16	17.85
7320	1-1/4"		30	.267		14.95	14.25		29.20	38
7330	1-1/2"		23	.348		21.50	18.55		40.05	51.50

26 27 73 – Door Chimes

26 27 73.10 Doorbell System

		Crew	Daily Output	Labor-Hours	Unit	Material	2014 Bare Costs Labor	Equipment	Total	Total Incl O&P
0010	DOORBELL SYSTEM, incl. transformer, button & signal									
0100	6" bell	1 Elec	4	2	Ea.	113	107		220	284
0200	Buzzer		4	2		104	107		211	274
1000	Door chimes, 2 notes, minimum		16	.500		24.50	26.50		51	67
1020	Maximum		12	.667		115	35.50		150.50	179
1100	Tube type, 3 tube system		12	.667		187	35.50		222.50	259
1180	4 tube system		10	.800		385	42.50		427.50	490
1900	For transformer & button, minimum add		5	1.600		16.30	85.50		101.80	146
1960	Maximum, add		4.50	1.778		78	95		173	228
3000	For push button only, minimum		24	.333		3.03	17.80		20.83	30
3100	Maximum		20	.400		21.50	21.50		43	55.50
3200	Bell transformer		16	.500		21	26.50		47.50	63

26 28 Low-Voltage Circuit Protective Devices

26 28 13 – Fuses

26 28 13.10 Fuse Elements

		Crew	Daily Output	Labor-Hours	Unit	Material	2014 Bare Costs Labor	Equipment	Total	Total Incl O&P
0010	FUSE ELEMENTS									
0020	Cartridge, nonrenewable									
0050	250 volt, 30 amp	1 Elec	50	.160	Ea.	2.32	8.55		10.87	15.35
0100	60 amp		50	.160		3.60	8.55		12.15	16.75
0150	100 amp		40	.200		15.10	10.65		25.75	32.50
0200	200 amp		36	.222		35.50	11.85		47.35	57
0250	400 amp		30	.267		70	14.25		84.25	98.50
0300	600 amp		24	.333		115	17.80		132.80	154
0400	600 volt, 30 amp		40	.200		8.30	10.65		18.95	25
0450	60 amp		40	.200		13.60	10.65		24.25	31
0500	100 amp		36	.222		27.50	11.85		39.35	48
0550	200 amp		30	.267		68	14.25		82.25	96.50
0600	400 amp		24	.333		141	17.80		158.80	182
0650	600 amp		20	.400		209	21.50		230.50	262
0800	Dual element, time delay, 250 volt, 30 amp		50	.160		8.25	8.55		16.80	22
0840	50 amp		50	.160		10.25	8.55		18.80	24

26 28 13 – Fuses

26 28 13.10 Fuse Elements		Crew	Daily Output	Labor-Hours	Unit	Material	2014 Bare Costs Labor	Equipment	Total	Total Incl O&P
0850	60 amp	1 Elec	50	.160	Ea.	12.10	8.55		20.65	26
0900	100 amp		40	.200		28	10.65		38.65	47
0950	200 amp		36	.222		61.50	11.85		73.35	85.50
1000	400 amp		30	.267		110	14.25		124.25	143
1050	600 amp		24	.333		181	17.80		198.80	226
1300	600 volt, 15 to 30 amp		40	.200		16.40	10.65		27.05	34
1350	35 to 60 amp		40	.200		28	10.65		38.65	46.50
1400	70 to 100 amp		36	.222		58	11.85		69.85	81.50
1450	110 to 200 amp		30	.267		117	14.25		131.25	150
1500	225 to 400 amp		24	.333		233	17.80		250.80	283
1550	600 amp		20	.400		335	21.50		356.50	400
1800	Class RK1, high capacity, 250 volt, 30 amp		50	.160		7.65	8.55		16.20	21.50
1850	60 amp		50	.160		14.30	8.55		22.85	28.50
1900	100 amp		40	.200		32	10.65		42.65	51
1950	200 amp		36	.222		70	11.85		81.85	95
2000	400 amp		30	.267		126	14.25		140.25	160
2050	600 amp		24	.333		175	17.80		192.80	219
2200	600 volt, 30 amp		40	.200		14.30	10.65		24.95	31.50
2250	60 amp		40	.200		30.50	10.65		41.15	49.50
2300	100 amp		36	.222		64.50	11.85		76.35	89
2350	200 amp		30	.267		127	14.25		141.25	161
2400	400 amp		24	.333		252	17.80		269.80	305
2450	600 amp		20	.400		270	21.50		291.50	330
2700	Class J, current limiting, 250 or 600 volt, 30 amp		40	.200		21.50	10.65		32.15	40
2750	60 amp		40	.200		35.50	10.65		46.15	55
2800	100 amp		36	.222		53.50	11.85		65.35	77
2850	200 amp		30	.267		100	14.25		114.25	132
2900	400 amp		24	.333		237	17.80		254.80	288
2950	600 amp		20	.400		355	21.50		376.50	420
3100	Class L, current limiting, 250 or 600 volt, 601 to 1200 amp		16	.500		605	26.50		631.50	705
3150	1500-1600 amp		13	.615		780	33		813	910
3200	1800-2000 amp		10	.800		1,050	42.50		1,092.50	1,225
3250	2500 amp		10	.800		1,175	42.50		1,217.50	1,375
3300	3000 amp		8	1		1,575	53.50		1,628.50	1,825
3350	3500-4000 amp		8	1		1,675	53.50		1,728.50	1,900
3400	4500-5000 amp		6.70	1.194		3,125	63.50		3,188.50	3,525
3450	6000 amp		5.70	1.404		4,200	75		4,275	4,700
3600	Plug, 120 volt, 1 to 10 amp		50	.160		2.72	8.55		11.27	15.80
3650	15 to 30 amp		50	.160		5.70	8.55		14.25	19.10
3700	Dual element 0.3 to 14 amp		50	.160		3.99	8.55		12.54	17.20
3750	15 to 30 amp		50	.160		7.80	8.55		16.35	21.50
3800	Fustat, 120 volt, 15 to 30 amp		50	.160		5.25	8.55		13.80	18.60
3850	0.3 to 14 amp		50	.160		8.40	8.55		16.95	22
3900	Adapters 0.3 to 10 amp		50	.160		5.80	8.55		14.35	19.20
3950	15 to 30 amp		50	.160		9.30	8.55		17.85	23

26 28 16 – Enclosed Switches and Circuit Breakers

26 28 16.10 Circuit Breakers

		Crew	Daily Output	Labor-Hours	Unit	Material	2014 Bare Costs Labor	Equipment	Total	Total Incl O&P
0010	**CIRCUIT BREAKERS** (in enclosure)									
0100	Enclosed (NEMA 1), 600 volt, 3 pole, 30 amp	1 Elec	3.20	2.500	Ea.	495	133		628	745
0200	60 amp		2.80	2.857		610	152		762	900
0400	100 amp		2.30	3.478		700	186		886	1,050
0500	200 amp		1.50	5.333		1,475	285		1,760	2,050

26 28 16 – Enclosed Switches and Circuit Breakers

26 28 16.10 Circuit Breakers

		Crew	Daily Output	Labor-Hours	Unit	Material	2014 Bare Costs Labor	Equipment	Total	Total Incl O&P
0600	225 amp	1 Elec	1.50	5.333	Ea.	1,625	285		1,910	2,200
0700	400 amp	2 Elec	1.60	10		2,750	535		3,285	3,825
0800	600 amp		1.20	13.333		4,000	710		4,710	5,475
1000	800 amp		.94	17.021		5,200	910		6,110	7,075
1200	1000 amp		.84	19.048		6,575	1,025		7,600	8,750
1220	1200 amp		.80	20		8,400	1,075		9,475	10,900
1240	1600 amp		.72	22.222		15,500	1,175		16,675	18,800
1260	2000 amp		.64	25		16,800	1,325		18,125	20,400
1400	1200 amp with ground fault		.80	20		13,400	1,075		14,475	16,300
1600	1600 amp with ground fault		.72	22.222		17,500	1,175		18,675	21,100
1800	2000 amp with ground fault		.64	25		18,800	1,325		20,125	22,700
2000	Disconnect, 240 volt 3 pole, 5 HP motor	1 Elec	3.20	2.500		430	133		563	670
2020	10 HP motor		3.20	2.500		430	133		563	670
2040	15 HP motor		2.80	2.857		430	152		582	700
2060	20 HP motor		2.30	3.478		520	186		706	850
2080	25 HP motor		2.30	3.478		520	186		706	850
2100	30 HP motor		2.30	3.478		520	186		706	850
2120	40 HP motor		2	4		890	213		1,103	1,300
2140	50 HP motor		1.50	5.333		890	285		1,175	1,400
2160	60 HP motor		1.50	5.333		2,050	285		2,335	2,675
2180	75 HP motor	2 Elec	2	8		2,050	425		2,475	2,900
2200	100 HP motor		1.60	10		2,050	535		2,585	3,050
2220	125 HP motor		1.60	10		2,050	535		2,585	3,050
2240	150 HP motor		1.20	13.333		4,000	710		4,710	5,475
2260	200 HP motor		1.20	13.333		5,200	710		5,910	6,800
2300	Enclosed (NEMA 7), explosion proof, 600 volt 3 pole, 50 amp	1 Elec	2.30	3.478		1,500	186		1,686	1,950
2350	100 amp		1.50	5.333		1,575	285		1,860	2,150
2400	150 amp		1	8		3,775	425		4,200	4,800
2450	250 amp	2 Elec	1.60	10		4,725	535		5,260	6,000
2500	400 amp	"	1.20	13.333		5,250	710		5,960	6,850

26 28 16.20 Safety Switches

		Crew	Daily Output	Labor-Hours	Unit	Material	2014 Bare Costs Labor	Equipment	Total	Total Incl O&P
0010	**SAFETY SWITCHES** R262816-80									
0100	General duty 240 volt, 3 pole NEMA 1, fusible, 30 amp	1 Elec	3.20	2.500	Ea.	86	133		219	295
0200	60 amp		2.30	3.478		146	186		332	440
0300	100 amp		1.90	4.211		250	225		475	610
0400	200 amp		1.30	6.154		535	330		865	1,075
0500	400 amp	2 Elec	1.80	8.889		1,350	475		1,825	2,200
0600	600 amp	"	1.20	13.333		2,550	710		3,260	3,875
0610	Nonfusible, 30 amp	1 Elec	3.20	2.500		69.50	133		202.50	277
0650	60 amp		2.30	3.478		92	186		278	380
0700	100 amp		1.90	4.211		215	225		440	570
0750	200 amp		1.30	6.154		395	330		725	925
0800	400 amp	2 Elec	1.80	8.889		955	475		1,430	1,750
0850	600 amp	"	1.20	13.333		1,825	710		2,535	3,075
1100	Heavy duty, 600 volt, 3 pole NEMA 1 nonfused									
1110	30 amp	1 Elec	3.20	2.500	Ea.	133	133		266	345
1500	60 amp		2.30	3.478		220	186		406	520
1700	100 amp		1.90	4.211		350	225		575	720
1900	200 amp		1.30	6.154		530	330		860	1,075
2100	400 amp	2 Elec	1.80	8.889		1,200	475		1,675	2,000
2300	600 amp		1.20	13.333		2,125	710		2,835	3,400
2500	800 amp		.94	17.021		4,300	910		5,210	6,100

26 28 16.20 Safety Switches		Crew	Daily Output	Labor-Hours	Unit	Material	2014 Bare Costs Labor	Equipment	Total	Total Incl O&P
2700	1200 amp	2 Elec	.80	20	Ea.	5,800	1,075		6,875	7,975
2900	Heavy duty, 240 volt, 3 pole NEMA 1 fusible									
2910	30 amp	1 Elec	3.20	2.500	Ea.	139	133		272	355
3000	60 amp		2.30	3.478		234	186		420	535
3300	100 amp		1.90	4.211		370	225		595	740
3500	200 amp		1.30	6.154		635	330		965	1,200
3700	400 amp	2 Elec	1.80	8.889		1,625	475		2,100	2,500
3900	600 amp		1.20	13.333		3,025	710		3,735	4,400
4100	800 amp		.94	17.021		6,575	910		7,485	8,575
4300	1200 amp		.80	20		8,650	1,075		9,725	11,100
4340	2 pole fusible, 30 amp	1 Elec	3.50	2.286		104	122		226	298
4350	600 volt, 3 pole, fusible, 30 amp		3.20	2.500		236	133		369	460
4380	60 amp		2.30	3.478		286	186		472	595
4400	100 amp		1.90	4.211		520	225		745	910
4420	200 amp		1.30	6.154		755	330		1,085	1,325
4440	400 amp	2 Elec	1.80	8.889		1,950	475		2,425	2,850
4450	600 amp		1.20	13.333		3,300	710		4,010	4,700
4460	800 amp		.94	17.021		6,575	910		7,485	8,575
4480	1200 amp		.80	20		8,650	1,075		9,725	11,100
4500	240 volt 3 pole NEMA 3R (no hubs), fusible									
4510	30 amp	1 Elec	3.10	2.581	Ea.	248	138		386	480
4700	60 amp		2.20	3.636		390	194		584	720
4900	100 amp		1.80	4.444		565	237		802	975
5100	200 amp		1.20	6.667		775	355		1,130	1,375
5300	400 amp	2 Elec	1.60	10		1,700	535		2,235	2,675
5500	600 amp	"	1	16		3,575	855		4,430	5,225
5510	Heavy duty, 600 volt, 3 pole 3ph. NEMA 3R fusible, 30 amp	1 Elec	3.10	2.581		395	138		533	640
5520	60 amp		2.20	3.636		465	194		659	800
5530	100 amp		1.80	4.444		725	237		962	1,150
5540	200 amp		1.20	6.667		1,000	355		1,355	1,625
5550	400 amp	2 Elec	1.60	10		2,325	535		2,860	3,375
5700	600 volt, 3 pole NEMA 3R nonfused									
5710	30 amp	1 Elec	3.10	2.581	Ea.	220	138		358	450
5900	60 amp		2.20	3.636		385	194		579	715
6100	100 amp		1.80	4.444		535	237		772	945
6300	200 amp		1.20	6.667		655	355		1,010	1,250
6500	400 amp	2 Elec	1.60	10		1,625	535		2,160	2,600
6700	600 amp	"	1	16		3,250	855		4,105	4,850
6900	600 volt, 6 pole NEMA 3R nonfused, 30 amp	1 Elec	2.70	2.963		1,400	158		1,558	1,775
7100	60 amp		2	4		1,625	213		1,838	2,100
7300	100 amp		1.50	5.333		2,000	285		2,285	2,625
7500	200 amp		1.20	6.667		6,850	355		7,205	8,050
7600	600 volt, 3 pole NEMA 7 explosion proof nonfused									
7610	30 amp	1 Elec	2.20	3.636	Ea.	1,300	194		1,494	1,725
7620	60 amp		1.80	4.444		1,575	237		1,812	2,075
7630	100 amp		1.20	6.667		1,900	355		2,255	2,625
7640	200 amp		.80	10		3,950	535		4,485	5,125
7710	600 volt 6 pole, NEMA 3R fusible, 30 amp		2.70	2.963		1,575	158		1,733	1,950
7900	60 amp		2	4		1,875	213		2,088	2,400
8100	100 amp		1.50	5.333		2,300	285		2,585	2,950
8110	240 volt 3 pole, NEMA 12 fusible, 30 amp		3.10	2.581		305	138		443	540
8120	60 amp		2.20	3.636		410	194		604	740
8130	100 amp		1.80	4.444		600	237		837	1,025

26 28 Low-Voltage Circuit Protective Devices

26 28 16 – Enclosed Switches and Circuit Breakers

26 28 16.20 Safety Switches		Crew	Daily Output	Labor-Hours	Unit	Material	2014 Bare Costs Labor	Equipment	Total	Total Incl O&P
8140	200 amp	1 Elec	1.20	6.667	Ea.	865	355		1,220	1,475
8150	400 amp	2 Elec	1.60	10		1,975	535		2,510	2,975
8160	600 amp	"	1	16		3,175	855		4,030	4,775
8180	600 volt 3 pole, NEMA 12 fusible, 30 amp	1 Elec	3.10	2.581		400	138		538	645
8190	60 amp		2.20	3.636		410	194		604	745
8200	100 amp		1.80	4.444		665	237		902	1,075
8210	200 amp		1.20	6.667		1,025	355		1,380	1,675
8220	400 amp	2 Elec	1.60	10		2,250	535		2,785	3,300
8230	600 amp	"	1	16		4,125	855		4,980	5,800
8240	600 volt 3 pole, NEMA 12 nonfused, 30 amp	1 Elec	3.10	2.581		281	138		419	515
8250	60 amp		2.20	3.636		365	194		559	690
8260	100 amp		1.80	4.444		520	237		757	925
8270	200 amp		1.20	6.667		690	355		1,045	1,275
8280	400 amp	2 Elec	1.60	10		1,700	535		2,235	2,675
8290	600 amp	"	1	16		3,050	855		3,905	4,625
8310	600 volt, 3 pole NEMA 4 fusible, 30 amp	1 Elec	3	2.667		1,100	142		1,242	1,425
8320	60 amp		2.20	3.636		1,225	194		1,419	1,650
8330	100 amp		1.80	4.444		2,700	237		2,937	3,325
8340	200 amp		1.20	6.667		3,425	355		3,780	4,275
8350	400 amp	2 Elec	1.60	10		6,775	535		7,310	8,250
8360	600 volt 3 pole NEMA 4 nonfused, 30 amp	1 Elec	3	2.667		905	142		1,047	1,225
8370	60 amp		2.20	3.636		1,075	194		1,269	1,475
8380	100 amp		1.80	4.444		2,200	237		2,437	2,775
8390	200 amp		1.20	6.667		3,000	355		3,355	3,825
8400	400 amp	2 Elec	1.60	10		6,100	535		6,635	7,500
8490	Motor starters, manual, single phase, NEMA 1	1 Elec	6.40	1.250		65.50	66.50		132	172
8500	NEMA 4		4	2		183	107		290	360
8700	NEMA 7		4	2		200	107		307	380
8900	NEMA 1 with pilot		6.40	1.250		90.50	66.50		157	200
8920	3 pole, NEMA 1, 230/460 volt, 5 HP, size 0		3.50	2.286		205	122		327	410
8940	10 HP, size 1		2	4		243	213		456	585
9010	Disc. switch, 600 V 3 pole fusible, 30 amp, to 10 HP motor		3.20	2.500		350	133		483	585
9050	60 amp, to 30 HP motor		2.30	3.478		805	186		991	1,175
9070	100 amp, to 60 HP motor		1.90	4.211		805	225		1,030	1,225
9100	200 amp, to 125 HP motor		1.30	6.154		1,200	330		1,530	1,825
9110	400 amp, to 200 HP motor	2 Elec	1.80	8.889		3,050	475		3,525	4,050

26 28 16.40 Time Switches

		Crew	Daily Output	Labor-Hours	Unit	Material	2014 Bare Costs Labor	Equipment	Total	Total Incl O&P
0010	**TIME SWITCHES**									
0100	Single pole, single throw, 24 hour dial	1 Elec	4	2	Ea.	138	107		245	310
0200	24 hour dial with reserve power		3.60	2.222		585	119		704	820
0300	Astronomic dial		3.60	2.222		180	119		299	375
0400	Astronomic dial with reserve power		3.30	2.424		665	129		794	925
0500	7 day calendar dial		3.30	2.424		158	129		287	370
0600	7 day calendar dial with reserve power		3.20	2.500		395	133		528	635
0700	Photo cell 2000 watt		8	1		27	53.50		80.50	110
1080	Load management device, 4 loads		2	4		870	213		1,083	1,275
1100	8 loads		1	8		2,200	425		2,625	3,075

275

26 29 13.10 Contactors, AC

26 29 13.10 Contactors, AC		Crew	Daily Output	Labor-Hours	Unit	Material	2014 Bare Costs Labor	Equipment	Total	Total Incl O&P
0010	**CONTACTORS, AC** Enclosed (NEMA 1)									
0050	Lighting, 600 volt 3 pole, electrically held									
0100	20 amp	1 Elec	4	2	Ea.	310	107		417	505
0200	30 amp		3.60	2.222		355	119		474	565
0300	60 amp		3	2.667		665	142		807	945
0400	100 amp		2.50	3.200		1,000	171		1,171	1,350
0500	200 amp		1.40	5.714		2,500	305		2,805	3,175
0600	300 amp	2 Elec	1.60	10		6,650	535		7,185	8,100
0800	600 volt 3 pole, mechanically held, 30 amp	1 Elec	3.60	2.222		450	119		569	670
0900	60 amp		3	2.667		885	142		1,027	1,200
1000	75 amp		2.80	2.857		1,200	152		1,352	1,550
1100	100 amp		2.50	3.200		1,250	171		1,421	1,625
1200	150 amp		2	4		3,300	213		3,513	3,975
1300	200 amp		1.40	5.714		3,450	305		3,755	4,250
1500	Magnetic with auxiliary contact, size 00, 9 amp		4	2		180	107		287	360
1600	Size 0, 18 amp		4	2		215	107		322	395
1700	Size 1, 27 amp		3.60	2.222		244	119		363	445
1800	Size 2, 45 amp		3	2.667		455	142		597	715
1900	Size 3, 90 amp		2.50	3.200		735	171		906	1,050
2000	Size 4, 135 amp		2.30	3.478		1,675	186		1,861	2,125
2100	Size 5, 270 amp	2 Elec	1.80	8.889		3,525	475		4,000	4,575
2200	Size 6, 540 amp		1.20	13.333		10,300	710		11,010	12,500
2300	Size 7, 810 amp		1	16		13,900	855		14,755	16,600
2310	Size 8, 1215 amp		.80	20		21,600	1,075		22,675	25,300
2500	Magnetic, 240 volt, 1-2 pole, .75 HP motor	1 Elec	4	2		146	107		253	320
2520	2 HP motor		3.60	2.222		163	119		282	355
2540	5 HP motor		2.50	3.200		395	171		566	690
2560	10 HP motor		1.40	5.714		650	305		955	1,175
2600	240 volt or less, 3 pole, .75 HP motor		4	2		146	107		253	320
2620	5 HP motor		3.60	2.222		180	119		299	375
2640	10 HP motor		3.60	2.222		210	119		329	405
2660	15 HP motor		2.50	3.200		420	171		591	715
2700	25 HP motor		2.50	3.200		420	171		591	715
2720	30 HP motor	2 Elec	2.80	5.714		700	305		1,005	1,225
2740	40 HP motor		2.80	5.714		700	305		1,005	1,225
2760	50 HP motor		1.60	10		700	535		1,235	1,575
2800	75 HP motor		1.60	10		1,650	535		2,185	2,600
2820	100 HP motor		1	16		1,650	855		2,505	3,075
2860	150 HP motor		1	16		3,500	855		4,355	5,125
2880	200 HP motor		1	16		3,500	855		4,355	5,125
3000	600 volt, 3 pole, 5 HP motor	1 Elec	4	2		180	107		287	360
3020	10 HP motor		3.60	2.222		210	119		329	405
3040	25 HP motor		3	2.667		420	142		562	675
3100	50 HP motor		2.50	3.200		700	171		871	1,025
3160	100 HP motor	2 Elec	2.80	5.714		1,650	305		1,955	2,250
3220	200 HP motor	"	1.60	10		3,500	535		4,035	4,650

26 29 13.20 Control Stations

26 29 13.20 Control Stations		Crew	Daily Output	Labor-Hours	Unit	Material	Labor	Equipment	Total	Total Incl O&P
0010	**CONTROL STATIONS**									
0050	NEMA 1, heavy duty, stop/start	1 Elec	8	1	Ea.	134	53.50		187.50	227
0100	Stop/start, pilot light		6.20	1.290		182	69		251	305
0200	Hand/off/automatic		6.20	1.290		99	69		168	212
0400	Stop/start/reverse		5.30	1.509		180	80.50		260.50	320

26 29 13 – Enclosed Controllers

26 29 13.20 Control Stations		Crew	Daily Output	Labor-Hours	Unit	Material	2014 Bare Costs Labor	Equipment	Total	Total Incl O&P
0500	NEMA 7, heavy duty, stop/start	1 Elec	6	1.333	Ea.	420	71		491	565
0600	Stop/start, pilot light		4	2		510	107		617	725
0700	NEMA 7 or 9, 1 element		6	1.333		340	71		411	480
0800	2 element		6	1.333		440	71		511	590
0900	3 element		4	2		875	107		982	1,125
0910	Selector switch, 2 position		6	1.333		340	71		411	480
0920	3 position		4	2		340	107		447	535
0930	Oiltight, 1 element		8	1		93	53.50		146.50	182
0940	2 element		6.20	1.290		134	69		203	250
0950	3 element		5.30	1.509		180	80.50		260.50	320
0960	Selector switch, 2 position		6.20	1.290		99	69		168	212
0970	3 position		5.30	1.509		99	80.50		179.50	230

26 29 13.30 Control Switches		Crew	Daily Output	Labor-Hours	Unit	Material	2014 Bare Costs Labor	Equipment	Total	Total Incl O&P
0010	**CONTROL SWITCHES** Field installed									
6000	Push button 600 V 10A, momentary contact									
6150	Standard operator with colored button	1 Elec	34	.235	Ea.	15.70	12.55		28.25	36
6160	With single block 1NO 1NC		18	.444		33	23.50		56.50	72
6170	With double block 2NO 2NC		15	.533		50.50	28.50		79	98
6180	Std operator w/mushroom button 1-9/16" diam.		34	.235		33	12.55		45.55	55.50
6190	Std operator w/mushroom button 2-1/4" diam.									
6200	With single block 1NO 1NC	1 Elec	18	.444	Ea.	50.50	23.50		74	91
6210	With double block 2NO 2NC		15	.533		68	28.50		96.50	118
6500	Maintained contact, selector operator		34	.235		50.50	12.55		63.05	74.50
6510	With single block 1NO 1NC		18	.444		68	23.50		91.50	111
6520	With double block 2NO 2NC		15	.533		85.50	28.50		114	137
6560	Spring-return selector operator		34	.235		50.50	12.55		63.05	74.50
6570	With single block 1NO 1NC		18	.444		68	23.50		91.50	111
6580	With double block 2NO 2NC		15	.533		85.50	28.50		114	137
6620	Transformer operator w/illuminated									
6630	button 6 V #12 lamp	1 Elec	32	.250	Ea.	85.50	13.35		98.85	114
6640	With single block 1NO 1NC w/guard		16	.500		103	26.50		129.50	153
6650	With double block 2NO 2NC w/guard		13	.615		120	33		153	182
6690	Combination operator		34	.235		50.50	12.55		63.05	74.50
6700	With single block 1NO 1NC		18	.444		68	23.50		91.50	111
6710	With double block 2NO 2NC		15	.533		85.50	28.50		114	137
9000	Indicating light unit, full voltage									
9010	110-125 V front mount	1 Elec	32	.250	Ea.	68	13.35		81.35	95
9020	130 V resistor type		32	.250		50.50	13.35		63.85	75.50
9030	6 V transformer type		32	.250		62.50	13.35		75.85	88.50

26 29 13.40 Relays		Crew	Daily Output	Labor-Hours	Unit	Material	2014 Bare Costs Labor	Equipment	Total	Total Incl O&P
0010	**RELAYS** Enclosed (NEMA 1)									
0050	600 volt AC, 1 pole, 12 amp	1 Elec	5.30	1.509	Ea.	87.50	80.50		168	217
0100	2 pole, 12 amp		5	1.600		87.50	85.50		173	224
0200	4 pole, 10 amp		4.50	1.778		116	95		211	270
0500	250 volt DC, 1 pole, 15 amp		5.30	1.509		114	80.50		194.50	247
0600	2 pole, 10 amp		5	1.600		110	85.50		195.50	249
0700	4 pole, 4 amp		4.50	1.778		145	95		240	300

26 29 Low-Voltage Controllers

26 29 23 – Variable-Frequency Motor Controllers

26 29 23.10 Variable Frequency Drives/Adj. Frequency Drives		Crew	Daily Output	Labor-Hours	Unit	Material	2014 Bare Costs Labor	Equipment	Total	Total Incl O&P
0010	**VARIABLE FREQUENCY DRIVES/ADJ. FREQUENCY DRIVES**									
0100	Enclosed (NEMA 1), 460 volt, for 3 HP motor size	G	1 Elec	.80	10	Ea.	1,475	535		2,010	2,425
0110	5 HP motor size	G		.80	10		1,675	535		2,210	2,625
0120	7.5 HP motor size	G		.67	11.940		2,000	635		2,635	3,150
0130	10 HP motor size	G		.67	11.940		2,325	635		2,960	3,500
0140	15 HP motor size	G	2 Elec	.89	17.978		2,825	960		3,785	4,525
0150	20 HP motor size	G		.89	17.978		3,325	960		4,285	5,075
0160	25 HP motor size	G		.67	23.881		4,100	1,275		5,375	6,400
0170	30 HP motor size	G		.67	23.881		5,100	1,275		6,375	7,525
0180	40 HP motor size	G		.67	23.881		6,125	1,275		7,400	8,650
0190	50 HP motor size	G		.53	30.189		7,575	1,600		9,175	10,700
0200	60 HP motor size	G	R-3	.56	35.714		9,125	1,900	250	11,275	13,100
0210	75 HP motor size	G		.56	35.714		10,500	1,900	250	12,650	14,700
0220	100 HP motor size	G		.50	40		12,200	2,125	280	14,605	16,900
0230	125 HP motor size	G		.50	40		13,600	2,125	280	16,005	18,500
0240	150 HP motor size	G		.50	40		15,600	2,125	280	18,005	20,700
0250	200 HP motor size	G		.42	47.619		20,300	2,525	335	23,160	26,400
1100	Custom-engineered, 460 volt, for 3 HP motor size	G	1 Elec	.56	14.286		2,700	760		3,460	4,125
1110	5 HP motor size	G		.56	14.286		2,700	760		3,460	4,125
1120	7.5 HP motor size	G		.47	17.021		2,825	910		3,735	4,475
1130	10 HP motor size	G		.47	17.021		2,975	910		3,885	4,625
1140	15 HP motor size	G	2 Elec	.62	25.806		3,725	1,375		5,100	6,125
1150	20 HP motor size	G		.62	25.806		4,175	1,375		5,550	6,650
1160	25 HP motor size	G		.47	34.043		4,900	1,825		6,725	8,100
1170	30 HP motor size	G		.47	34.043		6,100	1,825		7,925	9,450
1180	40 HP motor size	G		.47	34.043		7,225	1,825		9,050	10,700
1190	50 HP motor size	G		.37	43.243		8,300	2,300		10,600	12,600
1200	60 HP motor size	G	R-3	.39	51.282		12,500	2,725	360	15,585	18,300
1210	75 HP motor size	G		.39	51.282		13,300	2,725	360	16,385	19,100
1220	100 HP motor size	G		.35	57.143		14,600	3,025	400	18,025	21,000
1230	125 HP motor size	G		.35	57.143		15,600	3,025	400	19,025	22,100
1240	150 HP motor size	G		.35	57.143		17,800	3,025	400	21,225	24,600
1250	200 HP motor size	G		.29	68.966		23,300	3,650	485	27,435	31,600
2000	For complex & special design systems to meet specific										
2010	requirements, obtain quote from vendor.										

26 31 Photovoltaic Collectors

26 31 13 – Photovoltaics

26 31 13.50 Solar Energy - Photovoltaics		Crew	Daily Output	Labor-Hours	Unit	Material	2014 Bare Costs Labor	Equipment	Total	Total Incl O&P
0010	**SOLAR ENERGY - PHOTOVOLTAICS**									
0220	Alt. energy source, photovoltaic module, 6 watt, 15 volts	G	1 Elec	8	1	Ea.	67	53.50		120.50	154
0230	10 watt, 16.3 volts	G		8	1		107	53.50		160.50	198
0240	20 watt, 14.5 volts	G		8	1		161	53.50		214.50	257
0250	36 watt, 17 volts	G		8	1		217	53.50		270.50	320
0260	55 watt, 17 volts	G		8	1		300	53.50		353.50	410
0270	75 watt, 17 volts	G		8	1		390	53.50		443.50	505
0280	130 watt, 33 volts	G		8	1		560	53.50		613.50	695
0290	140 watt, 33 volts	G		8	1		590	53.50		643.50	725
0300	150 watt, 33 volts	G		8	1		645	53.50		698.50	790
0310	DC to AC inverter for, 12 V 2,000 watt	G		4	2		1,150	107		1,257	1,425

278

26 31 Photovoltaic Collectors

26 31 13 – Photovoltaics

26 31 13.50 Solar Energy - Photovoltaics		Crew	Daily Output	Labor-Hours	Unit	Material	2014 Bare Costs Labor	Equipment	Total	Total Incl O&P
0320	12 V, 2,500 watt	G 1 Elec	4	2	Ea.	1,400	107		1,507	1,700
0330	24 V, 2,500 watt	G	4	2		2,000	107		2,107	2,325
0340	12 V, 3,000 watt	G	3	2.667		1,650	142		1,792	2,025
0350	24 V, 3,000 watt	G	3	2.667		2,400	142		2,542	2,875
0360	24 V, 4,000 watt	G	2	4		3,300	213		3,513	3,950
0370	48 V, 4,000 watt	G	2	4		3,300	213		3,513	3,950
0380	48 V, 5,500 watt	G	2	4		3,750	213		3,963	4,450
0390	PV components, combiner box, 10 lug, NEMA 3R enclosure	G	4	2		189	107		296	370
0400	Fuse, 15 A for combiner box	G	40	.200		16.40	10.65		27.05	34
0410	Battery charger controller w/temperature sensor	G	4	2		485	107		592	695
0420	Digital readout panel, displays hours, volts, amps, etc.	G	4	2		129	107		236	300
0430	Deep cycle solar battery, 6 V, 180 Ah (C/20)	G	8	1		232	53.50		285.50	335
0440	Battery interconn, 15" AWG #2/0, sealed w/copper ring lugs	G	16	.500		12.70	26.50		39.20	54
0442	Battery interconn, 24" AWG #2/0, sealed w/copper ring lugs	G	16	.500		18.20	26.50		44.70	60
0444	Battery interconn, 60" AWG #2/0, sealed w/copper ring lugs	G	16	.500		43.50	26.50		70	87.50
0446	Batt temp computer probe, RJ11 jack, 15' cord	G	16	.500		19.50	26.50		46	61.50
0450	System disconnect, DC 175 amp circuit breaker	G	8	1		139	53.50		192.50	233
0460	Conduit box for inverter	G	8	1		44	53.50		97.50	128
0470	Low voltage disconnect	G	8	1		39.50	53.50		93	124
0480	Vented battery enclosure, wood	G 1 Carp	2	4		172	183		355	475
0490	PV rack system, roof, non-penetrating ballast, 1 panel	G R-1A	30.50	.525		885	23		908	1,000
0500	Penetrating surface mount, on steel framing, 1 panel		5.13	3.122		48.50	137		185.50	261
0510	On wood framing, 1 panel		11	1.455		47	64		111	149
0520	With standoff, 1 panel		11	1.455		55	64		119	158
0530	Ground, ballast, fixed, 3 panel		20.50	.780		1,150	34.50		1,184.50	1,300
0540	4 panel		20.50	.780		1,550	34.50		1,584.50	1,750
0550	5 panel		20.50	.780		2,050	34.50		2,084.50	2,300
0560	6 panel		20.50	.780		2,375	34.50		2,409.50	2,650
0570	Adjustable, 3 panel		20.50	.780		1,200	34.50		1,234.50	1,375
0580	4 panel		20.50	.780		1,650	34.50		1,684.50	1,875
0590	5 panel		20.50	.780		2,200	34.50		2,234.50	2,450
0600	6 panel		20.50	.780		2,500	34.50		2,534.50	2,800
0605	Top of Pole, see 32 31 13.30 6710+ for poles									
0610	Passive tracking, 1 panel	R-1A	20.50	.780	Ea.	695	34.50		729.50	815
0620	2 panel		20.50	.780		1,425	34.50		1,459.50	1,600
0630	3 panel		20.50	.780		1,975	34.50		2,009.50	2,225
0640	4 panel		20.50	.780		2,125	34.50		2,159.50	2,375
0650	6 panel		20.50	.780		2,400	34.50		2,434.50	2,675
0660	8 panel		20.50	.780		3,550	34.50		3,584.50	3,950

26 32 Packaged Generator Assemblies

26 32 13 – Engine Generators

26 32 13.13 Diesel-Engine-Driven Generator Sets

		Crew	Daily Output	Labor-Hours	Unit	Material	2014 Bare Costs Labor	Equipment	Total	Total Incl O&P
0010	**DIESEL-ENGINE-DRIVEN GENERATOR SETS**									
2000	Diesel engine, including battery, charger,									
2010	muffler, & day tank, 30 kW	R-3	.55	36.364	Ea.	16,400	1,925	255	18,580	21,200
2100	50 kW		.42	47.619		19,400	2,525	335	22,260	25,500
2110	60 kW		.39	51.282		21,800	2,725	360	24,885	28,500
2200	75 kW		.35	57.143		25,100	3,025	400	28,525	32,700
2300	100 kW		.31	64.516		28,000	3,425	450	31,875	36,400
2400	125 kW		.29	68.966		29,500	3,650	485	33,635	38,400

26 32 13 – Engine Generators

26 32 13.13 Diesel-Engine-Driven Generator Sets

		Crew	Daily Output	Labor-Hours	Unit	Material	2014 Bare Costs Labor	2014 Bare Costs Equipment	Total	Total Incl O&P
2500	150 kW	R-3	.26	76.923	Ea.	33,900	4,075	540	38,515	44,000
2600	175 kW		.25	80		36,900	4,225	560	41,685	47,600
2700	200 kW		.24	83.333		38,200	4,400	585	43,185	49,300
2800	250 kW		.23	86.957		44,800	4,600	610	50,010	57,000
2850	275 kW		.22	90.909		46,800	4,800	635	52,235	59,500
2900	300 kW		.22	90.909		48,600	4,800	635	54,035	61,500
3000	350 kW		.20	100		55,000	5,300	700	61,000	69,000
3100	400 kW		.19	105		68,000	5,575	735	74,310	83,500
3200	500 kW		.18	111		85,500	5,875	780	92,155	103,500
3220	600 kW	↓	.17	117		111,500	6,225	825	118,550	133,500
3230	650 kW	R-13	.38	110		137,000	5,675	500	143,175	160,000
3240	750 kW		.38	110		138,500	5,675	500	144,675	161,500
3250	800 kW		.36	116		146,000	6,000	525	152,525	170,000
3260	900 kW		.31	135		168,500	6,975	610	176,085	196,500
3270	1000 kW	↓	.27	155	↓	173,000	8,000	705	181,705	203,000

26 32 13.16 Gas-Engine-Driven Generator Sets

			Crew	Daily Output	Labor-Hours	Unit	Material	2014 Bare Costs Labor	2014 Bare Costs Equipment	Total	Total Incl O&P
0010	**GAS-ENGINE-DRIVEN GENERATOR SETS**	R263213-45									
0020	Gas or gasoline operated, includes battery,										
0050	charger, & muffler										
0200	3 phase 4 wire, 277/480 volt, 7.5 kW		R-3	.83	24.096	Ea.	7,275	1,275	169	8,719	10,100
0300	11.5 kW			.71	28.169		10,300	1,500	197	11,997	13,700
0400	20 kW			.63	31.746		12,200	1,675	222	14,097	16,200
0500	35 kW			.55	36.364		14,500	1,925	255	16,680	19,100
0520	60 kW			.50	40		19,100	2,125	280	21,505	24,500
0600	80 kW			.40	50		23,700	2,650	350	26,700	30,500
0700	100 kW			.33	60.606		26,000	3,200	425	29,625	33,900
0800	125 kW			.28	71.429		53,000	3,775	500	57,275	64,500
0900	185 kW		↓	.25	80	↓	70,500	4,225	560	75,285	84,500

26 33 43 – Battery Chargers

26 33 43.55 Electric Vehicle Charging

			Crew	Daily Output	Labor-Hours	Unit	Material	2014 Bare Costs Labor	2014 Bare Costs Equipment	Total	Total Incl O&P
0010	**ELECTRIC VEHICLE CHARGING**										
0020	Level 2, wall mounted										
2200	Heavy duty	G	R-1A	15.36	1.042	Ea.	3,525	46		3,571	3,975
2210	with RFID	G		12.29	1.302		3,800	57.50		3,857.50	4,275
2300	Free standing, single connector	G		10.24	1.563		4,050	69		4,119	4,550
2310	with RFID	G		8.78	1.822		4,650	80		4,730	5,225
2320	Double connector	G		7.68	2.083		6,050	91.50		6,141.50	6,825
2330	with RFID	G	↓	6.83	2.343	↓	6,675	103		6,778	7,475

26 33 53 – Static Uninterruptible Power Supply

26 33 53.10 Uninterruptible Power Supply/Conditioner Trans.

			Crew	Daily Output	Labor-Hours	Unit	Material	2014 Bare Costs Labor	2014 Bare Costs Equipment	Total	Total Incl O&P
0010	**UNINTERRUPTIBLE POWER SUPPLY/CONDITIONER TRANSFORMERS**										
0100	Volt. regulating, isolating transf., w/invert. & 10 min. battery pack										
0110	Single-phase, 120 V, 0.35 kVA	R263353-80	1 Elec	2.29	3.493	Ea.	1,050	186		1,236	1,425
0120	0.5 kVA			2	4		1,100	213		1,313	1,525
0130	For additional 55 min. battery, add to 0.35 kVA			2.29	3.493		615	186		801	955
0140	Add to 0.5 kVA			1.14	7.018		645	375		1,020	1,275
0150	Single-phase, 120 V, 0.75 kVA			.80	10		1,375	535		1,910	2,325
0160	1.0 kVA		↓	.80	10	↓	1,975	535		2,510	2,975

26 33 Battery Equipment

26 33 53 – Static Uninterruptible Power Supply

26 33 53.10 Uninterruptible Power Supply/Conditioner Trans.		Crew	Daily Output	Labor-Hours	Unit	Material	2014 Bare Costs Labor	Equipment	Total	Total Incl O&P
0170	1.5 kVA	2 Elec	1.14	14.035	Ea.	3,400	750		4,150	4,850
0180	2 kVA	"	.89	17.978		3,700	960		4,660	5,475
0190	3 kVA	R-3	.63	31.746		4,475	1,675	222	6,372	7,700
0200	5 kVA		.42	47.619		6,475	2,525	335	9,335	11,300
0210	7.5 kVA		.33	60.606		8,275	3,200	425	11,900	14,400
0220	10 kVA		.28	71.429		10,700	3,775	500	14,975	18,000
0230	15 kVA		.22	90.909		13,600	4,800	635	19,035	22,900
0240	3 phase, 120/208 V input 120/208 V output, 20 kVA, incl 17 min. battery		.21	95.238		25,100	5,050	665	30,815	35,900
0242	30 kVA, incl 11 min. battery		.20	100		28,100	5,300	700	34,100	39,600
0250	40 kVA, incl 15 min. battery		.20	100		37,500	5,300	700	43,500	49,900
0260	480 V input 277/480 V output, 60 kVA, incl 6 min. battery		.19	105		42,100	5,575	735	48,410	55,500
0262	80 kVA, incl 4 min. battery		.19	105		50,000	5,575	735	56,310	64,000
0400	For additional 34 min./15 min. battery, add to 40 kVA		.71	28.011		13,100	1,475	196	14,771	16,800
0500	For options & accessories add to above, minimum									10%
0520	Maximum									35%
0600	For complex & special design systems to meet specific									
0610	requirements, obtain quote from vendor									

26 35 Power Filters and Conditioners

26 35 13 – Capacitors

26 35 13.10 Capacitors Indoor

26 35 13.10 Capacitors Indoor		Crew	Daily Output	Labor-Hours	Unit	Material	2014 Bare Costs Labor	Equipment	Total	Total Incl O&P
0010	**CAPACITORS INDOOR**									
0020	240 volts, single & 3 phase, 0.5 kVAR	1 Elec	2.70	2.963	Ea.	425	158		583	700
0100	1.0 kVAR		2.70	2.963		510	158		668	795
0150	2.5 kVAR		2	4		575	213		788	955
0200	5.0 kVAR		1.80	4.444		690	237		927	1,100
0250	7.5 kVAR		1.60	5		800	267		1,067	1,275
0300	10 kVAR		1.50	5.333		965	285		1,250	1,475
0350	15 kVAR		1.30	6.154		1,325	330		1,655	1,950
0400	20 kVAR		1.10	7.273		1,650	390		2,040	2,400
0450	25 kVAR		1	8		1,950	425		2,375	2,800
1000	480 volts, single & 3 phase, 1 kVAR		2.70	2.963		385	158		543	660
1050	2 kVAR		2.70	2.963		445	158		603	725
1100	5 kVAR		2	4		560	213		773	935
1150	7.5 kVAR		2	4		605	213		818	985
1200	10 kVAR		2	4		675	213		888	1,075
1250	15 kVAR		2	4		835	213		1,048	1,225
1300	20 kVAR		1.60	5		925	267		1,192	1,425
1350	30 kVAR		1.50	5.333		1,150	285		1,435	1,700
1400	40 kVAR		1.20	6.667		1,475	355		1,830	2,150
1450	50 kVAR		1.10	7.273		1,675	390		2,065	2,425
2000	600 volts, single & 3 phase, 1 kVAR		2.70	2.963		400	158		558	675
2050	2 kVAR		2.70	2.963		455	158		613	740
2100	5 kVAR		2	4		560	213		773	935
2150	7.5 kVAR		2	4		605	213		818	985
2200	10 kVAR		2	4		675	213		888	1,075
2250	15 kVAR		1.60	5		835	267		1,102	1,325
2300	20 kVAR		1.60	5		925	267		1,192	1,425
2350	25 kVAR		1.50	5.333		1,050	285		1,335	1,575
2400	35 kVAR		1.40	5.714		1,325	305		1,630	1,900
2450	50 kVAR		1.30	6.154		1,775	330		2,105	2,450

26 35 Power Filters and Conditioners

26 35 26 – Harmonic Filters

26 35 26.10 Computer Isolation Transformer	Crew	Daily Output	Labor-Hours	Unit	Material	Labor	Equipment	Total	Total Incl O&P
0010 **COMPUTER ISOLATION TRANSFORMER**									
0100 Computer grade									
0110 Single-phase, 120/240 V, 0.5 kVA	1 Elec	4	2	Ea.	375	107		482	570
0120 1.0 kVA		2.67	2.996		530	160		690	825
0130 2.5 kVA		2	4		810	213		1,023	1,200
0140 5 kVA		1.14	7.018		920	375		1,295	1,575

26 35 26.20 Computer Regulator Transformer	Crew	Daily Output	Labor-Hours	Unit	Material	Labor	Equipment	Total	Total Incl O&P
0010 **COMPUTER REGULATOR TRANSFORMER**									
0100 Ferro-resonant, constant voltage, variable transformer									
0110 Single-phase, 240 V, 0.5 kVA	1 Elec	2.67	2.996	Ea.	385	160		545	665
0120 1.0 kVA		2	4		530	213		743	905
0130 2.0 kVA		1	8		910	425		1,335	1,650
0210 Plug-in unit 120 V, 0.14 kVA		8	1		224	53.50		277.50	325
0220 0.25 kVA		8	1		261	53.50		314.50	370
0230 0.5 kVA		8	1		385	53.50		438.50	505
0240 1.0 kVA		5.33	1.501		530	80		610	705
0250 2.0 kVA		4	2		910	107		1,017	1,150

26 35 26.30 Power Conditioner Transformer	Crew	Daily Output	Labor-Hours	Unit	Material	Labor	Equipment	Total	Total Incl O&P
0010 **POWER CONDITIONER TRANSFORMER**									
0100 Electronic solid state, buck-boost, transformer, w/tap switch									
0110 Single-phase, 115 V, 3.0 kVA, + or - 3% accuracy	2 Elec	1.60	10	Ea.	2,100	535		2,635	3,100
0120 208, 220, 230, or 240 V, 5.0 kVA, + or - 1.5% accuracy	3 Elec	1.60	15		2,700	800		3,500	4,175
0130 5.0 kVA, + or - 6% accuracy	2 Elec	1.14	14.035		2,425	750		3,175	3,775
0140 7.5 kVA, + or - 1.5% accuracy	3 Elec	1.50	16		3,425	855		4,280	5,050
0150 7.5 kVA, + or - 6% accuracy		1.60	15		2,850	800		3,650	4,350
0160 10.0 kVA, + or - 1.5% accuracy		1.33	18.045		4,575	965		5,540	6,475
0170 10.0 kVA, + or - 6% accuracy		1.41	17.021		3,900	910		4,810	5,625

26 35 26.40 Transient Voltage Suppressor Transformer	Crew	Daily Output	Labor-Hours	Unit	Material	Labor	Equipment	Total	Total Incl O&P
0010 **TRANSIENT VOLTAGE SUPPRESSOR TRANSFORMER**									
0110 Single-phase, 120 V, 1.8 kVA	1 Elec	4	2	Ea.	855	107		962	1,100
0120 3.6 kVA		4	2		1,675	107		1,782	2,000
0130 7.2 kVA		3.20	2.500		2,300	133		2,433	2,725
0150 240 V, 3.6 kVA		4	2		1,675	107		1,782	2,000
0160 7.2 kVA		4	2		2,175	107		2,282	2,550
0170 14.4 kVA		3.20	2.500		3,275	133		3,408	3,800
0210 Plug-in unit, 120 V, 1.8 kVA		8	1		650	53.50		703.50	795

26 35 53 – Voltage Regulators

26 35 53.10 Automatic Voltage Regulators

	Crew	Daily Output	Labor-Hours	Unit	Material	Labor	Equipment	Total	Total Incl O&P
0010 **AUTOMATIC VOLTAGE REGULATORS**									
0100 Computer grade, solid state, variable transf. volt. regulator									
0110 Single-phase, 120 V, 8.6 kVA	2 Elec	1.33	12.030	Ea.	4,850	640		5,490	6,275
0120 17.3 kVA		1.14	14.035		5,725	750		6,475	7,425
0130 208/240 V, 7.5/8.6 kVA		1.33	12.030		4,850	640		5,490	6,275
0140 13.5/15.6 kVA		1.33	12.030		5,725	640		6,365	7,250
0150 27.0/31.2 kVA		1.14	14.035		7,225	750		7,975	9,075
0210 Two-phase, single control, 208/240 V, 15.0/17.3 kVA		1.14	14.035		5,725	750		6,475	7,425
0220 Individual phase control, 15.0/17.3 kVA		1.14	14.035		5,725	750		6,475	7,425
0230 30.0/34.6 kVA	3 Elec	1.33	18.045		7,225	965		8,190	9,400
0310 Three-phase single control, 208/240 V, 26/30 kVA	2 Elec	1	16		5,725	855		6,580	7,575
0320 380/480 V, 24/30 kVA	"	1	16		5,725	855		6,580	7,575
0330 43/54 kVA	3 Elec	1.33	18.045		10,400	965		11,365	12,900

26 35 Power Filters and Conditioners

26 35 53 – Voltage Regulators

26 35 53.10 Automatic Voltage Regulators

		Crew	Daily Output	Labor-Hours	Unit	Material	2014 Bare Costs Labor	Equipment	Total	Total Incl O&P
0340	Individual phase control, 208 V, 26 kVA	3 Elec	1.33	18.045	Ea.	5,725	965		6,690	7,750
0350	52 kVA	R-3	.91	21.978		7,225	1,175	154	8,554	9,875
0360	340/480 V, 24/30 kVA	2 Elec	1	16		5,725	855		6,580	7,575
0370	43/54 kVA	"	1	16		7,225	855		8,080	9,225
0380	48/60 kVA	3 Elec	1.33	18.045		10,500	965		11,465	13,000
0390	86/108 kVA	R-3	.91	21.978	▼	11,600	1,175	154	12,929	14,700
0500	Standard grade, solid state, variable transformer volt. regulator									
0510	Single-phase, 115 V, 2.3 kVA	1 Elec	2.29	3.493	Ea.	1,875	186		2,061	2,325
0520	4.2 kVA		2	4		2,975	213		3,188	3,600
0530	6.6 kVA		1.14	7.018		3,650	375		4,025	4,575
0540	13.0 kVA	▼	1.14	7.018		6,325	375		6,700	7,500
0550	16.6 kVA	2 Elec	1.23	13.008		7,450	695		8,145	9,250
0610	230 V, 8.3 kVA		1.33	12.030		6,325	640		6,965	7,900
0620	21.4 kVA		1.23	13.008		7,450	695		8,145	9,250
0630	29.9 kVA		1.23	13.008		7,450	695		8,145	9,250
0710	460 V, 9.2 kVA		1.33	12.030		6,325	640		6,965	7,900
0720	20.7 kVA	▼	1.23	13.008		7,450	695		8,145	9,250
0810	Three-phase, 230 V, 13.1 kVA	3 Elec	1.41	17.021		6,325	910		7,235	8,300
0820	19.1 kVA		1.41	17.021		7,450	910		8,360	9,550
0830	25.1 kVA		1.23	19.512		7,450	1,050		8,500	9,750
0840	57.8 kVA	R-3	.95	21.053		13,500	1,125	147	14,772	16,700
0850	74.9 kVA	"	.91	21.978		13,500	1,175	154	14,829	16,800
0910	460 V, 14.3 kVA	3 Elec	1.41	17.021		6,325	910		7,235	8,300
0920	19.1 kVA		1.41	17.021		7,450	910		8,360	9,550
0930	27.9 kVA		1.23	19.512		7,450	1,050		8,500	9,750
0940	59.8 kVA	R-3	1	20		13,500	1,050	140	14,690	16,700
0950	79.7 kVA		.95	21.053		15,200	1,125	147	16,472	18,500
0960	118 kVA	▼	.95	21.053	▼	16,000	1,125	147	17,272	19,400
1000	Laboratory grade, precision, electronic voltage regulator									
1110	Single-phase, 115 V, 0.5 kVA	1 Elec	2.29	3.493	Ea.	1,200	186		1,386	1,575
1120	1.0 kVA		2	4		1,275	213		1,488	1,725
1130	3.0 kVA	▼	.80	10		1,775	535		2,310	2,750
1140	6.0 kVA	2 Elec	1.46	10.959		3,275	585		3,860	4,475
1150	10.0 kVA	3 Elec	1	24		4,250	1,275		5,525	6,600
1160	15.0 kVA	"	1.50	16		4,850	855		5,705	6,625
1210	230 V, 3.0 kVA	1 Elec	.80	10		2,000	535		2,535	3,000
1220	6.0 kVA	2 Elec	1.46	10.959		3,350	585		3,935	4,550
1230	10.0 kVA	3 Elec	1.71	14.035		4,450	750		5,200	6,025
1240	15.0 kVA	"	1.60	15	▼	5,025	800		5,825	6,725

26 35 53.30 Transient Suppressor/Voltage Regulator

		Crew	Daily Output	Labor-Hours	Unit	Material	2014 Bare Costs Labor	Equipment	Total	Total Incl O&P
0010	**TRANSIENT SUPPRESSOR/VOLTAGE REGULATOR** (without isolation)									
0110	Single-phase, 115 V, 1.0 kVA	1 Elec	2.67	2.996	Ea.	1,000	160		1,160	1,350
0120	2.0 kVA		2.29	3.493		1,375	186		1,561	1,775
0130	4.0 kVA		2.13	3.756		1,675	200		1,875	2,150
0140	220 V, 1.0 kVA		2.67	2.996		1,000	160		1,160	1,350
0150	2.0 kVA		2.29	3.493		1,375	186		1,561	1,775
0160	4.0 kVA		2.13	3.756		1,775	200		1,975	2,250
0210	Plug-in unit, 120 V, 1.0 kVA		8	1		955	53.50		1,008.50	1,125
0220	2.0 kVA	▼	8	1	▼	1,350	53.50		1,403.50	1,550

26 36 Transfer Switches

26 36 13 – Manual Transfer Switches

26 36 13.10 Non-Automatic Transfer Switches

		Crew	Daily Output	Labor-Hours	Unit	Material	2014 Bare Costs Labor	Equipment	Total	Total Incl O&P
0010	**NON-AUTOMATIC TRANSFER SWITCHES** enclosed									
0100	Manual operated, 480 volt 3 pole, 30 amp	1 Elec	2.30	3.478	Ea.	1,625	186		1,811	2,050
0150	60 amp		1.90	4.211		1,600	225		1,825	2,075
0200	100 amp	↓	1.30	6.154		1,600	330		1,930	2,250
0250	200 amp	2 Elec	2	8		3,175	425		3,600	4,125
0300	400 amp		1.60	10		4,625	535		5,160	5,875
0350	600 amp	↓	1	16		5,250	855		6,105	7,075
1000	250 volt 3 pole, 30 amp	1 Elec	2.30	3.478		1,175	186		1,361	1,575
1100	60 amp		1.90	4.211		1,350	225		1,575	1,800
1150	100 amp	↓	1.30	6.154		1,575	330		1,905	2,225
1200	200 amp	2 Elec	2	8		3,225	425		3,650	4,200
1300	600 amp	"	1	16		7,050	855		7,905	9,025
1500	Electrically operated, 480 volt 3 pole, 60 amp	1 Elec	1.90	4.211		2,350	225		2,575	2,925
1600	100 amp	"	1.30	6.154		2,350	330		2,680	3,100
1650	200 amp	2 Elec	2	8		3,900	425		4,325	4,925
1700	400 amp		1.60	10		5,425	535		5,960	6,775
1750	600 amp	↓	1	16		7,825	855		8,680	9,900
2000	250 volt 3 pole, 30 amp	1 Elec	2.30	3.478		2,350	186		2,536	2,875
2050	60 amp	"	1.90	4.211		2,350	225		2,575	2,925
2150	200 amp	2 Elec	2	8		3,900	425		4,325	4,925
2200	400 amp		1.60	10		5,425	535		5,960	6,775
2250	600 amp	↓	1	16		7,825	855		8,680	9,900
2500	NEMA 3R, 480 volt 3 pole, 60 amp	1 Elec	1.80	4.444		2,775	237		3,012	3,400
2550	100 amp	"	1.20	6.667		2,775	355		3,130	3,575
2600	200 amp	2 Elec	1.80	8.889		4,400	475		4,875	5,550
2650	400 amp	"	1.40	11.429		5,950	610		6,560	7,475
2800	NEMA 3R, 250 volt 3 pole solid state, 100 amp	1 Elec	1.20	6.667		2,775	355		3,130	3,575
2850	150 amp	2 Elec	1.80	8.889		3,700	475		4,175	4,750
2900	250 volt 2 pole solid state, 100 amp	1 Elec	1.30	6.154		2,700	330		3,030	3,475
2950	150 amp	2 Elec	2	8	↓	3,600	425		4,025	4,600

26 36 23 – Automatic Transfer Switches

26 36 23.10 Automatic Transfer Switch Devices

		Crew	Daily Output	Labor-Hours	Unit	Material	2014 Bare Costs Labor	Equipment	Total	Total Incl O&P
0010	**AUTOMATIC TRANSFER SWITCH DEVICES**　　R263623-60									
0015	Switches, enclosed 120/240 volt, 2 pole, 30 amp	1 Elec	2.40	3.333	Ea.	2,350	178		2,528	2,850
0020	70 amp		2	4		2,350	213		2,563	2,900
0030	100 amp	↓	1.35	5.926		2,350	315		2,665	3,050
0040	225 amp	2 Elec	2.10	7.619		3,475	405		3,880	4,400
0050	400 amp		1.70	9.412		5,275	500		5,775	6,550
0060	600 amp		1.06	15.094		7,725	805		8,530	9,700
0070	800 amp	↓	.84	19.048		9,075	1,025		10,100	11,500
0100	Switches, enclosed 480 volt, 3 pole, 30 amp	1 Elec	2.30	3.478		2,625	186		2,811	3,150
0200	60 amp		1.90	4.211		2,625	225		2,850	3,200
0300	100 amp	↓	1.30	6.154		2,625	330		2,955	3,375
0400	150 amp	2 Elec	2.40	6.667		3,225	355		3,580	4,075
0500	225 amp		2	8		4,125	425		4,550	5,175
0600	260 amp		2	8		4,750	425		5,175	5,875
0700	400 amp		1.60	10		6,200	535		6,735	7,625
0800	600 amp		1	16		8,600	855		9,455	10,700
0900	800 amp		.80	20		10,100	1,075		11,175	12,700
1000	1000 amp		.76	21.053		13,300	1,125		14,425	16,300
1100	1200 amp		.70	22.857		18,300	1,225		19,525	21,900
1200	1600 amp	↓	.60	26.667	↓	20,800	1,425		22,225	24,900

26 36 Transfer Switches

26 36 23 – Automatic Transfer Switches

26 36 23.10 Automatic Transfer Switch Devices	Crew	Daily Output	Labor-Hours	Unit	Material	2014 Bare Costs Labor	Equipment	Total	Total Incl O&P
1300	2000 amp	2 Elec	.50	32	Ea.	23,200	1,700		24,900	28,200
1600	Accessories, time delay on engine starting					204			204	224
1700	Adjustable time delay on retransfer					204			204	224
1800	Shunt trips for customer connections					365			365	400
1900	Maintenance select switch					83			83	91
2000	Auxiliary contact when normal fails					96			96	105
2100	Pilot light-emergency					83			83	91
2200	Pilot light-normal					83			83	91
2300	Auxiliary contact-closed on normal					96			96	105
2400	Auxiliary contact-closed on emergency					96			96	105
2500	Emergency source sensing, frequency relay					420			420	465

26 41 Facility Lightning Protection

26 41 13 – Lightning Protection for Structures

26 41 13.13 Lightning Protection for Buildings

		Crew	Daily Output	Labor-Hours	Unit	Material	2014 Bare Costs Labor	Equipment	Total	Total Incl O&P
0010	**LIGHTNING PROTECTION FOR BUILDINGS**									
0200	Air terminals & base, copper									
0400	3/8" diameter x 10" (to 75' high)	1 Elec	8	1	Ea.	24.50	53.50		78	107
0500	1/2" diameter x 12" (over 75' high)		8	1		28.50	53.50		82	111
0520	1/2" diameter x 24"		7.30	1.096		38	58.50		96.50	130
0540	1/2" diameter x 60"		6.70	1.194		65.50	63.50		129	168
1000	Aluminum, 1/2" diameter x 12" (to 75' high)		8	1		15.70	53.50		69.20	97.50
1020	1/2" diameter x 24"		7.30	1.096		18.20	58.50		76.70	108
1040	1/2" diameter x 60"		6.70	1.194		25.50	63.50		89	124
1100	5/8" diameter x 12" (over 75' high)		8	1		17.20	53.50		70.70	99
2000	Cable, copper, 220 lb. per thousand ft. (to 75' high)		320	.025	L.F.	3	1.33		4.33	5.30
2100	375 lb. per thousand ft. (over 75' high)		230	.035		5.45	1.86		7.31	8.80
2500	Aluminum, 101 lb. per thousand ft. (to 75' high)		280	.029		.93	1.52		2.45	3.30
2600	199 lb. per thousand ft. (over 75' high)		240	.033		1.38	1.78		3.16	4.18
3000	Arrester, 175 volt AC to ground		8	1	Ea.	88	53.50		141.50	177
3100	650 volt AC to ground		6.70	1.194	"	79	63.50		142.50	183
4000	Air terminals, copper									
4010	3/8" x 10"	1 Elec	50	.160	Ea.	8.40	8.55		16.95	22
4020	1/2" x 12"		38	.211		12.15	11.25		23.40	30
4030	5/8" x 12"		33	.242		18.10	12.95		31.05	39.50
4040	1/2" x 24"		30	.267		22	14.25		36.25	45.50
4050	1/2" x 60"		19	.421		49	22.50		71.50	87.50
4060	Air terminals, aluminum									
4070	3/8" x 10"	1 Elec	50	.160	Ea.	3.38	8.55		11.93	16.50
4080	1/2" x 12"		38	.211		3.95	11.25		15.20	21
4090	5/8" x 12"		33	.242		5.45	12.95		18.40	25.50
4100	1/2" x 24"		30	.267		6.45	14.25		20.70	28.50
4110	1/2" x 60"		19	.421		13.55	22.50		36.05	48.50
4200	Air terminal bases, copper									
4210	Adhesive bases, 1/2"	1 Elec	15	.533	Ea.	17.95	28.50		46.45	62
4215	Adhesive base, lightning air terminal					17.95			17.95	19.70
4220	Bolted bases	1 Elec	9	.889		16.15	47.50		63.65	89
4230	Hinged bases		7	1.143		29.50	61		90.50	124
4240	Side mounted bases		9	.889		16.15	47.50		63.65	89
4250	Offset point support		9	.889		26	47.50		73.50	99.50
4260	Concealed base assembly		5	1.600		34	85.50		119.50	166

285

26 41 13 – Lightning Protection for Structures

26 41 13.13 Lightning Protection for Buildings	Crew	Daily Output	Labor-Hours	Unit	Material	2014 Bare Costs Labor	Equipment	Total	Total Incl O&P
4270	Tee connector base	1 Elec	13	.615	Ea.	32.50	33		65.50	85
4280	Tripod base, 36" for 60" air terminal		7	1.143		47	61		108	143
4290	Intermediate base, 1/2"	↓	15	.533	↓	20.50	28.50		49	65
4300	Air terminal bases, aluminum									
4310	Adhesive bases, 1/2"	1 Elec	15	.533	Ea.	8.55	28.50		37.05	52
4320	Bolted bases		9	.889		11.75	47.50		59.25	84
4330	Hinged bases		7	1.143		21.50	61		82.50	115
4340	Side mounted bases		9	.889		11.45	47.50		58.95	83.50
4350	Offset point support		9	.889		15	47.50		62.50	87.50
4360	Concealed base assembly		5	1.600		20	85.50		105.50	150
4370	Tee connector base		13	.615		13.65	33		46.65	64
4380	Tripod base, 36" for 60" air terminal		7	1.143		33.50	61		94.50	129
4390	Intermediate base, 1/2"	↓	15	.533	↓	11.90	28.50		40.40	55.50
4400	Connector cable, copper									
4410	Through wall connector	1 Elec	7	1.143	Ea.	66	61		127	164
4420	Through roof connector		5	1.600		77	85.50		162.50	213
4430	Beam connector		7	1.143		28	61		89	123
4440	Double bolt connector		38	.211		13.95	11.25		25.20	32
4450	Bar, copper	↓	40	.200	↓	12.25	10.65		22.90	29.50
4500	Connector cable, aluminum									
4510	Through wall connector	1 Elec	7	1.143	Ea.	30	61		91	125
4520	Through roof connector		5	1.600		28	85.50		113.50	159
4530	Beam connector		7	1.143		16.35	61		77.35	110
4540	Double bolt connector	↓	38	.211		7.15	11.25		18.40	24.50
4600	Bonding plates, copper									
4610	I Beam, 8" square	1 Elec	7	1.143	Ea.	21.50	61		82.50	116
4620	Purlin, 8" square		7	1.143		28	61		89	123
4630	Large heavy duty, 16" square		5	1.600		28	85.50		113.50	159
4640	Aluminum, I Beam, 8" square		7	1.143		14.15	61		75.15	107
4650	Purlin, 8" square		7	1.143		13	61		74	106
4660	Large heavy duty, 16" square		5	1.600		16.90	85.50		102.40	147
4700	Cable support, copper, loop fastener		38	.211		.76	11.25		12.01	17.65
4710	Adhesive cable holder		38	.211		1.50	11.25		12.75	18.45
4720	Aluminum, loop fastener		38	.211		.27	11.25		11.52	17.10
4730	Adhesive cable holder		38	.211		.75	11.25		12	17.65
4740	Bonding strap, copper, 3/4" x 9-1/2"		15	.533		8.45	28.50		36.95	52
4750	Aluminum, 3/4" x 9-1/2"		15	.533		4.57	28.50		33.07	47.50
4760	Swivel adapter, copper, 1/2"		73	.110		17.50	5.85		23.35	28
4770	Aluminum, 1/2"	↓	73	.110	↓	9.85	5.85		15.70	19.60

26 42 Cathodic Protection

26 42 16 – Passive Cathodic Protection for Underground Storage Tank

26 42 16.50 Cathodic Protection Wiring Methods

		Crew	Daily Output	Labor-Hours	Unit	Material	Labor	Equipment	Total	Total Incl O&P
0010	**CATHODIC PROTECTION WIRING METHODS**									
1000	Anodes, magnesium type, 9 #	R-15	18.50	2.595	Ea.	39	136	17.55	192.55	266
1010	17 #		13	3.692		70.50	193	25	288.50	395
1020	32 #		10	4.800		123	251	32.50	406.50	545
1030	48 #	↓	7.20	6.667		162	350	45	557	755
1100	Graphite type w/epoxy cap, 3" x 60" (32 #)	R-22	8.40	4.438		127	201		328	445
1110	4" x 80" (68 #)		6	6.213		239	282		521	690
1120	6" x 72" (80 #)	↓	5.20	7.169		1,475	325		1,800	2,125

26 42 Cathodic Protection

26 42 16 – Passive Cathodic Protection for Underground Storage Tank

	26 42 16.50 Cathodic Protection Wiring Methods	Crew	Daily Output	Labor-Hours	Unit	Material	2014 Bare Costs Labor	2014 Bare Costs Equipment	Total	Total Incl O&P
1130	6" x 36" (45 #)	R-22	9.60	3.883	Ea.	750	176		926	1,100
2000	Rectifiers, silicon type, air cooled, 28 V/10 A	R-19	3.50	5.714		2,300	305		2,605	2,975
2010	20 V/20 A		3.50	5.714		2,350	305		2,655	3,025
2100	Oil immersed, 28 V/10 A		3	6.667		3,125	355		3,480	3,975
2110	20 V/20 A	↓	3	6.667	↓	3,150	355		3,505	4,000
3000	Anode backfill, coke breeze	R-22	3850	.010	Lb.	.22	.44		.66	.91
4000	Cable, HMWPE, No. 8		2.40	15.533	M.L.F.	495	705		1,200	1,625
4010	No. 6		2.40	15.533		740	705		1,445	1,875
4020	No. 4		2.40	15.533		1,125	705		1,830	2,300
4030	No. 2		2.40	15.533		1,750	705		2,455	3,000
4040	No. 1		2.20	16.945		2,375	770		3,145	3,800
4050	No. 1/0		2.20	16.945		2,950	770		3,720	4,425
4060	No. 2/0		2.20	16.945		4,575	770		5,345	6,225
4070	No. 4/0	↓	2	18.640	↓	5,950	845		6,795	7,800
5000	Test station, 7 terminal box, flush curb type w/lockable cover	R-19	12	1.667	Ea.	71	89		160	211
5010	Reference cell, 2" dia PVC conduit, cplg., plug, set flush	"	4.80	4.167	"	150	223		373	500

26 51 Interior Lighting

26 51 13 – Interior Lighting Fixtures, Lamps, and Ballasts

26 51 13.10 Fixture Hangers

		Crew	Daily Output	Labor-Hours	Unit	Material	2014 Bare Costs Labor	2014 Bare Costs Equipment	Total	Total Incl O&P
0010	**FIXTURE HANGERS**									
0220	Box hub cover	1 Elec	32	.250	Ea.	3.81	13.35		17.16	24
0240	Canopy		12	.667		9.10	35.50		44.60	63
0260	Connecting block		40	.200		2.97	10.65		13.62	19.20
0280	Cushion hanger		16	.500		22.50	26.50		49	65
0300	Box hanger, with mounting strap		8	1		9	53.50		62.50	90
0320	Connecting block		40	.200		2.15	10.65		12.80	18.30
0340	Flexible, 1/2" diameter, 4" long		12	.667		13.25	35.50		48.75	67.50
0360	6" long		12	.667		14.40	35.50		49.90	69
0380	8" long		12	.667		15.95	35.50		51.45	70.50
0400	10" long		12	.667		17.05	35.50		52.55	72
0420	12" long		12	.667		17.60	35.50		53.10	72.50
0440	15" long		12	.667		18.70	35.50		54.20	73.50
0460	18" long		12	.667		22	35.50		57.50	77
0480	3/4" diameter, 4" long		10	.800		16.35	42.50		58.85	82
0500	6" long		10	.800		18.15	42.50		60.65	84
0520	8" long		10	.800		19.80	42.50		62.30	86
0540	10" long		10	.800		19.90	42.50		62.40	86
0560	12" long		10	.800		21.50	42.50		64	87.50
0580	15" long		10	.800		24	42.50		66.50	90.50
0600	18" long	↓	10	.800	↓	27.50	42.50		70	94.50

26 51 13.40 Interior HID Fixtures

		Crew	Daily Output	Labor-Hours	Unit	Material	2014 Bare Costs Labor	2014 Bare Costs Equipment	Total	Total Incl O&P
0010	**INTERIOR HID FIXTURES** Incl. lamps, and mounting hardware									
0700	High pressure sodium, recessed, round, 70 watt	1 Elec	3.50	2.286	Ea.	425	122		547	655
0720	100 watt		3.50	2.286		455	122		577	685
0740	150 watt		3.20	2.500		510	133		643	760
0760	Square, 70 watt		3.60	2.222		430	119		549	645
0780	100 watt		3.60	2.222		455	119		574	675
0820	250 watt	↓	3	2.667		670	142		812	955
0840	1000 watt	2 Elec	4.80	3.333		1,200	178		1,378	1,600
0860	Surface, round, 70 watt	1 Elec	3	2.667	↓	610	142		752	885

26 51 13 – Interior Lighting Fixtures, Lamps, and Ballasts

26 51 13.40 Interior HID Fixtures		Crew	Daily Output	Labor-Hours	Unit	Material	2014 Bare Costs Labor	Equipment	Total	Total Incl O&P
0880	100 watt	1 Elec	3	2.667	Ea.	625	142		767	900
0900	150 watt		2.70	2.963		655	158		813	955
0920	Square, 70 watt		3	2.667		575	142		717	850
0940	100 watt		3	2.667		610	142		752	885
0980	250 watt		2.50	3.200		660	171		831	980
1040	Pendent, round, 70 watt		3	2.667		630	142		772	910
1060	100 watt		3	2.667		605	142		747	880
1080	150 watt		2.70	2.963		620	158		778	915
1100	Square, 70 watt		3	2.667		660	142		802	940
1120	100 watt		3	2.667		665	142		807	950
1140	150 watt		2.70	2.963		685	158		843	990
1160	250 watt		2.50	3.200		925	171		1,096	1,275
1180	400 watt		2.40	3.333		975	178		1,153	1,350
1220	Wall, round, 70 watt		3	2.667		555	142		697	825
1240	100 watt		3	2.667		565	142		707	835
1260	150 watt		2.70	2.963		580	158		738	875
1300	Square, 70 watt		3	2.667		590	142		732	860
1320	100 watt		3	2.667		620	142		762	895
1340	150 watt		2.70	2.963		640	158		798	940
1360	250 watt	↓	2.50	3.200		700	171		871	1,025
1380	400 watt	2 Elec	4.80	3.333		860	178		1,038	1,200
1400	1000 watt	"	3.60	4.444		1,225	237		1,462	1,700
1500	Metal halide, recessed, round, 175 watt	1 Elec	3.40	2.353		380	126		506	605
1520	250 watt	"	3.20	2.500		475	133		608	725
1540	400 watt	2 Elec	5.80	2.759		675	147		822	960
1580	Square, 175 watt	1 Elec	3.40	2.353		385	126		511	610
1640	Surface, round, 175 watt		2.90	2.759		530	147		677	800
1660	250 watt	↓	2.70	2.963		840	158		998	1,150
1680	400 watt	2 Elec	4.80	3.333		930	178		1,108	1,300
1720	Square, 175 watt	1 Elec	2.90	2.759		575	147		722	850
1800	Pendent, round, 175 watt		2.90	2.759		695	147		842	985
1820	250 watt	↓	2.70	2.963		910	158		1,068	1,225
1840	400 watt	2 Elec	4.80	3.333		995	178		1,173	1,375
1880	Square, 175 watt	1 Elec	2.90	2.759		500	147		647	770
1900	250 watt	"	2.70	2.963		600	158		758	895
1920	400 watt	2 Elec	4.80	3.333		965	178		1,143	1,350
1980	Wall, round, 175 watt	1 Elec	2.90	2.759		665	147		812	955
2000	250 watt	"	2.70	2.963		835	158		993	1,150
2020	400 watt	2 Elec	4.80	3.333		840	178		1,018	1,175
2060	Square, 175 watt	1 Elec	2.90	2.759		565	147		712	845
2080	250 watt	"	2.70	2.963		595	158		753	890
2100	400 watt	2 Elec	4.80	3.333		835	178		1,013	1,175
2800	High pressure sodium, vaporproof, recessed, 70 watt	1 Elec	3.50	2.286		560	122		682	800
2820	100 watt		3.50	2.286		575	122		697	815
2840	150 watt		3.20	2.500		590	133		723	845
2900	Surface, 70 watt		3	2.667		630	142		772	910
2920	100 watt		3	2.667		665	142		807	945
2940	150 watt		2.70	2.963		690	158		848	995
3000	Pendent, 70 watt		3	2.667		625	142		767	905
3020	100 watt		3	2.667		645	142		787	925
3040	150 watt		2.70	2.963		685	158		843	985
3100	Wall, 70 watt		3	2.667		675	142		817	955
3120	100 watt		3	2.667		705	142		847	990

26 51 Interior Lighting

26 51 13 – Interior Lighting Fixtures, Lamps, and Ballasts

26 51 13.40 Interior HID Fixtures		Crew	Daily Output	Labor-Hours	Unit	Material	2014 Bare Costs Labor	Equipment	Total	Total Incl O&P
3140	150 watt	1 Elec	2.70	2.963	Ea.	730	158		888	1,050
3200	Metal halide, vaporproof, recessed, 175 watt		3.40	2.353		500	126		626	740
3220	250 watt	↓	3.20	2.500		575	133		708	835
3240	400 watt	2 Elec	5.80	2.759		725	147		872	1,025
3260	1000 watt	"	4.80	3.333		1,325	178		1,503	1,725
3280	Surface, 175 watt	1 Elec	2.90	2.759		615	147		762	895
3300	250 watt	"	2.70	2.963		825	158		983	1,150
3320	400 watt	2 Elec	4.80	3.333		1,025	178		1,203	1,400
3340	1000 watt	"	3.60	4.444		1,500	237		1,737	2,000
3360	Pendent, 175 watt	1 Elec	2.90	2.759		640	147		787	920
3380	250 watt	"	2.70	2.963		845	158		1,003	1,175
3400	400 watt	2 Elec	4.80	3.333		1,000	178		1,178	1,400
3420	1000 watt	"	3.60	4.444		1,625	237		1,862	2,150
3440	Wall, 175 watt	1 Elec	2.90	2.759		695	147		842	985
3460	250 watt	"	2.70	2.963		895	158		1,053	1,225
3480	400 watt	2 Elec	4.80	3.333		1,050	178		1,228	1,450
3500	1000 watt	"	3.60	4.444		1,700	237		1,937	2,225

26 51 13.50 Interior Lighting Fixtures

		Crew	Daily Output	Labor-Hours	Unit	Material	2014 Bare Costs Labor	Equipment	Total	Total Incl O&P
0010	**INTERIOR LIGHTING FIXTURES** Including lamps, mounting	R265113-40								
0030	hardware and connections									
0100	Fluorescent, C.W. lamps, troffer, recess mounted in grid, RS									
0130	Grid ceiling mount									
0200	Acrylic lens, 1'W x 4'L, two 40 watt	1 Elec	5.70	1.404	Ea.	47.50	75		122.50	164
0210	1'W x 4'L, three 40 watt		5.40	1.481		54	79		133	177
0300	2'W x 2'L, two U40 watt		5.70	1.404		51	75		126	169
0400	2'W x 4'L, two 40 watt		5.30	1.509		50	80.50		130.50	176
0500	2'W x 4'L, three 40 watt		5	1.600		55	85.50		140.50	189
0600	2'W x 4'L, four 40 watt	↓	4.70	1.702		57.50	91		148.50	200
0700	4'W x 4'L, four 40 watt	2 Elec	6.40	2.500		293	133		426	520
0800	4'W x 4'L, six 40 watt		6.20	2.581		305	138		443	540
0900	4'W x 4'L, eight 40 watt	↓	5.80	2.759		315	147		462	565
0910	Acrylic lens, 1'W x 4'L, two 32 watt T8	G	1 Elec	5.70	1.404		52	75		127	169
0930	2'W x 2'L, two U32 watt T8	G		5.70	1.404		70.50	75		145.50	190
0940	2'W x 4'L, two 32 watt T8	G		5.30	1.509		58	80.50		138.50	185
0950	2'W x 4'L, three 32 watt T8	G		5	1.600		59	85.50		144.50	193
0960	2'W x 4'L, four 32 watt T8	G	↓	4.70	1.702	↓	64.50	91		155.50	207
1000	Surface mounted, RS									
1030	Acrylic lens with hinged & latched door frame									
1100	1'W x 4'L, two 40 watt	1 Elec	7	1.143	Ea.	63.50	61		124.50	162
1110	1'W x 4'L, three 40 watt		6.70	1.194		65.50	63.50		129	168
1200	2'W x 2'L, two U40 watt		7	1.143		68	61		129	167
1300	2'W x 4'L, two 40 watt		6.20	1.290		78	69		147	189
1400	2'W x 4'L, three 40 watt		5.70	1.404		79	75		154	199
1500	2'W x 4'L, four 40 watt	↓	5.30	1.509		81	80.50		161.50	210
1600	4'W x 4'L, four 40 watt	2 Elec	7.20	2.222		400	119		519	615
1700	4'W x 4'L, six 40 watt		6.60	2.424		435	129		564	670
1800	4'W x 4'L, eight 40 watt		6.20	2.581		450	138		588	700
1900	2'W x 8'L, four 40 watt	↓	6.40	2.500		159	133		292	375
2000	2'W x 8'L, eight 40 watt		6.20	2.581	↓	171	138		309	395
2010	Acrylic wrap around lens									
2020	6"W x 4'L, one 40 watt	1 Elec	8	1	Ea.	65	53.50		118.50	152
2030	6"W x 8'L, two 40 watt	2 Elec	8	2	↓	71.50	107		178.50	239

26 51 13 – Interior Lighting Fixtures, Lamps, and Ballasts

26 51 13.50 Interior Lighting Fixtures		Crew	Daily Output	Labor-Hours	Unit	Material	2014 Bare Costs Labor	Equipment	Total	Total Incl O&P
2040	11"W x 4'L, two 40 watt	1 Elec	7	1.143	Ea.	43.50	61		104.50	139
2050	11"W x 8'L, four 40 watt	2 Elec	6.60	2.424		71.50	129		200.50	273
2060	16"W x 4'L, four 40 watt	1 Elec	5.30	1.509		71.50	80.50		152	200
2070	16"W x 8'L, eight 40 watt	2 Elec	6.40	2.500		156	133		289	370
2080	2'W x 2'L, two U40 watt	1 Elec	7	1.143	↓	87.50	61		148.50	188
2100	Strip fixture									
2130	Surface mounted									
2200	4' long, one 40 watt, RS	1 Elec	8.50	.941	Ea.	27.50	50		77.50	105
2300	4' long, two 40 watt, RS		8	1		38.50	53.50		92	122
2310	4' long, two 32 watt T8, RS G		8	1		67	53.50		120.50	154
2400	4' long, one 40 watt, SL		8	1		46.50	53.50		100	131
2500	4' long, two 40 watt, SL	↓	7	1.143		63	61		124	161
2580	8' long, one 60 watt T8, SL G	2 Elec	13.40	1.194		83	63.50		146.50	187
2590	8' long, two 60 watt T8, SL G		12.40	1.290		85.50	69		154.50	197
2600	8' long, one 75 watt, SL		13.40	1.194		47.50	63.50		111	148
2700	8' long, two 75 watt, SL	↓	12.40	1.290		57.50	69		126.50	166
2800	4' long, two 60 watt, HO	1 Elec	6.70	1.194		92.50	63.50		156	198
2810	4' long, two 54 watt, T5HO G	"	6.70	1.194		160	63.50		223.50	272
2900	8' long, two 110 watt, HO	2 Elec	10.60	1.509		97.50	80.50		178	228
2910	4' long, two 115 watt, VHO	1 Elec	6.50	1.231		128	65.50		193.50	240
2920	8' long, two 215 watt, VHO	2 Elec	10.40	1.538		138	82		220	275
2950	High bay pendent mounted, 16" W x 4' L, four 54 watt, T5HO G		8.90	1.798		226	96		322	395
2952	2' W x 4' L, six 54 watt, T5HO G		8.50	1.882		305	100		405	485
2954	2' W x 4' L, six 32 watt, T8 G	↓	8.50	1.882	↓	177	100		277	345
3000	Strip, pendent mounted, industrial, white porcelain enamel									
3100	4' long, two 40 watt, RS	1 Elec	5.70	1.404	Ea.	52.50	75		127.50	170
3110	4' long, two 32 watt T8, RS G		5.70	1.404		73	75		148	193
3200	4' long, two 60 watt, HO	↓	5	1.600		82	85.50		167.50	218
3290	8' long, two 60 watt T8, SL G	2 Elec	8.80	1.818		114	97		211	271
3300	8' long, two 75 watt, SL		8.80	1.818		97	97		194	252
3400	8' long, two 110 watt, HO	↓	8	2		124	107		231	297
3410	Acrylic finish, 4' long, two 40 watt, RS	1 Elec	5.70	1.404		86	75		161	207
3420	4' long, two 60 watt, HO		5	1.600		159	85.50		244.50	300
3430	4' long, two 115 watt, VHO	↓	4.80	1.667		205	89		294	360
3440	8' long, two 75 watt, SL	2 Elec	8.80	1.818		168	97		265	330
3450	8' long, two 110 watt, HO		8	2		190	107		297	370
3460	8' long, two 215 watt, VHO	↓	7.60	2.105		269	112		381	465
3470	Troffer, air handling, 2'W x 4'L with four 32 watt T8 G	1 Elec	4	2		111	107		218	282
3480	2'W x 2'L with two U32 watt T8 G		5.50	1.455		107	77.50		184.50	234
3490	Air connector insulated, 5" diameter		20	.400		67	21.50		88.50	106
3500	6" diameter		20	.400		68	21.50		89.50	107
3502	Troffer, direct/indirect, 2'W x 4'L with two 32 W T8 G		5.30	1.509		271	80.50		351.50	420
3510	Troffer parabolic lay-in, 1'W x 4'L with one 32 W T8 G		5.70	1.404		113	75		188	236
3520	1'W x 4'L with two 32 W T8 G		5.30	1.509		135	80.50		215.50	270
3525	2'W x 2'L with two U32 W T8 G		5.70	1.404		114	75		189	238
3530	2'W x 4'L with three 32 W T8 G		5	1.600		126	85.50		211.50	267
3535	Downlight, recess mounted G		8	1		146	53.50		199.50	241
3540	Wall wash reflector, recess mounted G		8	1		104	53.50		157.50	194
3550	Direct/indirect, 4' long, stl., pendent mtd. G		5	1.600		161	85.50		246.50	305
3560	4' long, alum., pendent mtd. G		5	1.600		330	85.50		415.50	495
3565	Prefabricated cove, 4' long, stl. continuous row G		5	1.600		201	85.50		286.50	350
3570	4' long, alum. continuous row G		5	1.600		345	85.50		430.50	510
3580	Wet location, recess mounted, 2'W x 4'L with two 32 watt T8 G	↓	5.30	1.509	↓	234	80.50		314.50	380

26 51 13 – Interior Lighting Fixtures, Lamps, and Ballasts

26 51 13.50 Interior Lighting Fixtures		Crew	Daily Output	Labor-Hours	Unit	Material	2014 Bare Costs Labor	Equipment	Total	Total Incl O&P
3590	Pendent mounted, 2'W x 4'L with two 32 watt T8 [G]	1 Elec	5.70	1.404	Ea.	365	75		440	510
4000	Induction lamp, integral ballast, ceiling mounted									
4110	High bay, aluminum reflector, 160 watt	1 Elec	3.20	2.500	Ea.	945	133		1,078	1,250
4120	320 watt	"	3	2.667		1,700	142		1,842	2,100
4130	480 watt	2 Elec	5.80	2.759		2,575	147		2,722	3,050
4150	Low bay, aluminum reflector, 250 watt	1 Elec	3.20	2.500		745	133		878	1,025
4170	Garage, aluminum reflector, 80 watt		3.60	2.222		750	119		869	1,000
4180	Vandalproof, aluminum reflector, 100 watt		3.20	2.500		530	133		663	780
4220	Metal halide, integral ballast, ceiling, recess mounted									
4230	prismatic glass lens, floating door									
4240	2'W x 2'L, 250 watt	1 Elec	3.20	2.500	Ea.	315	133		448	545
4250	2'W x 2'L, 400 watt	2 Elec	5.80	2.759		360	147		507	620
4260	Surface mounted, 2'W x 2'L, 250 watt	1 Elec	2.70	2.963		340	158		498	610
4270	400 watt	2 Elec	4.80	3.333		400	178		578	705
4280	High bay, aluminum reflector,									
4290	Single unit, 400 watt	2 Elec	4.60	3.478	Ea.	405	186		591	730
4300	Single unit, 1000 watt		4	4		585	213		798	965
4310	Twin unit, 400 watt		3.20	5		815	267		1,082	1,300
4320	Low bay, aluminum reflector, 250W DX lamp	1 Elec	3.20	2.500		360	133		493	595
4330	400 watt lamp	2 Elec	5	3.200		520	171		691	825
4340	High pressure sodium integral ballast ceiling, recess mounted									
4350	prismatic glass lens, floating door									
4360	2'W x 2'L, 150 watt lamp	1 Elec	3.20	2.500	Ea.	380	133		513	620
4370	2'W x 2'L, 400 watt lamp	2 Elec	5.80	2.759		455	147		602	720
4380	Surface mounted, 2'W x 2'L, 150 watt lamp	1 Elec	2.70	2.963		465	158		623	750
4390	400 watt lamp	2 Elec	4.80	3.333		520	178		698	840
4400	High bay, aluminum reflector,									
4410	Single unit, 400 watt lamp	2 Elec	4.60	3.478	Ea.	375	186		561	695
4430	Single unit, 1000 watt lamp	"	4	4		540	213		753	915
4440	Low bay, aluminum reflector, 150 watt lamp	1 Elec	3.20	2.500		325	133		458	555
4445	High bay H.I.D. quartz restrike	"	16	.500		162	26.50		188.50	218
4450	Incandescent, high hat can, round alzak reflector, prewired									
4470	100 watt	1 Elec	8	1	Ea.	70.50	53.50		124	158
4480	150 watt		8	1		103	53.50		156.50	193
4500	300 watt		6.70	1.194		238	63.50		301.50	360
4520	Round with reflector and baffles, 150 watt		8	1		50.50	53.50		104	136
4540	Round with concentric louver, 150 watt PAR		8	1		78	53.50		131.50	166
4600	Square glass lens with metal trim, prewired									
4630	100 watt	1 Elec	6.70	1.194	Ea.	54.50	63.50		118	156
4700	200 watt		6.70	1.194		96	63.50		159.50	202
4800	300 watt		5.70	1.404		143	75		218	270
4810	500 watt		5	1.600		282	85.50		367.50	440
4900	Ceiling/wall, surface mounted, metal cylinder, 75 watt		10	.800		55.50	42.50		98	125
4920	150 watt		10	.800		80	42.50		122.50	152
4930	300 watt		8	1		166	53.50		219.50	262
5000	500 watt		6.70	1.194		360	63.50		423.50	490
5010	Square, 100 watt		8	1		111	53.50		164.50	202
5020	150 watt		8	1		122	53.50		175.50	214
5030	300 watt		7	1.143		335	61		396	455
5040	500 watt		6	1.333		340	71		411	475
5200	Ceiling, surface mounted, opal glass drum									
5300	8", one 60 watt lamp	1 Elec	10	.800	Ea.	44	42.50		86.50	113
5400	10", two 60 watt lamps		8	1		49.50	53.50		103	135

26 51 13 – Interior Lighting Fixtures, Lamps, and Ballasts

26 51 13.50 Interior Lighting Fixtures	Crew	Daily Output	Labor-Hours	Unit	Material	2014 Bare Costs Labor	Equipment	Total	Total Incl O&P
5500	12", four 60 watt lamps	1 Elec	6.70	1.194	Ea.	69.50	63.50		133	172
5510	Pendent, round, 100 watt		8	1		111	53.50		164.50	202
5520	150 watt		8	1		121	53.50		174.50	213
5530	300 watt		6.70	1.194		169	63.50		232.50	282
5540	500 watt		5.50	1.455		320	77.50		397.50	465
5550	Square, 100 watt		6.70	1.194		149	63.50		212.50	260
5560	150 watt		6.70	1.194		156	63.50		219.50	267
5570	300 watt		5.70	1.404		227	75		302	360
5580	500 watt		5	1.600		310	85.50		395.50	470
5600	Wall, round, 100 watt		8	1		64.50	53.50		118	151
5620	300 watt		8	1		113	53.50		166.50	204
5630	500 watt		6.70	1.194		375	63.50		438.50	505
5640	Square, 100 watt		8	1		102	53.50		155.50	192
5650	150 watt		8	1		104	53.50		157.50	194
5660	300 watt		7	1.143		164	61		225	272
5670	500 watt		6	1.333		288	71		359	420
6010	Vapor tight, incandescent, ceiling mounted, 200 watt		6.20	1.290		81.50	69		150.50	193
6020	Recessed, 200 watt		6.70	1.194		122	63.50		185.50	230
6030	Pendent, 200 watt		6.70	1.194		81.50	63.50		145	186
6040	Wall, 200 watt		8	1		79	53.50		132.50	167
6100	Fluorescent, surface mounted, 2 lamps, 4'L, RS, 40 watt		3.20	2.500		114	133		247	325
6110	Industrial, 2 lamps 4' long in tandem, 430 MA		2.20	3.636		210	194		404	520
6130	2 lamps 4' long, 800 MA		1.90	4.211		180	225		405	535
6160	Pendent, indust, 2 lamps 4'L in tandem, 430 MA		1.90	4.211		245	225		470	605
6170	2 lamps 4' long, 430 MA		2.30	3.478		165	186		351	460
6180	2 lamps 4' long, 800 MA		1.70	4.706		207	251		458	605
6850	Vandalproof, surface mounted, fluorescent, two 32 watt T8	G	3.20	2.500		250	133		383	475
6860	Incandescent, one 150 watt		8	1		85	53.50		138.50	174
6900	Mirror light, fluorescent, RS, acrylic enclosure, two 40 watt		8	1		126	53.50		179.50	219
6910	One 40 watt		8	1		105	53.50		158.50	195
6920	One 20 watt		12	.667		81	35.50		116.50	142
7000	Low bay, aluminum reflector. 70 watt, high pressure sodium		4	2		273	107		380	460
7010	250 watt		3.20	2.500		365	133		498	600
7020	400 watt	2 Elec	5	3.200		380	171		551	670
7500	Ballast replacement, by weight of ballast, to 15' high									
7520	Indoor fluorescent, less than 2 lb.	1 Elec	10	.800	Ea.	26	42.50		68.50	92.50
7540	Two 40W, watt reducer, 2 to 5 lb.		9.40	.851		41	45.50		86.50	113
7560	Two F96 slimline, over 5 lb.		8	1		76	53.50		129.50	164
7580	Vaportite ballast, less than 2 lb.		9.40	.851		26	45.50		71.50	96.50
7600	2 lb. to 5 lb.		8.90	.899		41	48		89	117
7620	Over 5 lb.		7.60	1.053		76	56		132	168
7630	Electronic ballast for two tubes		8	1		37	53.50		90.50	121
7640	Dimmable ballast one lamp	G	8	1		105	53.50		158.50	196
7650	Dimmable ballast two-lamp	G	7.60	1.053		103	56		159	197
7690	Emergency ballast (factory installed in fixture)					156			156	171
7990	Decorator									
8000	Pendent RLM in colors, shallow dome, 12" diam. 100 W	1 Elec	8	1	Ea.	80	53.50		133.50	168
8010	Regular dome, 12" diam., 100 watt		8	1		82.50	53.50		136	171
8020	16" diam., 200 watt		7	1.143		84	61		145	184
8030	18" diam., 300 watt		6	1.333		92	71		163	207
8100	Picture framing light, minimum		16	.500		93	26.50		119.50	142
8110	Maximum		16	.500		103	26.50		129.50	153
8150	Miniature low voltage, recessed, pinhole		8	1		160	53.50		213.50	256

26 51 Interior Lighting

26 51 13 – Interior Lighting Fixtures, Lamps, and Ballasts

26 51 13.50 Interior Lighting Fixtures		Crew	Daily Output	Labor-Hours	Unit	Material	2014 Bare Costs Labor	Equipment	Total	Total Incl O&P
8160	Star	1 Elec	8	1	Ea.	132	53.50		185.50	226
8170	Adjustable cone		8	1		179	53.50		232.50	277
8180	Eyeball		8	1		146	53.50		199.50	241
8190	Cone		8	1		129	53.50		182.50	221
8200	Coilex baffle		8	1		126	53.50		179.50	219
8210	Surface mounted, adjustable cylinder		8	1		130	53.50		183.50	223
8250	Chandeliers, incandescent									
8260	24" diam. x 42" high, 6 light candle	1 Elec	6	1.333	Ea.	440	71		511	590
8270	24" diam. x 42" high, 6 light candle w/glass shade		6	1.333		450	71		521	600
8280	17" diam. x 12" high, 8 light w/glass panels		8	1		279	53.50		332.50	385
8300	27" diam. x 29"H, 10 light bohemian lead crystal		4	2		585	107		692	800
8310	21" diam. x 9" high 6 light sculptured ice crystal		8	1		420	53.50		473.50	540
8500	Accent lights, on floor or edge, 0.5 W low volt incandescent									
8520	incl. transformer & fastenings, based on 100' lengths									
8550	Lights in clear tubing, 12" on center	1 Elec	230	.035	L.F.	8.70	1.86		10.56	12.40
8560	6" on center		160	.050		11.40	2.67		14.07	16.50
8570	4" on center		130	.062		17.40	3.28		20.68	24
8580	3" on center		125	.064		19.30	3.41		22.71	26
8590	2" on center		100	.080		28	4.27		32.27	37.50
8600	Carpet, lights both sides 6" OC, in alum. extrusion		270	.030		26.50	1.58		28.08	31.50
8610	In bronze extrusion		270	.030		29.50	1.58		31.08	35
8620	Carpet-bare floor, lights 18" OC, in alum. extrusion		270	.030		21	1.58		22.58	25.50
8630	In bronze extrusion		270	.030		24.50	1.58		26.08	29.50
8640	Carpet edge-wall, lights 6" OC in alum. extrusion		270	.030		26.50	1.58		28.08	31.50
8650	In bronze extrusion		270	.030		29.50	1.58		31.08	35
8660	Bare floor, lights 18" OC, in aluminum extrusion		300	.027		21	1.42		22.42	25
8670	In bronze extrusion		300	.027		24.50	1.42		25.92	29
8680	Bare floor conduit, aluminum extrusion		300	.027		6.95	1.42		8.37	9.80
8690	In bronze extrusion		300	.027		13.90	1.42		15.32	17.45
8700	Step edge to 36", lights 6" OC, in alum. extrusion		100	.080	Ea.	70	4.27		74.27	83.50
8710	In bronze extrusion		100	.080		73	4.27		77.27	87
8720	Step edge to 54", lights 6" OC, in alum. extrusion		100	.080		105	4.27		109.27	121
8730	In bronze extrusion		100	.080		111	4.27		115.27	128
8740	Step edge to 72", lights 6" OC, in alum. extrusion		100	.080		140	4.27		144.27	160
8750	In bronze extrusion		100	.080		153	4.27		157.27	174
8760	Connector, male		32	.250		2.59	13.35		15.94	23
8770	Female with pigtail		32	.250		5.45	13.35		18.80	26
8780	Clamps		400	.020		.51	1.07		1.58	2.16
8790	Transformers, 50 watt		8	1		78.50	53.50		132	167
8800	250 watt		4	2		260	107		367	445
8810	1000 watt		2.70	2.963		465	158		623	750

26 51 13.55 Interior LED Fixtures

			Crew	Daily Output	Labor-Hours	Unit	Material	2014 Bare Costs Labor	Equipment	Total	Total Incl O&P
0010	**INTERIOR LED FIXTURES** Incl. lamps, and mounting hardware										
0100	Downlight, recess mounted, 7.5" diameter, 25 watt	G	1 Elec	8	1	Ea.	770	53.50		823.50	930
0120	10" diameter, 36 watt	G		8	1		815	53.50		868.50	975
0140	12" x 12", 40 watt	G		6.70	1.194		855	63.50		918.50	1,025
0160	cylinder, 7.8 watts	G		8	1		440	53.50		493.50	565
0180	13.3 watts	G		8	1		585	53.50		638.50	725
1000	Troffer, recess mounted, 2' x 4', 3750 Lumens	G		5.30	1.509		194	80.50		274.50	335
1010	4000 Lumens	G		5	1.600		239	85.50		324.50	390
1020	4500 Lumens	G		4.70	1.702		325	91		416	495
1100	Troffer retrofit kit, 46", 37 watt	G		21	.381		405	20.50		425.50	475

293

26 51 13 – Interior Lighting Fixtures, Lamps, and Ballasts

26 51 13.55 Interior LED Fixtures		Crew	Daily Output	Labor-Hours	Unit	Material	2014 Bare Costs Labor	Equipment	Total	Total Incl O&P
1110	55 watt	G 1 Elec	20	.400	Ea.	560	21.50		581.50	645
1120	74 watt	G	18	.444		560	23.50		583.50	650
1200	Troffer, volumetric recess mounted, 2' x 2'	G	5.70	1.404		555	75		630	720
2000	Strip, surface mounted, one light bar 4' long, 40 watt	G	8.50	.941		720	50		770	865
2010	60 watt	G	8	1		745	53.50		798.50	900
2020	Two light bar 4' long, 80 watt	G	7	1.143		985	61		1,046	1,175
3000	Linear, suspended mounted, one light bar 4' long, 37 watt	G	6.70	1.194		330	63.50		393.50	460
3010	One light bar 8' long, 74 watt	G 2 Elec	12.20	1.311		615	70		685	780
3020	Two light bar 4' long, 74 watt	G 1 Elec	5.70	1.404		660	75		735	835
3030	Two light bar 8' long, 148 watt	G 2 Elec	8.80	1.818		1,225	97		1,322	1,500
4000	High bay, surface mounted, round, 150 watts	G	5.41	2.959		1,375	158		1,533	1,725
4010	2 bars,164 watts	G	5.41	2.959		970	158		1,128	1,300
4020	3 bars, 246 watts	G	5.01	3.197		1,250	171		1,421	1,625
4030	4 bars, 328 watts	G	4.60	3.478		1,525	186		1,711	1,950
4040	5 bars, 410 watts	G 3 Elec	4.20	5.716		1,775	305		2,080	2,425
4050	6 bars, 492 watts	G	3.80	6.324		2,025	335		2,360	2,725
4060	7 bars, 574 watts	G	3.39	7.075		2,300	375		2,675	3,100
4070	8 bars, 656 watts	G	2.99	8.029		2,550	430		2,980	3,450
5000	track, lighthead, 6 watt	G 1 Elec	32	.250		65.50	13.35		78.85	92
5010	9 watt	G "	32	.250		85.50	13.35		98.85	114
6000	Garage, surface mount, 103 watts	G 2 Elec	6.50	2.462		1,150	131		1,281	1,475
6100	pendent mount, 80 watts	G	6.50	2.462		830	131		961	1,100
6200	95 watts	G	6.50	2.462		940	131		1,071	1,225
6300	125 watts	G	6.50	2.462		1,050	131		1,181	1,350

26 51 13.70 Residential Fixtures

26 51 13.70 Residential Fixtures		Crew	Daily Output	Labor-Hours	Unit	Material	2014 Bare Costs Labor	Equipment	Total	Total Incl O&P
0010	**RESIDENTIAL FIXTURES**									
0400	Fluorescent, interior, surface, circline, 32 watt & 40 watt	1 Elec	20	.400	Ea.	118	21.50		139.50	161
0700	Shallow under cabinet, two 20 watt		16	.500		54.50	26.50		81	99.50
0900	Wall mounted, 4'L, two 32 watt T8, with baffle		10	.800		137	42.50		179.50	215
2000	Incandescent, exterior lantern, wall mounted, 60 watt		16	.500		48	26.50		74.50	93
2100	Post light, 150W, with 7' post		4	2		238	107		345	420
2500	Lamp holder, weatherproof with 150W PAR		16	.500		32	26.50		58.50	75
2550	With reflector and guard		12	.667		56	35.50		91.50	115
2600	Interior pendent, globe with shade, 150 watt		20	.400		191	21.50		212.50	242

26 51 13.90 Ballast, Replacement HID

26 51 13.90 Ballast, Replacement HID		Crew	Daily Output	Labor-Hours	Unit	Material	2014 Bare Costs Labor	Equipment	Total	Total Incl O&P
0010	**BALLAST, REPLACEMENT HID**									
7510	Multi-tap 120/208/240/277 volt									
7550	High pressure sodium, 70 watt	1 Elec	10	.800	Ea.	160	42.50		202.50	240
7560	100 watt		9.40	.851		167	45.50		212.50	251
7570	150 watt		9	.889		180	47.50		227.50	269
7580	250 watt		8.50	.941		268	50		318	370
7590	400 watt		7	1.143		305	61		366	425
7600	1000 watt		6	1.333		420	71		491	565
7610	Metal halide, 175 watt		8	1		98.50	53.50		152	188
7620	250 watt		8	1		127	53.50		180.50	220
7630	400 watt		7	1.143		159	61		220	267
7640	1000 watt		6	1.333		272	71		343	405
7650	1500 watt		5	1.600		335	85.50		420.50	500

26 52 Emergency Lighting

26 52 13 – Emergency Lighting Equipments

26 52 13.10 Emergency Lighting and Battery Units		Crew	Daily Output	Labor-Hours	Unit	Material	2014 Bare Costs Labor	Equipment	Total	Total Incl O&P
0010	**EMERGENCY LIGHTING AND BATTERY UNITS**									
0300	Emergency light units, battery operated									
0350	Twin sealed beam light, 25 watt, 6 volt each									
0500	Lead battery operated	1 Elec	4	2	Ea.	143	107		250	320
0700	Nickel cadmium battery operated		4	2		660	107		767	885
0780	Additional remote mount, sealed beam, 25 W 6 V		26.70	.300		31.50	16		47.50	59
0790	Twin sealed beam light, 25 W 6 V each		26.70	.300		55	16		71	84.50
0900	Self-contained fluorescent lamp pack		10	.800		159	42.50		201.50	239

26 53 Exit Signs

26 53 13 – Exit Lighting

26 53 13.10 Exit Lighting Fixtures

		Crew	Daily Output	Labor-Hours	Unit	Material	2014 Bare Costs Labor	Equipment	Total	Total Incl O&P
0010	**EXIT LIGHTING FIXTURES**									
0080	Exit light ceiling or wall mount, incandescent, single face	1 Elec	8	1	Ea.	39	53.50		92.50	123
0100	Double face		6.70	1.194		39	63.50		102.50	138
0120	Explosion proof		3.80	2.105		590	112		702	820
0150	Fluorescent, single face		8	1		66.50	53.50		120	153
0160	Double face		6.70	1.194		74.50	63.50		138	178
0200	LED standard, single face [G]		8	1		76	53.50		129.50	164
0220	Double face [G]		6.70	1.194		76	63.50		139.50	180
0230	LED vandal-resistant, single face [G]		7.27	1.100		227	58.50		285.50	340
0240	LED w/battery unit, single face [G]		4.40	1.818		140	97		237	298
0260	Double face [G]		4	2		142	107		249	315
0262	LED w/battery unit, vandal-resistant, single face [G]		4.40	1.818		380	97		477	560
0270	Combination emergency light units and exit sign		4	2		191	107		298	370
0290	LED retrofit kits [G]		60	.133		50.50	7.10		57.60	66
1780	With emergency battery, explosion proof	R-19	7.70	2.597		3,600	139		3,739	4,150

26 54 Classified Location Lighting

26 54 13 – Classified Lighting

26 54 13.20 Explosionproof

		Crew	Daily Output	Labor-Hours	Unit	Material	2014 Bare Costs Labor	Equipment	Total	Total Incl O&P
0010	**EXPLOSIONPROOF**, incl lamps, mounting hardware and connections									
6310	Metal halide with ballast, ceiling, surface mounted, 175 watt	1 Elec	2.90	2.759	Ea.	905	147		1,052	1,225
6320	250 watt	"	2.70	2.963		1,075	158		1,233	1,425
6330	400 watt	2 Elec	4.80	3.333		1,175	178		1,353	1,550
6340	Ceiling, pendent mounted, 175 watt	1 Elec	2.60	3.077		860	164		1,024	1,200
6350	250 watt	"	2.40	3.333		1,050	178		1,228	1,425
6360	400 watt	2 Elec	4.20	3.810		1,125	203		1,328	1,525
6370	Wall, surface mounted, 175 watt	1 Elec	2.90	2.759		970	147		1,117	1,300
6380	250 watt	"	2.70	2.963		1,150	158		1,308	1,500
6390	400 watt	2 Elec	4.80	3.333		1,225	178		1,403	1,625
6400	High pressure sodium, ceiling surface mounted, 70 watt	1 Elec	3	2.667		1,275	142		1,417	1,625
6410	100 watt		3	2.667		1,300	142		1,442	1,675
6420	150 watt		2.70	2.963		1,350	158		1,508	1,725
6430	Pendent mounted, 70 watt		2.70	2.963		1,200	158		1,358	1,550
6440	100 watt		2.70	2.963		1,250	158		1,408	1,600
6450	150 watt		2.40	3.333		1,300	178		1,478	1,700
6460	Wall mounted, 70 watt		3	2.667		1,350	142		1,492	1,725
6470	100 watt		3	2.667		1,400	142		1,542	1,775

26 54 Classified Location Lighting

26 54 13 – Classified Lighting

26 54 13.20 Explosionproof		Crew	Daily Output	Labor-Hours	Unit	Material	2014 Bare Costs Labor	Equipment	Total	Total Incl O&P
6480	150 watt	1 Elec	2.70	2.963	Ea.	1,450	158		1,608	1,800
6510	Incandescent, ceiling mounted, 200 watt		4	2		1,075	107		1,182	1,325
6520	Pendent mounted, 200 watt		3.50	2.286		915	122		1,037	1,175
6530	Wall mounted, 200 watt		4	2		1,050	107		1,157	1,325
6600	Fluorescent, RS, 4' long, ceiling mounted, two 40 watt		2.70	2.963		2,825	158		2,983	3,325
6610	Three 40 watt		2.20	3.636		4,075	194		4,269	4,800
6620	Four 40 watt		1.90	4.211		5,250	225		5,475	6,100
6630	Pendent mounted, two 40 watt		2.30	3.478		3,275	186		3,461	3,875
6640	Three 40 watt		1.90	4.211		4,650	225		4,875	5,450
6650	Four 40 watt	↓	1.70	4.706	↓	6,150	251		6,401	7,125

26 55 Special Purpose Lighting

26 55 59 – Display Lighting

26 55 59.10 Track Lighting

		Crew	Daily Output	Labor-Hours	Unit	Material	2014 Bare Costs Labor	Equipment	Total	Total Incl O&P
0010	**TRACK LIGHTING**									
0080	Track, 1 circuit, 4' section	1 Elec	6.70	1.194	Ea.	48	63.50		111.50	149
0100	8' section	2 Elec	10.60	1.509		74.50	80.50		155	203
0200	12' section	"	8.80	1.818		117	97		214	273
0300	3 circuits, 4' section	1 Elec	6.70	1.194		87	63.50		150.50	191
0400	8' section	2 Elec	10.60	1.509		109	80.50		189.50	241
0500	12' section	"	8.80	1.818		157	97		254	320
1000	Feed kit, surface mounting	1 Elec	16	.500		13	26.50		39.50	54.50
1100	End cover		24	.333		5.55	17.80		23.35	32.50
1200	Feed kit, stem mounting, 1 circuit		16	.500		42	26.50		68.50	86
1300	3 circuit		16	.500		42	26.50		68.50	86
2000	Electrical joiner, for continuous runs, 1 circuit		32	.250		18.05	13.35		31.40	40
2100	3 circuit		32	.250		51	13.35		64.35	76.50
2200	Fixtures, spotlight, 75W PAR halogen		16	.500		77	26.50		103.50	125
2210	50W MR16 halogen		16	.500		171	26.50		197.50	228
3000	Wall washer, 250 watt tungsten halogen		16	.500		130	26.50		156.50	183
3100	Low voltage, 25/50 watt, 1 circuit		16	.500		131	26.50		157.50	185
3120	3 circuit	↓	16	.500	↓	172	26.50		198.50	229

26 55 61 – Theatrical Lighting

26 55 61.10 Lights

		Crew	Daily Output	Labor-Hours	Unit	Material	2014 Bare Costs Labor	Equipment	Total	Total Incl O&P
0010	**LIGHTS**									
2000	Lights, border, quartz, reflector, vented,									
2100	colored or white	1 Elec	20	.400	L.F.	173	21.50		194.50	222
2500	Spotlight, follow spot, with transformer, 2,100 watt	"	4	2	Ea.	3,075	107		3,182	3,525
2600	For no transformer, deduct					910			910	1,000
3000	Stationary spot, fresnel quartz, 6" lens	1 Elec	4	2		131	107		238	305
3100	8" lens		4	2		231	107		338	415
3500	Ellipsoidal quartz, 1,000W, 6" lens		4	2		335	107		442	530
3600	12" lens		4	2		600	107		707	820
4000	Strobe light, 1 to 15 flashes per second, quartz		3	2.667		750	142		892	1,050
4500	Color wheel, portable, five hole, motorized	↓	4	2	↓	194	107		301	375

26 55 63 – Detention Lighting

26 55 63.10 Detention Lighting Fixtures

		Crew	Daily Output	Labor-Hours	Unit	Material	2014 Bare Costs Labor	Equipment	Total	Total Incl O&P
0010	**DETENTION LIGHTING FIXTURES**									
1000	Surface mounted, cold rolled steel, 14 ga., 1' x 4'	1 Elec	5.40	1.481	Ea.	595	79		674	775
1010	2' x 2'	↓	5.40	1.481	↓	705	79		784	895

26 55 Special Purpose Lighting

26 55 63 – Detention Lighting

26 55 63.10 Detention Lighting Fixtures	Crew	Daily Output	Labor-Hours	Unit	Material	2014 Bare Costs Labor	Equipment	Total	Total Incl O&P
1020	2' x 4'	1 Elec	4.80	1.667	Ea.	800	89		889	1,025
1100	12 ga., 1' x 4'		5.40	1.481		840	79		919	1,050
1110	2' x 2'		5.40	1.481		705	79		784	895
1120	2' x 4'		4.80	1.667		875	89		964	1,100

26 56 Exterior Lighting

26 56 13 – Lighting Poles and Standards

26 56 13.10 Lighting Poles

		Crew	Daily Output	Labor-Hours	Unit	Material	2014 Bare Costs Labor	Equipment	Total	Total Incl O&P
0010	**LIGHTING POLES**									
2800	Light poles, anchor base									
2820	not including concrete bases									
2840	Aluminum pole, 8' high	1 Elec	4	2	Ea.	695	107		802	925
2850	10' high		4	2		735	107		842	970
2860	12' high		3.80	2.105		765	112		877	1,025
2870	14' high		3.40	2.353		795	126		921	1,075
2880	16' high		3	2.667		875	142		1,017	1,175
3000	20' high	R-3	2.90	6.897		925	365	48.50	1,338.50	1,625
3200	30' high		2.60	7.692		1,750	405	54	2,209	2,600
3400	35' high		2.30	8.696		1,900	460	61	2,421	2,850
3600	40' high		2	10		2,175	530	70	2,775	3,275
3800	Bracket arms, 1 arm	1 Elec	8	1		119	53.50		172.50	211
4000	2 arms		8	1		240	53.50		293.50	345
4200	3 arms		5.30	1.509		360	80.50		440.50	515
4400	4 arms		5.30	1.509		480	80.50		560.50	650
4500	Steel pole, galvanized, 8' high		3.80	2.105		600	112		712	835
4510	10' high		3.70	2.162		630	115		745	870
4520	12' high		3.40	2.353		680	126		806	940
4530	14' high		3.10	2.581		725	138		863	1,000
4540	16' high		2.90	2.759		765	147		912	1,075
4550	18' high		2.70	2.963		810	158		968	1,125
4600	20' high	R-3	2.60	7.692		1,100	405	54	1,559	1,875
4800	30' high		2.30	8.696		1,300	460	61	1,821	2,175
5000	35' high		2.20	9.091		1,425	480	63.50	1,968.50	2,350
5200	40' high		1.70	11.765		1,750	625	82.50	2,457.50	2,950
5400	Bracket arms, 1 arm	1 Elec	8	1		179	53.50		232.50	276
5600	2 arms		8	1		277	53.50		330.50	385
5800	3 arms		5.30	1.509		300	80.50		380.50	450
6000	4 arms		5.30	1.509		415	80.50		495.50	580
6100	Fiberglass pole, 1 or 2 fixtures, 20' high	R-3	4	5		665	265	35	965	1,175
6200	30' high		3.60	5.556		825	294	39	1,158	1,400
6300	35' high		3.20	6.250		1,300	330	44	1,674	1,975
6400	40' high		2.80	7.143		1,500	380	50	1,930	2,275
6420	Wood pole, 4-1/2" x 5-1/8", 8' high	1 Elec	6	1.333		325	71		396	465
6430	10' high		6	1.333		375	71		446	515
6440	12' high		5.70	1.404		470	75		545	630
6450	15' high		5	1.600		550	85.50		635.50	735
6460	20' high		4	2		665	107		772	890
7300	Transformer bases, not including concrete bases									
7320	Maximum pole size, steel, 40' high	1 Elec	2	4	Ea.	1,400	213		1,613	1,875
7340	Cast aluminum, 30' high		3	2.667		740	142		882	1,025
7350	40' high		2.50	3.200		1,125	171		1,296	1,500

297

26 56 Exterior Lighting

26 56 16 – Parking Lighting

26 56 16.55 Parking LED Lighting

		Crew	Daily Output	Labor-Hours	Unit	Material	2014 Bare Costs Labor	Equipment	Total	Total Incl O&P
0010	**PARKING LED LIGHTING**									
0100	Round pole mounting, 88 lamp watts ⒢	1 Elec	2	4	Ea.	1,975	213		2,188	2,500
0110	Square pole mounting, 223 lamp watts ⒢	"	2	4	"	3,475	213		3,688	4,150

26 56 19 – Roadway Lighting

26 56 19.20 Roadway Luminaire

		Crew	Daily Output	Labor-Hours	Unit	Material	2014 Bare Costs Labor	Equipment	Total	Total Incl O&P
0010	**ROADWAY LUMINAIRE**									
2650	Roadway area luminaire, low pressure sodium, 135 watt	1 Elec	2	4	Ea.	645	213		858	1,025
2700	180 watt	"	2	4		690	213		903	1,075
2750	Metal halide, 400 watt	2 Elec	4.40	3.636		550	194		744	895
2760	1000 watt		4	4		620	213		833	1,000
2780	High pressure sodium, 400 watt		4.40	3.636		575	194		769	920
2790	1000 watt		4	4		655	213		868	1,050

26 56 19.55 Roadway LED Luminaire

		Crew	Daily Output	Labor-Hours	Unit	Material	2014 Bare Costs Labor	Equipment	Total	Total Incl O&P
0010	**ROADWAY LED LUMINAIRE**									
0100	LED fixture, 72 LEDs, 120 V AC or 12 V DC, equal to 60 watt ⒢	1 Elec	2.70	2.963	Ea.	585	158		743	880
0110	108 LEDs, 120 V AC or 12 V DC, equal to 90 watt ⒢		2.70	2.963		690	158		848	995
0120	144 LEDs, 120 V AC or 12 V DC, equal to 120 watt ⒢		2.70	2.963		845	158		1,003	1,175
0130	252 LEDs, 120 V AC or 12 V DC, equal to 210 watt ⒢	2 Elec	4.40	3.636		1,175	194		1,369	1,600
0140	Replaces high pressure sodium fixture, 75 watt ⒢	1 Elec	2.70	2.963		715	158		873	1,025
0150	125 watt ⒢		2.70	2.963		815	158		973	1,125
0160	150 watt ⒢		2.70	2.963		1,025	158		1,183	1,350
0170	175 watt ⒢		2.70	2.963		1,225	158		1,383	1,575
0180	200 watt ⒢		2.70	2.963		1,425	158		1,583	1,800
0190	250 watt ⒢	2 Elec	4.40	3.636		1,625	194		1,819	2,100
0200	320 watt ⒢	"	4.40	3.636		1,775	194		1,969	2,275

26 56 23 – Area Lighting

26 56 23.10 Exterior Fixtures

		Crew	Daily Output	Labor-Hours	Unit	Material	2014 Bare Costs Labor	Equipment	Total	Total Incl O&P
0010	**EXTERIOR FIXTURES** With lamps									
0200	Wall mounted, incandescent, 100 watt	1 Elec	8	1	Ea.	34.50	53.50		88	118
0400	Quartz, 500 watt		5.30	1.509		49.50	80.50		130	176
0420	1500 watt		4.20	1.905		102	102		204	264
1100	Wall pack, low pressure sodium, 35 watt		4	2		225	107		332	410
1150	55 watt		4	2		268	107		375	455
1160	High pressure sodium, 70 watt		4	2		221	107		328	405
1170	150 watt		4	2		242	107		349	425
1180	Metal Halide, 175 watt		4	2		245	107		352	430
1190	250 watt		4	2		279	107		386	465
1195	400 watt		4	2		330	107		437	525
1250	Induction lamp, 40 watt		4	2		340	107		447	535
1260	80 watt		4	2		575	107		682	795
1278	LED, poly lens, 26 watt		4	2		325	107		432	520
1280	110 watt		4	2		1,300	107		1,407	1,575
1500	LED, glass lens, 13 watt		4	2		325	107		432	520

26 56 23.55 Exterior LED Fixtures

		Crew	Daily Output	Labor-Hours	Unit	Material	2014 Bare Costs Labor	Equipment	Total	Total Incl O&P
0010	**EXTERIOR LED FIXTURES**									
0100	Wall mounted, indoor/outdoor, 12 watt ⒢	1 Elec	10	.800	Ea.	370	42.50		412.50	470
0110	32 watt ⒢		10	.800		495	42.50		537.50	610
0120	66 watt ⒢		10	.800		715	42.50		757.50	850
0200	outdoor, 110 watt ⒢		10	.800		1,100	42.50		1,142.50	1,275
0210	220 watt ⒢		10	.800		1,925	42.50		1,967.50	2,175

26 56 Exterior Lighting

26 56 23 – Area Lighting

26 56 23.55 Exterior LED Fixtures

		Crew	Daily Output	Labor-Hours	Unit	Material	2014 Bare Costs Labor	Equipment	Total	Total Incl O&P
0300	modular, type IV, 120 V, 50 lamp watts G	1 Elec	9	.889	Ea.	1,325	47.50		1,372.50	1,550
0310	101 lamp watts G		9	.889		1,500	47.50		1,547.50	1,725
0320	126 lamp watts G		9	.889		1,900	47.50		1,947.50	2,150
0330	202 lamp watts G		9	.889		2,150	47.50		2,197.50	2,425
0340	240 V, 50 lamp watts G		8	1		1,375	53.50		1,428.50	1,600
0350	101 lamp watts G		8	1		1,550	53.50		1,603.50	1,775
0360	126 lamp watts G		8	1		1,950	53.50		2,003.50	2,200
0370	202 lamp watts G		8	1		2,200	53.50		2,253.50	2,500
0400	wall pack, glass, 13 lamp watts G		4	2		515	107		622	725
0410	poly w/photocell, 26 lamp watts G		4	2		320	107		427	515
0420	50 lamp watts G		4	2		755	107		862	990
0430	replacement, 40 watts G		4	2		660	107		767	885
0440	60 watts G		4	2		830	107		937	1,075

26 56 26 – Landscape Lighting

26 56 26.20 Landscape Fixtures

		Crew	Daily Output	Labor-Hours	Unit	Material	2014 Bare Costs Labor	Equipment	Total	Total Incl O&P
0010	**LANDSCAPE FIXTURES**									
7380	Landscape recessed uplight, incl. housing, ballast, transformer									
7390	& reflector, not incl. conduit, wire, trench									
7420	Incandescent, 250 watt	1 Elec	5	1.600	Ea.	585	85.50		670.50	775
7440	Quartz, 250 watt		5	1.600		555	85.50		640.50	740
7460	500 watt		4	2		570	107		677	790

26 56 33 – Walkway Lighting

26 56 33.10 Walkway Luminaire

		Crew	Daily Output	Labor-Hours	Unit	Material	2014 Bare Costs Labor	Equipment	Total	Total Incl O&P
0010	**WALKWAY LUMINAIRE**									
6500	Bollard light, lamp & ballast, 42" high with polycarbonate lens									
6800	Metal halide, 175 watt	1 Elec	3	2.667	Ea.	805	142		947	1,100
6900	High pressure sodium, 70 watt		3	2.667		825	142		967	1,125
7000	100 watt		3	2.667		825	142		967	1,125
7100	150 watt		3	2.667		805	142		947	1,100
7200	Incandescent, 150 watt		3	2.667		590	142		732	860
7810	Walkway luminaire, square 16", metal halide 250 watt		2.70	2.963		630	158		788	930
7820	High pressure sodium, 70 watt		3	2.667		720	142		862	1,000
7830	100 watt		3	2.667		735	142		877	1,025
7840	150 watt		3	2.667		735	142		877	1,025
7850	200 watt		3	2.667		740	142		882	1,025
7910	Round 19", metal halide, 250 watt		2.70	2.963		920	158		1,078	1,225
7920	High pressure sodium, 70 watt		3	2.667		1,000	142		1,142	1,325
7930	100 watt		3	2.667		1,000	142		1,142	1,325
7940	150 watt		3	2.667		1,025	142		1,167	1,350
7950	250 watt		2.70	2.963		1,075	158		1,233	1,400
8000	Sphere 14" opal, incandescent, 200 watt		4	2		292	107		399	480
8020	Sphere 18" opal, incandescent, 300 watt		3.50	2.286		355	122		477	575
8040	Sphere 16" clear, high pressure sodium, 70 watt		3	2.667		610	142		752	885
8050	100 watt		3	2.667		650	142		792	930
8100	Cube 16" opal, incandescent, 300 watt		3.50	2.286		390	122		512	610
8120	High pressure sodium, 70 watt		3	2.667		565	142		707	840
8130	100 watt		3	2.667		580	142		722	855
8230	Lantern, high pressure sodium, 70 watt		3	2.667		505	142		647	775
8240	100 watt		3	2.667		545	142		687	815
8250	150 watt		3	2.667		510	142		652	775
8260	250 watt		2.70	2.963		715	158		873	1,025
8270	Incandescent, 300 watt		3.50	2.286		375	122		497	600

26 56 Exterior Lighting

26 56 33 – Walkway Lighting

26 56 33.10 Walkway Luminaire

		Crew	Daily Output	Labor-Hours	Unit	Material	2014 Bare Costs Labor	Equipment	Total	Total Incl O&P
8330	Reflector 22" w/globe, high pressure sodium, 70 watt	1 Elec	3	2.667	Ea.	485	142		627	750
8340	100 watt		3	2.667		495	142		637	760
8350	150 watt		3	2.667		500	142		642	765
8360	250 watt		2.70	2.963		640	158		798	940

26 56 33.55 Walkway LED Luminaire

			Crew	Daily Output	Labor-Hours	Unit	Material	Labor	Equipment	Total	Total Incl O&P
0010	**WALKWAY LED LUMINAIRE**										
0100	Pole mounted, 86 watts, 4350 lumens	G	1 Elec	3	2.667	Ea.	1,350	142		1,492	1,725
0110	4630 lumens	G		3	2.667		2,275	142		2,417	2,725
0120	80 watts, 4000 lumens	G		3	2.667		2,150	142		2,292	2,575

26 56 36 – Flood Lighting

26 56 36.20 Floodlights

			Crew	Daily Output	Labor-Hours	Unit	Material	Labor	Equipment	Total	Total Incl O&P
0010	**FLOODLIGHTS** with ballast and lamp,										
1290	floor mtd, mount with swivel bracket										
1300	Induction lamp, 40 watt		1 Elec	3	2.667	Ea.	445	142		587	705
1310	80 watt			3	2.667		680	142		822	960
1320	150 watt			3	2.667		1,225	142		1,367	1,575
1400	Pole mounted, pole not included										
1950	Metal halide, 175 watt		1 Elec	2.70	2.963	Ea.	335	158		493	605
2000	400 watt		2 Elec	4.40	3.636		415	194		609	750
2200	1000 watt			4	4		575	213		788	950
2210	1500 watt			3.70	4.324		595	231		826	1,000
2250	Low pressure sodium, 55 watt		1 Elec	2.70	2.963		580	158		738	870
2270	90 watt			2	4		640	213		853	1,025
2290	180 watt			2	4		810	213		1,023	1,225
2340	High pressure sodium, 70 watt			2.70	2.963		243	158		401	505
2360	100 watt			2.70	2.963		250	158		408	510
2380	150 watt			2.70	2.963		287	158		445	550
2400	400 watt		2 Elec	4.40	3.636		375	194		569	705
2600	1000 watt		"	4	4		645	213		858	1,025
9005	Solar powered floodlight, w/motion det, incl batt pack for cloudy days	G	1 Elec	8	1		118	53.50		171.50	210
9020	Battery pack for	G	"	8	1		13.45	53.50		66.95	95

26 56 36.55 LED Floodlights

			Crew	Daily Output	Labor-Hours	Unit	Material	Labor	Equipment	Total	Total Incl O&P
0010	**LED FLOODLIGHTS** with ballast and lamp,										
0020	Pole mounted, pole not included										
0100	11 watt	G	1 Elec	4	2	Ea.	345	107		452	540
0110	46 watt	G		4	2		1,025	107		1,132	1,300
0120	90 watt	G		4	2		1,500	107		1,607	1,800
0130	288 watt	G		4	2		2,325	107		2,432	2,700

26 61 Lighting Systems and Accessories

26 61 13 – Lighting Accessories

26 61 13.30 Fixture Whips

		Crew	Daily Output	Labor-Hours	Unit	Material	Labor	Equipment	Total	Total Incl O&P
0010	**FIXTURE WHIPS**									
0080	3/8" Greenfield, 2 connectors, 6' long									
0100	TFFN wire, three #18	1 Elec	32	.250	Ea.	8.60	13.35		21.95	29.50
0150	Four #18		28	.286		9.10	15.25		24.35	33
0200	Three #16		32	.250		8.65	13.35		22	29.50
0250	Four #16		28	.286		9.20	15.25		24.45	33
0300	THHN wire, three #14		32	.250		10.05	13.35		23.40	31
0350	Four #14		28	.286		11.45	15.25		26.70	35.50

26 61 13 – Lighting Accessories

26 61 13.30 Fixture Whips		Crew	Daily Output	Labor-Hours	Unit	Material	2014 Bare Costs Labor	Equipment	Total	Total Incl O&P
0360	Three #12	1 Elec	32	.250	Ea.	13	13.35		26.35	34.50

26 61 23 – Lamps Applications

26 61 23.10 Lamps

	26 61 23.10 Lamps		Crew	Daily Output	Labor-Hours	Unit	Material	2014 Bare Costs Labor	Equipment	Total	Total Incl O&P
0010	**LAMPS**	R265723-05									
0080	Fluorescent, rapid start, cool white, 2' long, 20 watt		1 Elec	1	8	C	310	425		735	980
0100	4' long, 40 watt			.90	8.889		236	475		711	970
0120	3' long, 30 watt			.90	8.889		410	475		885	1,150
0125	3' long, 25 watt energy saver	R265723-10	G		.90	8.889		1,200	475		1,675	2,000
0150	U-40 watt			.80	10		1,175	535		1,710	2,100
0155	U-34 watt energy saver		G		.80	10		1,200	535		1,735	2,100
0170	4' long, 34 watt energy saver		G		.90	8.889		425	475		900	1,175
0176	2' long, T8, 17 W energy saver	R265723-20	G		1	8		390	425		815	1,075
0178	3' long, T8, 25 W energy saver		G		.90	8.889		415	475		890	1,175
0180	4' long, T8, 32 watt energy saver		G		.90	8.889		365	475		840	1,100
0200	Slimline, 4' long, 40 watt			.90	8.889		1,175	475		1,650	2,000
0210	4' long, 30 watt energy saver	R265723-25	G		.90	8.889		1,175	475		1,650	2,000
0300	8' long, 75 watt			.80	10		1,050	535		1,585	1,950
0350	8' long, 60 watt energy saver		G		.80	10		450	535		985	1,300
0400	High output, 4' long, 60 watt			.90	8.889		740	475		1,215	1,525
0410	8' long, 95 watt energy saver		G		.80	10		750	535		1,285	1,625
0500	8' long, 110 watt			.80	10		750	535		1,285	1,625
0512	2' long, T5, 14 watt energy saver		G		1	8		1,000	425		1,425	1,775
0514	3' long, T5, 21 watt energy saver		G		.90	8.889		1,000	475		1,475	1,825
0516	4' long, T5, 28 watt energy saver		G		.90	8.889		870	475		1,345	1,675
0517	4' long, T5, 54 watt energy saver		G		.90	8.889		1,575	475		2,050	2,425
0520	Very high output, 4' long, 110 watt			.90	8.889		2,175	475		2,650	3,100
0525	8' long, 195 watt energy saver		G		.70	11.429		1,850	610		2,460	2,975
0550	8' long, 215 watt			.70	11.429		1,850	610		2,460	2,975
0554	Full spectrum, 4' long, 60 watt			.90	8.889		1,400	475		1,875	2,225
0556	6' long, 85 watt			.90	8.889		1,525	475		2,000	2,375
0558	8' long, 110 watt			.80	10		1,800	535		2,335	2,800
0560	Twin tube compact lamp		G		.90	8.889		425	475		900	1,175
0570	Double twin tube compact lamp		G		.80	10		945	535		1,480	1,850
0600	Mercury vapor, mogul base, deluxe white, 100 watt			.30	26.667		3,650	1,425		5,075	6,150
0650	175 watt			.30	26.667		1,725	1,425		3,150	4,000
0700	250 watt			.30	26.667		3,550	1,425		4,975	6,025
0800	400 watt			.30	26.667		5,100	1,425		6,525	7,725
0900	1000 watt			.20	40		7,350	2,125		9,475	11,300
1000	Metal halide, mogul base, 175 watt			.30	26.667		3,000	1,425		4,425	5,425
1100	250 watt			.30	26.667		3,400	1,425		4,825	5,875
1200	400 watt			.30	26.667		5,600	1,425		7,025	8,275
1300	1000 watt			.20	40		7,800	2,125		9,925	11,800
1320	1000 watt, 125,000 initial lumens			.20	40		22,100	2,125		24,225	27,500
1330	1500 watt			.20	40		20,400	2,125		22,525	25,700
1350	High pressure sodium, 70 watt			.30	26.667		3,650	1,425		5,075	6,150
1360	100 watt			.30	26.667		3,825	1,425		5,250	6,325
1370	150 watt			.30	26.667		3,925	1,425		5,350	6,450
1380	250 watt			.30	26.667		5,800	1,425		7,225	8,525
1400	400 watt			.30	26.667		4,275	1,425		5,700	6,825
1450	1000 watt			.20	40		11,800	2,125		13,925	16,100
1500	Low pressure sodium, 35 watt			.30	26.667		12,000	1,425		13,425	15,300
1550	55 watt			.30	26.667		11,900	1,425		13,325	15,200

26 61 23 – Lamps Applications

26 61 23.10 Lamps	Crew	Daily Output	Labor-Hours	Unit	Material	2014 Bare Costs Labor	Equipment	Total	Total Incl O&P
1600	90 watt	1 Elec	.30	26.667	C	13,300	1,425		14,725	16,700
1650	135 watt		.20	40		20,000	2,125		22,125	25,200
1700	180 watt		.20	40		23,100	2,125		25,225	28,600
1750	Quartz line, clear, 500 watt		1.10	7.273		1,375	390		1,765	2,100
1760	1500 watt		.20	40		3,575	2,125		5,700	7,150
1762	Spot, MR 16, 50 watt		1.30	6.154		1,050	330		1,380	1,675
1770	Tungsten halogen, T4, 400 watt		1.10	7.273		3,950	390		4,340	4,925
1775	T3, 1200 watt		.30	26.667		4,875	1,425		6,300	7,475
1778	PAR 30, 50 watt		1.30	6.154		1,125	330		1,455	1,750
1780	PAR 38, 90 watt		1.30	6.154		950	330		1,280	1,550
1800	Incandescent, interior, A21, 100 watt		1.60	5		233	267		500	655
1900	A21, 150 watt		1.60	5		166	267		433	585
2000	A23, 200 watt		1.60	5		330	267		597	765
2200	PS 35, 300 watt		1.60	5		860	267		1,127	1,350
2210	PS 35, 500 watt		1.60	5		1,425	267		1,692	1,950
2230	PS 52, 1000 watt		1.30	6.154		2,525	330		2,855	3,275
2240	PS 52, 1500 watt		1.30	6.154		6,775	330		7,105	7,950
2300	R30, 75 watt		1.30	6.154		610	330		940	1,175
2400	R40, 100 watt		1.30	6.154		610	330		940	1,175
2500	Exterior, PAR 38, 75 watt		1.30	6.154		1,850	330		2,180	2,550
2600	PAR 38, 150 watt		1.30	6.154		2,000	330		2,330	2,700
2700	PAR 46, 200 watt		1.10	7.273		3,875	390		4,265	4,825
2800	PAR 56, 300 watt		1.10	7.273		4,425	390		4,815	5,450
3000	Guards, fluorescent lamp, 4' long		1	8		1,400	425		1,825	2,200
3200	8' long		.90	8.889		2,800	475		3,275	3,800

26 61 23.55 LED Lamps

26 61 23.55 LED Lamps		Crew	Daily Output	Labor-Hours	Unit	Material	2014 Bare Costs Labor	Equipment	Total	Total Incl O&P
0010	**LED LAMPS**									
0100	LED lamp, interior, shape A60, equal to 60 watt	G	1 Elec	160	.050	Ea.	22	2.67		24.67	28.50
0200	Globe frosted A60, equal to 60 watt	G		160	.050		22	2.67		24.67	28.50
0300	Globe earth, equal to 100 watt	G		160	.050		73.50	2.67		76.17	84.50
1100	MR16, 3 watt, replacement of halogen lamp 25 watt	G		130	.062		21	3.28		24.28	28.50
1200	6 watt replacement of halogen lamp 45 watt	G		130	.062		46.50	3.28		49.78	56
2100	10 watt, PAR20, equal to 60 watt	G		130	.062		47.50	3.28		50.78	57
2200	15 watt, PAR30, equal to 100 watt	G		130	.062		91.50	3.28		94.78	106

26 71 Electrical Machines

26 71 13 – Motors Applications

26 71 13.10 Handling

		Crew	Daily Output	Labor-Hours	Unit	Material	2014 Bare Costs Labor	Equipment	Total	Total Incl O&P
0010	**HANDLING** Add to normal labor cost for restricted areas									
5000	Motors									
5100	1/2 HP, 23 pounds	1 Elec	4	2	Ea.		107		107	160
5110	3/4 HP, 28 pounds		4	2			107		107	160
5120	1 HP, 33 pounds		4	2			107		107	160
5130	1-1/2 HP, 44 pounds		3.20	2.500			133		133	200
5140	2 HP, 56 pounds		3	2.667			142		142	213
5150	3 HP, 71 pounds		2.30	3.478			186		186	278
5160	5 HP, 82 pounds		1.90	4.211			225		225	335
5170	7-1/2 HP, 124 pounds		1.50	5.333			285		285	425
5180	10 HP, 144 pounds		1.20	6.667			355		355	530
5190	15 HP, 185 pounds		1	8			425		425	640
5200	20 HP, 214 pounds	2 Elec	1.50	10.667			570		570	850

26 71 Electrical Machines

26 71 13 – Motors Applications

26 71 13.10 Handling		Crew	Daily Output	Labor-Hours	Unit	Material	2014 Bare Costs Labor	Equipment	Total	Total Incl O&P
5210	25 HP, 266 pounds	2 Elec	1.40	11.429	Ea.		610		610	915
5220	30 HP, 310 pounds		1.20	13.333			710		710	1,075
5230	40 HP, 400 pounds		1	16			855		855	1,275
5240	50 HP, 450 pounds		.90	17.778			950		950	1,425
5250	75 HP, 680 pounds		.80	20			1,075		1,075	1,600
5260	100 HP, 870 pounds	3 Elec	1	24			1,275		1,275	1,925
5270	125 HP, 940 pounds		.80	30			1,600		1,600	2,400
5280	150 HP, 1200 pounds		.70	34.286			1,825		1,825	2,750
5290	175 HP, 1300 pounds		.60	40			2,125		2,125	3,200
5300	200 HP, 1400 pounds		.50	48			2,550		2,550	3,825

26 71 13.20 Motors		Crew	Daily Output	Labor-Hours	Unit	Material	2014 Bare Costs Labor	Equipment	Total	Total Incl O&P
0010	**MOTORS** 230/460 volts, 60 HZ R263413-30									
0050	Dripproof, premium efficiency, 1.15 service factor									
0060	1800 RPM, 1/4 HP	1 Elec	5.33	1.501	Ea.	192	80		272	330
0070	1/3 HP		5.33	1.501		198	80		278	335
0080	1/2 HP		5.33	1.501		207	80		287	350
0090	3/4 HP		5.33	1.501		248	80		328	390
0100	1 HP		4.50	1.778		279	95		374	445
0150	2 HP		4.50	1.778		335	95		430	510
0200	3 HP		4.50	1.778		500	95		595	690
0250	5 HP		4.50	1.778		600	95		695	800
0300	7.5 HP		4.20	1.905		700	102		802	920
0350	10 HP		4	2		855	107		962	1,100
0400	15 HP		3.20	2.500		1,200	133		1,333	1,525
0450	20 HP	2 Elec	5.20	3.077		1,525	164		1,689	1,925
0500	25 HP		5	3.200		1,925	171		2,096	2,350
0550	30 HP		4.80	3.333		2,200	178		2,378	2,700
0600	40 HP		4	4		2,925	213		3,138	3,550
0650	50 HP		3.20	5		2,775	267		3,042	3,475
0700	60 HP		2.80	5.714		4,125	305		4,430	5,000
0750	75 HP		2.40	6.667		3,600	355		3,955	4,500
0800	100 HP	3 Elec	2.70	8.889		4,625	475		5,100	5,800
0850	125 HP		2.10	11.429		5,325	610		5,935	6,800
0900	150 HP		1.80	13.333		7,700	710		8,410	9,550
0950	200 HP		1.50	16		9,375	855		10,230	11,600
1000	1200 RPM, 1 HP	1 Elec	4.50	1.778		420	95		515	605
1050	2 HP		4.50	1.778		485	95		580	675
1100	3 HP		4.50	1.778		675	95		770	885
1150	5 HP		4.50	1.778		915	95		1,010	1,150
1200	3600 RPM, 2 HP		4.50	1.778		475	95		570	660
1250	3 HP		4.50	1.778		480	95		575	665
1300	5 HP		4.50	1.778		525	95		620	715
1350	Totally enclosed, premium efficiency 1.15 service factor									
1360	1800 RPM, 1/4 HP	1 Elec	5.33	1.501	Ea.	261	80		341	405
1370	1/3 HP		5.33	1.501		288	80		368	435
1380	1/2 HP		5.33	1.501		340	80		420	490
1390	3/4 HP		5.33	1.501		370	80		450	525
1400	1 HP		4.50	1.778		420	95		515	600
1450	2 HP		4.50	1.778		515	95		610	710
1500	3 HP		4.50	1.778		605	95		700	805
1550	5 HP		4.50	1.778		650	95		745	855
1600	7.5 HP		4.20	1.905		940	102		1,042	1,175

26 71 Electrical Machines

26 71 13 – Motors Applications

26 71 13.20 Motors		Crew	Daily Output	Labor-Hours	Unit	Material	2014 Bare Costs Labor	Equipment	Total	Total Incl O&P
1650	10 HP	1 Elec	4	2	Ea.	1,000	107		1,107	1,250
1700	15 HP	▼	3.20	2.500		1,425	133		1,558	1,750
1750	20 HP	2 Elec	5.20	3.077		1,950	164		2,114	2,400
1800	25 HP		5	3.200		2,250	171		2,421	2,750
1850	30 HP		4.80	3.333		2,375	178		2,553	2,875
1900	40 HP		4	4		2,875	213		3,088	3,500
1950	50 HP		3.20	5		3,100	267		3,367	3,800
2000	60 HP		2.80	5.714		5,200	305		5,505	6,175
2050	75 HP	▼	2.40	6.667		6,375	355		6,730	7,525
2100	100 HP	3 Elec	2.70	8.889		8,100	475		8,575	9,625
2150	125 HP		2.10	11.429		11,400	610		12,010	13,400
2200	150 HP		1.80	13.333		12,400	710		13,110	14,800
2250	200 HP	▼	1.50	16		15,800	855		16,655	18,600
2300	1200 RPM, 1 HP	1 Elec	4.50	1.778		405	95		500	585
2350	2 HP		4.50	1.778		465	95		560	655
2400	3 HP		4.50	1.778		635	95		730	840
2450	5 HP		4.50	1.778		865	95		960	1,100
2500	3600 RPM, 2 HP		4.50	1.778		365	95		460	540
2550	3 HP		4.50	1.778		450	95		545	635
2600	5 HP	▼	4.50	1.778	▼	560	95		655	760

Estimating Tips

27 20 00 Data Communications
27 30 00 Voice Communications
27 40 00 Audio-Video Communications

When estimating material costs for special systems, it is always prudent to obtain manufacturers' quotations for equipment prices and special installation requirements which will affect the total costs.

Reference Numbers

Reference numbers are shown in shaded boxes at the beginning of some major classifications. These numbers refer to related items in the Reference Section. The reference information may be an estimating procedure, an alternate pricing method, or technical information.

Note: Not all subdivisions listed here necessarily appear in this publication.

Note: **Trade Service,** *in part, has been used as a reference source for some of the material prices used in Division 27.*

27 01 Operation and Maintenance of Communications Systems

27 01 30 – Operation and Maintenance of Voice Communications

27 01 30.51 Operation and Maintenance of Voice Equipment	Crew	Daily Output	Labor-Hours	Unit	Material	2014 Bare Costs Labor	Equipment	Total	Total Incl O&P
0010 **OPERATION AND MAINTENANCE OF VOICE EQUIPMENT**									
3400 Remove and replace (reinstall), speaker	1 Elec	6	1.333	Ea.		71		71	106

27 05 Common Work Results for Communications

27 05 05 – Selective Communications Demolition

27 05 05.20 Electrical Demolition, Communications

	Crew	Daily Output	Labor-Hours	Unit	Material	2014 Bare Costs Labor	Equipment	Total	Total Incl O&P
0010 **ELECTRICAL DEMOLITION, COMMUNICATIONS** R024119-10									
0100 Fiber optics									
0120 Cable R260105-30	1 Elec	2400	.003	L.F.		.18		.18	.27
0160 Multi-channel rack enclosure		6	1.333	Ea.		71		71	106
0180 Patch panel	↓	18	.444	"		23.50		23.50	35.50
0200 Communication cables & fittings									
0220 Voice/data outlet	1 Elec	140	.057	Ea.		3.05		3.05	4.56
0240 Telephone cable		2800	.003	L.F.		.15		.15	.23
0260 Phone jack		135	.059	Ea.		3.16		3.16	4.73
0300 High performance cable, 2 pair		3000	.003	L.F.		.14		.14	.21
0320 4 pair		2100	.004			.20		.20	.30
0340 25 pair		900	.009	↓		.47		.47	.71
0400 Terminal cabinet	↓	5	1.600	Ea.		85.50		85.50	128
1000 Nurse call system									
1020 Station	1 Elec	24	.333	Ea.		17.80		17.80	26.50
1040 Standard call button		24	.333			17.80		17.80	26.50
1060 Corridor dome light or zone indictor	↓	24	.333			17.80		17.80	26.50
1080 Master control station	2 Elec	2	8	↓		425		425	640

27 05 05.30 Electrical Demolition, Sound and Video

	Crew	Daily Output	Labor-Hours	Unit	Material	2014 Bare Costs Labor	Equipment	Total	Total Incl O&P
0010 **ELECTRICAL DEMOLITION, SOUND & VIDEO** R024119-10									
0100 Cables									
0120 TV antenna lead-in cable R260105-30	1 Elec	2100	.004	L.F.		.20		.20	.30
0140 Sound cable		2400	.003			.18		.18	.27
0160 Microphone cable		2400	.003			.18		.18	.27
0180 Coaxial cable		2400	.003	↓		.18		.18	.27
0200 Doorbell system, not including wires, cables, and conduit		16	.500	Ea.		26.50		26.50	40
0220 Door chime or devices	↓	36	.222	"		11.85		11.85	17.75
0300 Public address system, not including wires, cables, and conduit									
0320 Conventional office	1 Elec	16	.500	Speaker		26.50		26.50	40
0340 Conventional industrial	"	8	1	"		53.50		53.50	80
0352 PA cabinet/panel	2 Elec	5	3.200	Ea.		171		171	256
0400 Sound system, not including wires, cables, and conduit									
0410 Components	1 Elec	24	.333	Ea.		17.80		17.80	26.50
0412 Speaker		24	.333			17.80		17.80	26.50
0416 Volume control		24	.333			17.80		17.80	26.50
0420 Intercom, master station		6	1.333			71		71	106
0440 Remote station		24	.333			17.80		17.80	26.50
0460 Intercom outlets		24	.333			17.80		17.80	26.50
0480 Handset	↓	12	.667	↓		35.50		35.50	53
0500 Emergency call system, not including wires, cables, and conduit									
0520 Annunciator	1 Elec	4	2	Ea.		107		107	160
0540 Devices	"	16	.500	↓		26.50		26.50	40
0600 Master door, buzzer type unit	2 Elec	1.60	10	↓		535		535	800
0800 TV System, not including wires, cables, and conduit									
0820 Master TV antenna system, per outlet	1 Elec	39	.205	Outlet		10.95		10.95	16.40

27 05 Common Work Results for Communications

27 05 05 – Selective Communications Demolition

27 05 05.30 Electrical Demolition, Sound and Video

		Crew	Daily Output	Labor-Hours	Unit	Material	2014 Bare Costs Labor	2014 Bare Costs Equipment	Total	Total Incl O&P
0840	School application, per outlet	1 Elec	16	.500	Outlet		26.50		26.50	40
0860	Amplifier		12	.667	Ea.		35.50		35.50	53
0880	Antenna	↓	6	1.333	"		71		71	106
0900	One camera & one monitor	2 Elec	7.80	2.051	Total		109		109	164
0920	One camera	1 Elec	8	1	Ea.		53.50		53.50	80

27 11 Communications Equipment Room Fittings

27 11 16 – Communications Cabinets, Racks, Frames and Enclosures

27 11 16.10 Public Phone

		Crew	Daily Output	Labor-Hours	Unit	Material	2014 Bare Costs Labor	2014 Bare Costs Equipment	Total	Total Incl O&P
0010	**PUBLIC PHONE**									
7600	Telephone with wood backboard									
7620	Single door, 12" H x 12" W x 4" D	1 Elec	5.30	1.509	Ea.	86	80.50		166.50	216
7650	18" H x 12" W x 4" D		4.70	1.702		101	91		192	247
7700	24" H x 12" W x 4" D		4.20	1.905		109	102		211	271
7720	18" H x 18" W x 4" D		4.20	1.905		120	102		222	285
7750	24" H x 18" W x 4" D		4	2		171	107		278	350
7780	36" H x 36" W x 4" D		3.60	2.222		191	119		310	385
7800	24" H x 24" W x 6" D		3.60	2.222		212	119		331	410
7820	30" H x 24" W x 6" D		3.20	2.500		253	133		386	480
7850	30" H x 30" W x 6" D		2.70	2.963		291	158		449	555
7880	36" H x 30" W x 6" D		2.50	3.200		360	171		531	650
7900	48" H x 36" W x 6" D		2.20	3.636		565	194		759	910
7920	Double door, 48" H x 36" W x 6" D	↓	2	4	↓	770	213		983	1,175

27 11 19 – Communications Termination Blocks and Patch Panels

27 11 19.10 Termination Blocks and Patch Panels

		Crew	Daily Output	Labor-Hours	Unit	Material	2014 Bare Costs Labor	2014 Bare Costs Equipment	Total	Total Incl O&P
0010	**TERMINATION BLOCKS AND PATCH PANELS**									
2960	Patch panel, RJ-45/110 type, 24 ports	2 Elec	6	2.667	Ea.	190	142		332	420
3000	48 ports	3 Elec	6	4		375	213		588	735
3040	96 ports	"	4	6		630	320		950	1,175
3100	Punch down termination per port	1 Elec	107	.075	↓		3.99		3.99	5.95

27 13 Communications Backbone Cabling

27 13 23 – Communications Optical Fiber Backbone Cabling

27 13 23.13 Communications Optical Fiber

			Crew	Daily Output	Labor-Hours	Unit	Material	2014 Bare Costs Labor	2014 Bare Costs Equipment	Total	Total Incl O&P
0010	**COMMUNICATIONS OPTICAL FIBER**										
0040	Specialized tools & techniques cause installation costs to vary.										
0070	Fiber optic, cable, bulk simplex, single mode	R271323-40	1 Elec	8	1	C.L.F.	23	53.50		76.50	106
0080	Multi mode			8	1		49.50	53.50		103	135
0090	4 strand, single mode			7.34	1.090		44	58		102	136
0095	Multi mode			7.34	1.090		50.50	58		108.50	143
0100	12 strand, single mode			6.67	1.199		105	64		169	212
0105	Multi mode		↓	6.67	1.199	↓	106	64		170	213
0150	Jumper					Ea.	33.50			33.50	37
0200	Pigtail						33.50			33.50	37
0300	Connector		1 Elec	24	.333		20.50	17.80		38.30	49
0350	Finger splice			32	.250		32.50	13.35		45.85	55.50
0400	Transceiver (low cost bi-directional)			8	1		415	53.50		468.50	540
0450	Rack housing, 4 rack spaces, 12 panels (144 fibers)			2	4		495	213		708	865
0500	Patch panel, 12 ports		↓	6	1.333	↓	245	71		316	375

27 15 10 – Special Communications Cabling

27 15 10.23 Sound and Video Cables and Fittings	Crew	Daily Output	Labor-Hours	Unit	Material	2014 Bare Costs Labor	Equipment	Total	Total Incl O&P
0010 **SOUND AND VIDEO CABLES & FITTINGS**									
0900 TV antenna lead-in, 300 ohm, #20-2 conductor	1 Elec	7	1.143	C.L.F.	23	61		84	117
0950 Coaxial, feeder outlet		7	1.143		20.50	61		81.50	114
1000 Coaxial, main riser		6	1.333		29	71		100	138
1100 Sound, shielded with drain, #22-2 conductor		8	1		15.90	53.50		69.40	97.50
1150 #22-3 conductor		7.50	1.067		20.50	57		77.50	108
1200 #22-4 conductor		6.50	1.231		25.50	65.50		91	127
1250 Nonshielded, #22-2 conductor		10	.800		14.25	42.50		56.75	79.50
1300 #22-3 conductor		9	.889		21.50	47.50		69	94.50
1350 #22-4 conductor		8	1		28.50	53.50		82	112
1400 Microphone cable		8	1		58.50	53.50		112	144

27 15 13 – Communications Copper Horizontal Cabling

27 15 13.13 Communication Cables

	Crew	Daily Output	Labor-Hours	Unit	Material	2014 Bare Costs Labor	Equipment	Total	Total Incl O&P
0010 **COMMUNICATION CABLES**									
2200 Telephone twisted, PVC insulation, #22-2 conductor	1 Elec	10	.800	C.L.F.	9.40	42.50		51.90	74.50
2250 #22-3 conductor		9	.889		12.30	47.50		59.80	84.50
2300 #22-4 conductor		8	1		18.65	53.50		72.15	101
2350 #18-2 conductor		9	.889		12.65	47.50		60.15	85
2370 Telephone jack, eight pins		32	.250	Ea.	4.84	13.35		18.19	25.50
5000 High performance unshielded twisted pair (UTP)									
5100 Cable, category 3, #24, 2 pair solid, PVC jacket R271513-75	1 Elec	10	.800	C.L.F.	8.10	42.50		50.60	73
5200 4 pair solid, PVC jacket		7	1.143		13.15	61		74.15	106
5300 25 pair solid, PVC jacket		3	2.667		68.50	142		210.50	289
5400 2 pair solid, plenum		10	.800		11.40	42.50		53.90	76.50
5500 4 pair solid, plenum		7	1.143		16.10	61		77.10	109
5600 25 pair solid, plenum		3	2.667		130	142		272	355
5700 4 pair stranded, PVC jacket		7	1.143		30.50	61		91.50	125
7000 Category 5, #24, 4 pair solid, PVC jacket		7	1.143		14.80	61		75.80	108
7100 4 pair solid, plenum		7	1.143		45.50	61		106.50	142
7200 4 pair stranded, PVC jacket		7	1.143		22	61		83	116
7210 Category 5e, #24, 4 pair solid, PVC jacket		7	1.143		14.15	61		75.15	107
7212 4 pair solid, plenum		7	1.143		23.50	61		84.50	118
7214 4 pair stranded, PVC jacket		7	1.143		25.50	61		86.50	120
7240 Category 6, #24, 4 pair solid, PVC jacket		7	1.143		20.50	61		81.50	114
7242 4 pair solid, plenum		7	1.143		24.50	61		85.50	119
7244 4 pair stranded, PVC jacket		7	1.143		25.50	61		86.50	120
7300 Connector, RJ-45, category 5		80	.100	Ea.	1.32	5.35		6.67	9.45
7302 Shielded RJ-45, category 5		72	.111		3.36	5.95		9.31	12.55
7310 Jack, UTP RJ-45, category 3		72	.111		2.33	5.95		8.28	11.40
7312 Category 5		65	.123		5	6.55		11.55	15.35
7314 Category 5e		65	.123		3.26	6.55		9.81	13.45
7316 Category 6		65	.123		3.31	6.55		9.86	13.50
7322 Jack, shielded RJ-45, category 5		60	.133		6	7.10		13.10	17.25
7324 Category 5e		60	.133		6	7.10		13.10	17.25
7326 Category 6		60	.133		6	7.10		13.10	17.25
7400 Voice/data expansion module, category 5e		8	1		46	53.50		99.50	131
8000 Multipair unshielded non-plenum cable, 150 V PVC jacket									
8002 #22, 2 pair	1 Elec	8.40	.952	C.L.F.	16.60	51		67.60	94.50
8003 3 pair		8	1		22	53.50		75.50	104
8004 4 pair		7.30	1.096		36	58.50		94.50	128
8006 6 pair		6.20	1.290		42	69		111	149
8008 8 pair		5.70	1.404		55	75		130	173

27 15 13 – Communications Copper Horizontal Cabling

27 15 13.13 Communication Cables	Crew	Daily Output	Labor-Hours	Unit	Material	2014 Bare Costs Labor	Equipment	Total	Total Incl O&P
8010	10 pair	1 Elec	5.30	1.509	C.L.F.	64	80.50		144.50	191
8012	12 pair		4.20	1.905		122	102		224	286
8015	15 pair		3.80	2.105		165	112		277	350
8020	20 pair	2 Elec	6.60	2.424		193	129		322	405
8025	25 pair		6	2.667		224	142		366	460
8030	30 pair		5.60	2.857		375	152		527	640
8040	40 pair		5.20	3.077		490	164		654	780
8050	50 pair		4.90	3.265		610	174		784	930
8100	Multipair unshielded non-plenum cable, 300 V PVC jacket									
8102	#20, 2 pair	1 Elec	7.30	1.096	C.L.F.	25.50	58.50		84	116
8103	3 pair		6.70	1.194		33.50	63.50		97	133
8104	4 pair		6	1.333		37.50	71		108.50	147
8106	6 pair		5.30	1.509		58.50	80.50		139	186
8108	8 pair		4.40	1.818		96	97		193	250
8110	10 pair	2 Elec	7.30	2.192		106	117		223	291
8112	12 pair		6.20	2.581		133	138		271	355
8115	15 pair		5.90	2.712		167	145		312	400
8201	#18, 1 pair	1 Elec	8	1		62.50	53.50		116	149
8202	2 pair		6.50	1.231		79.50	65.50		145	186
8203	3 pair		5.60	1.429		120	76		196	246
8204	4 pair		5	1.600		157	85.50		242.50	300
8206	6 pair		4.40	1.818		263	97		360	435
8208	8 pair		4	2		262	107		369	450
8215	15 pair	2 Elec	5	3.200		645	171		816	965
8300	Multipair shielded non-plenum cable, 300 V PVC jacket									
8303	#22, 3 pair	1 Elec	6.70	1.194	C.L.F.	61.50	63.50		125	164
8306	6 pair		5.70	1.404		123	75		198	248
8309	9 pair		5	1.600		229	85.50		314.50	380
8312	12 pair	2 Elec	8	2		485	107		592	690
8315	15 pair		7.30	2.192		500	117		617	725
8317	17 pair		6.80	2.353		680	126		806	940
8319	19 pair		6.40	2.500		685	133		818	950
8327	27 pair		5.90	2.712		1,225	145		1,370	1,575
8402	#20, 2 pair	1 Elec	6.70	1.194		77	63.50		140.50	180
8403	3 pair		6.20	1.290		96	69		165	208
8406	6 pair		4.40	1.818		242	97		339	410
8409	9 pair		3.60	2.222		305	119		424	510
8412	12 pair	2 Elec	5.90	2.712		560	145		705	830
8415	15 pair	"	5.70	2.807		550	150		700	830
8502	#18, 2 pair	1 Elec	6.20	1.290		131	69		200	247
8503	3 pair		5.30	1.509		205	80.50		285.50	345
8504	4 pair		4.70	1.702		245	91		336	405
8506	6 pair		4	2		380	107		487	580
8509	9 pair		3.40	2.353		585	126		711	830
8515	15 pair	2 Elec	4.80	3.333		980	178		1,158	1,350

27 15 33 – Communications Coaxial Horizontal Cabling

27 15 33.10 Coaxial Cable and Fittings

0010	COAXIAL CABLE & FITTINGS									
3500	Coaxial connectors, 50 ohm impedance quick disconnect									
3540	BNC plug, for RG A/U #58 cable	1 Elec	42	.190	Ea.	4.60	10.15		14.75	20.50
3550	RG A/U #59 cable		42	.190		4.60	10.15		14.75	20.50
3560	RG A/U #62 cable		42	.190		4.60	10.15		14.75	20.50

309

27 15 Communications Horizontal Cabling

27 15 33 – Communications Coaxial Horizontal Cabling

27 15 33.10 Coaxial Cable and Fittings	Crew	Daily Output	Labor-Hours	Unit	Material	2014 Bare Costs Labor	Equipment	Total	Total Incl O&P
3600	BNC jack, for RG A/U #58 cable	1 Elec	42	.190	Ea.	4.85	10.15		15	20.50
3610	RG A/U #59 cable		42	.190		4.85	10.15		15	20.50
3620	RG A/U #62 cable		42	.190		4.85	10.15		15	20.50
3660	BNC panel jack, for RG A/U #58 cable		40	.200		7.50	10.65		18.15	24
3670	RG A/U #59 cable		40	.200		7.50	10.65		18.15	24
3680	RG A/U #62 cable		40	.200		7.50	10.65		18.15	24
3720	BNC bulkhead jack, for RG A/U #58 cable		40	.200		7.75	10.65		18.40	24.50
3730	RG A/U #59 cable		40	.200		7.75	10.65		18.40	24.50
3740	RG A/U #62 cable		40	.200		7.75	10.65		18.40	24.50
3850	Coaxial cable, RG A/U 58, 50 ohm		8	1	C.L.F.	49	53.50		102.50	134
3860	RG A/U 59, 75 ohm		8	1		39	53.50		92.50	123
3870	RG A/U 62, 93 ohm		8	1		46	53.50		99.50	131
3875	RG 6/U, 75 ohm		8	1		28.50	53.50		82	111
3950	Fire rated, RG A/U 58, 50 ohm		8	1		93.50	53.50		147	183
3960	RG A/U 59, 75 ohm		8	1		123	53.50		176.50	216
3970	RG A/U 62, 93 ohm		8	1		109	53.50		162.50	200

27 15 43 – Communications Faceplates and Connectors

27 15 43.13 Communication Outlets

		Crew	Daily Output	Labor-Hours	Unit	Material	2014 Bare Costs Labor	Equipment	Total	Total Incl O&P
0010	**COMMUNICATION OUTLETS**									
0100	Voice/data devices not included									
0120	Voice/Data outlets, single opening	1 Elec	48	.167	Ea.	7.85	8.90		16.75	22
0140	Two jack openings		48	.167		2.56	8.90		11.46	16.10
0160	One jack & one 3/4" round opening		48	.167		6.95	8.90		15.85	21
0180	One jack & one twinaxial opening		48	.167		7.50	8.90		16.40	21.50
0200	One jack & one connector cabling opening		48	.167		6.95	8.90		15.85	21
0220	Two 3/8" coaxial openings		48	.167		6.95	8.90		15.85	21
0300	Data outlets, single opening		48	.167		6.95	8.90		15.85	21
0320	One 25-pin subminiature opening		48	.167		6.95	8.90		15.85	21
1000	Voice/Data wall plate plastic, 1 gang, 1-port		72	.111		2.03	5.95		7.98	11.10
1020	2-port		72	.111		2.03	5.95		7.98	11.10
1040	3-port		72	.111		2.03	5.95		7.98	11.10
1060	4-port		72	.111		2.03	5.95		7.98	11.10
1080	6-port		72	.111		2.03	5.95		7.98	11.10
1100	2 gang, 6-port		48	.167		6.05	8.90		14.95	19.95
1120	Voice/Data wall plate stainless steel, 1 gang, 1-port		72	.111		6.10	5.95		12.05	15.55
1140	2-port		72	.111		6.20	5.95		12.15	15.70
1160	3-port		72	.111		6.25	5.95		12.20	15.70
1180	4-port		72	.111		6.20	5.95		12.15	15.70
1200	2 gang, 6-port		48	.167		10.85	8.90		19.75	25.50

27 21 Data Communications Network Equipment

27 21 23 – Data Communications Switches and Hubs

27 21 23.10 Switching and Routing Equipment

		Crew	Daily Output	Labor-Hours	Unit	Material	2014 Bare Costs Labor	Equipment	Total	Total Incl O&P
0010	**SWITCHING AND ROUTING EQUIPMENT**									
1100	Network hub, dual speed, 24 ports, includes cabinet	3 Elec	.66	36.364	Ea.	1,700	1,950		3,650	4,775
2000	Network switch, 10/100/1000 Mbps, 24 ports		.75	32		1,850	1,700		3,550	4,575
2040	48 ports		.66	36.364		2,325	1,950		4,275	5,450

27 32 Voice Communications Terminal Equipment

27 32 36 – TTY Equipment

27 32 36.10 TTY Telephone Equipment	Crew	Daily Output	Labor-Hours	Unit	Material	2014 Bare Costs Labor	2014 Bare Costs Equipment	Total	Total Incl O&P
0010 **TTY TELEPHONE EQUIPMENT**									
1620 Telephone, TTY, compact, pocket type				Ea.	251			251	276
1630 Advanced, desk type	2 Elec	20	.800		505	42.50		547.50	620
1640 Full-featured public, wall type	"	4	4		720	213		933	1,100

27 41 Audio-Video Systems

27 41 33 – Master Antenna Television Systems

27 41 33.10 T.V. Systems

	Crew	Daily Output	Labor-Hours	Unit	Material	2014 Bare Costs Labor	2014 Bare Costs Equipment	Total	Total Incl O&P
0010 **T.V. SYSTEMS**, not including rough-in wires, cables & conduits									
0100 Master TV antenna system									
0200 VHF reception & distribution, 12 outlets	1 Elec	6	1.333	Outlet	158	71		229	280
0400 30 outlets		10	.800		140	42.50		182.50	218
0600 100 outlets		13	.615		142	33		175	206
0800 VHF & UHF reception & distribution, 12 outlets		6	1.333		211	71		282	340
1000 30 outlets		10	.800		140	42.50		182.50	218
1200 100 outlets		13	.615		142	33		175	206
1400 School and deluxe systems, 12 outlets		2.40	3.333		278	178		456	570
1600 30 outlets		4	2		244	107		351	430
1800 80 outlets		5.30	1.509		234	80.50		314.50	380
1900 Amplifier		4	2	Ea.	700	107		807	930
1910 Antenna		2	4		335	213		548	690
5000 T.V. Antenna, minimum		6	1.333		46.50	71		117.50	157
5100 Maximum		4	2		196	107		303	375
5110 Rotor unit		8	1		76.50	53.50		130	165
5120 Single booster		8	1		37.50	53.50		91	122
5130 Antenna pole, 10'		3.20	2.500		27.50	133		160.50	231
6100 Satellite TV system	2 Elec	1	16		2,125	855		2,980	3,600
6110 Dish, mesh, 10' diam.	"	2.40	6.667		1,150	355		1,505	1,800
6111 Two way RF/IF tapeoff	1 Elec	36	.222		3.52	11.85		15.37	21.50
6112 Two way RF/IF splitter		24	.333		14.60	17.80		32.40	42.50
6113 Line amplifier		24	.333		14.60	17.80		32.40	42.50
6114 Line splitters		36	.222		4.64	11.85		16.49	23
6115 Line multi switches		8	1		675	53.50		728.50	820
6120 Motor unit		2.40	3.333		365	178		543	670
7000 Home theater, widescreen, 42", high definition, TV		10.24	.781		505	41.50		546.50	620
7050 Flat wall mount bracket		10.24	.781		127	41.50		168.50	203
7100 7 channel home theater receiver		10.24	.781		375	41.50		416.50	480
7200 Home theater speakers		10.24	.781	Set	189	41.50		230.50	271
7300 Home theater programmable remote		10.24	.781	Ea.	256	41.50		297.50	345
8000 Main video splitter		4	2		2,425	107		2,532	2,825
8010 Video distribution units		4	2		122	107		229	295

27 51 Distributed Audio-Video Communications Systems

27 51 16 – Public Address and Mass Notification Systems

27 51 16.10 Public Address System		Crew	Daily Output	Labor-Hours	Unit	Material	2014 Bare Costs Labor	Equipment	Total	Total Incl O&P
0010	**PUBLIC ADDRESS SYSTEM**									
0100	Conventional, office	1 Elec	5.33	1.501	Speaker	134	80		214	267
0200	Industrial	"	2.70	2.963	"	258	158		416	520

27 51 19 – Sound Masking Systems

27 51 19.10 Sound System

		Crew	Daily Output	Labor-Hours	Unit	Material	2014 Bare Costs Labor	Equipment	Total	Total Incl O&P
0010	**SOUND SYSTEM**, not including rough-in wires, cables & conduits									
0100	Components, projector outlet	1 Elec	8	1	Ea.	44.50	53.50		98	129
0200	Microphone		4	2		79.50	107		186.50	248
0400	Speakers, ceiling or wall		8	1		116	53.50		169.50	208
0600	Trumpets		4	2		216	107		323	400
0800	Privacy switch		8	1		86.50	53.50		140	175
1000	Monitor panel		4	2		385	107		492	585
1200	Antenna, AM/FM		4	2		135	107		242	310
1400	Volume control		8	1		87.50	53.50		141	177
1600	Amplifier, 250 watts		1	8		1,250	425		1,675	2,025
1800	Cabinets		1	8		840	425		1,265	1,550
2000	Intercom, 30 station capacity, master station	2 Elec	2	8		2,300	425		2,725	3,175
2020	10 station capacity	"	4	4		1,275	213		1,488	1,725
2200	Remote station	1 Elec	8	1		162	53.50		215.50	258
2400	Intercom outlets		8	1		95	53.50		148.50	184
2600	Handset		4	2		315	107		422	505
2800	Emergency call system, 12 zones, annunciator		1.30	6.154		945	330		1,275	1,550
3000	Bell		5.30	1.509		98	80.50		178.50	229
3200	Light or relay		8	1		49	53.50		102.50	134
3400	Transformer		4	2		215	107		322	395
3600	House telephone, talking station		1.60	5		460	267		727	910
3800	Press to talk, release to listen		5.30	1.509		108	80.50		188.50	239
4000	System-on button					64.50			64.50	71
4200	Door release	1 Elec	4	2		115	107		222	287
4400	Combination speaker and microphone		8	1		196	53.50		249.50	296
4600	Termination box		3.20	2.500		61.50	133		194.50	268
4800	Amplifier or power supply		5.30	1.509		710	80.50		790.50	900
5000	Vestibule door unit		16	.500	Name	130	26.50		156.50	183
5200	Strip cabinet		27	.296	Ea.	246	15.80		261.80	295
5400	Directory		16	.500		116	26.50		142.50	167
6000	Master door, button buzzer type, 100 unit	2 Elec	.54	29.630		1,175	1,575		2,750	3,675
6020	200 unit		.30	53.333		2,225	2,850		5,075	6,700
6040	300 unit		.20	80		3,375	4,275		7,650	10,100
6060	Transformer	1 Elec	8	1		30	53.50		83.50	113
6080	Door opener		5.30	1.509		43	80.50		123.50	169
6100	Buzzer with door release and plate		4	2		43	107		150	208
6200	Intercom type, 100 unit	2 Elec	.54	29.630		1,475	1,575		3,050	3,975
6220	200 unit		.30	53.333		2,850	2,850		5,700	7,400
6240	300 unit		.20	80		4,325	4,275		8,600	11,200
6260	Amplifier	1 Elec	2	4		215	213		428	555
6280	Speaker with door release	"	4	2		64.50	107		171.50	231

27 52 Healthcare Communications and Monitoring Systems

27 52 23 – Nurse Call/Code Blue Systems

27 52 23.10 Nurse Call Systems

		Crew	Daily Output	Labor-Hours	Unit	Material	2014 Bare Costs Labor	Equipment	Total	Total Incl O&P
0010	**NURSE CALL SYSTEMS**									
0100	Single bedside call station	1 Elec	8	1	Ea.	228	53.50		281.50	330
0200	Ceiling speaker station		8	1		66.50	53.50		120	154
0400	Emergency call station		8	1		71.50	53.50		125	159
0600	Pillow speaker		8	1		176	53.50		229.50	273
0800	Double bedside call station		4	2		141	107		248	315
1000	Duty station		4	2		146	107		253	320
1200	Standard call button		8	1		86	53.50		139.50	175
1400	Lights, corridor, dome or zone indicator		8	1		48.50	53.50		102	134
1600	Master control station for 20 stations	2 Elec	.65	24.615	Total	3,950	1,325		5,275	6,300

27 53 Distributed Systems

27 53 13 – Clock Systems

27 53 13.50 Clock Equipments

		Crew	Daily Output	Labor-Hours	Unit	Material	2014 Bare Costs Labor	Equipment	Total	Total Incl O&P
0010	**CLOCK EQUIPMENTS**, not including wires & conduits									
0100	Time system components, master controller	1 Elec	.33	24.242	Ea.	1,850	1,300		3,150	3,950
0200	Program bell		8	1		86	53.50		139.50	175
0400	Combination clock & speaker		3.20	2.500		212	133		345	435
0600	Frequency generator		2	4		2,375	213		2,588	2,950
0800	Job time automatic stamp recorder		4	2		535	107		642	750
1600	Master time clock system, clocks & bells, 20 room	4 Elec	.20	160		6,100	8,525		14,625	19,500
1800	50 room	"	.08	400		12,500	21,300		33,800	45,700
1900	Time clock	1 Elec	3.20	2.500		440	133		573	685
2000	100 cards in & out, 1 color					9.05			9.05	10
2200	2 colors					9.05			9.05	10
2800	Metal rack for 25 cards	1 Elec	7	1.143		43.50	61		104.50	139
4000	Wireless time systems component, master controller	"	2	4		600	213		813	980
4010	For transceiver and antenna, see Section 28 39 10.10									
4100	Wireless analog clock w/battery operated	1 Elec	8	1	Ea.	112	53.50		165.50	203
4200	Wireless digital clock w/battery operated	"	8	1	"	365	53.50		418.50	480

Division Notes

	CREW	DAILY OUTPUT	LABOR-HOURS	UNIT	BARE COSTS				TOTAL INCL O&P
					MAT.	LABOR	EQUIP.	TOTAL	

Estimating Tips

- When estimating material costs for electronic safety and security systems, it is always prudent to obtain manufacturers' quotations for equipment prices and special installation requirements that affect the total cost.

- Fire alarm systems consist of control panels, annunciator panels, battery with rack, charger, and fire alarm actuating and indicating devices. Some fire alarm systems include speakers, telephone lines, door closer controls, and other components. Be careful not to overlook the costs related to installation for these items.

Also be aware of costs for integrated automation instrumentation and terminal devices, control equipment, control wiring, and programming.

- Security equipment includes items such as CCTV, access control, and other detection and identification systems to perform alert and alarm functions. Be sure to consider the costs related to installation for this security equipment, such as for integrated automation instrumentation and terminal devices, control equipment, control wiring, and programming.

Reference Numbers

Reference numbers are shown in shaded boxes at the beginning of some major classifications. These numbers refer to related items in the Reference Section. The reference information may be an estimating procedure, an alternate pricing method, or technical information.

Note: Not all subdivisions listed here necessarily appear in this publication.

28 01 Operation and Maint. of Electronic Safety and Security

28 01 30 – Operation and Maint. of Electronic Detection and Alarm

28 01 30.51 Maint. and Admin. of Elec. Detection and Alarm	Crew	Daily Output	Labor-Hours	Unit	Material	2014 Bare Costs Labor	Equipment	Total	Total Incl O&P
0010 **MAINT. AND ADMIN. OF ELEC. DETECTION AND ALARM**									
3300 Remove and replace (reinstall), fire alarm device	1 Elec	5.33	1.501	Ea.		80		80	120

28 05 Common Work Results for Electronic Safety and Security

28 05 05 – Selective Safety and Security Demolition

28 05 05.10 Safety and Security Demolition

		Crew	Daily Output	Labor-Hours	Unit	Material	2014 Bare Costs Labor	Equipment	Total	Total Incl O&P
0010	**SAFETY AND SECURITY DEMOLITION**									
1050	Finger print/card reader	1 Elec	8	1	Ea.		53.50		53.50	80
1060	Video camera		8	1			53.50		53.50	80
1070	Motion detector, multi-channel		7	1.143			61		61	91.50
1090	Infrared detector		7	1.143			61		61	91.50
1130	Video monitor, 19"		8	1			53.50		53.50	80
1210	Fire alarm horn and strobe light		16	.500			26.50		26.50	40
1220	Flame detector		23	.348			18.55		18.55	28
1230	Duct detector		10	.800			42.50		42.50	64
1240	Smoke detector		23	.348			18.55		18.55	28
1250	Fire alarm pull station		23	.348			18.55		18.55	28
1260	Fire alarm control panel, 4 to 8 zone	2 Elec	4	4			213		213	320
1270	12 to 16 zone	"	2.67	5.993			320		320	480
1280	Fire alarm annunciation panel, 4 to 8 zone	1 Elec	6	1.333			71		71	106
1290	12 to 16 zone	2 Elec	6	2.667			142		142	213

28 05 13 – Conductors and Cables for Electronic Safety and Security

28 05 13.23 Fire Alarm Communications Conductors and Cables

		Crew	Daily Output	Labor-Hours	Unit	Material	2014 Bare Costs Labor	Equipment	Total	Total Incl O&P
0010	**FIRE ALARM COMMUNICATIONS CONDUCTORS AND CABLES**									
1500	Fire alarm FEP teflon 150 volt to 200°C									
1550	#22, 1 pair	1 Elec	10	.800	C.L.F.	62	42.50		104.50	133
1600	2 pair		8	1		103	53.50		156.50	194
1650	4 pair		7	1.143		158	61		219	266
1700	6 pair		6	1.333		206	71		277	330
1750	8 pair		5.50	1.455		259	77.50		336.50	400
1800	10 pair		5	1.600		310	85.50		395.50	475
1850	#18, 1 pair		8	1		71	53.50		124.50	158
1900	2 pair		6.50	1.231		136	65.50		201.50	248
1950	4 pair		4.80	1.667		206	89		295	360
2000	6 pair		4	2		271	107		378	460
2050	8 pair		3.50	2.286		355	122		477	575
2100	10 pair		3	2.667		405	142		547	660

28 13 Access Control

28 13 53 – Security Access Detection

28 13 53.13 Security Access Metal Detectors

		Crew	Daily Output	Labor-Hours	Unit	Material	2014 Bare Costs Labor	Equipment	Total	Total Incl O&P
0010	**SECURITY ACCESS METAL DETECTORS**									
0240	Metal detector, hand-held, wand type, unit only				Ea.	89			89	98
0250	Metal detector, walk through portal type, single zone	1 Elec	2	4		3,675	213		3,888	4,350
0260	Multi-zone	"	2	4		4,650	213		4,863	5,450

28 13 Access Control

28 13 53 – Security Access Detection

28 13 53.16 Security Access X-Ray Equipment	Crew	Daily Output	Labor-Hours	Unit	Material	2014 Bare Costs Labor	2014 Bare Costs Equipment	Total	Total Incl O&P
0010 **SECURITY ACCESS X-RAY EQUIPMENT**									
0290 X-ray machine, desk top, for mail/small packages/letters	1 Elec	4	2	Ea.	3,400	107		3,507	3,875
0300 Conveyor type, incl monitor, minimum		2	4		15,800	213		16,013	17,700
0310 Maximum		2	4		28,300	213		28,513	31,400
0320 X-ray machine, large unit, for airports, incl monitor, min.	2 Elec	1	16		39,600	855		40,455	44,900
0330 Maximum	"	.50	32		68,000	1,700		69,700	77,000

28 13 53.23 Security Access Explosive Detection Equipment

	Crew	Daily Output	Labor-Hours	Unit	Material	2014 Bare Costs Labor	2014 Bare Costs Equipment	Total	Total Incl O&P
0010 **SECURITY ACCESS EXPLOSIVE DETECTION EQUIPMENT**									
0270 Explosives detector, walk through portal type	1 Elec	2	4	Ea.	43,700	213		43,913	48,400
0280 Hand-held, battery operated				"	25,500				28,100

28 16 Intrusion Detection

28 16 16 – Intrusion Detection Systems Infrastructure

28 16 16.50 Intrusion Detection

	Crew	Daily Output	Labor-Hours	Unit	Material	2014 Bare Costs Labor	2014 Bare Costs Equipment	Total	Total Incl O&P
0010 **INTRUSION DETECTION**, not including wires & conduits									
0100 Burglar alarm, battery operated, mechanical trigger	1 Elec	4	2	Ea.	275	107		382	460
0200 Electrical trigger		4	2		330	107		437	520
0400 For outside key control, add		8	1		83	53.50		136.50	172
0600 For remote signaling circuitry, add		8	1		124	53.50		177.50	216
0800 Card reader, flush type, standard		2.70	2.963		920	158		1,078	1,250
1000 Multi-code		2.70	2.963		1,175	158		1,333	1,525
1010 Card reader, proximity type		2.70	2.963		315	158		473	580
1200 Door switches, hinge switch		5.30	1.509		58	80.50		138.50	185
1400 Magnetic switch		5.30	1.509		68.50	80.50		149	197
1600 Exit control locks, horn alarm		4	2		274	107		381	460
1800 Flashing light alarm		4	2		300	107		407	490
2000 Indicating panels, 1 channel		2.70	2.963		365	158		523	635
2200 10 channel	2 Elec	3.20	5		1,250	267		1,517	1,775
2400 20 channel		2	8		2,425	425		2,850	3,325
2600 40 channel		1.14	14.035		4,425	750		5,175	5,975
2800 Ultrasonic motion detector, 12 volt	1 Elec	2.30	3.478		227	186		413	530
3000 Infrared photoelectric detector		4	2		188	107		295	365
3200 Passive infrared detector		4	2		280	107		387	470
3400 Glass break alarm switch		8	1		92.50	53.50		146	181
3420 Switchmats, 30" x 5'		5.30	1.509		84	80.50		164.50	214
3440 30" x 25'		4	2		201	107		308	380
3460 Police connect panel		4	2		242	107		349	425
3480 Telephone dialer		5.30	1.509		380	80.50		460.50	540
3500 Alarm bell		4	2		103	107		210	274
3520 Siren		4	2		145	107		252	320
3540 Microwave detector, 10' to 200'		2	4		660	213		873	1,050
3560 10' to 350'		2	4		1,925	213		2,138	2,450

28 23 Video Surveillance

28 23 13 – Video Surveillance Control and Management Systems

28 23 13.10 Closed Circuit Television System

28 23 13.10 Closed Circuit Television System	Crew	Daily Output	Labor-Hours	Unit	Material	2014 Bare Costs Labor	2014 Bare Costs Equipment	Total	Total Incl O&P
0010 **CLOSED CIRCUIT TELEVISION SYSTEM**									
2000 Surveillance, one station (camera & monitor)	2 Elec	2.60	6.154	Total	1,350	330		1,680	1,975
2200 For additional camera stations, add	1 Elec	2.70	2.963	Ea.	750	158		908	1,050
2400 Industrial quality, one station (camera & monitor)	2 Elec	2.60	6.154	Total	2,775	330		3,105	3,550
2600 For additional camera stations, add	1 Elec	2.70	2.963	Ea.	1,700	158		1,858	2,100
2610 For low light, add		2.70	2.963		1,375	158		1,533	1,725
2620 For very low light, add		2.70	2.963		10,100	158		10,258	11,300
2800 For weatherproof camera station, add		1.30	6.154		1,050	330		1,380	1,650
3000 For pan and tilt, add		1.30	6.154		2,700	330		3,030	3,475
3200 For zoom lens - remote control, add, minimum		2	4		2,500	213		2,713	3,075
3400 Maximum		2	4		9,125	213		9,338	10,300
3410 For automatic iris for low light, add		2	4		2,175	213		2,388	2,725
3600 Educational T.V. studio, basic 3 camera system, black & white,									
3800 electrical & electronic equip. only, minimum	4 Elec	.80	40	Total	13,000	2,125		15,125	17,500
4000 Maximum (full console)		.28	114		55,000	6,100		61,100	69,500
4100 As above, but color system, minimum		.28	114		73,000	6,100		79,100	89,500
4120 Maximum		.12	266		317,500	14,200		331,700	371,000
4200 For film chain, black & white, add	1 Elec	1	8	Ea.	14,800	425		15,225	16,900
4250 Color, add		.25	32		18,000	1,700		19,700	22,400
4400 For video recorders, add, minimum		1	8		3,125	425		3,550	4,075
4600 Maximum	4 Elec	.40	80		25,900	4,275		30,175	34,900

28 23 19 – Digital Video Recorders and Analog Recording Devices

28 23 19.10 Digital Video Recorder (DVR)

28 23 19.10 Digital Video Recorder (DVR)	Crew	Daily Output	Labor-Hours	Unit	Material	2014 Bare Costs Labor	2014 Bare Costs Equipment	Total	Total Incl O&P
0010 **DIGITAL VIDEO RECORDER (DVR)**									
0100 Pentaplex hybrid, internet protocol, and hard drive									
0200 4 channel	1 Elec	1.33	6.015	Ea.	1,450	320		1,770	2,050
0300 8 channel		1	8		2,250	425		2,675	3,125
0400 16 channel		1	8		2,175	425		2,600	3,050

28 23 23 – Video Surveillance Systems Infrastructure

28 23 23.50 Video Surveillance Equipments

28 23 23.50 Video Surveillance Equipments	Crew	Daily Output	Labor-Hours	Unit	Material	2014 Bare Costs Labor	2014 Bare Costs Equipment	Total	Total Incl O&P
0010 **VIDEO SURVEILLANCE EQUIPMENTS**									
0200 Video cameras, wireless, hidden in exit signs, clocks, etc., incl. receiver	1 Elec	3	2.667	Ea.	107	142		249	330
0210 Accessories for video recorder, single camera		3	2.667		181	142		323	410
0220 For multiple cameras		3	2.667		1,750	142		1,892	2,150
0230 Video cameras, wireless, for under vehicle searching, complete		2	4		10,100	213		10,313	11,400
0400 Internet protocol network camera, day/night, color & power supply		2.60	3.077		1,025	164		1,189	1,375
0500 Monitor, color flat screen, liquid crystal display (LCD), 15"		2.70	2.963		690	158		848	995
0520 17"		2.70	2.963		815	158		973	1,125
0540 19"		2.70	2.963		915	158		1,073	1,225

28 31 Fire Detection and Alarm

28 31 23 – Fire Detection and Alarm Annunciation Panels and Fire Stations

28 31 23.50 Alarm Panels and Devices	Crew	Daily Output	Labor-Hours	Unit	Material	2014 Bare Costs Labor	Equipment	Total	Total Incl O&P
0010 **ALARM PANELS AND DEVICES**, not including wires & conduits									
3594 Fire, alarm control panel									
3600 4 zone	2 Elec	2	8	Ea.	395	425		820	1,075
3800 8 zone		1	16		775	855		1,630	2,125
4000 12 zone		.67	23.988		2,375	1,275		3,650	4,525
4020 Alarm device	1 Elec	8	1		235	53.50		288.50	340
4050 Actuating device	"	8	1		335	53.50		388.50	445
4160 Alarm control panel, addressable w/o voice, up to 200 points	2 Elec	1.14	13.998		4,425	745		5,170	6,000
4170 addressable w/voice, up to 400 points	"	.73	22.008		9,325	1,175		10,500	12,100
4175 Addressable interface device	1 Elec	7.25	1.103		133	59		192	235
4200 Battery and rack		4	2		410	107		517	610
4400 Automatic charger		8	1		580	53.50		633.50	715
4600 Signal bell		8	1		78	53.50		131.50	166
4800 Trouble buzzer or manual station		8	1		82.50	53.50		136	171
5600 Strobe and horn		5.30	1.509		151	80.50		231.50	287
5610 Strobe and horn (ADA type)		5.30	1.509		151	80.50		231.50	287
5620 Visual alarm (ADA type)		6.70	1.194		102	63.50		165.50	208
5800 Fire alarm horn		6.70	1.194		60.50	63.50		124	162
6000 Door holder, electro-magnetic		4	2		102	107		209	272
6200 Combination holder and closer		3.20	2.500		122	133		255	335
6600 Drill switch		8	1		365	53.50		418.50	485
6800 Master box		2.70	2.963		6,350	158		6,508	7,200
7000 Break glass station		8	1		55	53.50		108.50	141
7010 Break glass station, addressable		7.25	1.103		146	59		205	249
7800 Remote annunciator, 8 zone lamp		1.80	4.444		207	237		444	580
8000 12 zone lamp	2 Elec	2.60	6.154		330	330		660	855
8200 16 zone lamp	"	2.20	7.273		415	390		805	1,025

28 31 43 – Fire Detection Sensors

28 31 43.50 Fire and Heat Detectors

	Crew	Daily Output	Labor-Hours	Unit	Material	2014 Bare Costs Labor	Equipment	Total	Total Incl O&P
0010 **FIRE & HEAT DETECTORS**									
5000 Detector, rate of rise	1 Elec	8	1	Ea.	50.50	53.50		104	136
5010 Heat addressable type		7.25	1.103		254	59		313	370
5100 Fixed temperature		8	1		50.50	53.50		104	136

28 31 46 – Smoke Detection Sensors

28 31 46.50 Smoke Detectors

	Crew	Daily Output	Labor-Hours	Unit	Material	2014 Bare Costs Labor	Equipment	Total	Total Incl O&P
0010 **SMOKE DETECTORS**									
5200 Smoke detector, ceiling type	1 Elec	6.20	1.290	Ea.	109	69		178	223
5240 Smoke detector addressable type		6	1.333		219	71		290	345
5400 Duct type		3.20	2.500		325	133		458	555
5420 Duct addressable type		3.20	2.500		505	133		638	755

28 39 Mass Notification Systems

28 39 10 – Notification Systems

28 39 10.10 Mass Notification System	Crew	Daily Output	Labor-Hours	Unit	Material	2014 Bare Costs Labor	Equipment	Total	Total Incl O&P
0010 **MASS NOTIFICATION SYSTEM**									
0100 Wireless command center, 10,000 devices	2 Elec	1.33	12.030	Ea.	3,575	640		4,215	4,875
0200 Option, email notification					1,750			1,750	1,925
0210 Remote device supervision & monitor					2,425			2,425	2,650
0300 Antenna VHF or UHF, for medium range	1 Elec	4	2		128	107		235	300
0310 For high-power transmitter		2	4		665	213		878	1,050
0400 Transmitter, 25 watt		4	2		2,000	107		2,107	2,350
0410 40 watt		2.66	3.008		2,675	160		2,835	3,200
0420 100 watt		1.33	6.015		6,700	320		7,020	7,850
0500 Wireless receiver/control module for speaker		8	1		262	53.50		315.50	370
0600 Desktop paging controller, stand alone		4	2		365	107		472	565

Estimating Tips

31 05 00 Common Work Results for Earthwork

- Estimating the actual cost of performing earthwork requires careful consideration of the variables involved. This includes items such as type of soil, whether water will be encountered, dewatering, whether banks need bracing, disposal of excavated earth, and length of haul to fill or spoil sites, etc. If the project has large quantities of cut or fill, consider raising or lowering the site to reduce costs, while paying close attention to the effect on site drainage and utilities.
- If the project has large quantities of fill, creating a borrow pit on the site can significantly lower the costs.
- It is very important to consider what time of year the project is scheduled for completion. Bad weather can create large cost overruns from dewatering, site repair, and lost productivity from cold weather.

Reference Numbers

Reference numbers are shown in shaded boxes at the beginning of some major classifications. These numbers refer to related items in the Reference Section. The reference information may be an estimating procedure, an alternate pricing method, or technical information.

Note: Not all subdivisions listed here necessarily appear in this publication.

31 23 Excavation and Fill

31 23 16 – Excavation

31 23 16.13 Excavating, Trench	Crew	Daily Output	Labor-Hours	Unit	Material	2014 Bare Costs Labor	2014 Bare Costs Equipment	Total	Total Incl O&P
0010 **EXCAVATING, TRENCH**									
0011 Or continuous footing									
0020 Common earth with no sheeting or dewatering included									
0050 1' to 4' deep, 3/8 C.Y. excavator	B-11C	150	.107	B.C.Y.		4.56	2.44	7	9.65
0060 1/2 C.Y. excavator	B-11M	200	.080			3.42	1.99	5.41	7.45
0062 3/4 C.Y. excavator	B-12F	270	.059			2.57	2.45	5.02	6.65
1400 By hand with pick and shovel 2' to 6' deep, light soil	1 Clab	8	1			36.50		36.50	56.50
1500 Heavy soil	"	4	2			73.50		73.50	113
3000 Backfill trench, F.E. loader, wheel mtd., 1 C.Y. bucket									
3020 Minimal haul	B-10R	400	.030	L.C.Y.		1.34	.74	2.08	2.86
5020 Loam & Sandy clay with no sheeting or dewatering included									
5050 1' to 4' deep, 3/8 C.Y. tractor loader/backhoe	B-11C	162	.099	B.C.Y.		4.23	2.25	6.48	8.95
5060 1/2 C.Y. excavator	B-11M	216	.074			3.17	1.84	5.01	6.85
5070 3/4 C.Y. excavator	B-12F	292	.055			2.38	2.27	4.65	6.15
6020 Sand & gravel with no sheeting or dewatering included									
6050 1' to 4' deep, 3/8 C.Y. excavator	B-11C	165	.097	B.C.Y.		4.15	2.21	6.36	8.80
6060 1/2 C.Y. excavator	B-11M	220	.073			3.11	1.81	4.92	6.75
6070 3/4 C.Y. excavator	B-12F	297	.054			2.34	2.23	4.57	6.05
7020 Dense hard clay with no sheeting or dewatering included									
7050 1' to 4' deep, 3/8 C.Y. excavator	B-11C	132	.121	B.C.Y.		5.20	2.77	7.97	10.95
7060 1/2 C.Y. excavator	B-11M	176	.091			3.89	2.26	6.15	8.45
7070 3/4 C.Y. excavator	B-12F	238	.067			2.92	2.78	5.70	7.50

31 23 16.14 Excavating, Utility Trench	Crew	Daily Output	Labor-Hours	Unit	Material	2014 Bare Costs Labor	2014 Bare Costs Equipment	Total	Total Incl O&P
0010 **EXCAVATING, UTILITY TRENCH**									
0011 Common earth									
0050 Trenching with chain trencher, 12 H.P., operator walking									
0100 4" wide trench, 12" deep	B-53	800	.010	L.F.		.47	.09	.56	.81
0150 18" deep		750	.011			.50	.10	.60	.86
0200 24" deep		700	.011			.54	.10	.64	.93
0300 6" wide trench, 12" deep		650	.012			.58	.11	.69	1
0350 18" deep		600	.013			.63	.12	.75	1.08
0400 24" deep		550	.015			.68	.13	.81	1.18
0450 36" deep		450	.018			.84	.16	1	1.44
0600 8" wide trench, 12" deep		475	.017			.79	.15	.94	1.36
0650 18" deep		400	.020			.94	.18	1.12	1.63
0700 24" deep		350	.023			1.08	.20	1.28	1.85
0750 36" deep		300	.027			1.25	.24	1.49	2.16
1000 Backfill by hand including compaction, add									
1050 4" wide trench, 12" deep	A-1G	800	.010	L.F.		.37	.07	.44	.65
1100 18" deep		530	.015			.55	.11	.66	.97
1150 24" deep		400	.020			.73	.14	.87	1.28
1300 6" wide trench, 12" deep		540	.015			.54	.10	.64	.95
1350 18" deep		405	.020			.72	.14	.86	1.27
1400 24" deep		270	.030			1.09	.21	1.30	1.91
1450 36" deep		180	.044			1.63	.31	1.94	2.85
1600 8" wide trench, 12" deep		400	.020			.73	.14	.87	1.28
1650 18" deep		265	.030			1.11	.21	1.32	1.94
1700 24" deep		200	.040			1.47	.28	1.75	2.57
1750 36" deep		135	.059			2.17	.41	2.58	3.81
2000 Chain trencher, 40 H.P. operator riding									
2050 6" wide trench and backfill, 12" deep	B-54	1200	.007	L.F.		.31	.28	.59	.79
2100 18" deep		1000	.008			.38	.34	.72	.94

31 23 Excavation and Fill

31 23 16 – Excavation

31 23 16.14 Excavating, Utility Trench

		Crew	Daily Output	Labor-Hours	Unit	Material	2014 Bare Costs Labor	Equipment	Total	Total Incl O&P
2150	24" deep	B-54	975	.008	L.F.		.39	.35	.74	.97
2200	36" deep		900	.009			.42	.38	.80	1.04
2250	48" deep		750	.011			.50	.45	.95	1.26
2300	60" deep		650	.012			.58	.52	1.10	1.45
2400	8" wide trench and backfill, 12" deep		1000	.008			.38	.34	.72	.94
2450	18" deep		950	.008			.40	.36	.76	.99
2500	24" deep		900	.009			.42	.38	.80	1.04
2550	36" deep		800	.010			.47	.42	.89	1.18
2600	48" deep		650	.012			.58	.52	1.10	1.45
2700	12" wide trench and backfill, 12" deep		975	.008			.39	.35	.74	.97
2750	18" deep		860	.009			.44	.39	.83	1.09
2800	24" deep		800	.010			.47	.42	.89	1.18
2850	36" deep		725	.011			.52	.47	.99	1.30
3000	16" wide trench and backfill, 12" deep		835	.010			.45	.41	.86	1.13
3050	18" deep		750	.011			.50	.45	.95	1.26
3100	24" deep	▼	700	.011	▼		.54	.49	1.03	1.35
3200	Compaction with vibratory plate, add								35%	35%
5100	Hand excavate and trim for pipe bells after trench excavation									
5200	8" pipe	1 Clab	155	.052	L.F.		1.89		1.89	2.92
5300	18" pipe	"	130	.062	"		2.26		2.26	3.48

31 23 16.15 Excavating, Utility Trench, Plow

		Crew	Daily Output	Labor-Hours	Unit	Material	2014 Bare Costs Labor	Equipment	Total	Total Incl O&P
0010	**EXCAVATING, UTILITY TRENCH, PLOW**									
0100	Single cable, plowed into fine material	B-54C	3800	.004	L.F.		.18	.32	.50	.63
0200	Two cable		3200	.005			.21	.38	.59	.74
0300	Single cable, plowed into coarse material	▼	2000	.008	▼		.34	.60	.94	1.18

31 23 16.16 Structural Excavation for Minor Structures

		Crew	Daily Output	Labor-Hours	Unit	Material	2014 Bare Costs Labor	Equipment	Total	Total Incl O&P
0010	**STRUCTURAL EXCAVATION FOR MINOR STRUCTURES**									
0015	Hand, pits to 6' deep, sandy soil	1 Clab	8	1	B.C.Y.		36.50		36.50	56.50
0100	Heavy soil or clay		4	2			73.50		73.50	113
0300	Pits 6' to 12' deep, sandy soil		5	1.600			58.50		58.50	90.50
0500	Heavy soil or clay		3	2.667			97.50		97.50	151
0700	Pits 12' to 18' deep, sandy soil		4	2			73.50		73.50	113
0900	Heavy soil or clay	▼	2	4			147		147	226
6030	Common earth, hydraulic backhoe, 1/2 C.Y. bucket	B-12E	55	.291			12.65	8.15	20.80	28.50
6035	3/4 C.Y. bucket	B-12F	90	.178			7.70	7.35	15.05	19.90
6040	1 C.Y. bucket	B-12A	108	.148			6.45	7.55	14	18.15
6050	1-1/2 C.Y. bucket	B-12B	144	.111			4.83	7.15	11.98	15.30
6060	2 C.Y. bucket	B-12C	200	.080			3.48	5.90	9.38	11.75
6070	Sand and gravel, 3/4 C.Y. bucket	B-12F	100	.160			6.95	6.60	13.55	17.90
6080	1 C.Y. bucket	B-12A	120	.133			5.80	6.80	12.60	16.30
6090	1-1/2 C.Y. bucket	B-12B	160	.100			4.35	6.45	10.80	13.75
6100	2 C.Y. bucket	B-12C	220	.073			3.16	5.35	8.51	10.75
6110	Clay, till, or blasted rock, 3/4 C.Y. bucket	B-12F	80	.200			8.70	8.30	17	22.50
6120	1 C.Y. bucket	B-12A	95	.168			7.30	8.60	15.90	20.50
6130	1-1/2 C.Y. bucket	B-12B	130	.123			5.35	7.95	13.30	16.85
6140	2 C.Y. bucket	B-12C	175	.091			3.97	6.70	10.67	13.45
6230	Sandy clay & loam, hydraulic backhoe, 1/2 C.Y. bucket	B-12E	60	.267			11.60	7.45	19.05	26
6235	3/4 C.Y. bucket	B-12F	98	.163			7.10	6.75	13.85	18.30
6240	1 C.Y. bucket	B-12A	116	.138			6	7	13	16.90
6250	1-1/2 C.Y. bucket	B-12B	156	.103	▼		4.46	6.60	11.06	14.05

31 23 Excavation and Fill

31 23 19 – Dewatering

31 23 19.20 Dewatering Systems		Crew	Daily Output	Labor-Hours	Unit	Material	2014 Bare Costs Labor	2014 Bare Costs Equipment	Total	Total Incl O&P
0010	**DEWATERING SYSTEMS**									
0020	Excavate drainage trench, 2' wide, 2' deep	B-11C	90	.178	C.Y.		7.60	4.06	11.66	16.05
0100	2' wide, 3' deep, with backhoe loader	"	135	.119			5.05	2.71	7.76	10.75
0200	Excavate sump pits by hand, light soil	1 Clab	7.10	1.127			41.50		41.50	63.50
0300	Heavy soil	"	3.50	2.286	↓		84		84	129
0500	Pumping 8 hr., attended 2 hrs. per day, including 20 L.F.									
0550	of suction hose & 100 L.F. discharge hose									
0600	2" diaphragm pump used for 8 hours	B-10H	4	3	Day		134	18.95	152.95	226
0650	4" diaphragm pump used for 8 hours	B-10I	4	3			134	31	165	239
0800	8 hrs. attended, 2" diaphragm pump	B-10H	1	12			540	75.50	615.50	905
0900	3" centrifugal pump	B-10J	1	12			540	84.50	624.50	915
1000	4" diaphragm pump	B-10I	1	12			540	124	664	955
1100	6" centrifugal pump	B-10K	1	12	↓		540	365	905	1,225

31 23 23 – Fill

31 23 23.13 Backfill

		Crew	Daily Output	Labor-Hours	Unit	Material	Labor	Equipment	Total	Total Incl O&P
0010	**BACKFILL**									
0015	By hand, no compaction, light soil	1 Clab	14	.571	L.C.Y.		21		21	32.50
0100	Heavy soil	↓	11	.727	"		26.50		26.50	41
0300	Compaction in 6" layers, hand tamp, add to above	↓	20.60	.388	E.C.Y.		14.25		14.25	22
0400	Roller compaction operator walking, add	B-10A	100	.120			5.40	1.83	7.23	10.20
0500	Air tamp, add	B-9D	190	.211			7.80	1.40	9.20	13.60
0600	Vibrating plate, add	A-1D	60	.133			4.89	.60	5.49	8.20
0800	Compaction in 12" layers, hand tamp, add to above	1 Clab	34	.235			8.60		8.60	13.30
0900	Roller compaction operator walking, add	B-10A	150	.080			3.59	1.22	4.81	6.80
1000	Air tamp, add	B-9	285	.140			5.20	.82	6.02	8.90
1100	Vibrating plate, add	A-1E	90	.089	↓		3.26	.51	3.77	5.60
1300	Dozer backfilling, bulk, up to 300' haul, no compaction	B-10B	1200	.010	L.C.Y.		.45	1.10	1.55	1.89
1400	Air tamped, add	B-11B	80	.200	E.C.Y.		8.35	3.78	12.13	16.95
1600	Compacting backfill, 6" to 12" lifts, vibrating roller	B-10C	800	.015			.67	2.17	2.84	3.41
1700	Sheepsfoot roller	B-10D	750	.016	↓		.72	2.36	3.08	3.68
1900	Dozer backfilling, trench, up to 300' haul, no compaction	B-10B	900	.013	L.C.Y.		.60	1.47	2.07	2.53
2000	Air tamped, add	B-11B	80	.200	E.C.Y.		8.35	3.78	12.13	16.95
2200	Compacting backfill, 6" to 12" lifts, vibrating roller	B-10C	700	.017			.77	2.48	3.25	3.90
2300	Sheepsfoot roller	B-10D	650	.018	↓		.83	2.72	3.55	4.25

31 23 23.16 Fill By Borrow and Utility Bedding

		Crew	Daily Output	Labor-Hours	Unit	Material	Labor	Equipment	Total	Total Incl O&P
0010	**FILL BY BORROW AND UTILITY BEDDING**									
0049	Utility bedding, for pipe & conduit, not incl. compaction									
0050	Crushed or screened bank run gravel	B-6	150	.160	L.C.Y.	25	6.40	2.44	33.84	40
0100	Crushed stone 3/4" to 1/2"		150	.160		28.50	6.40	2.44	37.34	43.50
0200	Sand, dead or bank	↓	150	.160	↓	17.55	6.40	2.44	26.39	32
0500	Compacting bedding in trench	A-1D	90	.089	E.C.Y.		3.26	.40	3.66	5.50
0610	See Section 31 23 23.20 for hauling mileage add.									

31 23 23.20 Hauling

		Crew	Daily Output	Labor-Hours	Unit	Material	Labor	Equipment	Total	Total Incl O&P
0010	**HAULING**									
0011	Excavated or borrow, loose cubic yards									
0012	no loading equipment, including hauling, waiting, loading/dumping									
0013	time per cycle (wait, load, travel, unload or dump & return)									
0014	8 C.Y. truck, 15 MPH ave, cycle 0.5 miles, 10 min. wait/Ld./Uld.	B-34A	320	.025	L.C.Y.		.94	1.30	2.24	2.86
0016	cycle 1 mile		272	.029			1.10	1.53	2.63	3.38
0018	cycle 2 miles		208	.038			1.44	2.01	3.45	4.41
0020	cycle 4 miles	↓	144	.056	↓		2.09	2.90	4.99	6.35

31 23 23.20 Hauling		Crew	Daily Output	Labor-Hours	Unit	Material	2014 Bare Costs Labor	2014 Bare Costs Equipment	Total	Total Incl O&P
0022	cycle 6 miles	B-34A	112	.071	L.C.Y.		2.68	3.73	6.41	8.20
0024	cycle 8 miles		88	.091			3.41	4.74	8.15	10.40
0026	20 MPH ave, cycle 0.5 mile		336	.024			.89	1.24	2.13	2.73
0028	cycle 1 mile		296	.027			1.02	1.41	2.43	3.10
0030	cycle 2 miles		240	.033			1.25	1.74	2.99	3.82
0032	cycle 4 miles		176	.045			1.71	2.37	4.08	5.20
0034	cycle 6 miles		136	.059			2.21	3.07	5.28	6.75
0036	cycle 8 miles		112	.071			2.68	3.73	6.41	8.20
0044	25 MPH ave, cycle 4 miles		192	.042			1.56	2.17	3.73	4.78
0046	cycle 6 miles		160	.050			1.88	2.61	4.49	5.75
0048	cycle 8 miles		128	.063			2.35	3.26	5.61	7.15
0050	30 MPH ave, cycle 4 miles		216	.037			1.39	1.93	3.32	4.25
0052	cycle 6 miles		176	.045			1.71	2.37	4.08	5.20
0054	cycle 8 miles		144	.056			2.09	2.90	4.99	6.35
0114	15 MPH ave, cycle 0.5 mile, 15 min. wait/Ld./Uld.		224	.036			1.34	1.86	3.20	4.10
0116	cycle 1 mile		200	.040			1.50	2.09	3.59	4.58
0118	cycle 2 miles		168	.048			1.79	2.48	4.27	5.45
0120	cycle 4 miles		120	.067			2.50	3.48	5.98	7.65
0122	cycle 6 miles		96	.083			3.13	4.35	7.48	9.55
0124	cycle 8 miles		80	.100			3.76	5.20	8.96	11.50
0126	20 MPH ave, cycle 0.5 mile		232	.034			1.29	1.80	3.09	3.96
0128	cycle 1 mile		208	.038			1.44	2.01	3.45	4.41
0130	cycle 2 miles		184	.043			1.63	2.27	3.90	4.98
0132	cycle 4 miles		144	.056			2.09	2.90	4.99	6.35
0134	cycle 6 miles		112	.071			2.68	3.73	6.41	8.20
0136	cycle 8 miles		96	.083			3.13	4.35	7.48	9.55
0144	25 MPH ave, cycle 4 miles		152	.053			1.98	2.74	4.72	6.05
0146	cycle 6 miles		128	.063			2.35	3.26	5.61	7.15
0148	cycle 8 miles		112	.071			2.68	3.73	6.41	8.20
0150	30 MPH ave, cycle 4 miles		168	.048			1.79	2.48	4.27	5.45
0152	cycle 6 miles		144	.056			2.09	2.90	4.99	6.35
0154	cycle 8 miles		120	.067			2.50	3.48	5.98	7.65
0214	15 MPH ave, cycle 0.5 mile, 20 min wait/Ld./Uld.		176	.045			1.71	2.37	4.08	5.20
0216	cycle 1 mile		160	.050			1.88	2.61	4.49	5.75
0218	cycle 2 miles		136	.059			2.21	3.07	5.28	6.75
0220	cycle 4 miles		104	.077			2.89	4.01	6.90	8.80
0222	cycle 6 miles		88	.091			3.41	4.74	8.15	10.40
0224	cycle 8 miles		72	.111			4.17	5.80	9.97	12.70
0226	20 MPH ave, cycle 0.5 mile		176	.045			1.71	2.37	4.08	5.20
0228	cycle 1 mile		168	.048			1.79	2.48	4.27	5.45
0230	cycle 2 miles		144	.056			2.09	2.90	4.99	6.35
0232	cycle 4 miles		120	.067			2.50	3.48	5.98	7.65
0234	cycle 6 miles		96	.083			3.13	4.35	7.48	9.55
0236	cycle 8 miles		88	.091			3.41	4.74	8.15	10.40
0244	25 MPH ave, cycle 4 miles		128	.063			2.35	3.26	5.61	7.15
0246	cycle 6 miles		112	.071			2.68	3.73	6.41	8.20
0248	cycle 8 miles		96	.083			3.13	4.35	7.48	9.55
0250	30 MPH ave, cycle 4 miles		136	.059			2.21	3.07	5.28	6.75
0252	cycle 6 miles		120	.067			2.50	3.48	5.98	7.65
0254	cycle 8 miles		104	.077			2.89	4.01	6.90	8.80
0314	15 MPH ave, cycle 0.5 mile, 25 min wait/Ld./Uld.		144	.056			2.09	2.90	4.99	6.35
0316	cycle 1 mile		128	.063			2.35	3.26	5.61	7.15
0318	cycle 2 miles		112	.071			2.68	3.73	6.41	8.20

31 23 23 – Fill

31 23 23.20 Hauling		Crew	Daily Output	Labor-Hours	Unit	Material	2014 Bare Costs Labor	2014 Bare Costs Equipment	Total	Total Incl O&P
0320	cycle 4 miles	B-34A	96	.083	L.C.Y.		3.13	4.35	7.48	9.55
0322	cycle 6 miles		80	.100			3.76	5.20	8.96	11.50
0324	cycle 8 miles		64	.125			4.69	6.50	11.19	14.30
0326	20 MPH ave, cycle 0.5 mile		144	.056			2.09	2.90	4.99	6.35
0328	cycle 1 mile		136	.059			2.21	3.07	5.28	6.75
0330	cycle 2 miles		120	.067			2.50	3.48	5.98	7.65
0332	cycle 4 miles		104	.077			2.89	4.01	6.90	8.80
0334	cycle 6 miles		88	.091			3.41	4.74	8.15	10.40
0336	cycle 8 miles		80	.100			3.76	5.20	8.96	11.50
0344	25 MPH ave, cycle 4 miles		112	.071			2.68	3.73	6.41	8.20
0346	cycle 6 miles		96	.083			3.13	4.35	7.48	9.55
0348	cycle 8 miles		88	.091			3.41	4.74	8.15	10.40
0350	30 MPH ave, cycle 4 miles		112	.071			2.68	3.73	6.41	8.20
0352	cycle 6 miles		104	.077			2.89	4.01	6.90	8.80
0354	cycle 8 miles		96	.083			3.13	4.35	7.48	9.55
0414	15 MPH ave, cycle 0.5 mile, 30 min wait/Ld./Uld.		120	.067			2.50	3.48	5.98	7.65
0416	cycle 1 mile		112	.071			2.68	3.73	6.41	8.20
0418	cycle 2 miles		96	.083			3.13	4.35	7.48	9.55
0420	cycle 4 miles		80	.100			3.76	5.20	8.96	11.50
0422	cycle 6 miles		72	.111			4.17	5.80	9.97	12.70
0424	cycle 8 miles		64	.125			4.69	6.50	11.19	14.30
0426	20 MPH ave, cycle 0.5 mile		120	.067			2.50	3.48	5.98	7.65
0428	cycle 1 mile		112	.071			2.68	3.73	6.41	8.20
0430	cycle 2 miles		104	.077			2.89	4.01	6.90	8.80
0432	cycle 4 miles		88	.091			3.41	4.74	8.15	10.40
0434	cycle 6 miles		80	.100			3.76	5.20	8.96	11.50
0436	cycle 8 miles		72	.111			4.17	5.80	9.97	12.70
0444	25 MPH ave, cycle 4 miles		96	.083			3.13	4.35	7.48	9.55
0446	cycle 6 miles		88	.091			3.41	4.74	8.15	10.40
0448	cycle 8 miles		80	.100			3.76	5.20	8.96	11.50
0450	30 MPH ave, cycle 4 miles		96	.083			3.13	4.35	7.48	9.55
0452	cycle 6 miles		88	.091			3.41	4.74	8.15	10.40
0454	cycle 8 miles		80	.100			3.76	5.20	8.96	11.50
0514	15 MPH ave, cycle 0.5 mile, 35 min wait/Ld./Uld.		104	.077			2.89	4.01	6.90	8.80
0516	cycle 1 mile		96	.083			3.13	4.35	7.48	9.55
0518	cycle 2 miles		88	.091			3.41	4.74	8.15	10.40
0520	cycle 4 miles		72	.111			4.17	5.80	9.97	12.70
0522	cycle 6 miles		64	.125			4.69	6.50	11.19	14.30
0524	cycle 8 miles		56	.143			5.35	7.45	12.80	16.40
0526	20 MPH ave, cycle 0.5 mile		104	.077			2.89	4.01	6.90	8.80
0528	cycle 1 mile		96	.083			3.13	4.35	7.48	9.55
0530	cycle 2 miles		96	.083			3.13	4.35	7.48	9.55
0532	cycle 4 miles		80	.100			3.76	5.20	8.96	11.50
0534	cycle 6 miles		72	.111			4.17	5.80	9.97	12.70
0536	cycle 8 miles		64	.125			4.69	6.50	11.19	14.30
0544	25 MPH ave, cycle 4 miles		88	.091			3.41	4.74	8.15	10.40
0546	cycle 6 miles		80	.100			3.76	5.20	8.96	11.50
0548	cycle 8 miles		72	.111			4.17	5.80	9.97	12.70
0550	30 MPH ave, cycle 4 miles		88	.091			3.41	4.74	8.15	10.40
0552	cycle 6 miles		80	.100			3.76	5.20	8.96	11.50
0554	cycle 8 miles		72	.111			4.17	5.80	9.97	12.70
1014	12 C.Y. truck, cycle 0.5 mile, 15 MPH ave, 15 min. wait/Ld./Uld.	B-34B	336	.024			.89	2.06	2.95	3.63
1016	cycle 1 mile		300	.027			1	2.31	3.31	4.07

31 23 Excavation and Fill

31 23 23 – Fill

31 23 23.20 Hauling	Crew	Daily Output	Labor-Hours	Unit	Material	Labor	Equipment	Total	Total Incl O&P
1018	cycle 2 miles	B-34B	252	.032	L.C.Y.		1.19	2.75	3.94	4.84
1020	cycle 4 miles		180	.044			1.67	3.85	5.52	6.80
1022	cycle 6 miles		144	.056			2.09	4.81	6.90	8.50
1024	cycle 8 miles		120	.067			2.50	5.75	8.25	10.15
1025	cycle 10 miles		96	.083			3.13	7.20	10.33	12.70
1026	20 MPH ave, cycle 0.5 mile		348	.023			.86	1.99	2.85	3.51
1028	cycle 1 mile		312	.026			.96	2.22	3.18	3.91
1030	cycle 2 miles		276	.029			1.09	2.51	3.60	4.42
1032	cycle 4 miles		216	.037			1.39	3.21	4.60	5.65
1034	cycle 6 miles		168	.048			1.79	4.12	5.91	7.25
1036	cycle 8 miles		144	.056			2.09	4.81	6.90	8.50
1038	cycle 10 miles		120	.067			2.50	5.75	8.25	10.15
1040	25 MPH ave, cycle 4 miles		228	.035			1.32	3.04	4.36	5.35
1042	cycle 6 miles		192	.042			1.56	3.61	5.17	6.35
1044	cycle 8 miles		168	.048			1.79	4.12	5.91	7.25
1046	cycle 10 miles		144	.056			2.09	4.81	6.90	8.50
1050	30 MPH ave, cycle 4 miles		252	.032			1.19	2.75	3.94	4.84
1052	cycle 6 miles		216	.037			1.39	3.21	4.60	5.65
1054	cycle 8 miles		180	.044			1.67	3.85	5.52	6.80
1056	cycle 10 miles		156	.051			1.93	4.44	6.37	7.80
1060	35 MPH ave, cycle 4 miles		264	.030			1.14	2.62	3.76	4.62
1062	cycle 6 miles		228	.035			1.32	3.04	4.36	5.35
1064	cycle 8 miles		204	.039			1.47	3.39	4.86	6
1066	cycle 10 miles		180	.044			1.67	3.85	5.52	6.80
1068	cycle 20 miles		120	.067			2.50	5.75	8.25	10.15
1069	cycle 30 miles		84	.095			3.58	8.25	11.83	14.50
1070	cycle 40 miles		72	.111			4.17	9.60	13.77	16.95
1072	40 MPH ave, cycle 6 miles		240	.033			1.25	2.88	4.13	5.10
1074	cycle 8 miles		216	.037			1.39	3.21	4.60	5.65
1076	cycle 10 miles		192	.042			1.56	3.61	5.17	6.35
1078	cycle 20 miles		120	.067			2.50	5.75	8.25	10.15
1080	cycle 30 miles		96	.083			3.13	7.20	10.33	12.70
1082	cycle 40 miles		72	.111			4.17	9.60	13.77	16.95
1084	cycle 50 miles		60	.133			5	11.55	16.55	20.50
1094	45 MPH ave, cycle 8 miles		216	.037			1.39	3.21	4.60	5.65
1096	cycle 10 miles		204	.039			1.47	3.39	4.86	6
1098	cycle 20 miles		132	.061			2.28	5.25	7.53	9.20
1100	cycle 30 miles		108	.074			2.78	6.40	9.18	11.30
1102	cycle 40 miles		84	.095			3.58	8.25	11.83	14.50
1104	cycle 50 miles		72	.111			4.17	9.60	13.77	16.95
1106	50 MPH ave, cycle 10 miles		216	.037			1.39	3.21	4.60	5.65
1108	cycle 20 miles		144	.056			2.09	4.81	6.90	8.50
1110	cycle 30 miles		108	.074			2.78	6.40	9.18	11.30
1112	cycle 40 miles		84	.095			3.58	8.25	11.83	14.50
1114	cycle 50 miles		72	.111			4.17	9.60	13.77	16.95
1214	15 MPH ave, cycle 0.5 mile, 20 min. wait/Ld./Uld.		264	.030			1.14	2.62	3.76	4.62
1216	cycle 1 mile		240	.033			1.25	2.88	4.13	5.10
1218	cycle 2 miles		204	.039			1.47	3.39	4.86	6
1220	cycle 4 miles		156	.051			1.93	4.44	6.37	7.80
1222	cycle 6 miles		132	.061			2.28	5.25	7.53	9.20
1224	cycle 8 miles		108	.074			2.78	6.40	9.18	11.30
1225	cycle 10 miles		96	.083			3.13	7.20	10.33	12.70
1226	20 MPH ave, cycle 0.5 mile		264	.030			1.14	2.62	3.76	4.62

31 23 23.20 Hauling		Crew	Daily Output	Labor-Hours	Unit	Material	2014 Bare Costs Labor	Equipment	Total	Total Incl O&P
1228	cycle 1 mile	B-34B	252	.032	L.C.Y.		1.19	2.75	3.94	4.84
1230	cycle 2 miles		216	.037			1.39	3.21	4.60	5.65
1232	cycle 4 miles		180	.044			1.67	3.85	5.52	6.80
1234	cycle 6 miles		144	.056			2.09	4.81	6.90	8.50
1236	cycle 8 miles		132	.061			2.28	5.25	7.53	9.20
1238	cycle 10 miles		108	.074			2.78	6.40	9.18	11.30
1240	25 MPH ave, cycle 4 miles		192	.042			1.56	3.61	5.17	6.35
1242	cycle 6 miles		168	.048			1.79	4.12	5.91	7.25
1244	cycle 8 miles		144	.056			2.09	4.81	6.90	8.50
1246	cycle 10 miles		132	.061			2.28	5.25	7.53	9.20
1250	30 MPH ave, cycle 4 miles		204	.039			1.47	3.39	4.86	6
1252	cycle 6 miles		180	.044			1.67	3.85	5.52	6.80
1254	cycle 8 miles		156	.051			1.93	4.44	6.37	7.80
1256	cycle 10 miles		144	.056			2.09	4.81	6.90	8.50
1260	35 MPH ave, cycle 4 miles		216	.037			1.39	3.21	4.60	5.65
1262	cycle 6 miles		192	.042			1.56	3.61	5.17	6.35
1264	cycle 8 miles		168	.048			1.79	4.12	5.91	7.25
1266	cycle 10 miles		156	.051			1.93	4.44	6.37	7.80
1268	cycle 20 miles		108	.074			2.78	6.40	9.18	11.30
1269	cycle 30 miles		72	.111			4.17	9.60	13.77	16.95
1270	cycle 40 miles		60	.133			5	11.55	16.55	20.50
1272	40 MPH ave, cycle 6 miles		192	.042			1.56	3.61	5.17	6.35
1274	cycle 8 miles		180	.044			1.67	3.85	5.52	6.80
1276	cycle 10 miles		156	.051			1.93	4.44	6.37	7.80
1278	cycle 20 miles		108	.074			2.78	6.40	9.18	11.30
1280	cycle 30 miles		84	.095			3.58	8.25	11.83	14.50
1282	cycle 40 miles		72	.111			4.17	9.60	13.77	16.95
1284	cycle 50 miles		60	.133			5	11.55	16.55	20.50
1294	45 MPH ave, cycle 8 miles		180	.044			1.67	3.85	5.52	6.80
1296	cycle 10 miles		168	.048			1.79	4.12	5.91	7.25
1298	cycle 20 miles		120	.067			2.50	5.75	8.25	10.15
1300	cycle 30 miles		96	.083			3.13	7.20	10.33	12.70
1302	cycle 40 miles		72	.111			4.17	9.60	13.77	16.95
1304	cycle 50 miles		60	.133			5	11.55	16.55	20.50
1306	50 MPH ave, cycle 10 miles		180	.044			1.67	3.85	5.52	6.80
1308	cycle 20 miles		132	.061			2.28	5.25	7.53	9.20
1310	cycle 30 miles		96	.083			3.13	7.20	10.33	12.70
1312	cycle 40 miles		84	.095			3.58	8.25	11.83	14.50
1314	cycle 50 miles		72	.111			4.17	9.60	13.77	16.95
1414	15 MPH ave, cycle 0.5 mile, 25 min. wait/Ld./Uld.		204	.039			1.47	3.39	4.86	6
1416	cycle 1 mile		192	.042			1.56	3.61	5.17	6.35
1418	cycle 2 miles		168	.048			1.79	4.12	5.91	7.25
1420	cycle 4 miles		132	.061			2.28	5.25	7.53	9.20
1422	cycle 6 miles		120	.067			2.50	5.75	8.25	10.15
1424	cycle 8 miles		96	.083			3.13	7.20	10.33	12.70
1425	cycle 10 miles		84	.095			3.58	8.25	11.83	14.50
1426	20 MPH ave, cycle 0.5 mile		216	.037			1.39	3.21	4.60	5.65
1428	cycle 1 mile		204	.039			1.47	3.39	4.86	6
1430	cycle 2 miles		180	.044			1.67	3.85	5.52	6.80
1432	cycle 4 miles		156	.051			1.93	4.44	6.37	7.80
1434	cycle 6 miles		132	.061			2.28	5.25	7.53	9.20
1436	cycle 8 miles		120	.067			2.50	5.75	8.25	10.15
1438	cycle 10 miles		96	.083			3.13	7.20	10.33	12.70

31 23 23 – Fill

31 23 23.20 Hauling	Crew	Daily Output	Labor-Hours	Unit	Material	2014 Bare Costs Labor	2014 Bare Costs Equipment	Total	Total Incl O&P
1440 25 MPH ave, cycle 4 miles	B-34B	168	.048	L.C.Y.		1.79	4.12	5.91	7.25
1442 cycle 6 miles		144	.056			2.09	4.81	6.90	8.50
1444 cycle 8 miles		132	.061			2.28	5.25	7.53	9.20
1446 cycle 10 miles		108	.074			2.78	6.40	9.18	11.30
1450 30 MPH ave, cycle 4 miles		168	.048			1.79	4.12	5.91	7.25
1452 cycle 6 miles		156	.051			1.93	4.44	6.37	7.80
1454 cycle 8 miles		132	.061			2.28	5.25	7.53	9.20
1456 cycle 10 miles		120	.067			2.50	5.75	8.25	10.15
1460 35 MPH ave, cycle 4 miles		180	.044			1.67	3.85	5.52	6.80
1462 cycle 6 miles		156	.051			1.93	4.44	6.37	7.80
1464 cycle 8 miles		144	.056			2.09	4.81	6.90	8.50
1466 cycle 10 miles		132	.061			2.28	5.25	7.53	9.20
1468 cycle 20 miles		96	.083			3.13	7.20	10.33	12.70
1469 cycle 30 miles		72	.111			4.17	9.60	13.77	16.95
1470 cycle 40 miles		60	.133			5	11.55	16.55	20.50
1472 40 MPH ave, cycle 6 miles		168	.048			1.79	4.12	5.91	7.25
1474 cycle 8 miles		156	.051			1.93	4.44	6.37	7.80
1476 cycle 10 miles		144	.056			2.09	4.81	6.90	8.50
1478 cycle 20 miles		96	.083			3.13	7.20	10.33	12.70
1480 cycle 30 miles		84	.095			3.58	8.25	11.83	14.50
1482 cycle 40 miles		60	.133			5	11.55	16.55	20.50
1484 cycle 50 miles		60	.133			5	11.55	16.55	20.50
1494 45 MPH ave, cycle 8 miles		156	.051			1.93	4.44	6.37	7.80
1496 cycle 10 miles		144	.056			2.09	4.81	6.90	8.50
1498 cycle 20 miles		108	.074			2.78	6.40	9.18	11.30
1500 cycle 30 miles		84	.095			3.58	8.25	11.83	14.50
1502 cycle 40 miles		72	.111			4.17	9.60	13.77	16.95
1504 cycle 50 miles		60	.133			5	11.55	16.55	20.50
1506 50 MPH ave, cycle 10 miles		156	.051			1.93	4.44	6.37	7.80
1508 cycle 20 miles		120	.067			2.50	5.75	8.25	10.15
1510 cycle 30 miles		96	.083			3.13	7.20	10.33	12.70
1512 cycle 40 miles		72	.111			4.17	9.60	13.77	16.95
1514 cycle 50 miles		60	.133			5	11.55	16.55	20.50
1614 15 MPH, cycle 0.5 mile, 30 min. wait/Ld./Uld.		180	.044			1.67	3.85	5.52	6.80
1616 cycle 1 mile		168	.048			1.79	4.12	5.91	7.25
1618 cycle 2 miles		144	.056			2.09	4.81	6.90	8.50
1620 cycle 4 miles		120	.067			2.50	5.75	8.25	10.15
1622 cycle 6 miles		108	.074			2.78	6.40	9.18	11.30
1624 cycle 8 miles		84	.095			3.58	8.25	11.83	14.50
1625 cycle 10 miles		84	.095			3.58	8.25	11.83	14.50
1626 20 MPH ave, cycle 0.5 mile		180	.044			1.67	3.85	5.52	6.80
1628 cycle 1 mile		168	.048			1.79	4.12	5.91	7.25
1630 cycle 2 miles		156	.051			1.93	4.44	6.37	7.80
1632 cycle 4 miles		132	.061			2.28	5.25	7.53	9.20
1634 cycle 6 miles		120	.067			2.50	5.75	8.25	10.15
1636 cycle 8 miles		108	.074			2.78	6.40	9.18	11.30
1638 cycle 10 miles		96	.083			3.13	7.20	10.33	12.70
1640 25 MPH ave, cycle 4 miles		144	.056			2.09	4.81	6.90	8.50
1642 cycle 6 miles		132	.061			2.28	5.25	7.53	9.20
1644 cycle 8 miles		108	.074			2.78	6.40	9.18	11.30
1646 cycle 10 miles		108	.074			2.78	6.40	9.18	11.30
1650 30 MPH ave, cycle 4 miles		144	.056			2.09	4.81	6.90	8.50
1652 cycle 6 miles		132	.061			2.28	5.25	7.53	9.20

31 23 23 – Fill

31 23 23.20 Hauling		Crew	Daily Output	Labor-Hours	Unit	Material	2014 Bare Costs Labor	2014 Bare Costs Equipment	Total	Total Incl O&P
1654	cycle 8 miles	B-34B	120	.067	L.C.Y.		2.50	5.75	8.25	10.15
1656	cycle 10 miles		108	.074			2.78	6.40	9.18	11.30
1660	35 MPH ave, cycle 4 miles		156	.051			1.93	4.44	6.37	7.80
1662	cycle 6 miles		144	.056			2.09	4.81	6.90	8.50
1664	cycle 8 miles		132	.061			2.28	5.25	7.53	9.20
1666	cycle 10 miles		120	.067			2.50	5.75	8.25	10.15
1668	cycle 20 miles		84	.095			3.58	8.25	11.83	14.50
1669	cycle 30 miles		72	.111			4.17	9.60	13.77	16.95
1670	cycle 40 miles		60	.133			5	11.55	16.55	20.50
1672	40 MPH, cycle 6 miles		144	.056			2.09	4.81	6.90	8.50
1674	cycle 8 miles		132	.061			2.28	5.25	7.53	9.20
1676	cycle 10 miles		120	.067			2.50	5.75	8.25	10.15
1678	cycle 20 miles		96	.083			3.13	7.20	10.33	12.70
1680	cycle 30 miles		72	.111			4.17	9.60	13.77	16.95
1682	cycle 40 miles		60	.133			5	11.55	16.55	20.50
1684	cycle 50 miles		48	.167			6.25	14.40	20.65	25.50
1694	45 MPH ave, cycle 8 miles		144	.056			2.09	4.81	6.90	8.50
1696	cycle 10 miles		132	.061			2.28	5.25	7.53	9.20
1698	cycle 20 miles		96	.083			3.13	7.20	10.33	12.70
1700	cycle 30 miles		84	.095			3.58	8.25	11.83	14.50
1702	cycle 40 miles		60	.133			5	11.55	16.55	20.50
1704	cycle 50 miles		60	.133			5	11.55	16.55	20.50
1706	50 MPH ave, cycle 10 miles		132	.061			2.28	5.25	7.53	9.20
1708	cycle 20 miles		108	.074			2.78	6.40	9.18	11.30
1710	cycle 30 miles		84	.095			3.58	8.25	11.83	14.50
1712	cycle 40 miles		72	.111			4.17	9.60	13.77	16.95
1714	cycle 50 miles		60	.133			5	11.55	16.55	20.50
2000	Hauling, 8 C.Y. truck, small project cost per hour	B-34A	8	1	Hr.		37.50	52	89.50	115
2100	12 C.Y. Truck	B-34B	8	1			37.50	86.50	124	153
2150	16.5 C.Y. Truck	B-34C	8	1			37.50	92.50	130	160
2175	18 C.Y. 8 wheel Truck	B-34I	8	1			37.50	108	145.50	177
2200	20 C.Y. Truck	B-34D	8	1			37.50	94	131.50	162
2300	Grading at dump, or embankment if required, by dozer	B-10B	1000	.012	L.C.Y.		.54	1.33	1.87	2.28
2310	Spotter at fill or cut, if required	1 Clab	8	1	Hr.		36.50		36.50	56.50
9014	18 C.Y. truck, 8 wheels,15 min. wait/Ld./Uld.,15 MPH, cycle 0.5 mi.	B-34I	504	.016	L.C.Y.		.60	1.72	2.32	2.80
9016	cycle 1 mile		450	.018			.67	1.93	2.60	3.14
9018	cycle 2 miles		378	.021			.79	2.29	3.08	3.73
9020	cycle 4 miles		270	.030			1.11	3.21	4.32	5.25
9022	cycle 6 miles		216	.037			1.39	4.02	5.41	6.55
9024	cycle 8 miles		180	.044			1.67	4.82	6.49	7.85
9025	cycle 10 miles		144	.056			2.09	6.05	8.14	9.85
9026	20 MPH ave, cycle 0.5 mile		522	.015			.58	1.66	2.24	2.71
9028	cycle 1 mile		468	.017			.64	1.85	2.49	3.02
9030	cycle 2 miles		414	.019			.73	2.10	2.83	3.41
9032	cycle 4 miles		324	.025			.93	2.68	3.61	4.36
9034	cycle 6 miles		252	.032			1.19	3.44	4.63	5.60
9036	cycle 8 miles		216	.037			1.39	4.02	5.41	6.55
9038	cycle 10 miles		180	.044			1.67	4.82	6.49	7.85
9040	25 MPH ave, cycle 4 miles		342	.023			.88	2.54	3.42	4.13
9042	cycle 6 miles		288	.028			1.04	3.01	4.05	4.90
9044	cycle 8 miles		252	.032			1.19	3.44	4.63	5.60
9046	cycle 10 miles		216	.037			1.39	4.02	5.41	6.55
9050	30 MPH ave, cycle 4 miles		378	.021			.79	2.29	3.08	3.73

31 23 23.20 Hauling		Crew	Daily Output	Labor-Hours	Unit	Material	2014 Bare Costs Labor	Equipment	Total	Total Incl O&P
9052	cycle 6 miles	B-34I	324	.025	L.C.Y.		.93	2.68	3.61	4.36
9054	cycle 8 miles		270	.030			1.11	3.21	4.32	5.25
9056	cycle 10 miles		234	.034			1.28	3.71	4.99	6.05
9060	35 MPH ave, cycle 4 miles		396	.020			.76	2.19	2.95	3.57
9062	cycle 6 miles		342	.023			.88	2.54	3.42	4.13
9064	cycle 8 miles		288	.028			1.04	3.01	4.05	4.90
9066	cycle 10 miles		270	.030			1.11	3.21	4.32	5.25
9068	cycle 20 miles		162	.049			1.85	5.35	7.20	8.75
9070	cycle 30 miles		126	.063			2.38	6.90	9.28	11.20
9072	cycle 40 miles		90	.089			3.34	9.65	12.99	15.70
9074	40 MPH ave, cycle 6 miles		360	.022			.83	2.41	3.24	3.92
9076	cycle 8 miles		324	.025			.93	2.68	3.61	4.36
9078	cycle 10 miles		288	.028			1.04	3.01	4.05	4.90
9080	cycle 20 miles		180	.044			1.67	4.82	6.49	7.85
9082	cycle 30 miles		144	.056			2.09	6.05	8.14	9.85
9084	cycle 40 miles		108	.074			2.78	8.05	10.83	13.10
9086	cycle 50 miles		90	.089			3.34	9.65	12.99	15.70
9094	45 MPH ave, cycle 8 miles		324	.025			.93	2.68	3.61	4.36
9096	cycle 10 miles		306	.026			.98	2.83	3.81	4.62
9098	cycle 20 miles		198	.040			1.52	4.38	5.90	7.15
9100	cycle 30 miles		144	.056			2.09	6.05	8.14	9.85
9102	cycle 40 miles		126	.063			2.38	6.90	9.28	11.20
9104	cycle 50 miles		108	.074			2.78	8.05	10.83	13.10
9106	50 MPH ave, cycle 10 miles		324	.025			.93	2.68	3.61	4.36
9108	cycle 20 miles		216	.037			1.39	4.02	5.41	6.55
9110	cycle 30 miles		162	.049			1.85	5.35	7.20	8.75
9112	cycle 40 miles		126	.063			2.38	6.90	9.28	11.20
9114	cycle 50 miles		108	.074			2.78	8.05	10.83	13.10
9214	20 min. wait/Ld./Uld.,15 MPH, cycle 0.5 mi.		396	.020			.76	2.19	2.95	3.57
9216	cycle 1 mile		360	.022			.83	2.41	3.24	3.92
9218	cycle 2 miles		306	.026			.98	2.83	3.81	4.62
9220	cycle 4 miles		234	.034			1.28	3.71	4.99	6.05
9222	cycle 6 miles		198	.040			1.52	4.38	5.90	7.15
9224	cycle 8 miles		162	.049			1.85	5.35	7.20	8.75
9225	cycle 10 miles		144	.056			2.09	6.05	8.14	9.85
9226	20 MPH ave, cycle 0.5 mile		396	.020			.76	2.19	2.95	3.57
9228	cycle 1 mile		378	.021			.79	2.29	3.08	3.73
9230	cycle 2 miles		324	.025			.93	2.68	3.61	4.36
9232	cycle 4 miles		270	.030			1.11	3.21	4.32	5.25
9234	cycle 6 miles		216	.037			1.39	4.02	5.41	6.55
9236	cycle 8 miles		198	.040			1.52	4.38	5.90	7.15
9238	cycle 10 miles		162	.049			1.85	5.35	7.20	8.75
9240	25 MPH ave, cycle 4 miles		288	.028			1.04	3.01	4.05	4.90
9242	cycle 6 miles		252	.032			1.19	3.44	4.63	5.60
9244	cycle 8 miles		216	.037			1.39	4.02	5.41	6.55
9246	cycle 10 miles		198	.040			1.52	4.38	5.90	7.15
9250	30 MPH ave, cycle 4 miles		306	.026			.98	2.83	3.81	4.62
9252	cycle 6 miles		270	.030			1.11	3.21	4.32	5.25
9254	cycle 8 miles		234	.034			1.28	3.71	4.99	6.05
9256	cycle 10 miles		216	.037			1.39	4.02	5.41	6.55
9260	35 MPH ave, cycle 4 miles		324	.025			.93	2.68	3.61	4.36
9262	cycle 6 miles		288	.028			1.04	3.01	4.05	4.90
9264	cycle 8 miles		252	.032			1.19	3.44	4.63	5.60

31 23 23.20 Hauling		Crew	Daily Output	Labor-Hours	Unit	Material	2014 Bare Costs Labor	2014 Bare Costs Equipment	Total	Total Incl O&P
9266	cycle 10 miles	B-34I	234	.034	L.C.Y.		1.28	3.71	4.99	6.05
9268	cycle 20 miles		162	.049			1.85	5.35	7.20	8.75
9270	cycle 30 miles		108	.074			2.78	8.05	10.83	13.10
9272	cycle 40 miles		90	.089			3.34	9.65	12.99	15.70
9274	40 MPH ave, cycle 6 miles		288	.028			1.04	3.01	4.05	4.90
9276	cycle 8 miles		270	.030			1.11	3.21	4.32	5.25
9278	cycle 10 miles		234	.034			1.28	3.71	4.99	6.05
9280	cycle 20 miles		162	.049			1.85	5.35	7.20	8.75
9282	cycle 30 miles		126	.063			2.38	6.90	9.28	11.20
9284	cycle 40 miles		108	.074			2.78	8.05	10.83	13.10
9286	cycle 50 miles		90	.089			3.34	9.65	12.99	15.70
9294	45 MPH ave, cycle 8 miles		270	.030			1.11	3.21	4.32	5.25
9296	cycle 10 miles		252	.032			1.19	3.44	4.63	5.60
9298	cycle 20 miles		180	.044			1.67	4.82	6.49	7.85
9300	cycle 30 miles		144	.056			2.09	6.05	8.14	9.85
9302	cycle 40 miles		108	.074			2.78	8.05	10.83	13.10
9304	cycle 50 miles		90	.089			3.34	9.65	12.99	15.70
9306	50 MPH ave, cycle 10 miles		270	.030			1.11	3.21	4.32	5.25
9308	cycle 20 miles		198	.040			1.52	4.38	5.90	7.15
9310	cycle 30 miles		144	.056			2.09	6.05	8.14	9.85
9312	cycle 40 miles		126	.063			2.38	6.90	9.28	11.20
9314	cycle 50 miles		108	.074			2.78	8.05	10.83	13.10
9414	25 min. wait/Ld./Uld.,15 MPH, cycle 0.5 mi.		306	.026			.98	2.83	3.81	4.62
9416	cycle 1 mile		288	.028			1.04	3.01	4.05	4.90
9418	cycle 2 miles		252	.032			1.19	3.44	4.63	5.60
9420	cycle 4 miles		198	.040			1.52	4.38	5.90	7.15
9422	cycle 6 miles		180	.044			1.67	4.82	6.49	7.85
9424	cycle 8 miles		144	.056			2.09	6.05	8.14	9.85
9425	cycle 10 miles		126	.063			2.38	6.90	9.28	11.20
9426	20 MPH ave, cycle 0.5 mile		324	.025			.93	2.68	3.61	4.36
9428	cycle 1 mile		306	.026			.98	2.83	3.81	4.62
9430	cycle 2 miles		270	.030			1.11	3.21	4.32	5.25
9432	cycle 4 miles		234	.034			1.28	3.71	4.99	6.05
9434	cycle 6 miles		198	.040			1.52	4.38	5.90	7.15
9436	cycle 8 miles		180	.044			1.67	4.82	6.49	7.85
9438	cycle 10 miles		144	.056			2.09	6.05	8.14	9.85
9440	25 MPH ave, cycle 4 miles		252	.032			1.19	3.44	4.63	5.60
9442	cycle 6 miles		216	.037			1.39	4.02	5.41	6.55
9444	cycle 8 miles		198	.040			1.52	4.38	5.90	7.15
9446	cycle 10 miles		180	.044			1.67	4.82	6.49	7.85
9450	30 MPH ave, cycle 4 miles		252	.032			1.19	3.44	4.63	5.60
9452	cycle 6 miles		234	.034			1.28	3.71	4.99	6.05
9454	cycle 8 miles		198	.040			1.52	4.38	5.90	7.15
9456	cycle 10 miles		180	.044			1.67	4.82	6.49	7.85
9460	35 MPH ave, cycle 4 miles		270	.030			1.11	3.21	4.32	5.25
9462	cycle 6 miles		234	.034			1.28	3.71	4.99	6.05
9464	cycle 8 miles		216	.037			1.39	4.02	5.41	6.55
9466	cycle 10 miles		198	.040			1.52	4.38	5.90	7.15
9468	cycle 20 miles		144	.056			2.09	6.05	8.14	9.85
9470	cycle 30 miles		108	.074			2.78	8.05	10.83	13.10
9472	cycle 40 miles		90	.089			3.34	9.65	12.99	15.70
9474	40 MPH ave, cycle 6 miles		252	.032			1.19	3.44	4.63	5.60
9476	cycle 8 miles		234	.034			1.28	3.71	4.99	6.05

31 23 23.20 Hauling		Crew	Daily Output	Labor-Hours	Unit	Material	2014 Bare Costs Labor	2014 Bare Costs Equipment	Total	Total Incl O&P
9478	cycle 10 miles	B-34I	216	.037	L.C.Y.		1.39	4.02	5.41	6.55
9480	cycle 20 miles		144	.056			2.09	6.05	8.14	9.85
9482	cycle 30 miles		126	.063			2.38	6.90	9.28	11.20
9484	cycle 40 miles		90	.089			3.34	9.65	12.99	15.70
9486	cycle 50 miles		90	.089			3.34	9.65	12.99	15.70
9494	45 MPH ave, cycle 8 miles		234	.034			1.28	3.71	4.99	6.05
9496	cycle 10 miles		216	.037			1.39	4.02	5.41	6.55
9498	cycle 20 miles		162	.049			1.85	5.35	7.20	8.75
9500	cycle 30 miles		126	.063			2.38	6.90	9.28	11.20
9502	cycle 40 miles		108	.074			2.78	8.05	10.83	13.10
9504	cycle 50 miles		90	.089			3.34	9.65	12.99	15.70
9506	50 MPH ave, cycle 10 miles		234	.034			1.28	3.71	4.99	6.05
9508	cycle 20 miles		180	.044			1.67	4.82	6.49	7.85
9510	cycle 30 miles		144	.056			2.09	6.05	8.14	9.85
9512	cycle 40 miles		108	.074			2.78	8.05	10.83	13.10
9514	cycle 50 miles		90	.089			3.34	9.65	12.99	15.70
9614	30 min. wait/Ld./Uld.,15 MPH, cycle 0.5 mi.		270	.030			1.11	3.21	4.32	5.25
9616	cycle 1 mile		252	.032			1.19	3.44	4.63	5.60
9618	cycle 2 miles		216	.037			1.39	4.02	5.41	6.55
9620	cycle 4 miles		180	.044			1.67	4.82	6.49	7.85
9622	cycle 6 miles		162	.049			1.85	5.35	7.20	8.75
9624	cycle 8 miles		126	.063			2.38	6.90	9.28	11.20
9625	cycle 10 miles		126	.063			2.38	6.90	9.28	11.20
9626	20 MPH ave, cycle 0.5 mile		270	.030			1.11	3.21	4.32	5.25
9628	cycle 1 mile		252	.032			1.19	3.44	4.63	5.60
9630	cycle 2 miles		234	.034			1.28	3.71	4.99	6.05
9632	cycle 4 miles		198	.040			1.52	4.38	5.90	7.15
9634	cycle 6 miles		180	.044			1.67	4.82	6.49	7.85
9636	cycle 8 miles		162	.049			1.85	5.35	7.20	8.75
9638	cycle 10 miles		144	.056			2.09	6.05	8.14	9.85
9640	25 MPH ave, cycle 4 miles		216	.037			1.39	4.02	5.41	6.55
9642	cycle 6 miles		198	.040			1.52	4.38	5.90	7.15
9644	cycle 8 miles		180	.044			1.67	4.82	6.49	7.85
9646	cycle 10 miles		162	.049			1.85	5.35	7.20	8.75
9650	30 MPH ave, cycle 4 miles		216	.037			1.39	4.02	5.41	6.55
9652	cycle 6 miles		198	.040			1.52	4.38	5.90	7.15
9654	cycle 8 miles		180	.044			1.67	4.82	6.49	7.85
9656	cycle 10 miles		162	.049			1.85	5.35	7.20	8.75
9660	35 MPH ave, cycle 4 miles		234	.034			1.28	3.71	4.99	6.05
9662	cycle 6 miles		216	.037			1.39	4.02	5.41	6.55
9664	cycle 8 miles		198	.040			1.52	4.38	5.90	7.15
9666	cycle 10 miles		180	.044			1.67	4.82	6.49	7.85
9668	cycle 20 miles		126	.063			2.38	6.90	9.28	11.20
9670	cycle 30 miles		108	.074			2.78	8.05	10.83	13.10
9672	cycle 40 miles		90	.089			3.34	9.65	12.99	15.70
9674	40 MPH ave, cycle 6 miles		216	.037			1.39	4.02	5.41	6.55
9676	cycle 8 miles		198	.040			1.52	4.38	5.90	7.15
9678	cycle 10 miles		180	.044			1.67	4.82	6.49	7.85
9680	cycle 20 miles		144	.056			2.09	6.05	8.14	9.85
9682	cycle 30 miles		108	.074			2.78	8.05	10.83	13.10
9684	cycle 40 miles		90	.089			3.34	9.65	12.99	15.70
9686	cycle 50 miles		72	.111			4.17	12.05	16.22	19.60
9694	45 MPH ave, cycle 8 miles		216	.037			1.39	4.02	5.41	6.55

31 23 23 – Fill

31 23 23.20 Hauling		Crew	Daily Output	Labor-Hours	Unit	Material	2014 Bare Costs		Total	Total Incl O&P
							Labor	Equipment		
9696	cycle 10 miles	B-34I	198	.040	L.C.Y.		1.52	4.38	5.90	7.15
9698	cycle 20 miles		144	.056			2.09	6.05	8.14	9.85
9700	cycle 30 miles		126	.063			2.38	6.90	9.28	11.20
9702	cycle 40 miles		108	.074			2.78	8.05	10.83	13.10
9704	cycle 50 miles		90	.089			3.34	9.65	12.99	15.70
9706	50 MPH ave, cycle 10 miles		198	.040			1.52	4.38	5.90	7.15
9708	cycle 20 miles		162	.049			1.85	5.35	7.20	8.75
9710	cycle 30 miles		126	.063			2.38	6.90	9.28	11.20
9712	cycle 40 miles		108	.074			2.78	8.05	10.83	13.10
9714	cycle 50 miles		90	.089			3.34	9.65	12.99	15.70

31 23 23.23 Compaction

		Crew	Daily Output	Labor-Hours	Unit	Material	Labor	Equipment	Total	Total Incl O&P
0010	**COMPACTION**									
7000	Walk behind, vibrating plate 18" wide, 6" lifts, 2 passes	A-1D	200	.040	E.C.Y.		1.47	.18	1.65	2.46
7020	3 passes		185	.043			1.58	.19	1.77	2.66
7040	4 passes		140	.057			2.09	.26	2.35	3.51

Estimating Tips

32 01 00 Operations and Maintenance of Exterior Improvements

- Recycling of asphalt pavement is becoming very popular and is an alternative to removal and replacement. It can be a good value engineering proposal if removed pavement can be recycled, either at the project site or at another site that is reasonably close to the project site. Sections on repair of flexible and rigid pavement are included.

32 10 00 Bases, Ballasts, and Paving

- When estimating paving, keep in mind the project schedule. Also note that prices for asphalt and concrete are generally higher in the cold seasons. Lines for pavement markings, including tactile warning systems and fence lines, are included.

32 90 00 Planting

- The timing of planting and guarantee specifications often dictate the costs for establishing tree and shrub growth and a stand of grass or ground cover. Establish the work performance schedule to coincide with the local planting season. Maintenance and growth guarantees can add from 20%–100% to the total landscaping cost and can be contractually cumbersome. The cost to replace trees and shrubs can be as high as 5% of the total cost, depending on the planting zone, soil conditions, and time of year.

Reference Numbers

Reference numbers are shown in shaded boxes at the beginning of some major classifications. These numbers refer to related items in the Reference Section. The reference information may be an estimating procedure, an alternate pricing method, or technical information.

Note: Not all subdivisions listed here necessarily appear in this publication.

32 31 Fences and Gates

32 31 13 – Chain Link Fences and Gates

32 31 13.30 Fence, Chain Link, Gates and Posts	Crew	Daily Output	Labor-Hours	Unit	Material	2014 Bare Costs Labor	Equipment	Total	Total Incl O&P
0010 **FENCE, CHAIN LINK, GATES & POSTS**									
0011 (1/3 post length in ground)									
6710 Corner post, galv. steel, 4" OD, set in conc., 4'	B-80	65	.492	Ea.	90	19.55	11.65	121.20	141
6715 6'		63	.508		99.50	20	12.05	131.55	153
6720 7'		61	.525		125	21	12.45	158.45	183
6725 8'		65	.492		129	19.55	11.65	160.20	185
7900 Auger fence post hole, 3' deep, medium soil, by hand	1 Clab	30	.267			9.75		9.75	15.10
7925 By machine	B-80	175	.183			7.25	4.33	11.58	15.85
7950 Rock, with jackhammer	B-9	32	1.250			46.50	7.30	53.80	79.50
7975 With rock drill	B-47C	65	.246			10.30	25	35.30	43.50

Estimating Tips

33 10 00 Water Utilities
33 30 00 Sanitary Sewerage Utilities
33 40 00 Storm Drainage Utilities

- Never assume that the water, sewer, and drainage lines will go in at the early stages of the project. Consider the site access needs before dividing the site in half with open trenches, loose pipe, and machinery obstructions. Always inspect the site to establish that the site drawings are complete. Check off all existing utilities on your drawings as you locate them. Be especially careful with underground utilities because appurtenances are sometimes buried during regrading or repaving operations. If you find any discrepancies, mark up the site plan for further research. Differing site conditions can be very costly if discovered later in the project.

- See also Section 33 01 00 for restoration of pipe where removal/replacement may be undesirable. Use of new types of piping materials can reduce the overall project cost. Owners/design engineers should consider the installing contractor as a valuable source of current information on utility products and local conditions that could lead to significant cost savings.

Reference Numbers

Reference numbers are shown in shaded boxes at the beginning of some major classifications. These numbers refer to related items in the Reference Section. The reference information may be an estimating procedure, an alternate pricing method, or technical information.

Note: Not all subdivisions listed here necessarily appear in this publication.

Division 33 - Utilities

Note: **Trade Service,** *in part, has been used as a reference source for some of the material prices used in Division 33.*

33 05 16 – Utility Structures

33 05 16.13 Precast Concrete Utility Boxes	Crew	Daily Output	Labor-Hours	Unit	Material	2014 Bare Costs Labor	2014 Bare Costs Equipment	Total	Total Incl O&P
0010 **PRECAST CONCRETE UTILITY BOXES**, 6" thick									
0050 5' x 10' x 6' high, I.D.	B-13	2	28	Ea.	3,675	1,125	375	5,175	6,125
0100 6' x 10' x 6' high, I.D.		2	28		3,800	1,125	375	5,300	6,300
0150 5' x 12' x 6' high, I.D.		2	28		4,025	1,125	375	5,525	6,525
0200 6' x 12' x 6' high, I.D.		1.80	31.111		4,525	1,250	415	6,190	7,325
0250 6' x 13' x 6' high, I.D.		1.50	37.333		5,925	1,500	495	7,920	9,350
0300 8' x 14' x 7' high, I.D.		1	56		6,400	2,225	745	9,370	11,300

33 05 23 – Trenchless Utility Installation

33 05 23.19 Microtunneling

	Crew	Daily Output	Labor-Hours	Unit	Material	2014 Bare Costs Labor	2014 Bare Costs Equipment	Total	Total Incl O&P
0010 **MICROTUNNELING**									
0011 Not including excavation, backfill, shoring,									
0020 or dewatering, average 50'/day, slurry method									
0100 24" to 48" outside diameter, minimum				L.F.				875	965
0110 Adverse conditions, add				%				50%	50%
1000 Rent microtunneling machine, average monthly lease				Month				97,500	107,000
1010 Operating technician				Day				630	705
1100 Mobilization and demobilization, minimum				Job				41,200	45,900
1110 Maximum				"				445,500	490,500

33 05 23.20 Horizontal Boring

	Crew	Daily Output	Labor-Hours	Unit	Material	2014 Bare Costs Labor	2014 Bare Costs Equipment	Total	Total Incl O&P
0010 **HORIZONTAL BORING**									
0011 Casing only, 100' minimum,									
0020 not incl. jacking pits or dewatering									
0100 Roadwork, 1/2" thick wall, 24" diameter casing	B-42	20	3.200	L.F.	121	132	68.50	321.50	415
0200 36" diameter		16	4		223	165	85.50	473.50	600
0300 48" diameter		15	4.267		310	176	91	577	715
0500 Railroad work, 24" diameter		15	4.267		121	176	91	388	510
0600 36" diameter		14	4.571		223	189	97.50	509.50	650
0700 48" diameter		12	5.333		310	220	114	644	810
0900 For ledge, add								20%	20%
1000 Small diameter boring, 3", sandy soil	B-82	900	.018		22	.74	.09	22.83	25
1040 Rocky soil	"	500	.032		22	1.34	.17	23.51	26
1100 Prepare jacking pits, incl. mobilization & demobilization, minimum				Ea.				3,225	3,700
1101 Maximum				"				22,000	25,500

33 05 23.22 Directional Drilling

	Crew	Daily Output	Labor-Hours	Unit	Material	2014 Bare Costs Labor	2014 Bare Costs Equipment	Total	Total Incl O&P
0010 **DIRECTIONAL DRILLING**									
0011 Excluding cost of conduit, 100' minimum									
0100 Small equipment to 300', not to exceed 12" diam.									
0102 small unit mobilization to site	B-82A	2	8	Ea.		335	550	885	1,125
0105 small unit setup per drill		4	4			167	276	443	560
0109 minimum charge gravel, sand & silt, up to 12" diam.		3.20	5		10.15	209	345	564.15	710
0110 gravel, sand & silt, up to 12" diam.		300	.053	L.F.	.07	2.23	3.68	5.98	7.55
0118 min charge , clay & soft sandstone, up to 10" diam.		3.20	5	Ea.	20.50	209	345	574.50	725
0120 Clay & soft sandstone, up to 10" diam.		300	.053	L.F.	.14	2.23	3.68	6.05	7.60
0128 min charge , hard clay & cobble, up to 8" diam.		1	16	Ea.	54	670	1,100	1,824	2,300
0130 Hard clay & cobble, up to 8" diam.		100	.160	L.F.	.54	6.70	11.05	18.29	23
0200 Medium equipment to 600', not to exceed 12" diam.									
0202 medium unit mobilization to site	B-82B	2	12	Ea.		480	690	1,170	1,500
0205 medium unit setup per drill		4	6			241	345	586	750
0209 minimum charge gravel, sand & silt, up to 12" diam.		3.20	7.500		10.15	300	430	740.15	945
0210 gravel, sand & silt, up to 12" diam.		350	.069	L.F.	.07	2.75	3.95	6.77	8.65
0218 min charge , clay & soft sandstone, up to 10" diam.		3.20	7.500	Ea.	20.50	300	430	750.50	960
0220 Clay & soft sandstone, up to 10" diam.		300	.080	L.F.	.14	3.21	4.61	7.96	10.10

33 05 Common Work Results for Utilities

33 05 23 – Trenchless Utility Installation

33 05 23.22 Directional Drilling		Crew	Daily Output	Labor-Hours	Unit	Material	2014 Bare Costs Labor	Equipment	Total	Total Incl O&P
0228	min charge , hard clay & cobble, up to 8" diam.	B-82B	1.78	13.483	Ea.	54	540	775	1,369	1,750
0230	Hard clay & cobble, up to 8" diam.	↓	150	.160	L.F.	.54	6.40	9.20	16.14	20.50
0300	Large equipment to 1000', not to exceed 12" diam.									
0302	large unit mobilization to site	B-82C	2	12	Ea.		480	845	1,325	1,675
0305	large unit setup per drill		4	6			241	425	666	835
0309	minimum charge gravel, sand & silt, up to 12" diam.		4	6	↓	10.15	241	425	676.15	845
0310	gravel, sand & silt, up to 12" diam.		400	.060	L.F.	.07	2.41	4.23	6.71	8.40
0318	min charge , clay & soft sandstone, up to 12" diam.		3.20	7.500	Ea.	20.50	300	530	850.50	1,075
0320	Clay & soft sandstone, up to 12" diam.		350	.069	L.F.	.14	2.75	4.84	7.73	9.65
0328	min charge , hard clay & cobble, up to 10" diam.		1.78	13.483	Ea.	54	540	950	1,544	1,950
0330	Rock & cobble, up to 10" diam.	↓	150	.160	L.F.	.54	6.40	11.30	18.24	23
0400	Directional drilling, mud trailer per day	B-82D	1	8	Day		375	375	750	980
1000	Additional charges for mobilization may apply at some locations									

33 05 26 – Utility Identification

33 05 26.10 Utility Accessories

		Crew	Daily Output	Labor-Hours	Unit	Material	2014 Bare Costs Labor	Equipment	Total	Total Incl O&P
0010	**UTILITY ACCESSORIES**									
0400	Underground tape, detectable, reinforced, alum. foil core, 2"	1 Clab	150	.053	C.L.F.	5.65	1.95		7.60	9.25
0500	6"	"	140	.057	"	27.50	2.09		29.59	33.50

33 44 Storm Utility Water Drains

33 44 13 – Utility Area Drains

33 44 13.13 Catchbasins

		Crew	Daily Output	Labor-Hours	Unit	Material	2014 Bare Costs Labor	Equipment	Total	Total Incl O&P
0010	**CATCHBASINS**									
0011	Not including footing & excavation									
1600	Frames & grates, C.I., 24" square, 500 lb.	B-6	7.80	3.077	Ea.	340	123	47	510	615
1700	26" D shape, 600 lb.		7	3.429		505	138	52	695	825
1800	Light traffic, 18" diameter, 100 lb.		10	2.400		122	96.50	36.50	255	320
1900	24" diameter, 300 lb.		8.70	2.759		196	111	42	349	430
2000	36" diameter, 900 lb.		5.80	4.138		570	166	63	799	950
2100	Heavy traffic, 24" diameter, 400 lb.		7.80	3.077		244	123	47	414	510
2200	36" diameter, 1150 lb.		3	8		795	320	122	1,237	1,500
2300	Mass. State standard, 26" diameter, 475 lb.		7	3.429		265	138	52	455	560
2400	30" diameter, 620 lb.		7	3.429		345	138	52	535	650
2500	Watertight, 24" diameter, 350 lb.		7.80	3.077		320	123	47	490	590
2600	26" diameter, 500 lb.		7	3.429		425	138	52	615	735
2700	32" diameter, 575 lb.	↓	6	4	↓	850	160	61	1,071	1,250
2800	3 piece cover & frame, 10" deep,									
2900	1200 lb., for heavy equipment	B-6	3	8	Ea.	1,050	320	122	1,492	1,775
3000	Raised for paving 1-1/4" to 2" high									
3100	4 piece expansion ring									
3200	20" to 26" diameter	1 Clab	3	2.667	Ea.	157	97.50		254.50	325
3300	30" to 36" diameter	"	3	2.667	"	216	97.50		313.50	390
3320	Frames and covers, existing, raised for paving, 2", including									
3340	row of brick, concrete collar, up to 12" wide frame	B-6	18	1.333	Ea.	44.50	53.50	20.50	118.50	154
3360	20" to 26" wide frame		11	2.182		66.50	87.50	33	187	244
3380	30" to 36" wide frame	↓	9	2.667		82.50	107	40.50	230	300
3400	Inverts, single channel brick	D-1	3	5.333		96	220		316	440
3500	Concrete		5	3.200		103	132		235	315
3600	Triple channel, brick		2	8		147	330		477	665
3700	Concrete	↓	3	5.333	↓	139	220		359	485

33 49 Storm Drainage Structures

33 49 13 – Storm Drainage Manholes, Frames, and Covers

33 49 13.10 Storm Drainage Manholes, Frames and Covers	Crew	Daily Output	Labor-Hours	Unit	Material	Labor	Equipment	Total	Total Incl O&P
0010	**STORM DRAINAGE MANHOLES, FRAMES & COVERS**								
0020	Excludes footing, excavation, backfill (See line items for frame & cover)								
0050	Brick, 4' inside diameter, 4' deep	D-1	1	16	Ea.	510	660		1,170	1,550
0100	6' deep		.70	22.857		725	945		1,670	2,250
0150	8' deep		.50	32		935	1,325		2,260	3,050
0200	For depths over 8', add		4	4	V.L.F.	83.50	165		248.50	345
0400	Concrete blocks (radial), 4' I.D., 4' deep		1.50	10.667	Ea.	370	440		810	1,075
0500	6' deep		1	16		495	660		1,155	1,550
0600	8' deep		.70	22.857		620	945		1,565	2,125
0700	For depths over 8', add		5.50	2.909	V.L.F.	65	120		185	255
0800	Concrete, cast in place, 4' x 4', 8" thick, 4' deep	C-14H	2	24	Ea.	505	1,075	16.55	1,596.55	2,250
0900	6' deep		1.50	32		725	1,450	22	2,197	3,050
1000	8' deep		1	48		1,050	2,175	33	3,258	4,500
1100	For depths over 8', add		8	6	V.L.F.	118	271	4.14	393.14	550
1110	Precast, 4' I.D., 4' deep	B-22	4.10	7.317	Ea.	775	315	51	1,141	1,375
1120	6' deep		3	10		970	425	70	1,465	1,800
1130	8' deep		2	15		1,100	640	105	1,845	2,325
1140	For depths over 8', add		16	1.875	V.L.F.	125	80	13.15	218.15	275
1150	5' I.D., 4' deep	B-6	3	8	Ea.	1,625	320	122	2,067	2,400
1160	6' deep		2	12		1,675	480	183	2,338	2,775
1170	8' deep		1.50	16		2,350	640	244	3,234	3,850
1180	For depths over 8', add		12	2	V.L.F.	277	80	30.50	387.50	460
1190	6' I.D., 4' deep		2	12	Ea.	2,125	480	183	2,788	3,275
1200	6' deep		1.50	16		2,550	640	244	3,434	4,050
1210	8' deep		1	24		3,150	965	365	4,480	5,325
1220	For depths over 8', add		8	3	V.L.F.	375	120	45.50	540.50	645
1250	Slab tops, precast, 8" thick									
1300	4' diameter manhole	B-6	8	3	Ea.	245	120	45.50	410.50	505
1400	5' diameter manhole		7.50	3.200		405	128	48.50	581.50	695
1500	6' diameter manhole		7	3.429		625	138	52	815	955
3800	Steps, heavyweight cast iron, 7" x 9"	1 Bric	40	.200		19.05	9.10		28.15	35
3900	8" x 9"		40	.200		23	9.10		32.10	39
3928	12" x 10-1/2"		40	.200		26.50	9.10		35.60	43.50
4000	Standard sizes, galvanized steel		40	.200		21.50	9.10		30.60	37.50
4100	Aluminum		40	.200		24	9.10		33.10	40
4150	Polyethylene		40	.200		24.50	9.10		33.60	41

33 71 Electrical Utility Transmission and Distribution

33 71 13 – Electrical Utility Towers

33 71 13.23 Steel Electrical Utility Towers

33 71 13.23 Steel Electrical Utility Towers	Crew	Daily Output	Labor-Hours	Unit	Material	Labor	Equipment	Total	Total Incl O&P
0010	**STEEL ELECTRICAL UTILITY TOWERS**								
0100	Excavation and backfill, earth	R-5	135.38	.650	C.Y.		30.50	11.20	41.70	58.50
0105	Rock		21.46	4.101	"		191	70.50	261.50	365
0200	Steel footings (grillage) in earth		3.91	22.506	Ton	2,175	1,050	385	3,610	4,400
0205	In rock		3.20	27.500	"	2,175	1,275	475	3,925	4,850
0290	See also Section 33 05 23.19									
0300	Rock anchors	R-5	5.87	14.991	Ea.	575	700	258	1,533	1,975
0400	Concrete foundations	"	12.85	6.848	C.Y.	131	320	118	569	755
0490	See also Section 03 30 53.40									
0500	Towers-material handling and spotting	R-7	22.56	2.128	Ton		80.50	9.95	90.45	135
0540	Steel tower erection	R-5	7.65	11.503		2,125	535	198	2,858	3,350

33 71 13 – Electrical Utility Towers

33 71 13.23 Steel Electrical Utility Towers	Crew	Daily Output	Labor-Hours	Unit	Material	2014 Bare Costs Labor	Equipment	Total	Total Incl O&P
0550	Lace and box	R-5	7.10	12.394	Ton	2,125	580	213	2,918	3,425
0560	Painting total structure	▼	1.47	59.864	Ea.	460	2,800	1,025	4,285	5,850
0570	Disposal of surplus material	R-7	20.87	2.300	Mile		87	10.75	97.75	146
0600	Special towers-material handling and spotting	"	12.31	3.899	Ton		148	18.25	166.25	247
0640	Special steel structure erection	R-6	6.52	13.497		2,650	630	525	3,805	4,450
0650	Special steel lace and box	"	6.29	13.990	▼	2,650	650	545	3,845	4,500
0670	Disposal of surplus material	R-7	7.87	6.099	Mile		231	28.50	259.50	385

33 71 13.80 Transmission Line Right of Way

		Crew	Daily Output	Labor-Hours	Unit	Material	Labor	Equipment	Total	Total Incl O&P
0010	**TRANSMISSION LINE RIGHT OF WAY**									
0100	Clearing right of way	B-87	6.67	5.997	Acre		279	450	729	920
0200	Restoration & seeding	B-10D	4	3	"	1,075	134	440	1,649	1,875

33 71 16 – Electrical Utility Poles

33 71 16.23 Steel Electrical Utility Poles

		Crew	Daily Output	Labor-Hours	Unit	Material	Labor	Equipment	Total	Total Incl O&P
0010	**STEEL ELECTRICAL UTILITY POLES**									
0100	Steel, square, tapered shaft, 20'	R-15A	6	8	Ea.	370	375	54	799	1,025
0120	25'		5.35	8.972		390	420	60.50	870.50	1,125
0140	30'		4.80	10		455	470	67.50	992.50	1,275
0160	35'		4.35	11.034		470	515	74.50	1,059.50	1,375
0180	40'		4	12		665	560	81	1,306	1,675
0300	Galvanized, round, tapered shaft, 40'		4	12		1,975	560	81	2,616	3,125
0320	45'		3.70	12.973		2,200	605	88	2,893	3,450
0340	50'		3.45	13.913		2,425	650	94	3,169	3,750
0360	55'		3	16		2,550	750	108	3,408	4,050
0380	60'	▼	2.65	18.113		2,850	850	123	3,823	4,525
0400	65'	R-13	2.20	19.091		3,525	980	86.50	4,591.50	5,450
0420	70'		1.90	22.105		3,850	1,125	100	5,075	6,025
0440	75'		1.70	24.706		5,050	1,275	112	6,437	7,575
0460	80'		1.50	28		4,550	1,450	127	6,127	7,325
0480	85'		1.40	30		4,950	1,550	136	6,636	7,925
0500	90'		1.25	33.600		5,300	1,725	152	7,177	8,625
0520	95'		1.15	36.522		5,725	1,875	165	7,765	9,300
0540	100'	▼	1	42		6,175	2,150	190	8,515	10,300
0700	12 sided, tapered shaft, 40'	R-15A	4	12		1,100	560	81	1,741	2,150
0720	45'		3.70	12.973		1,225	605	88	1,918	2,375
0740	50'		3.45	13.913		1,400	650	94	2,144	2,650
0760	55'		3	16		1,625	750	108	2,483	3,025
0780	60'	▼	2.65	18.113		1,675	850	123	2,648	3,250
0800	65'	R-13	2.20	19.091		1,975	980	86.50	3,041.50	3,750
0820	70'		1.90	22.105		2,150	1,125	100	3,375	4,150
0840	75'		1.70	24.706		2,325	1,275	112	3,712	4,575
0860	80'		1.50	28		3,950	1,450	127	5,527	6,650
0880	85'		1.40	30		6,400	1,550	136	8,086	9,525
0900	90'		1.25	33.600		8,375	1,725	152	10,252	12,000
0920	95'		1.15	36.522		10,500	1,875	165	12,540	14,600
0940	100'		1	42		12,800	2,150	190	15,140	17,600
0960	105'	▼	.90	46.667		14,200	2,400	211	16,811	19,400
0980	Ladder clips					226			226	249
1000	Galvanized steel, round, tapered, w/one 6' arm, 20'	R-15A	9.20	5.217		890	244	35.50	1,169.50	1,400
1020	25'		7.65	6.275		970	294	42.50	1,306.50	1,575
1040	30'		6.55	7.328		1,075	345	49.50	1,469.50	1,750
1060	35'		5.75	8.348		1,150	390	56.50	1,596.50	1,925
1080	40'	▼	5.15	9.320	▼	1,225	435	63	1,723	2,075

33 71 16 – Electrical Utility Poles

33 71 16.23 Steel Electrical Utility Poles

	Crew	Daily Output	Labor-Hours	Unit	Material	2014 Bare Costs Labor	Equipment	Total	Total Incl O&P
1200	Two 6' arms, 20'	R-15A	6.55	7.328	Ea.	1,150	345	49.50	1,544.50	1,850
1220	25'		5.75	8.348		1,175	390	56.50	1,621.50	1,950
1240	30'		5.15	9.320		1,300	435	63	1,798	2,150
1260	35'		4.60	10.435		1,400	490	70.50	1,960.50	2,350
1280	40'		4.20	11.429		1,600	535	77.50	2,212.50	2,675
1400	One 12' truss arm, 25'		7.65	6.275		1,100	294	42.50	1,436.50	1,725
1420	30'		6.55	7.328		1,150	345	49.50	1,544.50	1,825
1440	35'		5.75	8.348		1,250	390	56.50	1,696.50	2,025
1460	40'		5.15	9.320		1,400	435	63	1,898	2,250
1480	45'		4.60	10.435		1,600	490	70.50	2,160.50	2,600
1600	Two 12' truss arms, 25'		5.75	8.348		1,325	390	56.50	1,771.50	2,100
1620	30'		5.15	9.320		1,350	435	63	1,848	2,225
1640	35'		4.60	10.435		1,450	490	70.50	2,010.50	2,425
1660	40'		4.20	11.429		1,600	535	77.50	2,212.50	2,675
1680	45'		3.85	12.468		1,825	585	84.50	2,494.50	2,975
3400	Galvanized steel, tapered, 10'		15.50	3.097		277	145	21	443	545
3420	12'		11.50	4.174		405	195	28.50	628.50	770
3440	14'		9.20	5.217		299	244	35.50	578.50	740
3460	16'		8.45	5.680		310	266	38.50	614.50	785
3480	18'		7.65	6.275		320	294	42.50	656.50	840
3500	20'		7.10	6.761		300	315	46	661	860
5200	Galvanized steel, bracket type, 2", 6' long		18.50	2.595		138	121	17.55	276.55	355
5220	8' long		15.50	3.097		171	145	21	337	430
5240	10' long	↓	12.50	3.840		287	180	26	493	615
6000	Digging holes in earth, average	R-5	25.14	3.500			163	60	223	315
6010	In rock, average	"	4.51	19.512	↓		910	335	1,245	1,750
6020	Formed plate pole structure									
6030	Material handling and spotting	R-7	2.40	20	Ea.		755	93.50	848.50	1,275
6040	Erect steel plate pole	R-5	1.95	45.128		9,050	2,100	775	11,925	14,000
6050	Guys, anchors and hardware for pole, in earth		7.04	12.500		550	585	215	1,350	1,725
6060	In rock	↓	17.96	4.900	↓	655	228	84.50	967.50	1,150
6070	Foundations for line poles									
6080	Excavation, in earth	R-5	135.38	.650	C.Y.		30.50	11.20	41.70	58.50
6090	In rock		20	4.400			205	75.50	280.50	395
6110	Concrete foundations	↓	11	8		131	375	138	644	860

33 71 16.33 Wood Electrical Utility Poles

	Crew	Daily Output	Labor-Hours	Unit	Material	2014 Bare Costs Labor	Equipment	Total	Total Incl O&P
0010	**WOOD ELECTRICAL UTILITY POLES**									
0011	Excludes excavation, backfill and cast-in-place concrete									
1020	12" Ponderosa Pine Poles treated 0.40 ACQ, 16'	R-3	3.20	6.250	Ea.	880	330	44	1,254	1,500
5000	Wood, class 3 yellow pine, penta-treated, 25'	R-15A	8.60	5.581		199	261	38	498	655
5020	30'		7.70	6.234		221	292	42	555	730
5040	35'		5.80	8.276		292	385	56	733	965
5060	40'		5.30	9.057		365	425	61.50	851.50	1,100
5080	45'		4.70	10.213		405	480	69	954	1,250
5100	50'		4.20	11.429		505	535	77.50	1,117.50	1,450
5120	55'		3.80	12.632		605	590	85.50	1,280.50	1,650
5140	60'		3.50	13.714		840	640	93	1,573	2,000
5160	65'		3.20	15		1,125	700	102	1,927	2,400
5180	70'		3	16		1,600	750	108	2,458	3,000
5200	75'		2.80	17.143		2,075	805	116	2,996	3,625
6000	Wood, class 1 type C, CCA/ACA-treated, 25'		8.60	5.581		231	261	38	530	690
6020	30'	↓	7.70	6.234	↓	267	292	42	601	780

33 71 Electrical Utility Transmission and Distribution

33 71 16 – Electrical Utility Poles

33 71 16.33 Wood Electrical Utility Poles	Crew	Daily Output	Labor-Hours	Unit	Material	2014 Bare Costs Labor	Equipment	Total	Total Incl O&P
6040	35'	R-15A	5.80	8.276	Ea.	355	385	56	796	1,050
6060	40'		5.30	9.057		450	425	61.50	936.50	1,200
6080	45'		4.70	10.213		545	480	69	1,094	1,400
6100	50'		4.20	11.429		600	535	77.50	1,212.50	1,550
6120	55'		3.80	12.632		690	590	85.50	1,365.50	1,750
6200	Electric & tel sitework, 20' high, treated wd., see Section 26 56 13.10	R-3	3.10	6.452		223	340	45	608	805
6400	25' high		2.90	6.897		263	365	48.50	676.50	895
6600	30' high		2.60	7.692		430	405	54	889	1,150
6800	35' high		2.40	8.333		495	440	58.50	993.50	1,275
7000	40' high		2.30	8.696		710	460	61	1,231	1,525
7200	45' high		1.70	11.765		865	625	82.50	1,572.50	1,975
7400	Cross arms with hardware & insulators									
7600	4' long	1 Elec	2.50	3.200	Ea.	148	171		319	420
7800	5' long		2.40	3.333		170	178		348	455
8000	6' long		2.20	3.636		165	194		359	470
9000	Disposal of pole & hardware surplus material	R-7	20.87	2.300	Mile		87	10.75	97.75	146
9100	Disposal of crossarms & hardware surplus material	"	40	1.200	"		45.50	5.60	51.10	76

Note: 6400 row shows reference R337119-30

33 71 19 – Electrical Underground Ducts and Manholes

33 71 19.15 Underground Ducts and Manholes

		Crew	Daily Output	Labor-Hours	Unit	Material	2014 Bare Costs Labor	Equipment	Total	Total Incl O&P
0010	**UNDERGROUND DUCTS AND MANHOLES**									
0011	Not incl. excavation, backfill and concrete, in slab or duct bank									
1000	Direct burial									
1010	PVC, schedule 40, w/coupling, 1/2" diameter	1 Elec	340	.024	L.F.	.41	1.26		1.67	2.33
1020	3/4" diameter		290	.028		.55	1.47		2.02	2.81
1030	1" diameter		260	.031		.81	1.64		2.45	3.35
1040	1-1/2" diameter		210	.038		1.33	2.03		3.36	4.50
1050	2" diameter		180	.044		1.74	2.37		4.11	5.45
1060	3" diameter	2 Elec	240	.067		3.33	3.56		6.89	8.95
1070	4" diameter		160	.100		4.66	5.35		10.01	13.15
1080	5" diameter		120	.133		6.75	7.10		13.85	18.10
1090	6" diameter		90	.178		8.95	9.50		18.45	24
1110	Elbows, 1/2" diameter	1 Elec	48	.167	Ea.	1.06	8.90		9.96	14.45
1120	3/4" diameter		38	.211		1.04	11.25		12.29	17.95
1130	1" diameter		32	.250		1.57	13.35		14.92	21.50
1140	1-1/2" diameter		21	.381		3.11	20.50		23.61	34
1150	2" diameter		16	.500		4.22	26.50		30.72	44.50
1160	3" diameter		12	.667		12.70	35.50		48.20	67
1170	4" diameter		9	.889		20.50	47.50		68	93.50
1180	5" diameter		8	1		33	53.50		86.50	117
1190	6" diameter		5	1.600		76	85.50		161.50	212
1210	Adapters, 1/2" diameter		52	.154		.30	8.20		8.50	12.65
1220	3/4" diameter		43	.186		.49	9.95		10.44	15.40
1230	1" diameter		39	.205		.66	10.95		11.61	17.15
1240	1-1/2" diameter		35	.229		1.14	12.20		13.34	19.50
1250	2" diameter		26	.308		1.44	16.40		17.84	26
1260	3" diameter		20	.400		3.50	21.50		25	36
1270	4" diameter		14	.571		6.20	30.50		36.70	52.50
1280	5" diameter		12	.667		13.20	35.50		48.70	67.50
1290	6" diameter		9	.889		22	47.50		69.50	95
1340	Bell end & cap, 1-1/2" diameter		35	.229		8.20	12.20		20.40	27.50
1350	Bell end & plug, 2" diameter		26	.308		10.85	16.40		27.25	36.50
1360	3" diameter		20	.400		13.20	21.50		34.70	46.50

33 71 19.15 Underground Ducts and Manholes	Crew	Daily Output	Labor-Hours	Unit	Material	2014 Bare Costs Labor	Equipment	Total	Total Incl O&P
1370 4" diameter	1 Elec	14	.571	Ea.	16.65	30.50		47.15	64
1380 5" diameter		12	.667		18.80	35.50		54.30	73.50
1390 6" diameter		9	.889		22.50	47.50		70	96
1450 Base spacer, 2" diameter		56	.143		1.54	7.60		9.14	13.10
1460 3" diameter		46	.174		1.58	9.30		10.88	15.65
1470 4" diameter		41	.195		1.63	10.40		12.03	17.40
1480 5" diameter		37	.216		2.04	11.55		13.59	19.50
1490 6" diameter		34	.235		2.77	12.55		15.32	22
1550 Intermediate spacer, 2" diameter		60	.133		1.51	7.10		8.61	12.30
1560 3" diameter		46	.174		1.70	9.30		11	15.75
1570 4" diameter		41	.195		1.56	10.40		11.96	17.30
1580 5" diameter		37	.216		1.84	11.55		13.39	19.25
1590 6" diameter		34	.235	▼	2.69	12.55		15.24	22
4010 PVC, schedule 80, w/coupling, 1/2" diameter		215	.037	L.F.	.92	1.99		2.91	3.98
4020 3/4" diameter		180	.044		1.23	2.37		3.60	4.90
4030 1" diameter		145	.055		1.76	2.94		4.70	6.35
4040 1-1/2" diameter		120	.067		2.93	3.56		6.49	8.50
4050 2" diameter		100	.080		4.04	4.27		8.31	10.85
4060 3" diameter	2 Elec	130	.123		8.25	6.55		14.80	18.95
4070 4" diameter		90	.178		12	9.50		21.50	27.50
4080 5" diameter		70	.229		17.25	12.20		29.45	37
4090 6" diameter	▼	50	.320	▼	21.50	17.05		38.55	49.50
4110 Elbows, 1/2" diameter	1 Elec	29	.276	Ea.	1.32	14.70		16.02	23.50
4120 3/4" diameter		23	.348		2.56	18.55		21.11	31
4130 1" diameter		20	.400		3.19	21.50		24.69	35.50
4140 1-1/2" diameter		16	.500		6.90	26.50		33.40	47.50
4150 2" diameter		12	.667		8.30	35.50		43.80	62
4160 3" diameter		9	.889		23	47.50		70.50	96.50
4170 4" diameter		7	1.143		46.50	61		107.50	143
4180 5" diameter		6	1.333		195	71		266	320
4190 6" diameter		4	2		231	107		338	415
4210 Adapter, 1/2" diameter		39	.205		.30	10.95		11.25	16.75
4220 3/4" diameter		33	.242		.49	12.95		13.44	19.90
4230 1" diameter		29	.276		.66	14.70		15.36	22.50
4240 1-1/2" diameter		26	.308		1.14	16.40		17.54	26
4250 2" diameter		23	.348		1.44	18.55		19.99	29.50
4260 3" diameter		18	.444		3.50	23.50		27	39.50
4270 4" diameter		13	.615		6.20	33		39.20	56
4280 5" diameter		11	.727		13.20	39		52.20	72.50
4290 6" diameter		8	1		22	53.50		75.50	104
4310 Bell end & cap, 1-1/2" diameter		26	.308		8.20	16.40		24.60	33.50
4320 Bell end & plug, 2" diameter		23	.348		10.85	18.55		29.40	40
4330 3" diameter		18	.444		13.20	23.50		36.70	50
4340 4" diameter		13	.615		16.65	33		49.65	67.50
4350 5" diameter		11	.727		18.80	39		57.80	78.50
4360 6" diameter		8	1		22.50	53.50		76	105
4370 Base spacer, 2" diameter		42	.190		1.54	10.15		11.69	16.90
4380 3" diameter		33	.242		1.58	12.95		14.53	21
4390 4" diameter		29	.276		1.63	14.70		16.33	24
4400 5" diameter		26	.308		2.04	16.40		18.44	26.50
4410 6" diameter		25	.320		2.77	17.05		19.82	28.50
4420 Intermediate spacer, 2" diameter		45	.178		1.51	9.50		11.01	15.85
4430 3" diameter	▼	34	.235	▼	1.70	12.55		14.25	20.50

33 71 19.15 Underground Ducts and Manholes	Crew	Daily Output	Labor-Hours	Unit	Material	2014 Bare Costs Labor	Equipment	Total	Total Incl O&P
4440	4" diameter	1 Elec	31	.258	Ea.	1.56	13.75		15.31	22
4450	5" diameter		28	.286		1.84	15.25		17.09	25
4460	6" diameter	↓	25	.320	↓	2.69	17.05		19.74	28.50

33 71 19.17 Electric and Telephone Underground

		Crew	Daily Output	Labor-Hours	Unit	Material	2014 Bare Costs Labor	Equipment	Total	Total Incl O&P
0010	**ELECTRIC AND TELEPHONE UNDERGROUND**									
0011	Not including excavation									
0200	backfill and cast in place concrete									
0250	For bedding, see Section 31 23 23.16 R337119-30									
0400	Hand holes, precast concrete, with concrete cover									
0600	2' x 2' x 3' deep	R-3	2.40	8.333	Ea.	415	440	58.50	913.50	1,175
0800	3' x 3' x 3' deep		1.90	10.526		535	555	73.50	1,163.50	1,500
1000	4' x 4' x 4' deep	↓	1.40	14.286	↓	1,200	755	100	2,055	2,525
1200	Manholes, precast with iron racks & pulling irons, C.I. frame									
1400	and cover, 4' x 6' x 7' deep	B-13	2	28	Ea.	5,925	1,125	375	7,425	8,625
1600	6' x 8' x 7' deep		1.90	29.474		6,650	1,175	395	8,220	9,550
1800	6' x 10' x 7' deep	↓	1.80	31.111	↓	7,450	1,250	415	9,115	10,600
4200	Underground duct, banks ready for concrete fill, min. of 7.5"									
4400	between conduits, center to center									
4580	PVC, type EB, 1 @ 2" diameter	2 Elec	480	.033	L.F.	.72	1.78		2.50	3.46
4600	2 @ 2" diameter		240	.067		1.45	3.56		5.01	6.90
4800	4 @ 2" diameter		120	.133		2.90	7.10		10	13.85
4900	1 @ 3" diameter		400	.040		.99	2.13		3.12	4.28
5000	2 @ 3" diameter		200	.080		1.98	4.27		6.25	8.55
5200	4 @ 3" diameter		100	.160		3.95	8.55		12.50	17.15
5300	1 @ 4" diameter		320	.050		1.51	2.67		4.18	5.65
5400	2 @ 4" diameter		160	.100		3.03	5.35		8.38	11.35
5600	4 @ 4" diameter		80	.200		6.05	10.65		16.70	22.50
5800	6 @ 4" diameter		54	.296		9.10	15.80		24.90	33.50
5810	1 @ 5" diameter		260	.062		2.25	3.28		5.53	7.40
5820	2 @ 5" diameter		130	.123		4.51	6.55		11.06	14.80
5840	4 @ 5" diameter		70	.229		9	12.20		21.20	28
5860	6 @ 5" diameter		50	.320		13.50	17.05		30.55	40.50
5870	1 @ 6" diameter		200	.080		3.22	4.27		7.49	9.95
5880	2 @ 6" diameter		100	.160		6.45	8.55		15	19.90
5900	4 @ 6" diameter		50	.320		12.85	17.05		29.90	39.50
5920	6 @ 6" diameter		30	.533		19.30	28.50		47.80	63.50
6200	Rigid galvanized steel, 2 @ 2" diameter		180	.089		12.10	4.74		16.84	20.50
6400	4 @ 2" diameter		90	.178		24	9.50		33.50	40.50
6800	2 @ 3" diameter		100	.160		27.50	8.55		36.05	43
7000	4 @ 3" diameter		50	.320		55	17.05		72.05	86
7200	2 @ 4" diameter		70	.229		40	12.20		52.20	62.50
7400	4 @ 4" diameter		34	.471		80	25		105	126
7600	6 @ 4" diameter		22	.727		120	39		159	190
7620	2 @ 5" diameter		60	.267		83	14.25		97.25	113
7640	4 @ 5" diameter		30	.533		166	28.50		194.50	225
7660	6 @ 5" diameter		18	.889		248	47.50		295.50	345
7680	2 @ 6" diameter		40	.400		119	21.50		140.50	163
7700	4 @ 6" diameter		20	.800		238	42.50		280.50	325
7720	6 @ 6" diameter	↓	14	1.143	↓	360	61		421	485
7800	For cast-in-place concrete - Add									
7810	Under 1 C.Y.	C-6	16	3	C.Y.	158	115	4.14	277.14	355
7820	1 C.Y. - 5 C.Y.	↓	19.20	2.500	↓	144	95.50	3.45	242.95	310

33 71 19 – Electrical Underground Ducts and Manholes

33 71 19.17 Electric and Telephone Underground		Crew	Daily Output	Labor-Hours	Unit	Material	2014 Bare Costs Labor	Equipment	Total	Total Incl O&P
7830	Over 5 C.Y.	C-6	24	2	C.Y.	119	76.50	2.76	198.26	251
7850	For reinforcing rods - Add									
7860	#4 to #7	2 Rodm	1.10	14.545	Ton	1,000	735		1,735	2,250
7870	#8 to #14	"	1.50	10.667	"	1,000	540		1,540	1,950
8000	Fittings, PVC type EB, elbow, 2" diameter	1 Elec	16	.500	Ea.	15.90	26.50		42.40	57.50
8200	3" diameter		14	.571		20.50	30.50		51	68
8400	4" diameter		12	.667		28.50	35.50		64	84.50
8420	5" diameter		10	.800		152	42.50		194.50	231
8440	6" diameter		9	.889		173	47.50		220.50	261
8500	Coupling, 2" diameter					.70			.70	.77
8600	3" diameter					3.22			3.22	3.54
8700	4" diameter					3.52			3.52	3.87
8720	5" diameter					6.65			6.65	7.35
8740	6" diameter					22.50			22.50	24.50
8800	Adapter, 2" diameter	1 Elec	26	.308		1.03	16.40		17.43	25.50
9000	3" diameter		20	.400		2.88	21.50		24.38	35
9200	4" diameter		16	.500		4.25	26.50		30.75	44.50
9220	5" diameter		13	.615		10.25	33		43.25	60.50
9240	6" diameter		10	.800		13.30	42.50		55.80	78.50
9400	End bell, 2" diameter		16	.500		1.31	26.50		27.81	41.50
9600	3" diameter		14	.571		6.65	30.50		37.15	53
9800	4" diameter		12	.667		4.27	35.50		39.77	57.50
9810	5" diameter		10	.800		13	42.50		55.50	78.50
9820	6" diameter		8	1		66	53.50		119.50	153
9830	5° angle coupling, 2" diameter		26	.308		19.55	16.40		35.95	46
9840	3" diameter		20	.400		29.50	21.50		51	64.50
9850	4" diameter		16	.500		15.25	26.50		41.75	57
9860	5" diameter		13	.615		14.75	33		47.75	65
9870	6" diameter		10	.800		23	42.50		65.50	89
9880	Expansion joint, 2" diameter		16	.500		33.50	26.50		60	77
9890	3" diameter		18	.444		64	23.50		87.50	106
9900	4" diameter		12	.667		85.50	35.50		121	147
9910	5" diameter		10	.800		114	42.50		156.50	190
9920	6" diameter		8	1		143	53.50		196.50	237
9930	Heat bender, 2" diameter					545			545	600
9940	6" diameter					1,850			1,850	2,025
9950	Cement, quart					19.55			19.55	21.50
9960	Nylon polyethylene pull rope, 1/4"	2 Elec	2000	.008	L.F.	.17	.43		.60	.83

33 71 23 – Insulators and Fittings

33 71 23.16 Post Insulators

		Crew	Daily Output	Labor-Hours	Unit	Material	2014 Bare Costs Labor	Equipment	Total	Total Incl O&P
0010	**POST INSULATORS**									
7400	Insulators, pedestal type	R-11	112	.500	Ea.	25.50	7.95		33.45	46.50
7490	See also line 33 71 39.13 1000									

33 71 26 – Transmission and Distribution Equipment

33 71 26.13 Capacitor Banks

		Crew	Daily Output	Labor-Hours	Unit	Material	2014 Bare Costs Labor	Equipment	Total	Total Incl O&P
0010	**CAPACITOR BANKS**									
1300	Station capacitors									
1350	Synchronous, 13 to 26 kV	R-11	3.11	18.006	MVAR	6,650	910	285	7,845	9,025
1360	46 kV		3.33	16.817		8,500	850	267	9,617	10,900
1370	69 kV		3.81	14.698		8,350	745	233	9,328	10,600
1380	161 kV		6.51	8.602		7,800	435	136	8,371	9,400
1390	500 kV		10.37	5.400		6,800	273	85.50	7,158.50	7,975

33 71 26 – Transmission and Distribution Equipment

33 71 26.13 Capacitor Banks

		Crew	Daily Output	Labor-Hours	Unit	Material	2014 Bare Costs Labor	2014 Bare Costs Equipment	Total	Total Incl O&P
1450	Static, 13 to 26 kV	R-11	3.11	18.006	MVAR	5,650	910	285	6,845	7,900
1460	46 kV		3.01	18.605		7,125	940	295	8,360	9,600
1470	69 kV		3.81	14.698		6,925	745	233	7,903	9,000
1480	161 kV		6.51	8.602		6,425	435	136	6,996	7,875
1490	500 kV		10.37	5.400		5,875	273	85.50	6,233.50	6,950
1600	Voltage regulators, 13 to 26 kV		.75	74.667	Ea.	252,500	3,775	1,175	257,450	285,000

33 71 26.23 Current Transformers

		Crew	Daily Output	Labor-Hours	Unit	Material	2014 Bare Costs Labor	2014 Bare Costs Equipment	Total	Total Incl O&P
0010	**CURRENT TRANSFORMERS**									
4050	Current transformers, 13 to 26 kV	R-11	14	4	Ea.	3,100	202	63.50	3,365.50	3,775
4060	46 kV		9.33	6.002		9,075	305	95	9,475	10,500
4070	69 kV		7	8		9,400	405	127	9,932	11,100
4080	161 kV		1.87	29.947		30,500	1,525	475	32,500	36,300

33 71 26.26 Potential Transformers

		Crew	Daily Output	Labor-Hours	Unit	Material	2014 Bare Costs Labor	2014 Bare Costs Equipment	Total	Total Incl O&P
0010	**POTENTIAL TRANSFORMERS**									
4100	Potential transformers, 13 to 26 kV	R-11	11.20	5	Ea.	4,450	253	79.50	4,782.50	5,350
4110	46 kV		8	7		9,125	355	111	9,591	10,700
4120	69 kV		6.22	9.003		9,675	455	143	10,273	11,400
4130	161 kV		2.24	25		20,900	1,275	395	22,570	25,300
4140	500 kV		1.40	40		62,500	2,025	635	65,160	72,500

33 71 39 – High-Voltage Wiring

33 71 39.13 Overhead High-Voltage Wiring

		Crew	Daily Output	Labor-Hours	Unit	Material	2014 Bare Costs Labor	2014 Bare Costs Equipment	Total	Total Incl O&P
0010	**OVERHEAD HIGH-VOLTAGE WIRING** R337116-60									
0100	Conductors, primary circuits									
0110	Material handling and spotting	R-5	9.78	8.998	W.Mile		420	155	575	805
0120	For river crossing, add		11	8			375	138	513	715
0150	Conductors, per wire, 210 to 636 kcmil		1.96	44.898		10,300	2,100	775	13,175	15,300
0160	795 to 954 kcmil		1.87	47.059		20,500	2,200	810	23,510	26,700
0170	1000 to 1600 kcmil		1.47	59.864		35,100	2,800	1,025	38,925	44,000
0180	Over 1600 kcmil		1.35	65.185		49,100	3,050	1,125	53,275	60,000
0200	For river crossing, add, 210 to 636 kcmil		1.24	70.968			3,300	1,225	4,525	6,350
0220	795 to 954 kcmil		1.09	80.734			3,750	1,400	5,150	7,200
0230	1000 to 1600 kcmil		.97	90.722			4,225	1,550	5,775	8,125
0240	Over 1600 kcmil		.87	101			4,725	1,750	6,475	9,050
0300	Joints and dead ends	R-8	6	8	Ea.	1,400	380	63.50	1,843.50	2,200
0400	Sagging	R-5	7.33	12.001	W.Mile		560	207	767	1,075
0500	Clipping, per structure, 69 kV	R-10	9.60	5	Ea.		252	65	317	450
0510	161 kV		5.33	9.006			455	117	572	810
0520	345 to 500 kV		2.53	18.972			955	247	1,202	1,700
0600	Make and install jumpers, per structure, 69 kV	R-8	3.20	15		380	710	119	1,209	1,625
0620	161 kV		1.20	40		765	1,900	320	2,985	4,050
0640	345 to 500 kV		.32	150		1,275	7,075	1,200	9,550	13,500
0700	Spacers	R-10	68.57	.700		76	35	9.10	120.10	147
0720	For river crossings, add	"	60	.800			40.50	10.40	50.90	72
0800	Installing pulling line (500 kV only)	R-9	1.45	44.138	W.Mile	755	1,950	264	2,969	4,075
0810	Disposal of surplus material, high voltage conductors	R-7	6.96	6.897	Mile		261	32.50	293.50	435
0820	With trailer mounted reel stands	"	13.71	3.501	"		133	16.40	149.40	222
0900	Insulators and hardware, primary circuits									
0920	Material handling and spotting, 69 kV	R-7	480	.100	Ea.		3.79	.47	4.26	6.30
0930	161 kV		685.71	.070			2.65	.33	2.98	4.43
0950	345 to 500 kV		960	.050			1.89	.23	2.12	3.17
1000	Disk insulators, 69 kV	R-5	880	.100		77.50	4.66	1.72	83.88	94.50
1020	161 kV		977.78	.090		89	4.19	1.55	94.74	106

33 71 39 – High-Voltage Wiring

33 71 39.13 Overhead High-Voltage Wiring	Crew	Daily Output	Labor-Hours	Unit	Material	2014 Bare Costs Labor	Equipment	Total	Total Incl O&P
1040 345 to 500 kV	R-5	1100	.080	Ea.	89	3.73	1.38	94.11	105
1060 See Section 33 71 23.16 for pin or pedestal insulator									
1100 Install disk insulator at river crossing, add									
1110 69 kV	R-5	586.67	.150	Ea.		7	2.58	9.58	13.40
1120 161 kV		880	.100			4.66	1.72	6.38	8.95
1140 345 to 500 kV		880	.100			.4.66	1.72	6.38	8.95
1150 Disposal of surplus material, high voltage insulators	R-7	41.74	1.150	Mile		43.50	5.40	48.90	73
1300 Overhead ground wire installation									
1320 Material handling and spotting	R-7	5.65	8.496	W.Mile		320	40	360	540
1340 Overhead ground wire	R-5	1.76	50		2,975	2,325	860	6,160	7,750
1350 At river crossing, add		1.17	75.214			3,500	1,300	4,800	6,725
1360 Disposal of surplus material, grounding wire		41.74	2.108	Mile		98.50	36.50	135	188
1400 Installing conductors, underbuilt circuits									
1420 Material handling and spotting	R-7	5.65	8.496	W.Mile		320	40	360	540
1440 Conductors, per wire, 210 to 636 kcmil	R-5	1.96	44.898		10,300	2,100	775	13,175	15,300
1450 795 to 954 kcmil		1.87	47.059		20,500	2,200	810	23,510	26,700
1460 1000 to 1600 kcmil		1.47	59.864		35,100	2,800	1,025	38,925	44,000
1470 Over 1600 kcmil		1.35	65.185		49,100	3,050	1,125	53,275	60,000
1500 Joints and dead ends	R-8	6	8	Ea.	1,400	380	63.50	1,843.50	2,200
1550 Sagging	R-5	8.80	10	W.Mile		465	172	637	895
1600 Clipping, per structure, 69 kV	R-10	9.60	5	Ea.		252	65	317	450
1620 161 kV		5.33	9.006			455	117	572	810
1640 345 to 500 kV		2.53	18.972			955	247	1,202	1,700
1700 Making and installing jumpers, per structure, 69 kV	R-8	5.87	8.177		380	385	65	830	1,075
1720 161 kV		.96	50		765	2,350	400	3,515	4,850
1740 345 to 500 kV		.32	150		1,275	7,075	1,200	9,550	13,500
1800 Spacers	R-10	96	.500		76	25	6.50	107.50	129
1810 Disposal of surplus material, conductors & hardware	R-7	6.96	6.897	Mile		261	32.50	293.50	435
2000 Insulators and hardware for underbuilt circuits									
2100 Material handling and spotting	R-7	1200	.040	Ea.		1.51	.19	1.70	2.54
2150 Disk insulators, 69 kV	R-8	600	.080		77.50	3.78	.64	81.92	92
2160 161 kV		686	.070		89	3.30	.56	92.86	103
2170 345 to 500 kV		800	.060		89	2.83	.48	92.31	102
2180 Disposal of surplus material, insulators & hardware	R-7	41.74	1.150	Mile		43.50	5.40	48.90	73
2300 Sectionalizing switches, 69 kV	R-5	1.26	69.841	Ea.	20,000	3,250	1,200	24,450	28,300
2310 161 kV		.80	110		22,600	5,125	1,900	29,625	34,700
2500 Protective devices		5.50	16		6,550	745	275	7,570	8,625
2600 Clearance poles, 8 poles per mile									
2650 In earth, 69 kV	R-5	1.16	75.862	Mile	5,400	3,525	1,300	10,225	12,700
2660 161 kV	"	.64	137		8,850	6,400	2,375	17,625	22,000
2670 345 to 500 kV	R-6	.48	183		10,600	8,550	7,150	26,300	32,500
2800 In rock, 69 kV	R-5	.69	127		5,400	5,950	2,200	13,550	17,300
2820 161 kV	"	.35	251		8,850	11,700	4,325	24,875	32,200
2840 345 to 500 kV	R-6	.24	366		10,600	17,100	14,300	42,000	53,000

33 72 Utility Substations

33 72 26 – Substation Bus Assemblies

33 72 26.13 Aluminum Substation Bus Assemblies	Crew	Daily Output	Labor-Hours	Unit	Material	2014 Bare Costs Labor	Equipment	Total	Total Incl O&P
0010 **ALUMINUM SUBSTATION BUS ASSEMBLIES**									
7300 Bus	R-11	590	.095	Lb.	2.73	4.80	1.50	9.03	11.85

33 72 33 – Control House Equipment

33 72 33.33 Raceway/Boxes for Utility Substations

33 72 33.33 Raceway/Boxes for Utility Substations	Crew	Daily Output	Labor-Hours	Unit	Material	Labor	Equipment	Total	Total Incl O&P
0010 **RACEWAY/BOXES FOR UTILITY SUBSTATIONS**									
7000 Conduit, conductors, and insulators									
7100 Conduit, metallic	R-11	560	.100	Lb.	2.50	5.05	1.59	9.14	12.10
7110 Non-metallic	"	800	.070	"	1.75	3.54	1.11	6.40	8.50
7190 See Section 26 05 33									
7200 Wire and cable	R-11	700	.080	Lb.	8.20	4.05	1.27	13.52	16.50
7290 See Section 26 05 19									

33 72 33.36 Cable Trays for Utility Substations

33 72 33.36 Cable Trays for Utility Substations	Crew	Daily Output	Labor-Hours	Unit	Material	Labor	Equipment	Total	Total Incl O&P
0010 **CABLE TRAYS FOR UTILITY SUBSTATIONS**									
7700 Cable tray	R-11	40	1.400	L.F.	23.50	71	22	116.50	157
7790 See Section 26 05 36									

33 72 33.43 Substation Backup Batteries

33 72 33.43 Substation Backup Batteries	Crew	Daily Output	Labor-Hours	Unit	Material	Labor	Equipment	Total	Total Incl O&P
0010 **SUBSTATION BACKUP BATTERIES**									
9120 Battery chargers	R-11	11.20	5	Ea.	3,775	253	79.50	4,107.50	4,650
9200 Control batteries	"	14	4	K.A.H.	87	202	63.50	352.50	470

33 72 33.46 Substation Converter Stations

33 72 33.46 Substation Converter Stations	Crew	Daily Output	Labor-Hours	Unit	Material	Labor	Equipment	Total	Total Incl O&P
0010 **SUBSTATION CONVERTER STATIONS**									
9000 Station service equipment									
9100 Conversion equipment									
9110 Station service transformers	R-11	5.60	10	Ea.	91,500	505	159	92,164	101,500

33 73 Utility Transformers

33 73 23 – Dry-Type Utility Transformers

33 73 23.20 Primary Transformers

33 73 23.20 Primary Transformers	Crew	Daily Output	Labor-Hours	Unit	Material	Labor	Equipment	Total	Total Incl O&P
0010 **PRIMARY TRANSFORMERS**									
1000 Main conversion equipment									
1050 Power transformers, 13 to 26 kV	R-11	1.72	32.558	MVA	23,900	1,650	515	26,065	29,300
1060 46 kV		3.50	16		22,400	810	254	23,464	26,100
1070 69 kV		3.11	18.006		19,100	910	285	20,295	22,700
1080 110 kV		3.29	17.021		18,100	860	270	19,230	21,600
1090 161 kV		4.31	12.993		16,800	655	206	17,661	19,700
1100 500 kV		7	8		16,800	405	127	17,332	19,200
1200 Grounding transformers		3.11	18.006	Ea.	105,500	910	285	106,695	117,500

33 75 High-Voltage Switchgear and Protection Devices

33 75 13 – Air High-Voltage Circuit Breaker

33 75 13.13 Air High-Voltage Circuit Breaker

33 75 13.13 High-Voltage Circuit Breaker, Air		Crew	Daily Output	Labor-Hours	Unit	Material	2014 Bare Costs Labor	Equipment	Total	Total Incl O&P
0010	HIGH-VOLTAGE CIRCUIT BREAKER, AIR									
2100	Air circuit breakers, 13 to 26 kV	R-11	.56	100	Ea.	59,000	5,050	1,575	65,625	74,500
2110	161 kV	"	.24	233	"	253,500	11,800	3,700	269,000	301,000

33 75 16 – Oil High-Voltage Circuit Breaker

33 75 16.13 Oil Circuit Breaker

33 75 16.13 Oil Circuit Breaker		Crew	Daily Output	Labor-Hours	Unit	Material	2014 Bare Costs Labor	Equipment	Total	Total Incl O&P
0010	OIL CIRCUIT BREAKER									
2000	Power circuit breakers									
2050	Oil circuit breakers, 13 to 26 kV	R-11	1.12	50	Ea.	56,500	2,525	795	59,820	66,500
2060	46 kV		.75	74.667		82,500	3,775	1,175	87,450	97,500
2070	69 kV		.45	124		191,500	6,300	1,975	199,775	222,000
2080	161 kV		.16	350		292,500	17,700	5,550	315,750	354,000
2090	500 kV		.06	933		1,094,500	47,200	14,800	1,156,500	1,291,500

33 75 19 – Gas High-Voltage Circuit Breaker

33 75 19.13 Gas Circuit Breaker

33 75 19.13 Gas Circuit Breaker		Crew	Daily Output	Labor-Hours	Unit	Material	2014 Bare Costs Labor	Equipment	Total	Total Incl O&P
0010	GAS CIRCUIT BREAKER									
2000	Power circuit breakers									
2150	Gas circuit breakers, 13 to 26 kV	R-11	.56	100	Ea.	221,500	5,050	1,575	228,125	253,000
2160	161 kV		.08	700		291,000	35,400	11,100	337,500	385,500
2170	500 kV		.04	1400		1,024,500	71,000	22,200	1,117,700	1,258,000

33 75 23 – Vacuum High-Voltage Circuit Breaker

33 75 23.13 Vacuum Circuit Breaker

33 75 23.13 Vacuum Circuit Breaker		Crew	Daily Output	Labor-Hours	Unit	Material	2014 Bare Costs Labor	Equipment	Total	Total Incl O&P
0010	VACUUM CIRCUIT BREAKER									
2000	Power circuit breakers									
2200	Vacuum circuit breakers, 13 to 26 kV	R-11	.56	100	Ea.	48,300	5,050	1,575	54,925	62,500

33 75 36 – High-Voltage Utility Fuses

33 75 36.13 Fuses

33 75 36.13 Fuses		Crew	Daily Output	Labor-Hours	Unit	Material	2014 Bare Costs Labor	Equipment	Total	Total Incl O&P
0010	FUSES									
8000	Protective equipment									
8250	Fuses, 13 to 26 kV	R-11	18.67	2.999	Ea.	2,050	152	47.50	2,249.50	2,525
8260	46 kV		11.20	5		2,325	253	79.50	2,657.50	3,050
8270	69 kV		8	7		2,425	355	111	2,891	3,325
8280	161 kV		4.67	11.991		3,075	605	190	3,870	4,500

33 75 39 – High-Voltage Surge Arresters

33 75 39.13 Surge Arresters

33 75 39.13 Surge Arresters		Crew	Daily Output	Labor-Hours	Unit	Material	2014 Bare Costs Labor	Equipment	Total	Total Incl O&P
0010	SURGE ARRESTERS									
8000	Protective equipment									
8050	Lightning arresters, 13 to 26 kV	R-11	18.67	2.999	Ea.	1,525	152	47.50	1,724.50	1,950
8060	46 kV		14	4		4,125	202	63.50	4,390.50	4,925
8070	69 kV		11.20	5		5,300	253	79.50	5,632.50	6,300
8080	161 kV		5.60	10		7,300	505	159	7,964	8,975
8090	500 kV		1.40	40		24,800	2,025	635	27,460	31,000

33 75 43 – Shunt Reactors

33 75 43.13 Reactors

33 75 43.13 Reactors		Crew	Daily Output	Labor-Hours	Unit	Material	2014 Bare Costs Labor	Equipment	Total	Total Incl O&P
0010	REACTORS									
8000	Protective equipment									
8150	Reactors and resistors, 13 to 26 kV	R-11	28	2	Ea.	3,450	101	31.50	3,582.50	3,975
8160	46 kV		4.31	12.993		10,400	655	206	11,261	12,700
8170	69 kV		2.80	20		17,000	1,000	315	18,315	20,600
8180	161 kV		2.24	25		19,400	1,275	395	21,070	23,700

33 75 High-Voltage Switchgear and Protection Devices

33 75 43 – Shunt Reactors

33 75 43.13 Reactors		Crew	Daily Output	Labor-Hours	Unit	Material	2014 Bare Costs Labor	Equipment	Total	Total Incl O&P
8190	500 kV	R-11	.08	700	Ea.	78,000	35,400	11,100	124,500	151,000

33 75 53 – High-Voltage Switches

33 75 53.13 Switches

33 75 53.13 Switches		Crew	Daily Output	Labor-Hours	Unit	Material	2014 Bare Costs Labor	Equipment	Total	Total Incl O&P
0010	**SWITCHES**									
3000	Disconnecting switches									
3050	Gang operated switches									
3060	Manual operation, 13 to 26 kV	R-11	1.65	33.939	Ea.	13,600	1,725	540	15,865	18,100
3070	46 kV		1.12	50		22,600	2,525	795	25,920	29,500
3080	69 kV		.80	70		25,300	3,550	1,100	29,950	34,500
3090	161 kV		.56	100		30,500	5,050	1,575	37,125	42,900
3100	500 kV		.14	400		83,000	20,200	6,350	109,550	129,000
3110	Motor operation, 161 kV		.51	109		45,900	5,550	1,750	53,200	61,000
3120	500 kV		.28	200		127,000	10,100	3,175	140,275	158,000
3250	Circuit switches, 161 kV	▼	.41	136	▼	91,500	6,900	2,175	100,575	113,500
3300	Single pole switches									
3350	Disconnecting switches, 13 to 26 kV	R-11	28	2	Ea.	11,900	101	31.50	12,032.50	13,300
3360	46 kV		8	7		21,000	355	111	21,466	23,800
3370	69 kV		5.60	10		23,300	505	159	23,964	26,600
3380	161 kV		2.80	20		89,000	1,000	315	90,315	99,500
3390	500 kV		.22	254		253,000	12,900	4,025	269,925	302,500
3450	Grounding switches, 46 kV		5.60	10		32,100	505	159	32,764	36,200
3460	69 kV		3.73	15.013		33,300	760	238	34,298	38,100
3470	161 kV		2.24	25		35,300	1,275	395	36,970	41,100
3480	500 kV	▼	.62	90.323	▼	43,800	4,575	1,425	49,800	56,500

33 79 Site Grounding

33 79 83 – Site Grounding Conductors

33 79 83.13 Grounding Wire, Bar, and Rod

33 79 83.13 Grounding Wire, Bar, and Rod		Crew	Daily Output	Labor-Hours	Unit	Material	2014 Bare Costs Labor	Equipment	Total	Total Incl O&P
0010	**GROUNDING WIRE, BAR, AND ROD**									
7500	Grounding systems	R-11	280	.200	Lb.	10.80	10.10	3.17	24.07	30.50

33 81 Communications Structures

33 81 13 – Communications Transmission Towers

33 81 13.10 Radio Towers

33 81 13.10 Radio Towers		Crew	Daily Output	Labor-Hours	Unit	Material	2014 Bare Costs Labor	Equipment	Total	Total Incl O&P
0010	**RADIO TOWERS**									
0020	Guyed, 50' H, 40 lb. sect., 70 MPH basic wind spd.	2 Sswk	1	16	Ea.	2,750	820		3,570	4,475
0100	Wind load 90 MPH basic wind speed	"	1	16		3,700	820		4,520	5,525
0300	190' high, 40 lb. section, wind load 70 MPH basic wind speed	K-2	.33	72.727		9,550	3,400	1,000	13,950	17,400
0400	200' high, 70 lb. section, wind load 90 MPH basic wind speed		.33	72.727		15,900	3,400	1,000	20,300	24,400
0600	300' high, 70 lb. section, wind load 70 MPH basic wind speed		.20	120		25,500	5,625	1,675	32,800	39,400
0700	270' high, 90 lb. section, wind load 90 MPH basic wind speed		.20	120		27,600	5,625	1,675	34,900	41,700
0800	400' high, 100 lb. section, wind load 70 MPH basic wind speed		.14	171		38,100	8,050	2,375	48,525	58,000
0900	Self-supporting, 60' high, wind load 70 MPH basic wind speed		.80	30		4,500	1,400	415	6,315	7,800
0910	60' high, wind load 90 MPH basic wind speed		.45	53.333		5,275	2,500	740	8,515	10,900
1000	120' high, wind load 70 MPH basic wind speed		.40	60		10,000	2,825	835	13,660	16,700
1200	190' high, wind load 90 MPH basic wind speed	▼	.20	120		28,300	5,625	1,675	35,600	42,500
2000	For states west of Rocky Mountains, add for shipping				▼	10%				

351

Division Notes

	CREW	DAILY OUTPUT	LABOR-HOURS	UNIT	BARE COSTS				TOTAL INCL O&P
					MAT.	LABOR	EQUIP.	TOTAL	

Estimating Tips

34 11 00 Rail Tracks

This subdivision includes items that may involve either repair of existing, or construction of new, railroad tracks. Additional preparation work, such as the roadbed earthwork, would be found in Division 31. Additional new construction siding and turnouts are found in Subdivision 34 72. Maintenance of railroads is found under 34 01 23 Operation and Maintenance of Railways.

34 40 00 Traffic Signals

This subdivision includes traffic signal systems. Other traffic control devices such as traffic signs are found in Subdivision 10 14 53 Traffic Signage.

34 70 00 Vehicle Barriers

This subdivision includes security vehicle barriers, guide and guard rails, crash barriers, and delineators. The actual maintenance and construction of concrete and asphalt pavement is found in Division 32.

Reference Numbers

Reference numbers are shown in shaded boxes at the beginning of some major classifications. These numbers refer to related items in the Reference Section. The reference information may be an estimating procedure, an alternate pricing method, or technical information.

Note: Not all subdivisions listed here necessarily appear in this publication.

Division 34 - Transportation

34 43 13 – Airfield Signals

34 43 13.10 Airport Lighting	Crew	Daily Output	Labor-Hours	Unit	Material	2014 Bare Costs Labor	Equipment	Total	Total Incl O&P
0010 **AIRPORT LIGHTING**									
0100 Runway centerline, bidir., semi-flush, 200 W, w/shallow insert base	R-22	12.40	3.006	Ea.	1,625	136		1,761	1,975
0110 for mounting in base housing		18.60	2.004		815	91		906	1,050
0120 Flush, 200 W, w/shallow insert base		12.40	3.006		1,600	136		1,736	1,975
0130 for mounting in base housing		18.64	2		855	91		946	1,075
0140 45 W, for mounting in base housing		18.64	2		850	91		941	1,075
0150 Touchdown zone light, unidirectional, 200 W, w/shallow insert base		12.40	3.006		1,350	136		1,486	1,675
0160 115 W		12.40	3.006		1,300	136		1,436	1,625
0170 Bi-directional, 62 W		12.40	3.006		1,350	136		1,486	1,700
0180 Unidirectional, 200 W, for mounting in base housing		18.60	2.004		600	91		691	800
0190 115 W		18.60	2.004		645	91		736	850
0200 Bi-directional, 62 W		18.60	2.004		720	91		811	930
0210 Runway edge & threshold light, bidir., 200 W, for base housing		9.36	3.983		920	181		1,101	1,300
0220 300 W		9.36	3.983		1,300	181		1,481	1,700
0230 499 W		7.44	5.011		2,150	227		2,377	2,725
0240 Threshold & approach light, unidir., 200 W, for base housing		9.36	3.983		535	181		716	865
0250 499 W		7.44	5.011		650	227		877	1,050
0260 Runway edge, bi-directional, 2-115 W, for base housing		12.40	3.006		1,425	136		1,561	1,750
0270 2-185 W		12.40	3.006		1,575	136		1,711	1,925
0280 Runway threshold & end, bidir., 2-115 W, for base housing		12.40	3.006		1,550	136		1,686	1,925
0370 45 W, flush, for mounting in base housing		18.64	2		765	91		856	975
0380 115 W		18.64	2		780	91		871	995
0400 Transformer, 45 W		37.20	1.002		167	45.50		212.50	253
0410 65 W		37.20	1.002		172	45.50		217.50	258
0420 100 W		37.20	1.002		213	45.50		258.50	305
0430 200 W, 6.6/6.6 A		37.20	1.002		209	45.50		254.50	298
0440 20/6.6 A		24.80	1.503		223	68.50		291.50	350
0450 500 W		24.80	1.503		292	68.50		360.50	425
0460 Base housing, 24" deep, 12" diameter		18.64	2		470	91		561	655
0470 15" diameter		18.64	2		1,025	91		1,116	1,250
0480 Connection kit for 1 conductor cable		24.80	1.503		33.50	68.50		102	140
0500 Elevated runway marker, high intensity quartz, edge, 115 W		14.88	2.505		330	114		444	530
0510 175 W		14.88	2.505		340	114		454	540
0520 Threshold, 115 W		14.88	2.505		365	114		479	575
0530 Edge, 200 W, w/base and transformer		11.85	3.146		1,050	143		1,193	1,375
0540 Elevated runway light, quartz, runway edge, 45 W		14.88	2.505		222	114		336	415
0550 Threshold, 115 W		14.88	2.505		365	114		479	575
0560 Taxiway, 30 W		14.88	2.505		203	114		317	395
0570 Fixture, 200 W, w/base and transformer		11.85	3.146		1,050	143		1,193	1,375
0580 Elevated threshold marker, quartz, 115 W		14.88	2.505		465	114		579	680
0590 175 W		14.88	2.505		465	114		579	680

34 43 23 – Weather Observation Equipment

34 43 23.16 Airfield Wind Cones

	Crew	Daily Output	Labor-Hours	Unit	Material	2014 Bare Costs Labor	Equipment	Total	Total Incl O&P
0010 **AIRFIELD WIND CONES**									
1200 Wind cone, 12' lighted assembly, rigid, w/obstruction light	R-21	1.36	24.118	Ea.	9,025	1,275	44.50	10,344.50	11,900
1210 Without obstruction light		1.52	21.579		7,575	1,150	40	8,765	10,100
1220 Unlighted assembly, w/obstruction light		1.68	19.524		8,250	1,050	36	9,336	10,700
1230 Without obstruction light		1.84	17.826		7,575	950	33	8,558	9,775
1240 Wind cone slip fitter, 2-1/2" pipe		21.84	1.502		100	80	2.78	182.78	233
1250 Wind cone sock, 12' x 3', cotton		6.56	5		545	267	9.25	821.25	1,000
1260 Nylon		6.56	5		555	267	9.25	831.25	1,025

Estimating Tips

- When estimating costs for the installation of electrical power generation equipment, factors to review include access to the job site, access and setting at the installation site, required connections, uncrating pads, anchors, leveling, final assembly of the components, and temporary protection from physical damage, including from exposure to the environment.

- Be aware of the cost of equipment supports, concrete pads, and vibration isolators, and cross-reference to other trades' specifications. Also, review site and structural drawings for items that must be included in the estimates.

- It is important to include items that are not documented in the plans and specifications but must be priced. These items include, but are not limited to, testing, dust protection, roof penetration, core drilling concrete floors and walls, patching, cleanup, and final adjustments. Add a contingency or allowance for utility company fees for power hookups, if needed.

- The project size and scope of electrical power generation equipment will have a significant impact on cost. The intent of RSMeans cost data is to provide a benchmark cost so that owners, engineers, and electrical contractors will have a comfortable number with which to start a project. Additionally, there are many websites available to use for research and to obtain a vendor's quote to finalize costs.

Reference Numbers

Reference numbers are shown in shaded boxes at the beginning of some major classifications. These numbers refer to related items in the Reference Section. The reference information may be an estimating procedure, an alternate pricing method, or technical information.

Note: Not all subdivisions listed here necessarily appear in this publication.

48 15 Wind Energy Electrical Power Generation Equipment

48 15 13 – Wind Turbines

48 15 13.50 Wind Turbines and Components		Crew	Daily Output	Labor-Hours	Unit	Material	2014 Bare Costs Labor	Equipment	Total	Total Incl O&P
0010	**WIND TURBINES & COMPONENTS**									
0500	Complete system, grid connected									
1000	20 kW, 31' dia, incl. labor & material, min	G			System					49,900
1010	Maximum	G								92,000
1500	10 kW, 23' dia, incl. labor & material	G								74,000
2000	2.4 kW, 12' dia, incl. labor & material	G								18,000
2900	Component system									
3000	Turbine, 400 watt, 3' dia	G	1 Elec	3.41	2.346	Ea.	755	125		880	1,025
3100	600 watt, 3' dia	G		2.56	3.125		955	167		1,122	1,300
3200	1000 watt, 9' dia	G		2.05	3.902		3,825	208		4,033	4,525
3400	Mounting hardware									
3500	30' guyed tower kit	G	2 Clab	5.12	3.125	Ea.	455	115		570	675
3505	3' galvanized helical earth screw	G	1 Clab	8	1		53	36.50		89.50	115
3510	Attic mount kit	G	1 Rofc	2.56	3.125		133	122		255	355
3520	Roof mount kit	G	1 Clab	3.41	2.346		118	86		204	262
8900	Equipment									
9100	DC to AC inverter for, 48 V, 4,000 watt	G	1 Elec	2	4	Ea.	2,750	213		2,963	3,350

48 18 Fuel Cell Electrical Power Generation Equipment

48 18 13 – Electrical Power Generation Fuel Cells

48 18 13.10 Fuel Cells

		Crew	Daily Output	Labor-Hours	Unit	Material	2014 Bare Costs Labor	Equipment	Total	Total Incl O&P
0010	**FUEL CELLS**									
2001	Complete system, natural gas, installed, 200 kW	G			System					1,150,000
2002	400 kW	G								2,168,000
2003	600 kW	G								3,060,000
2004	800 kW	G								4,000,000
2005	1000 kW	G								5,005,000
2010	Comm type, for battery charging, uses hydrogen, 100 watt, 12 V	G			Ea.	2,075			2,075	2,300
2020	500 watt, 12 V	G				4,275			4,275	4,700
2030	1000 watt, 48 V	G				7,000			7,000	7,700
2040	Small, for demonstration or battery charging units, 12 V	G				188			188	207
2050	Spare anode set for	G				42.50			42.50	46.50
2060	3 watts, uses hydrogen gas	G				435			435	475
2070	6 watts, uses hydrogen gas	G				780			780	860
2080	10 watts, uses hydrogen gas	G				1,125			1,125	1,250
2090	One month rental of 261 C.F. hydrogen gas cylinder for, min	G				21.50			21.50	23.50
2100	Maximum	G				585			585	645

Assemblies Section

Table of Contents

Table No.	Page
D SERVICES	
D2020 Domestic Water Distribution	
D2020 210 Electric Water Heaters - Residential Systems	364
D2020 240 Electric Water Heaters - Commercial Systems	365
D3020 Heat Generating Systems	
D3020 102 Small Heating Systems, Hydronic, Electric Boilers	366
D3020 104 Large Heating Systems, Hydronic, Electric Boilers	367
D4090 Other Fire Protection Systems	
D4090 910 Fire Suppression Unit Components	368
D4090 920 FM200 Systems	368
D5010 Electrical Service/Distribution	
D5010 110 High Voltage Shielded Conductors	369
D5010 120 Overhead Electric Service, 3 Phase - 4 Wire	370
D5010 130 Underground Electric Service	372
D5010 230 Feeder Installation	374
D5010 240 Switchgear	375
D5010 250 Panelboard	376
D5020 Lighting and Branch Wiring	
D5020 110 Receptacle (by Wattage)	378
D5020 115 Receptacles, Floor	379
D5020 120 Receptacles & Wall Switches	380
D5020 125 Receptacles & Switches by Each	382
D5020 130 Wall Switch by Sq. Ft.	383
D5020 135 Miscellaneous Power	384
D5020 140 Central A. C. Power (by Wattage)	385
D5020 145 Motor Installation	386
D5020 155 Motor Feeder	388
D5020 160 Magnetic Starter	389
D5020 165 Safety Switches	390
D5020 170 Motor Connections	391
D5020 175 Motor & Starter	392
D5020 208 Fluorescent Fixtures (by Type)	398
D5020 210 Fluorescent Fixtures (by Wattage)	399
D5020 214 Incandescent Fixture (by Type)	400
D5020 216 Incandescent Fixture (by Wattage)	401
D5020 218 Fluorescent Fixture, High Bay, 8'-10' (by Wattage)	402
D5020 220 H.I.D. Fixture, High Bay, 8'-10' (by Type)	403
D5020 222 H.I.D. Fixture, High Bay, 8'-10' (by Wattage)	404
D5020 224 H.I.D. Fixture, High Bay, 16' (by Type)	405

Table No.	Page
D5020 226 H.I.D. Fixture, High Bay, 16' (by Wattage)	406
D5020 228 H.I.D. Fixture, High Bay, 20' (by Type)	407
D5020 230 H.I.D. Fixture, High Bay, 20' (by Wattage)	408
D5020 232 H.I.D. Fixture, High Bay, 30' (by Type)	409
D5020 234 H.I.D. Fixture, High Bay, 30' (by Wattage)	410
D5020 236 H.I.D. Fixture, Low Bay, 8'-10' (by Type)	411
D5020 238 H.I.D. Fixture, Low Bay, 8'-10' (by Wattage)	412
D5020 240 H.I.D. Fixture, Low Bay, 16' (by Type)	413
D5020 242 H.I.D. Fixture, Low Bay, 16' (by Wattage)	414
D5020 290 Daylight Dimming System	415
D5020 295 Lighting On/Off Control System	416
D5030 Communications and Security	
D5030 310 Telephone Systems	417
D5030 810 Security & Detection Systems	417
D5030 910 Communication & Alarm Systems	418
D5030 920 Data Communication	419
D5090 Other Electrical Systems	
D5090 210 Generators (by kW)	420
D5090 420 Photovoltaic Power System, Stand Alone	421
D5090 430 Photovoltaic Power System, Grid Connected	423
D5090 Energy Monitoring Systems	
D5090 480 Energy Monitoring Systems	424
D5090 Other Electrical Systems	
D5090 510 Electric Baseboard Radiation (Low Density)	425
D5090 520 Electric Baseboard Radiation (Medium Density)	426
G BUILDING SITEWORK	
G1030 Site Earthwork	
G1030 805 Trenching Common Earth	428
G1030 806 Trenching Loam & Sandy Clay	430
G1030 807 Trenching Sand & Gravel	432
G4010 Electrical Distribution	
G4010 312 Underground Power Feed	434
G4010 320 Underground Electrical Conduit	435
G4020 Site Lighting	
G4020 110 Site Lighting	438
G4020 210 Light Pole (Installed)	440

357

How RSMeans Assemblies Data Works

Assemblies estimating provides a fast and reasonably accurate way to develop construction costs. An assembly is the grouping of individual work items, with appropriate quantities, to provide a cost for a major construction component in a convenient unit of measure.

An assemblies estimate is often used during early stages of design development to compare the cost impact of various design alternatives on total building cost.

Assemblies estimates are also used as an efficient tool to verify construction estimates.

Assemblies estimates do not require a completed design or detailed drawings. Instead, they are based on the general size of the structure and other known parameters of the project. The degree of accuracy of an assemblies estimate is generally within +/- 15%.

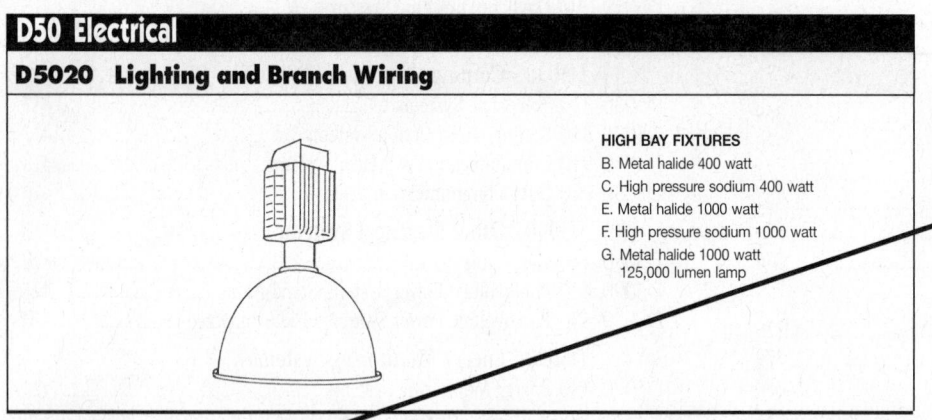

D50 Electrical

D5020 Lighting and Branch Wiring

HIGH BAY FIXTURES
B. Metal halide 400 watt
C. High pressure sodium 400 watt
E. Metal halide 1000 watt
F. High pressure sodium 1000 watt
G. Metal halide 1000 watt 125,000 lumen lamp

RSMeans assemblies are identified by a **unique 12-character identifier**. The assemblies are numbered using UNIFORMAT II, ASTM Standard E1557. The first 6 characters represent this system to Level 4. The last 6 characters represent further breakdown by RSMeans in order to arrange items in understandable groups of similar tasks. Line numbers are consistent across all RSMeans publications, so a line number in any RSMeans assemblies data set will always refer to the same work.

System Components	QUANTITY	UNIT	COST PER S.F.		
			MAT.	INST.	TOTAL
SYSTEM D5020 222 0240 HIGH INTENSITY DISCHARGE FIXTURE, 8'-10' ABOVE WORK PLANE 1 WATT/S.F., TYPE B, 29 FC, 2 FIXTURES/1000 S.F.					
Steel intermediate conduit, (IMC) 1/2" diam	.100	L.F.	.19	.64	.83
Wire, 600V, type THWN-THHN, copper, solid, #10	.002	C.L.F.	.05	.13	.18
Steel outlet box 4" concrete	.002	Ea.	.03	.06	.09
Steel outlet box plate with stud, 4" concrete	.002	Ea.	.02	.02	.04
Metal halide, hi bay, aluminum reflector, 400 W lamp	.002	Ea.	.90	.56	1.46
TOTAL			1.19	1.41	2.60

Information is available in the Reference Section to assist the estimator with estimating procedures, alternate pricing methods, and additional technical information.

The **Reference Box** indicates the exact location of this information in the Reference Section at the back of the book. The "R" stands for "reference," and the remaining characters are the RSMeans line numbers.

D5020 222	H.I.D. Fixture, High Bay, 8'-10' (by Wattage)		COST PER S.F.		
			MAT.	INST.	TOTAL
0190	High intensity discharge fixture, 8'-10' above work plane				
0240	1 watt/S.F., type B, 29 FC, 2 fixtures/1000 S.F.		1.19	1.41	2.60
0280	Type C, 54 FC, 2 fixtures/1000 S.F.		1.01	1.06	2.07
0400	2 watt/S.F., type B, 59 FC, 4 fixtures/1000 S.F.		2.34	2.72	5.06
0440	Type C, 108 FC, 4 fixtures/1000 S.F.	R265723 -05	2.01	2.07	4.08
0560	3 watt/S.F., type B, 103 FC, 7 fixtures/1000 S.F.		3.98	4.34	8.32
0600	Type C, 189 FC, 6 fixtures/1000 S.F.		3.24	2.68	5.92
0720	4 watt/S.F., type B, 133 FC, 9 fixtures/1000 S.F.		5.15	5.70	10.85
0760	Type C, 243 FC, 9 fixtures/1000 S.F.		4.51	4.61	9.12
0880	5 watt/S.F., type B, 162 FC, 11 fixtures/1000 S.F.		6.35	7.05	13.40
0920	Type C, 297 FC, 11 fixtures/1000 S.F.		5.55	5.70	11.25

RSMeans assemblies **Narrative descriptions** are shown in a hierarchical structure to make them readable. In order to read a complete description, read up through the indents to the top of the section. Include everything that is above and to the left that is not contradicted by information below.

For supplemental customizable square foot estimating forms, visit: **http://www.reedconstructiondata.com/rsmeans/assemblies**

Most assemblies consist of three major elements: a graphic, the system components, and the cost data itself. The **Graphic** is a visual representation showing the typical appearance of the assembly in question, frequently accompanied by additional explanatory technical information describing the class of items. The **System Components** is a listing of the individual tasks that make up the assembly, including the quantity and unit of measure for each item, along with the cost of material and installation. The **Assemblies Data** below lists prices for other similar systems with dimensional and/or size variations.

All RSMeans assemblies costs represent the cost for the installing contractor. An allowance for profit has been added to all material, labor, and equipment rental costs. A markup for labor burdens, including workers' compensation, fixed overhead, and business overhead, is included with installation costs.

The data published in RSMeans print books represents a "national average" cost. This data should be modified to the project location using the **City Cost Indexes** or **Location Factors** tables found in the Reference Section in the back of the book.

D50 Electrical

D5020 Lighting and Branch Wiring

HIGH BAY FIXTURES
B. Metal halide 400 watt
C. High pressure sodium 400 watt
E. Metal halide 1000 watt
F. High pressure sodium 1000 watt
G. Metal halide 1000 watt
125,000 lumen lamp

All RSMeans assemblies data includes a typical **Unit of Measure** used for estimating that item. For instance, while the unit of measure for the graphic shown here is Each, some A/C systems are estimated by the square foot (S.F.). The estimator needs to take special care that the unit in the data matches the unit in the takeoff. Abbreviations can be found in the Reference Section.

System Components	QUANTITY	UNIT	COST PER S.F.		
			MAT.	INST.	TOTAL
SYSTEM D5020 222 0240					
HIGH INTENSITY DISCHARGE FIXTURE, 8'-10' ABOVE WORK PLANE					
1 WATT/S.F., TYPE B, 29 FC, 2 FIXTURES/1000 S.F.					
Steel intermediate conduit, (IMC) 1/2" diam	.100	L.F.	.19	.64	.83
Wire, 600V, type THWN-THHN, copper, solid, #10	.002	C.L.F.	.05	.13	.18
Steel outlet box 4" concrete	.002	Ea.	.03	.06	.09
Steel outlet box plate with stud, 4" concrete	.002	Ea.	.02	.02	.04
Metal halide, hi bay, aluminum reflector, 400 W lamp	.002	Ea.	.90	.56	1.46
TOTAL			1.19	1.41	2.60

System components are listed separately to detail what is included in the development of the total system price.

D5020 222	H.I.D. Fixture, High Bay, 8'-10' (by Wattage)		COST PER S.F.		
			MAT.	INST.	TOTAL
0190	High intensity discharge fixture, 8'-10' above work plane				
0240	1 watt/S.F., type B, 29 FC, 2 fixtures/1000 S.F.		1.19	1.41	2.60
0280	Type C, 54 FC, 2 fixtures/1000 S.F.		1.01	1.06	2.07
0400	2 watt/S.F., type B, 59 FC, 4 fixtures/1000 S.F.		2.34	2.72	5.06
0440	Type C, 108 FC, 4 fixtures/1000 S.F.	R265723 -05	2.01	2.07	4.08
0560	3 watt/S.F., type B, 103 FC, 7 fixtures/1000 S.F.		3.98	4.34	8.32
0600	Type C, 189 FC, 6 fixtures/1000 S.F.		3.24	2.68	5.92
0720	4 watt/S.F., type B, 133 FC, 9 fixtures/1000 S.F.		5.15	5.70	10.85
0760	Type C, 243 FC, 9 fixtures/1000 S.F.		4.51	4.61	9.12
0880	5 watt/S.F., type B, 162 FC, 11 fixtures/1000 S.F.		6.35	7.05	13.40
0920	Type C, 297 FC, 11 fixtures/1000 S.F.		5.55	5.70	11.25

How RSMeans Assemblies Data Works (Continued)

Sample Estimate

This sample demonstrates the elements of an estimate, including a tally of the RSMeans data lines. Published assemblies costs include all markups for labor burden and profit for the installing contractor. This estimate adds a summary of the markups applied by a general contractor on the installing contractors' work. These figures represent the total cost to the owner. The RSMeans location factor is added at the bottom of the estimate to adjust the cost of the work to a specific location

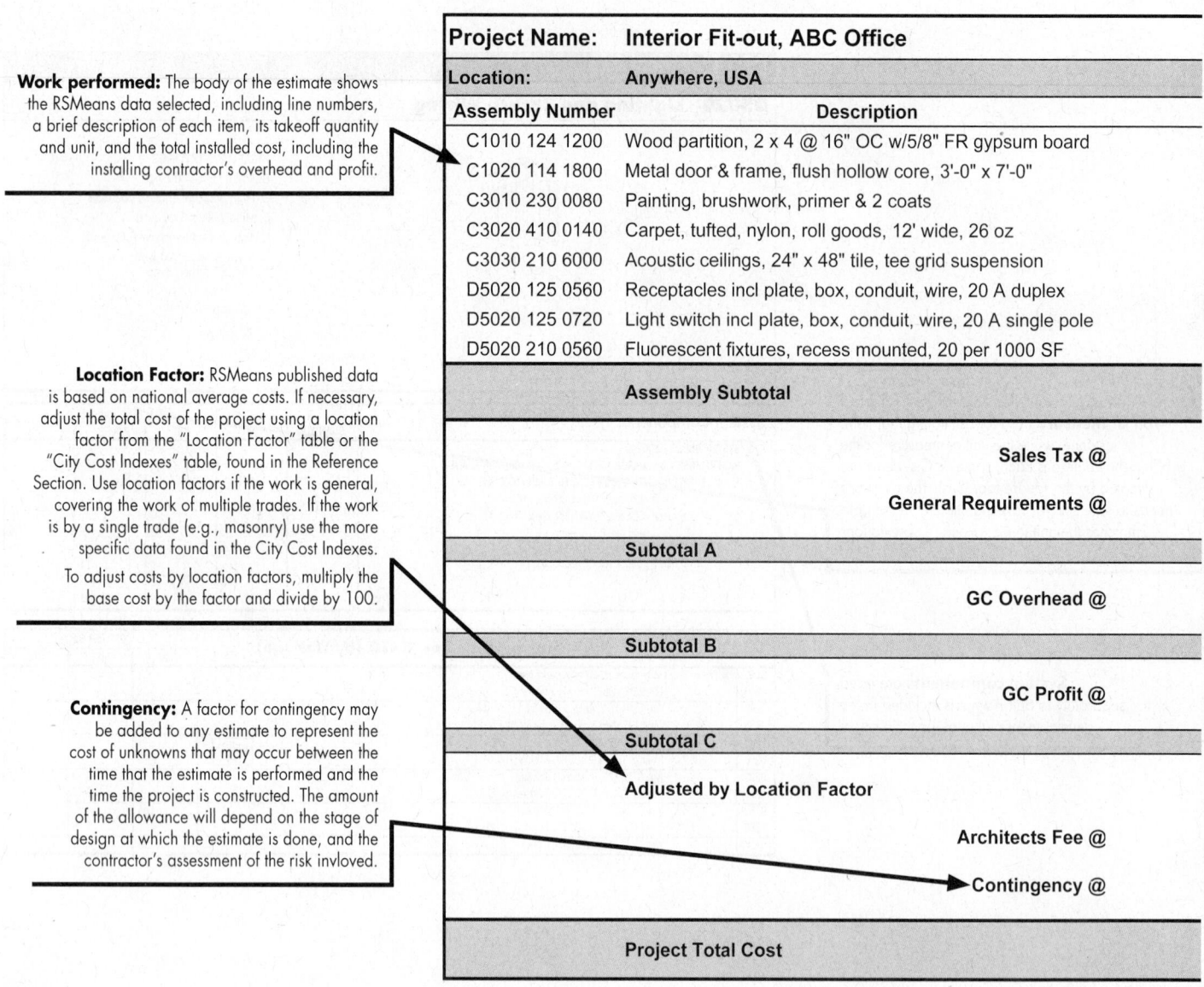

Work performed: The body of the estimate shows the RSMeans data selected, including line numbers, a brief description of each item, its takeoff quantity and unit, and the total installed cost, including the installing contractor's overhead and profit.

Location Factor: RSMeans published data is based on national average costs. If necessary, adjust the total cost of the project using a location factor from the "Location Factor" table or the "City Cost Indexes" table, found in the Reference Section. Use location factors if the work is general, covering the work of multiple trades. If the work is by a single trade (e.g., masonry) use the more specific data found in the City Cost Indexes.

To adjust costs by location factors, multiply the base cost by the factor and divide by 100.

Contingency: A factor for contingency may be added to any estimate to represent the cost of unknowns that may occur between the time that the estimate is performed and the time the project is constructed. The amount of the allowance will depend on the stage of design at which the estimate is done, and the contractor's assessment of the risk invloved.

Project Name:	Interior Fit-out, ABC Office
Location:	Anywhere, USA

Assembly Number	Description
C1010 124 1200	Wood partition, 2 x 4 @ 16" OC w/5/8" FR gypsum board
C1020 114 1800	Metal door & frame, flush hollow core, 3'-0" x 7'-0"
C3010 230 0080	Painting, brushwork, primer & 2 coats
C3020 410 0140	Carpet, tufted, nylon, roll goods, 12' wide, 26 oz
C3030 210 6000	Acoustic ceilings, 24" x 48" tile, tee grid suspension
D5020 125 0560	Receptacles incl plate, box, conduit, wire, 20 A duplex
D5020 125 0720	Light switch incl plate, box, conduit, wire, 20 A single pole
D5020 210 0560	Fluorescent fixtures, recess mounted, 20 per 1000 SF

Assembly Subtotal

Sales Tax @

General Requirements @

Subtotal A

GC Overhead @

Subtotal B

GC Profit @

Subtotal C

Adjusted by Location Factor

Architects Fee @

Contingency @

Project Total Cost

This estimate is based on an interactive spreadsheet.
A copy of this spreadsheet is located on the RSMeans website at
http://www.reedconstructiondata.com/rsmeans/extras/546011.
You are free to download it and adjust it to your methodology.

Date: 1/1/2014		STD
Qty.	Unit	Subtotal
560.000	S.F.	$2,592.80
2.000	Ea.	$2,610.00
1,120.000	S.F.	$1,321.60
240.000	S.F.	$1,147.20
200.000	S.F.	$914.00
8.000	Ea.	$2,104.00
2.000	Ea.	$527.00
200.000	S.F.	$2,092.00
		$13,308.60
5 %		$ 332.72
7 %		$ 931.60
		$14,572.92
5 %		$ 728.65
		$15,301.56
5 %		$ 765.08
		$16,066.64
118.1		$ 18,974.70
8 %		$ 1,517.98
15 %		$ 2,846.21
		$ 23,338.88

Sales Tax: If the work is subject to state or local sales taxes, the amount must be added to the estimate. In a conceptual estimate it can be assumed that one half of the total represents material costs. Therefore, apply the sales tax rate to 50% of the assembly subtotal.

General Requirements: This item covers project-wide needs provided by the General Contractor. These items vary by project, but may include temporary facilities and utilities, security, testing, project cleanup, etc. In assemblies estimates a percentage is used, typically between 5% and 15% of project cost.

General Contractor Overhead: This entry represents the General Contractor's markup on all work to cover project administration costs.

General Contractor Profit: This entry represents the GC's profit on all work performed. The value included here can vary widely by project, and is influenced by the GC's perception of the project's financial risk and market conditions.

Architects Fee: If appropriate, add the design cost to the project estimate. These fees vary based on project complexity and size. Typical design and engineering fees can be found in the reference section.

D2020 Domestic Water Distribution

Installation includes piping and fittings within 10' of heater. Electric water heaters do not require venting.

1 Kilowatt hour will raise:			
Gallons of Water	Degrees F	Gallons of Water	Degrees F
4.1	100°	6.8	60°
4.5	90°	8.2	50°
5.1	80°	10.0	40°
5.9	70°		

System Components	QUANTITY	UNIT	COST EACH		
			MAT.	INST.	TOTAL
SYSTEM D2020 210 1780					
ELECTRIC WATER HEATER, RESIDENTIAL, 100°F RISE					
10 GALLON TANK, 7 GPH					
Water heater, residential electric, glass lined tank, 10 gal.	1.000	Ea.	360	300	660
Copper tubing, type L, solder joint, hanger 10' OC 1/2″ diam.	30.000	L.F.	111.30	258	369.30
Wrought copper 90° elbow for solder joints 1/2″ diam.	4.000	Ea.	13.56	140	153.56
Wrought copper Tee for solder joints, 1/2″ diam.	2.000	Ea.	11.60	107	118.60
Union, wrought copper, 1/2″ diam.	2.000	Ea.	83	73	156
Valve, gate, bronze, 125 lb, NRS, soldered 1/2″ diam.	2.000	Ea.	103	58	161
Relief valve, bronze, press & temp, self-close, 3/4″ IPS	1.000	Ea.	278	25	303
Wrought copper adapter, CTS to MPT 3/4″ IPS	1.000	Ea.	11.90	41	52.90
Copper tubing, type L, solder joints, 3/4″ diam.	1.000	L.F.	5.65	9.15	14.80
Wrought copper 90° elbow for solder joints 3/4″ diam.	1.000	Ea.	7.60	36.50	44.10
TOTAL			985.61	1,047.65	2,033.26

D2020 210	Electric Water Heaters - Residential Systems		COST EACH		
			MAT.	INST.	TOTAL
1760	Electric water heater, residential, 100°F rise				
1780	10 gallon tank, 7 GPH		985	1,050	2,035
1820	20 gallon tank, 7 GPH	R224000 -20	1,200	1,125	2,325
1860	30 gallon tank, 7 GPH		1,525	1,175	2,700
1900	40 gallon tank, 8 GPH		2,175	1,300	3,475
1940	52 gallon tank, 10 GPH		2,300	1,300	3,600
1980	66 gallon tank, 13 GPH		3,625	1,475	5,100
2020	80 gallon tank, 16 GPH		3,800	1,550	5,350
2060	120 gallon tank, 23 GPH		5,200	1,825	7,025

D2020 Domestic Water Distribution

Systems below include piping and fittings within 10' of heater. Electric water heaters do not require venting.

System Components	QUANTITY	UNIT	COST EACH		
			MAT.	INST.	TOTAL
SYSTEM D2020 240 1820					
ELECTRIC WATER HEATER, COMMERCIAL, 100°F RISE					
50 GALLON TANK, 9 KW, 37 GPH					
Water heater, commercial, electric, 50 Gal, 9 KW, 37 GPH	1.000	Ea.	4,600	385	4,985
Copper tubing, type L, solder joint, hanger 10' OC, 3/4" diam	34.000	L.F.	192.10	311.10	503.20
Wrought copper 90° elbow for solder joints 3/4" diam	5.000	Ea.	38	182.50	220.50
Wrought copper Tee for solder joints, 3/4" diam	2.000	Ea.	27.90	116	143.90
Wrought copper union for soldered joints, 3/4" diam.	2.000	Ea.	105	77	182
Valve, gate, bronze, 125 lb, NRS, soldered 3/4" diam	2.000	Ea.	116	70	186
Relief valve, bronze, press & temp, self-close, 3/4" IPS	1.000	Ea.	278	25	303
Wrought copper adapter, copper tubing to male, 3/4" IPS	1.000	Ea.	11.90	41	52.90
TOTAL			5,368.90	1,207.60	6,576.50

D2020 240	Electric Water Heaters - Commercial Systems		COST EACH		
			MAT.	INST.	TOTAL
1800	Electric water heater, commercial, 100°F rise				
1820	50 gallon tank, 9 KW 37 GPH		5,375	1,200	6,575
1860	80 gal, 12 KW 49 GPH		7,150	1,500	8,650
1900	36 KW 147 GPH	R224000 -10	9,400	1,625	11,025
1940	120 gal, 36 KW 147 GPH		10,300	1,750	12,050
1980	150 gal, 120 KW 490 GPH	R224000 -20	29,900	1,875	31,775
2020	200 gal, 120 KW 490 GPH		31,400	1,925	33,325
2060	250 gal, 150 KW 615 GPH		38,700	2,250	40,950
2100	300 gal, 180 KW 738 GPH		44,300	2,350	46,650
2140	350 gal, 30 KW 123 GPH		29,000	2,550	31,550
2180	180 KW 738 GPH		39,700	2,550	42,250
2220	500 gal, 30 KW 123 GPH		37,400	3,000	40,400
2260	240 KW 984 GPH		53,500	3,000	56,500
2300	700 gal, 30 KW 123 GPH		31,100	3,425	34,525
2340	300 KW 1230 GPH		45,000	3,425	48,425
2380	1000 gal, 60 KW 245 GPH		39,200	4,750	43,950
2420	480 KW 1970 GPH		60,500	4,750	65,250
2460	1500 gal, 60 KW 245 GPH		79,500	5,875	85,375
2500	480 KW 1970 GPH		108,000	5,875	113,875

D3020 Heat Generating Systems

Boiler   Baseboard Radiation

Small Electric Boiler
System Considerations:
1. Terminal units are fin tube baseboard radiation rated at 720 BTU/hr with 200° water temperature or 820 BTU/hr steam.
2. Primary use being for residential or smaller supplementary areas, the floor levels are based on 7-1/2' ceiling heights.
3. All distribution piping is copper for boilers through 205 MBH. All piping for larger systems is steel pipe.

System Components	QUANTITY	UNIT	COST EACH		
			MAT.	INST.	TOTAL
SYSTEM D3020 102 1120					
SMALL HEATING SYSTEM, HYDRONIC, ELECTRIC BOILER					
1,480 S.F., 61 MBH, STEAM, 1 FLOOR					
Boiler, electric steam, std cntrls, trim, ftngs and valves, 18 KW, 61.4 MBH	1.000	Ea.	5,032.50	1,760	6,792.50
Copper tubing type L, solder joint, hanger 10'OC, 1-1/4" diam	160.000	L.F.	1,976	1,920	3,896
Radiation, 3/4" copper tube w/alum fin baseboard pkg 7" high	60.000	L.F.	486	1,320	1,806
Rough in baseboard panel or fin tube with valves & traps	10.000	Set	3,550	6,650	10,200
Pipe covering, calcium silicate w/cover, 1" wall 1-1/4" diam	160.000	L.F.	580.80	1,112	1,692.80
Low water cut-off, quick hookup, in gage glass tappings	1.000	Ea.	260	44	304
TOTAL			11,885.30	12,806	24,691.30
COST PER S.F.			8.03	8.65	16.68

D3020 102	Small Heating Systems, Hydronic, Electric Boilers	COST PER S.F.		
		MAT.	INST.	TOTAL
1100	Small heating systems, hydronic, electric boilers			
1120	Steam, 1 floor, 1480 S.F., 61 M.B.H.	8.04	8.65	16.69
1160	3,000 S.F., 123 M.B.H.	6.30	7.60	13.90
1200	5,000 S.F., 205 M.B.H.	5.55	7	12.55
1240	2 floors, 12,400 S.F., 512 M.B.H.	4.47	6.95	11.42
1280	3 floors, 24,800 S.F., 1023 M.B.H.	4.96	6.90	11.86
1320	34,750 S.F., 1,433 M.B.H.	4.63	6.70	11.33
1360	Hot water, 1 floor, 1,000 S.F., 41 M.B.H.	12.55	4.82	17.37
1400	2,500 S.F., 103 M.B.H.	9.55	8.65	18.20
1440	2 floors, 4,850 S.F., 205 M.B.H.	9.30	10.35	19.65
1480	3 floors, 9,700 S.F., 410 M.B.H.	9.90	10.70	20.60

D3020 Heat Generating Systems

Boiler

Unit Heater

Large Electric Boiler System Considerations:

1. Terminal units are all unit heaters of the same size. Quantities are varied to accommodate total requirements.
2. All air is circulated through the heaters a minimum of three times per hour.
3. As the capacities are adequate for commercial use, floor levels are based on 10' ceiling heights.
4. All distribution piping is black steel pipe.

System Components	QUANTITY	UNIT	COST EACH		
			MAT.	INST.	TOTAL
SYSTEM D3020 104 1240					
LARGE HEATING SYSTEM, HYDRONIC, ELECTRIC BOILER					
9,280 S.F., 135 KW, 461 MBH, 1 FLOOR					
Boiler, electric hot water, std ctrls, trim, ftngs, valves, 135 KW, 461 MBH	1.000	Ea.	12,225	3,825	16,050
Expansion tank, painted steel, 60 Gal capacity ASME	1.000	Ea.	4,175	212	4,387
Circulating pump, CI, close cpld, 50 GPM, 2 HP, 2" pipe conn	1.000	Ea.	2,875	415	3,290
Unit heater, 1 speed propeller, horizontal, 200° EWT, 72.7 MBH	7.000	Ea.	6,125	1,617	7,742
Unit heater piping hookup with controls	7.000	Set	5,250	10,500	15,750
Pipe, steel, black, schedule 40, welded, 2-1/2" diam	380.000	L.F.	5,282	10,560.20	15,842.20
Pipe covering, calcium silicate w/cover, 1" wall, 2-1/2" diam	380.000	L.F.	1,710	2,717	4,427
TOTAL			37,642	29,846.20	67,488.20
COST PER S.F.			4.06	3.22	7.28

D3020 104	Large Heating Systems, Hydronic, Electric Boilers	COST PER S.F.		
		MAT.	INST.	TOTAL
1230	Large heating systems, hydronic, electric boilers			
1240	9,280 S.F., 135 K.W., 461 M.B.H., 1 floor	4.05	3.22	7.27
1280	14,900 S.F., 240 K.W., 820 M.B.H., 2 floors	5.40	5.20	10.60
1320	18,600 S.F., 296 K.W., 1,010 M.B.H., 3 floors	5.30	5.65	10.95
1360	26,100 S.F., 420 K.W., 1,432 M.B.H., 4 floors	5.20	5.55	10.75
1400	39,100 S.F., 666 K.W., 2,273 M.B.H., 4 floors	4.45	4.64	9.09
1440	57,700 S.F., 900 K.W., 3,071 M.B.H., 5 floors	4.25	4.56	8.81
1480	111,700 S.F., 1,800 K.W., 6,148 M.B.H., 6 floors	3.79	3.93	7.72
1520	149,000 S.F., 2,400 K.W., 8,191 M.B.H., 8 floors	3.74	3.93	7.67
1560	223,300 S.F., 3,600 K.W., 12,283 M.B.H., 14 floors	3.98	4.44	8.42

D40 Fire Protection

D4090 Other Fire Protection Systems

General: Automatic fire protection (suppression) systems other than water sprinklers may be desired for special environments, high risk areas, isolated locations or unusual hazards. Some typical applications would include:

Paint dip tanks
Securities vaults
Electronic data processing
Tape and data storage
Transformer rooms
Spray booths
Petroleum storage
High rack storage

Piping and wiring costs are dependent on the individual application and must be added to the component costs shown below.

All areas are assumed to be open.

D4090 910	Fire Suppression Unit Components	COST EACH		
		MAT.	INST.	TOTAL
0020	Detectors with brackets			
0040	Fixed temperature heat detector	55.50	80	135.50
0060	Rate of temperature rise detector	55.50	80	135.50
0080	Ion detector (smoke) detector	120	103	223
0200	Extinguisher agent			
0240	200 lb FM200, container	8,300	315	8,615
0280	75 lb carbon dioxide cylinder	1,350	209	1,559
0320	Dispersion nozzle			
0340	FM200 1-1/2" dispersion nozzle	71.50	49.50	121
0380	Carbon dioxide 3" x 5" dispersion nozzle	71.50	38.50	110
0420	Control station			
0440	Single zone control station with batteries	1,850	640	2,490
0470	Multizone (4) control station with batteries	3,500	1,275	4,775
0490				
0500	Electric mechanical release	179	335	514
0520				
0550	Manual pull station	64.50	116	180.50
0570				
0640	Battery standby power 10" x 10" x 17"	450	160	610
0700				
0740	Bell signalling device	85.50	80	165.50

D4090 920	FM200 Systems	COST PER C.F.		
		MAT.	INST.	TOTAL
0820	Average FM200 system, minimum			1.88
0840	Maximum			3.74

D5010 Electrical Service/Distribution

System Components	QUANTITY	UNIT	COST PER L.F.		
			MAT.	INST.	TOTAL
SYSTEM D5010 110 0200					
HIGH VOLTAGE CABLE, NEUTRAL AND CONDUIT INCLUDED, COPPER #2, 5 kV					
Shielded cable, no splice/termn, copper, XLP shielding, 5 kV, #2	.030	C.L.F.	8.01	9.60	17.61
Wire, 600 volt, type THW, copper, stranded, #8	.010	C.L.F.	.42	.80	1.22
Rigid galv steel conduit to 15' H, 2" diam, w/term, ftng & support	1.000	L.F.	8.20	14.20	22.40
TOTAL			16.63	24.60	41.23

D5010 110	High Voltage Shielded Conductors	COST PER L.F.		
		MAT.	INST.	TOTAL
0200	High voltage cable, neutral & conduit included, copper #2, 5 kV	16.65	24.50	41.15
0240	Copper #1, 5 kV	24.50	29	53.50
0280	15 kV	26.50	29	55.50
0320	Copper 1/0, 5 kV	28	36.50	64.50
0360	15 kV	31	36.50	67.50
0400	25 kV	36	37	73
0440	35 kV	37.50	37.50	75
0480	Copper 2/0, 5 kV	31	37.50	68.50
0520	15 kV	33.50	37.50	71
0560	25 kV	38.50	38	76.50
0600	35 kV	45.50	42	87.50
0640	Copper 4/0, 5 kV	40	42.50	82.50
0680	15 kV	43.50	42.50	86
0720	25 kV	49	43	92
0760	35 kV	52	44	96
0800	Copper 250 kcmil, 5 kV	43.50	43	86.50
0840	15 kV	45.50	43	88.50
0880	25 kV	55	44	99
0920	35 kV	60.50	48	108.50
0960	Copper 350 kcmil, 5 kV	53	48	101
1000	15 kV	55	48	103
1040	25 kV	63.50	49.50	113
1080	35 kV	67.50	51	118.50
1120	Copper 500 kcmil, 5 kV	84.50	60.50	145
1160	15 kV	87	60.50	147.50
1200	25 kV	96	62	158
1240	35 kV	100	63.50	163.50

D5010 Electrical Service/Distribution

Service Entrance Cap

Meter Socket

Safety Switch

Switchboard or other

Ground Rod

System Components	QUANTITY	UNIT	COST EACH		
			MAT.	INST.	TOTAL
SYSTEM D5010 120 0220					
SERVICE INSTALLATION, INCLUDES BREAKERS, METERING, 20' CONDUIT & WIRE					
3 PHASE, 4 WIRE, 60 A					
Wire, 600 volt, copper type XHHW, stranded #6	.600	C.L.F.	48.60	59.10	107.70
Rigid galvanized steel conduit, 3/4" including fittings	20.000	L.F.	56.40	160	216.40
Service entrance cap, 3/4" diameter	1.000	Ea.	8.40	49	57.40
Conduit LB fitting with cover, 3/4" diameter	1.000	Ea.	11.40	49	60.40
Meter socket, three phase, 100 A	1.000	Ea.	158	228	386
Safety switches, heavy duty, 240 volt, 3 pole NEMA 1 fusible, 60 amp	1.000	Ea.	258	278	536
Grounding, wire ground bare armored, #8-1 conductor	.200	C.L.F.	19.70	64	83.70
Grounding, clamp, bronze, 3/4" diameter	1.000	Ea.	7.05	19.95	27
Grounding, rod, copper clad, 8' long, 3/4" diameter	1.000	Ea.	37.50	121	158.50
Wireway w/fittings, 2-1/2" x 2-1/2"	1.000	L.F.	11.40	14.20	25.60
TOTAL			616.45	1,042.25	1,658.70

D5010 120	Overhead Electric Service, 3 Phase - 4 Wire	COST EACH		
		MAT.	INST.	TOTAL
0200	Service installation, includes breakers, metering, 20' conduit & wire			
0220	3 phase, 4 wire, 120/208 volts, 60 A	615	1,050	1,665
0240	3 phase, 4 wire, 120/208 volts, 100 A	840	1,175	2,015
0245	100 A w/circuit breaker	1,600	1,450	3,050
0280	200 A	1,600	1,625	3,225
0285	200 A w/circuit breaker	3,225	2,050	5,275
0320	400 A	3,225	3,250	6,475
0325	400 A w/circuit breaker	6,250	4,050	10,300
0360	600 A	5,400	4,775	10,175
0365	600 A, w/switchboard	9,775	6,050	15,825
0400	800 A	7,725	5,725	13,450
0405	800 A, w/switchboard	12,100	7,175	19,275
0440	1000 A	9,375	7,050	16,425
0445	1000 A, w/switchboard	14,600	8,650	23,250
0480	1200 A	12,500	8,050	20,550
0485	1200 A, w/groundfault switchboard	35,200	9,975	45,175
0520	1600 A	15,500	10,500	26,000
0525	1600 A, w/groundfault switchboard	40,100	12,600	52,700

D5010 Electrical Service/Distribution

D5010 120	Overhead Electric Service, 3 Phase - 4 Wire	COST EACH		
		MAT.	INST.	TOTAL
0560	2000 A	20,600	12,800	33,400
0565	2000 A, w/groundfault switchboard	49,000	15,700	64,700
0610	1 phase, 3 wire, 120/240 volts, 100 A (no safety switch)	193	575	768
0615	100 A w/load center	555	1,100	1,655
0620	200 A	460	790	1,250
0625	200 A w/load center	1,250	1,600	2,850

D5010 Electrical Service/Distribution

Underground service conductor

Safety Switch

Meter and service equipment

Ground rod

System Components	QUANTITY	UNIT	COST EACH		
			MAT.	INST.	TOTAL
SYSTEM D5010 130 1000					
2000 AMP UNDERGROUND ELECTRIC SERVICE WITH GROUNDFAULT SWITCHBOARD					
INCLUDING EXCAVATION, BACKFILL, AND COMPACTION					
Excavate Trench	44.440	B.C.Y.		294.64	294.64
4 inch conduit bank	108.000	L.F.	1,080	2,538	3,618
4 inch fitting	8.000	Ea.	252	424	676
4 inch bells	4.000	Ea.	18.80	212	230.80
Concrete material	16.580	C.Y.	1,807.22		1,807.22
Concrete placement	16.580	C.Y.		399.58	399.58
Backfill trench	33.710	L.C.Y.		96.42	96.42
Compact fill material in trench	25.930	E.C.Y.		126.02	126.02
Dispose of excess fill material on-site	24.070	L.C.Y.		275.60	275.60
500 kcmil power cable	18.000	C.L.F.	22,950	7,200	30,150
Wire, 600 volt, type THW, copper, stranded, 1/0	6.000	C.L.F.	1,662	1,164	2,826
Saw cutting, concrete walls, plain, per inch of depth	64.000	L.F.	3.20	516.48	519.68
Meter centers and sockets, single pos, 4 terminal, 400 amp	5.000	Ea.	3,925	1,875	5,800
Safety switch, 400 amp	5.000	Ea.	9,000	3,550	12,550
Ground rod clamp	1.000	Ea.	7.05	19.95	27
Wireway	1.000	L.F.	11.40	14.20	25.60
600 volt stranded copper wire, 1/0	1.200	C.L.F.	332.40	232.80	565.20
Flexible metallic conduit	120.000	L.F.	103.20	478.80	582
Ground rod	6.000	Ea.	417	960	1,377
TOTAL			41,569.27	20,377.49	61,946.76

D5010 130	Underground Electric Service	COST EACH		
		MAT.	INST.	TOTAL
0950	Underground electric service including excavation, backfill, and compaction			
1000	3 phase, 4 wire, 277/480 volts, 2000 A	41,600	20,400	62,000
1050	2000 A w/groundfault switch	68,000	22,800	90,800
1100	1600 A	33,000	16,500	49,500
1150	1600 A w/groundfault switchboard	57,500	18,500	76,000
1200	1200 A	25,200	13,100	38,300
1250	1200 A w/groundfault switchboard	47,900	15,000	62,900
1400	800 A	18,300	10,900	29,200
1450	800 A w/ switchboard	22,600	12,300	34,900
1500	600 A	11,600	10,000	21,600

D5010 Electrical Service/Distribution

D5010 130	Underground Electric Service	COST EACH		
		MAT.	INST.	TOTAL
1550	600 A w/ switchboard	16,000	11,300	27,300
1600	1 phase, 3 wire, 120/240 volts, 200 A	3,325	3,475	6,800
1650	200 A w/load center	4,100	4,275	8,375
1700	100 A	2,250	2,825	5,075
1750	100 A w/load center	3,025	3,650	6,675

D5010 Electrical Service/Distribution

System Components			COST PER L.F.		
	QUANTITY	UNIT	MAT.	INST.	TOTAL
SYSTEM D5010 230 0200					
FEEDERS, INCLUDING STEEL CONDUIT & WIRE, 60 A					
Rigid galvanized steel conduit, 3/4", including fittings	1.000	L.F.	2.82	8	10.82
Wire 600 volt, type XHHW copper stranded #6	.030	C.L.F.	2.43	2.96	5.39
Wire 600 volt, type XHHW copper stranded #8	.010	C.L.F.	.49	.80	1.29
TOTAL			5.74	11.76	17.50

D5010 230	Feeder Installation	COST PER L.F.		
		MAT.	INST.	TOTAL
0200	Feeder installation 600 V, including RGS conduit and XHHW wire, 60 A	5.75	11.75	17.50
0240	100 A	10.50	15.30	25.80
0280	200 A	25	23	48
0320	400 A	50	46	96
0360	600 A	91	75	166
0400	800 A	119	89	208
0440	1000 A	147	117	264
0480	1200 A	182	150	332
0520	1600 A	238	178	416
0560	2000 A	293	234	527
1200	Branch installation 600 V, including EMT conduit and THW wire, 15 A	1.03	5.70	6.73
1240	20 A	1.26	6.10	7.36
1280	30 A	1.62	6.30	7.92
1320	50 A	3.29	8.75	12.04
1360	65 A	4.17	9.30	13.47
1400	85 A	6.45	10.85	17.30
1440	100 A	7.30	11.05	18.35
1480	130 A	10.85	12.90	23.75
1520	150 A	13.45	14.80	28.25
1560	200 A	18.60	16.90	35.50

D5010   Electrical Service/Distribution

System Components	QUANTITY	UNIT	COST EACH		
			MAT.	INST.	TOTAL
SYSTEM D5010 240 0240					
SWITCHGEAR INSTALLATION, INCL SWBD, PANELS & CIRC BREAKERS, 600 A					
Switchboards, 120/208 V, 600 amp	1.000	Ea.	4,375	1,275	5,650
Aluminum bus bars, 120/208 V, 600 amp	1.000	Ea.	1,775	1,275	3,050
Feeder section, circuit breakers, KA frame, 70 to 225 amp	1.000	Ea.	1,350	200	1,550
Feeder section, circuit breakers, LA frame, 125 to 400 amp	1.000	Ea.	3,050	278	3,328
TOTAL			10,550	3,028	13,578

D5010 240	Switchgear	COST EACH		
		MAT.	INST.	TOTAL
0190	Switchgear installation, including switchboard, panels, & circuit breaker			
0200	120/208 V, 1 phase, 400 A	8,075	2,700	10,775
0240	600 A	10,600	3,025	13,625
0280	800 A	13,500	3,500	17,000
0300	1000 A	16,500	3,900	20,400
0320	1200 A	17,300	4,400	21,700
0360	1600 A	26,800	5,050	31,850
0400	2000 A	31,900	5,500	37,400
0500	277/480 V, 3 phase, 400 A	10,900	5,000	15,900
0520	600 A	15,500	5,775	21,275
0540	800 A	18,100	6,400	24,500
0560	1000 A	22,500	7,100	29,600
0580	1200 A	24,400	7,950	32,350
0600	1600 A	33,500	8,700	42,200
0620	2000 A	41,000	9,600	50,600

D5010   Electrical Service/Distribution

System Components			COST EACH		
	QUANTITY	UNIT	MAT.	INST.	TOTAL
SYSTEM D5010 250 1020					
PANELBOARD INSTALLATION, INCLUDES PANELBOARD, CONDUCTOR, & CONDUIT					
Conduit, galvanized steel, 1-1/4"dia.	37.000	L.F.	197.95	394.05	592
Panelboards, NQOD, 4 wire, 120/208 volts, 100 amp main, 24 circuits	1.000	Ea.	1,275	1,350	2,625
Wire, 600 volt, type THW, copper, stranded, #3	1.480	C.L.F.	205.72	189.44	395.16
TOTAL			1,678.67	1,933.49	3,612.16

D5010 250	Panelboard	COST EACH		
		MAT.	INST.	TOTAL
0900	Panelboards, NQOD, 4 wire, 120/208 volts w/conductor & conduit			
1000	100 A, 0 stories, 0' horizontal	1,275	1,350	2,625
1020	1 stories, 25' horizontal	1,675	1,925	3,600
1040	5 stories, 50' horizontal	2,475	3,075	5,550
1060	10 stories, 75' horizontal	3,400	4,425	7,825
1080	225A, 0 stories, 0' horizontal	2,400	1,775	4,175
2000	1 stories, 25' horizontal	3,625	2,825	6,450
2020	5 stories, 50' horizontal	6,025	4,900	10,925
2040	10 stories, 75' horizontal	8,825	7,300	16,125
2060	400A, 0 stories, 0' horizontal	3,550	2,650	6,200
2080	1 stories, 25' horizontal	5,200	4,350	9,550
3000	5 stories, 50' horizontal	8,475	7,725	16,200
3020	10 stories, 75' horizontal	17,800	15,100	32,900
3040	600 A, 0 stories, 0' horizontal	5,250	3,200	8,450
3060	1 stories, 25' horizontal	8,750	5,975	14,725
3080	5 stories, 50' horizontal	15,600	11,500	27,100
4000	10 stories, 75' horizontal	35,400	22,600	58,000
4010	Panelboards, NEHB, 4 wire, 277/480 volts w/conductor, conduit, & safety switch			
4020	100 A, 0 stories, 0' horizontal, includes safety switch	3,000	1,850	4,850
4040	1 stories, 25' horizontal	3,400	2,450	5,850
4060	5 stories, 50' horizontal	4,200	3,600	7,800
4080	10 stories, 75' horizontal	5,125	4,925	10,050
5000	225 A, 0 stories, 0' horizontal	4,500	2,275	6,775
5020	1 stories, 25' horizontal	5,725	3,325	9,050
5040	5 stories, 50' horizontal	8,125	5,375	13,500
5060	10 stories, 75' horizontal	10,900	7,800	18,700

D50 Electrical

D5010 Electrical Service/Distribution

D5010 250	Panelboard	COST EACH		
		MAT.	INST.	TOTAL
5080	400 A, 0 stories, 0' horizontal	7,425	3,475	10,900
6000	1 stories, 25' horizontal	9,075	5,200	14,275
6020	5 stories, 50' horizontal	12,300	8,575	20,875
6040	10 stories, 75' horizontal	21,700	16,000	37,700
6060	600 A, 0 stories, 0' horizontal	11,100	4,425	15,525
6080	1 stories, 25' horizontal	14,500	7,200	21,700
7000	5 stories, 50' horizontal	21,400	12,700	34,100
7020	10 stories, 75' horizontal	41,200	23,800	65,000

D5020 Lighting and Branch Wiring

Duplex Receptacle

System Components	QUANTITY	UNIT	COST PER S.F.		
			MAT.	INST.	TOTAL
SYSTEM D5020 110 0200					
RECEPTACLES INCL. PLATE, BOX, CONDUIT, WIRE & TRANS. WHEN REQUIRED					
2.5 PER 1000 S.F., .3 WATTS PER S.F.					
Steel intermediate conduit, (IMC) 1/2" diam	167.000	L.F.	.32	1.07	1.39
Wire 600V type THWN-THHN, copper solid #12	3.340	C.L.F.	.05	.19	.24
Wiring device, receptacle, duplex, 120V grounded, 15 amp	2.500	Ea.		.04	.04
Wall plate, 1 gang, brown plastic	2.500	Ea.		.02	.02
Steel outlet box 4" square	2.500	Ea.	.01	.08	.09
Steel outlet box 4" plaster rings	2.500	Ea.	.01	.03	.04
TOTAL			.39	1.43	1.82

D5020 110	Receptacle (by Wattage)	COST PER S.F.		
		MAT.	INST.	TOTAL
0190	Receptacles include plate, box, conduit, wire & transformer when required			
0200	2.5 per 1000 S.F., .3 watts per S.F.	.39	1.43	1.82
0240	With transformer	.46	1.50	1.96
0280	4 per 1000 S.F., .5 watts per S.F.	.44	1.65	2.09
0320	With transformer	.55	1.76	2.31
0360	5 per 1000 S.F., .6 watts per S.F.	.51	1.96	2.47
0400	With transformer	.65	2.10	2.75
0440	8 per 1000 S.F., .9 watts per S.F.	.52	2.18	2.70
0480	With transformer	.74	2.41	3.15
0520	10 per 1000 S.F., 1.2 watts per S.F.	.56	2.36	2.92
0560	With transformer	.84	2.64	3.48
0600	16.5 per 1000 S.F., 2.0 watts per S.F.	.65	2.95	3.60
0640	With transformer	1.11	3.42	4.53
0680	20 per 1000 S.F., 2.4 watts per S.F.	.69	3.21	3.90
0720	With transformer	1.25	3.78	5.03

D50 Electrical

D5020 Lighting and Branch Wiring

Underfloor Receptacle System

Description: Table D5020 115 includes installed costs of raceways and copper wire from panel to and including receptacle.

National Electrical Code prohibits use of undercarpet system in residential, school or hospital buildings. Can only be used with carpet squares.

Low density = (1) Outlet per 259 S.F. of floor area.

High density = (1) Outlet per 127 S.F. of floor area.

System Components	QUANTITY	UNIT	COST PER S.F.		
			MAT.	INST.	TOTAL
SYSTEM D5020 115 0200					
RECEPTACLE SYSTEMS, UNDERFLOOR DUCT, 5′ ON CENTER, LOW DENSITY					
Underfloor duct 3-1/8″ x 7/8″ w/insert 24″ on center	.190	L.F.	3.99	1.74	5.73
Underfloor duct junction box, single duct, 3-1/8″	.003	Ea.	1.40	.48	1.88
Underfloor junction box carpet pan	.003	Ea.	1.16	.02	1.18
Underfloor duct outlet, high tension receptacle	.004	Ea.	.43	.32	.75
Wire 600V type THWN-THHN copper solid #12	.010	C.L.F.	.15	.58	.73
Vertical elbow for underfloor duct, 3-1/8″, included					
Underfloor duct conduit adapter, 2″ x 1-1/4″, included					
TOTAL			7.13	3.14	10.27

D5020 115	Receptacles, Floor	COST PER S.F.		
		MAT.	INST.	TOTAL
0200	Receptacle systems, underfloor duct, 5′ on center, low density	7.15	3.14	10.29
0240	High density	7.70	4.04	11.74
0280	7′ on center, low density	5.65	2.70	8.35
0320	High density	6.20	3.60	9.80
0400	Poke thru fittings, low density	1.23	1.54	2.77
0440	High density	2.47	3.07	5.54
0520	Telepoles, using Romex, low density	1.19	.95	2.14
0560	High density	2.39	1.89	4.28
0600	Using EMT, low density	1.25	1.26	2.51
0640	High density	2.51	2.50	5.01
0720	Conduit system with floor boxes, low density	1.15	1.08	2.23
0760	High density	2.31	2.15	4.46
0840	Undercarpet power system, 3 conductor with 5 conductor feeder, low density	1.52	.38	1.90
0880	High density	2.99	.79	3.78

D5020 Lighting and Branch Wiring

Duplex Receptacle

Wall Switch

System Components	QUANTITY	UNIT	COST PER S.F.		
			MAT.	INST.	TOTAL
SYSTEM D5020 120 0520					
RECEPTACLES AND WALL SWITCHES					
4 RECEPTACLES PER 400 S.F.					
Steel intermediate conduit, (IMC), 1/2" diam	.220	L.F.	.42	1.41	1.83
Wire, 600 volt, type THWN-THHN, copper, solid #12	.005	C.L.F.	.07	.29	.36
Steel outlet box 4" square	.010	Ea.	.02	.32	.34
Steel outlet box, 4" square, plaster rings	.010	Ea.	.03	.10	.13
Receptacle, duplex, 120 volt grounded, 15 amp	.010	Ea.	.01	.16	.17
Wall plate, 1 gang, brown plastic	.010	Ea.		.08	.08
TOTAL			.55	2.36	2.91

D5020 120	Receptacles & Wall Switches	COST PER S.F.		
		MAT.	INST.	TOTAL
0520	Receptacles and wall switches, 400 S.F., 4 receptacles	.55	2.36	2.91
0560	6 receptacles	.61	2.75	3.36
0600	8 receptacles	.68	3.20	3.88
0640	1 switch	.13	.54	.67
0680	600 S.F., 6 receptacles	.55	2.36	2.91
0720	8 receptacles	.62	2.68	3.30
0760	10 receptacles	.65	3	3.65
0800	2 switches	.18	.70	.88
0840	1000 S.F., 10 receptacles	.55	2.35	2.90
0880	12 receptacles	.58	2.48	3.06
0920	14 receptacles	.61	2.68	3.29
0960	2 switches	.09	.38	.47
1000	1600 S.F., 12 receptacles	.50	2.10	2.60
1040	14 receptacles	.54	2.28	2.82
1080	16 receptacles	.55	2.35	2.90
1120	4 switches	.11	.45	.56
1160	2000 S.F., 14 receptacles	.53	2.15	2.68
1200	16 receptacles	.52	2.16	2.68
1240	18 receptacles	.54	2.28	2.82
1280	4 switches	.09	.38	.47
1320	3000 S.F., 12 receptacles	.42	1.64	2.06
1360	18 receptacles	.51	2.03	2.54
1400	24 receptacles	.52	2.16	2.68
1440	6 switches	.09	.38	.47
1480	3600 S.F., 20 receptacles	.49	1.97	2.46
1520	24 receptacles	.50	2.03	2.53
1560	28 receptacles	.52	2.09	2.61
1600	8 switches	.11	.44	.55
1640	4000 S.F., 16 receptacles	.42	1.64	2.06
1680	24 receptacles	.51	2.03	2.54
1720	30 receptacles	.50	2.10	2.60
1760	8 switches	.09	.38	.47

D5020 Lighting and Branch Wiring

D5020 120	Receptacles & Wall Switches	COST PER S.F.		
		MAT.	INST.	TOTAL
1800	5000 S.F., 20 receptacles	.42	1.64	2.06
1840	26 receptacles	.51	1.96	2.47
1880	30 receptacles	.51	2.03	2.54
1920	10 switches	.09	.38	.47

D50 Electrical

D5020 Lighting and Branch Wiring

Duplex Receptacle

Wall Switch

System Components			COST PER EACH		
	QUANTITY	UNIT	MAT.	INST.	TOTAL
SYSTEM D5020 125 0520					
RECEPTACLES AND WALL SWITCHES, RECEPTICLE DUPLEX 120 V GROUNDED, 15 A					
Electric metallic tubing conduit, (EMT), 3/4" diam	22.000	L.F.	20.68	108.02	128.70
Wire, 600 volt, type THWN-THHN, copper, solid #12	.630	C.L.F.	9.39	36.54	45.93
Steel outlet box 4" square	1.000	Ea.	2.31	32	34.31
Steel outlet box, 4" square, plaster rings	1.000	Ea.	3.11	10	13.11
Receptacle, duplex, 120 volt grounded, 15 amp	1.000	Ea.	1.34	15.95	17.29
Wall plate, 1 gang, brown plastic	1.000	Ea.	.42	8	8.42
TOTAL			37.25	210.51	247.76

D5020 125	Receptacles & Switches by Each	COST PER EACH		
		MAT.	INST.	TOTAL
0460	Receptacles & Switches, with box, plate, 3/4" EMT conduit & wire			
0520	Receptacle duplex 120 V grounded, 15 A	37.50	211	248.50
0560	20 A	45	218	263
0600	Receptacle duplex ground fault interrupting, 15 A	77.50	218	295.50
0640	20 A	79.50	218	297.50
0680	Toggle switch single, 15 A	42	211	253
0720	20 A	45.50	218	263.50
0760	3 way switch, 15 A	42.50	223	265.50
0800	20 A	45	230	275
0840	4 way switch, 15 A	47.50	237	284.50
0880	20 A	82.50	253	335.50

D50 Electrical

D5020 Lighting and Branch Wiring

Description: Table D5020 130 includes the cost for switch, plate, box, conduit in slab or EMT exposed and copper wire. Add 20% for exposed conduit.

No power required for switches.

Federal energy guidelines recommend the maximum lighting area controlled per switch shall not exceed 1000 S.F. and that areas over 500 S.F. shall be so controlled that total illumination can be reduced by at least 50%.

System Components	QUANTITY	UNIT	COST PER S.F.		
			MAT.	INST.	TOTAL
SYSTEM D5020 130 0360					
WALL SWITCHES, 5.0 PER 1000 S.F.					
Steel, intermediate conduit (IMC), 1/2" diameter	88.000	L.F.	.17	.56	.73
Wire, 600V type THWN-THHN, copper solid #12	1.710	C.L.F.	.03	.10	.13
Toggle switch, single pole, 15 amp	5.000	Ea.	.03	.08	.11
Wall plate, 1 gang, brown plastic	5.000	Ea.		.04	.04
Steel outlet box 4" plaster rings	5.000	Ea.	.01	.16	.17
Plaster rings	5.000	Ea.	.02	.05	.07
TOTAL			.26	.99	1.25

D5020 130	Wall Switch by Sq. Ft.	COST PER S.F.		
		MAT.	INST.	TOTAL
0200	Wall switches, 1.0 per 1000 S.F.	.06	.23	.29
0240	1.2 per 1000 S.F.	.06	.26	.32
0280	2.0 per 1000 S.F.	.09	.37	.46
0320	2.5 per 1000 S.F.	.12	.47	.59
0360	5.0 per 1000 S.F.	.26	.99	1.25
0400	10.0 per 1000 S.F.	.50	2.01	2.51

D50 Electrical

D5020 Lighting and Branch Wiring

System D5020 135 includes all wiring and connections.

System Components	QUANTITY	UNIT	COST PER S.F.		
			MAT.	INST.	TOTAL
SYSTEM D5020 135 0200					
MISCELLANEOUS POWER, TO .5 WATTS					
Steel intermediate conduit, (IMC) 1/2″ diam	15.000	L.F.	.03	.10	.13
Wire 600V type THWN-THHN, copper solid #12	.325	C.L.F.		.02	.02
TOTAL			.03	.12	.15

D5020 135	Miscellaneous Power	COST PER S.F.		
		MAT.	INST.	TOTAL
0200	Miscellaneous power, to .5 watts	.03	.12	.15
0240	.8 watts	.05	.16	.21
0280	1 watt	.06	.22	.28
0320	1.2 watts	.07	.25	.32
0360	1.5 watts	.09	.30	.39
0400	1.8 watts	.10	.34	.44
0440	2 watts	.12	.41	.53
0480	2.5 watts	.15	.51	.66
0520	3 watts	.17	.60	.77

D50 Electrical

D5020 Lighting and Branch Wiring

System D5020 140 includes all wiring and connections for central air conditioning units.

System Components	QUANTITY	UNIT	COST PER S.F.		
			MAT.	INST.	TOTAL
SYSTEM D5020 140 0200					
CENTRAL AIR CONDITIONING POWER, 1 WATT					
Steel intermediate conduit, 1/2" diam.	.030	L.F.	.06	.19	.25
Wire 600V type THWN-THHN, copper solid #12	.001	C.L.F.	.01	.06	.07
TOTAL			.07	.25	.32

D5020 140	Central A. C. Power (by Wattage)	COST PER S.F.		
		MAT.	INST.	TOTAL
0200	Central air conditioning power, 1 watt	.07	.25	.32
0220	2 watts	.08	.29	.37
0240	3 watts	.11	.32	.43
0280	4 watts	.18	.43	.61
0320	6 watts	.34	.57	.91
0360	8 watts	.41	.62	1.03
0400	10 watts	.60	.72	1.32

D5020 Lighting and Branch Wiring

System D5020 145 installed cost of motor wiring using 50' of rigid conduit and copper wire. **Cost and setting of motor not included.**

System Components	QUANTITY	UNIT	COST EACH MAT.	COST EACH INST.	COST EACH TOTAL
SYSTEM D5020 145 0200					
MOTOR INST., SINGLE PHASE, 115V, TO AND INCLUDING 1/3 HP MOTOR SIZE					
Wire 600V type THWN-THHN, copper solid #12	1.250	C.L.F.	18.63	72.50	91.13
Steel intermediate conduit, (IMC) 1/2" diam.	50.000	L.F.	94.50	320	414.50
Magnetic FVNR, 115V, 1/3 HP, size 00 starter	1.000	Ea.	195	160	355
Safety switch, fused, heavy duty, 240V 2P 30 amp	1.000	Ea.	115	183	298
Safety switch, non fused, heavy duty, 600V, 3 phase, 30 A	1.000	Ea.	146	200	346
Flexible metallic conduit, Greenfield 1/2" diam.	1.500	L.F.	.95	4.79	5.74
Connectors for flexible metallic conduit Greenfield 1/2" diam.	1.000	Ea.	2.94	8	10.94
Coupling for Greenfield to conduit 1/2" diam. flexible metalic conduit	1.000	Ea.	1.93	12.80	14.73
Fuse cartridge nonrenewable, 250V 30 amp	1.000	Ea.	2.55	12.80	15.35
TOTAL			577.50	973.89	1,551.39

D5020 145	Motor Installation	COST EACH MAT.	COST EACH INST.	COST EACH TOTAL
0200	Motor installation, single phase, 115V, 1/3 HP motor size	580	975	1,555
0240	1 HP motor size	600	975	1,575
0280	2 HP motor size	650	1,025	1,675
0320	3 HP motor size	730	1,050	1,780
0360	230V, 1 HP motor size	580	985	1,565
0400	2 HP motor size	615	985	1,600
0440	3 HP motor size	695	1,050	1,745
0520	Three phase, 200V, 1-1/2 HP motor size	680	1,075	1,755
0560	3 HP motor size	725	1,175	1,900
0600	5 HP motor size	795	1,325	2,120
0640	7-1/2 HP motor size	825	1,350	2,175
0680	10 HP motor size	1,275	1,675	2,950
0720	15 HP motor size	1,675	1,875	3,550
0760	20 HP motor size	2,150	2,150	4,300
0800	25 HP motor size	2,200	2,150	4,350
0840	30 HP motor size	3,400	2,550	5,950
0880	40 HP motor size	4,300	3,025	7,325
0920	50 HP motor size	7,350	3,500	10,850
0960	60 HP motor size	7,675	3,700	11,375
1000	75 HP motor size	9,875	4,225	14,100
1040	100 HP motor size	26,900	5,000	31,900
1080	125 HP motor size	27,400	5,475	32,875
1120	150 HP motor size	31,900	6,450	38,350
1160	200 HP motor size	32,700	7,650	40,350
1240	230V, 1-1/2 HP motor size	650	1,075	1,725
1280	3 HP motor size	695	1,175	1,870
1320	5 HP motor size	760	1,300	2,060
1360	7-1/2 HP motor size	760	1,300	2,060
1400	10 HP motor size	1,175	1,575	2,750
1440	15 HP motor size	1,325	1,750	3,075
1480	20 HP motor size	2,025	2,100	4,125
1520	25 HP motor size	2,150	2,150	4,300

D5020 Lighting and Branch Wiring

D5020 145	Motor Installation	COST EACH		
		MAT.	INST.	TOTAL
1560	30 HP motor size	2,225	2,150	4,375
1600	40 HP motor size	4,100	2,950	7,050
1640	50 HP motor size	4,400	3,100	7,500
1680	60 HP motor size	7,350	3,525	10,875
1720	75 HP motor size	8,975	4,000	12,975
1760	100 HP motor size	10,500	4,450	14,950
1800	125 HP motor size	27,400	5,150	32,550
1840	150 HP motor size	29,100	5,875	34,975
1880	200 HP motor size	31,200	6,525	37,725
1960	460V, 2 HP motor size	800	1,075	1,875
2000	5 HP motor size	845	1,175	2,020
2040	10 HP motor size	875	1,300	2,175
2080	15 HP motor size	1,150	1,500	2,650
2120	20 HP motor size	1,225	1,600	2,825
2160	25 HP motor size	1,300	1,675	2,975
2200	30 HP motor size	1,650	1,800	3,450
2240	40 HP motor size	2,150	1,925	4,075
2280	50 HP motor size	2,425	2,150	4,575
2320	60 HP motor size	3,575	2,525	6,100
2360	75 HP motor size	4,200	2,800	7,000
2400	100 HP motor size	4,700	3,125	7,825
2440	125 HP motor size	7,675	3,525	11,200
2480	150 HP motor size	9,525	3,950	13,475
2520	200 HP motor size	11,300	4,450	15,750
2600	575V, 2 HP motor size	800	1,075	1,875
2640	5 HP motor size	845	1,175	2,020
2680	10 HP motor size	875	1,300	2,175
2720	20 HP motor size	1,150	1,500	2,650
2760	25 HP motor size	1,225	1,600	2,825
2800	30 HP motor size	1,650	1,800	3,450
2840	50 HP motor size	1,800	1,875	3,675
2880	60 HP motor size	3,500	2,525	6,025
2920	75 HP motor size	3,575	2,525	6,100
2960	100 HP motor size	4,200	2,800	7,000
3000	125 HP motor size	7,350	3,475	10,825
3040	150 HP motor size	7,675	3,525	11,200
3080	200 HP motor size	9,725	4,000	13,725

D50 Electrical

D5020 Lighting and Branch Wiring

System Components	QUANTITY	UNIT	COST PER L.F. MAT.	INST.	TOTAL
SYSTEM D5020 155 0200					
MOTOR FEEDER SYSTEMS, SINGLE PHASE, UP TO 115V, 1HP OR 230V, 2HP					
Steel intermediate conduit, (IMC) 1/2″ diam	1.000	L.F.	1.89	6.40	8.29
Wire 600V type THWN-THHN, copper solid #12	.020	C.L.F.	.30	1.16	1.46
TOTAL			2.19	7.56	9.75

D5020 155	Motor Feeder	MAT.	INST.	TOTAL
0200	Motor feeder systems, single phase, feed up to 115V 1HP or 230V 2 HP	2.19	7.55	9.74
0240	115V 2HP, 230V 3HP	2.36	7.70	10.06
0280	115V 3HP	2.72	8	10.72
0360	Three phase, feed to 200V 3HP, 230V 5HP, 460V 10HP, 575V 10HP	2.34	8.15	10.49
0440	200V 5HP, 230V 7.5HP, 460V 15HP, 575V 20HP	2.60	8.30	10.90
0520	200V 10HP, 230V 10HP, 460V 30HP, 575V 30HP	3.14	8.80	11.94
0600	200V 15HP, 230V 15HP, 460V 40HP, 575V 50HP	4.42	10.05	14.47
0680	200V 20HP, 230V 25HP, 460V 50HP, 575V 60HP	6.75	12.80	19.55
0760	200V 25HP, 230V 30HP, 460V 60HP, 575V 75HP	7.60	13	20.60
0840	200V 30HP	8.65	13.40	22.05
0920	230V 40HP, 460V 75HP, 575V 100HP	11.10	14.65	25.75
1000	200V 40HP	12.50	15.65	28.15
1080	230V 50HP, 460V 100HP, 575V 125HP	15.90	17.25	33.15
1160	200V 50HP, 230V 60HP, 460V 125HP, 575V 150HP	18.60	18.35	36.95
1240	200V 60HP, 460V 150HP	23	21.50	44.50
1320	230V 75HP, 575V 200HP	26.50	22.50	49
1400	200V 75HP	30.50	23	53.50
1480	230V 100HP, 460V 200HP	40	26.50	66.50
1560	200V 100HP	55	33.50	88.50
1640	230V 125HP	55	33.50	88.50
1720	200V 125HP, 230V 150HP	64.50	41.50	106
1800	200V 150HP	77.50	45	122.50
1880	200V 200HP	99	47.50	146.50
1960	230V 200HP	101	49.50	150.50

D5020 Lighting and Branch Wiring

Starters are full voltage, type NEMA 1 for general purpose indoor application with motor overload protection and include mounting and wire connections.

System Components	QUANTITY	UNIT	COST EACH		
			MAT.	INST.	TOTAL
SYSTEM D5020 160 0200					
MAGNETIC STARTER, SIZE 00 TO 1/3 HP, 1 PHASE 115V OR 1 HP 230V					
Magnetic starter, size 00, to 1/3 HP, 1 phase, 115V or 1 HP 230V	1.000	Ea.	195	160	355
TOTAL			195	160	355

D5020 160	Magnetic Starter	COST EACH		
		MAT.	INST.	TOTAL
0200	Magnetic starter, size 00, to 1/3 HP, 1 phase, 115V or 1 HP 230V	195	160	355
0280	Size 00, to 1-1/2 HP, 3 phase, 200-230V or 2 HP 460-575V	214	183	397
0360	Size 0, to 1 HP, 1 phase, 115V or 2 HP 230V	218	160	378
0440	Size 0, to 3 HP, 3 phase, 200-230V or 5 HP 460-575V	259	278	537
0520	Size 1, to 2 HP, 1 phase, 115V or 3 HP 230V	250	213	463
0600	Size 1, to 7-1/2 HP, 3 phase, 200-230V or 10 HP 460-575V	291	400	691
0680	Size 2, to 10 HP, 3 phase, 200V, 15 HP-230V or 25 HP 460-575V	545	580	1,125
0760	Size 3, to 25 HP, 3 phase, 200V, 30 HP-230V or 50 HP 460-575V	895	710	1,605
0840	Size 4, to 40 HP, 3 phase, 200V, 50 HP-230V or 100 HP 460-575V	1,975	1,075	3,050
0920	Size 5, to 75 HP, 3 phase, 200V, 100 HP-230V or 200 HP 460-575V	4,625	1,425	6,050
1000	Size 6, to 150 HP, 3 phase, 200V, 200 HP-230V or 400 HP 460-575V	20,100	1,600	21,700

D5020 Lighting and Branch Wiring

Safety switches are type NEMA 1 for general purpose indoor application, and include time delay fuses, insulation and wire terminations.

System Components				QUANTITY	UNIT	COST EACH		
						MAT.	INST.	TOTAL
SYSTEM D5020 165 0200								
SAFETY SWITCH, 30A FUSED, 1 PHASE, 115V OR 230V								
Safety switch fused, hvy duty, 240V 2p 30 amp				1.000	Ea.	115	183	298
Fuse, dual element time delay 250V, 30 amp				2.000	Ea.	18.20	25.60	43.80
			TOTAL			133.20	208.60	341.80

D5020 165	Safety Switches	COST EACH		
		MAT.	INST.	TOTAL
0200	Safety switch, 30 A fused, 1 phase, 2 HP 115 V or 3 HP, 230 V	133	209	342
0280	3 phase, 5 HP, 200 V or 7 1/2 HP, 230 V	180	238	418
0360	15 HP, 460 V or 20 HP, 575 V	315	248	563
0440	60 A fused, 3 phase, 15 HP 200 V or 15 HP 230 V	298	315	613
0520	30 HP 460 V or 40 HP 575 V	405	325	730
0600	100 A fused, 3 phase, 20 HP 200 V or 25 HP 230 V	500	385	885
0680	50 HP 460 V or 60 HP 575 V	765	390	1,155
0760	200 A fused, 3 phase, 50 HP 200 V or 60 HP 230 V	905	545	1,450
0840	125 HP 460 V or 150 HP 575 V	1,225	555	1,780
0920	400 A fused, 3 phase, 100 HP 200 V or 125 HP 230 V	2,175	775	2,950
1000	250 HP 460 V or 350 HP 575 V	2,925	790	3,715
1020	600 A fused, 3 phase, 150 HP 200 V or 200 HP 230 V	3,925	1,150	5,075
1040	400 HP 460 V	4,725	1,175	5,900

D5020 Lighting and Branch Wiring

Straight Connector

Angle Connector

Flexible Conduit

Table below includes costs for the flexible conduit. Not included are wire terminations and testing motor for correct rotation.

System Components	QUANTITY	UNIT	COST EACH		
			MAT.	INST.	TOTAL
SYSTEM D5020 170 0200 **MOTOR CONNECTIONS, SINGLE PHASE, 115V/230V UP TO 1 HP** Motor connection, flexible conduit & fittings, 1 HP motor 115V	1.000	Ea.	8	80	88
TOTAL			8	80	88

D5020 170	Motor Connections	COST EACH		
		MAT.	INST.	TOTAL
0200	Motor connections, single phase, 115/230V, up to 1 HP	8	80	88
0240	Up to 3 HP	15.15	98.50	113.65
0280	Three phase, 200/230/460/575V, up to 3 HP	9.55	94	103.55
0320	Up to 5 HP	8.15	117	125.15
0360	Up to 7-1/2 HP	11.55	139	150.55
0400	Up to 10 HP	26	152	178
0440	Up to 15 HP	26	194	220
0480	Up to 25 HP	36.50	237	273.50
0520	Up to 50 HP	74	290	364
0560	Up to 100 HP	174	425	599

D50 Electrical

D5020 Lighting and Branch Wiring

Manual Starter

Magnetic Starter

Induction Motor

For 230/460 Volt A.C., 3 phase, 60 cycle ball bearing squirrel cage induction motors, NEMA Class B standard line. Installation included.

No conduit, wire, or terminations included.

System Components			COST EACH		
	QUANTITY	UNIT	MAT.	INST.	TOTAL
SYSTEM D5020 175 0220					
MOTOR, DRIPPROOF CLASS B INSULATION, 1.15 SERVICE FACTOR, WITH STARTER					
1 H.P., 1200 RPM WITH MANUAL STARTER					
Motor, dripproof, class B insul, 1.15 serv fact, 1200 RPM, 1 HP	1.000	Ea.	465	142	607
Motor starter, manual, 3 phase, 1 HP motor	1.000	Ea.	225	183	408
TOTAL			690	325	1,015

D5020 175	Motor & Starter	COST EACH		
		MAT.	INST.	TOTAL
0190	Motor, dripproof, premium efficient, 1.15 service factor			
0200	1 HP, 1200 RPM, motor only	465	142	607
0220	With manual starter	690	325	1,015
0240	With magnetic starter	680	325	1,005
0260	1800 RPM, motor only	305	142	447
0280	With manual starter	530	325	855
0300	With magnetic starter	520	325	845
0320	2 HP, 1200 RPM, motor only	535	142	677
0340	With manual starter	760	325	1,085
0360	With magnetic starter	795	420	1,215
0380	1800 RPM, motor only	370	142	512
0400	With manual starter	595	325	920
0420	With magnetic starter	630	420	1,050
0440	3600 RPM, motor only	520	142	662
0460	With manual starter	745	325	1,070
0480	With magnetic starter	780	420	1,200
0500	3 HP, 1200 RPM, motor only	745	142	887
0520	With manual starter	970	325	1,295
0540	With magnetic starter	1,000	420	1,420
0560	1800 RPM, motor only	550	142	692
0580	With manual starter	775	325	1,100
0600	With magnetic starter	810	420	1,230
0620	3600 RPM, motor only	525	142	667
0640	With manual starter	750	325	1,075
0660	With magnetic starter	785	420	1,205
0680	5 HP, 1200 RPM, motor only	1,000	142	1,142
0700	With manual starter	1,275	460	1,735
0720	With magnetic starter	1,300	540	1,840
0740	1800 RPM, motor only	660	142	802
0760	With manual starter	925	460	1,385
0780	With magnetic starter	950	540	1,490
0800	3600 RPM, motor only	575	142	717

D5020 Lighting and Branch Wiring

D5020 175	Motor & Starter	COST EACH		
		MAT.	INST.	TOTAL
0820	With manual starter	840	460	1,300
0840	With magnetic starter	865	540	1,405
0860	7.5 HP, 1800 RPM, motor only	770	152	922
0880	With manual starter	1,025	470	1,495
0900	With magnetic starter	1,325	730	2,055
0920	10 HP, 1800 RPM, motor only	945	160	1,105
0940	With manual starter	1,200	480	1,680
0960	With magnetic starter	1,500	740	2,240
0980	15 HP, 1800 RPM, motor only	1,325	200	1,525
1000	With magnetic starter	1,875	780	2,655
1040	20 HP, 1800 RPM, motor only	1,675	246	1,921
1060	With magnetic starter	2,575	955	3,530
1100	25 HP, 1800 RPM, motor only	2,100	256	2,356
1120	With magnetic starter	3,000	965	3,965
1160	30 HP, 1800 RPM, motor only	2,425	266	2,691
1180	With magnetic starter	3,325	975	4,300
1220	40 HP, 1800 RPM, motor only	3,225	320	3,545
1240	With magnetic starter	5,200	1,400	6,600
1280	50 HP, 1800 RPM, motor only	3,075	400	3,475
1300	With magnetic starter	5,050	1,475	6,525
1340	60 HP, 1800 RPM, motor only	4,550	455	5,005
1360	With magnetic starter	9,175	1,875	11,050
1400	75 HP, 1800 RPM, motor only	3,975	530	4,505
1420	With magnetic starter	8,600	1,950	10,550
1460	100 HP, 1800 RPM, motor only	5,100	710	5,810
1480	With magnetic starter	9,725	2,125	11,850
1520	125 HP, 1800 RPM, motor only	5,875	915	6,790
1540	With magnetic starter	26,000	2,525	28,525
1580	150 HP, 1800 RPM, motor only	8,475	1,075	9,550
1600	With magnetic starter	28,600	2,675	31,275
1640	200 HP, 1800 RPM, motor only	10,300	1,275	11,575
1660	With magnetic starter	30,400	2,875	33,275
1680	Totally encl, premium efficient, 1.0 ser. fac., 1HP, 1200 RPM, motor only	445	142	587
1700	With manual starter	670	325	995
1720	With magnetic starter	660	325	985
1740	1800 RPM, motor only	460	142	602
1760	With manual starter	685	325	1,010
1780	With magnetic starter	675	325	1,000
1800	2 HP, 1200 RPM, motor only	515	142	657
1820	With manual starter	740	325	1,065
1840	With magnetic starter	775	420	1,195
1860	1800 RPM, motor only	570	142	712
1880	With manual starter	795	325	1,120
1900	With magnetic starter	830	420	1,250
1920	3600 RPM, motor only	400	142	542
1940	With manual starter	625	325	950
1960	With magnetic starter	660	420	1,080
1980	3 HP, 1200 RPM, motor only	700	142	842
2000	With manual starter	925	325	1,250
2020	With magnetic starter	960	420	1,380
2040	1800 RPM, motor only	665	142	807
2060	With manual starter	890	325	1,215
2080	With magnetic starter	925	420	1,345
2100	3600 RPM, motor only	495	142	637
2120	With manual starter	720	325	1,045
2140	With magnetic starter	755	420	1,175
2160	5 HP, 1200 RPM, motor only	950	142	1,092
2180	With manual starter	1,225	460	1,685

D5020 Lighting and Branch Wiring

D5020 175	Motor & Starter	COST EACH		
		MAT.	INST.	TOTAL
2200	With magnetic starter	1,250	540	1,790
2220	1800 RPM, motor only	715	142	857
2240	With manual starter	980	460	1,440
2260	With magnetic starter	1,000	540	1,540
2280	3600 RPM, motor only	620	142	762
2300	With manual starter	885	460	1,345
2320	With magnetic starter	910	540	1,450
2340	7.5 HP, 1800 RPM, motor only	1,025	152	1,177
2360	With manual starter	1,300	470	1,770
2380	With magnetic starter	1,575	730	2,305
2400	10 HP, 1800 RPM, motor only	1,100	160	1,260
2420	With manual starter	1,375	480	1,855
2440	With magnetic starter	1,650	740	2,390
2460	15 HP, 1800 RPM, motor only	1,550	200	1,750
2480	With magnetic starter	2,100	780	2,880
2500	20 HP, 1800 RPM, motor only	2,150	246	2,396
2520	With magnetic starter	3,050	955	4,005
2540	25 HP, 1800 RPM, motor only	2,500	256	2,756
2560	With magnetic starter	3,400	965	4,365
2580	30 HP, 1800 RPM, motor only	2,600	266	2,866
2600	With magnetic starter	3,500	975	4,475
2620	40 HP, 1800 RPM, motor only	3,175	320	3,495
2640	With magnetic starter	5,150	1,400	6,550
2660	50 HP, 1800 RPM, motor only	3,400	400	3,800
2680	With magnetic starter	5,375	1,475	6,850
2700	60 HP, 1800 RPM, motor only	5,725	455	6,180
2720	With magnetic starter	10,400	1,875	12,275
2740	75 HP, 1800 RPM, motor only	7,000	530	7,530
2760	With magnetic starter	11,600	1,950	13,550
2780	100 HP, 1800 RPM, motor only	8,925	710	9,635
2800	With magnetic starter	13,600	2,125	15,725
2820	125 HP, 1800 RPM, motor only	12,500	915	13,415
2840	With magnetic starter	32,600	2,525	35,125
2860	150 HP, 1800 RPM, motor only	13,700	1,075	14,775
2880	With magnetic starter	33,800	2,675	36,475
2900	200 HP, 1800 RPM, motor only	17,300	1,275	18,575
2920	With magnetic starter	37,400	2,875	40,275

General: The cost of the lighting portion of the electrical costs is dependent upon:
1. The footcandle requirement of the proposed building.
2. The type of fixtures required.
3. The ceiling heights of the building.
4. Reflectance value of ceilings, walls and floors.
5. Fixture efficiencies and spacing vs. mounting height ratios.

Footcandle Requirements: See Table D5020-204 for Footcandle and Watts per S.F. determination.

Table D5020-201   IESNA* Recommended Illumination Levels in Footcandles

Commercial Buildings			Industrial Buildings		
Type	Description	Footcandles	Type	Description	Footcandles
Bank	Lobby	50	Assembly Areas	Rough bench & machine work	50
	Customer Areas	70		Medium bench & machine work	100
	Teller Stations	150		Fine bench & machine work	500
	Accounting Areas	150	Inspection Areas	Ordinary	50
Offices	Routine Work	100		Difficult	100
	Accounting	150		Highly Difficult	200
	Drafting	200	Material Handling	Loading	20
	Corridors, Halls, Washrooms	30		Stock Picking	30
Schools	Reading or Writing	70		Packing, Wrapping	50
	Drafting, Labs, Shops	100	Stairways	Service Areas	20
	Libraries	70	Washrooms	Service Areas	20
	Auditoriums, Assembly	15	Storage Areas	Inactive	5
	Auditoriums, Exhibition	30		Active, Rough, Bulky	10
Stores	Circulation Areas	30		Active, Medium	20
	Stock Rooms	30		Active, Fine	50
	Merchandise Areas, Service	100	Garages	Active Traffic Areas	20
	Self-Service Areas	200		Service & Repair	100

*IESNA - Illuminating Engineering Society of North America

Table D5020-202 General Lighting Loads by Occupancies

Type of Occupancy	Unit Load per S.F. (Watts)
Armories and Auditoriums	1
Banks	5
Barber Shops and Beauty Parlors	3
Churches	1
Clubs	2
Court Rooms	2
*Dwelling Units	3
Garages — Commercial (storage)	½
Hospitals	2
*Hotels and Motels, including apartment houses without provisions for cooking by tenants	2
Industrial Commercial (Loft) Buildings	2
Lodge Rooms	1½
Office Buildings	5
Restaurants	2
Schools	3
Stores	3
Warehouses (storage)	¼
*In any of the above occupancies except one-family dwellings and individual dwelling units of multi-family dwellings:	
Assembly Halls and Auditoriums	1
Halls, Corridors, Closets	½
Storage Spaces	¼

Table D5020-203 Lighting Limit (Connected Load) for Listed Occupancies: New Building Proposed Energy Conservation Guideline

Type of Use	Maximum Watts per S.F.
Interior	3.00
Category A: Classrooms, office areas, automotive mechanical areas, museums, conference rooms, drafting rooms, clerical areas, laboratories, merchandising areas, kitchens, examining rooms, book stacks, athletic facilities.	
Category B: Auditoriums, waiting areas, spectator areas, restrooms, dining areas, transportation terminals, working corridors in prisons and hospitals, book storage areas, active inventory storage, hospital bedrooms, hotel and motel bedrooms, enclosed shopping mall concourse areas, stairways.	1.00
Category C: Corridors, lobbies, elevators, inactive storage areas.	0.50
Category D: Indoor parking.	0.25
Exterior	
Category E: Building perimeter: wall-wash, facade, canopy.	5.00 (per linear foot)
Category F: Outdoor parking.	0.10

Table D5020-204 Procedure for Calculating Footcandles and Watts Per Square Foot

1. Initial footcandles = No. of fixtures × lamps per fixture × lumens per lamp × coefficient of utilization ÷ square feet
2. Maintained footcandles = initial footcandles × maintenance factor
3. Watts per square foot = No. of fixtures × lamps × (lamp watts + ballast watts) ÷ square feet

Example: To find footcandles and watts per S.F. for an office 20′ x 20′ with 11 fluorescent fixtures each having 4–40 watt C.W. lamps.

Based on good reflectance and clean conditions:

Lumens per lamp = 40 watt cool white at 3150 lumens per lamp RD5020-250, Table RD5020-251

Coefficient of utilization = .42 (varies from .62 for light colored areas to .27 for dark)

Maintenance factor = .75 (varies from .80 for clean areas with good maintenance to .50 for poor)

Ballast loss = 8 watts per lamp. (Varies with manufacturer. See manufacturers' catalog.)

1. Initial footcandles:

$$\frac{11 \times 4 \times 3150 \times .42}{400} = \frac{58,212}{400} = 145 \text{ footcandles}$$

2. Maintained footcandles:

$$145 \times .75 = 109 \text{ footcandles}$$

3. Watts per S.F.

$$\frac{11 \times 4\,(40 + 8)}{400} = \frac{2,112}{400} = 5.3 \text{ watts per S.F.}$$

Table D5020-205 Approximate Watts Per Square Foot for Popular Fixture Types

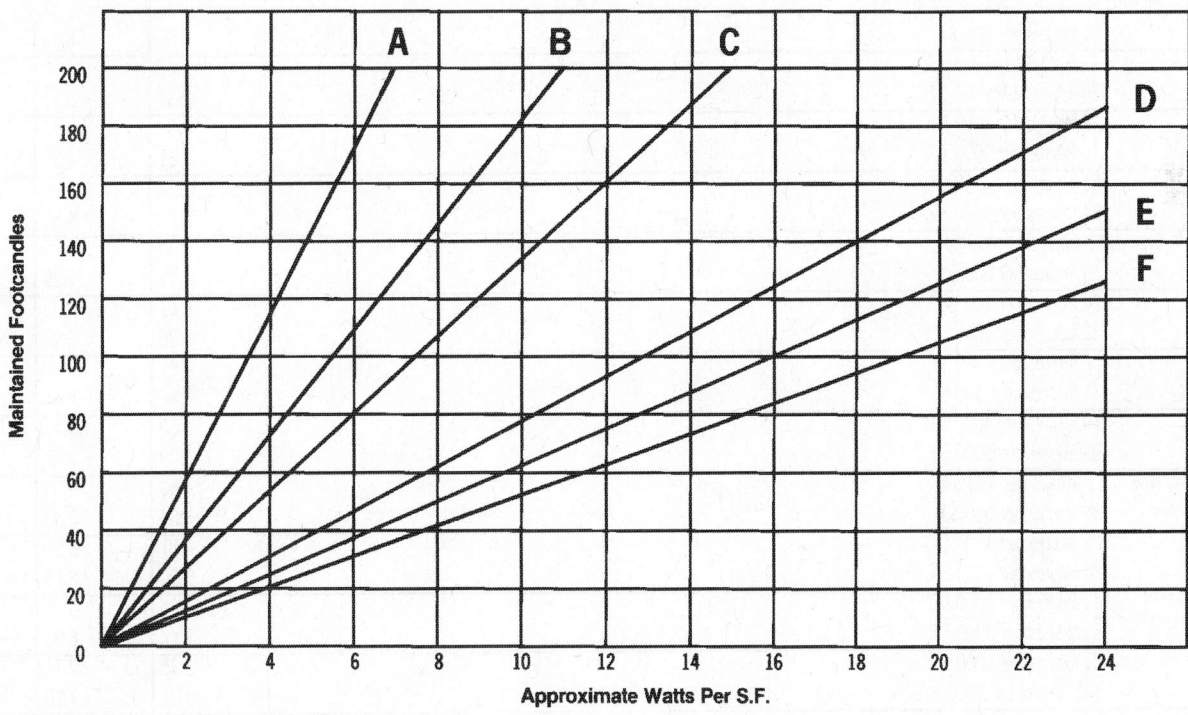

Due to the many variables involved, use for preliminary estimating only:
  a. Fluorescent – industrial System D5020 208
  b. Fluorescent – lens unit System D5020 208 Fixture types B & C
  c. Fluorescent – louvered unit
  d. Incandescent – open reflector System D5020 214, Type D
  e. Incandescent – lens unit System D5020 214, Type A
  f. Incandescent – down light System D5020 214, Type B

D5020 Lighting and Branch Wiring

A. Strip Fixture

B. Surface Mounted

C. Recessed

D. Pendent Mounted

Design Assumptions:

1. A 100 footcandle average maintained level of illumination.
2. Ceiling heights range from 9' to 11'.
3. Average reflectance values are assumed for ceilings, walls and floors.
4. Cool white (CW) fluorescent lamps with 3150 lumens for 40 watt lamps and 6300 lumens for 8' slimline lamps.
5. Four 40 watt lamps per 4' fixture and two 8' lamps per 8' fixture.
6. Average fixture efficiency values and spacing to mounting height ratios.
7. Installation labor is average U.S. rate as of January 1.

System Components	QUANTITY	UNIT	COST PER S.F.		
			MAT.	INST.	TOTAL
SYSTEM D5020 208 0520					
FLUORESCENT FIXTURES MOUNTED 9'-11" ABOVE FLOOR, 100 FC					
TYPE A, 8 FIXTURES PER 400 S.F.					
Conduit, steel intermediate, 1/2" diam.	.185	L.F.	.35	1.18	1.53
Wire, 600V, type THWN-THHN, copper, solid, #12	.004	C.L.F.	.06	.21	.27
Fluorescent strip fixture 8' long, surface mounted, two 75W SL	.020	Ea.	1.26	2.06	3.32
Steel outlet box 4" concrete	.020	Ea.	.32	.64	.96
Steel outlet box plate with stud, 4" concrete	.020	Ea.	.21	.16	.37
Fixture hangers, flexible, 1/2" diameter, 4" long	.040	Ea.	.58	2.12	2.70
Fixture whip, THHN wire, three #12, 3/8" Greenfield	.040	Ea.	.57	.80	1.37
TOTAL			3.35	7.17	10.52

D5020 208	Fluorescent Fixtures (by Type)	COST PER S.F.		
		MAT.	INST.	TOTAL
0520	Fluorescent fixtures, type A, 8 fixtures per 400 S.F.	3.35	7.15	10.50
0560	11 fixtures per 600 S.F.	3.05	6.55	9.60
0600	17 fixtures per 1000 S.F.	2.86	6.10	8.96
0640	23 fixtures per 1600 S.F.	2.37	5.15	7.52
0680	28 fixtures per 2000 S.F.	2.35	5	7.35
0720	41 fixtures per 3000 S.F.	2.31	4.93	7.24
0800	53 fixtures per 4000 S.F.	2.22	4.77	6.99
0840	64 fixtures per 5000 S.F.	2.18	4.66	6.84
0880	Type B, 11 fixtures per 400 S.F.	5.30	10.25	15.55
0920	15 fixtures per 600 S.F.	4.79	9.25	14.04
0960	24 fixtures per 1000 S.F.	4.61	8.90	13.51
1000	35 fixtures per 1600 S.F.	4.20	8.10	12.30
1040	42 fixtures per 2000 S.F.	4.01	7.75	11.76
1080	61 fixtures per 3000 S.F.	3.86	7.45	11.31
1160	80 fixtures per 4000 S.F.	2.79	4.75	7.54
1200	98 fixtures per 5000 S.F.	3.79	7.30	11.09
1240	Type C, 11 fixtures per 400 S.F.	4.60	10.65	15.25
1280	14 fixtures per 600 S.F.	3.84	8.90	12.74
1320	23 fixtures per 1000 S.F.	3.81	8.85	12.66
1360	34 fixtures per 1600 S.F.	3.50	8.15	11.65
1400	43 fixtures per 2000 S.F.	3.60	8.35	11.95
1440	63 fixtures per 3000 S.F.	3.47	8.05	11.52
1520	81 fixtures per 4000 S.F.	3.35	7.75	11.10
1560	101 fixtures per 5000 S.F.	3.31	7.70	11.01
1600	Type D, 8 fixtures per 400 S.F.	4.23	8	12.23
1640	12 fixtures per 600 S.F.	4.23	8	12.23
1680	19 fixtures per 1000 S.F.	4	7.60	11.60
1720	27 fixtures per 1600 S.F.	3.60	6.80	10.40
1760	34 fixtures per 2000 S.F.	3.61	6.80	10.41
1800	48 fixtures per 3000 S.F.	3.38	6.40	9.78
1880	64 fixtures per 4000 S.F.	3.38	6.40	9.78
1920	79 fixtures per 5000 S.F.	3.38	6.40	9.78

D5020 Lighting and Branch Wiring

Type C. Recessed, mounted on grid ceiling suspension system, 2' x 4', four 40 watt lamps, acrylic prismatic diffusers.

5.3 watts per S.F. for 100 footcandles.

3 watts per S.F. for 57 footcandles.

System Components	QUANTITY	UNIT	COST PER S.F.		
			MAT.	INST.	TOTAL
SYSTEM D5020 210 0200					
FLUORESCENT FIXTURES RECESS MOUNTED IN CEILING					
1 WATT PER S.F., 20 FC, 5 FIXTURES PER 1000 S.F.					
Steel intermediate conduit, (IMC) 1/2" diam.	.128	L.F.	.24	.82	1.06
Wire, 600 volt, type THW, copper, solid, #12	.003	C.L.F.	.04	.17	.21
Fluorescent fixture, recessed, 2'x 4', four 40W, w/lens, for grid ceiling	.005	Ea.	.32	.68	1
Steel outlet box 4" square	.005	Ea.	.08	.16	.24
Fixture whip, Greenfield w/#12 THHN wire	.005	Ea.	.05	.04	.09
TOTAL			.73	1.87	2.60

D5020 210	Fluorescent Fixtures (by Wattage)	COST PER S.F.		
		MAT.	INST.	TOTAL
0190	Fluorescent fixtures recess mounted in ceiling			
0195	T12, standard 40 watt lamps			
0200	1 watt per S.F., 20 FC, 5 fixtures @40 watts per 1000 S.F.	.73	1.87	2.60
0240	2 watt per S.F., 40 FC, 10 fixtures @40 watt per 1000 S.F.	1.46	3.69	5.15
0280	3 watt per S.F., 60 FC, 15 fixtures @40 watt per 1000 S.F	2.19	5.55	7.74
0320	4 watt per S.F., 80 FC, 20 fixtures @40 watt per 1000 S.F.	2.91	7.40	10.31
0400	5 watt per S.F., 100 FC, 25 fixtures @40 watt per 1000 S.F.	3.65	9.25	12.90
0450	T8, energy saver 32 watt lamps			
0500	0.8 watt per S.F., 20 FC, 5 fixtures @32 watt per 1000 S.F.	.76	1.87	2.63
0520	1.6 watt per S.F., 40 FC, 10 fixtures @32 watt per 1000 S.F.	1.53	3.69	5.22
0540	2.4 watt per S.F., 60 FC, 15 fixtures @ 32 watt per 1000 S.F	2.31	5.55	7.86
0560	3.2 watt per S.F., 80 FC, 20 fixtures @32 watt per 1000 S.F.	3.06	7.40	10.46
0580	4 watt per S.F., 100 FC, 25 fixtures @32 watt per 1000 S.F.	3.82	9.25	13.07

D5020 Lighting and Branch Wiring

Type A. Recessed wide distribution reflector with flat glass lens 150 W.

Maximum spacing = 1.2 x mounting height.

13 watts per S.F. for 100 footcandles.

Type B. Recessed reflector down light with baffles 150 W.

Maximum spacing = 0.8 x mounting height.

18 watts per S.F. for 100 footcandles.

Type C. Recessed PAR–38 flood lamp with concentric louver 150 W.

Maximum spacing = 0.5 x mounting height.

19 watts per S.F. for 100 footcandles.

Type D. Recessed R–40 flood lamp with reflector skirt.

Maximum spacing = 0.7 x mounting height.

15 watts per S.F. for 100 footcandles.

System Components	QUANTITY	UNIT	COST PER S.F.		
			MAT.	INST.	TOTAL
SYSTEM D5020 214 0400					
INCANDESCENT FIXTURE RECESS MOUNTED, 100 FC					
TYPE A, 34 FIXTURES PER 400 S.F.					
Steel intermediate conduit, (IMC) 1/2" diam	1.060	L.F.	2	6.78	8.78
Wire, 600V, type THWN-THHN, copper, solid, #12	.033	C.L.F.	.49	1.91	2.40
Steel outlet box 4" square	.085	Ea.	1.35	2.72	4.07
Fixture whip, Greenfield w/#12 THHN wire	.085	Ea.	.89	.68	1.57
Incandescent fixture, recessed, w/lens, prewired, square trim, 200W	.085	Ea.	9.01	8.12	17.13
TOTAL			13.74	20.21	33.95

D5020 214	Incandescent Fixture (by Type)	COST PER S.F.		
		MAT.	INST.	TOTAL
0380	Incandescent fixture recess mounted, 100 FC			
0400	Type A, 34 fixtures per 400 S.F.	13.75	20	33.75
0440	49 fixtures per 600 S.F.	13.35	19.90	33.25
0480	63 fixtures per 800 S.F.	13	19.65	32.65
0520	90 fixtures per 1200 S.F.	12.55	19.25	31.80
0560	116 fixtures per 1600 S.F.	12.30	19.10	31.40
0600	143 fixtures per 2000 S.F.	12.20	19	31.20
0640	Type B, 47 fixtures per 400 S.F.	12.95	25.50	38.45
0680	66 fixtures per 600 S.F.	12.35	25	37.35
0720	88 fixtures per 800 S.F.	12.35	25	37.35
0760	127 fixtures per 1200 S.F.	12.10	24.50	36.60
0800	160 fixtures per 1600 S.F.	11.90	24.50	36.40
0840	206 fixtures per 2000 S.F.	11.90	24.50	36.40
0880	Type C, 51 fixtures per 400 S.F.	17.55	26.50	44.05
0920	74 fixtures per 600 S.F.	17.05	26	43.05
0960	97 fixtures per 800 S.F.	16.85	26	42.85
1000	142 fixtures per 1200 S.F.	16.55	26	42.55
1040	186 fixtures per 1600 S.F.	16.35	25.50	41.85
1080	230 fixtures per 2000 S.F.	16.25	25.50	41.75
1120	Type D, 39 fixtures per 400 S.F.	16.30	21	37.30
1160	57 fixtures per 600 S.F.	15.90	20.50	36.40
1200	75 fixtures per 800 S.F.	15.80	20.50	36.30
1240	109 fixtures per 1200 S.F.	15.45	20.50	35.95
1280	143 fixtures per 1600 S.F.	15.15	20.50	35.65
1320	176 fixtures per 2000 S.F.	15	20	35

D5020 Lighting and Branch Wiring

Type A. Recessed, wide distribution reflector with flat glass lens.

150 watt inside frost—2500 lumens per lamp.

PS–25 extended service lamp.

Maximum spacing = 1.2 x mounting height.

13 watts per S.F. for 100 footcandles.

System Components	QUANTITY	UNIT	COST PER S.F.		
			MAT.	INST.	TOTAL
SYSTEM D5020 216 0200					
INCANDESCENT FIXTURE RECESS MOUNTED, TYPE A					
1 WATT PER S.F., 8 FC, 6 FIXT PER 1000 S.F.					
Steel intermediate conduit, (IMC) 1/2" diam	.091	L.F.	.17	.58	.75
Wire, 600V, type THWN-THHN, copper, solid, #12	.002	C.L.F.	.03	.12	.15
Incandescent fixture, recessed, w/lens, prewired, square trim, 200W	.006	Ea.	.64	.57	1.21
Steel outlet box 4" square	.006	Ea.	.10	.19	.29
Fixture whip, Greenfield w/#12 THHN wire	.006	Ea.	.06	.05	.11
TOTAL			1	1.51	2.51

D5020 216	Incandescent Fixture (by Wattage)	COST PER S.F.		
		MAT.	INST.	TOTAL
0190	Incandescent fixture recess mounted, type A			
0200	1 watt per S.F., 8 FC, 6 fixtures per 1000 S.F.	1	1.51	2.51
0240	2 watt per S.F., 16 FC, 12 fixtures per 1000 S.F.	1.99	3.02	5.01
0280	3 watt per S.F., 24 FC, 18 fixtures, per 1000 S.F.	2.98	4.48	7.46
0320	4 watt per S.F., 32 FC, 24 fixtures per 1000 S.F.	3.96	6	9.96
0400	5 watt per S.F., 40 FC, 30 fixtures per 1000 S.F.	4.97	7.50	12.47

D5020 Lighting and Branch Wiring

High Bay Fluorescent Fixtures
Four T5 HO/54 watt lamps

System Components			COST PER S.F.		
	QUANTITY	UNIT	MAT.	INST.	TOTAL
SYSTEM D5020 218 0200					
FLUORESCENT HIGH BAY FIXTURE, 4 LAMP, 8'-10' ABOVE WORK PLANE					
0.5 WATT/S.F., 29 FC, 2 FIXTURES/1000 S.F.					
Steel intermediate conduit, (IMC) 1/2" diam	.100	L.F.	.19	.64	.83
Wire, 600V, type THWN-THHN, copper, solid, #10	.002	C.L.F.	.05	.13	.18
Steel outlet box 4" concrete	.002	Ea.	.03	.06	.09
Steel outlet box plate with stud, 4" concrete	.002	Ea.	.02	.02	.04
Flourescent , hi bay, 4 lamp, T-5 fixture	.002	Ea.	.50	.29	.79
TOTAL			.79	1.14	1.93

D5020 218	Fluorescent Fixture, High Bay, 8'-10' (by Wattage)	COST PER S.F.		
		MAT.	INST.	TOTAL
0190	Fluorescent high bay-4 lamp fixture, 8'-10' above work plane			
0200	.5 watt/SF, 29 FC, 2 fixtures per 1000 S.F.	.79	1.14	1.93
0400	1 watt/SF, 59 FC, 4 fixtures per 1000 S.F.	1.54	2.19	3.73
0600	1.5 watt/SF, 103 FC, 7 fixtures per 1000 S.F.	2.57	3.40	5.97
0800	2 watt/SF, 133 FC, 9 fixtures per 1000 S.F.	3.33	4.49	7.82
1000	2.5 watt/SF, 162 FC, 11 fixtures per 1000 S.F.	4.12	5.60	9.72

D5020 Lighting and Branch Wiring

HIGH BAY FIXTURES

B. Metal halide 400 watt

C. High pressure sodium 400 watt

E. Metal halide 1000 watt

F. High pressure sodium 1000 watt

G. Metal halide 1000 watt
125,000 lumen lamp

System Components	QUANTITY	UNIT	COST PER S.F.		
			MAT.	INST.	TOTAL
SYSTEM D5020 220 0880					
HIGH INTENSITY DISCHARGE FIXTURE, 8'-10' ABOVE WORK PLANE, 100 FC					
TYPE B, 8 FIXTURES PER 900 S.F.					
Steel intermediate conduit, (IMC) 1/2" diam	.460	L.F.	.87	2.94	3.81
Wire, 600V, type THWN-THHN, copper, solid, #10	.009	C.L.F.	.21	.58	.79
Steel outlet box 4" concrete	.009	Ea.	.14	.29	.43
Steel outlet box plate with stud 4" concrete	.009	Ea.	.09	.07	.16
Metal halide, hi bay, aluminum reflector, 400 W lamp	.009	Ea.	4.05	2.50	6.55
TOTAL			5.36	6.38	11.74

D5020 220	H.I.D. Fixture, High Bay, 8'-10' (by Type)	COST PER S.F.		
		MAT.	INST.	TOTAL
0500	High intensity discharge fixture, 8'-10' above work plane, 100 FC			
0880	Type B, 8 fixtures per 900 S.F.	5.35	6.40	11.75
0920	15 fixtures per 1800 S.F.	4.92	6.10	11.02
0960	24 fixtures per 3000 S.F.	4.92	6.10	11.02
1000	31 fixtures per 4000 S.F.	4.91	6.15	11.06
1040	38 fixtures per 5000 S.F.	4.91	6.15	11.06
1080	60 fixtures per 8000 S.F.	4.91	6.15	11.06
1120	72 fixtures per 10000 S.F.	4.39	5.70	10.09
1160	115 fixtures per 16000 S.F.	4.39	5.70	10.09
1200	230 fixtures per 32000 S.F.	4.39	5.70	10.09
1240	Type C, 4 fixtures per 900 S.F.	2.57	3.90	6.47
1280	8 fixtures per 1800 S.F.	2.59	3.92	6.51
1320	13 fixtures per 3000 S.F.	2.57	3.90	6.47
1360	17 fixtures per 4000 S.F.	2.57	3.90	6.47
1400	21 fixtures per 5000 S.F.	2.47	3.57	6.04
1440	33 fixtures per 8000 S.F.	2.45	3.51	5.96
1480	40 fixtures per 10000 S.F.	2.39	3.32	5.71
1520	63 fixtures per 16000 S.F.	2.39	3.32	5.71
1560	126 fixtures per 32000 S.F.	2.39	3.32	5.71

D5020 Lighting and Branch Wiring

HIGH BAY FIXTURES

B. Metal halide 400 watt

C. High pressure sodium 400 watt

E. Metal halide 1000 watt

F. High pressure sodium 1000 watt

G. Metal halide 1000 watt
125,000 lumen lamp

System Components			COST PER S.F.		
	QUANTITY	UNIT	MAT.	INST.	TOTAL
SYSTEM D5020 222 0240					
HIGH INTENSITY DISCHARGE FIXTURE, 8'-10' ABOVE WORK PLANE					
1 WATT/S.F., TYPE B, 29 FC, 2 FIXTURES/1000 S.F.					
Steel intermediate conduit, (IMC) 1/2" diam	.100	L.F.	.19	.64	.83
Wire, 600V, type THWN-THHN, copper, solid, #10	.002	C.L.F.	.05	.13	.18
Steel outlet box 4" concrete	.002	Ea.	.03	.06	.09
Steel outlet box plate with stud, 4" concrete	.002	Ea.	.02	.02	.04
Metal halide, hi bay, aluminum reflector, 400 W lamp	.002	Ea.	.90	.56	1.46
TOTAL			1.19	1.41	2.60

D5020 222	H.I.D. Fixture, High Bay, 8'-10' (by Wattage)		COST PER S.F.		
			MAT.	INST.	TOTAL
0190	High intensity discharge fixture, 8'-10' above work plane				
0240	1 watt/S.F., type B, 29 FC, 2 fixtures/1000 S.F.		1.19	1.41	2.60
0280	Type C, 54 FC, 2 fixtures/1000 S.F.		1.01	1.06	2.07
0400	2 watt/S.F., type B, 59 FC, 4 fixtures/1000 S.F.		2.34	2.72	5.06
0440	Type C, 108 FC, 4 fixtures/1000 S.F.	R265723 -05	2.01	2.07	4.08
0560	3 watt/S.F., type B, 103 FC, 7 fixtures/1000 S.F.		3.98	4.34	8.32
0600	Type C, 189 FC, 6 fixtures/1000 S.F.		3.24	2.68	5.92
0720	4 watt/S.F., type B, 133 FC, 9 fixtures/1000 S.F.		5.15	5.70	10.85
0760	Type C, 243 FC, 9 fixtures/1000 S.F.		4.51	4.61	9.12
0880	5 watt/S.F., type B, 162 FC, 11 fixtures/1000 S.F.		6.35	7.05	13.40
0920	Type C, 297 FC, 11 fixtures/1000 S.F.		5.55	5.70	11.25

D5020 Lighting and Branch Wiring

D5020 Lighting and Branch Wiring

HIGH BAY FIXTURES

B. Metal halide 400 watt

C. High pressure sodium 400 watt

E. Metal halide 1000 watt

F. High pressure sodium 1000 watt

G. Metal halide 1000 watt
125,000 lumen lamp

System Components	QUANTITY	UNIT	COST PER S.F.		
			MAT.	INST.	TOTAL
SYSTEM D5020 224 1240					
HIGH INTENSITY DISCHARGE FIXTURE, 16′ ABOVE WORK PLANE, 100 FC					
TYPE C, 5 FIXTURES PER 900 S.F.					
Steel intermediate conduit, (IMC) 1/2″ diam	.260	L.F.	.49	1.66	2.15
Wire, 600V, type THWN-THHN, copper, solid, #10	.007	C.L.F.	.16	.45	.61
Steel outlet box 4″ concrete	.006	Ea.	.10	.19	.29
Steel outlet box plate with stud, 4″ concrete	.006	Ea.	.06	.05	.11
High pressure sodium, hi bay, aluminum reflector, 400 W lamp	.006	Ea.	2.49	1.67	4.16
TOTAL			3.30	4.02	7.32

D5020 224	H.I.D. Fixture, High Bay, 16′ (by Type)	COST PER S.F.		
		MAT.	INST.	TOTAL
0510	High intensity discharge fixture, 16′ above work plane, 100 FC			
1240	Type C, 5 fixtures per 900 S.F.	3.30	4.02	7.32
1280	9 fixtures per 1800 S.F.	3.02	4.21	7.23
1320	15 fixtures per 3000 S.F.	3.02	4.21	7.23
1360	18 fixtures per 4000 S.F.	2.89	3.76	6.65
1400	22 fixtures per 5000 S.F.	2.89	3.76	6.65
1440	36 fixtures per 8000 S.F.	2.89	3.76	6.65
1480	42 fixtures per 10,000 S.F.	2.57	3.89	6.46
1520	65 fixtures per 16,000 S.F.	2.57	3.89	6.46
1600	Type G, 4 fixtures per 900 S.F.	4.20	6.35	10.55
1640	6 fixtures per 1800 S.F.	3.53	5.95	9.48
1720	9 fixtures per 4000 S.F.	2.90	5.80	8.70
1760	11 fixtures per 5000 S.F.	2.90	5.80	8.70
1840	21 fixtures per 10,000 S.F.	2.73	5.20	7.93
1880	33 fixtures per 16,000 S.F.	2.73	5.20	7.93

D5020 Lighting and Branch Wiring

HIGH BAY FIXTURES

B. Metal halide 400 watt

C. High pressure sodium 400 watt

E. Metal halide 1000 watt

F. High pressure sodium 1000 watt

G. Metal halide 1000 watt
125,000 lumen lamp

System Components	QUANTITY	UNIT	COST PER S.F.		
			MAT.	INST.	TOTAL
SYSTEM D5020 226 0240					
HIGH INTENSITY DISCHARGE FIXTURE, 16′ ABOVE WORK PLANE					
1 WATT/S.F., TYPE E, 42 FC, 1 FIXTURE/1000 S.F.					
Steel intermediate conduit, (IMC) 1/2″ diam	.160	L.F.	.30	1.02	1.32
Wire, 600V, type THWN-THHN, copper, solid, #10	.003	C.L.F.	.07	.19	.26
Steel outlet box 4″ concrete	.001	Ea.	.02	.03	.05
Steel outlet box plate with stud, 4″ concrete	.001	Ea.	.01	.01	.02
Metal halide, hi bay, aluminum reflector, 1000 W lamp	.001	Ea.	.65	.32	.97
TOTAL			1.05	1.57	2.62

D5020 226	H.I.D. Fixture, High Bay, 16′ (by Wattage)		COST PER S.F.		
			MAT.	INST.	TOTAL
0190	High intensity discharge fixture, 16′ above work plane				
0240	1 watt/S.F., type E, 42 FC, 1 fixture/1000 S.F.		1.05	1.57	2.62
0280	Type G, 52 FC, 1 fixture/1000 S.F.		1.05	1.57	2.62
0320	Type C, 54 FC, 2 fixture/1000 S.F.	R265723 -05	1.23	1.76	2.99
0440	2 watt/S.F., type E, 84 FC, 2 fixture/1000 S.F.		2.10	3.22	5.32
0480	Type G, 105 FC, 2 fixture/1000 S.F.		2.10	3.22	5.32
0520	Type C, 108 FC, 4 fixture/1000 S.F.		2.45	3.52	5.97
0640	3 watt/S.F., type E, 126 FC, 3 fixture/1000 S.F.		3.17	4.79	7.96
0680	Type G, 157 FC, 3 fixture/1000 S.F.		3.17	4.79	7.96
0720	Type C, 162 FC, 6 fixture/1000 S.F.		3.69	5.30	8.99
0840	4 watt/S.F., type E, 168 FC, 4 fixture/1000 S.F.		4.22	6.45	10.67
0880	Type G, 210 FC, 4 fixture/1000 S.F.		4.22	6.45	10.67
0920	Type C, 243 FC, 9 fixture/1000 S.F.		5.35	7.35	12.70
1040	5 watt/S.F., type E, 210 FC, 5 fixture/1000 S.F.		5.25	8	13.25
1080	Type G, 262 FC, 5 fixture/1000 S.F.		5.25	8	13.25
1120	Type C, 297 FC, 11 fixture/1000 S.F.		6.60	9.15	15.75

D5020   Lighting and Branch Wiring

HIGH BAY FIXTURES

B. Metal halide 400 watt

C. High pressure sodium 400 watt

E. Metal halide 1000 watt

F. High pressure sodium 1000 watt

G. Metal halide 1000 watt
   125,000 lumen lamp

System Components	QUANTITY	UNIT	COST PER S.F.		
			MAT.	INST.	TOTAL
SYSTEM D5020 228 1240					
HIGH INTENSITY DISCHARGE FIXTURE, 20′ ABOVE WORK PLANE, 100 FC					
TYPE C, 6 FIXTURES PER 900 S.F.					
Steel intermediate conduit, (IMC) 1/2″ diam	.350	L.F.	.66	2.24	2.90
Wire, 600V, type THWN-THHN, copper, solid, #10.	.011	C.L.F.	.26	.70	.96
Steel outlet box 4″ concrete	.007	Ea.	.11	.22	.33
Steel outlet box plate with stud, 4″ concrete	.007	Ea.	.07	.06	.13
High pressure sodium, hi bay, aluminum reflector, 400 W lamp	.007	Ea.	2.91	1.95	4.86
TOTAL			4.01	5.17	9.18

D5020 228	H.I.D. Fixture, High Bay, 20′ (by Type)	COST PER S.F.		
		MAT.	INST.	TOTAL
0510	High intensity discharge fixture 20′ above work plane, 100 FC			
1240	Type C, 6 fixtures per 900 S.F.	4.01	5.15	9.16
1280	10 fixtures per 1800 S.F.	3.57	4.85	8.42
1320	16 fixtures per 3000 S.F.	3.17	4.66	7.83
1360	20 fixtures per 4000 S.F.	3.17	4.66	7.83
1400	24 fixtures per 5000 S.F.	3.17	4.66	7.83
1440	38 fixtures per 8000 S.F.	3.11	4.47	7.58
1520	68 fixtures per 16000 S.F.	2.89	4.91	7.80
1560	132 fixtures per 32000 S.F.	2.89	4.91	7.80
1600	Type G, 4 fixtures per 900 S.F.	4.18	6.25	10.43
1640	6 fixtures per 1800 S.F.	3.54	5.95	9.49
1680	7 fixtures per 3000 S.F.	3.48	5.75	9.23
1720	10 fixtures per 4000 S.F.	3.46	5.70	9.16
1760	11 fixtures per 5000 S.F.	3.01	6.10	9.11
1800	18 fixtures per 8000 S.F.	3.01	6.10	9.11
1840	22 fixtures per 10000 S.F.	3.01	6.10	9.11
1880	34 fixtures per 16000 S.F.	2.95	5.90	8.85
1920	66 fixtures per 32000 S.F.	2.80	5.40	8.20

Note: Between rows 1280 and 1320 there is a box reading "R265723 -05".

D5020 Lighting and Branch Wiring

HIGH BAY FIXTURES

B. Metal halide 400 watt

C. High pressure sodium 400 watt

E. Metal halide 1000 watt

F. High pressure sodium 1000 watt

G. Metal halide 1000 watt
   125,000 lumen lamp

System Components	QUANTITY	UNIT	COST PER S.F.		
			MAT.	INST.	TOTAL
SYSTEM D5020 230 0240					
HIGH INTENSITY DISCHARGE FIXTURE, 20' ABOVE WORK PLANE					
1 WATT/S.F., TYPE E, 40 FC, 1 FIXTURE 1000 S.F.					
Steel intermediate conduit, (IMC) 1/2" diam	.160	L.F.	.30	1.02	1.32
Wire, 600V, type THWN-THHN, copper, solid, #10	.005	C.L.F.	.12	.32	.44
Steel outlet box 4" concrete	.001	Ea.	.02	.03	.05
Steel outlet box plate with stud, 4" concrete	.001	Ea.	.01	.01	.02
Metal halide, hi bay, aluminum reflector, 1000 W lamp	.001	Ea.	.65	.32	.97
TOTAL			1.10	1.70	2.80

D5020 230	H.I.D. Fixture, High Bay, 20' (by Wattage)	COST PER S.F.		
		MAT.	INST.	TOTAL
0190	High intensity discharge fixture, 20' above work plane			
0240	1 watt/S.F., type E, 40 FC, 1 fixture/1000 S.F.	1.10	1.70	2.80
0280	Type G, 50 FC, 1 fixture/1000 S.F.	1.10	1.70	2.80
0320	Type C, 52 FC, 2 fixtures/1000 S.F.	1.30	1.97	3.27
0440	2 watt/S.F., type E, 81 FC, 2 fixtures/1000 S.F.	2.18	3.41	5.59
0480	Type G, 101 FC, 2 fixtures/1000 S.F.	2.18	3.41	5.59
0520	Type C, 104 FC, 4 fixtures/1000 S.F.	2.57	3.87	6.44
0640	3 watt/S.F., type E, 121 FC, 3 fixtures/1000 S.F.	3.28	5.10	8.38
0680	Type G, 151 FC, 3 fixtures/1000 S.F.	3.28	5.10	8.38
0720	Type C, 155 FC, 6 fixtures/1000 S.F.	3.88	5.85	9.73
0840	4 watt/S.F., type E, 161 FC, 4 fixtures/1000 S.F.	4.36	6.80	11.16
0880	Type G, 202 FC, 4 fixtures/1000 S.F.	4.36	6.80	11.16
0920	Type C, 233 FC, 9 fixtures/1000 S.F.	5.60	8.05	13.65
1040	5 watt/S.F., type E, 202 FC, 5 fixtures/1000 S.F.	5.45	8.50	13.95
1080	Type G, 252 FC, 5 fixtures/1000 S.F.	5.45	8.50	13.95
1120	Type C, 285 FC, 11 fixtures/1000 S.F.	6.90	10.05	16.95

R265723 -05 (in row 0320)

D5020 Lighting and Branch Wiring

HIGH BAY FIXTURES

B. Metal halide 400 watt

C. High pressure sodium 400 watt

E. Metal halide 1000 watt

F. High pressure sodium 1000 watt

G. Metal halide 1000 watt
125,000 lumen lamp

System Components	QUANTITY	UNIT	COST PER S.F.		
			MAT.	INST.	TOTAL
SYSTEM D5020 232 1240					
HIGH INTENSITY DISCHARGE FIXTURE, 30' ABOVE WORK PLANE, 100 FC					
TYPE F, 4 FIXTURES PER 900 S.F.					
Steel intermediate conduit, (IMC) 1/2" diam	.580	L.F.	1.10	3.71	4.81
Wire, 600V, type THWN-THHN, copper, solid, #10	.018	C.L.F.	.42	1.15	1.57
Steel outlet box 4" concrete	.004	Ea.	.06	.13	.19
Steel outlet box plate with stud, 4" concrete	.004	Ea.	.04	.03	.07
High pressure sodium, hi bay, aluminum, 1000 W lamp	.004	Ea.	2.38	1.28	3.66
TOTAL			4	6.30	10.30

D5020 232	H.I.D. Fixture, High Bay, 30' (by Type)		COST PER S.F.		
			MAT.	INST.	TOTAL
0510	High intensity discharge fixture, 30' above work plane, 100 FC				
1240	Type F, 4 fixtures per 900 S.F.		4	6.30	10.30
1280	6 fixtures per 1800 S.F.	R265723 -05	3.41	6	9.41
1320	8 fixtures per 3000 S.F.		3.41	6	9.41
1360	9 fixtures per 4000 S.F.		2.79	5.70	8.49
1400	10 fixtures per 5000 S.F.		2.79	5.70	8.49
1440	17 fixtures per 8000 S.F.		2.79	5.70	8.49
1480	18 fixtures per 10,000 S.F.		2.66	5.25	7.91
1520	27 fixtures per 16,000 S.F.		2.66	5.25	7.91
1560	52 fixtures per 32000 S.F.		2.64	5.20	7.84
1600	Type G, 4 fixtures per 900 S.F.		4.31	6.70	11.01
1640	6 fixtures per 1800 S.F.		3.57	6.05	9.62
1680	9 fixtures per 3000 S.F.		3.52	5.90	9.42
1720	11 fixtures per 4000 S.F.		3.46	5.70	9.16
1760	13 fixtures per 5000 S.F.		3.46	5.70	9.16
1800	21 fixtures per 8000 S.F.		3.46	5.70	9.16
1840	23 fixtures per 10,000 S.F.		3.04	6.05	9.09
1880	36 fixtures per 16,000 S.F.		3.04	6.05	9.09
1920	70 fixtures per 32,000 S.F.		3.04	6.05	9.09

D5020   Lighting and Branch Wiring

HIGH BAY FIXTURES

B. Metal halide 400 watt

C. High pressure sodium 400 watt

E. Metal halide 1000 watt

F. High pressure sodium 1000 watt

G. Metal halide 1000 watt
   125,000 lumen lamp

System Components	QUANTITY	UNIT	COST PER S.F.		
			MAT.	INST.	TOTAL
SYSTEM D5020 234 0240					
HIGH INTENSITY DISCHARGE FIXTURE, 30′ ABOVE WORK PLANE					
1 WATT/S.F., TYPE E, 37 FC, 1 FIXTURE/1000 S.F.					
Steel intermediate conduit, (IMC) 1/2″ diam	.196	L.F.	.37	1.25	1.62
Wire, 600V type THWN-THHN, copper, solid, #10	.006	C.L.F.	.14	.38	.52
Steel outlet box 4″ concrete	.001	Ea.	.02	.03	.05
Steel outlet box plate with stud, 4″ concrete	.001	Ea.	.01	.01	.02
Metal halide, hi bay, aluminum reflector, 1000 W lamp	.001	Ea.	.65	.32	.97
TOTAL			1.19	1.99	3.18

D5020 234	H.I.D. Fixture, High Bay, 30′ (by Wattage)		COST PER S.F.		
			MAT.	INST.	TOTAL
0190	High intensity discharge fixture, 30′ above work plane				
0240	1 watt/S.F., type E, 37 FC, 1 fixture/1000 S.F.		1.19	1.99	3.18
0280	Type G, 45 FC., 1 fixture/1000 S.F.	R265723	1.19	1.99	3.18
0320	Type F, 50 FC, 1 fixture/1000 S.F.	-05	1	1.57	2.57
0440	2 watt/S.F., type E, 74 FC, 2 fixtures/1000 S.F.		2.36	4	6.36
0480	Type G, 92 FC, 2 fixtures/1000 S.F.		2.36	4	6.36
0520	Type F, 100 FC, 2 fixtures/1000 S.F.		2.01	3.19	5.20
0640	3 watt/S.F., type E, 110 FC, 3 fixtures/1000 S.F.		3.58	6.05	9.63
0680	Type G, 138 FC, 3 fixtures/1000 S.F.		3.58	6.05	9.63
0720	Type F, 150 FC, 3 fixtures/1000 S.F.		3.02	4.75	7.77
0840	4 watt/S.F., type E, 148 FC, 4 fixtures/1000 S.F.		4.75	8.05	12.80
0880	Type G, 185 FC, 4 fixtures/1000 S.F.		4.75	8.05	12.80
0920	Type F, 200 FC, 4 fixtures/1000 S.F.		4.02	6.40	10.42
1040	5 watt/S.F., type E, 185 FC, 5 fixtures/1000 S.F.		5.95	10.05	16
1080	Type G, 230 FC, 5 fixtures/1000 S.F.		5.95	10.05	16
1120	Type F, 250 FC, 5 fixtures/1000 S.F.		5.05	7.95	13

D5020 Lighting and Branch Wiring

LOW BAY FIXTURES

J. Metal halide 250 watt

K. High pressure sodium 150 watt

System Components	QUANTITY	UNIT	COST PER S.F.		
			MAT.	INST.	TOTAL
SYSTEM D5020 236 0920					
HIGH INTENSITY DISCHARGE FIXTURE, 8'-10' ABOVE WORK PLANE, 50 FC					
TYPE J, 13 FIXTURES PER 1800 S.F.					
Steel intermediate conduit, (IMC) 1/2" diam	.550	L.F.	1.04	3.52	4.56
Wire, 600V, type THWN-THHN, copper, solid, #10	.012	C.L.F.	.28	.77	1.05
Steel outlet box 4" concrete	.007	Ea.	.11	.22	.33
Steel outlet box plate with stud, 4" concrete	.007	Ea.	.07	.06	.13
Metal halide, lo bay, aluminum reflector, 250 W DX lamp	.007	Ea.	2.77	1.40	4.17
TOTAL			4.27	5.97	10.24

D5020 236	H.I.D. Fixture, Low Bay, 8'-10' (by Type)		COST PER S.F.		
			MAT.	INST.	TOTAL
0510	High intensity discharge fixture, 8'-10' above work plane, 50 FC				
0880	Type J, 7 fixtures per 900 S.F.		4.63	6	10.63
0920	13 fixtures per 1800 S.F.	R265723 -05	4.27	5.95	10.22
0960	21 fixtures per 3000 S.F.		4.27	5.95	10.22
1000	28 fixtures per 4000 S.F.		4.27	5.95	10.22
1040	35 fixtures per 5000 S.F.		4.27	5.95	10.22
1120	62 fixtures per 10,000 S.F.		3.85	5.75	9.60
1160	99 fixtures per 16,000 S.F.		3.85	5.75	9.60
1200	199 fixtures per 32,000 S.F.		3.85	5.75	9.60
1240	Type K, 9 fixtures per 900 S.F.		4.67	5.15	9.82
1280	16 fixtures per 1800 S.F.		4.30	4.97	9.27
1320	26 fixtures per 3000 S.F.		4.28	4.91	9.19
1360	31 fixtures per 4000 S.F.		3.94	4.80	8.74
1400	39 fixtures per 5000 S.F.		3.94	4.80	8.74
1440	62 fixtures per 8000 S.F.		3.94	4.80	8.74
1480	78 fixtures per 10,000 S.F.		3.94	4.80	8.74
1520	124 fixtures per 16,000 S.F.		3.89	4.67	8.56
1560	248 fixtures per 32,000 S.F.		3.75	5.20	8.95

D5020 Lighting and Branch Wiring

LOW BAY FIXTURES
J. Metal halide 250 watt
K. High pressure sodium 150 watt

System Components	QUANTITY	UNIT	COST PER S.F.		
			MAT.	INST.	TOTAL
SYSTEM D5020 238 0240					
HIGH INTENSITY DISCHARGE FIXTURE, 8'-10' ABOVE WORK PLANE					
1 WATT/S.F., TYPE J, 30 FC, 4 FIXTURES/1000 S.F.					
Steel intermediate conduit, (IMC) 1/2" diam	.280	L.F.	.53	1.79	2.32
Wire, 600V, type THWN-THHN, copper, solid, #10	.008	C.L.F.	.19	.51	.70
Steel outlet box 4" concrete	.004	Ea.	.06	.13	.19
Steel outlet box plate with stud, 4" concrete	.004	Ea.	.04	.03	.07
Metal halide, lo bay, aluminum reflector, 250 W DX lamp	.004	Ea.	1.58	.80	2.38
TOTAL			2.40	3.26	5.66

D5020 238	H.I.D. Fixture, Low Bay, 8'-10' (by Wattage)	COST PER S.F.		
		MAT.	INST.	TOTAL
0190	High intensity discharge fixture, 8'-10' above work plane			
0240	1 watt/S.F., type J, 30 FC, 4 fixtures/1000 S.F.	2.40	3.26	5.66
0280	Type K, 29 FC, 5 fixtures/1000 S.F.	2.45	2.92	5.37
0400	2 watt/S.F., type J, 52 FC, 7 fixtures/1000 S.F.	4.30	6	10.30
0440	Type K, 63 FC, 11 fixtures/1000 S.F.	5.30	6.15	11.45
0560	3 watt/S.F., type J, 81 FC, 11 fixtures/1000 S.F.	6.65	9.05	15.70
0600	Type K, 92 FC, 16 fixtures/1000 S.F.	7.75	9.10	16.85
0720	4 watt/S.F., type J, 103 FC, 14 fixtures/1000 S.F.	8.55	11.90	20.45
0760	Type K, 127 FC, 22 fixtures/1000 S.F.	10.60	12.30	22.90
0880	5 watt/S.F., type J, 133 FC, 18 fixtures/1000 S.F.	10.95	15	25.95
0920	Type K, 155 FC, 27 fixtures/1000 S.F.	13.05	15.25	28.30

R265723 -05

D5020 Lighting and Branch Wiring

LOW BAY FIXTURES

J. Metal halide 250 watt

K. High pressure sodium 150 watt

System Components	QUANTITY	UNIT	COST PER S.F.		
			MAT.	INST.	TOTAL
SYSTEM D5020 240 0880					
HIGH INTENSITY DISCHARGE FIXTURE, 16′ ABOVE WORK PLANE, 50 FC					
TYPE J, 9 FIXTURES PER 900 S.F.					
Steel intermediate conduit, (IMC) 1/2″ diam	.630	L.F.	1.19	4.03	5.22
Wire, 600V type, THWN-THHN, copper, solid, #10	.012	C.L.F.	.28	.77	1.05
Steel outlet box 4″ concrete	.010	Ea.	.16	.32	.48
Steel outlet box plate with stud, 4″ concrete	.010	Ea.	.11	.08	.19
Metal halide, lo bay, aluminum reflector, 250 W DX lamp	.010	Ea.	3.95	2	5.95
TOTAL			5.69	7.20	12.89

D5020 240	H.I.D. Fixture, Low Bay, 16′ (by Type)		COST PER S.F.		
			MAT.	INST.	TOTAL
0510	High intensity discharge fixture, 16′ above work plane, 50 FC				
0880	Type J, 9 fixtures per 900 S.F.		5.70	7.20	12.90
0920	14 fixtures per 1800 S.F.	R265723 -05	4.87	6.80	11.67
0960	24 fixtures per 3000 S.F.		4.91	6.90	11.81
1000	32 fixtures per 4000 S.F.		4.91	6.90	11.81
1040	35 fixtures per 5000 S.F.		4.51	6.75	11.26
1080	56 fixtures per 8000 S.F.		4.51	6.75	11.26
1120	70 fixtures per 10,000 S.F.		4.51	6.75	11.26
1160	111 fixtures per 16,000 S.F.		4.51	6.75	11.26
1200	222 fixtures per 32,000 S.F.		4.51	6.75	11.26
1240	Type K, 11 fixtures per 900 S.F.		5.75	6.70	12.45
1280	20 fixtures per 1800 S.F.		5.30	6.25	11.55
1320	29 fixtures per 3000 S.F.		4.97	6.10	11.07
1360	39 fixtures per 4000 S.F.		4.95	6.05	11
1400	44 fixtures per 5000 S.F.		4.60	5.95	10.55
1440	62 fixtures per 8000 S.F.		4.60	5.95	10.55
1480	87 fixtures per 10,000 S.F.		4.60	5.95	10.55
1520	138 fixtures per 16,000 S.F.		4.60	5.95	10.55

D5020 Lighting and Branch Wiring

LOW BAY FIXTURES

J. Metal halide 250 watt

K. High pressure sodium 150 watt

System Components	QUANTITY	UNIT	COST PER S.F.		
			MAT.	INST.	TOTAL
SYSTEM D5020 242 0240					
HIGH INTENSITY DISCHARGE FIXTURE, 16' ABOVE WORK PLANE					
1 WATT/S.F., TYPE J, 28 FC, 4 FIXTURES/1000 S.F.					
Steel intermediate conduit, (IMC) 1/2" diam	.328	L.F.	.62	2.10	2.72
Wire, 600V, type THWN-THHN, copper, solid, #10	.010	C.L.F.	.24	.64	.88
Steel outlet box 4" concrete	.004	Ea.	.06	.13	.19
Steel outlet box plate with stud, 4" concrete	.004	Ea.	.04	.03	.07
Metal halide, lo bay, aluminum reflector, 250 W DX lamp	.004	Ea.	1.58	.80	2.38
TOTAL			2.54	3.70	6.24

D5020 242	H.I.D. Fixture, Low Bay, 16' (by Wattage)		COST PER S.F.		
			MAT.	INST.	TOTAL
0190	High intensity discharge fixture, mounted 16' above work plane				
0240	1 watt/S.F., type J, 28 FC, 4 fixt./1000 S.F.		2.54	3.70	6.24
0280	Type K, 27 FC, 5 fixt./1000 S.F.	R265723	2.83	4.14	6.97
0400	2 watt/S.F., type J, 48 FC, 7 fixt/1000 S.F.	-05	4.66	7.15	11.81
0440	Type K, 58 FC, 11 fixt/1000 S.F.		6	8.45	14.45
0560	3 watt/S.F., type J, 75 FC, 11 fixt/1000 S.F.		7.20	10.85	18.05
0600	Type K, 85 FC, 16 fixt/1000 S.F.		8.85	12.60	21.45
0720	4 watt/S.F., type J, 95 FC, 14 fixt/1000 S.F.		9.30	14.30	23.60
0760	Type K, 117 FC, 22 fixt/1000 S.F.		12.05	16.95	29
0880	5 watt/S.F., type J, 122 FC, 18 fixt/1000 S.F.		11.85	18	29.85
0920	Type K, 143 FC, 27 fixt/1000 S.F.		14.85	21	35.85

D5020 Lighting and Branch Wiring

Light level sensor

Wall switch

To building control system

0-10V signal to dimmable ballasts (up to 50)

Daylighting controller

Daylight Dimming System

System Components	QUANTITY	UNIT	COST PER S.F.		
			MAT.	INST.	TOTAL
SYSTEM D5020 290 0800					
DAYLIGHT DIMMING CONTROL SYSTEM					
5 FIXTURES PER 1000 M.S.F.					
Tray cable, type TC, copper #16-4 conductor	.300	C.L.F.	.01	.03	.04
Wire, 600 volt, type THWN-THHN, copper, solid, #12	.320	C.L.F.		.02	.02
Conduit (EMT) , to 15′ H, incl 2 termn,2 elb&11 bm clp per 100′, 3/4″	10.000	L.F.	.01	.05	.06
Cabinet, hinged, sheet metal, NEMA 1, 12″W x 12″H x 4″D	.200	Ea.	.01	.02	.03
Lighting control module	.200	Ea.	.07	.06	.13
Dimmable ballast three-lamp	5.000	Ea.	.67	.46	1.13
Daylight level sensor, wall mounted, on/off or dimming	.200	Ea.	.03	.02	.05
Automatic wall switches	.200	Ea.	.01	.01	.02
Remote power pack	.100	Ea.		.01	.01
TOTAL			.81	.68	1.49

D5020 290	Daylight Dimming System	COST PER S.F.		
		MAT.	INST.	TOTAL
0500	Daylight Dimming System (no fixtures or fixture power)			
0800	5 fixtures per 1000 S.F.	.81	.68	1.49
1000	10 fixtures per 1000 S.F.	1.61	1.27	2.88
2000	12 fixtures per 1000 S.F.	1.93	1.51	3.44
3000	15 fixtures per 1000 S.F.	2.41	1.88	4.29
4000	20 fixtures per 1000 S.F.	3.21	2.48	5.69
5000	25 fixtures per 1000 S.F.	4.02	3.08	7.10
6000	50 fixtures per 1000 S.F.	8.05	6.10	14.15

D5020 Lighting and Branch Wiring

Wall switch

Time switch

Occupancy sensor

To building control system

Relay for power to level 1 lights

Relay for power to level 2 lights

Lighting controller

Daylight On/Off Control System

System Components			COST PER S.F.		
	QUANTITY	UNIT	MAT.	INST.	TOTAL
SYSTEM D5020 295 0800					
LIGHTING ON/OFF CONTROL SYSTEM					
5 FIXTURES PER 1000 S.F.					
Tray cable, type TC, copper #16-4 conductor	.600	C.L.F.	.02	.05	.07
Wire, 600 volt, type THWN-THHN, copper, solid, #12	.320	C.L.F.		.02	.02
Conduit (EMT) , to 15' H, incl 2 termn,2 elb&11 bm clp per 100', 3/4"	10.000	L.F.	.01	.05	.06
Cabinet, hinged, sheet metal, NEMA 1, 12"W x 12"H x 4"D	.400	Ea.	.01	.04	.05
Relays, 120 V or 277 V standard	.800	Ea.	.04	.04	.08
24 hour dial with reserve power	.400	Ea.	.26	.07	.33
Lighting control module	.200	Ea.	.07	.06	.13
Occupancy sensors, passive infrared ceiling mounted	.400	Ea.	.04	.04	.08
Automatic wall switches	.400	Ea.	.03	.01	.04
Remote power pack	.200	Ea.	.01	.01	.02
TOTAL			.49	.39	.88

D5020 295	Lighting On/Off Control System	COST PER S.F.		
		MAT.	INST.	TOTAL
0500	Includes occupancy and time switching (no fixtures or fixture power)			
0800	5 fixtures per 1000 SF	.49	.39	.88
1000	10 fixtures per 1000 S.F.	.55	.46	1.01
2000	12 fixtures per 1000 S.F.	.65	.53	1.18
3000	15 fixtures per 1000 S.F.	.81	.66	1.47
4000	20 fixtures per 1000 S.F.	1.08	.86	1.94
5000	25 fixtures per 1000 S.F.	1.37	1.07	2.44
6000	50 fixtures per 1000 S.F.	2.71	2.03	4.74

D5030 Communications and Security

Description: System below includes telephone fitting installed. Does not include cable.

When poke thru fittings and telepoles are used for power, they can also be used for telephones at a negligible additional cost.

System Components	QUANTITY	UNIT	COST PER S.F.		
			MAT.	INST.	TOTAL
SYSTEM D5030 310 0200					
TELEPHONE SYSTEMS, UNDERFLOOR DUCT, 5' ON CENTER, LOW DENSITY					
Underfloor duct 7-1/4" w/insert 2' O.C. 1-3/8" x 7-1/4" super duct	.190	L.F.	6.94	2.43	9.37
Vertical elbow for underfloor superduct, 7-1/4", included					
Underfloor duct conduit adapter, 2" x 1-1/4", included					
Underfloor duct junction box, single duct, 7-1/4" x 3 1/8"	.003	Ea.	1.62	.48	2.10
Underfloor junction box carpet pan	.003	Ea.	1.16	.02	1.18
Underfloor duct outlet, low tension	.004	Ea.	.43	.32	.75
TOTAL			10.15	3.25	13.40

D5030 310	Telephone Systems	COST PER S.F.		
		MAT.	INST.	TOTAL
0200	Telephone systems, underfloor duct, 5' on center, low density	10.15	3.25	13.40
0240	5' on center, high density	10.60	3.57	14.17
0280	7' on center, low density	8.10	2.71	10.81
0320	7' on center, high density	8.55	3.03	11.58
0400	Poke thru fittings, low density	1.16	1.16	2.32
0440	High density	2.33	2.31	4.64
0520	Telepoles, low density	1.15	.69	1.84
0560	High density	2.30	1.38	3.68
0640	Conduit system with floor boxes, low density	1.38	1.14	2.52
0680	High density	2.74	2.25	4.99
1020	Telephone wiring for offices & laboratories, 8 jacks/M.S.F.	.41	1.77	2.18

D5030 810	Security & Detection Systems	COST EACH		
		MAT.	INST.	TOTAL
0200	Security system, head end equipment	29,300	4,250	33,550
0240	Security system, door /window contact biased, box, conduit & cable	258	480	738
0280	Security system, door /window contact balanced, box, conduit & cable	258	480	738
0440	Security system, proximity card reader, box, conduit & cable	460	640	1,100
1600	Card control entrance system, to 6 zone, including hardware for 100 doors	42,000	29,600	71,600

Antenna
Speaker
Volume Control
Amplifier

Sound System Includes AM–FM antenna, outlets, rigid conduit, and copper wire.

Fire Detection System Includes pull stations, signals, smoke and heat detectors, rigid conduit, and copper wire.

Intercom System Includes master and remote stations, rigid conduit, and copper wire.

Master Clock System Includes clocks, bells, rigid conduit, and copper wire.

Master TV Antenna Includes antenna, VHF–UHF reception and distribution, rigid conduit, and copper wire.

System Components	QUANTITY	UNIT	COST EACH		
			MAT.	INST.	TOTAL
SYSTEM D5030 910 0220					
SOUND SYSTEM, INCLUDES OUTLETS, BOXES, CONDUIT & WIRE					
Steel intermediate conduit, (IMC) 1/2" diam	1200.000	L.F.	2,268	7,680	9,948
Wire sound shielded w/drain, #22-2 conductor	15.500	C.L.F.	271.25	1,240	1,511.25
Sound system speakers ceiling or wall	12.000	Ea.	1,536	960	2,496
Sound system volume control	12.000	Ea.	1,158	960	2,118
Sound system amplifier, 250 Watts	1.000	Ea.	1,375	640	2,015
Sound system antenna, AM FM	1.000	Ea.	148	160	308
Sound system monitor panel	1.000	Ea.	425	160	585
Sound system cabinet	1.000	Ea.	920	640	1,560
Steel outlet box 4" square	12.000	Ea.	27.72	384	411.72
Steel outlet box 4" plaster rings	12.000	Ea.	37.32	120	157.32
TOTAL			8,166.29	12,944	21,110.29

D5030 910	Communication & Alarm Systems	COST EACH		
		MAT.	INST.	TOTAL
0200	Communication & alarm systems, includes outlets, boxes, conduit & wire			
0210	Sound system, 6 outlets	5,700	8,100	13,800
0220	12 outlets	8,175	12,900	21,075
0240	30 outlets	14,600	24,500	39,100
0280	100 outlets	44,200	82,000	126,200
0320	Fire detection systems, non-addressable, 12 detectors	3,275	6,700	9,975
0360	25 detectors	5,725	11,300	17,025
0400	50 detectors	11,200	22,500	33,700
0440	100 detectors	21,000	40,900	61,900
0450	Addressable type, 12 detectors	4,925	6,750	11,675
0452	25 detectors	8,975	11,400	20,375
0454	50 detectors	17,300	22,400	39,700
0456	100 detectors	33,500	41,200	74,700
0458	Fire alarm control panel, 8 zone, excluding wire and conduit	850	1,275	2,125
0459	12 zone	2,600	1,925	4,525
0460	Fire alarm command center, addressable without voice, excl. wire & conduit	4,875	1,125	6,000
0462	Addressable with voice	10,300	1,750	12,050
0480	Intercom systems, 6 stations	3,875	5,525	9,400
0520	12 stations	7,475	11,000	18,475
0560	25 stations	12,500	21,200	33,700
0600	50 stations	24,200	39,900	64,100
0640	100 stations	47,800	78,000	125,800
0680	Master clock systems, 6 rooms	5,100	9,025	14,125
0720	12 rooms	7,975	15,400	23,375
0760	20 rooms	11,100	21,700	32,800
0800	30 rooms	18,000	40,100	58,100

D50 Electrical

D5030 Communications and Security

D5030 910	Communication & Alarm Systems	COST EACH		
		MAT.	INST.	TOTAL
0840	50 rooms	29,200	67,500	96,700
0880	100 rooms	56,500	134,000	190,500
0920	Master TV antenna systems, 6 outlets	2,550	5,700	8,250
0960	12 outlets	4,800	10,600	15,400
1000	30 outlets	11,100	24,600	35,700
1040	100 outlets	36,800	80,500	117,300

D5030 920	Data Communication	COST PER M.S.F.		
		MAT.	INST.	TOTAL
0100	Data communication system, incl data/voice outlets, boxes, conduit, & cable			
0102	Data and voice system, 2 data/voice outlets per 1000 S.F.	174	475	649
0104	4 data/voice outlets per 1000 S.F.	325	930	1,255
0106	6 data/voice outlets per 1000 S.F.	460	1,375	1,835
0110	8 data/voice outlets per 1000 S.F.	590	1,825	2,415

D5090   Other Electrical Systems

Description: System below tabulates the installed cost for generators by kW. Included in costs are battery, charger, muffler, and transfer switch.

No conduit, wire, or terminations included.

System Components	QUANTITY	UNIT	COST PER kW		
			MAT.	INST.	TOTAL
SYSTEM D5090 210 0200 GENERATOR SET, INCL. BATTERY, CHARGER, MUFFLER & TRANSFER SWITCH GAS/GASOLINE OPER., 3 PHASE, 4 WIRE, 277/480V, 7.5 kW					
Generator set, gas or gasoline operated, 3 ph 4 W, 277/480 V, 7.5 kW	.133	Ea.	1,066.67	281.47	1,348.14
TOTAL			1,066.67	281.47	1,348.14

D5090 210	Generators (by kW)	COST PER kW		
		MAT.	INST.	TOTAL
0190	Generator sets, include battery, charger, muffler & transfer switch			
0200	Gas/gasoline operated, 3 phase, 4 wire, 277/480 volt, 7.5 kW	1,075	282	1,357
0240	11.5 kW	985	212	1,197
0280	20 kW	670	138	808
0320	35 kW	455	91	546
0360	80 kW	325	54.50	379.50
0400	100 kW	286	53	339
0440	125 kW	470	50	520
0480	185 kW	420	38	458
0560	Diesel engine with fuel tank, 30 kW	600	106	706
0600	50 kW	430	83	513
0640	75 kW	370	66.50	436.50
0680	100 kW	310	56.50	366.50
0720	125 kW	259	48	307
0760	150 kW	249	44.50	293.50
0800	175 kW	232	40	272
0840	200 kW	210	36	246
0880	250 kW	197	30	227
0920	300 kW	178	26.50	204.50
0960	350 kW	173	24.50	197.50
1000	400 kW	186	23	209
1040	500 kW	188	19.35	207.35
1200	750 kW	203	12.10	215.10
1400	1000 kW	190	12.80	202.80

D5090 Other Electrical Systems

PV power system, stand alone, AC and DC loads

Combiner box 15amp pull out fuses

PV array disconnect 50 amp 2-pole, added lightning protection

Battery disconnect 175 amp

DC to AC inverter

To AC loads

Low voltage disconnect

AC load center

DC loads

PV modules

PV Charge battery temp sensor

C40 charge controller 40 amp, 48 V

Batteries

DC load center

Grounding rod

System Components	QUANTITY	UNIT	COST EACH		
			MAT.	INST.	TOTAL
SYSTEM D5090 420 0100					
PHOTOVOLTAIC POWER SYSTEM, STAND ALONE					
Alternative energy sources, photovoltaic module, 75 watt, 17 volts	12.000	Ea.	5,100	960	6,060
PV rack system, roof, pentrating surface mount, on wood framing, 1 panel	12.000	Ea.	618	1,164	1,782
DC to AC inverter for, 24 V, 2,500 watt	1.000	Ea.	2,175	160	2,335
PV components, combiner box, 10 lug, NEMA 3R enclosure	1.000	Ea.	208	160	368
Alternative energy sources, PV components, fuse, 15 A for combiner box	10.000	Ea.	180.50	159.50	340
Battery charger controller w/temperature sensor	1.000	Ea.	535	160	695
Digital readout panel, displays hours, volts, amps, etc.	1.000	Ea.	142	160	302
Deep cycle solar battery, 6 V, 180 Ah (C/20)	4.000	Ea.	1,024	320	1,344
Battery interconnection, 15" AWG #2/0, sealed w/copper ring lugs	3.000	Ea.	41.85	120	161.85
Battery interconn, 24" AWG #2/0, sealed w/copper ring lugs	2.000	Ea.	40	80	120
Battery interconn, 60" AWG #2/0, sealed w/copper ring lugs	2.000	Ea.	95	80	175
Batt temp computer probe, RJ11 jack, 15' cord	1.000	Ea.	21.50	40	61.50
System disconnect, DC 175 amp circuit breaker	1.000	Ea.	153	80	233
Conduit box for inverter	1.000	Ea.	48	80	128
Low voltage disconnect	1.000	Ea.	43.50	80	123.50
Vented battery enclosure, wood	1.000	Ea.	190	283	473
Grounding, rod, copper clad, 8' long, 5/8" diameter	1.000	Ea.	21.50	116	137.50
Grounding, clamp, bronze, 5/8" dia	1.000	Ea.	5.90	19.95	25.85
Bare copper wire, stranded, #8	1.000	C.L.F.	38.50	58	96.50
Wire, 600 volt, type THW, copper, stranded, #12	3.600	C.L.F.	57.24	208.80	266.04
Wire, 600 volt, type THW, copper, stranded, #10	1.050	C.L.F.	26.25	67.20	93.45
Cond to 15' H,incl 2 termn,2 elb&11 bm CLP per 100',galv stl,1/2" dia	120.000	L.F.	316.80	852	1,168.80
Conduit,to 15' H,incl 2 termn,2 elb&11 bm clp per 100',(EMT), 1" dia	30.000	L.F.	48.90	166.50	215.40
Lightning surge suppressor	1.000	Ea.	56	19.95	75.95
General duty 240 volt, 2 pole, nonfusible, NEMA 3R, 60 amp	1.000	Ea.	300	278	578
Load centers, 3 wire, 120/240V, 100 amp main lugs, indoor, 8 circuits	2.000	Ea.	382	910	1,292
Circuit breaker, Plug-in, 120/240 volt, to 60 amp, 1 pole	5.000	Ea.	61.50	265	326.50
Fuses, dual element, time delay, 250 volt, 50 amp	2.000	Ea.	22.60	25.60	48.20
TOTAL			11,952.54	7,073.50	19,026.04

D5090 420	Photovoltaic Power System, Stand Alone	COST EACH		
		MAT.	INST.	TOTAL
0050	24 V capacity, 900 W (~102 SF)			
0100	Roof mounted, on wood framing	12,000	7,075	19,075

D5090 Other Electrical Systems

D5090 420	Photovoltaic Power System, Stand Alone	COST EACH		
		MAT.	INST.	TOTAL
0150	with standoff	12,100	7,075	19,175
0200	Ground placement, ballast, fixed	17,400	7,675	25,075
0250	Adjustable	17,700	7,675	25,375
0300	Top of pole, passive tracking	17,500	7,675	25,175
0350	1.8 kW (~204 SF)			
0400	Roof mounted, on wood framing	18,800	11,200	30,000
0450	with standoff	19,100	11,200	30,300
0500	Ground placement, ballast, fixed	28,900	10,800	39,700
0550	Adjustable	29,500	10,800	40,300
0600	Top of pole, passive tracking	30,200	10,700	40,900
0650	3.2 kW (~408 SF)			
0700	Roof mounted, on wood framing	33,900	19,100	53,000
0750	with standoff	34,300	19,100	53,400
0800	Ground placement, ballast, fixed	53,000	16,600	69,600
0850	Adjustable	54,500	16,600	71,100
0900	Top of pole, passive tracking	56,000	16,500	72,500

D5090 Other Electrical Systems

Photovoltaic power system, grid connected 10 kW

System Components	QUANTITY	UNIT	COST EACH		
			MAT.	INST.	TOTAL
SYSTEM D5090 430 0100					
PHOTOVOLTAIC POWER SYSTEM, GRID CONNECTED 10 kW					
Photovoltaic module, 150 watt, 33 volts	60.000	Ea.	42,600	4,800	47,400
PV rack system, roof, non-penetrating ballast, 1 panel	60.000	Ea.	58,500	2,100	60,600
DC to AC inverter for, 24 V, 4,000 watt	3.000	Ea.	10,875	960	11,835
Combiner box 10 lug, NEMA 3R	1.000	Ea.	208	160	368
15 amp fuses	10.000	Ea.	180.50	159.50	340
Safety switch, 60 amp	1.000	Ea.	430	290	720
Safety switch, 100 amp	1.000	Ea.	800	355	1,155
Utility connection, 3 pole breaker	1.000	Ea.	250	103	353
Fuse, 60 A	3.000	Ea.	39.90	38.40	78.30
Fuse, 100 A	3.000	Ea.	93	47.85	140.85
10 kVA isolation tranformer	1.000	Ea.	1,875	800	2,675
EMT Conduit w/fittings & support	1.000	L.F.	293.40	999	1,292.40
RGS Conduit w/fittings & support	267.000	L.F.	1,084.02	2,629.95	3,713.97
Enclosure 24″ x 24″ x 10″, NEMA 4	1.000	Ea.	5,425	1,275	6,700
Wire, 600 volt, type THWN-THHN, copper, stranded, #6	8.000	C.L.F.	572	788	1,360
Wire, 600 volt, copper type XLPE-USE(RHW), stranded, #12	18.000	C.L.F.	486	1,044	1,530
Grounding, bare copper wire stranded, 4/0	2.000	C.L.F.	1,010	448	1,458
Grounding, exothermic weld, 4/0 wire to building steel	4.000	Ea.	40.60	366	406.60
Grounding, brazed connections, #6 wire	2.000	Ea.	35.20	106	141.20
Insulated ground wire, copper, #6	.400	C.L.F.	28.60	39.40	68
TOTAL			124,826.22	17,509.10	142,335.32

D5090 430	Photovoltaic Power System, Grid Connected	COST EACH		
		MAT.	INST.	TOTAL
0050	10 kW			
0100	Roof mounted, non-penetrating ballast	125,000	17,500	142,500
0200	Penetrating surface mount, on steel framing	69,500	27,900	97,400
0300	on wood framing	69,500	21,200	90,700
0400	with standoff	70,000	21,200	91,200
0500	Ground placement, ballast, fixed	93,000	17,500	110,500
0600	adjustable	94,500	17,500	112,000
0700	Top of pole, passive tracking	95,000	18,000	113,000
1050	20 kW			
1100	Roof mounted, non-penetrating ballast	243,000	29,300	272,300
1200	Penetrating surface mount, on steel framing	132,500	50,000	182,500
1300	on wood framing	132,500	36,700	169,200
1400	with standoff	133,500	36,700	170,200
1500	Ground placement, ballast, fixed	180,000	28,400	208,400
1600	adjustable	183,000	28,400	211,400
1700	Top of pole, passive tracking	183,000	29,300	212,300

D5090 Energy Monitoring Systems

System Components	QUANTITY	UNIT	COST EACH		
			MAT.	INST.	TOTAL
SYSTEM D5090 480 1100					
ELECTRICAL, SINGLE PHASE, 1 METER					
#18 twisted shielded pair in 1/2″ EMT conduit	.050	C.L.F.	7.80	10.65	18.45
Wire, 600 volt, type THW, copper, solid, #12	.150	C.L.F.	2.06	8.70	10.76
Conduit (EMT) , to 15′ H, incl 2 termn,2 elb&11 bm clp per 100′, 3/4″	5.000	L.F.	4.70	24.55	29.25
Outlet boxes, pressed steel, handy box	1.000	Ea.	3.73	23.50	27.23
Outlet boxes, pressed steel, handy box, covers, device	1.000	Ea.	1.31	10	11.31
Wiring devices, receptacle, duplex, 120 volt, ground, 20 amp	1.000	Ea.	9.20	23.50	32.70
Single phase, 277 volt, 200 amp	1.000	Ea.	475	73	548
Data recorder, 8 meters	1.000	Ea.	1,500	58	1,558
Software package, per meter, premium	1.000	Ea.	665		665
TOTAL			2,668.80	231.90	2,900.70

D5090 480	Energy Monitoring Systems	COST EACH		
		MAT.	INST.	TOTAL
1000	Electrical			
1100	Single phase, 1 meter	2,675	232	2,907
1110	4 meters	6,875	1,750	8,625
1120	8 meters	13,800	3,625	17,425
1200	Three phase, 1 meter	2,975	295	3,270
1210	5 meters	9,650	2,750	12,400
1220	10 meters	19,600	5,625	25,225
1230	25 meters	44,500	11,200	55,700
2000	Mechanical			
2100	BTU, 1 meter	3,550	855	4,405
2110	w/1 duct sensor	3,750	1,050	4,800
2120	& 1 space sensor	4,250	1,200	5,450
2130	& 5 space sensors	6,225	1,775	8,000
2140	& 10 space sensors	9,325	2,500	11,825
2200	BTU, 3 meters	7,675	2,450	10,125
2210	w/3 duct sensors	8,275	3,050	11,325
2220	& 3 space sensors	9,750	3,475	13,225
2230	& 15 space sensors	15,700	5,175	20,875
2240	& 30 space sensors	23,700	7,325	31,025
9000	Front end display	695	125	820
9100	Computer workstation	7,750	1,650	9,400

D50 Electrical

D5090   Other Electrical Systems

Low Density and Medium Density Baseboard Radiation

The costs shown in Table below are based on the following system considerations:

1. The heat loss per square foot is based on approximately 34 BTU/hr. per S.F. of floor or 10 watts per S.F. of floor.
2. Baseboard radiation is based on the low watt density type rated 187 watts per L.F. and the medium density type rated 250 watts per L.F.
3. Thermostat is not included.
4. Wiring costs include branch circuit wiring.

System Components	QUANTITY	UNIT	COST PER S.F.		
			MAT.	INST.	TOTAL
SYSTEM D5090 510 1000					
ELECTRIC BASEBOARD RADIATION, LOW DENSITY, 900 S.F., 31 MBH, 9 kW					
Electric baseboard radiator, 5' long 935 watt	.011	Ea.	.62	1.23	1.85
Steel intermediate conduit, (IMC) 1/2" diam	.170	L.F.	.32	1.09	1.41
Wire 600 volt, type THW, copper, solid, #12	.005	C.L.F.	.07	.29	.36
TOTAL			1.01	2.61	3.62

D5090 510	Electric Baseboard Radiation (Low Density)	COST PER S.F.		
		MAT.	INST.	TOTAL
1000	Electric baseboard radiation, low density, 900 S.F., 31 MBH, 9 kW	1.01	2.61	3.62
1200	1500 S.F., 51 MBH, 15 kW	.99	2.54	3.53
1400	2100 S.F., 72 MBH, 21 kW	.89	2.28	3.17
1600	3000 S.F., 102 MBH, 30 kW	.81	2.08	2.89
2000	Medium density, 900 S.F., 31 MBH, 9 kW	.89	2.39	3.28
2200	1500 S.F., 51 MBH, 15 kW	.87	2.32	3.19
2400	2100 S.F., 72 MBH, 21 kW	.83	2.17	3
2600	3000 S.F., 102 MBH, 30 kW	.76	1.97	2.73

D5090   Other Electrical Systems

Commercial Duty Baseboard Radiation
The costs shown in Table below are based on the following system considerations:

1. The heat loss per square foot is based on approximately 41 BTU/hr. per S.F. of floor or 12 watts per S.F.
2. The baseboard radiation is of the commercial duty type rated 250 watts per L.F. served by 277 volt, single phase power.
3. Thermostat is not included.
4. Wiring costs include branch circuit wiring.

System Components	QUANTITY	UNIT	COST PER S.F.		
			MAT.	INST.	TOTAL
SYSTEM D5090 520 1000					
ELECTRIC BASEBOARD RADIATION, MEDIUM DENSITY, 1230 S.F., 51 MBH, 15 kW					
Electric baseboard radiator, 5' long	.013	Ea.	.73	1.46	2.19
Steel intermediate conduit, (IMC) 1/2" diam	.154	L.F.	.29	.99	1.28
Wire 600 volt, type THW, copper, solid, #12	.004	C.L.F.	.06	.23	.29
TOTAL			1.08	2.68	3.76

D5090 520	Electric Baseboard Radiation (Medium Density)	COST PER S.F.		
		MAT.	INST.	TOTAL
1000	Electric baseboard radiation, medium density, 1230 SF, 51 MBH, 15 kW	1.08	2.68	3.76
1200	2500 S.F. floor area, 106 MBH, 31 kW	1	2.50	3.50
1400	3700 S.F. floor area, 157 MBH, 46 kW	.98	2.43	3.41
1600	4800 S.F. floor area, 201 MBH, 59 kW	.91	2.25	3.16
1800	11,300 S.F. floor area, 464 MBH, 136 kW	.85	2.06	2.91
2000	30,000 S.F. floor area, 1229 MBH, 360 kW	.84	2.01	2.85

G1030 Site Earthwork

Trenching Systems are shown on a cost per linear foot basis. The systems include: excavation; backfill and removal of spoil; and compaction for various depths and trench bottom widths. The backfill has been reduced to accommodate a pipe of suitable diameter and bedding.

The slope for trench sides varies from none to 1:1.

The Expanded System Listing shows Trenching Systems that range from 2' to 12' in width. Depths range from 2' to 25'.

System Components	QUANTITY	UNIT	COST PER L.F. EQUIP.	LABOR	TOTAL
SYSTEM G1030 805 1310					
TRENCHING COMMON EARTH, NO SLOPE, 2' WIDE, 2' DP, 3/8 C.Y. BUCKET					
Excavation, trench, hyd. backhoe, track mtd., 3/8 C.Y. bucket	.148	B.C.Y.	.40	1.03	1.43
Backfill and load spoil, from stockpile	.153	L.C.Y.	.12	.31	.43
Compaction by vibrating plate, 6" lifts, 4 passes	.118	E.C.Y.	.03	.38	.41
Remove excess spoil, 8 C.Y. dump truck, 2 mile roundtrip	.040	L.C.Y.	.15	.15	.30
TOTAL			.70	1.87	2.57

G1030 805	Trenching Common Earth	COST PER L.F. EQUIP.	LABOR	TOTAL
1310	Trenching, common earth, no slope, 2' wide, 2' deep, 3/8 C.Y. bucket	.70	1.87	2.57
1320	3' deep, 3/8 C.Y. bucket	.99	2.82	3.81
1330	4' deep, 3/8 C.Y. bucket	1.29	3.78	5.07
1340	6' deep, 3/8 C.Y. bucket	1.68	4.92	6.60
1350	8' deep, 1/2 C.Y. bucket	2.22	6.55	8.77
1360	10' deep, 1 C.Y. bucket	3.48	7.75	11.23
1400	4' wide, 2' deep, 3/8 C.Y. bucket	1.61	3.70	5.31
1410	3' deep, 3/8 C.Y. bucket	2.21	5.60	7.81
1420	4' deep, 1/2 C.Y. bucket	2.57	6.30	8.87
1430	6' deep, 1/2 C.Y. bucket	4.15	10.15	14.30
1440	8' deep, 1/2 C.Y. bucket	6.65	13.05	19.70
1450	10' deep, 1 C.Y. bucket	8.05	16.20	24.25
1460	12' deep, 1 C.Y. bucket	10.35	20.50	30.85
1470	15' deep, 1-1/2 C.Y. bucket	9.40	18.50	27.90
1480	18' deep, 2-1/2 C.Y. bucket	13.05	26	39.05
1520	6' wide, 6' deep, 5/8 C.Y. bucket w/trench box	8.60	14.95	23.55
1530	8' deep, 3/4 C.Y. bucket	11.40	19.75	31.15
1540	10' deep, 1 C.Y. bucket	11.55	20.50	32.05
1550	12' deep, 1-1/2 C.Y. bucket	12.40	22	34.40
1560	16' deep, 2-1/2 C.Y. bucket	17.10	27.50	44.60
1570	20' deep, 3-1/2 C.Y. bucket	22	33	55
1580	24' deep, 3-1/2 C.Y. bucket	26	39.50	65.50
1640	8' wide, 12' deep, 1-1/2 C.Y. bucket w/trench box	17.40	28	45.40
1650	15' deep, 1-1/2 C.Y. bucket	22.50	36.50	59
1660	18' deep, 2-1/2 C.Y. bucket	25	36.50	61.50
1680	24' deep, 3-1/2 C.Y. bucket	35.50	51	86.50
1730	10' wide, 20' deep, 3-1/2 C.Y. bucket w/trench box	28.50	48.50	77
1740	24' deep, 3-1/2 C.Y. bucket	42.50	58.50	101
1780	12' wide, 20' deep, 3-1/2 C.Y. bucket w/trench box	45	61.50	106.50
1790	25' deep, bucket	56	78.50	134.50
1800	1/2 to 1 slope, 2' wide, 2' deep, 3/8 C.Y. bucket	.99	2.82	3.81
1810	3' deep, 3/8 C.Y. bucket	1.67	4.97	6.64
1820	4' deep, 3/8 C.Y. bucket	2.49	7.60	10.09
1840	6' deep, 3/8 C.Y. bucket	4.02	12.35	16.37

G1030 Site Earthwork

G1030 805	Trenching Common Earth	COST PER L.F.		
		EQUIP.	LABOR	TOTAL
1860	8' deep, 1/2 C.Y. bucket	6.40	19.75	26.15
1880	10' deep, 1 C.Y. bucket	11.95	27.50	39.45
2300	4' wide, 2' deep, 3/8 C.Y. bucket	1.91	4.65	6.56
2310	3' deep, 3/8 C.Y. bucket	2.88	7.75	10.63
2320	4' deep, 1/2 C.Y. bucket	3.63	9.65	13.28
2340	6' deep, 1/2 C.Y. bucket	6.95	18	24.95
2360	8' deep, 1/2 C.Y. bucket	12.85	26.50	39.35
2380	10' deep, 1 C.Y. bucket	17.80	37	54.80
2400	12' deep, 1 C.Y. bucket	24.50	50.50	75
2430	15' deep, 1-1/2 C.Y. bucket	26.50	54	80.50
2460	18' deep, 2-1/2 C.Y. bucket	46.50	84.50	131
2840	6' wide, 6' deep, 5/8 C.Y. bucket w/trench box	12.70	22.50	35.20
2860	8' deep, 3/4 C.Y. bucket	18.35	33.50	51.85
2880	10' deep, 1 C.Y. bucket	18.60	34	52.60
2900	12' deep, 1-1/2 C.Y. bucket	23.50	44.50	68
2940	16' deep, 2-1/2 C.Y. bucket	39	65.50	104.50
2980	20' deep, 3-1/2 C.Y. bucket	55	88.50	143.50
3020	24' deep, 3-1/2 C.Y. bucket	77.50	121	198.50
3100	8' wide, 12' deep, 1-1/2 C.Y. bucket w/trench box	29	51	80
3120	15' deep, 1-1/2 C.Y. bucket	42.50	74	116.50
3140	18' deep, 2-1/2 C.Y. bucket	54	88	142
3180	24' deep, 3-1/2 C.Y. bucket	87	133	220
3270	10' wide, 20' deep, 3-1/2 C.Y. bucket w/trench box	56	102	158
3280	24' deep, 3-1/2 C.Y. bucket	96	145	241
3370	12' wide, 20' deep, 3-1/2 C.Y. bucket w/trench box	81	118	199
3380	25' deep, 3-1/2 C.Y. bucket	112	169	281
3500	1 to 1 slope, 2' wide, 2' deep, 3/8 C.Y. bucket	1.29	3.78	5.07
3520	3' deep, 3/8 C.Y. bucket	3.72	8.50	12.22
3540	4' deep, 3/8 C.Y. bucket	3.68	11.40	15.08
3560	6' deep, 1/2 C.Y. bucket	4.02	12.35	16.37
3580	8' deep, 1/2 C.Y. bucket	7.95	24.50	32.45
3600	10' deep, 1 C.Y. bucket	20.50	47	67.50
3800	4' wide, 2' deep, 3/8 C.Y. bucket	2.21	5.60	7.81
3820	3' deep, 3/8 C.Y. bucket	3.55	9.90	13.45
3840	4' deep, 1/2 C.Y. bucket	4.67	12.90	17.57
3860	6' deep, 1/2 C.Y. bucket	9.75	26	35.75
3880	8' deep, 1/2 C.Y. bucket	19.10	40	59.10
3900	10' deep, 1 C.Y. bucket	27.50	58	85.50
3920	12' deep, 1 C.Y. bucket	40.50	84	124.50
3940	15' deep, 1-1/2 C.Y. bucket	43.50	89.50	133
3960	18' deep, 2-1/2 C.Y. bucket	64	116	180
4030	6' wide, 6' deep, 5/8 C.Y. bucket w/trench box	16.60	30	46.60
4040	8' deep, 3/4 C.Y. bucket	24	41.50	65.50
4050	10' deep, 1 C.Y. bucket	27	50	77
4060	12' deep, 1-1/2 C.Y. bucket	36	68	104
4070	16' deep, 2-1/2 C.Y. bucket	61	104	165
4080	20' deep, 3-1/2 C.Y. bucket	88.50	144	232.50
4090	24' deep, 3-1/2 C.Y. bucket	129	203	332
4500	8' wide, 12' deep, 1-1/2 C.Y. bucket w/trench box	41	74	115
4550	15' deep, 1-1/2 C.Y. bucket	62	111	173
4600	18' deep, 2-1/2 C.Y. bucket	81.50	136	217.50
4650	24' deep, 3-1/2 C.Y. bucket	139	214	353
4800	10' wide, 20' deep, 3-1/2 C.Y. bucket w/trench box	83	155	238
4850	24' deep, 3-1/2 C.Y. bucket	148	226	374
4950	12' wide, 20' deep, 3-1/2 C.Y. bucket w/trench box	117	175	292
4980	25' deep, 3-1/2 C.Y. bucket	168	256	424

G1030 Site Earthwork

Trenching Systems are shown on a cost per linear foot basis. The systems include: excavation; backfill and removal of spoil; and compaction for various depths and trench bottom widths. The backfill has been reduced to accommodate a pipe of suitable diameter and bedding.

The slope for trench sides varies from none to 1:1.

The Expanded System Listing shows Trenching Systems that range from 2' to 12' in width. Depths range from 2' to 25'.

System Components	QUANTITY	UNIT	COST PER L.F.		
			EQUIP.	LABOR	TOTAL
SYSTEM G1030 806 1310					
TRENCHING LOAM & SANDY CLAY, NO SLOPE, 2' WIDE, 2' DP, 3/8 C.Y. BUCKET					
Excavation, trench, hyd. backhoe, track mtd., 3/8 C.Y. bucket	.148	B.C.Y.	.37	.95	1.32
Backfill and load spoil, from stockpile	.165	L.C.Y.	.13	.34	.47
Compaction by vibrating plate 18" wide, 6" lifts, 4 passes	.118	E.C.Y.	.03	.38	.41
Remove excess spoil, 8 C.Y. dump truck, 2 mile roundtrip	.042	L.C.Y.	.16	.16	.32
TOTAL			.70	1.87	2.57

G1030 806	Trenching Loam & Sandy Clay	COST PER L.F.		
		EQUIP.	LABOR	TOTAL
1310	Trenching, loam & sandy clay, no slope, 2' wide, 2' deep, 3/8 C.Y. bucket	.69	1.83	2.52
1320	3' deep, 3/8 C.Y. bucket	1.08	3.03	4.11
1330	4' deep, 3/8 C.Y. bucket	1.26	3.69	4.95
1340	6' deep, 3/8 C.Y. bucket	1.80	4.40	6.20
1350	8' deep, 1/2 C.Y. bucket	2.39	5.80	8.20
1360	10' deep, 1 C.Y. bucket	2.69	6.25	8.95
1400	4' wide, 2' deep, 3/8 C.Y. bucket	1.60	3.63	5.25
1410	3' deep, 3/8 C.Y. bucket	2.18	5.50	7.70
1420	4' deep, 1/2 C.Y. bucket	2.56	6.20	8.75
1430	6' deep, 1/2 C.Y. bucket	4.42	9.05	13.45
1440	8' deep, 1/2 C.Y. bucket	6.50	12.85	19.35
1450	10' deep, 1 C.Y. bucket	6.45	13.20	19.65
1460	12' deep, 1 C.Y. bucket	8.10	16.40	24.50
1470	15' deep, 1-1/2 C.Y. bucket	9.80	19.10	29
1480	18' deep, 2-1/2 C.Y. bucket	11.60	21	32.50
1520	6' wide, 6' deep, 5/8 C.Y. bucket w/trench box	8.35	14.75	23
1530	8' deep, 3/4 C.Y. bucket	11	19.35	30.50
1540	10' deep, 1 C.Y. bucket	10.60	19.70	30.50
1550	12' deep, 1-1/2 C.Y. bucket	12.10	22	34
1560	16' deep, 2-1/2 C.Y. bucket	16.70	28	44.50
1570	20' deep, 3-1/2 C.Y. bucket	20	33	53
1580	24' deep, 3-1/2 C.Y. bucket	25.50	40	65.50
1640	8' wide, 12' deep, 1-1/4 C.Y. bucket w/trench box	17.15	28	45
1650	15' deep, 1-1/2 C.Y. bucket	21	35.50	56.50
1660	18' deep, 2-1/2 C.Y. bucket	25.50	40	65.50
1680	24' deep, 3-1/2 C.Y. bucket	34.50	52	86.50
1730	10' wide, 20' deep, 3-1/2 C.Y. bucket w/trench box	35	52	87
1740	24' deep, 3-1/2 C.Y. bucket	43.50	64	108
1780	12' wide, 20' deep, 3-1/2 C.Y. bucket w/trench box	42	61.50	104
1790	25' deep, 3-1/2 C.Y. bucket	54.50	79.50	134
1800	1/2:1 slope, 2' wide, 2' deep, 3/8 C.Y. bucket	.98	2.76	3.74
1810	3' deep, 3/8 C.Y. bucket	1.63	4.86	6.50
1820	4' deep, 3/8 C.Y. bucket	2.43	7.40	9.85
1840	6' deep, 3/8 C.Y. bucket	4.32	11	15.30

G1030 Site Earthwork

G1030 806	Trenching Loam & Sandy Clay	COST PER L.F.		
		EQUIP.	LABOR	TOTAL
1860	8' deep, 1/2 C.Y. bucket	6.85	17.65	24.50
1880	10' deep, 1 C.Y. bucket	9.15	22	31
2300	4' wide, 2' deep, 3/8 C.Y. bucket	1.90	4.56	6.45
2310	3' deep, 3/8 C.Y. bucket	2.84	7.60	10.45
2320	4' deep, 1/2 C.Y. bucket	3.59	9.45	13.05
2340	6' deep, 1/2 C.Y. bucket	7.40	16.15	23.50
2360	8' deep, 1/2 C.Y. bucket	12.50	26	38.50
2380	10' deep, 1 C.Y. bucket	14.20	30.50	44.50
2400	12' deep, 1 C.Y. bucket	23.50	50	73.50
2430	15' deep, 1-1/2 C.Y. bucket	27.50	56	83.50
2460	18' deep, 2-1/2 C.Y. bucket	46	85.50	132
2840	6' wide, 6' deep, 5/8 C.Y. bucket w/trench box	12.10	22.50	34.50
2860	8' deep, 3/4 C.Y. bucket	17.70	33	50.50
2880	10' deep, 1 C.Y. bucket	18.95	37	56
2900	12' deep, 1-1/2 C.Y. bucket	23.50	45	68.50
2940	16' deep, 2-1/2 C.Y. bucket	38	66.50	105
2980	20' deep, 3-1/2 C.Y. bucket	53	89.50	143
3020	24' deep, 3-1/2 C.Y. bucket	75	123	198
3100	8' wide, 12' deep, 1-1/2 C.Y. bucket w/trench box	28.50	51	79.50
3120	15' deep, 1-1/2 C.Y. bucket	39	72	111
3140	18' deep, 2-1/2 C.Y. bucket	52.50	89	142
3180	24' deep, 3-1/2 C.Y. bucket	84.50	135	220
3270	10' wide, 20' deep, 3-1/2 C.Y. bucket w/trench box	68	109	177
3280	24' deep, 3-1/2 C.Y. bucket	93.50	147	241
3320	12' wide, 20' deep, 3-1/2 C.Y. bucket w/trench box	75	118	193
3380	25' deep, 3-1/2 C.Y. bucket w/trench box	102	159	261
3500	1:1 slope, 2' wide, 2' deep, 3/8 C.Y. bucket	1.26	3.69	4.95
3520	3' deep, 3/8 C.Y. bucket	2.29	6.95	9.25
3540	4' deep, 3/8 C.Y. bucket	3.57	11.10	14.65
3560	6' deep, 1/2 C.Y. bucket	4.32	11	15.30
3580	8' deep, 1/2 C.Y. bucket	11.35	29.50	41
3600	10' deep, 1 C.Y. bucket	15.65	38	53.50
3800	4' wide, 2' deep, 3/8 C.Y. bucket	2.18	5.50	7.70
3820	3' deep, 1/2 C.Y. bucket	3.49	9.70	13.20
3840	4' deep, 1/2 C.Y. bucket	4.61	12.65	17.25
3860	6' deep, 1/2 C.Y. bucket	10.35	23	33.50
3880	8' deep, 1/2 C.Y. bucket	18.55	39.50	58
3900	10' deep, 1 C.Y. bucket	22	47.50	69.50
3920	12' deep, 1 C.Y. bucket	31.50	67	98.50
3940	15' deep, 1-1/2 C.Y. bucket	45.50	92.50	138
3960	18' deep, 2-1/2 C.Y. bucket	63	118	181
4030	6' wide, 6' deep, 5/8 C.Y. bucket w/trench box	15.85	30	46
4040	8' deep, 3/4 C.Y. bucket	24.50	46.50	71
4050	10' deep, 1 C.Y. bucket	27.50	54.50	82
4060	12' deep, 1-1/2 C.Y. bucket	35	68.50	104
4070	16' deep, 2-1/2 C.Y. bucket	59.50	105	165
4080	20' deep, 3-1/2 C.Y. bucket	86	146	232
4090	24' deep, 3-1/2 C.Y. bucket	125	206	330
4500	8' wide, 12' deep, 1-1/4 C.Y. bucket w/trench box	40	74	114
4550	15' deep, 1-1/2 C.Y. bucket	57	108	165
4600	18' deep, 2-1/2 C.Y. bucket	79.50	138	218
4650	24' deep, 3-1/2 C.Y. bucket	134	217	350
4800	10' wide, 20' deep, 3-1/2 C.Y. bucket w/trench box	101	166	267
4850	24' deep, 3-1/2 C.Y. bucket	143	230	375
4950	12' wide, 20' deep, 3-1/2 C.Y. bucket w/trench box	108	175	283
4980	25' deep, 3-1/2 C.Y. bucket	163	259	420

G1030 Site Earthwork

Trenching Systems are shown on a cost per linear foot basis. The systems include: excavation; backfill and removal of spoil; and compaction for various depths and trench bottom widths. The backfill has been reduced to accommodate a pipe of suitable diameter and bedding.

The slope for trench sides varies from none to 1:1.

The Expanded System Listing shows Trenching Systems that range from 2' to 12' in width. Depths range from 2' to 25'.

System Components			COST PER L.F.		
	QUANTITY	UNIT	EQUIP.	LABOR	TOTAL
SYSTEM G1030 807 1310					
TRENCHING SAND & GRAVEL, NO SLOPE, 2' WIDE, 2' DEEP, 3/8 C.Y. BUCKET					
Excavation, trench, hyd. backhoe, track mtd., 3/8 C.Y. bucket	.148	B.C.Y.	.36	.94	1.30
Backfill and load spoil, from stockpile	.140	L.C.Y.	.11	.29	.40
Compaction by vibrating plate 18" wide, 6" lifts, 4 passes	.118	E.C.Y.	.03	.38	.41
Remove excess spoil, 8 C.Y. dump truck, 2 mile roundtrip	.035	L.C.Y.	.14	.14	.28
TOTAL			.64	1.75	2.39

G1030 807	Trenching Sand & Gravel	COST PER L.F.		
		EQUIP.	LABOR	TOTAL
1310	Trenching, sand & gravel, no slope, 2' wide, 2' deep, 3/8 C.Y. bucket	.64	1.75	2.39
1320	3' deep, 3/8 C.Y. bucket	1.02	2.93	3.95
1330	4' deep, 3/8 C.Y. bucket	1.18	3.52	4.70
1340	6' deep, 3/8 C.Y. bucket	1.72	4.19	5.90
1350	8' deep, 1/2 C.Y. bucket	2.26	5.55	7.80
1360	10' deep, 1 C.Y. bucket	2.51	5.90	8.40
1400	4' wide, 2' deep, 3/8 C.Y. bucket	1.46	3.42	4.88
1410	3' deep, 3/8 C.Y. bucket	2	5.20	7.20
1420	4' deep, 1/2 C.Y. bucket	2.35	5.85	8.20
1430	6' deep, 1/2 C.Y. bucket	4.18	8.65	12.85
1440	8' deep, 1/2 C.Y. bucket	6.15	12.20	18.35
1450	10' deep, 1 C.Y. bucket	6.10	12.45	18.55
1460	12' deep, 1 C.Y. bucket	7.65	15.55	23
1470	15' deep, 1-1/2 C.Y. bucket	9.25	18	27.50
1480	18' deep, 2-1/2 C.Y. bucket	11	19.70	30.50
1520	6' wide, 6' deep, 5/8 C.Y. bucket w/trench box	7.85	13.95	22
1530	8' deep, 3/4 C.Y. bucket	10.40	18.40	29
1540	10' deep, 1 C.Y. bucket	10	18.60	28.50
1550	12' deep, 1-1/2 C.Y. bucket	11.40	21	32.50
1560	16' deep, 2 C.Y. bucket	16	26	42
1570	20' deep, 3-1/2 C.Y. bucket	19.15	31	50
1580	24' deep, 3-1/2 C.Y. bucket	24	37.50	61.50
1640	8' wide, 12' deep, 1-1/2 C.Y. bucket w/trench box	16.10	26.50	42.50
1650	15' deep, 1-1/2 C.Y. bucket	19.55	33.50	53
1660	18' deep, 2-1/2 C.Y. bucket	24	37.50	61.50
1680	24' deep, 3-1/2 C.Y. bucket	32.50	48.50	81
1730	10' wide, 20' deep, 3-1/2 C.Y. bucket w/trench box	32.50	48.50	81
1740	24' deep, 3-1/2 C.Y. bucket	41	60	101
1780	12' wide, 20' deep, 3-1/2 C.Y. bucket w/trench box	39.50	57.50	97
1790	25' deep, 3-1/2 C.Y. bucket	51.50	74.50	126
1800	1/2:1 slope, 2' wide, 2' deep, 3/8 C.Y. bucket	.91	2.63	3.54
1810	3' deep, 3/8 C.Y. bucket	1.53	4.64	6.15
1820	4' deep, 3/8 C.Y. bucket	2.28	7.10	9.40
1840	6' deep, 3/8 C.Y. bucket	4.11	10.55	14.65

G1030   Site Earthwork

G1030 807	Trenching Sand & Gravel	COST PER L.F.		
		EQUIP.	LABOR	TOTAL
1860	8' deep, 1/2 C.Y. bucket	6.55	16.85	23.50
1880	10' deep, 1 C.Y. bucket	8.60	21	29.50
2300	4' wide, 2' deep, 3/8 C.Y. bucket	1.73	4.31	6.05
2310	3' deep, 3/8 C.Y. bucket	2.61	7.20	9.80
2320	4' deep, 1/2 C.Y. bucket	3.31	8.95	12.25
2340	6' deep, 1/2 C.Y. bucket	7.05	15.45	22.50
2360	8' deep, 1/2 C.Y. bucket	11.85	24.50	36.50
2380	10' deep, 1 C.Y. bucket	13.40	28.50	42
2400	12' deep, 1 C.Y. bucket	22.50	47.50	70
2430	15' deep, 1-1/2 C.Y. bucket	26	52.50	78.50
2460	18' deep, 2-1/2 C.Y. bucket	43.50	80.50	124
2840	6' wide, 6' deep, 5/8 C.Y. bucket w/trench box	11.45	21	32.50
2860	8' deep, 3/4 C.Y. bucket	16.80	31	48
2880	10' deep, 1 C.Y. bucket	17.95	35	53
2900	12' deep, 1-1/2 C.Y. bucket	22.50	42.50	65
2940	16' deep, 2 C.Y. bucket	36.50	62.50	99
2980	20' deep, 3-1/2 C.Y. bucket	50.50	84	135
3020	24' deep, 3-1/2 C.Y. bucket	71.50	115	187
3100	8' wide, 12' deep, 1-1/4 C.Y. bucket w/trench box	27	48.50	75.50
3120	15' deep, 1-1/2 C.Y. bucket	36.50	67.50	104
3140	18' deep, 2-1/2 C.Y. bucket	50	83.50	134
3180	24' deep, 3-1/2 C.Y. bucket	80	126	206
3270	10' wide, 20' deep, 3-1/2 C.Y. bucket w/trench box	64	102	166
3280	24' deep, 3-1/2 C.Y. bucket	88.50	138	227
3370	12' wide, 20' deep, 3-1/2 C.Y. bucket w/trench box	71.50	113	185
3380	25' deep, 3-1/2 C.Y. bucket	103	159	262
3500	1:1 slope, 2' wide, 2' deep, 3/8 C.Y. bucket	2.05	4.29	6.35
3520	3' deep, 3/8 C.Y. bucket	2.14	6.65	8.80
3540	4' deep, 3/8 C.Y. bucket	3.36	10.65	14
3560	6' deep, 3/8 C.Y. bucket	4.11	10.55	14.65
3580	8' deep, 1/2 C.Y. bucket	10.80	28	39
3600	10' deep, 1 C.Y. bucket	14.65	35.50	50
3800	4' wide, 2' deep, 3/8 C.Y. bucket	2	5.20	7.20
3820	3' deep, 3/8 C.Y. bucket	3.23	9.20	12.45
3840	4' deep, 1/2 C.Y. bucket	4.26	12	16.25
3860	6' deep, 1/2 C.Y. bucket	9.90	22	32
3880	8' deep, 1/2 C.Y. bucket	17.60	37.50	55
3900	10' deep, 1 C.Y. bucket	21	45	66
3920	12' deep, 1 C.Y. bucket	30	63.50	93.50
3940	15' deep, 1-1/2 C.Y. bucket	43	87.50	131
3960	18' deep, 2-1/2 C.Y. bucket	59.50	111	171
4030	6' wide, 6' deep, 5/8 C.Y. bucket w/trench box	15.05	28.50	43.50
4040	8' deep, 3/4 C.Y. bucket	23	44	67
4050	10' deep, 1 C.Y. bucket	26	51.50	77.50
4060	12' deep, 1-1/2 C.Y. bucket	33	64.50	97.50
4070	16' deep, 2 C.Y. bucket	57	99	156
4080	20' deep, 3-1/2 C.Y. bucket	81.50	137	219
4090	24' deep, 3-1/2 C.Y. bucket	119	193	310
4500	8' wide, 12' deep, 1-1/2 C.Y. bucket w/trench box	38	70	108
4550	15' deep, 1-1/2 C.Y. bucket	53.50	102	156
4600	18' deep, 2-1/2 C.Y. bucket	76	129	205
4650	24' deep, 3-1/2 C.Y. bucket	127	204	330
4800	10' wide, 20' deep, 3-1/2 C.Y. bucket w/trench box	95.50	155	251
4850	24' deep, 3-1/2 C.Y. bucket	136	216	350
4950	12' wide, 20' deep, 3-1/2 C.Y. bucket w/trench box	102	164	266
4980	25' deep, 3-1/2 C.Y. bucket	154	243	395

G4010 Electrical Distribution

UNDERGROUND ELECTRICAL CONDUIT

System Components			COST EACH		
	QUANTITY	UNIT	MAT.	INST.	TOTAL
SYSTEM G4010 312 1000					
2000 AMP UNDERGROUND SERVICE INCLUDING COMMON EARTH EXCAVATION,					
CONCRETE, BACKFILL AND COMPACTION					
Excavate Trench	44.440	B.C.Y.		294.64	294.64
4 inch conduit bank	108.000	L.F.	1,080	2,538	3,618
4 inch fitting	8.000	Ea.	252	424	676
4 inch bells	4.000	Ea.	18.80	212	230.80
Concrete material	16.580	C.Y.	1,807.22		1,807.22
Concrete placement	16.580	C.Y.		399.58	399.58
Backfill trench	33.710	L.C.Y.		96.42	96.42
Compact fill material in trench	25.930	E.C.Y.		126.02	126.02
Dispose of excess fill material on-site	24.070	L.C.Y.		275.60	275.60
500 kcmil power cable	18.000	C.L.F.	22,950	7,200	30,150
Wire, 600 volt stranded copper wire, type THW, 1/0	6.000	C.L.F.	1,662	1,164	2,826
TOTAL			27,770.02	12,730.26	40,500.28

G4010 312		Underground Power Feed	COST EACH		
			MAT.	INST.	TOTAL
0950	Underground electrical including common earth excavation,concrete, backfill and				
1000	2000 Amp service, 100' length, 4' depth		27,800	12,800	40,600
1100	1600 Amp service		21,800	9,825	31,625
1200	1200 Amp service		16,900	8,150	25,050
1300	1000 Amp service		14,700	7,475	22,175
1400	800 Amp service		12,600	7,025	19,625
1500	600 Amp service		8,325	6,525	14,850

G4010 Electrical Distribution

System Components	QUANTITY	UNIT	COST L.F.		
			MAT.	INST.	TOTAL
SYSTEM G4010 320 1012					
UNDERGROUND ELECTRICAL CONDUIT, 2″ DIAMETER, INCLUDING EXCAVATION,					
CONCRETE, BEDDING, BACKFILL, AND COMPACTION					
Excavate Trench	.150	B.C.Y.		1	1
Base spacer	.200	Ea.	.34	2.28	2.62
Conduit	1.000	L.F.	1.91	3.55	5.46
Concrete material	.050	C.Y.	5.45		5.45
Concrete placement	.050	C.Y.		1.21	1.21
Backfill trench	.120	L.C.Y.		.35	.35
Compact fill material in trench	.100	E.C.Y.		.49	.49
Dispose of excess fill material on-site	.120	L.C.Y.		1.37	1.37
Marking tape	.010	C.L.F.	.31	.03	.34
TOTAL			8.01	10.28	18.29

G4010 320	Underground Electrical Conduit	COST L.F.		
		MAT.	INST.	TOTAL
0950	Underground electrical conduit, including common earth excavation,			
0952	Concrete, bedding, backfill and compaction			
1012	2″ dia. Schedule 40 PVC, 2′ deep	8	10.30	18.30
1014	4′ deep	8	14.70	22.70
1016	6′ deep	8	18.90	26.90
1022	2 @ 2″ dia. Schedule 40 PVC, 2′ deep	10.25	16.15	26.40
1024	4′ deep	10.25	20.50	30.75
1026	6′ deep	10.25	24.50	34.75
1032	3 @ 2″ dia. Schedule 40 PVC, 2′ deep	12.50	22	34.50
1034	4′ deep	12.50	26.50	39
1036	6′ deep	12.50	30.50	43
1042	4 @ 2″ dia. Schedule 40 PVC, 2′ deep	16.90	27.50	44.40
1044	4′ deep	16.90	31.50	48.40
1046	6′ deep	16.90	36	52.90

G4010 Electrical Distribution

G4010 320	Underground Electrical Conduit	COST L.F.		
		MAT.	INST.	TOTAL
1062	6 @ 2" dia. Schedule 40 PVC, 2' deep	21.50	39	60.50
1064	4' deep	21.50	43	64.50
1066	6' deep	21.50	47.50	69
1082	8 @ 2" dia. Schedule 40 PVC, 2' deep	26	50.50	76.50
1084	4' deep	26	54.50	80.50
1086	6' deep	26	59	85
1092	9 @ 2" dia. Schedule 40 PVC, 2' deep	28	56	84
1094	4' deep	30.50	60	90.50
1096	6' deep	30.50	64.50	95
1212	3" dia. Schedule 40 PVC, 2' deep	10.85	12.60	23.45
1214	4' deep	10.85	17	27.85
1216	6' deep	10.85	21.50	32.35
1222	2 @ 3" dia. Schedule 40 PVC, 2' deep	13.80	20.50	34.30
1224	4' deep	13.80	25	38.80
1226	6' deep	13.80	29	42.80
1232	3 @ 3" dia. Schedule 40 PVC, 2' deep	17.80	28.50	46.30
1234	4' deep	17.80	33	50.80
1236	6' deep	17.80	37	54.80
1242	4 @ 3" dia. Schedule 40 PVC, 2' deep	25	37	62
1244	4' deep	25	41	66
1246	6' deep	25	45.50	70.50
1262	6 @ 3" dia. Schedule 40 PVC, 2' deep	32	52.50	84.50
1264	4' deep	32	57	89
1266	6' deep	32	61.50	93.50
1282	8 @ 3" dia. Schedule 40 PVC, 2' deep	40	69	109
1284	4' deep	40	73	113
1286	6' deep	40	78	118
1292	9 @ 3" dia. Schedule 40 PVC, 2' deep	46.50	78	124.50
1294	4' deep	47.50	81	128.50
1296	6' deep	47.50	85	132.50
1312	4" dia. Schedule 40 PVC, 2' deep	12.35	15.45	27.80
1314	4' deep	12.35	19.85	32.20
1316	6' deep	12.35	24	36.35
1322	2 @ 4" dia. Schedule 40 PVC, 2' deep	17.85	26.50	44.35
1324	4' deep	17.85	30.50	48.35
1326	6' deep	17.85	35	52.85
1332	3 @ 4" dia. Schedule 40 PVC, 2' deep	22.50	37.50	60
1334	4' deep	22.50	41.50	64
1336	6' deep	22.50	46	68.50
1342	4 @ 4" dia. Schedule 40 PVC, 2' deep	32	49	81
1344	4' deep	32	53	85
1346	6' deep	32	57.50	89.50
1362	6 @ 4" dia. Schedule 40 PVC, 2' deep	42	71	113
1364	4' deep	42	75	117
1366	6' deep	42	79.50	121.50
1382	8 @ 4" dia. Schedule 40 PVC, 2' deep	54	93.50	147.50
1384	4' deep	54	99	153
1386	6' deep	54	104	158
1392	9 @ 4" dia. Schedule 40 PVC, 2' deep	60.50	105	165.50
1394	4' deep	60.50	108	168.50
1396	6' deep	60.50	112	172.50
1412	5" dia. Schedule 40 PVC, 2' deep	14.75	18.40	33.15
1414	4' deep	14.75	22.50	37.25
1416	6' deep	14.75	27	41.75
1422	2 @ 5" dia. Schedule 40 PVC, 2' deep	22.50	32.50	55
1424	4' deep	22.50	36.50	59
1426	6' deep	22.50	41	63.50
1432	3 @ 5" dia. Schedule 40 PVC, 2' deep	30.50	46.50	77

G4010 Electrical Distribution

G4010 320	Underground Electrical Conduit	COST L.F.		
		MAT.	INST.	TOTAL
1434	4' deep	30.50	51.50	82
1436	6' deep	30.50	56	86.50
1442	4 @ 5" dia. Schedule 40 PVC, 2' deep	41.50	60	101.50
1444	4' deep	41.50	65	106.50
1446	6' deep	41.50	69	110.50
1462	6 @ 5" dia. Schedule 40 PVC, 2' deep	56.50	88.50	145
1464	4' deep	56.50	93	149.50
1466	6' deep	56.50	97.50	154
1482	8 @ 5" dia. Schedule 40 PVC, 2' deep	74	118	192
1484	4' deep	74	124	198
1486	6' deep	74	129	203
1492	9 @ 5" dia. Schedule 40 PVC, 2' deep	83	132	215
1494	4' deep	83	135	218
1496	6' deep	83	139	222
1512	6" dia. Schedule 40 PVC, 2' deep	18.40	22.50	40.90
1514	4' deep	18.40	26.50	44.90
1516	6' deep	18.40	31	49.40
1522	2 @ 6" dia. Schedule 40 PVC, 2' deep	28	40	68
1524	4' deep	28	44	72
1526	6' deep	28	48.50	76.50
1532	3 @ 6" dia. Schedule 40 PVC, 2' deep	39.50	59	98.50
1534	4' deep	39.50	64	103.50
1536	6' deep	39.50	69	108.50
1542	4 @ 6" dia. Schedule 40 PVC, 2' deep	52	75.50	127.50
1544	4' deep	52	80	132
1546	6' deep	52	84	136
1562	6 @ 6" dia. Schedule 40 PVC, 2' deep	74	112	186
1564	4' deep	74	117	191
1566	6' deep	74	123	197
1582	8 @ 6" dia. Schedule 40 PVC, 2' deep	98	149	247
1584	4' deep	98	156	254
1586	6' deep	98	163	261
1592	9 @ 6" dia. Schedule 40 PVC, 2' deep	110	168	278
1594	4' deep	110	171	281
1596	6' deep	110	175	285

G4020 Site Lighting

The Site Lighting System includes the complete unit from foundation to electrical fixtures. Each system includes: excavation; concrete base; backfill by hand; compaction with a plate compacter; pole of specified material; all fixtures; and lamps.

The Expanded System Listing shows Site Lighting Systems that use one of two types of lamps: high pressure sodium; and metal halide. Systems are listed for 400-watt and 1000-watt lamps. Pole height varies from 20' to 40'. There are four types of poles possibly listed: aluminum, fiberglass, steel and wood.

System Components	QUANTITY	UNIT	COST EACH		
			MAT.	INST.	TOTAL
SYSTEM G4020 110 5820					
SITE LIGHTING, METAL HALIDE, 400 WATT, ALUMINUM POLE, 20' HIGH					
Excavating, by hand, pits to 6' deep, heavy soil	2.368	C.Y.		267.58	267.58
Concrete in place incl. forms and reinf. stl. spread footings under 1 CY	.465	C.Y.	84.17	126.62	210.79
Backfill	1.903	C.Y.		78.02	78.02
Compact, vibrating plate	1.903	C.Y.		10.68	10.68
Aluminum light pole, 20', no concrete base	1.000	Ea.	1,025	603	1,628
Road area luminaire wire	.300	C.L.F.	12.45	24	36.45
Roadway area luminaire, metal halide 400 W	1.000	Ea.	605	290	895
TOTAL			1,726.62	1,399.90	3,126.52

G4020 110	Site Lighting	COST EACH		
		MAT.	INST.	TOTAL
2320	Site lighting, high pressure sodium, 400 watt, aluminum pole, 20' high	1,750	1,400	3,150
2340	30' high	2,675	1,700	4,375
2360	40' high	3,175	2,200	5,375
2520	Fiberglass pole, 20' high	1,450	1,250	2,700
2540	30' high	1,650	1,525	3,175
2560	40' high	2,425	1,950	4,375
2720	Steel pole, 20' high	1,925	1,450	3,375
2740	30' high	2,175	1,800	3,975
2760	40' high	2,700	2,375	5,075
2920	Wood pole, 20' high	970	1,350	2,320
2940	30' high	1,225	1,700	2,925
2960	40' high	1,550	2,100	3,650
3120	1000 watt, aluminum pole, 20' high	1,850	1,425	3,275
3140	30' high	2,775	1,725	4,500
3160	40' high	3,275	2,225	5,500
3320	Fiberglass pole, 20' high	1,550	1,275	2,825
3340	30' high	1,750	1,550	3,300
3360	40' high	2,525	1,975	4,500
3420	Steel pole, 20' high	2,025	1,475	3,500
3440	30' high	2,275	1,825	4,100

G4020 Site Lighting

G4020 110	Site Lighting	COST EACH		
		MAT.	INST.	TOTAL
3460	40' high	2,800	2,400	5,200
3520	Wood pole, 20' high	1,050	1,375	2,425
3540	30' high	1,325	1,725	3,050
3560	40' high	1,650	2,125	3,775
5820	Metal halide, 400 watt, aluminum pole, 20' high	1,725	1,400	3,125
5840	30' high	2,650	1,700	4,350
5860	40' high	3,150	2,200	5,350
6120	Fiberglass pole, 20' high	1,425	1,250	2,675
6140	30' high	1,625	1,525	3,150
6160	40' high	2,400	1,950	4,350
6320	Steel pole, 20' high	1,900	1,450	3,350
6340	30' high	2,150	1,800	3,950
6360	40' high	2,675	2,375	5,050
7620	1000 watt, aluminum pole, 20' high	1,800	1,425	3,225
7640	30' high	2,725	1,725	4,450
7660	40' high	3,225	2,225	5,450
7900	Fiberglass pole, 20' high	1,500	1,275	2,775
7920	30' high	1,700	1,550	3,250
7940	40' high	2,475	1,975	4,450
8120	Steel pole, 20' high	1,975	1,475	3,450
8140	30' high	2,225	1,825	4,050
8160	40' high	2,750	2,400	5,150

Table G4020 210 Procedure for Calculating Floodlights Required for Various Footcandles

Poles should not be spaced more than 4 times the fixture mounting height for good light distribution.

Estimating Chart

Select Lamp type.

Determine total square feet.

Chart will show quantity of fixtures to provide 1 footcandle initial, at intersection of lines. Multiply fixture quantity by desired footcandle level.

Chart based on use of wide beam luminaires in an area whose dimensions are large compared to mounting height and is approximate only.

To maintain 1 footcandle over a large area use these watts per square foot:

Incandescent	0.15
Metal Halide	0.032
Mercury Vapor	0.05
High Pressure Sodium	0.024

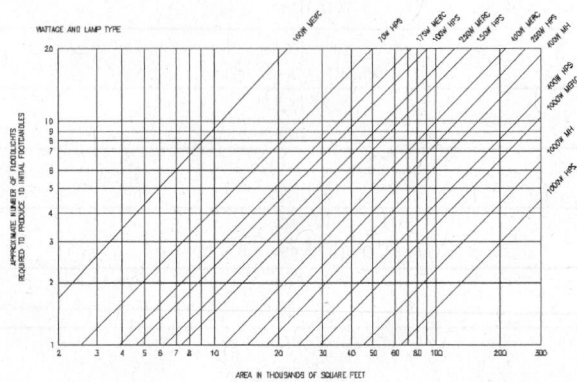

System Components			COST EACH		
	QUANTITY	UNIT	MAT.	INST.	TOTAL
SYSTEM G4020 210 0200					
LIGHT POLES, ALUMINUM, 20' HIGH, 1 ARM BRACKET					
Aluminum light pole, 20', no concrete base	1.000	Ea.	1,025	603	1,628
Bracket arm for Aluminum light pole	1.000	Ea.	131	80	211
Excavation by hand, pits to 6' deep, heavy soil or clay	2.368	C.Y.		267.58	267.58
Footing, concrete incl forms, reinforcing, spread, under 1 C.Y.	.465	C.Y.	84.17	126.62	210.79
Backfill by hand	1.903	C.Y.		78.02	78.02
Compaction vibrating plate	1.903	C.Y.		10.68	10.68
TOTAL			1,240.17	1,165.90	2,406.07

G4020 210	Light Pole (Installed)	COST EACH		
		MAT.	INST.	TOTAL
0200	Light pole, aluminum, 20' high, 1 arm bracket	1,250	1,150	2,400
0240	2 arm brackets	1,375	1,150	2,525
0280	3 arm brackets	1,500	1,200	2,700
0320	4 arm brackets	1,650	1,200	2,850
0360	30' high, 1 arm bracket	2,150	1,450	3,600
0400	2 arm brackets	2,300	1,450	3,750
0440	3 arm brackets	2,425	1,500	3,925
0480	4 arm brackets	2,550	1,500	4,050
0680	40' high, 1 arm bracket	2,650	1,950	4,600
0720	2 arm brackets	2,800	1,950	4,750
0760	3 arm brackets	2,925	2,000	4,925
0800	4 arm brackets	3,050	2,000	5,050
0840	Steel, 20' high, 1 arm bracket	1,475	1,225	2,700
0880	2 arm brackets	1,600	1,225	2,825
0920	3 arm brackets	1,625	1,250	2,875
0960	4 arm brackets	1,750	1,250	3,000
1000	30' high, 1 arm bracket	1,725	1,575	3,300
1040	2 arm brackets	1,825	1,575	3,400
1080	3 arm brackets	1,850	1,600	3,450
1120	4 arm brackets	2,000	1,600	3,600
1320	40' high, 1 arm bracket	2,250	2,125	4,375
1360	2 arm brackets	2,350	2,125	4,475
1400	3 arm brackets	2,375	2,150	4,525
1440	4 arm brackets	2,500	2,150	4,650

Reference Section

All the reference information is in one section, making it easy to find what you need to know . . . and easy to use the book on a daily basis. This section is visually identified by a vertical gray bar on the page edges.

In this Reference Section, we've included Equipment Rental Costs, a listing of rental and operating costs; Crew Listings, a full listing of all crews, equipment, and their costs; Historical Cost Indexes for cost comparisons over time; City Cost Indexes and Location Factors for adjusting costs to the region you are in; Reference Tables, where you will find explanations, estimating information and procedures, or technical data; Square Foot Costs that allow you to make a rough estimate for the overall cost of a project; and an explanation of all the Abbreviations in the book.

Table of Contents

Construction Equipment Rental Costs — 443

Crew Listings — 455

Historical Cost Indexes — 491

City Cost Indexes — 492

Location Factors — 535

Reference Tables — 541
R01 General Requirements 541
R02 Existing Conditions 555
R05 Metals 556
R07 Thermal & Moisture Protection 556

Reference Tables (cont.)
R22 Plumbing 557
R23 Heating, Ventilating & Air Conditioning................................... 559
R26 Electrical.. 561
R27 Communications 611
R33 Utilities.. 612

Change Orders — 614

Square Foot Costs — 619

Abbreviations — 629

Estimating Tips

- This section contains the average costs to rent and operate hundreds of pieces of construction equipment. This is useful information when estimating the time and material requirements of any particular operation in order to establish a unit or total cost. Equipment costs include not only rental, but also operating costs for equipment under normal use.

Rental Costs

- Equipment rental rates are obtained from industry sources throughout North America–contractors, suppliers, dealers, manufacturers, and distributors.
- Rental rates vary throughout the country, with larger cities generally having lower rates. Lease plans for new equipment are available for periods in excess of six months, with a percentage of payments applying toward purchase.
- Monthly rental rates vary from 2% to 5% of the purchase price of the equipment depending on the anticipated life of the equipment and its wearing parts.
- Weekly rental rates are about 1/3 the monthly rates, and daily rental rates are about 1/3 the weekly rate.
- Rental rates can also be treated as reimbursement costs for contractor-owned equipment. Owned equipment costs include depreciation, loan payments, interest, taxes, insurance, storage, and major repairs.

Operating Costs

- The operating costs include parts and labor for routine servicing, such as repair and replacement of pumps, filters and worn lines. Normal operating expendables, such as fuel, lubricants, tires and electricity (where applicable), are also included.
- Extraordinary operating expendables with highly variable wear patterns, such as diamond bits and blades, are excluded. These costs can be found as material costs in the Unit Price section.
- The hourly operating costs listed do not include the operator's wages.

Equipment Cost/Day

- Any power equipment required by a crew is shown in the Crew Listings with a daily cost.
- The daily cost of equipment needed by a crew is based on dividing the weekly rental rate by 5 (number of working days in the week), and then adding the hourly operating cost times 8 (the number of hours in a day). This "Equipment Cost/ Day" is shown in the far right column of the Equipment Rental pages.
- If equipment is needed for only one or two days, it is best to develop your own cost by including components for daily rent and hourly operating cost. This is important when the listed Crew for a task does not contain the equipment needed, such as a crane for lifting mechanical heating/cooling equipment up onto a roof.
- If the quantity of work is less than the crew's Daily Output shown for a Unit Price line item that includes a bare unit equipment cost, it is recommended to estimate one day's rental cost and operating cost for equipment shown in the Crew Listing for that line item.

Mobilization/ Demobilization

- The cost to move construction equipment from an equipment yard or rental company to the jobsite and back again is not included in equipment rental costs listed in the Reference section, nor in the bare equipment cost of any Unit Price line item, nor in any equipment costs shown in the Crew listings.
- Mobilization (to the site) and demobilization (from the site) costs can be found in the Unit Price section.
- If a piece of equipment is already at the jobsite, it is not appropriate to utilize mobil./ demob. costs again in an estimate.

01 54 33 | Equipment Rental

			UNIT	HOURLY OPER. COST	RENT PER DAY	RENT PER WEEK	RENT PER MONTH	EQUIPMENT COST/DAY	
10	0010	**CONCRETE EQUIPMENT RENTAL** without operators							10
	0200	Bucket, concrete lightweight, 1/2 C.Y.	Ea.	.80	23	69	207	20.20	
	0300	1 C.Y.		.85	27	81	243	23	
	0400	1-1/2 C.Y.		1.10	36.50	110	330	30.80	
	0500	2 C.Y.		1.20	43.50	130	390	35.60	
	0580	8 C.Y.		5.95	252	755	2,275	198.60	
	0600	Cart, concrete, self-propelled, operator walking, 10 C.F.		3.40	55	165	495	60.20	
	0700	Operator riding, 18 C.F.		5.65	93.50	280	840	101.20	
	0800	Conveyer for concrete, portable, gas, 16" wide, 26' long		13.50	122	365	1,100	181	
	0900	46' long		13.90	147	440	1,325	199.20	
	1000	56' long		14.05	155	465	1,400	205.40	
	1100	Core drill, electric, 2-1/2 H.P., 1" to 8" bit diameter		1.50	57	171	515	46.20	
	1150	11 H.P., 8" to 18" cores		5.90	115	345	1,025	116.20	
	1200	Finisher, concrete floor, gas, riding trowel, 96" wide		14.05	142	425	1,275	197.40	
	1300	Gas, walk-behind, 3 blade, 36" trowel		2.35	20	60	180	30.80	
	1400	4 blade, 48" trowel		4.70	26.50	80	240	53.60	
	1500	Float, hand-operated (Bull float) 48" wide		.08	13.35	40	120	8.65	
	1570	Curb builder, 14 H.P., gas, single screw		14.65	250	750	2,250	267.20	
	1590	Double screw		15.30	287	860	2,575	294.40	
	1600	Floor grinder, concrete and terrazzo, electric, 22" path		2.19	134	402	1,200	97.90	
	1700	Edger, concrete, electric, 7" path		1.02	51.50	155	465	39.15	
	1750	Vacuum pick-up system for floor grinders, wet/dry		1.47	81.50	245	735	60.75	
	1800	Mixer, powered, mortar and concrete, gas, 6 C.F., 18 H.P.		9.20	117	350	1,050	143.60	
	1900	10 C.F., 25 H.P.		11.60	140	420	1,250	176.80	
	2000	16 C.F.		11.95	162	485	1,450	192.60	
	2100	Concrete, stationary, tilt drum, 2 C.Y.		6.80	228	685	2,050	191.40	
	2120	Pump, concrete, truck mounted 4" line 80' boom		24.55	855	2,570	7,700	710.40	
	2140	5" line, 110' boom		31.75	1,125	3,400	10,200	934	
	2160	Mud jack, 50 C.F. per hr.		7.35	127	380	1,150	134.80	
	2180	225 C.F. per hr.		9.65	145	435	1,300	164.20	
	2190	Shotcrete pump rig, 12 C.Y./hr.		15.10	220	660	1,975	252.80	
	2200	35 C.Y./hr.		17.80	237	710	2,125	284.40	
	2600	Saw, concrete, manual, gas, 18 H.P.		7.25	43.50	130	390	84	
	2650	Self-propelled, gas, 30 H.P.		14.40	100	300	900	175.20	
	2675	V-groove crack chaser, manual, gas, 6 H.P.		2.55	17.35	52	156	30.80	
	2700	Vibrators, concrete, electric, 60 cycle, 2 H.P.		.43	8.65	26	78	8.65	
	2800	3 H.P.		.61	11.65	35	105	11.90	
	2900	Gas engine, 5 H.P.		2.20	16	48	144	27.20	
	3000	8 H.P.		3.00	15	45	135	33	
	3050	Vibrating screed, gas engine, 8 H.P.		2.78	68.50	205	615	63.25	
	3120	Concrete transit mixer, 6 x 4, 250 H.P., 8 C.Y., rear discharge		62.50	565	1,690	5,075	838	
	3200	Front discharge		73.60	690	2,065	6,200	1,002	
	3300	6 x 6, 285 H.P., 12 C.Y., rear discharge		72.65	650	1,950	5,850	971.20	
	3400	Front discharge		75.95	695	2,085	6,250	1,025	
20	0010	**EARTHWORK EQUIPMENT RENTAL** without operators							20
	0040	Aggregate spreader, push type 8' to 12' wide	Ea.	3.25	25.50	76	228	41.20	
	0045	Tailgate type, 8' wide		3.10	32.50	98	294	44.40	
	0055	Earth auger, truck-mounted, for fence & sign posts, utility poles		19.75	445	1,335	4,000	425	
	0060	For borings and monitoring wells		47.30	680	2,040	6,125	786.40	
	0070	Portable, trailer mounted		3.15	32.50	98	294	44.80	
	0075	Truck-mounted, for caissons, water wells		100.90	2,900	8,720	26,200	2,551	
	0080	Horizontal boring machine, 12" to 36" diameter, 45 H.P.		24.60	193	580	1,750	312.80	
	0090	12" to 48" diameter, 65 H.P.		34.85	335	1,005	3,025	479.80	
	0095	Auger, for fence posts, gas engine, hand held		.55	6	18	54	8	
	0100	Excavator, diesel hydraulic, crawler mounted, 1/2 C.Y. cap.		25.00	415	1,240	3,725	448	
	0120	5/8 C.Y. capacity		33.05	550	1,650	4,950	594.40	
	0140	3/4 C.Y. capacity		36.00	625	1,870	5,600	662	
	0150	1 C.Y. capacity		50.60	685	2,050	6,150	814.80	

R015433 -10

01 54 33 | Equipment Rental

		UNIT	HOURLY OPER. COST	RENT PER DAY	RENT PER WEEK	RENT PER MONTH	EQUIPMENT COST/DAY	
20	0200	1-1/2 C.Y. capacity	Ea.	60.80	910	2,725	8,175	1,031	20
	0300	2 C.Y. capacity		68.50	1,050	3,135	9,400	1,175	
	0320	2-1/2 C.Y. capacity		104.75	1,325	3,995	12,000	1,637	
	0325	3-1/2 C.Y. capacity		147.35	2,100	6,315	18,900	2,442	
	0330	4-1/2 C.Y. capacity		176.65	2,725	8,150	24,500	3,043	
	0335	6 C.Y. capacity		224.40	2,875	8,640	25,900	3,523	
	0340	7 C.Y. capacity		226.60	2,975	8,915	26,700	3,596	
	0342	Excavator attachments, bucket thumbs		3.20	245	735	2,200	172.60	
	0345	Grapples		2.75	193	580	1,750	138	
	0347	Hydraulic hammer for boom mounting, 5000 ft lb.		14.30	425	1,280	3,850	370.40	
	0349	11,000 ft lb.		23.20	750	2,250	6,750	635.60	
	0350	Gradall type, truck mounted, 3 ton @ 15' radius, 5/8 C.Y.		43.45	890	2,670	8,000	881.60	
	0370	1 C.Y. capacity		47.65	1,050	3,120	9,350	1,005	
	0400	Backhoe-loader, 40 to 45 H.P., 5/8 C.Y. capacity		14.95	233	700	2,100	259.60	
	0450	45 H.P. to 60 H.P., 3/4 C.Y. capacity		23.90	290	870	2,600	365.20	
	0460	80 H.P., 1-1/4 C.Y. capacity		26.20	315	940	2,825	397.60	
	0470	112 H.P., 1-1/2 C.Y. capacity		41.65	645	1,935	5,800	720.20	
	0482	Backhoe-loader attachment, compactor, 20,000 lb.		5.75	142	425	1,275	131	
	0485	Hydraulic hammer, 750 ft lb.		3.25	96.50	290	870	84	
	0486	Hydraulic hammer, 1200 ft lb.		6.30	217	650	1,950	180.40	
	0500	Brush chipper, gas engine, 6" cutter head, 35 H.P.		11.95	107	320	960	159.60	
	0550	Diesel engine, 12" cutter head, 130 H.P.		28.20	285	855	2,575	396.60	
	0600	15" cutter head, 165 H.P.		33.90	330	985	2,950	468.20	
	0750	Bucket, clamshell, general purpose, 3/8 C.Y.		1.25	38.50	115	345	33	
	0800	1/2 C.Y.		1.35	45	135	405	37.80	
	0850	3/4 C.Y.		1.50	53.50	160	480	44	
	0900	1 C.Y.		1.55	58.50	175	525	47.40	
	0950	1-1/2 C.Y.		2.55	80	240	720	68.40	
	1000	2 C.Y.		2.65	88.50	265	795	74.20	
	1010	Bucket, dragline, medium duty, 1/2 C.Y.		.75	23.50	70	210	20	
	1020	3/4 C.Y.		.75	24.50	74	222	20.80	
	1030	1 C.Y.		.80	26	78	234	22	
	1040	1-1/2 C.Y.		1.20	40	120	360	33.60	
	1050	2 C.Y.		1.25	45	135	405	37	
	1070	3 C.Y.		1.95	61.50	185	555	52.60	
	1200	Compactor, manually guided 2-drum vibratory smooth roller, 7.5 H.P.		7.35	207	620	1,850	182.80	
	1250	Rammer/tamper, gas, 8"		2.70	46.50	140	420	49.60	
	1260	15"		3.00	53.50	160	480	56	
	1300	Vibratory plate, gas, 18" plate, 3000 lb. blow		2.65	24.50	73	219	35.80	
	1350	21" plate, 5000 lb. blow		3.30	32.50	98	294	46	
	1370	Curb builder/extruder, 14 H.P., gas, single screw		14.65	250	750	2,250	267.20	
	1390	Double screw		15.30	287	860	2,575	294.40	
	1500	Disc harrow attachment, for tractor		.43	72.50	217	650	46.85	
	1810	Feller buncher, shearing & accumulating trees, 100 H.P.		42.50	650	1,945	5,825	729	
	1860	Grader, self-propelled, 25,000 lb.		39.55	595	1,780	5,350	672.40	
	1910	30,000 lb.		43.25	600	1,805	5,425	707	
	1920	40,000 lb.		64.65	1,000	3,015	9,050	1,120	
	1930	55,000 lb.		84.65	1,600	4,800	14,400	1,637	
	1950	Hammer, pavement breaker, self-propelled, diesel, 1000 to 1250 lb.		30.25	355	1,060	3,175	454	
	2000	1300 to 1500 lb.		45.38	705	2,120	6,350	787.05	
	2050	Pile driving hammer, steam or air, 4150 ft lb. @ 225 bpm		10.35	475	1,425	4,275	367.80	
	2100	8750 ft lb. @ 145 bpm		12.35	660	1,980	5,950	494.80	
	2150	15,000 ft lb. @ 60 bpm		13.95	795	2,380	7,150	587.60	
	2200	24,450 ft lb. @ 111 bpm		14.95	880	2,640	7,925	647.60	
	2250	Leads, 60' high for pile driving hammers up to 20,000 ft lb.		3.30	80.50	242	725	74.80	
	2300	90' high for hammers over 20,000 ft lb.		4.95	141	424	1,275	124.40	
	2350	Diesel type hammer, 22,400 ft lb.		26.93	535	1,610	4,825	537.45	
	2400	41,300 ft lb.		36.30	585	1,760	5,275	642.40	

01 54 33 | Equipment Rental

		UNIT	HOURLY OPER. COST	RENT PER DAY	RENT PER WEEK	RENT PER MONTH	EQUIPMENT COST/DAY
20	2450	141,000 ft lb.	Ea.	55.55	1,050	3,160	9,475	1,076
	2500	Vib. elec. hammer/extractor, 200 kW diesel generator, 34 H.P.		56.25	635	1,910	5,725	832
	2550	80 H.P.		103.45	930	2,785	8,350	1,385
	2600	150 H.P.		196.65	1,775	5,335	16,000	2,640
	2800	Log chipper, up to 22" diameter, 600 H.P.		64.95	630	1,895	5,675	898.60
	2850	Logger, for skidding & stacking logs, 150 H.P.		53.45	820	2,460	7,375	919.60
	2860	Mulcher, diesel powered, trailer mounted		24.70	213	640	1,925	325.60
	2900	Rake, spring tooth, with tractor		18.31	335	1,007	3,025	347.90
	3000	Roller, vibratory, tandem, smooth drum, 20 H.P.		8.75	148	445	1,325	159
	3050	35 H.P.		11.25	253	760	2,275	242
	3100	Towed type vibratory compactor, smooth drum, 50 H.P.		26.25	335	1,010	3,025	412
	3150	Sheepsfoot, 50 H.P.		27.60	370	1,115	3,350	443.80
	3170	Landfill compactor, 220 H.P.		91.85	1,500	4,485	13,500	1,632
	3200	Pneumatic tire roller, 80 H.P.		17.00	355	1,065	3,200	349
	3250	120 H.P.		25.45	605	1,815	5,450	566.60
	3300	Sheepsfoot vibratory roller, 240 H.P.		69.65	1,100	3,305	9,925	1,218
	3320	340 H.P.		101.85	1,600	4,830	14,500	1,781
	3350	Smooth drum vibratory roller, 75 H.P.		25.15	595	1,790	5,375	559.20
	3400	125 H.P.		32.85	715	2,150	6,450	692.80
	3410	Rotary mower, brush, 60", with tractor		22.65	305	915	2,750	364.20
	3420	Rototiller, walk-behind, gas, 5 H.P.		2.23	66	198	595	57.45
	3422	8 H.P.		3.30	91.50	275	825	81.40
	3440	Scrapers, towed type, 7 C.Y. capacity		5.70	110	330	990	111.60
	3450	10 C.Y. capacity		6.45	150	450	1,350	141.60
	3500	15 C.Y. capacity		6.85	173	520	1,550	158.80
	3525	Self-propelled, single engine, 14 C.Y. capacity		113.60	1,525	4,600	13,800	1,829
	3550	Dual engine, 21 C.Y. capacity		174.50	2,000	6,015	18,000	2,599
	3600	31 C.Y. capacity		232.10	2,850	8,545	25,600	3,566
	3640	44 C.Y. capacity		287.55	3,650	10,960	32,900	4,492
	3650	Elevating type, single engine, 11 C.Y. capacity		71.20	1,025	3,050	9,150	1,180
	3700	22 C.Y. capacity		138.15	2,225	6,690	20,100	2,443
	3710	Screening plant 110 H.P. w/5' x 10' screen		31.63	435	1,300	3,900	513.05
	3720	5' x 16' screen		34.85	540	1,615	4,850	601.80
	3850	Shovel, crawler-mounted, front-loading, 7 C.Y. capacity		246.75	2,900	8,700	26,100	3,714
	3855	12 C.Y. capacity		354.50	3,950	11,840	35,500	5,204
	3860	Shovel/backhoe bucket, 1/2 C.Y.		2.45	66.50	200	600	59.60
	3870	3/4 C.Y.		2.50	73.50	220	660	64
	3880	1 C.Y.		2.60	83.50	250	750	70.80
	3890	1-1/2 C.Y.		2.75	96.50	290	870	80
	3910	3 C.Y.		3.10	130	390	1,175	102.80
	3950	Stump chipper, 18" deep, 30 H.P.		7.06	176	528	1,575	162.10
	4110	Dozer, crawler, torque converter, diesel 80 H.P.		29.75	405	1,215	3,650	481
	4150	105 H.P.		36.10	505	1,515	4,550	591.80
	4200	140 H.P.		52.10	780	2,335	7,000	883.80
	4260	200 H.P.		77.35	1,175	3,530	10,600	1,325
	4310	300 H.P.		100.55	1,700	5,065	15,200	1,817
	4360	410 H.P.		135.65	2,200	6,620	19,900	2,409
	4370	500 H.P.		174.30	2,875	8,610	25,800	3,116
	4380	700 H.P.		261.35	4,000	11,970	35,900	4,485
	4400	Loader, crawler, torque conv., diesel, 1-1/2 C.Y., 80 H.P.		31.35	455	1,360	4,075	522.80
	4450	1-1/2 to 1-3/4 C.Y., 95 H.P.		34.15	585	1,760	5,275	625.20
	4510	1-3/4 to 2-1/4 C.Y., 130 H.P.		54.60	870	2,615	7,850	959.80
	4530	2-1/2 to 3-1/4 C.Y., 190 H.P.		67.00	1,100	3,275	9,825	1,191
	4560	3-1/2 to 5 C.Y., 275 H.P.		86.10	1,400	4,215	12,600	1,532
	4610	Front end loader, 4WD, articulated frame, diesel, 1 to 1-1/4 C.Y., 70 H.P.		19.90	227	680	2,050	295.20
	4620	1-1/2 to 1-3/4 C.Y., 95 H.P.		25.30	285	855	2,575	373.40
	4650	1-3/4 to 2 C.Y., 130 H.P.		29.45	365	1,100	3,300	455.60
	4710	2-1/2 to 3-1/2 C.Y., 145 H.P.		33.35	425	1,280	3,850	522.80

20

01 54 33 | Equipment Rental

		UNIT	HOURLY OPER. COST	RENT PER DAY	RENT PER WEEK	RENT PER MONTH	EQUIPMENT COST/DAY	
20	4730	3 to 4-1/2 C.Y., 185 H.P.	Ea.	42.85	550	1,655	4,975	673.80	**20**
	4760	5-1/4 to 5-3/4 C.Y., 270 H.P.		67.80	845	2,535	7,600	1,049	
	4810	7 to 9 C.Y., 475 H.P.		116.25	1,775	5,295	15,900	1,989	
	4870	9 - 11 C.Y., 620 H.P.		155.75	2,425	7,285	21,900	2,703	
	4880	Skid steer loader, wheeled, 10 C.F., 30 H.P. gas		10.90	145	435	1,300	174.20	
	4890	1 C.Y., 78 H.P., diesel		20.35	243	730	2,200	308.80	
	4892	Skid-steer attachment, auger		.50	83.50	251	755	54.20	
	4893	Backhoe		.66	111	332	995	71.70	
	4894	Broom		.74	124	372	1,125	80.30	
	4895	Forks		.23	38.50	115	345	24.85	
	4896	Grapple		.56	92.50	278	835	60.10	
	4897	Concrete hammer		1.06	177	531	1,600	114.70	
	4898	Tree spade		1.03	172	515	1,550	111.25	
	4899	Trencher		.67	112	335	1,000	72.35	
	4900	Trencher, chain, boom type, gas, operator walking, 12 H.P.		5.40	46.50	140	420	71.20	
	4910	Operator riding, 40 H.P.		20.55	292	875	2,625	339.40	
	5000	Wheel type, diesel, 4' deep, 12" wide		87.50	845	2,530	7,600	1,206	
	5100	6' deep, 20" wide		93.25	1,925	5,760	17,300	1,898	
	5150	Chain type, diesel, 5' deep, 8" wide		38.85	560	1,685	5,050	647.80	
	5200	Diesel, 8' deep, 16" wide		156.45	3,550	10,620	31,900	3,376	
	5202	Rock trencher, wheel type, 6" wide x 18" deep		21.50	330	985	2,950	369	
	5206	Chain type, 18" wide x 7' deep		112.60	2,825	8,445	25,300	2,590	
	5210	Tree spade, self-propelled		14.00	267	800	2,400	272	
	5250	Truck, dump, 2-axle, 12 ton, 8 C.Y. payload, 220 H.P.		35.15	227	680	2,050	417.20	
	5300	Three axle dump, 16 ton, 12 C.Y. payload, 400 H.P.		62.40	320	965	2,900	692.20	
	5310	Four axle dump, 25 ton, 18 C.Y. payload, 450 H.P.		73.20	470	1,410	4,225	867.60	
	5350	Dump trailer only, rear dump, 16-1/2 C.Y.		5.40	138	415	1,250	126.20	
	5400	20 C.Y.		5.85	157	470	1,400	140.80	
	5450	Flatbed, single axle, 1-1/2 ton rating		28.65	68.50	205	615	270.20	
	5500	3 ton rating		34.25	98.50	295	885	333	
	5550	Off highway rear dump, 25 ton capacity		74.55	1,200	3,590	10,800	1,314	
	5600	35 ton capacity		83.10	1,350	4,025	12,100	1,470	
	5610	50 ton capacity		103.40	1,600	4,800	14,400	1,787	
	5620	65 ton capacity		106.30	1,575	4,735	14,200	1,797	
	5630	100 ton capacity		154.35	2,775	8,315	24,900	2,898	
	6000	Vibratory plow, 25 H.P., walking		9.25	60	180	540	110	
40	0010	**GENERAL EQUIPMENT RENTAL** without operators							**40**
	0150	Aerial lift, scissor type, to 15' high, 1000 lb. cap., electric	Ea.	3.10	51.50	155	465	55.80	
	0160	To 25' high, 2000 lb. capacity		3.55	66.50	200	600	68.40	
	0170	Telescoping boom to 40' high, 500 lb. capacity, diesel		17.00	315	945	2,825	325	
	0180	To 45' high, 500 lb. capacity		18.85	355	1,070	3,200	364.80	
	0190	To 60' high, 600 lb. capacity		21.25	475	1,430	4,300	456	
	0195	Air compressor, portable, 6.5 CFM, electric		.56	12.65	38	114	12.10	
	0196	Gasoline		.79	19	57	171	17.70	
	0200	Towed type, gas engine, 60 CFM		13.55	48.50	145	435	137.40	
	0300	160 CFM		15.80	51.50	155	465	157.40	
	0400	Diesel engine, rotary screw, 250 CFM		17.05	110	330	990	202.40	
	0500	365 CFM		23.10	133	400	1,200	264.80	
	0550	450 CFM		29.35	167	500	1,500	334.80	
	0600	600 CFM		51.70	232	695	2,075	552.60	
	0700	750 CFM		51.90	240	720	2,150	559.20	
	0800	For silenced models, small sizes, add to rent		3%	5%	5%	5%		
	0900	Large sizes, add to rent		5%	7%	7%	7%		
	0930	Air tools, breaker, pavement, 60 lb.	Ea.	.50	9.65	29	87	9.80	
	0940	80 lb.		.50	10.35	31	93	10.20	
	0950	Drills, hand (jackhammer) 65 lb.		.60	17.35	52	156	15.20	
	0960	Track or wagon, swing boom, 4" drifter		63.45	885	2,660	7,975	1,040	
	0970	5" drifter		75.50	1,075	3,255	9,775	1,255	

R015433 -10

01 54 33 | Equipment Rental

		UNIT	HOURLY OPER. COST	RENT PER DAY	RENT PER WEEK	RENT PER MONTH	EQUIPMENT COST/DAY	
40	0975	Track mounted quarry drill, 6" diameter drill	Ea.	129.75	1,600	4,830	14,500	2,004	40
	0980	Dust control per drill		.97	24	72	216	22.15	
	0990	Hammer, chipping, 12 lb.		.55	26	78	234	20	
	1000	Hose, air with couplings, 50' long, 3/4" diameter		.03	5	15	45	3.25	
	1100	1" diameter		.04	6.35	19	57	4.10	
	1200	1-1/2" diameter		.05	9	27	81	5.80	
	1300	2" diameter		.07	12	36	108	7.75	
	1400	2-1/2" diameter		.11	19	57	171	12.30	
	1410	3" diameter		.14	23	69	207	14.90	
	1450	Drill, steel, 7/8" x 2'		.05	8.65	26	78	5.60	
	1460	7/8" x 6'		.05	9	27	81	5.80	
	1520	Moil points		.02	3.33	10	30	2.15	
	1525	Pneumatic nailer w/accessories		.47	31	93	279	22.35	
	1530	Sheeting driver for 60 lb. breaker		.03	5.65	17	51	3.65	
	1540	For 90 lb. breaker		.11	7.35	22	66	5.30	
	1550	Spade, 25 lb.		.45	7	21	63	7.80	
	1560	Tamper, single, 35 lb.		.55	36.50	109	325	26.20	
	1570	Triple, 140 lb.		.82	54.50	164	490	39.35	
	1580	Wrenches, impact, air powered, up to 3/4" bolt		.40	12.65	38	114	10.80	
	1590	Up to 1-1/4" bolt		.50	23.50	70	210	18	
	1600	Barricades, barrels, reflectorized, 1 to 99 barrels		.03	4.60	13.80	41.50	3	
	1610	100 to 200 barrels		.02	3.53	10.60	32	2.30	
	1620	Barrels with flashers, 1 to 99 barrels		.03	5.25	15.80	47.50	3.40	
	1630	100 to 200 barrels		.03	4.20	12.60	38	2.75	
	1640	Barrels with steady burn type C lights		.04	7	21	63	4.50	
	1650	Illuminated board, trailer mounted, with generator		3.55	130	390	1,175	106.40	
	1670	Portable barricade, stock, with flashers, 1 to 6 units		.03	5.25	15.80	47.50	3.40	
	1680	25 to 50 units		.03	4.90	14.70	44	3.20	
	1685	Butt fusion machine, wheeled, 1.5 HP electric, 2" - 8" diameter pipe		2.62	167	500	1,500	120.95	
	1690	Tracked, 20 HP diesel, 4"-12" diameter pipe		11.10	490	1,465	4,400	381.80	
	1695	83 HP diesel, 8" - 24" diameter pipe		30.30	975	2,930	8,800	828.40	
	1700	Carts, brick, hand powered, 1000 lb. capacity		.44	72.50	218	655	47.10	
	1800	Gas engine, 1500 lb., 7-1/2' lift		4.10	115	345	1,025	101.80	
	1822	Dehumidifier, medium, 6 lb./hr., 150 CFM		.99	62	186	560	45.10	
	1824	Large, 18 lb./hr., 600 CFM		2.01	126	378	1,125	91.70	
	1830	Distributor, asphalt, trailer mounted, 2000 gal., 38 H.P. diesel		10.15	330	985	2,950	278.20	
	1840	3000 gal., 38 H.P. diesel		11.65	360	1,075	3,225	308.20	
	1850	Drill, rotary hammer, electric		1.00	26.50	80	240	24	
	1860	Carbide bit, 1-1/2" diameter, add to electric rotary hammer		.02	3.61	10.84	32.50	2.35	
	1865	Rotary, crawler, 250 H.P.		150.45	2,100	6,285	18,900	2,461	
	1870	Emulsion sprayer, 65 gal., 5 H.P. gas engine		2.90	98	294	880	82	
	1880	200 gal., 5 H.P. engine		7.85	163	490	1,475	160.80	
	1900	Floor auto-scrubbing machine, walk-behind, 28" path		4.52	293	880	2,650	212.15	
	1930	Floodlight, mercury vapor, or quartz, on tripod, 1000 watt		.41	19.35	58	174	14.90	
	1940	2000 watt		.78	39	117	350	29.65	
	1950	Floodlights, trailer mounted with generator, 1 - 300 watt light		3.65	71.50	215	645	72.20	
	1960	2 - 1000 watt lights		4.90	98.50	295	885	98.20	
	2000	4 - 300 watt lights		4.60	93.50	280	840	92.80	
	2005	Foam spray rig, incl. box trailer, compressor, generator, proportioner		36.53	490	1,465	4,400	585.25	
	2020	Forklift, straight mast, 12' lift, 5000 lb., 2 wheel drive, gas		25.35	203	610	1,825	324.80	
	2040	21' lift, 5000 lb., 4 wheel drive, diesel		21.15	242	725	2,175	314.20	
	2050	For rough terrain, 42' lift, 35' reach, 9000 lb., 110 H.P.		29.95	485	1,460	4,375	531.60	
	2060	For plant, 4 ton capacity, 80 H.P., 2 wheel drive, gas		15.50	93.50	280	840	180	
	2080	10 ton capacity, 120 H.P., 2 wheel drive, diesel		24.30	162	485	1,450	291.40	
	2100	Generator, electric, gas engine, 1.5 kW to 3 kW		3.95	11.35	34	102	38.40	
	2200	5 kW		5.10	15.35	46	138	50	
	2300	10 kW		9.65	36.50	110	330	99.20	
	2400	25 kW		11.00	86.50	260	780	140	

01 54 33 | Equipment Rental

		UNIT	HOURLY OPER. COST	RENT PER DAY	RENT PER WEEK	RENT PER MONTH	EQUIPMENT COST/DAY	
40	2500	Diesel engine, 20 kW	Ea.	12.55	68.50	205	615	141.40	40
	2600	50 kW		24.35	103	310	930	256.80	
	2700	100 kW		45.15	130	390	1,175	439.20	
	2800	250 kW		88.85	237	710	2,125	852.80	
	2850	Hammer, hydraulic, for mounting on boom, to 500 ft lb.		2.55	75	225	675	65.40	
	2860	1000 ft lb.		4.35	127	380	1,150	110.80	
	2900	Heaters, space, oil or electric, 50 MBH		1.93	7.65	23	69	20.05	
	3000	100 MBH		3.61	10.65	32	96	35.30	
	3100	300 MBH		10.58	38.50	115	345	107.65	
	3150	500 MBH		17.40	45	135	405	166.20	
	3200	Hose, water, suction with coupling, 20' long, 2" diameter		.02	3	9	27	1.95	
	3210	3" diameter		.03	4.33	13	39	2.85	
	3220	4" diameter		.03	5	15	45	3.25	
	3230	6" diameter		.11	17.65	53	159	11.50	
	3240	8" diameter		.20	33.50	100	300	21.60	
	3250	Discharge hose with coupling, 50' long, 2" diameter		.01	1.33	4	12	.90	
	3260	3" diameter		.01	2.33	7	21	1.50	
	3270	4" diameter		.02	3.67	11	33	2.35	
	3280	6" diameter		.06	9.35	28	84	6.10	
	3290	8" diameter		.20	33.50	100	300	21.60	
	3295	Insulation blower		.77	6	18	54	9.75	
	3300	Ladders, extension type, 16' to 36' long		.14	22.50	68	204	14.70	
	3400	40' to 60' long		.19	31	93	279	20.10	
	3405	Lance for cutting concrete		2.52	89	267	800	73.55	
	3407	Lawn mower, rotary, 22", 5 H.P.		2.10	59	177	530	52.20	
	3408	48" self propelled		2.92	75	225	675	68.35	
	3410	Level, electronic, automatic, with tripod and leveling rod		1.49	99	297	890	71.30	
	3430	Laser type, for pipe and sewer line and grade		.73	48.50	145	435	34.85	
	3440	Rotating beam for interior control		.97	64	192	575	46.15	
	3460	Builder's optical transit, with tripod and rod		.10	16.65	50	150	10.80	
	3500	Light towers, towable, with diesel generator, 2000 watt		4.60	93.50	280	840	92.80	
	3600	4000 watt		4.90	98.50	295	885	98.20	
	3700	Mixer, powered, plaster and mortar, 6 C.F., 7 H.P.		2.90	19	57	171	34.60	
	3800	10 C.F., 9 H.P.		3.05	30.50	91	273	42.60	
	3850	Nailer, pneumatic		.47	31	93	279	22.35	
	3900	Paint sprayers complete, 8 CFM		.94	62.50	187	560	44.90	
	4000	17 CFM		1.57	104	313	940	75.15	
	4020	Pavers, bituminous, rubber tires, 8' wide, 50 H.P., diesel		31.85	500	1,500	4,500	554.80	
	4030	10' wide, 150 H.P.		107.95	1,825	5,460	16,400	1,956	
	4050	Crawler, 8' wide, 100 H.P., diesel		90.45	1,825	5,500	16,500	1,824	
	4060	10' wide, 150 H.P.		115.10	2,250	6,715	20,100	2,264	
	4070	Concrete paver, 12' to 24' wide, 250 H.P.		102.05	1,525	4,600	13,800	1,736	
	4080	Placer-spreader-trimmer, 24' wide, 300 H.P.		147.05	2,475	7,405	22,200	2,657	
	4100	Pump, centrifugal gas pump, 1-1/2" diam., 65 GPM		3.95	50	150	450	61.60	
	4200	2" diameter, 130 GPM		5.30	55	165	495	75.40	
	4300	3" diameter, 250 GPM		5.60	56.50	170	510	78.80	
	4400	6" diameter, 1500 GPM		29.35	177	530	1,600	340.80	
	4500	Submersible electric pump, 1-1/4" diameter, 55 GPM		.38	16.65	50	150	13.05	
	4600	1-1/2" diameter, 83 GPM		.42	19	57	171	14.75	
	4700	2" diameter, 120 GPM		1.43	23.50	71	213	25.65	
	4800	3" diameter, 300 GPM		2.55	41.50	125	375	45.40	
	4900	4" diameter, 560 GPM		10.65	160	480	1,450	181.20	
	5000	6" diameter, 1590 GPM		15.75	215	645	1,925	255	
	5100	Diaphragm pump, gas, single, 1-1/2" diameter		1.18	50.50	152	455	39.85	
	5200	2" diameter		4.25	63.50	190	570	72	
	5300	3" diameter		4.25	63.50	190	570	72	
	5400	Double, 4" diameter		6.35	108	325	975	115.80	
	5450	Pressure washer 5 GPM, 3000 psi		4.80	51.50	155	465	69.40	

01 54 33 | Equipment Rental

		UNIT	HOURLY OPER. COST	RENT PER DAY	RENT PER WEEK	RENT PER MONTH	EQUIPMENT COST/DAY	
40	5460	7 GPM, 3000 psi	Ea.	6.25	60	180	540	86	40
	5500	Trash pump, self-priming, gas, 2" diameter		4.50	21.50	64	192	48.80	
	5600	Diesel, 4" diameter		9.30	90	270	810	128.40	
	5650	Diesel, 6" diameter		25.55	150	450	1,350	294.40	
	5655	Grout Pump		26.80	260	780	2,350	370.40	
	5700	Salamanders, L.P. gas fired, 100,000 Btu		3.81	13.65	41	123	38.70	
	5705	50,000 Btu		2.39	10.35	31	93	25.30	
	5720	Sandblaster, portable, open top, 3 C.F. capacity		.55	26.50	80	240	20.40	
	5730	6 C.F. capacity		.95	40	120	360	31.60	
	5740	Accessories for above		.13	21.50	64	192	13.85	
	5750	Sander, floor		.71	17	51	153	15.90	
	5760	Edger		.57	18.35	55	165	15.55	
	5800	Saw, chain, gas engine, 18" long		2.25	22	66	198	31.20	
	5900	Hydraulic powered, 36" long		.75	65	195	585	45	
	5950	60" long		.75	66.50	200	600	46	
	6000	Masonry, table mounted, 14" diameter, 5 H.P.		1.31	56.50	170	510	44.50	
	6050	Portable cut-off, 8 H.P.		2.40	33.50	100	300	39.20	
	6100	Circular, hand held, electric, 7-1/4" diameter		.22	4.67	14	42	4.55	
	6200	12" diameter		.22	8	24	72	6.55	
	6250	Wall saw, w/hydraulic power, 10 H.P.		10.45	60	180	540	119.60	
	6275	Shot blaster, walk-behind, 20" wide		4.80	293	880	2,650	214.40	
	6280	Sidewalk broom, walk-behind		2.32	70	210	630	60.55	
	6300	Steam cleaner, 100 gallons per hour		3.70	76.50	230	690	75.60	
	6310	200 gallons per hour		5.25	95	285	855	99	
	6340	Tar Kettle/Pot, 400 gallons		10.50	76.50	230	690	130	
	6350	Torch, cutting, acetylene-oxygen, 150' hose, excludes gases		.30	15	45	135	11.40	
	6360	Hourly operating cost includes tips and gas		19.00				152	
	6410	Toilet, portable chemical		.12	20.50	62	186	13.35	
	6420	Recycle flush type		.15	24.50	74	222	16	
	6430	Toilet, fresh water flush, garden hose,		.18	29.50	89	267	19.25	
	6440	Hoisted, non-flush, for high rise		.15	24.50	73	219	15.80	
	6465	Tractor, farm with attachment		21.90	277	830	2,500	341.20	
	6500	Trailers, platform, flush deck, 2 axle, 25 ton capacity		5.45	117	350	1,050	113.60	
	6600	40 ton capacity		7.00	160	480	1,450	152	
	6700	3 axle, 50 ton capacity		7.55	177	530	1,600	166.40	
	6800	75 ton capacity		9.35	232	695	2,075	213.80	
	6810	Trailer mounted cable reel for high voltage line work		5.35	255	764	2,300	195.60	
	6820	Trailer mounted cable tensioning rig		10.64	505	1,520	4,550	389.10	
	6830	Cable pulling rig		71.33	2,825	8,510	25,500	2,273	
	6900	Water tank trailer, engine driven discharge, 5000 gallons		6.95	143	430	1,300	141.60	
	6925	10,000 gallons		9.50	198	595	1,775	195	
	6950	Water truck, off highway, 6000 gallons		89.80	775	2,330	7,000	1,184	
	7010	Tram car for high voltage line work, powered, 2 conductor		6.27	138	415	1,250	133.15	
	7020	Transit (builder's level) with tripod		.10	16.65	50	150	10.80	
	7030	Trench box, 3000 lb., 6' x 8'		.56	93.50	280	840	60.50	
	7040	7200 lb., 6' x 20'		.78	130	389	1,175	84.05	
	7050	8000 lb., 8' x 16'		1.08	180	540	1,625	116.65	
	7060	9500 lb., 8' x 20'		1.21	201	603	1,800	130.30	
	7065	11,000 lb., 8' x 24'		1.27	211	633	1,900	136.75	
	7070	12,000 lb., 10' x 20'		1.36	227	680	2,050	146.90	
	7100	Truck, pickup, 3/4 ton, 2 wheel drive		15.25	58.50	175	525	157	
	7200	4 wheel drive		15.55	73.50	220	660	168.40	
	7250	Crew carrier, 9 passenger		21.60	86.50	260	780	224.80	
	7290	Flat bed truck, 20,000 lb. GVW		22.30	125	375	1,125	253.40	
	7300	Tractor, 4 x 2, 220 H.P.		31.05	197	590	1,775	366.40	
	7410	330 H.P.		46.05	272	815	2,450	531.40	
	7500	6 x 4, 380 H.P.		52.85	315	950	2,850	612.80	
	7600	450 H.P.		64.05	385	1,150	3,450	742.40	

01 54 33 | Equipment Rental

		UNIT	HOURLY OPER. COST	RENT PER DAY	RENT PER WEEK	RENT PER MONTH	EQUIPMENT COST/DAY	
40	7610	Tractor, with A frame, boom and winch, 225 H.P.	Ea.	33.95	272	815	2,450	434.60	40
	7620	Vacuum truck, hazardous material, 2500 gallons		12.00	292	875	2,625	271	
	7625	5,000 gallons		14.27	410	1,230	3,700	360.15	
	7650	Vacuum, HEPA, 16 gallon, wet/dry		.83	18	54	162	17.45	
	7655	55 gallon, wet/dry		.77	27	81	243	22.35	
	7660	Water tank, portable		.16	26.50	80	240	17.30	
	7690	Sewer/catch basin vacuum, 14 C.Y., 1500 gallons		18.61	615	1,850	5,550	518.90	
	7700	Welder, electric, 200 amp		3.64	16.35	49	147	38.90	
	7800	300 amp		5.41	20	60	180	55.30	
	7900	Gas engine, 200 amp		13.90	23.50	70	210	125.20	
	8000	300 amp		15.90	24.50	74	222	142	
	8100	Wheelbarrow, any size		.08	13	39	117	8.45	
	8200	Wrecking ball, 4000 lb.		2.30	68.50	205	615	59.40	
50	0010	**HIGHWAY EQUIPMENT RENTAL** without operators R015433 -10	Ea.						50
	0050	Asphalt batch plant, portable drum mixer, 100 ton/hr.		79.15	1,450	4,320	13,000	1,497	
	0060	200 ton/hr.		90.32	1,525	4,590	13,800	1,641	
	0070	300 ton/hr.		107.39	1,800	5,390	16,200	1,937	
	0100	Backhoe attachment, long stick, up to 185 H.P., 10.5' long		.35	23	69	207	16.60	
	0140	Up to 250 H.P., 12' long		.37	24.50	74	222	17.75	
	0180	Over 250 H.P., 15' long		.52	34.50	103	310	24.75	
	0200	Special dipper arm, up to 100 H.P., 32' long		1.06	70.50	211	635	50.70	
	0240	Over 100 H.P., 33' long		1.32	87.50	263	790	63.15	
	0280	Catch basin/sewer cleaning truck, 3 ton, 9 C.Y., 1000 gal.		46.25	390	1,170	3,500	604	
	0300	Concrete batch plant, portable, electric, 200 C.Y./hr.		22.08	510	1,525	4,575	481.65	
	0520	Grader/dozer attachment, ripper/scarifier, rear mounted, up to 135 H.P.		3.45	65	195	585	66.60	
	0540	Up to 180 H.P.		4.05	81.50	245	735	81.40	
	0580	Up to 250 H.P.		4.45	93.50	280	840	91.60	
	0700	Pvmt. removal bucket, for hyd. excavator, up to 90 H.P.		1.90	51.50	155	465	46.20	
	0740	Up to 200 H.P.		2.10	73.50	220	660	60.80	
	0780	Over 200 H.P.		2.25	85	255	765	69	
	0900	Aggregate spreader, self-propelled, 187 H.P.		53.75	685	2,055	6,175	841	
	1000	Chemical spreader, 3 C.Y.		3.40	43.50	130	390	53.20	
	1900	Hammermill, traveling, 250 H.P.		77.73	2,050	6,120	18,400	1,846	
	2000	Horizontal borer, 3" diameter, 13 H.P. gas driven		6.25	55	165	495	83	
	2150	Horizontal directional drill, 20,000 lb. thrust, 78 H.P. diesel		31.55	675	2,030	6,100	658.40	
	2160	30,000 lb. thrust, 115 H.P.		39.40	1,025	3,105	9,325	936.20	
	2170	50,000 lb. thrust, 170 H.P.		56.45	1,325	3,970	11,900	1,246	
	2190	Mud trailer for HDD, 1500 gallons, 175 H.P., gas		35.35	153	460	1,375	374.80	
	2200	Hydromulcher, diesel, 3000 gallon, for truck mounting		24.05	247	740	2,225	340.40	
	2300	Gas, 600 gallon		9.20	98.50	295	885	132.60	
	2400	Joint & crack cleaner, walk behind, 25 H.P.		4.35	50	150	450	64.80	
	2500	Filler, trailer mounted, 400 gallons, 20 H.P.		9.60	212	635	1,900	203.80	
	3000	Paint striper, self-propelled, 40 gallon, 22 H.P.		7.60	157	470	1,400	154.80	
	3100	120 gallon, 120 H.P.		24.80	400	1,195	3,575	437.40	
	3200	Post drivers, 6" I-Beam frame, for truck mounting		19.65	380	1,145	3,425	386.20	
	3400	Road sweeper, self-propelled, 8' wide, 90 H.P.		40.30	570	1,715	5,150	665.40	
	3450	Road sweeper, vacuum assisted, 4 C.Y., 220 gallons		77.55	625	1,870	5,600	994.40	
	4000	Road mixer, self-propelled, 130 H.P.		47.80	775	2,320	6,950	846.40	
	4100	310 H.P.		84.40	2,100	6,335	19,000	1,942	
	4220	Cold mix paver, incl. pug mill and bitumen tank, 165 H.P.		98.80	2,325	6,965	20,900	2,183	
	4240	Pavement brush, towed		3.20	93.50	280	840	81.60	
	4250	Paver, asphalt, wheel or crawler, 130 H.P., diesel		98.45	2,300	6,905	20,700	2,169	
	4300	Paver, road widener, gas 1' to 6', 67 H.P.		49.40	905	2,715	8,150	938.20	
	4400	Diesel, 2' to 14', 88 H.P.		63.70	1,050	3,170	9,500	1,144	
	4600	Slipform pavers, curb and gutter, 2 track, 75 H.P.		52.00	805	2,410	7,225	898	
	4700	4 track, 165 H.P.		41.65	730	2,185	6,550	770.20	
	4800	Median barrier, 215 H.P.		54.00	890	2,670	8,000	966	

01 54 33 | Equipment Rental

			UNIT	HOURLY OPER. COST	RENT PER DAY	RENT PER WEEK	RENT PER MONTH	EQUIPMENT COST/DAY	
50	4901	Trailer, low bed, 75 ton capacity	Ea.	10.10	232	695	2,075	219.80	50
	5000	Road planer, walk behind, 10" cutting width, 10 H.P.		4.00	33.50	100	300	52	
	5100	Self-propelled, 12" cutting width, 64 H.P.		11.45	113	340	1,025	159.60	
	5120	Traffic line remover, metal ball blaster, truck mounted, 115 H.P.		48.75	750	2,245	6,725	839	
	5140	Grinder, truck mounted, 115 H.P.		55.45	810	2,435	7,300	930.60	
	5160	Walk-behind, 11 H.P.		4.55	53.50	160	480	68.40	
	5200	Pavement profiler, 4' to 6' wide, 450 H.P.		258.10	3,350	10,045	30,100	4,074	
	5300	8' to 10' wide, 750 H.P.		405.10	4,400	13,180	39,500	5,877	
	5400	Roadway plate, steel, 1" x 8' x 20'		.08	13	39	117	8.45	
	5600	Stabilizer, self-propelled, 150 H.P.		50.10	610	1,835	5,500	767.80	
	5700	310 H.P.		100.25	1,700	5,125	15,400	1,827	
	5800	Striper, truck mounted, 120 gallon paint, 460 H.P.		71.05	485	1,460	4,375	860.40	
	5900	Thermal paint heating kettle, 115 gallons		5.95	25.50	77	231	63	
	6000	Tar kettle, 330 gallon, trailer mounted		8.70	58.50	175	525	104.60	
	7000	Tunnel locomotive, diesel, 8 to 12 ton		32.40	585	1,760	5,275	611.20	
	7005	Electric, 10 ton		26.55	670	2,010	6,025	614.40	
	7010	Muck cars, 1/2 C.Y. capacity		2.05	24.50	74	222	31.20	
	7020	1 C.Y. capacity		2.30	32.50	98	294	38	
	7030	2 C.Y. capacity		2.45	36.50	110	330	41.60	
	7040	Side dump, 2 C.Y. capacity		2.65	45	135	405	48.20	
	7050	3 C.Y. capacity		3.55	51.50	155	465	59.40	
	7060	5 C.Y. capacity		5.10	65	195	585	79.80	
	7100	Ventilating blower for tunnel, 7-1/2 H.P.		2.05	51.50	155	465	47.40	
	7110	10 H.P.		2.22	53.50	160	480	49.75	
	7120	20 H.P.		3.43	69.50	208	625	69.05	
	7140	40 H.P.		5.59	98.50	295	885	103.70	
	7160	60 H.P.		8.54	152	455	1,375	159.30	
	7175	75 H.P.		11.25	207	620	1,850	214	
	7180	200 H.P.		22.56	305	910	2,725	362.50	
	7800	Windrow loader, elevating		55.95	1,325	3,960	11,900	1,240	
60	0010	**LIFTING AND HOISTING EQUIPMENT RENTAL** without operators R015433 -10							60
	0120	Aerial lift truck, 2 person, to 80'	Ea.	26.75	715	2,145	6,425	643	
	0140	Boom work platform, 40' snorkel R015433 -15		16.10	277	830	2,500	294.80	
	0150	Crane, flatbed mounted, 3 ton capacity		16.55	193	580	1,750	248.40	
	0200	Crane, climbing, 106' jib, 6000 lb. capacity, 410 fpm R312316 -45		37.76	1,625	4,860	14,600	1,274	
	0300	101' jib, 10,250 lb. capacity, 270 fpm		44.26	2,050	6,160	18,500	1,586	
	0500	Tower, static, 130' high, 106' jib, 6200 lb. capacity at 400 fpm		41.56	1,875	5,620	16,900	1,456	
	0600	Crawler mounted, lattice boom, 1/2 C.Y., 15 tons at 12' radius		37.52	640	1,920	5,750	684.15	
	0700	3/4 C.Y., 20 tons at 12' radius		50.03	800	2,400	7,200	880.25	
	0800	1 C.Y., 25 tons at 12' radius		66.70	1,075	3,195	9,575	1,173	
	0900	1-1/2 C.Y., 40 tons at 12' radius		66.70	1,075	3,225	9,675	1,179	
	1000	2 C.Y., 50 tons at 12' radius		70.70	1,250	3,765	11,300	1,319	
	1100	3 C.Y., 75 tons at 12' radius		75.65	1,475	4,435	13,300	1,492	
	1200	100 ton capacity, 60' boom		85.65	1,700	5,090	15,300	1,703	
	1300	165 ton capacity, 60' boom		109.15	2,000	5,980	17,900	2,069	
	1400	200 ton capacity, 70' boom		132.10	2,500	7,480	22,400	2,553	
	1500	350 ton capacity, 80' boom		184.60	3,725	11,210	33,600	3,719	
	1600	Truck mounted, lattice boom, 6 x 4, 20 tons at 10' radius		37.31	1,100	3,310	9,925	960.50	
	1700	25 tons at 10' radius		40.35	1,200	3,610	10,800	1,045	
	1800	8 x 4, 30 tons at 10' radius		43.85	1,275	3,840	11,500	1,119	
	1900	40 tons at 12' radius		46.92	1,325	4,010	12,000	1,177	
	2000	60 tons at 15' radius		53.24	1,425	4,240	12,700	1,274	
	2050	82 tons at 15' radius		60.06	1,525	4,540	13,600	1,388	
	2100	90 tons at 15' radius		67.57	1,650	4,940	14,800	1,529	
	2200	115 tons at 15' radius		76.34	1,850	5,520	16,600	1,715	
	2300	150 tons at 18' radius		84.40	1,950	5,815	17,400	1,838	
	2350	165 tons at 18' radius		90.25	2,050	6,160	18,500	1,954	
	2400	Truck mounted, hydraulic, 12 ton capacity		43.00	530	1,595	4,775	663	

01 54 33 | Equipment Rental

			UNIT	HOURLY OPER. COST	RENT PER DAY	RENT PER WEEK	RENT PER MONTH	EQUIPMENT COST/DAY
60	2500	25 ton capacity	Ea.	45.35	640	1,915	5,750	745.80
	2550	33 ton capacity		45.90	655	1,960	5,875	759.20
	2560	40 ton capacity		59.35	765	2,290	6,875	932.80
	2600	55 ton capacity		77.50	860	2,585	7,750	1,137
	2700	80 ton capacity		101.35	1,400	4,205	12,600	1,652
	2720	100 ton capacity		94.95	1,450	4,325	13,000	1,625
	2740	120 ton capacity		109.95	1,550	4,665	14,000	1,813
	2760	150 ton capacity		128.55	2,050	6,135	18,400	2,255
	2800	Self-propelled, 4 x 4, with telescoping boom, 5 ton		17.75	230	690	2,075	280
	2900	12-1/2 ton capacity		32.65	365	1,100	3,300	481.20
	3000	15 ton capacity		33.35	385	1,160	3,475	498.80
	3050	20 ton capacity		36.30	450	1,350	4,050	560.40
	3100	25 ton capacity		37.90	505	1,520	4,550	607.20
	3150	40 ton capacity		46.50	570	1,705	5,125	713
	3200	Derricks, guy, 20 ton capacity, 60' boom, 75' mast		27.18	395	1,190	3,575	455.45
	3300	100' boom, 115' mast		42.63	685	2,050	6,150	751.05
	3400	Stiffleg, 20 ton capacity, 70' boom, 37' mast		29.70	515	1,550	4,650	547.60
	3500	100' boom, 47' mast		45.64	825	2,480	7,450	861.10
	3550	Helicopter, small, lift to 1250 lb. maximum, w/pilot		100.87	3,200	9,610	28,800	2,729
	3600	Hoists, chain type, overhead, manual, 3/4 ton		.10	.33	1	3	1
	3900	10 ton		.75	6	18	54	9.60
	4000	Hoist and tower, 5000 lb. cap., portable electric, 40' high		4.77	228	685	2,050	175.15
	4100	For each added 10' section, add		.11	18	54	162	11.70
	4200	Hoist and single tubular tower, 5000 lb. electric, 100' high		6.47	320	957	2,875	243.15
	4300	For each added 6'-6" section, add		.19	31	93	279	20.10
	4400	Hoist and double tubular tower, 5000 lb., 100' high		6.95	350	1,054	3,150	266.40
	4500	For each added 6'-6" section, add		.21	34.50	103	310	22.30
	4550	Hoist and tower, mast type, 6000 lb., 100' high		7.48	365	1,093	3,275	278.45
	4570	For each added 10' section, add		.13	21.50	64	192	13.85
	4600	Hoist and tower, personnel, electric, 2000 lb., 100' @ 125 fpm		15.90	970	2,910	8,725	709.20
	4700	3000 lb., 100' @ 200 fpm		18.13	1,100	3,290	9,875	803.05
	4800	3000 lb., 150' @ 300 fpm		20.13	1,225	3,690	11,100	899.05
	4900	4000 lb., 100' @ 300 fpm		20.82	1,250	3,760	11,300	918.55
	5000	6000 lb., 100' @ 275 fpm		22.44	1,325	3,950	11,900	969.50
	5100	For added heights up to 500', add	L.F.	.01	1.67	5	15	1.10
	5200	Jacks, hydraulic, 20 ton	Ea.	.05	2	6	18	1.60
	5500	100 ton		.40	11.65	35	105	10.20
	6100	Jacks, hydraulic, climbing w/50' jackrods, control console, 30 ton cap.		1.97	131	394	1,175	94.55
	6150	For each added 10' jackrod section, add		.05	3.33	10	30	2.40
	6300	50 ton capacity		3.17	211	633	1,900	151.95
	6350	For each added 10' jackrod section, add		.06	4	12	36	2.90
	6500	125 ton capacity		8.30	555	1,660	4,975	398.40
	6550	For each added 10' jackrod section, add		.57	37.50	113	340	27.15
	6600	Cable jack, 10 ton capacity with 200' cable		1.65	110	329	985	79
	6650	For each added 50' of cable, add		.20	13	39	117	9.40
70	0010	**WELLPOINT EQUIPMENT RENTAL** without operators						
	0020	Based on 2 months rental						
	0100	Combination jetting & wellpoint pump, 60 H.P. diesel	Ea.	18.14	325	976	2,925	340.30
	0200	High pressure gas jet pump, 200 H.P., 300 psi	"	43.54	278	834	2,500	515.10
	0300	Discharge pipe, 8" diameter	L.F.	.01	.53	1.59	4.77	.40
	0350	12" diameter		.01	.78	2.33	7	.55
	0400	Header pipe, flows up to 150 GPM, 4" diameter		.01	.48	1.44	4.32	.35
	0500	400 GPM, 6" diameter		.01	.56	1.69	5.05	.40
	0600	800 GPM, 8" diameter		.01	.78	2.33	7	.55
	0700	1500 GPM, 10" diameter		.01	.82	2.46	7.40	.55
	0800	2500 GPM, 12" diameter		.02	1.55	4.65	13.95	1.10
	0900	4500 GPM, 16" diameter		.03	1.98	5.95	17.85	1.45

R015433 -10

01 54 33 | Equipment Rental

			UNIT	HOURLY OPER. COST	RENT PER DAY	RENT PER WEEK	RENT PER MONTH	EQUIPMENT COST/DAY	
70	0950	For quick coupling aluminum and plastic pipe, add	L.F.	.03	2.05	6.16	18.50	1.45	70
	1100	Wellpoint, 25' long, with fittings & riser pipe, 1-1/2" or 2" diameter		.06	4.10	12.29	37	2.95	
	1200	Wellpoint pump, diesel powered, 4" suction, 20 H.P.		7.71	188	563	1,700	174.30	
	1300	6" suction, 30 H.P.		10.54	233	698	2,100	223.90	
	1400	8" suction, 40 H.P.		14.24	320	957	2,875	305.30	
	1500	10" suction, 75 H.P.		21.97	375	1,119	3,350	399.55	
	1600	12" suction, 100 H.P.		31.24	590	1,770	5,300	603.90	
	1700	12" suction, 175 H.P.		46.78	655	1,970	5,900	768.25	
80	0010	**MARINE EQUIPMENT RENTAL** without operators R015433 -10							80
	0200	Barge, 400 Ton, 30' wide x 90' long	Ea.	17.50	1,075	3,205	9,625	781	
	0240	800 Ton, 45' wide x 90' long		21.20	1,300	3,900	11,700	949.60	
	2000	Tugboat, diesel, 100 H.P.		38.95	217	650	1,950	441.60	
	2040	250 H.P.		81.55	395	1,185	3,550	889.40	
	2080	380 H.P.		160.00	1,175	3,545	10,600	1,989	
	3000	Small work boat, gas, 16-foot, 50 H.P.		19.15	60	180	540	189.20	
	4000	Large, diesel, 48-foot, 200 H.P.		90.80	1,250	3,745	11,200	1,475	

Crew No.	Bare Costs		Incl. Subs O&P		Cost Per Labor-Hour	
Crew A-1	Hr.	Daily	Hr.	Daily	Bare Costs	Incl. O&P
1 Building Laborer	$36.65	$293.20	$56.55	$452.40	$36.65	$56.55
1 Concrete Saw, Gas Manual		84.00		92.40	10.50	11.55
8 L.H., Daily Totals		$377.20		$544.80	$47.15	$68.10
Crew A-1A	Hr.	Daily	Hr.	Daily	Bare Costs	Incl. O&P
1 Skilled Worker	$47.30	$378.40	$73.25	$586.00	$47.30	$73.25
1 Shot Blaster, 20"		214.40		235.84	26.80	29.48
8 L.H., Daily Totals		$592.80		$821.84	$74.10	$102.73
Crew A-1B	Hr.	Daily	Hr.	Daily	Bare Costs	Incl. O&P
1 Building Laborer	$36.65	$293.20	$56.55	$452.40	$36.65	$56.55
1 Concrete Saw		175.20		192.72	21.90	24.09
8 L.H., Daily Totals		$468.40		$645.12	$58.55	$80.64
Crew A-1C	Hr.	Daily	Hr.	Daily	Bare Costs	Incl. O&P
1 Building Laborer	$36.65	$293.20	$56.55	$452.40	$36.65	$56.55
1 Chain Saw, Gas, 18"		31.20		34.32	3.90	4.29
8 L.H., Daily Totals		$324.40		$486.72	$40.55	$60.84
Crew A-1D	Hr.	Daily	Hr.	Daily	Bare Costs	Incl. O&P
1 Building Laborer	$36.65	$293.20	$56.55	$452.40	$36.65	$56.55
1 Vibrating Plate, Gas, 18"		35.80		39.38	4.47	4.92
8 L.H., Daily Totals		$329.00		$491.78	$41.13	$61.47
Crew A-1E	Hr.	Daily	Hr.	Daily	Bare Costs	Incl. O&P
1 Building Laborer	$36.65	$293.20	$56.55	$452.40	$36.65	$56.55
1 Vibrating Plate, Gas, 21"		46.00		50.60	5.75	6.33
8 L.H., Daily Totals		$339.20		$503.00	$42.40	$62.88
Crew A-1F	Hr.	Daily	Hr.	Daily	Bare Costs	Incl. O&P
1 Building Laborer	$36.65	$293.20	$56.55	$452.40	$36.65	$56.55
1 Rammer/Tamper, Gas, 8"		49.60		54.56	6.20	6.82
8 L.H., Daily Totals		$342.80		$506.96	$42.85	$63.37
Crew A-1G	Hr.	Daily	Hr.	Daily	Bare Costs	Incl. O&P
1 Building Laborer	$36.65	$293.20	$56.55	$452.40	$36.65	$56.55
1 Rammer/Tamper, Gas, 15"		56.00		61.60	7.00	7.70
8 L.H., Daily Totals		$349.20		$514.00	$43.65	$64.25
Crew A-1H	Hr.	Daily	Hr.	Daily	Bare Costs	Incl. O&P
1 Building Laborer	$36.65	$293.20	$56.55	$452.40	$36.65	$56.55
1 Exterior Steam Cleaner		75.60		83.16	9.45	10.40
8 L.H., Daily Totals		$368.80		$535.56	$46.10	$66.94
Crew A-1J	Hr.	Daily	Hr.	Daily	Bare Costs	Incl. O&P
1 Building Laborer	$36.65	$293.20	$56.55	$452.40	$36.65	$56.55
1 Cultivator, Walk-Behind, 5 H.P.		57.45		63.20	7.18	7.90
8 L.H., Daily Totals		$350.65		$515.60	$43.83	$64.45
Crew A-1K	Hr.	Daily	Hr.	Daily	Bare Costs	Incl. O&P
1 Building Laborer	$36.65	$293.20	$56.55	$452.40	$36.65	$56.55
1 Cultivator, Walk-Behind, 8 H.P.		81.40		89.54	10.18	11.19
8 L.H., Daily Totals		$374.60		$541.94	$46.83	$67.74
Crew A-1M	Hr.	Daily	Hr.	Daily	Bare Costs	Incl. O&P
1 Building Laborer	$36.65	$293.20	$56.55	$452.40	$36.65	$56.55
1 Snow Blower, Walk-Behind		60.55		66.61	7.57	8.33
8 L.H., Daily Totals		$353.75		$519.01	$44.22	$64.88

Crew No.	Bare Costs		Incl. Subs O&P		Cost Per Labor-Hour	
Crew A-2	Hr.	Daily	Hr.	Daily	Bare Costs	Incl. O&P
2 Laborers	$36.65	$586.40	$56.55	$904.80	$36.60	$56.27
1 Truck Driver (light)	36.50	292.00	55.70	445.60		
1 Flatbed Truck, Gas, 1.5 Ton		270.20		297.22	11.26	12.38
24 L.H., Daily Totals		$1148.60		$1647.62	$47.86	$68.65
Crew A-2A	Hr.	Daily	Hr.	Daily	Bare Costs	Incl. O&P
2 Laborers	$36.65	$586.40	$56.55	$904.80	$36.60	$56.27
1 Truck Driver (light)	36.50	292.00	55.70	445.60		
1 Flatbed Truck, Gas, 1.5 Ton		270.20		297.22		
1 Concrete Saw		175.20		192.72	18.56	20.41
24 L.H., Daily Totals		$1323.80		$1840.34	$55.16	$76.68
Crew A-2B	Hr.	Daily	Hr.	Daily	Bare Costs	Incl. O&P
1 Truck Driver (light)	$36.50	$292.00	$55.70	$445.60	$36.50	$55.70
1 Flatbed Truck, Gas, 1.5 Ton		270.20		297.22	33.77	37.15
8 L.H., Daily Totals		$562.20		$742.82	$70.28	$92.85
Crew A-3A	Hr.	Daily	Hr.	Daily	Bare Costs	Incl. O&P
1 Truck Driver (light)	$36.50	$292.00	$55.70	$445.60	$36.50	$55.70
1 Pickup Truck, 4 x 4, 3/4 Ton		168.40		185.24	21.05	23.16
8 L.H., Daily Totals		$460.40		$630.84	$57.55	$78.86
Crew A-3B	Hr.	Daily	Hr.	Daily	Bare Costs	Incl. O&P
1 Equip. Oper. (medium)	$48.90	$391.20	$74.15	$593.20	$43.23	$65.72
1 Truck Driver (heavy)	37.55	300.40	57.30	458.40		
1 Dump Truck, 12 C.Y., 400 H.P.	.	692.20		761.42		
1 F.E. Loader, W.M.,2.5 C.Y.		522.80		575.08	75.94	83.53
16 L.H., Daily Totals		$1906.60		$2388.10	$119.16	$149.26
Crew A-3C	Hr.	Daily	Hr.	Daily	Bare Costs	Incl. O&P
1 Equip. Oper. (light)	$47.05	$376.40	$71.35	$570.80	$47.05	$71.35
1 Loader, Skid Steer, 78 H.P.		308.80		339.68	38.60	42.46
8 L.H., Daily Totals		$685.20		$910.48	$85.65	$113.81
Crew A-3D	Hr.	Daily	Hr.	Daily	Bare Costs	Incl. O&P
1 Truck Driver (light)	$36.50	$292.00	$55.70	$445.60	$36.50	$55.70
1 Pickup Truck, 4 x 4, 3/4 Ton		168.40		185.24		
1 Flatbed Trailer, 25 Ton		113.60		124.96	35.25	38.77
8 L.H., Daily Totals		$574.00		$755.80	$71.75	$94.47
Crew A-3E	Hr.	Daily	Hr.	Daily	Bare Costs	Incl. O&P
1 Equip. Oper. (crane)	$50.25	$402.00	$76.20	$609.60	$43.90	$66.75
1 Truck Driver (heavy)	37.55	300.40	57.30	458.40		
1 Pickup Truck, 4 x 4, 3/4 Ton		168.40		185.24	10.53	11.58
16 L.H., Daily Totals		$870.80		$1253.24	$54.42	$78.33
Crew A-3F	Hr.	Daily	Hr.	Daily	Bare Costs	Incl. O&P
1 Equip. Oper. (crane)	$50.25	$402.00	$76.20	$609.60	$43.90	$66.75
1 Truck Driver (heavy)	37.55	300.40	57.30	458.40		
1 Pickup Truck, 4 x 4, 3/4 Ton		168.40		185.24		
1 Truck Tractor, 6x4, 380 H.P.		612.80		674.08		
1 Lowbed Trailer, 75 Ton		219.80		241.78	62.56	68.82
16 L.H., Daily Totals		$1703.40		$2169.10	$106.46	$135.57

Crew A-3G

Crew No.	Bare Costs Hr.	Daily	Incl. Subs O&P Hr.	Daily	Cost Per Labor-Hour Bare Costs	Incl. O&P
1 Equip. Oper. (crane)	$50.25	$402.00	$76.20	$609.60	$43.90	$66.75
1 Truck Driver (heavy)	37.55	300.40	57.30	458.40		
1 Pickup Truck, 4 x 4, 3/4 Ton		168.40		185.24		
1 Truck Tractor, 6x4, 450 H.P.		742.40		816.64		
1 Lowbed Trailer, 75 Ton		219.80		241.78	70.66	77.73
16 L.H., Daily Totals		$1833.00		$2311.66	$114.56	$144.48

Crew A-3H

Crew No.	Hr.	Daily	Hr.	Daily	Bare Costs	Incl. O&P
1 Equip. Oper. (crane)	$50.25	$402.00	$76.20	$609.60	$50.25	$76.20
1 Hyd. Crane, 12 Ton (Daily)		874.00		961.40	109.25	120.18
8 L.H., Daily Totals		$1276.00		$1571.00	$159.50	$196.38

Crew A-3I

Crew No.	Hr.	Daily	Hr.	Daily	Bare Costs	Incl. O&P
1 Equip. Oper. (crane)	$50.25	$402.00	$76.20	$609.60	$50.25	$76.20
1 Hyd. Crane, 25 Ton (Daily)		1003.00		1103.30	125.38	137.91
8 L.H., Daily Totals		$1405.00		$1712.90	$175.63	$214.11

Crew A-3J

Crew No.	Hr.	Daily	Hr.	Daily	Bare Costs	Incl. O&P
1 Equip. Oper. (crane)	$50.25	$402.00	$76.20	$609.60	$50.25	$76.20
1 Hyd. Crane, 40 Ton (Daily)		1240.00		1364.00	155.00	170.50
8 L.H., Daily Totals		$1642.00		$1973.60	$205.25	$246.70

Crew A-3K

Crew No.	Hr.	Daily	Hr.	Daily	Bare Costs	Incl. O&P
1 Equip. Oper. (crane)	$50.25	$402.00	$76.20	$609.60	$46.90	$71.10
1 Equip. Oper. (oiler)	43.55	348.40	66.00	528.00		
1 Hyd. Crane, 55 Ton (Daily)		1480.00		1628.00		
1 P/U Truck, 3/4 Ton (Daily)		182.00		200.20	103.88	114.26
16 L.H., Daily Totals		$2412.40		$2965.80	$150.78	$185.36

Crew A-3L

Crew No.	Hr.	Daily	Hr.	Daily	Bare Costs	Incl. O&P
1 Equip. Oper. (crane)	$50.25	$402.00	$76.20	$609.60	$46.90	$71.10
1 Equip. Oper. (oiler)	43.55	348.40	66.00	528.00		
1 Hyd. Crane, 80 Ton (Daily)		2211.00		2432.10		
1 P/U Truck, 3/4 Ton (Daily)		182.00		200.20	149.56	164.52
16 L.H., Daily Totals		$3143.40		$3769.90	$196.46	$235.62

Crew A-3M

Crew No.	Hr.	Daily	Hr.	Daily	Bare Costs	Incl. O&P
1 Equip. Oper. (crane)	$50.25	$402.00	$76.20	$609.60	$46.90	$71.10
1 Equip. Oper. (oiler)	43.55	348.40	66.00	528.00		
1 Hyd. Crane, 100 Ton (Daily)		2200.00		2420.00		
1 P/U Truck, 3/4 Ton (Daily)		182.00		200.20	148.88	163.76
16 L.H., Daily Totals		$3132.40		$3757.80	$195.78	$234.86

Crew A-3N

Crew No.	Hr.	Daily	Hr.	Daily	Bare Costs	Incl. O&P
1 Equip. Oper. (crane)	$50.25	$402.00	$76.20	$609.60	$50.25	$76.20
1 Tower Cane (monthly)		1100.00		1210.00	137.50	151.25
8 L.H., Daily Totals		$1502.00		$1819.60	$187.75	$227.45

Crew A-3P

Crew No.	Hr.	Daily	Hr.	Daily	Bare Costs	Incl. O&P
1 Equip. Oper. (light)	$47.05	$376.40	$71.35	$570.80	$47.05	$71.35
1 A.T. Forklift, 42' lift		531.60		584.76	66.45	73.09
8 L.H., Daily Totals		$908.00		$1155.56	$113.50	$144.44

Crew A-4

Crew No.	Hr.	Daily	Hr.	Daily	Bare Costs	Incl. O&P
2 Carpenters	$45.85	$733.60	$70.75	$1132.00	$43.75	$67.00
1 Painter, Ordinary	39.55	316.40	59.50	476.00		
24 L.H., Daily Totals		$1050.00		$1608.00	$43.75	$67.00

Crew A-5

Crew No.	Hr.	Daily	Hr.	Daily	Bare Costs	Incl. O&P
2 Laborers	$36.65	$586.40	$56.55	$904.80	$36.63	$56.46
.25 Truck Driver (light)	36.50	73.00	55.70	111.40		
.25 Flatbed Truck, Gas, 1.5 Ton		67.55		74.31	3.75	4.13
18 L.H., Daily Totals		$726.95		$1090.51	$40.39	$60.58

Crew A-6

Crew No.	Hr.	Daily	Hr.	Daily	Bare Costs	Incl. O&P
1 Instrument Man	$47.30	$378.40	$73.25	$586.00	$45.67	$70.13
1 Rodman/Chainman	44.05	352.40	67.00	536.00		
1 Level, Electronic		71.30		78.43	4.46	4.90
16 L.H., Daily Totals		$802.10		$1200.43	$50.13	$75.03

Crew A-7

Crew No.	Hr.	Daily	Hr.	Daily	Bare Costs	Incl. O&P
1 Chief of Party	$59.90	$479.20	$92.15	$737.20	$50.42	$77.47
1 Instrument Man	47.30	378.40	73.25	586.00		
1 Rodman/Chainman	44.05	352.40	67.00	536.00		
1 Level, Electronic		71.30		78.43	2.97	3.27
24 L.H., Daily Totals		$1281.30		$1937.63	$53.39	$80.73

Crew A-8

Crew No.	Hr.	Daily	Hr.	Daily	Bare Costs	Incl. O&P
1 Chief of Party	$59.90	$479.20	$92.15	$737.20	$48.83	$74.85
1 Instrument Man	47.30	378.40	73.25	586.00		
2 Rodmen/Chainmen	44.05	704.80	67.00	1072.00		
1 Level, Electronic		71.30		78.43	2.23	2.45
32 L.H., Daily Totals		$1633.70		$2473.63	$51.05	$77.30

Crew A-9

Crew No.	Hr.	Daily	Hr.	Daily	Bare Costs	Incl. O&P
1 Asbestos Foreman	$51.65	$413.20	$80.35	$642.80	$51.21	$79.69
7 Asbestos Workers	51.15	2864.40	79.60	4457.60		
64 L.H., Daily Totals		$3277.60		$5100.40	$51.21	$79.69

Crew A-10A

Crew No.	Hr.	Daily	Hr.	Daily	Bare Costs	Incl. O&P
1 Asbestos Foreman	$51.65	$413.20	$80.35	$642.80	$51.32	$79.85
2 Asbestos Workers	51.15	818.40	79.60	1273.60		
24 L.H., Daily Totals		$1231.60		$1916.40	$51.32	$79.85

Crew A-10B

Crew No.	Hr.	Daily	Hr.	Daily	Bare Costs	Incl. O&P
1 Asbestos Foreman	$51.65	$413.20	$80.35	$642.80	$51.27	$79.79
3 Asbestos Workers	51.15	1227.60	79.60	1910.40		
32 L.H., Daily Totals		$1640.80		$2553.20	$51.27	$79.79

Crew A-10C

Crew No.	Hr.	Daily	Hr.	Daily	Bare Costs	Incl. O&P
3 Asbestos Workers	$51.15	$1227.60	$79.60	$1910.40	$51.15	$79.60
1 Flatbed Truck, Gas, 1.5 Ton		270.20		297.22	11.26	12.38
24 L.H., Daily Totals		$1497.80		$2207.62	$62.41	$91.98

Crew A-10D

Crew No.	Hr.	Daily	Hr.	Daily	Bare Costs	Incl. O&P
2 Asbestos Workers	$51.15	$818.40	$79.60	$1273.60	$49.02	$75.35
1 Equip. Oper. (crane)	50.25	402.00	76.20	609.60		
1 Equip. Oper. (oiler)	43.55	348.40	66.00	528.00		
1 Hydraulic Crane, 33 Ton		759.20		835.12	23.73	26.10
32 L.H., Daily Totals		$2328.00		$3246.32	$72.75	$101.45

Crew A-11

Crew No.	Hr.	Daily	Hr.	Daily	Bare Costs	Incl. O&P
1 Asbestos Foreman	$51.65	$413.20	$80.35	$642.80	$51.21	$79.69
7 Asbestos Workers	51.15	2864.40	79.60	4457.60		
2 Chip. Hammers, 12 Lb., Elec.		40.00		44.00	.63	.69
64 L.H., Daily Totals		$3317.60		$5144.40	$51.84	$80.38

Crew No.	Bare Costs		Incl. Subs O&P		Cost Per Labor-Hour	

Crew A-12

Crew A-12	Hr.	Daily	Hr.	Daily	Bare Costs	Incl. O&P
1 Asbestos Foreman	$51.65	$413.20	$80.35	$642.80	$51.21	$79.69
7 Asbestos Workers	51.15	2864.40	79.60	4457.60		
1 Trk-Mtd Vac, 14 CY, 1500 Gal.		518.90		570.79		
1 Flatbed Truck, 20,000 GVW		253.40		278.74	12.07	13.27
64 L.H., Daily Totals		$4049.90		$5949.93	$63.28	$92.97

Crew A-13	Hr.	Daily	Hr.	Daily	Bare Costs	Incl. O&P
1 Equip. Oper. (light)	$47.05	$376.40	$71.35	$570.80	$47.05	$71.35
1 Trk-Mtd Vac, 14 CY, 1500 Gal.		518.90		570.79		
1 Flatbed Truck, 20,000 GVW		253.40		278.74	96.54	106.19
8 L.H., Daily Totals		$1148.70		$1420.33	$143.59	$177.54

Crew B-1	Hr.	Daily	Hr.	Daily	Bare Costs	Incl. O&P
1 Labor Foreman (outside)	$38.65	$309.20	$59.65	$477.20	$37.32	$57.58
2 Laborers	36.65	586.40	56.55	904.80		
24 L.H., Daily Totals		$895.60		$1382.00	$37.32	$57.58

Crew B-1A	Hr.	Daily	Hr.	Daily	Bare Costs	Incl. O&P
1 Labor Foreman (outside)	$38.65	$309.20	$59.65	$477.20	$37.32	$57.58
2 Laborers	36.65	586.40	56.55	904.80		
2 Cutting Torches		22.80		25.08		
2 Sets of Gases		304.00		334.40	13.62	14.98
24 L.H., Daily Totals		$1222.40		$1741.48	$50.93	$72.56

Crew B-1B	Hr.	Daily	Hr.	Daily	Bare Costs	Incl. O&P
1 Labor Foreman (outside)	$38.65	$309.20	$59.65	$477.20	$40.55	$62.24
2 Laborers	36.65	586.40	56.55	904.80		
1 Equip. Oper. (crane)	50.25	402.00	76.20	609.60		
2 Cutting Torches		22.80		25.08		
2 Sets of Gases		304.00		334.40		
1 Hyd. Crane, 12 Ton		663.00		729.30	30.93	34.02
32 L.H., Daily Totals		$2287.40		$3080.38	$71.48	$96.26

Crew B-1C	Hr.	Daily	Hr.	Daily	Bare Costs	Incl. O&P
1 Labor Foreman (outside)	$38.65	$309.20	$59.65	$477.20	$37.32	$57.58
2 Laborers	36.65	586.40	56.55	904.80		
1 Aerial Lift Truck, 60' Boom		456.00		501.60	19.00	20.90
24 L.H., Daily Totals		$1351.60		$1883.60	$56.32	$78.48

Crew B-1D	Hr.	Daily	Hr.	Daily	Bare Costs	Incl. O&P
2 Laborers	$36.65	$586.40	$56.55	$904.80	$36.65	$56.55
1 Small Work Boat, Gas, 50 H.P.		189.20		208.12		
1 Pressure Washer, 7 GPM		86.00		94.60	17.20	18.92
16 L.H., Daily Totals		$861.60		$1207.52	$53.85	$75.47

Crew B-1E	Hr.	Daily	Hr.	Daily	Bare Costs	Incl. O&P
1 Labor Foreman (outside)	$38.65	$309.20	$59.65	$477.20	$37.15	$57.33
3 Laborers	36.65	879.60	56.55	1357.20		
1 Work Boat, Diesel, 200 H.P.		1475.00		1622.50		
2 Pressure Washer, 7 GPM		172.00		189.20	51.47	56.62
32 L.H., Daily Totals		$2835.80		$3646.10	$88.62	$113.94

Crew B-1F	Hr.	Daily	Hr.	Daily	Bare Costs	Incl. O&P
2 Skilled Workers	$47.30	$756.80	$73.25	$1172.00	$43.75	$67.68
1 Laborer	36.65	293.20	56.55	452.40		
1 Small Work Boat, Gas, 50 H.P.		189.20		208.12		
1 Pressure Washer, 7 GPM		86.00		94.60	11.47	12.61
24 L.H., Daily Totals		$1325.20		$1927.12	$55.22	$80.30

Crew B-1G	Hr.	Daily	Hr.	Daily	Bare Costs	Incl. O&P
2 Laborers	$36.65	$586.40	$56.55	$904.80	$36.65	$56.55
1 Small Work Boat, Gas, 50 H.P.		189.20		208.12	11.82	13.01
16 L.H., Daily Totals		$775.60		$1112.92	$48.48	$69.56

Crew B-1H	Hr.	Daily	Hr.	Daily	Bare Costs	Incl. O&P
2 Skilled Workers	$47.30	$756.80	$73.25	$1172.00	$43.75	$67.68
1 Laborer	36.65	293.20	56.55	452.40		
1 Small Work Boat, Gas, 50 H.P.		189.20		208.12	7.88	8.67
24 L.H., Daily Totals		$1239.20		$1832.52	$51.63	$76.36

Crew B-1J	Hr.	Daily	Hr.	Daily	Bare Costs	Incl. O&P
1 Labor Foreman (inside)	$37.15	$297.20	$57.30	$458.40	$36.90	$56.92
1 Laborer	36.65	293.20	56.55	452.40		
16 L.H., Daily Totals		$590.40		$910.80	$36.90	$56.92

Crew B-1K	Hr.	Daily	Hr.	Daily	Bare Costs	Incl. O&P
1 Carpenter Foreman (inside)	$46.35	$370.80	$71.50	$572.00	$46.10	$71.13
1 Carpenter	45.85	366.80	70.75	566.00		
16 L.H., Daily Totals		$737.60		$1138.00	$46.10	$71.13

Crew B-2	Hr.	Daily	Hr.	Daily	Bare Costs	Incl. O&P
1 Labor Foreman (outside)	$38.65	$309.20	$59.65	$477.20	$37.05	$57.17
4 Laborers	36.65	1172.80	56.55	1809.60		
40 L.H., Daily Totals		$1482.00		$2286.80	$37.05	$57.17

Crew B-2A	Hr.	Daily	Hr.	Daily	Bare Costs	Incl. O&P
1 Labor Foreman (outside)	$38.65	$309.20	$59.65	$477.20	$37.32	$57.58
2 Laborers	36.65	586.40	56.55	904.80		
1 Aerial Lift Truck, 60' Boom		456.00		501.60	19.00	20.90
24 L.H., Daily Totals		$1351.60		$1883.60	$56.32	$78.48

Crew B-3	Hr.	Daily	Hr.	Daily	Bare Costs	Incl. O&P
1 Labor Foreman (outside)	$38.65	$309.20	$59.65	$477.20	$39.33	$60.25
2 Laborers	36.65	586.40	56.55	904.80		
1 Equip. Oper. (medium)	48.90	391.20	74.15	593.20		
2 Truck Drivers (heavy)	37.55	600.80	57.30	916.80		
1 Crawler Loader, 3 C.Y.		1191.00		1310.10		
2 Dump Trucks, 12 C.Y., 400 H.P.		1384.40		1522.84	53.65	59.02
48 L.H., Daily Totals		$4463.00		$5724.94	$92.98	$119.27

Crew B-3A	Hr.	Daily	Hr.	Daily	Bare Costs	Incl. O&P
4 Laborers	$36.65	$1172.80	$56.55	$1809.60	$39.10	$60.07
1 Equip. Oper. (medium)	48.90	391.20	74.15	593.20		
1 Hyd. Excavator, 1.5 C.Y.		1031.00		1134.10	25.77	28.35
40 L.H., Daily Totals		$2595.00		$3536.90	$64.88	$88.42

Crew B-3B	Hr.	Daily	Hr.	Daily	Bare Costs	Incl. O&P
2 Laborers	$36.65	$586.40	$56.55	$904.80	$39.94	$61.14
1 Equip. Oper. (medium)	48.90	391.20	74.15	593.20		
1 Truck Driver (heavy)	37.55	300.40	57.30	458.40		
1 Backhoe Loader, 80 H.P.		397.60		437.36		
1 Dump Truck, 12 C.Y., 400 H.P.		692.20		761.42	34.06	37.46
32 L.H., Daily Totals		$2367.80		$3155.18	$73.99	$98.60

Crew B-3C	Hr.	Daily	Hr.	Daily	Bare Costs	Incl. O&P
3 Laborers	$36.65	$879.60	$56.55	$1357.20	$39.71	$60.95
1 Equip. Oper. (medium)	48.90	391.20	74.15	593.20		
1 Crawler Loader, 4 C.Y.		1532.00		1685.20	47.88	52.66
32 L.H., Daily Totals		$2802.80		$3635.60	$87.59	$113.61

Crew No.	Bare Costs		Incl. Subs O&P		Cost Per Labor-Hour	
Crew B-4	Hr.	Daily	Hr.	Daily	Bare Costs	Incl. O&P
1 Labor Foreman (outside)	$38.65	$309.20	$59.65	$477.20	$37.13	$57.19
4 Laborers	36.65	1172.80	56.55	1809.60		
1 Truck Driver (heavy)	37.55	300.40	57.30	458.40		
1 Truck Tractor, 220 H.P.		366.40		403.04		
1 Flatbed Trailer, 40 Ton		152.00		167.20	10.80	11.88
48 L.H., Daily Totals		$2300.80		$3315.44	$47.93	$69.07

Crew No.	Bare Costs		Incl. Subs O&P		Cost Per Labor-Hour	
Crew B-5	Hr.	Daily	Hr.	Daily	Bare Costs	Incl. O&P
1 Labor Foreman (outside)	$38.65	$309.20	$59.65	$477.20	$40.44	$62.02
4 Laborers	36.65	1172.80	56.55	1809.60		
2 Equip. Oper. (medium)	48.90	782.40	74.15	1186.40		
1 Air Compressor, 250 cfm		202.40		222.64		
2 Breakers, Pavement, 60 lb.		19.60		21.56		
2 -50' Air Hoses, 1.5"		11.60		12.76		
1 Crawler Loader, 3 C.Y.		1191.00		1310.10	25.44	27.98
56 L.H., Daily Totals		$3689.00		$5040.26	$65.88	$90.00

Crew No.	Bare Costs		Incl. Subs O&P		Cost Per Labor-Hour	
Crew B-5A	Hr.	Daily	Hr.	Daily	Bare Costs	Incl. O&P
1 Labor Foreman (outside)	$38.65	$309.20	$59.65	$477.20	$39.88	$61.10
6 Laborers	36.65	1759.20	56.55	2714.40		
2 Equip. Oper. (medium)	48.90	782.40	74.15	1186.40		
1 Equip. Oper. (light)	47.05	376.40	71.35	570.80		
2 Truck Drivers (heavy)	37.55	600.80	57.30	916.80		
1 Air Compressor, 365 cfm		264.80		291.28		
2 Breakers, Pavement, 60 lb.		19.60		21.56		
8 -50' Air Hoses, 1"		32.80		36.08		
2 Dump Trucks, 8 C.Y., 220 H.P.		834.40		917.84	12.00	13.20
96 L.H., Daily Totals		$4979.60		$7132.36	$51.87	$74.30

Crew No.	Bare Costs		Incl. Subs O&P		Cost Per Labor-Hour	
Crew B-5B	Hr.	Daily	Hr.	Daily	Bare Costs	Incl. O&P
1 Powderman	$47.30	$378.40	$73.25	$586.00	$42.96	$65.58
2 Equip. Oper. (medium)	48.90	782.40	74.15	1186.40		
3 Truck Drivers (heavy)	37.55	901.20	57.30	1375.20		
1 F.E. Loader, W.M.,2.5 C.Y.		522.80		575.08		
3 Dump Trucks, 12 C.Y., 400 H.P.		2076.60		2284.26		
1 Air Compressor, 365 CFM		264.80		291.28	59.67	65.64
48 L.H., Daily Totals		$4926.20		$6298.22	$102.63	$131.21

Crew No.	Bare Costs		Incl. Subs O&P		Cost Per Labor-Hour	
Crew B-5C	Hr.	Daily	Hr.	Daily	Bare Costs	Incl. O&P
3 Laborers	$36.65	$879.60	$56.55	$1357.20	$40.97	$62.58
1 Equip. Oper. (medium)	48.90	391.20	74.15	593.20		
2 Truck Drivers (heavy)	37.55	600.80	57.30	916.80		
1 Equip. Oper. (crane)	50.25	402.00	76.20	609.60		
1 Equip. Oper. (oiler)	43.55	348.40	66.00	528.00		
2 Dump Trucks, 12 C.Y., 400 H.P.		1384.40		1522.84		
1 Crawler Loader, 4 C.Y.		1532.00		1685.20		
1 S.P. Crane, 4x4, 25 Ton		607.20		667.92	55.06	60.56
64 L.H., Daily Totals		$6145.60		$7880.76	$96.03	$123.14

Crew No.	Bare Costs		Incl. Subs O&P		Cost Per Labor-Hour	
Crew B-5D	Hr.	Daily	Hr.	Daily	Bare Costs	Incl. O&P
1 Labor Foreman (outside)	$38.65	$309.20	$59.65	$477.20	$40.08	$61.43
4 Laborers	36.65	1172.80	56.55	1809.60		
2 Equip. Oper. (medium)	48.90	782.40	74.15	1186.40		
1 Truck Driver (heavy)	37.55	300.40	57.30	458.40		
1 Air Compressor, 250 cfm		202.40		222.64		
2 Breakers, Pavement, 60 lb.		19.60		21.56		
2 -50' Air Hoses, 1.5"		11.60		12.76		
1 Crawler Loader, 3 C.Y.		1191.00		1310.10		
1 Dump Truck, 12 C.Y., 400 H.P.		692.20		761.42	33.08	36.38
64 L.H., Daily Totals		$4681.60		$6260.08	$73.15	$97.81

Crew No.	Bare Costs		Incl. Subs O&P		Cost Per Labor-Hour	
Crew B-6	Hr.	Daily	Hr.	Daily	Bare Costs	Incl. O&P
2 Laborers	$36.65	$586.40	$56.55	$904.80	$40.12	$61.48
1 Equip. Oper. (light)	47.05	376.40	71.35	570.80		
1 Backhoe Loader, 48 H.P.		365.20		401.72	15.22	16.74
24 L.H., Daily Totals		$1328.00		$1877.32	$55.33	$78.22

Crew No.	Bare Costs		Incl. Subs O&P		Cost Per Labor-Hour	
Crew B-6A	Hr.	Daily	Hr.	Daily	Bare Costs	Incl. O&P
.5 Labor Foreman (outside)	$38.65	$154.60	$59.65	$238.60	$41.95	$64.21
1 Laborer	36.65	293.20	56.55	452.40		
1 Equip. Oper. (medium)	48.90	391.20	74.15	593.20		
1 Vacuum Truck, 5000 Gal.		360.15		396.17	18.01	19.81
20 L.H., Daily Totals		$1199.15		$1680.37	$59.96	$84.02

Crew No.	Bare Costs		Incl. Subs O&P		Cost Per Labor-Hour	
Crew B-6B	Hr.	Daily	Hr.	Daily	Bare Costs	Incl. O&P
2 Labor Foremen (outside)	$38.65	$618.40	$59.65	$954.40	$37.32	$57.58
4 Laborers	36.65	1172.80	56.55	1809.60		
1 S.P. Crane, 4x4, 5 Ton		280.00		308.00		
1 Flatbed Truck, Gas, 1.5 Ton		270.20		297.22		
1 Butt Fusion Mach., 4"-12" diam.		381.80		419.98	19.42	21.36
48 L.H., Daily Totals		$2723.20		$3789.20	$56.73	$78.94

Crew No.	Bare Costs		Incl. Subs O&P		Cost Per Labor-Hour	
Crew B-6C	Hr.	Daily	Hr.	Daily	Bare Costs	Incl. O&P
2 Labor Foremen (outside)	$38.65	$618.40	$59.65	$954.40	$37.32	$57.58
4 Laborers	36.65	1172.80	56.55	1809.60		
1 S.P. Crane, 4x4, 12 Ton		481.20		529.32		
1 Flatbed Truck, Gas, 3 Ton		333.00		366.30		
1 Butt Fusion Mach., 8"-24" diam.		828.40		911.24	34.22	37.64
48 L.H., Daily Totals		$3433.80		$4570.86	$71.54	$95.23

Crew No.	Bare Costs		Incl. Subs O&P		Cost Per Labor-Hour	
Crew B-7	Hr.	Daily	Hr.	Daily	Bare Costs	Incl. O&P
1 Labor Foreman (outside)	$38.65	$309.20	$59.65	$477.20	$39.02	$60.00
4 Laborers	36.65	1172.80	56.55	1809.60		
1 Equip. Oper. (medium)	48.90	391.20	74.15	593.20		
1 Brush Chipper, 12", 130 H.P.		396.60		436.26		
1 Crawler Loader, 3 C.Y.		1191.00		1310.10		
2 Chain Saws, Gas, 36" Long		90.00		99.00	34.95	38.45
48 L.H., Daily Totals		$3550.80		$4725.36	$73.97	$98.44

Crew No.	Bare Costs		Incl. Subs O&P		Cost Per Labor-Hour	
Crew B-7A	Hr.	Daily	Hr.	Daily	Bare Costs	Incl. O&P
2 Laborers	$36.65	$586.40	$56.55	$904.80	$40.12	$61.48
1 Equip. Oper. (light)	47.05	376.40	71.35	570.80		
1 Rake w/Tractor		347.90		382.69		
2 Chain Saw, Gas, 18"		62.40		68.64	17.10	18.81
24 L.H., Daily Totals		$1373.10		$1926.93	$57.21	$80.29

Crew No.	Bare Costs		Incl. Subs O&P		Cost Per Labor-Hour	
Crew B-7B	Hr.	Daily	Hr.	Daily	Bare Costs	Incl. O&P
1 Labor Foreman (outside)	$38.65	$309.20	$59.65	$477.20	$38.81	$59.61
4 Laborers	36.65	1172.80	56.55	1809.60		
1 Equip. Oper. (medium)	48.90	391.20	74.15	593.20		
1 Truck Driver (heavy)	37.55	300.40	57.30	458.40		
1 Brush Chipper, 12", 130 H.P.		396.60		436.26		
1 Crawler Loader, 3 C.Y.		1191.00		1310.10		
2 Chain Saws, Gas, 36" Long		90.00		99.00		
1 Dump Truck, 8 C.Y., 220 H.P.		417.20		458.92	37.41	41.15
56 L.H., Daily Totals		$4268.40		$5642.68	$76.22	$100.76

Crew No.	Bare Costs		Incl. Subs O&P		Cost Per Labor-Hour	
Crew B-7C	Hr.	Daily	Hr.	Daily	Bare Costs	Incl. O&P
1 Labor Foreman (outside)	$38.65	$309.20	$59.65	$477.20	$38.81	$59.61
4 Laborers	36.65	1172.80	56.55	1809.60		
1 Equip. Oper. (medium)	48.90	391.20	74.15	593.20		
1 Truck Driver (heavy)	37.55	300.40	57.30	458.40		
1 Brush Chipper, 12", 130 H.P.		396.60		436.26		
1 Crawler Loader, 3 C.Y.		1191.00		1310.10		
2 Chain Saws, Gas, 36" Long		90.00		99.00		
1 Dump Truck, 12 C.Y., 400 H.P.		692.20		761.42	42.32	46.55
56 L.H., Daily Totals		$4543.40		$5945.18	$81.13	$106.16

Crew No.	Bare Costs		Incl. Subs O&P		Cost Per Labor-Hour	
Crew B-8	Hr.	Daily	Hr.	Daily	Bare Costs	Incl. O&P
1 Labor Foreman (outside)	$38.65	$309.20	$59.65	$477.20	$41.05	$62.71
2 Laborers	36.65	586.40	56.55	904.80		
2 Equip. Oper. (medium)	48.90	782.40	74.15	1186.40		
1 Equip. Oper. (oiler)	43.55	348.40	66.00	528.00		
2 Truck Drivers (heavy)	37.55	600.80	57.30	916.80		
1 Hyd. Crane, 25 Ton		745.80		820.38		
1 Crawler Loader, 3 C.Y.		1191.00		1310.10		
2 Dump Trucks, 12 C.Y., 400 H.P.		1384.40		1522.84	51.89	57.08
64 L.H., Daily Totals		$5948.40		$7666.52	$92.94	$119.79

Crew No.	Bare Costs		Incl. Subs O&P		Cost Per Labor-Hour	
Crew B-9	Hr.	Daily	Hr.	Daily	Bare Costs	Incl. O&P
1 Labor Foreman (outside)	$38.65	$309.20	$59.65	$477.20	$37.05	$57.17
4 Laborers	36.65	1172.80	56.55	1809.60		
1 Air Compressor, 250 cfm		202.40		222.64		
2 Breakers, Pavement, 60 lb.		19.60		21.56		
2 -50' Air Hoses, 1.5"		11.60		12.76	5.84	6.42
40 L.H., Daily Totals		$1715.60		$2543.76	$42.89	$63.59

Crew No.	Bare Costs		Incl. Subs O&P		Cost Per Labor-Hour	
Crew B-9A	Hr.	Daily	Hr.	Daily	Bare Costs	Incl. O&P
2 Laborers	$36.65	$586.40	$56.55	$904.80	$36.95	$56.80
1 Truck Driver (heavy)	37.55	300.40	57.30	458.40		
1 Water Tank Trailer, 5000 Gal.		141.60		155.76		
1 Truck Tractor, 220 H.P.		366.40		403.04		
2 -50' Discharge Hoses, 3"		3.00		3.30	21.29	23.42
24 L.H., Daily Totals		$1397.80		$1925.30	$58.24	$80.22

Crew No.	Bare Costs		Incl. Subs O&P		Cost Per Labor-Hour	
Crew B-9B	Hr.	Daily	Hr.	Daily	Bare Costs	Incl. O&P
2 Laborers	$36.65	$586.40	$56.55	$904.80	$36.95	$56.80
1 Truck Driver (heavy)	37.55	300.40	57.30	458.40		
2 -50' Discharge Hoses, 3"		3.00		3.30		
1 Water Tank Trailer, 5000 Gal.		141.60		155.76		
1 Truck Tractor, 220 H.P.		366.40		403.04		
1 Pressure Washer		69.40		76.34	24.18	26.60
24 L.H., Daily Totals		$1467.20		$2001.64	$61.13	$83.40

Crew No.	Bare Costs		Incl. Subs O&P		Cost Per Labor-Hour	
Crew B-9D	Hr.	Daily	Hr.	Daily	Bare Costs	Incl. O&P
1 Labor Foreman (outside)	$38.65	$309.20	$59.65	$477.20	$37.05	$57.17
4 Common Laborers	36.65	1172.80	56.55	1809.60		
1 Air Compressor, 250 cfm		202.40		222.64		
2 -50' Air Hoses, 1.5"		11.60		12.76		
2 Air Powered Tampers		52.40		57.64	6.66	7.33
40 L.H., Daily Totals		$1748.40		$2579.84	$43.71	$64.50

Crew No.	Bare Costs		Incl. Subs O&P		Cost Per Labor-Hour	
Crew B-10	Hr.	Daily	Hr.	Daily	Bare Costs	Incl. O&P
1 Equip. Oper. (medium)	$48.90	$391.20	$74.15	$593.20	$44.82	$68.28
.5 Laborer	36.65	146.60	56.55	226.20		
12 L.H., Daily Totals		$537.80		$819.40	$44.82	$68.28

Crew No.	Bare Costs		Incl. Subs O&P		Cost Per Labor-Hour	
Crew B-10A	Hr.	Daily	Hr.	Daily	Bare Costs	Incl. O&P
1 Equip. Oper. (medium)	$48.90	$391.20	$74.15	$593.20	$44.82	$68.28
.5 Laborer	36.65	146.60	56.55	226.20		
1 Roller, 2-Drum, W.B., 7.5 H.P.		182.80		201.08	15.23	16.76
12 L.H., Daily Totals		$720.60		$1020.48	$60.05	$85.04

Crew No.	Bare Costs		Incl. Subs O&P		Cost Per Labor-Hour	
Crew B-10B	Hr.	Daily	Hr.	Daily	Bare Costs	Incl. O&P
1 Equip. Oper. (medium)	$48.90	$391.20	$74.15	$593.20	$44.82	$68.28
.5 Laborer	36.65	146.60	56.55	226.20		
1 Dozer, 200 H.P.		1325.00		1457.50	110.42	121.46
12 L.H., Daily Totals		$1862.80		$2276.90	$155.23	$189.74

Crew No.	Bare Costs		Incl. Subs O&P		Cost Per Labor-Hour	
Crew B-10C	Hr.	Daily	Hr.	Daily	Bare Costs	Incl. O&P
1 Equip. Oper. (medium)	$48.90	$391.20	$74.15	$593.20	$44.82	$68.28
.5 Laborer	36.65	146.60	56.55	226.20		
1 Dozer, 200 H.P.		1325.00		1457.50		
1 Vibratory Roller, Towed, 23 Ton		412.00		453.20	144.75	159.22
12 L.H., Daily Totals		$2274.80		$2730.10	$189.57	$227.51

Crew No.	Bare Costs		Incl. Subs O&P		Cost Per Labor-Hour	
Crew B-10D	Hr.	Daily	Hr.	Daily	Bare Costs	Incl. O&P
1 Equip. Oper. (medium)	$48.90	$391.20	$74.15	$593.20	$44.82	$68.28
.5 Laborer	36.65	146.60	56.55	226.20		
1 Dozer, 200 H.P.		1325.00		1457.50		
1 Sheepsft. Roller, Towed		443.80		488.18	147.40	162.14
12 L.H., Daily Totals		$2306.60		$2765.08	$192.22	$230.42

Crew No.	Bare Costs		Incl. Subs O&P		Cost Per Labor-Hour	
Crew B-10E	Hr.	Daily	Hr.	Daily	Bare Costs	Incl. O&P
1 Equip. Oper. (medium)	$48.90	$391.20	$74.15	$593.20	$44.82	$68.28
.5 Laborer	36.65	146.60	56.55	226.20		
1 Tandem Roller, 5 Ton		159.00		174.90	13.25	14.57
12 L.H., Daily Totals		$696.80		$994.30	$58.07	$82.86

Crew No.	Bare Costs		Incl. Subs O&P		Cost Per Labor-Hour	
Crew B-10F	Hr.	Daily	Hr.	Daily	Bare Costs	Incl. O&P
1 Equip. Oper. (medium)	$48.90	$391.20	$74.15	$593.20	$44.82	$68.28
.5 Laborer	36.65	146.60	56.55	226.20		
1 Tandem Roller, 10 Ton		242.00		266.20	20.17	22.18
12 L.H., Daily Totals		$779.80		$1085.60	$64.98	$90.47

Crew No.	Bare Costs		Incl. Subs O&P		Cost Per Labor-Hour	
Crew B-10G	Hr.	Daily	Hr.	Daily	Bare Costs	Incl. O&P
1 Equip. Oper. (medium)	$48.90	$391.20	$74.15	$593.20	$44.82	$68.28
.5 Laborer	36.65	146.60	56.55	226.20		
1 Sheepsfoot Roller, 240 H.P.		1218.00		1339.80	101.50	111.65
12 L.H., Daily Totals		$1755.80		$2159.20	$146.32	$179.93

Crew No.	Bare Costs		Incl. Subs O&P		Cost Per Labor-Hour	
Crew B-10H	Hr.	Daily	Hr.	Daily	Bare Costs	Incl. O&P
1 Equip. Oper. (medium)	$48.90	$391.20	$74.15	$593.20	$44.82	$68.28
.5 Laborer	36.65	146.60	56.55	226.20		
1 Diaphragm Water Pump, 2"		72.00		79.20		
1 -20' Suction Hose, 2"		1.95		2.15		
2 -50' Discharge Hoses, 2"		1.80		1.98	6.31	6.94
12 L.H., Daily Totals		$613.55		$902.73	$51.13	$75.23

Crew No.	Bare Costs		Incl. Subs O&P		Cost Per Labor-Hour	
Crew B-10I	Hr.	Daily	Hr.	Daily	Bare Costs	Incl. O&P
1 Equip. Oper. (medium)	$48.90	$391.20	$74.15	$593.20	$44.82	$68.28
.5 Laborer	36.65	146.60	56.55	226.20		
1 Diaphragm Water Pump, 4"		115.80		127.38		
1 -20' Suction Hose, 4"		3.25		3.58		
2 -50' Discharge Hoses, 4"		4.70		5.17	10.31	11.34
12 L.H., Daily Totals		$661.55		$955.52	$55.13	$79.63

| Crew No. | Bare Costs | | Incl. Subs O&P | | Cost Per Labor-Hour | |

Crew B-10J

	Hr.	Daily	Hr.	Daily	Bare Costs	Incl. O&P
1 Equip. Oper. (medium)	$48.90	$391.20	$74.15	$593.20	$44.82	$68.28
.5 Laborer	36.65	146.60	56.55	226.20		
1 Centrifugal Water Pump, 3"		78.80		86.68		
1 -20' Suction Hose, 3"		2.85		3.13		
2 -50' Discharge Hoses, 3"		3.00		3.30	7.05	7.76
12 L.H., Daily Totals		$622.45		$912.51	$51.87	$76.04

Crew B-10K

	Hr.	Daily	Hr.	Daily	Bare Costs	Incl. O&P
1 Equip. Oper. (medium)	$48.90	$391.20	$74.15	$593.20	$44.82	$68.28
.5 Laborer	36.65	146.60	56.55	226.20		
1 Centr. Water Pump, 6"		340.80		374.88		
1 -20' Suction Hose, 6"		11.50		12.65		
2 -50' Discharge Hoses, 6"		12.20		13.42	30.38	33.41
12 L.H., Daily Totals		$902.30		$1220.35	$75.19	$101.70

Crew B-10L

	Hr.	Daily	Hr.	Daily	Bare Costs	Incl. O&P
1 Equip. Oper. (medium)	$48.90	$391.20	$74.15	$593.20	$44.82	$68.28
.5 Laborer	36.65	146.60	56.55	226.20		
1 Dozer, 80 H.P.		481.00		529.10	40.08	44.09
12 L.H., Daily Totals		$1018.80		$1348.50	$84.90	$112.38

Crew B-10M

	Hr.	Daily	Hr.	Daily	Bare Costs	Incl. O&P
1 Equip. Oper. (medium)	$48.90	$391.20	$74.15	$593.20	$44.82	$68.28
.5 Laborer	36.65	146.60	56.55	226.20		
1 Dozer, 300 H.P.		1817.00		1998.70	151.42	166.56
12 L.H., Daily Totals		$2354.80		$2818.10	$196.23	$234.84

Crew B-10N

	Hr.	Daily	Hr.	Daily	Bare Costs	Incl. O&P
1 Equip. Oper. (medium)	$48.90	$391.20	$74.15	$593.20	$44.82	$68.28
.5 Laborer	36.65	146.60	56.55	226.20		
1 F.E. Loader, T.M., 1.5 C.Y		522.80		575.08	43.57	47.92
12 L.H., Daily Totals		$1060.60		$1394.48	$88.38	$116.21

Crew B-10O

	Hr.	Daily	Hr.	Daily	Bare Costs	Incl. O&P
1 Equip. Oper. (medium)	$48.90	$391.20	$74.15	$593.20	$44.82	$68.28
.5 Laborer	36.65	146.60	56.55	226.20		
1 F.E. Loader, T.M., 2.25 C.Y.		959.80		1055.78	79.98	87.98
12 L.H., Daily Totals		$1497.60		$1875.18	$124.80	$156.26

Crew B-10P

	Hr.	Daily	Hr.	Daily	Bare Costs	Incl. O&P
1 Equip. Oper. (medium)	$48.90	$391.20	$74.15	$593.20	$44.82	$68.28
.5 Laborer	36.65	146.60	56.55	226.20		
1 Crawler Loader, 3 C.Y.		1191.00		1310.10	99.25	109.18
12 L.H., Daily Totals		$1728.80		$2129.50	$144.07	$177.46

Crew B-10Q

	Hr.	Daily	Hr.	Daily	Bare Costs	Incl. O&P
1 Equip. Oper. (medium)	$48.90	$391.20	$74.15	$593.20	$44.82	$68.28
.5 Laborer	36.65	146.60	56.55	226.20		
1 Crawler Loader, 4 C.Y.		1532.00		1685.20	127.67	140.43
12 L.H., Daily Totals		$2069.80		$2504.60	$172.48	$208.72

Crew B-10R

	Hr.	Daily	Hr.	Daily	Bare Costs	Incl. O&P
1 Equip. Oper. (medium)	$48.90	$391.20	$74.15	$593.20	$44.82	$68.28
.5 Laborer	36.65	146.60	56.55	226.20		
1 F.E. Loader, W.M., 1 C.Y.		295.20		324.72	24.60	27.06
12 L.H., Daily Totals		$833.00		$1144.12	$69.42	$95.34

Crew B-10S

	Hr.	Daily	Hr.	Daily	Bare Costs	Incl. O&P
1 Equip. Oper. (medium)	$48.90	$391.20	$74.15	$593.20	$44.82	$68.28
.5 Laborer	36.65	146.60	56.55	226.20		
1 F.E. Loader, W.M., 1.5 C.Y.		373.40		410.74	31.12	34.23
12 L.H., Daily Totals		$911.20		$1230.14	$75.93	$102.51

Crew B-10T

	Hr.	Daily	Hr.	Daily	Bare Costs	Incl. O&P
1 Equip. Oper. (medium)	$48.90	$391.20	$74.15	$593.20	$44.82	$68.28
.5 Laborer	36.65	146.60	56.55	226.20		
1 F.E. Loader, W.M.,2.5 C.Y.		522.80		575.08	43.57	47.92
12 L.H., Daily Totals		$1060.60		$1394.48	$88.38	$116.21

Crew B-10U

	Hr.	Daily	Hr.	Daily	Bare Costs	Incl. O&P
1 Equip. Oper. (medium)	$48.90	$391.20	$74.15	$593.20	$44.82	$68.28
.5 Laborer	36.65	146.60	56.55	226.20		
1 F.E. Loader, W.M., 5.5 C.Y.		1049.00		1153.90	87.42	96.16
12 L.H., Daily Totals		$1586.80		$1973.30	$132.23	$164.44

Crew B-10V

	Hr.	Daily	Hr.	Daily	Bare Costs	Incl. O&P
1 Equip. Oper. (medium)	$48.90	$391.20	$74.15	$593.20	$44.82	$68.28
.5 Laborer	36.65	146.60	56.55	226.20		
1 Dozer, 700 H.P.		4485.00		4933.50	373.75	411.13
12 L.H., Daily Totals		$5022.80		$5752.90	$418.57	$479.41

Crew B-10W

	Hr.	Daily	Hr.	Daily	Bare Costs	Incl. O&P
1 Equip. Oper. (medium)	$48.90	$391.20	$74.15	$593.20	$44.82	$68.28
.5 Laborer	36.65	146.60	56.55	226.20		
1 Dozer, 105 H.P.		591.80		650.98	49.32	54.25
12 L.H., Daily Totals		$1129.60		$1470.38	$94.13	$122.53

Crew B-10X

	Hr.	Daily	Hr.	Daily	Bare Costs	Incl. O&P
1 Equip. Oper. (medium)	$48.90	$391.20	$74.15	$593.20	$44.82	$68.28
.5 Laborer	36.65	146.60	56.55	226.20		
1 Dozer, 410 H.P.		2409.00		2649.90	200.75	220.82
12 L.H., Daily Totals		$2946.80		$3469.30	$245.57	$289.11

Crew B-10Y

	Hr.	Daily	Hr.	Daily	Bare Costs	Incl. O&P
1 Equip. Oper. (medium)	$48.90	$391.20	$74.15	$593.20	$44.82	$68.28
.5 Laborer	36.65	146.60	56.55	226.20		
1 Vibr. Roller, Towed, 12 Ton		559.20		615.12	46.60	51.26
12 L.H., Daily Totals		$1097.00		$1434.52	$91.42	$119.54

Crew B-11A

	Hr.	Daily	Hr.	Daily	Bare Costs	Incl. O&P
1 Equipment Oper. (med.)	$48.90	$391.20	$74.15	$593.20	$42.77	$65.35
1 Laborer	36.65	293.20	56.55	452.40		
1 Dozer, 200 H.P.		1325.00		1457.50	82.81	91.09
16 L.H., Daily Totals		$2009.40		$2503.10	$125.59	$156.44

Crew B-11B

	Hr.	Daily	Hr.	Daily	Bare Costs	Incl. O&P
1 Equipment Oper. (light)	$47.05	$376.40	$71.35	$570.80	$41.85	$63.95
1 Laborer	36.65	293.20	56.55	452.40		
1 Air Powered Tamper		26.20		28.82		
1 Air Compressor, 365 cfm		264.80		291.28		
2 -50' Air Hoses, 1.5"		11.60		12.76	18.91	20.80
16 L.H., Daily Totals		$972.20		$1356.06	$60.76	$84.75

Crew B-11C

	Hr.	Daily	Hr.	Daily	Bare Costs	Incl. O&P
1 Equipment Oper. (med.)	$48.90	$391.20	$74.15	$593.20	$42.77	$65.35
1 Laborer	36.65	293.20	56.55	452.40		
1 Backhoe Loader, 48 H.P.		365.20		401.72	22.82	25.11
16 L.H., Daily Totals		$1049.60		$1447.32	$65.60	$90.46

Crew No.	Bare Costs		Incl. Subs O&P		Cost Per Labor-Hour	
Crew B-11J	Hr.	Daily	Hr.	Daily	Bare Costs	Incl. O&P
1 Equipment Oper. (med.)	$48.90	$391.20	$74.15	$593.20	$42.77	$65.35
1 Laborer	36.65	293.20	56.55	452.40		
1 Grader, 30,000 Lbs.		707.00		777.70		
1 Ripper, Beam & 1 Shank		81.40		89.54	49.27	54.20
16 L.H., Daily Totals		$1472.80		$1912.84	$92.05	$119.55

Crew No.	Bare Costs		Incl. Subs O&P		Cost Per Labor-Hour	
Crew B-11K	Hr.	Daily	Hr.	Daily	Bare Costs	Incl. O&P
1 Equipment Oper. (med.)	$48.90	$391.20	$74.15	$593.20	$42.77	$65.35
1 Laborer	36.65	293.20	56.55	452.40		
1 Trencher, Chain Type, 8' D		3376.00		3713.60	211.00	232.10
16 L.H., Daily Totals		$4060.40		$4759.20	$253.78	$297.45

Crew No.	Bare Costs		Incl. Subs O&P		Cost Per Labor-Hour	
Crew B-11L	Hr.	Daily	Hr.	Daily	Bare Costs	Incl. O&P
1 Equipment Oper. (med.)	$48.90	$391.20	$74.15	$593.20	$42.77	$65.35
1 Laborer	36.65	293.20	56.55	452.40		
1 Grader, 30,000 Lbs.		707.00		777.70	44.19	48.61
16 L.H., Daily Totals		$1391.40		$1823.30	$86.96	$113.96

Crew No.	Bare Costs		Incl. Subs O&P		Cost Per Labor-Hour	
Crew B-11M	Hr.	Daily	Hr.	Daily	Bare Costs	Incl. O&P
1 Equipment Oper. (med.)	$48.90	$391.20	$74.15	$593.20	$42.77	$65.35
1 Laborer	36.65	293.20	56.55	452.40		
1 Backhoe Loader, 80 H.P.		397.60		437.36	24.85	27.34
16 L.H., Daily Totals		$1082.00		$1482.96	$67.63	$92.69

Crew No.	Bare Costs		Incl. Subs O&P		Cost Per Labor-Hour	
Crew B-11N	Hr.	Daily	Hr.	Daily	Bare Costs	Incl. O&P
1 Labor Foreman (outside)	$38.65	$309.20	$59.65	$477.20	$40.19	$61.31
2 Equipment Operators (med.)	48.90	782.40	74.15	1186.40		
6 Truck Drivers (heavy)	37.55	1802.40	57.30	2750.40		
1 F.E. Loader, W.M., 5.5 C.Y.		1049.00		1153.90		
1 Dozer, 410 H.P.		2409.00		2649.90		
6 Dump Trucks, Off Hwy., 50 Ton		10722.00		11794.20	196.94	216.64
72 L.H., Daily Totals		$17074.00		$20012.00	$237.14	$277.94

Crew No.	Bare Costs		Incl. Subs O&P		Cost Per Labor-Hour	
Crew B-11Q	Hr.	Daily	Hr.	Daily	Bare Costs	Incl. O&P
1 Equipment Operator (med.)	$48.90	$391.20	$74.15	$593.20	$44.82	$68.28
.5 Laborer	36.65	146.60	56.55	226.20		
1 Dozer, 140 H.P.		883.80		972.18	73.65	81.02
12 L.H., Daily Totals		$1421.60		$1791.58	$118.47	$149.30

Crew No.	Bare Costs		Incl. Subs O&P		Cost Per Labor-Hour	
Crew B-11R	Hr.	Daily	Hr.	Daily	Bare Costs	Incl. O&P
1 Equipment Operator (med.)	$48.90	$391.20	$74.15	$593.20	$44.82	$68.28
.5 Laborer	36.65	146.60	56.55	226.20		
1 Dozer, 200 H.P.		1325.00		1457.50	110.42	121.46
12 L.H., Daily Totals		$1862.80		$2276.90	$155.23	$189.74

Crew No.	Bare Costs		Incl. Subs O&P		Cost Per Labor-Hour	
Crew B-11S	Hr.	Daily	Hr.	Daily	Bare Costs	Incl. O&P
1 Equipment Operator (med.)	$48.90	$391.20	$74.15	$593.20	$44.82	$68.28
.5 Laborer	36.65	146.60	56.55	226.20		
1 Dozer, 300 H.P.		1817.00		1998.70		
1 Ripper, Beam & 1 Shank		81.40		89.54	158.20	174.02
12 L.H., Daily Totals		$2436.20		$2907.64	$203.02	$242.30

Crew No.	Bare Costs		Incl. Subs O&P		Cost Per Labor-Hour	
Crew B-11T	Hr.	Daily	Hr.	Daily	Bare Costs	Incl. O&P
1 Equipment Operator (med.)	$48.90	$391.20	$74.15	$593.20	$44.82	$68.28
.5 Laborer	36.65	146.60	56.55	226.20		
1 Dozer, 410 H.P.		2409.00		2649.90		
1 Ripper, Beam & 2 Shanks		91.60		100.76	208.38	229.22
12 L.H., Daily Totals		$3038.40		$3570.06	$253.20	$297.51

Crew No.	Bare Costs		Incl. Subs O&P		Cost Per Labor-Hour	
Crew B-11U	Hr.	Daily	Hr.	Daily	Bare Costs	Incl. O&P
1 Equipment Operator (med.)	$48.90	$391.20	$74.15	$593.20	$44.82	$68.28
.5 Laborer	36.65	146.60	56.55	226.20		
1 Dozer, 520 H.P.		3116.00		3427.60	259.67	285.63
12 L.H., Daily Totals		$3653.80		$4247.00	$304.48	$353.92

Crew No.	Bare Costs		Incl. Subs O&P		Cost Per Labor-Hour	
Crew B-11V	Hr.	Daily	Hr.	Daily	Bare Costs	Incl. O&P
3 Laborers	$36.65	$879.60	$56.55	$1357.20	$36.65	$56.55
1 Roller, 2-Drum, W.B., 7.5 H.P.		182.80		201.08	7.62	8.38
24 L.H., Daily Totals		$1062.40		$1558.28	$44.27	$64.93

Crew No.	Bare Costs		Incl. Subs O&P		Cost Per Labor-Hour	
Crew B-11W	Hr.	Daily	Hr.	Daily	Bare Costs	Incl. O&P
1 Equipment Operator (med.)	$48.90	$391.20	$74.15	$593.20	$38.42	$58.64
1 Common Laborer	36.65	293.20	56.55	452.40		
10 Truck Drivers (heavy)	37.55	3004.00	57.30	4584.00		
1 Dozer, 200 H.P.		1325.00		1457.50		
1 Vibratory Roller, Towed, 23 Ton		412.00		453.20		
10 Dump Trucks, 8 C.Y., 220 H.P.		4172.00		4589.20	61.55	67.71
96 L.H., Daily Totals		$9597.40		$12129.50	$99.97	$126.35

Crew No.	Bare Costs		Incl. Subs O&P		Cost Per Labor-Hour	
Crew B-11Y	Hr.	Daily	Hr.	Daily	Bare Costs	Incl. O&P
1 Labor Foreman (outside)	$38.65	$309.20	$59.65	$477.20	$40.96	$62.76
5 Common Laborers	36.65	1466.00	56.55	2262.00		
3 Equipment Operators (med.)	48.90	1173.60	74.15	1779.60		
1 Dozer, 80 H.P.		481.00		529.10		
2 Roller, 2-Drum, W.B., 7.5 H.P.		365.60		402.16		
4 Vibrating Plate, Gas, 21"		184.00		202.40	14.31	15.75
72 L.H., Daily Totals		$3979.40		$5652.46	$55.27	$78.51

Crew No.	Bare Costs		Incl. Subs O&P		Cost Per Labor-Hour	
Crew B-12A	Hr.	Daily	Hr.	Daily	Bare Costs	Incl. O&P
1 Equip. Oper. (crane)	$50.25	$402.00	$76.20	$609.60	$43.45	$66.38
1 Laborer	36.65	293.20	56.55	452.40		
1 Hyd. Excavator, 1 C.Y.		814.80		896.28	50.92	56.02
16 L.H., Daily Totals		$1510.00		$1958.28	$94.38	$122.39

Crew No.	Bare Costs		Incl. Subs O&P		Cost Per Labor-Hour	
Crew B-12B	Hr.	Daily	Hr.	Daily	Bare Costs	Incl. O&P
1 Equip. Oper. (crane)	$50.25	$402.00	$76.20	$609.60	$43.45	$66.38
1 Laborer	36.65	293.20	56.55	452.40		
1 Hyd. Excavator, 1.5 C.Y.		1031.00		1134.10	64.44	70.88
16 L.H., Daily Totals		$1726.20		$2196.10	$107.89	$137.26

Crew No.	Bare Costs		Incl. Subs O&P		Cost Per Labor-Hour	
Crew B-12C	Hr.	Daily	Hr.	Daily	Bare Costs	Incl. O&P
1 Equip. Oper. (crane)	$50.25	$402.00	$76.20	$609.60	$43.45	$66.38
1 Laborer	36.65	293.20	56.55	452.40		
1 Hyd. Excavator, 2 C.Y.		1175.00		1292.50	73.44	80.78
16 L.H., Daily Totals		$1870.20		$2354.50	$116.89	$147.16

Crew No.	Bare Costs		Incl. Subs O&P		Cost Per Labor-Hour	
Crew B-12D	Hr.	Daily	Hr.	Daily	Bare Costs	Incl. O&P
1 Equip. Oper. (crane)	$50.25	$402.00	$76.20	$609.60	$43.45	$66.38
1 Laborer	36.65	293.20	56.55	452.40		
1 Hyd. Excavator, 3.5 C.Y.		2442.00		2686.20	152.63	167.89
16 L.H., Daily Totals		$3137.20		$3748.20	$196.07	$234.26

Crew No.	Bare Costs		Incl. Subs O&P		Cost Per Labor-Hour	
Crew B-12E	Hr.	Daily	Hr.	Daily	Bare Costs	Incl. O&P
1 Equip. Oper. (crane)	$50.25	$402.00	$76.20	$609.60	$43.45	$66.38
1 Laborer	36.65	293.20	56.55	452.40		
1 Hyd. Excavator, .5 C.Y.		448.00		492.80	28.00	30.80
16 L.H., Daily Totals		$1143.20		$1554.80	$71.45	$97.17

Crew No.	Bare Costs Hr.	Daily	Incl. Subs O&P Hr.	Daily	Cost Per Labor-Hour Bare Costs	Incl. O&P
Crew B-12F	Hr.	Daily	Hr.	Daily	Bare Costs	Incl. O&P
1 Equip. Oper. (crane)	$50.25	$402.00	$76.20	$609.60	$43.45	$66.38
1 Laborer	36.65	293.20	56.55	452.40		
1 Hyd. Excavator, .75 C.Y.		662.00		728.20	41.38	45.51
16 L.H., Daily Totals		$1357.20		$1790.20	$84.83	$111.89
Crew B-12G	Hr.	Daily	Hr.	Daily	Bare Costs	Incl. O&P
1 Equip. Oper. (crane)	$50.25	$402.00	$76.20	$609.60	$43.45	$66.38
1 Laborer	36.65	293.20	56.55	452.40		
1 Crawler Crane, 15 Ton		684.15		752.57		
1 Clamshell Bucket, .5 C.Y.		37.80		41.58	45.12	49.63
16 L.H., Daily Totals		$1417.15		$1856.15	$88.57	$116.01
Crew B-12H	Hr.	Daily	Hr.	Daily	Bare Costs	Incl. O&P
1 Equip. Oper. (crane)	$50.25	$402.00	$76.20	$609.60	$43.45	$66.38
1 Laborer	36.65	293.20	56.55	452.40		
1 Crawler Crane, 25 Ton		1173.00		1290.30		
1 Clamshell Bucket, 1 C.Y.		47.40		52.14	76.28	83.90
16 L.H., Daily Totals		$1915.60		$2404.44	$119.72	$150.28
Crew B-12I	Hr.	Daily	Hr.	Daily	Bare Costs	Incl. O&P
1 Equip. Oper. (crane)	$50.25	$402.00	$76.20	$609.60	$43.45	$66.38
1 Laborer	36.65	293.20	56.55	452.40		
1 Crawler Crane, 20 Ton		880.25		968.27		
1 Dragline Bucket, .75 C.Y.		20.80		22.88	56.32	61.95
16 L.H., Daily Totals		$1596.25		$2053.16	$99.77	$128.32
Crew B-12J	Hr.	Daily	Hr.	Daily	Bare Costs	Incl. O&P
1 Equip. Oper. (crane)	$50.25	$402.00	$76.20	$609.60	$43.45	$66.38
1 Laborer	36.65	293.20	56.55	452.40		
1 Gradall, 5/8 C.Y.		881.60		969.76	55.10	60.61
16 L.H., Daily Totals		$1576.80		$2031.76	$98.55	$126.99
Crew B-12K	Hr.	Daily	Hr.	Daily	Bare Costs	Incl. O&P
1 Equip. Oper. (crane)	$50.25	$402.00	$76.20	$609.60	$43.45	$66.38
1 Laborer	36.65	293.20	56.55	452.40		
1 Gradall, 3 Ton, 1 C.Y.		1005.00		1105.50	62.81	69.09
16 L.H., Daily Totals		$1700.20		$2167.50	$106.26	$135.47
Crew B-12L	Hr.	Daily	Hr.	Daily	Bare Costs	Incl. O&P
1 Equip. Oper. (crane)	$50.25	$402.00	$76.20	$609.60	$43.45	$66.38
1 Laborer	36.65	293.20	56.55	452.40		
1 Crawler Crane, 15 Ton		684.15		752.57		
1 F.E. Attachment, .5 C.Y.		59.60		65.56	46.48	51.13
16 L.H., Daily Totals		$1438.95		$1880.13	$89.93	$117.51
Crew B-12M	Hr.	Daily	Hr.	Daily	Bare Costs	Incl. O&P
1 Equip. Oper. (crane)	$50.25	$402.00	$76.20	$609.60	$43.45	$66.38
1 Laborer	36.65	293.20	56.55	452.40		
1 Crawler Crane, 20 Ton		880.25		968.27		
1 F.E. Attachment, .75 C.Y.		64.00		70.40	59.02	64.92
16 L.H., Daily Totals		$1639.45		$2100.68	$102.47	$131.29
Crew B-12N	Hr.	Daily	Hr.	Daily	Bare Costs	Incl. O&P
1 Equip. Oper. (crane)	$50.25	$402.00	$76.20	$609.60	$43.45	$66.38
1 Laborer	36.65	293.20	56.55	452.40		
1 Crawler Crane, 25 Ton		1173.00		1290.30		
1 F.E. Attachment, 1 C.Y.		70.80		77.88	77.74	85.51
16 L.H., Daily Totals		$1939.00		$2430.18	$121.19	$151.89

Crew No.	Bare Costs Hr.	Daily	Incl. Subs O&P Hr.	Daily	Cost Per Labor-Hour Bare Costs	Incl. O&P
Crew B-12O	Hr.	Daily	Hr.	Daily	Bare Costs	Incl. O&P
1 Equip. Oper. (crane)	$50.25	$402.00	$76.20	$609.60	$43.45	$66.38
1 Laborer	36.65	293.20	56.55	452.40		
1 Crawler Crane, 40 Ton		1179.00		1296.90		
1 F.E. Attachment, 1.5 C.Y.		80.00		88.00	78.69	86.56
16 L.H., Daily Totals		$1954.20		$2446.90	$122.14	$152.93
Crew B-12P	Hr.	Daily	Hr.	Daily	Bare Costs	Incl. O&P
1 Equip. Oper. (crane)	$50.25	$402.00	$76.20	$609.60	$43.45	$66.38
1 Laborer	36.65	293.20	56.55	452.40		
1 Crawler Crane, 40 Ton		1179.00		1296.90		
1 Dragline Bucket, 1.5 C.Y.		33.60		36.96	75.79	83.37
16 L.H., Daily Totals		$1907.80		$2395.86	$119.24	$149.74
Crew B-12Q	Hr.	Daily	Hr.	Daily	Bare Costs	Incl. O&P
1 Equip. Oper. (crane)	$50.25	$402.00	$76.20	$609.60	$43.45	$66.38
1 Laborer	36.65	293.20	56.55	452.40		
1 Hyd. Excavator, 5/8 C.Y.		594.40		653.84	37.15	40.87
16 L.H., Daily Totals		$1289.60		$1715.84	$80.60	$107.24
Crew B-12S	Hr.	Daily	Hr.	Daily	Bare Costs	Incl. O&P
1 Equip. Oper. (crane)	$50.25	$402.00	$76.20	$609.60	$43.45	$66.38
1 Laborer	36.65	293.20	56.55	452.40		
1 Hyd. Excavator, 2.5 C.Y.		1637.00		1800.70	102.31	112.54
16 L.H., Daily Totals		$2332.20		$2862.70	$145.76	$178.92
Crew B-12T	Hr.	Daily	Hr.	Daily	Bare Costs	Incl. O&P
1 Equip. Oper. (crane)	$50.25	$402.00	$76.20	$609.60	$43.45	$66.38
1 Laborer	36.65	293.20	56.55	452.40		
1 Crawler Crane, 75 Ton		1492.00		1641.20		
1 F.E. Attachment, 3 C.Y.		102.80		113.08	99.67	109.64
16 L.H., Daily Totals		$2290.00		$2816.28	$143.13	$176.02
Crew B-12V	Hr.	Daily	Hr.	Daily	Bare Costs	Incl. O&P
1 Equip. Oper. (crane)	$50.25	$402.00	$76.20	$609.60	$43.45	$66.38
1 Laborer	36.65	293.20	56.55	452.40		
1 Crawler Crane, 75 Ton		1492.00		1641.20		
1 Dragline Bucket, 3 C.Y.		52.60		57.86	96.54	106.19
16 L.H., Daily Totals		$2239.80		$2761.06	$139.99	$172.57
Crew B-12Y	Hr.	Daily	Hr.	Daily	Bare Costs	Incl. O&P
1 Equip. Oper. (crane)	$50.25	$402.00	$76.20	$609.60	$41.18	$63.10
2 Laborers	36.65	586.40	56.55	904.80		
1 Hyd. Excavator, 3.5 C.Y.		2442.00		2686.20	101.75	111.93
24 L.H., Daily Totals		$3430.40		$4200.60	$142.93	$175.03
Crew B-12Z	Hr.	Daily	Hr.	Daily	Bare Costs	Incl. O&P
1 Equip. Oper. (crane)	$50.25	$402.00	$76.20	$609.60	$41.18	$63.10
2 Laborers	36.65	586.40	56.55	904.80		
1 Hyd. Excavator, 2.5 C.Y.		1637.00		1800.70	68.21	75.03
24 L.H., Daily Totals		$2625.40		$3315.10	$109.39	$138.13
Crew B-13	Hr.	Daily	Hr.	Daily	Bare Costs	Incl. O&P
1 Labor Foreman (outside)	$38.65	$309.20	$59.65	$477.20	$39.86	$61.15
4 Laborers	36.65	1172.80	56.55	1809.60		
1 Equip. Oper. (crane)	50.25	402.00	76.20	609.60		
1 Equip. Oper. (oiler)	43.55	348.40	66.00	528.00		
1 Hyd. Crane, 25 Ton		745.80		820.38	13.32	14.65
56 L.H., Daily Totals		$2978.20		$4244.78	$53.18	$75.80

Crew No.	Bare Costs		Incl. Subs O&P		Cost Per Labor-Hour	
Crew B-13A	Hr.	Daily	Hr.	Daily	Bare Costs	Incl. O&P
1 Labor Foreman (outside)	$38.65	$309.20	$59.65	$477.20	$40.69	$62.24
2 Laborers	36.65	586.40	56.55	904.80		
2 Equipment Operators (med.)	48.90	782.40	74.15	1186.40		
2 Truck Drivers (heavy)	37.55	600.80	57.30	916.80		
1 Crawler Crane, 75 Ton		1492.00		1641.20		
1 Crawler Loader, 4 C.Y.		1532.00		1685.20		
2 Dump Trucks, 8 C.Y., 220 H.P.		834.40		917.84	68.90	75.79
56 L.H., Daily Totals		$6137.20		$7729.44	$109.59	$138.03

Crew No.	Bare Costs		Incl. Subs O&P		Cost Per Labor-Hour	
Crew B-13B	Hr.	Daily	Hr.	Daily	Bare Costs	Incl. O&P
1 Labor Foreman (outside)	$38.65	$309.20	$59.65	$477.20	$39.86	$61.15
4 Laborers	36.65	1172.80	56.55	1809.60		
1 Equip. Oper. (crane)	50.25	402.00	76.20	609.60		
1 Equip. Oper. (oiler)	43.55	348.40	66.00	528.00		
1 Hyd. Crane, 55 Ton		1137.00		1250.70	20.30	22.33
56 L.H., Daily Totals		$3369.40		$4675.10	$60.17	$83.48

Crew No.	Bare Costs		Incl. Subs O&P		Cost Per Labor-Hour	
Crew B-13C	Hr.	Daily	Hr.	Daily	Bare Costs	Incl. O&P
1 Labor Foreman (outside)	$38.65	$309.20	$59.65	$477.20	$39.86	$61.15
4 Laborers	36.65	1172.80	56.55	1809.60		
1 Equip. Oper. (crane)	50.25	402.00	76.20	609.60		
1 Equip. Oper. (oiler)	43.55	348.40	66.00	528.00		
1 Crawler Crane, 100 Ton		1703.00		1873.30	30.41	33.45
56 L.H., Daily Totals		$3935.40		$5297.70	$70.28	$94.60

Crew No.	Bare Costs		Incl. Subs O&P		Cost Per Labor-Hour	
Crew B-13D	Hr.	Daily	Hr.	Daily	Bare Costs	Incl. O&P
1 Laborer	$36.65	$293.20	$56.55	$452.40	$43.45	$66.38
1 Equip. Oper. (crane)	50.25	402.00	76.20	609.60		
1 Hyd. Excavator, 1 C.Y.		814.80		896.28		
1 Trench Box		84.05		92.45	56.18	61.80
16 L.H., Daily Totals		$1594.05		$2050.74	$99.63	$128.17

Crew No.	Bare Costs		Incl. Subs O&P		Cost Per Labor-Hour	
Crew B-13E	Hr.	Daily	Hr.	Daily	Bare Costs	Incl. O&P
1 Laborer	$36.65	$293.20	$56.55	$452.40	$43.45	$66.38
1 Equip. Oper. (crane)	50.25	402.00	76.20	609.60		
1 Hyd. Excavator, 1.5 C.Y.		1031.00		1134.10		
1 Trench Box		84.05		92.45	69.69	76.66
16 L.H., Daily Totals		$1810.25		$2288.55	$113.14	$143.03

Crew No.	Bare Costs		Incl. Subs O&P		Cost Per Labor-Hour	
Crew B-13F	Hr.	Daily	Hr.	Daily	Bare Costs	Incl. O&P
1 Laborer	$36.65	$293.20	$56.55	$452.40	$43.45	$66.38
1 Equip. Oper. (crane)	50.25	402.00	76.20	609.60		
1 Hyd. Excavator, 3.5 C.Y.		2442.00		2686.20		
1 Trench Box		84.05		92.45	157.88	173.67
16 L.H., Daily Totals		$3221.25		$3840.66	$201.33	$240.04

Crew No.	Bare Costs		Incl. Subs O&P		Cost Per Labor-Hour	
Crew B-13G	Hr.	Daily	Hr.	Daily	Bare Costs	Incl. O&P
1 Laborer	$36.65	$293.20	$56.55	$452.40	$43.45	$66.38
1 Equip. Oper. (crane)	50.25	402.00	76.20	609.60		
1 Hyd. Excavator, .75 C.Y.		662.00		728.20		
1 Trench Box		84.05		92.45	46.63	51.29
16 L.H., Daily Totals		$1441.25		$1882.66	$90.08	$117.67

Crew No.	Bare Costs		Incl. Subs O&P		Cost Per Labor-Hour	
Crew B-13H	Hr.	Daily	Hr.	Daily	Bare Costs	Incl. O&P
1 Laborer	$36.65	$293.20	$56.55	$452.40	$43.45	$66.38
1 Equip. Oper. (crane)	50.25	402.00	76.20	609.60		
1 Gradall, 5/8 C.Y.		881.60		969.76		
1 Trench Box		84.05		92.45	60.35	66.39
16 L.H., Daily Totals		$1660.85		$2124.22	$103.80	$132.76

Crew No.	Bare Costs		Incl. Subs O&P		Cost Per Labor-Hour	
Crew B-13I	Hr.	Daily	Hr.	Daily	Bare Costs	Incl. O&P
1 Laborer	$36.65	$293.20	$56.55	$452.40	$43.45	$66.38
1 Equip. Oper. (crane)	50.25	402.00	76.20	609.60		
1 Gradall, 3 Ton, 1 C.Y.		1005.00		1105.50		
1 Trench Box		84.05		92.45	68.07	74.87
16 L.H., Daily Totals		$1784.25		$2259.95	$111.52	$141.25

Crew No.	Bare Costs		Incl. Subs O&P		Cost Per Labor-Hour	
Crew B-13J	Hr.	Daily	Hr.	Daily	Bare Costs	Incl. O&P
1 Laborer	$36.65	$293.20	$56.55	$452.40	$43.45	$66.38
1 Equip. Oper. (crane)	50.25	402.00	76.20	609.60		
1 Hyd. Excavator, 2.5 C.Y.		1637.00		1800.70		
1 Trench Box		84.05		92.45	107.57	118.32
16 L.H., Daily Totals		$2416.25		$2955.16	$151.02	$184.70

Crew No.	Bare Costs		Incl. Subs O&P		Cost Per Labor-Hour	
Crew B-14	Hr.	Daily	Hr.	Daily	Bare Costs	Incl. O&P
1 Labor Foreman (outside)	$38.65	$309.20	$59.65	$477.20	$38.72	$59.53
4 Laborers	36.65	1172.80	56.55	1809.60		
1 Equip. Oper. (light)	47.05	376.40	71.35	570.80		
1 Backhoe Loader, 48 H.P.		365.20		401.72	7.61	8.37
48 L.H., Daily Totals		$2223.60		$3259.32	$46.33	$67.90

Crew No.	Bare Costs		Incl. Subs O&P		Cost Per Labor-Hour	
Crew B-14A	Hr.	Daily	Hr.	Daily	Bare Costs	Incl. O&P
1 Equip. Oper. (crane)	$50.25	$402.00	$76.20	$609.60	$45.72	$69.65
.5 Laborer	36.65	146.60	56.55	226.20		
1 Hyd. Excavator, 4.5 C.Y.		3043.00		3347.30	253.58	278.94
12 L.H., Daily Totals		$3591.60		$4183.10	$299.30	$348.59

Crew No.	Bare Costs		Incl. Subs O&P		Cost Per Labor-Hour	
Crew B-14B	Hr.	Daily	Hr.	Daily	Bare Costs	Incl. O&P
1 Equip. Oper. (crane)	$50.25	$402.00	$76.20	$609.60	$45.72	$69.65
.5 Laborer	36.65	146.60	56.55	226.20		
1 Hyd. Excavator, 6 C.Y.		3523.00		3875.30	293.58	322.94
12 L.H., Daily Totals		$4071.60		$4711.10	$339.30	$392.59

Crew No.	Bare Costs		Incl. Subs O&P		Cost Per Labor-Hour	
Crew B-14C	Hr.	Daily	Hr.	Daily	Bare Costs	Incl. O&P
1 Equip. Oper. (crane)	$50.25	$402.00	$76.20	$609.60	$45.72	$69.65
.5 Laborer	36.65	146.60	56.55	226.20		
1 Hyd. Excavator, 7 C.Y.		3596.00		3955.60	299.67	329.63
12 L.H., Daily Totals		$4144.60		$4791.40	$345.38	$399.28

Crew No.	Bare Costs		Incl. Subs O&P		Cost Per Labor-Hour	
Crew B-14F	Hr.	Daily	Hr.	Daily	Bare Costs	Incl. O&P
1 Equip. Oper. (crane)	$50.25	$402.00	$76.20	$609.60	$45.72	$69.65
.5 Laborer	36.65	146.60	56.55	226.20		
1 Hyd. Shovel, 7 C.Y.		3714.00		4085.40	309.50	340.45
12 L.H., Daily Totals		$4262.60		$4921.20	$355.22	$410.10

Crew No.	Bare Costs		Incl. Subs O&P		Cost Per Labor-Hour	
Crew B-14G	Hr.	Daily	Hr.	Daily	Bare Costs	Incl. O&P
1 Equip. Oper. (crane)	$50.25	$402.00	$76.20	$609.60	$45.72	$69.65
.5 Laborer	36.65	146.60	56.55	226.20		
1 Hyd. Shovel, 12 C.Y.		5204.00		5724.40	433.67	477.03
12 L.H., Daily Totals		$5752.60		$6560.20	$479.38	$546.68

Crew No.	Bare Costs		Incl. Subs O&P		Cost Per Labor-Hour	
Crew B-14J	Hr.	Daily	Hr.	Daily	Bare Costs	Incl. O&P
1 Equip. Oper. (medium)	$48.90	$391.20	$74.15	$593.20	$44.82	$68.28
.5 Laborer	36.65	146.60	56.55	226.20		
1 F.E. Loader, 8 C.Y.		1989.00		2187.90	165.75	182.32
12 L.H., Daily Totals		$2526.80		$3007.30	$210.57	$250.61

Crew B-14K

	Bare Costs Hr.	Bare Costs Daily	Incl. Subs O&P Hr.	Incl. Subs O&P Daily	Cost Per Labor-Hour Bare Costs	Cost Per Labor-Hour Incl. O&P
1 Equip. Oper. (medium)	$48.90	$391.20	$74.15	$593.20	$44.82	$68.28
.5 Laborer	36.65	146.60	56.55	226.20		
1 F.E. Loader, 10 C.Y.		2703.00		2973.30	225.25	247.78
12 L.H., Daily Totals		$3240.80		$3792.70	$270.07	$316.06

Crew B-15

	Bare Costs Hr.	Bare Costs Daily	Incl. Subs O&P Hr.	Incl. Subs O&P Daily	Cost Per Labor-Hour Bare Costs	Cost Per Labor-Hour Incl. O&P
1 Equipment Oper. (med.)	$48.90	$391.20	$74.15	$593.20	$40.66	$62.01
.5 Laborer	36.65	146.60	56.55	226.20		
2 Truck Drivers (heavy)	37.55	600.80	57.30	916.80		
2 Dump Trucks, 12 C.Y., 400 H.P.		1384.40		1522.84		
1 Dozer, 200 H.P.		1325.00		1457.50	96.76	106.44
28 L.H., Daily Totals		$3848.00		$4716.54	$137.43	$168.45

Crew B-16

	Bare Costs Hr.	Bare Costs Daily	Incl. Subs O&P Hr.	Incl. Subs O&P Daily	Cost Per Labor-Hour Bare Costs	Cost Per Labor-Hour Incl. O&P
1 Labor Foreman (outside)	$38.65	$309.20	$59.65	$477.20	$37.38	$57.51
2 Laborers	36.65	586.40	56.55	904.80		
1 Truck Driver (heavy)	37.55	300.40	57.30	458.40		
1 Dump Truck, 12 C.Y., 400 H.P.		692.20		761.42	21.63	23.79
32 L.H., Daily Totals		$1888.20		$2601.82	$59.01	$81.31

Crew B-17

	Bare Costs Hr.	Bare Costs Daily	Incl. Subs O&P Hr.	Incl. Subs O&P Daily	Cost Per Labor-Hour Bare Costs	Cost Per Labor-Hour Incl. O&P
2 Laborers	$36.65	$586.40	$56.55	$904.80	$39.48	$60.44
1 Equip. Oper. (light)	47.05	376.40	71.35	570.80		
1 Truck Driver (heavy)	37.55	300.40	57.30	458.40		
1 Backhoe Loader, 48 H.P.		365.20		401.72		
1 Dump Truck, 8 C.Y., 220 H.P.		417.20		458.92	24.45	26.90
32 L.H., Daily Totals		$2045.60		$2794.64	$63.92	$87.33

Crew B-17A

	Bare Costs Hr.	Bare Costs Daily	Incl. Subs O&P Hr.	Incl. Subs O&P Daily	Cost Per Labor-Hour Bare Costs	Cost Per Labor-Hour Incl. O&P
2 Labor Foremen (outside)	$38.65	$618.40	$59.65	$954.40	$39.38	$60.82
6 Laborers	36.65	1759.20	56.55	2714.40		
1 Skilled Worker Foreman (out)	49.30	394.40	76.35	610.80		
1 Skilled Worker	47.30	378.40	73.25	586.00		
80 L.H., Daily Totals		$3150.40		$4865.60	$39.38	$60.82

Crew B-17B

	Bare Costs Hr.	Bare Costs Daily	Incl. Subs O&P Hr.	Incl. Subs O&P Daily	Cost Per Labor-Hour Bare Costs	Cost Per Labor-Hour Incl. O&P
2 Laborers	$36.65	$586.40	$56.55	$904.80	$39.48	$60.44
1 Equip. Oper. (light)	47.05	376.40	71.35	570.80		
1 Truck Driver (heavy)	37.55	300.40	57.30	458.40		
1 Backhoe Loader, 48 H.P.		365.20		401.72		
1 Dump Truck, 12 C.Y., 400 H.P.		692.20		761.42	33.04	36.35
32 L.H., Daily Totals		$2320.60		$3097.14	$72.52	$96.79

Crew B-18

	Bare Costs Hr.	Bare Costs Daily	Incl. Subs O&P Hr.	Incl. Subs O&P Daily	Cost Per Labor-Hour Bare Costs	Cost Per Labor-Hour Incl. O&P
1 Labor Foreman (outside)	$38.65	$309.20	$59.65	$477.20	$37.32	$57.58
2 Laborers	36.65	586.40	56.55	904.80		
1 Vibrating Plate, Gas, 21"		46.00		50.60	1.92	2.11
24 L.H., Daily Totals		$941.60		$1432.60	$39.23	$59.69

Crew B-19

	Bare Costs Hr.	Bare Costs Daily	Incl. Subs O&P Hr.	Incl. Subs O&P Daily	Cost Per Labor-Hour Bare Costs	Cost Per Labor-Hour Incl. O&P
1 Pile Driver Foreman (outside)	$46.40	$371.20	$73.60	$588.80	$46.01	$71.70
4 Pile Drivers	44.40	1420.80	70.40	2252.80		
2 Equip. Oper. (crane)	50.25	804.00	76.20	1219.20		
1 Equip. Oper. (oiler)	43.55	348.40	66.00	528.00		
1 Crawler Crane, 40 Ton		1179.00		1296.90		
1 Lead, 90' High		124.40		136.84		
1 Hammer, Diesel, 22k ft-lb		537.45		591.20	28.76	31.64
64 L.H., Daily Totals		$4785.25		$6613.73	$74.77	$103.34

Crew B-19A

	Bare Costs Hr.	Bare Costs Daily	Incl. Subs O&P Hr.	Incl. Subs O&P Daily	Cost Per Labor-Hour Bare Costs	Cost Per Labor-Hour Incl. O&P
1 Pile Driver Foreman (outside)	$46.40	$371.20	$73.60	$588.80	$46.01	$71.70
4 Pile Drivers	44.40	1420.80	70.40	2252.80		
2 Equip. Oper. (crane)	50.25	804.00	76.20	1219.20		
1 Equip. Oper. (oiler)	43.55	348.40	66.00	528.00		
1 Crawler Crane, 75 Ton		1492.00		1641.20		
1 Lead, 90' high		124.40		136.84		
1 Hammer, Diesel, 41k ft-lb		642.40		706.64	35.29	38.82
64 L.H., Daily Totals		$5203.20		$7073.48	$81.30	$110.52

Crew B-19B

	Bare Costs Hr.	Bare Costs Daily	Incl. Subs O&P Hr.	Incl. Subs O&P Daily	Cost Per Labor-Hour Bare Costs	Cost Per Labor-Hour Incl. O&P
1 Pile Driver Foreman (outside)	$46.40	$371.20	$73.60	$588.80	$46.01	$71.70
4 Pile Drivers	44.40	1420.80	70.40	2252.80		
2 Equip. Oper. (crane)	50.25	804.00	76.20	1219.20		
1 Equip. Oper. (oiler)	43.55	348.40	66.00	528.00		
1 Crawler Crane, 40 Ton		1179.00		1296.90		
1 Lead, 90' High		124.40		136.84		
1 Hammer, Diesel, 22k ft-lb		537.45		591.20		
1 Barge, 400 Ton		781.00		859.10	40.97	45.06
64 L.H., Daily Totals		$5566.25		$7472.84	$86.97	$116.76

Crew B-19C

	Bare Costs Hr.	Bare Costs Daily	Incl. Subs O&P Hr.	Incl. Subs O&P Daily	Cost Per Labor-Hour Bare Costs	Cost Per Labor-Hour Incl. O&P
1 Pile Driver Foreman (outside)	$46.40	$371.20	$73.60	$588.80	$46.01	$71.70
4 Pile Drivers	44.40	1420.80	70.40	2252.80		
2 Equip. Oper. (crane)	50.25	804.00	76.20	1219.20		
1 Equip. Oper. (oiler)	43.55	348.40	66.00	528.00		
1 Crawler Crane, 75 Ton		1492.00		1641.20		
1 Lead, 90' High		124.40		136.84		
1 Hammer, Diesel, 41k ft-lb		642.40		706.64		
1 Barge, 400 Ton		781.00		859.10	47.50	52.25
64 L.H., Daily Totals		$5984.20		$7932.58	$93.50	$123.95

Crew B-20

	Bare Costs Hr.	Bare Costs Daily	Incl. Subs O&P Hr.	Incl. Subs O&P Daily	Cost Per Labor-Hour Bare Costs	Cost Per Labor-Hour Incl. O&P
1 Labor Foreman (outside)	$38.65	$309.20	$59.65	$477.20	$40.87	$63.15
1 Skilled Worker	47.30	378.40	73.25	586.00		
1 Laborer	36.65	293.20	56.55	452.40		
24 L.H., Daily Totals		$980.80		$1515.60	$40.87	$63.15

Crew B-20A

	Bare Costs Hr.	Bare Costs Daily	Incl. Subs O&P Hr.	Incl. Subs O&P Daily	Cost Per Labor-Hour Bare Costs	Cost Per Labor-Hour Incl. O&P
1 Labor Foreman (outside)	$38.65	$309.20	$59.65	$477.20	$44.73	$68.16
1 Laborer	36.65	293.20	56.55	452.40		
1 Plumber	57.55	460.40	86.90	695.20		
1 Plumber Apprentice	46.05	368.40	69.55	556.40		
32 L.H., Daily Totals		$1431.20		$2181.20	$44.73	$68.16

Crew B-21

	Bare Costs Hr.	Bare Costs Daily	Incl. Subs O&P Hr.	Incl. Subs O&P Daily	Cost Per Labor-Hour Bare Costs	Cost Per Labor-Hour Incl. O&P
1 Labor Foreman (outside)	$38.65	$309.20	$59.65	$477.20	$42.21	$65.01
1 Skilled Worker	47.30	378.40	73.25	586.00		
1 Laborer	36.65	293.20	56.55	452.40		
.5 Equip. Oper. (crane)	50.25	201.00	76.20	304.80		
.5 S.P. Crane, 4x4, 5 Ton		140.00		154.00	5.00	5.50
28 L.H., Daily Totals		$1321.80		$1974.40	$47.21	$70.51

Crew B-21A

	Bare Costs Hr.	Bare Costs Daily	Incl. Subs O&P Hr.	Incl. Subs O&P Daily	Cost Per Labor-Hour Bare Costs	Cost Per Labor-Hour Incl. O&P
1 Labor Foreman (outside)	$38.65	$309.20	$59.65	$477.20	$45.83	$69.77
1 Laborer	36.65	293.20	56.55	452.40		
1 Plumber	57.55	460.40	86.90	695.20		
1 Plumber Apprentice	46.05	368.40	69.55	556.40		
1 Equip. Oper. (crane)	50.25	402.00	76.20	609.60		
1 S.P. Crane, 4x4, 12 Ton		481.20		529.32	12.03	13.23
40 L.H., Daily Totals		$2314.40		$3320.12	$57.86	$83.00

Crew B-21B

Crew No.	Bare Costs Hr.	Daily	Incl. Subs O&P Hr.	Daily	Cost Per Labor-Hour Bare Costs	Incl. O&P
1 Labor Foreman (outside)	$38.65	$309.20	$59.65	$477.20	$39.77	$61.10
3 Laborers	36.65	879.60	56.55	1357.20		
1 Equip. Oper. (crane)	50.25	402.00	76.20	609.60		
1 Hyd. Crane, 12 Ton		663.00		729.30	16.57	18.23
40 L.H., Daily Totals		$2253.80		$3173.30	$56.34	$79.33

Crew B-21C

Crew No.	Bare Costs Hr.	Daily	Incl. Subs O&P Hr.	Daily	Cost Per Labor-Hour Bare Costs	Incl. O&P
1 Labor Foreman (outside)	$38.65	$309.20	$59.65	$477.20	$39.86	$61.15
4 Laborers	36.65	1172.80	56.55	1809.60		
1 Equip. Oper. (crane)	50.25	402.00	76.20	609.60		
1 Equip. Oper. (oiler)	43.55	348.40	66.00	528.00		
2 Cutting Torches		22.80		25.08		
2 Sets of Gases		304.00		334.40		
1 Lattice Boom Crane, 90 Ton		1529.00		1681.90	33.14	36.45
56 L.H., Daily Totals		$4088.20		$5465.78	$73.00	$97.60

Crew B-22

Crew No.	Bare Costs Hr.	Daily	Incl. Subs O&P Hr.	Daily	Cost Per Labor-Hour Bare Costs	Incl. O&P
1 Labor Foreman (outside)	$38.65	$309.20	$59.65	$477.20	$42.74	$65.76
1 Skilled Worker	47.30	378.40	73.25	586.00		
1 Laborer	36.65	293.20	56.55	452.40		
.75 Equip. Oper. (crane)	50.25	301.50	76.20	457.20		
.75 S.P. Crane, 4x4, 5 Ton		210.00		231.00	7.00	7.70
30 L.H., Daily Totals		$1492.30		$2203.80	$49.74	$73.46

Crew B-22A

Crew No.	Bare Costs Hr.	Daily	Incl. Subs O&P Hr.	Daily	Cost Per Labor-Hour Bare Costs	Incl. O&P
1 Labor Foreman (outside)	$38.65	$309.20	$59.65	$477.20	$41.90	$64.44
1 Skilled Worker	47.30	378.40	73.25	586.00		
2 Laborers	36.65	586.40	56.55	904.80		
1 Equipment Operator, Crane	50.25	402.00	76.20	609.60		
1 S.P. Crane, 4x4, 5 Ton		280.00		308.00		
1 Butt Fusion Mach., 4"-12" diam.		381.80		419.98	16.55	18.20
40 L.H., Daily Totals		$2337.80		$3305.58	$58.45	$82.64

Crew B-22B

Crew No.	Bare Costs Hr.	Daily	Incl. Subs O&P Hr.	Daily	Cost Per Labor-Hour Bare Costs	Incl. O&P
1 Labor Foreman (outside)	$38.65	$309.20	$59.65	$477.20	$41.90	$64.44
1 Skilled Worker	47.30	378.40	73.25	586.00		
2 Laborers	36.65	586.40	56.55	904.80		
1 Equip. Oper. (crane)	50.25	402.00	76.20	609.60		
1 S.P. Crane, 4x4, 5 Ton		280.00		308.00		
1 Butt Fusion Mach., 8"-24" diam.		828.40		911.24	27.71	30.48
40 L.H., Daily Totals		$2784.40		$3796.84	$69.61	$94.92

Crew B-22C

Crew No.	Bare Costs Hr.	Daily	Incl. Subs O&P Hr.	Daily	Cost Per Labor-Hour Bare Costs	Incl. O&P
1 Skilled Worker	$47.30	$378.40	$73.25	$586.00	$41.98	$64.90
1 Laborer	36.65	293.20	56.55	452.40		
1 Butt Fusion Mach., 2"-8" diam.		120.95		133.04	7.56	8.32
16 L.H., Daily Totals		$792.55		$1171.44	$49.53	$73.22

Crew B-23

Crew No.	Bare Costs Hr.	Daily	Incl. Subs O&P Hr.	Daily	Cost Per Labor-Hour Bare Costs	Incl. O&P
1 Labor Foreman (outside)	$38.65	$309.20	$59.65	$477.20	$37.05	$57.17
4 Laborers	36.65	1172.80	56.55	1809.60		
1 Drill Rig, Truck-Mounted		2551.00		2806.10		
1 Flatbed Truck, Gas, 3 Ton		333.00		366.30	72.10	79.31
40 L.H., Daily Totals		$4366.00		$5459.20	$109.15	$136.48

Crew B-23A

Crew No.	Bare Costs Hr.	Daily	Incl. Subs O&P Hr.	Daily	Cost Per Labor-Hour Bare Costs	Incl. O&P
1 Labor Foreman (outside)	$38.65	$309.20	$59.65	$477.20	$41.40	$63.45
1 Laborer	36.65	293.20	56.55	452.40		
1 Equip. Oper. (medium)	48.90	391.20	74.15	593.20		
1 Drill Rig, Truck-Mounted		2551.00		2806.10		
1 Pickup Truck, 3/4 Ton		157.00		172.70	112.83	124.12
24 L.H., Daily Totals		$3701.60		$4501.60	$154.23	$187.57

Crew B-23B

Crew No.	Bare Costs Hr.	Daily	Incl. Subs O&P Hr.	Daily	Cost Per Labor-Hour Bare Costs	Incl. O&P
1 Labor Foreman (outside)	$38.65	$309.20	$59.65	$477.20	$41.40	$63.45
1 Laborer	36.65	293.20	56.55	452.40		
1 Equip. Oper. (medium)	48.90	391.20	74.15	593.20		
1 Drill Rig, Truck-Mounted		2551.00		2806.10		
1 Pickup Truck, 3/4 Ton		157.00		172.70		
1 Centr. Water Pump, 6"		340.80		374.88	127.03	139.74
24 L.H., Daily Totals		$4042.40		$4876.48	$168.43	$203.19

Crew B-24

Crew No.	Bare Costs Hr.	Daily	Incl. Subs O&P Hr.	Daily	Cost Per Labor-Hour Bare Costs	Incl. O&P
1 Cement Finisher	$44.05	$352.40	$65.10	$520.80	$42.18	$64.13
1 Laborer	36.65	293.20	56.55	452.40		
1 Carpenter	45.85	366.80	70.75	566.00		
24 L.H., Daily Totals		$1012.40		$1539.20	$42.18	$64.13

Crew B-25

Crew No.	Bare Costs Hr.	Daily	Incl. Subs O&P Hr.	Daily	Cost Per Labor-Hour Bare Costs	Incl. O&P
1 Labor Foreman (outside)	$38.65	$309.20	$59.65	$477.20	$40.17	$61.63
7 Laborers	36.65	2052.40	56.55	3166.80		
3 Equip. Oper. (medium)	48.90	1173.60	74.15	1779.60		
1 Asphalt Paver, 130 H.P.		2169.00		2385.90		
1 Tandem Roller, 10 Ton		242.00		266.20		
1 Roller, Pneum. Whl., 12 Ton		349.00		383.90	31.36	34.50
88 L.H., Daily Totals		$6295.20		$8459.60	$71.54	$96.13

Crew B-25B

Crew No.	Bare Costs Hr.	Daily	Incl. Subs O&P Hr.	Daily	Cost Per Labor-Hour Bare Costs	Incl. O&P
1 Labor Foreman (outside)	$38.65	$309.20	$59.65	$477.20	$40.90	$62.67
7 Laborers	36.65	2052.40	56.55	3166.80		
4 Equip. Oper. (medium)	48.90	1564.80	74.15	2372.80		
1 Asphalt Paver, 130 H.P.		2169.00		2385.90		
2 Tandem Rollers, 10 Ton		484.00		532.40		
1 Roller, Pneum. Whl., 12 Ton		349.00		383.90	31.27	34.40
96 L.H., Daily Totals		$6928.40		$9319.00	$72.17	$97.07

Crew B-25C

Crew No.	Bare Costs Hr.	Daily	Incl. Subs O&P Hr.	Daily	Cost Per Labor-Hour Bare Costs	Incl. O&P
1 Labor Foreman (outside)	$38.65	$309.20	$59.65	$477.20	$41.07	$62.93
3 Laborers	36.65	879.60	56.55	1357.20		
2 Equip. Oper. (medium)	48.90	782.40	74.15	1186.40		
1 Asphalt Paver, 130 H.P.		2169.00		2385.90		
1 Tandem Roller, 10 Ton		242.00		266.20	50.23	55.25
48 L.H., Daily Totals		$4382.20		$5672.90	$91.30	$118.19

Crew B-25D

Crew No.	Bare Costs Hr.	Daily	Incl. Subs O&P Hr.	Daily	Cost Per Labor-Hour Bare Costs	Incl. O&P
1 Labor Foreman (outside)	$38.65	$309.20	$59.65	$477.20	$41.15	$63.05
3 Laborers	36.65	879.60	56.55	1357.20		
2.125 Equip. Oper. (medium)	48.90	831.30	74.15	1260.55		
.125 Truck Driver (heavy)	37.55	37.55	57.30	57.30		
.125 Truck Tractor, 6x4, 380 H.P.		76.60		84.26		
.125 Dist. Tanker, 3000 Gallon		38.52		42.38		
1 Asphalt Paver, 130 H.P.		2169.00		2385.90		
1 Tandem Roller, 10 Ton		242.00		266.20	50.52	55.57
50 L.H., Daily Totals		$4583.77		$5930.99	$91.68	$118.62

Crew No.	Bare Costs Hr.	Daily	Incl. Subs O&P Hr.	Daily	Cost Per Labor-Hour Bare Costs	Incl. O&P
Crew B-25E	**Hr.**	**Daily**	**Hr.**	**Daily**	**Bare Costs**	**Incl. O&P**
1 Labor Foreman (outside)	$38.65	$309.20	$59.65	$477.20	$41.23	$63.15
3 Laborers	36.65	879.60	56.55	1357.20		
2.250 Equip. Oper. (medium)	48.90	880.20	74.15	1334.70		
.25 Truck Driver (heavy)	37.55	75.10	57.30	114.60		
.25 Truck Tractor, 6x4, 380 H.P.		153.20		168.52		
.25 Dist. Tanker, 3000 Gallon		77.05		84.75		
1 Asphalt Paver, 130 H.P.		2169.00		2385.90		
1 Tandem Roller, 10 Ton		242.00		266.20	50.79	55.87
52 L.H., Daily Totals		$4785.35		$6189.07	$92.03	$119.02
Crew B-26	**Hr.**	**Daily**	**Hr.**	**Daily**	**Bare Costs**	**Incl. O&P**
1 Labor Foreman (outside)	$38.65	$309.20	$59.65	$477.20	$41.00	$62.90
6 Laborers	36.65	1759.20	56.55	2714.40		
2 Equip. Oper. (medium)	48.90	782.40	74.15	1186.40		
1 Rodman (reinf.)	50.65	405.20	79.55	636.40		
1 Cement Finisher	44.05	352.40	65.10	520.80		
1 Grader, 30,000 Lbs.		707.00		777.70		
1 Paving Mach. & Equip.		2657.00		2922.70	38.23	42.05
88 L.H., Daily Totals		$6972.40		$9235.60	$79.23	$104.95
Crew B-26A	**Hr.**	**Daily**	**Hr.**	**Daily**	**Bare Costs**	**Incl. O&P**
1 Labor Foreman (outside)	$38.65	$309.20	$59.65	$477.20	$41.00	$62.90
6 Laborers	36.65	1759.20	56.55	2714.40		
2 Equip. Oper. (medium)	48.90	782.40	74.15	1186.40		
1 Rodman (reinf.)	50.65	405.20	79.55	636.40		
1 Cement Finisher	44.05	352.40	65.10	520.80		
1 Grader, 30,000 Lbs.		707.00		777.70		
1 Paving Mach. & Equip.		2657.00		2922.70		
1 Concrete Saw		175.20		192.72	40.22	44.24
88 L.H., Daily Totals		$7147.60		$9428.32	$81.22	$107.14
Crew B-26B	**Hr.**	**Daily**	**Hr.**	**Daily**	**Bare Costs**	**Incl. O&P**
1 Labor Foreman (outside)	$38.65	$309.20	$59.65	$477.20	$41.66	$63.84
6 Laborers	36.65	1759.20	56.55	2714.40		
3 Equip. Oper. (medium)	48.90	1173.60	74.15	1779.60		
1 Rodman (reinf.)	50.65	405.20	79.55	636.40		
1 Cement Finisher	44.05	352.40	65.10	520.80		
1 Grader, 30,000 Lbs.		707.00		777.70		
1 Paving Mach. & Equip.		2657.00		2922.70		
1 Concrete Pump, 110' Boom		934.00		1027.40	44.77	49.25
96 L.H., Daily Totals		$8297.60		$10856.20	$86.43	$113.09
Crew B-26C	**Hr.**	**Daily**	**Hr.**	**Daily**	**Bare Costs**	**Incl. O&P**
1 Labor Foreman (outside)	$38.65	$309.20	$59.65	$477.20	$40.22	$61.77
6 Laborers	36.65	1759.20	56.55	2714.40		
1 Equip. Oper. (medium)	48.90	391.20	74.15	593.20		
1 Rodman (reinf.)	50.65	405.20	79.55	636.40		
1 Cement Finisher	44.05	352.40	65.10	520.80		
1 Paving Mach. & Equip.		2657.00		2922.70		
1 Concrete Saw		175.20		192.72	35.40	38.94
80 L.H., Daily Totals		$6049.40		$8057.42	$75.62	$100.72
Crew B-27	**Hr.**	**Daily**	**Hr.**	**Daily**	**Bare Costs**	**Incl. O&P**
1 Labor Foreman (outside)	$38.65	$309.20	$59.65	$477.20	$37.15	$57.33
3 Laborers	36.65	879.60	56.55	1357.20		
1 Berm Machine		294.40		323.84	9.20	10.12
32 L.H., Daily Totals		$1483.20		$2158.24	$46.35	$67.44

Crew No.	Bare Costs Hr.	Daily	Incl. Subs O&P Hr.	Daily	Cost Per Labor-Hour Bare Costs	Incl. O&P
Crew B-28	**Hr.**	**Daily**	**Hr.**	**Daily**	**Bare Costs**	**Incl. O&P**
2 Carpenters	$45.85	$733.60	$70.75	$1132.00	$42.78	$66.02
1 Laborer	36.65	293.20	56.55	452.40		
24 L.H., Daily Totals		$1026.80		$1584.40	$42.78	$66.02
Crew B-29	**Hr.**	**Daily**	**Hr.**	**Daily**	**Bare Costs**	**Incl. O&P**
1 Labor Foreman (outside)	$38.65	$309.20	$59.65	$477.20	$39.86	$61.15
4 Laborers	36.65	1172.80	56.55	1809.60		
1 Equip. Oper. (crane)	50.25	402.00	76.20	609.60		
1 Equip. Oper. (oiler)	43.55	348.40	66.00	528.00		
1 Gradall, 5/8 C.Y.		881.60		969.76	15.74	17.32
56 L.H., Daily Totals		$3114.00		$4394.16	$55.61	$78.47
Crew B-30	**Hr.**	**Daily**	**Hr.**	**Daily**	**Bare Costs**	**Incl. O&P**
1 Equip. Oper. (medium)	$48.90	$391.20	$74.15	$593.20	$41.33	$62.92
2 Truck Drivers (heavy)	37.55	600.80	57.30	916.80		
1 Hyd. Excavator, 1.5 C.Y.		1031.00		1134.10		
2 Dump Trucks, 12 C.Y., 400 H.P.		1384.40		1522.84	100.64	110.71
24 L.H., Daily Totals		$3407.40		$4166.94	$141.97	$173.62
Crew B-31	**Hr.**	**Daily**	**Hr.**	**Daily**	**Bare Costs**	**Incl. O&P**
1 Labor Foreman (outside)	$38.65	$309.20	$59.65	$477.20	$38.89	$60.01
3 Laborers	36.65	879.60	56.55	1357.20		
1 Carpenter	45.85	366.80	70.75	566.00		
1 Air Compressor, 250 cfm		202.40		222.64		
1 Sheeting Driver		5.30		5.83		
2 -50' Air Hoses, 1.5"		11.60		12.76	5.48	6.03
40 L.H., Daily Totals		$1774.90		$2641.63	$44.37	$66.04
Crew B-32	**Hr.**	**Daily**	**Hr.**	**Daily**	**Bare Costs**	**Incl. O&P**
1 Laborer	$36.65	$293.20	$56.55	$452.40	$45.84	$69.75
3 Equip. Oper. (medium)	48.90	1173.60	74.15	1779.60		
1 Grader, 30,000 Lbs.		707.00		777.70		
1 Tandem Roller, 10 Ton		242.00		266.20		
1 Dozer, 200 H.P.		1325.00		1457.50	71.06	78.17
32 L.H., Daily Totals		$3740.80		$4733.40	$116.90	$147.92
Crew B-32A	**Hr.**	**Daily**	**Hr.**	**Daily**	**Bare Costs**	**Incl. O&P**
1 Laborer	$36.65	$293.20	$56.55	$452.40	$44.82	$68.28
2 Equip. Oper. (medium)	48.90	782.40	74.15	1186.40		
1 Grader, 30,000 Lbs.		707.00		777.70		
1 Roller, Vibratory, 25 Ton		692.80		762.08	58.33	64.16
24 L.H., Daily Totals		$2475.40		$3178.58	$103.14	$132.44
Crew B-32B	**Hr.**	**Daily**	**Hr.**	**Daily**	**Bare Costs**	**Incl. O&P**
1 Laborer	$36.65	$293.20	$56.55	$452.40	$44.82	$68.28
2 Equip. Oper. (medium)	48.90	782.40	74.15	1186.40		
1 Dozer, 200 H.P.		1325.00		1457.50		
1 Roller, Vibratory, 25 Ton		692.80		762.08	84.08	92.48
24 L.H., Daily Totals		$3093.40		$3858.38	$128.89	$160.77
Crew B-32C	**Hr.**	**Daily**	**Hr.**	**Daily**	**Bare Costs**	**Incl. O&P**
1 Labor Foreman (outside)	$38.65	$309.20	$59.65	$477.20	$43.11	$65.87
2 Laborers	36.65	586.40	56.55	904.80		
3 Equip. Oper. (medium)	48.90	1173.60	74.15	1779.60		
1 Grader, 30,000 Lbs.		707.00		777.70		
1 Tandem Roller, 10 Ton		242.00		266.20		
1 Dozer, 200 H.P.		1325.00		1457.50	47.38	52.11
48 L.H., Daily Totals		$4343.20		$5663.00	$90.48	$117.98

Crew B-33A

Crew B-33A	Hr.	Daily	Hr.	Daily	Bare Costs	Incl. O&P
1 Equip. Oper. (medium)	$48.90	$391.20	$74.15	$593.20	$45.40	$69.12
.5 Laborer	36.65	146.60	56.55	226.20		
.25 Equip. Oper. (medium)	48.90	97.80	74.15	148.30		
1 Scraper, Towed, 7 C.Y.		111.60		122.76		
1.25 Dozers, 300 H.P.		2271.25		2498.38	170.20	187.22
14 L.H., Daily Totals		$3018.45		$3588.84	$215.60	$256.35

Crew B-33B

Crew B-33B	Hr.	Daily	Hr.	Daily	Bare Costs	Incl. O&P
1 Equip. Oper. (medium)	$48.90	$391.20	$74.15	$593.20	$45.40	$69.12
.5 Laborer	36.65	146.60	56.55	226.20		
.25 Equip. Oper. (medium)	48.90	97.80	74.15	148.30		
1 Scraper, Towed, 10 C.Y.		141.60		155.76		
1.25 Dozers, 300 H.P.		2271.25		2498.38	172.35	189.58
14 L.H., Daily Totals		$3048.45		$3621.84	$217.75	$258.70

Crew B-33C

Crew B-33C	Hr.	Daily	Hr.	Daily	Bare Costs	Incl. O&P
1 Equip. Oper. (medium)	$48.90	$391.20	$74.15	$593.20	$45.40	$69.12
.5 Laborer	36.65	146.60	56.55	226.20		
.25 Equip. Oper. (medium)	48.90	97.80	74.15	148.30		
1 Scraper, Towed, 15 C.Y.		158.80		174.68		
1.25 Dozers, 300 H.P.		2271.25		2498.38	173.57	190.93
14 L.H., Daily Totals		$3065.65		$3640.76	$218.97	$260.05

Crew B-33D

Crew B-33D	Hr.	Daily	Hr.	Daily	Bare Costs	Incl. O&P
1 Equip. Oper. (medium)	$48.90	$391.20	$74.15	$593.20	$45.40	$69.12
.5 Laborer	36.65	146.60	56.55	226.20		
.25 Equip. Oper. (medium)	48.90	97.80	74.15	148.30		
1 S.P. Scraper, 14 C.Y.		1829.00		2011.90		
.25 Dozer, 300 H.P.		454.25		499.68	163.09	179.40
14 L.H., Daily Totals		$2918.85		$3479.28	$208.49	$248.52

Crew B-33E

Crew B-33E	Hr.	Daily	Hr.	Daily	Bare Costs	Incl. O&P
1 Equip. Oper. (medium)	$48.90	$391.20	$74.15	$593.20	$45.40	$69.12
.5 Laborer	36.65	146.60	56.55	226.20		
.25 Equip. Oper. (medium)	48.90	97.80	74.15	148.30		
1 S.P. Scraper, 21 C.Y.		2599.00		2858.90		
.25 Dozer, 300 H.P.		454.25		499.68	218.09	239.90
14 L.H., Daily Totals		$3688.85		$4326.27	$263.49	$309.02

Crew B-33F

Crew B-33F	Hr.	Daily	Hr.	Daily	Bare Costs	Incl. O&P
1 Equip. Oper. (medium)	$48.90	$391.20	$74.15	$593.20	$45.40	$69.12
.5 Laborer	36.65	146.60	56.55	226.20		
.25 Equip. Oper. (medium)	48.90	97.80	74.15	148.30		
1 Elev. Scraper, 11 C.Y.		1180.00		1298.00		
.25 Dozer, 300 H.P.		454.25		499.68	116.73	128.41
14 L.H., Daily Totals		$2269.85		$2765.38	$162.13	$197.53

Crew B-33G

Crew B-33G	Hr.	Daily	Hr.	Daily	Bare Costs	Incl. O&P
1 Equip. Oper. (medium)	$48.90	$391.20	$74.15	$593.20	$45.40	$69.12
.5 Laborer	36.65	146.60	56.55	226.20		
.25 Equip. Oper. (medium)	48.90	97.80	74.15	148.30		
1 Elev. Scraper, 22 C.Y.		2443.00		2687.30		
.25 Dozer, 300 H.P.		454.25		499.68	206.95	227.64
14 L.H., Daily Totals		$3532.85		$4154.68	$252.35	$296.76

Crew B-33H

Crew B-33H	Hr.	Daily	Hr.	Daily	Bare Costs	Incl. O&P
.5 Laborer	$36.65	$146.60	$56.55	$226.20	$45.40	$69.12
1 Equipment Operator (med.)	48.90	391.20	74.15	593.20		
.25 Equipment Operator (med.)	48.90	97.80	74.15	148.30		
1 S.P. Scraper, 44 C.Y.		4492.00		4941.20		
.25 Dozer, 410 H.P.		602.25		662.48	363.88	400.26
14 L.H., Daily Totals		$5729.85		$6571.38	$409.27	$469.38

Crew B-33J

Crew B-33J	Hr.	Daily	Hr.	Daily	Bare Costs	Incl. O&P
1 Equipment Operator (med.)	$48.90	$391.20	$74.15	$593.20	$48.90	$74.15
1 S.P. Scraper, 14 C.Y.		1829.00		2011.90	228.63	251.49
8 L.H., Daily Totals		$2220.20		$2605.10	$277.52	$325.64

Crew B-33K

Crew B-33K	Hr.	Daily	Hr.	Daily	Bare Costs	Incl. O&P
1 Equipment Operator (med.)	$48.90	$391.20	$74.15	$593.20	$45.40	$69.12
.25 Equipment Operator (med.)	48.90	97.80	74.15	148.30		
.5 Laborer	36.65	146.60	56.55	226.20		
1 S.P. Scraper, 31 C.Y.		3566.00		3922.60		
.25 Dozer, 410 H.P.		602.25		662.48	297.73	327.51
14 L.H., Daily Totals		$4803.85		$5552.77	$343.13	$396.63

Crew B-34A

Crew B-34A	Hr.	Daily	Hr.	Daily	Bare Costs	Incl. O&P
1 Truck Driver (heavy)	$37.55	$300.40	$57.30	$458.40	$37.55	$57.30
1 Dump Truck, 8 C.Y., 220 H.P.		417.20		458.92	52.15	57.37
8 L.H., Daily Totals		$717.60		$917.32	$89.70	$114.67

Crew B-34B

Crew B-34B	Hr.	Daily	Hr.	Daily	Bare Costs	Incl. O&P
1 Truck Driver (heavy)	$37.55	$300.40	$57.30	$458.40	$37.55	$57.30
1 Dump Truck, 12 C.Y., 400 H.P.		692.20		761.42	86.53	95.18
8 L.H., Daily Totals		$992.60		$1219.82	$124.08	$152.48

Crew B-34C

Crew B-34C	Hr.	Daily	Hr.	Daily	Bare Costs	Incl. O&P
1 Truck Driver (heavy)	$37.55	$300.40	$57.30	$458.40	$37.55	$57.30
1 Truck Tractor, 6x4, 380 H.P.		612.80		674.08		
1 Dump Trailer, 16.5 C.Y.		126.20		138.82	92.38	101.61
8 L.H., Daily Totals		$1039.40		$1271.30	$129.93	$158.91

Crew B-34D

Crew B-34D	Hr.	Daily	Hr.	Daily	Bare Costs	Incl. O&P
1 Truck Driver (heavy)	$37.55	$300.40	$57.30	$458.40	$37.55	$57.30
1 Truck Tractor, 6x4, 380 H.P.		612.80		674.08		
1 Dump Trailer, 20 C.Y.		140.80		154.88	94.20	103.62
8 L.H., Daily Totals		$1054.00		$1287.36	$131.75	$160.92

Crew B-34E

Crew B-34E	Hr.	Daily	Hr.	Daily	Bare Costs	Incl. O&P
1 Truck Driver (heavy)	$37.55	$300.40	$57.30	$458.40	$37.55	$57.30
1 Dump Truck, Off Hwy., 25 Ton		1314.00		1445.40	164.25	180.68
8 L.H., Daily Totals		$1614.40		$1903.80	$201.80	$237.97

Crew B-34F

Crew B-34F	Hr.	Daily	Hr.	Daily	Bare Costs	Incl. O&P
1 Truck Driver (heavy)	$37.55	$300.40	$57.30	$458.40	$37.55	$57.30
1 Dump Truck, Off Hwy., 35 Ton		1470.00		1617.00	183.75	202.13
8 L.H., Daily Totals		$1770.40		$2075.40	$221.30	$259.43

Crew B-34G

Crew B-34G	Hr.	Daily	Hr.	Daily	Bare Costs	Incl. O&P
1 Truck Driver (heavy)	$37.55	$300.40	$57.30	$458.40	$37.55	$57.30
1 Dump Truck, Off Hwy., 50 Ton		1787.00		1965.70	223.38	245.71
8 L.H., Daily Totals		$2087.40		$2424.10	$260.93	$303.01

Crew B-34H

Crew No.	Bare Costs Hr.	Daily	Incl. Subs O&P Hr.	Daily	Bare Costs	Incl. O&P
1 Truck Driver (heavy)	$37.55	$300.40	$57.30	$458.40	$37.55	$57.30
1 Dump Truck, Off Hwy., 65 Ton		1797.00		1976.70	224.63	247.09
8 L.H., Daily Totals		$2097.40		$2435.10	$262.18	$304.39

Crew B-34I

Crew No.	Bare Costs Hr.	Daily	Incl. Subs O&P Hr.	Daily	Bare Costs	Incl. O&P
1 Truck Driver (heavy)	$37.55	$300.40	$57.30	$458.40	$37.55	$57.30
1 Dump Truck, 18 C.Y., 450 H.P.		867.60		954.36	108.45	119.30
8 L.H., Daily Totals		$1168.00		$1412.76	$146.00	$176.60

Crew B-34J

Crew No.	Bare Costs Hr.	Daily	Incl. Subs O&P Hr.	Daily	Bare Costs	Incl. O&P
1 Truck Driver (heavy)	$37.55	$300.40	$57.30	$458.40	$37.55	$57.30
1 Dump Truck, Off Hwy., 100 Ton		2898.00		3187.80	362.25	398.48
8 L.H., Daily Totals		$3198.40		$3646.20	$399.80	$455.77

Crew B-34K

Crew No.	Bare Costs Hr.	Daily	Incl. Subs O&P Hr.	Daily	Bare Costs	Incl. O&P
1 Truck Driver (heavy)	$37.55	$300.40	$57.30	$458.40	$37.55	$57.30
1 Truck Tractor, 6x4, 450 H.P.		742.40		816.64		
1 Lowbed Trailer, 75 Ton		219.80		241.78	120.28	132.30
8 L.H., Daily Totals		$1262.60		$1516.82	$157.82	$189.60

Crew B-34L

Crew No.	Bare Costs Hr.	Daily	Incl. Subs O&P Hr.	Daily	Bare Costs	Incl. O&P
1 Equip. Oper. (light)	$47.05	$376.40	$71.35	$570.80	$47.05	$71.35
1 Flatbed Truck, Gas, 1.5 Ton		270.20		297.22	33.77	37.15
8 L.H., Daily Totals		$646.60		$868.02	$80.83	$108.50

Crew B-34M

Crew No.	Bare Costs Hr.	Daily	Incl. Subs O&P Hr.	Daily	Bare Costs	Incl. O&P
1 Equip. Oper. (light)	$47.05	$376.40	$71.35	$570.80	$47.05	$71.35
1 Flatbed Truck, Gas, 3 Ton		333.00		366.30	41.63	45.79
8 L.H., Daily Totals		$709.40		$937.10	$88.67	$117.14

Crew B-34N

Crew No.	Bare Costs Hr.	Daily	Incl. Subs O&P Hr.	Daily	Bare Costs	Incl. O&P
1 Truck Driver (heavy)	$37.55	$300.40	$57.30	$458.40	$37.55	$57.30
1 Dump Truck, 8 C.Y., 220 H.P.		417.20		458.92		
1 Flatbed Trailer, 40 Ton		152.00		167.20	71.15	78.27
8 L.H., Daily Totals		$869.60		$1084.52	$108.70	$135.57

Crew B-34P

Crew No.	Bare Costs Hr.	Daily	Incl. Subs O&P Hr.	Daily	Bare Costs	Incl. O&P
1 Pipe Fitter	$58.50	$468.00	$88.35	$706.80	$47.97	$72.73
1 Truck Driver (light)	36.50	292.00	55.70	445.60		
1 Equip. Oper. (medium)	48.90	391.20	74.15	593.20		
1 Flatbed Truck, Gas, 3 Ton		333.00		366.30		
1 Backhoe Loader, 48 H.P.		365.20		401.72	29.09	32.00
24 L.H., Daily Totals		$1849.40		$2513.62	$77.06	$104.73

Crew B-34Q

Crew No.	Bare Costs Hr.	Daily	Incl. Subs O&P Hr.	Daily	Bare Costs	Incl. O&P
1 Pipe Fitter	$58.50	$468.00	$88.35	$706.80	$48.42	$73.42
1 Truck Driver (light)	36.50	292.00	55.70	445.60		
1 Equip. Oper. (crane)	50.25	402.00	76.20	609.60		
1 Flatbed Trailer, 25 Ton		113.60		124.96		
1 Dump Truck, 8 C.Y., 220 H.P.		417.20		458.92		
1 Hyd. Crane, 25 Ton		745.80		820.38	53.19	58.51
24 L.H., Daily Totals		$2438.60		$3166.26	$101.61	$131.93

Crew B-34R

Crew No.	Bare Costs Hr.	Daily	Incl. Subs O&P Hr.	Daily	Bare Costs	Incl. O&P
1 Pipe Fitter	$58.50	$468.00	$88.35	$706.80	$48.42	$73.42
1 Truck Driver (light)	36.50	292.00	55.70	445.60		
1 Equip. Oper. (crane)	50.25	402.00	76.20	609.60		
1 Flatbed Trailer, 25 Ton		113.60		124.96		
1 Dump Truck, 8 C.Y., 220 H.P.		417.20		458.92		
1 Hyd. Crane, 25 Ton		745.80		820.38		
1 Hyd. Excavator, 1 C.Y.		814.80		896.28	87.14	95.86
24 L.H., Daily Totals		$3253.40		$4062.54	$135.56	$169.27

Crew B-34S

Crew No.	Bare Costs Hr.	Daily	Incl. Subs O&P Hr.	Daily	Bare Costs	Incl. O&P
2 Pipe Fitters	$58.50	$936.00	$88.35	$1413.60	$51.20	$77.55
1 Truck Driver (heavy)	37.55	300.40	57.30	458.40		
1 Equip. Oper. (crane)	50.25	402.00	76.20	609.60		
1 Flatbed Trailer, 40 Ton		152.00		167.20		
1 Truck Tractor, 6x4, 380 H.P.		612.80		674.08		
1 Hyd. Crane, 80 Ton		1652.00		1817.20		
1 Hyd. Excavator, 2 C.Y.		1175.00		1292.50	112.24	123.47
32 L.H., Daily Totals		$5230.20		$6432.58	$163.44	$201.02

Crew B-34T

Crew No.	Bare Costs Hr.	Daily	Incl. Subs O&P Hr.	Daily	Bare Costs	Incl. O&P
2 Pipe Fitters	$58.50	$936.00	$88.35	$1413.60	$51.20	$77.55
1 Truck Driver (heavy)	37.55	300.40	57.30	458.40		
1 Equip. Oper. (crane)	50.25	402.00	76.20	609.60		
1 Flatbed Trailer, 40 Ton		152.00		167.20		
1 Truck Tractor, 6x4, 380 H.P.		612.80		674.08		
1 Hyd. Crane, 80 Ton		1652.00		1817.20	75.53	83.08
32 L.H., Daily Totals		$4055.20		$5140.08	$126.72	$160.63

Crew B-35

Crew No.	Bare Costs Hr.	Daily	Incl. Subs O&P Hr.	Daily	Bare Costs	Incl. O&P
1 Labor Foreman (outside)	$38.65	$309.20	$59.65	$477.20	$45.66	$69.76
1 Skilled Worker	47.30	378.40	73.25	586.00		
1 Welder (plumber)	57.55	460.40	86.90	695.20		
1 Laborer	36.65	293.20	56.55	452.40		
1 Equip. Oper. (crane)	50.25	402.00	76.20	609.60		
1 Equip. Oper. (oiler)	43.55	348.40	66.00	528.00		
1 Welder, Electric, 300 amp		55.30		60.83		
1 Hyd. Excavator, .75 C.Y.		662.00		728.20	14.94	16.44
48 L.H., Daily Totals		$2908.90		$4137.43	$60.60	$86.20

Crew B-35A

Crew No.	Bare Costs Hr.	Daily	Incl. Subs O&P Hr.	Daily	Bare Costs	Incl. O&P
1 Labor Foreman (outside)	$38.65	$309.20	$59.65	$477.20	$44.37	$67.87
2 Laborers	36.65	586.40	56.55	904.80		
1 Skilled Worker	47.30	378.40	73.25	586.00		
1 Welder (plumber)	57.55	460.40	86.90	695.20		
1 Equip. Oper. (crane)	50.25	402.00	76.20	609.60		
1 Equip. Oper. (oiler)	43.55	348.40	66.00	528.00		
1 Welder, Gas Engine, 300 amp		142.00		156.20		
1 Crawler Crane, 75 Ton		1492.00		1641.20	29.18	32.10
56 L.H., Daily Totals		$4118.80		$5598.20	$73.55	$99.97

Crew B-36

Crew No.	Bare Costs Hr.	Daily	Incl. Subs O&P Hr.	Daily	Bare Costs	Incl. O&P
1 Labor Foreman (outside)	$38.65	$309.20	$59.65	$477.20	$41.95	$64.21
2 Laborers	36.65	586.40	56.55	904.80		
2 Equip. Oper. (medium)	48.90	782.40	74.15	1186.40		
1 Dozer, 200 H.P.		1325.00		1457.50		
1 Aggregate Spreader		41.20		45.32		
1 Tandem Roller, 10 Ton		242.00		266.20	40.20	44.23
40 L.H., Daily Totals		$3286.20		$4337.42	$82.16	$108.44

Crews

Crew B-36A

Crew No.	Bare Costs Hr.	Daily	Incl. Subs O&P Hr.	Daily	Cost Per Labor-Hour Bare Costs	Incl. O&P
1 Labor Foreman (outside)	$38.65	$309.20	$59.65	$477.20	$43.94	$67.05
2 Laborers	36.65	586.40	56.55	904.80		
4 Equip. Oper. (medium)	48.90	1564.80	74.15	2372.80		
1 Dozer, 200 H.P.		1325.00		1457.50		
1 Aggregate Spreader		41.20		45.32		
1 Tandem Roller, 10 Ton		242.00		266.20		
1 Roller, Pneum. Whl., 12 Ton		349.00		383.90	34.95	38.45
56 L.H., Daily Totals		$4417.60		$5907.72	$78.89	$105.50

Crew B-36B

Crew No.	Bare Costs Hr.	Daily	Incl. Subs O&P Hr.	Daily	Cost Per Labor-Hour Bare Costs	Incl. O&P
1 Labor Foreman (outside)	$38.65	$309.20	$59.65	$477.20	$43.14	$65.83
2 Laborers	36.65	586.40	56.55	904.80		
4 Equip. Oper. (medium)	48.90	1564.80	74.15	2372.80		
1 Truck Driver (heavy)	37.55	300.40	57.30	458.40		
1 Grader, 30,000 Lbs.		707.00		777.70		
1 F.E. Loader, Crl. 1.5 C.Y.		625.20		687.72		
1 Dozer, 300 H.P.		1817.00		1998.70		
1 Roller, Vibratory, 25 Ton		692.80		762.08		
1 Truck Tractor, 6x4, 450 H.P.		742.40		816.64		
1 Water Tank Trailer, 5000 Gal.		141.60		155.76	73.84	81.23
64 L.H., Daily Totals		$7486.80		$9411.80	$116.98	$147.06

Crew B-36C

Crew No.	Bare Costs Hr.	Daily	Incl. Subs O&P Hr.	Daily	Cost Per Labor-Hour Bare Costs	Incl. O&P
1 Labor Foreman (outside)	$38.65	$309.20	$59.65	$477.20	$44.58	$67.88
3 Equip. Oper. (medium)	48.90	1173.60	74.15	1779.60		
1 Truck Driver (heavy)	37.55	300.40	57.30	458.40		
1 Grader, 30,000 Lbs.		707.00		777.70		
1 Dozer, 300 H.P.		1817.00		1998.70		
1 Roller, Vibratory, 25 Ton		692.80		762.08		
1 Truck Tractor, 6x4, 450 H.P.		742.40		816.64		
1 Water Tank Trailer, 5000 Gal.		141.60		155.76	102.52	112.77
40 L.H., Daily Totals		$5884.00		$7226.08	$147.10	$180.65

Crew B-36D

Crew No.	Bare Costs Hr.	Daily	Incl. Subs O&P Hr.	Daily	Cost Per Labor-Hour Bare Costs	Incl. O&P
1 Labor Foreman (outside)	$38.65	$309.20	$59.65	$477.20	$46.34	$70.53
3 Equip. Oper. (medium)	48.90	1173.60	74.15	1779.60		
1 Grader, 30,000 Lbs.		707.00		777.70		
1 Dozer, 300 H.P.		1817.00		1998.70		
1 Roller, Vibratory, 25 Ton		692.80		762.08	100.53	110.58
32 L.H., Daily Totals		$4699.60		$5795.28	$146.86	$181.10

Crew B-37

Crew No.	Bare Costs Hr.	Daily	Incl. Subs O&P Hr.	Daily	Cost Per Labor-Hour Bare Costs	Incl. O&P
1 Labor Foreman (outside)	$38.65	$309.20	$59.65	$477.20	$38.72	$59.53
4 Laborers	36.65	1172.80	56.55	1809.60		
1 Equip. Oper. (light)	47.05	376.40	71.35	570.80		
1 Tandem Roller, 5 Ton		159.00		174.90	3.31	3.64
48 L.H., Daily Totals		$2017.40		$3032.50	$42.03	$63.18

Crew B-37A

Crew No.	Bare Costs Hr.	Daily	Incl. Subs O&P Hr.	Daily	Cost Per Labor-Hour Bare Costs	Incl. O&P
2 Laborers	$36.65	$586.40	$56.55	$904.80	$36.60	$56.27
1 Truck Driver (light)	36.50	292.00	55.70	445.60		
1 Flatbed Truck, Gas, 1.5 Ton		270.20		297.22		
1 Tar Kettle, T.M.		104.60		115.06	15.62	17.18
24 L.H., Daily Totals		$1253.20		$1762.68	$52.22	$73.44

Crew B-37B

Crew No.	Bare Costs Hr.	Daily	Incl. Subs O&P Hr.	Daily	Cost Per Labor-Hour Bare Costs	Incl. O&P
3 Laborers	$36.65	$879.60	$56.55	$1357.20	$36.61	$56.34
1 Truck Driver (light)	36.50	292.00	55.70	445.60		
1 Flatbed Truck, Gas, 1.5 Ton		270.20		297.22		
1 Tar Kettle, T.M.		104.60		115.06	11.71	12.88
32 L.H., Daily Totals		$1546.40		$2215.08	$48.33	$69.22

Crew B-37C

Crew No.	Bare Costs Hr.	Daily	Incl. Subs O&P Hr.	Daily	Cost Per Labor-Hour Bare Costs	Incl. O&P
2 Laborers	$36.65	$586.40	$56.55	$904.80	$36.58	$56.13
2 Truck Drivers (light)	36.50	584.00	55.70	891.20		
2 Flatbed Trucks, Gas, 1.5 Ton		540.40		594.44		
1 Tar Kettle, T.M.		104.60		115.06	20.16	22.17
32 L.H., Daily Totals		$1815.40		$2505.50	$56.73	$78.30

Crew B-37D

Crew No.	Bare Costs Hr.	Daily	Incl. Subs O&P Hr.	Daily	Cost Per Labor-Hour Bare Costs	Incl. O&P
1 Laborer	$36.65	$293.20	$56.55	$452.40	$36.58	$56.13
1 Truck Driver (light)	36.50	292.00	55.70	445.60		
1 Pickup Truck, 3/4 Ton		157.00		172.70	9.81	10.79
16 L.H., Daily Totals		$742.20		$1070.70	$46.39	$66.92

Crew B-37E

Crew No.	Bare Costs Hr.	Daily	Incl. Subs O&P Hr.	Daily	Cost Per Labor-Hour Bare Costs	Incl. O&P
3 Laborers	$36.65	$879.60	$56.55	$1357.20	$39.84	$60.94
1 Equip. Oper. (light)	47.05	376.40	71.35	570.80		
1 Equip. Oper. (medium)	48.90	391.20	74.15	593.20		
2 Truck Drivers (light)	36.50	584.00	55.70	891.20		
4 Barrels w/ Flasher		13.60		14.96		
1 Concrete Saw		175.20		192.72		
1 Rotary Hammer Drill		24.00		26.40		
1 Hammer Drill Bit		2.35		2.59		
1 Loader, Skid Steer, 30 H.P.		174.20		191.62		
1 Conc. Hammer Attach.		114.70		126.17		
1 Vibrating Plate, Gas, 18"		35.80		39.38		
2 Flatbed Trucks, Gas, 1.5 Ton		540.40		594.44	19.29	21.22
56 L.H., Daily Totals		$3311.45		$4600.68	$59.13	$82.15

Crew B-37F

Crew No.	Bare Costs Hr.	Daily	Incl. Subs O&P Hr.	Daily	Cost Per Labor-Hour Bare Costs	Incl. O&P
3 Laborers	$36.65	$879.60	$56.55	$1357.20	$36.61	$56.34
1 Truck Driver (light)	36.50	292.00	55.70	445.60		
4 Barrels w/ Flasher		13.60		14.96		
1 Concrete Mixer, 10 C.F.		176.80		194.48		
1 Air Compressor, 60 cfm		137.40		151.14		
1 -50' Air Hose, 3/4"		3.25		3.58		
1 Spade (Chipper)		7.80		8.58		
1 Flatbed Truck, Gas, 1.5 Ton		270.20		297.22	19.03	20.94
32 L.H., Daily Totals		$1780.65		$2472.76	$55.65	$77.27

Crew B-37G

Crew No.	Bare Costs Hr.	Daily	Incl. Subs O&P Hr.	Daily	Cost Per Labor-Hour Bare Costs	Incl. O&P
1 Labor Foreman (outside)	$38.65	$309.20	$59.65	$477.20	$38.72	$59.53
4 Laborers	36.65	1172.80	56.55	1809.60		
1 Equip. Oper. (light)	47.05	376.40	71.35	570.80		
1 Berm Machine		294.40		323.84		
1 Tandem Roller, 5 Ton		159.00		174.90	9.45	10.39
48 L.H., Daily Totals		$2311.80		$3356.34	$48.16	$69.92

Crew B-37H

Crew No.	Bare Costs Hr.	Daily	Incl. Subs O&P Hr.	Daily	Cost Per Labor-Hour Bare Costs	Incl. O&P
1 Labor Foreman (outside)	$38.65	$309.20	$59.65	$477.20	$38.72	$59.53
4 Laborers	36.65	1172.80	56.55	1809.60		
1 Equip. Oper. (light)	47.05	376.40	71.35	570.80		
1 Tandem Roller, 5 Ton		159.00		174.90		
1 Flatbed Trucks, Gas, 1.5 Ton		270.20		297.22		
1 Tar Kettle, T.M.		104.60		115.06	11.12	12.23
48 L.H., Daily Totals		$2392.20		$3444.78	$49.84	$71.77

Crew No.	Bare Costs Hr.	Daily	Incl. Subs O&P Hr.	Daily	Cost Per Labor-Hour Bare Costs	Incl. O&P
Crew B-37I	Hr.	Daily	Hr.	Daily	Bare Costs	Incl. O&P
3 Laborers	$36.65	$879.60	$56.55	$1357.20	$39.84	$60.94
1 Equip. Oper. (light)	47.05	376.40	71.35	570.80		
1 Equip. Oper. (medium)	48.90	391.20	74.15	593.20		
2 Truck Drivers (light)	36.50	584.00	55.70	891.20		
4 Barrels w/ Flasher		13.60		14.96		
1 Concrete Saw		175.20		192.72		
1 Rotary Hammer Drill		24.00		26.40		
1 Hammer Drill Bit		2.35		2.59		
1 Air Compressor, 60 cfm		137.40		151.14		
1 -50' Air Hose, 3/4"		3.25		3.58		
1 Spade (Chipper)		7.80		8.58		
1 Loader, Skid Steer, 30 H.P.		174.20		191.62		
1 Conc. Hammer Attach.		114.70		126.17		
1 Concrete Mixer, 10 C.F.		176.80		194.48		
1 Vibrating Plate, Gas, 18"		35.80		39.38		
2 Flatbed Trucks, Gas, 1.5 Ton		540.40		594.44	25.10	27.61
56 L.H., Daily Totals		$3636.70		$4958.45	$64.94	$88.54

Crew B-37J	Hr.	Daily	Hr.	Daily	Bare Costs	Incl. O&P
1 Labor Foreman (outside)	$38.65	$309.20	$59.65	$477.20	$38.72	$59.53
4 Laborers	36.65	1172.80	56.55	1809.60		
1 Equip. Oper. (light)	47.05	376.40	71.35	570.80		
1 Air Compressor, 60 cfm		137.40		151.14		
1 -50' Air Hose, 3/4"		3.25		3.58		
2 Concrete Mixer, 10 C.F.		353.60		388.96		
2 Flatbed Trucks, Gas, 1.5 Ton		540.40		594.44		
1 Shot Blaster, 20"		214.40		235.84	26.02	28.62
48 L.H., Daily Totals		$3107.45		$4231.56	$64.74	$88.16

Crew B-37K	Hr.	Daily	Hr.	Daily	Bare Costs	Incl. O&P
1 Labor Foreman (outside)	$38.65	$309.20	$59.65	$477.20	$38.72	$59.53
4 Laborers	36.65	1172.80	56.55	1809.60		
1 Equip. Oper. (light)	47.05	376.40	71.35	570.80		
1 Air Compressor, 60 cfm		137.40		151.14		
1 -50' Air Hose, 3/4"		3.25		3.58		
2 Flatbed Trucks, Gas, 1.5 Ton		540.40		594.44		
1 Shot Blaster, 20"		214.40		235.84	18.66	20.52
48 L.H., Daily Totals		$2753.85		$3842.59	$57.37	$80.05

Crew B-38	Hr.	Daily	Hr.	Daily	Bare Costs	Incl. O&P
1 Labor Foreman (outside)	$38.65	$309.20	$59.65	$477.20	$41.58	$63.65
2 Laborers	36.65	586.40	56.55	904.80		
1 Equip. Oper. (light)	47.05	376.40	71.35	570.80		
1 Equip. Oper. (medium)	48.90	391.20	74.15	593.20		
1 Backhoe Loader, 48 H.P.		365.20		401.72		
1 Hyd. Hammer, (1200 lb.)		180.40		198.44		
1 F.E. Loader, W.M., 4 C.Y.		673.80		741.18		
1 Pvmt. Rem. Bucket		60.80		66.88	32.01	35.21
40 L.H., Daily Totals		$2943.40		$3954.22	$73.58	$98.86

Crew B-39	Hr.	Daily	Hr.	Daily	Bare Costs	Incl. O&P
1 Labor Foreman (outside)	$38.65	$309.20	$59.65	$477.20	$38.72	$59.53
4 Laborers	36.65	1172.80	56.55	1809.60		
1 Equip. Oper. (light)	47.05	376.40	71.35	570.80		
1 Air Compressor, 250 cfm		202.40		222.64		
2 Breakers, Pavement, 60 lb.		19.60		21.56		
2 -50' Air Hoses, 1.5"		11.60		12.76	4.87	5.35
48 L.H., Daily Totals		$2092.00		$3114.56	$43.58	$64.89

Crew B-40	Hr.	Daily	Hr.	Daily	Bare Costs	Incl. O&P
1 Pile Driver Foreman (outside)	$46.40	$371.20	$73.60	$588.80	$46.01	$71.70
4 Pile Drivers	44.40	1420.80	70.40	2252.80		
2 Equip. Oper. (crane)	50.25	804.00	76.20	1219.20		
1 Equip. Oper. (oiler)	43.55	348.40	66.00	528.00		
1 Crawler Crane, 40 Ton		1179.00		1296.90		
1 Vibratory Hammer & Gen.		2640.00		2904.00	59.67	65.64
64 L.H., Daily Totals		$6763.40		$8789.70	$105.68	$137.34

Crew B-40B	Hr.	Daily	Hr.	Daily	Bare Costs	Incl. O&P
1 Labor Foreman (outside)	$38.65	$309.20	$59.65	$477.20	$40.40	$61.92
3 Laborers	36.65	879.60	56.55	1357.20		
1 Equip. Oper. (crane)	50.25	402.00	76.20	609.60		
1 Equip. Oper. (oiler)	43.55	348.40	66.00	528.00		
1 Lattice Boom Crane, 40 Ton		1177.00		1294.70	24.52	26.97
48 L.H., Daily Totals		$3116.20		$4266.70	$64.92	$88.89

Crew B-41	Hr.	Daily	Hr.	Daily	Bare Costs	Incl. O&P
1 Labor Foreman (outside)	$38.65	$309.20	$59.65	$477.20	$37.95	$58.44
4 Laborers	36.65	1172.80	56.55	1809.60		
.25 Equip. Oper. (crane)	50.25	100.50	76.20	152.40		
.25 Equip. Oper. (oiler)	43.55	87.10	66.00	132.00		
.25 Crawler Crane, 40 Ton		294.75		324.23	6.70	7.37
44 L.H., Daily Totals		$1964.35		$2895.43	$44.64	$65.81

Crew B-42	Hr.	Daily	Hr.	Daily	Bare Costs	Incl. O&P
1 Labor Foreman (outside)	$38.65	$309.20	$59.65	$477.20	$41.27	$64.78
4 Laborers	36.65	1172.80	56.55	1809.60		
1 Equip. Oper. (crane)	50.25	402.00	76.20	609.60		
1 Equip. Oper. (oiler)	43.55	348.40	66.00	528.00		
1 Welder	51.10	408.80	90.20	721.60		
1 Hyd. Crane, 25 Ton		745.80		820.38		
1 Welder, Gas Engine, 300 amp		142.00		156.20		
1 Horz. Boring Csg. Mch.		479.80		527.78	21.37	23.51
64 L.H., Daily Totals		$4008.80		$5650.36	$62.64	$88.29

Crew B-43	Hr.	Daily	Hr.	Daily	Bare Costs	Incl. O&P
1 Labor Foreman (outside)	$38.65	$309.20	$59.65	$477.20	$40.40	$61.92
3 Laborers	36.65	879.60	56.55	1357.20		
1 Equip. Oper. (crane)	50.25	402.00	76.20	609.60		
1 Equip. Oper. (oiler)	43.55	348.40	66.00	528.00		
1 Drill Rig, Truck-Mounted		2551.00		2806.10	53.15	58.46
48 L.H., Daily Totals		$4490.20		$5778.10	$93.55	$120.38

Crew B-44	Hr.	Daily	Hr.	Daily	Bare Costs	Incl. O&P
1 Pile Driver Foreman (outside)	$46.40	$371.20	$73.60	$588.80	$45.14	$70.52
4 Pile Drivers	44.40	1420.80	70.40	2252.80		
2 Equip. Oper. (crane)	50.25	804.00	76.20	1219.20		
1 Laborer	36.65	293.20	56.55	452.40		
1 Crawler Crane, 40 Ton		1179.00		1296.90		
1 Lead, 60' High		74.80		82.28		
1 Hammer, Diesel, 15K ft.-lbs.		587.60		646.36	28.77	31.65
64 L.H., Daily Totals		$4730.60		$6538.74	$73.92	$102.17

Crew B-45	Hr.	Daily	Hr.	Daily	Bare Costs	Incl. O&P
1 Equip. Oper. (medium)	$48.90	$391.20	$74.15	$593.20	$43.23	$65.72
1 Truck Driver (heavy)	37.55	300.40	57.30	458.40		
1 Dist. Tanker, 3000 Gallon		308.20		339.02		
1 Truck Tractor, 6x4, 380 H.P.		612.80		674.08	57.56	63.32
16 L.H., Daily Totals		$1612.60		$2064.70	$100.79	$129.04

Crew No.	Bare Costs		Incl. Subs O&P		Cost Per Labor-Hour	
Crew B-46	Hr.	Daily	Hr.	Daily	Bare Costs	Incl. O&P
1 Pile Driver Foreman (outside)	$46.40	$371.20	$73.60	$588.80	$40.86	$64.01
2 Pile Drivers	44.40	710.40	70.40	1126.40		
3 Laborers	36.65	879.60	56.55	1357.20		
1 Chain Saw, Gas, 36" Long		45.00		49.50	.94	1.03
48 L.H., Daily Totals		$2006.20		$3121.90	$41.80	$65.04
Crew B-47	Hr.	Daily	Hr.	Daily	Bare Costs	Incl. O&P
1 Blast Foreman (outside)	$38.65	$309.20	$59.65	$477.20	$40.78	$62.52
1 Driller	36.65	293.20	56.55	452.40		
1 Equip. Oper. (light)	47.05	376.40	71.35	570.80		
1 Air Track Drill, 4"		1040.00		1144.00		
1 Air Compressor, 600 cfm		552.60		607.86		
2 -50' Air Hoses, 3"		29.80		32.78	67.60	74.36
24 L.H., Daily Totals		$2601.20		$3285.04	$108.38	$136.88
Crew B-47A	Hr.	Daily	Hr.	Daily	Bare Costs	Incl. O&P
1 Drilling Foreman (outside)	$38.65	$309.20	$59.65	$477.20	$44.15	$67.28
1 Equip. Oper. (heavy)	50.25	402.00	76.20	609.60		
1 Equip. Oper. (oiler)	43.55	348.40	66.00	528.00		
1 Air Track Drill, 5"		1255.00		1380.50	52.29	57.52
24 L.H., Daily Totals		$2314.60		$2995.30	$96.44	$124.80
Crew B-47C	Hr.	Daily	Hr.	Daily	Bare Costs	Incl. O&P
1 Laborer	$36.65	$293.20	$56.55	$452.40	$41.85	$63.95
1 Equip. Oper. (light)	47.05	376.40	71.35	570.80		
1 Air Compressor, 750 cfm		559.20		615.12		
2 -50' Air Hoses, 3"		29.80		32.78		
1 Air Track Drill, 4"		1040.00		1144.00	101.81	111.99
16 L.H., Daily Totals		$2298.60		$2815.10	$143.66	$175.94
Crew B-47E	Hr.	Daily	Hr.	Daily	Bare Costs	Incl. O&P
1 Labor Foreman (outside)	$38.65	$309.20	$59.65	$477.20	$37.15	$57.33
3 Laborers	36.65	879.60	56.55	1357.20		
1 Flatbed Truck, Gas, 3 Ton		333.00		366.30	10.41	11.45
32 L.H., Daily Totals		$1521.80		$2200.70	$47.56	$68.77
Crew B-47G	Hr.	Daily	Hr.	Daily	Bare Costs	Incl. O&P
1 Labor Foreman (outside)	$38.65	$309.20	$59.65	$477.20	$39.75	$61.02
2 Laborers	36.65	586.40	56.55	904.80		
1 Equip. Oper. (light)	47.05	376.40	71.35	570.80		
1 Air Track Drill, 4"		1040.00		1144.00		
1 Air Compressor, 600 cfm		552.60		607.86		
2 -50' Air Hoses, 3"		29.80		32.78		
1 Gunite Pump Rig		370.40		407.44	62.27	68.50
32 L.H., Daily Totals		$3264.80		$4144.88	$102.03	$129.53
Crew B-47H	Hr.	Daily	Hr.	Daily	Bare Costs	Incl. O&P
1 Skilled Worker Foreman (out)	$49.30	$394.40	$76.35	$610.80	$47.80	$74.03
3 Skilled Workers	47.30	1135.20	73.25	1758.00		
1 Flatbed Truck, Gas, 3 Ton		333.00		366.30	10.41	11.45
32 L.H., Daily Totals		$1862.60		$2735.10	$58.21	$85.47

Crew No.	Bare Costs		Incl. Subs O&P		Cost Per Labor-Hour	
Crew B-48	Hr.	Daily	Hr.	Daily	Bare Costs	Incl. O&P
1 Labor Foreman (outside)	$38.65	$309.20	$59.65	$477.20	$41.35	$63.26
3 Laborers	36.65	879.60	56.55	1357.20		
1 Equip. Oper. (crane)	50.25	402.00	76.20	609.60		
1 Equip. Oper. (oiler)	43.55	348.40	66.00	528.00		
1 Equip. Oper. (light)	47.05	376.40	71.35	570.80		
1 Centr. Water Pump, 6"		340.80		374.88		
1 -20' Suction Hose, 6"		11.50		12.65		
1 -50' Discharge Hose, 6"		6.10		6.71		
1 Drill Rig, Truck-Mounted		2551.00		2806.10	51.95	57.15
56 L.H., Daily Totals		$5225.00		$6743.14	$93.30	$120.41
Crew B-49	Hr.	Daily	Hr.	Daily	Bare Costs	Incl. O&P
1 Labor Foreman (outside)	$38.65	$309.20	$59.65	$477.20	$42.91	$65.99
3 Laborers	36.65	879.60	56.55	1357.20		
2 Equip. Oper. (crane)	50.25	804.00	76.20	1219.20		
2 Equip. Oper. (oilers)	43.55	696.80	66.00	1056.00		
1 Equip. Oper. (light)	47.05	376.40	71.35	570.80		
2 Pile Drivers	44.40	710.40	70.40	1126.40		
1 Hyd. Crane, 25 Ton		745.80		820.38		
1 Centr. Water Pump, 6"		340.80		374.88		
1 -20' Suction Hose, 6"		11.50		12.65		
1 -50' Discharge Hose, 6"		6.10		6.71		
1 Drill Rig, Truck-Mounted		2551.00		2806.10	41.54	45.69
88 L.H., Daily Totals		$7431.60		$9827.52	$84.45	$111.68
Crew B-50	Hr.	Daily	Hr.	Daily	Bare Costs	Incl. O&P
2 Pile Driver Foremen (outside)	$46.40	$742.40	$73.60	$1177.60	$43.80	$68.40
6 Pile Drivers	44.40	2131.20	70.40	3379.20		
2 Equip. Oper. (crane)	50.25	804.00	76.20	1219.20		
1 Equip. Oper. (oiler)	43.55	348.40	66.00	528.00		
3 Laborers	36.65	879.60	56.55	1357.20		
1 Crawler Crane, 40 Ton		1179.00		1296.90		
1 Lead, 60' High		74.80		82.28		
1 Hammer, Diesel, 15K ft.-lbs.		587.60		646.36		
1 Air Compressor, 600 cfm		552.60		607.86		
2 -50' Air Hoses, 3"		29.80		32.78		
1 Chain Saw, Gas, 36" Long		45.00		49.50	22.04	24.25
112 L.H., Daily Totals		$7374.40		$10376.88	$65.84	$92.65
Crew B-51	Hr.	Daily	Hr.	Daily	Bare Costs	Incl. O&P
1 Labor Foreman (outside)	$38.65	$309.20	$59.65	$477.20	$36.96	$56.92
4 Laborers	36.65	1172.80	56.55	1809.60		
1 Truck Driver (light)	36.50	292.00	55.70	445.60		
1 Flatbed Truck, Gas, 1.5 Ton		270.20		297.22	5.63	6.19
48 L.H., Daily Totals		$2044.20		$3029.62	$42.59	$63.12
Crew B-52	Hr.	Daily	Hr.	Daily	Bare Costs	Incl. O&P
1 Carpenter Foreman (outside)	$47.85	$382.80	$73.85	$590.80	$42.50	$65.17
1 Carpenter	45.85	366.80	70.75	566.00		
3 Laborers	36.65	879.60	56.55	1357.20		
1 Cement Finisher	44.05	352.40	65.10	520.80		
.5 Rodman (reinf.)	50.65	202.60	79.55	318.20		
.5 Equip. Oper. (medium)	48.90	195.60	74.15	296.60		
.5 Crawler Loader, 3 C.Y.		595.50		655.05	10.63	11.70
56 L.H., Daily Totals		$2975.30		$4304.65	$53.13	$76.87
Crew B-53	Hr.	Daily	Hr.	Daily	Bare Costs	Incl. O&P
1 Equip. Oper. (light)	$47.05	$376.40	$71.35	$570.80	$47.05	$71.35
1 Trencher, Chain, 12 H.P.		71.20		78.32	8.90	9.79
8 L.H., Daily Totals		$447.60		$649.12	$55.95	$81.14

Crew No.	Bare Costs		Incl. Subs O&P		Cost Per Labor-Hour	

Crew B-54

	Hr.	Daily	Hr.	Daily	Bare Costs	Incl. O&P
1 Equip. Oper. (light)	$47.05	$376.40	$71.35	$570.80	$47.05	$71.35
1 Trencher, Chain, 40 H.P.		339.40		373.34	42.42	46.67
8 L.H., Daily Totals		$715.80		$944.14	$89.47	$118.02

Crew B-54A

	Hr.	Daily	Hr.	Daily	Bare Costs	Incl. O&P
.17 Labor Foreman (outside)	$38.65	$52.56	$59.65	$81.12	$47.41	$72.04
1 Equipment Operator (med.)	48.90	391.20	74.15	593.20		
1 Wheel Trencher, 67 H.P.		1206.00		1326.60	128.85	141.73
9.36 L.H., Daily Totals		$1649.76		$2000.92	$176.26	$213.77

Crew B-54B

	Hr.	Daily	Hr.	Daily	Bare Costs	Incl. O&P
.25 Labor Foreman (outside)	$38.65	$77.30	$59.65	$119.30	$46.85	$71.25
1 Equipment Operator (med.)	48.90	391.20	74.15	593.20		
1 Wheel Trencher, 150 H.P.		1898.00		2087.80	189.80	208.78
10 L.H., Daily Totals		$2366.50		$2800.30	$236.65	$280.03

Crew B-54C

	Hr.	Daily	Hr.	Daily	Bare Costs	Incl. O&P
1 Laborer	$36.65	$293.20	$56.55	$452.40	$42.77	$65.35
1 Equipment Operator (med.)	48.90	391.20	74.15	593.20		
1 Wheel Trencher, 67 H.P.		1206.00		1326.60	75.38	82.91
16 L.H., Daily Totals		$1890.40		$2372.20	$118.15	$148.26

Crew B-54D

	Hr.	Daily	Hr.	Daily	Bare Costs	Incl. O&P
1 Laborer	$36.65	$293.20	$56.55	$452.40	$42.77	$65.35
1 Equipment Operator (med.)	48.90	391.20	74.15	593.20		
1 Rock Trencher, 6" Width		369.00		405.90	23.06	25.37
16 L.H., Daily Totals		$1053.40		$1451.50	$65.84	$90.72

Crew B-54E

	Hr.	Daily	Hr.	Daily	Bare Costs	Incl. O&P
1 Laborer	$36.65	$293.20	$56.55	$452.40	$42.77	$65.35
1 Equipment Operator (med.)	48.90	391.20	74.15	593.20		
1 Rock Trencher, 18" Width		2590.00		2849.00	161.88	178.06
16 L.H., Daily Totals		$3274.40		$3894.60	$204.65	$243.41

Crew B-55

	Hr.	Daily	Hr.	Daily	Bare Costs	Incl. O&P
2 Laborers	$36.65	$586.40	$56.55	$904.80	$36.60	$56.27
1 Truck Driver (light)	36.50	292.00	55.70	445.60		
1 Truck-Mounted Earth Auger		786.40		865.04		
1 Flatbed Truck, Gas, 3 Ton		333.00		366.30	46.64	51.31
24 L.H., Daily Totals		$1997.80		$2581.74	$83.24	$107.57

Crew B-56

	Hr.	Daily	Hr.	Daily	Bare Costs	Incl. O&P
1 Laborer	$36.65	$293.20	$56.55	$452.40	$41.85	$63.95
1 Equip. Oper. (light)	47.05	376.40	71.35	570.80		
1 Air Track Drill, 4"		1040.00		1144.00		
1 Air Compressor, 600 cfm		552.60		607.86		
1 -50' Air Hose, 3"		14.90		16.39	100.47	110.52
16 L.H., Daily Totals		$2277.10		$2791.45	$142.32	$174.47

Crew B-57

	Hr.	Daily	Hr.	Daily	Bare Costs	Incl. O&P
1 Labor Foreman (outside)	$38.65	$309.20	$59.65	$477.20	$42.13	$64.38
2 Laborers	36.65	586.40	56.55	904.80		
1 Equip. Oper. (crane)	50.25	402.00	76.20	609.60		
1 Equip. Oper. (light)	47.05	376.40	71.35	570.80		
1 Equip. Oper. (oiler)	43.55	348.40	66.00	528.00		
1 Crawler Crane, 25 Ton		1173.00		1290.30		
1 Clamshell Bucket, 1 C.Y.		47.40		52.14		
1 Centr. Water Pump, 6"		340.80		374.88		
1 -20' Suction Hose, 6"		11.50		12.65		
20 -50' Discharge Hoses, 6"		122.00		134.20	35.31	38.84
48 L.H., Daily Totals		$3717.10		$4954.57	$77.44	$103.22

Crew B-58

	Hr.	Daily	Hr.	Daily	Bare Costs	Incl. O&P
2 Laborers	$36.65	$586.40	$56.55	$904.80	$40.12	$61.48
1 Equip. Oper. (light)	47.05	376.40	71.35	570.80		
1 Backhoe Loader, 48 H.P.		365.20		401.72		
1 Small Helicopter, w/ Pilot		2729.00		3001.90	128.93	141.82
24 L.H., Daily Totals		$4057.00		$4879.22	$169.04	$203.30

Crew B-59

	Hr.	Daily	Hr.	Daily	Bare Costs	Incl. O&P
1 Truck Driver (heavy)	$37.55	$300.40	$57.30	$458.40	$37.55	$57.30
1 Truck Tractor, 220 H.P.		366.40		403.04		
1 Water Tank Trailer, 5000 Gal.		141.60		155.76	63.50	69.85
8 L.H., Daily Totals		$808.40		$1017.20	$101.05	$127.15

Crew B-59A

	Hr.	Daily	Hr.	Daily	Bare Costs	Incl. O&P
2 Laborers	$36.65	$586.40	$56.55	$904.80	$36.95	$56.80
1 Truck Driver (heavy)	37.55	300.40	57.30	458.40		
1 Water Tank Trailer, 5000 Gal.		141.60		155.76		
1 Truck Tractor, 220 H.P.		366.40		403.04	21.17	23.28
24 L.H., Daily Totals		$1394.80		$1922.00	$58.12	$80.08

Crew B-60

	Hr.	Daily	Hr.	Daily	Bare Costs	Incl. O&P
1 Labor Foreman (outside)	$38.65	$309.20	$59.65	$477.20	$42.84	$65.38
2 Laborers	36.65	586.40	56.55	904.80		
1 Equip. Oper. (crane)	50.25	402.00	76.20	609.60		
2 Equip. Oper. (light)	47.05	752.80	71.35	1141.60		
1 Equip. Oper. (oiler)	43.55	348.40	66.00	528.00		
1 Crawler Crane, 40 Ton		1179.00		1296.90		
1 Lead, 60' High		74.80		82.28		
1 Hammer, Diesel, 15K ft.-lbs.		587.60		646.36		
1 Backhoe Loader, 48 H.P.		365.20		401.72	39.40	43.34
56 L.H., Daily Totals		$4605.40		$6088.46	$82.24	$108.72

Crew B-61

	Hr.	Daily	Hr.	Daily	Bare Costs	Incl. O&P
1 Labor Foreman (outside)	$38.65	$309.20	$59.65	$477.20	$39.13	$60.13
3 Laborers	36.65	879.60	56.55	1357.20		
1 Equip. Oper. (light)	47.05	376.40	71.35	570.80		
1 Cement Mixer, 2 C.Y.		191.40		210.54		
1 Air Compressor, 160 cfm		157.40		173.14	8.72	9.59
40 L.H., Daily Totals		$1914.00		$2788.88	$47.85	$69.72

Crew B-62

	Hr.	Daily	Hr.	Daily	Bare Costs	Incl. O&P
2 Laborers	$36.65	$586.40	$56.55	$904.80	$40.12	$61.48
1 Equip. Oper. (light)	47.05	376.40	71.35	570.80		
1 Loader, Skid Steer, 30 H.P.		174.20		191.62	7.26	7.98
24 L.H., Daily Totals		$1137.00		$1667.22	$47.38	$69.47

Crew B-63

Crew No.	Bare Costs Hr.	Bare Costs Daily	Incl. Subs O&P Hr.	Incl. Subs O&P Daily	Cost Per Labor-Hour Bare Costs	Cost Per Labor-Hour Incl. O&P
4 Laborers	$36.65	$1172.80	$56.55	$1809.60	$38.73	$59.51
1 Equip. Oper. (light)	47.05	376.40	71.35	570.80		
1 Loader, Skid Steer, 30 H.P.		174.20		191.62	4.36	4.79
40 L.H., Daily Totals		$1723.40		$2572.02	$43.09	$64.30

Crew B-63B

Crew No.	Bare Costs Hr.	Bare Costs Daily	Incl. Subs O&P Hr.	Incl. Subs O&P Daily	Cost Per Labor-Hour Bare Costs	Cost Per Labor-Hour Incl. O&P
1 Labor Foreman (inside)	$37.15	$297.20	$57.30	$458.40	$39.38	$60.44
2 Laborers	36.65	586.40	56.55	904.80		
1 Equip. Oper. (light)	47.05	376.40	71.35	570.80		
1 Loader, Skid Steer, 78 H.P.		308.80		339.68	9.65	10.62
32 L.H., Daily Totals		$1568.80		$2273.68	$49.02	$71.05

Crew B-64

Crew No.	Bare Costs Hr.	Bare Costs Daily	Incl. Subs O&P Hr.	Incl. Subs O&P Daily	Cost Per Labor-Hour Bare Costs	Cost Per Labor-Hour Incl. O&P
1 Laborer	$36.65	$293.20	$56.55	$452.40	$36.58	$56.13
1 Truck Driver (light)	36.50	292.00	55.70	445.60		
1 Power Mulcher (small)		159.60		175.56		
1 Flatbed Truck, Gas, 1.5 Ton		270.20		297.22	26.86	29.55
16 L.H., Daily Totals		$1015.00		$1370.78	$63.44	$85.67

Crew B-65

Crew No.	Bare Costs Hr.	Bare Costs Daily	Incl. Subs O&P Hr.	Incl. Subs O&P Daily	Cost Per Labor-Hour Bare Costs	Cost Per Labor-Hour Incl. O&P
1 Laborer	$36.65	$293.20	$56.55	$452.40	$36.58	$56.13
1 Truck Driver (light)	36.50	292.00	55.70	445.60		
1 Power Mulcher (Large)		325.60		358.16		
1 Flatbed Truck, Gas, 1.5 Ton		270.20		297.22	37.24	40.96
16 L.H., Daily Totals		$1181.00		$1553.38	$73.81	$97.09

Crew B-66

Crew No.	Bare Costs Hr.	Bare Costs Daily	Incl. Subs O&P Hr.	Incl. Subs O&P Daily	Cost Per Labor-Hour Bare Costs	Cost Per Labor-Hour Incl. O&P
1 Equip. Oper. (light)	$47.05	$376.40	$71.35	$570.80	$47.05	$71.35
1 Loader-Backhoe, 40 H.P.		259.60		285.56	32.45	35.70
8 L.H., Daily Totals		$636.00		$856.36	$79.50	$107.05

Crew B-67

Crew No.	Bare Costs Hr.	Bare Costs Daily	Incl. Subs O&P Hr.	Incl. Subs O&P Daily	Cost Per Labor-Hour Bare Costs	Cost Per Labor-Hour Incl. O&P
1 Millwright	$48.10	$384.80	$70.85	$566.80	$47.58	$71.10
1 Equip. Oper. (light)	47.05	376.40	71.35	570.80		
1 Forklift, R/T, 4,000 Lb.		314.20		345.62	19.64	21.60
16 L.H., Daily Totals		$1075.40		$1483.22	$67.21	$92.70

Crew B-67B

Crew No.	Bare Costs Hr.	Bare Costs Daily	Incl. Subs O&P Hr.	Incl. Subs O&P Daily	Cost Per Labor-Hour Bare Costs	Cost Per Labor-Hour Incl. O&P
1 Millwright Foreman (inside)	$48.60	$388.80	$71.60	$572.80	$48.35	$71.22
1 Millwright	48.10	384.80	70.85	566.80		
16 L.H., Daily Totals		$773.60		$1139.60	$48.35	$71.22

Crew B-68

Crew No.	Bare Costs Hr.	Bare Costs Daily	Incl. Subs O&P Hr.	Incl. Subs O&P Daily	Cost Per Labor-Hour Bare Costs	Cost Per Labor-Hour Incl. O&P
2 Millwrights	$48.10	$769.60	$70.85	$1133.60	$47.75	$71.02
1 Equip. Oper. (light)	47.05	376.40	71.35	570.80		
1 Forklift, R/T, 4,000 Lb.		314.20		345.62	13.09	14.40
24 L.H., Daily Totals		$1460.20		$2050.02	$60.84	$85.42

Crew B-68A

Crew No.	Bare Costs Hr.	Bare Costs Daily	Incl. Subs O&P Hr.	Incl. Subs O&P Daily	Cost Per Labor-Hour Bare Costs	Cost Per Labor-Hour Incl. O&P
1 Millwright Foreman (inside)	$48.60	$388.80	$71.60	$572.80	$48.27	$71.10
2 Millwrights	48.10	769.60	70.85	1133.60		
1 Forklift, 8,000 Lb.		180.00		198.00	7.50	8.25
24 L.H., Daily Totals		$1338.40		$1904.40	$55.77	$79.35

Crew B-68B

Crew No.	Bare Costs Hr.	Bare Costs Daily	Incl. Subs O&P Hr.	Incl. Subs O&P Daily	Cost Per Labor-Hour Bare Costs	Cost Per Labor-Hour Incl. O&P
1 Millwright Foreman (inside)	$48.60	$388.80	$71.60	$572.80	$52.37	$78.11
2 Millwrights	48.10	769.60	70.85	1133.60		
2 Electricians	53.35	853.60	79.85	1277.60		
2 Plumbers	57.55	920.80	86.90	1390.40		
1 Forklift, 5,000 Lb.		324.80		357.28	5.80	6.38
56 L.H., Daily Totals		$3257.60		$4731.68	$58.17	$84.49

Crew B-68C

Crew No.	Bare Costs Hr.	Bare Costs Daily	Incl. Subs O&P Hr.	Incl. Subs O&P Daily	Cost Per Labor-Hour Bare Costs	Cost Per Labor-Hour Incl. O&P
1 Millwright Foreman (inside)	$48.60	$388.80	$71.60	$572.80	$51.90	$77.30
1 Millwright	48.10	384.80	70.85	566.80		
1 Electrician	53.35	426.80	79.85	638.80		
1 Plumber	57.55	460.40	86.90	695.20		
1 Forklift, 5,000 Lb.		324.80		357.28	10.15	11.16
32 L.H., Daily Totals		$1985.60		$2830.88	$62.05	$88.47

Crew B-68D

Crew No.	Bare Costs Hr.	Bare Costs Daily	Incl. Subs O&P Hr.	Incl. Subs O&P Daily	Cost Per Labor-Hour Bare Costs	Cost Per Labor-Hour Incl. O&P
1 Labor Foreman (inside)	$37.15	$297.20	$57.30	$458.40	$40.28	$61.73
1 Laborer	36.65	293.20	56.55	452.40		
1 Equip. Oper. (light)	47.05	376.40	71.35	570.80		
1 Forklift, 5,000 Lb.		324.80		357.28	13.53	14.89
24 L.H., Daily Totals		$1291.60		$1838.88	$53.82	$76.62

Crew B-68E

Crew No.	Bare Costs Hr.	Bare Costs Daily	Incl. Subs O&P Hr.	Incl. Subs O&P Daily	Cost Per Labor-Hour Bare Costs	Cost Per Labor-Hour Incl. O&P
1 Struc. Steel Foreman (inside)	$51.60	$412.80	$91.05	$728.40	$51.20	$90.37
3 Struc. Steel Workers	51.10	1226.40	90.20	2164.80		
1 Welder	51.10	408.80	90.20	721.60		
1 Forklift, 8,000 Lb.		180.00		198.00	4.50	4.95
40 L.H., Daily Totals		$2228.00		$3812.80	$55.70	$95.32

Crew B-68F

Crew No.	Bare Costs Hr.	Bare Costs Daily	Incl. Subs O&P Hr.	Incl. Subs O&P Daily	Cost Per Labor-Hour Bare Costs	Cost Per Labor-Hour Incl. O&P
1 Skilled Worker Foreman (out)	$49.30	$394.40	$76.35	$610.80	$47.97	$74.28
2 Skilled Workers	47.30	756.80	73.25	1172.00		
1 Forklift, 5,000 Lb.		324.80		357.28	13.53	14.89
24 L.H., Daily Totals		$1476.00		$2140.08	$61.50	$89.17

Crew B-69

Crew No.	Bare Costs Hr.	Bare Costs Daily	Incl. Subs O&P Hr.	Incl. Subs O&P Daily	Cost Per Labor-Hour Bare Costs	Cost Per Labor-Hour Incl. O&P
1 Labor Foreman (outside)	$38.65	$309.20	$59.65	$477.20	$40.40	$61.92
3 Laborers	36.65	879.60	56.55	1357.20		
1 Equip. Oper. (crane)	50.25	402.00	76.20	609.60		
1 Equip. Oper. (oiler)	43.55	348.40	66.00	528.00		
1 Hyd. Crane, 80 Ton		1652.00		1817.20	34.42	37.86
48 L.H., Daily Totals		$3591.20		$4789.20	$74.82	$99.78

Crew B-69A

Crew No.	Bare Costs Hr.	Bare Costs Daily	Incl. Subs O&P Hr.	Incl. Subs O&P Daily	Cost Per Labor-Hour Bare Costs	Cost Per Labor-Hour Incl. O&P
1 Labor Foreman (outside)	$38.65	$309.20	$59.65	$477.20	$40.26	$61.42
3 Laborers	36.65	879.60	56.55	1357.20		
1 Equip. Oper. (medium)	48.90	391.20	74.15	593.20		
1 Concrete Finisher	44.05	352.40	65.10	520.80		
1 Curb/Gutter Paver, 2-Track		898.00		987.80	18.71	20.58
48 L.H., Daily Totals		$2830.40		$3936.20	$58.97	$82.00

Crew B-69B

Crew No.	Bare Costs Hr.	Bare Costs Daily	Incl. Subs O&P Hr.	Incl. Subs O&P Daily	Cost Per Labor-Hour Bare Costs	Cost Per Labor-Hour Incl. O&P
1 Labor Foreman (outside)	$38.65	$309.20	$59.65	$477.20	$40.26	$61.42
3 Laborers	36.65	879.60	56.55	1357.20		
1 Equip. Oper. (medium)	48.90	391.20	74.15	593.20		
1 Cement Finisher	44.05	352.40	65.10	520.80		
1 Curb/Gutter Paver, 4-Track		770.20		847.22	16.05	17.65
48 L.H., Daily Totals		$2702.60		$3795.62	$56.30	$79.08

Crew B-70

Crew B-70	Hr.	Daily	Hr.	Daily	Bare Costs	Incl. O&P
1 Labor Foreman (outside)	$38.65	$309.20	$59.65	$477.20	$42.19	$64.54
3 Laborers	36.65	879.60	56.55	1357.20		
3 Equip. Oper. (medium)	48.90	1173.60	74.15	1779.60		
1 Grader, 30,000 Lbs.		707.00		777.70		
1 Ripper, Beam & 1 Shank		81.40		89.54		
1 Road Sweeper, S.P., 8' wide		665.40		731.94		
1 F.E. Loader, W.M., 1.5 C.Y.		373.40		410.74	32.63	35.89
56 L.H., Daily Totals		$4189.60		$5623.92	$74.81	$100.43

Crew B-70A

Crew B-70A	Hr.	Daily	Hr.	Daily	Bare Costs	Incl. O&P
1 Laborer	$36.65	$293.20	$56.55	$452.40	$46.45	$70.63
4 Equip. Oper. (medium)	48.90	1564.80	74.15	2372.80		
1 Grader, 40,000 Lbs.		1120.00		1232.00		
1 F.E. Loader, W.M.,2.5 C.Y.		522.80		575.08		
1 Dozer, 80 H.P.		481.00		529.10		
1 Roller, Pneum. Whl., 12 Ton		349.00		383.90	61.82	68.00
40 L.H., Daily Totals		$4330.80		$5545.28	$108.27	$138.63

Crew B-71

Crew B-71	Hr.	Daily	Hr.	Daily	Bare Costs	Incl. O&P
1 Labor Foreman (outside)	$38.65	$309.20	$59.65	$477.20	$42.19	$64.54
3 Laborers	36.65	879.60	56.55	1357.20		
3 Equip. Oper. (medium)	48.90	1173.60	74.15	1779.60		
1 Pvmt. Profiler, 750 H.P.		5877.00		6464.70		
1 Road Sweeper, S.P., 8' wide		665.40		731.94		
1 F.E. Loader, W.M., 1.5 C.Y.		373.40		410.74	123.50	135.85
56 L.H., Daily Totals		$9278.20		$11221.38	$165.68	$200.38

Crew B-72

Crew B-72	Hr.	Daily	Hr.	Daily	Bare Costs	Incl. O&P
1 Labor Foreman (outside)	$38.65	$309.20	$59.65	$477.20	$43.02	$65.74
3 Laborers	36.65	879.60	56.55	1357.20		
4 Equip. Oper. (medium)	48.90	1564.80	74.15	2372.80		
1 Pvmt. Profiler, 750 H.P.		5877.00		6464.70		
1 Hammermill, 250 H.P.		1846.00		2030.60		
1 Windrow Loader		1240.00		1364.00		
1 Mix Paver 165 H.P.		2183.00		2401.30		
1 Roller, Pneum. Whl., 12 Ton		349.00		383.90	179.61	197.57
64 L.H., Daily Totals		$14248.60		$16851.70	$222.63	$263.31

Crew B-73

Crew B-73	Hr.	Daily	Hr.	Daily	Bare Costs	Incl. O&P
1 Labor Foreman (outside)	$38.65	$309.20	$59.65	$477.20	$44.56	$67.94
2 Laborers	36.65	586.40	56.55	904.80		
5 Equip. Oper. (medium)	48.90	1956.00	74.15	2966.00		
1 Road Mixer, 310 H.P.		1942.00		2136.20		
1 Tandem Roller, 10 Ton		242.00		266.20		
1 Hammermill, 250 H.P.		1846.00		2030.60		
1 Grader, 30,000 Lbs.		707.00		777.70		
.5 F.E. Loader, W.M., 1.5 C.Y.		186.70		205.37		
.5 Truck Tractor, 220 H.P.		183.20		201.52		
.5 Water Tank Trailer, 5000 Gal.		70.80		77.88	80.90	88.99
64 L.H., Daily Totals		$8029.30		$10043.47	$125.46	$156.93

Crew B-74

Crew B-74	Hr.	Daily	Hr.	Daily	Bare Costs	Incl. O&P
1 Labor Foreman (outside)	$38.65	$309.20	$59.65	$477.20	$43.25	$65.92
1 Laborer	36.65	293.20	56.55	452.40		
4 Equip. Oper. (medium)	48.90	1564.80	74.15	2372.80		
2 Truck Drivers (heavy)	37.55	600.80	57.30	916.80		
1 Grader, 30,000 Lbs.		707.00		777.70		
1 Ripper, Beam & 1 Shank		81.40		89.54		
2 Stabilizers, 310 H.P.		3654.00		4019.40		
1 Flatbed Truck, Gas, 3 Ton		333.00		366.30		
1 Chem. Spreader, Towed		53.20		58.52		
1 Roller, Vibratory, 25 Ton		692.80		762.08		
1 Water Tank Trailer, 5000 Gal.		141.60		155.76		
1 Truck Tractor, 220 H.P.		366.40		403.04	94.21	103.63
64 L.H., Daily Totals		$8797.40		$10851.54	$137.46	$169.56

Crew B-75

Crew B-75	Hr.	Daily	Hr.	Daily	Bare Costs	Incl. O&P
1 Labor Foreman (outside)	$38.65	$309.20	$59.65	$477.20	$44.06	$67.16
1 Laborer	36.65	293.20	56.55	452.40		
4 Equip. Oper. (medium)	48.90	1564.80	74.15	2372.80		
1 Truck Driver (heavy)	37.55	300.40	57.30	458.40		
1 Grader, 30,000 Lbs.		707.00		777.70		
1 Ripper, Beam & 1 Shank		81.40		89.54		
2 Stabilizers, 310 H.P.		3654.00		4019.40		
1 Dist. Tanker, 3000 Gallon		308.20		339.02		
1 Truck Tractor, 6x4, 380 H.P.		612.80		674.08		
1 Roller, Vibratory, 25 Ton		692.80		762.08	108.15	118.96
56 L.H., Daily Totals		$8523.80		$10422.62	$152.21	$186.12

Crew B-76

Crew B-76	Hr.	Daily	Hr.	Daily	Bare Costs	Incl. O&P
1 Dock Builder Foreman (outside)	$46.40	$371.20	$73.60	$588.80	$45.83	$71.56
5 Dock Builders	44.40	1776.00	70.40	2816.00		
2 Equip. Oper. (crane)	50.25	804.00	76.20	1219.20		
1 Equip. Oper. (oiler)	43.55	348.40	66.00	528.00		
1 Crawler Crane, 50 Ton		1319.00		1450.90		
1 Barge, 400 Ton		781.00		859.10		
1 Hammer, Diesel, 15K ft.-lbs.		587.60		646.36		
1 Lead, 60' High		74.80		82.28		
1 Air Compressor, 600 cfm		552.60		607.86		
2 -50' Air Hoses, 3"		29.80		32.78	46.46	51.10
72 L.H., Daily Totals		$6644.40		$8831.28	$92.28	$122.66

Crew B-76A

Crew B-76A	Hr.	Daily	Hr.	Daily	Bare Costs	Incl. O&P
1 Labor Foreman (outside)	$38.65	$309.20	$59.65	$477.20	$39.46	$60.58
5 Laborers	36.65	1466.00	56.55	2262.00		
1 Equip. Oper. (crane)	50.25	402.00	76.20	609.60		
1 Equip. Oper. (oiler)	43.55	348.40	66.00	528.00		
1 Crawler Crane, 50 Ton		1319.00		1450.90		
1 Barge, 400 Ton		781.00		859.10	32.81	36.09
64 L.H., Daily Totals		$4625.60		$6186.80	$72.28	$96.67

Crew B-77

Crew B-77	Hr.	Daily	Hr.	Daily	Bare Costs	Incl. O&P
1 Labor Foreman (outside)	$38.65	$309.20	$59.65	$477.20	$37.02	$57.00
3 Laborers	36.65	879.60	56.55	1357.20		
1 Truck Driver (light)	36.50	292.00	55.70	445.60		
1 Crack Cleaner, 25 H.P.		64.80		71.28		
1 Crack Filler, Trailer Mtd.		203.80		224.18		
1 Flatbed Truck, Gas, 3 Ton		333.00		366.30	15.04	16.54
40 L.H., Daily Totals		$2082.40		$2941.76	$52.06	$73.54

Crew B-78

Crew B-78	Hr.	Daily	Hr.	Daily	Bare Costs	Incl. O&P
1 Labor Foreman (outside)	$38.65	$309.20	$59.65	$477.20	$36.96	$56.92
4 Laborers	36.65	1172.80	56.55	1809.60		
1 Truck Driver (light)	36.50	292.00	55.70	445.60		
1 Paint Striper, S.P., 40 Gallon		154.80		170.28		
1 Flatbed Truck, Gas, 3 Ton		333.00		366.30		
1 Pickup Truck, 3/4 Ton		157.00		172.70	13.43	14.78
48 L.H., Daily Totals		$2418.80		$3441.68	$50.39	$71.70

Crew B-78A

Crew B-78A	Hr.	Daily	Hr.	Daily	Bare Costs	Incl. O&P
1 Equip. Oper. (light)	$47.05	$376.40	$71.35	$570.80	$47.05	$71.35
1 Line Rem. (Metal Balls) 115 H.P.		839.00		922.90	104.88	115.36
8 L.H., Daily Totals		$1215.40		$1493.70	$151.93	$186.71

Crew B-78B

Crew B-78B	Hr.	Daily	Hr.	Daily	Bare Costs	Incl. O&P
2 Laborers	$36.65	$586.40	$56.55	$904.80	$37.81	$58.19
.25 Equip. Oper. (light)	47.05	94.10	71.35	142.70		
1 Pickup Truck, 3/4 Ton		157.00		172.70		
1 Line Rem.,11 H.P.,Walk Behind		68.40		75.24		
.25 Road Sweeper, S.P., 8' wide		166.35		182.99	21.76	23.94
18 L.H., Daily Totals		$1072.25		$1478.43	$59.57	$82.13

Crew B-78C

Crew B-78C	Hr.	Daily	Hr.	Daily	Bare Costs	Incl. O&P
1 Labor Foreman (outside)	$38.65	$309.20	$59.65	$477.20	$36.96	$56.92
4 Laborers	36.65	1172.80	56.55	1809.60		
1 Truck Driver (light)	36.50	292.00	55.70	445.60		
1 Paint Striper, T.M., 120 Gal.		860.40		946.44		
1 Flatbed Truck, Gas, 3 Ton		333.00		366.30		
1 Pickup Truck, 3/4 Ton		157.00		172.70	28.13	30.95
48 L.H., Daily Totals		$3124.40		$4217.84	$65.09	$87.87

Crew B-78D

Crew B-78D	Hr.	Daily	Hr.	Daily	Bare Costs	Incl. O&P
2 Labor Foremen (outside)	$38.65	$618.40	$59.65	$954.40	$37.03	$57.09
7 Laborers	36.65	2052.40	56.55	3166.80		
1 Truck Driver (light)	36.50	292.00	55.70	445.60		
1 Paint Striper, T.M., 120 Gal.		860.40		946.44		
1 Flatbed Truck, Gas, 3 Ton		333.00		366.30		
3 Pickup Trucks, 3/4 Ton		471.00		518.10		
1 Air Compressor, 60 cfm		137.40		151.14		
1 -50' Air Hose, 3/4"		3.25		3.58		
1 Breakers, Pavement, 60 lb.		9.80		10.78	22.69	24.95
80 L.H., Daily Totals		$4777.65		$6563.14	$59.72	$82.04

Crew B-78E

Crew B-78E	Hr.	Daily	Hr.	Daily	Bare Costs	Incl. O&P
2 Labor Foremen (outside)	$38.65	$618.40	$59.65	$954.40	$36.97	$57.00
9 Laborers	36.65	2638.80	56.55	4071.60		
1 Truck Driver (light)	36.50	292.00	55.70	445.60		
1 Paint Striper, T.M., 120 Gal.		860.40		946.44		
1 Flatbed Truck, Gas, 3 Ton		333.00		366.30		
4 Pickup Trucks, 3/4 Ton		628.00		690.80		
2 Air Compressor, 60 cfm		274.80		302.28		
2 -50' Air Hose, 3/4"		6.50		7.15		
2 Breakers, Pavement, 60 lb.		19.60		21.56	22.11	24.32
96 L.H., Daily Totals		$5671.50		$7806.13	$59.08	$81.31

Crew B-78F

Crew B-78F	Hr.	Daily	Hr.	Daily	Bare Costs	Incl. O&P
2 Labor Foremen (outside)	$38.65	$618.40	$59.65	$954.40	$36.92	$56.93
11 Laborers	36.65	3225.20	56.55	4976.40		
1 Truck Driver (light)	36.50	292.00	55.70	445.60		
1 Paint Striper, T.M., 120 Gal.		860.40		946.44		
1 Flatbed Truck, Gas, 3 Ton		333.00		366.30		
7 Pickup Trucks, 3/4 Ton		1099.00		1208.90		
3 Air Compressor, 60 cfm		412.20		453.42		
3 -50' Air Hose, 3/4"		9.75		10.73		
3 Breakers, Pavement, 60 lb.		29.40		32.34	24.50	26.95
112 L.H., Daily Totals		$6879.35		$9394.52	$61.42	$83.88

Crew B-79

Crew B-79	Hr.	Daily	Hr.	Daily	Bare Costs	Incl. O&P
1 Labor Foreman (outside)	$38.65	$309.20	$59.65	$477.20	$37.02	$57.00
3 Laborers	36.65	879.60	56.55	1357.20		
1 Truck Driver (light)	36.50	292.00	55.70	445.60		
1 Paint Striper, T.M., 120 Gal.		860.40		946.44		
1 Heating Kettle, 115 Gallon		63.00		69.30		
1 Flatbed Truck, Gas, 3 Ton		333.00		366.30		
2 Pickup Trucks, 3/4 Ton		314.00		345.40	39.26	43.19
40 L.H., Daily Totals		$3051.20		$4007.44	$76.28	$100.19

Crew B-79A

Crew B-79A	Hr.	Daily	Hr.	Daily	Bare Costs	Incl. O&P
1.5 Equip. Oper. (light)	$47.05	$564.60	$71.35	$856.20	$47.05	$71.35
.5 Line Remov. (Grinder) 115 H.P.		465.30		511.83		
1 Line Rem. (Metal Balls) 115 H.P.		839.00		922.90	108.69	119.56
12 L.H., Daily Totals		$1868.90		$2290.93	$155.74	$190.91

Crew B-79B

Crew B-79B	Hr.	Daily	Hr.	Daily	Bare Costs	Incl. O&P
1 Laborer	$36.65	$293.20	$56.55	$452.40	$36.65	$56.55
1 Set of Gases		152.00		167.20	19.00	20.90
8 L.H., Daily Totals		$445.20		$619.60	$55.65	$77.45

Crew B-79C

Crew B-79C	Hr.	Daily	Hr.	Daily	Bare Costs	Incl. O&P
1 Labor Foreman (outside)	$38.65	$309.20	$59.65	$477.20	$36.91	$56.87
5 Laborers	36.65	1466.00	56.55	2262.00		
1 Truck Driver (light)	36.50	292.00	55.70	445.60		
1 Paint Striper, T.M., 120 Gal.		860.40		946.44		
1 Heating Kettle, 115 Gallon		63.00		69.30		
1 Flatbed Truck, Gas, 3 Ton		333.00		366.30		
3 Pickup Trucks, 3/4 Ton		471.00		518.10		
1 Air Compressor, 60 cfm		137.40		151.14		
1 -50' Air Hose, 3/4"		3.25		3.58		
1 Breakers, Pavement, 60 lb.		9.80		10.78	33.53	36.89
56 L.H., Daily Totals		$3945.05		$5250.44	$70.45	$93.76

Crew B-79D

Crew B-79D	Hr.	Daily	Hr.	Daily	Bare Costs	Incl. O&P
2 Labor Foremen (outside)	$38.65	$618.40	$59.65	$954.40	$37.13	$57.22
5 Laborers	36.65	1466.00	56.55	2262.00		
1 Truck Driver (light)	36.50	292.00	55.70	445.60		
1 Paint Striper, T.M., 120 Gal.		860.40		946.44		
1 Heating Kettle, 115 Gallon		63.00		69.30		
1 Flatbed Truck, Gas, 3 Ton		333.00		366.30		
4 Pickup Trucks, 3/4 Ton		628.00		690.80		
1 Air Compressor, 60 cfm		137.40		151.14		
1 -50' Air Hose, 3/4"		3.25		3.58		
1 Breakers, Pavement, 60 lb.		9.80		10.78	31.79	34.97
64 L.H., Daily Totals		$4411.25		$5900.34	$68.93	$92.19

Crew B-79E

	Bare Costs Hr.	Daily	Incl. Subs O&P Hr.	Daily	Cost Per Labor-Hour Bare Costs	Incl. O&P
2 Labor Foremen (outside)	$38.65	$618.40	$59.65	$954.40	$37.03	$57.09
7 Laborers	36.65	2052.40	56.55	3166.80		
1 Truck Driver (light)	36.50	292.00	55.70	445.60		
1 Paint Striper, T.M., 120 Gal.		860.40		946.44		
1 Heating Kettle, 115 Gallon		63.00		69.30		
1 Flatbed Truck, Gas, 3 Ton		333.00		366.30		
5 Pickup Trucks, 3/4 Ton		785.00		863.50		
2 Air Compressors, 60 cfm		274.80		302.28		
2 -50' Air Hoses, 3/4"		6.50		7.15		
2 Breakers, Pavement, 60 lb.		19.60		21.56	29.28	32.21
80 L.H., Daily Totals		$5305.10		$7143.33	$66.31	$89.29

Crew B-80

	Bare Costs Hr.	Daily	Incl. Subs O&P Hr.	Daily	Cost Per Labor-Hour Bare Costs	Incl. O&P
1 Labor Foreman (outside)	$38.65	$309.20	$59.65	$477.20	$39.71	$60.81
1 Laborer	36.65	293.20	56.55	452.40		
1 Truck Driver (light)	36.50	292.00	55.70	445.60		
1 Equip. Oper. (light)	47.05	376.40	71.35	570.80		
1 Flatbed Truck, Gas, 3 Ton		333.00		366.30		
1 Earth Auger, Truck-Mtd.		425.00		467.50	23.69	26.06
32 L.H., Daily Totals		$2028.80		$2779.80	$63.40	$86.87

Crew B-80A

	Bare Costs Hr.	Daily	Incl. Subs O&P Hr.	Daily	Cost Per Labor-Hour Bare Costs	Incl. O&P
3 Laborers	$36.65	$879.60	$56.55	$1357.20	$36.65	$56.55
1 Flatbed Truck, Gas, 3 Ton		333.00		366.30	13.88	15.26
24 L.H., Daily Totals		$1212.60		$1723.50	$50.52	$71.81

Crew B-80B

	Bare Costs Hr.	Daily	Incl. Subs O&P Hr.	Daily	Cost Per Labor-Hour Bare Costs	Incl. O&P
3 Laborers	$36.65	$879.60	$56.55	$1357.20	$39.25	$60.25
1 Equip. Oper. (light)	47.05	376.40	71.35	570.80		
1 Crane, Flatbed Mounted, 3 Ton		248.40		273.24	7.76	8.54
32 L.H., Daily Totals		$1504.40		$2201.24	$47.01	$68.79

Crew B-80C

	Bare Costs Hr.	Daily	Incl. Subs O&P Hr.	Daily	Cost Per Labor-Hour Bare Costs	Incl. O&P
2 Laborers	$36.65	$586.40	$56.55	$904.80	$36.60	$56.27
1 Truck Driver (light)	36.50	292.00	55.70	445.60		
1 Flatbed Truck, Gas, 1.5 Ton		270.20		297.22		
1 Manual Fence Post Auger, Gas		8.00		8.80	11.59	12.75
24 L.H., Daily Totals		$1156.60		$1656.42	$48.19	$69.02

Crew B-81

	Bare Costs Hr.	Daily	Incl. Subs O&P Hr.	Daily	Cost Per Labor-Hour Bare Costs	Incl. O&P
1 Laborer	$36.65	$293.20	$56.55	$452.40	$41.03	$62.67
1 Equip. Oper. (medium)	48.90	391.20	74.15	593.20		
1 Truck Driver (heavy)	37.55	300.40	57.30	458.40		
1 Hydromulcher, T.M., 3000 Gal.		340.40		374.44		
1 Truck Tractor, 220 H.P.		366.40		403.04	29.45	32.40
24 L.H., Daily Totals		$1691.60		$2281.48	$70.48	$95.06

Crew B-81A

	Bare Costs Hr.	Daily	Incl. Subs O&P Hr.	Daily	Cost Per Labor-Hour Bare Costs	Incl. O&P
1 Laborer	$36.65	$293.20	$56.55	$452.40	$36.58	$56.13
1 Truck Driver (light)	36.50	292.00	55.70	445.60		
1 Hydromulcher, T.M., 600 Gal.		132.60		145.86		
1 Flatbed Truck, Gas, 3 Ton		333.00		366.30	29.10	32.01
16 L.H., Daily Totals		$1050.80		$1410.16	$65.67	$88.14

Crew B-82

	Bare Costs Hr.	Daily	Incl. Subs O&P Hr.	Daily	Cost Per Labor-Hour Bare Costs	Incl. O&P
1 Laborer	$36.65	$293.20	$56.55	$452.40	$41.85	$63.95
1 Equip. Oper. (light)	47.05	376.40	71.35	570.80		
1 Horiz. Borer, 6 H.P.		83.00		91.30	5.19	5.71
16 L.H., Daily Totals		$752.60		$1114.50	$47.04	$69.66

Crew B-82A

	Bare Costs Hr.	Daily	Incl. Subs O&P Hr.	Daily	Cost Per Labor-Hour Bare Costs	Incl. O&P
1 Laborer	$36.65	$293.20	$56.55	$452.40	$41.85	$63.95
1 Equip. Oper. (light)	47.05	376.40	71.35	570.80		
1 Flatbed Truck, Gas, 3 Ton		333.00		366.30		
1 Flatbed Trailer, 25 Ton		113.60		124.96		
1 Horiz. Dir. Drill, 20k lb. Thrust		658.40		724.24	69.06	75.97
16 L.H., Daily Totals		$1774.60		$2238.70	$110.91	$139.92

Crew B-82B

	Bare Costs Hr.	Daily	Incl. Subs O&P Hr.	Daily	Cost Per Labor-Hour Bare Costs	Incl. O&P
2 Laborers	$36.65	$586.40	$56.55	$904.80	$40.12	$61.48
1 Equip. Oper. (light)	47.05	376.40	71.35	570.80		
1 Flatbed Truck, Gas, 3 Ton		333.00		366.30		
1 Flatbed Trailer, 25 Ton		113.60		124.96		
1 Horiz. Dir. Drill, 30k lb. Thrust		936.20		1029.82	57.62	63.38
24 L.H., Daily Totals		$2345.60		$2996.68	$97.73	$124.86

Crew B-82C

	Bare Costs Hr.	Daily	Incl. Subs O&P Hr.	Daily	Cost Per Labor-Hour Bare Costs	Incl. O&P
2 Laborers	$36.65	$586.40	$56.55	$904.80	$40.12	$61.48
1 Equip. Oper. (light)	47.05	376.40	71.35	570.80		
1 Flatbed Truck, Gas, 3 Ton		333.00		366.30		
1 Flatbed Trailer, 25 Ton		113.60		124.96		
1 Horiz. Dir. Drill, 50k lb. Thrust		1246.00		1370.60	70.53	77.58
24 L.H., Daily Totals		$2655.40		$3337.46	$110.64	$139.06

Crew B-82D

	Bare Costs Hr.	Daily	Incl. Subs O&P Hr.	Daily	Cost Per Labor-Hour Bare Costs	Incl. O&P
1 Equip. Oper. (light)	$47.05	$376.40	$71.35	$570.80	$47.05	$71.35
1 Mud Trailer for HDD, 1500 Gal.		374.80		412.28	46.85	51.53
8 L.H., Daily Totals		$751.20		$983.08	$93.90	$122.89

Crew B-83

	Bare Costs Hr.	Daily	Incl. Subs O&P Hr.	Daily	Cost Per Labor-Hour Bare Costs	Incl. O&P
1 Tugboat Captain	$48.90	$391.20	$74.15	$593.20	$42.77	$65.35
1 Tugboat Hand	36.65	293.20	56.55	452.40		
1 Tugboat, 250 H.P.		889.40		978.34	55.59	61.15
16 L.H., Daily Totals		$1573.80		$2023.94	$98.36	$126.50

Crew B-84

	Bare Costs Hr.	Daily	Incl. Subs O&P Hr.	Daily	Cost Per Labor-Hour Bare Costs	Incl. O&P
1 Equip. Oper. (medium)	$48.90	$391.20	$74.15	$593.20	$48.90	$74.15
1 Rotary Mower/Tractor		364.20		400.62	45.52	50.08
8 L.H., Daily Totals		$755.40		$993.82	$94.42	$124.23

Crew B-85

	Bare Costs Hr.	Daily	Incl. Subs O&P Hr.	Daily	Cost Per Labor-Hour Bare Costs	Incl. O&P
3 Laborers	$36.65	$879.60	$56.55	$1357.20	$39.28	$60.22
1 Equip. Oper. (medium)	48.90	391.20	74.15	593.20		
1 Truck Driver (heavy)	37.55	300.40	57.30	458.40		
1 Aerial Lift Truck, 80'		643.00		707.30		
1 Brush Chipper, 12", 130 H.P.		396.60		436.26		
1 Pruning Saw, Rotary		6.55		7.21	26.15	28.77
40 L.H., Daily Totals		$2617.35		$3559.57	$65.43	$88.99

Crew B-86

	Bare Costs Hr.	Daily	Incl. Subs O&P Hr.	Daily	Cost Per Labor-Hour Bare Costs	Incl. O&P
1 Equip. Oper. (medium)	$48.90	$391.20	$74.15	$593.20	$48.90	$74.15
1 Stump Chipper, S.P.		162.10		178.31	20.26	22.29
8 L.H., Daily Totals		$553.30		$771.51	$69.16	$96.44

Crew B-86A

	Bare Costs Hr.	Daily	Incl. Subs O&P Hr.	Daily	Cost Per Labor-Hour Bare Costs	Incl. O&P
1 Equip. Oper. (medium)	$48.90	$391.20	$74.15	$593.20	$48.90	$74.15
1 Grader, 30,000 Lbs.		707.00		777.70	88.38	97.21
8 L.H., Daily Totals		$1098.20		$1370.90	$137.28	$171.36

Crew No.	Bare Costs		Incl. Subs O&P		Cost Per Labor-Hour	
Crew B-86B	Hr.	Daily	Hr.	Daily	Bare Costs	Incl. O&P
1 Equip. Oper. (medium)	$48.90	$391.20	$74.15	$593.20	$48.90	$74.15
1 Dozer, 200 H.P.		1325.00		1457.50	165.63	182.19
8 L.H., Daily Totals		$1716.20		$2050.70	$214.53	$256.34

Crew No.	Bare Costs		Incl. Subs O&P		Cost Per Labor-Hour	
Crew B-87	Hr.	Daily	Hr.	Daily	Bare Costs	Incl. O&P
1 Laborer	$36.65	$293.20	$56.55	$452.40	$46.45	$70.63
4 Equip. Oper. (medium)	48.90	1564.80	74.15	2372.80		
2 Feller Bunchers, 100 H.P.		1458.00		1603.80		
1 Log Chipper, 22" Tree		898.60		988.46		
1 Dozer, 105 H.P.		591.80		650.98		
1 Chain Saw, Gas, 36" Long		45.00		49.50	74.83	82.32
40 L.H., Daily Totals		$4851.40		$6117.94	$121.29	$152.95

Crew No.	Bare Costs		Incl. Subs O&P		Cost Per Labor-Hour	
Crew B-88	Hr.	Daily	Hr.	Daily	Bare Costs	Incl. O&P
1 Laborer	$36.65	$293.20	$56.55	$452.40	$47.15	$71.64
6 Equip. Oper. (medium)	48.90	2347.20	74.15	3559.20		
2 Feller Bunchers, 100 H.P.		1458.00		1603.80		
1 Log Chipper, 22" Tree		898.60		988.46		
2 Log Skidders, 50 H.P.		1839.20		2023.12		
1 Dozer, 105 H.P.		591.80		650.98		
1 Chain Saw, Gas, 36" Long		45.00		49.50	86.30	94.93
56 L.H., Daily Totals		$7473.00		$9327.46	$133.45	$166.56

Crew No.	Bare Costs		Incl. Subs O&P		Cost Per Labor-Hour	
Crew B-89	Hr.	Daily	Hr.	Daily	Bare Costs	Incl. O&P
1 Equip. Oper. (light)	$47.05	$376.40	$71.35	$570.80	$41.77	$63.52
1 Truck Driver (light)	36.50	292.00	55.70	445.60		
1 Flatbed Truck, Gas, 3 Ton		333.00		366.30		
1 Concrete Saw		175.20		192.72		
1 Water Tank, 65 Gal.		17.30		19.03	32.84	36.13
16 L.H., Daily Totals		$1193.90		$1594.45	$74.62	$99.65

Crew No.	Bare Costs		Incl. Subs O&P		Cost Per Labor-Hour	
Crew B-89A	Hr.	Daily	Hr.	Daily	Bare Costs	Incl. O&P
1 Skilled Worker	$47.30	$378.40	$73.25	$586.00	$41.98	$64.90
1 Laborer	36.65	293.20	56.55	452.40		
1 Core Drill (Large)		116.20		127.82	7.26	7.99
16 L.H., Daily Totals		$787.80		$1166.22	$49.24	$72.89

Crew No.	Bare Costs		Incl. Subs O&P		Cost Per Labor-Hour	
Crew B-89B	Hr.	Daily	Hr.	Daily	Bare Costs	Incl. O&P
1 Equip. Oper. (light)	$47.05	$376.40	$71.35	$570.80	$41.77	$63.52
1 Truck Driver (light)	36.50	292.00	55.70	445.60		
1 Wall Saw, Hydraulic, 10 H.P.		119.60		131.56		
1 Generator, Diesel, 100 kW		439.20		483.12		
1 Water Tank, 65 Gal.		17.30		19.03		
1 Flatbed Truck, Gas, 3 Ton		333.00		366.30	56.82	62.50
16 L.H., Daily Totals		$1577.50		$2016.41	$98.59	$126.03

Crew No.	Bare Costs		Incl. Subs O&P		Cost Per Labor-Hour	
Crew B-90	Hr.	Daily	Hr.	Daily	Bare Costs	Incl. O&P
1 Labor Foreman (outside)	$38.65	$309.20	$59.65	$477.20	$39.73	$60.83
3 Laborers	36.65	879.60	56.55	1357.20		
2 Equip. Oper. (light)	47.05	752.80	71.35	1141.60		
2 Truck Drivers (heavy)	37.55	600.80	57.30	916.80		
1 Road Mixer, 310 H.P.		1942.00		2136.20		
1 Dist. Truck, 2000 Gal.		278.20		306.02	34.69	38.16
64 L.H., Daily Totals		$4762.60		$6335.02	$74.42	$98.98

Crew No.	Bare Costs		Incl. Subs O&P		Cost Per Labor-Hour	
Crew B-90A	Hr.	Daily	Hr.	Daily	Bare Costs	Incl. O&P
1 Labor Foreman (outside)	$38.65	$309.20	$59.65	$477.20	$43.94	$67.05
2 Laborers	36.65	586.40	56.55	904.80		
4 Equip. Oper. (medium)	48.90	1564.80	74.15	2372.80		
2 Graders, 30,000 Lbs.		1414.00		1555.40		
1 Tandem Roller, 10 Ton		242.00		266.20		
1 Roller, Pneum. Whl., 12 Ton		349.00		383.90	35.80	39.38
56 L.H., Daily Totals		$4465.40		$5960.30	$79.74	$106.43

Crew No.	Bare Costs		Incl. Subs O&P		Cost Per Labor-Hour	
Crew B-90B	Hr.	Daily	Hr.	Daily	Bare Costs	Incl. O&P
1 Labor Foreman (outside)	$38.65	$309.20	$59.65	$477.20	$43.11	$65.87
2 Laborers	36.65	586.40	56.55	904.80		
3 Equip. Oper. (medium)	48.90	1173.60	74.15	1779.60		
1 Roller, Pneum. Whl., 12 Ton		349.00		383.90		
1 Road Mixer, 310 H.P.		1942.00		2136.20	47.73	52.50
48 L.H., Daily Totals		$4360.20		$5681.70	$90.84	$118.37

Crew No.	Bare Costs		Incl. Subs O&P		Cost Per Labor-Hour	
Crew B-90C	Hr.	Daily	Hr.	Daily	Bare Costs	Incl. O&P
1 Labor Foreman (outside)	$38.65	$309.20	$59.65	$477.20	$40.42	$61.84
4 Laborers	36.65	1172.80	56.55	1809.60		
3 Equip. Oper. (medium)	48.90	1173.60	74.15	1779.60		
3 Truck Drivers (heavy)	37.55	901.20	57.30	1375.20		
3 Road Mixers, 310 H.P.		5826.00		6408.60	66.20	72.83
88 L.H., Daily Totals		$9382.80		$11850.20	$106.62	$134.66

Crew No.	Bare Costs		Incl. Subs O&P		Cost Per Labor-Hour	
Crew B-90D	Hr.	Daily	Hr.	Daily	Bare Costs	Incl. O&P
1 Labor Foreman (outside)	$38.65	$309.20	$59.65	$477.20	$39.84	$61.02
6 Laborers	36.65	1759.20	56.55	2714.40		
3 Equip. Oper. (medium)	48.90	1173.60	74.15	1779.60		
3 Truck Drivers (heavy)	37.55	901.20	57.30	1375.20		
3 Road Mixers, 310 H.P.		5826.00		6408.60	56.02	61.62
104 L.H., Daily Totals		$9969.20		$12755.00	$95.86	$122.64

Crew No.	Bare Costs		Incl. Subs O&P		Cost Per Labor-Hour	
Crew B-90E	Hr.	Daily	Hr.	Daily	Bare Costs	Incl. O&P
1 Labor Foreman (outside)	$38.65	$309.20	$59.65	$477.20	$41.06	$62.84
4 Laborers	36.65	1172.80	56.55	1809.60		
3 Equip. Oper. (medium)	48.90	1173.60	74.15	1779.60		
1 Truck Driver (heavy)	37.55	300.40	57.30	458.40		
1 Road Mixers, 310 H.P.		1942.00		2136.20	26.97	29.67
72 L.H., Daily Totals		$4898.00		$6661.00	$68.03	$92.51

Crew No.	Bare Costs		Incl. Subs O&P		Cost Per Labor-Hour	
Crew B-91	Hr.	Daily	Hr.	Daily	Bare Costs	Incl. O&P
1 Labor Foreman (outside)	$38.65	$309.20	$59.65	$477.20	$43.14	$65.83
2 Laborers	36.65	586.40	56.55	904.80		
4 Equip. Oper. (medium)	48.90	1564.80	74.15	2372.80		
1 Truck Driver (heavy)	37.55	300.40	57.30	458.40		
1 Dist. Tanker, 3000 Gallon		308.20		339.02		
1 Truck Tractor, 6x4, 380 H.P.		612.80		674.08		
1 Aggreg. Spreader, S.P.		841.00		925.10		
1 Roller, Pneum. Whl., 12 Ton		349.00		383.90		
1 Tandem Roller, 10 Ton		242.00		266.20	36.77	40.44
64 L.H., Daily Totals		$5113.80		$6801.50	$79.90	$106.27

Crew No.	Bare Costs		Incl. Subs O&P		Cost Per Labor-Hour	
Crew B-91B	Hr.	Daily	Hr.	Daily	Bare Costs	Incl. O&P
1 Laborer	$36.65	$293.20	$56.55	$452.40	$42.77	$65.35
1 Equipment Oper. (med.)	48.90	391.20	74.15	593.20		
1 Road Sweeper, Vac. Assist.		994.40		1093.84	62.15	68.36
16 L.H., Daily Totals		$1678.80		$2139.44	$104.93	$133.72

Crew B-91C

Crew No.	Bare Costs Hr.	Daily	Incl. Subs O&P Hr.	Daily	Cost Per Labor-Hour Bare Costs	Incl. O&P
1 Laborer	$36.65	$293.20	$56.55	$452.40	$36.58	$56.13
1 Truck Driver (light)	36.50	292.00	55.70	445.60		
1 Catch Basin Cleaning Truck		604.00		664.40	37.75	41.52
16 L.H., Daily Totals		$1189.20		$1562.40	$74.33	$97.65

Crew B-91D

Crew No.	Bare Costs Hr.	Daily	Incl. Subs O&P Hr.	Daily	Cost Per Labor-Hour Bare Costs	Incl. O&P
1 Labor Foreman (outside)	$38.65	$309.20	$59.65	$477.20	$41.65	$63.67
5 Laborers	36.65	1466.00	56.55	2262.00		
5 Equip. Oper. (medium)	48.90	1956.00	74.15	2966.00		
2 Truck Drivers (heavy)	37.55	600.80	57.30	916.80		
1 Aggreg. Spreader, S.P.		841.00		925.10		
2 Truck Tractor, 6x4, 380 H.P.		1225.60		1348.16		
2 Dist. Tanker, 3000 Gallon		616.40		678.04		
2 Pavement Brush, Towed		163.20		179.52		
2 Roller, Pneum. Whl., 12 Ton		698.00		767.80	34.08	37.49
104 L.H., Daily Totals		$7876.20		$10520.62	$75.73	$101.16

Crew B-92

Crew No.	Bare Costs Hr.	Daily	Incl. Subs O&P Hr.	Daily	Cost Per Labor-Hour Bare Costs	Incl. O&P
1 Labor Foreman (outside)	$38.65	$309.20	$59.65	$477.20	$37.15	$57.33
3 Laborers	36.65	879.60	56.55	1357.20		
1 Crack Cleaner, 25 H.P.		64.80		71.28		
1 Air Compressor, 60 cfm		137.40		151.14		
1 Tar Kettle, T.M.		104.60		115.06		
1 Flatbed Truck, Gas, 3 Ton		333.00		366.30	19.99	21.99
32 L.H., Daily Totals		$1828.60		$2538.18	$57.14	$79.32

Crew B-93

Crew No.	Bare Costs Hr.	Daily	Incl. Subs O&P Hr.	Daily	Cost Per Labor-Hour Bare Costs	Incl. O&P
1 Equip. Oper. (medium)	$48.90	$391.20	$74.15	$593.20	$48.90	$74.15
1 Feller Buncher, 100 H.P.		729.00		801.90	91.13	100.24
8 L.H., Daily Totals		$1120.20		$1395.10	$140.03	$174.39

Crew B-94A

Crew No.	Bare Costs Hr.	Daily	Incl. Subs O&P Hr.	Daily	Cost Per Labor-Hour Bare Costs	Incl. O&P
1 Laborer	$36.65	$293.20	$56.55	$452.40	$36.65	$56.55
1 Diaphragm Water Pump, 2"		72.00		79.20		
1 -20' Suction Hose, 2"		1.95		2.15		
2 -50' Discharge Hoses, 2"		1.80		1.98	9.47	10.42
8 L.H., Daily Totals		$368.95		$535.73	$46.12	$66.97

Crew B-94B

Crew No.	Bare Costs Hr.	Daily	Incl. Subs O&P Hr.	Daily	Cost Per Labor-Hour Bare Costs	Incl. O&P
1 Laborer	$36.65	$293.20	$56.55	$452.40	$36.65	$56.55
1 Diaphragm Water Pump, 4"		115.80		127.38		
1 -20' Suction Hose, 4"		3.25		3.58		
2 -50' Discharge Hoses, 4"		4.70		5.17	15.47	17.02
8 L.H., Daily Totals		$416.95		$588.52	$52.12	$73.57

Crew B-94C

Crew No.	Bare Costs Hr.	Daily	Incl. Subs O&P Hr.	Daily	Cost Per Labor-Hour Bare Costs	Incl. O&P
1 Laborer	$36.65	$293.20	$56.55	$452.40	$36.65	$56.55
1 Centrifugal Water Pump, 3"		78.80		86.68		
1 -20' Suction Hose, 3"		2.85		3.13		
2 -50' Discharge Hoses, 3"		3.00		3.30	10.58	11.64
8 L.H., Daily Totals		$377.85		$545.51	$47.23	$68.19

Crew B-94D

Crew No.	Bare Costs Hr.	Daily	Incl. Subs O&P Hr.	Daily	Cost Per Labor-Hour Bare Costs	Incl. O&P
1 Laborer	$36.65	$293.20	$56.55	$452.40	$36.65	$56.55
1 Centr. Water Pump, 6"		340.80		374.88		
1 -20' Suction Hose, 6"		11.50		12.65		
2 -50' Discharge Hoses, 6"		12.20		13.42	45.56	50.12
8 L.H., Daily Totals		$657.70		$853.35	$82.21	$106.67

Crew C-1

Crew No.	Bare Costs Hr.	Daily	Incl. Subs O&P Hr.	Daily	Cost Per Labor-Hour Bare Costs	Incl. O&P
3 Carpenters	$45.85	$1100.40	$70.75	$1698.00	$43.55	$67.20
1 Laborer	36.65	293.20	56.55	452.40		
32 L.H., Daily Totals		$1393.60		$2150.40	$43.55	$67.20

Crew C-2

Crew No.	Bare Costs Hr.	Daily	Incl. Subs O&P Hr.	Daily	Cost Per Labor-Hour Bare Costs	Incl. O&P
1 Carpenter Foreman (outside)	$47.85	$382.80	$73.85	$590.80	$44.65	$68.90
4 Carpenters	45.85	1467.20	70.75	2264.00		
1 Laborer	36.65	293.20	56.55	452.40		
48 L.H., Daily Totals		$2143.20		$3307.20	$44.65	$68.90

Crew C-2A

Crew No.	Bare Costs Hr.	Daily	Incl. Subs O&P Hr.	Daily	Cost Per Labor-Hour Bare Costs	Incl. O&P
1 Carpenter Foreman (outside)	$47.85	$382.80	$73.85	$590.80	$44.35	$67.96
3 Carpenters	45.85	1100.40	70.75	1698.00		
1 Cement Finisher	44.05	352.40	65.10	520.80		
1 Laborer	36.65	293.20	56.55	452.40		
48 L.H., Daily Totals		$2128.80		$3262.00	$44.35	$67.96

Crew C-3

Crew No.	Bare Costs Hr.	Daily	Incl. Subs O&P Hr.	Daily	Cost Per Labor-Hour Bare Costs	Incl. O&P
1 Rodman Foreman (outside)	$52.65	$421.20	$82.70	$661.60	$46.95	$73.17
4 Rodmen (reinf.)	50.65	1620.80	79.55	2545.60		
1 Equip. Oper. (light)	47.05	376.40	71.35	570.80		
2 Laborers	36.65	586.40	56.55	904.80		
3 Stressing Equipment		30.60		33.66		
.5 Grouting Equipment		82.10		90.31	1.76	1.94
64 L.H., Daily Totals		$3117.50		$4806.77	$48.71	$75.11

Crew C-4

Crew No.	Bare Costs Hr.	Daily	Incl. Subs O&P Hr.	Daily	Cost Per Labor-Hour Bare Costs	Incl. O&P
1 Rodman Foreman (outside)	$52.65	$421.20	$82.70	$661.60	$51.15	$80.34
3 Rodmen (reinf.)	50.65	1215.60	79.55	1909.20		
3 Stressing Equipment		30.60		33.66	.96	1.05
32 L.H., Daily Totals		$1667.40		$2604.46	$52.11	$81.39

Crew C-4A

Crew No.	Bare Costs Hr.	Daily	Incl. Subs O&P Hr.	Daily	Cost Per Labor-Hour Bare Costs	Incl. O&P
2 Rodmen (reinf.)	$50.65	$810.40	$79.55	$1272.80	$50.65	$79.55
4 Stressing Equipment		40.80		44.88	2.55	2.81
16 L.H., Daily Totals		$851.20		$1317.68	$53.20	$82.36

Crew C-5

Crew No.	Bare Costs Hr.	Daily	Incl. Subs O&P Hr.	Daily	Cost Per Labor-Hour Bare Costs	Incl. O&P
1 Rodman Foreman (outside)	$52.65	$421.20	$82.70	$661.60	$49.86	$77.59
4 Rodmen (reinf.)	50.65	1620.80	79.55	2545.60		
1 Equip. Oper. (crane)	50.25	402.00	76.20	609.60		
1 Equip. Oper. (oiler)	43.55	348.40	66.00	528.00		
1 Hyd. Crane, 25 Ton		745.80		820.38	13.32	14.65
56 L.H., Daily Totals		$3538.20		$5165.18	$63.18	$92.24

Crew C-6

Crew No.	Bare Costs Hr.	Daily	Incl. Subs O&P Hr.	Daily	Cost Per Labor-Hour Bare Costs	Incl. O&P
1 Labor Foreman (outside)	$38.65	$309.20	$59.65	$477.20	$38.22	$58.49
4 Laborers	36.65	1172.80	56.55	1809.60		
1 Cement Finisher	44.05	352.40	65.10	520.80		
2 Gas Engine Vibrators		66.00		72.60	1.38	1.51
48 L.H., Daily Totals		$1900.40		$2880.20	$39.59	$60.00

Crew No.	Bare Costs		Incl. Subs O&P		Cost Per Labor-Hour	
Crew C-7	Hr.	Daily	Hr.	Daily	Bare Costs	Incl. O&P
1 Labor Foreman (outside)	$38.65	$309.20	$59.65	$477.20	$39.82	$60.85
5 Laborers	36.65	1466.00	56.55	2262.00		
1 Cement Finisher	44.05	352.40	65.10	520.80		
1 Equip. Oper. (medium)	48.90	391.20	74.15	593.20		
1 Equip. Oper. (oiler)	43.55	348.40	66.00	528.00		
2 Gas Engine Vibrators		66.00		72.60		
1 Concrete Bucket, 1 C.Y.		23.00		25.30		
1 Hyd. Crane, 55 Ton		1137.00		1250.70	17.03	18.73
72 L.H., Daily Totals		$4093.20		$5729.80	$56.85	$79.58

Crew C-7A	Hr.	Daily	Hr.	Daily	Bare Costs	Incl. O&P
1 Labor Foreman (outside)	$38.65	$309.20	$59.65	$477.20	$37.13	$57.13
5 Laborers	36.65	1466.00	56.55	2262.00		
2 Truck Drivers (heavy)	37.55	600.80	57.30	916.80		
2 Conc. Transit Mixers		2050.00		2255.00	32.03	35.23
64 L.H., Daily Totals		$4426.00		$5911.00	$69.16	$92.36

Crew C-7B	Hr.	Daily	Hr.	Daily	Bare Costs	Incl. O&P
1 Labor Foreman (outside)	$38.65	$309.20	$59.65	$477.20	$39.46	$60.58
5 Laborers	36.65	1466.00	56.55	2262.00		
1 Equipment Operator, Crane	50.25	402.00	76.20	609.60		
1 Equipment Oiler	43.55	348.40	66.00	528.00		
1 Conc. Bucket, 2 C.Y.		35.60		39.16		
1 Lattice Boom Crane, 165 Ton		1954.00		2149.40	31.09	34.20
64 L.H., Daily Totals		$4515.20		$6065.36	$70.55	$94.77

Crew C-7C	Hr.	Daily	Hr.	Daily	Bare Costs	Incl. O&P
1 Labor Foreman (outside)	$38.65	$309.20	$59.65	$477.20	$39.96	$61.34
5 Laborers	36.65	1466.00	56.55	2262.00		
2 Equipment Operators (med.)	48.90	782.40	74.15	1186.40		
2 F.E. Loaders, W.M., 4 C.Y.		1347.60		1482.36	21.06	23.16
64 L.H., Daily Totals		$3905.20		$5407.96	$61.02	$84.50

Crew C-7D	Hr.	Daily	Hr.	Daily	Bare Costs	Incl. O&P
1 Labor Foreman (outside)	$38.65	$309.20	$59.65	$477.20	$38.69	$59.51
5 Laborers	36.65	1466.00	56.55	2262.00		
1 Equip. Oper. (medium)	48.90	391.20	74.15	593.20		
1 Concrete Conveyer		205.40		225.94	3.67	4.03
56 L.H., Daily Totals		$2371.80		$3558.34	$42.35	$63.54

Crew C-8	Hr.	Daily	Hr.	Daily	Bare Costs	Incl. O&P
1 Labor Foreman (outside)	$38.65	$309.20	$59.65	$477.20	$40.80	$61.95
3 Laborers	36.65	879.60	56.55	1357.20		
2 Cement Finishers	44.05	704.80	65.10	1041.60		
1 Equip. Oper. (medium)	48.90	391.20	74.15	593.20		
1 Concrete Pump (Small)		710.40		781.44	12.69	13.95
56 L.H., Daily Totals		$2995.20		$4250.64	$53.49	$75.90

Crew C-8A	Hr.	Daily	Hr.	Daily	Bare Costs	Incl. O&P
1 Labor Foreman (outside)	$38.65	$309.20	$59.65	$477.20	$39.45	$59.92
3 Laborers	36.65	879.60	56.55	1357.20		
2 Cement Finishers	44.05	704.80	65.10	1041.60		
48 L.H., Daily Totals		$1893.60		$2876.00	$39.45	$59.92

Crew C-8B	Hr.	Daily	Hr.	Daily	Bare Costs	Incl. O&P
1 Labor Foreman (outside)	$38.65	$309.20	$59.65	$477.20	$39.50	$60.69
3 Laborers	36.65	879.60	56.55	1357.20		
1 Equip. Oper. (medium)	48.90	391.20	74.15	593.20		
1 Vibrating Power Screed		63.25		69.58		
1 Roller, Vibratory, 25 Ton		692.80		762.08		
1 Dozer, 200 H.P.		1325.00		1457.50	52.03	57.23
40 L.H., Daily Totals		$3661.05		$4716.76	$91.53	$117.92

Crew C-8C	Hr.	Daily	Hr.	Daily	Bare Costs	Incl. O&P
1 Labor Foreman (outside)	$38.65	$309.20	$59.65	$477.20	$40.26	$61.42
3 Laborers	36.65	879.60	56.55	1357.20		
1 Cement Finisher	44.05	352.40	65.10	520.80		
1 Equip. Oper. (medium)	48.90	391.20	74.15	593.20		
1 Shotcrete Rig, 12 C.Y./hr		252.80		278.08		
1 Air Compressor, 160 cfm		157.40		173.14		
4 -50' Air Hoses, 1"		16.40		18.04		
4 -50' Air Hoses, 2"		31.00		34.10	9.53	10.49
48 L.H., Daily Totals		$2390.00		$3451.76	$49.79	$71.91

Crew C-8D	Hr.	Daily	Hr.	Daily	Bare Costs	Incl. O&P
1 Labor Foreman (outside)	$38.65	$309.20	$59.65	$477.20	$41.60	$63.16
1 Laborer	36.65	293.20	56.55	452.40		
1 Cement Finisher	44.05	352.40	65.10	520.80		
1 Equipment Oper. (light)	47.05	376.40	71.35	570.80		
1 Air Compressor, 250 cfm		202.40		222.64		
2 -50' Air Hoses, 1"		8.20		9.02	6.58	7.24
32 L.H., Daily Totals		$1541.80		$2252.86	$48.18	$70.40

Crew C-8E	Hr.	Daily	Hr.	Daily	Bare Costs	Incl. O&P
1 Labor Foreman (outside)	$38.65	$309.20	$59.65	$477.20	$39.95	$60.96
3 Laborers	36.65	879.60	56.55	1357.20		
1 Cement Finisher	44.05	352.40	65.10	520.80		
1 Equipment Oper. (light)	47.05	376.40	71.35	570.80		
1 Shotcrete Rig, 35 C.Y./hr		284.40		312.84		
1 Air Compressor, 250 cfm		202.40		222.64		
4 -50' Air Hoses, 1"		16.40		18.04		
4 -50' Air Hoses, 2"		31.00		34.10	11.13	12.24
48 L.H., Daily Totals		$2451.80		$3513.62	$51.08	$73.20

Crew C-10	Hr.	Daily	Hr.	Daily	Bare Costs	Incl. O&P
1 Laborer	$36.65	$293.20	$56.55	$452.40	$41.58	$62.25
2 Cement Finishers	44.05	704.80	65.10	1041.60		
24 L.H., Daily Totals		$998.00		$1494.00	$41.58	$62.25

Crew C-10B	Hr.	Daily	Hr.	Daily	Bare Costs	Incl. O&P
3 Laborers	$36.65	$879.60	$56.55	$1357.20	$39.61	$59.97
2 Cement Finishers	44.05	704.80	65.10	1041.60		
1 Concrete Mixer, 10 C.F.		176.80		194.48		
2 Trowels, 48" Walk-Behind		107.20		117.92	7.10	7.81
40 L.H., Daily Totals		$1868.40		$2711.20	$46.71	$67.78

Crew C-10C	Hr.	Daily	Hr.	Daily	Bare Costs	Incl. O&P
1 Laborer	$36.65	$293.20	$56.55	$452.40	$41.58	$62.25
2 Cement Finishers	44.05	704.80	65.10	1041.60		
1 Trowel, 48" Walk-Behind		53.60		58.96	2.23	2.46
24 L.H., Daily Totals		$1051.60		$1552.96	$43.82	$64.71

Crew No.	Bare Costs		Incl. Subs O&P		Cost Per Labor-Hour	
Crew C-10D	Hr.	Daily	Hr.	Daily	Bare Costs	Incl. O&P
1 Laborer	$36.65	$293.20	$56.55	$452.40	$41.58	$62.25
2 Cement Finishers	44.05	704.80	65.10	1041.60		
1 Vibrating Power Screed		63.25		69.58		
1 Trowel, 48" Walk-Behind		53.60		58.96	4.87	5.36
24 L.H., Daily Totals		$1114.85		$1622.54	$46.45	$67.61

Crew No.	Bare Costs		Incl. Subs O&P		Cost Per Labor-Hour	
Crew C-10E	Hr.	Daily	Hr.	Daily	Bare Costs	Incl. O&P
1 Laborer	$36.65	$293.20	$56.55	$452.40	$41.58	$62.25
2 Cement Finishers	44.05	704.80	65.10	1041.60		
1 Vibrating Power Screed		63.25		69.58		
1 Cement Trowel, 96" Ride-On		197.40		217.14	10.86	11.95
24 L.H., Daily Totals		$1258.65		$1780.71	$52.44	$74.20

Crew No.	Bare Costs		Incl. Subs O&P		Cost Per Labor-Hour	
Crew C-10F	Hr.	Daily	Hr.	Daily	Bare Costs	Incl. O&P
1 Laborer	$36.65	$293.20	$56.55	$452.40	$41.58	$62.25
2 Cement Finishers	44.05	704.80	65.10	1041.60		
1 Aerial Lift Truck, 60' Boom		456.00		501.60	19.00	20.90
24 L.H., Daily Totals		$1454.00		$1995.60	$60.58	$83.15

Crew No.	Bare Costs		Incl. Subs O&P		Cost Per Labor-Hour	
Crew C-11	Hr.	Daily	Hr.	Daily	Bare Costs	Incl. O&P
1 Struc. Steel Foreman (outside)	$53.10	$424.80	$93.70	$749.60	$50.39	$86.34
6 Struc. Steel Workers	51.10	2452.80	90.20	4329.60		
1 Equip. Oper. (crane)	50.25	402.00	76.20	609.60		
1 Equip. Oper. (oiler)	43.55	348.40	66.00	528.00		
1 Lattice Boom Crane, 150 Ton		1838.00		2021.80	25.53	28.08
72 L.H., Daily Totals		$5466.00		$8238.60	$75.92	$114.43

Crew No.	Bare Costs		Incl. Subs O&P		Cost Per Labor-Hour	
Crew C-12	Hr.	Daily	Hr.	Daily	Bare Costs	Incl. O&P
1 Carpenter Foreman (outside)	$47.85	$382.80	$73.85	$590.80	$45.38	$69.81
3 Carpenters	45.85	1100.40	70.75	1698.00		
1 Laborer	36.65	293.20	56.55	452.40		
1 Equip. Oper. (crane)	50.25	402.00	76.20	609.60		
1 Hyd. Crane, 12 Ton		663.00		729.30	13.81	15.19
48 L.H., Daily Totals		$2841.40		$4080.10	$59.20	$85.00

Crew No.	Bare Costs		Incl. Subs O&P		Cost Per Labor-Hour	
Crew C-13	Hr.	Daily	Hr.	Daily	Bare Costs	Incl. O&P
1 Struc. Steel Worker	$51.10	$408.80	$90.20	$721.60	$49.35	$83.72
1 Welder	51.10	408.80	90.20	721.60		
1 Carpenter	45.85	366.80	70.75	566.00		
1 Welder, Gas Engine, 300 amp		142.00		156.20	5.92	6.51
24 L.H., Daily Totals		$1326.40		$2165.40	$55.27	$90.22

Crew No.	Bare Costs		Incl. Subs O&P		Cost Per Labor-Hour	
Crew C-14	Hr.	Daily	Hr.	Daily	Bare Costs	Incl. O&P
1 Carpenter Foreman (outside)	$47.85	$382.80	$73.85	$590.80	$44.90	$69.13
5 Carpenters	45.85	1834.00	70.75	2830.00		
4 Laborers	36.65	1172.80	56.55	1809.60		
4 Rodmen (reinf.)	50.65	1620.80	79.55	2545.60		
2 Cement Finishers	44.05	704.80	65.10	1041.60		
1 Equip. Oper. (crane)	50.25	402.00	76.20	609.60		
1 Equip. Oper. (oiler)	43.55	348.40	66.00	528.00		
1 Hyd. Crane, 80 Ton		1652.00		1817.20	11.47	12.62
144 L.H., Daily Totals		$8117.60		$11772.40	$56.37	$81.75

Crew No.	Bare Costs		Incl. Subs O&P		Cost Per Labor-Hour	
Crew C-14A	Hr.	Daily	Hr.	Daily	Bare Costs	Incl. O&P
1 Carpenter Foreman (outside)	$47.85	$382.80	$73.85	$590.80	$46.01	$71.06
16 Carpenters	45.85	5868.80	70.75	9056.00		
4 Rodmen (reinf.)	50.65	1620.80	79.55	2545.60		
2 Laborers	36.65	586.40	56.55	904.80		
1 Cement Finisher	44.05	352.40	65.10	520.80		
1 Equip. Oper. (medium)	48.90	391.20	74.15	593.20		
1 Gas Engine Vibrator		33.00		36.30		
1 Concrete Pump (Small)		710.40		781.44	3.72	4.09
200 L.H., Daily Totals		$9945.80		$15028.94	$49.73	$75.14

Crew No.	Bare Costs		Incl. Subs O&P		Cost Per Labor-Hour	
Crew C-14B	Hr.	Daily	Hr.	Daily	Bare Costs	Incl. O&P
1 Carpenter Foreman (outside)	$47.85	$382.80	$73.85	$590.80	$45.94	$70.83
16 Carpenters	45.85	5868.80	70.75	9056.00		
4 Rodmen (reinf.)	50.65	1620.80	79.55	2545.60		
2 Laborers	36.65	586.40	56.55	904.80		
2 Cement Finishers	44.05	704.80	65.10	1041.60		
1 Equip. Oper. (medium)	48.90	391.20	74.15	593.20		
1 Gas Engine Vibrator		33.00		36.30		
1 Concrete Pump (Small)		710.40		781.44	3.57	3.93
208 L.H., Daily Totals		$10298.20		$15549.74	$49.51	$74.76

Crew No.	Bare Costs		Incl. Subs O&P		Cost Per Labor-Hour	
Crew C-14C	Hr.	Daily	Hr.	Daily	Bare Costs	Incl. O&P
1 Carpenter Foreman (outside)	$47.85	$382.80	$73.85	$590.80	$43.92	$67.77
6 Carpenters	45.85	2200.80	70.75	3396.00		
2 Rodmen (reinf.)	50.65	810.40	79.55	1272.80		
4 Laborers	36.65	1172.80	56.55	1809.60		
1 Cement Finisher	44.05	352.40	65.10	520.80		
1 Gas Engine Vibrator		33.00		36.30	.29	.32
112 L.H., Daily Totals		$4952.20		$7626.30	$44.22	$68.09

Crew No.	Bare Costs		Incl. Subs O&P		Cost Per Labor-Hour	
Crew C-14D	Hr.	Daily	Hr.	Daily	Bare Costs	Incl. O&P
1 Carpenter Foreman (outside)	$47.85	$382.80	$73.85	$590.80	$45.63	$70.35
18 Carpenters	45.85	6602.40	70.75	10188.00		
2 Rodmen (reinf.)	50.65	810.40	79.55	1272.80		
2 Laborers	36.65	586.40	56.55	904.80		
1 Cement Finisher	44.05	352.40	65.10	520.80		
1 Equip. Oper. (medium)	48.90	391.20	74.15	593.20		
1 Gas Engine Vibrator		33.00		36.30		
1 Concrete Pump (Small)		710.40		781.44	3.72	4.09
200 L.H., Daily Totals		$9869.00		$14888.14	$49.34	$74.44

Crew No.	Bare Costs		Incl. Subs O&P		Cost Per Labor-Hour	
Crew C-14E	Hr.	Daily	Hr.	Daily	Bare Costs	Incl. O&P
1 Carpenter Foreman (outside)	$47.85	$382.80	$73.85	$590.80	$45.10	$69.85
2 Carpenters	45.85	733.60	70.75	1132.00		
4 Rodmen (reinf.)	50.65	1620.80	79.55	2545.60		
3 Laborers	36.65	879.60	56.55	1357.20		
1 Cement Finisher	44.05	352.40	65.10	520.80		
1 Gas Engine Vibrator		33.00		36.30	.38	.41
88 L.H., Daily Totals		$4002.20		$6182.70	$45.48	$70.26

Crew No.	Bare Costs		Incl. Subs O&P		Cost Per Labor-Hour	
Crew C-14F	Hr.	Daily	Hr.	Daily	Bare Costs	Incl. O&P
1 Labor Foreman (outside)	$38.65	$309.20	$59.65	$477.20	$41.81	$62.59
2 Laborers	36.65	586.40	56.55	904.80		
6 Cement Finishers	44.05	2114.40	65.10	3124.80		
1 Gas Engine Vibrator		33.00		36.30	.46	.50
72 L.H., Daily Totals		$3043.00		$4543.10	$42.26	$63.10

Crew No.	Bare Costs		Incl. Subs O&P		Cost Per Labor-Hour	

Crew C-14G

	Hr.	Daily	Hr.	Daily	Bare Costs	Incl. O&P
1 Labor Foreman (outside)	$38.65	$309.20	$59.65	$477.20	$41.16	$61.88
2 Laborers	36.65	586.40	56.55	904.80		
4 Cement Finishers	44.05	1409.60	65.10	2083.20		
1 Gas Engine Vibrator		33.00		36.30	.59	.65
56 L.H., Daily Totals		$2338.20		$3501.50	$41.75	$62.53

Crew C-14H

	Hr.	Daily	Hr.	Daily	Bare Costs	Incl. O&P
1 Carpenter Foreman (outside)	$47.85	$382.80	$73.85	$590.80	$45.15	$69.42
2 Carpenters	45.85	733.60	70.75	1132.00		
1 Rodman (reinf.)	50.65	405.20	79.55	636.40		
1 Laborer	36.65	293.20	56.55	452.40		
1 Cement Finisher	44.05	352.40	65.10	520.80		
1 Gas Engine Vibrator		33.00		36.30	.69	.76
48 L.H., Daily Totals		$2200.20		$3368.70	$45.84	$70.18

Crew C-14L

	Hr.	Daily	Hr.	Daily	Bare Costs	Incl. O&P
1 Carpenter Foreman (outside)	$47.85	$382.80	$73.85	$590.80	$42.80	$65.80
6 Carpenters	45.85	2200.80	70.75	3396.00		
4 Laborers	36.65	1172.80	56.55	1809.60		
1 Cement Finisher	44.05	352.40	65.10	520.80		
1 Gas Engine Vibrator		33.00		36.30	.34	.38
96 L.H., Daily Totals		$4141.80		$6353.50	$43.14	$66.18

Crew C-14M

	Hr.	Daily	Hr.	Daily	Bare Costs	Incl. O&P
1 Carpenter Foreman (outside)	$47.85	$382.80	$73.85	$590.80	$44.56	$68.41
2 Carpenters	45.85	733.60	70.75	1132.00		
1 Rodman (reinf.)	50.65	405.20	79.55	636.40		
2 Laborers	36.65	586.40	56.55	904.80		
1 Cement Finisher	44.05	352.40	65.10	520.80		
1 Equip. Oper. (medium)	48.90	391.20	74.15	593.20		
1 Gas Engine Vibrator		33.00		36.30		
1 Concrete Pump (Small)		710.40		781.44	11.62	12.78
64 L.H., Daily Totals		$3595.00		$5195.74	$56.17	$81.18

Crew C-15

	Hr.	Daily	Hr.	Daily	Bare Costs	Incl. O&P
1 Carpenter Foreman (outside)	$47.85	$382.80	$73.85	$590.80	$43.14	$66.08
2 Carpenters	45.85	733.60	70.75	1132.00		
3 Laborers	36.65	879.60	56.55	1357.20		
2 Cement Finishers	44.05	704.80	65.10	1041.60		
1 Rodman (reinf.)	50.65	405.20	79.55	636.40		
72 L.H., Daily Totals		$3106.00		$4758.00	$43.14	$66.08

Crew C-16

	Hr.	Daily	Hr.	Daily	Bare Costs	Incl. O&P
1 Labor Foreman (outside)	$38.65	$309.20	$59.65	$477.20	$40.80	$61.95
3 Laborers	36.65	879.60	56.55	1357.20		
2 Cement Finishers	44.05	704.80	65.10	1041.60		
1 Equip. Oper. (medium)	48.90	391.20	74.15	593.20		
1 Gunite Pump Rig		370.40		407.44		
2 -50' Air Hoses, 3/4"		6.50		7.15		
2 -50' Air Hoses, 2"		15.50		17.05	7.01	7.71
56 L.H., Daily Totals		$2677.20		$3900.84	$47.81	$69.66

Crew C-16A

	Hr.	Daily	Hr.	Daily	Bare Costs	Incl. O&P
1 Laborer	$36.65	$293.20	$56.55	$452.40	$43.41	$65.22
2 Cement Finishers	44.05	704.80	65.10	1041.60		
1 Equip. Oper. (medium)	48.90	391.20	74.15	593.20		
1 Gunite Pump Rig		370.40		407.44		
2 -50' Air Hoses, 3/4"		6.50		7.15		
2 -50' Air Hoses, 2"		15.50		17.05		
1 Aerial Lift Truck, 60' Boom		456.00		501.60	26.51	29.16
32 L.H., Daily Totals		$2237.60		$3020.44	$69.92	$94.39

Crew No.	Bare Costs		Incl. Subs O&P		Cost Per Labor-Hour	

Crew C-17

	Hr.	Daily	Hr.	Daily	Bare Costs	Incl. O&P
2 Skilled Worker Foremen (out)	$49.30	$788.80	$76.35	$1221.60	$47.70	$73.87
8 Skilled Workers	47.30	3027.20	73.25	4688.00		
80 L.H., Daily Totals		$3816.00		$5909.60	$47.70	$73.87

Crew C-17A

	Hr.	Daily	Hr.	Daily	Bare Costs	Incl. O&P
2 Skilled Worker Foremen (out)	$49.30	$788.80	$76.35	$1221.60	$47.73	$73.90
8 Skilled Workers	47.30	3027.20	73.25	4688.00		
.125 Equip. Oper. (crane)	50.25	50.25	76.20	76.20		
.125 Hyd. Crane, 80 Ton		206.50		227.15	2.55	2.80
81 L.H., Daily Totals		$4072.75		$6212.95	$50.28	$76.70

Crew C-17B

	Hr.	Daily	Hr.	Daily	Bare Costs	Incl. O&P
2 Skilled Worker Foremen (out)	$49.30	$788.80	$76.35	$1221.60	$47.76	$73.93
8 Skilled Workers	47.30	3027.20	73.25	4688.00		
.25 Equip. Oper. (crane)	50.25	100.50	76.20	152.40		
.25 Hyd. Crane, 80 Ton		413.00		454.30		
.25 Trowel, 48" Walk-Behind		13.40		14.74	5.20	5.72
82 L.H., Daily Totals		$4342.90		$6531.04	$52.96	$79.65

Crew C-17C

	Hr.	Daily	Hr.	Daily	Bare Costs	Incl. O&P
2 Skilled Worker Foremen (out)	$49.30	$788.80	$76.35	$1221.60	$47.79	$73.95
8 Skilled Workers	47.30	3027.20	73.25	4688.00		
.375 Equip. Oper. (crane)	50.25	150.75	76.20	228.60		
.375 Hyd. Crane, 80 Ton		619.50		681.45	7.46	8.21
83 L.H., Daily Totals		$4586.25		$6819.65	$55.26	$82.16

Crew C-17D

	Hr.	Daily	Hr.	Daily	Bare Costs	Incl. O&P
2 Skilled Worker Foremen (out)	$49.30	$788.80	$76.35	$1221.60	$47.82	$73.98
8 Skilled Workers	47.30	3027.20	73.25	4688.00		
.5 Equip. Oper. (crane)	50.25	201.00	76.20	304.80		
.5 Hyd. Crane, 80 Ton		826.00		908.60	9.83	10.82
84 L.H., Daily Totals		$4843.00		$7123.00	$57.65	$84.80

Crew C-17E

	Hr.	Daily	Hr.	Daily	Bare Costs	Incl. O&P
2 Skilled Worker Foremen (out)	$49.30	$788.80	$76.35	$1221.60	$47.70	$73.87
8 Skilled Workers	47.30	3027.20	73.25	4688.00		
1 Hyd. Jack with Rods		94.55		104.01	1.18	1.30
80 L.H., Daily Totals		$3910.55		$6013.60	$48.88	$75.17

Crew C-18

	Hr.	Daily	Hr.	Daily	Bare Costs	Incl. O&P
.125 Labor Foreman (outside)	$38.65	$38.65	$59.65	$59.65	$36.87	$56.89
1 Laborer	36.65	293.20	56.55	452.40		
1 Concrete Cart, 10 C.F.		60.20		66.22	6.69	7.36
9 L.H., Daily Totals		$392.05		$578.27	$43.56	$64.25

Crew C-19

	Hr.	Daily	Hr.	Daily	Bare Costs	Incl. O&P
.125 Labor Foreman (outside)	$38.65	$38.65	$59.65	$59.65	$36.87	$56.89
1 Laborer	36.65	293.20	56.55	452.40		
1 Concrete Cart, 18 C.F.		101.20		111.32	11.24	12.37
9 L.H., Daily Totals		$433.05		$623.37	$48.12	$69.26

Crew C-20

	Hr.	Daily	Hr.	Daily	Bare Costs	Incl. O&P
1 Labor Foreman (outside)	$38.65	$309.20	$59.65	$477.20	$39.36	$60.21
5 Laborers	36.65	1466.00	56.55	2262.00		
1 Cement Finisher	44.05	352.40	65.10	520.80		
1 Equip. Oper. (medium)	48.90	391.20	74.15	593.20		
2 Gas Engine Vibrators		66.00		72.60		
1 Concrete Pump (Small)		710.40		781.44	12.13	13.34
64 L.H., Daily Totals		$3295.20		$4707.24	$51.49	$73.55

Crew C-21

Crew No.	Bare Costs Hr.	Daily	Incl. Subs O&P Hr.	Daily	Cost Per Labor-Hour Bare Costs	Incl. O&P
1 Labor Foreman (outside)	$38.65	$309.20	$59.65	$477.20	$39.36	$60.21
5 Laborers	36.65	1466.00	56.55	2262.00		
1 Cement Finisher	44.05	352.40	65.10	520.80		
1 Equip. Oper. (medium)	48.90	391.20	74.15	593.20		
2 Gas Engine Vibrators		66.00		72.60		
1 Concrete Conveyer		205.40		225.94	4.24	4.66
64 L.H., Daily Totals		$2790.20		$4151.74	$43.60	$64.87

Crew C-22

Crew No.	Hr.	Daily	Hr.	Daily	Bare Costs	Incl. O&P
1 Rodman Foreman (outside)	$52.65	$421.20	$82.70	$661.60	$50.85	$79.75
4 Rodmen (reinf.)	50.65	1620.80	79.55	2545.60		
.125 Equip. Oper. (crane)	50.25	50.25	76.20	76.20		
.125 Equip. Oper. (oiler)	43.55	43.55	66.00	66.00		
.125 Hyd. Crane, 25 Ton		93.22		102.55	2.22	2.44
42 L.H., Daily Totals		$2229.03		$3451.95	$53.07	$82.19

Crew C-23

Crew No.	Hr.	Daily	Hr.	Daily	Bare Costs	Incl. O&P
2 Skilled Worker Foremen (out)	$49.30	$788.80	$76.35	$1221.60	$47.62	$73.44
6 Skilled Workers	47.30	2270.40	73.25	3516.00		
1 Equip. Oper. (crane)	50.25	402.00	76.20	609.60		
1 Equip. Oper. (oiler)	43.55	348.40	66.00	528.00		
1 Lattice Boom Crane, 90 Ton		1529.00		1681.90	19.11	21.02
80 L.H., Daily Totals		$5338.60		$7557.10	$66.73	$94.46

Crew C-23A

Crew No.	Hr.	Daily	Hr.	Daily	Bare Costs	Incl. O&P
1 Labor Foreman (outside)	$38.65	$309.20	$59.65	$477.20	$41.15	$62.99
2 Laborers	36.65	586.40	56.55	904.80		
1 Equip. Oper. (crane)	50.25	402.00	76.20	609.60		
1 Equip. Oper. (oiler)	43.55	348.40	66.00	528.00		
1 Crawler Crane, 100 Ton		1703.00		1873.30		
3 Conc. Buckets, 8 C.Y.		595.80		655.38	57.47	63.22
40 L.H., Daily Totals		$3944.80		$5048.28	$98.62	$126.21

Crew C-24

Crew No.	Hr.	Daily	Hr.	Daily	Bare Costs	Incl. O&P
2 Skilled Worker Foremen (out)	$49.30	$788.80	$76.35	$1221.60	$47.62	$73.44
6 Skilled Workers	47.30	2270.40	73.25	3516.00		
1 Equip. Oper. (crane)	50.25	402.00	76.20	609.60		
1 Equip. Oper. (oiler)	43.55	348.40	66.00	528.00		
1 Lattice Boom Crane, 150 Ton		1838.00		2021.80	22.98	25.27
80 L.H., Daily Totals		$5647.60		$7897.00	$70.59	$98.71

Crew C-25

Crew No.	Hr.	Daily	Hr.	Daily	Bare Costs	Incl. O&P
2 Rodmen (reinf.)	$50.65	$810.40	$79.55	$1272.80	$39.90	$64.58
2 Rodmen Helpers	29.15	466.40	49.60	793.60		
32 L.H., Daily Totals		$1276.80		$2066.40	$39.90	$64.58

Crew C-27

Crew No.	Hr.	Daily	Hr.	Daily	Bare Costs	Incl. O&P
2 Cement Finishers	$44.05	$704.80	$65.10	$1041.60	$44.05	$65.10
1 Concrete Saw		175.20		192.72	10.95	12.05
16 L.H., Daily Totals		$880.00		$1234.32	$55.00	$77.14

Crew C-28

Crew No.	Hr.	Daily	Hr.	Daily	Bare Costs	Incl. O&P
1 Cement Finisher	$44.05	$352.40	$65.10	$520.80	$44.05	$65.10
1 Portable Air Compressor, Gas		17.70		19.47	2.21	2.43
8 L.H., Daily Totals		$370.10		$540.27	$46.26	$67.53

Crew C-29

Crew No.	Hr.	Daily	Hr.	Daily	Bare Costs	Incl. O&P
1 Laborer	$36.65	$293.20	$56.55	$452.40	$36.65	$56.55
1 Pressure Washer		69.40		76.34	8.68	9.54
8 L.H., Daily Totals		$362.60		$528.74	$45.33	$66.09

Crew C-30

Crew No.	Hr.	Daily	Hr.	Daily	Bare Costs	Incl. O&P
1 Laborer	$36.65	$293.20	$56.55	$452.40	$36.65	$56.55
1 Concrete Mixer, 10 C.F.		176.80		194.48	22.10	24.31
8 L.H., Daily Totals		$470.00		$646.88	$58.75	$80.86

Crew C-31

Crew No.	Hr.	Daily	Hr.	Daily	Bare Costs	Incl. O&P
1 Cement Finisher	$44.05	$352.40	$65.10	$520.80	$44.05	$65.10
1 Grout Pump		370.40		407.44	46.30	50.93
8 L.H., Daily Totals		$722.80		$928.24	$90.35	$116.03

Crew C-32

Crew No.	Hr.	Daily	Hr.	Daily	Bare Costs	Incl. O&P
1 Cement Finisher	$44.05	$352.40	$65.10	$520.80	$40.35	$60.83
1 Laborer	36.65	293.20	56.55	452.40		
1 Crack Chaser Saw, Gas, 6 H.P.		30.80		33.88		
1 Vacuum Pick-Up System		60.75		66.83	5.72	6.29
16 L.H., Daily Totals		$737.15		$1073.91	$46.07	$67.12

Crew D-1

Crew No.	Hr.	Daily	Hr.	Daily	Bare Costs	Incl. O&P
1 Bricklayer	$45.60	$364.80	$69.50	$556.00	$41.30	$62.95
1 Bricklayer Helper	37.00	296.00	56.40	451.20		
16 L.H., Daily Totals		$660.80		$1007.20	$41.30	$62.95

Crew D-2

Crew No.	Hr.	Daily	Hr.	Daily	Bare Costs	Incl. O&P
3 Bricklayers	$45.60	$1094.40	$69.50	$1668.00	$42.50	$64.85
2 Bricklayer Helpers	37.00	592.00	56.40	902.40		
.5 Carpenter	45.85	183.40	70.75	283.00		
44 L.H., Daily Totals		$1869.80		$2853.40	$42.50	$64.85

Crew D-3

Crew No.	Hr.	Daily	Hr.	Daily	Bare Costs	Incl. O&P
3 Bricklayers	$45.60	$1094.40	$69.50	$1668.00	$42.34	$64.57
2 Bricklayer Helpers	37.00	592.00	56.40	902.40		
.25 Carpenter	45.85	91.70	70.75	141.50		
42 L.H., Daily Totals		$1778.10		$2711.90	$42.34	$64.57

Crew D-4

Crew No.	Hr.	Daily	Hr.	Daily	Bare Costs	Incl. O&P
1 Bricklayer	$45.60	$364.80	$69.50	$556.00	$41.66	$63.41
2 Bricklayer Helpers	37.00	592.00	56.40	902.40		
1 Equip. Oper. (light)	47.05	376.40	71.35	570.80		
1 Grout Pump, 50 C.F./hr.		134.80		148.28	4.21	4.63
32 L.H., Daily Totals		$1468.00		$2177.48	$45.88	$68.05

Crew D-5

Crew No.	Hr.	Daily	Hr.	Daily	Bare Costs	Incl. O&P
1 Bricklayer	$45.60	$364.80	$69.50	$556.00	$45.60	$69.50
8 L.H., Daily Totals		$364.80		$556.00	$45.60	$69.50

Crew D-6

Crew No.	Hr.	Daily	Hr.	Daily	Bare Costs	Incl. O&P
3 Bricklayers	$45.60	$1094.40	$69.50	$1668.00	$41.48	$63.26
3 Bricklayer Helpers	37.00	888.00	56.40	1353.60		
.25 Carpenter	45.85	91.70	70.75	141.50		
50 L.H., Daily Totals		$2074.10		$3163.10	$41.48	$63.26

Crew D-7

Crew No.	Hr.	Daily	Hr.	Daily	Bare Costs	Incl. O&P
1 Tile Layer	$41.95	$335.60	$61.85	$494.80	$37.60	$55.42
1 Tile Layer Helper	33.25	266.00	49.00	392.00		
16 L.H., Daily Totals		$601.60		$886.80	$37.60	$55.42

Crew D-8

Crew No.	Hr.	Daily	Hr.	Daily	Bare Costs	Incl. O&P
3 Bricklayers	$45.60	$1094.40	$69.50	$1668.00	$42.16	$64.26
2 Bricklayer Helpers	37.00	592.00	56.40	902.40		
40 L.H., Daily Totals		$1686.40		$2570.40	$42.16	$64.26

Crew No.	Bare Costs Hr.	Bare Costs Daily	Incl. Subs O&P Hr.	Incl. Subs O&P Daily	Cost Per Labor-Hour Bare Costs	Cost Per Labor-Hour Incl. O&P
Crew D-9						
3 Bricklayers	$45.60	$1094.40	$69.50	$1668.00	$41.30	$62.95
3 Bricklayer Helpers	37.00	888.00	56.40	1353.60		
48 L.H., Daily Totals		$1982.40		$3021.60	$41.30	$62.95
Crew D-10						
1 Bricklayer Foreman (outside)	$47.60	$380.80	$72.55	$580.40	$45.11	$68.66
1 Bricklayer	45.60	364.80	69.50	556.00		
1 Bricklayer Helper	37.00	296.00	56.40	451.20		
1 Equip. Oper. (crane)	50.25	402.00	76.20	609.60		
1 S.P. Crane, 4x4, 12 Ton		481.20		529.32	15.04	16.54
32 L.H., Daily Totals		$1924.80		$2726.52	$60.15	$85.20
Crew D-11						
1 Bricklayer Foreman (outside)	$47.60	$380.80	$72.55	$580.40	$43.40	$66.15
1 Bricklayer	45.60	364.80	69.50	556.00		
1 Bricklayer Helper	37.00	296.00	56.40	451.20		
24 L.H., Daily Totals		$1041.60		$1587.60	$43.40	$66.15
Crew D-12						
1 Bricklayer Foreman (outside)	$47.60	$380.80	$72.55	$580.40	$41.80	$63.71
1 Bricklayer	45.60	364.80	69.50	556.00		
2 Bricklayer Helpers	37.00	592.00	56.40	902.40		
32 L.H., Daily Totals		$1337.60		$2038.80	$41.80	$63.71
Crew D-13						
1 Bricklayer Foreman (outside)	$47.60	$380.80	$72.55	$580.40	$43.88	$66.97
1 Bricklayer	45.60	364.80	69.50	556.00		
2 Bricklayer Helpers	37.00	592.00	56.40	902.40		
1 Carpenter	45.85	366.80	70.75	566.00		
1 Equip. Oper. (crane)	50.25	402.00	76.20	609.60		
1 S.P. Crane, 4x4, 12 Ton		481.20		529.32	10.03	11.03
48 L.H., Daily Totals		$2587.60		$3743.72	$53.91	$77.99
Crew E-1						
1 Welder Foreman (outside)	$53.10	$424.80	$93.70	$749.60	$50.42	$85.08
1 Welder	51.10	408.80	90.20	721.60		
1 Equip. Oper. (light)	47.05	376.40	71.35	570.80		
1 Welder, Gas Engine, 300 amp		142.00		156.20	5.92	6.51
24 L.H., Daily Totals		$1352.00		$2198.20	$56.33	$91.59
Crew E-2						
1 Struc. Steel Foreman (outside)	$53.10	$424.80	$93.70	$749.60	$50.19	$85.24
4 Struc. Steel Workers	51.10	1635.20	90.20	2886.40		
1 Equip. Oper. (crane)	50.25	402.00	76.20	609.60		
1 Equip. Oper. (oiler)	43.55	348.40	66.00	528.00		
1 Lattice Boom Crane, 90 Ton		1529.00		1681.90	27.30	30.03
56 L.H., Daily Totals		$4339.40		$6455.50	$77.49	$115.28
Crew E-3						
1 Struc. Steel Foreman (outside)	$53.10	$424.80	$93.70	$749.60	$51.77	$91.37
1 Struc. Steel Worker	51.10	408.80	90.20	721.60		
1 Welder	51.10	408.80	90.20	721.60		
1 Welder, Gas Engine, 300 amp		142.00		156.20	5.92	6.51
24 L.H., Daily Totals		$1384.40		$2349.00	$57.68	$97.88

Crew No.	Bare Costs Hr.	Bare Costs Daily	Incl. Subs O&P Hr.	Incl. Subs O&P Daily	Cost Per Labor-Hour Bare Costs	Cost Per Labor-Hour Incl. O&P
Crew E-3A						
1 Struc. Steel Foreman (outside)	$53.10	$424.80	$93.70	$749.60	$51.77	$91.37
1 Struc. Steel Worker	51.10	408.80	90.20	721.60		
1 Welder	51.10	408.80	90.20	721.60		
1 Welder, Gas Engine, 300 amp		142.00		156.20		
1 Aerial Lift Truck, 40' Boom		325.00		357.50	19.46	21.40
24 L.H., Daily Totals		$1709.40		$2706.50	$71.22	$112.77
Crew E-4						
1 Struc. Steel Foreman (outside)	$53.10	$424.80	$93.70	$749.60	$51.60	$91.08
3 Struc. Steel Workers	51.10	1226.40	90.20	2164.80		
1 Welder, Gas Engine, 300 amp		142.00		156.20	4.44	4.88
32 L.H., Daily Totals		$1793.20		$3070.60	$56.04	$95.96
Crew E-5						
2 Struc. Steel Foremen (outside)	$53.10	$849.60	$93.70	$1499.20	$50.66	$87.08
5 Struc. Steel Workers	51.10	2044.00	90.20	3608.00		
1 Equip. Oper. (crane)	50.25	402.00	76.20	609.60		
1 Welder	51.10	408.80	90.20	721.60		
1 Equip. Oper. (oiler)	43.55	348.40	66.00	528.00		
1 Lattice Boom Crane, 90 Ton		1529.00		1681.90		
1 Welder, Gas Engine, 300 amp		142.00		156.20	20.89	22.98
80 L.H., Daily Totals		$5723.80		$8804.50	$71.55	$110.06
Crew E-6						
3 Struc. Steel Foremen (outside)	$53.10	$1274.40	$93.70	$2248.80	$50.70	$87.29
9 Struc. Steel Workers	51.10	3679.20	90.20	6494.40		
1 Equip. Oper. (crane)	50.25	402.00	76.20	609.60		
1 Welder	51.10	408.80	90.20	721.60		
1 Equip. Oper. (oiler)	43.55	348.40	66.00	528.00		
1 Equip. Oper. (light)	47.05	376.40	71.35	570.80		
1 Lattice Boom Crane, 90 Ton		1529.00		1681.90		
1 Welder, Gas Engine, 300 amp		142.00		156.20		
1 Air Compressor, 160 cfm		157.40		173.14		
2 Impact Wrenches		36.00		39.60	14.57	16.02
128 L.H., Daily Totals		$8353.60		$13224.04	$65.26	$103.31
Crew E-7						
1 Struc. Steel Foreman (outside)	$53.10	$424.80	$93.70	$749.60	$50.66	$87.08
4 Struc. Steel Workers	51.10	1635.20	90.20	2886.40		
1 Equip. Oper. (crane)	50.25	402.00	76.20	609.60		
1 Equip. Oper. (oiler)	43.55	348.40	66.00	528.00		
1 Welder Foreman (outside)	53.10	424.80	93.70	749.60		
2 Welders	51.10	817.60	90.20	1443.20		
1 Lattice Boom Crane, 90 Ton		1529.00		1681.90		
2 Welder, Gas Engine, 300 amp		284.00		312.40	22.66	24.93
80 L.H., Daily Totals		$5865.80		$8960.70	$73.32	$112.01
Crew E-8						
1 Struc. Steel Foreman (outside)	$53.10	$424.80	$93.70	$749.60	$50.45	$86.35
4 Struc. Steel Workers	51.10	1635.20	90.20	2886.40		
1 Welder Foreman (outside)	53.10	424.80	93.70	749.60		
4 Welders	51.10	1635.20	90.20	2886.40		
1 Equip. Oper. (crane)	50.25	402.00	76.20	609.60		
1 Equip. Oper. (oiler)	43.55	348.40	66.00	528.00		
1 Equip. Oper. (light)	47.05	376.40	71.35	570.80		
1 Lattice Boom Crane, 90 Ton		1529.00		1681.90		
4 Welder, Gas Engine, 300 amp		568.00		624.80	20.16	22.18
104 L.H., Daily Totals		$7343.80		$11287.10	$70.61	$108.53

Crew No.	Bare Costs		Incl. Subs O&P		Cost Per Labor-Hour	

Crew E-9

	Hr.	Daily	Hr.	Daily	Bare Costs	Incl. O&P
2 Struc. Steel Foremen (outside)	$53.10	$849.60	$93.70	$1499.20	$50.70	$87.29
5 Struc. Steel Workers	51.10	2044.00	90.20	3608.00		
1 Welder Foreman (outside)	53.10	424.80	93.70	749.60		
5 Welders	51.10	2044.00	90.20	3608.00		
1 Equip. Oper. (crane)	50.25	402.00	76.20	609.60		
1 Equip. Oper. (oiler)	43.55	348.40	66.00	528.00		
1 Equip. Oper. (light)	47.05	376.40	71.35	570.80		
1 Lattice Boom Crane, 90 Ton		1529.00		1681.90		
5 Welder, Gas Engine, 300 amp		710.00		781.00	17.49	19.24
128 L.H., Daily Totals		$8728.20		$13636.10	$68.19	$106.53

Crew E-10

	Hr.	Daily	Hr.	Daily	Bare Costs	Incl. O&P
1 Welder Foreman (outside)	$53.10	$424.80	$93.70	$749.60	$52.10	$91.95
1 Welder	51.10	408.80	90.20	721.60		
1 Welder, Gas Engine, 300 amp		142.00		156.20		
1 Flatbed Truck, Gas, 3 Ton		333.00		366.30	29.69	32.66
16 L.H., Daily Totals		$1308.60		$1993.70	$81.79	$124.61

Crew E-11

	Hr.	Daily	Hr.	Daily	Bare Costs	Incl. O&P
2 Painters, Struc. Steel	$40.50	$648.00	$73.05	$1168.80	$41.17	$68.50
1 Building Laborer	36.65	293.20	56.55	452.40		
1 Equip. Oper. (light)	47.05	376.40	71.35	570.80		
1 Air Compressor, 250 cfm		202.40		222.64		
1 Sandblaster, Portable, 3 C.F.		20.40		22.44		
1 Set Sand Blasting Accessories		13.85		15.23	7.40	8.13
32 L.H., Daily Totals		$1554.25		$2452.32	$48.57	$76.63

Crew E-11A

	Hr.	Daily	Hr.	Daily	Bare Costs	Incl. O&P
2 Painters, Struc. Steel	$40.50	$648.00	$73.05	$1168.80	$41.17	$68.50
1 Building Laborer	36.65	293.20	56.55	452.40		
1 Equip. Oper. (light)	47.05	376.40	71.35	570.80		
1 Air Compressor, 250 cfm		202.40		222.64		
1 Sandblaster, Portable, 3 C.F.		20.40		22.44		
1 Set Sand Blasting Accessories		13.85		15.23		
1 Aerial Lift Truck, 60' Boom		456.00		501.60	21.65	23.81
32 L.H., Daily Totals		$2010.25		$2953.92	$62.82	$92.31

Crew E-11B

	Hr.	Daily	Hr.	Daily	Bare Costs	Incl. O&P
2 Painters, Struc. Steel	$40.50	$648.00	$73.05	$1168.80	$39.22	$67.55
1 Building Laborer	36.65	293.20	56.55	452.40		
2 Paint Sprayer, 8 C.F.M.		89.80		98.78		
1 Aerial Lift Truck, 60' Boom		456.00		501.60	22.74	25.02
24 L.H., Daily Totals		$1487.00		$2221.58	$61.96	$92.57

Crew E-12

	Hr.	Daily	Hr.	Daily	Bare Costs	Incl. O&P
1 Welder Foreman (outside)	$53.10	$424.80	$93.70	$749.60	$50.08	$82.53
1 Equip. Oper. (light)	47.05	376.40	71.35	570.80		
1 Welder, Gas Engine, 300 amp		142.00		156.20	8.88	9.76
16 L.H., Daily Totals		$943.20		$1476.60	$58.95	$92.29

Crew E-13

	Hr.	Daily	Hr.	Daily	Bare Costs	Incl. O&P
1 Welder Foreman (outside)	$53.10	$424.80	$93.70	$749.60	$51.08	$86.25
.5 Equip. Oper. (light)	47.05	188.20	71.35	285.40		
1 Welder, Gas Engine, 300 amp		142.00		156.20	11.83	13.02
12 L.H., Daily Totals		$755.00		$1191.20	$62.92	$99.27

Crew E-14

	Hr.	Daily	Hr.	Daily	Bare Costs	Incl. O&P
1 Welder Foreman (outside)	$53.10	$424.80	$93.70	$749.60	$53.10	$93.70
1 Welder, Gas Engine, 300 amp		142.00		156.20	17.75	19.52
8 L.H., Daily Totals		$566.80		$905.80	$70.85	$113.22

Crew E-16

	Hr.	Daily	Hr.	Daily	Bare Costs	Incl. O&P
1 Welder Foreman (outside)	$53.10	$424.80	$93.70	$749.60	$52.10	$91.95
1 Welder	51.10	408.80	90.20	721.60		
1 Welder, Gas Engine, 300 amp		142.00		156.20	8.88	9.76
16 L.H., Daily Totals		$975.60		$1627.40	$60.98	$101.71

Crew E-17

	Hr.	Daily	Hr.	Daily	Bare Costs	Incl. O&P
1 Struc. Steel Foreman (outside)	$53.10	$424.80	$93.70	$749.60	$52.10	$91.95
1 Structural Steel Worker	51.10	408.80	90.20	721.60		
16 L.H., Daily Totals		$833.60		$1471.20	$52.10	$91.95

Crew E-18

	Hr.	Daily	Hr.	Daily	Bare Costs	Incl. O&P
1 Struc. Steel Foreman (outside)	$53.10	$424.80	$93.70	$749.60	$51.06	$87.69
3 Structural Steel Workers	51.10	1226.40	90.20	2164.80		
1 Equipment Operator (med.)	48.90	391.20	74.15	593.20		
1 Lattice Boom Crane, 20 Ton		960.50		1056.55	24.01	26.41
40 L.H., Daily Totals		$3002.90		$4564.15	$75.07	$114.10

Crew E-19

	Hr.	Daily	Hr.	Daily	Bare Costs	Incl. O&P
1 Struc. Steel Foreman (outside)	$53.10	$424.80	$93.70	$749.60	$50.42	$85.08
1 Structural Steel Worker	51.10	408.80	90.20	721.60		
1 Equip. Oper. (light)	47.05	376.40	71.35	570.80		
1 Lattice Boom Crane, 20 Ton		960.50		1056.55	40.02	44.02
24 L.H., Daily Totals		$2170.50		$3098.55	$90.44	$129.11

Crew E-20

	Hr.	Daily	Hr.	Daily	Bare Costs	Incl. O&P
1 Struc. Steel Foreman (outside)	$53.10	$424.80	$93.70	$749.60	$50.30	$85.86
5 Structural Steel Workers	51.10	2044.00	90.20	3608.00		
1 Equip. Oper. (crane)	50.25	402.00	76.20	609.60		
1 Equip. Oper. (oiler)	43.55	348.40	66.00	528.00		
1 Lattice Boom Crane, 40 Ton		1177.00		1294.70	18.39	20.23
64 L.H., Daily Totals		$4396.20		$6789.90	$68.69	$106.09

Crew E-22

	Hr.	Daily	Hr.	Daily	Bare Costs	Incl. O&P
1 Skilled Worker Foreman (out)	$49.30	$394.40	$76.35	$610.80	$47.97	$74.28
2 Skilled Workers	47.30	756.80	73.25	1172.00		
24 L.H., Daily Totals		$1151.20		$1782.80	$47.97	$74.28

Crew E-24

	Hr.	Daily	Hr.	Daily	Bare Costs	Incl. O&P
3 Structural Steel Workers	$51.10	$1226.40	$90.20	$2164.80	$50.55	$86.19
1 Equipment Operator (med.)	48.90	391.20	74.15	593.20		
1 Hyd. Crane, 25 Ton		745.80		820.38	23.31	25.64
32 L.H., Daily Totals		$2363.40		$3578.38	$73.86	$111.82

Crew E-25

	Hr.	Daily	Hr.	Daily	Bare Costs	Incl. O&P
1 Welder Foreman (outside)	$53.10	$424.80	$93.70	$749.60	$53.10	$93.70
1 Cutting Torch		11.40		12.54	1.43	1.57
8 L.H., Daily Totals		$436.20		$762.14	$54.52	$95.27

Crew F-3

	Hr.	Daily	Hr.	Daily	Bare Costs	Incl. O&P
4 Carpenters	$45.85	$1467.20	$70.75	$2264.00	$46.73	$71.84
1 Equip. Oper. (crane)	50.25	402.00	76.20	609.60		
1 Hyd. Crane, 12 Ton		663.00		729.30	16.57	18.23
40 L.H., Daily Totals		$2532.20		$3602.90	$63.31	$90.07

Crew F-4

	Hr.	Daily	Hr.	Daily	Bare Costs	Incl. O&P
4 Carpenters	$45.85	$1467.20	$70.75	$2264.00	$46.20	$70.87
1 Equip. Oper. (crane)	50.25	402.00	76.20	609.60		
1 Equip. Oper. (oiler)	43.55	348.40	66.00	528.00		
1 Hyd. Crane, 55 Ton		1137.00		1250.70	23.69	26.06
48 L.H., Daily Totals		$3354.60		$4652.30	$69.89	$96.92

Crew F-5

Crew F-5	Hr.	Daily	Hr.	Daily	Bare Costs	Incl. O&P
1 Carpenter Foreman (outside)	$47.85	$382.80	$73.85	$590.80	$46.35	$71.53
3 Carpenters	45.85	1100.40	70.75	1698.00		
32 L.H., Daily Totals		$1483.20		$2288.80	$46.35	$71.53

Crew F-6

Crew F-6	Hr.	Daily	Hr.	Daily	Bare Costs	Incl. O&P
2 Carpenters	$45.85	$733.60	$70.75	$1132.00	$43.05	$66.16
2 Building Laborers	36.65	586.40	56.55	904.80		
1 Equip. Oper. (crane)	50.25	402.00	76.20	609.60		
1 Hyd. Crane, 12 Ton		663.00		729.30	16.57	18.23
40 L.H., Daily Totals		$2385.00		$3375.70	$59.63	$84.39

Crew F-7

Crew F-7	Hr.	Daily	Hr.	Daily	Bare Costs	Incl. O&P
2 Carpenters	$45.85	$733.60	$70.75	$1132.00	$41.25	$63.65
2 Building Laborers	36.65	586.40	56.55	904.80		
32 L.H., Daily Totals		$1320.00		$2036.80	$41.25	$63.65

Crew G-1

Crew G-1	Hr.	Daily	Hr.	Daily	Bare Costs	Incl. O&P
1 Roofer Foreman (outside)	$41.15	$329.20	$70.05	$560.40	$36.58	$62.26
4 Roofers Composition	39.15	1252.80	66.65	2132.80		
2 Roofer Helpers	29.15	466.40	49.60	793.60		
1 Application Equipment		199.20		219.12		
1 Tar Kettle/Pot		130.00		143.00		
1 Crew Truck		224.80		247.28	9.89	10.88
56 L.H., Daily Totals		$2602.40		$4096.20	$46.47	$73.15

Crew G-2

Crew G-2	Hr.	Daily	Hr.	Daily	Bare Costs	Incl. O&P
1 Plasterer	$41.95	$335.60	$63.15	$505.20	$38.60	$58.57
1 Plasterer Helper	37.20	297.60	56.00	448.00		
1 Building Laborer	36.65	293.20	56.55	452.40		
1 Grout Pump, 50 C.F./hr.		134.80		148.28	5.62	6.18
24 L.H., Daily Totals		$1061.20		$1553.88	$44.22	$64.75

Crew G-2A

Crew G-2A	Hr.	Daily	Hr.	Daily	Bare Costs	Incl. O&P
1 Roofer Composition	$39.15	$313.20	$66.65	$533.20	$34.98	$57.60
1 Roofer Helper	29.15	233.20	49.60	396.80		
1 Building Laborer	36.65	293.20	56.55	452.40		
1 Foam Spray Rig, Trailer-Mtd.		585.25		643.77		
1 Pickup Truck, 3/4 Ton		157.00		172.70	30.93	34.02
24 L.H., Daily Totals		$1581.85		$2198.88	$65.91	$91.62

Crew G-3

Crew G-3	Hr.	Daily	Hr.	Daily	Bare Costs	Incl. O&P
2 Sheet Metal Workers	$54.70	$875.20	$83.55	$1336.80	$45.67	$70.05
2 Building Laborers	36.65	586.40	56.55	904.80		
32 L.H., Daily Totals		$1461.60		$2241.60	$45.67	$70.05

Crew G-4

Crew G-4	Hr.	Daily	Hr.	Daily	Bare Costs	Incl. O&P
1 Labor Foreman (outside)	$38.65	$309.20	$59.65	$477.20	$37.32	$57.58
2 Building Laborers	36.65	586.40	56.55	904.80		
1 Flatbed Truck, Gas, 1.5 Ton		270.20		297.22		
1 Air Compressor, 160 cfm		157.40		173.14	17.82	19.60
24 L.H., Daily Totals		$1323.20		$1852.36	$55.13	$77.18

Crew G-5

Crew G-5	Hr.	Daily	Hr.	Daily	Bare Costs	Incl. O&P
1 Roofer Foreman (outside)	$41.15	$329.20	$70.05	$560.40	$35.55	$60.51
2 Roofers Composition	39.15	626.40	66.65	1066.40		
2 Roofer Helpers	29.15	466.40	49.60	793.60		
1 Application Equipment		199.20		219.12	4.98	5.48
40 L.H., Daily Totals		$1621.20		$2639.52	$40.53	$65.99

Crew G-6A

Crew G-6A	Hr.	Daily	Hr.	Daily	Bare Costs	Incl. O&P
2 Roofers Composition	$39.15	$626.40	$66.65	$1066.40	$39.15	$66.65
1 Small Compressor, Electric		12.10		13.31		
2 Pneumatic Nailers		44.70		49.17	3.55	3.90
16 L.H., Daily Totals		$683.20		$1128.88	$42.70	$70.56

Crew G-7

Crew G-7	Hr.	Daily	Hr.	Daily	Bare Costs	Incl. O&P
1 Carpenter	$45.85	$366.80	$70.75	$566.00	$45.85	$70.75
1 Small Compressor, Electric		12.10		13.31		
1 Pneumatic Nailer		22.35		24.59	4.31	4.74
8 L.H., Daily Totals		$401.25		$603.89	$50.16	$75.49

Crew H-1

Crew H-1	Hr.	Daily	Hr.	Daily	Bare Costs	Incl. O&P
2 Glaziers	$44.05	$704.80	$67.00	$1072.00	$47.58	$78.60
2 Struc. Steel Workers	51.10	817.60	90.20	1443.20		
32 L.H., Daily Totals		$1522.40		$2515.20	$47.58	$78.60

Crew H-2

Crew H-2	Hr.	Daily	Hr.	Daily	Bare Costs	Incl. O&P
2 Glaziers	$44.05	$704.80	$67.00	$1072.00	$41.58	$63.52
1 Building Laborer	36.65	293.20	56.55	452.40		
24 L.H., Daily Totals		$998.00		$1524.40	$41.58	$63.52

Crew H-3

Crew H-3	Hr.	Daily	Hr.	Daily	Bare Costs	Incl. O&P
1 Glazier	$44.05	$352.40	$67.00	$536.00	$39.35	$60.35
1 Helper	34.65	277.20	53.70	429.60		
16 L.H., Daily Totals		$629.60		$965.60	$39.35	$60.35

Crew H-4

Crew H-4	Hr.	Daily	Hr.	Daily	Bare Costs	Incl. O&P
1 Carpenter	$45.85	$366.80	$70.75	$566.00	$42.87	$65.75
1 Carpenter Helper	34.65	277.20	53.70	429.60		
.5 Electrician	53.35	213.40	79.85	319.40		
20 L.H., Daily Totals		$857.40		$1315.00	$42.87	$65.75

Crew J-1

Crew J-1	Hr.	Daily	Hr.	Daily	Bare Costs	Incl. O&P
3 Plasterers	$41.95	$1006.80	$63.15	$1515.60	$40.05	$60.29
2 Plasterer Helpers	37.20	595.20	56.00	896.00		
1 Mixing Machine, 6 C.F.		143.60		157.96	3.59	3.95
40 L.H., Daily Totals		$1745.60		$2569.56	$43.64	$64.24

Crew J-2

Crew J-2	Hr.	Daily	Hr.	Daily	Bare Costs	Incl. O&P
3 Plasterers	$41.95	$1006.80	$63.15	$1515.60	$40.14	$60.19
2 Plasterer Helpers	37.20	595.20	56.00	896.00		
1 Lather	40.60	324.80	59.70	477.60		
1 Mixing Machine, 6 C.F.		143.60		157.96	2.99	3.29
48 L.H., Daily Totals		$2070.40		$3047.16	$43.13	$63.48

Crew J-3

Crew J-3	Hr.	Daily	Hr.	Daily	Bare Costs	Incl. O&P
1 Terrazzo Worker	$41.65	$333.20	$61.40	$491.20	$38.23	$56.35
1 Terrazzo Helper	34.80	278.40	51.30	410.40		
1 Floor Grinder, 22" Path		97.90		107.69		
1 Terrazzo Mixer		192.60		211.86	18.16	19.97
16 L.H., Daily Totals		$902.10		$1221.15	$56.38	$76.32

Crew J-4

Crew J-4	Hr.	Daily	Hr.	Daily	Bare Costs	Incl. O&P
2 Cement Finishers	$44.05	$704.80	$65.10	$1041.60	$41.58	$62.25
1 Laborer	36.65	293.20	56.55	452.40		
1 Floor Grinder, 22" Path		97.90		107.69		
1 Floor Edger, 7" Path		39.15		43.06		
1 Vacuum Pick-Up System		60.75		66.83	8.24	9.07
24 L.H., Daily Totals		$1195.80		$1711.58	$49.83	$71.32

Crew J-4A

Crew No.	Bare Costs Hr.	Bare Costs Daily	Incl. Subs O&P Hr.	Incl. Subs O&P Daily	Cost Per L-H Bare Costs	Cost Per L-H Incl. O&P
2 Cement Finishers	$44.05	$704.80	$65.10	$1041.60	$40.35	$60.83
2 Laborers	36.65	586.40	56.55	904.80		
1 Floor Grinder, 22" Path		97.90		107.69		
1 Floor Edger, 7" Path		39.15		43.06		
1 Vacuum Pick-Up System		60.75		66.83		
1 Floor Auto Scrubber		212.15		233.37	12.81	14.09
32 L.H., Daily Totals		$1701.15		$2397.34	$53.16	$74.92

Crew J-4B

Crew No.	Bare Costs Hr.	Bare Costs Daily	Incl. Subs O&P Hr.	Incl. Subs O&P Daily	Cost Per L-H Bare Costs	Cost Per L-H Incl. O&P
1 Laborer	$36.65	$293.20	$56.55	$452.40	$36.65	$56.55
1 Floor Auto Scrubber		212.15		233.37	26.52	29.17
8 L.H., Daily Totals		$505.35		$685.76	$63.17	$85.72

Crew J-6

Crew No.	Bare Costs Hr.	Bare Costs Daily	Incl. Subs O&P Hr.	Incl. Subs O&P Daily	Cost Per L-H Bare Costs	Cost Per L-H Incl. O&P
2 Painters	$39.55	$632.80	$59.50	$952.00	$40.70	$61.73
1 Building Laborer	36.65	293.20	56.55	452.40		
1 Equip. Oper. (light)	47.05	376.40	71.35	570.80		
1 Air Compressor, 250 cfm		202.40		222.64		
1 Sandblaster, Portable, 3 C.F.		20.40		22.44		
1 Set Sand Blasting Accessories		13.85		15.23	7.40	8.13
32 L.H., Daily Totals		$1539.05		$2235.51	$48.10	$69.86

Crew J-7

Crew No.	Bare Costs Hr.	Bare Costs Daily	Incl. Subs O&P Hr.	Incl. Subs O&P Daily	Cost Per L-H Bare Costs	Cost Per L-H Incl. O&P
2 Painters	$39.55	$632.80	$59.50	$952.00	$39.55	$59.50
1 Floor Belt Sander		15.90		17.49		
1 Floor Sanding Edger		15.55		17.11	1.97	2.16
16 L.H., Daily Totals		$664.25		$986.60	$41.52	$61.66

Crew K-1

Crew No.	Bare Costs Hr.	Bare Costs Daily	Incl. Subs O&P Hr.	Incl. Subs O&P Daily	Cost Per L-H Bare Costs	Cost Per L-H Incl. O&P
1 Carpenter	$45.85	$366.80	$70.75	$566.00	$41.17	$63.23
1 Truck Driver (light)	36.50	292.00	55.70	445.60		
1 Flatbed Truck, Gas, 3 Ton		333.00		366.30	20.81	22.89
16 L.H., Daily Totals		$991.80		$1377.90	$61.99	$86.12

Crew K-2

Crew No.	Bare Costs Hr.	Bare Costs Daily	Incl. Subs O&P Hr.	Incl. Subs O&P Daily	Cost Per L-H Bare Costs	Cost Per L-H Incl. O&P
1 Struc. Steel Foreman (outside)	$53.10	$424.80	$93.70	$749.60	$46.90	$79.87
1 Struc. Steel Worker	51.10	408.80	90.20	721.60		
1 Truck Driver (light)	36.50	292.00	55.70	445.60		
1 Flatbed Truck, Gas, 3 Ton		333.00		366.30	13.88	15.26
24 L.H., Daily Totals		$1458.60		$2283.10	$60.77	$95.13

Crew L-1

Crew No.	Bare Costs Hr.	Bare Costs Daily	Incl. Subs O&P Hr.	Incl. Subs O&P Daily	Cost Per L-H Bare Costs	Cost Per L-H Incl. O&P
1 Electrician	$53.35	$426.80	$79.85	$638.80	$55.45	$83.38
1 Plumber	57.55	460.40	86.90	695.20		
16 L.H., Daily Totals		$887.20		$1334.00	$55.45	$83.38

Crew L-2

Crew No.	Bare Costs Hr.	Bare Costs Daily	Incl. Subs O&P Hr.	Incl. Subs O&P Daily	Cost Per L-H Bare Costs	Cost Per L-H Incl. O&P
1 Carpenter	$45.85	$366.80	$70.75	$566.00	$40.25	$62.23
1 Carpenter Helper	34.65	277.20	53.70	429.60		
16 L.H., Daily Totals		$644.00		$995.60	$40.25	$62.23

Crew L-3

Crew No.	Bare Costs Hr.	Bare Costs Daily	Incl. Subs O&P Hr.	Incl. Subs O&P Daily	Cost Per L-H Bare Costs	Cost Per L-H Incl. O&P
1 Carpenter	$45.85	$366.80	$70.75	$566.00	$49.94	$76.22
.5 Electrician	53.35	213.40	79.85	319.40		
.5 Sheet Metal Worker	54.70	218.80	83.55	334.20		
16 L.H., Daily Totals		$799.00		$1219.60	$49.94	$76.22

Crew L-3A

Crew No.	Bare Costs Hr.	Bare Costs Daily	Incl. Subs O&P Hr.	Incl. Subs O&P Daily	Cost Per L-H Bare Costs	Cost Per L-H Incl. O&P
1 Carpenter Foreman (outside)	$47.85	$382.80	$73.85	$590.80	$50.13	$77.08
.5 Sheet Metal Worker	54.70	218.80	83.55	334.20		
12 L.H., Daily Totals		$601.60		$925.00	$50.13	$77.08

Crew L-4

Crew No.	Bare Costs Hr.	Bare Costs Daily	Incl. Subs O&P Hr.	Incl. Subs O&P Daily	Cost Per L-H Bare Costs	Cost Per L-H Incl. O&P
2 Skilled Workers	$47.30	$756.80	$73.25	$1172.00	$43.08	$66.73
1 Helper	34.65	277.20	53.70	429.60		
24 L.H., Daily Totals		$1034.00		$1601.60	$43.08	$66.73

Crew L-5

Crew No.	Bare Costs Hr.	Bare Costs Daily	Incl. Subs O&P Hr.	Incl. Subs O&P Daily	Cost Per L-H Bare Costs	Cost Per L-H Incl. O&P
1 Struc. Steel Foreman (outside)	$53.10	$424.80	$93.70	$749.60	$51.26	$88.70
5 Struc. Steel Workers	51.10	2044.00	90.20	3608.00		
1 Equip. Oper. (crane)	50.25	402.00	76.20	609.60		
1 Hyd. Crane, 25 Ton		745.80		820.38	13.32	14.65
56 L.H., Daily Totals		$3616.60		$5787.58	$64.58	$103.35

Crew L-5A

Crew No.	Bare Costs Hr.	Bare Costs Daily	Incl. Subs O&P Hr.	Incl. Subs O&P Daily	Cost Per L-H Bare Costs	Cost Per L-H Incl. O&P
1 Struc. Steel Foreman (outside)	$53.10	$424.80	$93.70	$749.60	$51.39	$87.58
2 Structural Steel Workers	51.10	817.60	90.20	1443.20		
1 Equip. Oper. (crane)	50.25	402.00	76.20	609.60		
1 S.P. Crane, 4x4, 25 Ton		607.20		667.92	18.98	20.87
32 L.H., Daily Totals		$2251.60		$3470.32	$70.36	$108.45

Crew L-5B

Crew No.	Bare Costs Hr.	Bare Costs Daily	Incl. Subs O&P Hr.	Incl. Subs O&P Daily	Cost Per L-H Bare Costs	Cost Per L-H Incl. O&P
1 Struc. Steel Foreman (outside)	$53.10	$424.80	$93.70	$749.60	$52.53	$83.63
2 Structural Steel Workers	51.10	817.60	90.20	1443.20		
2 Electricians	53.35	853.60	79.85	1277.60		
2 Steamfitters/Pipefitters	58.50	936.00	88.35	1413.60		
1 Equip. Oper. (crane)	50.25	402.00	76.20	609.60		
1 Equip. Oper. (oiler)	43.55	348.40	66.00	528.00		
1 Hyd. Crane, 80 Ton		1652.00		1817.20	22.94	25.24
72 L.H., Daily Totals		$5434.40		$7838.80	$75.48	$108.87

Crew L-6

Crew No.	Bare Costs Hr.	Bare Costs Daily	Incl. Subs O&P Hr.	Incl. Subs O&P Daily	Cost Per L-H Bare Costs	Cost Per L-H Incl. O&P
1 Plumber	$57.55	$460.40	$86.90	$695.20	$56.15	$84.55
.5 Electrician	53.35	213.40	79.85	319.40		
12 L.H., Daily Totals		$673.80		$1014.60	$56.15	$84.55

Crew L-7

Crew No.	Bare Costs Hr.	Bare Costs Daily	Incl. Subs O&P Hr.	Incl. Subs O&P Daily	Cost Per L-H Bare Costs	Cost Per L-H Incl. O&P
2 Carpenters	$45.85	$733.60	$70.75	$1132.00	$44.29	$67.99
1 Building Laborer	36.65	293.20	56.55	452.40		
.5 Electrician	53.35	213.40	79.85	319.40		
28 L.H., Daily Totals		$1240.20		$1903.80	$44.29	$67.99

Crew L-8

Crew No.	Bare Costs Hr.	Bare Costs Daily	Incl. Subs O&P Hr.	Incl. Subs O&P Daily	Cost Per L-H Bare Costs	Cost Per L-H Incl. O&P
2 Carpenters	$45.85	$733.60	$70.75	$1132.00	$48.19	$73.98
.5 Plumber	57.55	230.20	86.90	347.60		
20 L.H., Daily Totals		$963.80		$1479.60	$48.19	$73.98

Crew L-9

Crew No.	Bare Costs Hr.	Bare Costs Daily	Incl. Subs O&P Hr.	Incl. Subs O&P Daily	Cost Per L-H Bare Costs	Cost Per L-H Incl. O&P
1 Labor Foreman (inside)	$37.15	$297.20	$57.30	$458.40	$41.83	$66.78
2 Building Laborers	36.65	586.40	56.55	904.80		
1 Struc. Steel Worker	51.10	408.80	90.20	721.60		
.5 Electrician	53.35	213.40	79.85	319.40		
36 L.H., Daily Totals		$1505.80		$2404.20	$41.83	$66.78

Crew No.	Bare Costs Hr.	Daily	Incl. Subs O&P Hr.	Daily	Cost Per Labor-Hour Bare Costs	Incl. O&P
Crew L-10						
1 Struc. Steel Foreman (outside)	$53.10	$424.80	$93.70	$749.60	$51.48	$86.70
1 Structural Steel Worker	51.10	408.80	90.20	721.60		
1 Equip. Oper. (crane)	50.25	402.00	76.20	609.60		
1 Hyd. Crane, 12 Ton		663.00		729.30	27.63	30.39
24 L.H., Daily Totals		$1898.60		$2810.10	$79.11	$117.09
Crew L-11						
2 Wreckers	$36.65	$586.40	$61.60	$985.60	$42.65	$67.69
1 Equip. Oper. (crane)	50.25	402.00	76.20	609.60		
1 Equip. Oper. (light)	47.05	376.40	71.35	570.80		
1 Hyd. Excavator, 2.5 C.Y.		1637.00		1800.70		
1 Loader, Skid Steer, 78 H.P.		308.80		339.68	60.81	66.89
32 L.H., Daily Totals		$3310.60		$4306.38	$103.46	$134.57
Crew M-1						
3 Elevator Constructors	$74.15	$1779.60	$110.95	$2662.80	$70.44	$105.39
1 Elevator Apprentice	59.30	474.40	88.70	709.60		
5 Hand Tools		48.00		52.80	1.50	1.65
32 L.H., Daily Totals		$2302.00		$3425.20	$71.94	$107.04
Crew M-3						
1 Electrician Foreman (outside)	$55.35	$442.80	$82.85	$662.80	$55.92	$84.14
1 Common Laborer	36.65	293.20	56.55	452.40		
.25 Equipment Operator (med.)	48.90	97.80	74.15	148.30		
1 Elevator Constructor	74.15	593.20	110.95	887.60		
1 Elevator Apprentice	59.30	474.40	88.70	709.60		
.25 S.P. Crane, 4x4, 20 Ton		140.10		154.11	4.12	4.53
34 L.H., Daily Totals		$2041.50		$3014.81	$60.04	$88.67
Crew M-4						
1 Electrician Foreman (outside)	$55.35	$442.80	$82.85	$662.80	$55.31	$83.24
1 Common Laborer	36.65	293.20	56.55	452.40		
.25 Equipment Operator, Crane	50.25	100.50	76.20	152.40		
.25 Equip. Oper. (oiler)	43.55	87.10	66.00	132.00		
1 Elevator Constructor	74.15	593.20	110.95	887.60		
1 Elevator Apprentice	59.30	474.40	88.70	709.60		
.25 S.P. Crane, 4x4, 40 Ton		178.25		196.07	4.95	5.45
36 L.H., Daily Totals		$2169.45		$3192.88	$60.26	$88.69
Crew Q-1						
1 Plumber	$57.55	$460.40	$86.90	$695.20	$51.80	$78.22
1 Plumber Apprentice	46.05	368.40	69.55	556.40		
16 L.H., Daily Totals		$828.80		$1251.60	$51.80	$78.22
Crew Q-1A						
.25 Plumber Foreman (outside)	$59.55	$119.10	$89.90	$179.80	$57.95	$87.50
1 Plumber	57.55	460.40	86.90	695.20		
10 L.H., Daily Totals		$579.50		$875.00	$57.95	$87.50
Crew Q-1C						
1 Plumber	$57.55	$460.40	$86.90	$695.20	$50.83	$76.87
1 Plumber Apprentice	46.05	368.40	69.55	556.40		
1 Equip. Oper. (medium)	48.90	391.20	74.15	593.20		
1 Trencher, Chain Type, 8' D		3376.00		3713.60	140.67	154.73
24 L.H., Daily Totals		$4596.00		$5558.40	$191.50	$231.60
Crew Q-2						
2 Plumbers	$57.55	$920.80	$86.90	$1390.40	$53.72	$81.12
1 Plumber Apprentice	46.05	368.40	69.55	556.40		
24 L.H., Daily Totals		$1289.20		$1946.80	$53.72	$81.12
Crew Q-3						
1 Plumber Foreman (inside)	$58.05	$464.40	$87.65	$701.20	$54.80	$82.75
2 Plumbers	57.55	920.80	86.90	1390.40		
1 Plumber Apprentice	46.05	368.40	69.55	556.40		
32 L.H., Daily Totals		$1753.60		$2648.00	$54.80	$82.75
Crew Q-4						
1 Plumber Foreman (inside)	$58.05	$464.40	$87.65	$701.20	$54.80	$82.75
1 Plumber	57.55	460.40	86.90	695.20		
1 Welder (plumber)	57.55	460.40	86.90	695.20		
1 Plumber Apprentice	46.05	368.40	69.55	556.40		
1 Welder, Electric, 300 amp		55.30		60.83	1.73	1.90
32 L.H., Daily Totals		$1808.90		$2708.83	$56.53	$84.65
Crew Q-5						
1 Steamfitter	$58.50	$468.00	$88.35	$706.80	$52.65	$79.50
1 Steamfitter Apprentice	46.80	374.40	70.65	565.20		
16 L.H., Daily Totals		$842.40		$1272.00	$52.65	$79.50
Crew Q-6						
2 Steamfitters	$58.50	$936.00	$88.35	$1413.60	$54.60	$82.45
1 Steamfitter Apprentice	46.80	374.40	70.65	565.20		
24 L.H., Daily Totals		$1310.40		$1978.80	$54.60	$82.45
Crew Q-7						
1 Steamfitter Foreman (inside)	$59.00	$472.00	$89.10	$712.80	$55.70	$84.11
2 Steamfitters	58.50	936.00	88.35	1413.60		
1 Steamfitter Apprentice	46.80	374.40	70.65	565.20		
32 L.H., Daily Totals		$1782.40		$2691.60	$55.70	$84.11
Crew Q-8						
1 Steamfitter Foreman (inside)	$59.00	$472.00	$89.10	$712.80	$55.70	$84.11
1 Steamfitter	58.50	468.00	88.35	706.80		
1 Welder (steamfitter)	58.50	468.00	88.35	706.80		
1 Steamfitter Apprentice	46.80	374.40	70.65	565.20		
1 Welder, Electric, 300 amp		55.30		60.83	1.73	1.90
32 L.H., Daily Totals		$1837.70		$2752.43	$57.43	$86.01
Crew Q-9						
1 Sheet Metal Worker	$54.70	$437.60	$83.55	$668.40	$49.23	$75.17
1 Sheet Metal Apprentice	43.75	350.00	66.80	534.40		
16 L.H., Daily Totals		$787.60		$1202.80	$49.23	$75.17
Crew Q-10						
2 Sheet Metal Workers	$54.70	$875.20	$83.55	$1336.80	$51.05	$77.97
1 Sheet Metal Apprentice	43.75	350.00	66.80	534.40		
24 L.H., Daily Totals		$1225.20		$1871.20	$51.05	$77.97
Crew Q-11						
1 Sheet Metal Foreman (inside)	$55.20	$441.60	$84.30	$674.40	$52.09	$79.55
2 Sheet Metal Workers	54.70	875.20	83.55	1336.80		
1 Sheet Metal Apprentice	43.75	350.00	66.80	534.40		
32 L.H., Daily Totals		$1666.80		$2545.60	$52.09	$79.55
Crew Q-12						
1 Sprinkler Installer	$55.40	$443.20	$83.70	$669.60	$49.85	$75.33
1 Sprinkler Apprentice	44.30	354.40	66.95	535.60		
16 L.H., Daily Totals		$797.60		$1205.20	$49.85	$75.33

Crew No.	Bare Costs Hr.	Daily	Incl. Subs O&P Hr.	Daily	Cost Per Labor-Hour Bare Costs	Incl. O&P
Crew Q-13	Hr.	Daily	Hr.	Daily	Bare Costs	Incl. O&P
1 Sprinkler Foreman (inside)	$55.90	$447.20	$84.45	$675.60	$52.75	$79.70
2 Sprinkler Installers	55.40	886.40	83.70	1339.20		
1 Sprinkler Apprentice	44.30	354.40	66.95	535.60		
32 L.H., Daily Totals		$1688.00		$2550.40	$52.75	$79.70
Crew Q-14	Hr.	Daily	Hr.	Daily	Bare Costs	Incl. O&P
1 Asbestos Worker	$51.15	$409.20	$79.60	$636.80	$46.02	$71.63
1 Asbestos Apprentice	40.90	327.20	63.65	509.20		
16 L.H., Daily Totals		$736.40		$1146.00	$46.02	$71.63
Crew Q-15	Hr.	Daily	Hr.	Daily	Bare Costs	Incl. O&P
1 Plumber	$57.55	$460.40	$86.90	$695.20	$51.80	$78.22
1 Plumber Apprentice	46.05	368.40	69.55	556.40		
1 Welder, Electric, 300 amp		55.30		60.83	3.46	3.80
16 L.H., Daily Totals		$884.10		$1312.43	$55.26	$82.03
Crew Q-16	Hr.	Daily	Hr.	Daily	Bare Costs	Incl. O&P
2 Plumbers	$57.55	$920.80	$86.90	$1390.40	$53.72	$81.12
1 Plumber Apprentice	46.05	368.40	69.55	556.40		
1 Welder, Electric, 300 amp		55.30		60.83	2.30	2.53
24 L.H., Daily Totals		$1344.50		$2007.63	$56.02	$83.65
Crew Q-17	Hr.	Daily	Hr.	Daily	Bare Costs	Incl. O&P
1 Steamfitter	$58.50	$468.00	$88.35	$706.80	$52.65	$79.50
1 Steamfitter Apprentice	46.80	374.40	70.65	565.20		
1 Welder, Electric, 300 amp		55.30		60.83	3.46	3.80
16 L.H., Daily Totals		$897.70		$1332.83	$56.11	$83.30
Crew Q-17A	Hr.	Daily	Hr.	Daily	Bare Costs	Incl. O&P
1 Steamfitter	$58.50	$468.00	$88.35	$706.80	$51.85	$78.40
1 Steamfitter Apprentice	46.80	374.40	70.65	565.20		
1 Equip. Oper. (crane)	50.25	402.00	76.20	609.60		
1 Hyd. Crane, 12 Ton		663.00		729.30		
1 Welder, Electric, 300 amp		55.30		60.83	29.93	32.92
24 L.H., Daily Totals		$1962.70		$2671.73	$81.78	$111.32
Crew Q-18	Hr.	Daily	Hr.	Daily	Bare Costs	Incl. O&P
2 Steamfitters	$58.50	$936.00	$88.35	$1413.60	$54.60	$82.45
1 Steamfitter Apprentice	46.80	374.40	70.65	565.20		
1 Welder, Electric, 300 amp		55.30		60.83	2.30	2.53
24 L.H., Daily Totals		$1365.70		$2039.63	$56.90	$84.98
Crew Q-19	Hr.	Daily	Hr.	Daily	Bare Costs	Incl. O&P
1 Steamfitter	$58.50	$468.00	$88.35	$706.80	$52.88	$79.62
1 Steamfitter Apprentice	46.80	374.40	70.65	565.20		
1 Electrician	53.35	426.80	79.85	638.80		
24 L.H., Daily Totals		$1269.20		$1910.80	$52.88	$79.62
Crew Q-20	Hr.	Daily	Hr.	Daily	Bare Costs	Incl. O&P
1 Sheet Metal Worker	$54.70	$437.60	$83.55	$668.40	$50.05	$76.11
1 Sheet Metal Apprentice	43.75	350.00	66.80	534.40		
.5 Electrician	53.35	213.40	79.85	319.40		
20 L.H., Daily Totals		$1001.00		$1522.20	$50.05	$76.11
Crew Q-21	Hr.	Daily	Hr.	Daily	Bare Costs	Incl. O&P
2 Steamfitters	$58.50	$936.00	$88.35	$1413.60	$54.29	$81.80
1 Steamfitter Apprentice	46.80	374.40	70.65	565.20		
1 Electrician	53.35	426.80	79.85	638.80		
32 L.H., Daily Totals		$1737.20		$2617.60	$54.29	$81.80

Crew No.	Bare Costs Hr.	Daily	Incl. Subs O&P Hr.	Daily	Cost Per Labor-Hour Bare Costs	Incl. O&P
Crew Q-22	Hr.	Daily	Hr.	Daily	Bare Costs	Incl. O&P
1 Plumber	$57.55	$460.40	$86.90	$695.20	$51.80	$78.22
1 Plumber Apprentice	46.05	368.40	69.55	556.40		
1 Hyd. Crane, 12 Ton		663.00		729.30	41.44	45.58
16 L.H., Daily Totals		$1491.80		$1980.90	$93.24	$123.81
Crew Q-22A	Hr.	Daily	Hr.	Daily	Bare Costs	Incl. O&P
1 Plumber	$57.55	$460.40	$86.90	$695.20	$47.63	$72.30
1 Plumber Apprentice	46.05	368.40	69.55	556.40		
1 Laborer	36.65	293.20	56.55	452.40		
1 Equip. Oper. (crane)	50.25	402.00	76.20	609.60		
1 Hyd. Crane, 12 Ton		663.00		729.30	20.72	22.79
32 L.H., Daily Totals		$2187.00		$3042.90	$68.34	$95.09
Crew Q-23	Hr.	Daily	Hr.	Daily	Bare Costs	Incl. O&P
1 Plumber Foreman (outside)	$59.55	$476.40	$89.90	$719.20	$55.33	$83.65
1 Plumber	57.55	460.40	86.90	695.20		
1 Equip. Oper. (medium)	48.90	391.20	74.15	593.20		
1 Lattice Boom Crane, 20 Ton		960.50		1056.55	40.02	44.02
24 L.H., Daily Totals		$2288.50		$3064.15	$95.35	$127.67
Crew R-1	Hr.	Daily	Hr.	Daily	Bare Costs	Incl. O&P
1 Electrician Foreman	$53.85	$430.80	$80.60	$644.80	$47.20	$71.26
3 Electricians	53.35	1280.40	79.85	1916.40		
2 Helpers	34.65	554.40	53.70	859.20		
48 L.H., Daily Totals		$2265.60		$3420.40	$47.20	$71.26
Crew R-1A	Hr.	Daily	Hr.	Daily	Bare Costs	Incl. O&P
1 Electrician	$53.35	$426.80	$79.85	$638.80	$44.00	$66.78
1 Helper	34.65	277.20	53.70	429.60		
16 L.H., Daily Totals		$704.00		$1068.40	$44.00	$66.78
Crew R-2	Hr.	Daily	Hr.	Daily	Bare Costs	Incl. O&P
1 Electrician Foreman	$53.85	$430.80	$80.60	$644.80	$47.64	$71.96
3 Electricians	53.35	1280.40	79.85	1916.40		
2 Helpers	34.65	554.40	53.70	859.20		
1 Equip. Oper. (crane)	50.25	402.00	76.20	609.60		
1 S.P. Crane, 4x4, 5 Ton		280.00		308.00	5.00	5.50
56 L.H., Daily Totals		$2947.60		$4338.00	$52.64	$77.46
Crew R-3	Hr.	Daily	Hr.	Daily	Bare Costs	Incl. O&P
1 Electrician Foreman	$53.85	$430.80	$80.60	$644.80	$52.93	$79.42
1 Electrician	53.35	426.80	79.85	638.80		
.5 Equip. Oper. (crane)	50.25	201.00	76.20	304.80		
.5 S.P. Crane, 4x4, 5 Ton		140.00		154.00	7.00	7.70
20 L.H., Daily Totals		$1198.60		$1742.40	$59.93	$87.12
Crew R-4	Hr.	Daily	Hr.	Daily	Bare Costs	Incl. O&P
1 Struc. Steel Foreman (outside)	$53.10	$424.80	$93.70	$749.60	$51.95	$88.83
3 Struc. Steel Workers	51.10	1226.40	90.20	2164.80		
1 Electrician	53.35	426.80	79.85	638.80		
1 Welder, Gas Engine, 300 amp		142.00		156.20	3.55	3.90
40 L.H., Daily Totals		$2220.00		$3709.40	$55.50	$92.73

Crew No.	Bare Costs		Incl. Subs O&P		Cost Per Labor-Hour	
Crew R-5	Hr.	Daily	Hr.	Daily	Bare Costs	Incl. O&P
1 Electrician Foreman	$53.85	$430.80	$80.60	$644.80	$46.60	$70.41
4 Electrician Linemen	53.35	1707.20	79.85	2555.20		
2 Electrician Operators	53.35	853.60	79.85	1277.60		
4 Electrician Groundmen	34.65	1108.80	53.70	1718.40		
1 Crew Truck		224.80		247.28		
1 Flatbed Truck, 20,000 GVW		253.40		278.74		
1 Pickup Truck, 3/4 Ton		157.00		172.70		
.2 Hyd. Crane, 55 Ton		227.40		250.14		
.2 Hyd. Crane, 12 Ton		132.60		145.86		
.2 Earth Auger, Truck-Mtd.		85.00		93.50		
1 Tractor w/Winch		434.60		478.06	17.21	18.93
88 L.H., Daily Totals		$5615.20		$7862.28	$63.81	$89.34

	Bare Costs		Incl. Subs O&P			
Crew R-6	Hr.	Daily	Hr.	Daily	Bare Costs	Incl. O&P
1 Electrician Foreman	$53.85	$430.80	$80.60	$644.80	$46.60	$70.41
4 Electrician Linemen	53.35	1707.20	79.85	2555.20		
2 Electrician Operators	53.35	853.60	79.85	1277.60		
4 Electrician Groundmen	34.65	1108.80	53.70	1718.40		
1 Crew Truck		224.80		247.28		
1 Flatbed Truck, 20,000 GVW		253.40		278.74		
1 Pickup Truck, 3/4 Ton		157.00		172.70		
.2 Hyd. Crane, 55 Ton		227.40		250.14		
.2 Hyd. Crane, 12 Ton		132.60		145.86		
.2 Earth Auger, Truck-Mtd.		85.00		93.50		
1 Tractor w/Winch		434.60		478.06		
3 Cable Trailers		586.80		645.48		
.5 Tensioning Rig		194.55		214.01		
.5 Cable Pulling Rig		1136.50		1250.15	39.01	42.91
88 L.H., Daily Totals		$7533.05		$9971.92	$85.60	$113.32

	Bare Costs		Incl. Subs O&P			
Crew R-7	Hr.	Daily	Hr.	Daily	Bare Costs	Incl. O&P
1 Electrician Foreman	$53.85	$430.80	$80.60	$644.80	$37.85	$58.18
5 Electrician Groundmen	34.65	1386.00	53.70	2148.00		
1 Crew Truck		224.80		247.28	4.68	5.15
48 L.H., Daily Totals		$2041.60		$3040.08	$42.53	$63.34

	Bare Costs		Incl. Subs O&P			
Crew R-8	Hr.	Daily	Hr.	Daily	Bare Costs	Incl. O&P
1 Electrician Foreman	$53.85	$430.80	$80.60	$644.80	$47.20	$71.26
3 Electrician Linemen	53.35	1280.40	79.85	1916.40		
2 Electrician Groundmen	34.65	554.40	53.70	859.20		
1 Pickup Truck, 3/4 Ton		157.00		172.70		
1 Crew Truck		224.80		247.28	7.95	8.75
48 L.H., Daily Totals		$2647.40		$3840.38	$55.15	$80.01

	Bare Costs		Incl. Subs O&P			
Crew R-9	Hr.	Daily	Hr.	Daily	Bare Costs	Incl. O&P
1 Electrician Foreman	$53.85	$430.80	$80.60	$644.80	$44.06	$66.87
1 Electrician Lineman	53.35	426.80	79.85	638.80		
2 Electrician Operators	53.35	853.60	79.85	1277.60		
4 Electrician Groundmen	34.65	1108.80	53.70	1718.40		
1 Pickup Truck, 3/4 Ton		157.00		172.70		
1 Crew Truck		224.80		247.28	5.97	6.56
64 L.H., Daily Totals		$3201.80		$4699.58	$50.03	$73.43

	Bare Costs		Incl. Subs O&P			
Crew R-10	Hr.	Daily	Hr.	Daily	Bare Costs	Incl. O&P
1 Electrician Foreman	$53.85	$430.80	$80.60	$644.80	$50.32	$75.62
4 Electrician Linemen	53.35	1707.20	79.85	2555.20		
1 Electrician Groundman	34.65	277.20	53.70	429.60		
1 Crew Truck		224.80		247.28		
3 Tram Cars		399.45		439.39	13.01	14.31
48 L.H., Daily Totals		$3039.45		$4316.27	$63.32	$89.92

Crew No.	Bare Costs		Incl. Subs O&P		Cost Per Labor-Hour	
Crew R-11	Hr.	Daily	Hr.	Daily	Bare Costs	Incl. O&P
1 Electrician Foreman	$53.85	$430.80	$80.60	$644.80	$50.59	$76.11
4 Electricians	53.35	1707.20	79.85	2555.20		
1 Equip. Oper. (crane)	50.25	402.00	76.20	609.60		
1 Common Laborer	36.65	293.20	56.55	452.40		
1 Crew Truck		224.80		247.28		
1 Hyd. Crane, 12 Ton		663.00		729.30	15.85	17.44
56 L.H., Daily Totals		$3721.00		$5238.58	$66.45	$93.55

	Bare Costs		Incl. Subs O&P			
Crew R-12	Hr.	Daily	Hr.	Daily	Bare Costs	Incl. O&P
1 Carpenter Foreman (inside)	$46.35	$370.80	$71.50	$572.00	$43.30	$67.73
4 Carpenters	45.85	1467.20	70.75	2264.00		
4 Common Laborers	36.65	1172.80	56.55	1809.60		
1 Equip. Oper. (medium)	48.90	391.20	74.15	593.20		
1 Steel Worker	51.10	408.80	90.20	721.60		
1 Dozer, 200 H.P.		1325.00		1457.50		
1 Pickup Truck, 3/4 Ton		157.00		172.70	16.84	18.52
88 L.H., Daily Totals		$5292.80		$7590.60	$60.15	$86.26

	Bare Costs		Incl. Subs O&P			
Crew R-13	Hr.	Daily	Hr.	Daily	Bare Costs	Incl. O&P
1 Electrician Foreman	$53.85	$430.80	$80.60	$644.80	$51.43	$77.18
3 Electricians	53.35	1280.40	79.85	1916.40		
.25 Equip. Oper. (crane)	50.25	100.50	76.20	152.40		
1 Equipment Oiler	43.55	348.40	66.00	528.00		
.25 Hydraulic Crane, 33 Ton		189.80		208.78	4.52	4.97
42 L.H., Daily Totals		$2349.90		$3450.38	$55.95	$82.15

	Bare Costs		Incl. Subs O&P			
Crew R-15	Hr.	Daily	Hr.	Daily	Bare Costs	Incl. O&P
1 Electrician Foreman	$53.85	$430.80	$80.60	$644.80	$52.38	$78.56
4 Electricians	53.35	1707.20	79.85	2555.20		
1 Equipment Oper. (light)	47.05	376.40	71.35	570.80		
1 Aerial Lift Truck, 40' Boom		325.00		357.50	6.77	7.45
48 L.H., Daily Totals		$2839.40		$4128.30	$59.15	$86.01

	Bare Costs		Incl. Subs O&P			
Crew R-18	Hr.	Daily	Hr.	Daily	Bare Costs	Incl. O&P
.25 Electrician Foreman	$53.85	$107.70	$80.60	$161.20	$41.88	$63.82
1 Electrician	53.35	426.80	79.85	638.80		
2 Helpers	34.65	554.40	53.70	859.20		
26 L.H., Daily Totals		$1088.90		$1659.20	$41.88	$63.82

	Bare Costs		Incl. Subs O&P			
Crew R-19	Hr.	Daily	Hr.	Daily	Bare Costs	Incl. O&P
.5 Electrician Foreman	$53.85	$215.40	$80.60	$322.40	$53.45	$80.00
2 Electricians	53.35	853.60	79.85	1277.60		
20 L.H., Daily Totals		$1069.00		$1600.00	$53.45	$80.00

	Bare Costs		Incl. Subs O&P			
Crew R-21	Hr.	Daily	Hr.	Daily	Bare Costs	Incl. O&P
1 Electrician Foreman	$53.85	$430.80	$80.60	$644.80	$53.36	$79.89
3 Electricians	53.35	1280.40	79.85	1916.40		
.1 Equip. Oper. (medium)	48.90	39.12	74.15	59.32		
.1 S.P. Crane, 4x4, 25 Ton		60.72		66.79	1.85	2.04
32.8 L.H., Daily Totals		$1811.04		$2687.31	$55.21	$81.93

	Bare Costs		Incl. Subs O&P			
Crew R-22	Hr.	Daily	Hr.	Daily	Bare Costs	Incl. O&P
.66 Electrician Foreman	$53.85	$284.33	$80.60	$425.57	$45.40	$68.73
2 Electricians	53.35	853.60	79.85	1277.60		
2 Helpers	34.65	554.40	53.70	859.20		
37.28 L.H., Daily Totals		$1692.33		$2562.37	$45.40	$68.73

Crew No.	Bare Costs		Incl. Subs O & P		Cost Per Labor-Hour	
Crew R-30	Hr.	Daily	Hr.	Daily	Bare Costs	Incl. O&P
.25 Electrician Foreman (outside)	$55.35	$110.70	$82.85	$165.70	$43.23	$65.74
1 Electrician	53.35	426.80	79.85	638.80		
2 Laborers, (Semi-Skilled)	36.65	586.40	56.55	904.80		
26 L.H., Daily Totals		$1123.90		$1709.30	$43.23	$65.74
Crew R-31	Hr.	Daily	Hr.	Daily	Bare Costs	Incl. O&P
1 Electrician	$53.35	$426.80	$79.85	$638.80	$53.35	$79.85
1 Core Drill, Electric, 2.5 H.P.		46.20		50.82	5.78	6.35
8 L.H., Daily Totals		$473.00		$689.62	$59.13	$86.20
Crew W-41E	Hr.	Daily	Hr.	Daily	Bare Costs	Incl. O&P
.5 Plumber Foreman (outside)	$59.55	$238.20	$89.90	$359.60	$49.59	$75.36
1 Plumber	57.55	460.40	86.90	695.20		
1 Laborer	36.65	293.20	56.55	452.40		
20 L.H., Daily Totals		$991.80		$1507.20	$49.59	$75.36

Historical Cost Indexes

The table below lists both the RSMeans® historical cost index based on Jan. 1, 1993 = 100 as well as the computed value of an index based on Jan. 1, 2014 costs. Since the Jan. 1, 2014 figure is estimated, space is left to write in the actual index figures as they become available through either the quarterly *RSMeans Construction Cost Indexes* or as printed in the *Engineering News-Record.* To compute the actual index based on Jan. 1, 2014 = 100, divide the historical cost index for a particular year by the actual Jan. 1, 2014 construction cost index. Space has been left to advance the index figures as the year progresses.

Year	Historical Cost Index Jan. 1, 1993 = 100		Current Index Based on Jan. 1, 2014 = 100		Year	Historical Cost Index Jan. 1, 1993 = 100	Current Index Based on Jan. 1, 2014 = 100		Year	Historical Cost Index Jan. 1, 1993 = 100	Current Index Based on Jan. 1, 2014 = 100	
	Est.	Actual	Est.	Actual		Actual	Est.	Actual		Actual	Est.	Actual
Oct 2014*					July 1999	117.6	58.0		July 1981	70.0	34.5	
July 2014*					1998	115.1	56.8		1980	62.9	31.0	
April 2014*					1997	112.8	55.6		1979	57.8	28.5	
Jan 2014*	202.7		100.0	100.0	1996	110.2	54.4		1978	53.5	26.4	
July 2013		201.2	99.3		1995	107.6	53.1		1977	49.5	24.4	
2012		194.6	96.0		1994	104.4	51.5		1976	46.9	23.1	
2011		191.2	94.3		1993	101.7	50.2		1975	44.8	22.1	
2010		183.5	90.5		1992	99.4	49.1		1974	41.4	20.4	
2009		180.1	88.9		1991	96.8	47.8		1973	37.7	18.6	
2008		180.4	89.0		1990	94.3	46.5		1972	34.8	17.2	
2007		169.4	83.6		1989	92.1	45.5		1971	32.1	15.8	
2006		162.0	79.9		1988	89.9	44.3		1970	28.7	14.2	
2005		151.6	74.8		1987	87.7	43.3		1969	26.9	13.3	
2004		143.7	70.9		1986	84.2	41.6		1968	24.9	12.3	
2003		132.0	65.1		1985	82.6	40.8		1967	23.5	11.6	
2002		128.7	63.5		1984	82.0	40.4		1966	22.7	11.2	
2001		125.1	61.7		1983	80.2	39.5		1965	21.7	10.7	
2000		120.9	59.6		1982	76.1	37.6		1964	21.2	10.5	

Adjustments to Costs

The "Historical Cost Index" can be used to convert national average building costs at a particular time to the approximate building costs for some other time.

Example:

Estimate and compare construction costs for different years in the same city.

To estimate the national average construction cost of a building in 1970, knowing that it cost $900,000 in 2014:

INDEX in 1970 = 28.7

INDEX in 2014 = 202.7

Note: The city cost indexes for Canada can be used to convert U.S. national averages to local costs in Canadian dollars.

*Historical Cost Index updates and other resources are provided on the following website.
http://www.reedconstructiondata.com/rsmeans/chgnotice/456321

Time Adjustment Using the Historical Cost Indexes:

$$\frac{\text{Index for Year A}}{\text{Index for Year B}} \times \text{Cost in Year B} = \text{Cost in Year A}$$

$$\frac{\text{INDEX 1970}}{\text{INDEX 2014}} \times \text{Cost 2014} = \text{Cost 1970}$$

$$\frac{28.7}{202.7} \times \$900{,}000 = .142 \times \$900{,}000 = \$128{,}000$$

The construction cost of the building in 1970 is $128,000.

How to Use the City Cost Indexes

What you should know before you begin

RSMeans City Cost Indexes (CCI) are an extremely useful tool to use when you want to compare costs from city to city and region to region.

This publication contains average construction cost indexes for 731 U.S. and Canadian cities covering over 930 three-digit zip code locations, as listed directly under each city.

Keep in mind that a City Cost Index number is a percentage ratio of a specific city's cost to the national average cost of the same item at a stated time period.

In other words, these index figures represent relative construction factors (or, if you prefer, multipliers) for Material and Installation costs, as well as the weighted average for Total In Place costs for each CSI MasterFormat division. Installation costs include both labor and equipment rental costs. When estimating equipment rental rates only, for a specific location, use 01 54 33 EQUIPMENT RENTAL COSTS in the Reference Section at the back of the book.

The 30 City Average Index is the average of 30 major U.S. cities and serves as a National Average.

Index figures for both material and installation are based on the 30 major city average of 100 and represent the cost relationship as of July 1, 2013. The index for each division is computed from representative material and labor quantities for that division. The weighted average for each city is a weighted total of the components listed above it, but does not include relative productivity between trades or cities.

As changes occur in local material prices, labor rates, and equipment rental rates (including fuel costs), the impact of these changes should be accurately measured by the change in the City Cost Index for each particular city (as compared to the 30 City Average).

Therefore, if you know (or have estimated) building costs in one city today, you can easily convert those costs to expected building costs in another city.

In addition, by using the Historical Cost Index, you can easily convert National Average building costs at a particular time to the approximate building costs for some other time. The City Cost Indexes can then be applied to calculate the costs for a particular city.

Quick Calculations

Location Adjustment Using the City Cost Indexes:

$$\frac{\text{Index for City A}}{\text{Index for City B}} \times \text{Cost in City B} = \text{Cost in City A}$$

Time Adjustment for the National Average
Using the Historical Cost Index:

$$\frac{\text{Index for Year A}}{\text{Index for Year B}} \times \text{Cost in Year B} = \text{Cost in Year A}$$

Adjustment from the National Average:

$$\frac{\text{Index for City A}}{100} \times \text{National Average Cost} = \text{Cost in City A}$$

Since each of the other RSMeans publications contains many different items, any *one* item multiplied by the particular city index may give incorrect results. However, the larger the number of items compiled, the closer the results should be to actual costs for that particular city.

The City Cost Indexes for Canadian cities are calculated using Canadian material and equipment prices and labor rates, in Canadian dollars. Therefore, indexes for Canadian cities can be used to convert U.S. National Average prices to local costs in Canadian dollars.

How to use this section

1. **Compare costs from city to city.**

In using the RSMeans Indexes, remember that an index number is not a fixed number but a ratio: It's a percentage ratio of a building component's cost at any stated time to the National Average cost of that same component at the same time period. Put in the form of an equation:

$$\frac{\text{Specific City Cost}}{\text{National Average Cost}} \times 100 = \text{City Index Number}$$

Therefore, when making cost comparisons between cities, do not subtract one city's index number from the index number of another city and read the result as a percentage difference. Instead, divide one city's index number by that of the other city. The resulting number may then be used as a multiplier to calculate cost differences from city to city.

The formula used to find cost differences between cities for the purpose of comparison is as follows:

$$\frac{\text{City A Index}}{\text{City B Index}} \times \text{City B Cost (Known)} = \text{City A Cost (Unknown)}$$

In addition, you can use RSMeans CCI to calculate and compare costs division by division between cities using the same basic formula. (Just be sure that you're comparing similar divisions.)

2. **Compare a specific city's construction costs with the National Average.**

When you're studying construction location feasibility, it's advisable to compare a prospective project's cost index with an index of the National Average cost.

For example, divide the weighted average index of construction costs of a specific city by that of the 30 City Average, which = 100.

$$\frac{\text{City Index}}{100} = \% \text{ of National Average}$$

As a result, you get a ratio that indicates the relative cost of construction in that city in comparison with the National Average.

3. **Convert U.S. National Average to actual costs in Canadian City.**

$$\frac{\text{Index for Canadian City}}{100} \times \text{National Average Cost} = \text{Cost in Canadian City in \$ CAN}$$

4. Adjust construction cost data based on a National Average.

When you use a source of construction cost data which is based on a National Average (such as RSMeans cost data publications), it is necessary to adjust those costs to a specific location.

$$\frac{\text{City Index}}{100} \times \text{``Book'' Cost Based on National Average Costs} = \text{City Cost (Unknown)}$$

5. When applying the City Cost Indexes to demolition projects, use the appropriate division installation index. For example, for removal of existing doors and windows, use Division 8 (Openings) index.

What you might like to know about how we developed the Indexes

The information presented in the CCI is organized according to the Construction Specifications Institute (CSI) MasterFormat 2012 classification system.

To create a reliable index, RSMeans researched the building type most often constructed in the United States and Canada. Because it was concluded that no one type of building completely represented the building construction industry, nine different types of buildings were combined to create a composite model.

The exact material, labor, and equipment quantities are based on detailed analyses of these nine building types, and then each quantity is weighted in proportion to expected usage. These various material items, labor hours, and equipment rental rates are thus combined to form a composite building representing as closely as possible the actual usage of materials, labor, and equipment used in the North American building construction industry.

The following structures were chosen to make up that composite model:

1. Factory, 1 story
2. Office, 2–4 story
3. Store, Retail
4. Town Hall, 2–3 story
5. High School, 2–3 story
6. Hospital, 4–8 story
7. Garage, Parking
8. Apartment, 1–3 story
9. Hotel/Motel, 2–3 story

For the purposes of ensuring the timeliness of the data, the components of the index for the composite model have been streamlined. They currently consist of:

- specific quantities of 66 commonly used construction materials;
- specific labor-hours for 21 building construction trades; and
- specific days of equipment rental for 6 types of construction equipment (normally used to install the 66 material items by the 21 trades.) Fuel costs and routine maintenance costs are included in the equipment cost.

A sophisticated computer program handles the updating of all costs for each city on a quarterly basis. Material and equipment price quotations are gathered quarterly from 731 cities in the United States and Canada. These prices and the latest negotiated labor wage rates for 21 different building trades are used to compile the quarterly update of the City Cost Index.

The 30 major U.S. cities used to calculate the National Average are:

Atlanta, GA	Memphis, TN
Baltimore, MD	Milwaukee, WI
Boston, MA	Minneapolis, MN
Buffalo, NY	Nashville, TN
Chicago, IL	New Orleans, LA
Cincinnati, OH	New York, NY
Cleveland, OH	Philadelphia, PA
Columbus, OH	Phoenix, AZ
Dallas, TX	Pittsburgh, PA
Denver, CO	St. Louis, MO
Detroit, MI	San Antonio, TX
Houston, TX	San Diego, CA
Indianapolis, IN	San Francisco, CA
Kansas City, MO	Seattle, WA
Los Angeles, CA	Washington, DC

What the CCI does not indicate

The weighted average for each city is a total of the divisional components weighted to reflect typical usage, but it does not include the productivity variations between trades or cities.

In addition, the CCI does not take into consideration factors such as the following:

- managerial efficiency
- competitive conditions
- automation
- restrictive union practices
- unique local requirements
- regional variations due to specific building codes

UNITED STATES / ALABAMA

DIVISION		30 CITY AVERAGE			ANNISTON 362			BIRMINGHAM 350 - 352			BUTLER 369			DECATUR 356			DOTHAN 363		
		MAT.	INST.	TOTAL	MAT.	INST.	TOTAL	MAT.	INST.	TOTAL	MAT.	INST.	TOTAL	MAT.	INST.	TOTAL	MAT.	INST.	TOTAL
015433	CONTRACTOR EQUIPMENT		100.0	100.0		101.6	101.6		101.7	101.7		98.8	98.8		101.6	101.6		98.8	98.8
0241, 31 - 34	SITE & INFRASTRUCTURE, DEMOLITION	100.0	100.0	100.0	87.7	94.3	92.3	88.8	94.9	93.1	101.0	88.3	92.1	81.3	92.8	89.3	98.7	88.3	91.4
0310	Concrete Forming & Accessories	100.0	100.0	100.0	92.7	48.2	54.3	93.3	70.2	73.4	89.0	43.6	49.8	94.7	46.4	53.0	97.9	43.0	50.5
0320	Concrete Reinforcing	100.0	100.0	100.0	87.0	87.4	87.2	92.9	87.8	90.4	92.0	47.4	69.7	87.0	79.5	83.3	92.0	46.8	69.4
0330	Cast-in-Place Concrete	100.0	100.0	100.0	102.0	50.5	80.6	99.7	74.7	89.3	99.4	55.7	81.3	93.7	65.1	81.9	99.4	47.0	77.7
03	CONCRETE	100.0	100.0	100.0	99.8	57.9	79.1	95.1	76.1	85.7	100.6	50.4	75.8	91.4	60.7	76.3	100.0	47.0	73.9
04	MASONRY	100.0	100.0	100.0	90.5	67.9	76.5	89.9	76.2	81.4	95.3	49.4	66.9	88.4	52.1	66.0	96.6	40.4	61.8
05	METALS	100.0	100.0	100.0	96.9	92.5	95.5	97.0	94.7	96.3	95.8	78.1	90.4	99.0	87.7	95.5	95.9	76.9	90.0
06	WOOD, PLASTICS & COMPOSITES	100.0	100.0	100.0	92.7	43.7	65.1	96.3	68.3	80.5	87.3	43.4	62.5	98.0	43.9	67.5	99.6	43.4	67.9
07	THERMAL & MOISTURE PROTECTION	100.0	100.0	100.0	100.1	53.8	81.3	95.9	81.4	90.0	100.1	57.6	82.9	93.1	59.3	79.4	100.1	49.9	79.7
08	OPENINGS	100.0	100.0	100.0	93.9	50.9	83.8	95.8	72.3	90.3	94.0	45.2	82.4	98.8	51.6	87.6	94.0	45.2	82.5
0920	Plaster & Gypsum Board	100.0	100.0	100.0	87.5	42.4	56.8	97.0	67.8	77.1	84.9	42.0	55.7	98.1	42.6	60.4	95.3	42.0	59.1
0950, 0980	Ceilings & Acoustic Treatment	100.0	100.0	100.0	87.8	42.4	57.6	98.4	67.8	78.0	87.8	42.0	57.3	95.7	42.6	60.4	87.8	42.0	57.3
0960	Flooring	100.0	100.0	100.0	104.1	41.0	85.2	109.0	76.2	99.2	109.2	57.0	93.6	106.0	51.5	89.7	114.9	28.0	88.9
0970, 0990	Wall Finishes & Painting/Coating	100.0	100.0	100.0	91.4	32.1	55.8	93.5	56.8	71.4	91.4	54.5	69.2	89.3	68.2	76.7	91.4	54.5	69.2
09	FINISHES	100.0	100.0	100.0	92.7	44.0	66.6	99.4	69.3	82.6	95.9	46.7	68.5	95.9	48.8	69.7	98.7	41.0	66.6
COVERS	DIVS. 10 - 14, 25, 28, 41, 43, 44, 46	100.0	100.0	100.0	100.0	70.1	94.0	100.0	87.4	97.5	100.0	42.5	88.5	100.0	44.4	88.9	100.0	42.6	88.5
21, 22, 23	FIRE SUPPRESSION, PLUMBING & HVAC	100.0	100.0	100.0	99.9	56.6	82.6	100.0	62.8	85.1	97.3	34.0	71.9	100.0	41.7	76.6	97.3	33.9	71.9
26, 27, 3370	ELECTRICAL, COMMUNICATIONS & UTIL.	100.0	100.0	100.0	90.7	58.7	74.0	99.7	62.8	80.4	92.6	41.2	65.7	94.5	66.5	79.9	91.5	58.8	74.4
MF2010	WEIGHTED AVERAGE	100.0	100.0	100.0	96.5	63.0	81.9	97.5	74.1	87.3	96.8	50.5	76.6	96.6	59.1	80.3	96.9	50.5	76.7

ALABAMA

DIVISION		EVERGREEN 364			GADSDEN 359			HUNTSVILLE 357 - 358			JASPER 355			MOBILE 365 - 366			MONTGOMERY 360 - 361		
		MAT.	INST.	TOTAL	MAT.	INST.	TOTAL	MAT.	INST.	TOTAL	MAT.	INST.	TOTAL	MAT.	INST.	TOTAL	MAT.	INST.	TOTAL
015433	CONTRACTOR EQUIPMENT		98.8	98.8		101.6	101.6		101.6	101.6		101.6	101.6		98.8	98.8		98.8	98.8
0241, 31 - 34	SITE & INFRASTRUCTURE, DEMOLITION	101.5	88.6	92.4	86.7	93.8	91.7	81.1	93.8	90.0	86.3	93.6	91.4	94.0	89.3	90.7	94.0	89.2	90.6
0310	Concrete Forming & Accessories	85.6	45.2	50.8	87.0	43.3	49.3	94.7	71.1	74.3	92.2	35.3	43.0	97.1	55.0	60.7	96.9	45.0	52.1
0320	Concrete Reinforcing	92.0	47.4	69.7	92.2	87.7	90.0	87.0	80.5	83.8	87.0	86.5	86.8	89.8	79.4	84.6	94.9	87.1	91.0
0330	Cast-in-Place Concrete	99.5	58.3	82.4	93.8	66.7	82.5	91.3	69.6	82.3	103.9	44.8	79.3	104.3	67.4	88.9	105.4	57.4	85.5
03	CONCRETE	100.9	52.0	76.8	95.7	61.4	78.8	90.3	73.3	81.9	98.9	50.1	74.8	97.0	65.3	81.4	98.3	58.9	78.8
04	MASONRY	95.4	54.1	69.8	86.7	62.3	71.6	89.7	67.7	76.1	84.2	42.0	58.1	93.8	55.1	69.9	90.4	50.5	65.7
05	METALS	95.9	78.1	90.4	96.9	94.0	96.0	99.0	90.5	96.4	96.9	91.1	95.1	97.9	90.6	95.6	97.0	92.8	95.7
06	WOOD, PLASTICS & COMPOSITES	83.6	43.4	60.9	87.8	39.6	60.6	98.0	74.3	84.6	94.8	32.3	59.6	98.2	54.1	73.3	97.7	43.4	67.0
07	THERMAL & MOISTURE PROTECTION	100.1	57.9	83.0	93.1	72.4	84.7	93.0	78.5	87.1	93.2	46.1	74.1	99.8	71.3	88.2	100.2	67.8	87.0
08	OPENINGS	94.0	45.2	82.4	95.4	49.8	84.6	98.8	68.2	91.5	95.3	48.7	84.3	97.2	58.6	88.1	98.3	54.0	87.8
0920	Plaster & Gypsum Board	84.2	42.0	55.5	89.8	38.2	54.7	98.1	73.9	81.7	94.2	30.7	51.0	92.1	53.1	65.5	91.5	42.0	57.8
0950, 0980	Ceilings & Acoustic Treatment	87.8	42.0	57.3	92.1	38.2	56.2	97.6	73.9	81.9	92.1	30.7	51.2	93.3	53.1	66.5	95.9	42.0	60.0
0960	Flooring	107.0	57.0	92.1	102.0	76.2	94.3	106.0	76.2	97.1	104.3	41.0	85.3	114.6	59.5	98.1	110.8	59.5	95.5
0970, 0990	Wall Finishes & Painting/Coating	91.4	54.5	69.2	89.3	62.0	72.9	89.3	59.7	71.5	89.3	39.4	59.3	95.0	55.4	71.2	91.2	54.5	69.1
09	FINISHES	95.3	47.9	68.9	93.3	49.4	68.8	96.4	71.1	82.3	94.5	35.2	61.5	99.0	54.7	74.3	99.0	47.1	70.1
COVERS	DIVS. 10 - 14, 25, 28, 41, 43, 44, 46	100.0	44.1	88.8	100.0	79.5	95.9	100.0	85.3	97.0	100.0	40.8	88.1	100.0	83.9	96.8	100.0	79.4	95.9
21, 22, 23	FIRE SUPPRESSION, PLUMBING & HVAC	97.3	36.3	72.9	102.0	36.9	75.9	100.0	52.9	81.1	102.0	53.8	82.6	100.0	61.5	84.5	100.1	34.8	73.9
26, 27, 3370	ELECTRICAL, COMMUNICATIONS & UTIL.	90.1	41.2	64.6	94.5	62.8	78.0	95.7	66.5	80.4	94.1	62.8	77.7	93.4	58.5	75.2	93.9	62.3	77.4
MF2010	WEIGHTED AVERAGE	96.5	51.9	77.1	96.7	60.8	81.1	96.7	70.8	85.4	97.0	56.7	79.4	97.8	65.9	83.9	97.9	58.1	80.5

ALABAMA / ALASKA

DIVISION		PHENIX CITY 368			SELMA 367			TUSCALOOSA 354			ANCHORAGE 995 - 996			FAIRBANKS 997			JUNEAU 998		
		MAT.	INST.	TOTAL	MAT.	INST.	TOTAL	MAT.	INST.	TOTAL	MAT.	INST.	TOTAL	MAT.	INST.	TOTAL	MAT.	INST.	TOTAL
015433	CONTRACTOR EQUIPMENT		98.8	98.8		98.8	98.8		101.6	101.6		113.5	113.5		113.5	113.5		113.5	113.5
0241, 31 - 34	SITE & INFRASTRUCTURE, DEMOLITION	105.2	89.3	94.0	98.6	89.1	92.0	81.6	94.0	90.3	125.5	128.8	127.8	116.4	128.8	125.1	132.0	128.8	129.8
0310	Concrete Forming & Accessories	92.6	39.9	47.1	90.3	43.5	49.9	94.6	48.8	55.1	127.8	119.0	120.2	133.0	118.7	120.7	129.2	118.9	120.3
0320	Concrete Reinforcing	91.9	63.3	77.7	92.0	86.2	89.1	87.0	87.7	87.3	148.0	111.5	129.8	142.1	111.5	126.8	122.7	111.5	117.1
0330	Cast-in-Place Concrete	99.5	57.9	82.2	99.4	47.2	77.8	95.1	68.5	84.1	131.0	117.4	125.3	129.2	117.7	124.4	136.1	117.4	128.3
03	CONCRETE	103.6	52.4	78.3	99.4	54.5	77.3	92.1	64.5	78.5	143.0	116.3	129.8	126.5	116.3	121.4	139.9	116.2	128.2
04	MASONRY	95.3	41.8	62.2	98.7	39.3	62.0	88.6	65.5	74.3	180.9	122.2	144.6	182.2	122.2	145.1	167.6	122.2	139.5
05	METALS	95.8	83.5	92.0	95.8	90.3	94.1	98.2	94.1	96.9	114.6	102.5	110.9	118.8	102.6	113.8	117.6	102.4	112.9
06	WOOD, PLASTICS & COMPOSITES	92.5	36.3	60.8	89.3	43.4	63.4	98.0	45.4	68.3	117.1	117.9	117.6	125.0	117.4	120.7	119.8	117.9	118.7
07	THERMAL & MOISTURE PROTECTION	100.4	64.9	86.0	100.0	53.0	80.9	93.1	74.4	85.5	169.9	115.5	147.8	169.2	116.6	147.9	163.4	115.5	143.9
08	OPENINGS	93.9	44.6	82.3	93.9	54.0	84.5	98.8	59.8	89.6	119.7	115.2	118.7	116.2	115.2	116.0	116.6	115.2	116.3
0920	Plaster & Gypsum Board	89.0	34.7	52.1	86.8	42.0	56.3	98.1	44.2	61.4	126.3	118.3	120.9	148.9	117.8	127.7	131.2	118.3	122.4
0950, 0980	Ceilings & Acoustic Treatment	87.8	34.7	52.5	87.8	42.0	57.3	97.6	44.2	62.1	126.5	118.3	121.0	121.7	117.8	119.1	123.5	118.3	120.0
0960	Flooring	111.3	59.5	95.8	109.8	29.0	85.6	106.0	76.2	97.1	150.1	128.3	143.6	140.7	128.3	137.0	142.1	128.3	138.0
0970, 0990	Wall Finishes & Painting/Coating	91.4	54.5	69.2	91.4	54.5	69.2	89.3	46.9	63.8	146.5	113.3	126.5	143.3	121.9	130.7	143.3	113.3	125.2
09	FINISHES	97.5	43.2	67.2	96.0	41.2	65.5	96.3	52.0	71.6	143.1	120.5	130.5	139.4	121.2	129.2	139.6	120.5	129.0
COVERS	DIVS. 10 - 14, 25, 28, 41, 43, 44, 46	100.0	79.0	95.8	100.0	42.5	88.5	100.0	81.3	96.2	100.0	112.5	102.5	100.0	112.5	102.5	100.0	112.5	102.5
21, 22, 23	FIRE SUPPRESSION, PLUMBING & HVAC	97.3	35.1	72.4	97.3	34.3	72.1	100.0	34.6	73.8	100.2	104.3	101.9	100.2	106.7	102.8	100.1	100.6	100.3
26, 27, 3370	ELECTRICAL, COMMUNICATIONS & UTIL.	92.1	71.2	81.2	91.2	41.2	65.1	95.1	62.8	78.2	120.5	115.6	117.9	134.7	115.6	124.7	125.5	115.6	120.3
MF2010	WEIGHTED AVERAGE	97.3	55.7	79.2	96.7	50.8	76.7	96.7	62.0	81.6	121.3	114.3	118.3	120.7	115.0	118.2	120.6	113.5	117.5

		ALASKA			ARIZONA														
		KETCHIKAN			CHAMBERS			FLAGSTAFF			GLOBE			KINGMAN			MESA/TEMPE		
	DIVISION	999			865			860			855			864			852		
		MAT.	INST.	TOTAL	MAT.	INST.	TOTAL	MAT.	INST.	TOTAL	MAT.	INST.	TOTAL	MAT.	INST.	TOTAL	MAT.	INST.	TOTAL
015433	CONTRACTOR EQUIPMENT		113.5	113.5		92.6	92.6		92.6	92.6		94.2	94.2		92.6	92.6		94.2	94.2
0241, 31 - 34	SITE & INFRASTRUCTURE, DEMOLITION	170.9	128.8	141.3	67.2	96.6	87.9	84.8	96.8	93.3	100.7	98.4	99.1	67.2	96.8	88.0	91.4	98.6	96.4
0310	Concrete Forming & Accessories	124.5	118.9	119.7	98.2	59.2	64.5	103.7	66.4	71.5	96.3	59.3	64.3	96.5	66.4	70.5	99.3	69.6	73.6
0320	Concrete Reinforcing	111.6	111.5	111.5	100.0	83.5	91.7	99.8	83.5	91.7	110.1	83.5	96.8	100.1	83.5	91.8	110.8	83.5	97.2
0330	Cast-in-Place Concrete	261.0	117.4	201.4	90.3	73.5	83.3	90.4	73.8	83.5	93.8	73.6	85.4	90.0	73.8	83.3	94.6	73.9	86.0
03	CONCRETE	214.6	116.2	166.0	96.5	68.9	82.8	116.1	72.2	94.4	109.8	68.9	89.7	96.1	72.2	84.3	101.7	73.7	87.8
04	MASONRY	191.0	122.2	148.4	90.6	62.7	73.3	90.7	62.8	73.4	110.3	62.7	80.9	90.6	62.8	73.4	110.6	62.8	81.0
05	METALS	119.0	102.4	113.9	95.1	75.0	88.9	95.6	75.7	89.5	92.4	75.5	87.2	95.7	75.8	89.6	92.7	76.4	87.7
06	WOOD, PLASTICS & COMPOSITES	115.7	117.9	117.0	96.1	56.3	73.6	102.2	65.7	81.6	90.0	56.4	71.0	91.6	65.7	77.0	93.4	69.9	80.2
07	THERMAL & MOISTURE PROTECTION	172.3	115.5	149.2	95.1	65.1	82.9	96.7	68.2	85.1	99.3	64.6	85.2	95.0	65.7	83.1	98.6	66.6	85.6
08	OPENINGS	118.1	115.2	117.4	108.1	65.7	98.1	108.2	70.8	99.4	102.2	65.8	93.6	108.3	70.8	99.4	102.3	73.2	95.4
0920	Plaster & Gypsum Board	137.4	118.3	124.4	85.6	55.0	64.8	89.1	64.7	72.5	94.2	55.0	67.5	79.1	64.7	69.3	96.4	69.0	77.7
0950, 0980	Ceilings & Acoustic Treatment	114.7	118.3	117.1	107.8	55.0	72.7	108.7	64.7	79.4	96.4	55.0	68.8	108.7	64.7	79.4	96.4	69.0	78.1
0960	Flooring	140.7	128.3	137.0	92.1	40.3	76.6	94.1	40.4	78.1	100.0	40.3	82.1	91.1	54.7	80.2	101.1	50.4	86.0
0970, 0990	Wall Finishes & Painting/Coating	143.9	113.3	125.5	97.8	56.3	72.9	97.8	56.3	72.9	108.9	56.3	77.3	97.8	56.3	72.9	108.9	56.3	77.3
09	FINISHES	140.7	120.5	129.5	95.7	54.0	72.4	98.8	59.6	76.9	97.6	54.1	73.4	94.7	62.0	76.5	97.1	64.0	78.7
COVERS	DIVS. 10 - 14, 25, 28, 41, 43, 44, 46	100.0	112.5	102.5	100.0	82.3	96.5	100.0	83.3	96.7	100.0	82.5	96.5	100.0	83.3	96.7	100.0	84.0	96.8
21, 22, 23	FIRE SUPPRESSION, PLUMBING & HVAC	98.4	100.6	99.3	97.0	78.9	89.7	100.2	80.0	92.1	95.2	78.9	88.7	97.0	80.0	90.2	100.1	79.0	91.6
26, 27, 3370	ELECTRICAL, COMMUNICATIONS & UTIL.	134.6	115.6	124.7	102.2	72.6	86.7	101.1	65.8	82.7	98.7	65.8	81.6	102.2	65.8	83.2	95.3	65.8	79.9
MF2010	WEIGHTED AVERAGE	132.0	113.5	124.0	97.3	71.9	86.3	101.0	72.9	88.8	99.0	71.2	86.9	97.3	73.2	86.8	98.8	73.8	87.9

		ARIZONA												ARKANSAS					
		PHOENIX			PRESCOTT			SHOW LOW			TUCSON			BATESVILLE			CAMDEN		
	DIVISION	850,853			863			859			856 - 857			725			717		
		MAT.	INST.	TOTAL	MAT.	INST.	TOTAL	MAT.	INST.	TOTAL	MAT.	INST.	TOTAL	MAT.	INST.	TOTAL	MAT.	INST.	TOTAL
015433	CONTRACTOR EQUIPMENT		94.7	94.7		92.6	92.6		94.2	94.2		94.2	94.2		89.0	89.0		89.0	89.0
0241, 31 - 34	SITE & INFRASTRUCTURE, DEMOLITION	91.9	98.8	96.7	73.5	96.6	89.7	102.8	98.4	99.7	87.1	98.6	95.1	73.4	84.9	81.5	73.3	84.5	81.2
0310	Concrete Forming & Accessories	100.2	69.7	73.9	99.8	53.5	59.8	103.2	69.3	74.0	99.8	69.5	73.6	85.2	45.9	51.3	83.9	31.4	38.6
0320	Concrete Reinforcing	109.0	83.7	96.3	99.8	83.5	91.6	110.8	83.5	97.2	91.8	83.5	87.6	86.7	65.8	76.3	88.2	65.7	76.9
0330	Cast-in-Place Concrete	94.7	74.0	86.1	90.3	73.5	83.3	93.9	73.6	85.5	97.4	73.9	87.6	81.1	46.1	66.6	81.2	39.0	63.7
03	CONCRETE	101.3	73.8	87.7	101.7	66.3	84.2	112.1	73.4	93.0	99.7	73.6	86.8	85.2	50.5	68.0	86.9	41.5	64.5
04	MASONRY	98.0	64.6	77.3	90.7	62.7	73.4	110.4	62.7	80.9	95.9	62.8	75.4	98.7	41.5	63.3	111.6	31.9	62.3
05	METALS	94.2	77.2	88.9	95.6	74.7	89.2	92.2	75.7	87.1	93.5	76.3	88.2	98.0	66.9	88.4	97.3	66.5	87.8
06	WOOD, PLASTICS & COMPOSITES	94.4	69.9	80.6	97.5	48.5	69.8	98.0	69.9	82.1	93.7	69.9	80.3	86.3	46.5	63.9	89.8	30.4	56.3
07	THERMAL & MOISTURE PROTECTION	98.2	68.7	86.2	95.5	64.2	82.8	99.5	66.5	86.1	99.6	65.8	85.9	100.2	44.8	77.7	96.1	36.4	71.9
08	OPENINGS	104.2	73.2	96.9	108.2	61.5	97.2	101.3	73.2	94.7	98.3	73.2	92.4	94.6	46.0	83.1	93.5	40.6	81.0
0920	Plaster & Gypsum Board	98.5	69.0	78.4	85.7	47.0	59.4	98.6	69.0	78.4	101.9	69.0	79.5	81.0	45.4	56.7	80.2	28.7	45.2
0950, 0980	Ceilings & Acoustic Treatment	103.5	69.0	80.5	106.9	47.0	67.1	96.4	69.0	78.1	97.3	69.0	78.4	90.6	45.4	60.5	86.2	28.7	47.9
0960	Flooring	101.4	52.7	86.8	92.9	40.3	77.1	102.5	45.9	85.5	91.9	45.0	77.9	104.6	60.1	91.3	111.2	40.3	90.0
0970, 0990	Wall Finishes & Painting/Coating	108.9	62.0	80.7	97.8	56.3	72.9	108.9	56.3	77.3	110.1	56.3	77.8	99.8	42.0	65.1	99.5	50.8	70.2
09	FINISHES	99.1	65.0	80.1	96.2	49.4	70.1	99.2	63.0	79.0	95.4	62.9	77.3	89.8	47.8	66.4	92.1	34.2	59.8
COVERS	DIVS. 10 - 14, 25, 28, 41, 43, 44, 46	100.0	84.0	96.8	100.0	81.4	96.3	100.0	84.0	96.8	100.0	84.0	96.8	100.0	40.6	88.1	100.0	36.5	87.3
21, 22, 23	FIRE SUPPRESSION, PLUMBING & HVAC	100.0	80.0	92.0	100.2	78.9	91.7	95.7	79.0	88.7	100.1	79.0	91.6	95.3	52.1	77.9	95.3	53.5	78.5
26, 27, 3370	ELECTRICAL, COMMUNICATIONS & UTIL.	102.2	68.3	84.5	100.7	65.8	82.5	95.7	65.8	80.1	97.6	65.8	81.0	95.0	64.0	78.8	91.5	62.0	76.1
MF2010	WEIGHTED AVERAGE	99.4	74.8	88.7	98.9	69.7	86.2	99.1	73.5	88.0	97.6	73.6	87.1	94.2	55.1	77.2	94.5	50.4	75.3

		ARKANSAS																	
		FAYETTEVILLE			FORT SMITH			HARRISON			HOT SPRINGS			JONESBORO			LITTLE ROCK		
	DIVISION	727			729			726			719			724			720 - 722		
		MAT.	INST.	TOTAL	MAT.	INST.	TOTAL	MAT.	INST.	TOTAL	MAT.	INST.	TOTAL	MAT.	INST.	TOTAL	MAT.	INST.	TOTAL
015433	CONTRACTOR EQUIPMENT		89.0	89.0		89.0	89.0		89.0	89.0		89.0	89.0		107.5	107.5		89.0	89.0
0241, 31 - 34	SITE & INFRASTRUCTURE, DEMOLITION	72.8	86.4	82.4	77.7	86.3	83.8	77.9	84.9	82.8	76.4	85.9	83.0	96.9	100.3	99.3	87.9	86.3	86.8
0310	Concrete Forming & Accessories	80.6	39.4	45.1	100.2	57.7	63.5	89.7	45.8	51.8	81.2	36.5	42.6	88.7	49.2	54.6	95.3	60.4	65.2
0320	Concrete Reinforcing	86.7	71.6	79.2	87.7	72.3	80.0	86.3	65.7	76.0	86.5	65.8	76.2	83.9	64.8	74.4	92.7	66.5	79.6
0330	Cast-in-Place Concrete	81.1	50.2	68.3	92.9	70.4	83.5	90.0	44.2	71.0	83.0	39.5	64.9	88.3	55.9	74.9	91.5	70.5	82.8
03	CONCRETE	84.9	50.1	67.7	93.1	65.3	79.4	92.6	49.8	71.4	90.7	44.0	67.6	89.8	55.9	73.1	94.7	65.4	80.2
04	MASONRY	89.1	42.1	60.0	94.1	50.0	66.8	99.0	39.7	62.3	83.9	34.1	53.1	90.5	42.6	60.9	93.9	50.0	66.7
05	METALS	98.0	69.3	89.1	100.3	71.1	91.3	99.1	66.6	89.1	97.3	66.9	87.9	94.6	78.8	89.7	100.5	69.2	90.8
06	WOOD, PLASTICS & COMPOSITES	81.9	36.3	56.2	103.9	60.4	79.4	92.3	46.5	66.5	86.8	36.5	58.4	90.1	49.4	67.1	96.3	63.9	78.1
07	THERMAL & MOISTURE PROTECTION	101.0	46.2	78.8	101.4	54.5	82.4	100.5	43.5	77.4	96.3	38.6	72.9	105.2	50.0	82.8	97.7	54.9	80.3
08	OPENINGS	94.6	45.8	83.0	95.4	57.1	86.4	95.4	46.7	83.9	93.5	42.4	81.4	97.4	53.5	87.0	96.2	58.9	87.4
0920	Plaster & Gypsum Board	78.7	34.9	48.9	86.6	59.7	68.3	85.5	45.4	58.2	78.8	35.1	49.0	95.4	48.0	63.1	92.1	63.3	72.5
0950, 0980	Ceilings & Acoustic Treatment	90.6	34.9	53.5	92.4	59.7	70.6	92.4	45.4	61.1	86.2	35.1	52.2	93.8	48.0	63.3	95.8	63.3	74.2
0960	Flooring	102.0	60.1	89.5	111.0	61.9	96.3	106.9	60.1	92.9	110.0	60.1	95.1	74.5	54.4	68.5	109.1	61.9	95.0
0970, 0990	Wall Finishes & Painting/Coating	99.8	30.0	57.8	99.8	55.0	72.8	99.8	42.0	65.1	99.5	50.8	70.2	88.3	49.4	64.9	100.0	56.4	74.2
09	FINISHES	88.9	40.9	62.2	93.0	58.4	73.8	91.8	47.8	67.3	91.8	41.3	63.7	86.5	49.6	66.0	95.9	60.7	76.3
COVERS	DIVS. 10 - 14, 25, 28, 41, 43, 44, 46	100.0	50.0	90.0	100.0	76.8	95.4	100.0	50.5	90.1	100.0	37.5	87.5	100.0	46.3	89.3	100.0	77.2	95.4
21, 22, 23	FIRE SUPPRESSION, PLUMBING & HVAC	95.3	50.6	77.4	100.2	50.1	80.1	95.3	49.8	77.0	95.3	49.1	76.8	100.4	52.9	81.4	100.0	56.4	82.5
26, 27, 3370	ELECTRICAL, COMMUNICATIONS & UTIL.	88.8	52.4	69.8	92.3	66.1	78.6	93.5	38.0	64.5	93.5	69.4	81.0	99.0	64.0	80.7	99.3	69.5	83.7
MF2010	WEIGHTED AVERAGE	93.0	52.9	75.6	96.8	61.7	81.5	95.5	51.1	76.1	93.8	52.3	75.8	96.3	59.4	80.2	98.0	63.8	83.1

495

ARKANSAS / CALIFORNIA

	DIVISION	PINE BLUFF 716 MAT.	INST.	TOTAL	RUSSELLVILLE 728 MAT.	INST.	TOTAL	TEXARKANA 718 MAT.	INST.	TOTAL	WEST MEMPHIS 723 MAT.	INST.	TOTAL	ALHAMBRA 917-918 MAT.	INST.	TOTAL	ANAHEIM 928 MAT.	INST.	TOTAL
015433	CONTRACTOR EQUIPMENT		89.0	89.0		89.0	89.0		89.7	89.7		107.5	107.5		98.9	98.9		100.0	100.0
0241, 31 - 34	SITE & INFRASTRUCTURE, DEMOLITION	78.5	86.3	84.0	74.5	84.9	81.8	90.2	86.9	87.9	103.2	100.3	101.2	96.4	109.7	105.7	96.5	107.7	104.4
0310	Concrete Forming & Accessories	80.9	60.3	63.2	85.9	56.0	60.1	87.6	41.0	47.4	94.4	49.5	55.6	115.1	116.9	116.6	105.5	123.1	120.7
0320	Concrete Reinforcing	88.1	66.5	77.3	87.3	65.6	76.5	87.7	65.8	76.8	83.9	64.9	74.4	107.4	114.9	111.1	93.2	114.8	103.9
0330	Cast-in-Place Concrete	83.0	70.4	77.8	85.0	46.0	68.8	90.5	44.0	71.2	92.6	56.0	77.4	94.7	119.1	104.8	91.3	121.7	104.0
03	CONCRETE	91.5	65.4	78.6	88.4	54.9	71.9	89.1	47.6	68.6	97.3	56.0	76.9	104.0	116.4	110.1	103.1	120.0	111.4
04	MASONRY	120.2	50.0	76.8	93.3	38.0	59.1	98.3	33.7	58.2	76.8	42.6	55.7	122.3	118.6	120.0	77.4	116.2	101.4
05	METALS	98.0	69.1	89.1	98.0	66.5	88.3	90.8	67.0	83.5	93.6	79.1	89.2	84.0	99.9	88.9	103.1	100.2	102.2
06	WOOD, PLASTICS & COMPOSITES	86.3	63.9	73.7	87.7	60.4	72.3	95.2	43.0	65.7	96.3	49.4	69.9	96.6	113.1	105.9	98.5	121.4	111.4
07	THERMAL & MOISTURE PROTECTION	96.4	54.9	79.6	101.2	44.7	78.3	97.1	45.3	76.1	105.6	50.0	83.1	95.2	117.2	104.2	97.9	121.3	107.4
08	OPENINGS	93.5	58.9	85.3	94.6	56.2	85.5	98.3	46.7	86.1	97.4	53.5	87.0	89.8	115.5	95.9	100.3	120.2	105.0
0920	Plaster & Gypsum Board	78.4	63.3	68.1	81.0	59.7	66.5	82.2	41.7	54.6	98.0	48.0	64.0	102.1	113.7	110.0	100.6	122.0	115.2
0950, 0980	Ceilings & Acoustic Treatment	86.2	63.3	71.0	90.6	59.7	70.0	89.7	41.7	57.8	91.8	48.0	62.7	102.5	113.7	109.9	106.7	122.0	116.9
0960	Flooring	109.7	61.9	95.4	104.3	60.1	91.1	111.9	51.9	94.0	76.7	54.4	70.0	93.0	109.2	97.8	105.1	114.1	107.8
0970, 0990	Wall Finishes & Painting/Coating	99.5	56.4	73.6	99.8	35.7	61.3	99.5	31.0	58.3	88.3	50.8	65.8	101.7	108.4	105.7	100.4	106.2	103.9
09	FINISHES	91.8	60.7	74.5	90.0	55.3	70.7	94.3	41.3	64.7	87.7	49.8	66.6	98.0	114.0	106.9	102.0	119.6	111.8
COVERS	DIVS. 10 - 14, 25, 28, 41, 43, 44, 46	100.0	77.2	95.4	100.0	42.1	88.4	100.0	33.5	86.7	100.0	46.4	89.3	100.0	110.1	102.0	100.0	111.3	102.3
21, 22, 23	FIRE SUPPRESSION, PLUMBING & HVAC	100.2	52.4	81.1	95.3	51.9	77.9	100.2	54.6	81.9	95.5	66.5	83.9	95.1	113.7	102.6	100.0	107.2	102.7
26, 27, 3370	ELECTRICAL, COMMUNICATIONS & UTIL.	91.6	69.5	80.1	92.2	45.7	68.0	93.4	36.8	63.9	100.6	65.8	82.4	120.2	119.9	120.0	91.4	105.9	99.0
MF2010	WEIGHTED AVERAGE	96.9	62.9	82.1	94.1	54.4	76.8	95.7	49.9	75.7	95.6	62.7	81.2	98.2	113.9	105.0	98.9	114.1	105.6

CALIFORNIA

	DIVISION	BAKERSFIELD 932-933 MAT.	INST.	TOTAL	BERKELEY 947 MAT.	INST.	TOTAL	EUREKA 955 MAT.	INST.	TOTAL	FRESNO 936-938 MAT.	INST.	TOTAL	INGLEWOOD 903-905 MAT.	INST.	TOTAL	LONG BEACH 906-908 MAT.	INST.	TOTAL
015433	CONTRACTOR EQUIPMENT		97.9	97.9		98.5	98.5		97.6	97.6		97.9	97.9		95.7	95.7		95.7	95.7
0241, 31 - 34	SITE & INFRASTRUCTURE, DEMOLITION	97.5	105.3	103.0	119.7	106.0	110.1	108.5	101.6	103.7	99.4	104.8	103.2	92.6	102.2	99.3	99.8	102.2	101.5
0310	Concrete Forming & Accessories	104.6	122.7	120.2	118.3	148.0	143.9	114.9	130.6	128.4	103.9	131.7	127.9	113.1	116.9	116.4	108.0	116.9	115.7
0320	Concrete Reinforcing	108.4	114.7	111.5	92.5	116.2	104.3	101.9	115.4	108.6	84.7	115.4	100.0	108.6	114.9	111.7	107.6	114.9	111.2
0330	Cast-in-Place Concrete	92.1	120.7	104.0	127.8	122.0	125.4	99.1	116.3	106.2	97.5	117.3	105.7	87.6	120.2	101.1	99.7	120.2	108.2
03	CONCRETE	101.1	119.4	110.1	110.4	131.2	120.6	114.4	121.4	117.8	101.5	122.4	111.8	95.7	116.8	106.1	105.0	116.8	110.8
04	MASONRY	93.2	115.7	107.1	119.2	131.9	127.0	99.4	123.4	114.3	96.3	121.3	111.7	75.8	118.7	102.4	84.5	118.7	105.7
05	METALS	105.8	99.4	103.8	106.5	100.3	104.6	102.9	97.8	101.3	105.9	100.4	104.2	93.8	100.6	95.9	93.7	100.6	95.8
06	WOOD, PLASTICS & COMPOSITES	93.8	121.5	109.4	110.4	154.0	135.0	112.3	135.9	125.6	105.3	135.9	122.5	106.0	113.2	110.1	99.5	113.2	107.2
07	THERMAL & MOISTURE PROTECTION	104.0	111.9	107.2	103.2	135.9	116.5	101.9	116.0	107.6	94.0	113.6	101.9	100.9	118.5	108.1	101.1	118.5	108.2
08	OPENINGS	96.5	116.2	101.2	96.8	140.5	107.1	99.8	112.7	102.8	96.7	124.0	103.1	91.0	115.6	96.8	91.0	115.5	96.8
0920	Plaster & Gypsum Board	98.9	122.0	114.7	102.8	155.2	138.4	106.0	136.9	127.0	97.5	136.9	124.3	103.9	113.7	110.5	99.8	113.7	109.2
0950, 0980	Ceilings & Acoustic Treatment	109.6	122.0	117.9	104.1	155.2	138.1	112.1	136.9	128.6	106.0	136.9	126.5	101.7	113.7	109.7	101.7	113.7	109.7
0960	Flooring	104.9	109.2	106.2	108.7	126.4	114.0	109.3	118.7	112.1	110.4	135.9	118.0	98.9	109.2	102.0	96.5	109.2	100.3
0970, 0990	Wall Finishes & Painting/Coating	100.8	104.2	102.8	102.7	142.1	126.4	102.0	47.2	69.1	115.0	108.6	111.1	102.1	108.4	105.9	102.1	108.4	105.9
09	FINISHES	102.9	118.7	111.7	103.9	145.5	127.1	107.9	122.2	115.9	104.3	131.9	119.7	101.4	114.1	108.4	100.6	114.1	108.1
COVERS	DIVS. 10 - 14, 25, 28, 41, 43, 44, 46	100.0	111.3	102.3	100.0	125.1	105.0	100.0	121.4	104.3	100.0	121.4	104.3	100.0	110.2	102.0	100.0	110.2	102.0
21, 22, 23	FIRE SUPPRESSION, PLUMBING & HVAC	100.1	108.3	103.4	95.2	144.7	115.0	95.1	99.1	96.7	100.2	109.6	104.0	94.7	113.7	102.4	94.7	113.7	102.4
26, 27, 3370	ELECTRICAL, COMMUNICATIONS & UTIL.	102.0	102.3	102.2	109.2	142.6	126.6	98.8	112.6	106.0	91.7	102.9	97.6	105.5	119.9	113.0	105.2	119.9	112.9
MF2010	WEIGHTED AVERAGE	100.8	110.6	105.1	103.4	133.1	116.4	101.6	111.6	106.0	100.0	114.5	106.3	95.7	113.4	103.4	97.1	113.4	104.2

CALIFORNIA

	DIVISION	LOS ANGELES 900-902 MAT.	INST.	TOTAL	MARYSVILLE 959 MAT.	INST.	TOTAL	MODESTO 953 MAT.	INST.	TOTAL	MOJAVE 935 MAT.	INST.	TOTAL	OAKLAND 946 MAT.	INST.	TOTAL	OXNARD 930 MAT.	INST.	TOTAL
015433	CONTRACTOR EQUIPMENT		99.2	99.2		97.6	97.6		97.6	97.6		97.9	97.9		98.5	98.5		96.8	96.8
0241, 31 - 34	SITE & INFRASTRUCTURE, DEMOLITION	100.3	105.7	104.1	104.8	104.5	104.6	99.6	104.5	103.0	93.1	105.3	101.7	126.2	106.0	112.0	99.7	103.4	102.3
0310	Concrete Forming & Accessories	109.9	123.2	121.4	105.0	131.8	128.2	101.3	131.8	127.7	116.3	113.5	113.9	107.2	148.0	142.4	107.7	123.1	121.0
0320	Concrete Reinforcing	109.4	115.0	112.2	101.9	115.3	108.6	105.5	115.3	110.5	102.0	114.6	108.3	94.6	116.2	105.4	100.1	114.8	107.4
0330	Cast-in-Place Concrete	94.6	120.9	105.5	110.7	117.6	113.6	99.1	117.6	106.8	84.4	120.7	99.4	121.2	122.0	121.5	98.2	121.0	107.6
03	CONCRETE	100.9	119.9	110.3	115.1	122.6	118.8	105.9	122.6	114.1	93.9	115.4	104.5	109.9	131.2	120.4	102.1	119.8	110.8
04	MASONRY	90.0	119.9	108.5	100.3	118.2	111.4	99.3	122.3	113.5	94.7	115.7	107.7	126.6	131.9	129.9	97.5	116.5	109.2
05	METALS	100.6	102.0	101.0	102.4	101.1	102.0	99.6	101.4	100.2	102.6	99.3	101.6	100.8	100.3	100.7	100.7	100.0	100.5
06	WOOD, PLASTICS & COMPOSITES	105.7	121.4	114.5	99.5	135.9	120.0	95.3	135.9	118.2	105.0	109.2	107.4	98.4	154.0	129.8	99.6	121.5	111.9
07	THERMAL & MOISTURE PROTECTION	100.3	121.1	108.7	101.4	119.3	108.6	101.0	118.5	108.1	100.1	110.2	104.2	101.2	135.9	115.3	103.3	119.4	109.8
08	OPENINGS	97.8	120.2	103.1	99.1	124.7	105.2	98.0	124.7	104.3	92.1	109.5	96.2	96.8	140.5	107.2	94.5	120.2	100.6
0920	Plaster & Gypsum Board	103.5	122.0	116.1	98.3	136.9	124.5	101.0	136.9	125.4	108.9	109.4	109.2	97.4	155.2	136.7	101.8	122.0	115.6
0950, 0980	Ceilings & Acoustic Treatment	111.6	122.0	118.5	111.2	136.9	128.3	106.7	136.9	126.8	107.4	109.4	108.7	107.0	155.2	139.1	109.9	122.0	118.0
0960	Flooring	96.8	114.1	101.9	105.2	111.3	107.0	105.6	113.5	107.9	110.4	109.2	110.0	104.4	126.4	111.0	102.9	114.1	106.2
0970, 0990	Wall Finishes & Painting/Coating	101.1	108.4	105.5	102.0	118.3	111.8	102.0	118.3	111.8	100.5	104.2	102.7	102.7	142.1	126.4	100.5	101.2	100.9
09	FINISHES	102.9	119.8	112.3	105.0	128.9	118.3	104.0	129.2	118.1	104.6	111.3	108.3	103.1	145.5	126.7	102.3	119.1	111.6
COVERS	DIVS. 10 - 14, 25, 28, 41, 43, 44, 46	100.0	111.3	102.3	100.0	121.4	104.3	100.0	121.4	104.3	100.0	110.0	102.0	100.0	125.1	105.0	100.0	111.5	102.3
21, 22, 23	FIRE SUPPRESSION, PLUMBING & HVAC	100.0	118.0	107.2	95.1	108.9	101.0	100.0	108.8	103.5	95.2	108.3	100.4	100.1	144.7	118.0	100.1	118.0	107.3
26, 27, 3370	ELECTRICAL, COMMUNICATIONS & UTIL.	103.6	120.6	112.5	95.4	109.9	103.0	97.9	106.5	102.4	90.5	102.3	96.6	108.3	142.6	126.2	96.2	109.9	103.3
MF2010	WEIGHTED AVERAGE	100.1	116.5	107.3	100.8	115.1	107.0	100.4	114.9	106.7	96.8	108.6	101.9	103.9	133.1	116.6	99.6	114.2	105.9

City Cost Indexes

CALIFORNIA

DIVISION		PALM SPRINGS 922			PALO ALTO 943			PASADENA 910 - 912			REDDING 960			RICHMOND 948			RIVERSIDE 925		
		MAT.	INST.	TOTAL	MAT.	INST.	TOTAL	MAT.	INST.	TOTAL	MAT.	INST.	TOTAL	MAT.	INST.	TOTAL	MAT.	INST.	TOTAL
015433	CONTRACTOR EQUIPMENT		98.8	98.8		98.5	98.5		98.9	98.9		97.6	97.6		98.5	98.5		98.8	98.8
0241, 31 - 34	SITE & INFRASTRUCTURE, DEMOLITION	88.2	105.7	100.5	115.6	106.0	108.8	93.4	109.7	104.8	109.5	104.5	106.0	125.4	106.0	111.8	95.0	105.7	102.5
0310	Concrete Forming & Accessories	102.1	116.9	114.9	105.4	144.5	139.2	104.9	116.9	115.2	105.0	129.6	126.2	120.9	147.8	144.2	105.8	123.1	120.7
0320	Concrete Reinforcing	107.1	114.7	110.9	92.5	115.9	104.2	108.3	114.9	111.6	104.7	115.3	110.0	92.5	116.0	104.2	104.1	114.7	109.4
0330	Cast-in-Place Concrete	87.1	121.6	101.4	108.1	122.0	113.8	89.9	119.1	102.0	110.9	117.6	113.7	124.3	122.0	123.4	94.7	121.7	105.9
03	CONCRETE	97.7	117.2	107.3	99.3	129.6	114.2	99.4	116.4	107.8	115.2	121.5	118.3	112.0	131.0	121.4	103.8	120.0	111.8
04	MASONRY	74.9	115.9	100.3	101.7	127.8	117.8	106.8	118.6	114.1	110.3	118.2	115.2	119.0	127.8	124.4	75.9	115.9	100.7
05	METALS	103.8	99.9	102.6	98.3	99.8	98.8	84.1	99.9	88.9	100.5	101.1	100.7	98.4	99.8	98.8	103.4	100.1	102.4
06	WOOD, PLASTICS & COMPOSITES	93.2	113.3	104.5	95.8	149.6	126.1	83.3	113.1	100.1	97.8	132.8	117.5	114.2	154.0	136.7	98.5	121.4	111.4
07	THERMAL & MOISTURE PROTECTION	97.6	119.0	106.3	100.8	133.4	114.0	95.0	117.2	104.0	108.6	118.9	112.8	101.4	134.2	114.7	98.1	121.2	107.5
08	OPENINGS	96.6	115.5	101.1	96.8	137.5	106.5	89.8	115.5	95.9	102.0	123.0	107.0	96.9	139.9	107.1	99.1	120.2	104.1
0920	Plaster & Gypsum Board	95.5	113.7	107.9	95.6	150.6	133.0	95.0	113.7	107.7	104.5	133.7	124.4	104.9	155.2	139.1	99.9	122.0	115.0
0950, 0980	Ceilings & Acoustic Treatment	104.1	113.7	110.5	105.0	150.6	135.4	102.5	113.7	109.9	131.8	133.7	133.1	105.0	155.2	138.4	111.0	122.0	118.4
0960	Flooring	107.7	109.2	108.1	103.4	126.4	110.3	88.9	109.2	103.2	104.5	111.3	107.2	110.3	126.4	115.1	109.1	114.1	110.6
0970, 0990	Wall Finishes & Painting/Coating	98.7	113.7	107.7	102.7	142.1	126.4	101.7	108.4	105.7	104.1	118.3	112.6	102.7	142.1	126.4	98.7	106.2	103.2
09	FINISHES	100.7	114.7	108.5	101.4	142.9	124.5	95.7	114.0	105.9	110.9	127.1	119.9	105.4	145.5	127.7	103.5	119.6	112.5
COVERS	DIVS. 10 - 14, 25, 28, 41, 43, 44, 46	100.0	110.4	102.1	100.0	124.7	104.9	100.0	110.1	102.0	100.0	121.1	104.2	100.0	125.1	105.0	100.0	111.3	102.3
21, 22, 23	FIRE SUPPRESSION, PLUMBING & HVAC	95.1	113.6	102.5	95.2	143.3	114.5	95.1	113.7	102.6	100.1	109.8	104.0	95.2	144.7	115.0	100.0	118.0	107.2
26, 27, 3370	ELECTRICAL, COMMUNICATIONS & UTIL.	95.0	106.0	100.8	108.1	153.8	132.0	116.8	119.9	118.4	97.8	109.9	104.1	108.8	132.4	121.1	91.6	106.0	99.1
MF2010	WEIGHTED AVERAGE	96.7	111.6	103.2	99.5	133.1	114.2	96.3	113.9	104.0	103.6	114.6	108.4	102.5	131.2	115.0	98.9	114.0	105.5

CALIFORNIA

DIVISION		SACRAMENTO 942,956 - 958			SALINAS 939			SAN BERNARDINO 923 - 924			SAN DIEGO 919 - 921			SAN FRANCISCO 940 - 941			SAN JOSE 951		
		MAT.	INST.	TOTAL	MAT.	INST.	TOTAL	MAT.	INST.	TOTAL	MAT.	INST.	TOTAL	MAT.	INST.	TOTAL	MAT.	INST.	TOTAL
015433	CONTRACTOR EQUIPMENT		98.0	98.0		97.9	97.9		98.8	98.8		99.0	99.0		108.8	108.8		98.5	98.5
0241, 31 - 34	SITE & INFRASTRUCTURE, DEMOLITION	101.0	111.5	108.3	112.3	105.1	107.2	75.1	105.7	96.6	99.7	103.3	102.2	128.6	111.8	116.8	132.0	99.7	109.3
0310	Concrete Forming & Accessories	105.6	132.4	128.7	110.9	132.8	129.8	109.5	116.9	115.9	103.8	112.0	110.9	106.9	149.0	143.2	107.4	147.9	142.3
0320	Concrete Reinforcing	87.6	115.6	101.6	100.6	115.9	108.3	104.1	114.7	109.4	104.9	114.7	109.8	107.8	116.6	112.2	92.8	116.2	104.5
0330	Cast-in-Place Concrete	102.5	118.4	109.1	97.0	117.9	105.7	65.5	121.6	88.8	98.5	108.7	102.7	124.3	123.5	124.0	115.1	121.6	117.8
03	CONCRETE	99.8	122.7	111.1	112.4	123.2	117.7	77.7	117.2	97.2	104.6	110.7	107.6	113.5	132.7	123.0	111.8	131.4	121.5
04	MASONRY	103.4	122.3	115.1	94.4	127.4	114.8	82.4	115.9	103.1	97.1	114.2	107.7	127.2	136.2	132.8	126.8	131.9	130.0
05	METALS	96.4	95.7	96.2	105.3	102.5	104.4	103.3	99.9	102.3	99.6	101.3	100.1	107.1	110.1	108.0	97.8	107.1	100.6
06	WOOD, PLASTICS & COMPOSITES	92.3	136.1	117.0	104.4	136.1	122.3	102.3	113.3	108.5	97.0	108.8	103.6	98.4	154.2	129.9	107.5	153.8	133.6
07	THERMAL & MOISTURE PROTECTION	108.5	121.1	113.6	100.6	125.9	110.9	96.8	119.0	105.8	99.7	106.6	102.5	102.9	139.0	117.6	97.4	138.4	114.1
08	OPENINGS	110.7	124.8	114.1	95.6	130.7	103.9	96.6	115.5	101.1	96.8	111.4	100.3	101.4	140.6	110.7	89.5	140.3	101.5
0920	Plaster & Gypsum Board	92.5	136.9	122.7	102.9	137.1	126.2	101.7	113.7	109.9	102.3	108.9	106.8	99.8	155.2	137.5	96.2	155.2	136.3
0950, 0980	Ceilings & Acoustic Treatment	105.0	136.9	126.2	107.4	137.1	127.2	106.7	113.7	111.4	109.6	108.9	109.1	114.8	155.2	141.7	104.8	155.2	138.3
0960	Flooring	103.3	119.7	108.2	105.1	123.3	110.6	111.0	109.2	110.4	96.2	114.1	101.5	104.4	126.4	111.0	98.7	126.4	107.0
0970, 0990	Wall Finishes & Painting/Coating	100.3	118.3	111.1	101.6	142.1	126.0	98.7	106.2	103.2	98.3	106.4	103.2	102.7	152.6	132.7	102.3	142.1	126.2
09	FINISHES	99.8	130.4	116.9	104.2	133.7	120.6	101.8	113.9	108.6	101.0	111.3	106.7	105.2	146.8	128.4	102.7	145.3	126.4
COVERS	DIVS. 10 - 14, 25, 28, 41, 43, 44, 46	100.0	121.9	104.4	100.0	121.5	104.3	100.0	110.2	102.1	100.0	109.1	101.8	100.0	125.6	105.1	100.0	124.7	105.0
21, 22, 23	FIRE SUPPRESSION, PLUMBING & HVAC	100.0	121.0	108.4	95.2	115.3	103.2	95.1	113.8	102.6	100.0	116.3	106.5	100.0	171.6	128.7	100.0	143.3	117.3
26, 27, 3370	ELECTRICAL, COMMUNICATIONS & UTIL.	103.4	111.3	107.5	91.7	122.8	107.9	95.0	103.7	99.6	102.2	98.9	100.5	108.3	161.4	136.1	101.0	153.8	128.6
MF2010	WEIGHTED AVERAGE	101.2	118.6	108.8	100.2	120.3	108.9	94.6	111.2	101.9	100.2	109.1	104.1	106.1	143.8	122.5	102.2	134.6	116.3

CALIFORNIA

DIVISION		SAN LUIS OBISPO 934			SAN MATEO 944			SAN RAFAEL 949			SANTA ANA 926 - 927			SANTA BARBARA 931			SANTA CRUZ 950		
		MAT.	INST.	TOTAL	MAT.	INST.	TOTAL	MAT.	INST.	TOTAL	MAT.	INST.	TOTAL	MAT.	INST.	TOTAL	MAT.	INST.	TOTAL
015433	CONTRACTOR EQUIPMENT		97.9	97.9		98.5	98.5		98.5	98.5		98.8	98.8		97.9	97.9		98.5	98.5
0241, 31 - 34	SITE & INFRASTRUCTURE, DEMOLITION	104.7	105.3	105.1	122.8	106.0	111.0	113.7	111.6	112.2	86.7	105.7	100.0	99.6	105.3	103.6	131.6	99.5	109.0
0310	Concrete Forming & Accessories	118.0	116.9	117.0	111.2	147.9	142.9	115.2	147.8	143.4	109.7	116.9	115.9	108.6	123.1	121.1	107.5	133.0	129.5
0320	Concrete Reinforcing	102.0	114.8	108.4	92.5	116.1	104.3	93.1	116.1	104.6	107.6	114.7	111.2	100.1	114.8	107.4	114.3	119.8	116.6
0330	Cast-in-Place Concrete	104.2	120.8	111.1	120.4	122.0	121.0	140.1	121.2	132.2	83.6	121.6	99.4	97.8	120.9	107.4	115.0	124.1	119.5
03	CONCRETE	110.1	116.9	113.4	108.4	131.1	119.6	129.7	130.6	130.2	95.2	117.2	106.1	102.0	119.7	110.7	115.0	124.1	119.5
04	MASONRY	96.1	117.4	109.3	118.7	130.8	126.2	98.0	130.8	118.3	72.6	116.2	99.6	94.9	117.4	108.8	130.4	127.5	128.6
05	METALS	103.2	99.7	102.1	98.2	100.1	98.8	99.6	97.8	99.1	103.4	99.9	102.3	101.1	100.0	100.8	105.1	105.8	105.3
06	WOOD, PLASTICS & COMPOSITES	107.5	113.4	110.8	103.5	154.0	132.0	100.4	153.8	130.5	103.9	113.3	109.2	99.6	121.5	111.9	107.5	136.2	123.7
07	THERMAL & MOISTURE PROTECTION	100.8	117.5	107.6	101.2	136.2	115.4	105.2	132.9	116.5	97.9	119.1	106.5	100.2	119.0	107.9	97.4	128.5	110.0
08	OPENINGS	93.9	111.8	98.1	96.8	139.9	107.0	108.2	139.8	115.7	96.0	115.5	100.6	95.2	120.2	101.1	90.7	130.8	100.2
0920	Plaster & Gypsum Board	110.4	113.7	112.6	100.8	155.2	137.8	102.5	155.2	138.4	102.4	113.7	110.1	101.8	122.0	115.6	103.7	137.1	126.4
0950, 0980	Ceilings & Acoustic Treatment	107.4	113.7	111.6	105.0	155.2	138.4	113.0	155.2	141.1	106.7	113.7	111.4	109.9	122.0	118.0	108.5	137.1	127.5
0960	Flooring	111.3	109.2	110.7	106.0	126.4	112.1	115.1	121.3	117.0	111.5	109.2	110.8	104.4	109.2	105.8	102.8	123.3	108.9
0970, 0990	Wall Finishes & Painting/Coating	100.5	101.1	100.9	102.7	142.1	126.4	99.0	142.1	124.9	98.7	106.2	103.2	100.5	101.2	100.9	102.5	142.1	126.3
09	FINISHES	106.1	112.6	109.7	103.4	145.5	126.8	105.8	144.4	127.3	103.3	113.9	109.2	102.9	118.2	111.4	105.7	133.8	121.3
COVERS	DIVS. 10 - 14, 25, 28, 41, 43, 44, 46	100.0	120.4	104.1	100.0	125.2	105.0	100.0	124.5	104.9	100.0	110.4	102.1	100.0	111.5	102.3	100.0	121.8	104.4
21, 22, 23	FIRE SUPPRESSION, PLUMBING & HVAC	95.2	113.8	102.6	95.2	142.8	114.3	95.1	164.7	123.0	95.1	113.8	102.6	100.1	118.0	107.3	100.0	115.3	106.1
26, 27, 3370	ELECTRICAL, COMMUNICATIONS & UTIL.	90.5	106.0	98.6	108.1	143.6	126.7	105.1	116.4	111.0	95.0	105.9	100.7	89.4	111.8	101.1	99.9	122.8	111.9
MF2010	WEIGHTED AVERAGE	99.3	111.6	104.7	101.7	132.7	115.2	104.2	133.6	117.0	96.5	111.6	103.1	98.9	114.6	105.7	104.1	120.4	111.2

City Cost Indexes

		CALIFORNIA																COLORADO		
	DIVISION	SANTA ROSA			STOCKTON			SUSANVILLE			VALLEJO			VAN NUYS			ALAMOSA		
		954			952			961			945			913 - 916			811		
		MAT.	INST.	TOTAL	MAT.	INST.	TOTAL	MAT.	INST.	TOTAL	MAT.	INST.	TOTAL	MAT.	INST.	TOTAL	MAT.	INST.	TOTAL
015433	CONTRACTOR EQUIPMENT		98.2	98.2		97.6	97.6		97.6	97.6		98.5	98.5		98.9	98.9		93.8	93.8
0241, 31 - 34	SITE & INFRASTRUCTURE, DEMOLITION	100.1	104.5	103.2	99.4	104.5	103.0	115.9	104.5	107.9	100.5	111.4	108.2	109.9	109.7	109.7	132.1	89.0	101.9
0310	Concrete Forming & Accessories	103.1	146.8	140.8	105.3	131.8	128.2	106.2	129.6	126.4	105.6	146.5	140.9	110.9	116.9	116.1	105.2	69.0	73.9
0320	Concrete Reinforcing	102.7	116.4	109.6	105.5	115.5	110.5	104.7	115.3	110.0	94.3	116.3	105.3	108.3	114.9	111.6	108.0	78.4	93.2
0330	Cast-in-Place Concrete	108.7	119.2	113.1	96.6	117.6	105.3	100.9	117.6	107.8	111.6	119.6	114.9	94.8	119.1	104.9	100.6	79.3	91.8
03	CONCRETE	114.9	130.0	122.4	104.9	122.6	113.6	117.6	121.5	119.5	105.2	129.5	117.2	113.8	116.4	115.1	115.5	74.6	95.3
04	MASONRY	98.6	133.7	120.3	99.2	122.3	113.5	108.3	118.2	114.4	74.6	132.4	110.4	122.3	118.6	120.0	123.2	72.7	92.0
05	METALS	103.7	104.3	103.9	99.8	101.4	100.3	99.6	101.1	100.1	99.6	97.6	99.0	83.2	99.9	88.3	97.2	81.1	92.2
06	WOOD, PLASTICS & COMPOSITES	94.6	153.6	127.9	100.8	135.9	120.6	99.5	132.8	118.3	89.8	153.8	125.9	91.6	113.1	103.8	93.8	69.0	79.8
07	THERMAL & MOISTURE PROTECTION	98.3	134.5	113.0	101.1	118.5	108.2	109.2	118.9	113.2	103.0	134.0	115.6	95.8	117.2	104.5	105.1	78.4	94.2
08	OPENINGS	97.4	140.2	107.6	98.0	124.7	104.3	101.9	123.0	106.9	110.0	140.3	117.2	89.7	115.5	95.8	98.4	75.3	92.9
0920	Plaster & Gypsum Board	97.2	155.2	136.7	101.0	136.9	125.4	105.4	133.7	124.7	97.3	155.2	136.7	99.9	113.7	109.3	75.8	67.9	70.4
0950, 0980	Ceilings & Acoustic Treatment	106.7	155.2	139.0	113.7	136.9	129.1	123.9	133.7	130.4	115.0	155.2	141.8	99.8	113.7	109.1	104.2	67.9	80.1
0960	Flooring	108.1	117.0	110.8	105.6	113.5	107.9	105.9	111.3	107.5	111.2	126.4	115.7	91.1	109.2	96.5	112.4	55.8	95.5
0970, 0990	Wall Finishes & Painting/Coating	98.7	142.1	124.8	102.0	118.3	111.8	104.1	118.3	112.6	99.9	142.1	125.3	101.7	108.4	105.7	113.0	24.1	59.6
09	FINISHES	102.6	143.0	125.1	100.5	129.2	118.7	110.4	127.1	119.7	102.8	144.7	126.1	97.7	114.0	106.8	104.9	61.2	80.5
COVERS	DIVS. 10 - 14, 25, 28, 41, 43, 44, 46	100.0	123.2	104.7	100.0	121.4	104.3	100.0	121.1	104.2	100.0	123.7	104.7	100.0	110.1	102.0	100.0	89.1	97.8
21, 22, 23	FIRE SUPPRESSION, PLUMBING & HVAC	95.1	163.4	122.5	100.0	108.8	103.5	95.2	109.8	101.0	100.0	128.8	111.6	95.1	113.7	102.6	95.2	72.6	86.1
26, 27, 3370	ELECTRICAL, COMMUNICATIONS & UTIL.	95.4	116.6	106.4	97.9	110.9	104.7	98.1	109.9	104.3	100.6	125.1	113.4	116.8	119.9	118.4	96.8	74.2	85.0
MF2010	WEIGHTED AVERAGE	100.3	133.4	114.7	100.5	115.5	107.0	102.6	114.6	107.8	100.6	127.2	112.2	99.1	113.9	105.5	102.0	74.6	90.1

		COLORADO																		
	DIVISION	BOULDER			COLORADO SPRINGS			DENVER			DURANGO			FORT COLLINS			FORT MORGAN		
		803			808 - 809			800 - 802			813			805			807		
		MAT.	INST.	TOTAL	MAT.	INST.	TOTAL	MAT.	INST.	TOTAL	MAT.	INST.	TOTAL	MAT.	INST.	TOTAL	MAT.	INST.	TOTAL
015433	CONTRACTOR EQUIPMENT		94.9	94.9		93.4	93.4		97.7	97.7		93.8	93.8		94.9	94.9		94.9	94.9
0241, 31 - 34	SITE & INFRASTRUCTURE, DEMOLITION	92.4	95.1	94.3	94.5	93.1	93.5	93.7	101.2	99.0	125.8	89.0	100.0	104.3	94.7	97.5	95.3	94.6	94.8
0310	Concrete Forming & Accessories	102.0	81.7	84.4	92.9	77.4	79.5	98.5	77.7	80.5	111.4	69.2	74.9	99.3	76.4	79.5	102.5	76.6	80.2
0320	Concrete Reinforcing	102.0	78.5	90.3	101.2	82.7	92.0	101.2	82.7	92.0	108.0	78.4	93.2	102.1	78.6	90.3	102.2	78.4	90.3
0330	Cast-in-Place Concrete	102.4	79.8	93.0	105.2	82.9	95.9	99.6	82.6	92.5	115.7	79.4	100.6	115.6	78.6	100.2	100.5	78.6	91.4
03	CONCRETE	104.7	80.6	92.8	108.1	80.6	94.6	102.4	80.6	91.7	117.8	74.8	96.5	115.3	77.8	96.8	103.2	77.8	90.7
04	MASONRY	93.1	73.2	80.8	94.1	73.0	81.0	96.1	73.3	82.0	110.9	72.4	87.1	110.4	76.9	89.7	107.5	72.9	86.1
05	METALS	97.7	84.4	93.6	100.8	86.5	96.4	103.4	86.7	98.2	97.2	81.2	92.2	98.9	81.3	93.5	97.5	81.1	92.4
06	WOOD, PLASTICS & COMPOSITES	93.3	84.9	88.6	83.8	79.1	81.2	91.0	79.0	84.3	102.6	69.0	83.7	90.8	79.0	84.1	93.3	79.0	85.2
07	THERMAL & MOISTURE PROTECTION	104.4	80.5	94.7	105.3	80.3	95.1	103.7	74.6	91.9	105.0	78.3	94.2	104.8	72.2	91.6	104.4	79.9	94.4
08	OPENINGS	101.5	83.9	97.3	105.8	82.0	100.2	106.5	81.9	100.7	105.3	75.3	98.2	101.4	80.7	96.5	101.4	80.7	96.5
0920	Plaster & Gypsum Board	104.0	84.6	90.8	86.1	78.6	81.0	99.6	78.6	85.3	88.1	67.9	74.3	98.0	78.6	84.8	104.0	78.6	86.7
0950, 0980	Ceilings & Acoustic Treatment	98.0	84.6	89.1	105.1	78.6	87.4	108.2	78.6	88.5	104.2	67.9	80.1	98.0	78.6	85.0	98.0	78.6	85.0
0960	Flooring	103.7	85.3	98.2	95.8	69.5	87.9	100.5	86.3	96.3	116.8	55.8	98.5	100.4	55.8	87.1	104.1	55.8	89.7
0970, 0990	Wall Finishes & Painting/Coating	100.8	68.0	81.1	100.5	41.4	65.0	100.8	77.3	86.7	113.0	24.1	59.6	100.8	41.1	64.9	100.8	54.7	73.1
09	FINISHES	102.6	81.2	90.7	99.4	72.0	84.2	103.1	79.0	89.7	106.9	61.2	81.5	101.4	68.9	83.3	102.6	70.4	84.7
COVERS	DIVS. 10 - 14, 25, 28, 41, 43, 44, 46	100.0	90.0	98.0	100.0	89.7	97.9	100.0	89.4	97.9	100.0	89.1	97.8	100.0	89.4	97.9	100.0	89.4	97.9
21, 22, 23	FIRE SUPPRESSION, PLUMBING & HVAC	95.2	77.6	88.1	100.2	72.7	89.2	100.0	79.9	91.9	95.2	84.9	91.0	100.1	77.5	91.0	95.2	77.5	88.1
26, 27, 3370	ELECTRICAL, COMMUNICATIONS & UTIL.	96.9	85.2	90.8	100.5	83.0	91.4	102.1	85.2	93.3	96.2	71.1	83.1	96.9	85.1	90.7	97.2	85.1	90.9
MF2010	WEIGHTED AVERAGE	98.6	82.0	91.3	101.3	79.3	91.7	101.6	82.6	93.3	102.4	76.8	91.2	102.1	79.6	92.3	99.1	79.6	90.6

		COLORADO																		
	DIVISION	GLENWOOD SPRINGS			GOLDEN			GRAND JUNCTION			GREELEY			MONTROSE			PUEBLO		
		816			804			815			806			814			810		
		MAT.	INST.	TOTAL	MAT.	INST.	TOTAL	MAT.	INST.	TOTAL	MAT.	INST.	TOTAL	MAT.	INST.	TOTAL	MAT.	INST.	TOTAL
015433	CONTRACTOR EQUIPMENT		96.5	96.5		94.9	94.9		96.5	96.5		94.9	94.9		95.1	95.1		93.8	93.8
0241, 31 - 34	SITE & INFRASTRUCTURE, DEMOLITION	140.8	95.9	109.3	104.4	94.9	97.7	125.5	95.5	104.5	91.7	94.1	93.4	134.6	92.1	104.8	117.7	91.5	99.3
0310	Concrete Forming & Accessories	102.3	76.7	80.2	95.2	76.5	79.1	110.1	76.2	80.8	97.5	80.2	82.5	101.8	76.4	79.9	107.3	77.8	81.8
0320	Concrete Reinforcing	106.9	78.4	92.7	102.2	78.4	90.3	107.3	78.3	92.8	102.0	77.0	89.5	106.8	78.3	92.6	103.3	82.7	93.1
0330	Cast-in-Place Concrete	100.6	78.8	91.5	100.5	78.6	91.4	111.4	78.1	97.5	96.6	58.5	80.8	100.6	78.7	91.5	99.9	83.7	93.2
03	CONCRETE	120.6	77.9	99.6	113.8	77.8	96.0	114.1	77.4	96.0	99.9	72.2	86.3	111.6	77.7	94.9	104.3	81.1	92.8
04	MASONRY	97.5	72.9	82.3	110.3	72.9	87.2	129.6	72.7	94.4	104.6	49.7	70.6	103.9	72.7	84.6	94.1	73.2	81.2
05	METALS	96.9	81.8	92.2	97.7	81.2	92.6	98.4	80.5	92.9	98.9	78.5	92.6	96.0	80.8	91.3	100.0	87.4	96.1
06	WOOD, PLASTICS & COMPOSITES	89.1	79.1	83.4	86.2	79.0	82.1	100.4	79.1	88.4	88.4	84.9	86.4	90.2	79.2	84.0	96.3	79.4	86.8
07	THERMAL & MOISTURE PROTECTION	105.0	79.5	94.6	105.3	74.2	92.7	104.1	68.8	89.8	104.3	65.0	88.3	105.1	79.5	94.7	103.6	80.0	94.0
08	OPENINGS	104.3	80.8	98.7	101.4	80.7	96.5	105.0	80.8	99.3	101.4	83.9	97.3	105.5	80.8	99.7	100.1	82.1	95.9
0920	Plaster & Gypsum Board	111.5	78.6	89.1	95.8	78.6	84.1	123.4	78.6	92.9	96.5	84.6	88.4	75.0	78.6	77.4	80.2	78.6	79.1
0950, 0980	Ceilings & Acoustic Treatment	103.4	78.6	86.9	98.0	78.6	85.0	103.4	78.6	86.9	98.0	84.6	89.1	104.2	78.6	87.2	113.1	78.6	90.1
0960	Flooring	111.7	51.4	93.7	98.3	55.8	85.6	116.2	55.8	98.1	99.5	55.8	86.4	114.6	46.2	94.2	113.4	86.3	105.3
0970, 0990	Wall Finishes & Painting/Coating	112.9	54.7	77.9	100.8	68.0	81.1	112.9	68.0	85.9	100.8	25.6	55.6	113.0	24.1	59.6	113.0	38.4	68.1
09	FINISHES	109.9	69.8	87.5	101.3	72.6	85.3	110.8	72.0	89.2	100.1	70.7	83.7	105.4	65.3	83.1	105.8	74.9	88.6
COVERS	DIVS. 10 - 14, 25, 28, 41, 43, 44, 46	100.0	89.5	97.9	100.0	89.4	97.9	100.0	89.5	97.9	100.0	90.0	98.0	100.0	89.8	98.0	100.0	90.3	98.1
21, 22, 23	FIRE SUPPRESSION, PLUMBING & HVAC	95.2	84.8	91.0	95.2	77.0	87.9	100.0	84.3	93.7	100.1	77.5	91.0	95.2	84.8	91.0	100.0	72.8	89.1
26, 27, 3370	ELECTRICAL, COMMUNICATIONS & UTIL.	93.5	71.1	81.8	97.2	85.1	90.9	95.9	55.1	74.6	96.9	85.1	90.7	95.9	57.7	76.0	96.9	74.3	85.1
MF2010	WEIGHTED AVERAGE	102.2	79.3	92.2	100.6	79.6	91.4	104.4	76.7	92.4	99.7	76.1	89.4	101.2	76.5	90.5	100.9	78.6	91.2

498

City Cost Indexes

DIVISION		COLORADO SALIDA 812 MAT.	INST.	TOTAL	CONNECTICUT BRIDGEPORT 066 MAT.	INST.	TOTAL	BRISTOL 060 MAT.	INST.	TOTAL	HARTFORD 061 MAT.	INST.	TOTAL	MERIDEN 064 MAT.	INST.	TOTAL	NEW BRITAIN 060 MAT.	INST.	TOTAL
015433	CONTRACTOR EQUIPMENT		95.1	95.1		100.6	100.6		100.6	100.6		100.6	100.6		101.1	101.1		100.6	100.6
0241, 31 - 34	SITE & INFRASTRUCTURE, DEMOLITION	125.5	92.4	102.3	111.7	104.0	106.3	110.9	104.0	106.1	106.0	104.0	104.6	108.9	104.7	105.9	111.1	104.0	106.1
0310	Concrete Forming & Accessories	110.0	76.4	81.0	98.2	123.8	120.3	98.2	123.7	120.3	95.1	123.7	119.8	97.9	123.7	120.2	98.6	123.7	120.3
0320	Concrete Reinforcing	106.6	78.3	92.5	110.0	129.3	119.7	110.0	129.3	119.7	109.5	129.3	119.4	110.0	129.3	119.7	110.0	129.3	119.7
0330	Cast-in-Place Concrete	115.2	78.7	100.1	103.7	127.5	113.6	102.7	127.4	110.8	99.0	127.4	110.8	93.5	127.4	110.7	103.0	127.4	114.7
03	CONCRETE	112.8	77.7	95.5	105.3	125.8	115.4	102.2	125.7	113.8	102.8	125.8	114.1	100.5	125.7	112.9	103.0	125.7	114.2
04	MASONRY	131.2	72.7	95.0	98.1	134.8	120.8	90.8	134.8	118.0	94.7	134.8	119.5	90.4	134.8	117.9	92.1	134.8	118.6
05	METALS	95.7	80.8	91.1	102.0	125.1	109.1	102.0	125.1	109.1	107.8	125.1	113.1	99.5	125.1	107.4	98.6	125.1	106.8
06	WOOD, PLASTICS & COMPOSITES	97.4	79.2	87.2	99.1	122.9	112.5	99.1	122.9	112.5	92.6	122.9	109.7	99.1	122.9	112.5	99.1	122.9	112.5
07	THERMAL & MOISTURE PROTECTION	104.1	79.5	94.1	100.3	129.2	112.0	100.4	126.9	111.2	104.8	126.9	113.8	100.4	126.9	111.2	100.4	126.9	111.2
08	OPENINGS	98.5	80.8	94.3	107.0	132.2	113.0	107.0	132.2	113.0	106.1	132.2	112.2	109.4	132.2	114.8	107.0	132.2	113.0
0920	Plaster & Gypsum Board	75.4	78.6	77.5	99.9	123.0	115.6	99.9	123.0	115.6	95.4	123.0	116.1	101.5	123.0	116.1	99.9	123.0	115.6
0950, 0980	Ceilings & Acoustic Treatment	104.2	78.6	87.2	97.9	123.0	114.6	97.9	123.0	114.6	101.3	123.0	115.7	101.2	123.0	115.7	97.9	123.0	114.6
0960	Flooring	119.1	46.2	97.3	99.3	133.6	109.6	99.3	133.6	109.6	97.8	133.6	108.5	99.3	133.6	109.6	99.3	133.6	109.6
0970, 0990	Wall Finishes & Painting/Coating	113.0	24.1	59.6	96.4	122.3	112.0	96.4	122.3	112.0	97.3	122.3	112.3	96.4	122.3	112.0	96.4	122.3	112.0
09	FINISHES	105.6	65.3	83.2	100.8	125.1	114.3	100.8	125.1	114.3	99.7	125.1	113.8	101.9	125.1	114.8	100.8	125.1	114.3
COVERS	DIVS. 10 - 14, 25, 28, 41, 43, 44, 46	100.0	89.9	98.0	100.0	110.4	102.1	100.0	110.5	102.1	100.0	110.5	102.1	100.0	110.5	102.1	100.0	110.5	102.1
21, 22, 23	FIRE SUPPRESSION, PLUMBING & HVAC	95.2	72.6	86.1	100.0	116.0	106.4	100.0	116.0	106.4	100.0	116.0	106.4	95.1	116.0	103.5	100.0	116.0	106.4
26, 27, 3370	ELECTRICAL, COMMUNICATIONS & UTIL.	96.1	74.2	84.7	99.2	108.7	104.2	99.2	111.3	105.5	98.6	112.3	105.7	99.2	111.3	105.5	99.3	111.3	105.6
MF2010	WEIGHTED AVERAGE	101.7	76.2	90.6	101.8	120.2	109.8	101.1	120.4	109.5	102.0	120.6	110.1	99.6	120.5	108.7	100.8	120.4	109.3

DIVISION		CONNECTICUT NEW HAVEN 065 MAT.	INST.	TOTAL	NEW LONDON 063 MAT.	INST.	TOTAL	NORWALK 068 MAT.	INST.	TOTAL	STAMFORD 069 MAT.	INST.	TOTAL	WATERBURY 067 MAT.	INST.	TOTAL	WILLIMANTIC 062 MAT.	INST.	TOTAL
015433	CONTRACTOR EQUIPMENT		101.1	101.1		101.1	101.1		100.6	100.6		100.6	100.6		100.6	100.6		100.6	100.6
0241, 31 - 34	SITE & INFRASTRUCTURE, DEMOLITION	110.9	104.7	106.5	103.1	104.7	104.2	111.5	104.0	106.2	112.2	104.0	106.4	111.4	104.0	106.2	111.6	104.0	106.3
0310	Concrete Forming & Accessories	98.0	123.7	120.2	97.9	123.7	120.2	98.2	124.2	120.6	98.2	124.2	120.7	98.2	123.7	120.2	98.2	123.7	120.2
0320	Concrete Reinforcing	110.0	129.3	119.7	86.3	129.3	107.7	110.0	129.4	119.7	110.0	129.4	119.7	110.0	129.3	119.7	110.0	129.3	119.7
0330	Cast-in-Place Concrete	100.5	127.4	111.7	85.7	127.4	103.0	102.0	128.9	113.2	103.7	128.9	114.2	103.7	127.4	113.6	96.9	127.4	109.6
03	CONCRETE	117.2	125.7	121.4	90.2	125.7	107.7	104.5	126.5	115.3	105.3	126.5	115.8	105.3	125.7	115.4	102.1	125.7	113.7
04	MASONRY	91.0	134.8	118.1	89.4	134.8	117.5	90.6	134.8	118.0	91.3	134.8	118.2	91.3	134.8	118.2	90.6	134.8	118.0
05	METALS	98.8	125.1	106.9	98.5	125.1	106.7	102.0	125.5	109.3	102.0	125.6	109.3	102.0	125.0	109.1	101.8	125.0	108.9
06	WOOD, PLASTICS & COMPOSITES	99.1	122.9	112.5	99.1	122.9	112.5	99.1	122.9	112.5	99.1	122.9	112.5	99.1	122.9	112.5	99.1	122.9	112.5
07	THERMAL & MOISTURE PROTECTION	100.5	126.7	111.1	100.4	126.9	111.1	100.4	129.4	112.2	100.4	129.4	112.2	100.4	126.7	111.1	100.6	125.7	110.8
08	OPENINGS	107.0	132.2	113.0	109.9	132.2	115.2	107.0	132.2	113.0	107.0	132.2	113.0	107.0	132.2	113.0	110.0	132.2	115.2
0920	Plaster & Gypsum Board	99.9	123.0	115.6	99.9	123.0	115.6	99.9	123.0	115.6	99.9	123.0	115.6	99.9	123.0	115.6	99.9	123.0	115.6
0950, 0980	Ceilings & Acoustic Treatment	97.9	123.0	114.6	95.9	123.0	113.9	97.9	123.0	114.6	97.9	123.0	114.6	97.9	123.0	114.6	95.9	123.0	113.9
0960	Flooring	99.3	133.6	109.6	99.3	133.6	109.6	99.3	133.6	109.6	99.3	133.6	109.6	99.3	133.6	109.6	99.3	135.7	110.2
0970, 0990	Wall Finishes & Painting/Coating	96.4	122.3	112.0	96.4	122.3	112.0	96.4	122.3	112.0	96.4	122.3	112.0	96.4	122.3	112.0	96.4	122.3	112.0
09	FINISHES	100.8	125.1	114.4	99.8	125.1	113.9	100.8	125.1	114.4	100.9	125.1	114.4	100.7	125.1	114.3	100.5	125.4	114.4
COVERS	DIVS. 10 - 14, 25, 28, 41, 43, 44, 46	100.0	110.5	102.1	100.0	110.5	102.1	100.0	110.6	102.1	100.0	110.7	102.1	100.0	110.4	102.1	100.0	110.5	102.1
21, 22, 23	FIRE SUPPRESSION, PLUMBING & HVAC	100.0	116.0	106.4	95.1	116.0	103.5	100.0	116.1	106.4	100.0	116.1	106.4	100.0	116.0	106.4	100.0	116.0	106.4
26, 27, 3370	ELECTRICAL, COMMUNICATIONS & UTIL.	99.2	111.3	105.5	95.9	111.3	103.9	99.2	156.4	129.1	99.2	163.8	132.9	98.7	108.7	103.9	99.2	112.2	106.0
MF2010	WEIGHTED AVERAGE	102.2	120.5	110.2	97.7	120.5	107.6	101.4	126.9	112.5	101.5	127.9	113.0	101.4	120.1	109.6	101.4	120.6	109.7

DIVISION		D.C. WASHINGTON 200 - 205 MAT.	INST.	TOTAL	DELAWARE DOVER 199 MAT.	INST.	TOTAL	NEWARK 197 MAT.	INST.	TOTAL	WILMINGTON 198 MAT.	INST.	TOTAL	FLORIDA DAYTONA BEACH 321 MAT.	INST.	TOTAL	FORT LAUDERDALE 333 MAT.	INST.	TOTAL
015433	CONTRACTOR EQUIPMENT		104.4	104.4		117.0	117.0		117.0	117.0		117.2	117.2		98.8	98.8		91.7	91.7
0241, 31 - 34	SITE & INFRASTRUCTURE, DEMOLITION	106.8	94.0	97.8	99.7	112.3	108.6	99.1	112.3	108.4	92.4	112.6	106.6	97.3	69.7	73.5	95.5	70.1	73.6
0310	Concrete Forming & Accessories	99.5	79.2	81.9	97.1	101.5	100.9	97.2	101.5	100.9	98.3	101.5	101.1	92.8	79.0	85.9	89.8	74.4	82.1
0320	Concrete Reinforcing	109.0	91.3	100.2	98.5	105.2	101.9	93.8	105.2	99.5	96.4	105.2	100.8	92.8	79.0	85.9	90.9	78.4	85.7
0330	Cast-in-Place Concrete	121.4	88.1	107.5	99.1	100.6	99.7	84.4	100.6	91.1	88.9	100.6	93.8	86.6	72.0	80.5	90.9	78.4	85.7
03	CONCRETE	109.6	85.8	97.9	99.9	102.8	101.3	92.1	102.8	97.4	94.8	102.8	98.7	88.2	73.4	80.9	90.7	75.0	82.9
04	MASONRY	95.9	80.0	86.1	103.3	97.2	99.5	99.9	97.2	98.2	103.6	97.2	99.6	92.3	66.8	76.5	89.9	69.4	77.3
05	METALS	100.1	107.3	102.3	103.1	117.0	107.4	103.7	117.0	107.8	102.9	117.0	107.3	96.4	92.8	95.3	95.5	91.5	94.2
06	WOOD, PLASTICS & COMPOSITES	103.5	77.2	88.7	95.0	101.1	98.4	97.6	101.1	99.6	98.2	101.1	99.8	99.7	70.8	83.4	96.0	68.7	80.6
07	THERMAL & MOISTURE PROTECTION	102.4	86.0	95.8	101.9	110.2	105.3	104.2	110.2	106.7	102.2	110.2	105.5	99.6	74.9	89.6	96.9	82.3	91.0
08	OPENINGS	100.3	87.6	97.4	94.3	110.5	98.1	93.4	110.5	97.4	96.6	110.5	99.9	98.9	69.1	91.8	96.9	67.1	89.9
0920	Plaster & Gypsum Board	105.8	76.5	85.9	105.9	101.0	102.6	103.4	101.0	101.8	104.8	101.0	102.2	95.1	70.3	78.3	98.5	68.2	77.9
0950, 0980	Ceilings & Acoustic Treatment	107.8	76.5	87.0	104.2	101.0	102.0	95.0	101.0	99.0	102.2	101.0	101.4	90.7	70.3	77.2	91.3	68.2	75.9
0960	Flooring	104.7	95.3	101.9	99.1	109.0	102.1	95.5	109.0	99.6	101.4	109.0	103.7	115.2	74.9	103.2	115.4	70.6	102.0
0970, 0990	Wall Finishes & Painting/Coating	104.6	84.0	92.2	97.3	105.3	102.1	99.5	105.3	103.0	97.3	105.3	102.1	99.4	75.7	85.2	96.5	71.7	81.6
09	FINISHES	101.6	81.7	90.5	101.4	102.8	102.2	97.4	102.8	100.4	102.0	102.8	102.4	100.8	70.9	84.1	99.4	69.6	82.8
COVERS	DIVS. 10 - 14, 25, 28, 41, 43, 44, 46	100.0	98.4	99.7	100.0	89.7	97.9	100.0	89.7	97.9	100.0	89.7	97.9	100.0	83.5	96.7	100.0	86.5	97.3
21, 22, 23	FIRE SUPPRESSION, PLUMBING & HVAC	100.1	92.9	97.2	100.0	115.8	106.3	100.2	115.8	106.5	100.2	115.8	106.4	100.0	77.8	91.1	100.0	68.0	87.2
26, 27, 3370	ELECTRICAL, COMMUNICATIONS & UTIL.	102.7	102.6	102.6	95.2	109.0	102.4	97.7	109.0	103.6	96.6	109.0	103.1	94.4	56.2	74.5	92.6	72.7	82.2
MF2010	WEIGHTED AVERAGE	101.6	91.6	97.2	99.7	108.1	103.3	98.8	108.1	102.8	99.4	108.1	103.2	97.4	74.4	87.3	96.4	73.9	86.6

499

FLORIDA

DIVISION		FORT MYERS 339,341			GAINESVILLE 326,344			JACKSONVILLE 320,322			LAKELAND 338			MELBOURNE 329			MIAMI 330 - 332,340		
		MAT.	INST.	TOTAL	MAT.	INST.	TOTAL	MAT.	INST.	TOTAL	MAT.	INST.	TOTAL	MAT.	INST.	TOTAL	MAT.	INST.	TOTAL
015433	CONTRACTOR EQUIPMENT		98.8	98.8		98.8	98.8		98.8	98.8		98.8	98.8		98.8	98.8		91.7	91.7
0241, 31 - 34	SITE & INFRASTRUCTURE, DEMOLITION	104.3	89.5	93.9	115.4	89.5	97.2	107.0	89.8	94.9	106.2	89.9	94.7	113.9	89.7	96.9	96.7	78.1	83.6
0310	Concrete Forming & Accessories	91.1	76.6	78.6	92.3	55.5	60.5	97.1	55.9	61.5	87.4	77.1	78.5	93.5	71.4	74.4	100.7	70.5	74.6
0320	Concrete Reinforcing	90.8	95.7	93.3	98.5	66.9	82.7	92.8	67.0	79.9	93.1	96.6	94.8	93.9	79.0	86.5	95.9	74.5	85.2
0330	Cast-in-Place Concrete	94.8	69.4	84.3	99.4	63.6	84.5	87.4	69.5	80.0	97.0	70.8	86.1	104.3	74.3	91.9	91.7	79.4	86.6
03	CONCRETE	91.3	78.6	85.1	98.7	62.0	80.6	88.6	64.2	76.6	93.0	79.5	86.3	99.0	75.0	87.1	92.4	75.5	84.0
04	MASONRY	84.1	62.9	71.0	106.7	62.3	79.2	92.3	62.3	73.7	100.3	77.1	85.9	91.2	71.0	78.7	90.3	73.2	79.7
05	METALS	97.6	98.6	97.9	95.3	86.6	92.7	95.0	87.0	92.5	97.5	99.5	98.1	104.4	93.0	100.9	96.3	90.3	94.1
06	WOOD, PLASTICS & COMPOSITES	92.5	79.2	85.0	93.2	53.0	70.5	99.7	53.0	73.4	87.5	79.2	82.8	94.9	70.8	81.3	103.4	68.7	83.8
07	THERMAL & MOISTURE PROTECTION	96.8	81.3	90.5	99.9	62.5	84.7	99.8	63.1	84.9	96.7	86.0	92.4	100.0	77.3	90.8	98.1	76.5	89.4
08	OPENINGS	97.9	76.2	92.8	97.1	54.0	86.9	98.9	57.2	89.0	97.9	76.9	92.9	98.1	72.4	92.0	99.1	67.0	91.5
0920	Plaster & Gypsum Board	93.9	79.0	83.8	90.8	51.9	64.4	95.1	51.9	65.8	90.5	79.0	82.7	90.8	70.3	76.9	96.2	68.2	77.1
0950, 0980	Ceilings & Acoustic Treatment	85.8	79.0	81.3	86.1	51.9	63.4	90.7	51.9	64.9	85.8	79.0	81.3	88.1	70.3	76.3	95.9	68.2	77.4
0960	Flooring	112.1	56.1	95.4	112.6	43.9	92.1	115.2	65.7	100.4	109.8	57.6	94.2	112.8	74.9	101.5	118.5	74.3	105.3
0970, 0990	Wall Finishes & Painting/Coating	101.6	62.1	77.8	99.4	62.1	77.0	99.4	66.3	79.5	101.6	62.1	77.8	99.4	93.9	96.1	93.5	71.7	80.4
09	FINISHES	98.9	71.5	83.6	99.4	52.9	73.5	100.9	57.7	76.8	99.7	71.8	83.3	99.6	73.9	85.3	101.9	70.8	84.5
COVERS	DIVS. 10 - 14, 25, 28, 41, 43, 44, 46	100.0	74.6	94.9	100.0	81.9	96.4	100.0	81.4	96.3	100.0	74.6	94.9	100.0	84.9	97.0	100.0	87.1	97.4
21, 22, 23	FIRE SUPPRESSION, PLUMBING & HVAC	97.4	65.7	84.7	98.8	65.5	85.5	100.0	65.5	86.2	97.4	82.5	91.5	100.0	79.9	92.0	100.0	67.9	87.1
26, 27, 3370	ELECTRICAL, COMMUNICATIONS & UTIL.	94.7	64.3	78.9	94.7	73.5	83.6	94.1	63.7	78.2	92.8	63.1	77.3	95.7	69.5	82.0	96.6	75.5	85.6
MF2010	WEIGHTED AVERAGE	96.4	74.1	86.7	98.7	68.0	85.3	97.2	67.7	84.3	97.1	79.3	89.4	99.9	77.9	90.3	97.7	74.6	87.6

FLORIDA

DIVISION		ORLANDO 327 - 328,347			PANAMA CITY 324			PENSACOLA 325			SARASOTA 342			ST. PETERSBURG 337			TALLAHASSEE 323		
		MAT.	INST.	TOTAL	MAT.	INST.	TOTAL	MAT.	INST.	TOTAL	MAT.	INST.	TOTAL	MAT.	INST.	TOTAL	MAT.	INST.	TOTAL
015433	CONTRACTOR EQUIPMENT		98.8	98.8		98.8	98.8		98.8	98.8		98.8	98.8		98.8	98.8		98.8	98.8
0241, 31 - 34	SITE & INFRASTRUCTURE, DEMOLITION	108.9	89.6	95.4	119.4	88.6	97.8	118.9	89.1	98.0	111.3	89.6	96.1	107.8	89.3	94.8	105.5	88.9	93.8
0310	Concrete Forming & Accessories	99.8	73.7	77.3	96.4	44.9	52.0	94.4	53.0	58.6	96.2	76.7	79.4	95.0	52.2	58.1	100.4	45.1	52.7
0320	Concrete Reinforcing	95.9	76.2	86.1	97.0	74.9	85.9	99.4	75.4	87.4	92.8	96.5	94.6	93.1	88.8	91.0	102.5	66.8	84.6
0330	Cast-in-Place Concrete	107.4	71.8	92.6	91.9	57.7	77.7	113.4	67.0	94.2	104.0	70.6	90.1	98.0	65.5	84.5	93.5	57.3	78.5
03	CONCRETE	99.0	74.6	86.9	96.8	56.7	77.0	106.3	63.6	85.2	97.2	79.2	88.3	94.6	65.2	80.1	94.3	55.2	75.0
04	MASONRY	97.1	66.8	78.4	97.0	48.1	66.7	114.9	55.4	78.1	92.6	77.1	83.0	138.8	49.9	83.8	93.8	54.8	69.6
05	METALS	95.2	91.3	94.0	96.2	88.2	93.7	97.3	89.6	94.9	99.8	99.1	99.6	98.4	94.6	97.2	93.2	86.2	91.1
06	WOOD, PLASTICS & COMPOSITES	99.3	76.9	86.7	98.5	42.7	67.0	96.1	52.7	71.6	99.4	79.2	88.0	97.4	51.3	71.4	98.3	42.3	66.7
07	THERMAL & MOISTURE PROTECTION	98.9	75.5	89.4	100.1	57.5	82.8	100.0	64.3	85.1	98.4	86.0	93.4	96.9	58.7	81.4	105.5	72.7	92.2
08	OPENINGS	100.0	70.9	93.1	96.8	47.6	85.1	96.7	58.3	87.6	99.6	76.0	94.0	97.9	62.4	89.5	102.0	48.5	89.3
0920	Plaster & Gypsum Board	99.8	76.7	84.0	93.7	41.4	58.1	94.8	51.7	65.5	95.7	79.0	84.3	96.5	50.2	65.0	105.5	40.9	61.6
0950, 0980	Ceilings & Acoustic Treatment	99.6	76.7	84.3	88.1	41.4	57.0	88.1	51.7	63.9	91.3	79.0	83.1	87.8	50.2	62.7	99.5	40.9	60.5
0960	Flooring	112.1	74.9	101.0	114.8	43.8	93.5	110.4	64.1	96.6	117.1	58.9	99.7	114.2	56.2	96.8	112.1	63.1	97.4
0970, 0990	Wall Finishes & Painting/Coating	96.2	71.7	81.5	99.4	66.3	79.5	99.4	66.3	79.5	102.2	62.1	78.1	101.6	62.1	77.8	96.4	66.3	78.3
09	FINISHES	103.1	74.1	86.9	101.3	45.8	70.4	99.8	55.7	75.3	102.6	72.0	85.6	100.4	52.9	74.0	103.5	49.3	73.3
COVERS	DIVS. 10 - 14, 25, 28, 41, 43, 44, 46	100.0	84.2	96.8	100.0	47.4	89.5	100.0	47.5	89.5	100.0	74.6	94.9	100.0	57.3	91.4	100.0	67.6	93.5
21, 22, 23	FIRE SUPPRESSION, PLUMBING & HVAC	100.0	57.8	83.1	100.0	53.4	81.3	100.0	53.6	81.4	99.9	64.1	85.5	100.0	59.7	83.9	100.0	39.8	75.9
26, 27, 3370	ELECTRICAL, COMMUNICATIONS & UTIL.	98.9	61.4	79.3	93.1	60.7	76.2	97.0	56.8	76.0	95.7	63.0	78.6	92.8	63.0	77.3	99.2	60.8	79.1
MF2010	WEIGHTED AVERAGE	99.3	71.4	87.1	98.5	59.1	81.3	100.8	62.4	84.1	99.3	75.3	88.8	100.2	64.7	84.7	98.7	58.0	81.0

FLORIDA / GEORGIA

DIVISION		TAMPA 335 - 336,346			WEST PALM BEACH 334,349			ALBANY 317,398			ATHENS 306			ATLANTA 300 - 303,399			AUGUSTA 308 - 309		
		MAT.	INST.	TOTAL	MAT.	INST.	TOTAL	MAT.	INST.	TOTAL	MAT.	INST.	TOTAL	MAT.	INST.	TOTAL	MAT.	INST.	TOTAL
015433	CONTRACTOR EQUIPMENT		98.8	98.8		91.7	91.7		92.5	92.5		93.8	93.8		94.3	94.3		93.8	93.8
0241, 31 - 34	SITE & INFRASTRUCTURE, DEMOLITION	108.3	89.8	95.3	90.5	78.4	82.0	96.0	79.8	84.6	93.9	93.2	93.4	90.8	94.7	93.5	87.9	92.9	91.4
0310	Concrete Forming & Accessories	98.0	77.3	80.1	99.2	69.7	73.7	90.9	44.5	50.9	90.6	47.0	52.9	94.1	75.0	77.6	91.6	67.0	70.4
0320	Concrete Reinforcing	89.8	96.6	93.2	92.3	74.2	83.3	89.6	81.4	85.5	98.6	80.4	89.5	97.9	82.4	90.2	99.0	74.1	86.6
0330	Cast-in-Place Concrete	95.8	70.9	85.5	86.4	75.5	81.9	92.4	55.3	77.0	107.4	56.5	86.3	107.4	71.1	92.3	101.4	50.6	80.3
03	CONCRETE	93.2	79.6	86.5	87.8	73.7	80.9	91.0	56.7	74.1	103.8	57.3	80.8	101.3	71.5	88.4	101.4	63.0	80.1
04	MASONRY	90.9	77.1	82.4	89.5	67.7	76.0	93.8	49.9	66.6	77.7	54.7	63.5	90.9	66.9	76.1	91.2	44.0	62.0
05	METALS	97.4	99.7	98.1	94.6	90.9	93.5	96.6	86.4	93.5	90.8	74.6	85.8	91.6	77.9	87.4	90.5	72.7	85.0
06	WOOD, PLASTICS & COMPOSITES	101.7	79.2	89.0	101.3	68.7	82.9	93.0	38.9	62.5	90.3	41.9	63.0	94.4	77.5	84.9	91.7	72.6	80.9
07	THERMAL & MOISTURE PROTECTION	97.1	86.0	92.6	96.7	71.4	86.4	96.5	61.4	82.3	93.6	53.8	77.5	93.5	73.0	85.2	93.3	58.1	79.0
08	OPENINGS	99.1	79.2	94.4	96.2	67.1	89.3	98.8	45.0	86.0	90.4	46.9	80.1	95.8	73.4	90.5	90.4	63.7	84.1
0920	Plaster & Gypsum Board	99.2	79.0	85.5	103.2	68.2	79.4	100.1	37.4	57.5	96.9	40.4	58.4	99.1	77.2	84.2	98.0	72.1	80.4
0950, 0980	Ceilings & Acoustic Treatment	91.3	79.0	83.1	85.8	68.2	74.1	93.5	37.4	56.2	94.1	40.4	58.4	94.1	77.2	82.8	95.1	72.1	79.8
0960	Flooring	115.4	57.6	98.1	117.4	67.7	102.5	116.2	48.7	96.0	94.5	54.9	82.7	95.8	66.4	87.0	94.8	47.4	80.6
0970, 0990	Wall Finishes & Painting/Coating	101.6	62.1	77.8	96.5	71.7	81.6	97.6	53.9	71.3	96.1	46.9	66.6	96.1	85.9	89.9	96.1	46.9	64.6
09	FINISHES	102.0	71.8	85.2	99.1	69.0	82.4	101.6	44.2	69.6	93.5	46.8	67.5	93.8	74.8	83.2	93.3	61.9	75.8
COVERS	DIVS. 10 - 14, 25, 28, 41, 43, 44, 46	100.0	84.9	97.0	100.0	86.5	97.3	100.0	80.3	96.1	100.0	79.4	95.9	100.0	85.7	97.1	100.0	80.6	96.1
21, 22, 23	FIRE SUPPRESSION, PLUMBING & HVAC	100.0	82.6	93.0	97.4	64.0	84.0	100.0	69.5	87.7	95.1	71.1	85.5	100.0	71.9	88.7	100.0	62.6	85.0
26, 27, 3370	ELECTRICAL, COMMUNICATIONS & UTIL.	92.6	63.1	77.2	93.8	72.7	82.8	96.1	60.1	77.3	99.5	71.2	84.7	98.8	72.7	85.2	100.1	63.2	80.9
MF2010	WEIGHTED AVERAGE	97.9	79.8	90.0	95.3	72.3	85.3	97.5	62.5	82.3	94.7	65.1	81.8	96.8	75.3	87.5	95.6	64.8	82.2

GEORGIA

DIVISION		COLUMBUS 318-319			DALTON 307			GAINESVILLE 305			MACON 310-312			SAVANNAH 313-314			STATESBORO 304		
		MAT.	INST.	TOTAL	MAT.	INST.	TOTAL	MAT.	INST.	TOTAL	MAT.	INST.	TOTAL	MAT.	INST.	TOTAL	MAT.	INST.	TOTAL
015433	CONTRACTOR EQUIPMENT		92.5	92.5		105.5	105.5		93.8	93.8		102.8	102.8		93.5	93.5		93.2	93.2
0241, 31-34	SITE & INFRASTRUCTURE, DEMOLITION	95.9	79.9	84.6	94.2	97.5	96.5	93.7	93.2	93.3	96.6	94.8	95.3	97.6	81.2	86.1	95.0	77.6	82.7
0310	Concrete Forming & Accessories	90.8	55.6	60.4	83.8	47.8	52.7	93.8	43.9	50.7	90.3	53.1	58.2	92.5	51.0	56.7	78.7	52.6	56.2
0320	Concrete Reinforcing	89.8	81.8	85.8	98.1	76.3	87.2	98.4	80.3	89.4	91.0	81.5	86.2	96.8	74.2	85.5	97.7	43.1	70.4
0330	Cast-in-Place Concrete	92.1	54.8	76.6	104.3	51.0	82.2	112.8	53.8	88.3	90.9	66.7	80.8	100.2	55.8	81.8	107.1	60.7	87.9
03	CONCRETE	90.8	61.5	76.4	102.9	55.8	79.7	105.7	54.9	80.6	90.4	64.4	77.6	95.9	58.4	77.4	103.2	55.3	79.5
04	MASONRY	93.7	56.2	70.5	78.8	37.3	53.2	86.3	55.1	67.0	105.6	45.3	68.3	90.3	51.4	66.2	80.2	41.3	56.1
05	METALS	96.1	87.5	93.5	91.6	84.3	89.3	90.0	73.8	85.0	92.2	87.1	90.6	93.3	83.8	90.3	95.0	74.0	88.5
06	WOOD, PLASTICS & COMPOSITES	93.0	53.3	70.6	74.1	48.5	59.7	94.2	38.8	62.9	101.2	52.4	73.7	105.9	47.0	72.7	67.6	55.1	60.6
07	THERMAL & MOISTURE PROTECTION	96.5	64.6	83.5	95.6	52.5	78.1	93.6	55.8	78.2	94.9	63.2	82.1	95.6	57.4	80.1	94.2	52.1	77.1
08	OPENINGS	98.8	60.2	89.6	91.2	50.7	81.6	90.4	40.9	78.7	97.6	53.7	87.2	103.0	49.6	90.4	92.2	41.6	80.2
0920	Plaster & Gypsum Board	100.1	52.3	67.6	84.0	47.3	59.0	99.1	37.2	57.0	105.5	51.3	68.6	98.1	45.8	62.5	84.1	54.1	63.7
0950, 0980	Ceilings & Acoustic Treatment	93.5	52.3	66.1	106.9	47.3	67.2	94.1	37.2	56.2	88.4	51.3	63.7	101.5	45.8	64.4	102.2	54.1	70.2
0960	Flooring	116.2	51.3	96.8	94.7	47.8	80.7	95.7	47.4	81.3	91.7	48.7	78.8	113.7	47.2	93.8	112.7	45.7	92.7
0970, 0990	Wall Finishes & Painting/Coating	97.6	67.9	79.7	87.3	61.1	71.5	96.1	46.9	66.6	99.8	53.9	72.2	96.0	59.5	74.1	94.0	40.1	61.6
09	FINISHES	101.5	54.7	75.4	101.7	48.1	71.9	94.0	42.6	65.3	89.7	51.0	68.2	101.7	49.9	72.9	104.7	50.1	74.3
COVERS	DIVS. 10-14, 25, 28, 41, 43, 44, 46	100.0	81.9	96.4	100.0	23.4	84.6	100.0	40.5	88.1	100.0	80.3	96.1	100.0	80.0	96.0	100.0	44.7	88.9
21, 22, 23	FIRE SUPPRESSION, PLUMBING & HVAC	100.0	64.7	85.8	95.2	58.2	80.4	95.1	71.0	85.5	100.0	68.2	87.2	100.0	62.9	85.2	95.7	58.0	80.5
26, 27, 3370	ELECTRICAL, COMMUNICATIONS & UTIL.	96.2	71.3	83.2	109.8	69.3	88.7	99.5	71.2	84.7	95.2	62.7	78.2	101.0	58.7	78.9	99.6	58.7	78.3
MF2010	WEIGHTED AVERAGE	97.4	66.6	84.0	96.4	60.1	80.6	95.2	62.8	81.1	96.2	65.9	83.0	98.4	62.0	82.6	96.5	56.9	79.2

DIVISION		GEORGIA VALDOSTA 316			WAYCROSS 315			HAWAII HILO 967			HONOLULU 968			STATES & POSS., GUAM 969			IDAHO BOISE 836-837		
		MAT.	INST.	TOTAL	MAT.	INST.	TOTAL	MAT.	INST.	TOTAL	MAT.	INST.	TOTAL	MAT.	INST.	TOTAL	MAT.	INST.	TOTAL
015433	CONTRACTOR EQUIPMENT		92.5	92.5		92.5	92.5		98.6	98.6		98.6	98.6		162.1	162.1		97.9	97.9
0241, 31-34	SITE & INFRASTRUCTURE, DEMOLITION	104.6	79.9	87.3	101.5	78.6	85.4	126.9	106.4	112.5	136.4	106.4	115.3	165.7	103.6	122.1	82.4	97.1	92.7
0310	Concrete Forming & Accessories	81.4	45.0	50.0	83.1	66.3	68.6	108.1	135.3	131.5	119.2	135.3	133.1	110.3	64.7	70.9	100.7	79.2	82.2
0320	Concrete Reinforcing	91.7	78.8	85.3	91.7	75.0	83.3	109.0	121.6	115.3	116.9	121.6	119.2	188.7	31.1	110.0	101.9	81.2	91.6
0330	Cast-in-Place Concrete	90.5	56.5	76.4	102.2	49.0	80.1	186.3	126.3	161.4	153.8	126.3	142.4	161.5	106.6	138.7	91.1	89.4	90.4
03	CONCRETE	95.4	56.9	76.4	98.7	62.8	81.0	148.2	128.4	138.4	141.5	128.4	135.0	149.6	73.5	112.0	100.7	83.2	92.1
04	MASONRY	99.3	51.0	69.4	100.0	39.9	62.8	127.9	128.5	128.3	128.1	128.5	128.4	182.7	44.3	97.1	118.9	82.4	96.3
05	METALS	95.7	85.4	92.5	94.8	79.3	90.0	107.6	107.9	107.7	120.8	107.9	116.6	138.8	76.3	119.5	101.1	81.2	95.0
06	WOOD, PLASTICS & COMPOSITES	81.0	38.5	57.0	82.8	74.1	77.9	101.3	137.9	122.0	119.2	137.9	129.7	113.8	68.6	88.3	88.9	77.7	82.6
07	THERMAL & MOISTURE PROTECTION	96.7	63.3	83.1	96.5	51.2	78.1	108.8	124.8	114.8	123.3	124.8	123.9	124.5	67.7	101.5	95.0	81.2	89.4
08	OPENINGS	94.7	43.5	82.6	94.7	59.2	86.3	99.7	132.7	107.5	108.8	132.7	114.4	104.3	55.2	92.6	99.1	72.5	92.8
0920	Plaster & Gypsum Board	92.0	37.1	54.6	92.4	73.7	79.7	107.7	138.9	128.9	143.9	138.9	140.5	213.3	56.8	106.8	84.9	76.9	79.5
0950, 0980	Ceilings & Acoustic Treatment	90.8	37.1	55.1	88.9	73.7	78.8	121.3	138.9	133.0	132.4	138.9	136.7	240.6	56.8	118.3	107.2	76.9	87.0
0960	Flooring	110.1	48.7	91.7	111.2	30.4	87.1	125.5	138.2	129.3	144.2	138.2	142.4	147.9	47.2	117.8	95.5	85.2	92.4
0970, 0990	Wall Finishes & Painting/Coating	97.6	53.3	70.9	97.6	46.9	67.1	99.1	145.5	127.0	107.4	145.5	130.3	104.1	36.7	63.6	102.9	40.3	65.3
09	FINISHES	99.0	44.5	68.7	98.6	58.7	76.4	115.6	138.1	128.1	130.3	138.1	134.6	192.9	60.0	118.8	97.8	76.5	85.9
COVERS	DIVS. 10-14, 25, 28, 41, 43, 44, 46	100.0	79.1	95.8	100.0	52.9	90.6	100.0	116.4	103.3	100.0	116.4	103.3	100.0	76.3	95.3	100.0	89.8	98.0
21, 22, 23	FIRE SUPPRESSION, PLUMBING & HVAC	100.0	71.4	88.5	96.8	58.3	81.4	100.1	109.4	103.8	100.2	109.4	103.8	100.2	38.1	76.4	100.0	72.5	89.0
26, 27, 3370	ELECTRICAL, COMMUNICATIONS & UTIL.	94.2	57.6	75.1	98.7	58.7	77.8	105.2	124.0	115.0	106.5	124.0	115.7	151.3	42.2	94.3	96.6	74.6	85.1
MF2010	WEIGHTED AVERAGE	97.5	62.6	82.3	97.3	60.7	81.3	110.4	121.0	115.0	114.9	121.0	117.6	131.2	58.9	99.7	99.8	79.4	90.9

DIVISION		IDAHO COEUR D'ALENE 838			IDAHO FALLS 834			LEWISTON 835			POCATELLO 832			TWIN FALLS 833			ILLINOIS BLOOMINGTON 617		
		MAT.	INST.	TOTAL	MAT.	INST.	TOTAL	MAT.	INST.	TOTAL	MAT.	INST.	TOTAL	MAT.	INST.	TOTAL	MAT.	INST.	TOTAL
015433	CONTRACTOR EQUIPMENT		92.5	92.5		97.9	97.9		92.5	92.5		97.9	97.9		97.9	97.9		101.4	101.4
0241, 31-34	SITE & INFRASTRUCTURE, DEMOLITION	79.9	92.3	88.6	80.3	96.9	91.9	86.3	93.1	91.1	83.1	97.1	92.9	89.4	98.0	95.5	96.3	97.8	97.3
0310	Concrete Forming & Accessories	113.8	82.2	86.5	94.9	79.0	81.1	118.9	83.1	88.0	100.9	78.9	81.9	101.8	56.0	62.3	85.5	119.2	114.6
0320	Concrete Reinforcing	109.5	96.9	103.2	103.9	81.0	92.4	109.5	97.1	103.3	102.3	81.2	91.8	104.2	81.0	92.6	95.8	111.2	103.5
0330	Cast-in-Place Concrete	98.5	86.4	93.5	86.8	75.0	81.9	111.1	86.8	95.9	93.7	89.2	91.8	96.1	64.4	82.9	98.7	113.8	104.9
03	CONCRETE	107.5	86.3	97.1	92.9	78.0	85.6	111.1	86.9	99.2	99.9	83.0	91.5	107.5	64.2	86.1	97.2	115.8	106.4
04	MASONRY	120.4	85.3	98.7	114.1	82.4	94.5	120.8	85.3	98.8	116.6	82.4	95.4	119.0	82.4	96.4	113.4	118.3	116.4
05	METALS	95.2	87.1	92.7	108.9	79.6	99.9	94.6	88.0	92.6	109.0	80.9	100.3	109.0	79.5	99.9	97.3	113.3	102.2
06	WOOD, PLASTICS & COMPOSITES	92.7	81.8	86.5	83.0	79.4	80.9	97.9	81.8	88.8	88.9	77.7	82.6	89.8	48.3	66.4	88.1	118.1	105.0
07	THERMAL & MOISTURE PROTECTION	147.5	81.5	120.7	94.6	71.7	85.3	147.7	81.5	120.8	95.0	73.9	86.4	95.7	75.2	87.4	100.3	113.3	105.6
08	OPENINGS	117.8	73.5	107.4	102.8	68.6	94.7	117.8	76.1	107.9	99.8	67.5	92.1	102.8	49.6	90.2	93.1	106.6	96.3
0920	Plaster & Gypsum Board	152.1	81.2	103.9	75.3	78.7	77.6	153.2	81.2	104.3	77.6	76.9	77.1	78.5	46.7	56.8	90.3	118.6	109.5
0950, 0980	Ceilings & Acoustic Treatment	138.9	81.2	100.5	105.1	78.7	87.5	138.9	81.2	100.5	113.1	76.9	89.0	107.8	46.7	67.1	94.3	118.6	110.5
0960	Flooring	137.5	46.7	110.3	95.7	43.9	80.2	140.2	92.5	126.0	98.7	85.2	94.6	99.7	43.9	83.0	98.0	120.9	104.8
0970, 0990	Wall Finishes & Painting/Coating	122.3	66.3	88.7	102.9	41.2	65.8	122.3	66.3	88.7	102.8	42.0	66.2	102.9	38.1	63.9	95.3	127.8	114.8
09	FINISHES	166.2	74.0	114.8	95.5	69.5	81.0	167.3	83.0	120.4	98.9	76.7	86.5	98.9	50.8	72.1	97.4	121.1	110.6
COVERS	DIVS. 10-14, 25, 28, 41, 43, 44, 46	100.0	89.7	97.9	100.0	48.9	89.8	100.0	89.7	97.9	100.0	89.8	98.0	100.0	45.4	89.1	100.0	103.9	100.8
21, 22, 23	FIRE SUPPRESSION, PLUMBING & HVAC	99.5	84.3	93.4	101.0	72.3	89.5	100.7	87.2	95.3	100.0	72.4	88.9	100.0	70.8	88.3	95.0	107.8	100.1
26, 27, 3370	ELECTRICAL, COMMUNICATIONS & UTIL.	88.7	79.4	83.9	88.7	70.9	79.4	86.7	80.8	83.6	94.1	70.9	82.0	90.2	61.6	75.2	93.5	96.8	95.3
MF2010	WEIGHTED AVERAGE	107.6	83.3	97.0	99.6	75.5	89.1	108.3	85.5	98.4	100.8	78.4	91.1	101.9	68.5	87.4	96.9	109.9	102.6

ILLINOIS

DIVISION		CARBONDALE 629			CENTRALIA 628			CHAMPAIGN 618 - 619			CHICAGO 606 - 608			DECATUR 625			EAST ST. LOUIS 620 - 622		
		MAT.	INST.	TOTAL	MAT.	INST.	TOTAL	MAT.	INST.	TOTAL	MAT.	INST.	TOTAL	MAT.	INST.	TOTAL	MAT.	INST.	TOTAL
015433	CONTRACTOR EQUIPMENT		107.7	107.7		107.7	107.7		102.1	102.1		93.9	93.9		102.1	102.1		107.7	107.7
0241, 31 - 34	SITE & INFRASTRUCTURE, DEMOLITION	100.2	98.6	99.0	100.5	99.2	99.6	105.2	98.4	100.4	105.3	95.4	98.4	92.2	98.6	96.7	102.6	99.2	100.2
0310	Concrete Forming & Accessories	95.5	107.9	106.2	97.4	114.8	112.4	91.6	117.1	113.7	96.6	159.8	151.1	97.7	116.9	114.2	92.8	115.0	112.0
0320	Concrete Reinforcing	89.8	100.4	95.1	89.8	113.3	101.5	95.8	105.1	100.5	95.5	163.2	129.3	87.9	105.9	96.9	89.7	113.4	101.5
0330	Cast-in-Place Concrete	91.8	101.1	95.7	92.3	118.5	103.2	114.4	109.2	112.2	103.4	149.6	122.6	100.2	111.9	105.1	93.8	118.5	104.1
03	CONCRETE	87.0	104.8	95.8	87.5	116.3	101.7	109.6	112.1	110.8	99.2	155.5	127.0	97.5	113.1	105.2	88.4	116.5	102.3
04	MASONRY	79.4	107.7	96.9	79.5	117.4	103.0	136.2	118.4	125.2	101.8	158.0	136.6	75.9	115.5	100.4	79.7	117.4	103.0
05	METALS	94.3	113.7	100.3	94.4	120.8	102.5	97.3	108.7	100.8	95.0	135.9	107.6	97.9	109.6	101.5	95.4	121.0	103.3
06	WOOD, PLASTICS & COMPOSITES	106.0	106.1	106.1	108.7	112.3	110.7	95.0	116.5	107.1	98.2	159.6	132.8	106.2	116.5	112.0	103.2	112.3	108.3
07	THERMAL & MOISTURE PROTECTION	98.0	101.6	99.5	98.0	112.3	103.8	100.9	114.5	106.5	99.2	147.5	118.8	104.0	109.6	106.3	98.0	111.9	103.6
08	OPENINGS	88.3	110.5	93.6	88.4	117.3	95.2	93.7	112.1	98.1	101.3	162.6	115.8	99.3	112.4	102.4	88.4	117.3	95.3
0920	Plaster & Gypsum Board	95.7	106.2	102.9	96.8	112.6	107.6	92.5	116.9	109.1	89.7	161.3	138.4	98.4	116.9	111.0	94.6	112.6	106.9
0950, 0980	Ceilings & Acoustic Treatment	96.2	106.2	102.9	96.2	112.6	107.1	94.3	116.9	109.4	101.5	161.3	141.3	102.4	116.9	112.1	96.2	112.6	107.1
0960	Flooring	121.3	122.4	121.6	122.3	117.9	120.9	100.7	122.4	107.1	93.1	150.7	110.3	108.6	118.8	111.7	120.2	117.9	119.5
0970, 0990	Wall Finishes & Painting/Coating	104.5	99.5	101.5	104.5	105.2	104.9	95.3	111.1	104.8	93.8	156.1	131.2	96.1	108.4	103.5	104.5	105.2	104.9
09	FINISHES	102.4	108.9	106.0	102.8	114.4	109.3	99.3	118.0	109.7	98.7	158.7	132.1	102.0	117.3	110.9	102.0	114.5	109.0
COVERS	DIVS. 10 - 14, 25, 28, 41, 43, 44, 46	100.0	104.1	100.8	100.0	106.5	101.3	100.0	103.1	100.6	100.0	125.0	105.0	100.0	103.0	100.6	100.0	106.5	101.3
21, 22, 23	FIRE SUPPRESSION, PLUMBING & HVAC	95.0	104.3	98.7	95.0	94.4	94.8	95.0	105.3	99.1	99.9	136.1	114.4	100.0	98.9	99.5	99.9	98.4	99.3
26, 27, 3370	ELECTRICAL, COMMUNICATIONS & UTIL.	94.0	106.7	100.6	95.5	104.3	100.1	96.8	94.5	95.6	96.1	137.1	117.5	98.4	91.3	94.7	95.0	104.3	99.9
MF2010	WEIGHTED AVERAGE	93.9	106.2	99.2	94.1	108.4	100.4	100.2	108.0	103.6	98.9	141.9	117.6	98.2	105.9	101.6	95.5	109.3	101.5

ILLINOIS

DIVISION		EFFINGHAM 624			GALESBURG 614			JOLIET 604			KANKAKEE 609			LA SALLE 613			NORTH SUBURBAN 600 - 603		
		MAT.	INST.	TOTAL	MAT.	INST.	TOTAL	MAT.	INST.	TOTAL	MAT.	INST.	TOTAL	MAT.	INST.	TOTAL	MAT.	INST.	TOTAL
015433	CONTRACTOR EQUIPMENT		102.1	102.1		101.4	101.4		92.0	92.0		92.0	92.0		101.4	101.4		92.0	92.0
0241, 31 - 34	SITE & INFRASTRUCTURE, DEMOLITION	97.5	98.2	98.0	98.7	97.7	98.0	105.2	94.7	97.8	98.9	93.9	95.4	98.1	98.4	98.3	104.4	93.7	96.8
0310	Concrete Forming & Accessories	102.2	115.5	113.7	91.5	119.0	115.3	98.4	163.3	154.4	92.4	141.3	134.6	105.1	125.1	122.4	97.8	153.4	145.8
0320	Concrete Reinforcing	90.7	101.0	95.9	95.4	111.1	103.2	95.5	155.7	125.6	96.3	146.5	121.4	95.5	143.7	119.6	95.5	157.3	126.4
0330	Cast-in-Place Concrete	99.9	107.3	103.0	101.6	107.9	104.2	103.3	149.4	122.4	96.4	132.5	111.4	101.5	123.6	110.7	103.4	143.5	120.0
03	CONCRETE	98.4	109.9	104.1	100.1	113.7	106.8	99.3	155.5	127.1	93.7	138.2	115.6	100.9	127.9	114.3	99.3	149.4	124.0
04	MASONRY	83.8	107.6	98.5	113.5	119.7	117.4	105.0	151.1	133.5	101.3	139.8	125.1	113.5	125.6	121.0	101.8	146.6	129.5
05	METALS	95.1	106.1	98.5	97.3	112.8	102.1	92.9	130.6	104.6	92.9	124.7	102.7	97.4	130.3	107.5	94.1	130.1	105.2
06	WOOD, PLASTICS & COMPOSITES	108.8	116.5	113.1	94.8	118.2	108.0	99.7	164.8	136.4	92.8	140.7	119.8	110.0	123.5	117.6	98.2	154.1	129.7
07	THERMAL & MOISTURE PROTECTION	103.6	107.6	105.2	100.5	109.9	104.3	99.1	144.5	117.5	98.4	135.6	113.5	100.6	119.4	108.2	99.5	140.8	116.3
08	OPENINGS	93.8	111.2	97.9	93.1	112.2	97.6	99.0	163.5	114.2	91.9	147.0	105.0	93.1	128.3	101.5	99.0	157.1	112.8
0920	Plaster & Gypsum Board	98.3	116.9	111.0	92.5	118.7	110.3	86.7	166.8	141.2	84.1	142.0	123.5	99.2	124.2	116.2	89.7	155.8	134.6
0950, 0980	Ceilings & Acoustic Treatment	96.2	116.9	110.0	94.3	118.7	110.5	101.5	166.8	144.9	101.5	142.0	128.4	94.3	124.2	114.2	101.5	155.8	137.6
0960	Flooring	109.7	122.4	113.5	100.6	120.9	106.7	92.7	142.9	107.7	90.1	134.3	103.3	106.5	124.8	112.0	93.1	142.9	108.0
0970, 0990	Wall Finishes & Painting/Coating	96.1	104.2	101.0	95.3	96.0	95.7	92.0	151.4	127.7	92.0	127.8	113.5	95.3	127.8	114.8	93.8	151.4	128.4
09	FINISHES	101.9	116.8	110.2	98.6	117.8	109.3	98.1	160.0	132.6	96.6	137.1	119.2	101.2	124.7	114.3	98.7	152.5	128.7
COVERS	DIVS. 10 - 14, 25, 28, 41, 43, 44, 46	100.0	71.7	94.3	100.0	103.8	100.8	100.0	125.4	105.1	100.0	117.6	103.5	100.0	104.4	100.9	100.0	120.4	104.1
21, 22, 23	FIRE SUPPRESSION, PLUMBING & HVAC	95.1	103.5	98.4	95.0	105.7	99.3	100.0	132.7	113.1	95.1	126.6	107.7	95.0	124.8	107.7	99.9	130.9	112.3
26, 27, 3370	ELECTRICAL, COMMUNICATIONS & UTIL.	96.0	106.6	101.6	94.5	87.3	90.7	95.4	141.2	119.3	90.1	140.4	114.1	91.6	140.4	117.1	95.2	131.5	114.2
MF2010	WEIGHTED AVERAGE	96.3	106.3	100.6	97.6	107.6	102.0	98.4	140.7	116.8	94.8	130.8	110.5	97.7	125.1	109.6	98.5	135.9	114.8

ILLINOIS

DIVISION		PEORIA 615 - 616			QUINCY 623			ROCK ISLAND 612			ROCKFORD 610 - 611			SOUTH SUBURBAN 605			SPRINGFIELD 626 - 627		
		MAT.	INST.	TOTAL	MAT.	INST.	TOTAL	MAT.	INST.	TOTAL	MAT.	INST.	TOTAL	MAT.	INST.	TOTAL	MAT.	INST.	TOTAL
015433	CONTRACTOR EQUIPMENT		101.4	101.4		102.1	102.1		101.4	101.4		101.4	101.4		92.0	92.0		102.1	102.1
0241, 31 - 34	SITE & INFRASTRUCTURE, DEMOLITION	99.2	97.8	98.2	96.3	98.1	97.6	96.8	96.6	96.8	98.5	98.8	98.7	104.4	93.7	96.8	99.2	98.6	98.8
0310	Concrete Forming & Accessories	94.5	119.9	116.4	100.0	113.8	111.9	93.2	101.3	100.2	98.9	130.1	125.8	97.8	153.4	145.8	98.1	117.3	114.6
0320	Concrete Reinforcing	92.9	111.2	102.0	90.4	105.6	98.0	95.4	104.2	99.8	87.8	138.2	113.0	95.5	157.3	126.4	92.8	106.1	99.4
0330	Cast-in-Place Concrete	98.5	116.0	105.8	98.0	104.2	101.8	99.4	98.4	99.0	100.9	126.1	111.5	103.4	143.5	120.0	98.2	108.5	100.7
03	CONCRETE	97.3	116.9	106.9	98.0	109.0	103.4	98.1	101.2	99.6	97.8	130.1	113.7	99.3	149.4	124.0	95.6	112.1	103.7
04	MASONRY	113.5	120.0	117.5	104.2	104.6	104.4	113.4	96.2	102.7	88.8	134.3	117.0	101.8	146.6	129.5	86.1	119.0	106.5
05	METALS	100.0	113.6	104.2	95.1	108.4	99.2	97.3	108.3	100.7	100.0	127.8	108.6	94.1	130.1	105.2	95.5	109.7	99.9
06	WOOD, PLASTICS & COMPOSITES	102.5	118.2	111.4	106.1	116.5	112.0	96.4	100.6	98.8	102.5	127.8	116.8	98.2	154.1	129.7	103.0	116.5	110.6
07	THERMAL & MOISTURE PROTECTION	101.2	113.8	106.3	103.6	104.1	103.8	100.4	98.5	99.7	103.5	128.4	113.6	99.5	140.8	116.3	105.2	112.8	108.3
08	OPENINGS	99.2	116.7	103.4	94.6	112.4	98.8	93.1	101.0	95.0	99.2	132.5	107.1	99.0	157.1	112.8	101.3	112.4	103.9
0920	Plaster & Gypsum Board	96.3	118.7	111.5	96.8	116.9	110.5	92.5	100.5	97.9	96.3	128.5	118.2	89.7	155.8	134.6	97.1	116.9	110.6
0950, 0980	Ceilings & Acoustic Treatment	100.5	118.7	112.6	94.3	116.9	109.0	94.3	100.5	98.4	100.5	128.5	119.2	101.5	155.8	137.6	101.5	116.9	113.5
0960	Flooring	103.8	120.9	108.9	108.6	110.1	109.1	101.7	105.5	103.4	103.8	124.8	110.1	93.1	142.9	108.0	113.1	109.8	112.1
0970, 0990	Wall Finishes & Painting/Coating	95.3	127.8	114.8	96.1	105.1	101.5	95.3	96.0	95.7	95.3	135.3	119.3	93.8	151.4	128.4	94.6	105.1	100.9
09	FINISHES	101.4	121.4	112.6	101.3	113.6	108.2	98.8	101.9	100.6	101.4	129.8	117.2	98.7	152.5	128.7	106.1	115.4	111.3
COVERS	DIVS. 10 - 14, 25, 28, 41, 43, 44, 46	100.0	104.3	100.9	100.0	72.7	94.5	100.0	97.2	99.4	100.0	111.6	102.3	100.0	120.4	104.1	100.0	103.3	100.7
21, 22, 23	FIRE SUPPRESSION, PLUMBING & HVAC	99.9	105.8	102.3	95.1	100.8	97.4	95.0	99.8	96.9	100.0	114.9	106.0	99.9	130.9	112.3	99.9	103.4	101.3
26, 27, 3370	ELECTRICAL, COMMUNICATIONS & UTIL.	95.5	97.4	96.5	93.2	81.3	87.0	86.5	95.7	91.3	95.8	127.0	112.1	95.2	131.5	114.2	101.0	92.8	96.8
MF2010	WEIGHTED AVERAGE	99.9	110.4	104.5	96.9	101.6	99.0	96.6	99.8	98.0	98.9	123.5	109.6	98.5	135.9	114.8	99.0	107.2	102.5

INDIANA

DIVISION		ANDERSON 460 MAT.	INST.	TOTAL	BLOOMINGTON 474 MAT.	INST.	TOTAL	COLUMBUS 472 MAT.	INST.	TOTAL	EVANSVILLE 476 - 477 MAT.	INST.	TOTAL	FORT WAYNE 467 - 468 MAT.	INST.	TOTAL	GARY 463 - 464 MAT.	INST.	TOTAL
015433	CONTRACTOR EQUIPMENT		96.7	96.7		87.2	87.2		87.2	87.2		116.1	116.1		96.7	96.7		96.7	96.7
0241, 31 - 34	SITE & INFRASTRUCTURE, DEMOLITION	90.5	96.0	94.4	82.7	94.8	91.2	79.4	94.6	90.1	87.8	123.5	112.9	91.5	96.0	94.6	91.0	99.2	96.8
0310	Concrete Forming & Accessories	98.8	80.8	83.3	101.9	81.1	83.9	95.9	78.9	81.2	95.4	81.9	83.7	97.1	75.4	78.3	98.9	115.2	113.0
0320	Concrete Reinforcing	98.0	83.7	90.9	87.1	83.4	85.2	87.5	83.4	85.4	95.4	76.7	86.1	98.0	76.8	87.4	98.0	110.1	104.0
0330	Cast-in-Place Concrete	104.8	78.1	93.7	100.8	78.0	91.3	100.4	72.5	88.8	96.3	87.7	92.8	111.3	83.7	99.9	109.5	112.4	110.7
03	CONCRETE	96.3	80.9	88.7	102.3	80.3	91.4	101.6	77.3	89.6	102.7	83.1	93.0	99.3	79.1	89.3	98.5	113.1	105.7
04	MASONRY	84.8	79.7	81.7	90.5	76.4	81.8	90.4	76.2	81.6	86.4	83.3	84.5	88.2	77.3	81.4	86.1	112.8	102.7
05	METALS	92.9	89.8	92.0	95.9	78.7	90.6	95.9	78.0	90.4	89.2	84.4	87.8	92.9	86.4	90.9	92.9	108.0	97.6
06	WOOD, PLASTICS & COMPOSITES	108.2	80.8	92.8	118.0	81.1	97.2	112.5	78.3	93.2	97.4	80.8	88.0	107.9	74.8	89.2	105.7	115.0	110.9
07	THERMAL & MOISTURE PROTECTION	105.4	77.1	93.9	95.3	79.7	88.9	94.9	79.4	88.6	99.3	84.7	93.4	105.2	79.0	94.6	104.0	109.0	106.0
08	OPENINGS	95.0	81.9	91.9	105.0	82.0	99.6	100.8	80.5	96.0	98.3	79.8	93.9	95.0	74.7	90.2	95.0	119.4	100.8
0920	Plaster & Gypsum Board	94.3	80.6	85.0	98.1	81.2	86.6	94.7	78.3	83.6	93.1	79.6	83.9	93.5	74.4	80.5	87.6	115.8	106.8
0950, 0980	Ceilings & Acoustic Treatment	89.6	80.6	83.7	83.9	81.2	82.1	83.9	78.3	80.2	88.8	79.6	82.7	89.6	74.4	79.5	89.6	115.8	107.1
0960	Flooring	99.9	86.6	96.0	105.8	76.4	97.0	101.4	76.4	93.9	100.8	82.2	95.2	99.9	80.2	94.0	99.9	123.6	107.0
0970, 0990	Wall Finishes & Painting/Coating	97.3	70.9	81.4	89.6	83.0	85.7	89.6	83.0	85.7	94.7	86.7	89.9	97.3	74.4	83.5	97.3	126.6	114.9
09	FINISHES	95.0	80.9	87.1	96.2	80.4	87.4	94.5	78.7	85.7	95.2	82.4	88.1	94.8	76.1	84.4	94.0	118.3	107.5
COVERS	DIVS. 10 - 14, 25, 28, 41, 43, 44, 46	100.0	91.5	98.3	100.0	89.1	97.8	100.0	88.7	97.7	100.0	94.1	98.8	100.0	89.5	97.9	100.0	105.3	101.1
21, 22, 23	FIRE SUPPRESSION, PLUMBING & HVAC	100.0	75.8	90.3	99.8	79.0	91.4	94.9	79.0	88.5	100.0	80.2	92.0	100.0	72.4	88.9	100.0	105.9	102.3
26, 27, 3370	ELECTRICAL, COMMUNICATIONS & UTIL.	85.9	88.4	87.2	98.2	83.5	90.5	97.3	85.2	91.0	94.5	86.3	90.2	86.6	78.4	82.3	97.7	110.2	104.2
MF2010	WEIGHTED AVERAGE	95.5	83.1	90.1	98.7	81.5	91.2	96.6	80.9	89.7	96.5	86.6	92.1	96.0	79.2	88.7	96.8	110.1	102.6

INDIANA

DIVISION		INDIANAPOLIS 461 - 462 MAT.	INST.	TOTAL	KOKOMO 469 MAT.	INST.	TOTAL	LAFAYETTE 479 MAT.	INST.	TOTAL	LAWRENCEBURG 470 MAT.	INST.	TOTAL	MUNCIE 473 MAT.	INST.	TOTAL	NEW ALBANY 471 MAT.	INST.	TOTAL
015433	CONTRACTOR EQUIPMENT		93.3	93.3		96.7	96.7		87.2	87.2		105.0	105.0		95.7	95.7		95.1	95.1
0241, 31 - 34	SITE & INFRASTRUCTURE, DEMOLITION	90.3	99.3	96.6	87.1	95.8	93.2	80.2	94.5	90.3	78.0	111.0	101.2	82.5	95.0	91.3	75.3	97.4	90.8
0310	Concrete Forming & Accessories	98.0	85.1	86.9	102.4	77.8	81.2	93.5	82.8	84.3	92.4	77.2	79.3	93.5	80.5	82.3	90.2	75.1	77.2
0320	Concrete Reinforcing	100.0	85.4	92.7	88.7	83.7	86.2	87.1	83.6	85.4	86.3	76.6	81.5	96.3	83.6	90.0	87.7	79.6	83.6
0330	Cast-in-Place Concrete	103.3	84.7	95.6	103.7	81.0	94.3	100.9	80.7	92.5	94.4	76.6	87.0	105.9	77.7	94.2	97.4	75.7	88.4
03	CONCRETE	99.5	84.8	92.3	93.1	80.5	86.9	101.8	81.9	92.0	95.0	77.4	86.3	101.1	80.6	91.0	100.2	76.5	88.5
04	MASONRY	90.3	81.3	84.7	84.4	80.0	81.7	95.5	79.8	85.8	75.6	76.0	75.9	92.4	79.8	84.6	82.3	69.5	74.4
05	METALS	93.6	81.7	89.9	89.5	89.3	89.4	94.3	78.6	89.5	91.1	85.2	89.3	97.7	89.9	95.3	93.0	82.7	89.8
06	WOOD, PLASTICS & COMPOSITES	105.3	85.3	94.0	111.6	76.6	91.9	109.5	83.2	94.7	95.8	77.0	85.2	111.4	80.6	94.0	97.9	76.0	85.5
07	THERMAL & MOISTURE PROTECTION	100.9	82.6	93.5	105.0	77.1	93.7	94.9	81.5	89.5	99.2	78.3	90.7	97.9	78.0	89.8	96.8	71.4	80.6
08	OPENINGS	103.1	84.8	98.8	90.0	79.6	87.5	99.2	83.1	95.4	100.4	76.8	94.8	98.5	81.7	94.5	97.9	78.2	93.2
0920	Plaster & Gypsum Board	88.5	85.1	86.2	99.8	76.3	83.8	91.8	83.3	86.0	72.6	76.8	75.5	93.1	80.6	84.6	91.2	75.6	80.6
0950, 0980	Ceilings & Acoustic Treatment	91.3	85.1	87.2	89.6	76.3	80.8	79.4	83.3	82.0	93.2	76.8	82.3	84.8	80.6	82.0	88.8	75.6	80.0
0960	Flooring	101.7	86.6	97.2	103.6	89.9	99.5	100.4	88.1	96.8	74.8	86.6	78.3	100.7	86.6	96.5	98.3	64.0	88.1
0970, 0990	Wall Finishes & Painting/Coating	99.5	83.0	89.6	97.3	73.3	82.9	89.6	95.1	92.9	90.6	73.2	80.1	89.6	70.9	78.4	94.7	84.5	88.6
09	FINISHES	95.4	85.4	89.8	96.6	79.3	87.0	92.9	85.2	88.6	85.8	78.9	82.0	94.0	80.7	86.6	94.5	74.3	83.2
COVERS	DIVS. 10 - 14, 25, 28, 41, 43, 44, 46	100.0	93.0	98.6	100.0	89.1	97.8	100.0	91.3	98.3	100.0	44.9	89.0	100.0	90.9	98.2	100.0	44.3	88.8
21, 22, 23	FIRE SUPPRESSION, PLUMBING & HVAC	99.9	80.1	92.0	95.1	79.2	88.7	94.9	79.0	89.0	95.6	75.7	87.6	99.8	75.8	90.2	95.1	76.9	87.8
26, 27, 3370	ELECTRICAL, COMMUNICATIONS & UTIL.	100.7	88.3	94.2	90.4	78.8	84.4	96.8	82.6	89.4	92.3	75.3	83.4	90.2	80.3	85.0	92.8	77.6	84.9
MF2010	WEIGHTED AVERAGE	98.3	85.1	92.6	93.3	82.0	88.4	96.3	82.9	90.5	93.3	79.3	87.2	97.3	81.8	90.6	94.4	77.0	86.8

INDIANA / IOWA

DIVISION		SOUTH BEND 465 - 466 MAT.	INST.	TOTAL	TERRE HAUTE 478 MAT.	INST.	TOTAL	WASHINGTON 475 MAT.	INST.	TOTAL	BURLINGTON 526 MAT.	INST.	TOTAL	CARROLL 514 MAT.	INST.	TOTAL	CEDAR RAPIDS 522 - 524 MAT.	INST.	TOTAL
015433	CONTRACTOR EQUIPMENT		105.0	105.0		116.1	116.1		116.1	116.1		100.2	100.2		100.2	100.2		96.8	96.8
0241, 31 - 34	SITE & INFRASTRUCTURE, DEMOLITION	92.9	95.9	95.0	89.7	124.2	113.9	89.2	121.5	111.9	96.5	97.3	97.1	85.8	97.3	93.9	98.5	96.0	96.8
0310	Concrete Forming & Accessories	102.6	79.8	82.9	96.3	78.1	80.6	97.1	80.6	82.9	97.3	77.3	80.1	85.2	51.6	56.2	103.2	81.0	84.1
0320	Concrete Reinforcing	101.1	79.1	90.1	95.4	83.8	89.6	88.2	50.3	69.3	94.3	85.4	89.9	95.1	81.1	88.1	109.9	81.3	98.0
0330	Cast-in-Place Concrete	96.3	80.7	89.8	93.3	87.0	90.7	101.5	86.6	95.3	109.6	56.3	87.5	109.6	61.0	89.4	101.8	82.8	92.4
03	CONCRETE	90.8	81.3	86.1	105.6	82.6	94.2	111.1	77.1	94.3	101.7	72.5	87.3	100.5	61.7	81.3	101.8	82.8	92.4
04	MASONRY	90.7	79.1	83.5	93.6	80.5	85.5	86.4	82.3	83.9	97.2	66.0	77.9	99.0	73.5	83.2	102.8	77.5	87.1
05	METALS	93.0	99.8	95.1	89.9	88.1	89.4	84.7	69.6	80.0	87.0	94.9	89.4	87.0	90.4	88.1	89.3	93.8	90.7
06	WOOD, PLASTICS & COMPOSITES	105.0	79.2	90.4	99.5	76.4	86.5	100.0	80.7	89.1	95.3	78.4	85.7	81.8	45.9	61.6	101.8	80.5	89.8
07	THERMAL & MOISTURE PROTECTION	97.5	81.9	91.2	99.4	81.9	92.3	99.5	84.4	93.4	103.8	75.6	92.4	104.1	70.5	90.5	104.7	79.9	94.6
08	OPENINGS	91.0	78.9	88.1	98.9	79.4	94.3	95.6	69.2	89.3	92.4	72.1	87.6	96.6	54.7	86.7	97.6	81.7	93.8
0920	Plaster & Gypsum Board	87.0	78.9	81.5	93.1	75.0	80.8	93.0	79.5	83.8	100.7	77.7	85.1	95.9	44.3	60.8	105.3	80.1	88.2
0950, 0980	Ceilings & Acoustic Treatment	93.4	78.9	83.8	88.8	75.0	79.6	82.5	79.5	80.5	110.6	77.7	88.7	110.6	44.3	66.5	114.1	80.1	91.5
0960	Flooring	97.4	89.9	95.1	100.8	86.6	96.6	101.7	77.8	94.6	111.2	38.5	89.5	105.6	33.8	84.2	127.9	79.7	113.5
0970, 0990	Wall Finishes & Painting/Coating	87.6	87.4	87.5	94.7	83.7	88.1	94.7	85.2	89.0	99.0	80.2	87.7	99.0	79.9	87.5	100.4	73.7	84.3
09	FINISHES	95.3	82.4	88.1	95.2	79.9	86.6	94.3	81.4	87.1	107.2	70.4	86.7	103.3	49.4	73.3	113.2	80.0	94.7
COVERS	DIVS. 10 - 14, 25, 28, 41, 43, 44, 46	100.0	90.7	98.1	100.0	93.2	98.6	100.0	94.0	98.8	100.0	89.2	97.8	100.0	66.9	93.4	100.0	90.2	98.0
21, 22, 23	FIRE SUPPRESSION, PLUMBING & HVAC	99.9	77.4	90.9	100.0	80.4	92.1	95.1	79.3	88.7	95.3	76.9	87.9	95.3	74.2	86.8	100.2	80.7	92.4
26, 27, 3370	ELECTRICAL, COMMUNICATIONS & UTIL.	97.6	86.6	91.9	92.6	83.6	87.9	93.1	86.3	89.5	100.7	69.3	84.3	101.4	81.3	90.9	98.2	81.3	89.3
MF2010	WEIGHTED AVERAGE	95.7	84.2	90.7	97.2	85.8	92.2	95.0	83.3	89.9	96.6	76.9	88.0	96.4	72.2	85.9	99.4	83.4	92.4

IOWA

DIVISION		COUNCIL BLUFFS 515			CRESTON 508			DAVENPORT 527 - 528			DECORAH 521			DES MOINES 500 - 503,509			DUBUQUE 520		
		MAT.	INST.	TOTAL	MAT.	INST.	TOTAL	MAT.	INST.	TOTAL	MAT.	INST.	TOTAL	MAT.	INST.	TOTAL	MAT.	INST.	TOTAL
015433	CONTRACTOR EQUIPMENT		96.3	96.3		100.2	100.2		100.2	100.2		100.2	100.2		101.8	101.8		95.7	95.7
0241, 31 - 34	SITE & INFRASTRUCTURE, DEMOLITION	103.6	92.3	95.7	86.7	96.4	93.5	96.8	99.3	98.6	95.1	96.3	95.9	95.5	99.7	98.5	96.4	93.3	94.2
0310	Concrete Forming & Accessories	84.6	70.7	72.6	84.5	67.2	69.6	102.5	91.2	92.7	94.9	46.5	53.1	103.2	78.9	82.2	85.9	73.8	75.4
0320	Concrete Reinforcing	97.0	79.3	88.2	93.3	81.2	87.3	95.1	100.0	97.5	94.3	79.0	86.7	102.7	81.4	92.0	93.7	86.6	90.2
0330	Cast-in-Place Concrete	114.3	75.1	98.0	102.1	65.4	86.9	105.9	91.7	100.0	106.7	59.3	87.0	89.8	88.0	89.1	107.6	100.4	104.6
03	CONCRETE	104.1	74.7	89.6	96.9	70.2	83.7	99.9	93.5	96.7	99.6	58.4	79.3	95.0	83.2	89.2	98.6	86.2	92.4
04	MASONRY	104.5	75.9	86.8	96.9	81.4	87.3	100.0	85.5	91.1	117.7	72.7	89.8	90.7	80.4	84.3	103.7	73.7	85.1
05	METALS	94.3	90.2	93.1	89.4	90.6	89.8	89.3	103.9	93.8	87.1	87.7	87.3	95.2	92.9	94.5	88.0	93.5	89.7
06	WOOD, PLASTICS & COMPOSITES	80.6	69.7	74.4	79.1	63.1	70.1	101.9	90.3	95.3	92.3	37.9	61.6	98.8	77.4	86.8	82.2	71.7	76.3
07	THERMAL & MOISTURE PROTECTION	104.2	68.4	89.6	106.2	78.8	95.1	104.2	87.4	97.4	104.0	54.7	84.0	100.6	79.5	92.0	104.4	71.3	91.0
08	OPENINGS	96.6	75.5	91.6	106.6	64.4	96.6	97.6	91.4	96.1	95.2	49.8	84.5	101.9	82.8	97.4	96.6	79.3	92.5
0920	Plaster & Gypsum Board	95.9	69.0	77.6	94.1	62.0	72.2	105.3	90.0	94.9	99.6	36.0	56.4	91.9	76.8	81.6	95.9	71.1	79.0
0950, 0980	Ceilings & Acoustic Treatment	110.6	69.0	82.9	111.5	62.0	78.5	114.1	90.0	98.1	110.6	36.0	61.0	115.0	76.8	89.6	110.6	71.1	84.3
0960	Flooring	104.3	82.1	97.6	104.2	33.8	83.2	113.5	98.7	109.1	111.1	48.0	92.2	109.4	87.6	102.9	118.8	77.6	106.5
0970, 0990	Wall Finishes & Painting/Coating	95.3	66.9	78.2	95.8	79.9	86.3	99.0	97.1	98.1	99.0	33.6	59.7	91.2	79.9	84.4	99.4	65.1	78.8
09	FINISHES	104.1	72.0	86.2	102.4	61.5	79.6	109.1	93.2	100.2	106.9	43.6	71.6	105.5	80.4	91.5	108.4	72.9	88.6
COVERS	DIVS. 10 - 14, 25, 28, 41, 43, 44, 46	100.0	88.9	97.8	100.0	71.4	94.3	100.0	93.1	98.6	100.0	85.2	97.0	100.0	90.3	98.1	100.0	88.6	97.7
21, 22, 23	FIRE SUPPRESSION, PLUMBING & HVAC	100.2	75.8	90.4	95.3	80.3	89.2	100.2	93.2	97.4	95.3	74.5	86.9	100.0	79.8	91.9	100.2	75.4	90.2
26, 27, 3370	ELECTRICAL, COMMUNICATIONS & UTIL.	103.7	81.6	92.1	94.8	81.3	87.8	95.5	91.2	93.2	98.1	45.4	70.6	106.9	81.3	93.6	102.2	78.4	89.8
MF2010	WEIGHTED AVERAGE	100.2	78.8	90.9	96.7	77.9	88.5	98.4	93.4	96.2	97.4	65.6	83.5	99.4	83.9	92.7	98.5	80.4	90.6

IOWA

DIVISION		FORT DODGE 505			MASON CITY 504			OTTUMWA 525			SHENANDOAH 516			SIBLEY 512			SIOUX CITY 510 - 511		
		MAT.	INST.	TOTAL	MAT.	INST.	TOTAL	MAT.	INST.	TOTAL	MAT.	INST.	TOTAL	MAT.	INST.	TOTAL	MAT.	INST.	TOTAL
015433	CONTRACTOR EQUIPMENT		100.2	100.2		100.2	100.2		95.7	95.7		96.3	96.3		100.2	100.2		100.2	100.2
0241, 31 - 34	SITE & INFRASTRUCTURE, DEMOLITION	94.6	95.0	94.9	94.6	96.2	95.8	96.8	91.4	93.0	102.0	91.4	94.6	106.0	95.0	98.3	107.6	95.8	99.3
0310	Concrete Forming & Accessories	85.1	46.0	51.3	89.6	46.4	52.3	92.9	71.9	74.8	86.2	57.8	61.7	86.6	38.7	45.3	103.2	67.9	72.7
0320	Concrete Reinforcing	93.3	67.4	80.4	93.2	80.6	86.9	94.3	85.6	90.0	97.0	68.6	82.8	97.0	67.3	82.2	95.1	78.2	86.7
0330	Cast-in-Place Concrete	95.8	45.0	74.7	95.8	58.7	80.4	110.3	54.8	87.3	110.5	61.1	90.0	108.3	45.7	82.3	109.0	56.7	87.3
03	CONCRETE	92.6	51.1	72.1	92.9	58.5	75.9	101.2	69.4	85.5	101.5	62.1	82.1	100.5	48.1	74.6	101.2	66.9	84.3
04	MASONRY	95.9	38.5	60.4	108.7	69.4	84.4	100.3	57.1	73.6	104.0	75.9	86.6	122.1	38.8	70.6	97.2	55.6	71.4
05	METALS	89.5	81.9	87.2	89.5	88.8	89.3	86.9	92.0	88.5	93.3	84.3	90.5	87.2	81.4	85.4	89.3	88.2	89.0
06	WOOD, PLASTICS & COMPOSITES	79.6	46.2	60.8	83.9	37.9	58.0	89.4	78.2	83.1	82.2	54.0	66.3	83.1	36.7	56.9	101.8	68.2	82.9
07	THERMAL & MOISTURE PROTECTION	105.6	58.5	86.5	105.2	64.6	88.7	104.6	65.2	88.6	103.5	66.6	88.5	103.8	49.2	81.6	104.3	64.9	88.3
08	OPENINGS	100.8	50.1	88.8	93.3	50.2	83.1	96.6	76.2	91.8	88.1	55.5	80.4	93.4	44.9	81.9	97.6	66.9	90.3
0920	Plaster & Gypsum Board	94.1	44.6	60.4	94.1	36.0	54.6	97.0	77.7	83.9	95.9	52.8	66.6	95.9	34.7	54.3	105.3	67.2	79.4
0950, 0980	Ceilings & Acoustic Treatment	111.5	44.6	67.0	111.5	36.0	61.3	110.6	77.7	88.7	110.6	52.8	72.1	110.6	34.7	60.1	114.1	67.2	82.9
0960	Flooring	105.8	48.0	88.5	107.9	48.0	90.0	121.8	51.1	100.6	104.9	34.9	84.0	106.4	34.6	85.0	113.6	54.7	96.0
0970, 0990	Wall Finishes & Painting/Coating	95.8	63.1	76.1	95.8	30.7	56.7	99.4	79.9	87.7	95.3	66.9	78.2	99.0	63.1	77.4	99.0	63.1	77.4
09	FINISHES	104.4	47.1	72.5	105.0	42.9	70.4	109.5	69.2	87.0	104.2	53.4	75.9	106.7	38.9	68.9	110.7	64.8	85.1
COVERS	DIVS. 10 - 14, 25, 28, 41, 43, 44, 46	100.0	79.8	96.0	100.0	84.8	97.0	100.0	84.2	96.8	100.0	67.9	93.6	100.0	78.9	95.8	100.0	86.9	97.4
21, 22, 23	FIRE SUPPRESSION, PLUMBING & HVAC	95.3	65.8	83.4	95.3	71.9	85.9	95.3	71.3	85.7	95.3	75.1	87.2	95.3	69.5	84.9	100.2	76.3	90.6
26, 27, 3370	ELECTRICAL, COMMUNICATIONS & UTIL.	101.9	42.9	71.1	100.9	57.0	77.9	100.5	67.8	83.4	98.1	80.6	89.0	98.1	42.4	69.1	98.2	74.7	85.9
MF2010	WEIGHTED AVERAGE	96.6	58.8	80.2	96.5	66.6	83.5	97.3	73.0	86.7	97.1	72.1	86.2	97.7	57.5	80.2	99.1	73.6	88.0

IOWA / KANSAS

DIVISION		SPENCER 513			WATERLOO 506 - 507			BELLEVILLE 669			COLBY 677			DODGE CITY 678			EMPORIA 668		
		MAT.	INST.	TOTAL	MAT.	INST.	TOTAL	MAT.	INST.	TOTAL	MAT.	INST.	TOTAL	MAT.	INST.	TOTAL	MAT.	INST.	TOTAL
015433	CONTRACTOR EQUIPMENT		100.2	100.2		100.2	100.2		103.1	103.1		103.1	103.1		103.1	103.1		101.4	101.4
0241, 31 - 34	SITE & INFRASTRUCTURE, DEMOLITION	106.1	95.0	98.3	99.5	96.1	97.1	111.0	94.7	99.5	109.4	94.8	99.1	111.9	94.5	99.7	102.8	92.2	95.3
0310	Concrete Forming & Accessories	92.8	38.7	46.1	101.6	48.1	55.4	95.0	55.3	60.7	102.2	59.6	65.4	95.3	59.5	64.4	86.5	66.7	69.4
0320	Concrete Reinforcing	97.0	67.3	82.2	94.0	86.5	90.2	99.6	56.1	77.9	101.2	56.1	78.7	98.8	56.1	77.5	98.2	56.7	77.5
0330	Cast-in-Place Concrete	108.3	45.7	82.3	102.6	55.0	82.9	118.9	56.9	93.2	114.3	57.0	90.6	116.4	56.7	91.6	114.9	54.8	90.0
03	CONCRETE	100.9	48.1	74.9	98.4	59.1	79.0	117.7	57.4	88.0	114.9	59.3	87.5	116.2	59.1	88.0	110.0	61.8	86.2
04	MASONRY	122.1	38.8	70.6	96.6	73.0	82.0	94.9	59.8	73.2	96.9	59.8	73.9	104.4	60.0	76.9	100.5	68.4	80.7
05	METALS	87.1	81.4	85.4	91.8	92.8	92.1	97.8	78.4	91.9	98.2	78.7	92.2	99.7	77.8	92.9	97.6	79.9	92.1
06	WOOD, PLASTICS & COMPOSITES	89.3	36.7	59.6	98.6	37.9	64.4	96.4	53.3	72.1	103.9	59.2	78.7	95.3	59.2	74.9	87.7	68.2	76.7
07	THERMAL & MOISTURE PROTECTION	104.6	49.2	82.1	105.4	71.4	91.6	98.4	62.3	83.7	101.9	62.9	86.1	101.9	62.3	85.8	97.0	77.4	89.1
08	OPENINGS	104.3	44.9	90.2	94.2	53.5	84.5	96.4	49.4	85.3	99.5	52.6	88.4	99.5	52.6	88.4	94.4	57.5	85.6
0920	Plaster & Gypsum Board	97.0	34.7	54.7	103.9	36.0	57.7	98.0	51.8	66.6	100.7	57.9	71.6	94.8	57.9	69.7	95.1	67.1	76.0
0950, 0980	Ceilings & Acoustic Treatment	110.6	34.7	60.1	115.0	36.0	62.5	91.6	51.8	65.1	100.0	57.9	70.1	94.3	57.9	70.1	91.6	67.1	75.3
0960	Flooring	109.1	34.6	86.8	113.3	76.9	102.4	106.4	39.9	86.5	105.7	39.9	86.0	102.2	39.9	83.6	101.8	37.1	82.4
0970, 0990	Wall Finishes & Painting/Coating	99.0	63.1	77.4	95.8	79.4	86.0	97.4	39.8	62.8	97.4	39.8	62.8	97.4	39.8	62.8	97.4	39.8	62.8
09	FINISHES	107.6	38.9	69.3	108.8	54.4	78.5	101.3	50.4	72.9	101.5	53.9	75.0	99.8	53.9	74.2	98.6	58.6	76.3
COVERS	DIVS. 10 - 14, 25, 28, 41, 43, 44, 46	100.0	78.9	95.8	100.0	85.3	97.1	100.0	41.8	88.3	100.0	42.5	88.5	100.0	42.5	88.5	100.0	41.5	88.3
21, 22, 23	FIRE SUPPRESSION, PLUMBING & HVAC	95.3	69.5	84.9	100.2	80.0	92.1	95.1	73.1	86.3	95.1	68.7	84.5	100.0	68.7	87.4	95.1	73.4	86.4
26, 27, 3370	ELECTRICAL, COMMUNICATIONS & UTIL.	99.9	42.4	69.9	97.0	57.0	76.1	93.3	66.0	79.1	97.9	66.0	81.3	94.7	73.4	83.6	90.8	73.1	81.5
MF2010	WEIGHTED AVERAGE	99.2	57.5	81.0	98.3	70.9	86.4	99.4	65.6	84.7	100.2	65.6	85.1	101.7	66.5	86.4	97.8	70.0	85.7

KANSAS

DIVISION		FORT SCOTT 667			HAYS 676			HUTCHINSON 675			INDEPENDENCE 673			KANSAS CITY 660 - 662			LIBERAL 679		
		MAT.	INST.	TOTAL	MAT.	INST.	TOTAL	MAT.	INST.	TOTAL	MAT.	INST.	TOTAL	MAT.	INST.	TOTAL	MAT.	INST.	TOTAL
015433	CONTRACTOR EQUIPMENT		102.3	102.3		103.1	103.1		103.1	103.1		103.1	103.1		99.7	99.7		103.1	103.1
0241, 31 - 34	SITE & INFRASTRUCTURE, DEMOLITION	99.4	93.0	94.9	115.5	94.8	100.9	93.4	94.6	94.3	113.1	94.6	100.1	93.5	92.9	93.1	115.2	94.6	100.7
0310	Concrete Forming & Accessories	102.6	78.6	81.9	99.7	59.6	65.1	89.9	59.5	63.6	110.8	67.8	73.7	99.3	96.3	96.7	95.8	59.5	64.5
0320	Concrete Reinforcing	97.5	96.3	96.9	98.8	56.1	77.5	98.8	56.1	77.5	98.2	63.0	80.6	94.6	97.8	96.2	100.2	56.1	78.2
0330	Cast-in-Place Concrete	106.6	55.7	85.5	90.2	57.0	76.5	83.6	54.2	71.4	116.8	54.5	91.0	91.4	99.3	94.7	90.3	54.2	75.3
03	CONCRETE	105.2	74.6	90.1	106.7	59.3	83.3	89.1	58.3	73.9	117.4	63.3	90.7	97.3	97.9	97.6	108.9	58.3	83.9
04	MASONRY	101.0	61.4	76.5	104.3	59.8	76.8	96.4	59.6	73.6	94.0	65.7	76.5	103.6	99.5	101.1	102.4	59.6	75.9
05	METALS	97.5	94.4	96.5	97.8	78.7	91.9	97.6	77.7	91.5	97.5	81.3	92.5	104.9	100.6	103.5	98.1	77.7	91.8
06	WOOD, PLASTICS & COMPOSITES	106.5	85.4	94.6	100.6	59.2	77.2	89.9	59.2	72.6	114.7	69.6	89.3	102.4	96.7	99.1	95.9	59.2	75.2
07	THERMAL & MOISTURE PROTECTION	97.8	82.8	91.7	102.1	62.9	86.2	100.9	62.2	85.2	101.9	82.4	94.0	97.5	99.8	98.4	102.3	62.2	86.0
08	OPENINGS	94.4	80.1	91.0	99.4	52.6	88.3	99.4	52.6	88.3	97.2	59.8	88.4	95.6	90.5	94.4	99.5	52.6	88.4
0920	Plaster & Gypsum Board	100.6	84.9	89.9	98.1	57.9	70.7	93.6	57.9	69.3	107.8	68.6	81.1	93.5	96.5	95.5	95.5	57.9	69.9
0950, 0980	Ceilings & Acoustic Treatment	91.6	84.9	87.2	94.3	57.9	70.1	94.3	57.9	70.1	94.3	68.6	77.2	91.6	96.5	94.9	94.3	57.9	70.1
0960	Flooring	117.6	40.1	94.4	104.7	39.9	85.3	99.5	39.9	81.7	109.9	39.9	89.0	93.5	100.5	95.6	102.4	39.9	83.7
0970, 0990	Wall Finishes & Painting/Coating	99.2	39.8	63.5	97.4	39.8	62.8	97.4	39.8	62.8	97.4	39.8	62.8	105.6	69.2	83.7	97.4	39.8	62.8
09	FINISHES	104.0	68.4	84.2	101.4	53.9	74.9	97.3	53.9	73.1	103.8	60.0	79.4	98.5	93.1	95.4	100.7	53.9	74.6
COVERS	DIVS. 10 - 14, 25, 28, 41, 43, 44, 46	100.0	47.4	89.5	100.0	42.5	88.5	100.0	42.5	88.5	100.0	43.7	88.7	100.0	63.5	92.7	100.0	42.5	88.5
21, 22, 23	FIRE SUPPRESSION, PLUMBING & HVAC	95.1	69.3	84.7	95.1	68.7	84.5	95.1	68.7	84.5	95.1	71.1	85.5	99.9	96.8	98.6	95.1	68.7	84.5
26, 27, 3370	ELECTRICAL, COMMUNICATIONS & UTIL.	90.1	73.1	81.2	96.8	66.0	80.7	91.7	66.0	78.3	94.0	73.1	83.0	95.5	97.3	96.4	94.7	73.4	83.6
MF2010	WEIGHTED AVERAGE	97.7	74.3	87.5	99.6	65.6	84.8	95.8	65.4	82.5	99.9	70.3	87.0	99.4	95.7	97.7	99.5	66.4	85.1

KANSAS / KENTUCKY

DIVISION		SALINA 674			TOPEKA 664 - 666			WICHITA 670 - 672			ASHLAND 411 - 412			BOWLING GREEN 421 - 422			CAMPTON 413 - 414		
		MAT.	INST.	TOTAL	MAT.	INST.	TOTAL	MAT.	INST.	TOTAL	MAT.	INST.	TOTAL	MAT.	INST.	TOTAL	MAT.	INST.	TOTAL
015433	CONTRACTOR EQUIPMENT		103.1	103.1		101.4	101.4		103.1	103.1		98.0	98.0		95.1	95.1		101.5	101.5
0241, 31 - 34	SITE & INFRASTRUCTURE, DEMOLITION	102.2	94.8	97.0	97.9	91.2	93.2	98.7	93.3	94.9	108.9	87.0	93.5	75.6	97.7	91.1	83.8	98.8	94.3
0310	Concrete Forming & Accessories	91.8	55.7	60.6	97.2	41.8	49.4	97.8	51.7	58.0	89.0	105.7	103.4	86.5	84.0	84.4	90.4	84.6	85.4
0320	Concrete Reinforcing	98.2	62.6	80.4	97.7	97.2	97.5	97.5	75.8	86.7	89.1	106.1	97.6	86.4	89.5	88.0	87.3	105.0	96.1
0330	Cast-in-Place Concrete	101.0	54.7	81.8	96.1	46.5	75.6	94.7	52.7	77.3	88.7	102.9	94.6	88.0	97.5	91.9	98.1	75.2	88.6
03	CONCRETE	103.0	58.0	80.8	98.3	55.4	77.1	97.6	57.8	78.0	96.1	105.2	100.6	94.4	89.7	92.1	98.2	85.2	91.8
04	MASONRY	118.4	59.8	82.1	97.6	56.9	72.4	94.9	51.1	67.8	92.2	105.1	100.2	94.0	81.6	86.3	91.6	70.2	78.4
05	METALS	99.5	81.5	93.9	101.6	96.4	100.0	101.6	83.7	96.1	92.1	109.8	97.6	93.7	86.9	91.6	93.0	93.7	93.2
06	WOOD, PLASTICS & COMPOSITES	91.5	53.3	70.0	95.4	38.1	63.1	95.3	52.5	71.2	79.0	105.4	93.9	92.3	83.9	87.5	90.8	91.1	90.9
07	THERMAL & MOISTURE PROTECTION	101.4	62.6	85.7	100.0	69.1	87.8	100.0	57.6	82.8	91.3	99.3	94.5	86.7	82.1	84.8	99.6	69.4	87.3
08	OPENINGS	99.4	49.7	87.6	98.6	55.3	88.4	101.6	56.6	91.0	97.0	99.3	97.5	97.9	79.4	93.5	99.3	89.8	97.1
0920	Plaster & Gypsum Board	93.6	51.8	65.2	97.6	36.1	55.8	92.4	51.0	64.2	61.2	105.6	91.4	86.4	83.7	84.6	86.4	90.2	89.0
0950, 0980	Ceilings & Acoustic Treatment	94.3	51.8	66.0	97.6	36.1	56.7	99.4	51.0	67.2	83.8	105.6	98.3	88.8	83.7	85.4	88.8	90.2	89.7
0960	Flooring	100.7	63.1	89.5	106.9	75.0	97.3	108.3	64.1	95.1	80.1	104.2	87.3	96.5	87.6	93.9	98.6	42.0	81.7
0970, 0990	Wall Finishes & Painting/Coating	97.4	39.8	62.8	99.2	53.7	71.9	96.3	49.0	67.9	96.9	103.4	100.8	94.7	74.3	82.4	94.7	66.8	77.9
09	FINISHES	98.5	54.8	74.2	103.1	47.0	71.8	102.7	53.5	75.3	82.5	105.7	95.4	93.1	84.0	88.0	93.9	75.7	83.8
COVERS	DIVS. 10 - 14, 25, 28, 41, 43, 44, 46	100.0	54.2	90.8	100.0	48.4	89.7	100.0	51.1	90.2	100.0	96.3	99.3	100.0	59.8	91.9	100.0	57.3	91.4
21, 22, 23	FIRE SUPPRESSION, PLUMBING & HVAC	100.0	68.7	87.5	100.0	69.0	87.5	99.0	64.3	85.6	95.0	92.9	94.1	100.0	84.3	93.7	95.1	81.5	89.7
26, 27, 3370	ELECTRICAL, COMMUNICATIONS & UTIL.	94.5	73.3	83.4	94.2	73.1	83.2	99.0	73.3	85.6	90.7	99.6	95.4	93.3	82.1	87.4	90.6	67.7	78.6
MF2010	WEIGHTED AVERAGE	100.4	67.0	85.9	99.4	66.5	85.1	99.9	65.1	84.8	94.0	100.1	96.6	95.5	84.8	90.8	95.1	80.2	88.6

KENTUCKY

DIVISION		CORBIN 407 - 409			COVINGTON 410			ELIZABETHTOWN 427			FRANKFORT 406			HAZARD 417 - 418			HENDERSON 424		
		MAT.	INST.	TOTAL	MAT.	INST.	TOTAL	MAT.	INST.	TOTAL	MAT.	INST.	TOTAL	MAT.	INST.	TOTAL	MAT.	INST.	TOTAL
015433	CONTRACTOR EQUIPMENT		101.5	101.5		105.0	105.0		95.1	95.1		101.5	101.5		101.5	101.5		116.1	116.1
0241, 31 - 34	SITE & INFRASTRUCTURE, DEMOLITION	83.0	98.5	93.9	79.4	112.3	102.5	70.4	97.8	89.6	87.1	99.7	95.9	81.8	100.1	94.6	78.4	124.1	110.5
0310	Concrete Forming & Accessories	88.9	72.3	74.6	86.0	92.9	92.0	81.6	82.4	82.3	99.5	73.4	77.0	87.1	85.4	85.6	93.4	83.5	84.8
0320	Concrete Reinforcing	87.0	72.1	79.5	85.9	94.0	90.0	86.9	93.8	90.3	99.8	93.7	96.8	87.7	105.0	96.3	86.6	87.6	87.1
0330	Cast-in-Place Concrete	92.2	59.9	78.8	93.9	99.6	96.3	79.6	74.6	77.5	87.2	70.6	80.3	94.4	84.2	90.2	77.8	92.5	83.9
03	CONCRETE	92.6	68.4	80.7	96.5	95.5	96.0	86.6	82.0	84.3	93.0	76.6	84.9	95.0	88.7	91.9	91.6	87.4	89.6
04	MASONRY	89.8	63.0	73.2	105.4	103.3	104.1	78.5	76.8	77.4	87.2	80.8	83.3	89.3	72.8	79.1	97.8	94.5	95.8
05	METALS	93.4	79.8	89.2	91.0	91.7	91.2	92.8	88.8	91.6	95.8	89.5	93.9	93.0	93.7	93.2	84.4	88.6	85.7
06	WOOD, PLASTICS & COMPOSITES	88.5	77.7	82.4	88.7	86.0	87.2	87.4	83.9	85.4	105.5	70.3	85.6	87.7	91.1	89.6	95.1	80.8	87.1
07	THERMAL & MOISTURE PROTECTION	98.8	65.6	85.3	99.4	95.4	97.8	86.4	78.8	83.3	98.8	77.1	90.0	99.5	71.3	88.1	99.0	93.6	96.8
08	OPENINGS	99.3	67.2	91.7	101.3	85.7	97.6	97.9	85.3	94.9	102.1	78.2	96.4	99.7	80.3	95.1	96.0	81.8	92.7
0920	Plaster & Gypsum Board	90.6	76.4	80.9	69.4	86.1	80.8	85.6	83.7	84.3	100.9	68.8	79.0	85.6	90.2	88.7	89.7	79.7	82.9
0950, 0980	Ceilings & Acoustic Treatment	86.1	76.4	79.6	92.3	86.1	88.2	88.8	83.7	85.4	96.8	68.8	78.1	88.8	90.2	89.7	82.5	79.7	80.6
0960	Flooring	100.6	42.0	83.1	72.7	86.2	76.7	94.1	87.6	92.2	108.0	55.4	92.3	97.1	42.7	80.9	100.1	87.6	96.4
0970, 0990	Wall Finishes & Painting/Coating	94.2	50.5	67.9	90.6	84.1	86.7	94.7	77.8	84.6	93.9	81.5	86.5	94.7	66.8	77.9	94.7	98.1	96.7
09	FINISHES	93.9	65.0	77.8	84.8	90.5	87.9	92.0	82.5	86.7	101.6	70.3	84.2	93.2	76.5	83.9	92.6	85.7	88.8
COVERS	DIVS. 10 - 14, 25, 28, 41, 43, 44, 46	100.0	49.7	89.9	100.0	102.8	100.6	100.0	86.6	97.3	100.0	67.2	93.4	100.0	58.1	91.6	100.0	67.1	93.4
21, 22, 23	FIRE SUPPRESSION, PLUMBING & HVAC	95.1	72.9	86.2	95.6	90.6	93.6	95.4	81.9	90.0	100.1	84.1	93.7	95.1	82.4	90.0	95.4	79.3	88.9
26, 27, 3370	ELECTRICAL, COMMUNICATIONS & UTIL.	90.5	82.1	86.1	94.6	79.6	86.8	90.4	82.1	86.1	99.0	89.2	93.9	90.6	59.7	74.5	92.7	82.2	87.2
MF2010	WEIGHTED AVERAGE	94.4	73.3	85.2	95.2	93.2	94.3	92.2	83.7	88.5	97.9	82.4	91.1	94.6	79.9	88.2	93.0	87.9	90.8

505

City Cost Indexes

KENTUCKY

DIVISION		LEXINGTON 403 - 405			LOUISVILLE 400 - 402			OWENSBORO 423			PADUCAH 420			PIKEVILLE 415 - 416			SOMERSET 425 - 426		
		MAT.	INST.	TOTAL	MAT.	INST.	TOTAL	MAT.	INST.	TOTAL	MAT.	INST.	TOTAL	MAT.	INST.	TOTAL	MAT.	INST.	TOTAL
015433	CONTRACTOR EQUIPMENT		101.5	101.5		95.1	95.1		116.1	116.1		116.1	116.1		98.0	98.0		101.5	101.5
0241, 31 - 34	SITE & INFRASTRUCTURE, DEMOLITION	85.0	100.9	96.2	75.7	97.8	91.3	87.9	124.1	113.3	80.9	123.6	110.9	119.6	86.0	96.0	75.1	99.3	92.1
0310	Concrete Forming & Accessories	101.5	74.3	78.0	100.6	83.0	85.4	91.8	83.4	84.5	89.9	81.8	82.9	97.7	90.9	91.9	88.0	77.6	79.0
0320	Concrete Reinforcing	95.5	93.5	94.5	97.1	93.8	95.4	86.6	90.2	88.4	87.1	89.0	88.0	89.6	105.9	97.7	86.9	93.7	90.3
0330	Cast-in-Place Concrete	94.3	93.9	94.2	95.8	75.4	87.4	90.7	96.1	92.9	82.9	88.3	85.1	97.5	98.7	98.0	77.8	100.3	87.1
03	CONCRETE	95.8	84.9	90.5	96.7	82.5	89.7	103.4	89.1	96.4	96.0	85.5	90.8	109.6	97.1	103.5	81.8	88.6	85.2
04	MASONRY	88.4	73.4	79.1	88.9	78.2	82.3	90.5	92.6	91.8	93.5	90.0	91.3	90.0	98.5	95.2	84.7	78.1	80.6
05	METALS	95.8	89.9	93.9	97.6	88.9	94.9	85.9	90.2	87.2	83.0	88.4	84.6	92.0	108.5	97.1	92.9	89.3	91.8
06	WOOD, PLASTICS & COMPOSITES	105.1	70.3	85.5	103.1	83.9	92.3	93.0	80.8	86.1	90.6	80.9	85.2	88.6	90.4	89.6	88.3	77.7	82.3
07	THERMAL & MOISTURE PROTECTION	99.1	85.7	93.6	97.9	79.5	90.4	99.3	93.9	97.1	99.0	80.9	91.7	91.9	81.5	87.7	99.0	71.8	87.9
08	OPENINGS	99.5	75.3	93.8	99.0	85.3	95.8	96.0	83.9	93.1	95.2	79.4	91.5	97.7	88.4	95.5	98.7	79.6	94.2
0920	Plaster & Gypsum Board	100.6	68.8	79.0	93.2	83.7	86.7	88.2	79.7	82.4	87.1	79.8	82.1	64.5	90.2	82.0	85.6	76.4	79.4
0950, 0980	Ceilings & Acoustic Treatment	90.5	68.8	76.1	94.3	83.7	87.2	82.5	79.7	80.6	82.5	79.8	80.7	83.8	90.2	88.0	88.8	76.4	80.5
0960	Flooring	105.0	71.6	95.0	102.9	87.6	98.3	99.5	87.6	96.0	98.5	60.4	87.1	83.8	104.2	89.9	97.4	42.0	80.8
0970, 0990	Wall Finishes & Painting/Coating	94.2	82.8	87.3	93.9	77.8	84.2	94.7	98.1	96.7	94.7	81.4	86.7	96.9	83.0	88.5	94.7	77.8	84.6
09	FINISHES	97.5	73.6	84.2	97.9	83.5	89.9	99.2	85.2	88.6	92.1	77.8	84.1	84.9	93.0	89.4	92.6	70.9	80.5
COVERS	DIVS. 10 - 14, 25, 28, 41, 43, 44, 46	100.0	88.4	97.7	100.0	87.1	97.4	100.0	100.9	100.2	100.0	63.4	92.7	100.0	57.8	91.5	100.0	60.0	92.0
21, 22, 23	FIRE SUPPRESSION, PLUMBING & HVAC	100.0	80.6	92.2	100.0	82.6	93.0	100.0	78.4	91.3	95.4	82.9	90.4	95.0	89.0	92.6	95.4	79.1	88.8
26, 27, 3370	ELECTRICAL, COMMUNICATIONS & UTIL.	93.4	82.1	87.5	100.0	82.1	90.6	92.7	82.2	87.3	95.1	81.8	88.2	93.7	75.1	84.0	91.0	82.1	86.4
MF2010	WEIGHTED AVERAGE	97.0	82.4	90.7	97.8	84.2	91.9	95.6	88.9	92.7	93.2	86.2	90.2	96.2	90.0	93.5	92.7	81.5	87.8

LOUISIANA

DIVISION		ALEXANDRIA 713 - 714			BATON ROUGE 707 - 708			HAMMOND 704			LAFAYETTE 705			LAKE CHARLES 706			MONROE 712		
		MAT.	INST.	TOTAL	MAT.	INST.	TOTAL	MAT.	INST.	TOTAL	MAT.	INST.	TOTAL	MAT.	INST.	TOTAL	MAT.	INST.	TOTAL
015433	CONTRACTOR EQUIPMENT		89.7	89.7		89.3	89.3		89.9	89.9		89.9	89.9		89.3	89.3		89.7	89.7
0241, 31 - 34	SITE & INFRASTRUCTURE, DEMOLITION	96.3	86.7	89.5	96.0	86.0	89.0	99.1	86.9	90.5	100.1	87.0	90.9	100.8	85.9	90.3	96.3	85.9	89.0
0310	Concrete Forming & Accessories	83.0	44.2	49.5	94.1	61.7	66.1	78.1	46.9	51.2	95.2	54.9	60.4	96.1	57.5	62.8	82.5	43.9	49.2
0320	Concrete Reinforcing	89.4	56.0	72.7	95.1	59.0	77.1	93.3	59.4	76.4	94.6	59.0	76.8	94.6	59.2	76.9	88.4	55.6	72.0
0330	Cast-in-Place Concrete	94.6	50.3	76.2	91.8	58.7	78.0	93.9	43.9	73.2	93.5	47.6	74.4	98.4	67.4	85.5	94.6	56.7	78.8
03	CONCRETE	94.7	49.7	72.5	95.9	60.7	78.5	93.3	49.1	71.5	94.4	53.8	74.4	96.7	61.8	79.5	94.5	51.5	73.3
04	MASONRY	116.0	54.2	77.8	97.6	54.7	71.0	99.0	51.8	69.8	99.0	53.0	70.5	98.3	59.5	74.3	110.9	48.4	72.2
05	METALS	89.5	71.1	83.9	106.0	70.1	94.9	97.6	69.5	88.9	96.7	70.0	88.5	96.7	70.5	88.7	89.5	68.5	83.0
06	WOOD, PLASTICS & COMPOSITES	89.8	42.7	63.2	98.5	65.3	79.8	84.8	48.1	64.1	106.3	57.0	78.5	104.4	59.1	78.8	89.1	43.2	63.2
07	THERMAL & MOISTURE PROTECTION	97.5	54.1	79.9	95.8	62.4	82.2	96.9	57.0	80.7	97.4	60.8	82.5	96.5	59.5	81.5	97.5	54.7	80.1
08	OPENINGS	99.8	45.8	87.0	98.6	60.0	89.5	92.7	52.5	83.2	96.3	53.9	86.3	96.3	55.7	86.7	99.7	48.8	87.7
0920	Plaster & Gypsum Board	79.0	41.4	53.4	99.4	64.6	75.7	100.1	46.8	63.8	109.6	56.0	73.2	109.6	58.1	74.6	78.6	41.9	53.6
0950, 0980	Ceilings & Acoustic Treatment	87.1	41.4	56.7	103.1	64.6	77.4	98.7	46.8	64.2	97.0	56.0	69.7	98.0	58.1	71.5	87.1	41.9	57.0
0960	Flooring	110.0	66.9	97.1	106.0	66.9	94.3	98.6	66.9	89.1	106.8	66.9	94.9	106.8	74.3	97.1	109.5	58.1	94.2
0970, 0990	Wall Finishes & Painting/Coating	99.5	65.8	79.2	96.1	47.3	66.8	103.4	49.3	70.9	103.4	59.2	76.8	103.4	51.0	71.9	99.5	49.1	69.2
09	FINISHES	93.2	50.0	69.1	100.6	61.6	78.9	97.8	50.9	71.7	101.1	57.5	76.8	101.3	59.9	78.2	93.0	46.5	67.1
COVERS	DIVS. 10 - 14, 25, 28, 41, 43, 44, 46	100.0	51.4	90.3	100.0	80.9	96.2	100.0	46.6	89.3	100.0	79.4	95.9	100.0	80.4	96.1	100.0	47.0	89.4
21, 22, 23	FIRE SUPPRESSION, PLUMBING & HVAC	100.2	56.5	82.7	100.0	59.5	83.8	95.2	43.7	74.5	100.1	60.7	84.3	100.1	61.7	84.7	100.2	54.1	81.7
26, 27, 3370	ELECTRICAL, COMMUNICATIONS & UTIL.	90.5	58.2	73.6	103.0	60.8	81.0	98.2	57.3	76.8	99.5	66.6	82.3	98.9	66.6	82.0	92.2	59.6	75.2
MF2010	WEIGHTED AVERAGE	96.8	57.8	79.8	100.4	63.6	84.4	96.3	54.9	78.3	98.5	62.7	82.9	98.6	65.0	84.0	96.7	56.4	79.2

LOUISIANA / MAINE

DIVISION		NEW ORLEANS 700 - 701			SHREVEPORT 710 - 711			THIBODAUX 703			AUGUSTA 043			BANGOR 044			BATH 045		
		MAT.	INST.	TOTAL	MAT.	INST.	TOTAL	MAT.	INST.	TOTAL	MAT.	INST.	TOTAL	MAT.	INST.	TOTAL	MAT.	INST.	TOTAL
015433	CONTRACTOR EQUIPMENT		90.2	90.2		89.7	89.7		89.9	89.9		100.6	100.6		100.6	100.6		100.6	100.6
0241, 31 - 34	SITE & INFRASTRUCTURE, DEMOLITION	105.2	88.4	93.4	100.8	86.0	90.4	101.4	87.0	91.3	89.2	100.8	97.3	91.0	101.0	98.0	88.9	100.8	97.2
0310	Concrete Forming & Accessories	97.7	66.5	70.8	98.8	44.7	52.2	90.1	64.3	67.8	99.2	97.0	97.3	93.6	98.8	98.1	89.7	97.2	96.2
0320	Concrete Reinforcing	96.7	60.1	78.4	92.6	56.1	74.4	93.3	59.2	76.3	96.4	110.1	103.2	90.1	111.5	100.8	89.2	111.2	100.2
0330	Cast-in-Place Concrete	99.7	70.1	87.4	99.4	53.4	80.3	100.9	51.3	80.3	97.9	60.3	82.3	80.2	113.0	93.8	80.2	60.3	72.0
03	CONCRETE	97.8	67.1	82.6	99.6	50.9	75.6	98.4	59.3	79.1	104.6	86.2	95.5	94.4	105.3	99.8	94.5	86.4	90.5
04	MASONRY	101.8	60.7	76.4	102.0	51.0	70.4	124.8	49.3	78.1	96.8	64.8	77.0	106.0	103.9	104.7	112.3	85.8	95.9
05	METALS	109.5	73.0	98.3	93.3	69.2	85.9	97.6	69.3	88.9	97.8	84.0	94.8	88.7	90.4	89.2	87.2	89.6	87.9
06	WOOD, PLASTICS & COMPOSITES	99.7	67.9	81.8	101.7	43.7	69.0	92.9	69.7	79.8	98.6	107.7	103.7	95.1	98.7	97.1	89.7	107.7	99.9
07	THERMAL & MOISTURE PROTECTION	96.1	69.0	85.1	97.1	52.2	78.9	96.7	54.6	79.6	104.3	66.8	89.1	100.2	86.0	94.5	100.2	73.2	89.2
08	OPENINGS	96.2	66.4	89.2	98.8	46.4	86.4	97.1	64.2	89.3	104.2	90.0	100.8	102.5	85.0	98.4	102.5	90.0	99.5
0920	Plaster & Gypsum Board	108.2	67.2	80.3	90.0	42.5	57.7	102.7	69.1	79.8	100.8	107.4	105.2	98.8	98.0	98.2	95.1	107.4	103.4
0950, 0980	Ceilings & Acoustic Treatment	98.7	67.2	77.8	91.3	42.5	58.8	98.7	69.1	79.0	101.5	107.4	105.4	91.6	98.0	95.9	89.6	107.4	101.4
0960	Flooring	105.3	66.9	93.8	113.4	62.4	98.1	104.3	44.9	86.5	103.1	56.3	89.1	98.9	113.0	103.1	97.3	56.3	85.0
0970, 0990	Wall Finishes & Painting/Coating	108.8	65.5	82.8	95.5	45.5	65.4	104.7	50.6	72.2	101.0	64.7	79.2	95.3	45.1	65.1	95.3	45.1	65.1
09	FINISHES	102.4	66.6	82.4	98.2	47.4	69.9	100.1	60.1	77.8	101.4	87.7	93.8	98.6	96.3	97.3	97.1	85.6	90.7
COVERS	DIVS. 10 - 14, 25, 28, 41, 43, 44, 46	100.0	83.2	96.6	100.0	77.3	95.5	100.0	81.3	96.3	100.0	103.3	100.7	100.0	110.0	102.0	100.0	103.3	100.7
21, 22, 23	FIRE SUPPRESSION, PLUMBING & HVAC	100.0	65.8	86.3	100.0	58.9	83.5	95.2	61.7	81.7	99.9	65.6	86.2	100.1	75.9	90.4	95.2	65.7	83.4
26, 27, 3370	ELECTRICAL, COMMUNICATIONS & UTIL.	103.2	72.9	87.4	97.7	67.2	81.8	96.7	72.8	84.2	99.3	80.0	89.2	98.5	77.2	87.4	96.4	80.0	87.9
MF2010	WEIGHTED AVERAGE	101.5	69.8	87.7	98.4	59.6	81.5	98.7	64.9	84.0	100.3	80.8	91.8	97.6	90.8	94.7	96.1	83.0	90.4

City Cost Indexes

MAINE

DIVISION		HOULTON 047			KITTERY 039			LEWISTON 042			MACHIAS 046			PORTLAND 040-041			ROCKLAND 048		
		MAT.	INST.	TOTAL	MAT.	INST.	TOTAL	MAT.	INST.	TOTAL	MAT.	INST.	TOTAL	MAT.	INST.	TOTAL	MAT.	INST.	TOTAL
015433	CONTRACTOR EQUIPMENT		100.6	100.6		100.6	100.6		100.6	100.6		100.6	100.6		100.6	100.6		100.6	100.6
0241, 31 - 34	SITE & INFRASTRUCTURE, DEMOLITION	90.7	101.8	98.5	85.6	100.8	96.3	88.2	101.0	97.2	90.1	101.8	98.3	86.9	101.0	96.8	86.3	101.8	97.2
0310	Concrete Forming & Accessories	97.1	103.6	102.7	90.4	97.7	96.7	98.9	98.9	98.9	94.4	103.5	102.3	95.8	98.9	98.4	95.4	103.5	102.4
0320	Concrete Reinforcing	90.1	109.9	100.0	84.5	111.3	97.8	110.8	111.5	111.1	90.1	109.9	100.0	107.5	111.5	109.5	90.1	109.9	100.0
0330	Cast-in-Place Concrete	80.3	72.6	77.1	81.4	61.2	73.0	81.9	113.0	94.8	80.2	72.0	76.8	95.5	113.0	102.7	81.9	72.0	77.8
03	CONCRETE	95.5	93.3	94.4	90.1	87.0	88.6	95.1	105.3	100.1	94.9	93.1	94.0	100.8	105.3	103.0	92.2	93.1	92.6
04	MASONRY	91.0	76.3	81.9	101.0	86.8	92.2	91.7	103.9	99.2	91.0	76.3	81.9	101.3	103.9	102.9	85.9	76.3	80.0
05	METALS	87.4	87.9	87.5	87.1	89.8	87.9	91.7	90.4	91.3	87.4	87.9	87.6	99.4	90.4	96.6	87.3	87.9	87.5
06	WOOD, PLASTICS & COMPOSITES	99.1	107.7	104.0	88.7	107.7	99.4	101.2	98.7	99.8	96.1	107.7	102.6	96.1	98.7	97.5	96.9	107.7	103.0
07	THERMAL & MOISTURE PROTECTION	100.3	74.0	89.6	103.2	72.5	90.7	100.1	86.0	94.4	100.3	73.0	89.2	101.4	86.0	95.2	100.0	73.0	89.1
08	OPENINGS	102.6	88.2	99.2	102.7	90.0	99.6	105.6	85.0	100.7	102.6	88.2	99.2	105.7	85.0	100.8	102.5	88.2	99.1
0920	Plaster & Gypsum Board	101.0	107.4	105.3	94.7	107.4	103.3	104.5	98.0	100.1	99.5	107.4	104.8	101.0	98.0	99.0	99.5	107.4	104.8
0950, 0980	Ceilings & Acoustic Treatment	89.6	107.4	101.4	90.6	107.4	101.8	101.4	98.0	99.1	89.6	107.4	101.4	104.4	98.0	100.1	89.6	107.4	101.4
0960	Flooring	100.1	52.7	85.9	97.5	58.7	85.9	101.6	113.0	105.0	99.3	52.7	85.4	101.9	113.0	105.2	99.6	52.7	85.6
0970, 0990	Wall Finishes & Painting/Coating	95.3	140.1	122.2	95.4	39.3	61.7	95.3	45.1	65.1	95.3	140.1	122.2	93.4	45.1	64.4	95.3	140.1	122.2
09	FINISHES	98.8	99.7	99.3	96.2	85.6	90.3	101.8	96.3	98.7	98.3	99.7	99.1	102.5	96.3	99.0	98.0	99.7	98.9
COVERS	DIVS. 10 - 14, 25, 28, 41, 43, 44, 46	100.0	108.9	101.8	100.0	103.6	100.7	100.0	110.0	102.0	100.0	108.9	101.8	100.0	110.0	102.0	100.0	108.9	101.8
21, 22, 23	FIRE SUPPRESSION, PLUMBING & HVAC	95.2	74.1	86.8	95.2	77.3	88.0	100.1	75.9	90.4	95.2	74.1	86.8	100.0	75.9	90.3	95.2	74.1	86.8
26, 27, 3370	ELECTRICAL, COMMUNICATIONS & UTIL.	100.6	80.0	89.9	98.0	80.0	88.6	100.6	80.1	89.9	100.6	80.0	89.9	103.3	80.1	91.2	100.5	80.0	89.8
MF2010	WEIGHTED AVERAGE	95.9	86.6	91.8	95.1	85.7	91.0	98.3	91.2	95.2	95.7	86.5	91.7	100.8	91.2	96.6	95.0	86.5	91.3

MAINE / MARYLAND

DIVISION		WATERVILLE 049			ANNAPOLIS 214			BALTIMORE 210-212			COLLEGE PARK 207-208			CUMBERLAND 215			EASTON 216		
		MAT.	INST.	TOTAL	MAT.	INST.	TOTAL	MAT.	INST.	TOTAL	MAT.	INST.	TOTAL	MAT.	INST.	TOTAL	MAT.	INST.	TOTAL
015433	CONTRACTOR EQUIPMENT		100.6	100.6		99.3	99.3		102.8	102.8		104.3	104.3		99.3	99.3		99.3	99.3
0241, 31 - 34	SITE & INFRASTRUCTURE, DEMOLITION	90.8	100.8	97.8	99.8	90.9	93.1	103.7	94.1	96.9	91.4	90.9	91.1	98.2	88.0	91.0	98.2	88.0	91.0
0310	Concrete Forming & Accessories	89.2	97.0	96.0	98.1	70.3	74.1	100.3	72.6	76.4	86.2	71.2	73.2	91.6	80.8	82.3	89.5	71.7	74.2
0320	Concrete Reinforcing	90.1	110.1	100.1	104.0	82.1	93.0	104.5	82.1	93.3	109.0	80.5	94.7	87.7	74.6	81.2	87.0	80.6	83.8
0330	Cast-in-Place Concrete	80.3	60.3	72.0	110.8	77.1	96.8	111.3	77.9	97.4	122.5	78.0	104.0	93.6	85.7	90.3	104.0	49.2	81.2
03	CONCRETE	96.0	86.2	91.1	103.7	75.9	90.0	104.2	77.2	90.9	109.4	76.6	93.2	89.4	82.2	85.9	97.0	66.6	82.0
04	MASONRY	100.7	64.8	78.5	96.1	72.9	81.7	98.4	72.9	82.6	108.4	69.4	84.3	94.6	84.4	88.3	108.1	43.7	68.2
05	METALS	87.3	88.0	87.5	104.7	93.8	101.3	101.5	94.3	99.3	87.0	97.5	90.2	99.2	91.2	96.7	99.4	88.1	95.9
06	WOOD, PLASTICS & COMPOSITES	89.2	107.7	99.6	97.6	70.6	82.4	97.3	73.7	84.0	87.6	70.8	78.1	86.3	79.8	82.6	84.1	79.1	81.3
07	THERMAL & MOISTURE PROTECTION	100.3	66.8	86.7	101.1	79.6	92.4	99.1	80.2	91.4	102.6	78.5	92.9	99.5	80.4	91.7	99.6	60.4	83.7
08	OPENINGS	102.6	90.0	99.6	99.3	78.1	94.3	99.7	79.8	93.5	94.7	75.2	90.1	94.0	78.7	91.9	94.4	72.9	89.3
0920	Plaster & Gypsum Board	95.1	107.4	103.4	100.5	70.0	79.7	103.4	73.0	82.8	97.1	70.0	78.6	100.7	79.5	86.3	100.3	78.8	85.7
0950, 0980	Ceilings & Acoustic Treatment	89.6	107.4	101.4	100.6	70.0	80.2	98.1	73.0	81.4	98.1	70.0	79.4	100.1	79.5	86.4	100.1	78.8	85.9
0960	Flooring	97.0	56.3	84.8	100.7	78.6	94.1	100.0	78.6	93.6	97.9	80.5	92.7	96.2	92.5	95.1	95.4	52.6	82.6
0970, 0990	Wall Finishes & Painting/Coating	95.3	64.7	76.9	91.7	80.9	85.2	94.6	80.9	86.4	104.6	77.5	88.3	94.4	75.0	82.7	94.4	77.5	84.3
09	FINISHES	97.2	87.7	91.9	97.2	72.1	83.2	99.9	73.9	85.4	96.3	72.8	83.2	97.6	82.3	89.1	97.8	70.5	82.6
COVERS	DIVS. 10 - 14, 25, 28, 41, 43, 44, 46	100.0	103.3	100.7	100.0	85.6	97.1	100.0	86.3	97.3	100.0	83.8	96.8	100.0	90.2	98.0	100.0	77.2	95.4
21, 22, 23	FIRE SUPPRESSION, PLUMBING & HVAC	95.2	65.6	83.3	100.0	81.7	92.6	100.0	81.7	92.7	95.3	85.0	91.2	95.1	73.0	86.2	95.1	68.6	84.5
26, 27, 3370	ELECTRICAL, COMMUNICATIONS & UTIL.	100.6	80.0	89.8	96.7	91.7	94.1	101.4	91.7	96.3	102.2	99.8	100.9	96.6	81.6	88.8	96.1	64.4	79.5
MF2010	WEIGHTED AVERAGE	96.2	80.8	89.5	100.3	81.8	92.3	100.5	82.8	92.8	97.6	83.9	91.7	96.0	81.9	89.9	97.4	83.3	85.2

MARYLAND / MASSACHUSETTS

DIVISION		ELKTON 219			HAGERSTOWN 217			SALISBURY 218			SILVER SPRING 209			WALDORF 206			BOSTON 020-022, 024		
		MAT.	INST.	TOTAL	MAT.	INST.	TOTAL	MAT.	INST.	TOTAL	MAT.	INST.	TOTAL	MAT.	INST.	TOTAL	MAT.	INST.	TOTAL
015433	CONTRACTOR EQUIPMENT		99.3	99.3		99.3	99.3		99.3	99.3		97.2	97.2		97.2	97.2		106.3	106.3
0241, 31 - 34	SITE & INFRASTRUCTURE, DEMOLITION	85.8	88.7	87.8	89.8	91.6	91.0	98.1	88.0	91.0	91.3	87.5	88.6	98.0	87.1	90.3	100.8	108.2	106.0
0310	Concrete Forming & Accessories	95.5	82.6	84.4	90.6	76.3	78.3	103.5	52.9	59.8	94.8	71.4	74.6	102.4	69.6	74.1	100.9	142.0	136.4
0320	Concrete Reinforcing	87.0	101.6	94.3	87.7	74.6	81.2	87.0	65.6	76.3	107.6	80.4	94.0	108.3	80.4	94.4	108.7	157.1	132.8
0330	Cast-in-Place Concrete	84.2	76.5	81.0	89.2	85.7	87.8	104.0	47.9	80.7	125.4	79.8	106.5	140.5	76.9	114.1	104.6	152.2	124.3
03	CONCRETE	82.6	84.7	83.7	86.0	80.2	83.1	97.9	55.0	76.7	108.0	77.1	92.7	118.7	75.3	97.3	108.3	147.2	127.5
04	MASONRY	94.1	59.9	73.0	100.3	84.4	90.4	107.7	48.1	70.9	107.3	72.7	85.9	92.7	67.6	77.2	108.6	162.1	141.7
05	METALS	99.5	99.2	99.4	99.3	91.3	96.8	99.5	82.4	94.2	91.2	93.9	92.1	91.2	93.8	92.0	100.8	132.1	110.4
06	WOOD, PLASTICS & COMPOSITES	91.3	88.8	89.9	85.4	73.5	78.7	101.5	56.3	76.0	94.5	70.1	80.7	102.3	70.1	84.1	100.3	141.7	123.6
07	THERMAL & MOISTURE PROTECTION	99.2	75.1	89.4	99.2	78.6	90.8	99.8	65.2	85.7	105.6	84.4	97.0	106.1	82.2	96.4	104.0	151.9	123.5
08	OPENINGS	94.4	81.4	91.3	94.4	74.5	89.6	94.6	62.8	87.1	86.8	74.7	84.0	87.5	74.7	84.5	99.2	144.4	109.9
0920	Plaster & Gypsum Board	103.6	88.8	93.6	100.3	73.0	81.8	110.0	55.3	72.8	102.9	70.0	80.5	106.2	70.0	81.6	104.2	142.4	130.2
0950, 0980	Ceilings & Acoustic Treatment	100.1	88.8	92.6	101.0	73.0	82.4	100.1	55.3	70.3	107.8	70.0	82.6	107.8	70.0	82.6	101.8	142.4	128.8
0960	Flooring	97.6	59.5	86.2	95.8	92.5	94.8	101.0	65.9	90.5	104.0	80.5	97.0	107.5	80.5	99.4	98.6	183.1	123.9
0970, 0990	Wall Finishes & Painting/Coating	94.4	77.5	84.3	94.4	77.5	84.3	94.4	77.5	84.3	111.6	77.5	91.1	111.6	77.5	91.1	98.6	150.8	130.0
09	FINISHES	98.0	78.4	87.1	97.5	78.9	87.1	100.7	57.3	76.5	96.8	72.3	83.2	98.5	71.7	83.6	102.1	151.0	129.3
COVERS	DIVS. 10 - 14, 25, 28, 41, 43, 44, 46	100.0	61.4	92.3	100.0	89.5	97.9	100.0	39.7	87.9	100.0	83.1	96.6	100.0	81.4	96.3	100.0	118.6	103.7
21, 22, 23	FIRE SUPPRESSION, PLUMBING & HVAC	95.1	79.3	88.8	100.0	81.4	92.5	95.1	61.2	81.5	95.3	86.3	91.7	95.3	83.7	90.6	100.1	127.0	110.9
26, 27, 3370	ELECTRICAL, COMMUNICATIONS & UTIL.	97.9	91.7	94.6	96.4	81.6	88.7	94.8	68.7	81.2	99.4	99.8	99.6	96.6	99.8	98.3	100.9	138.6	120.6
MF2010	WEIGHTED AVERAGE	95.1	81.9	89.4	96.9	82.8	90.7	97.8	63.3	82.8	96.9	83.8	91.2	97.5	82.2	90.9	101.8	138.1	117.6

City Cost Indexes

MASSACHUSETTS

DIVISION		BROCKTON 023			BUZZARDS BAY 025			FALL RIVER 027			FITCHBURG 014			FRAMINGHAM 017			GREENFIELD 013		
		MAT.	INST.	TOTAL	MAT.	INST.	TOTAL	MAT.	INST.	TOTAL	MAT.	INST.	TOTAL	MAT.	INST.	TOTAL	MAT.	INST.	TOTAL
015433	CONTRACTOR EQUIPMENT		102.4	102.4		102.4	102.4		103.3	103.3		100.6	100.6		101.8	101.8		100.6	100.6
0241, 31 - 34	SITE & INFRASTRUCTURE, DEMOLITION	95.7	104.8	102.1	85.8	104.8	99.1	94.8	104.9	101.9	85.8	104.7	99.1	82.5	104.5	97.9	89.6	103.2	99.2
0310	Concrete Forming & Accessories	102.6	131.5	127.5	100.2	131.4	127.1	102.6	131.7	127.7	94.8	126.0	121.7	102.0	131.2	127.2	93.2	112.1	109.5
0320	Concrete Reinforcing	106.8	157.0	131.9	85.6	165.2	125.4	106.8	165.2	136.0	85.4	156.0	120.6	85.3	156.7	121.0	88.6	126.0	107.3
0330	Cast-in-Place Concrete	98.6	152.4	120.9	82.0	152.4	111.2	95.4	152.9	119.3	85.7	148.3	111.7	85.8	140.8	108.6	88.2	129.2	105.2
03	CONCRETE	104.8	142.5	123.4	88.7	144.0	115.9	103.3	144.3	123.5	86.8	138.3	112.2	89.7	138.4	113.7	90.6	119.9	105.1
04	MASONRY	102.4	159.6	137.8	95.0	159.6	134.9	103.3	159.5	138.1	93.0	155.2	131.5	98.6	155.3	133.7	96.9	133.2	119.4
05	METALS	99.4	129.7	108.8	94.4	132.9	106.3	99.4	133.3	109.9	95.6	120.0	105.0	95.6	120.0	105.9	97.8	109.2	101.3
06	WOOD, PLASTICS & COMPOSITES	102.0	129.4	117.5	98.9	129.4	116.1	102.0	129.6	117.6	95.7	122.7	111.0	102.2	129.1	117.4	93.5	109.8	102.7
07	THERMAL & MOISTURE PROTECTION	101.7	148.3	120.6	101.0	146.7	119.5	101.6	145.9	119.6	100.8	140.1	116.8	100.9	145.2	118.8	100.9	121.2	109.1
08	OPENINGS	98.1	137.7	107.4	94.0	135.4	103.8	98.1	135.5	106.9	104.5	133.9	111.5	94.7	137.6	104.8	104.5	114.7	106.9
0920	Plaster & Gypsum Board	95.4	129.7	118.7	91.1	129.7	117.3	95.4	129.7	118.7	97.3	122.8	114.6	99.9	129.7	120.1	98.0	109.4	105.8
0950, 0980	Ceilings & Acoustic Treatment	100.9	129.7	120.1	90.0	129.7	116.4	100.9	129.7	120.1	90.6	122.8	112.0	90.6	129.7	116.6	99.5	109.4	106.1
0960	Flooring	100.4	183.1	125.1	98.9	183.1	124.1	99.8	183.1	124.7	98.2	183.1	123.6	99.6	183.1	124.5	97.5	149.9	113.2
0970, 0990	Wall Finishes & Painting/Coating	96.0	135.1	119.5	96.0	135.1	119.5	96.0	135.1	119.5	95.1	135.1	119.2	96.1	135.1	119.6	95.1	108.5	103.1
09	FINISHES	100.0	141.4	123.0	96.0	141.4	121.3	99.9	141.6	123.1	96.1	137.4	119.1	96.7	141.2	121.5	98.3	118.7	109.7
COVERS	DIVS. 10 - 14, 25, 28, 41, 43, 44, 46	100.0	116.4	103.3	100.0	116.4	103.3	100.0	117.0	103.4	100.0	102.5	100.5	100.0	115.9	103.2	100.0	105.1	101.0
21, 22, 23	FIRE SUPPRESSION, PLUMBING & HVAC	100.1	111.2	104.5	95.2	111.2	101.6	100.1	111.3	104.6	95.8	113.6	103.0	95.8	120.4	105.7	95.8	100.8	97.8
26, 27, 3370	ELECTRICAL, COMMUNICATIONS & UTIL.	100.7	100.7	100.7	97.6	100.7	99.2	100.6	100.7	100.6	100.4	104.2	102.4	96.7	129.2	113.7	100.4	92.8	96.4
MF2010	WEIGHTED AVERAGE	100.4	126.3	111.7	94.9	126.7	108.8	100.2	126.8	111.8	96.3	124.6	108.6	95.5	131.0	111.0	97.5	110.2	103.0

MASSACHUSETTS

DIVISION		HYANNIS 026			LAWRENCE 019			LOWELL 018			NEW BEDFORD 027			PITTSFIELD 012			SPRINGFIELD 010 - 011		
		MAT.	INST.	TOTAL	MAT.	INST.	TOTAL	MAT.	INST.	TOTAL	MAT.	INST.	TOTAL	MAT.	INST.	TOTAL	MAT.	INST.	TOTAL
015433	CONTRACTOR EQUIPMENT		102.4	102.4		102.4	102.4		100.6	100.6		103.3	103.3		100.6	100.6		100.6	100.6
0241, 31 - 34	SITE & INFRASTRUCTURE, DEMOLITION	91.9	104.8	101.0	94.5	104.8	101.7	93.7	104.7	101.5	93.4	104.9	101.5	94.7	103.1	100.6	94.2	103.2	100.5
0310	Concrete Forming & Accessories	94.6	131.4	126.3	103.2	131.7	127.8	100.0	131.5	127.2	102.6	131.7	127.7	99.9	111.5	109.9	102.2	111.7	110.1
0320	Concrete Reinforcing	85.6	165.2	125.4	105.6	146.6	126.0	106.5	146.3	126.4	106.8	165.2	136.0	87.9	126.0	106.9	106.5	126.0	116.2
0330	Cast-in-Place Concrete	90.0	152.4	115.9	99.2	149.8	120.2	90.2	149.7	114.9	84.1	152.9	112.6	98.4	128.3	110.8	93.9	128.5	108.3
03	CONCRETE	95.0	144.0	119.2	104.9	139.8	122.1	96.3	139.5	117.6	97.9	144.3	120.8	97.2	119.3	108.1	98.0	119.4	108.6
04	MASONRY	101.4	159.6	137.4	103.3	159.6	138.1	92.4	155.2	131.3	101.3	159.5	137.3	93.0	131.5	116.8	92.7	131.9	117.0
05	METALS	95.9	132.9	107.3	98.2	125.8	106.7	98.1	123.0	105.8	99.4	133.3	109.9	98.0	109.2	101.4	100.6	109.2	103.2
06	WOOD, PLASTICS & COMPOSITES	91.9	129.4	113.1	102.7	129.4	117.7	101.9	129.4	117.4	102.0	129.6	117.6	101.8	109.8	106.4	101.9	109.8	106.4
07	THERMAL & MOISTURE PROTECTION	101.2	146.7	119.7	101.3	147.9	120.2	101.2	145.8	119.3	101.6	145.9	119.6	101.2	120.4	109.0	101.2	120.6	109.1
08	OPENINGS	94.6	135.4	104.3	98.4	134.9	107.1	105.7	134.9	112.6	98.1	135.5	106.8	105.7	114.7	107.8	105.7	114.7	107.8
0920	Plaster & Gypsum Board	86.7	129.7	115.9	102.8	129.7	121.1	102.8	129.7	121.1	95.4	129.7	118.7	102.8	109.4	107.3	102.8	109.4	107.3
0950, 0980	Ceilings & Acoustic Treatment	92.9	129.7	117.4	101.5	129.7	120.2	101.5	129.7	120.2	100.9	129.7	120.1	101.5	109.4	106.8	101.5	109.4	106.8
0960	Flooring	96.8	183.1	122.6	100.2	183.1	125.0	100.2	183.1	125.0	99.8	183.1	124.7	100.4	149.9	115.2	99.9	149.9	114.8
0970, 0990	Wall Finishes & Painting/Coating	96.0	135.1	119.5	95.2	135.1	119.2	95.1	135.1	119.2	96.0	135.1	119.5	95.1	108.5	103.1	96.5	108.5	103.7
09	FINISHES	95.8	141.4	121.2	100.4	141.4	123.2	100.3	141.4	123.2	99.7	141.6	123.0	100.4	118.3	110.4	100.3	118.4	110.4
COVERS	DIVS. 10 - 14, 25, 28, 41, 43, 44, 46	100.0	116.4	103.3	100.0	116.5	103.3	100.0	116.5	103.3	100.0	117.0	103.4	100.0	104.5	100.9	100.0	104.6	100.9
21, 22, 23	FIRE SUPPRESSION, PLUMBING & HVAC	100.1	111.2	104.5	100.0	120.4	108.2	100.0	118.1	107.3	100.1	111.3	104.6	100.0	100.0	100.0	100.0	100.2	100.1
26, 27, 3370	ELECTRICAL, COMMUNICATIONS & UTIL.	98.1	100.7	99.4	99.4	129.2	115.0	99.9	129.2	115.2	101.5	100.7	101.1	99.9	92.8	96.2	99.9	92.8	96.2
MF2010	WEIGHTED AVERAGE	97.6	126.7	110.3	100.1	131.3	113.7	99.4	130.1	112.8	99.6	126.8	111.4	99.6	109.6	103.9	100.0	109.8	104.3

MASSACHUSETTS / MICHIGAN

DIVISION		WORCESTER 015 - 016			ANN ARBOR 481			BATTLE CREEK 490			BAY CITY 487			DEARBORN 481			DETROIT 482		
		MAT.	INST.	TOTAL	MAT.	INST.	TOTAL	MAT.	INST.	TOTAL	MAT.	INST.	TOTAL	MAT.	INST.	TOTAL	MAT.	INST.	TOTAL
015433	CONTRACTOR EQUIPMENT		100.6	100.6		110.7	110.7		102.6	102.6		110.7	110.7		110.7	110.7		98.7	98.7
0241, 31 - 34	SITE & INFRASTRUCTURE, DEMOLITION	94.1	104.7	101.5	78.7	97.2	91.6	89.4	87.9	88.4	70.2	96.4	88.6	78.4	97.4	91.7	92.0	99.0	96.9
0310	Concrete Forming & Accessories	100.5	126.1	122.6	97.8	111.4	109.6	97.5	87.1	88.5	97.9	89.0	90.2	97.7	116.2	113.6	99.5	116.2	113.9
0320	Concrete Reinforcing	106.5	155.2	130.8	95.7	123.7	109.7	92.2	96.0	94.1	95.7	122.8	109.3	95.7	123.8	109.8	97.2	123.8	110.5
0330	Cast-in-Place Concrete	93.3	148.3	116.2	86.2	109.5	95.9	95.1	100.0	97.1	82.5	92.0	86.5	84.3	112.5	96.0	91.4	112.5	100.1
03	CONCRETE	97.8	138.2	117.7	92.7	113.2	102.8	96.0	92.6	94.3	91.0	97.0	94.0	91.8	116.4	103.9	95.5	115.7	105.2
04	MASONRY	92.2	155.2	131.2	97.2	108.2	104.0	99.1	86.5	91.3	96.8	87.3	90.9	97.1	113.6	107.3	95.5	113.6	106.7
05	METALS	100.6	126.0	108.4	93.7	118.2	101.2	95.8	88.6	93.6	94.3	115.0	100.7	93.7	118.5	101.4	96.3	101.2	97.8
06	WOOD, PLASTICS & COMPOSITES	102.4	122.7	113.9	94.9	112.4	104.8	100.3	86.3	92.4	94.9	88.7	91.4	94.9	116.9	107.3	97.2	116.9	108.3
07	THERMAL & MOISTURE PROTECTION	101.2	140.1	117.0	100.0	108.5	103.4	96.7	85.5	92.1	97.8	94.2	96.3	98.5	113.5	104.6	97.3	113.5	103.9
08	OPENINGS	105.7	133.7	112.3	97.7	110.0	100.6	94.2	82.7	91.5	97.7	94.8	97.0	97.7	112.4	101.1	100.0	112.1	102.8
0920	Plaster & Gypsum Board	102.8	122.8	116.4	106.2	111.8	110.0	93.0	82.0	85.6	106.2	87.3	93.4	106.2	116.4	113.2	109.0	116.4	114.0
0950, 0980	Ceilings & Acoustic Treatment	101.6	122.8	115.7	93.1	111.8	105.5	96.5	82.0	86.9	94.1	87.3	89.6	93.1	116.4	108.6	96.8	116.4	109.8
0960	Flooring	100.2	183.1	125.0	96.1	117.6	102.5	103.7	93.6	100.7	96.1	81.9	91.8	95.7	116.0	101.8	97.1	116.0	102.7
0970, 0990	Wall Finishes & Painting/Coating	95.1	135.1	119.2	86.5	100.8	95.1	97.2	81.8	87.9	86.5	83.7	84.9	86.5	99.0	94.0	88.5	99.0	94.8
09	FINISHES	100.3	137.4	121.0	94.3	111.7	104.0	97.7	87.5	92.0	94.1	86.5	89.9	94.2	114.8	105.7	96.8	114.8	106.8
COVERS	DIVS. 10 - 14, 25, 28, 41, 43, 44, 46	100.0	108.6	101.7	100.0	106.1	101.2	100.0	98.2	99.6	100.0	92.6	98.5	100.0	107.7	101.5	100.0	107.7	101.5
21, 22, 23	FIRE SUPPRESSION, PLUMBING & HVAC	100.0	113.1	105.3	100.0	99.9	100.0	100.0	87.7	95.1	100.0	84.6	93.8	100.0	108.8	103.5	100.0	109.0	103.6
26, 27, 3370	ELECTRICAL, COMMUNICATIONS & UTIL.	99.9	104.2	102.1	96.0	110.0	103.4	90.3	80.4	85.1	95.0	91.1	92.9	96.0	105.2	100.8	97.2	105.1	101.3
MF2010	WEIGHTED AVERAGE	100.0	124.6	110.7	96.4	107.9	101.4	96.8	87.4	92.7	95.8	92.5	94.4	96.2	110.9	102.6	97.9	109.3	102.8

MICHIGAN

DIVISION		FLINT 484 - 485			GAYLORD 497			GRAND RAPIDS 493,495			IRON MOUNTAIN 498 - 499			JACKSON 492			KALAMAZOO 491		
		MAT.	INST.	TOTAL	MAT.	INST.	TOTAL	MAT.	INST.	TOTAL	MAT.	INST.	TOTAL	MAT.	INST.	TOTAL	MAT.	INST.	TOTAL
015433	CONTRACTOR EQUIPMENT		110.7	110.7		105.1	105.1		102.6	102.6		94.2	94.2		105.1	105.1		102.6	102.6
0241, 31 - 34	SITE & INFRASTRUCTURE, DEMOLITION	68.2	96.5	88.1	83.4	85.7	85.0	89.5	88.0	88.4	92.1	94.1	93.5	105.8	88.0	93.3	89.7	87.9	88.4
0310	Concrete Forming & Accessories	100.6	91.4	92.7	95.4	77.7	80.1	97.0	84.7	86.4	87.5	86.1	86.3	92.4	88.3	88.9	97.5	86.8	88.2
0320	Concrete Reinforcing	95.7	123.2	109.5	85.9	109.3	97.6	98.6	95.9	97.2	85.7	95.4	90.6	83.5	123.1	103.3	92.2	93.6	92.9
0330	Cast-in-Place Concrete	86.8	94.7	90.1	94.9	87.8	92.0	100.1	97.6	99.1	112.3	87.2	101.9	94.8	93.8	94.4	97.0	99.8	98.2
03	CONCRETE	93.2	99.1	96.1	92.4	88.2	90.3	99.3	90.7	95.1	101.8	88.4	95.2	87.2	97.3	92.2	99.4	92.0	95.8
04	MASONRY	97.3	96.8	97.0	109.6	80.5	91.6	96.1	86.6	90.2	95.6	87.6	90.7	89.7	92.8	91.7	95.8	86.5	90.8
05	METALS	93.7	115.7	100.5	97.2	111.7	101.6	93.0	88.2	91.5	96.6	91.1	94.9	97.3	113.6	102.4	95.8	87.5	93.3
06	WOOD, PLASTICS & COMPOSITES	98.2	89.9	93.5	92.8	76.9	83.8	101.5	83.1	91.1	87.5	86.1	86.7	91.5	86.6	88.7	100.3	86.3	92.4
07	THERMAL & MOISTURE PROTECTION	97.8	97.9	97.9	95.1	76.2	87.4	99.5	77.4	90.5	98.6	83.1	92.3	94.5	94.5	94.5	96.7	85.5	92.1
08	OPENINGS	97.7	95.5	97.1	94.8	76.0	90.3	100.6	85.6	97.0	101.3	77.3	95.7	93.8	93.9	93.8	94.2	82.1	91.3
0920	Plaster & Gypsum Board	107.7	88.6	94.7	92.8	74.9	80.6	94.9	78.8	83.9	53.7	86.1	75.7	91.0	84.9	86.9	93.0	82.0	85.6
0950, 0980	Ceilings & Acoustic Treatment	93.1	88.6	90.1	95.7	74.9	81.9	105.1	78.8	87.6	94.6	86.1	88.9	95.7	84.9	88.5	96.5	82.0	86.9
0960	Flooring	96.1	94.9	95.7	95.7	93.9	95.1	103.6	84.3	97.8	119.6	95.6	112.4	94.4	84.3	91.4	103.7	77.1	95.8
0970, 0990	Wall Finishes & Painting/Coating	86.5	87.3	87.0	93.1	83.7	87.5	98.9	80.0	87.1	111.4	71.8	87.6	93.1	98.8	96.5	97.2	81.8	87.9
09	FINISHES	93.6	90.7	92.0	97.5	79.8	87.7	101.7	83.8	91.7	99.9	86.4	92.3	98.6	88.0	92.7	97.7	84.2	90.2
COVERS	DIVS. 10 - 14, 25, 28, 41, 43, 44, 46	100.0	93.9	98.8	100.0	90.7	98.1	100.0	98.2	99.6	100.0	89.5	97.9	100.0	93.7	98.7	100.0	98.2	99.6
21, 22, 23	FIRE SUPPRESSION, PLUMBING & HVAC	100.0	90.6	96.2	95.5	80.1	89.3	100.0	83.8	93.5	95.4	86.0	91.6	95.5	88.8	92.8	100.0	81.6	92.6
26, 27, 3370	ELECTRICAL, COMMUNICATIONS & UTIL.	96.0	98.3	97.2	88.4	78.9	83.4	95.2	72.3	83.3	94.5	86.3	90.2	92.2	110.0	101.5	90.2	83.7	86.8
MF2010	WEIGHTED AVERAGE	96.1	96.8	96.4	95.5	84.4	90.7	98.1	84.5	92.2	97.5	87.4	93.1	95.0	95.9	95.4	97.1	85.9	92.2

MICHIGAN / MINNESOTA

DIVISION		LANSING 488 - 489			MUSKEGON 494			ROYAL OAK 480,483			SAGINAW 486			TRAVERSE CITY 496			BEMIDJI 566		
		MAT.	INST.	TOTAL	MAT.	INST.	TOTAL	MAT.	INST.	TOTAL	MAT.	INST.	TOTAL	MAT.	INST.	TOTAL	MAT.	INST.	TOTAL
015433	CONTRACTOR EQUIPMENT		110.7	110.7		102.6	102.6		96.2	96.2		110.7	110.7		94.2	94.2		99.0	99.0
0241, 31 - 34	SITE & INFRASTRUCTURE, DEMOLITION	90.5	96.6	94.8	87.5	87.9	87.8	83.0	96.8	92.7	71.2	96.4	88.9	78.4	93.6	89.0	94.0	98.3	97.0
0310	Concrete Forming & Accessories	98.0	89.2	90.4	97.9	85.6	87.3	93.6	112.7	110.1	97.8	88.8	90.0	87.5	75.7	77.3	88.7	91.4	91.0
0320	Concrete Reinforcing	99.0	123.0	111.0	92.8	96.1	94.5	86.8	123.8	105.2	95.7	122.8	109.3	87.0	92.9	90.0	97.0	108.0	102.5
0330	Cast-in-Place Concrete	100.5	92.9	97.3	94.8	95.8	95.2	75.5	108.3	89.1	85.2	91.9	88.0	87.6	80.0	84.5	104.1	104.2	104.1
03	CONCRETE	100.0	97.4	98.7	94.3	90.6	92.5	79.9	112.2	95.8	92.2	96.9	94.5	83.9	80.8	82.4	96.9	100.0	98.4
04	MASONRY	95.3	93.3	94.1	96.3	89.4	92.0	92.0	112.1	104.5	98.7	87.3	91.6	93.7	82.3	86.7	97.2	104.8	101.9
05	METALS	91.4	115.1	98.7	93.6	88.5	92.0	96.8	99.4	97.6	93.8	114.8	100.3	96.5	89.7	94.4	90.2	121.0	99.7
06	WOOD, PLASTICS & COMPOSITES	96.4	88.0	91.6	96.5	85.2	90.1	90.4	113.1	103.2	91.2	88.7	89.8	87.5	75.2	80.6	71.5	87.5	80.6
07	THERMAL & MOISTURE PROTECTION	98.6	90.6	95.3	95.8	80.4	89.5	96.5	112.3	102.9	98.6	94.2	96.9	97.8	72.8	87.6	105.6	97.1	102.1
08	OPENINGS	98.7	94.4	97.7	93.4	86.5	91.8	97.6	110.4	100.6	95.6	94.8	95.4	101.3	71.4	94.3	97.5	109.2	100.2
0920	Plaster & Gypsum Board	92.2	86.6	88.4	74.3	80.9	78.8	103.2	112.5	109.5	106.2	87.3	93.4	53.7	74.9	68.1	100.6	87.4	91.6
0950, 0980	Ceilings & Acoustic Treatment	96.8	86.6	90.0	97.4	80.9	86.4	92.3	112.5	105.8	93.1	87.3	89.3	94.6	74.9	81.5	137.1	87.4	104.0
0960	Flooring	102.8	84.3	97.2	102.5	84.3	97.1	93.7	116.0	100.4	96.1	81.9	91.8	119.6	93.9	111.9	107.8	123.0	112.3
0970, 0990	Wall Finishes & Painting/Coating	96.3	81.5	87.4	95.6	80.0	86.3	87.9	98.8	94.5	86.5	83.7	84.9	111.4	45.6	71.9	95.3	97.9	96.9
09	FINISHES	98.2	87.0	91.9	94.3	84.0	88.6	93.3	112.2	103.8	93.9	86.5	89.8	98.6	74.3	85.1	111.3	97.1	103.4
COVERS	DIVS. 10 - 14, 25, 28, 41, 43, 44, 46	100.0	93.3	98.7	100.0	98.0	99.6	100.0	104.6	100.9	100.0	92.6	98.5	100.0	86.8	97.4	100.0	99.3	99.9
21, 22, 23	FIRE SUPPRESSION, PLUMBING & HVAC	99.9	88.9	95.5	99.9	83.1	93.1	95.4	108.0	100.5	100.0	84.2	93.7	95.4	79.4	89.0	95.5	84.8	91.2
26, 27, 3370	ELECTRICAL, COMMUNICATIONS & UTIL.	99.2	91.4	95.1	90.7	77.2	83.6	98.3	104.5	101.5	93.5	91.1	92.2	89.9	78.9	84.2	104.4	105.2	104.8
MF2010	WEIGHTED AVERAGE	97.7	94.1	96.1	95.6	85.5	91.2	94.2	107.5	100.0	95.7	92.4	94.2	94.6	81.0	88.6	97.7	99.5	98.5

MINNESOTA

DIVISION		BRAINERD 564			DETROIT LAKES 565			DULUTH 556 - 558			MANKATO 560			MINNEAPOLIS 553 - 555			ROCHESTER 559		
		MAT.	INST.	TOTAL	MAT.	INST.	TOTAL	MAT.	INST.	TOTAL	MAT.	INST.	TOTAL	MAT.	INST.	TOTAL	MAT.	INST.	TOTAL
015433	CONTRACTOR EQUIPMENT		101.6	101.6		99.0	99.0		100.4	100.4		101.6	101.6		103.8	103.8		100.4	100.4
0241, 31 - 34	SITE & INFRASTRUCTURE, DEMOLITION	94.2	103.0	100.4	92.2	98.6	96.7	96.3	100.4	99.2	91.1	103.7	99.9	97.9	105.6	103.3	97.0	100.0	99.1
0310	Concrete Forming & Accessories	90.6	93.6	93.2	85.7	93.5	92.5	100.4	112.2	110.6	98.4	101.1	100.7	101.2	127.4	123.8	102.8	106.6	106.1
0320	Concrete Reinforcing	95.9	108.2	102.0	97.0	108.0	102.5	102.6	108.5	105.5	95.8	118.2	107.0	97.2	118.5	107.8	94.8	118.2	106.5
0330	Cast-in-Place Concrete	113.1	108.2	111.1	101.1	107.5	103.7	107.4	107.0	107.2	104.2	109.0	106.2	110.5	117.3	113.3	100.7	107.0	104.1
03	CONCRETE	100.8	102.4	101.6	94.5	102.1	98.2	101.3	110.3	105.7	96.5	107.9	102.1	103.4	122.6	112.9	100.7	107.7	104.1
04	MASONRY	119.6	113.5	115.8	119.9	110.8	114.2	104.6	117.0	112.3	108.6	110.5	109.8	114.7	125.3	121.3	107.0	112.5	110.4
05	METALS	91.3	121.0	100.5	90.2	120.9	99.7	93.4	123.3	102.6	91.2	126.7	102.1	98.6	130.0	108.3	95.2	128.4	105.4
06	WOOD, PLASTICS & COMPOSITES	89.5	87.3	88.3	68.5	87.5	79.3	103.6	111.9	108.3	99.0	98.7	98.8	109.0	126.0	118.6	109.0	105.2	106.9
07	THERMAL & MOISTURE PROTECTION	104.7	109.9	106.8	105.4	110.1	107.3	100.1	116.9	106.9	105.0	100.6	103.2	102.1	128.6	112.9	103.5	102.0	102.9
08	OPENINGS	84.8	109.0	90.5	97.4	109.2	100.2	98.3	118.2	103.0	89.1	117.7	95.9	94.9	133.8	104.1	93.5	122.5	100.4
0920	Plaster & Gypsum Board	84.6	87.4	86.5	99.5	87.4	91.3	91.1	112.6	105.8	89.5	99.1	96.0	99.3	127.0	103.8	99.5	105.8	103.8
0950, 0980	Ceilings & Acoustic Treatment	66.7	87.4	80.5	137.1	87.4	104.0	98.7	112.6	108.0	66.7	99.1	88.3	98.5	127.0	117.5	95.0	105.8	102.2
0960	Flooring	106.7	123.0	111.6	106.6	123.0	111.5	107.1	124.3	112.3	108.3	123.0	112.7	98.3	123.0	105.7	103.2	123.0	109.1
0970, 0990	Wall Finishes & Painting/Coating	89.6	97.9	94.6	95.3	97.9	96.9	91.0	111.3	103.2	101.0	105.3	103.6	96.5	128.1	115.5	91.3	105.3	99.7
09	FINISHES	93.1	98.5	96.1	110.7	98.6	104.0	101.4	114.7	108.8	94.8	105.3	100.7	100.5	127.2	115.4	99.4	109.5	105.0
COVERS	DIVS. 10 - 14, 25, 28, 41, 43, 44, 46	100.0	100.9	100.2	100.0	101.2	100.2	100.0	100.1	100.0	100.0	101.1	100.2	100.0	109.1	101.8	100.0	102.6	100.5
21, 22, 23	FIRE SUPPRESSION, PLUMBING & HVAC	94.8	87.8	92.0	95.5	87.6	92.3	99.9	97.2	98.8	94.8	88.5	92.3	99.9	116.9	106.7	100.0	96.0	98.4
26, 27, 3370	ELECTRICAL, COMMUNICATIONS & UTIL.	101.9	104.9	103.4	104.1	69.1	85.8	98.4	104.9	101.8	108.8	89.2	98.5	101.8	109.0	105.6	98.2	89.2	93.5
MF2010	WEIGHTED AVERAGE	96.3	102.3	98.9	98.3	96.7	97.6	99.0	108.6	103.2	96.6	102.4	99.1	100.6	119.9	109.0	98.8	104.9	101.5

		MINNESOTA																MISSISSIPPI		
	DIVISION	SAINT PAUL			ST. CLOUD			THIEF RIVER FALLS			WILLMAR			WINDOM			BILOXI		
		550 - 551			563			567			562			561			395		
		MAT.	INST.	TOTAL	MAT.	INST.	TOTAL	MAT.	INST.	TOTAL	MAT.	INST.	TOTAL	MAT.	INST.	TOTAL	MAT.	INST.	TOTAL
015433	CONTRACTOR EQUIPMENT		100.4	100.4		101.6	101.6		99.0	99.0		101.6	101.6		101.6	101.6		99.3	99.3
0241, 31 - 34	SITE & INFRASTRUCTURE, DEMOLITION	99.1	101.1	100.5	89.7	105.1	100.5	92.9	98.2	96.7	89.1	102.8	98.7	83.3	101.7	96.3	98.8	89.6	92.3
0310	Concrete Forming & Accessories	102.0	124.9	121.8	87.7	120.4	116.0	89.4	90.8	90.6	87.5	93.1	92.3	92.0	86.8	87.5	95.6	54.1	59.8
0320	Concrete Reinforcing	98.5	118.5	108.5	95.9	118.4	107.1	97.4	107.9	102.7	95.6	118.0	106.8	95.6	117.1	106.4	95.8	58.4	77.1
0330	Cast-in-Place Concrete	120.2	116.7	118.7	99.9	115.8	106.5	103.2	89.5	97.5	101.4	86.9	95.4	87.7	67.8	79.4	108.7	54.8	86.3
03	CONCRETE	110.6	121.3	115.9	92.3	118.9	105.4	95.6	94.6	95.1	92.2	96.7	94.4	83.2	87.2	85.2	100.0	56.8	78.7
04	MASONRY	115.6	125.3	121.6	105.3	122.1	115.7	97.2	104.8	101.9	109.7	113.5	112.0	119.4	91.2	102.0	90.3	45.1	62.4
05	METALS	94.1	129.8	105.1	92.0	128.0	103.1	90.3	120.3	99.6	91.1	125.8	101.8	91.0	123.5	101.0	91.9	82.4	89.0
06	WOOD, PLASTICS & COMPOSITES	102.8	122.7	114.0	86.8	117.6	104.2	72.5	87.5	81.0	86.6	87.5	87.1	90.8	84.8	87.4	102.3	57.4	77.0
07	THERMAL & MOISTURE PROTECTION	101.8	128.1	112.5	104.8	120.1	111.0	106.5	100.0	103.9	104.6	110.8	107.1	104.6	91.0	99.1	97.0	55.4	80.1
08	OPENINGS	97.1	132.0	105.3	89.1	129.2	98.6	97.5	109.2	100.2	86.7	93.5	88.3	90.2	92.0	90.6	97.2	58.6	88.0
0920	Plaster & Gypsum Board	87.3	123.8	112.1	84.6	118.6	107.7	100.2	87.4	91.5	84.6	87.6	86.7	84.6	84.8	84.7	104.4	56.5	71.8
0950, 0980	Ceilings & Acoustic Treatment	98.5	123.8	115.4	66.7	118.6	101.2	137.1	87.4	104.0	66.7	87.6	80.6	66.7	84.8	78.7	95.9	56.5	69.7
0960	Flooring	99.5	123.0	106.5	103.3	123.0	109.2	107.5	123.0	112.1	105.1	123.0	110.4	107.3	123.0	112.0	114.4	56.1	97.0
0970, 0990	Wall Finishes & Painting/Coating	99.9	127.1	116.3	101.0	128.1	117.3	95.3	97.9	96.9	95.3	97.9	96.9	95.3	105.3	101.3	94.7	44.4	64.5
09	FINISHES	99.5	125.1	113.8	92.6	121.9	108.9	111.1	97.1	103.3	92.7	96.2	94.6	92.8	93.2	93.0	100.8	53.6	74.5
COVERS	DIVS. 10 - 14, 25, 28, 41, 43, 44, 46	100.0	108.5	101.7	100.0	107.2	101.4	100.0	99.2	99.8	100.0	100.8	100.2	100.0	95.6	99.1	100.0	56.9	91.4
21, 22, 23	FIRE SUPPRESSION, PLUMBING & HVAC	99.9	114.9	105.9	99.7	116.3	106.3	95.5	84.5	91.1	94.8	108.4	100.2	94.8	85.3	91.0	100.0	50.1	80.0
26, 27, 3370	ELECTRICAL, COMMUNICATIONS & UTIL.	95.2	111.4	103.7	101.9	111.4	106.9	101.4	69.1	84.6	101.9	87.2	94.2	108.8	89.1	98.5	98.5	57.6	77.1
MF2010	WEIGHTED AVERAGE	100.2	118.8	108.3	96.3	117.8	105.7	97.3	93.8	95.7	94.9	103.0	98.4	95.3	93.3	94.4	97.7	59.1	80.9

		MISSISSIPPI																		
	DIVISION	CLARKSDALE			COLUMBUS			GREENVILLE			GREENWOOD			JACKSON			LAUREL		
		386			397			387			389			390 - 392			394		
		MAT.	INST.	TOTAL	MAT.	INST.	TOTAL	MAT.	INST.	TOTAL	MAT.	INST.	TOTAL	MAT.	INST.	TOTAL	MAT.	INST.	TOTAL
015433	CONTRACTOR EQUIPMENT		99.3	99.3		99.3	99.3		99.3	99.3		99.3	99.3		99.3	99.3		99.3	99.3
0241, 31 - 34	SITE & INFRASTRUCTURE, DEMOLITION	95.6	89.4	91.2	97.1	89.6	91.9	101.3	89.6	93.1	98.3	89.1	91.8	95.9	89.6	91.5	102.9	89.5	93.5
0310	Concrete Forming & Accessories	85.3	50.4	55.2	84.4	52.4	56.8	82.0	65.2	67.5	94.3	50.4	56.4	94.6	58.9	63.8	84.4	54.2	58.3
0320	Concrete Reinforcing	100.0	42.9	71.5	102.7	43.9	73.3	100.6	62.2	81.4	100.0	54.2	77.1	101.2	57.9	79.4	103.3	40.3	71.9
0330	Cast-in-Place Concrete	103.0	54.4	82.8	110.8	59.5	89.5	106.0	56.5	85.5	110.7	54.1	87.2	98.9	57.8	81.8	108.4	56.1	86.7
03	CONCRETE	95.9	52.1	74.3	101.6	54.9	78.6	101.1	63.0	82.3	102.1	53.9	78.3	96.1	59.7	78.2	104.0	53.9	79.3
04	MASONRY	93.5	45.8	64.0	114.7	52.5	76.2	138.0	50.3	83.8	94.1	45.6	64.1	95.7	50.3	67.6	110.9	49.0	72.6
05	METALS	90.5	74.9	85.7	88.9	75.8	84.9	91.5	84.1	89.3	90.5	76.4	86.2	98.5	82.3	93.5	89.0	74.8	84.6
06	WOOD, PLASTICS & COMPOSITES	85.6	52.4	66.9	87.2	53.4	68.1	82.0	69.7	75.1	98.7	52.4	72.6	99.6	61.0	77.8	88.3	55.9	70.0
07	THERMAL & MOISTURE PROTECTION	95.9	48.0	76.5	97.0	52.3	78.8	96.2	61.0	81.9	96.3	47.9	76.6	97.1	60.3	82.1	97.1	54.0	79.6
08	OPENINGS	96.8	50.6	85.8	96.5	50.7	85.7	96.8	62.9	88.8	96.8	51.9	86.2	100.1	56.8	90.5	93.5	51.5	83.6
0920	Plaster & Gypsum Board	87.7	51.4	63.0	92.7	52.4	65.3	87.3	69.2	75.0	98.9	51.4	66.6	97.4	60.2	72.1	92.7	54.9	67.0
0950, 0980	Ceilings & Acoustic Treatment	87.8	51.4	63.6	90.3	52.4	65.1	89.7	69.2	76.1	87.8	51.4	63.6	98.4	60.2	73.0	90.3	54.9	66.8
0960	Flooring	111.6	53.8	94.3	108.2	60.3	93.9	109.9	53.8	93.1	117.1	53.8	98.2	111.9	56.1	95.2	106.8	53.8	91.0
0970, 0990	Wall Finishes & Painting/Coating	97.4	47.1	67.2	94.7	54.1	70.3	97.4	67.3	79.3	97.4	47.1	67.2	91.7	67.3	77.0	94.7	62.2	75.2
09	FINISHES	96.5	51.0	71.1	96.3	54.4	73.0	97.0	64.3	78.8	99.9	51.0	72.7	100.6	59.6	77.8	96.4	55.5	73.6
COVERS	DIVS. 10 - 14, 25, 28, 41, 43, 44, 46	100.0	57.2	91.4	100.0	58.4	91.7	100.0	60.0	92.0	100.0	57.2	91.4	100.0	59.1	91.8	100.0	38.5	87.7
21, 22, 23	FIRE SUPPRESSION, PLUMBING & HVAC	97.9	42.2	75.6	97.3	38.3	73.7	100.0	48.6	79.4	97.9	40.2	74.7	100.0	58.3	83.3	97.4	46.8	77.1
26, 27, 3370	ELECTRICAL, COMMUNICATIONS & UTIL.	95.6	47.5	70.5	95.7	50.3	72.0	95.6	64.2	79.2	95.6	44.5	68.9	100.1	64.2	81.3	97.3	58.3	77.0
MF2010	WEIGHTED AVERAGE	95.8	53.8	77.5	97.1	55.1	78.8	99.4	63.1	83.6	97.0	53.3	78.0	99.1	63.3	83.6	97.2	57.1	79.7

		MISSISSIPPI									MISSOURI								
	DIVISION	MCCOMB			MERIDIAN			TUPELO			BOWLING GREEN			CAPE GIRARDEAU			CHILLICOTHE		
		396			393			388			633			637			646		
		MAT.	INST.	TOTAL	MAT.	INST.	TOTAL	MAT.	INST.	TOTAL	MAT.	INST.	TOTAL	MAT.	INST.	TOTAL	MAT.	INST.	TOTAL
015433	CONTRACTOR EQUIPMENT		99.3	99.3		99.3	99.3		99.3	99.3		106.6	106.6		106.6	106.6		102.1	102.1
0241, 31 - 34	SITE & INFRASTRUCTURE, DEMOLITION	90.3	89.2	89.6	94.3	89.9	91.2	93.1	89.3	90.4	90.1	95.1	93.6	91.7	95.0	94.1	105.0	94.6	97.7
0310	Concrete Forming & Accessories	84.4	52.2	56.6	81.7	64.8	67.1	82.5	52.5	56.6	94.2	88.8	89.5	87.3	75.1	76.8	89.9	95.2	94.5
0320	Concrete Reinforcing	103.9	42.3	73.1	102.7	57.5	80.1	97.9	54.2	76.1	103.2	100.8	102.0	104.5	88.8	96.7	97.4	110.0	103.7
0330	Cast-in-Place Concrete	96.2	53.9	78.7	102.9	60.1	85.2	103.0	55.8	83.4	92.6	85.4	89.6	91.6	87.3	89.8	97.9	87.8	93.7
03	CONCRETE	91.6	52.5	72.3	95.8	63.2	79.7	95.4	55.4	75.7	95.9	91.0	93.5	95.0	83.4	89.3	102.1	95.9	99.1
04	MASONRY	116.2	45.1	72.2	90.0	54.5	68.0	127.9	48.7	78.9	122.1	97.9	107.1	118.5	79.4	94.1	103.6	100.6	101.7
05	METALS	89.1	73.1	84.2	90.0	82.5	87.7	90.4	76.6	86.2	96.1	113.8	101.6	97.2	107.3	100.3	96.9	110.5	101.1
06	WOOD, PLASTICS & COMPOSITES	87.2	55.3	69.2	84.6	66.7	74.5	82.8	53.4	66.2	95.3	88.0	91.2	88.1	71.5	78.7	96.0	95.4	95.7
07	THERMAL & MOISTURE PROTECTION	96.6	47.2	76.6	96.8	62.6	82.9	95.9	53.8	78.8	103.2	98.8	101.4	103.1	83.5	95.1	97.8	97.0	97.4
08	OPENINGS	96.6	52.2	86.1	96.5	65.5	89.2	96.8	51.8	86.1	97.0	98.5	97.3	97.0	75.5	91.9	96.0	98.1	96.5
0920	Plaster & Gypsum Board	92.7	54.3	66.6	92.7	66.1	74.6	87.3	52.4	63.6	97.9	87.6	90.9	96.4	70.6	78.8	97.4	95.1	95.8
0950, 0980	Ceilings & Acoustic Treatment	90.3	54.3	66.4	92.3	66.1	74.8	87.8	52.4	64.2	97.1	87.6	90.8	97.1	70.6	79.5	92.2	95.1	94.1
0960	Flooring	108.2	53.8	91.9	106.7	56.1	91.6	110.1	53.8	93.3	98.4	98.5	98.4	95.5	85.6	92.3	96.0	104.4	98.5
0970, 0990	Wall Finishes & Painting/Coating	94.7	43.2	63.8	94.7	77.8	84.5	97.4	60.0	74.9	102.3	102.8	102.6	102.3	74.8	85.8	96.2	91.3	93.3
09	FINISHES	95.7	52.1	71.4	96.0	65.2	78.8	96.0	53.8	72.5	100.8	91.2	95.4	99.6	75.4	86.1	99.9	96.4	97.9
COVERS	DIVS. 10 - 14, 25, 28, 41, 43, 44, 46	100.0	60.7	92.1	100.0	61.1	92.2	100.0	58.4	91.7	100.0	90.2	98.0	100.0	92.3	98.5	100.0	90.9	98.2
21, 22, 23	FIRE SUPPRESSION, PLUMBING & HVAC	97.3	35.3	72.5	100.0	60.4	84.1	98.1	44.2	76.5	95.0	97.7	96.1	99.9	96.6	98.6	95.2	99.2	96.8
26, 27, 3370	ELECTRICAL, COMMUNICATIONS & UTIL.	94.1	50.6	71.4	97.3	58.2	76.9	95.4	50.4	71.9	98.0	80.8	89.1	98.0	103.9	101.1	93.7	80.5	86.8
MF2010	WEIGHTED AVERAGE	95.7	52.9	77.1	96.2	65.3	82.7	97.3	56.1	79.4	98.1	94.7	96.6	99.1	90.6	95.4	97.7	96.2	97.1

MISSOURI

DIVISION		COLUMBIA 652			FLAT RIVER 636			HANNIBAL 634			HARRISONVILLE 647			JEFFERSON CITY 650 - 651			JOPLIN 648		
		MAT.	INST.	TOTAL	MAT.	INST.	TOTAL	MAT.	INST.	TOTAL	MAT.	INST.	TOTAL	MAT.	INST.	TOTAL	MAT.	INST.	TOTAL
015433	CONTRACTOR EQUIPMENT		107.7	107.7		106.6	106.6		106.6	106.6		102.1	102.1		107.7	107.7		105.8	105.8
0241, 31 - 34	SITE & INFRASTRUCTURE, DEMOLITION	100.0	96.6	97.6	92.8	95.2	94.5	88.0	95.1	93.0	96.1	95.6	95.8	99.2	96.6	97.4	105.6	99.4	101.3
0310	Concrete Forming & Accessories	85.5	80.0	80.7	100.1	86.2	88.1	92.4	79.3	81.1	87.3	100.9	99.1	95.6	80.0	82.1	101.6	73.7	77.5
0320	Concrete Reinforcing	98.1	116.3	107.2	104.5	108.0	106.3	102.7	100.8	101.7	97.0	110.1	103.5	100.4	113.1	106.7	100.6	101.7	101.1
0330	Cast-in-Place Concrete	89.8	87.1	88.7	95.6	94.2	95.0	87.6	85.2	86.6	100.3	106.1	102.7	95.3	87.1	91.9	106.0	79.2	94.9
03	CONCRETE	86.8	90.6	88.7	98.8	94.2	96.6	92.2	86.8	89.5	98.5	104.8	101.6	91.9	90.0	91.0	101.7	81.7	91.8
04	MASONRY	133.2	88.7	105.7	119.0	80.5	95.2	113.5	97.9	103.8	98.3	105.3	102.6	97.0	88.7	91.9	96.6	86.8	90.5
05	METALS	97.1	119.7	104.1	96.0	115.8	102.1	96.1	113.5	101.5	97.3	111.5	101.7	96.6	118.4	103.3	100.0	102.2	100.7
06	WOOD, PLASTICS & COMPOSITES	95.4	76.3	84.6	103.2	84.5	92.7	93.3	75.5	83.2	92.5	99.8	96.6	102.0	76.3	87.5	107.5	71.8	87.4
07	THERMAL & MOISTURE PROTECTION	95.5	88.1	92.5	103.4	93.8	99.5	103.0	96.9	100.5	97.1	105.8	100.6	100.8	88.1	95.7	97.1	79.6	90.0
08	OPENINGS	94.7	93.3	94.4	96.9	98.8	97.4	97.0	84.3	94.0	96.1	102.9	97.7	94.2	92.5	93.8	97.2	78.2	92.7
0920	Plaster & Gypsum Board	85.3	75.5	78.7	103.8	84.1	90.4	96.7	74.7	81.8	92.5	99.6	97.3	89.2	75.5	79.9	103.5	70.7	81.2
0950, 0980	Ceilings & Acoustic Treatment	97.9	75.5	83.0	97.1	84.1	88.4	97.1	74.7	82.2	92.2	99.6	97.1	102.1	75.5	84.4	93.0	70.7	78.2
0960	Flooring	109.7	97.5	106.1	101.3	85.6	96.6	97.7	98.5	98.0	91.7	104.4	95.5	116.0	97.5	110.5	121.0	76.3	107.6
0970, 0990	Wall Finishes & Painting/Coating	102.4	78.8	88.2	102.3	80.0	88.9	102.3	102.8	102.6	101.1	105.3	103.6	99.6	78.8	87.1	95.8	76.5	84.2
09	FINISHES	97.5	81.6	88.6	102.5	84.1	92.3	100.2	83.8	91.1	97.5	101.7	99.9	101.7	81.6	90.5	106.1	74.5	88.4
COVERS	DIVS. 10 - 14, 25, 28, 41, 43, 44, 46	100.0	95.6	99.1	100.0	94.3	98.8	100.0	88.8	97.8	100.0	93.1	98.6	100.0	95.6	99.1	100.0	89.6	97.9
21, 22, 23	FIRE SUPPRESSION, PLUMBING & HVAC	99.9	97.8	99.0	95.0	97.6	96.1	95.0	97.6	96.1	95.1	100.5	97.2	99.9	98.7	99.4	100.1	72.7	89.1
26, 27, 3370	ELECTRICAL, COMMUNICATIONS & UTIL.	98.4	86.6	92.2	102.4	104.0	103.2	96.8	82.9	89.6	100.4	99.7	100.1	104.3	86.6	95.1	91.5	69.7	80.1
MF2010	WEIGHTED AVERAGE	98.6	93.5	96.4	99.0	95.8	97.6	97.1	92.6	95.1	97.3	102.2	99.5	98.4	93.4	96.2	99.5	81.0	91.5

MISSOURI

DIVISION		KANSAS CITY 640 - 641			KIRKSVILLE 635			POPLAR BLUFF 639			ROLLA 654 - 655			SEDALIA 653			SIKESTON 638		
		MAT.	INST.	TOTAL	MAT.	INST.	TOTAL	MAT.	INST.	TOTAL	MAT.	INST.	TOTAL	MAT.	INST.	TOTAL	MAT.	INST.	TOTAL
015433	CONTRACTOR EQUIPMENT		103.4	103.4		99.0	99.0		101.3	101.3		107.7	107.7		99.7	99.7		101.3	101.3
0241, 31 - 34	SITE & INFRASTRUCTURE, DEMOLITION	98.2	98.0	98.0	90.9	91.2	91.1	78.4	95.0	90.1	99.0	96.6	97.3	94.9	92.5	93.2	81.7	95.1	91.1
0310	Concrete Forming & Accessories	100.3	107.8	106.8	85.2	79.3	80.1	85.3	75.0	76.4	92.7	94.0	93.8	97.1	109.4	103.2	105.9	101.4	103.6
0320	Concrete Reinforcing	95.5	117.1	106.3	103.4	101.9	102.6	106.5	97.6	102.0	98.6	97.7	98.2	97.1	109.4	103.2	105.9	101.4	103.6
0330	Cast-in-Place Concrete	98.6	107.8	102.4	95.5	85.1	91.2	73.5	87.8	79.5	91.8	96.4	93.7	95.9	84.5	91.2	78.6	87.8	82.4
03	CONCRETE	97.8	109.7	103.7	110.2	86.4	98.5	86.0	84.6	85.3	88.6	96.5	92.5	101.1	87.6	94.4	89.6	85.3	87.5
04	MASONRY	100.4	109.1	105.8	125.6	88.7	102.8	117.1	79.1	93.6	109.4	89.4	97.0	115.2	86.6	97.5	116.8	79.1	93.5
05	METALS	107.7	114.9	109.9	95.7	105.2	98.6	96.3	102.9	98.3	96.5	111.9	101.3	95.1	109.3	99.5	96.6	104.4	99.0
06	WOOD, PLASTICS & COMPOSITES	106.7	107.5	107.1	81.1	76.3	78.4	80.3	71.5	75.3	103.0	94.9	98.5	95.4	76.3	84.6	81.9	71.5	76.0
07	THERMAL & MOISTURE PROTECTION	97.2	108.6	101.8	109.8	89.1	101.4	108.3	85.3	99.0	95.6	95.5	95.6	101.8	93.7	98.5	108.5	83.9	98.5
08	OPENINGS	103.5	109.8	105.0	102.0	85.6	98.1	103.0	77.9	97.1	94.7	90.4	93.7	99.0	87.3	96.2	103.0	78.9	97.3
0920	Plaster & Gypsum Board	100.2	107.5	105.1	91.8	75.5	80.7	92.3	70.6	77.5	87.2	94.7	92.3	80.4	75.5	77.1	94.1	70.6	78.1
0950, 0980	Ceilings & Acoustic Treatment	99.3	107.5	104.7	95.3	75.5	82.1	97.1	70.6	79.5	97.9	94.7	95.8	97.9	75.5	83.0	97.1	70.6	79.5
0960	Flooring	96.6	109.3	100.4	76.5	97.5	82.8	90.2	83.5	88.2	113.0	97.5	108.3	86.4	97.5	89.7	90.6	83.5	88.5
0970, 0990	Wall Finishes & Painting/Coating	101.1	112.6	108.0	97.6	80.4	87.2	97.1	74.8	83.7	102.4	102.8	102.7	102.4	102.4	102.4	97.1	74.8	83.7
09	FINISHES	101.4	108.5	105.4	100.0	81.7	89.8	99.6	75.1	85.9	98.8	95.5	96.9	95.7	83.7	89.0	100.3	75.1	86.2
COVERS	DIVS. 10 - 14, 25, 28, 41, 43, 44, 46	100.0	101.3	100.3	100.0	89.0	97.8	100.0	92.4	98.5	100.0	91.4	98.3	100.0	91.5	98.3	100.0	92.4	98.5
21, 22, 23	FIRE SUPPRESSION, PLUMBING & HVAC	100.0	106.7	102.7	95.1	97.1	95.9	95.1	96.0	95.4	95.0	98.0	96.2	94.9	93.8	94.5	95.1	96.1	95.5
26, 27, 3370	ELECTRICAL, COMMUNICATIONS & UTIL.	102.0	100.1	101.0	97.0	82.9	89.6	97.2	103.9	100.7	96.6	86.6	91.4	98.2	99.7	99.0	96.3	103.9	100.3
MF2010	WEIGHTED AVERAGE	101.6	106.7	103.8	100.3	90.0	95.8	97.0	90.3	94.1	96.3	95.6	96.0	98.1	92.6	95.7	97.5	90.6	94.5

MISSOURI / MONTANA

DIVISION		SPRINGFIELD 656 - 658			ST. JOSEPH 644 - 645			ST. LOUIS 630 - 631			BILLINGS 590 - 591			BUTTE 597			GREAT FALLS 594		
		MAT.	INST.	TOTAL	MAT.	INST.	TOTAL	MAT.	INST.	TOTAL	MAT.	INST.	TOTAL	MAT.	INST.	TOTAL	MAT.	INST.	TOTAL
015433	CONTRACTOR EQUIPMENT		102.3	102.3		102.1	102.1		107.6	107.6		99.3	99.3		99.0	99.0		99.0	99.0
0241, 31 - 34	SITE & INFRASTRUCTURE, DEMOLITION	97.2	94.4	95.2	99.8	93.7	95.5	91.9	98.4	96.5	95.9	97.1	96.8	103.3	96.9	98.8	107.1	96.9	100.0
0310	Concrete Forming & Accessories	98.2	74.7	77.9	100.2	91.1	92.3	100.2	103.6	103.1	101.3	71.6	75.7	88.4	73.3	75.4	101.1	71.6	75.6
0320	Concrete Reinforcing	94.3	116.0	105.1	94.4	109.8	102.1	95.5	114.0	104.7	93.9	82.8	88.3	101.9	82.9	92.4	94.0	82.8	88.4
0330	Cast-in-Place Concrete	97.5	78.8	89.7	98.5	103.6	100.6	91.6	105.0	97.2	120.2	68.9	98.9	132.3	70.3	106.6	139.8	65.0	108.7
03	CONCRETE	97.7	84.6	91.2	97.5	99.5	98.5	94.5	106.8	100.5	106.9	73.6	90.5	113.6	74.9	93.0	115.9	72.2	94.3
04	MASONRY	89.5	87.2	88.1	100.1	96.7	98.0	99.7	112.5	107.6	117.9	75.0	91.4	113.3	77.9	91.4	117.5	76.7	92.3
05	METALS	101.2	107.5	103.1	103.7	110.7	105.9	101.9	120.5	107.7	107.5	91.0	102.4	100.6	91.1	97.7	103.8	91.1	99.9
06	WOOD, PLASTICS & COMPOSITES	103.4	72.8	86.1	107.5	89.7	97.4	103.1	100.8	101.8	99.5	70.0	82.8	86.6	72.5	78.6	100.7	70.0	83.4
07	THERMAL & MOISTURE PROTECTION	99.9	81.1	92.2	97.7	97.4	97.5	103.2	107.6	105.0	108.1	71.2	93.1	108.0	72.5	93.6	108.5	69.8	92.8
08	OPENINGS	101.6	81.2	96.8	101.7	98.3	100.9	97.2	110.5	100.4	93.9	68.7	88.0	92.3	71.3	87.3	95.0	68.1	88.6
0920	Plaster & Gypsum Board	88.3	71.9	77.1	104.8	89.1	94.1	104.7	100.8	102.1	99.9	69.3	79.1	100.3	71.9	81.0	109.7	69.3	82.2
0950, 0980	Ceilings & Acoustic Treatment	97.9	71.9	80.6	98.4	89.1	92.2	102.4	100.8	101.3	99.8	69.3	79.5	107.0	71.9	83.6	108.8	69.3	82.5
0960	Flooring	110.5	76.3	100.3	99.8	104.4	101.2	101.1	99.4	100.6	106.0	71.5	95.7	103.9	75.5	95.4	111.0	75.5	100.4
0970, 0990	Wall Finishes & Painting/Coating	96.2	76.5	84.4	96.2	103.5	100.6	102.3	106.5	104.8	99.0	88.5	92.7	97.2	52.3	70.2	97.2	88.5	92.0
09	FINISHES	101.0	75.1	86.6	102.5	92.7	97.0	103.7	102.5	103.0	102.7	73.0	86.2	103.4	71.2	85.5	107.4	73.7	88.6
COVERS	DIVS. 10 - 14, 25, 28, 41, 43, 44, 46	100.0	93.0	98.6	100.0	96.9	99.4	100.0	102.3	100.5	100.0	96.1	99.2	100.0	96.4	99.3	100.0	96.1	99.2
21, 22, 23	FIRE SUPPRESSION, PLUMBING & HVAC	99.9	73.0	89.1	100.1	89.5	95.8	100.0	107.6	103.0	100.1	74.9	90.0	100.2	70.7	88.4	100.2	72.2	89.0
26, 27, 3370	ELECTRICAL, COMMUNICATIONS & UTIL.	102.9	70.3	85.9	100.5	80.5	90.1	100.5	107.1	103.9	96.8	73.9	84.8	104.4	72.2	87.6	96.9	72.2	84.0
MF2010	WEIGHTED AVERAGE	99.9	82.1	92.1	100.7	93.9	97.8	99.6	107.6	103.1	102.2	77.9	91.6	102.0	77.2	91.2	103.4	77.1	92.0

511

City Cost Indexes

MONTANA

DIVISION		HAVRE 595 MAT.	INST.	TOTAL	HELENA 596 MAT.	INST.	TOTAL	KALISPELL 599 MAT.	INST.	TOTAL	MILES CITY 593 MAT.	INST.	TOTAL	MISSOULA 598 MAT.	INST.	TOTAL	WOLF POINT 592 MAT.	INST.	TOTAL
015433	CONTRACTOR EQUIPMENT		99.0	99.0		99.0	99.0		99.0	99.0		99.0	99.0		99.0	99.0		99.0	99.0
0241, 31 - 34	SITE & INFRASTRUCTURE, DEMOLITION	111.0	96.7	100.9	100.1	96.6	97.7	93.4	96.8	95.8	99.9	96.3	97.4	85.8	96.6	93.4	118.0	96.4	102.9
0310	Concrete Forming & Accessories	81.6	69.1	70.8	104.4	66.7	71.9	91.6	71.5	74.3	99.6	67.6	72.0	91.6	71.5	74.2	92.2	67.8	71.2
0320	Concrete Reinforcing	102.7	82.8	92.8	106.8	82.8	94.8	104.6	83.4	94.0	102.3	82.8	92.6	103.6	83.4	93.5	103.7	82.8	93.3
0330	Cast-in-Place Concrete	142.5	61.8	109.0	106.9	64.3	89.2	114.9	64.0	93.8	125.8	61.6	99.1	97.5	66.2	84.5	140.8	62.0	108.1
03	CONCRETE	119.1	70.0	94.9	104.5	69.8	87.3	100.0	72.0	86.2	107.4	69.3	88.6	88.0	72.7	80.4	123.0	69.6	96.6
04	MASONRY	114.2	71.1	87.5	110.7	77.1	89.9	112.2	78.3	91.2	119.4	66.5	86.7	135.6	77.9	99.9	120.5	67.2	87.5
05	METALS	96.8	90.9	95.0	102.7	90.0	98.7	96.6	91.4	95.0	95.9	91.1	94.4	97.2	91.3	95.4	96.0	91.1	94.5
06	WOOD, PLASTICS & COMPOSITES	78.4	70.0	73.7	103.7	63.8	81.2	89.9	70.0	78.7	97.8	70.0	82.1	89.9	70.0	78.7	89.5	70.0	78.5
07	THERMAL & MOISTURE PROTECTION	108.3	69.4	92.5	105.4	72.2	91.9	107.7	74.3	94.1	108.0	66.8	91.3	107.4	80.6	96.5	108.7	67.1	91.8
08	OPENINGS	92.4	68.1	86.6	96.2	66.3	89.2	92.3	68.3	86.6	91.8	68.0	86.2	92.3	68.0	86.6	91.9	67.8	86.2
0920	Plaster & Gypsum Board	96.2	69.3	77.9	97.9	62.9	74.1	100.3	69.3	79.2	109.0	69.3	82.0	100.3	69.3	79.2	104.2	69.3	80.4
0950, 0980	Ceilings & Acoustic Treatment	107.0	69.3	81.9	109.7	62.9	78.5	107.0	69.3	81.9	106.1	69.3	81.6	107.0	69.3	81.9	106.1	69.3	81.6
0960	Flooring	101.3	75.5	93.6	106.1	60.2	92.4	105.7	75.5	96.7	110.7	71.5	99.0	105.7	75.5	96.7	107.1	71.5	96.5
0970, 0990	Wall Finishes & Painting/Coating	97.2	49.7	68.7	93.5	48.2	66.3	97.2	48.2	67.8	97.2	48.2	67.8	97.2	63.3	76.8	97.2	49.7	68.7
09	FINISHES	102.8	67.9	83.4	104.7	63.0	81.5	103.3	69.4	84.4	106.2	65.9	83.8	102.7	71.0	85.1	105.9	66.2	83.8
COVERS	DIVS. 10 - 14, 25, 28, 41, 43, 44, 46	100.0	72.6	94.5	100.0	95.6	99.1	100.0	96.2	99.2	100.0	92.6	98.5	100.0	96.1	99.2	100.0	92.9	98.6
21, 22, 23	FIRE SUPPRESSION, PLUMBING & HVAC	95.3	69.3	84.9	100.1	70.9	88.4	95.3	70.3	85.3	95.3	69.7	85.0	100.2	69.9	88.0	95.3	70.0	85.2
26, 27, 3370	ELECTRICAL, COMMUNICATIONS & UTIL.	96.9	72.7	84.3	103.8	72.7	87.6	101.2	71.0	85.4	96.9	77.4	86.7	102.2	69.3	85.0	96.9	77.4	86.7
MF2010	WEIGHTED AVERAGE	100.6	74.2	89.1	102.0	75.1	90.2	98.4	76.3	88.8	99.5	74.7	88.7	99.4	76.4	89.4	101.7	74.9	90.0

NEBRASKA

DIVISION		ALLIANCE 693 MAT.	INST.	TOTAL	COLUMBUS 686 MAT.	INST.	TOTAL	GRAND ISLAND 688 MAT.	INST.	TOTAL	HASTINGS 689 MAT.	INST.	TOTAL	LINCOLN 683 - 685 MAT.	INST.	TOTAL	MCCOOK 690 MAT.	INST.	TOTAL
015433	CONTRACTOR EQUIPMENT		97.5	97.5		101.4	101.4		101.4	101.4		101.4	101.4		101.4	101.4		101.4	101.4
0241, 31 - 34	SITE & INFRASTRUCTURE, DEMOLITION	98.8	98.4	98.6	99.1	92.0	94.1	103.8	92.8	96.1	102.6	92.0	95.2	93.6	92.8	93.0	102.1	91.9	95.0
0310	Concrete Forming & Accessories	88.3	57.7	61.9	95.9	68.1	71.9	95.4	74.0	76.9	98.5	73.6	77.0	97.5	68.9	72.8	93.9	57.3	62.3
0320	Concrete Reinforcing	110.1	85.0	97.6	101.7	76.9	89.3	101.1	76.8	89.0	101.1	77.5	89.3	99.2	76.3	87.7	103.1	76.5	89.8
0330	Cast-in-Place Concrete	109.4	61.3	89.4	114.9	61.3	92.6	121.6	68.6	99.4	121.6	64.6	97.9	99.8	71.0	87.9	117.6	60.6	94.0
03	CONCRETE	120.2	64.5	92.7	106.9	68.2	87.8	111.7	73.4	92.8	111.9	71.9	92.1	98.0	71.9	85.1	109.5	63.1	86.6
04	MASONRY	109.7	74.9	88.2	112.6	79.0	91.8	105.7	77.4	88.2	114.2	89.5	98.9	95.9	68.4	78.9	105.5	74.7	86.5
05	METALS	101.2	79.1	94.4	91.9	85.2	89.8	93.6	87.0	91.5	94.4	86.5	91.9	95.6	86.2	92.7	96.0	84.2	92.4
06	WOOD, PLASTICS & COMPOSITES	88.3	53.4	68.6	99.7	67.2	81.4	99.0	73.9	84.8	102.7	73.9	86.4	98.3	67.2	80.7	97.1	53.4	72.5
07	THERMAL & MOISTURE PROTECTION	106.4	69.5	91.4	104.4	70.7	90.7	104.5	74.2	92.2	104.5	85.4	96.8	100.4	70.1	88.1	100.8	67.4	87.2
08	OPENINGS	92.8	59.4	84.9	94.1	66.7	87.6	94.2	70.9	88.7	94.2	70.3	88.5	103.2	63.5	93.8	93.2	57.5	84.7
0920	Plaster & Gypsum Board	81.0	51.9	61.2	88.7	66.1	73.3	88.0	73.0	77.8	89.9	73.0	78.4	90.1	66.1	73.8	92.5	51.9	64.9
0950, 0980	Ceilings & Acoustic Treatment	98.5	51.9	66.5	93.3	66.1	75.2	93.3	73.0	79.8	93.3	73.0	79.8	99.3	66.1	77.2	94.3	51.9	66.1
0960	Flooring	104.5	83.6	98.2	102.5	102.2	102.4	102.3	109.2	104.4	103.4	102.2	103.1	107.6	85.1	100.9	101.6	83.6	96.2
0970, 0990	Wall Finishes & Painting/Coating	174.1	56.0	103.1	91.9	65.1	75.8	91.9	69.1	78.2	91.9	65.1	75.8	100.0	79.6	87.7	95.3	48.8	67.3
09	FINISHES	102.8	60.8	79.4	98.5	72.9	84.3	98.7	78.8	87.6	99.2	76.9	86.8	102.3	72.7	85.8	99.6	59.9	77.5
COVERS	DIVS. 10 - 14, 25, 28, 41, 43, 44, 46	100.0	67.0	93.4	100.0	88.3	97.7	100.0	89.3	97.9	100.0	89.1	97.8	100.0	88.6	97.7	100.0	66.3	93.2
21, 22, 23	FIRE SUPPRESSION, PLUMBING & HVAC	95.3	72.0	85.9	95.1	77.1	87.9	100.0	78.5	91.4	95.1	77.2	87.9	99.9	78.5	91.3	95.0	77.1	87.8
26, 27, 3370	ELECTRICAL, COMMUNICATIONS & UTIL.	91.7	71.0	80.9	92.0	82.4	87.0	90.7	65.9	77.8	90.1	85.6	87.7	103.2	65.9	83.7	93.9	71.0	81.9
MF2010	WEIGHTED AVERAGE	100.4	71.6	87.9	97.5	77.9	88.9	99.1	77.8	89.9	98.5	81.2	90.9	99.5	75.3	89.0	98.1	72.1	86.8

NEBRASKA / NEVADA

DIVISION		NORFOLK 687 MAT.	INST.	TOTAL	NORTH PLATTE 691 MAT.	INST.	TOTAL	OMAHA 680 - 681 MAT.	INST.	TOTAL	VALENTINE 692 MAT.	INST.	TOTAL	CARSON CITY 897 MAT.	INST.	TOTAL	ELKO 898 MAT.	INST.	TOTAL
015433	CONTRACTOR EQUIPMENT		92.7	92.7		101.4	101.4		92.7	92.7		95.6	95.6		97.9	97.9		97.9	97.9
0241, 31 - 34	SITE & INFRASTRUCTURE, DEMOLITION	82.3	91.1	88.5	103.4	92.0	95.4	86.4	91.8	90.2	86.4	95.9	93.1	78.7	99.6	93.4	61.9	98.4	87.5
0310	Concrete Forming & Accessories	82.4	72.7	74.0	96.3	73.1	76.3	93.5	72.9	75.7	84.5	57.1	60.8	105.3	94.6	96.1	111.2	83.2	87.0
0320	Concrete Reinforcing	101.8	66.6	84.3	102.5	77.4	90.0	104.1	76.9	90.5	103.1	66.3	84.8	107.5	118.0	112.7	106.9	113.9	110.4
0330	Cast-in-Place Concrete	116.1	63.8	94.4	117.6	65.9	96.2	92.6	77.1	86.2	104.3	58.6	85.3	104.1	86.2	96.7	98.7	80.9	91.3
03	CONCRETE	105.7	68.7	87.5	109.6	72.1	91.1	94.6	75.4	85.1	107.0	60.0	83.8	104.1	96.0	100.1	98.6	88.3	93.5
04	MASONRY	118.7	78.7	93.9	93.2	89.5	90.9	98.9	79.9	87.1	105.3	74.8	86.4	116.6	89.3	99.7	114.9	73.1	89.0
05	METALS	95.2	74.1	88.7	95.1	86.1	92.3	95.4	79.5	90.5	107.3	74.1	97.1	90.1	104.0	94.4	92.8	99.5	94.8
06	WOOD, PLASTICS & COMPOSITES	82.8	73.4	77.5	99.1	73.9	84.9	91.2	72.1	80.5	82.8	52.9	65.9	91.6	96.6	94.4	99.4	83.7	90.6
07	THERMAL & MOISTURE PROTECTION	104.4	74.9	92.4	100.8	84.1	94.0	100.0	79.8	91.8	101.6	69.1	88.4	102.7	91.2	98.0	99.7	76.6	89.9
08	OPENINGS	95.9	67.7	89.2	92.5	70.3	87.3	101.4	71.9	94.4	95.0	56.5	85.9	99.6	109.4	101.9	100.6	87.3	97.4
0920	Plaster & Gypsum Board	88.7	73.0	78.0	92.5	73.0	79.3	99.5	71.7	80.6	94.7	51.9	65.6	91.3	96.4	94.8	93.3	83.1	86.4
0950, 0980	Ceilings & Acoustic Treatment	106.9	73.0	84.4	94.3	73.0	80.1	112.8	71.7	85.5	113.1	51.9	72.4	101.8	96.4	98.2	102.3	83.1	89.5
0960	Flooring	131.7	102.2	122.9	102.4	102.2	102.3	129.9	83.4	116.0	133.3	81.3	117.7	100.8	109.9	103.6	103.7	52.1	88.2
0970, 0990	Wall Finishes & Painting/Coating	164.9	65.1	104.9	95.3	65.1	77.1	148.8	68.2	100.4	171.4	67.5	108.9	102.7	82.5	90.6	104.2	97.9	100.4
09	FINISHES	120.0	76.9	96.0	99.9	76.9	87.1	119.6	74.0	94.2	123.7	61.1	88.8	100.4	96.2	98.1	99.4	78.9	88.0
COVERS	DIVS. 10 - 14, 25, 28, 41, 43, 44, 46	100.0	87.9	97.6	100.0	68.6	93.7	100.0	88.1	97.6	100.0	92.0	98.4	100.0	64.3	92.8	100.0	64.5	92.9
21, 22, 23	FIRE SUPPRESSION, PLUMBING & HVAC	94.9	76.9	87.7	99.9	77.1	90.8	99.8	78.0	91.1	94.7	76.9	87.6	100.0	81.4	92.6	99.7	81.2	91.1
26, 27, 3370	ELECTRICAL, COMMUNICATIONS & UTIL.	90.9	82.4	86.5	92.0	78.9	85.2	97.9	82.4	89.8	89.0	88.6	88.8	100.4	97.4	98.8	98.1	97.3	97.7
MF2010	WEIGHTED AVERAGE	99.2	77.5	89.8	98.5	79.6	90.2	99.6	79.3	90.7	100.6	73.6	88.9	99.1	93.8	96.8	97.6	86.1	92.6

City Cost Indexes

NEVADA / NEW HAMPSHIRE

DIVISION		ELY 893 MAT.	INST.	TOTAL	LAS VEGAS 889-891 MAT.	INST.	TOTAL	RENO 894-895 MAT.	INST.	TOTAL	CHARLESTON 036 MAT.	INST.	TOTAL	CLAREMONT 037 MAT.	INST.	TOTAL	CONCORD 032-033 MAT.	INST.	TOTAL
015433	CONTRACTOR EQUIPMENT		97.9	97.9		97.9	97.9		97.9	97.9		100.6	100.6		100.6	100.6		100.6	100.6
0241, 31 - 34	SITE & INFRASTRUCTURE, DEMOLITION	67.0	99.9	90.1	69.7	101.6	92.1	67.0	99.6	89.9	87.4	99.2	95.7	81.2	99.2	93.9	94.5	101.4	99.3
0310	Concrete Forming & Accessories	104.2	105.9	105.6	105.4	111.0	110.3	100.5	94.5	95.3	88.1	82.9	83.6	93.8	82.9	84.4	95.9	94.3	94.5
0320	Concrete Reinforcing	105.7	115.2	110.4	97.5	121.8	109.6	99.9	121.4	110.6	84.5	95.2	89.8	84.5	95.2	89.8	92.9	95.8	94.4
0330	Cast-in-Place Concrete	105.9	106.7	106.2	102.6	107.5	104.6	112.0	86.2	101.3	97.4	72.1	86.9	89.5	72.1	82.3	111.5	92.2	103.5
03	CONCRETE	106.6	107.6	107.1	101.8	111.4	106.6	106.3	96.6	101.5	99.5	81.6	90.6	91.5	81.6	86.6	106.0	93.7	100.0
04	MASONRY	119.0	104.2	109.9	108.1	102.8	104.8	114.2	89.3	98.8	87.8	82.3	84.4	87.3	82.3	84.2	97.6	102.1	100.4
05	METALS	92.7	104.8	96.4	99.8	109.3	102.7	94.2	105.3	97.6	95.2	90.6	93.8	95.2	90.6	93.8	101.2	94.0	99.0
06	WOOD, PLASTICS & COMPOSITES	90.3	106.3	99.3	89.2	110.4	101.2	85.2	96.6	91.6	86.9	92.3	90.0	93.3	92.3	92.7	94.8	93.3	93.9
07	THERMAL & MOISTURE PROTECTION	99.6	95.9	98.1	112.5	102.0	108.2	99.1	91.2	95.9	102.8	72.0	90.3	102.7	72.0	90.2	104.9	91.4	99.4
08	OPENINGS	100.5	99.6	100.3	99.6	117.9	104.0	98.6	104.2	99.9	102.6	78.0	96.8	103.8	78.0	97.7	104.1	88.4	100.4
0920	Plaster & Gypsum Board	90.0	106.4	101.1	87.8	110.7	103.4	82.0	96.4	91.8	94.7	91.4	92.5	95.8	91.4	92.8	98.4	92.4	94.4
0950, 0980	Ceilings & Acoustic Treatment	102.3	106.4	105.0	111.1	110.7	110.8	107.6	96.4	100.1	90.6	91.4	91.2	90.6	91.4	91.2	95.9	92.4	93.6
0960	Flooring	101.5	58.2	88.6	93.3	109.9	98.3	98.8	109.9	102.1	96.3	32.7	77.3	98.2	32.7	78.6	97.1	112.2	101.6
0970, 0990	Wall Finishes & Painting/Coating	104.2	123.6	115.9	107.2	123.6	117.0	104.2	82.5	91.2	95.4	46.7	66.2	95.4	46.9	66.3	95.1	97.5	96.5
09	FINISHES	98.9	99.4	99.2	98.4	112.2	106.1	98.0	96.2	97.0	94.9	71.3	81.8	95.1	71.3	81.9	95.9	98.0	97.0
COVERS	DIVS. 10 - 14, 25, 28, 41, 43, 44, 46	100.0	65.0	93.0	100.0	106.3	101.3	100.0	92.0	98.4	100.0	90.0	98.0	100.0	90.0	98.0	100.0	104.8	101.0
21, 22, 23	FIRE SUPPRESSION, PLUMBING & HVAC	97.7	106.7	101.3	100.2	105.1	102.1	100.0	81.5	92.6	95.2	40.4	73.2	95.2	40.5	73.3	99.9	86.1	94.4
26, 27, 3370	ELECTRICAL, COMMUNICATIONS & UTIL.	98.4	109.7	104.3	102.7	120.6	112.0	98.8	97.4	98.0	99.6	54.1	75.8	99.6	54.1	75.8	99.6	80.4	89.6
MF2010	WEIGHTED AVERAGE	98.7	103.4	100.8	100.2	109.4	104.2	99.0	93.8	96.7	96.9	70.1	85.2	96.0	70.1	84.7	100.8	92.3	97.1

NEW HAMPSHIRE / NEW JERSEY

DIVISION		KEENE 034 MAT.	INST.	TOTAL	LITTLETON 035 MAT.	INST.	TOTAL	MANCHESTER 031 MAT.	INST.	TOTAL	NASHUA 030 MAT.	INST.	TOTAL	PORTSMOUTH 038 MAT.	INST.	TOTAL	ATLANTIC CITY 082,084 MAT.	INST.	TOTAL
015433	CONTRACTOR EQUIPMENT		100.6	100.6		100.6	100.6		100.6	100.6		100.6	100.6		100.6	100.6		98.8	98.8
0241, 31 - 34	SITE & INFRASTRUCTURE, DEMOLITION	95.2	99.5	98.2	81.3	99.5	94.1	93.6	101.4	99.1	96.5	101.4	100.0	90.1	101.5	98.1	98.0	104.4	102.5
0310	Concrete Forming & Accessories	92.6	84.9	85.9	103.4	84.9	87.4	98.4	94.8	95.3	100.3	94.8	95.5	89.4	95.1	94.3	110.1	128.4	125.9
0320	Concrete Reinforcing	84.5	95.3	89.9	85.2	95.3	90.3	106.7	95.8	101.3	105.5	95.8	100.7	84.5	95.9	90.2	79.1	121.6	100.3
0330	Cast-in-Place Concrete	97.9	74.4	88.1	87.9	74.4	82.3	111.9	115.3	113.3	92.6	115.3	102.0	87.9	115.4	99.3	84.2	132.2	104.1
03	CONCRETE	99.2	83.2	91.3	91.0	83.3	87.2	108.6	101.9	105.3	99.5	101.9	100.6	91.0	102.1	96.4	92.9	127.3	109.9
04	MASONRY	91.7	85.9	88.1	96.9	85.9	90.1	94.9	102.1	99.4	91.4	102.1	98.0	87.7	102.1	96.6	99.2	125.4	115.4
05	METALS	95.8	91.2	94.4	95.9	91.2	94.4	103.8	94.5	100.9	100.8	94.5	98.9	97.3	95.1	96.7	95.6	106.5	99.0
06	WOOD, PLASTICS & COMPOSITES	91.6	92.3	92.0	102.9	92.3	96.9	95.1	93.3	94.1	101.4	93.3	96.8	88.2	93.3	91.1	115.6	130.2	123.9
07	THERMAL & MOISTURE PROTECTION	103.2	75.0	91.8	102.8	73.5	90.9	106.9	94.9	102.0	103.6	94.9	100.1	103.3	115.6	108.3	104.6	123.2	112.2
08	OPENINGS	101.1	82.9	96.8	104.8	78.7	98.6	106.9	88.4	102.6	105.8	88.4	101.6	106.5	85.1	101.4	102.3	125.4	107.7
0920	Plaster & Gypsum Board	95.1	91.4	92.6	108.4	91.4	96.9	98.7	92.4	94.4	103.8	92.4	96.1	94.7	92.4	93.2	107.3	130.5	123.1
0950, 0980	Ceilings & Acoustic Treatment	90.6	91.4	91.2	90.6	91.4	91.2	98.6	92.4	94.5	101.4	92.4	95.4	91.6	92.4	92.2	88.9	130.5	116.6
0960	Flooring	97.9	53.6	84.7	106.2	32.7	84.2	98.2	112.2	102.4	100.9	112.2	104.2	96.5	112.2	101.2	102.2	150.6	116.6
0970, 0990	Wall Finishes & Painting/Coating	95.4	46.7	66.2	95.4	61.2	74.9	98.2	97.5	97.8	95.4	97.5	96.7	95.4	97.5	96.7	93.0	117.3	107.6
09	FINISHES	96.7	76.2	85.3	99.0	73.8	85.0	98.5	98.0	98.2	101.0	98.0	99.3	95.9	98.0	97.1	99.4	132.5	117.8
COVERS	DIVS. 10 - 14, 25, 28, 41, 43, 44, 46	100.0	91.2	98.2	100.0	99.2	99.8	100.0	104.8	101.0	100.0	104.8	101.0	100.0	104.8	101.0	100.0	110.5	102.1
21, 22, 23	FIRE SUPPRESSION, PLUMBING & HVAC	95.2	44.2	74.7	95.2	65.2	83.2	99.9	86.2	94.4	100.1	86.2	94.5	100.1	86.2	94.5	99.6	119.6	107.6
26, 27, 3370	ELECTRICAL, COMMUNICATIONS & UTIL.	99.6	64.7	81.4	100.8	57.3	78.1	101.6	80.4	90.6	102.2	80.4	90.8	100.1	80.4	89.8	92.7	141.4	118.1
MF2010	WEIGHTED AVERAGE	97.4	74.0	87.2	97.1	77.2	88.4	102.1	93.6	98.4	100.6	93.6	97.6	98.2	94.2	96.4	98.1	123.5	109.2

NEW JERSEY

DIVISION		CAMDEN 081 MAT.	INST.	TOTAL	DOVER 078 MAT.	INST.	TOTAL	ELIZABETH 072 MAT.	INST.	TOTAL	HACKENSACK 076 MAT.	INST.	TOTAL	JERSEY CITY 073 MAT.	INST.	TOTAL	LONG BRANCH 077 MAT.	INST.	TOTAL
015433	CONTRACTOR EQUIPMENT		98.8	98.8		100.6	100.6		100.6	100.6		100.6	100.6		98.8	98.8		98.4	98.4
0241, 31 - 34	SITE & INFRASTRUCTURE, DEMOLITION	99.1	104.7	103.0	108.2	105.4	106.3	112.7	105.4	107.6	108.9	105.5	106.5	98.0	105.4	103.2	102.6	105.3	104.5
0310	Concrete Forming & Accessories	101.7	128.4	124.7	97.4	129.0	124.7	109.3	129.0	126.3	97.4	128.9	124.6	101.1	129.0	125.2	101.7	128.7	125.0
0320	Concrete Reinforcing	103.8	121.8	112.8	80.0	128.6	104.3	80.0	128.6	104.3	80.0	128.6	104.3	103.8	128.6	116.2	80.0	128.6	104.3
0330	Cast-in-Place Concrete	81.6	132.1	102.6	102.2	127.3	112.6	87.8	127.3	104.2	99.9	127.3	111.3	79.8	127.4	99.5	88.7	132.1	106.7
03	CONCRETE	94.1	127.3	110.4	99.7	127.4	113.3	95.3	127.4	111.1	97.8	127.3	112.4	93.2	127.2	110.0	97.1	128.7	112.7
04	MASONRY	89.7	125.4	111.8	87.3	125.9	111.2	101.2	125.9	116.5	90.8	125.9	112.5	80.8	125.9	108.7	94.4	125.4	113.6
05	METALS	100.7	106.6	102.5	95.6	113.6	101.2	97.1	113.6	102.2	95.7	113.4	101.2	100.8	111.2	104.0	95.7	111.1	100.5
06	WOOD, PLASTICS & COMPOSITES	104.2	130.2	118.9	101.5	130.2	117.7	117.1	130.2	124.5	101.5	130.2	117.7	102.3	130.2	118.0	103.4	130.1	118.5
07	THERMAL & MOISTURE PROTECTION	104.5	122.7	111.9	101.5	130.6	113.4	101.7	130.6	113.5	101.3	123.3	110.3	101.1	130.6	113.1	101.2	123.4	110.3
08	OPENINGS	104.6	125.4	109.5	108.8	127.9	113.4	106.9	127.9	111.9	106.2	127.9	111.4	104.5	127.9	110.0	100.7	127.9	107.1
0920	Plaster & Gypsum Board	102.8	130.5	121.6	99.9	130.5	120.7	107.7	130.5	123.2	99.9	130.5	120.7	103.2	130.5	121.8	101.7	130.5	121.3
0950, 0980	Ceilings & Acoustic Treatment	99.7	130.5	120.2	88.6	130.5	116.5	90.6	130.5	117.2	88.6	130.5	116.5	99.5	130.5	120.1	88.6	130.5	116.5
0960	Flooring	98.7	150.6	114.2	96.8	172.3	119.4	101.2	172.3	122.5	96.8	172.3	119.4	97.7	172.3	120.0	97.9	172.3	120.1
0970, 0990	Wall Finishes & Painting/Coating	93.0	117.3	107.6	94.6	119.3	109.4	94.6	119.3	109.4	94.6	119.3	109.4	94.7	119.3	109.5	94.7	117.3	108.3
09	FINISHES	100.0	132.5	118.1	97.1	135.5	118.5	100.4	135.5	119.9	96.9	135.5	118.4	99.8	135.5	119.7	97.9	136.0	119.2
COVERS	DIVS. 10 - 14, 25, 28, 41, 43, 44, 46	100.0	110.5	102.1	100.0	113.6	102.7	100.0	113.6	102.7	100.0	113.6	102.7	100.0	113.6	102.7	100.0	110.4	102.1
21, 22, 23	FIRE SUPPRESSION, PLUMBING & HVAC	100.0	119.7	107.9	99.6	123.4	109.1	100.0	122.8	109.1	99.6	123.3	109.1	100.0	123.4	109.3	99.6	122.6	108.8
26, 27, 3370	ELECTRICAL, COMMUNICATIONS & UTIL.	97.9	141.4	120.6	90.9	137.6	115.3	91.5	137.6	115.6	90.9	139.7	116.4	95.5	139.7	118.6	90.6	131.5	111.9
MF2010	WEIGHTED AVERAGE	99.4	123.6	109.9	98.7	125.4	110.3	99.5	125.3	110.7	98.4	125.5	110.2	98.5	125.5	110.2	97.8	124.1	109.2

513

NEW JERSEY

DIVISION		NEW BRUNSWICK 088 - 089			NEWARK 070 - 071			PATERSON 074 - 075			POINT PLEASANT 087			SUMMIT 079			TRENTON 085 - 086		
		MAT.	INST.	TOTAL	MAT.	INST.	TOTAL	MAT.	INST.	TOTAL	MAT.	INST.	TOTAL	MAT.	INST.	TOTAL	MAT.	INST.	TOTAL
015433	CONTRACTOR EQUIPMENT		98.4	98.4		100.6	100.6		100.6	100.6		98.4	98.4		100.6	100.6		98.4	98.4
0241, 31 - 34	SITE & INFRASTRUCTURE, DEMOLITION	111.8	105.3	107.3	114.4	105.4	108.1	110.9	105.5	107.1	113.5	105.3	107.7	110.1	105.4	106.8	96.6	105.3	102.7
0310	Concrete Forming & Accessories	104.6	128.9	125.6	97.2	129.0	124.6	99.4	128.9	124.8	99.3	120.3	117.4	100.2	129.0	125.0	99.9	128.6	124.6
0320	Concrete Reinforcing	80.0	128.6	104.3	103.3	128.6	116.0	103.8	128.6	116.2	80.0	128.5	104.2	80.0	128.6	104.3	103.3	115.7	109.5
0330	Cast-in-Place Concrete	103.9	132.3	115.7	109.6	127.3	117.0	101.6	127.3	112.3	103.9	128.4	114.1	84.9	127.3	102.5	99.5	132.0	113.0
03	CONCRETE	109.8	128.9	119.3	106.9	127.4	117.0	103.4	127.3	115.2	109.5	123.7	116.5	92.3	127.4	109.6	102.3	126.3	114.1
04	MASONRY	96.6	125.9	114.7	93.4	125.9	113.5	87.8	125.9	111.4	86.4	124.6	110.0	89.3	125.9	112.0	92.1	125.4	112.7
05	METALS	95.7	111.2	100.4	102.8	113.6	106.1	95.7	113.5	101.2	95.7	110.7	100.3	95.6	113.6	101.1	100.4	105.7	102.0
06	WOOD, PLASTICS & COMPOSITES	108.9	130.1	120.9	99.0	130.2	116.6	104.1	130.2	118.8	101.6	119.1	111.5	105.3	130.2	119.3	101.4	130.1	117.6
07	THERMAL & MOISTURE PROTECTION	104.8	129.9	115.0	102.0	130.6	113.6	101.6	123.3	110.4	104.8	122.1	111.8	101.9	130.6	113.6	104.2	123.1	111.9
08	OPENINGS	97.2	127.9	104.5	107.5	127.9	112.3	111.6	127.9	115.4	99.1	122.2	104.6	113.6	127.9	117.0	106.7	124.4	110.9
0920	Plaster & Gypsum Board	103.9	130.5	122.0	99.9	130.5	120.7	103.2	130.5	121.8	99.4	119.1	112.8	101.7	130.5	121.3	99.9	130.5	120.7
0950, 0980	Ceilings & Acoustic Treatment	88.9	130.5	116.6	102.4	130.5	121.1	99.5	130.5	120.1	88.9	119.1	109.0	88.6	130.5	116.5	101.4	130.5	120.8
0960	Flooring	100.0	172.3	121.6	98.5	172.3	120.6	97.7	172.3	120.0	97.8	150.6	113.6	98.0	172.3	120.2	99.3	172.3	121.1
0970, 0990	Wall Finishes & Painting/Coating	93.0	119.3	108.8	96.0	119.3	110.0	94.6	119.3	109.4	93.0	117.3	107.6	94.6	119.3	109.4	98.3	117.3	109.7
09	FINISHES	99.7	135.4	119.6	98.6	135.5	119.1	100.0	135.5	119.8	98.5	125.9	113.7	98.0	135.5	118.9	100.1	136.0	120.1
COVERS	DIVS. 10 - 14, 25, 28, 41, 43, 44, 46	100.0	113.4	102.7	100.0	113.6	102.7	100.0	113.6	102.7	100.0	104.2	100.8	100.0	113.6	102.7	100.0	110.4	102.1
21, 22, 23	FIRE SUPPRESSION, PLUMBING & HVAC	99.6	122.9	108.9	100.0	122.8	109.1	100.0	123.4	109.4	99.6	122.5	108.7	99.6	122.8	108.9	100.0	122.5	109.0
26, 27, 3370	ELECTRICAL, COMMUNICATIONS & UTIL.	93.4	137.6	116.5	99.2	139.7	120.3	95.5	137.6	117.5	92.7	131.5	112.9	91.5	137.6	115.6	101.4	138.7	120.8
MF2010	WEIGHTED AVERAGE	99.7	125.2	110.8	101.9	125.6	112.2	100.2	125.2	111.1	99.2	121.4	108.8	98.7	125.3	110.3	100.8	124.0	110.9

NEW JERSEY / NEW MEXICO

DIVISION		VINELAND (NJ) 080,083			ALBUQUERQUE 870 - 872			CARRIZOZO 883			CLOVIS 881			FARMINGTON 874			GALLUP 873		
		MAT.	INST.	TOTAL	MAT.	INST.	TOTAL	MAT.	INST.	TOTAL	MAT.	INST.	TOTAL	MAT.	INST.	TOTAL	MAT.	INST.	TOTAL
015433	CONTRACTOR EQUIPMENT		98.8	98.8		109.6	109.6		109.6	109.6		109.6	109.6		109.6	109.6		109.6	109.6
0241, 31 - 34	SITE & INFRASTRUCTURE, DEMOLITION	102.7	104.7	104.1	82.2	104.1	97.6	102.9	104.1	103.7	91.2	104.1	100.2	88.4	104.1	99.4	96.0	104.1	101.7
0310	Concrete Forming & Accessories	96.8	128.4	124.1	101.4	65.0	70.0	99.2	65.0	69.7	99.1	64.9	69.6	101.4	65.0	70.0	101.5	65.0	70.0
0320	Concrete Reinforcing	79.1	118.5	98.7	100.5	71.1	85.8	109.9	71.1	90.5	111.2	71.1	91.2	109.9	71.1	90.5	105.2	71.1	88.2
0330	Cast-in-Place Concrete	90.7	132.2	107.9	96.9	71.2	86.2	95.4	71.2	85.3	95.3	71.1	85.3	97.8	71.2	86.8	92.1	71.2	83.4
03	CONCRETE	97.6	126.7	111.9	101.9	69.4	85.9	119.1	69.4	94.6	107.6	69.3	88.7	105.7	69.4	87.8	112.5	69.4	91.2
04	MASONRY	87.9	125.4	111.1	99.5	60.6	75.4	101.6	60.6	76.2	101.7	60.6	76.2	106.3	60.6	78.0	95.4	60.6	73.8
05	METALS	95.6	106.0	98.8	103.8	87.9	98.9	98.4	87.9	95.2	98.1	87.8	94.9	101.4	87.9	97.2	100.6	87.9	96.6
06	WOOD, PLASTICS & COMPOSITES	98.5	130.2	116.4	92.8	65.5	77.4	88.9	65.5	75.7	88.9	65.5	75.7	92.9	65.5	77.4	92.9	65.5	77.4
07	THERMAL & MOISTURE PROTECTION	104.4	123.2	112.1	99.4	71.8	88.2	100.7	71.8	89.0	99.6	71.8	88.3	99.6	71.8	88.3	100.5	71.8	88.8
08	OPENINGS	98.6	125.0	104.9	101.4	68.5	93.6	98.6	68.5	91.5	98.8	68.5	91.6	104.0	68.5	95.6	104.1	68.5	95.7
0920	Plaster & Gypsum Board	98.0	130.5	120.1	88.3	64.1	71.8	75.4	64.1	67.7	75.4	64.1	67.7	81.6	64.1	69.7	81.6	64.1	69.7
0950, 0980	Ceilings & Acoustic Treatment	88.9	130.5	116.6	106.3	64.1	78.2	104.2	64.1	77.5	104.2	64.1	77.5	103.1	64.1	77.2	103.1	64.1	77.2
0960	Flooring	97.1	150.6	113.1	100.2	67.2	90.3	100.2	67.2	90.4	100.2	67.2	90.4	101.7	67.2	91.4	101.7	67.2	91.4
0970, 0990	Wall Finishes & Painting/Coating	93.0	117.3	107.6	109.1	68.1	84.4	102.9	68.1	82.0	102.9	68.1	82.0	102.9	68.1	82.0	102.9	68.1	82.0
09	FINISHES	97.2	132.5	116.8	98.3	65.5	80.0	99.0	65.5	80.3	97.6	65.5	79.7	96.9	65.5	79.4	98.1	65.5	79.9
COVERS	DIVS. 10 - 14, 25, 28, 41, 43, 44, 46	100.0	110.5	102.1	100.0	76.3	95.2	100.0	76.3	95.2	100.0	76.3	95.2	100.0	76.3	95.2	100.0	76.3	95.2
21, 22, 23	FIRE SUPPRESSION, PLUMBING & HVAC	99.6	119.6	107.6	100.2	70.7	88.4	97.2	70.7	86.6	97.2	70.5	86.5	100.1	70.7	88.3	97.1	70.7	86.6
26, 27, 3370	ELECTRICAL, COMMUNICATIONS & UTIL.	92.7	141.4	118.1	88.5	72.9	80.4	90.6	72.9	81.4	88.1	72.9	80.2	86.8	72.9	79.6	86.0	72.9	79.2
MF2010	WEIGHTED AVERAGE	97.5	123.4	108.8	99.2	73.6	88.0	100.1	73.6	88.5	98.1	73.5	87.4	99.7	73.6	88.3	99.3	73.6	88.1

NEW MEXICO

DIVISION		LAS CRUCES 880			LAS VEGAS 877			ROSWELL 882			SANTA FE 875			SOCORRO 878			TRUTH/CONSEQUENCES 879		
		MAT.	INST.	TOTAL	MAT.	INST.	TOTAL	MAT.	INST.	TOTAL	MAT.	INST.	TOTAL	MAT.	INST.	TOTAL	MAT.	INST.	TOTAL
015433	CONTRACTOR EQUIPMENT		86.0	86.0		109.6	109.6		109.6	109.6		109.6	109.6		109.6	109.6		86.1	86.1
0241, 31 - 34	SITE & INFRASTRUCTURE, DEMOLITION	91.6	83.8	86.1	87.8	104.1	99.2	93.3	104.1	100.9	92.9	104.1	100.7	84.3	104.1	98.2	101.9	83.8	89.2
0310	Concrete Forming & Accessories	95.6	63.8	68.2	101.4	65.0	70.0	99.2	65.0	69.7	100.1	65.0	69.8	101.5	65.0	70.0	98.7	63.8	68.6
0320	Concrete Reinforcing	107.3	71.0	89.1	106.9	71.1	89.0	111.2	71.1	91.2	106.0	71.1	88.6	109.1	71.1	90.1	102.8	71.0	86.9
0330	Cast-in-Place Concrete	89.9	63.7	79.0	95.2	71.2	85.2	95.3	71.2	85.3	103.2	71.2	89.9	93.2	71.2	84.1	102.1	63.7	86.1
03	CONCRETE	86.1	65.9	76.1	103.1	69.4	86.4	108.3	69.4	89.1	105.6	69.4	87.7	102.0	69.4	85.9	95.1	65.9	80.7
04	MASONRY	97.5	60.2	74.4	95.6	60.6	73.9	112.2	60.6	80.3	99.5	60.6	75.4	95.5	60.6	73.9	93.1	60.2	72.8
05	METALS	97.1	81.3	92.2	100.3	87.9	96.4	99.3	87.9	95.8	97.7	87.9	94.7	100.6	87.9	96.6	100.1	81.4	94.3
06	WOOD, PLASTICS & COMPOSITES	78.8	64.4	70.7	92.9	65.5	77.4	88.9	65.5	75.7	94.6	65.5	78.2	92.9	65.5	77.4	84.6	64.4	73.2
07	THERMAL & MOISTURE PROTECTION	86.2	67.2	78.5	99.2	71.8	88.1	99.7	71.8	88.4	101.7	71.8	89.5	99.1	71.8	88.0	86.9	67.2	78.9
08	OPENINGS	91.5	67.9	85.9	100.3	68.5	92.8	98.6	68.5	91.5	102.4	68.5	94.4	100.1	68.5	92.7	93.3	67.9	87.3
0920	Plaster & Gypsum Board	74.2	64.1	67.3	81.6	64.1	69.7	75.4	64.1	67.7	91.6	64.1	72.9	81.6	64.1	69.7	83.4	64.1	70.3
0950, 0980	Ceilings & Acoustic Treatment	91.4	64.1	73.2	103.1	64.1	77.2	104.2	64.1	77.5	103.6	64.1	77.3	103.1	64.1	77.2	90.4	64.1	72.9
0960	Flooring	129.3	67.2	110.7	101.7	67.2	91.4	100.2	67.2	90.4	108.8	67.2	96.4	101.7	67.2	91.4	132.2	67.2	112.7
0970, 0990	Wall Finishes & Painting/Coating	90.1	68.1	76.8	102.9	68.1	82.0	102.9	68.1	82.0	107.7	68.1	83.9	102.9	68.1	82.0	93.7	68.1	78.3
09	FINISHES	107.1	64.6	83.4	96.8	65.5	79.3	97.7	65.5	79.8	101.9	65.5	81.6	96.7	65.5	79.3	109.0	64.6	84.3
COVERS	DIVS. 10 - 14, 25, 28, 41, 43, 44, 46	100.0	73.6	94.7	100.0	76.3	95.2	100.0	76.3	95.2	100.0	76.3	95.2	100.0	76.3	95.2	100.0	73.6	94.7
21, 22, 23	FIRE SUPPRESSION, PLUMBING & HVAC	100.4	70.5	88.4	97.1	70.7	86.6	100.0	70.7	88.3	100.1	70.7	88.3	97.1	70.7	86.6	97.0	70.5	86.4
26, 27, 3370	ELECTRICAL, COMMUNICATIONS & UTIL.	89.9	72.9	81.0	88.5	72.9	80.4	89.6	72.9	80.9	100.8	72.9	86.3	86.5	72.9	79.4	90.1	72.9	81.1
MF2010	WEIGHTED AVERAGE	95.8	70.4	84.7	97.7	73.6	87.2	99.8	73.6	88.4	100.5	73.6	88.8	97.3	73.6	87.0	97.0	70.4	85.4

		NEW MEXICO			NEW YORK												
	DIVISION	TUCUMCARI			ALBANY			BINGHAMTON			BRONX			BROOKLYN			BUFFALO	
		884			120 - 122			137 - 139			104			112			140 - 142	
		MAT.	INST.	TOTAL	MAT.	INST.	TOTAL	MAT.	INST.	TOTAL	MAT.	INST.	TOTAL	MAT.	INST.	TOTAL	MAT.	INST.	TOTAL
015433	CONTRACTOR EQUIPMENT		109.6	109.6		113.5	113.5		114.4	114.4		112.8	112.8		114.1	114.1		97.0	97.0
0241, 31 - 34	SITE & INFRASTRUCTURE, DEMOLITION	90.8	104.1	100.1	75.0	107.1	97.5	92.8	94.3	93.9	110.9	123.1	119.5	117.0	128.4	125.0	98.8	98.3	98.4
0310	Concrete Forming & Accessories	99.1	64.9	69.6	100.7	96.1	96.7	102.2	83.3	85.9	103.3	171.7	162.3	108.6	171.2	162.6	97.4	115.5	113.1
0320	Concrete Reinforcing	109.0	71.1	90.1	105.6	106.3	106.0	93.5	96.4	94.9	96.7	184.6	140.6	94.9	184.6	139.7	102.1	103.0	102.5
0330	Cast-in-Place Concrete	95.3	71.1	85.3	88.1	100.9	93.4	100.9	127.9	112.1	94.6	173.8	127.5	103.0	172.0	131.7	105.5	119.2	111.2
03	CONCRETE	106.7	69.3	88.3	101.0	100.3	100.6	96.2	102.7	99.4	96.5	173.2	134.4	108.3	172.0	139.7	104.6	113.7	109.1
04	MASONRY	112.9	60.6	80.5	90.9	100.3	96.7	105.8	126.6	118.7	94.3	176.4	145.1	116.4	176.4	153.5	107.3	119.6	114.9
05	METALS	98.1	87.8	94.9	101.3	109.5	103.9	96.0	118.2	102.9	100.8	149.3	115.8	104.4	140.3	115.4	99.5	95.6	98.3
06	WOOD, PLASTICS & COMPOSITES	88.9	65.5	75.7	97.7	93.6	95.4	107.3	79.3	91.5	100.1	171.4	140.4	109.3	171.1	144.1	98.7	115.2	108.0
07	THERMAL & MOISTURE PROTECTION	99.5	71.8	88.3	99.4	96.7	98.3	106.9	96.3	102.6	108.5	163.2	130.7	108.9	162.7	130.8	101.5	110.4	105.1
08	OPENINGS	98.5	68.5	91.4	99.9	92.0	98.0	90.7	81.6	88.5	87.0	172.0	107.1	88.0	166.5	106.6	91.8	103.3	94.6
0920	Plaster & Gypsum Board	75.4	64.1	67.7	105.0	93.1	96.9	111.6	78.2	88.9	105.4	173.3	151.6	106.9	173.3	152.1	103.2	115.4	111.5
0950, 0980	Ceilings & Acoustic Treatment	104.2	64.1	77.5	95.9	93.1	94.0	94.6	78.2	83.7	82.8	173.3	143.0	90.2	173.3	145.5	101.6	115.4	110.8
0960	Flooring	100.2	67.2	90.4	97.0	105.2	99.4	106.2	93.1	102.3	96.6	185.6	123.2	113.2	185.6	134.8	101.0	120.1	106.7
0970, 0990	Wall Finishes & Painting/Coating	102.9	68.1	82.0	96.8	89.4	92.4	93.3	95.3	94.5	97.9	148.0	128.0	124.2	148.0	138.5	102.4	112.5	108.5
09	FINISHES	97.6	65.5	79.7	97.8	96.7	97.2	96.7	84.9	90.1	95.3	172.0	138.1	111.3	171.8	145.0	101.0	116.9	109.9
COVERS	DIVS. 10 - 14, 25, 28, 41, 43, 44, 46	100.0	76.3	95.2	100.0	96.0	99.2	100.0	93.5	98.7	100.0	132.7	106.6	100.0	132.0	106.4	100.0	106.1	101.2
21, 22, 23	FIRE SUPPRESSION, PLUMBING & HVAC	97.2	70.5	86.5	100.0	98.8	99.5	100.0	87.3	95.1	100.2	164.7	126.1	99.8	164.4	125.7	100.0	97.1	98.8
26, 27, 3370	ELECTRICAL, COMMUNICATIONS & UTIL.	90.6	72.9	81.4	96.0	99.3	97.7	98.9	107.9	103.6	96.1	175.0	137.3	98.7	175.0	138.5	97.5	101.9	99.8
MF2010	WEIGHTED AVERAGE	98.8	73.5	87.8	98.6	100.2	99.3	98.1	99.4	98.6	98.0	163.9	126.7	102.6	162.9	128.9	99.7	105.8	102.4

		NEW YORK																
	DIVISION	ELMIRA			FAR ROCKAWAY			FLUSHING			GLENS FALLS			HICKSVILLE			JAMAICA	
		148 - 149			116			113			128			115,117,118			114	
		MAT.	INST.	TOTAL	MAT.	INST.	TOTAL	MAT.	INST.	TOTAL	MAT.	INST.	TOTAL	MAT.	INST.	TOTAL	MAT.	INST.	TOTAL
015433	CONTRACTOR EQUIPMENT		116.0	116.0		114.1	114.1		114.1	114.1		113.5	113.5		114.1	114.1		114.1	114.1
0241, 31 - 34	SITE & INFRASTRUCTURE, DEMOLITION	93.3	94.3	94.0	120.2	128.4	125.9	120.2	128.4	126.0	65.2	106.7	94.3	110.1	127.3	122.2	114.5	128.4	124.3
0310	Concrete Forming & Accessories	83.9	86.3	86.0	94.9	171.2	160.8	98.7	171.2	161.3	87.1	86.9	86.9	91.5	144.5	137.3	98.7	171.2	161.3
0320	Concrete Reinforcing	100.5	90.6	95.6	94.9	184.6	139.7	96.5	184.6	140.5	97.0	94.1	95.5	94.9	184.4	139.6	94.9	184.6	139.7
0330	Cast-in-Place Concrete	94.7	96.4	95.4	111.4	172.0	136.6	111.4	172.0	136.6	81.7	96.9	88.0	94.7	165.6	124.1	103.0	172.0	131.7
03	CONCRETE	95.0	92.3	93.7	114.3	172.0	142.8	114.9	172.0	143.1	89.9	92.6	91.3	100.1	157.8	128.6	107.6	172.0	139.4
04	MASONRY	101.4	93.7	96.6	119.7	176.4	154.8	113.9	176.4	152.6	90.1	93.2	92.0	110.5	166.0	144.9	118.1	176.4	154.2
05	METALS	95.4	114.7	101.3	104.4	140.3	115.5	104.4	140.3	115.5	94.8	105.1	98.0	105.9	138.2	115.8	104.4	140.3	115.5
06	WOOD, PLASTICS & COMPOSITES	86.3	84.9	85.5	92.2	171.1	136.7	97.1	171.1	138.8	86.1	84.9	85.4	88.7	141.5	118.5	97.1	171.1	138.8
07	THERMAL & MOISTURE PROTECTION	102.2	84.1	94.8	108.8	162.7	130.7	108.9	162.7	130.7	93.0	92.8	92.9	108.5	154.3	127.1	108.7	162.7	130.6
08	OPENINGS	92.4	83.8	90.3	86.7	166.5	105.6	86.7	166.5	105.6	91.3	84.5	89.7	86.7	150.3	101.8	86.7	166.5	105.6
0920	Plaster & Gypsum Board	100.2	84.2	89.3	95.1	173.3	148.3	97.8	173.3	149.1	95.2	84.2	87.7	94.8	142.7	127.4	97.8	173.3	149.1
0950, 0980	Ceilings & Acoustic Treatment	93.4	84.2	87.2	78.7	173.3	141.6	78.7	173.3	141.6	85.9	84.2	84.8	77.7	142.7	121.0	78.7	173.3	141.6
0960	Flooring	92.9	93.1	93.0	108.5	185.6	131.6	110.0	185.6	132.6	86.1	105.2	91.8	107.6	145.3	118.9	110.0	185.6	132.6
0970, 0990	Wall Finishes & Painting/Coating	101.7	87.6	93.2	124.2	148.0	138.5	124.2	148.0	138.5	94.1	89.3	91.2	124.2	148.0	138.5	124.2	148.0	138.5
09	FINISHES	96.1	87.4	91.3	106.4	171.8	142.8	107.1	171.8	143.1	89.1	89.8	89.5	104.9	143.7	126.5	106.6	171.8	142.9
COVERS	DIVS. 10 - 14, 25, 28, 41, 43, 44, 46	100.0	93.4	98.7	100.0	132.0	106.4	100.0	132.0	106.4	100.0	92.7	98.5	100.0	125.1	105.0	100.0	132.0	106.4
21, 22, 23	FIRE SUPPRESSION, PLUMBING & HVAC	95.3	87.8	92.3	94.9	164.4	122.8	94.9	164.4	122.8	95.4	92.0	94.0	99.8	151.8	120.6	94.9	164.4	122.8
26, 27, 3370	ELECTRICAL, COMMUNICATIONS & UTIL.	95.5	96.9	96.2	106.1	175.0	142.1	106.1	175.0	142.1	91.1	99.3	95.4	98.0	135.8	117.7	96.9	175.0	137.7
MF2010	WEIGHTED AVERAGE	95.9	93.1	94.7	102.4	162.9	128.7	102.3	162.9	128.7	92.6	95.0	93.6	100.7	146.6	120.7	100.6	162.9	127.7

		NEW YORK																
	DIVISION	JAMESTOWN			KINGSTON			LONG ISLAND CITY			MONTICELLO			MOUNT VERNON			NEW ROCHELLE	
		147			124			111			127			105			108	
		MAT.	INST.	TOTAL	MAT.	INST.	TOTAL	MAT.	INST.	TOTAL	MAT.	INST.	TOTAL	MAT.	INST.	TOTAL	MAT.	INST.	TOTAL
015433	CONTRACTOR EQUIPMENT		94.0	94.0		114.1	114.1		114.1	114.1		114.1	114.1		112.8	112.8		112.8	112.8
0241, 31 - 34	SITE & INFRASTRUCTURE, DEMOLITION	94.5	93.8	94.0	122.1	124.3	123.7	118.1	128.4	125.3	117.5	124.2	122.2	117.9	118.8	118.6	117.1	118.8	118.3
0310	Concrete Forming & Accessories	83.9	81.2	81.6	88.7	106.0	103.6	103.1	171.2	161.9	95.7	105.0	103.7	93.5	128.5	123.7	108.6	128.5	125.8
0320	Concrete Reinforcing	100.7	98.6	99.6	97.4	140.6	119.0	94.9	184.6	139.7	96.6	140.4	118.5	95.2	183.2	139.2	95.3	183.2	139.2
0330	Cast-in-Place Concrete	98.3	91.3	95.4	110.4	131.3	119.1	106.3	172.0	133.6	103.4	126.8	113.1	105.7	128.9	115.3	105.7	128.9	115.3
03	CONCRETE	98.0	88.0	93.0	112.5	120.8	116.6	110.7	172.0	140.9	107.2	118.8	112.9	105.9	137.8	121.6	105.4	137.8	121.4
04	MASONRY	109.1	91.1	98.0	104.0	136.8	124.3	113.3	176.4	152.3	98.0	131.0	118.4	100.0	126.1	116.2	100.0	126.1	116.2
05	METALS	92.7	90.9	92.2	103.3	116.5	107.4	104.4	140.3	115.4	103.3	116.1	107.2	100.6	134.6	111.1	100.9	134.6	111.3
06	WOOD, PLASTICS & COMPOSITES	84.8	78.6	81.3	87.3	98.6	93.7	103.3	171.1	141.5	95.2	98.6	97.1	89.3	131.7	113.2	107.6	131.7	121.2
07	THERMAL & MOISTURE PROTECTION	101.6	84.0	94.5	114.3	136.3	123.3	108.8	162.7	130.7	114.0	134.0	122.1	109.4	135.8	120.1	109.4	135.8	120.2
08	OPENINGS	92.2	82.5	89.9	92.8	117.5	98.7	86.7	166.5	105.6	88.5	117.5	95.4	87.0	145.0	100.7	87.0	145.0	100.7
0920	Plaster & Gypsum Board	93.2	77.7	82.6	97.2	98.6	98.1	102.6	173.3	150.7	99.5	98.6	98.9	100.1	132.3	122.0	112.8	132.3	126.1
0950, 0980	Ceilings & Acoustic Treatment	90.7	77.7	82.0	74.3	98.6	90.5	78.7	173.3	141.6	74.3	98.6	90.5	81.1	132.3	115.2	81.1	132.3	115.2
0960	Flooring	95.9	89.7	94.1	103.1	73.8	94.4	111.4	185.6	133.6	105.4	73.8	95.9	89.0	180.7	116.4	95.2	180.7	120.8
0970, 0990	Wall Finishes & Painting/Coating	103.0	91.3	96.0	125.9	113.8	118.6	124.2	148.0	138.5	125.9	113.8	118.6	96.1	148.0	127.3	96.1	148.0	127.3
09	FINISHES	95.6	83.0	88.6	103.7	98.4	100.7	107.9	171.8	143.5	104.2	97.8	100.7	92.8	139.7	118.9	96.0	139.7	120.3
COVERS	DIVS. 10 - 14, 25, 28, 41, 43, 44, 46	100.0	94.8	99.0	100.0	109.7	101.9	100.0	132.0	106.4	100.0	108.9	101.8	100.0	117.8	103.6	100.0	117.8	103.6
21, 22, 23	FIRE SUPPRESSION, PLUMBING & HVAC	95.2	84.6	91.0	95.3	121.1	105.6	99.8	164.4	125.7	95.3	117.9	104.4	95.4	118.6	104.7	95.4	118.6	104.7
26, 27, 3370	ELECTRICAL, COMMUNICATIONS & UTIL.	94.3	90.4	92.3	92.7	113.7	103.6	97.5	175.0	137.9	92.7	113.7	103.6	94.4	158.9	128.1	94.4	158.9	128.1
MF2010	WEIGHTED AVERAGE	96.0	87.8	92.4	100.7	118.3	108.4	102.2	162.9	128.6	99.4	116.5	106.8	97.9	133.3	113.3	98.2	133.3	113.5

515

NEW YORK

DIVISION		NEW YORK 100 - 102			NIAGARA FALLS 143			PLATTSBURGH 129			POUGHKEEPSIE 125 - 126			QUEENS 110			RIVERHEAD 119		
		MAT.	INST.	TOTAL	MAT.	INST.	TOTAL	MAT.	INST.	TOTAL	MAT.	INST.	TOTAL	MAT.	INST.	TOTAL	MAT.	INST.	TOTAL
015433	CONTRACTOR EQUIPMENT		113.2	113.2		94.0	94.0		99.2	99.2		114.1	114.1		114.1	114.1		114.1	114.1
0241, 31 - 34	SITE & INFRASTRUCTURE, DEMOLITION	119.6	123.8	122.5	96.6	95.4	95.8	89.9	102.7	98.9	118.3	124.0	122.3	113.4	128.4	123.9	111.3	127.3	122.5
0310	Concrete Forming & Accessories	107.7	179.3	169.5	83.9	113.2	109.2	91.9	83.8	84.9	88.7	148.2	140.0	91.7	171.2	160.3	96.0	144.5	137.9
0320	Concrete Reinforcing	102.4	184.7	143.5	99.3	99.3	99.3	101.3	93.6	97.5	97.4	140.9	119.1	96.5	184.6	140.5	96.7	184.4	140.5
0330	Cast-in-Place Concrete	106.4	179.2	136.7	101.6	118.7	108.7	100.1	92.7	97.0	106.9	127.3	115.4	97.9	172.0	128.7	96.3	165.6	125.0
03	CONCRETE	106.7	178.5	142.1	100.2	111.7	105.9	104.1	88.4	96.3	109.6	138.2	123.7	103.3	172.0	137.2	100.8	157.8	128.9
04	MASONRY	104.1	178.4	150.0	117.3	118.8	118.2	85.9	87.3	86.8	97.8	128.1	116.6	107.8	176.4	150.2	115.8	166.0	146.9
05	METALS	114.1	149.5	125.0	95.4	92.3	94.5	99.2	85.5	95.0	103.3	118.5	108.0	104.4	140.3	115.4	106.3	138.2	116.1
06	WOOD, PLASTICS & COMPOSITES	104.6	180.6	147.5	84.7	110.9	99.5	94.1	81.9	87.2	87.3	158.2	127.3	88.8	171.1	135.3	94.0	141.5	120.8
07	THERMAL & MOISTURE PROTECTION	108.5	165.3	131.5	101.7	108.0	104.3	108.4	89.8	100.9	114.3	138.8	124.2	108.5	162.7	130.5	109.2	154.3	127.5
08	OPENINGS	92.6	177.5	112.7	92.2	100.2	94.1	98.8	80.1	94.4	92.8	149.1	106.2	86.7	166.5	105.6	86.7	150.3	101.8
0920	Plaster & Gypsum Board	112.2	182.7	160.2	93.2	111.0	105.3	119.4	80.6	93.0	97.2	160.0	139.9	94.8	173.3	148.2	96.1	142.7	127.8
0950, 0980	Ceilings & Acoustic Treatment	101.5	182.7	155.5	90.7	111.0	104.2	100.8	80.6	87.4	74.3	160.0	131.3	78.7	173.3	141.6	78.5	142.7	121.3
0960	Flooring	97.6	185.6	123.9	95.9	113.8	101.3	109.9	105.2	108.5	103.1	148.0	116.6	107.6	185.6	130.9	108.5	145.3	119.5
0970, 0990	Wall Finishes & Painting/Coating	97.9	156.4	133.1	103.0	109.7	107.0	123.3	76.4	95.1	125.9	113.8	118.6	124.2	148.0	138.5	124.2	148.0	138.5
09	FINISHES	101.2	178.9	144.5	95.7	113.6	105.7	100.9	86.4	92.9	103.5	146.3	127.3	105.3	171.8	142.3	105.5	143.7	126.8
COVERS	DIVS. 10 - 14, 25, 28, 41, 43, 44, 46	100.0	137.1	107.4	100.0	104.7	100.9	100.0	48.9	89.8	100.0	113.3	102.7	100.0	132.0	106.4	100.0	125.1	105.0
21, 22, 23	FIRE SUPPRESSION, PLUMBING & HVAC	100.1	165.7	126.4	95.2	97.1	95.9	95.3	92.0	94.0	95.3	117.9	104.4	99.8	164.4	125.7	100.0	151.8	120.7
26, 27, 3370	ELECTRICAL, COMMUNICATIONS & UTIL.	104.2	175.0	141.1	92.9	99.3	96.3	89.2	90.4	89.8	92.7	113.7	103.6	98.0	175.0	138.2	99.7	135.8	118.5
MF2010	WEIGHTED AVERAGE	103.8	166.4	131.1	96.9	103.9	100.0	97.3	88.4	93.4	100.0	127.5	112.0	100.8	162.9	127.8	101.4	146.6	121.1

NEW YORK

DIVISION		ROCHESTER 144 - 146			SCHENECTADY 123			STATEN ISLAND 103			SUFFERN 109			SYRACUSE 130 - 132			UTICA 133 - 135		
		MAT.	INST.	TOTAL	MAT.	INST.	TOTAL	MAT.	INST.	TOTAL	MAT.	INST.	TOTAL	MAT.	INST.	TOTAL	MAT.	INST.	TOTAL
015433	CONTRACTOR EQUIPMENT		116.4	116.4		113.5	113.5		112.8	112.8		112.8	112.8		113.5	113.5		113.5	113.5
0241, 31 - 34	SITE & INFRASTRUCTURE, DEMOLITION	77.8	109.6	100.2	74.9	107.1	97.5	122.7	123.1	123.0	113.8	119.1	117.5	91.8	106.6	102.2	70.8	104.5	94.4
0310	Concrete Forming & Accessories	104.8	92.4	94.1	103.1	96.1	97.1	93.0	171.7	160.9	102.1	131.2	127.2	101.1	91.2	92.5	102.4	82.2	85.0
0320	Concrete Reinforcing	101.9	90.7	96.3	96.0	106.3	101.2	96.7	184.6	140.6	95.3	140.9	118.1	94.4	94.8	94.6	94.4	91.9	93.2
0330	Cast-in-Place Concrete	97.6	98.9	98.1	98.6	100.9	99.5	105.7	173.8	134.0	102.5	134.8	115.9	93.8	105.4	98.6	85.8	98.3	91.0
03	CONCRETE	107.3	95.2	101.3	104.5	100.3	102.4	107.6	173.2	140.0	102.4	133.3	117.6	99.5	97.5	98.6	97.8	90.6	94.2
04	MASONRY	100.2	98.6	99.2	88.1	100.3	95.7	106.7	176.4	149.8	99.4	138.2	123.4	98.1	106.3	103.2	90.0	97.7	94.8
05	METALS	101.4	104.5	102.3	98.8	109.5	102.1	98.7	149.3	114.3	98.7	118.1	104.7	99.5	105.0	101.2	97.5	103.8	99.5
06	WOOD, PLASTICS & COMPOSITES	103.8	90.6	96.4	106.0	93.6	99.0	87.8	171.4	135.0	99.7	131.7	117.7	103.7	87.6	94.6	103.7	79.4	90.0
07	THERMAL & MOISTURE PROTECTION	100.5	98.0	99.5	94.2	96.7	95.3	108.8	163.2	130.9	109.3	140.9	122.1	102.5	97.9	100.6	90.9	94.4	92.3
08	OPENINGS	96.7	86.9	94.4	96.9	92.0	95.7	87.0	172.2	107.1	87.0	135.5	98.5	92.5	80.4	89.6	95.2	79.2	91.4
0920	Plaster & Gypsum Board	111.1	90.3	96.9	106.6	93.1	97.4	100.2	173.3	149.9	104.6	132.3	123.5	101.9	87.0	91.7	101.9	78.6	86.0
0950, 0980	Ceilings & Acoustic Treatment	99.4	90.3	93.3	92.0	93.1	92.8	82.8	173.3	143.0	81.1	132.3	115.2	94.6	87.0	89.5	94.6	78.6	83.9
0960	Flooring	92.3	104.9	96.1	92.6	105.2	96.3	92.9	185.6	120.6	92.0	145.3	107.9	93.6	94.9	94.0	90.9	95.0	92.1
0970, 0990	Wall Finishes & Painting/Coating	102.9	97.4	99.6	94.1	89.4	91.3	97.9	148.0	128.0	96.1	121.7	111.5	98.3	97.0	97.5	90.8	97.0	94.5
09	FINISHES	103.1	95.2	98.7	94.5	96.7	95.8	94.7	172.0	137.8	93.8	130.2	114.1	96.0	92.1	93.8	94.1	85.4	89.2
COVERS	DIVS. 10 - 14, 25, 28, 41, 43, 44, 46	100.0	97.0	99.4	100.0	96.0	99.2	100.0	132.7	106.6	100.0	121.8	104.4	100.0	96.3	99.3	100.0	91.9	98.4
21, 22, 23	FIRE SUPPRESSION, PLUMBING & HVAC	100.0	87.9	95.1	100.2	98.8	99.6	100.2	164.7	126.1	95.4	119.6	105.1	100.2	91.2	96.6	100.2	88.8	95.6
26, 27, 3370	ELECTRICAL, COMMUNICATIONS & UTIL.	99.2	94.3	96.7	95.5	99.3	97.5	96.1	175.0	137.3	102.3	121.5	112.3	99.1	97.7	98.4	97.3	97.7	97.5
MF2010	WEIGHTED AVERAGE	100.3	95.6	98.3	97.8	100.2	98.8	99.7	163.9	127.6	98.0	126.2	110.3	98.5	97.0	97.9	96.7	93.2	95.1

DIVISION		NEW YORK — WATERTOWN 136			WHITE PLAINS 106			YONKERS 107			NORTH CAROLINA — ASHEVILLE 287 - 288			CHARLOTTE 281 - 282			DURHAM 277		
		MAT.	INST.	TOTAL	MAT.	INST.	TOTAL	MAT.	INST.	TOTAL	MAT.	INST.	TOTAL	MAT.	INST.	TOTAL	MAT.	INST.	TOTAL
015433	CONTRACTOR EQUIPMENT		113.5	113.5		112.8	112.8		112.8	112.8		95.6	95.6		95.6	95.6		100.8	100.8
0241, 31 - 34	SITE & INFRASTRUCTURE, DEMOLITION	78.1	106.1	97.8	110.6	118.8	116.4	119.1	118.8	118.9	102.4	76.6	84.3	106.0	76.7	85.4	102.6	85.3	90.4
0310	Concrete Forming & Accessories	87.5	87.4	87.4	107.1	128.5	125.6	107.4	128.5	125.7	95.4	42.1	49.4	98.6	43.7	51.2	100.0	45.1	52.7
0320	Concrete Reinforcing	95.0	94.8	94.9	95.3	183.2	139.2	99.1	183.2	141.1	94.4	63.0	78.7	100.3	58.3	79.3	94.8	57.6	76.2
0330	Cast-in-Place Concrete	99.8	91.8	96.5	93.8	128.9	108.4	105.0	128.9	114.9	110.7	51.9	86.3	115.3	50.4	88.4	104.1	47.6	80.7
03	CONCRETE	110.2	91.2	100.8	96.1	137.8	116.7	105.6	137.9	121.5	102.3	51.2	77.1	104.9	50.5	78.1	98.8	50.1	74.8
04	MASONRY	91.0	100.2	96.7	98.9	126.1	115.7	103.6	126.1	117.5	81.7	44.5	58.7	88.9	52.1	66.1	85.5	38.6	56.4
05	METALS	97.6	104.8	99.8	100.3	134.6	110.9	109.9	134.6	117.5	94.2	82.7	90.7	95.0	81.3	90.8	108.6	80.2	99.9
06	WOOD, PLASTICS & COMPOSITES	85.0	86.3	85.8	105.6	131.7	120.3	105.4	131.7	120.3	96.4	41.2	65.2	101.6	42.8	68.4	99.5	45.7	69.1
07	THERMAL & MOISTURE PROTECTION	91.2	94.9	92.7	109.2	135.8	120.0	109.5	135.8	120.2	103.8	44.5	79.7	98.5	47.5	77.8	106.6	46.6	82.2
08	OPENINGS	95.2	83.2	92.4	87.0	145.0	100.7	90.0	145.3	103.1	94.7	44.6	82.9	99.9	45.5	87.1	97.8	47.7	85.9
0920	Plaster & Gypsum Board	92.2	85.7	87.7	107.6	132.3	124.4	111.8	132.3	125.8	97.5	39.2	57.8	97.8	40.9	59.1	108.8	43.8	64.6
0950, 0980	Ceilings & Acoustic Treatment	94.6	85.7	88.6	81.1	132.3	115.2	99.7	132.3	121.4	91.6	39.2	56.7	96.0	40.9	59.3	95.1	43.8	61.0
0960	Flooring	84.8	95.0	87.9	93.8	180.7	119.8	93.4	180.7	119.5	101.5	43.5	84.1	101.1	44.2	84.1	106.2	43.5	87.4
0970, 0990	Wall Finishes & Painting/Coating	90.8	91.4	91.2	96.1	148.0	127.3	96.1	148.0	127.3	107.1	41.1	67.4	107.3	50.4	73.1	103.4	38.2	64.2
09	FINISHES	91.7	88.8	90.1	94.3	139.7	119.5	99.4	139.7	121.8	96.8	41.5	66.0	97.0	43.8	67.4	100.2	43.8	68.8
COVERS	DIVS. 10 - 14, 25, 28, 41, 43, 44, 46	100.0	93.8	98.8	100.0	117.8	103.6	100.0	120.5	104.1	100.0	78.2	95.6	100.0	78.6	95.7	100.0	73.8	94.8
21, 22, 23	FIRE SUPPRESSION, PLUMBING & HVAC	100.2	81.5	92.7	100.4	118.6	107.7	100.4	118.6	107.7	100.4	53.5	81.6	100.0	55.3	82.1	100.3	53.1	81.4
26, 27, 3370	ELECTRICAL, COMMUNICATIONS & UTIL.	99.1	90.5	94.6	94.4	158.9	128.1	102.4	158.9	131.9	96.8	57.0	76.0	95.7	59.9	77.1	94.4	45.6	68.9
MF2010	WEIGHTED AVERAGE	98.1	91.9	95.4	98.0	133.3	113.4	102.5	133.4	116.0	97.6	55.9	79.4	98.7	57.6	80.8	100.2	54.4	80.2

		NORTH CAROLINA																
	DIVISION	ELIZABETH CITY			FAYETTEVILLE			GASTONIA			GREENSBORO			HICKORY			KINSTON		
		279			283			280			270,272 - 274			286			285		
		MAT.	INST.	TOTAL	MAT.	INST.	TOTAL	MAT.	INST.	TOTAL	MAT.	INST.	TOTAL	MAT.	INST.	TOTAL	MAT.	INST.	TOTAL
015433	CONTRACTOR EQUIPMENT		105.0	105.0		100.8	100.8		95.6	95.6		100.8	100.8		100.8	100.8		100.8	100.8
0241, 31 - 34	SITE & INFRASTRUCTURE, DEMOLITION	107.1	87.0	93.0	101.6	85.2	90.1	102.2	76.6	84.3	102.5	85.3	90.4	101.3	85.1	90.0	100.4	85.2	89.7
0310	Concrete Forming & Accessories	85.5	43.3	49.1	95.0	60.4	65.1	101.9	39.5	48.0	99.7	45.4	52.8	91.6	37.9	45.2	87.9	43.0	49.2
0320	Concrete Reinforcing	92.8	47.3	70.1	98.0	57.7	77.9	94.8	58.3	76.6	93.7	57.8	75.8	94.4	57.9	76.1	93.9	47.2	70.6
0330	Cast-in-Place Concrete	104.3	47.4	80.7	116.1	48.9	88.2	108.2	53.4	85.5	103.4	48.2	80.5	110.7	48.7	84.9	106.9	45.4	81.3
03	CONCRETE	98.8	47.3	73.4	104.4	57.3	81.2	100.9	49.7	75.6	98.3	50.4	74.7	102.0	47.3	75.0	98.9	46.4	73.0
04	MASONRY	97.5	48.8	67.4	84.7	39.6	56.8	85.8	51.7	64.7	82.9	42.1	57.6	72.2	44.4	55.0	77.5	49.0	59.8
05	METALS	96.1	76.9	90.2	113.2	80.4	103.1	94.9	81.2	90.7	101.9	80.4	95.3	94.3	79.9	89.9	93.2	76.0	87.9
06	WOOD, PLASTICS & COMPOSITES	82.0	44.5	60.9	95.9	66.6	79.4	105.5	37.4	67.1	99.2	45.9	69.1	90.9	36.0	59.9	87.7	43.9	63.0
07	THERMAL & MOISTURE PROTECTION	105.9	44.1	80.8	103.7	46.6	80.5	104.0	46.0	80.5	106.3	43.7	80.9	104.1	42.5	79.1	103.9	43.7	79.5
08	OPENINGS	95.0	39.3	81.8	94.8	59.1	86.4	98.4	42.0	85.1	97.8	47.8	86.0	94.8	37.6	81.2	94.9	44.2	82.9
0920	Plaster & Gypsum Board	100.1	42.0	60.6	102.8	65.4	77.3	105.1	35.3	57.6	110.5	44.0	65.3	97.5	33.9	54.2	97.6	42.0	59.8
0950, 0980	Ceilings & Acoustic Treatment	95.1	42.0	59.7	92.5	65.4	74.5	95.2	35.3	55.3	95.1	44.0	61.1	91.6	33.9	53.2	95.2	42.0	59.8
0960	Flooring	98.3	23.6	76.0	101.7	43.5	84.3	104.8	43.5	86.5	106.2	40.4	86.5	101.4	33.5	81.1	98.7	23.0	76.1
0970, 0990	Wall Finishes & Painting/Coating	103.4	42.9	67.0	107.1	33.9	63.1	107.1	41.1	67.4	103.4	32.3	60.7	107.1	41.1	67.4	107.1	40.1	66.8
09	FINISHES	97.1	39.5	65.0	97.7	55.5	74.2	99.5	39.4	66.0	100.5	42.8	68.3	96.9	36.2	63.1	96.8	38.8	64.5
COVERS	DIVS. 10 - 14, 25, 28, 41, 43, 44, 46	100.0	72.2	94.4	100.0	76.0	95.2	100.0	77.9	95.6	100.0	77.8	95.6	100.0	77.6	95.5	100.0	73.4	94.7
21, 22, 23	FIRE SUPPRESSION, PLUMBING & HVAC	95.3	52.2	78.0	100.3	52.8	81.3	100.4	53.5	81.6	100.2	53.3	81.4	95.5	53.3	78.6	95.5	52.3	78.2
26, 27, 3370	ELECTRICAL, COMMUNICATIONS & UTIL.	94.1	35.3	63.4	95.6	51.9	72.8	96.2	58.7	76.6	93.4	57.0	74.4	94.2	58.7	75.7	94.0	47.2	69.6
MF2010	WEIGHTED AVERAGE	96.9	52.2	77.4	100.9	58.5	82.5	98.4	56.1	80.0	98.8	56.3	80.3	95.6	54.9	77.9	95.3	53.7	77.2

		NORTH CAROLINA															NORTH DAKOTA		
	DIVISION	MURPHY			RALEIGH			ROCKY MOUNT			WILMINGTON			WINSTON-SALEM			BISMARCK		
		289			275 - 276			278			284			271			585		
		MAT.	INST.	TOTAL	MAT.	INST.	TOTAL	MAT.	INST.	TOTAL	MAT.	INST.	TOTAL	MAT.	INST.	TOTAL	MAT.	INST.	TOTAL
015433	CONTRACTOR EQUIPMENT		95.6	95.6		100.8	100.8		100.8	100.8		95.6	95.6		100.8	100.8		99.0	99.0
0241, 31 - 34	SITE & INFRASTRUCTURE, DEMOLITION	103.6	76.5	84.5	103.7	85.3	90.8	105.0	85.3	91.2	103.6	76.8	84.8	102.8	85.3	90.5	100.7	97.6	98.5
0310	Concrete Forming & Accessories	102.6	40.2	48.7	98.7	47.6	54.6	92.1	45.2	51.6	96.8	49.9	56.3	101.6	49.0	56.2	103.6	41.3	49.8
0320	Concrete Reinforcing	93.9	46.9	70.4	100.0	57.8	78.9	92.8	57.6	75.2	95.1	57.6	76.4	93.7	57.6	75.7	105.6	83.2	94.4
0330	Cast-in-Place Concrete	114.5	40.0	85.7	109.3	52.0	85.5	102.0	48.3	79.7	110.3	50.2	85.5	105.9	50.6	83.0	109.5	48.4	84.1
03	CONCRETE	105.3	45.0	75.5	102.0	52.8	77.7	99.5	50.3	75.2	102.2	53.1	77.9	99.4	52.8	76.5	103.6	53.0	78.6
04	MASONRY	74.5	41.9	54.4	88.1	42.9	60.2	76.2	40.2	53.9	72.4	42.4	53.9	83.0	39.3	56.0	110.9	57.4	77.8
05	METALS	92.1	75.3	87.0	94.8	80.4	90.4	95.3	79.6	90.5	93.9	80.2	89.7	99.4	80.2	93.5	97.7	85.4	93.9
06	WOOD, PLASTICS & COMPOSITES	106.2	40.5	69.2	97.9	48.6	70.1	89.5	45.7	64.8	98.8	50.1	71.3	99.2	50.4	71.7	87.3	35.6	58.1
07	THERMAL & MOISTURE PROTECTION	104.0	41.6	78.6	100.6	47.9	79.2	106.3	41.6	80.1	103.8	47.9	81.1	106.3	44.3	81.2	109.3	50.8	85.6
08	OPENINGS	94.7	39.8	81.7	97.0	48.5	85.5	94.3	42.9	82.1	94.9	51.7	84.6	97.8	50.3	86.5	100.9	47.9	88.4
0920	Plaster & Gypsum Board	104.2	38.5	59.5	102.5	46.8	64.6	102.1	43.8	62.5	100.3	48.3	65.0	110.5	48.7	68.5	94.3	33.8	53.1
0950, 0980	Ceilings & Acoustic Treatment	91.6	38.5	56.3	95.9	46.8	63.2	92.4	43.8	60.1	92.5	48.3	63.1	95.1	48.7	64.2	136.9	33.8	68.3
0960	Flooring	105.2	24.5	81.1	103.4	43.5	85.4	101.8	22.2	78.0	102.3	45.3	85.2	106.2	43.5	87.4	109.8	73.4	98.9
0970, 0990	Wall Finishes & Painting/Coating	107.1	40.2	66.9	102.1	37.7	63.4	103.4	39.3	64.9	107.1	39.5	66.5	103.4	37.5	63.8	94.1	30.0	55.6
09	FINISHES	99.0	36.8	64.3	99.7	45.7	69.6	97.9	40.2	65.8	97.6	47.8	69.9	100.5	46.7	70.5	111.7	43.8	73.9
COVERS	DIVS. 10 - 14, 25, 28, 41, 43, 44, 46	100.0	77.5	95.5	100.0	74.5	94.9	100.0	74.4	94.9	100.0	75.7	95.1	100.0	79.3	95.9	100.0	84.8	97.0
21, 22, 23	FIRE SUPPRESSION, PLUMBING & HVAC	95.5	52.0	78.0	100.0	53.6	81.4	95.3	53.9	78.7	100.4	55.1	82.2	100.2	53.5	81.5	100.2	61.3	84.6
26, 27, 3370	ELECTRICAL, COMMUNICATIONS & UTIL.	97.8	30.0	62.4	96.5	41.3	67.7	96.1	40.6	67.1	97.0	51.9	73.4	93.4	57.0	74.4	102.2	73.3	87.1
MF2010	WEIGHTED AVERAGE	96.4	49.2	75.8	98.2	55.1	79.4	96.0	53.2	77.4	97.2	56.6	79.5	98.5	57.1	80.5	102.0	64.0	85.5

		NORTH DAKOTA																	
	DIVISION	DEVILS LAKE			DICKINSON			FARGO			GRAND FORKS			JAMESTOWN			MINOT		
		583			586			580 - 581			582			584			587		
		MAT.	INST.	TOTAL	MAT.	INST.	TOTAL	MAT.	INST.	TOTAL	MAT.	INST.	TOTAL	MAT.	INST.	TOTAL	MAT.	INST.	TOTAL
015433	CONTRACTOR EQUIPMENT		99.0	99.0		99.0	99.0		99.0	99.0		99.0	99.0		99.0	99.0		99.0	99.0
0241, 31 - 34	SITE & INFRASTRUCTURE, DEMOLITION	103.3	95.2	97.6	111.5	93.7	99.0	98.4	97.7	97.9	107.3	93.7	97.8	102.3	93.7	96.3	104.8	97.6	99.8
0310	Concrete Forming & Accessories	102.6	36.0	45.1	91.7	34.9	42.7	102.5	42.2	50.5	95.6	34.8	43.1	93.2	34.6	42.6	91.4	66.0	69.5
0320	Concrete Reinforcing	101.9	83.5	92.7	102.8	41.0	71.9	94.5	83.3	88.9	100.3	83.3	91.8	102.5	48.3	75.4	103.8	83.6	93.7
0330	Cast-in-Place Concrete	122.9	47.0	91.4	111.2	45.2	83.8	100.9	50.3	79.9	111.2	45.1	83.8	121.4	45.0	89.7	111.2	47.5	84.7
03	CONCRETE	110.8	50.2	80.9	110.1	41.0	76.0	104.7	54.0	79.7	107.3	48.7	78.4	109.3	42.2	76.2	105.8	63.7	85.0
04	MASONRY	111.7	64.4	82.4	113.4	59.2	79.8	97.9	57.4	72.9	106.2	63.8	80.0	123.8	33.4	67.8	105.6	64.2	80.0
05	METALS	95.1	84.6	91.9	95.1	64.1	85.5	99.0	86.0	95.0	95.1	79.3	90.2	95.1	65.1	85.8	95.4	86.3	92.6
06	WOOD, PLASTICS & COMPOSITES	88.7	32.8	57.1	76.6	32.8	51.9	86.8	36.1	58.2	80.8	32.8	53.7	78.5	32.8	52.7	76.4	68.7	72.0
07	THERMAL & MOISTURE PROTECTION	106.7	49.4	83.4	107.1	47.3	82.8	108.7	51.8	85.6	106.9	49.4	83.5	106.6	41.5	80.1	106.6	56.0	86.1
08	OPENINGS	96.6	40.8	83.4	96.5	30.9	81.0	98.8	48.2	86.8	96.6	40.8	83.4	96.6	33.0	81.5	96.7	66.0	89.4
0920	Plaster & Gypsum Board	115.2	30.9	57.9	105.2	30.9	54.7	91.5	34.3	52.6	106.7	30.9	55.1	106.3	30.9	55.0	105.2	67.9	79.8
0950, 0980	Ceilings & Acoustic Treatment	142.1	30.9	68.1	142.1	30.9	68.1	138.7	34.3	69.2	142.1	30.9	68.1	142.1	30.9	68.1	142.1	67.9	92.7
0960	Flooring	117.0	36.5	92.9	110.5	36.5	88.4	109.8	73.4	98.9	112.4	36.5	89.7	111.2	36.5	88.9	110.3	86.4	103.1
0970, 0990	Wall Finishes & Painting/Coating	94.9	23.0	51.7	94.9	23.0	51.7	91.0	72.2	79.7	94.9	28.9	55.2	94.9	23.0	51.7	94.9	25.2	53.0
09	FINISHES	117.6	32.9	70.4	115.3	32.9	69.4	112.3	48.8	76.9	115.5	33.6	69.9	114.7	32.9	69.1	114.5	65.4	87.1
COVERS	DIVS. 10 - 14, 25, 28, 41, 43, 44, 46	100.0	33.3	86.6	100.0	33.5	86.7	100.0	84.9	97.0	100.0	33.4	86.7	100.0	82.8	96.6	100.0	88.5	97.7
21, 22, 23	FIRE SUPPRESSION, PLUMBING & HVAC	95.5	74.6	87.1	95.5	67.1	84.1	100.1	77.6	91.1	100.4	33.9	73.8	95.5	35.8	71.5	100.4	61.4	84.7
26, 27, 3370	ELECTRICAL, COMMUNICATIONS & UTIL.	99.3	36.7	66.6	109.1	74.2	90.9	104.8	67.7	85.5	103.3	54.8	78.0	99.3	36.6	66.6	106.9	77.5	91.5
MF2010	WEIGHTED AVERAGE	101.0	58.6	82.5	101.9	57.9	82.7	101.7	67.6	86.9	101.8	51.6	79.9	101.0	45.1	76.7	101.8	71.0	88.4

NORTH DAKOTA / OHIO

DIVISION		WILLISTON 588			AKRON 442 - 443			ATHENS 457			CANTON 446 - 447			CHILLICOTHE 456			CINCINNATI 451 - 452		
		MAT.	INST.	TOTAL	MAT.	INST.	TOTAL	MAT.	INST.	TOTAL	MAT.	INST.	TOTAL	MAT.	INST.	TOTAL	MAT.	INST.	TOTAL
015433	CONTRACTOR EQUIPMENT		99.0	99.0		95.3	95.3		90.7	90.7		95.3	95.3		100.3	100.3		100.0	100.0
0241, 31 - 34	SITE & INFRASTRUCTURE, DEMOLITION	105.3	93.7	97.2	96.8	102.4	100.7	111.6	92.6	98.3	96.9	102.0	100.4	96.5	104.0	101.8	93.6	103.5	100.5
0310	Concrete Forming & Accessories	97.3	34.9	43.5	100.2	92.2	93.3	94.1	85.8	86.9	100.2	82.3	84.7	96.2	92.1	92.6	98.1	80.6	83.0
0320	Concrete Reinforcing	104.8	41.0	72.9	93.9	93.3	93.6	90.1	86.6	88.4	93.9	76.5	85.2	87.0	81.0	84.0	92.2	81.7	87.0
0330	Cast-in-Place Concrete	111.2	45.2	83.8	92.4	97.4	94.5	111.9	92.2	103.7	93.3	95.2	94.1	101.6	96.1	99.3	93.6	93.6	93.6
03	CONCRETE	107.3	41.0	74.6	92.9	93.5	93.2	106.4	87.6	97.2	93.4	85.3	89.4	100.8	91.3	96.1	95.3	85.5	90.4
04	MASONRY	100.0	59.2	74.7	92.1	94.1	93.3	82.7	89.5	87.0	92.8	84.6	87.7	90.1	96.7	94.2	89.8	85.5	87.1
05	METALS	95.2	64.2	85.7	94.8	82.2	90.9	101.6	78.6	94.5	94.8	75.1	88.7	93.7	84.9	91.0	95.9	84.6	92.4
06	WOOD, PLASTICS & COMPOSITES	82.2	32.8	54.3	99.1	91.4	94.7	83.4	86.8	85.3	99.5	80.8	88.9	96.0	90.6	92.9	98.5	78.5	87.2
07	THERMAL & MOISTURE PROTECTION	106.8	47.3	82.7	111.2	95.8	104.9	99.7	91.9	96.6	112.3	91.6	103.9	101.2	94.3	98.4	99.5	89.9	95.6
08	OPENINGS	96.7	30.9	81.1	115.2	92.5	109.8	99.9	83.6	96.1	108.5	77.2	101.1	91.9	84.0	90.0	99.9	78.9	94.9
0920	Plaster & Gypsum Board	106.7	30.9	55.1	102.0	90.8	94.4	93.9	86.1	88.6	103.1	79.9	87.3	96.6	90.5	92.5	98.3	78.1	84.6
0950, 0980	Ceilings & Acoustic Treatment	142.1	30.9	68.1	98.7	90.8	93.5	102.0	86.1	91.5	98.7	79.9	86.2	96.3	90.5	92.5	97.2	78.1	84.5
0960	Flooring	113.2	36.5	90.3	98.6	92.5	96.7	124.5	100.8	117.4	98.7	82.0	93.7	101.5	100.8	101.3	102.5	90.0	98.7
0970, 0990	Wall Finishes & Painting/Coating	94.9	32.2	57.2	97.1	103.9	101.2	104.9	93.1	97.8	97.1	82.8	88.5	102.2	89.8	94.7	102.2	84.8	91.7
09	FINISHES	115.6	34.0	70.1	100.1	93.3	96.4	102.9	90.1	95.8	100.4	81.8	90.0	100.4	93.7	96.6	100.8	82.2	90.4
COVERS	DIVS. 10 - 14, 25, 28, 41, 43, 44, 46	100.0	33.5	86.7	100.0	95.3	99.1	100.0	53.3	90.6	100.0	93.1	98.6	100.0	93.1	98.6	100.0	91.4	98.3
21, 22, 23	FIRE SUPPRESSION, PLUMBING & HVAC	95.5	67.1	84.1	100.0	94.3	97.7	95.2	52.4	78.0	100.0	82.5	93.0	95.6	94.0	94.9	100.0	81.3	92.5
26, 27, 3370	ELECTRICAL, COMMUNICATIONS & UTIL.	103.7	74.2	88.3	99.3	92.8	95.9	97.8	98.9	98.3	98.5	87.8	92.9	97.3	85.4	91.1	96.1	79.7	87.6
MF2010	WEIGHTED AVERAGE	100.3	58.0	81.9	99.8	93.4	97.0	99.1	80.6	91.1	99.2	85.0	93.0	96.4	92.2	94.5	97.8	84.8	92.1

OHIO

DIVISION		CLEVELAND 441			COLUMBUS 430 - 432			DAYTON 453 - 454			HAMILTON 450			LIMA 458			LORAIN 440		
		MAT.	INST.	TOTAL	MAT.	INST.	TOTAL	MAT.	INST.	TOTAL	MAT.	INST.	TOTAL	MAT.	INST.	TOTAL	MAT.	INST.	TOTAL
015433	CONTRACTOR EQUIPMENT		95.6	95.6		94.9	94.9		95.3	95.3		100.3	100.3		93.1	93.1		95.3	95.3
0241, 31 - 34	SITE & INFRASTRUCTURE, DEMOLITION	96.7	102.9	101.1	95.2	99.3	98.1	92.4	103.4	100.1	92.3	103.6	100.2	105.2	92.5	96.3	96.3	101.3	99.8
0310	Concrete Forming & Accessories	100.3	97.9	98.2	99.1	85.9	87.7	98.1	79.0	81.6	98.1	80.1	82.5	94.0	84.3	85.6	100.3	85.3	87.3
0320	Concrete Reinforcing	94.3	93.7	94.0	101.6	81.7	91.7	92.2	80.6	86.4	92.2	81.7	87.0	90.1	79.1	84.6	93.9	93.6	93.7
0330	Cast-in-Place Concrete	90.7	105.5	96.8	90.6	95.9	92.8	87.0	86.0	86.6	93.3	92.8	93.1	102.8	97.2	100.5	88.0	103.5	94.4
03	CONCRETE	92.2	99.0	95.8	95.8	88.3	92.1	92.2	81.5	86.9	95.1	85.0	90.1	99.3	87.6	93.5	90.8	92.6	91.7
04	MASONRY	96.6	105.1	101.9	97.9	94.8	96.0	89.3	81.7	84.6	89.6	84.0	86.1	112.9	81.7	93.6	88.8	98.5	94.8
05	METALS	96.3	85.2	92.9	95.9	80.5	91.2	95.2	77.8	89.8	95.2	85.1	92.1	101.6	80.4	95.1	95.4	83.3	91.6
06	WOOD, PLASTICS & COMPOSITES	98.2	94.8	96.2	97.5	83.7	89.7	99.7	77.3	87.1	98.5	78.5	87.2	83.3	84.0	83.7	99.1	82.0	89.4
07	THERMAL & MOISTURE PROTECTION	109.8	108.1	109.1	101.8	95.7	99.3	104.5	86.4	97.2	101.3	88.7	96.2	99.3	92.5	96.5	112.2	101.1	107.7
08	OPENINGS	104.5	94.3	102.1	103.0	80.4	97.7	99.3	78.0	94.2	96.9	78.9	92.6	99.9	78.3	94.8	108.5	87.4	103.5
0920	Plaster & Gypsum Board	101.2	94.3	96.5	97.6	83.2	87.8	98.3	76.8	83.7	98.3	78.1	84.6	93.9	83.2	86.6	102.0	81.2	87.8
0950, 0980	Ceilings & Acoustic Treatment	97.0	94.3	95.2	100.4	83.2	89.0	98.2	76.8	84.0	97.2	78.1	84.5	101.0	83.2	89.2	98.7	81.2	87.1
0960	Flooring	98.4	105.1	100.4	94.7	87.9	92.6	105.2	80.7	97.9	102.5	90.0	98.7	123.6	88.3	113.1	98.4	105.1	100.6
0970, 0990	Wall Finishes & Painting/Coating	97.1	105.2	102.0	100.9	93.3	96.3	102.2	85.2	92.0	102.2	84.8	91.7	104.9	83.8	92.2	97.1	105.2	102.0
09	FINISHES	99.7	99.6	99.6	97.5	86.6	91.4	101.7	79.5	89.3	100.7	81.9	90.2	101.9	84.7	92.3	100.2	90.4	94.7
COVERS	DIVS. 10 - 14, 25, 28, 41, 43, 44, 46	100.0	102.1	100.4	100.0	93.8	98.8	100.0	90.9	98.2	100.0	90.9	98.2	100.0	94.1	98.8	100.0	98.3	99.7
21, 22, 23	FIRE SUPPRESSION, PLUMBING & HVAC	100.0	99.8	99.9	100.0	92.3	96.9	100.8	81.7	93.1	100.5	80.5	92.5	95.2	84.7	91.0	100.0	85.2	94.0
26, 27, 3370	ELECTRICAL, COMMUNICATIONS & UTIL.	98.8	106.2	102.6	97.1	88.1	92.4	94.7	82.7	88.4	95.1	81.0	87.8	98.1	78.4	87.8	98.7	86.7	92.4
MF2010	WEIGHTED AVERAGE	98.9	100.0	99.4	98.6	89.7	94.7	97.5	83.2	91.3	97.4	84.5	91.8	99.6	84.4	93.0	98.8	90.4	95.2

OHIO

DIVISION		MANSFIELD 448 - 449			MARION 433			SPRINGFIELD 455			STEUBENVILLE 439			TOLEDO 434 - 436			YOUNGSTOWN 444 - 445		
		MAT.	INST.	TOTAL	MAT.	INST.	TOTAL	MAT.	INST.	TOTAL	MAT.	INST.	TOTAL	MAT.	INST.	TOTAL	MAT.	INST.	TOTAL
015433	CONTRACTOR EQUIPMENT		95.3	95.3		94.5	94.5		95.3	95.3		99.3	99.3		97.3	97.3		95.3	95.3
0241, 31 - 34	SITE & INFRASTRUCTURE, DEMOLITION	92.7	102.2	99.3	91.1	97.8	95.8	92.7	102.1	99.3	134.7	107.7	115.8	94.4	99.8	98.2	96.7	103.0	101.1
0310	Concrete Forming & Accessories	90.3	83.5	84.4	95.7	81.3	83.3	98.1	83.3	85.4	97.6	87.9	89.2	99.0	92.5	93.4	100.2	85.9	87.9
0320	Concrete Reinforcing	85.4	75.6	80.5	93.7	81.5	87.6	92.2	78.9	85.6	91.1	85.9	88.5	101.6	87.2	94.4	93.9	86.1	90.0
0330	Cast-in-Place Concrete	85.6	94.5	89.3	82.8	92.9	87.0	89.4	85.9	87.9	89.9	94.7	91.9	90.6	100.2	94.6	91.5	97.1	93.9
03	CONCRETE	85.6	85.4	85.5	88.0	85.1	86.5	93.3	83.1	88.3	91.7	89.3	90.5	95.8	93.9	94.9	92.5	89.3	90.9
04	MASONRY	91.2	93.7	92.7	100.0	92.4	95.3	89.5	81.7	84.6	86.5	93.8	91.0	106.6	96.8	100.6	92.4	91.1	91.6
05	METALS	95.6	75.5	89.4	95.0	78.1	89.8	95.1	77.0	89.5	91.6	79.9	87.9	95.7	86.9	93.0	94.8	79.3	90.0
06	WOOD, PLASTICS & COMPOSITES	86.8	82.0	84.1	93.5	81.7	86.9	101.1	83.4	91.1	89.3	86.1	87.5	97.5	91.5	94.1	99.1	84.1	90.6
07	THERMAL & MOISTURE PROTECTION	111.6	93.6	104.3	101.4	82.0	93.5	104.4	87.0	97.4	113.5	96.6	106.6	103.4	100.9	102.4	112.4	92.8	104.4
08	OPENINGS	108.8	76.7	101.2	97.1	77.6	92.5	97.4	78.6	92.9	97.5	82.9	94.0	100.4	88.8	97.6	108.5	84.2	102.7
0920	Plaster & Gypsum Board	95.2	81.2	85.7	95.4	81.2	85.7	98.3	83.2	88.0	95.5	85.1	88.4	97.6	91.2	93.2	102.0	83.3	89.3
0950, 0980	Ceilings & Acoustic Treatment	99.6	81.2	87.4	100.4	81.2	87.6	98.2	83.2	88.2	97.3	85.1	89.2	100.4	91.2	94.3	98.7	83.3	88.5
0960	Flooring	94.6	104.7	97.6	93.5	71.0	86.8	105.2	80.7	97.9	122.7	99.8	115.9	93.9	101.5	96.2	98.7	92.1	96.7
0970, 0990	Wall Finishes & Painting/Coating	97.1	88.5	91.9	100.9	48.5	69.4	102.2	85.2	92.0	114.2	102.6	107.2	100.9	102.1	101.6	97.1	92.6	94.4
09	FINISHES	98.0	87.7	92.3	96.4	76.1	85.1	101.7	83.1	91.4	114.3	91.2	101.4	97.2	94.8	95.9	100.3	87.3	93.0
COVERS	DIVS. 10 - 14, 25, 28, 41, 43, 44, 46	100.0	93.2	98.6	100.0	54.9	91.0	100.0	91.6	98.3	100.0	92.5	98.5	100.0	96.3	99.3	100.0	94.0	98.8
21, 22, 23	FIRE SUPPRESSION, PLUMBING & HVAC	95.1	83.4	90.4	95.2	89.7	93.0	100.8	81.7	93.1	95.5	91.9	94.0	100.1	99.2	99.7	100.0	88.4	95.3
26, 27, 3370	ELECTRICAL, COMMUNICATIONS & UTIL.	96.2	77.7	86.5	91.2	77.7	84.1	94.7	85.6	90.0	86.4	112.7	100.1	97.2	101.6	99.5	98.7	82.9	90.5
MF2010	WEIGHTED AVERAGE	96.6	85.6	91.8	95.0	83.7	90.1	97.5	84.2	91.7	96.8	94.5	95.8	98.7	96.4	97.7	99.1	88.3	94.4

City Cost Indexes

DIVISION		OHIO			OKLAHOMA														
		ZANESVILLE			ARDMORE			CLINTON			DURANT			ENID			GUYMON		
		437 - 438			734			736			747			737			739		
		MAT.	INST.	TOTAL	MAT.	INST.	TOTAL	MAT.	INST.	TOTAL	MAT.	INST.	TOTAL	MAT.	INST.	TOTAL	MAT.	INST.	TOTAL
015433	CONTRACTOR EQUIPMENT		94.5	94.5		82.5	82.5		81.6	81.6		81.6	81.6		81.6	81.6		81.6	81.6
0241, 31 - 34	SITE & INFRASTRUCTURE, DEMOLITION	93.9	99.5	97.9	94.7	91.4	92.4	96.1	90.0	91.8	92.5	89.7	90.6	97.6	90.0	92.2	99.9	89.7	92.7
0310	Concrete Forming & Accessories	92.8	82.6	84.0	94.4	41.3	48.6	92.9	47.7	53.9	85.7	45.1	50.6	96.6	34.8	43.3	100.3	45.9	53.3
0320	Concrete Reinforcing	93.1	85.3	89.2	87.6	80.2	83.9	88.1	80.2	84.2	92.3	80.6	86.4	87.5	80.2	83.9	88.1	80.2	84.1
0330	Cast-in-Place Concrete	87.2	93.3	89.8	99.3	44.2	76.4	95.9	45.9	75.2	92.1	43.5	71.9	95.9	47.2	75.7	96.0	43.9	74.3
03	CONCRETE	91.5	86.5	89.0	97.5	50.0	74.1	97.1	53.5	75.6	92.9	51.5	72.5	97.7	48.2	73.3	100.7	51.9	76.7
04	MASONRY	96.8	83.4	88.5	93.4	58.1	71.5	118.8	58.1	81.2	92.4	63.4	74.5	100.6	58.1	74.3	95.9	54.6	70.3
05	METALS	96.4	80.3	91.4	93.3	70.6	86.3	93.4	70.6	86.4	91.4	71.3	85.2	94.8	70.5	87.3	93.9	70.3	86.6
06	WOOD, PLASTICS & COMPOSITES	89.6	81.7	85.2	100.7	38.7	65.7	99.7	47.3	70.1	89.5	44.0	63.8	103.6	30.0	62.0	107.5	47.0	73.4
07	THERMAL & MOISTURE PROTECTION	101.5	91.1	97.3	106.1	61.0	87.8	106.1	62.0	88.2	99.9	62.3	84.6	106.2	60.3	87.6	106.5	58.4	87.0
08	OPENINGS	97.1	80.2	93.1	93.9	49.6	83.4	93.9	54.0	84.4	93.9	52.7	84.1	93.9	44.4	82.1	94.0	49.8	83.5
0920	Plaster & Gypsum Board	92.8	81.2	84.9	84.3	37.5	52.4	84.0	46.3	58.3	79.1	42.9	54.5	85.1	28.4	46.5	85.7	46.0	58.7
0950, 0980	Ceilings & Acoustic Treatment	100.4	81.2	87.6	86.2	37.5	53.8	86.2	46.3	59.7	85.4	42.9	57.1	86.2	28.4	47.8	88.9	46.0	60.4
0960	Flooring	92.0	86.9	90.5	110.5	43.9	90.6	109.4	41.9	89.2	106.9	63.1	93.8	111.2	41.9	90.5	112.9	24.8	86.5
0970, 0990	Wall Finishes & Painting/Coating	100.9	76.0	85.9	96.5	52.4	70.0	96.5	52.4	70.0	100.1	52.4	71.4	96.5	52.4	70.0	96.5	33.8	58.8
09	FINISHES	95.9	82.2	88.3	93.8	41.0	64.4	93.6	45.8	67.0	91.8	47.7	67.2	94.3	35.4	61.5	95.8	39.7	64.6
COVERS	DIVS. 10 - 14, 25, 28, 41, 43, 44, 46	100.0	90.0	98.0	100.0	78.0	95.6	100.0	79.0	95.8	100.0	78.3	95.7	100.0	77.1	95.4	100.0	77.8	95.5
21, 22, 23	FIRE SUPPRESSION, PLUMBING & HVAC	95.2	89.7	93.0	95.4	66.3	83.7	95.4	66.3	83.7	95.4	65.8	83.5	100.3	66.3	86.6	95.4	64.2	82.9
26, 27, 3370	ELECTRICAL, COMMUNICATIONS & UTIL.	91.5	78.2	84.6	89.7	70.4	79.6	90.7	70.4	80.1	94.7	70.4	82.0	90.7	70.4	80.1	92.3	63.0	77.0
MF2010	WEIGHTED AVERAGE	95.5	85.6	91.2	95.2	62.3	80.9	96.5	63.6	82.2	94.3	64.0	81.1	97.2	60.9	81.4	96.3	60.5	80.7

DIVISION		OKLAHOMA																	
		LAWTON			MCALESTER			MIAMI			MUSKOGEE			OKLAHOMA CITY			PONCA CITY		
		735			745			743			744			730 - 731			746		
		MAT.	INST.	TOTAL	MAT.	INST.	TOTAL	MAT.	INST.	TOTAL	MAT.	INST.	TOTAL	MAT.	INST.	TOTAL	MAT.	INST.	TOTAL
015433	CONTRACTOR EQUIPMENT		82.5	82.5		81.6	81.6		89.7	89.7		89.7	89.7		82.7	82.7		81.6	81.6
0241, 31 - 34	SITE & INFRASTRUCTURE, DEMOLITION	94.3	91.4	92.2	86.3	89.9	88.9	87.8	87.2	87.4	88.0	87.0	87.3	93.7	91.8	92.3	93.0	89.9	90.8
0310	Concrete Forming & Accessories	100.2	45.3	52.8	83.6	44.6	49.9	96.8	67.7	71.7	101.3	34.1	43.3	98.2	55.5	61.3	92.2	47.3	53.4
0320	Concrete Reinforcing	87.7	80.2	84.0	91.9	80.2	86.1	90.5	80.3	85.4	91.3	79.7	85.5	92.8	80.2	86.5	91.3	80.3	85.8
0330	Cast-in-Place Concrete	92.8	47.3	73.9	80.8	46.0	66.3	84.6	48.3	69.6	85.7	46.6	69.4	94.5	47.7	75.1	94.6	45.5	74.2
03	CONCRETE	93.8	52.9	73.6	83.4	52.1	68.0	88.2	64.0	76.2	90.0	48.3	69.4	97.3	57.5	77.7	95.0	53.1	74.3
04	MASONRY	95.9	58.1	72.5	110.7	57.1	77.5	95.2	57.2	71.7	113.0	46.0	71.5	98.7	57.4	73.1	87.8	57.1	68.8
05	METALS	98.5	70.6	89.9	91.3	70.5	84.9	91.3	82.8	88.6	92.7	81.0	89.1	92.0	70.7	85.4	91.3	70.8	84.9
06	WOOD, PLASTICS & COMPOSITES	106.4	44.1	71.3	86.9	44.0	62.7	102.4	74.6	86.7	107.0	31.4	64.3	100.3	58.2	76.5	97.6	47.0	69.1
07	THERMAL & MOISTURE PROTECTION	106.1	61.7	88.1	99.6	61.0	83.9	99.9	64.9	85.7	100.0	49.5	79.5	98.8	62.8	84.2	100.1	66.7	86.5
08	OPENINGS	95.4	52.1	85.1	93.9	52.5	84.1	93.9	69.3	88.1	93.9	42.8	81.8	94.8	59.7	86.5	93.9	53.8	84.4
0920	Plaster & Gypsum Board	87.1	43.0	57.1	77.9	42.9	54.1	85.0	74.3	77.7	87.2	29.7	48.1	92.9	57.5	68.8	83.5	46.0	58.0
0950, 0980	Ceilings & Acoustic Treatment	94.2	43.0	60.1	85.4	42.9	57.1	85.4	74.3	78.0	94.3	29.7	51.3	97.5	57.5	70.9	85.4	46.0	59.2
0960	Flooring	113.1	41.9	91.8	105.8	41.9	86.7	112.8	63.1	97.9	115.3	40.5	92.9	109.5	41.9	89.3	109.9	41.9	89.6
0970, 0990	Wall Finishes & Painting/Coating	96.5	52.4	70.0	100.1	38.6	63.1	100.1	79.1	87.5	100.1	34.9	60.9	97.9	52.4	70.6	100.1	52.4	71.4
09	FINISHES	96.4	43.8	67.1	90.8	42.0	63.6	93.7	68.8	79.8	96.7	33.7	61.6	98.6	52.0	72.6	93.4	45.3	66.6
COVERS	DIVS. 10 - 14, 25, 28, 41, 43, 44, 46	100.0	78.6	95.7	100.0	78.3	95.6	100.0	82.1	96.4	100.0	76.5	95.3	100.0	80.0	96.0	100.0	78.6	95.7
21, 22, 23	FIRE SUPPRESSION, PLUMBING & HVAC	100.3	66.3	86.7	95.4	61.5	81.8	95.4	61.8	81.9	100.3	60.5	84.3	100.0	66.0	86.4	95.4	61.7	81.9
26, 27, 3370	ELECTRICAL, COMMUNICATIONS & UTIL.	92.3	70.4	80.9	93.1	66.1	79.0	94.6	66.1	79.7	92.7	53.7	72.3	97.3	70.4	83.3	92.7	70.0	80.8
MF2010	WEIGHTED AVERAGE	97.5	63.3	82.6	93.7	61.1	79.5	94.0	68.3	82.8	96.6	56.3	79.0	97.3	65.4	83.4	94.3	62.5	80.5

DIVISION		OKLAHOMA												OREGON					
		POTEAU			SHAWNEE			TULSA			WOODWARD			BEND			EUGENE		
		749			748			740 - 741			738			977			974		
		MAT.	INST.	TOTAL	MAT.	INST.	TOTAL	MAT.	INST.	TOTAL	MAT.	INST.	TOTAL	MAT.	INST.	TOTAL	MAT.	INST.	TOTAL
015433	CONTRACTOR EQUIPMENT		89.0	89.0		81.6	81.6		89.7	89.7		81.6	81.6		97.9	97.9		97.9	97.9
0241, 31 - 34	SITE & INFRASTRUCTURE, DEMOLITION	74.3	85.9	82.4	96.0	89.9	91.7	94.2	87.4	89.4	96.3	90.0	91.9	106.8	102.8	103.9	96.9	102.7	101.0
0310	Concrete Forming & Accessories	90.0	41.9	48.5	85.5	44.7	50.3	101.4	41.1	49.4	93.1	47.8	54.0	111.2	100.6	102.0	107.6	100.4	101.4
0320	Concrete Reinforcing	92.4	80.3	86.3	91.3	80.2	85.8	91.6	80.2	85.9	87.5	80.3	83.9	95.7	101.5	98.6	99.7	101.5	100.6
0330	Cast-in-Place Concrete	84.6	45.7	68.5	97.6	43.7	75.2	93.3	47.0	74.1	95.9	46.1	75.2	94.2	104.1	98.3	91.2	104.0	96.5
03	CONCRETE	90.6	51.6	71.3	96.6	51.4	74.3	95.3	51.7	73.8	97.4	53.6	75.8	104.8	101.6	103.2	96.3	101.5	98.9
04	MASONRY	95.5	57.2	71.8	111.9	57.1	78.0	96.1	61.0	74.4	89.5	58.1	70.1	98.7	104.5	102.3	96.0	104.5	101.3
05	METALS	91.3	82.4	88.5	91.2	70.4	84.8	95.9	82.0	91.6	93.5	71.0	86.5	90.7	96.6	92.5	91.4	96.4	93.0
06	WOOD, PLASTICS & COMPOSITES	94.3	40.0	63.7	89.3	44.0	63.7	106.2	37.5	67.4	99.8	47.0	70.0	103.3	100.3	101.6	98.8	100.3	99.6
07	THERMAL & MOISTURE PROTECTION	100.0	61.0	84.2	100.0	60.3	83.9	100.0	62.6	84.8	106.2	67.1	90.3	104.6	96.3	101.2	103.9	93.4	99.6
08	OPENINGS	93.9	50.5	83.6	93.9	52.5	84.1	95.4	48.1	84.2	93.9	53.8	84.4	93.8	103.8	96.2	94.2	103.8	96.5
0920	Plaster & Gypsum Board	82.0	38.7	52.5	79.1	42.9	54.5	87.2	36.0	52.4	84.6	46.0	58.4	103.7	100.2	101.3	101.1	100.2	100.5
0950, 0980	Ceilings & Acoustic Treatment	85.4	38.7	54.3	85.4	42.9	57.1	94.3	36.0	55.5	88.9	46.0	60.4	92.3	100.2	97.5	93.3	100.2	97.9
0960	Flooring	109.3	63.1	95.4	106.4	33.3	84.9	114.0	43.6	92.9	109.4	43.9	89.8	107.6	105.3	106.9	106.1	105.3	105.9
0970, 0990	Wall Finishes & Painting/Coating	100.1	52.4	71.4	100.1	34.9	60.9	100.1	41.4	64.8	96.5	52.4	70.0	104.6	72.3	85.2	104.6	72.3	85.2
09	FINISHES	91.5	45.3	65.8	92.1	40.1	63.1	96.5	39.7	64.9	94.4	45.9	67.4	103.3	98.1	100.4	101.7	98.1	99.7
COVERS	DIVS. 10 - 14, 25, 28, 41, 43, 44, 46	100.0	78.1	95.6	100.0	78.3	95.7	100.0	79.3	95.8	100.0	79.0	95.8	100.0	101.9	100.4	100.0	101.9	100.4
21, 22, 23	FIRE SUPPRESSION, PLUMBING & HVAC	95.4	61.7	81.9	95.4	65.8	83.5	100.3	63.5	85.5	95.4	66.3	83.7	95.1	102.6	98.1	100.0	102.6	101.0
26, 27, 3370	ELECTRICAL, COMMUNICATIONS & UTIL.	92.8	66.1	78.9	94.8	70.4	82.1	94.7	66.1	79.8	92.2	70.4	80.8	95.8	96.3	96.1	94.5	96.3	95.4
MF2010	WEIGHTED AVERAGE	93.5	62.1	79.8	95.7	62.2	81.1	97.4	62.2	82.1	95.3	63.8	81.6	97.3	100.5	98.7	97.1	100.4	98.5

OREGON

DIVISION		KLAMATH FALLS 976			MEDFORD 975			PENDLETON 978			PORTLAND 970 - 972			SALEM 973			VALE 979		
		MAT.	INST.	TOTAL	MAT.	INST.	TOTAL	MAT.	INST.	TOTAL	MAT.	INST.	TOTAL	MAT.	INST.	TOTAL	MAT.	INST.	TOTAL
015433	CONTRACTOR EQUIPMENT		97.9	97.9		97.9	97.9		95.4	95.4		97.9	97.9		97.9	97.9		95.4	95.4
0241, 31 - 34	SITE & INFRASTRUCTURE, DEMOLITION	110.8	102.7	105.1	104.5	102.7	103.3	103.2	96.4	98.4	99.4	102.8	101.7	92.9	102.8	99.8	91.4	96.3	94.9
0310	Concrete Forming & Accessories	104.1	100.3	100.9	103.2	100.3	100.7	104.4	100.7	101.2	108.8	100.6	101.7	107.1	100.6	101.5	110.9	99.7	101.2
0320	Concrete Reinforcing	95.7	101.5	98.6	97.3	101.5	99.4	95.1	101.5	98.3	100.5	101.5	101.0	106.1	101.5	103.8	92.7	101.4	97.0
0330	Cast-in-Place Concrete	94.2	104.0	98.3	94.2	104.0	98.3	95.0	105.4	99.3	93.7	104.1	98.0	88.4	104.1	94.9	74.9	105.0	87.4
03	CONCRETE	107.6	101.5	104.6	102.2	101.5	101.8	88.4	102.1	95.2	97.7	101.6	99.7	96.0	101.6	98.8	74.7	101.5	88.0
04	MASONRY	110.8	104.5	106.9	93.1	104.5	100.2	101.7	104.6	103.5	96.5	104.5	101.5	103.3	104.5	104.1	100.2	104.6	102.9
05	METALS	90.7	96.3	92.5	91.0	96.3	92.6	97.1	97.1	97.1	92.4	96.6	93.7	97.8	96.6	97.4	97.0	96.0	96.7
06	WOOD, PLASTICS & COMPOSITES	93.8	100.3	97.5	92.6	100.3	97.0	95.9	100.4	98.4	99.8	100.3	100.1	96.7	100.3	98.7	104.8	100.4	102.3
07	THERMAL & MOISTURE PROTECTION	104.7	94.8	100.7	104.4	94.8	100.5	97.4	95.4	96.6	103.8	98.1	101.5	101.7	96.3	99.5	96.9	94.8	96.1
08	OPENINGS	93.9	103.8	96.2	96.3	103.8	98.1	89.5	103.9	92.9	92.2	103.8	95.0	95.9	103.8	97.8	89.4	95.8	90.9
0920	Plaster & Gypsum Board	98.2	100.2	99.6	97.5	100.2	99.3	85.2	100.2	95.4	100.9	100.2	100.4	97.3	100.2	99.3	91.9	100.2	97.5
0950, 0980	Ceilings & Acoustic Treatment	100.3	100.2	100.2	107.3	100.2	102.6	64.6	100.2	88.3	95.3	100.2	98.5	100.7	100.2	100.4	64.6	100.2	88.3
0960	Flooring	105.0	105.3	105.1	104.4	105.3	104.7	72.5	105.3	82.3	103.6	105.3	104.1	104.3	105.3	104.6	74.2	105.3	83.5
0970, 0990	Wall Finishes & Painting/Coating	104.6	59.4	77.4	104.6	59.4	77.4	96.1	66.9	78.5	104.3	66.9	81.8	102.8	72.3	84.5	96.1	72.3	81.8
09	FINISHES	104.0	96.7	99.9	104.4	96.7	100.1	71.8	97.6	86.2	101.4	97.5	99.2	100.5	98.1	99.2	72.3	98.2	86.7
COVERS	DIVS. 10 - 14, 25, 28, 41, 43, 44, 46	100.0	101.8	100.4	100.0	101.8	100.4	100.0	92.2	98.4	100.0	101.9	100.4	100.0	101.9	100.4	100.0	102.1	100.4
21, 22, 23	FIRE SUPPRESSION, PLUMBING & HVAC	95.1	102.6	98.1	100.0	102.6	101.0	96.9	115.1	104.3	100.0	102.6	101.0	100.0	102.6	101.1	96.9	71.5	86.7
26, 27, 3370	ELECTRICAL, COMMUNICATIONS & UTIL.	94.6	81.3	87.6	97.9	81.3	89.2	87.3	96.5	92.1	94.7	103.2	99.1	102.1	96.3	99.1	87.3	73.9	80.3
MF2010	WEIGHTED AVERAGE	98.2	98.2	98.2	98.4	98.2	98.3	93.1	102.5	97.2	97.2	101.4	99.1	99.0	100.5	99.7	91.3	89.8	90.7

PENNSYLVANIA

DIVISION		ALLENTOWN 181			ALTOONA 166			BEDFORD 155			BRADFORD 167			BUTLER 160			CHAMBERSBURG 172		
		MAT.	INST.	TOTAL	MAT.	INST.	TOTAL	MAT.	INST.	TOTAL	MAT.	INST.	TOTAL	MAT.	INST.	TOTAL	MAT.	INST.	TOTAL
015433	CONTRACTOR EQUIPMENT		113.5	113.5		113.5	113.5		110.5	110.5		113.5	113.5		113.5	113.5		112.7	112.7
0241, 31 - 34	SITE & INFRASTRUCTURE, DEMOLITION	90.7	105.4	101.0	94.2	105.5	102.1	103.7	102.2	102.7	89.5	104.8	100.3	85.7	106.9	100.6	85.1	102.7	97.4
0310	Concrete Forming & Accessories	100.5	114.2	112.3	85.7	80.4	81.1	85.2	82.7	83.0	87.9	82.5	83.2	87.2	95.7	94.5	89.5	81.8	82.9
0320	Concrete Reinforcing	94.4	111.5	103.0	91.5	106.0	98.7	89.7	83.5	86.6	93.5	106.4	99.9	92.2	112.5	102.3	88.7	106.3	97.5
0330	Cast-in-Place Concrete	85.0	101.5	91.8	94.7	87.5	91.7	109.6	71.2	93.6	90.4	92.9	91.5	83.7	98.1	89.6	94.9	71.0	85.0
03	CONCRETE	94.3	110.0	102.1	89.5	89.1	89.3	103.4	80.2	91.9	96.0	92.0	94.0	81.7	100.8	91.1	103.3	84.1	93.8
04	MASONRY	94.2	102.2	99.2	96.9	65.6	77.5	114.6	88.5	98.5	94.0	88.5	90.6	98.9	98.5	98.7	94.7	88.9	91.1
05	METALS	99.9	123.1	107.0	93.7	118.3	101.3	97.5	105.6	100.0	97.8	118.2	104.1	93.4	123.4	102.7	99.5	116.5	104.7
06	WOOD, PLASTICS & COMPOSITES	103.0	117.3	111.1	81.2	82.7	82.0	83.7	82.6	83.1	87.6	80.3	83.5	82.6	94.8	89.5	92.5	82.7	87.0
07	THERMAL & MOISTURE PROTECTION	102.5	120.4	109.7	101.8	92.0	97.8	104.0	88.6	97.8	102.4	92.3	98.3	101.5	100.5	101.1	102.8	74.0	91.1
08	OPENINGS	92.5	116.3	98.1	86.4	90.1	87.3	92.1	83.6	90.1	92.6	92.3	92.5	86.4	105.7	91.0	89.7	84.8	88.6
0920	Plaster & Gypsum Board	99.7	117.6	111.8	91.1	81.9	84.8	91.6	81.9	85.0	91.9	79.4	83.4	91.1	94.4	93.3	99.4	81.9	87.5
0950, 0980	Ceilings & Acoustic Treatment	85.7	117.6	106.9	90.1	81.9	84.6	95.7	81.9	86.5	88.4	79.4	82.4	91.1	94.4	93.3	90.3	81.9	84.7
0960	Flooring	93.6	97.6	94.8	87.4	99.3	91.0	96.9	89.8	94.7	88.5	105.8	93.6	88.2	82.5	86.5	93.7	47.0	79.7
0970, 0990	Wall Finishes & Painting/Coating	98.3	72.1	82.5	94.6	108.9	103.2	104.7	83.3	91.9	98.3	92.6	94.9	94.6	108.9	103.2	101.9	83.3	90.7
09	FINISHES	93.8	107.1	101.2	92.2	85.9	88.7	98.3	82.6	89.6	91.9	85.7	88.5	92.1	93.5	92.9	93.9	75.4	83.6
COVERS	DIVS. 10 - 14, 25, 28, 41, 43, 44, 46	100.0	104.5	100.9	100.0	95.3	99.1	100.0	98.6	99.7	100.0	99.3	99.9	100.0	102.3	100.5	100.0	95.2	99.0
21, 22, 23	FIRE SUPPRESSION, PLUMBING & HVAC	100.2	110.6	104.4	99.8	84.3	93.6	95.0	89.1	92.7	95.4	90.9	93.6	95.0	93.9	94.5	95.4	89.0	92.8
26, 27, 3370	ELECTRICAL, COMMUNICATIONS & UTIL.	98.2	98.9	98.6	88.3	111.1	100.2	93.7	111.1	102.8	91.7	111.1	101.8	88.9	110.1	100.0	89.4	88.3	88.9
MF2010	WEIGHTED AVERAGE	97.6	108.8	102.4	94.3	92.7	93.6	98.0	92.6	95.7	95.3	96.8	96.0	92.1	102.2	96.5	96.0	89.7	93.2

PENNSYLVANIA

DIVISION		DOYLESTOWN 189			DUBOIS 158			ERIE 164 - 165			GREENSBURG 156			HARRISBURG 170 - 171			HAZLETON 182		
		MAT.	INST.	TOTAL	MAT.	INST.	TOTAL	MAT.	INST.	TOTAL	MAT.	INST.	TOTAL	MAT.	INST.	TOTAL	MAT.	INST.	TOTAL
015433	CONTRACTOR EQUIPMENT		94.6	94.6		110.5	110.5		113.5	113.5		110.5	110.5		112.7	112.7		113.5	113.5
0241, 31 - 34	SITE & INFRASTRUCTURE, DEMOLITION	103.5	91.4	95.1	108.7	102.5	104.4	91.1	105.9	101.5	99.8	104.7	103.2	85.2	104.7	98.7	84.1	105.7	99.3
0310	Concrete Forming & Accessories	84.7	129.1	123.0	84.7	83.9	84.0	99.9	88.3	89.9	91.5	95.6	95.1	99.8	89.4	90.8	82.3	90.2	89.1
0320	Concrete Reinforcing	91.3	135.0	113.1	89.1	106.7	97.9	93.5	106.5	100.0	89.1	112.5	100.8	98.1	106.7	102.4	91.7	110.5	101.0
0330	Cast-in-Place Concrete	80.3	87.6	83.4	105.7	93.5	100.6	93.1	82.6	88.7	101.7	97.5	100.0	97.1	96.8	97.0	80.3	92.7	85.5
03	CONCRETE	90.0	115.6	102.6	104.8	92.8	98.9	88.6	90.9	89.7	98.4	100.5	99.4	99.2	96.5	97.9	87.0	96.1	91.5
04	MASONRY	97.2	127.9	116.2	114.6	89.6	99.2	86.7	92.8	90.5	125.3	98.5	108.7	95.0	90.6	92.3	106.2	96.5	100.2
05	METALS	97.4	125.6	106.1	97.5	117.6	103.7	93.9	117.0	101.0	97.5	121.9	105.0	106.1	120.3	110.5	99.6	121.4	106.3
06	WOOD, PLASTICS & COMPOSITES	83.1	131.5	110.4	82.6	82.4	82.5	99.2	86.5	92.1	90.5	94.7	92.9	100.1	88.7	93.7	81.6	88.6	85.6
07	THERMAL & MOISTURE PROTECTION	100.6	132.8	113.7	104.3	95.5	100.7	102.3	93.4	98.7	103.9	100.1	102.4	106.2	112.2	108.6	102.0	107.6	104.2
08	OPENINGS	94.7	140.0	105.4	92.1	93.5	92.4	86.6	93.0	88.1	92.1	105.7	95.3	95.8	95.4	95.7	93.0	94.5	93.4
0920	Plaster & Gypsum Board	89.6	132.2	118.6	90.5	81.7	84.5	99.7	86.0	90.4	92.9	94.4	93.9	103.4	88.1	93.0	90.0	88.0	88.7
0950, 0980	Ceilings & Acoustic Treatment	84.9	132.2	116.4	95.7	81.7	86.4	85.7	86.0	85.9	94.8	94.4	94.5	99.4	88.1	91.9	86.7	88.0	87.6
0960	Flooring	78.6	136.8	96.0	96.7	105.8	99.4	92.3	93.5	92.7	99.8	69.3	90.7	98.7	93.4	97.1	86.2	96.3	89.2
0970, 0990	Wall Finishes & Painting/Coating	97.8	70.5	81.4	104.7	108.9	107.2	103.3	92.6	96.9	104.7	108.9	107.2	102.3	91.5	95.8	98.3	105.6	102.7
09	FINISHES	85.1	123.5	106.5	98.6	89.3	93.4	94.3	89.2	91.5	98.7	91.9	94.9	98.8	89.9	93.8	90.2	91.0	90.6
COVERS	DIVS. 10 - 14, 25, 28, 41, 43, 44, 46	100.0	71.9	94.4	100.0	98.6	99.7	100.0	100.9	100.2	100.0	102.1	100.4	100.0	96.4	99.3	100.0	99.1	99.8
21, 22, 23	FIRE SUPPRESSION, PLUMBING & HVAC	95.0	125.4	107.2	95.0	89.5	92.8	99.8	93.1	97.1	95.0	92.8	94.1	100.0	90.7	96.3	95.4	89.0	92.8
26, 27, 3370	ELECTRICAL, COMMUNICATIONS & UTIL.	91.1	126.9	109.8	94.3	111.1	103.1	90.1	87.4	88.7	94.3	111.1	103.1	96.8	88.4	92.4	92.6	91.5	92.0
MF2010	WEIGHTED AVERAGE	94.6	120.8	106.0	98.4	97.1	97.8	94.1	95.0	94.5	98.0	101.6	99.5	99.6	95.9	98.0	95.0	98.8	96.7

City Cost Indexes

PENNSYLVANIA

DIVISION		INDIANA 157			JOHNSTOWN 159			KITTANNING 162			LANCASTER 175 - 176			LEHIGH VALLEY 180			MONTROSE 188		
		MAT.	INST.	TOTAL	MAT.	INST.	TOTAL	MAT.	INST.	TOTAL	MAT.	INST.	TOTAL	MAT.	INST.	TOTAL	MAT.	INST.	TOTAL
015433	CONTRACTOR EQUIPMENT		110.5	110.5		110.5	110.5		113.5	113.5		112.7	112.7		113.5	113.5		113.5	113.5
0241, 31 - 34	SITE & INFRASTRUCTURE, DEMOLITION	97.9	103.3	101.7	104.2	103.9	104.0	88.2	106.8	101.3	77.7	104.4	96.5	87.9	105.1	99.9	86.6	103.4	98.4
0310	Concrete Forming & Accessories	85.8	86.3	86.2	84.7	84.2	84.3	87.2	95.8	94.6	91.6	88.9	89.3	94.2	113.6	110.9	83.3	88.5	87.8
0320	Concrete Reinforcing	88.4	112.6	100.5	89.7	112.2	100.9	92.2	112.6	102.4	88.4	106.6	97.5	91.7	111.4	101.5	96.0	108.4	102.2
0330	Cast-in-Place Concrete	99.7	97.3	98.7	110.6	92.9	103.2	86.9	98.0	91.5	80.8	97.7	87.8	86.9	100.9	92.7	85.2	90.9	87.6
03	CONCRETE	95.9	96.3	96.1	104.2	93.7	99.0	84.1	100.8	92.4	90.9	96.5	93.7	93.3	109.4	101.2	92.0	94.1	93.1
04	MASONRY	110.3	100.2	104.1	111.1	88.0	96.8	101.7	100.2	100.8	99.9	90.8	94.3	94.2	101.7	98.8	94.1	96.0	95.3
05	METALS	97.6	121.5	105.0	97.5	120.1	104.5	93.5	123.4	102.7	99.5	119.7	105.7	99.5	121.7	106.4	97.9	115.3	103.2
06	WOOD, PLASTICS & COMPOSITES	84.5	82.4	83.3	82.6	82.6	82.6	82.6	94.8	89.5	92.5	88.7	91.5	94.4	117.3	107.3	82.4	88.3	85.8
07	THERMAL & MOISTURE PROTECTION	103.8	97.9	101.4	104.0	93.7	99.8	101.6	100.9	101.3	102.4	98.5	100.8	102.3	103.6	102.8	102.0	92.4	98.1
08	OPENINGS	92.1	95.2	92.8	92.1	91.8	92.0	86.4	102.0	90.1	89.7	99.3	92.0	93.0	116.3	98.5	89.8	93.8	90.8
0920	Plaster & Gypsum Board	92.0	81.7	85.0	90.3	81.9	84.6	91.1	94.4	93.3	101.3	88.1	92.3	93.0	117.6	109.7	90.4	87.7	88.6
0950, 0980	Ceilings & Acoustic Treatment	95.7	81.7	86.4	94.8	81.9	86.2	91.1	94.4	93.3	90.3	88.1	88.8	86.7	117.6	107.2	88.4	87.7	88.0
0960	Flooring	97.4	105.8	99.9	96.7	100.0	97.7	88.2	105.8	93.5	94.5	93.4	94.2	90.9	97.6	92.9	86.7	58.9	78.4
0970, 0990	Wall Finishes & Painting/Coating	104.7	108.9	107.2	104.7	108.9	107.2	94.6	108.9	103.2	101.9	58.4	75.7	98.3	67.9	80.0	98.3	105.6	102.7
09	FINISHES	98.0	90.2	93.7	97.9	88.7	92.8	92.3	97.5	95.2	93.9	86.2	89.6	92.2	105.5	99.6	91.0	84.8	87.5
COVERS	DIVS. 10 - 14, 25, 28, 41, 43, 44, 46	100.0	100.7	100.1	100.0	98.8	99.8	100.0	102.3	100.5	100.0	96.3	99.3	100.0	104.1	100.8	100.0	99.1	99.8
21, 22, 23	FIRE SUPPRESSION, PLUMBING & HVAC	95.0	92.7	94.1	95.0	89.8	92.9	95.0	96.1	95.4	95.4	90.8	93.5	95.4	110.3	101.4	95.4	98.0	96.4
26, 27, 3370	ELECTRICAL, COMMUNICATIONS & UTIL.	94.3	111.1	103.1	94.3	111.1	103.1	88.3	111.1	100.2	90.9	45.1	67.0	92.6	133.5	113.9	91.7	97.2	94.5
MF2010	WEIGHTED AVERAGE	96.9	100.1	98.3	97.9	97.3	97.7	92.5	103.4	97.3	94.8	89.3	92.4	95.4	112.5	102.9	94.4	97.1	95.6

PENNSYLVANIA

DIVISION		NEW CASTLE 161			NORRISTOWN 194			OIL CITY 163			PHILADELPHIA 190 - 191			PITTSBURGH 150 - 152			POTTSVILLE 179		
		MAT.	INST.	TOTAL	MAT.	INST.	TOTAL	MAT.	INST.	TOTAL	MAT.	INST.	TOTAL	MAT.	INST.	TOTAL	MAT.	INST.	TOTAL
015433	CONTRACTOR EQUIPMENT		113.5	113.5		98.0	98.0		113.5	113.5		97.5	97.5		111.8	111.8		112.7	112.7
0241, 31 - 34	SITE & INFRASTRUCTURE, DEMOLITION	86.1	106.9	100.7	92.4	99.8	97.6	84.7	105.0	99.0	99.9	99.2	99.4	103.4	106.7	105.7	80.4	104.6	97.4
0310	Concrete Forming & Accessories	87.2	97.4	96.0	83.5	130.8	124.3	87.2	83.5	84.0	98.7	139.9	134.3	99.0	98.7	98.7	83.2	90.0	89.1
0320	Concrete Reinforcing	91.0	93.3	92.2	92.5	142.2	117.3	92.2	93.1	92.6	100.5	142.2	121.3	90.1	112.9	101.5	87.7	103.8	95.7
0330	Cast-in-Place Concrete	84.4	96.9	89.6	79.7	127.5	99.6	82.0	96.5	88.1	94.8	131.7	110.1	105.7	97.7	102.4	86.0	97.9	91.0
03	CONCRETE	82.0	97.6	89.7	86.9	131.4	108.9	80.5	91.2	85.7	98.2	136.9	117.3	101.7	102.0	101.9	94.6	96.6	95.6
04	MASONRY	98.1	98.5	98.4	106.5	125.7	118.4	98.2	95.2	96.3	95.5	130.2	117.0	103.3	103.3	103.3	94.5	92.9	93.5
05	METALS	93.5	114.0	99.8	99.4	130.9	109.1	93.5	112.1	99.2	100.7	131.2	110.1	98.9	122.9	106.3	99.7	118.9	105.6
06	WOOD, PLASTICS & COMPOSITES	82.6	98.1	91.4	81.5	131.4	109.7	82.6	80.3	81.3	97.2	141.5	122.2	100.5	98.0	99.1	84.9	88.6	87.0
07	THERMAL & MOISTURE PROTECTION	101.5	97.0	99.7	103.9	132.6	115.6	101.4	94.9	98.8	102.7	135.7	116.1	104.1	102.0	103.3	102.5	107.1	104.3
08	OPENINGS	86.4	98.0	89.2	88.6	142.0	101.2	86.4	80.7	85.1	94.4	147.5	106.9	95.5	107.5	98.3	89.8	92.5	90.4
0920	Plaster & Gypsum Board	91.1	97.8	95.7	88.2	132.2	118.1	91.1	79.4	83.2	97.2	142.6	128.1	99.8	97.8	98.5	95.7	88.0	90.5
0950, 0980	Ceilings & Acoustic Treatment	91.1	97.8	95.6	90.5	132.2	118.3	91.1	79.4	83.3	91.6	142.6	125.6	95.7	97.8	97.1	90.3	88.0	88.8
0960	Flooring	88.2	58.5	79.4	91.6	139.5	105.9	88.2	105.8	93.5	98.5	139.5	110.7	102.8	108.0	104.3	91.1	93.4	91.8
0970, 0990	Wall Finishes & Painting/Coating	94.6	108.9	103.2	100.8	147.4	128.8	94.6	108.9	103.2	105.9	154.0	134.8	104.7	121.2	114.6	101.9	105.6	104.1
09	FINISHES	92.1	91.2	91.6	92.9	133.9	115.7	92.0	88.3	89.9	98.2	141.6	122.4	100.9	102.1	101.6	92.4	91.9	92.1
COVERS	DIVS. 10 - 14, 25, 28, 41, 43, 44, 46	100.0	102.6	100.5	100.0	111.7	102.3	100.0	100.4	100.1	100.0	120.5	104.1	100.0	102.5	100.5	100.0	96.7	99.3
21, 22, 23	FIRE SUPPRESSION, PLUMBING & HVAC	95.0	93.5	94.4	95.2	125.8	107.4	95.0	92.6	94.0	100.0	130.5	112.2	99.9	101.2	100.4	95.4	98.2	96.5
26, 27, 3370	ELECTRICAL, COMMUNICATIONS & UTIL.	88.9	88.7	88.8	92.0	148.7	121.6	90.8	110.1	100.9	98.6	148.7	124.7	96.7	111.1	104.2	89.0	91.1	90.1
MF2010	WEIGHTED AVERAGE	92.1	97.2	94.4	94.9	129.5	110.0	92.1	97.0	94.3	98.9	133.4	113.9	99.7	105.4	102.3	94.7	98.0	96.1

PENNSYLVANIA

DIVISION		READING 195 - 196			SCRANTON 184 - 185			STATE COLLEGE 168			STROUDSBURG 183			SUNBURY 178			UNIONTOWN 154		
		MAT.	INST.	TOTAL	MAT.	INST.	TOTAL	MAT.	INST.	TOTAL	MAT.	INST.	TOTAL	MAT.	INST.	TOTAL	MAT.	INST.	TOTAL
015433	CONTRACTOR EQUIPMENT		117.0	117.0		113.5	113.5		112.7	112.7		113.5	113.5		113.5	113.5		110.5	110.5
0241, 31 - 34	SITE & INFRASTRUCTURE, DEMOLITION	96.9	111.5	107.1	91.1	105.9	101.5	81.6	104.2	97.5	85.6	103.5	98.2	90.9	105.6	101.2	98.5	104.7	102.8
0310	Concrete Forming & Accessories	99.0	90.5	91.7	100.7	90.1	91.5	85.4	80.1	80.9	88.7	89.5	89.4	95.8	88.9	89.9	79.0	95.7	93.4
0320	Concrete Reinforcing	93.8	103.9	98.8	94.4	108.7	101.6	92.7	106.7	99.7	94.7	115.2	104.9	90.1	106.7	98.4	89.1	112.5	100.8
0330	Cast-in-Place Concrete	71.5	98.2	82.6	88.7	92.9	90.5	85.6	66.6	77.7	83.6	73.4	79.4	94.0	96.5	95.0	99.7	97.5	98.8
03	CONCRETE	86.2	96.9	91.5	96.1	95.8	96.0	96.3	81.9	89.2	90.8	89.8	90.3	98.2	96.2	97.2	95.5	100.6	98.0
04	MASONRY	96.3	94.0	94.9	94.6	96.5	95.8	99.5	79.2	87.0	91.8	101.3	97.7	93.9	90.4	91.7	127.7	98.5	109.6
05	METALS	99.7	119.2	105.7	101.9	121.2	107.9	97.6	119.0	104.2	99.6	117.8	105.2	99.4	120.0	105.8	97.3	122.0	104.9
06	WOOD, PLASTICS & COMPOSITES	100.3	88.5	93.7	103.0	88.2	94.7	89.7	82.7	85.7	88.6	88.3	88.4	93.2	88.6	90.6	76.4	94.7	86.7
07	THERMAL & MOISTURE PROTECTION	104.3	113.6	108.1	102.4	96.3	99.9	101.6	92.1	97.7	102.1	85.8	95.5	103.5	105.8	104.4	103.7	100.2	102.3
08	OPENINGS	93.1	98.3	94.3	92.5	93.7	92.8	89.6	90.1	89.7	93.0	89.1	92.1	89.9	93.4	90.7	92.0	105.7	95.3
0920	Plaster & Gypsum Board	100.4	88.0	92.0	101.9	87.6	92.2	93.1	81.9	85.5	91.4	87.7	88.9	95.2	88.0	90.3	88.4	94.4	92.5
0950, 0980	Ceilings & Acoustic Treatment	81.5	88.0	85.8	94.6	87.6	89.9	85.8	81.9	83.2	84.9	87.7	86.8	86.8	88.0	87.6	94.8	94.4	94.5
0960	Flooring	95.5	93.4	94.9	93.6	108.9	98.2	91.4	99.3	93.8	89.0	52.3	78.0	91.8	89.3	91.0	94.4	105.8	97.8
0970, 0990	Wall Finishes & Painting/Coating	99.5	105.6	103.1	98.3	105.6	102.7	98.3	108.9	104.7	98.3	66.0	78.9	101.9	105.6	104.1	104.7	108.9	107.2
09	FINISHES	94.0	91.5	92.6	96.0	94.6	95.2	91.2	85.9	88.3	90.9	80.3	85.0	93.1	90.6	91.7	96.5	97.3	96.9
COVERS	DIVS. 10 - 14, 25, 28, 41, 43, 44, 46	100.0	99.0	99.8	100.0	99.1	99.8	100.0	91.8	98.4	100.0	63.5	92.7	100.0	95.9	99.2	100.0	102.1	100.4
21, 22, 23	FIRE SUPPRESSION, PLUMBING & HVAC	100.2	107.3	103.0	100.2	98.8	99.6	95.4	84.6	91.1	95.4	100.5	97.4	95.4	90.4	93.4	95.0	92.8	94.1
26, 27, 3370	ELECTRICAL, COMMUNICATIONS & UTIL.	98.7	91.1	94.7	98.2	97.2	97.7	90.8	111.1	101.4	92.6	142.5	118.6	89.3	93.0	91.2	91.3	111.1	101.6
MF2010	WEIGHTED AVERAGE	97.1	101.1	98.8	98.3	99.7	98.9	94.9	93.0	94.1	94.9	102.0	98.0	95.5	96.3	95.8	97.1	102.2	99.4

521

PENNSYLVANIA

DIVISION		WASHINGTON 153 MAT.	INST.	TOTAL	WELLSBORO 169 MAT.	INST.	TOTAL	WESTCHESTER 193 MAT.	INST.	TOTAL	WILKES-BARRE 186-187 MAT.	INST.	TOTAL	WILLIAMSPORT 177 MAT.	INST.	TOTAL	YORK 173-174 MAT.	INST.	TOTAL
015433	CONTRACTOR EQUIPMENT		110.5	110.5		113.5	113.5		98.0	98.0		113.5	113.5		113.5	113.5		112.7	112.7
0241, 31 - 34	SITE & INFRASTRUCTURE, DEMOLITION	98.6	104.7	102.9	93.1	103.3	100.3	98.0	97.2	97.4	83.7	105.7	99.2	82.7	104.2	97.8	81.1	104.5	97.5
0310	Concrete Forming & Accessories	85.9	95.8	94.5	87.3	87.1	87.1	90.1	129.3	123.9	91.3	90.4	90.5	92.3	61.5	65.7	86.6	89.5	89.1
0320	Concrete Reinforcing	89.1	112.6	100.9	92.7	108.3	100.5	91.7	109.7	100.7	93.5	110.5	102.0	89.4	66.6	78.0	90.1	106.7	98.4
0330	Cast-in-Place Concrete	99.7	97.6	98.8	89.7	88.8	89.3	88.2	126.7	104.2	80.3	92.6	85.4	79.6	74.3	77.4	86.4	97.9	91.2
03	CONCRETE	96.0	100.6	98.3	98.5	92.8	95.7	94.3	124.4	109.1	87.9	96.2	92.0	86.0	68.8	77.5	95.8	96.9	96.4
04	MASONRY	109.8	100.2	103.9	100.2	90.4	94.2	101.2	125.7	116.3	106.6	96.1	100.1	86.4	94.2	91.2	94.9	90.8	92.4
05	METALS	97.2	122.3	105.0	99.7	115.4	103.1	99.4	114.9	104.2	97.8	121.8	105.2	99.5	100.2	99.7	101.1	120.3	107.0
06	WOOD, PLASTICS & COMPOSITES	84.6	94.7	90.3	87.1	88.6	87.9	88.8	131.4	112.8	91.0	88.6	89.7	89.6	52.2	68.5	88.7	88.7	88.7
07	THERMAL & MOISTURE PROTECTION	103.8	100.6	102.5	102.6	90.3	97.6	104.3	131.7	115.4	102.0	107.4	104.2	103.0	103.3	103.1	102.5	112.2	106.5
08	OPENINGS	92.0	105.7	95.3	92.5	94.0	92.9	88.6	125.7	97.4	89.8	100.4	92.3	89.9	54.8	81.6	89.8	95.4	91.1
0920	Plaster & Gypsum Board	91.8	94.4	93.5	91.3	88.0	89.1	90.0	132.2	118.7	92.3	88.0	89.4	95.7	50.5	64.9	96.8	88.1	90.9
0950, 0980	Ceilings & Acoustic Treatment	94.8	94.4	94.5	85.8	88.0	87.3	90.5	132.2	118.3	88.4	88.0	88.2	90.3	50.5	63.8	89.3	88.1	88.5
0960	Flooring	97.5	105.8	100.0	88.2	51.7	77.3	94.4	139.5	107.9	89.8	96.3	91.7	90.8	50.6	78.8	92.3	93.4	92.6
0970, 0990	Wall Finishes & Painting/Coating	104.7	106.8	106.0	98.3	105.6	102.7	100.8	147.4	128.8	98.3	105.6	102.7	101.9	105.6	104.1	101.9	91.5	95.6
09	FINISHES	97.8	97.3	97.6	91.6	82.6	86.6	94.4	131.9	115.3	91.9	92.3	92.1	92.9	63.1	76.3	92.7	89.9	91.1
COVERS	DIVS. 10 - 14, 25, 28, 41, 43, 44, 46	100.0	102.1	100.4	100.0	95.8	99.2	100.0	115.5	103.1	100.0	99.0	99.8	100.0	93.2	98.6	100.0	96.5	99.3
21, 22, 23	FIRE SUPPRESSION, PLUMBING & HVAC	95.0	96.0	95.4	95.4	89.6	93.0	95.2	125.5	107.9	95.4	98.7	96.7	95.4	91.4	93.8	100.2	90.8	96.4
26, 27, 3370	ELECTRICAL, COMMUNICATIONS & UTIL.	93.7	111.1	102.8	91.7	83.3	87.3	91.9	117.6	105.3	92.6	91.5	92.0	89.8	47.7	67.8	90.9	88.4	89.6
MF2010	WEIGHTED AVERAGE	96.7	103.2	99.5	95.9	92.3	94.3	95.8	121.7	107.1	94.7	99.2	96.7	93.6	79.3	87.4	96.5	96.0	96.3

DIVISION		PUERTO RICO SAN JUAN 009 MAT.	INST.	TOTAL	RHODE ISLAND NEWPORT 028 MAT.	INST.	TOTAL	PROVIDENCE 029 MAT.	INST.	TOTAL	SOUTH CAROLINA AIKEN 298 MAT.	INST.	TOTAL	BEAUFORT 299 MAT.	INST.	TOTAL	CHARLESTON 294 MAT.	INST.	TOTAL
015433	CONTRACTOR EQUIPMENT		90.2	90.2		102.1	102.1		102.1	102.1		100.5	100.5		100.5	100.5		100.5	100.5
0241, 31 - 34	SITE & INFRASTRUCTURE, DEMOLITION	132.3	90.5	103.0	90.6	104.2	100.2	92.0	104.2	100.6	118.8	87.1	96.5	113.9	85.3	93.8	97.9	86.0	89.5
0310	Concrete Forming & Accessories	93.7	18.1	28.4	102.6	120.6	118.1	103.9	120.6	118.3	98.6	68.9	73.0	97.5	38.8	46.8	96.5	64.4	68.8
0320	Concrete Reinforcing	191.6	13.1	102.5	106.8	153.9	130.3	103.4	153.9	128.6	94.8	67.3	81.0	93.9	26.6	60.3	93.7	61.2	77.5
0330	Cast-in-Place Concrete	103.7	31.2	73.6	80.3	124.1	98.5	96.9	124.1	108.2	80.9	70.8	76.7	80.9	47.8	67.2	94.9	50.2	76.3
03	CONCRETE	110.5	22.7	67.2	96.1	127.4	111.6	103.6	127.4	115.3	101.3	70.4	86.0	98.6	41.7	70.5	94.1	60.2	77.3
04	MASONRY	83.6	17.2	42.5	96.4	128.9	116.5	100.6	128.9	118.1	73.9	62.5	66.8	87.4	33.8	54.3	88.6	41.6	59.6
05	METALS	117.2	34.9	91.8	99.4	126.8	107.8	106.5	126.8	112.8	92.9	85.1	90.5	92.9	69.3	85.6	94.8	81.6	90.7
06	WOOD, PLASTICS & COMPOSITES	97.0	17.9	52.4	101.9	119.8	112.0	104.2	119.8	113.0	99.2	70.1	82.8	97.6	38.7	64.3	96.4	70.0	81.5
07	THERMAL & MOISTURE PROTECTION	129.7	21.3	85.7	101.5	120.4	109.2	102.3	120.4	109.6	105.3	67.7	90.0	105.0	39.6	78.5	104.2	48.2	81.5
08	OPENINGS	150.1	15.1	118.2	98.1	128.0	105.1	100.9	128.0	107.3	92.5	66.3	86.3	92.6	37.5	79.5	96.0	64.8	88.6
0920	Plaster & Gypsum Board	154.5	15.3	59.8	94.7	119.8	111.8	97.0	119.8	112.5	104.6	69.1	80.4	108.1	36.6	59.4	109.8	68.9	82.0
0950, 0980	Ceilings & Acoustic Treatment	230.8	15.3	87.4	95.3	119.8	111.6	95.2	119.8	111.6	96.0	69.1	78.1	99.5	36.6	57.7	99.5	68.9	79.1
0960	Flooring	225.0	13.4	161.7	99.8	140.3	111.9	100.4	140.3	112.4	104.8	69.7	94.3	106.1	52.1	90.0	105.8	59.3	91.9
0970, 0990	Wall Finishes & Painting/Coating	203.3	17.2	91.4	96.0	124.2	112.9	94.4	124.2	112.3	105.7	71.9	85.4	105.7	35.4	63.5	105.7	68.3	83.2
09	FINISHES	214.4	17.4	104.6	98.6	124.7	113.2	98.0	124.7	112.9	102.4	69.5	84.0	103.5	40.7	68.5	101.6	64.9	81.2
COVERS	DIVS. 10 - 14, 25, 28, 41, 43, 44, 46	100.0	16.5	83.3	100.0	105.8	101.2	100.0	105.8	101.2	100.0	73.6	94.7	100.0	72.1	94.4	100.0	70.7	94.1
21, 22, 23	FIRE SUPPRESSION, PLUMBING & HVAC	102.8	14.2	67.3	100.1	108.4	103.4	99.9	108.4	103.3	95.5	64.5	83.0	95.5	37.3	72.1	100.4	54.9	82.1
26, 27, 3370	ELECTRICAL, COMMUNICATIONS & UTIL.	126.9	13.5	67.7	101.5	100.4	100.9	102.5	100.4	101.4	96.8	67.4	81.5	100.7	34.4	66.1	99.0	90.9	94.8
MF2010	WEIGHTED AVERAGE	122.1	24.6	79.7	99.0	116.6	106.6	101.5	116.6	108.1	96.4	70.4	85.1	97.1	45.7	74.7	97.7	66.4	84.1

DIVISION		SOUTH CAROLINA COLUMBIA 290-292 MAT.	INST.	TOTAL	FLORENCE 295 MAT.	INST.	TOTAL	GREENVILLE 296 MAT.	INST.	TOTAL	ROCK HILL 297 MAT.	INST.	TOTAL	SPARTANBURG 293 MAT.	INST.	TOTAL	SOUTH DAKOTA ABERDEEN 574 MAT.	INST.	TOTAL
015433	CONTRACTOR EQUIPMENT		100.5	100.5		100.5	100.5		100.5	100.5		100.5	100.5		100.5	100.5		99.0	99.0
0241, 31 - 34	SITE & INFRASTRUCTURE, DEMOLITION	98.0	85.9	89.5	107.9	86.0	92.5	102.9	85.6	90.7	100.4	84.6	89.3	102.7	85.6	90.7	94.3	94.5	94.4
0310	Concrete Forming & Accessories	95.4	46.1	52.9	84.6	46.3	51.5	96.0	46.1	52.9	94.5	39.6	47.1	99.0	46.2	53.5	98.9	37.5	45.9
0320	Concrete Reinforcing	96.8	61.0	78.9	93.3	61.2	77.3	93.2	45.9	69.6	94.1	45.6	69.8	93.2	58.9	76.1	100.5	39.3	69.9
0330	Cast-in-Place Concrete	97.9	50.7	78.3	80.9	50.0	68.1	80.9	49.9	68.0	80.9	44.5	65.8	80.9	50.0	68.1	110.5	44.5	83.1
03	CONCRETE	95.9	52.2	74.3	93.3	52.0	72.9	92.2	49.1	70.9	90.3	44.2	67.5	92.4	51.6	72.3	103.7	41.7	73.1
04	MASONRY	85.5	38.3	56.3	74.0	41.6	54.0	72.0	41.6	53.2	94.1	35.2	57.7	74.0	41.6	54.0	104.8	55.8	74.5
05	METALS	92.1	80.8	88.6	93.7	81.2	89.8	93.7	75.7	88.1	92.9	72.4	86.6	93.7	80.4	89.6	98.4	65.2	88.2
06	WOOD, PLASTICS & COMPOSITES	100.4	45.8	69.6	82.1	45.8	61.6	96.0	45.8	67.7	94.3	39.8	63.5	100.3	45.8	69.6	96.3	36.8	62.8
07	THERMAL & MOISTURE PROTECTION	100.0	44.2	77.4	104.5	45.7	80.6	104.4	45.7	80.6	104.3	40.5	78.4	104.5	45.7	80.6	104.3	48.8	81.8
08	OPENINGS	97.5	51.7	86.7	92.6	51.7	82.9	92.5	48.1	82.0	92.6	40.3	80.2	92.5	51.1	82.7	93.9	37.6	80.6
0920	Plaster & Gypsum Board	102.7	44.0	62.8	97.4	44.0	61.0	103.1	44.0	62.9	102.4	37.7	58.4	106.1	44.0	63.8	106.1	35.1	57.8
0950, 0980	Ceilings & Acoustic Treatment	101.2	44.0	63.1	96.9	44.0	61.7	96.0	44.0	61.4	96.0	37.7	57.2	96.0	44.0	61.4	108.2	35.1	59.6
0960	Flooring	99.7	44.4	83.1	98.4	44.4	82.3	103.7	58.2	90.1	102.9	45.3	85.7	105.0	58.2	91.0	111.1	51.2	93.2
0970, 0990	Wall Finishes & Painting/Coating	102.4	68.3	81.9	105.7	68.3	83.2	105.7	68.3	83.2	105.7	40.0	66.2	105.7	68.3	83.2	95.6	37.9	60.9
09	FINISHES	99.7	47.7	70.7	98.5	48.1	70.4	100.2	50.4	72.5	99.6	40.2	66.5	101.0	50.4	72.8	106.0	39.7	69.5
COVERS	DIVS. 10 - 14, 25, 28, 41, 43, 44, 46	100.0	67.9	93.6	100.0	67.9	93.6	100.0	67.9	93.6	100.0	66.2	93.2	100.0	67.9	93.6	100.0	41.7	88.3
21, 22, 23	FIRE SUPPRESSION, PLUMBING & HVAC	100.0	53.2	81.2	100.4	53.3	81.5	100.4	53.1	81.4	95.5	45.2	75.3	100.4	53.3	81.5	100.2	40.6	76.3
26, 27, 3370	ELECTRICAL, COMMUNICATIONS & UTIL.	99.0	61.1	79.3	96.8	61.1	78.2	99.0	58.0	77.6	99.0	57.9	77.6	99.0	58.0	77.6	99.2	52.4	74.7
MF2010	WEIGHTED AVERAGE	97.2	57.3	79.8	96.2	57.8	79.4	96.2	56.5	78.9	95.6	51.2	76.3	96.4	57.4	79.5	100.1	50.6	78.6

SOUTH DAKOTA

	DIVISION	MITCHELL 573			MOBRIDGE 576			PIERRE 575			RAPID CITY 577			SIOUX FALLS 570 - 571			WATERTOWN 572		
		MAT.	INST.	TOTAL	MAT.	INST.	TOTAL	MAT.	INST.	TOTAL	MAT.	INST.	TOTAL	MAT.	INST.	TOTAL	MAT.	INST.	TOTAL
015433	CONTRACTOR EQUIPMENT		99.0	99.0		99.0	99.0		99.0	99.0		99.0	99.0		100.0	100.0		99.0	99.0
0241, 31 - 34	SITE & INFRASTRUCTURE, DEMOLITION	91.2	94.5	93.5	91.2	94.5	93.5	93.0	94.6	94.1	92.8	94.6	94.1	92.4	96.2	95.1	91.1	94.5	93.5
0310	Concrete Forming & Accessories	98.1	37.8	46.1	88.8	37.7	44.7	101.3	39.3	47.8	106.7	37.5	47.0	104.1	41.4	50.0	85.4	37.5	44.1
0320	Concrete Reinforcing	99.9	47.8	73.9	102.5	39.3	70.9	104.3	71.5	87.9	94.0	71.6	82.8	101.7	71.5	86.6	97.1	39.4	68.3
0330	Cast-in-Place Concrete	107.4	43.3	80.8	107.4	44.5	81.3	96.8	43.2	74.6	106.5	43.8	80.5	102.8	44.0	78.4	107.4	47.0	82.3
03	CONCRETE	101.3	42.9	72.5	101.1	41.8	71.8	97.3	48.1	73.0	100.6	47.6	74.4	99.2	49.3	74.6	100.0	42.6	71.7
04	MASONRY	93.9	54.2	69.4	102.5	55.8	73.6	103.2	54.4	73.0	102.6	57.4	74.6	97.5	54.6	70.9	127.0	58.5	84.6
05	METALS	97.4	65.4	87.6	97.5	65.2	87.5	99.7	80.1	93.6	100.2	80.7	94.2	101.4	80.3	94.9	97.4	65.8	87.7
06	WOOD, PLASTICS & COMPOSITES	95.2	37.4	62.6	83.9	37.1	57.5	98.2	37.6	64.0	100.1	34.1	62.9	96.2	40.2	64.6	80.3	36.8	55.8
07	THERMAL & MOISTURE PROTECTION	104.1	46.8	80.9	104.2	48.8	81.7	105.1	47.1	81.5	104.7	49.0	82.1	109.3	49.4	85.0	104.0	49.5	81.9
08	OPENINGS	92.7	38.2	79.8	95.1	37.2	81.4	101.6	47.3	88.7	97.5	45.4	85.2	98.9	48.7	87.0	92.7	37.3	79.6
0920	Plaster & Gypsum Board	104.5	35.7	57.7	97.2	35.4	55.1	97.2	35.9	55.5	105.6	32.3	55.8	97.3	38.6	57.4	95.2	35.1	54.3
0950, 0980	Ceilings & Acoustic Treatment	104.6	35.7	58.7	108.2	35.4	59.7	106.4	35.9	59.5	110.8	32.3	58.6	107.3	38.6	61.6	104.6	35.1	58.4
0960	Flooring	110.7	51.2	93.0	106.4	51.2	89.9	110.5	36.5	88.4	110.4	80.1	101.4	105.9	76.3	97.1	105.1	51.2	89.0
0970, 0990	Wall Finishes & Painting/Coating	95.6	41.5	63.1	95.6	42.6	63.7	99.1	45.5	66.9	95.6	45.5	65.5	93.9	45.5	64.8	95.6	37.9	60.9
09	FINISHES	104.8	40.4	68.9	103.3	40.3	68.2	105.5	38.2	68.0	106.3	44.9	72.1	103.8	47.5	72.4	101.9	39.7	67.2
COVERS	DIVS. 10 - 14, 25, 28, 41, 43, 44, 46	100.0	38.7	87.7	100.0	41.7	88.3	100.0	77.9	95.6	100.0	77.8	95.6	100.0	78.3	95.7	100.0	41.7	88.3
21, 22, 23	FIRE SUPPRESSION, PLUMBING & HVAC	95.3	40.3	73.2	95.3	40.8	73.4	100.0	64.6	85.8	100.2	65.1	86.1	100.1	39.4	75.8	95.3	40.3	73.4
26, 27, 3370	ELECTRICAL, COMMUNICATIONS & UTIL.	97.5	45.0	70.1	99.2	45.9	71.4	104.6	52.4	77.4	95.7	52.5	73.1	97.4	76.9	86.7	96.7	45.9	70.2
MF2010	WEIGHTED AVERAGE	97.5	49.5	76.6	98.1	49.8	77.1	100.8	59.2	82.7	100.0	60.3	82.7	100.0	58.8	82.1	98.5	50.2	77.5

TENNESSEE

	DIVISION	CHATTANOOGA 373 - 374			COLUMBIA 384			COOKEVILLE 385			JACKSON 383			JOHNSON CITY 376			KNOXVILLE 377 - 379		
		MAT.	INST.	TOTAL	MAT.	INST.	TOTAL	MAT.	INST.	TOTAL	MAT.	INST.	TOTAL	MAT.	INST.	TOTAL	MAT.	INST.	TOTAL
015433	CONTRACTOR EQUIPMENT		104.8	104.8		98.8	98.8		98.8	98.8		104.9	104.9		98.7	98.7		98.7	98.7
0241, 31 - 34	SITE & INFRASTRUCTURE, DEMOLITION	103.4	99.6	100.7	86.7	88.1	87.7	92.0	87.6	88.9	95.1	97.9	97.1	110.7	88.4	95.1	89.5	88.9	89.1
0310	Concrete Forming & Accessories	97.9	56.6	62.3	82.3	63.5	66.1	82.5	36.3	42.6	89.0	46.0	51.9	84.6	39.4	45.6	96.4	42.4	49.8
0320	Concrete Reinforcing	89.9	63.2	76.6	87.7	63.2	75.4	87.7	63.2	75.4	87.6	64.0	75.8	90.5	62.1	76.3	89.9	62.2	76.1
0330	Cast-in-Place Concrete	100.0	60.4	83.5	91.8	52.4	75.5	104.0	44.3	79.2	101.6	50.1	80.2	80.4	61.0	72.4	93.9	48.8	75.2
03	CONCRETE	93.8	60.8	77.5	93.6	61.2	77.6	103.2	46.3	75.1	94.7	52.7	74.0	100.2	53.2	77.0	91.2	50.3	71.1
04	MASONRY	108.3	49.7	72.1	118.0	55.5	79.3	112.3	44.5	70.4	118.2	48.0	74.8	123.3	45.6	75.2	85.2	53.0	65.3
05	METALS	97.7	87.4	94.5	91.5	87.6	90.3	91.6	87.5	90.3	93.8	87.6	91.9	94.9	86.4	92.3	98.3	86.7	94.7
06	WOOD, PLASTICS & COMPOSITES	107.1	58.6	79.8	73.0	65.9	69.0	73.2	34.4	51.3	88.9	46.4	65.0	79.1	37.8	55.8	93.4	37.8	62.0
07	THERMAL & MOISTURE PROTECTION	100.1	57.6	82.8	93.7	60.2	80.1	94.1	46.5	74.8	96.0	55.0	79.3	95.2	53.5	78.3	93.3	58.2	79.0
08	OPENINGS	101.7	57.1	91.2	92.5	59.3	84.6	92.5	43.2	80.8	100.1	52.9	88.9	98.0	46.2	85.7	95.3	46.2	83.7
0920	Plaster & Gypsum Board	82.0	57.8	65.6	85.1	65.3	71.6	85.1	32.8	49.5	86.8	45.2	58.5	99.3	36.3	56.5	106.9	36.3	58.9
0950, 0980	Ceilings & Acoustic Treatment	97.1	57.8	71.0	81.5	65.3	70.7	81.5	32.8	49.1	90.3	45.2	60.3	93.3	36.3	55.4	94.2	36.3	55.7
0960	Flooring	98.2	57.7	86.1	94.5	20.9	72.5	94.5	59.7	84.1	91.6	41.5	76.6	97.2	41.4	80.5	101.3	56.8	88.0
0970, 0990	Wall Finishes & Painting/Coating	97.0	53.9	71.1	87.7	34.1	55.5	87.7	37.3	57.4	89.2	50.6	66.0	94.3	43.0	63.5	94.3	49.6	67.4
09	FINISHES	94.5	56.6	73.4	92.3	53.2	70.5	92.8	39.3	63.0	91.9	44.7	65.6	99.1	38.9	65.6	92.1	44.6	65.6
COVERS	DIVS. 10 - 14, 25, 28, 41, 43, 44, 46	100.0	42.4	88.5	100.0	49.2	89.8	100.0	42.1	88.4	100.0	44.2	88.8	100.0	44.3	88.8	100.0	46.8	89.3
21, 22, 23	FIRE SUPPRESSION, PLUMBING & HVAC	100.2	61.9	84.9	97.4	79.8	90.3	97.4	74.2	88.1	100.1	69.5	87.8	100.0	60.4	84.1	100.0	65.7	86.2
26, 27, 3370	ELECTRICAL, COMMUNICATIONS & UTIL.	101.9	69.8	85.1	92.6	58.8	75.0	94.4	63.1	78.0	99.5	59.9	78.8	91.9	47.4	68.7	97.8	57.3	76.6
MF2010	WEIGHTED AVERAGE	99.5	65.5	84.7	95.3	67.5	83.2	96.4	60.4	80.7	98.3	62.5	82.7	99.2	56.7	80.7	96.2	60.5	80.6

TENNESSEE / TEXAS

	DIVISION	TENNESSEE									TEXAS								
		MCKENZIE 382			MEMPHIS 375,380 - 381			NASHVILLE 370 - 372			ABILENE 795 - 796			AMARILLO 790 - 791			AUSTIN 786 - 787		
		MAT.	INST.	TOTAL	MAT.	INST.	TOTAL	MAT.	INST.	TOTAL	MAT.	INST.	TOTAL	MAT.	INST.	TOTAL	MAT.	INST.	TOTAL
015433	CONTRACTOR EQUIPMENT		98.8	98.8		103.2	103.2		105.2	105.2		89.7	89.7		89.7	89.7		89.0	89.0
0241, 31 - 34	SITE & INFRASTRUCTURE, DEMOLITION	91.7	87.7	88.9	93.0	94.6	94.1	97.7	100.3	99.6	94.0	86.7	88.8	92.7	87.7	89.2	95.8	86.8	89.5
0310	Concrete Forming & Accessories	90.1	39.0	46.0	95.4	64.7	68.9	99.3	65.7	70.3	100.3	38.9	47.3	101.2	52.5	59.2	96.9	54.3	60.2
0320	Concrete Reinforcing	87.8	63.9	75.9	95.5	64.5	80.0	97.2	63.6	80.4	89.4	51.7	70.6	93.2	51.7	72.5	91.1	49.4	70.2
0330	Cast-in-Place Concrete	101.8	58.3	83.7	92.4	61.9	79.8	89.4	66.0	79.7	100.1	47.4	78.2	92.5	53.9	76.5	93.3	61.7	80.2
03	CONCRETE	101.9	52.4	77.5	91.6	65.2	78.6	90.6	66.9	78.9	98.3	45.5	72.2	97.5	53.7	75.9	92.7	56.7	74.9
04	MASONRY	116.5	47.0	73.5	100.1	63.2	77.3	91.0	59.3	71.4	98.7	56.9	72.9	100.5	57.7	74.0	99.5	45.8	66.3
05	METALS	91.6	87.8	90.4	99.0	89.3	96.0	99.5	88.7	96.2	100.6	68.5	90.7	96.1	68.9	87.7	97.3	64.8	87.3
06	WOOD, PLASTICS & COMPOSITES	82.1	37.1	56.7	96.4	66.6	79.6	106.2	66.5	83.8	104.2	37.3	66.4	98.2	54.7	73.7	93.4	58.1	73.5
07	THERMAL & MOISTURE PROTECTION	94.1	49.5	76.0	95.5	64.0	82.7	97.1	61.7	82.7	101.2	49.6	80.2	101.8	53.5	82.2	101.1	49.6	80.2
08	OPENINGS	92.5	45.1	81.3	101.3	64.6	92.6	101.5	64.7	92.8	92.8	41.0	80.5	94.3	52.8	84.5	101.9	55.1	90.8
0920	Plaster & Gypsum Board	88.0	35.6	52.4	92.8	65.9	74.5	95.0	65.9	75.2	86.2	35.8	51.9	92.4	53.8	66.2	88.6	57.3	67.3
0950, 0980	Ceilings & Acoustic Treatment	81.5	35.6	51.0	95.4	65.9	75.8	92.1	65.9	74.6	93.3	35.8	55.1	102.3	53.8	70.0	99.4	57.3	71.4
0960	Flooring	97.2	40.1	80.1	98.9	36.7	80.3	97.4	65.4	87.9	113.4	77.9	102.8	108.3	77.9	99.2	105.5	40.2	86.0
0970, 0990	Wall Finishes & Painting/Coating	87.7	50.6	65.4	91.9	56.0	70.3	95.3	73.6	82.2	98.8	50.2	69.6	98.9	49.5	69.2	102.8	40.8	65.6
09	FINISHES	93.9	39.1	63.3	96.5	58.4	75.3	94.8	66.4	79.0	96.2	46.4	68.5	99.3	56.9	75.6	100.1	50.3	72.4
COVERS	DIVS. 10 - 14, 25, 28, 41, 43, 44, 46	100.0	28.2	85.6	100.0	82.1	96.4	100.0	82.6	96.5	100.0	76.1	95.2	100.0	67.7	93.5	100.0	78.3	95.7
21, 22, 23	FIRE SUPPRESSION, PLUMBING & HVAC	97.4	69.8	86.3	100.0	74.9	90.0	100.0	83.2	93.3	100.3	46.0	78.5	100.1	50.8	80.3	100.1	56.1	82.5
26, 27, 3370	ELECTRICAL, COMMUNICATIONS & UTIL.	94.1	64.4	78.6	100.9	64.9	82.1	97.6	63.1	79.6	94.9	48.7	70.7	98.7	64.6	80.9	97.0	62.1	78.8
MF2010	WEIGHTED AVERAGE	96.6	60.4	80.9	98.5	71.3	86.7	97.9	74.1	87.5	98.3	53.6	78.8	98.2	60.0	81.6	98.6	59.1	81.4

TEXAS

DIVISION		BEAUMONT 776 - 777			BROWNWOOD 768			BRYAN 778			CHILDRESS 792			CORPUS CHRISTI 783 - 784			DALLAS 752 - 753		
		MAT.	INST.	TOTAL	MAT.	INST.	TOTAL	MAT.	INST.	TOTAL	MAT.	INST.	TOTAL	MAT.	INST.	TOTAL	MAT.	INST.	TOTAL
015433	CONTRACTOR EQUIPMENT		91.2	91.2		89.7	89.7		91.2	91.2		89.7	89.7		96.2	96.2		99.1	99.1
0241, 31 - 34	SITE & INFRASTRUCTURE, DEMOLITION	89.0	87.6	88.0	100.5	86.1	90.4	80.7	90.3	87.5	103.1	87.0	91.8	138.3	82.8	99.3	107.7	88.2	94.0
0310	Concrete Forming & Accessories	106.7	48.4	56.4	97.7	33.0	41.8	84.5	45.5	50.8	98.7	61.9	66.9	98.8	39.8	47.9	97.5	64.1	68.7
0320	Concrete Reinforcing	89.0	41.5	65.3	91.7	50.7	71.3	91.3	49.5	70.4	89.5	51.0	70.3	84.6	47.2	65.9	99.1	52.8	76.0
0330	Cast-in-Place Concrete	86.8	51.7	72.2	105.3	42.7	79.3	69.6	65.7	68.0	102.7	49.4	80.6	109.1	46.9	83.3	96.4	54.8	79.1
03	CONCRETE	96.2	49.3	73.1	105.2	40.8	73.4	80.7	54.5	67.8	107.8	56.2	82.3	99.0	45.5	72.6	99.6	60.1	80.1
04	MASONRY	99.8	57.3	73.5	130.0	48.3	79.5	137.2	56.3	87.2	102.9	49.8	70.0	88.6	51.4	65.6	101.9	56.5	73.8
05	METALS	95.2	64.6	85.8	97.3	62.8	86.6	94.8	70.5	87.3	98.1	65.7	88.1	92.7	75.0	87.2	100.6	79.6	94.1
06	WOOD, PLASTICS & COMPOSITES	114.0	48.4	77.0	101.5	29.2	60.7	79.0	39.6	56.8	103.5	67.5	83.2	116.8	38.2	72.4	102.3	67.6	82.7
07	THERMAL & MOISTURE PROTECTION	103.5	55.5	84.0	100.1	44.1	77.4	95.4	56.2	79.5	101.7	48.8	80.3	102.3	48.4	80.4	92.1	62.5	80.1
08	OPENINGS	94.4	45.2	82.8	91.6	37.0	78.7	97.4	45.9	85.2	90.0	56.6	82.1	106.7	39.5	90.8	103.9	58.4	93.2
0920	Plaster & Gypsum Board	97.4	47.3	63.4	87.8	27.5	46.8	84.7	38.3	53.1	85.8	66.9	73.0	93.9	36.7	54.9	96.7	66.9	76.4
0950, 0980	Ceilings & Acoustic Treatment	95.5	47.3	63.5	91.7	27.5	49.0	89.0	38.3	55.3	91.5	66.9	75.2	91.5	36.7	55.0	103.4	66.9	79.1
0960	Flooring	109.1	74.3	98.7	109.2	45.5	90.1	84.2	62.9	77.9	111.6	41.6	90.7	115.4	64.2	100.1	107.8	48.6	90.1
0970, 0990	Wall Finishes & Painting/Coating	96.1	48.6	67.6	96.3	29.8	56.3	92.6	62.5	74.5	98.8	38.6	62.7	115.0	57.7	80.6	105.5	51.5	73.1
09	FINISHES	90.2	52.7	69.4	95.3	34.0	61.2	79.5	48.8	62.4	96.4	56.8	74.4	103.7	45.6	71.3	100.4	60.4	78.1
COVERS	DIVS. 10 - 14, 25, 28, 41, 43, 44, 46	100.0	79.4	95.9	100.0	37.3	87.4	100.0	79.8	96.0	100.0	70.0	94.0	100.0	78.3	95.7	100.0	82.0	96.4
21, 22, 23	FIRE SUPPRESSION, PLUMBING & HVAC	100.2	59.7	83.9	95.3	49.1	76.8	95.3	65.9	83.5	95.4	52.3	78.1	100.2	40.0	76.0	100.0	62.1	84.8
26, 27, 3370	ELECTRICAL, COMMUNICATIONS & UTIL.	92.1	70.8	81.0	93.2	43.7	67.4	90.2	67.7	78.4	94.9	43.4	68.0	91.8	61.4	76.0	95.7	65.2	79.7
MF2010	WEIGHTED AVERAGE	96.7	61.2	81.2	98.5	48.4	76.7	94.1	63.0	80.5	97.9	56.9	80.0	99.7	53.7	79.6	100.1	65.8	85.2

TEXAS

DIVISION		DEL RIO 788			DENTON 762			EASTLAND 764			EL PASO 798 - 799,885			FORT WORTH 760 - 761			GALVESTON 775		
		MAT.	INST.	TOTAL	MAT.	INST.	TOTAL	MAT.	INST.	TOTAL	MAT.	INST.	TOTAL	MAT.	INST.	TOTAL	MAT.	INST.	TOTAL
015433	CONTRACTOR EQUIPMENT		89.0	89.0		95.7	95.7		89.7	89.7		89.8	89.8		89.7	89.7		100.1	100.1
0241, 31 - 34	SITE & INFRASTRUCTURE, DEMOLITION	119.1	85.4	95.4	100.5	79.3	85.6	103.1	86.0	91.1	95.1	86.1	88.8	97.4	87.1	90.1	108.0	88.1	94.1
0310	Concrete Forming & Accessories	95.3	36.5	44.6	104.6	32.3	42.2	98.5	35.8	44.4	97.2	43.8	51.1	96.9	63.5	68.1	94.2	65.6	69.5
0320	Concrete Reinforcing	85.2	46.7	66.0	93.0	51.0	72.1	92.0	50.7	71.3	97.5	45.8	71.7	97.3	52.7	75.0	90.8	62.8	76.8
0330	Cast-in-Place Concrete	117.4	42.8	86.4	81.5	44.4	66.1	111.3	43.3	83.1	96.6	41.3	73.7	98.6	50.7	78.7	92.3	66.9	81.8
03	CONCRETE	119.8	41.7	81.3	82.8	42.0	62.7	110.0	42.4	76.6	99.9	44.3	72.5	100.7	57.7	79.5	97.2	67.1	82.3
04	MASONRY	103.6	50.0	70.4	138.8	47.1	82.1	99.9	48.2	67.9	96.5	46.9	65.8	98.4	56.4	72.4	95.9	56.4	71.5
05	METALS	92.5	60.8	82.7	96.9	75.5	90.3	97.1	63.2	86.6	95.3	62.6	85.2	98.8	68.5	89.5	96.3	88.8	94.0
06	WOOD, PLASTICS & COMPOSITES	97.0	34.6	61.8	112.9	31.9	67.2	108.3	32.8	65.7	99.9	45.8	69.4	99.5	67.5	81.4	96.4	66.4	79.4
07	THERMAL & MOISTURE PROTECTION	99.2	43.6	76.6	98.3	44.1	76.3	100.5	44.2	77.6	98.9	54.5	80.9	98.4	52.6	79.8	94.5	65.3	82.6
08	OPENINGS	99.3	33.9	83.9	107.8	37.5	91.1	72.9	39.0	64.9	91.2	43.9	80.0	98.7	58.3	89.2	101.8	64.0	92.9
0920	Plaster & Gypsum Board	89.7	33.0	51.2	92.3	30.2	50.0	87.8	31.2	49.3	93.7	44.6	60.3	92.0	66.9	74.9	92.0	65.7	74.1
0950, 0980	Ceilings & Acoustic Treatment	87.8	33.0	51.4	96.2	30.2	52.2	91.7	31.2	51.4	95.9	44.6	61.7	102.3	66.9	78.8	91.6	65.7	74.4
0960	Flooring	97.3	34.7	78.6	102.5	43.5	84.8	141.2	45.5	112.6	114.0	63.5	98.9	137.1	41.4	108.5	98.1	62.9	87.6
0970, 0990	Wall Finishes & Painting/Coating	101.3	38.4	63.5	106.2	29.6	60.2	97.8	30.5	57.4	93.2	34.9	58.2	95.2	51.1	68.7	104.3	47.0	69.9
09	FINISHES	95.8	36.0	62.5	92.0	33.5	59.4	104.8	36.3	66.6	99.2	46.5	69.9	107.4	58.8	80.3	88.8	63.0	74.4
COVERS	DIVS. 10 - 14, 25, 28, 41, 43, 44, 46	100.0	37.1	87.4	100.0	36.0	87.2	100.0	38.4	87.7	100.0	77.3	95.4	100.0	81.7	96.3	100.0	84.3	96.9
21, 22, 23	FIRE SUPPRESSION, PLUMBING & HVAC	95.2	46.9	75.8	95.3	37.9	72.3	95.3	49.4	76.9	100.0	34.2	73.6	100.0	57.7	83.1	95.2	66.0	83.5
26, 27, 3370	ELECTRICAL, COMMUNICATIONS & UTIL.	94.0	30.1	60.6	95.6	43.8	68.6	93.1	43.8	67.3	92.7	54.9	73.0	94.5	62.4	77.8	91.7	69.5	80.1
MF2010	WEIGHTED AVERAGE	99.4	46.2	76.3	98.2	46.6	75.7	96.5	49.1	75.9	97.2	50.6	76.9	99.6	62.5	83.5	96.3	69.7	84.7

TEXAS

DIVISION		GIDDINGS 789			GREENVILLE 754			HOUSTON 770 - 772			HUNTSVILLE 773			LAREDO 780			LONGVIEW 756		
		MAT.	INST.	TOTAL	MAT.	INST.	TOTAL	MAT.	INST.	TOTAL	MAT.	INST.	TOTAL	MAT.	INST.	TOTAL	MAT.	INST.	TOTAL
015433	CONTRACTOR EQUIPMENT		89.0	89.0		96.7	96.7		100.0	100.0		91.2	91.2		89.0	89.0		91.5	91.5
0241, 31 - 34	SITE & INFRASTRUCTURE, DEMOLITION	105.0	85.4	91.3	101.0	83.5	88.7	106.7	87.9	93.5	95.8	87.9	90.3	99.1	86.8	90.5	97.3	90.3	92.4
0310	Concrete Forming & Accessories	93.0	55.5	60.7	89.1	27.3	35.7	96.3	65.8	69.9	91.8	55.4	60.4	95.3	40.0	47.5	85.5	28.1	35.9
0320	Concrete Reinforcing	85.8	46.8	66.3	99.6	24.9	62.3	90.6	63.3	77.0	91.5	50.5	71.0	85.2	47.6	66.4	98.4	22.7	60.6
0330	Cast-in-Place Concrete	99.5	42.9	76.0	87.8	37.7	67.0	89.5	66.9	80.1	95.6	48.2	75.9	84.1	59.3	73.8	102.1	35.0	74.2
03	CONCRETE	96.5	50.2	73.7	92.2	32.6	62.8	94.9	67.2	81.3	103.8	52.8	78.7	91.4	49.1	70.6	108.4	30.8	70.1
04	MASONRY	111.8	50.0	73.5	159.0	46.1	89.1	95.8	66.6	77.7	135.8	51.6	83.7	97.3	58.6	73.3	154.6	42.0	85.0
05	METALS	92.0	61.8	82.6	97.9	63.6	87.3	99.2	89.7	96.3	94.7	66.3	85.9	95.0	64.0	85.5	91.2	50.5	78.7
06	WOOD, PLASTICS & COMPOSITES	96.0	59.4	75.4	92.3	26.3	55.1	98.8	66.4	80.5	88.3	58.2	71.3	97.0	38.1	63.7	86.2	28.4	53.6
07	THERMAL & MOISTURE PROTECTION	99.9	47.7	78.7	92.2	39.5	70.8	94.1	68.1	83.6	96.3	48.2	76.8	98.2	51.9	79.4	93.5	34.9	69.7
08	OPENINGS	98.5	49.0	86.7	102.1	26.0	84.1	104.4	64.0	94.8	97.4	48.5	85.8	99.2	39.4	85.0	91.7	26.0	76.1
0920	Plaster & Gypsum Board	88.6	58.7	68.3	90.4	24.5	45.5	94.5	65.7	74.9	89.5	57.4	67.7	90.8	36.7	54.0	88.7	26.8	46.6
0950, 0980	Ceilings & Acoustic Treatment	87.8	58.7	68.4	99.0	24.5	49.4	97.0	65.7	76.2	89.0	57.4	67.9	92.2	36.7	55.2	95.4	26.8	49.7
0960	Flooring	97.8	34.7	79.0	103.7	47.8	87.0	99.1	62.9	88.3	87.4	34.7	71.7	97.1	64.2	87.3	108.1	35.1	86.2
0970, 0990	Wall Finishes & Painting/Coating	101.3	30.5	58.8	105.5	25.1	57.2	104.3	62.5	79.2	92.6	31.3	55.8	101.3	54.4	73.1	95.1	29.3	55.6
09	FINISHES	94.7	50.2	69.9	96.9	30.4	59.8	96.0	64.7	78.5	81.9	49.9	64.1	95.2	45.1	67.3	101.0	29.4	61.1
COVERS	DIVS. 10 - 14, 25, 28, 41, 43, 44, 46	100.0	42.9	88.6	100.0	34.4	86.9	100.0	84.3	96.9	100.0	43.3	88.6	100.0	76.3	95.2	100.0	21.9	84.4
21, 22, 23	FIRE SUPPRESSION, PLUMBING & HVAC	95.3	62.5	82.1	95.2	43.5	74.4	100.0	66.1	86.4	95.3	63.3	82.5	100.1	38.8	75.5	95.1	27.5	68.0
26, 27, 3370	ELECTRICAL, COMMUNICATIONS & UTIL.	90.8	43.4	66.1	92.3	27.1	58.3	93.7	67.8	80.2	90.2	43.4	65.8	94.2	59.9	76.3	92.7	53.2	72.0
MF2010	WEIGHTED AVERAGE	96.4	55.6	78.6	99.4	42.2	74.5	98.7	70.9	86.6	97.2	56.9	79.6	97.1	53.7	78.2	99.0	40.5	73.5

TEXAS

DIVISION		LUBBOCK 793 - 794			LUFKIN 759			MCALLEN 785			MCKINNEY 750			MIDLAND 797			ODESSA 797		
		MAT.	INST.	TOTAL	MAT.	INST.	TOTAL	MAT.	INST.	TOTAL	MAT.	INST.	TOTAL	MAT.	INST.	TOTAL	MAT.	INST.	TOTAL
015433	CONTRACTOR EQUIPMENT		98.2	98.2		91.5	91.5		96.4	96.4		96.7	96.7		98.2	98.2		89.7	89.7
0241, 31 - 34	SITE & INFRASTRUCTURE, DEMOLITION	116.1	85.0	94.3	92.5	91.8	92.0	142.5	82.9	100.6	97.4	83.6	87.7	118.7	85.0	95.1	94.2	87.0	89.1
0310	Concrete Forming & Accessories	99.1	51.0	57.6	88.6	56.8	61.2	99.9	35.8	44.6	88.2	36.3	43.4	103.0	51.0	58.2	100.2	50.9	57.7
0320	Concrete Reinforcing	90.5	51.7	71.1	100.2	62.4	81.3	84.8	47.1	66.0	99.6	51.1	75.4	91.4	50.9	71.2	89.4	50.9	70.2
0330	Cast-in-Place Concrete	100.3	48.4	78.8	91.3	44.7	72.0	118.5	45.3	88.1	82.4	45.2	66.9	106.6	48.6	82.5	100.1	47.5	78.3
03	CONCRETE	96.9	52.0	74.7	100.8	54.3	77.8	106.6	43.1	75.3	87.6	44.0	66.0	101.6	51.9	77.1	98.3	50.7	74.8
04	MASONRY	98.1	55.7	71.8	118.7	52.3	77.6	103.2	51.4	71.1	171.0	48.4	95.1	114.8	42.0	69.7	98.7	42.0	63.6
05	METALS	104.1	80.1	96.7	98.0	69.1	89.1	92.4	74.5	86.9	97.8	74.4	90.6	102.4	79.0	95.2	99.9	67.1	89.8
06	WOOD, PLASTICS & COMPOSITES	104.1	54.9	76.3	94.3	59.4	74.6	115.4	33.6	69.2	91.0	36.4	60.2	109.1	54.9	78.5	104.2	54.7	76.3
07	THERMAL & MOISTURE PROTECTION	90.8	52.7	75.3	93.3	48.0	74.9	102.6	45.0	79.2	92.0	45.9	73.3	91.0	47.4	73.3	101.2	46.6	79.0
08	OPENINGS	103.6	49.4	90.8	69.6	53.4	65.8	103.4	36.8	87.6	102.0	39.8	87.3	102.5	49.1	89.9	92.8	49.0	82.4
0920	Plaster & Gypsum Board	86.7	53.8	64.3	87.7	58.7	68.0	94.9	31.9	52.0	89.6	34.8	52.4	88.3	53.8	64.8	86.2	53.8	64.2
0950, 0980	Ceilings & Acoustic Treatment	95.1	53.8	67.6	88.3	58.7	68.6	92.2	31.9	52.1	99.0	34.8	56.3	92.4	53.8	66.7	93.3	53.8	67.0
0960	Flooring	107.0	39.3	86.7	143.4	38.2	112.0	114.9	65.0	100.0	103.3	35.6	83.1	108.2	48.6	90.4	113.4	48.6	94.0
0970, 0990	Wall Finishes & Painting/Coating	110.2	49.5	73.7	95.1	37.8	60.6	115.0	28.9	63.2	105.5	41.2	66.9	110.2	49.5	73.7	98.8	49.5	69.2
09	FINISHES	98.8	48.8	70.9	109.1	52.0	77.3	104.3	39.8	68.4	96.4	36.7	63.1	99.2	50.7	72.1	96.2	50.6	70.8
COVERS	DIVS. 10 - 14, 25, 28, 41, 43, 44, 46	100.0	77.8	95.6	100.0	70.8	94.2	100.0	76.0	95.2	100.0	39.7	87.9	100.0	78.6	95.7	100.0	78.3	95.6
21, 22, 23	FIRE SUPPRESSION, PLUMBING & HVAC	99.8	45.2	77.9	95.1	63.5	82.4	95.3	30.8	69.4	95.2	41.4	73.6	94.9	45.3	75.0	100.3	45.3	78.2
26, 27, 3370	ELECTRICAL, COMMUNICATIONS & UTIL.	93.6	49.6	70.6	94.0	43.4	67.6	91.6	34.2	61.6	92.4	45.2	67.8	93.5	40.4	65.8	95.0	40.3	66.4
MF2010	WEIGHTED AVERAGE	100.0	56.3	81.0	95.9	59.1	79.9	99.7	46.6	76.6	99.3	48.9	77.3	99.9	53.6	79.7	98.2	52.5	78.3

TEXAS

DIVISION		PALESTINE 758			SAN ANGELO 769			SAN ANTONIO 781 - 782			TEMPLE 765			TEXARKANA 755			TYLER 757		
		MAT.	INST.	TOTAL	MAT.	INST.	TOTAL	MAT.	INST.	TOTAL	MAT.	INST.	TOTAL	MAT.	INST.	TOTAL	MAT.	INST.	TOTAL
015433	CONTRACTOR EQUIPMENT		91.5	91.5		89.7	89.7		91.5	91.5		89.7	89.7		91.5	91.5		91.5	91.5
0241, 31 - 34	SITE & INFRASTRUCTURE, DEMOLITION	98.6	91.1	93.3	97.0	88.1	90.8	101.8	91.4	94.5	86.0	85.9	85.9	87.0	91.1	89.9	96.6	91.5	93.1
0310	Concrete Forming & Accessories	80.1	55.8	59.1	98.0	36.5	44.9	94.5	54.8	60.3	101.3	39.7	48.1	95.4	34.0	42.4	90.3	33.1	40.9
0320	Concrete Reinforcing	97.7	50.7	74.2	91.6	50.9	71.3	91.8	49.1	70.5	91.8	47.4	69.6	97.6	52.4	75.0	98.4	53.0	75.7
0330	Cast-in-Place Concrete	83.5	42.6	66.5	99.3	46.6	77.4	84.5	65.4	76.6	81.4	48.4	67.7	84.1	43.4	67.2	102.0	42.5	76.3
03	CONCRETE	102.9	50.9	77.2	100.4	43.9	72.5	92.6	58.3	75.7	86.0	45.2	65.9	93.6	42.0	68.1	108.0	41.4	75.1
04	MASONRY	113.5	48.5	73.3	126.6	41.6	74.0	98.9	66.5	78.8	137.9	48.4	82.5	175.4	44.7	94.5	164.8	54.4	96.5
05	METALS	97.7	62.7	86.9	97.4	66.5	87.9	96.1	68.7	87.6	97.1	62.8	86.5	91.1	66.9	83.6	97.6	68.0	88.5
06	WOOD, PLASTICS & COMPOSITES	84.7	59.4	70.4	101.9	36.1	64.7	96.7	52.8	71.9	111.0	39.4	70.6	98.1	32.3	61.0	96.0	30.5	59.0
07	THERMAL & MOISTURE PROTECTION	93.7	46.4	74.5	100.0	42.1	76.5	95.9	64.3	83.1	99.6	45.0	77.4	93.2	47.2	74.5	93.6	53.2	77.2
08	OPENINGS	69.5	50.7	65.1	91.7	38.8	79.2	99.9	51.7	88.5	69.6	41.8	63.0	91.7	37.9	78.9	69.5	36.9	61.8
0920	Plaster & Gypsum Board	84.4	58.7	66.9	87.8	34.6	51.6	88.8	51.8	63.6	87.8	38.0	53.9	93.2	30.7	50.7	87.7	28.9	47.7
0950, 0980	Ceilings & Acoustic Treatment	88.3	58.7	68.6	91.7	34.6	53.7	91.2	51.8	65.0	91.7	38.0	56.0	95.4	30.7	52.4	88.3	28.9	48.7
0960	Flooring	135.3	58.9	112.5	109.2	34.7	86.9	97.6	64.3	87.6	143.4	41.7	113.0	115.6	59.4	98.7	145.3	37.0	112.9
0970, 0990	Wall Finishes & Painting/Coating	95.1	30.5	56.3	96.3	49.5	68.2	98.5	54.4	72.0	97.8	38.5	62.2	95.1	39.4	61.6	95.1	39.4	61.6
09	FINISHES	106.9	54.4	77.7	95.0	36.4	62.4	94.8	55.4	72.8	104.1	39.6	68.2	103.1	38.1	66.9	109.9	32.6	66.9
COVERS	DIVS. 10 - 14, 25, 28, 41, 43, 44, 46	100.0	40.7	88.1	100.0	75.0	95.0	100.0	80.6	96.1	100.0	36.8	87.3	100.0	74.9	95.0	100.0	74.7	94.9
21, 22, 23	FIRE SUPPRESSION, PLUMBING & HVAC	95.1	58.5	80.4	95.3	44.6	75.0	100.0	64.5	85.8	95.3	48.2	76.4	95.1	32.9	70.2	95.1	57.7	80.1
26, 27, 3370	ELECTRICAL, COMMUNICATIONS & UTIL.	90.1	43.4	65.7	97.2	45.8	70.4	94.3	60.0	76.4	94.3	64.9	79.0	93.8	61.1	76.8	92.7	55.0	73.0
MF2010	WEIGHTED AVERAGE	95.4	55.8	78.2	98.2	49.5	77.0	97.5	64.5	83.1	95.0	52.7	76.6	98.4	49.7	77.2	98.9	54.6	79.6

TEXAS / UTAH

DIVISION		VICTORIA 779			WACO 766 - 767			WAXAHACHIE 751			WHARTON 774			WICHITA FALLS 763			LOGAN 843		
		MAT.	INST.	TOTAL	MAT.	INST.	TOTAL	MAT.	INST.	TOTAL	MAT.	INST.	TOTAL	MAT.	INST.	TOTAL	MAT.	INST.	TOTAL
015433	CONTRACTOR EQUIPMENT		99.0	99.0		89.7	89.7		96.7	96.7		100.1	100.1		89.7	89.7		97.2	97.2
0241, 31 - 34	SITE & INFRASTRUCTURE, DEMOLITION	112.7	82.3	91.4	94.2	87.0	89.2	99.0	84.4	88.8	118.4	85.5	95.3	94.9	86.7	89.1	93.4	96.2	95.4
0310	Concrete Forming & Accessories	94.0	35.1	43.2	99.9	39.6	47.9	88.2	59.5	63.5	88.7	36.7	43.9	99.9	37.2	45.8	104.8	58.1	64.5
0320	Concrete Reinforcing	87.2	30.3	58.8	91.5	49.4	70.5	99.6	51.1	75.4	90.7	50.2	70.5	91.5	51.7	71.6	102.3	80.2	91.3
0330	Cast-in-Place Concrete	103.6	41.0	77.7	88.1	65.8	78.8	86.1	47.1	68.1	106.5	51.2	83.5	94.1	44.7	73.6	86.9	72.8	81.0
03	CONCRETE	104.5	38.5	71.9	93.0	51.7	72.6	91.3	53.2	72.5	108.8	46.4	78.0	95.8	43.8	70.1	107.9	67.9	88.2
04	MASONRY	112.4	34.1	64.0	98.3	53.7	70.7	159.7	48.5	90.9	97.0	44.6	64.6	98.8	56.9	72.9	103.9	59.7	76.5
05	METALS	94.9	70.3	87.3	99.4	67.0	89.4	97.9	77.0	91.4	96.3	78.5	90.8	99.4	68.5	89.8	101.0	78.2	94.0
06	WOOD, PLASTICS & COMPOSITES	99.9	35.9	63.8	109.2	35.1	67.4	91.0	67.5	77.8	89.4	32.6	57.3	109.2	35.0	67.3	81.5	56.2	67.2
07	THERMAL & MOISTURE PROTECTION	98.4	39.5	74.5	100.5	46.9	78.8	92.1	46.3	73.5	94.8	49.3	76.3	100.5	50.5	80.2	97.9	61.8	83.3
08	OPENINGS	101.7	35.2	85.9	81.5	42.5	72.2	102.1	56.8	91.3	101.8	37.4	86.6	81.5	40.3	71.7	94.3	57.1	85.5
0920	Plaster & Gypsum Board	89.3	34.3	51.9	88.3	33.6	51.1	90.1	66.9	74.3	86.8	30.9	48.8	88.3	33.4	51.0	75.6	54.8	61.4
0950, 0980	Ceilings & Acoustic Treatment	92.5	34.3	53.8	93.5	33.6	53.6	100.7	66.9	78.2	91.6	30.9	51.2	93.5	33.4	53.5	105.1	54.8	71.6
0960	Flooring	97.8	38.2	80.0	142.3	45.1	113.3	103.3	38.2	83.8	96.1	45.5	80.9	143.5	77.9	123.9	100.2	48.6	84.8
0970, 0990	Wall Finishes & Painting/Coating	104.5	37.8	64.4	97.8	49.8	69.0	105.5	37.8	64.8	104.3	31.3	60.5	100.5	49.5	69.9	102.9	62.0	78.3
09	FINISHES	87.0	36.0	58.6	104.7	40.2	68.8	97.1	54.8	73.5	88.3	36.7	59.5	105.3	45.0	71.7	98.5	56.0	74.8
COVERS	DIVS. 10 - 14, 25, 28, 41, 43, 44, 46	100.0	36.6	87.3	100.0	78.2	95.6	100.0	40.6	88.1	100.0	38.7	87.7	100.0	65.2	93.0	100.0	54.1	90.8
21, 22, 23	FIRE SUPPRESSION, PLUMBING & HVAC	95.3	31.2	69.6	100.2	54.2	81.8	95.2	40.5	73.2	95.2	63.4	82.5	100.2	48.4	79.5	100.0	69.0	87.6
26, 27, 3370	ELECTRICAL, COMMUNICATIONS & UTIL.	96.5	39.7	66.9	97.4	65.1	80.5	92.4	65.2	78.2	95.3	43.8	68.4	99.1	61.0	79.2	95.4	74.3	84.4
MF2010	WEIGHTED AVERAGE	98.2	42.7	74.1	97.2	57.3	79.8	99.3	56.4	80.6	98.1	53.8	78.8	97.8	55.0	79.2	99.7	68.8	86.2

UTAH / VERMONT

	DIVISION	OGDEN 842,844 MAT.	INST.	TOTAL	PRICE 845 MAT.	INST.	TOTAL	PROVO 846 - 847 MAT.	INST.	TOTAL	SALT LAKE CITY 840 - 841 MAT.	INST.	TOTAL	BELLOWS FALLS 051 MAT.	INST.	TOTAL	BENNINGTON 052 MAT.	INST.	TOTAL
015433	CONTRACTOR EQUIPMENT		97.2	97.2		96.3	96.3		96.3	96.3		97.2	97.2		100.6	100.6		100.6	100.6
0241, 31 - 34	SITE & INFRASTRUCTURE, DEMOLITION	82.2	96.2	92.1	90.8	94.6	93.5	89.8	94.6	93.2	81.8	96.2	91.9	82.6	100.1	94.9	82.0	100.1	94.7
0310	Concrete Forming & Accessories	104.8	58.1	64.5	107.1	49.4	57.3	106.3	58.0	64.6	107.3	58.0	64.7	100.2	87.5	89.3	97.8	109.1	107.6
0320	Concrete Reinforcing	102.0	80.2	91.1	109.8	80.1	95.0	110.7	80.2	95.5	104.3	80.2	92.3	83.0	87.3	85.2	83.0	87.3	85.2
0330	Cast-in-Place Concrete	88.2	72.8	81.8	87.0	60.0	75.8	87.0	72.7	81.1	96.4	72.7	86.6	89.5	113.1	99.3	89.5	113.1	99.3
03	CONCRETE	97.5	67.9	82.9	109.3	59.6	84.8	107.8	67.8	88.1	116.9	67.8	92.6	93.5	96.2	94.8	93.3	105.8	99.5
04	MASONRY	97.8	59.7	74.2	109.1	59.7	78.5	109.3	59.7	78.6	111.0	59.7	79.3	94.6	96.9	96.0	102.1	96.9	98.9
05	METALS	101.5	78.2	94.3	98.3	77.6	91.9	99.3	78.0	92.7	106.1	78.0	97.4	95.6	89.3	93.6	95.5	89.2	93.6
06	WOOD, PLASTICS & COMPOSITES	81.5	56.2	67.2	84.4	45.3	62.3	82.9	56.2	67.8	83.2	56.2	68.0	102.2	85.5	92.8	99.2	115.1	108.1
07	THERMAL & MOISTURE PROTECTION	96.9	61.8	82.6	99.5	59.4	83.3	99.5	61.8	84.2	103.8	61.8	86.8	98.0	84.8	92.7	98.0	83.1	92.0
08	OPENINGS	94.3	57.1	85.5	98.3	50.1	86.9	98.3	57.1	88.5	96.1	57.1	86.9	103.5	88.1	99.9	103.5	104.2	103.7
0920	Plaster & Gypsum Board	75.6	54.8	61.4	77.9	43.5	54.5	76.0	54.8	61.6	83.0	54.8	63.8	99.8	84.4	89.3	97.9	114.9	109.5
0950, 0980	Ceilings & Acoustic Treatment	105.1	54.8	71.6	105.1	43.5	64.2	105.1	54.8	71.6	99.2	54.8	69.6	92.3	84.4	87.1	92.3	114.9	107.3
0960	Flooring	98.0	48.6	83.3	101.2	36.1	81.8	100.9	48.6	85.3	102.2	48.6	86.2	101.6	106.5	103.1	100.8	106.5	102.5
0970, 0990	Wall Finishes & Painting/Coating	102.9	62.0	78.3	102.9	38.3	64.1	102.9	49.0	70.5	106.5	49.0	71.9	95.7	106.8	102.3	95.7	106.8	102.3
09	FINISHES	96.5	56.0	74.0	99.5	44.5	68.9	99.1	54.6	74.3	98.2	54.6	73.9	97.2	92.6	94.6	96.7	110.1	104.2
COVERS	DIVS. 10 - 14, 25, 28, 41, 43, 44, 46	100.0	54.1	90.8	100.0	47.0	89.4	100.0	54.0	90.8	100.0	54.0	90.8	100.0	100.3	100.1	100.0	103.5	100.7
21, 22, 23	FIRE SUPPRESSION, PLUMBING & HVAC	100.0	69.0	87.6	97.5	68.6	85.9	100.0	69.0	87.6	100.1	69.0	87.6	95.2	98.5	96.5	95.2	98.4	96.5
26, 27, 3370	ELECTRICAL, COMMUNICATIONS & UTIL.	95.7	74.3	84.5	101.2	74.3	87.1	96.2	61.2	77.9	98.6	61.2	79.1	98.6	91.5	94.9	98.6	62.6	79.8
MF2010	WEIGHTED AVERAGE	97.9	68.8	85.2	100.1	65.3	84.9	100.1	66.7	85.6	102.2	66.8	86.8	96.6	94.7	95.8	96.9	95.3	96.2

VERMONT

	DIVISION	BRATTLEBORO 053 MAT.	INST.	TOTAL	BURLINGTON 054 MAT.	INST.	TOTAL	GUILDHALL 059 MAT.	INST.	TOTAL	MONTPELIER 056 MAT.	INST.	TOTAL	RUTLAND 057 MAT.	INST.	TOTAL	ST. JOHNSBURY 058 MAT.	INST.	TOTAL
015433	CONTRACTOR EQUIPMENT		100.6	100.6		100.6	100.6		100.6	100.6		100.6	100.6		100.6	100.6		100.6	100.6
0241, 31 - 34	SITE & INFRASTRUCTURE, DEMOLITION	83.4	100.1	95.1	86.2	100.0	95.9	81.7	98.7	93.6	86.6	99.8	95.8	85.9	99.8	95.6	81.7	98.7	93.6
0310	Concrete Forming & Accessories	100.5	87.4	89.2	98.0	89.6	90.8	98.1	80.1	82.6	101.8	86.1	88.2	100.8	89.6	91.1	96.5	80.1	82.4
0320	Concrete Reinforcing	82.2	87.3	84.7	105.7	87.2	96.4	83.8	87.2	85.5	89.0	87.2	88.1	103.7	87.2	95.5	82.2	87.2	84.7
0330	Cast-in-Place Concrete	92.3	113.1	100.9	102.6	112.0	106.5	86.8	103.2	93.6	105.0	111.9	107.9	87.7	111.9	97.8	86.8	103.2	93.6
03	CONCRETE	95.5	96.2	95.9	104.0	96.7	100.4	91.0	89.5	90.3	102.8	95.2	99.0	96.9	96.7	96.8	90.7	89.5	90.1
04	MASONRY	101.1	96.9	98.5	104.8	95.4	99.0	101.3	79.6	87.8	100.1	95.4	97.2	85.8	95.4	91.7	124.9	79.6	96.9
05	METALS	95.5	89.2	93.6	104.0	88.4	99.2	95.6	88.3	93.3	100.2	88.3	96.5	101.0	88.4	97.1	95.6	88.3	93.3
06	WOOD, PLASTICS & COMPOSITES	102.6	85.5	92.9	98.3	90.3	93.8	98.5	85.5	91.2	96.4	85.5	90.3	102.7	90.3	95.7	93.6	85.5	89.0
07	THERMAL & MOISTURE PROTECTION	98.1	80.0	90.8	104.3	82.5	95.4	97.9	72.6	87.6	105.0	82.0	95.6	98.2	82.5	91.8	97.8	72.6	87.6
08	OPENINGS	103.5	88.1	99.9	107.5	86.4	102.5	103.5	83.8	98.8	103.8	83.8	99.1	106.6	86.4	101.8	103.5	83.8	98.8
0920	Plaster & Gypsum Board	99.8	84.4	89.3	100.5	89.3	92.9	106.1	84.4	91.3	105.4	84.4	91.1	100.5	89.3	92.9	108.0	84.4	91.9
0950, 0980	Ceilings & Acoustic Treatment	92.3	84.4	87.1	96.0	89.3	91.6	92.3	84.4	87.1	102.5	84.4	90.5	97.0	89.3	91.9	92.3	84.4	87.1
0960	Flooring	101.7	106.5	103.2	103.1	106.5	104.1	104.6	106.5	105.2	104.0	106.5	104.8	101.6	106.5	103.1	107.7	106.5	107.3
0970, 0990	Wall Finishes & Painting/Coating	95.7	106.8	102.3	100.4	85.3	91.4	95.7	85.3	89.4	100.5	85.3	91.4	95.7	85.3	89.4	95.7	85.3	89.4
09	FINISHES	97.3	92.6	94.7	99.1	92.7	95.5	98.8	85.9	91.6	101.0	89.8	94.8	98.4	92.7	95.2	99.9	85.9	92.1
COVERS	DIVS. 10 - 14, 25, 28, 41, 43, 44, 46	100.0	100.3	100.1	100.0	100.2	100.0	100.0	94.3	98.9	100.0	99.7	99.9	100.0	100.2	100.0	100.0	94.3	98.9
21, 22, 23	FIRE SUPPRESSION, PLUMBING & HVAC	95.2	98.4	96.5	99.9	73.3	89.3	95.2	65.3	83.2	95.0	73.3	86.3	100.1	73.3	89.3	95.2	65.3	83.2
26, 27, 3370	ELECTRICAL, COMMUNICATIONS & UTIL.	98.6	91.5	94.9	98.0	62.6	79.5	98.6	62.6	79.8	98.0	62.6	79.5	98.6	62.6	79.8	98.6	62.6	79.8
MF2010	WEIGHTED AVERAGE	97.2	94.5	96.0	101.5	85.1	94.4	96.8	79.2	89.1	99.2	84.3	92.7	99.2	85.1	93.0	97.9	79.2	89.8

VERMONT / VIRGINIA

	DIVISION	WHITE RIVER JCT. 050 MAT.	INST.	TOTAL	ALEXANDRIA 223 MAT.	INST.	TOTAL	ARLINGTON 222 MAT.	INST.	TOTAL	BRISTOL 242 MAT.	INST.	TOTAL	CHARLOTTESVILLE 229 MAT.	INST.	TOTAL	CULPEPER 227 MAT.	INST.	TOTAL
015433	CONTRACTOR EQUIPMENT		100.6	100.6		102.1	102.1		100.8	100.8		100.8	100.8		105.1	105.1		100.8	100.8
0241, 31 - 34	SITE & INFRASTRUCTURE, DEMOLITION	85.7	98.8	94.9	113.5	89.3	96.5	123.7	87.3	98.2	107.8	85.7	92.3	112.6	87.7	95.1	111.2	87.2	94.3
0310	Concrete Forming & Accessories	95.0	80.7	82.7	93.3	73.6	76.3	92.5	73.0	75.7	88.4	43.4	49.6	86.8	49.6	54.7	83.9	72.3	73.9
0320	Concrete Reinforcing	83.0	87.2	85.1	85.0	87.0	86.0	95.9	87.0	91.5	95.9	67.3	81.6	95.3	73.0	84.2	95.9	86.9	91.4
0330	Cast-in-Place Concrete	92.3	104.1	97.2	107.5	82.2	97.0	104.6	81.4	95.0	104.1	49.1	81.3	108.3	56.1	86.6	107.1	68.8	91.2
03	CONCRETE	97.3	90.1	93.7	101.2	80.1	90.8	105.6	79.6	92.8	102.1	51.7	77.2	102.8	57.9	80.7	100.1	74.9	87.6
04	MASONRY	112.6	81.2	93.1	86.0	72.9	77.8	99.1	71.3	81.9	89.1	52.3	66.3	112.3	54.2	76.3	100.9	71.2	82.5
05	METALS	95.6	88.3	93.3	102.5	97.0	100.8	101.1	97.4	99.9	99.9	85.8	95.6	100.2	90.6	97.2	100.3	95.4	98.8
06	WOOD, PLASTICS & COMPOSITES	95.9	85.5	90.0	98.1	71.9	83.3	94.4	71.9	81.7	86.6	40.9	60.8	85.1	46.6	63.4	83.7	71.9	77.1
07	THERMAL & MOISTURE PROTECTION	98.2	73.3	88.1	103.2	82.4	94.7	105.1	81.2	95.4	104.7	61.3	87.1	104.3	69.0	90.0	104.5	78.7	94.0
08	OPENINGS	103.5	83.8	98.8	96.4	75.6	91.5	94.6	75.6	90.1	97.4	46.8	85.5	95.7	52.7	85.5	96.0	75.6	91.2
0920	Plaster & Gypsum Board	96.5	84.4	88.3	108.4	70.9	82.9	104.4	70.9	81.6	100.1	38.9	58.4	100.1	44.1	62.0	100.3	70.9	80.3
0950, 0980	Ceilings & Acoustic Treatment	92.3	84.4	87.1	97.7	70.9	79.9	95.1	70.9	79.0	94.2	38.9	57.4	94.2	44.1	60.8	95.1	70.9	79.0
0960	Flooring	99.7	106.5	101.8	106.7	84.5	100.0	105.2	82.7	98.5	101.9	67.7	91.7	100.6	67.7	90.7	100.6	82.7	95.2
0970, 0990	Wall Finishes & Painting/Coating	95.7	85.3	89.4	120.5	79.9	96.1	120.5	79.9	96.1	106.3	55.5	75.7	106.3	77.7	89.1	120.5	77.7	94.8
09	FINISHES	96.5	86.3	90.8	102.9	75.5	87.6	102.6	74.7	87.0	99.2	48.2	70.8	98.8	54.6	74.2	99.6	74.3	85.5
COVERS	DIVS. 10 - 14, 25, 28, 41, 43, 44, 46	100.0	94.9	99.0	100.0	88.5	97.7	100.0	86.0	97.2	100.0	69.5	93.9	100.0	78.0	95.6	100.0	86.0	97.2
21, 22, 23	FIRE SUPPRESSION, PLUMBING & HVAC	95.2	66.1	83.5	100.3	87.7	95.2	100.3	86.8	94.9	95.4	51.6	77.8	95.4	68.7	84.7	95.4	73.2	86.5
26, 27, 3370	ELECTRICAL, COMMUNICATIONS & UTIL.	98.6	62.6	79.8	95.8	99.2	97.6	93.4	99.2	96.5	95.5	36.2	64.6	95.4	70.8	82.6	98.0	99.2	98.6
MF2010	WEIGHTED AVERAGE	97.9	79.7	90.0	99.8	85.4	93.5	100.6	84.7	93.6	98.1	55.6	79.6	99.2	67.3	85.3	98.6	80.8	90.9

VIRGINIA

DIVISION		FAIRFAX 220 - 221			FARMVILLE 239			FREDERICKSBURG 224 - 225			GRUNDY 246			HARRISONBURG 228			LYNCHBURG 245		
		MAT.	INST.	TOTAL	MAT.	INST.	TOTAL	MAT.	INST.	TOTAL	MAT.	INST.	TOTAL	MAT.	INST.	TOTAL	MAT.	INST.	TOTAL
015433	CONTRACTOR EQUIPMENT		100.8	100.8		105.1	105.1		100.8	100.8		100.8	100.8		100.8	100.8		100.8	100.8
0241, 31 - 34	SITE & INFRASTRUCTURE, DEMOLITION	122.4	87.3	97.8	107.2	87.2	93.2	110.8	87.2	94.2	105.6	84.4	90.7	119.4	85.9	95.9	106.5	85.9	92.1
0310	Concrete Forming & Accessories	86.8	72.8	74.7	100.4	44.6	52.3	86.8	70.8	73.0	91.5	37.2	44.6	82.9	43.5	48.9	88.4	61.6	65.3
0320	Concrete Reinforcing	95.9	87.0	91.5	94.7	51.5	73.1	96.7	86.9	91.8	94.6	52.0	73.3	95.9	66.9	81.5	95.3	67.3	81.3
0330	Cast-in-Place Concrete	104.6	81.3	94.9	106.8	54.1	85.0	106.2	64.7	89.0	104.1	48.1	80.9	104.6	58.8	85.6	104.1	58.4	85.2
03	CONCRETE	105.3	79.4	92.5	101.4	50.9	76.5	99.8	72.8	86.5	100.9	45.3	73.5	103.2	54.9	79.3	100.8	63.0	82.1
04	MASONRY	99.0	71.3	81.8	95.6	45.6	64.7	99.9	71.2	82.1	90.9	48.1	64.4	97.4	51.2	68.8	104.9	54.1	73.5
05	METALS	100.4	97.0	99.3	100.0	78.2	93.3	100.3	96.2	99.0	99.9	71.0	91.0	100.2	85.1	95.6	100.1	86.3	95.9
06	WOOD, PLASTICS & COMPOSITES	86.6	71.9	78.3	101.9	45.0	69.8	86.6	69.6	77.0	89.6	35.8	59.3	82.7	39.8	58.5	86.6	64.0	73.9
07	THERMAL & MOISTURE PROTECTION	105.0	73.9	92.4	105.3	50.2	83.0	104.5	79.1	94.2	104.7	46.1	80.9	104.9	64.5	88.5	104.5	65.1	88.5
08	OPENINGS	94.6	75.6	90.1	94.9	44.0	82.9	95.7	74.0	90.6	97.4	35.2	82.7	96.0	50.6	85.3	96.0	59.4	87.3
0920	Plaster & Gypsum Board	100.3	70.9	80.3	109.3	42.5	63.8	100.3	68.5	78.7	100.1	33.7	54.9	100.1	37.8	57.7	100.1	62.7	74.7
0950, 0980	Ceilings & Acoustic Treatment	95.1	70.9	79.0	92.5	42.5	59.2	95.1	68.5	77.4	94.2	33.7	53.9	94.2	37.8	56.7	94.2	62.7	73.3
0960	Flooring	102.2	82.7	96.4	108.6	61.4	94.5	102.2	82.7	96.4	103.2	34.3	82.6	100.3	75.3	92.9	101.9	67.7	91.7
0970, 0990	Wall Finishes & Painting/Coating	120.5	79.9	96.1	107.7	39.6	66.7	120.5	77.7	94.8	106.3	35.9	64.0	120.5	55.5	81.4	106.3	55.5	75.7
09	FINISHES	101.2	74.7	86.4	101.8	47.0	71.3	100.1	72.9	85.0	99.3	35.9	64.0	100.0	48.9	71.5	99.0	61.5	78.1
COVERS	DIVS. 10 - 14, 25, 28, 41, 43, 44, 46	100.0	86.0	97.2	100.0	46.8	89.3	100.0	81.9	96.4	100.0	44.8	88.9	100.0	69.3	93.8	100.0	72.7	94.5
21, 22, 23	FIRE SUPPRESSION, PLUMBING & HVAC	95.4	86.8	91.9	95.3	44.3	74.9	95.4	86.5	91.8	95.4	63.5	82.6	95.4	67.1	84.1	95.4	68.4	84.6
26, 27, 3370	ELECTRICAL, COMMUNICATIONS & UTIL.	96.6	99.2	98.0	90.4	49.4	69.0	93.6	99.2	96.6	95.5	45.7	69.5	95.7	98.7	97.3	96.6	53.9	74.3
MF2010	WEIGHTED AVERAGE	99.3	84.4	92.8	97.8	53.3	78.5	98.2	83.0	91.6	98.0	53.4	78.6	98.8	68.1	85.5	98.6	66.1	84.5

VIRGINIA

DIVISION		NEWPORT NEWS 236			NORFOLK 233 - 235			PETERSBURG 238			PORTSMOUTH 237			PULASKI 243			RICHMOND 230 - 232		
		MAT.	INST.	TOTAL	MAT.	INST.	TOTAL	MAT.	INST.	TOTAL	MAT.	INST.	TOTAL	MAT.	INST.	TOTAL	MAT.	INST.	TOTAL
015433	CONTRACTOR EQUIPMENT		105.1	105.1		105.7	105.7		105.1	105.1		105.0	105.0		100.8	100.8		105.0	105.0
0241, 31 - 34	SITE & INFRASTRUCTURE, DEMOLITION	106.2	88.4	93.7	105.7	89.4	94.3	109.7	88.6	94.9	104.7	88.2	93.1	104.9	85.1	91.0	103.5	88.6	93.0
0310	Concrete Forming & Accessories	99.7	65.4	70.1	99.5	65.6	70.2	93.0	58.3	63.1	88.9	54.2	58.9	91.5	39.5	46.6	100.7	58.3	64.1
0320	Concrete Reinforcing	94.4	72.8	83.6	99.5	72.8	86.2	94.1	73.2	83.6	94.1	72.8	83.4	94.6	66.7	80.7	100.7	73.2	87.0
0330	Cast-in-Place Concrete	103.8	66.8	88.5	111.9	66.9	93.2	110.3	54.2	87.0	102.8	66.7	87.8	104.1	48.9	81.2	98.8	54.2	80.3
03	CONCRETE	98.6	68.6	83.8	103.3	68.7	86.2	103.6	61.2	82.7	97.3	63.5	80.7	100.9	49.6	75.6	97.3	61.2	79.5
04	MASONRY	90.6	54.9	68.5	99.1	54.9	71.7	103.3	53.6	72.5	96.0	54.9	70.6	85.9	49.1	63.1	92.1	53.5	68.2
05	METALS	102.2	90.4	98.6	102.0	90.5	98.4	100.1	91.1	97.3	101.2	89.7	97.7	100.0	83.8	95.0	106.2	91.1	101.5
06	WOOD, PLASTICS & COMPOSITES	100.6	67.7	82.0	97.0	67.7	80.5	91.4	59.8	73.6	87.6	52.6	67.8	89.6	37.6	60.3	101.6	59.8	78.0
07	THERMAL & MOISTURE PROTECTION	105.3	69.7	90.8	101.9	69.7	88.8	105.3	69.6	90.8	105.3	67.0	89.7	104.7	51.0	82.9	103.7	69.6	89.8
08	OPENINGS	95.3	63.3	87.7	94.2	63.3	86.9	94.6	59.9	86.4	95.4	55.0	85.8	97.5	43.2	84.6	99.0	59.9	89.8
0920	Plaster & Gypsum Board	110.2	65.8	80.0	106.3	65.8	78.8	101.7	57.7	71.8	102.0	50.2	66.8	100.1	35.5	56.1	104.7	57.7	72.8
0950, 0980	Ceilings & Acoustic Treatment	96.1	65.8	75.9	97.7	65.8	76.5	93.4	57.7	69.7	96.1	50.2	65.6	94.2	35.5	55.1	100.4	57.7	72.0
0960	Flooring	108.6	67.7	96.4	103.5	67.7	92.8	104.7	72.5	95.1	101.7	67.7	91.5	103.2	67.7	92.6	109.4	72.5	98.4
0970, 0990	Wall Finishes & Painting/Coating	107.7	42.9	68.8	105.8	77.7	88.9	107.7	77.7	89.7	107.7	77.7	89.7	106.3	55.5	75.7	107.0	77.7	89.4
09	FINISHES	102.5	63.5	80.8	100.7	67.4	82.2	100.1	62.5	79.1	99.5	57.9	76.3	99.3	44.7	68.9	104.4	62.5	81.1
COVERS	DIVS. 10 - 14, 25, 28, 41, 43, 44, 46	100.0	75.4	95.1	100.0	75.4	95.1	100.0	78.2	95.6	100.0	73.7	94.7	100.0	45.0	89.0	100.0	78.2	95.6
21, 22, 23	FIRE SUPPRESSION, PLUMBING & HVAC	100.2	64.1	85.7	100.0	64.7	85.8	95.3	66.9	83.9	100.2	64.7	86.0	95.4	63.4	82.5	100.0	66.9	86.7
26, 27, 3370	ELECTRICAL, COMMUNICATIONS & UTIL.	93.1	60.3	76.0	96.9	59.7	77.4	93.3	70.8	81.5	91.4	59.7	74.9	95.5	55.7	74.7	96.2	70.8	83.0
MF2010	WEIGHTED AVERAGE	99.2	68.2	85.7	100.0	68.8	86.4	98.6	68.9	85.6	98.6	66.1	84.5	97.7	58.3	80.6	100.4	68.9	86.7

VIRGINIA / WASHINGTON

DIVISION		ROANOKE 240 - 241			STAUNTON 244			WINCHESTER 226			CLARKSTON 994			EVERETT 982			OLYMPIA 985		
		MAT.	INST.	TOTAL	MAT.	INST.	TOTAL	MAT.	INST.	TOTAL	MAT.	INST.	TOTAL	MAT.	INST.	TOTAL	MAT.	INST.	TOTAL
015433	CONTRACTOR EQUIPMENT		100.8	100.8		105.1	105.1		100.8	100.8		90.4	90.4		102.1	102.1		102.1	102.1
0241, 31 - 34	SITE & INFRASTRUCTURE, DEMOLITION	105.2	85.9	91.7	109.1	87.3	93.8	118.1	87.2	96.4	97.1	90.0	92.1	89.4	110.3	104.1	90.2	110.4	104.4
0310	Concrete Forming & Accessories	97.9	61.8	66.8	91.1	51.1	56.6	85.2	70.8	72.8	114.8	68.8	75.1	114.9	99.6	101.7	103.1	99.8	100.2
0320	Concrete Reinforcing	95.6	67.4	81.5	95.3	50.5	72.9	95.3	86.3	90.8	105.7	87.4	96.6	110.2	100.2	105.2	116.1	100.2	108.2
0330	Cast-in-Place Concrete	118.3	58.5	93.5	108.3	52.0	84.9	104.6	64.7	88.1	97.0	83.2	91.3	99.8	103.8	101.5	89.2	105.4	96.0
03	CONCRETE	105.5	63.1	84.6	123.0	53.0	78.0	102.6	72.7	87.9	110.7	77.4	94.3	98.7	102.0	99.7	96.9	101.3	99.1
04	MASONRY	92.1	54.1	68.6	100.6	52.3	70.7	95.1	60.0	73.3	97.2	80.7	87.0	111.9	99.0	103.9	103.1	99.9	101.1
05	METALS	102.2	86.5	97.4	100.2	81.7	94.5	100.3	94.8	98.6	88.1	81.2	86.0	104.3	91.7	100.4	104.2	91.6	100.3
06	WOOD, PLASTICS & COMPOSITES	99.0	64.0	79.3	89.6	51.0	67.8	85.1	69.6	76.4	103.3	65.2	81.8	112.1	99.3	104.9	94.6	99.3	97.2
07	THERMAL & MOISTURE PROTECTION	104.3	67.2	89.3	104.3	53.0	83.5	105.0	76.2	93.3	142.7	76.9	116.0	100.5	100.5	100.5	101.6	97.8	100.1
08	OPENINGS	96.4	59.4	87.6	96.0	49.2	84.9	97.5	74.7	92.1	106.7	69.7	98.0	101.9	99.8	101.4	103.5	99.8	102.6
0920	Plaster & Gypsum Board	108.4	62.7	77.3	100.1	48.6	65.1	100.3	68.5	78.7	130.6	64.0	85.3	110.7	99.2	102.9	103.1	99.2	100.5
0950, 0980	Ceilings & Acoustic Treatment	97.7	62.7	74.5	94.2	48.6	63.9	95.1	68.5	77.4	97.9	64.0	75.3	99.9	99.2	99.5	106.3	99.2	101.6
0960	Flooring	106.7	67.7	95.0	102.3	53.0	87.9	101.6	82.7	96.0	98.3	48.0	83.2	117.9	93.3	110.6	109.2	93.4	104.5
0970, 0990	Wall Finishes & Painting/Coating	106.3	55.5	75.7	106.3	35.3	63.6	120.5	89.1	101.6	106.4	59.8	78.4	104.8	90.1	96.0	110.3	90.1	98.1
09	FINISHES	101.8	62.4	79.8	99.2	47.6	70.4	100.5	74.2	85.9	114.8	62.6	85.7	109.0	97.1	102.4	104.6	97.4	100.6
COVERS	DIVS. 10 - 14, 25, 28, 41, 43, 44, 46	100.0	72.7	94.5	100.0	72.6	94.5	100.0	85.7	97.1	100.0	77.3	95.5	100.0	99.6	99.9	100.0	100.1	100.0
21, 22, 23	FIRE SUPPRESSION, PLUMBING & HVAC	100.3	65.8	86.5	95.4	60.1	81.2	95.4	86.8	91.9	95.4	84.7	91.1	100.1	100.3	100.2	100.0	100.8	100.3
26, 27, 3370	ELECTRICAL, COMMUNICATIONS & UTIL.	95.5	55.7	74.7	94.3	74.6	84.0	94.1	99.2	96.8	94.9	94.4	94.6	103.0	98.5	100.7	101.0	99.7	100.3
MF2010	WEIGHTED AVERAGE	100.2	66.0	85.3	98.4	62.7	82.9	98.7	82.1	91.5	100.6	80.7	92.0	102.1	99.6	101.0	101.0	100.0	100.6

City Cost Indexes

WASHINGTON

DIVISION		RICHLAND 993			SEATTLE 980 - 981,987			SPOKANE 990 - 992			TACOMA 983 - 984			VANCOUVER 986			WENATCHEE 988		
		MAT.	INST.	TOTAL	MAT.	INST.	TOTAL	MAT.	INST.	TOTAL	MAT.	INST.	TOTAL	MAT.	INST.	TOTAL	MAT.	INST.	TOTAL
015433	CONTRACTOR EQUIPMENT		90.4	90.4		102.0	102.0		90.4	90.4		102.1	102.1		97.1	97.1		102.1	102.1
0241, 31 - 34	SITE & INFRASTRUCTURE, DEMOLITION	99.6	90.8	93.4	94.5	108.6	104.4	98.7	90.8	93.2	92.4	110.4	105.0	102.3	98.1	99.3	101.4	109.2	106.9
0310	Concrete Forming & Accessories	114.9	78.4	83.4	106.9	100.3	101.2	119.8	78.0	83.7	105.5	99.8	100.6	106.2	92.1	94.0	107.2	77.5	81.6
0320	Concrete Reinforcing	101.4	87.6	94.5	112.2	100.3	106.3	102.1	87.6	94.9	109.0	100.2	104.6	109.9	100.0	104.9	109.8	89.3	99.6
0330	Cast-in-Place Concrete	97.2	85.5	92.4	103.6	105.6	104.5	101.0	85.3	94.5	102.6	105.4	103.8	114.6	99.5	108.3	104.8	77.4	93.4
03	CONCRETE	110.3	82.6	96.6	102.0	101.7	101.8	112.6	82.3	97.6	100.5	101.3	100.9	110.5	95.9	103.3	108.4	79.8	94.3
04	MASONRY	99.3	83.2	89.3	109.2	99.9	103.4	99.9	83.8	89.9	106.7	99.9	102.5	107.3	96.7	100.7	109.1	91.5	98.2
05	METALS	88.5	84.1	87.1	104.5	93.4	101.1	90.8	83.7	88.6	106.0	91.6	101.6	103.4	92.1	99.9	103.5	86.1	98.1
06	WOOD, PLASTICS & COMPOSITES	103.4	75.7	87.8	102.2	99.3	100.6	112.1	75.7	91.6	101.3	99.3	100.2	93.1	91.9	92.5	102.9	75.9	87.6
07	THERMAL & MOISTURE PROTECTION	144.1	79.7	117.9	97.6	100.9	99.0	140.6	79.1	115.6	100.2	98.7	99.6	100.6	92.5	97.3	100.8	82.2	93.3
08	OPENINGS	109.2	71.2	100.2	105.2	99.8	103.9	109.7	71.7	100.7	102.7	99.8	102.0	98.5	94.4	97.5	102.2	71.5	94.9
0920	Plaster & Gypsum Board	130.6	74.8	92.6	106.9	99.2	101.7	121.5	74.8	89.7	111.5	99.2	103.1	109.4	91.9	97.5	114.0	75.1	87.5
0950, 0980	Ceilings & Acoustic Treatment	104.8	74.8	84.8	101.8	99.2	100.1	99.8	74.8	83.2	103.5	99.2	100.6	98.4	91.9	94.1	95.3	75.1	81.8
0960	Flooring	98.7	81.6	93.5	110.9	101.0	108.0	98.2	86.9	94.8	111.4	93.4	106.0	118.1	104.0	113.9	114.1	63.8	99.1
0970, 0990	Wall Finishes & Painting/Coating	106.4	66.2	82.2	104.2	90.1	95.7	106.6	72.4	86.0	104.8	90.1	96.0	107.2	68.4	83.9	104.8	66.2	81.6
09	FINISHES	116.6	76.7	94.4	108.1	98.9	102.9	114.3	78.5	94.3	108.2	97.4	102.2	105.6	91.9	98.0	108.5	72.9	88.7
COVERS	DIVS. 10 - 14, 25, 28, 41, 43, 44, 46	100.0	87.2	97.4	100.0	100.1	100.0	100.0	87.1	97.4	100.0	100.1	100.0	100.0	62.0	92.4	100.0	78.5	95.7
21, 22, 23	FIRE SUPPRESSION, PLUMBING & HVAC	100.3	109.7	104.1	100.0	117.1	106.8	100.2	85.9	94.5	100.1	100.8	100.4	100.2	88.6	95.5	95.2	87.9	92.3
26, 27, 3370	ELECTRICAL, COMMUNICATIONS & UTIL.	92.1	94.4	93.3	102.4	107.0	104.8	90.5	79.1	84.5	102.8	99.7	101.2	107.5	101.0	104.1	103.7	98.5	101.0
MF2010	WEIGHTED AVERAGE	102.2	89.6	96.7	102.5	104.8	103.5	102.5	82.7	93.8	102.3	100.0	101.3	103.1	93.2	98.8	102.0	87.1	95.5

DIVISION		WASHINGTON YAKIMA 989			WEST VIRGINIA BECKLEY 258 - 259			BLUEFIELD 247 - 248			BUCKHANNON 262			CHARLESTON 250 - 253			CLARKSBURG 263 - 264		
		MAT.	INST.	TOTAL	MAT.	INST.	TOTAL	MAT.	INST.	TOTAL	MAT.	INST.	TOTAL	MAT.	INST.	TOTAL	MAT.	INST.	TOTAL
015433	CONTRACTOR EQUIPMENT		102.1	102.1		100.8	100.8		100.8	100.8		100.8	100.8		100.8	100.8		100.8	100.8
0241, 31 - 34	SITE & INFRASTRUCTURE, DEMOLITION	94.7	109.8	105.3	97.9	89.2	91.8	99.6	89.2	92.3	105.8	89.2	94.1	99.0	90.0	92.7	106.3	89.2	94.3
0310	Concrete Forming & Accessories	106.0	95.3	96.8	86.3	92.0	91.2	88.3	91.9	91.4	87.7	92.0	90.9	99.6	93.4	94.2	85.3	92.1	91.1
0320	Concrete Reinforcing	109.5	88.8	99.2	92.3	89.4	90.9	94.2	84.9	89.6	94.8	85.0	89.9	100.6	89.5	95.1	94.8	84.4	89.6
0330	Cast-in-Place Concrete	109.6	84.4	99.2	102.0	102.1	102.0	101.9	102.1	102.0	101.6	100.3	101.0	100.6	102.7	101.4	111.3	98.4	105.9
03	CONCRETE	105.1	90.0	97.7	96.5	95.5	96.0	96.9	94.7	95.8	99.7	94.2	97.0	98.1	96.4	97.2	103.7	93.4	98.6
04	MASONRY	100.4	81.6	88.8	90.8	93.4	92.4	87.5	93.4	91.1	97.4	92.3	94.2	89.2	95.3	93.0	100.7	92.3	95.5
05	METALS	104.2	85.7	98.5	101.6	98.2	100.3	100.2	96.5	99.1	100.4	97.3	99.4	99.8	98.9	99.5	100.4	96.9	99.3
06	WOOD, PLASTICS & COMPOSITES	101.7	99.3	100.4	85.1	92.8	89.4	88.6	92.8	91.0	87.9	92.8	90.6	101.1	92.8	96.4	84.4	92.8	89.1
07	THERMAL & MOISTURE PROTECTION	100.3	81.6	92.7	107.1	89.4	99.9	104.4	89.4	98.4	104.7	90.0	98.7	103.6	90.2	98.2	104.6	88.9	98.3
08	OPENINGS	102.2	81.0	97.2	96.1	85.0	93.5	97.9	83.9	94.6	97.9	83.9	94.6	98.1	85.0	95.0	97.9	85.1	94.9
0920	Plaster & Gypsum Board	111.1	99.2	103.0	93.3	92.4	92.7	99.3	92.4	94.6	99.7	92.4	94.7	97.5	92.4	94.0	97.5	92.4	94.0
0950, 0980	Ceilings & Acoustic Treatment	97.3	99.2	98.6	93.4	92.4	92.7	92.4	92.4	92.4	94.2	92.4	93.0	102.2	92.4	95.7	94.2	92.4	93.0
0960	Flooring	112.4	59.1	96.4	100.5	114.1	104.6	99.8	114.1	104.1	99.5	114.1	103.9	104.4	114.1	107.3	98.6	114.1	103.2
0970, 0990	Wall Finishes & Painting/Coating	104.8	72.4	85.3	107.1	93.5	98.9	106.3	93.5	98.6	106.3	93.8	98.8	106.6	93.5	98.7	106.3	93.8	98.8
09	FINISHES	107.2	86.5	95.7	97.8	96.9	97.3	97.5	96.9	97.2	98.4	96.7	97.4	102.4	97.4	99.6	97.7	96.7	97.1
COVERS	DIVS. 10 - 14, 25, 28, 41, 43, 44, 46	100.0	97.3	99.5	100.0	56.6	91.3	100.0	56.6	91.3	100.0	96.7	99.3	100.0	97.7	99.5	100.0	96.7	99.3
21, 22, 23	FIRE SUPPRESSION, PLUMBING & HVAC	100.1	109.1	103.7	95.5	91.5	93.9	95.4	91.5	93.8	95.4	92.3	94.2	100.1	92.9	97.2	95.4	92.2	94.1
26, 27, 3370	ELECTRICAL, COMMUNICATIONS & UTIL.	105.9	94.4	99.9	91.2	91.9	91.6	94.4	91.9	93.1	95.8	96.8	96.3	96.3	91.9	94.0	95.8	96.8	96.3
MF2010	WEIGHTED AVERAGE	102.5	94.4	99.0	96.9	92.0	94.8	97.1	91.7	94.7	98.2	93.6	96.2	99.0	94.1	96.9	98.8	93.5	96.5

DIVISION		WEST VIRGINIA GASSAWAY 266			HUNTINGTON 255 - 257			LEWISBURG 249			MARTINSBURG 254			MORGANTOWN 265			PARKERSBURG 261		
		MAT.	INST.	TOTAL	MAT.	INST.	TOTAL	MAT.	INST.	TOTAL	MAT.	INST.	TOTAL	MAT.	INST.	TOTAL	MAT.	INST.	TOTAL
015433	CONTRACTOR EQUIPMENT		100.8	100.8		100.8	100.8		100.8	100.8		100.8	100.8		100.8	100.8		100.8	100.8
0241, 31 - 34	SITE & INFRASTRUCTURE, DEMOLITION	103.2	89.2	93.3	102.4	90.4	94.0	115.3	89.2	97.0	101.4	89.7	93.2	100.6	89.8	93.0	108.8	89.9	95.5
0310	Concrete Forming & Accessories	87.2	92.2	91.5	98.2	96.0	96.3	85.6	91.8	91.0	86.3	82.4	83.0	85.6	91.8	91.0	89.8	90.0	90.0
0320	Concrete Reinforcing	94.8	85.0	89.9	93.7	90.6	92.2	94.8	84.9	89.9	92.3	83.0	87.6	94.8	84.3	89.6	94.2	84.8	89.5
0330	Cast-in-Place Concrete	106.3	101.0	104.1	111.1	103.5	108.0	101.9	102.1	102.0	106.8	96.0	102.3	101.5	100.1	101.0	103.8	97.1	101.0
03	CONCRETE	100.2	94.4	97.3	101.8	98.0	99.9	105.8	94.6	100.3	99.9	88.0	94.1	96.6	93.8	95.3	101.6	92.1	96.9
04	MASONRY	101.5	92.3	95.8	89.8	99.9	96.0	90.3	93.4	92.2	91.7	85.6	87.9	117.9	92.3	102.0	78.3	90.7	86.0
05	METALS	100.3	97.2	99.4	103.6	99.5	102.3	100.3	96.5	99.1	101.6	94.9	99.5	100.4	96.7	99.3	101.0	96.8	99.7
06	WOOD, PLASTICS & COMPOSITES	87.0	92.8	90.2	97.2	95.4	96.2	84.7	92.8	89.3	85.1	81.3	82.9	84.7	92.8	89.3	88.0	88.8	88.4
07	THERMAL & MOISTURE PROTECTION	104.4	90.0	98.6	107.3	91.4	100.8	105.2	89.4	98.8	107.4	78.2	95.5	104.5	88.9	98.2	104.4	89.6	98.4
08	OPENINGS	96.1	83.9	93.2	95.4	86.6	93.4	97.9	83.9	94.6	98.1	73.9	92.4	99.1	85.1	95.8	96.9	81.8	93.3
0920	Plaster & Gypsum Board	99.0	92.4	94.5	102.1	95.1	97.4	97.5	92.4	94.0	93.9	80.5	84.8	97.5	92.4	94.0	100.1	88.3	92.0
0950, 0980	Ceilings & Acoustic Treatment	94.2	92.4	93.0	96.1	95.1	95.4	94.2	92.4	93.0	96.1	80.5	85.7	94.2	92.4	93.0	94.2	88.3	90.3
0960	Flooring	99.3	114.1	103.8	107.9	120.3	111.6	98.7	114.1	103.3	100.5	114.1	104.6	98.7	114.1	103.3	102.4	114.1	105.9
0970, 0990	Wall Finishes & Painting/Coating	106.3	93.5	98.6	107.1	93.5	98.9	106.3	77.0	88.7	107.1	42.2	68.1	106.3	93.8	98.8	106.3	93.8	98.8
09	FINISHES	97.9	96.6	97.2	101.7	100.5	101.0	98.8	95.1	96.7	98.7	82.3	89.6	97.3	96.7	96.9	99.3	94.8	96.8
COVERS	DIVS. 10 - 14, 25, 28, 41, 43, 44, 46	100.0	96.7	99.3	100.0	98.6	99.7	100.0	56.6	91.3	100.0	95.4	99.1	100.0	65.4	93.1	100.0	97.1	99.4
21, 22, 23	FIRE SUPPRESSION, PLUMBING & HVAC	95.4	91.4	93.8	100.4	89.3	95.9	95.4	91.5	93.8	95.5	86.7	91.9	95.4	92.2	94.1	100.2	88.1	95.4
26, 27, 3370	ELECTRICAL, COMMUNICATIONS & UTIL.	95.8	91.9	93.8	94.9	96.1	95.5	91.8	91.9	91.8	96.9	82.8	89.6	96.0	96.8	96.4	96.0	92.8	94.3
MF2010	WEIGHTED AVERAGE	98.2	92.8	95.8	99.8	95.2	97.8	98.4	91.4	95.4	98.3	86.1	93.0	98.8	92.6	96.1	98.9	91.4	95.6

		WEST VIRGINIA									WISCONSIN								
		PETERSBURG			ROMNEY			WHEELING			BELOIT			EAU CLAIRE			GREEN BAY		
	DIVISION	268			267			260			535			547			541 - 543		
		MAT.	INST.	TOTAL	MAT.	INST.	TOTAL	MAT.	INST.	TOTAL	MAT.	INST.	TOTAL	MAT.	INST.	TOTAL	MAT.	INST.	TOTAL
015433	CONTRACTOR EQUIPMENT		100.8	100.8		100.8	100.8		100.8	100.8		102.6	102.6		101.2	101.2		99.0	99.0
0241, 31 - 34	SITE & INFRASTRUCTURE, DEMOLITION	99.9	89.8	92.8	102.6	89.8	93.6	109.5	89.9	95.7	95.8	108.6	104.8	94.6	103.6	100.9	98.2	99.6	99.2
0310	Concrete Forming & Accessories	88.7	91.6	91.2	84.8	91.6	90.7	91.5	92.6	92.4	100.1	102.1	101.9	99.7	101.3	101.1	107.8	100.5	101.5
0320	Concrete Reinforcing	94.2	84.9	89.6	94.8	76.3	85.6	93.6	84.3	89.0	96.6	117.1	106.9	93.7	100.9	97.3	91.9	93.2	92.6
0330	Cast-in-Place Concrete	101.5	96.3	99.4	106.3	96.3	102.1	103.8	100.9	102.6	103.7	101.9	102.9	102.1	99.5	101.0	105.6	100.9	103.7
03	CONCRETE	96.8	92.5	94.7	100.0	91.0	95.5	101.6	94.4	98.1	99.9	104.9	102.3	98.8	100.7	99.7	101.9	99.5	100.7
04	MASONRY	92.9	92.3	92.5	90.7	92.3	91.7	100.2	91.7	94.9	98.5	104.8	102.4	89.8	102.7	97.8	120.3	101.9	108.9
05	METALS	100.5	96.6	99.3	100.5	93.7	98.4	101.2	96.9	99.9	94.2	108.4	98.6	93.1	102.7	96.0	95.4	99.1	96.5
06	WOOD, PLASTICS & COMPOSITES	88.9	92.8	91.1	83.8	92.8	88.9	89.6	92.8	91.4	98.2	101.6	100.1	105.2	101.6	103.2	110.0	101.6	105.3
07	THERMAL & MOISTURE PROTECTION	104.5	84.5	96.4	104.6	81.2	95.1	104.7	89.5	98.5	103.4	97.9	101.2	103.2	88.1	97.1	105.0	87.1	97.7
08	OPENINGS	99.1	83.9	95.5	99.0	82.0	95.0	97.6	85.1	94.7	103.6	110.2	105.1	101.0	98.1	100.3	97.4	97.8	97.5
0920	Plaster & Gypsum Board	99.7	92.4	94.7	97.1	92.4	93.9	100.1	92.4	94.8	92.8	101.9	99.0	101.4	101.9	101.8	96.6	101.9	100.2
0950, 0980	Ceilings & Acoustic Treatment	94.2	92.4	93.0	94.2	92.4	93.0	94.2	92.4	93.0	92.5	101.9	98.8	101.7	101.9	101.9	92.7	101.9	98.9
0960	Flooring	100.3	114.1	104.5	98.5	114.1	103.2	103.2	114.1	106.4	101.9	120.3	107.4	94.2	116.2	100.8	113.3	116.2	114.2
0970, 0990	Wall Finishes & Painting/Coating	106.3	93.8	98.8	106.3	93.8	98.8	106.3	93.8	98.8	91.8	99.3	96.3	88.0	88.3	88.2	98.1	85.0	90.2
09	FINISHES	98.1	96.7	97.3	97.4	96.7	97.0	99.5	96.9	98.1	99.7	105.1	102.7	99.2	103.2	101.4	103.3	102.6	102.9
COVERS	DIVS. 10 - 14, 25, 28, 41, 43, 44, 46	100.0	65.3	93.1	100.0	96.7	99.3	100.0	97.1	99.4	100.0	98.2	99.6	100.0	98.5	99.7	100.0	98.1	99.6
21, 22, 23	FIRE SUPPRESSION, PLUMBING & HVAC	95.4	91.2	93.7	95.4	90.8	93.6	100.3	92.8	97.3	100.0	100.0	100.0	100.1	87.5	95.1	100.4	84.3	93.9
26, 27, 3370	ELECTRICAL, COMMUNICATIONS & UTIL.	99.2	82.9	90.7	98.5	82.9	90.3	93.1	96.8	95.0	102.8	87.3	94.7	103.9	86.9	95.0	98.2	82.9	90.2
MF2010	WEIGHTED AVERAGE	98.0	90.1	94.6	98.2	90.4	94.8	99.8	93.8	97.2	99.6	101.9	100.6	98.7	96.4	97.7	100.5	94.1	97.7

		WISCONSIN																	
		KENOSHA			LA CROSSE			LANCASTER			MADISON			MILWAUKEE			NEW RICHMOND		
	DIVISION	531			546			538			537			530,532			540		
		MAT.	INST.	TOTAL	MAT.	INST.	TOTAL	MAT.	INST.	TOTAL	MAT.	INST.	TOTAL	MAT.	INST.	TOTAL	MAT.	INST.	TOTAL
015433	CONTRACTOR EQUIPMENT		100.5	100.5		101.2	101.2		102.6	102.6		102.6	102.6		90.8	90.8		101.6	101.6
0241, 31 - 34	SITE & INFRASTRUCTURE, DEMOLITION	101.8	105.5	104.4	88.5	103.6	99.1	95.0	108.5	104.4	91.3	108.6	103.4	98.0	98.7	98.5	91.7	104.0	100.3
0310	Concrete Forming & Accessories	106.8	102.5	103.1	86.7	100.8	98.9	99.5	99.7	99.7	102.1	101.7	102.0	102.2	120.1	117.6	95.2	102.8	101.8
0320	Concrete Reinforcing	96.4	98.6	97.5	93.4	92.9	93.1	97.9	92.8	95.4	102.3	93.1	97.7	100.5	98.9	99.7	91.0	100.7	95.9
0330	Cast-in-Place Concrete	112.8	106.0	109.9	91.8	97.0	94.0	103.0	101.0	102.2	94.3	101.8	97.4	102.3	110.6	105.7	106.3	83.6	96.9
03	CONCRETE	104.6	103.0	103.8	90.1	98.1	94.1	99.6	98.9	99.2	97.1	100.1	98.6	100.3	111.9	106.0	96.6	95.9	96.3
04	MASONRY	95.9	115.5	108.1	89.0	102.7	97.5	98.6	104.9	102.5	99.4	104.8	102.8	102.2	120.5	113.5	115.4	102.8	107.6
05	METALS	95.0	101.6	97.0	93.0	98.9	94.8	91.8	96.1	93.1	93.5	98.1	94.9	98.1	94.9	97.1	93.2	101.5	95.8
06	WOOD, PLASTICS & COMPOSITES	101.8	98.7	100.1	90.2	101.6	96.7	97.5	101.6	99.8	97.0	101.6	99.6	101.1	121.3	112.5	94.0	104.9	100.2
07	THERMAL & MOISTURE PROTECTION	103.4	108.4	105.4	102.7	87.2	96.4	103.1	83.5	95.2	106.5	94.0	101.4	99.9	113.0	105.2	104.9	93.8	100.4
08	OPENINGS	97.6	102.8	98.9	101.0	86.7	97.6	99.2	87.5	96.4	105.1	101.6	104.3	106.2	115.1	108.3	86.9	99.9	90.0
0920	Plaster & Gypsum Board	83.2	99.0	93.9	95.9	101.9	100.0	91.5	101.9	98.6	94.2	101.9	99.5	99.9	122.1	115.0	85.5	105.5	99.1
0950, 0980	Ceilings & Acoustic Treatment	92.5	99.0	96.8	100.8	101.9	101.6	87.2	101.9	97.0	102.4	101.9	102.1	96.1	122.1	113.4	65.8	105.5	92.2
0960	Flooring	120.9	112.4	118.4	88.3	116.2	96.7	101.6	112.6	104.9	103.5	112.6	106.2	101.6	116.1	105.9	106.5	116.2	109.4
0970, 0990	Wall Finishes & Painting/Coating	101.5	115.2	109.7	88.0	75.5	80.5	91.8	67.8	77.4	89.8	99.3	95.5	93.2	125.6	112.7	101.0	88.3	93.4
09	FINISHES	104.6	105.5	105.1	96.1	101.8	99.3	98.3	98.4	98.3	101.0	103.9	102.6	102.5	121.0	112.8	93.7	103.3	99.0
COVERS	DIVS. 10 - 14, 25, 28, 41, 43, 44, 46	100.0	101.6	100.3	100.0	98.5	99.7	100.0	50.6	90.1	100.0	98.2	99.6	100.0	104.8	101.0	100.0	98.4	99.7
21, 22, 23	FIRE SUPPRESSION, PLUMBING & HVAC	100.2	96.9	98.9	100.1	87.4	95.0	95.1	87.2	92.0	100.0	96.3	98.5	100.0	102.5	101.0	94.8	86.8	91.6
26, 27, 3370	ELECTRICAL, COMMUNICATIONS & UTIL.	103.4	99.6	101.4	104.2	87.1	95.3	102.6	87.1	94.5	103.6	94.3	98.8	102.9	101.9	102.4	102.0	87.1	94.3
MF2010	WEIGHTED AVERAGE	100.2	102.9	101.4	97.2	95.0	96.3	97.4	93.5	95.5	99.5	99.8	99.6	100.9	107.9	104.0	96.2	95.9	96.1

		WISCONSIN																	
		OSHKOSH			PORTAGE			RACINE			RHINELANDER			SUPERIOR			WAUSAU		
	DIVISION	549			539			534			545			548			544		
		MAT.	INST.	TOTAL	MAT.	INST.	TOTAL	MAT.	INST.	TOTAL	MAT.	INST.	TOTAL	MAT.	INST.	TOTAL	MAT.	INST.	TOTAL
015433	CONTRACTOR EQUIPMENT		99.0	99.0		102.6	102.6		102.6	102.6		99.0	99.0		101.6	101.6		99.0	99.0
0241, 31 - 34	SITE & INFRASTRUCTURE, DEMOLITION	90.0	99.6	96.7	85.8	108.4	101.6	95.6	109.3	105.2	102.1	99.6	100.3	88.6	104.0	99.4	85.9	99.6	95.5
0310	Concrete Forming & Accessories	91.0	99.9	98.7	92.1	100.7	99.5	100.4	102.5	102.2	88.4	99.9	98.3	93.4	94.0	94.0	90.3	100.5	99.1
0320	Concrete Reinforcing	92.0	93.1	92.6	98.0	93.0	95.5	96.6	98.6	97.6	92.2	92.2	92.2	91.0	92.7	91.8	92.2	92.9	92.5
0330	Cast-in-Place Concrete	97.9	100.8	99.1	88.5	103.0	94.5	101.7	105.7	103.3	111.0	101.0	106.9	100.1	100.0	100.1	91.3	100.9	95.3
03	CONCRETE	91.7	99.1	95.3	87.5	100.1	93.7	98.9	102.9	100.9	103.6	99.0	101.3	91.2	96.1	93.6	86.8	99.4	93.0
04	MASONRY	102.5	102.2	102.3	97.5	104.8	102.1	98.6	115.5	109.0	118.7	102.7	108.8	114.7	106.3	109.5	102.1	101.9	102.0
05	METALS	93.4	98.4	94.9	92.5	97.1	93.9	95.9	101.5	97.6	93.3	97.8	94.7	94.1	99.4	95.7	93.2	98.9	94.9
06	WOOD, PLASTICS & COMPOSITES	89.8	101.6	96.5	88.4	101.6	95.8	98.4	98.7	98.6	87.4	101.6	95.4	92.4	92.2	92.3	89.3	101.6	96.3
07	THERMAL & MOISTURE PROTECTION	104.2	87.2	97.3	102.6	93.8	99.0	103.4	107.9	105.2	105.0	85.2	96.9	104.5	94.3	100.4	104.1	85.8	96.6
08	OPENINGS	93.4	97.8	94.5	99.3	94.2	98.1	103.6	102.8	103.4	93.5	88.8	92.4	86.4	92.5	87.8	93.6	97.7	94.6
0920	Plaster & Gypsum Board	83.9	101.9	96.2	85.4	101.9	96.7	92.8	99.0	97.0	83.9	101.9	96.2	85.4	92.4	90.2	83.9	101.9	96.2
0950, 0980	Ceilings & Acoustic Treatment	92.7	101.9	98.9	89.9	101.9	97.9	92.5	99.0	96.8	92.7	101.9	98.9	66.7	92.4	83.8	92.7	101.9	98.9
0960	Flooring	105.1	116.2	108.4	98.2	120.3	104.8	101.9	112.6	105.0	104.3	116.2	107.9	107.8	124.7	112.9	104.9	116.2	108.3
0970, 0990	Wall Finishes & Painting/Coating	95.3	85.0	89.1	91.8	67.8	77.4	91.8	115.3	105.9	95.3	61.7	75.1	89.6	103.7	98.1	95.3	85.0	89.1
09	FINISHES	98.4	100.9	99.8	96.4	100.5	98.7	99.7	105.5	103.0	99.3	99.1	99.2	93.1	100.7	97.3	98.1	102.6	100.6
COVERS	DIVS. 10 - 14, 25, 28, 41, 43, 44, 46	100.0	66.1	93.2	100.0	64.8	92.9	100.0	101.6	100.3	100.0	56.4	91.3	100.0	96.7	99.3	100.0	98.1	99.6
21, 22, 23	FIRE SUPPRESSION, PLUMBING & HVAC	95.5	83.4	90.6	95.1	96.1	95.5	100.0	96.9	98.8	95.5	87.0	92.1	94.8	90.4	93.0	95.5	86.3	91.8
26, 27, 3370	ELECTRICAL, COMMUNICATIONS & UTIL.	102.6	81.0	91.3	106.8	94.3	100.3	102.6	95.2	98.7	101.9	82.6	91.8	107.2	99.3	103.1	103.8	82.6	92.7
MF2010	WEIGHTED AVERAGE	96.3	92.4	94.6	96.1	97.9	96.9	99.7	102.6	101.0	98.7	92.4	95.9	96.1	97.6	96.7	95.7	94.4	95.2

529

WYOMING

DIVISION		CASPER 826			CHEYENNE 820			NEWCASTLE 827			RAWLINS 823			RIVERTON 825			ROCK SPRINGS 829 - 831		
		MAT.	INST.	TOTAL	MAT.	INST.	TOTAL	MAT.	INST.	TOTAL	MAT.	INST.	TOTAL	MAT.	INST.	TOTAL	MAT.	INST.	TOTAL
015433	CONTRACTOR EQUIPMENT		97.9	97.9		97.9	97.9		97.9	97.9		97.9	97.9		97.9	97.9		97.9	97.9
0241, 31 - 34	SITE & INFRASTRUCTURE, DEMOLITION	97.2	95.2	95.8	90.1	95.2	93.7	82.0	94.8	91.0	95.8	94.8	95.1	89.3	94.8	93.2	85.8	94.7	92.1
0310	Concrete Forming & Accessories	104.4	43.5	51.9	106.6	54.7	61.8	97.1	63.5	68.1	101.3	63.5	68.6	96.0	63.0	67.5	103.5	62.9	68.4
0320	Concrete Reinforcing	108.6	70.5	89.6	100.8	71.3	86.1	108.4	71.7	90.1	108.1	71.7	89.9	109.1	71.4	90.2	109.1	71.4	90.3
0330	Cast-in-Place Concrete	98.6	72.6	87.8	91.1	72.7	83.5	92.0	62.1	79.6	92.1	62.1	79.6	92.0	52.6	75.7	92.0	52.3	75.5
03	CONCRETE	103.0	59.4	81.5	100.1	64.6	82.6	100.6	64.8	82.9	114.3	64.8	89.9	108.9	61.3	85.4	101.1	61.2	81.4
04	MASONRY	100.0	34.5	59.5	95.8	47.6	65.9	92.6	55.7	69.7	92.6	55.7	69.7	92.6	49.9	66.1	147.5	48.7	86.4
05	METALS	98.1	71.0	89.7	101.5	72.4	92.5	96.6	72.4	89.2	96.7	72.4	89.2	96.8	71.6	89.0	97.5	71.9	89.6
06	WOOD, PLASTICS & COMPOSITES	96.3	41.0	65.1	95.6	55.6	73.0	85.6	68.4	75.9	89.6	68.4	77.6	84.5	68.4	75.5	93.6	68.4	79.4
07	THERMAL & MOISTURE PROTECTION	105.2	49.4	82.5	99.7	54.4	81.3	101.2	55.9	82.8	102.7	55.9	83.7	102.1	59.7	84.9	101.3	59.3	84.2
08	OPENINGS	102.1	48.9	89.5	102.3	56.9	91.6	106.2	60.8	95.5	105.9	60.8	95.2	106.1	60.8	95.3	106.5	60.8	95.7
0920	Plaster & Gypsum Board	93.1	39.1	56.3	84.6	54.2	63.9	80.3	67.4	71.5	80.7	67.4	71.6	80.3	67.4	71.5	90.0	67.4	74.6
0950, 0980	Ceilings & Acoustic Treatment	108.9	39.1	62.4	100.3	54.2	69.6	102.3	67.4	79.0	102.3	67.4	79.0	102.3	67.4	79.0	102.3	67.4	79.0
0960	Flooring	109.6	33.6	86.8	107.4	54.0	91.5	101.0	52.6	86.5	103.8	52.6	88.5	100.6	52.6	86.2	105.9	52.6	89.9
0970, 0990	Wall Finishes & Painting/Coating	100.0	44.0	66.3	106.9	44.0	69.1	102.6	56.6	74.9	102.6	56.6	74.9	102.6	56.6	74.9	102.6	56.6	74.9
09	FINISHES	106.4	40.1	69.5	103.2	52.8	75.1	97.9	60.6	77.1	100.2	60.6	78.2	98.7	60.6	77.5	100.6	60.6	78.2
COVERS	DIVS. 10 - 14, 25, 28, 41, 43, 44, 46	100.0	88.9	97.8	100.0	90.6	98.1	100.0	91.5	98.3	100.0	91.5	98.3	100.0	58.0	91.6	100.0	82.7	96.5
21, 22, 23	FIRE SUPPRESSION, PLUMBING & HVAC	100.0	65.4	86.1	100.0	65.5	86.2	97.5	64.8	84.3	97.5	64.8	84.3	97.5	64.7	84.3	100.0	64.2	85.6
26, 27, 3370	ELECTRICAL, COMMUNICATIONS & UTIL.	96.0	65.6	80.1	99.1	70.2	84.0	96.8	70.6	83.2	96.8	70.6	83.2	96.8	64.4	79.9	95.0	64.4	79.0
MF2010	WEIGHTED AVERAGE	100.4	60.7	83.1	100.1	65.7	85.2	98.2	67.8	84.9	100.2	67.8	86.1	99.3	64.9	84.3	101.8	65.4	85.9

WYOMING / CANADA

DIVISION		SHERIDAN 828			WHEATLAND 822			WORLAND 824			YELLOWSTONE NAT'L PA 821			BARRIE, ONTARIO			BATHURST, NEW BRUNSWICK		
		MAT.	INST.	TOTAL	MAT.	INST.	TOTAL	MAT.	INST.	TOTAL	MAT.	INST.	TOTAL	MAT.	INST.	TOTAL	MAT.	INST.	TOTAL
015433	CONTRACTOR EQUIPMENT		97.9	97.9		97.9	97.9		97.9	97.9		97.9	97.9		101.6	101.6		101.9	101.9
0241, 31 - 34	SITE & INFRASTRUCTURE, DEMOLITION	89.8	95.2	93.6	86.4	94.8	92.3	83.9	94.7	91.5	84.0	94.9	91.7	115.3	101.0	105.3	98.4	97.3	97.6
0310	Concrete Forming & Accessories	104.3	51.6	58.8	99.1	51.1	57.7	99.2	62.7	67.7	99.2	64.0	68.8	125.1	91.5	96.1	103.7	65.1	70.4
0320	Concrete Reinforcing	109.1	71.8	90.4	108.4	71.7	90.1	109.1	71.1	90.1	111.0	71.1	91.1	174.2	86.5	130.4	139.2	58.3	98.8
0330	Cast-in-Place Concrete	95.1	72.7	85.8	96.0	61.8	81.8	92.0	52.2	75.5	92.0	54.2	76.3	166.0	90.4	134.6	136.1	62.8	105.7
03	CONCRETE	108.8	63.3	86.3	105.4	59.2	82.6	100.8	61.0	81.2	101.1	62.3	81.9	151.0	90.3	121.1	130.9	63.7	97.7
04	MASONRY	92.8	57.2	70.8	92.9	55.2	69.6	92.6	44.9	63.1	92.6	48.5	65.3	165.2	99.8	124.7	165.6	66.1	104.0
05	METALS	100.3	72.9	91.9	96.6	72.2	89.1	96.8	71.2	88.9	97.4	71.2	89.3	108.2	91.4	103.0	104.4	72.6	94.6
06	WOOD, PLASTICS & COMPOSITES	97.0	51.6	71.4	87.5	52.1	67.5	87.5	68.4	76.8	87.5	68.4	76.8	120.6	90.1	103.4	97.8	65.4	79.5
07	THERMAL & MOISTURE PROTECTION	102.4	56.5	83.7	101.5	55.0	82.6	101.3	56.0	82.9	100.8	57.5	83.2	112.4	91.0	103.7	104.6	63.6	87.9
08	OPENINGS	106.8	52.5	93.9	104.8	54.7	93.0	106.4	60.8	95.6	99.8	60.8	90.6	92.5	88.6	91.5	89.2	57.2	81.6
0920	Plaster & Gypsum Board	105.6	50.0	67.8	80.3	50.5	60.0	80.3	67.4	71.5	80.5	67.4	71.6	151.2	89.8	109.4	151.8	64.2	92.2
0950, 0980	Ceilings & Acoustic Treatment	105.2	50.0	68.5	102.3	50.5	67.8	102.3	67.4	79.0	103.1	67.4	79.3	95.2	89.8	91.6	100.9	64.2	76.5
0960	Flooring	105.0	52.6	89.3	102.5	51.7	87.3	102.5	52.6	87.6	102.5	52.6	87.6	130.7	100.7	121.7	117.8	48.0	96.9
0970, 0990	Wall Finishes & Painting/Coating	105.1	44.0	68.4	102.6	35.2	62.1	102.6	56.6	74.9	102.6	56.6	74.9	111.9	91.8	99.8	108.4	52.7	74.9
09	FINISHES	106.2	49.7	74.7	98.6	48.3	70.6	98.3	60.5	77.2	98.5	61.4	77.8	117.3	93.5	104.0	113.4	60.9	84.2
COVERS	DIVS. 10 - 14, 25, 28, 41, 43, 44, 46	100.0	90.1	98.0	100.0	82.0	96.4	100.0	57.8	91.5	100.0	59.0	91.8	139.2	76.2	126.6	131.1	66.9	118.3
21, 22, 23	FIRE SUPPRESSION, PLUMBING & HVAC	97.5	65.5	84.6	97.5	64.5	84.2	97.5	64.4	84.2	97.5	66.2	84.9	102.2	101.5	101.9	102.5	71.3	90.0
26, 27, 3370	ELECTRICAL, COMMUNICATIONS & UTIL.	99.5	64.5	81.2	96.8	70.2	82.9	96.8	64.4	79.9	95.7	64.4	79.4	116.5	92.6	104.0	114.2	63.3	87.6
MF2010	WEIGHTED AVERAGE	100.9	65.2	85.4	98.7	64.6	83.9	98.3	64.1	83.4	97.7	65.2	83.5	117.3	94.9	107.5	112.1	68.6	93.1

CANADA

DIVISION		BRANDON, MANITOBA			BRANTFORD, ONTARIO			BRIDGEWATER, NOVA SCOTIA			CALGARY, ALBERTA			CAP-DE-LA-MADELEINE, QUEBEC			CHARLESBOURG, QUEBEC		
		MAT.	INST.	TOTAL	MAT.	INST.	TOTAL	MAT.	INST.	TOTAL	MAT.	INST.	TOTAL	MAT.	INST.	TOTAL	MAT.	INST.	TOTAL
015433	CONTRACTOR EQUIPMENT		103.9	103.9		101.6	101.6		101.5	101.5		106.4	106.4		102.5	102.5		102.5	102.5
0241, 31 - 34	SITE & INFRASTRUCTURE, DEMOLITION	117.9	99.9	105.3	114.9	101.3	105.4	99.9	98.9	99.2	124.6	105.7	111.3	95.9	100.6	99.2	95.9	100.6	99.2
0310	Concrete Forming & Accessories	126.8	74.1	81.3	124.9	99.1	102.6	97.1	73.2	76.5	127.7	96.8	101.0	130.3	90.0	95.5	130.3	90.0	95.5
0320	Concrete Reinforcing	189.5	54.9	122.3	167.9	85.2	126.6	142.9	48.2	95.6	130.9	74.1	102.5	142.9	81.4	112.2	142.9	81.4	112.2
0330	Cast-in-Place Concrete	146.1	77.7	117.7	162.0	112.0	141.3	166.5	73.2	127.8	201.5	106.2	162.0	130.8	99.3	117.7	130.8	99.3	117.7
03	CONCRETE	138.9	72.4	106.1	147.1	100.9	124.3	144.7	69.3	107.5	167.1	95.9	132.0	128.7	91.7	110.5	128.7	91.7	110.5
04	MASONRY	186.5	68.8	113.7	167.9	104.4	128.6	163.2	73.2	107.5	203.4	90.7	133.7	163.7	88.6	117.2	163.7	88.6	117.2
05	METALS	117.8	77.8	105.5	105.5	92.2	101.4	104.6	75.3	95.6	130.4	89.2	117.7	103.8	88.8	99.2	103.8	88.8	99.2
06	WOOD, PLASTICS & COMPOSITES	124.4	75.1	96.6	119.8	97.9	107.4	89.2	72.6	79.9	107.8	96.5	101.4	131.6	89.9	108.1	131.6	89.9	108.1
07	THERMAL & MOISTURE PROTECTION	109.1	73.3	94.6	112.6	96.6	106.1	107.5	71.4	92.9	121.9	94.1	110.6	106.2	91.8	100.3	106.2	91.8	100.3
08	OPENINGS	97.0	66.3	89.7	93.6	94.9	93.9	86.6	66.4	81.8	94.2	86.1	92.3	94.7	83.0	91.9	94.7	83.0	91.9
0920	Plaster & Gypsum Board	129.5	73.9	91.7	142.7	97.8	112.1	141.8	71.7	94.1	140.9	95.9	110.3	172.3	89.4	115.9	172.3	89.4	115.9
0950, 0980	Ceilings & Acoustic Treatment	102.5	73.9	83.5	97.4	97.8	97.6	97.4	71.7	80.3	145.9	95.9	112.6	97.4	89.4	92.1	97.4	89.4	92.1
0960	Flooring	136.3	71.5	116.9	130.5	100.7	121.6	113.1	68.5	99.8	133.3	94.3	121.7	130.5	101.4	121.8	130.5	101.4	121.8
0970, 0990	Wall Finishes & Painting/Coating	113.1	60.0	81.2	109.0	100.8	104.1	109.0	65.5	82.9	114.2	112.4	113.1	109.0	93.7	99.8	109.0	93.7	99.8
09	FINISHES	118.8	72.5	93.0	115.6	100.1	107.0	110.9	71.8	88.7	129.1	98.5	112.0	118.9	92.8	104.4	118.9	92.8	104.4
COVERS	DIVS. 10 - 14, 25, 28, 41, 43, 44, 46	131.1	69.1	118.7	131.1	78.4	120.6	131.1	68.2	118.5	131.1	98.4	124.6	131.1	87.3	122.4	131.1	87.3	122.4
21, 22, 23	FIRE SUPPRESSION, PLUMBING & HVAC	102.5	85.9	95.9	102.5	104.5	103.3	102.5	86.3	96.0	101.3	92.2	97.6	102.9	93.0	98.9	102.9	93.0	98.9
26, 27, 3370	ELECTRICAL, COMMUNICATIONS & UTIL.	120.0	71.3	94.6	112.9	92.1	102.0	118.7	65.9	91.1	114.2	100.8	107.2	112.6	74.3	92.6	112.6	74.3	92.6
MF2010	WEIGHTED AVERAGE	118.7	77.3	100.7	115.5	98.8	108.2	113.4	76.1	97.2	124.7	95.4	111.9	112.8	89.4	102.6	112.8	89.4	102.6

City Cost Indexes

	DIVISION	CHARLOTTETOWN, PRINCE EDWARD ISLAND			CHICOUTIMI, QUEBEC			CORNER BROOK, NEWFOUNDLAND			CORNWALL, ONTARIO			DALHOUSIE, NEW BRUNSWICK			DARTMOUTH, NOVA SCOTIA		
		MAT.	INST.	TOTAL	MAT.	INST.	TOTAL	MAT.	INST.	TOTAL	MAT.	INST.	TOTAL	MAT.	INST.	TOTAL	MAT.	INST.	TOTAL
015433	CONTRACTOR EQUIPMENT		101.5	101.5		102.5	102.5		103.1	103.1		101.6	101.6		101.9	101.9		101.5	101.5
0241, 31 - 34	SITE & INFRASTRUCTURE, DEMOLITION	115.8	96.4	102.2	96.1	99.7	98.6	120.7	97.9	104.7	113.0	100.8	104.5	97.6	97.3	97.4	111.0	98.9	102.5
0310	Concrete Forming & Accessories	105.6	59.5	65.8	131.4	92.8	98.1	106.9	63.0	69.0	122.8	91.6	95.9	103.2	65.3	70.5	98.9	73.2	76.8
0320	Concrete Reinforcing	146.3	48.3	97.3	105.0	96.4	100.7	173.8	50.1	112.0	167.9	84.9	126.5	144.1	58.4	101.3	181.7	48.2	115.0
0330	Cast-in-Place Concrete	164.6	62.0	122.0	128.2	99.2	116.1	170.1	71.7	129.2	145.7	102.2	127.7	137.6	62.8	106.6	164.2	73.2	126.4
03	CONCRETE	153.1	59.1	106.7	120.1	95.6	108.0	167.6	64.5	116.8	139.3	94.1	117.0	135.2	63.8	100.0	152.9	69.3	111.6
04	MASONRY	182.6	63.7	109.0	167.4	94.1	122.0	183.2	65.3	110.2	166.7	95.7	122.7	158.4	66.1	101.3	196.5	73.2	120.2
05	METALS	127.0	68.6	108.9	104.2	92.3	100.6	118.3	74.0	104.6	105.5	90.8	101.0	101.7	72.8	92.8	118.3	75.3	105.0
06	WOOD, PLASTICS & COMPOSITES	88.5	59.1	71.9	132.6	93.2	110.4	104.3	62.1	80.5	118.6	91.0	103.0	97.7	65.4	79.5	94.6	72.6	82.2
07	THERMAL & MOISTURE PROTECTION	119.5	61.7	96.1	105.0	95.9	101.3	112.7	63.2	92.6	112.4	91.1	103.8	110.8	63.6	91.6	110.4	71.4	94.6
08	OPENINGS	93.4	51.1	83.4	93.8	80.5	90.6	102.9	58.5	92.4	94.7	88.2	93.1	90.3	57.2	82.4	88.3	66.4	83.1
0920	Plaster & Gypsum Board	144.0	57.7	85.3	176.1	92.8	119.4	160.5	60.8	92.7	208.0	90.7	128.2	148.2	64.2	91.1	156.0	71.7	98.6
0950, 0980	Ceilings & Acoustic Treatment	121.3	57.7	79.0	100.1	92.8	95.2	103.4	60.8	75.0	100.9	90.7	94.1	100.2	64.2	76.3	101.5	71.7	84.7
0960	Flooring	116.1	64.0	100.5	132.3	101.4	123.0	122.1	58.3	103.0	130.5	99.2	121.1	116.3	73.4	103.5	118.0	68.5	103.2
0970, 0990	Wall Finishes & Painting/Coating	112.9	43.9	71.4	108.4	105.7	106.7	113.0	63.1	83.0	109.0	93.9	99.9	109.9	52.7	75.5	113.0	65.5	84.4
09	FINISHES	118.9	59.1	85.6	120.3	96.5	107.0	119.5	62.2	87.6	125.3	93.6	107.6	112.7	65.9	86.6	118.0	71.8	92.3
COVERS	DIVS. 10 - 14, 25, 28, 41, 43, 44, 46	131.1	66.4	118.2	131.1	88.0	122.5	131.1	67.2	118.3	131.1	75.8	120.1	131.1	66.9	118.3	131.1	68.2	118.5
21, 22, 23	FIRE SUPPRESSION, PLUMBING & HVAC	102.7	65.2	87.7	102.5	93.7	99.0	102.5	73.4	90.8	102.9	102.3	102.6	102.5	71.3	90.0	102.5	86.3	96.0
26, 27, 3370	ELECTRICAL, COMMUNICATIONS & UTIL.	111.9	53.6	81.5	111.0	86.5	98.2	117.4	59.7	87.2	113.9	93.1	103.0	115.3	59.8	86.3	122.2	65.9	92.8
MF2010	WEIGHTED AVERAGE	120.5	64.0	95.9	111.9	93.1	103.7	122.2	69.0	99.0	115.5	95.1	106.7	112.1	68.8	93.2	119.7	76.1	100.7

	DIVISION	EDMONTON, ALBERTA			FORT MCMURRAY, ALBERTA			FREDERICTON, NEW BRUNSWICK			GATINEAU, QUEBEC			GRANBY, QUEBEC			HALIFAX, NOVA SCOTIA		
		MAT.	INST.	TOTAL	MAT.	INST.	TOTAL	MAT.	INST.	TOTAL	MAT.	INST.	TOTAL	MAT.	INST.	TOTAL	MAT.	INST.	TOTAL
015433	CONTRACTOR EQUIPMENT		106.4	106.4		103.7	103.7		101.9	101.9		102.5	102.5		102.5	102.5		101.5	101.5
0241, 31 - 34	SITE & INFRASTRUCTURE, DEMOLITION	146.0	105.7	117.7	120.4	101.9	107.4	100.8	97.3	98.3	95.7	100.5	99.1	96.2	100.5	99.2	105.2	98.8	100.7
0310	Concrete Forming & Accessories	127.2	96.8	101.0	123.0	94.3	98.2	122.7	65.6	73.5	130.3	89.8	95.4	130.3	89.8	95.4	104.9	78.3	81.9
0320	Concrete Reinforcing	140.3	74.1	107.2	155.3	74.1	114.7	136.1	58.5	97.4	151.1	81.3	116.3	151.1	81.3	116.3	147.6	62.8	105.3
0330	Cast-in-Place Concrete	201.0	106.2	161.7	215.2	104.8	169.4	134.8	62.9	104.9	129.0	99.3	116.7	133.3	99.2	119.2	165.6	81.0	130.5
03	CONCRETE	168.4	95.9	132.6	170.0	94.3	132.7	133.1	64.0	99.0	129.2	91.7	110.7	131.2	91.6	111.7	146.0	76.9	111.9
04	MASONRY	193.2	90.7	129.8	207.6	89.6	134.6	192.4	67.6	115.2	163.6	88.6	117.2	163.9	88.6	117.3	193.7	83.4	125.5
05	METALS	130.4	89.2	117.7	130.8	89.0	117.9	123.0	73.3	107.7	103.8	88.7	99.2	103.8	88.6	99.1	130.3	81.6	115.3
06	WOOD, PLASTICS & COMPOSITES	108.6	96.5	101.8	113.7	93.8	102.5	112.9	65.4	86.1	131.6	89.9	108.1	131.6	89.9	108.1	88.7	77.4	82.3
07	THERMAL & MOISTURE PROTECTION	127.2	94.1	113.8	120.1	92.6	108.9	117.3	64.6	95.9	106.2	91.8	100.3	106.2	90.2	99.7	118.6	77.9	102.1
08	OPENINGS	94.4	86.1	92.4	94.7	84.6	92.3	89.3	56.1	81.5	94.7	78.2	90.8	94.7	78.2	90.8	86.5	72.0	83.1
0920	Plaster & Gypsum Board	145.5	95.9	111.8	140.9	93.3	108.5	153.7	64.2	92.8	140.5	89.4	105.8	143.0	89.4	106.6	140.8	76.6	97.1
0950, 0980	Ceilings & Acoustic Treatment	150.5	95.9	114.2	107.2	93.3	97.9	113.3	64.2	80.6	97.4	89.4	92.1	97.4	89.4	92.1	116.1	76.6	89.8
0960	Flooring	135.0	94.3	122.8	130.5	94.3	119.7	128.5	76.7	113.0	130.5	101.4	121.8	130.5	101.4	121.8	118.2	90.0	109.8
0970, 0990	Wall Finishes & Painting/Coating	114.6	112.4	113.3	109.1	97.5	102.2	112.6	67.2	85.3	109.0	93.7	99.8	109.0	93.7	99.8	113.9	82.0	94.7
09	FINISHES	133.3	98.5	113.9	118.9	95.0	105.6	121.4	68.1	91.7	114.5	92.8	102.4	114.9	92.8	102.6	117.2	81.2	97.2
COVERS	DIVS. 10 - 14, 25, 28, 41, 43, 44, 46	131.1	98.4	124.6	131.1	97.5	124.4	131.1	66.9	118.3	131.1	87.3	122.4	131.1	87.3	122.4	131.1	69.6	118.8
21, 22, 23	FIRE SUPPRESSION, PLUMBING & HVAC	101.2	92.2	97.6	102.9	101.0	102.2	102.6	81.1	94.0	102.9	93.0	98.9	102.5	93.0	98.7	100.7	77.7	91.5
26, 27, 3370	ELECTRICAL, COMMUNICATIONS & UTIL.	110.3	100.8	105.4	107.1	87.5	96.8	117.2	78.9	97.2	112.6	74.3	92.6	113.2	74.3	92.9	120.1	80.5	99.4
MF2010	WEIGHTED AVERAGE	125.0	95.4	112.1	124.1	94.2	111.1	118.1	74.0	98.9	112.5	89.2	102.3	112.8	89.1	102.4	119.9	80.6	102.8

	DIVISION	HAMILTON, ONTARIO			HULL, QUEBEC			JOLIETTE, QUEBEC			KAMLOOPS, BRITISH COLUMBIA			KINGSTON, ONTARIO			KITCHENER, ONTARIO		
		MAT.	INST.	TOTAL	MAT.	INST.	TOTAL	MAT.	INST.	TOTAL	MAT.	INST.	TOTAL	MAT.	INST.	TOTAL	MAT.	INST.	TOTAL
015433	CONTRACTOR EQUIPMENT		108.3	108.3		102.5	102.5		102.5	102.5		105.6	105.6		103.9	103.9		103.8	103.8
0241, 31 - 34	SITE & INFRASTRUCTURE, DEMOLITION	117.0	112.7	114.0	95.7	100.5	99.1	96.3	100.6	99.3	118.4	104.0	108.3	113.0	104.5	107.1	101.2	105.1	104.0
0310	Concrete Forming & Accessories	127.9	94.2	98.8	130.3	89.8	95.4	130.3	90.0	95.4	123.5	92.8	97.0	122.9	91.7	96.0	118.0	86.6	90.9
0320	Concrete Reinforcing	153.0	93.1	123.1	151.1	81.3	116.3	142.9	81.4	112.2	112.0	79.2	95.6	167.9	84.9	126.5	103.3	93.0	98.2
0330	Cast-in-Place Concrete	151.7	100.2	130.3	129.0	99.3	116.7	134.4	99.3	119.8	116.3	103.1	110.8	145.7	102.2	127.6	140.7	90.1	119.7
03	CONCRETE	140.5	96.0	118.6	129.2	91.7	110.7	130.4	91.7	111.3	136.9	93.9	115.7	141.2	94.1	118.0	118.8	89.2	104.2
04	MASONRY	193.0	99.7	135.3	163.6	88.6	117.2	164.0	88.6	117.3	170.7	96.6	124.9	173.6	95.7	125.4	159.0	95.3	119.6
05	METALS	121.5	92.7	112.6	103.8	88.7	99.2	103.8	88.8	99.2	106.2	88.7	100.8	107.2	90.7	102.1	113.9	92.5	107.3
06	WOOD, PLASTICS & COMPOSITES	105.6	93.7	98.9	131.6	89.9	108.1	131.6	89.9	108.1	101.3	91.3	95.7	118.6	91.1	103.1	109.7	85.3	96.0
07	THERMAL & MOISTURE PROTECTION	124.3	95.0	112.4	106.2	91.8	100.3	106.2	91.8	100.3	121.2	88.6	108.0	112.4	92.2	104.2	112.1	91.6	103.8
08	OPENINGS	91.9	92.1	91.9	94.7	78.2	90.8	94.7	83.0	91.9	91.0	87.7	90.2	94.7	87.9	93.1	85.7	85.8	85.7
0920	Plaster & Gypsum Board	172.8	93.5	118.9	140.5	89.4	105.8	172.3	89.4	115.9	124.3	90.6	101.4	211.3	90.8	129.4	141.4	84.9	103.0
0950, 0980	Ceilings & Acoustic Treatment	130.0	93.5	105.7	97.4	89.4	92.1	97.4	89.4	92.1	97.4	90.6	92.9	114.3	90.8	98.7	105.3	84.9	91.7
0960	Flooring	132.6	104.1	124.1	130.5	101.4	121.8	130.5	101.4	121.8	129.8	56.6	107.9	130.5	99.2	121.1	124.4	104.1	118.3
0970, 0990	Wall Finishes & Painting/Coating	114.2	102.5	107.2	109.0	93.7	99.8	109.0	93.7	99.8	109.0	85.5	95.0	109.0	86.8	95.6	107.9	92.5	98.6
09	FINISHES	129.3	97.1	111.3	114.5	92.8	102.4	118.9	92.8	104.4	115.0	86.0	98.9	128.7	92.9	108.7	114.3	90.1	100.8
COVERS	DIVS. 10 - 14, 25, 28, 41, 43, 44, 46	131.1	100.4	125.0	131.1	87.3	122.4	131.1	87.3	122.4	131.1	97.9	124.5	131.1	75.8	120.1	131.1	98.3	124.6
21, 22, 23	FIRE SUPPRESSION, PLUMBING & HVAC	101.4	85.5	95.1	102.5	93.0	98.7	102.5	93.0	98.7	102.5	96.5	100.1	102.9	102.5	102.7	100.5	87.6	95.3
26, 27, 3370	ELECTRICAL, COMMUNICATIONS & UTIL.	111.6	96.0	103.5	114.5	74.3	93.5	113.2	74.3	92.9	116.7	83.9	99.6	113.9	91.7	102.3	114.4	93.6	103.6
MF2010	WEIGHTED AVERAGE	119.3	95.3	108.8	112.6	89.2	102.4	113.0	89.4	102.7	114.9	92.4	105.1	116.6	95.2	107.3	111.6	92.0	103.1

CANADA

DIVISION		LAVAL, QUEBEC			LETHBRIDGE, ALBERTA			LLOYDMINSTER, ALBERTA			LONDON, ONTARIO			MEDICINE HAT, ALBERTA			MONCTON, NEW BRUNSWICK		
		MAT.	INST.	TOTAL	MAT.	INST.	TOTAL	MAT.	INST.	TOTAL	MAT.	INST.	TOTAL	MAT.	INST.	TOTAL	MAT.	INST.	TOTAL
015433	CONTRACTOR EQUIPMENT		102.5	102.5		103.7	103.7		103.7	103.7		103.9	103.9		103.7	103.7		101.9	101.9
0241, 31 - 34	SITE & INFRASTRUCTURE, DEMOLITION	96.1	100.5	99.2	113.0	102.5	105.6	112.9	101.9	105.2	116.2	105.3	108.5	111.6	101.9	104.8	97.8	97.5	97.6
0310	Concrete Forming & Accessories	130.5	89.8	95.4	124.2	94.4	98.5	122.4	84.4	89.6	126.1	88.0	93.2	124.1	84.3	89.8	103.7	65.9	71.0
0320	Concrete Reinforcing	151.1	81.3	116.3	155.3	74.1	114.8	155.3	74.0	114.7	130.9	93.0	112.0	155.3	74.0	114.7	139.2	61.5	100.4
0330	Cast-in-Place Concrete	133.3	99.3	119.2	161.4	104.8	137.9	149.7	101.0	129.5	161.6	98.1	135.2	149.7	101.0	129.5	131.1	79.8	109.8
03	CONCRETE	131.2	91.7	111.7	144.7	94.4	119.9	139.1	88.6	114.2	141.5	92.5	117.3	139.2	88.6	114.2	128.5	70.8	100.1
04	MASONRY	163.8	88.6	117.3	181.6	89.6	124.7	163.2	82.9	113.5	196.0	97.0	134.7	163.2	82.9	113.5	165.2	66.2	103.9
05	METALS	103.9	88.7	99.2	124.2	89.0	113.4	106.4	88.9	101.0	122.9	92.0	113.4	106.4	88.8	100.9	104.4	81.5	97.4
06	WOOD, PLASTICS & COMPOSITES	131.8	89.9	108.2	117.3	93.8	104.1	113.7	83.8	96.9	108.5	86.4	96.0	117.3	83.8	98.4	97.8	65.4	79.5
07	THERMAL & MOISTURE PROTECTION	106.8	91.8	100.7	117.4	92.6	107.3	114.3	88.0	103.7	125.2	92.8	112.0	120.9	88.0	107.6	108.9	65.8	91.4
08	OPENINGS	94.7	78.2	90.8	94.7	84.6	92.3	94.7	79.2	91.0	91.1	87.0	90.1	94.7	79.2	91.0	89.2	62.4	82.8
0920	Plaster & Gypsum Board	143.3	89.4	106.7	131.4	93.3	105.5	126.2	83.0	96.8	173.0	86.0	113.8	129.2	83.0	97.7	151.8	64.2	92.2
0950, 0980	Ceilings & Acoustic Treatment	97.4	89.4	92.1	102.1	93.3	96.2	97.4	83.0	87.8	130.1	86.0	100.7	97.4	83.0	87.8	100.9	64.2	76.5
0960	Flooring	130.5	101.4	121.8	130.5	94.3	119.7	130.5	94.3	119.7	132.6	104.1	124.1	130.5	94.3	119.7	117.8	73.4	104.5
0970, 0990	Wall Finishes & Painting/Coating	109.0	93.7	99.8	109.0	106.5	107.5	109.1	83.0	93.4	112.6	99.3	104.6	109.0	83.0	93.4	108.4	52.7	74.9
09	FINISHES	114.9	92.8	102.6	116.5	95.9	105.0	113.9	85.7	98.2	129.0	91.9	108.3	114.2	85.7	98.3	113.4	65.9	86.9
COVERS	DIVS. 10 - 14, 25, 28, 41, 43, 44, 46	131.1	87.3	122.4	131.1	97.5	124.4	131.1	94.1	123.7	131.1	99.0	124.7	131.1	94.1	123.7	131.1	66.9	118.3
21, 22, 23	FIRE SUPPRESSION, PLUMBING & HVAC	100.5	93.0	97.5	102.7	97.6	100.7	102.9	97.6	100.8	101.5	82.8	94.0	102.5	94.2	99.2	102.5	71.6	90.1
26, 27, 3370	ELECTRICAL, COMMUNICATIONS & UTIL.	114.4	74.3	93.5	108.4	87.5	97.5	106.1	87.5	96.4	109.7	93.6	101.3	106.1	87.5	96.4	118.8	63.3	89.8
MF2010	WEIGHTED AVERAGE	112.4	89.2	102.3	118.7	93.7	107.8	113.9	90.3	103.6	119.5	91.9	107.5	114.0	89.6	103.4	112.3	71.4	94.5

CANADA

DIVISION		MONTREAL, QUEBEC			MOOSE JAW, SASKATCHEWAN			NEW GLASGOW, NOVA SCOTIA			NEWCASTLE, NEW BRUNSWICK			NORTH BAY, ONTARIO			OSHAWA, ONTARIO		
		MAT.	INST.	TOTAL	MAT.	INST.	TOTAL	MAT.	INST.	TOTAL	MAT.	INST.	TOTAL	MAT.	INST.	TOTAL	MAT.	INST.	TOTAL
015433	CONTRACTOR EQUIPMENT		104.0	104.0		100.2	100.2		101.5	101.5		101.9	101.9		101.6	101.6		103.8	103.8
0241, 31 - 34	SITE & INFRASTRUCTURE, DEMOLITION	107.7	100.0	102.3	112.7	96.5	101.3	104.4	98.9	100.5	98.4	97.3	97.6	119.5	100.4	106.1	112.9	104.5	107.0
0310	Concrete Forming & Accessories	131.4	93.1	98.4	107.3	61.9	68.1	98.8	73.2	76.8	103.7	65.3	70.6	127.4	89.0	94.3	123.8	91.1	95.6
0320	Concrete Reinforcing	135.8	96.4	116.1	109.6	63.8	86.7	173.8	48.2	111.1	139.2	58.4	98.8	205.4	84.4	145.0	163.7	87.1	125.4
0330	Cast-in-Place Concrete	174.5	100.5	143.8	145.3	72.1	114.9	164.2	73.2	126.4	136.1	62.9	105.7	158.1	88.4	129.1	162.7	89.2	132.2
03	CONCRETE	148.7	96.2	122.8	122.3	66.5	94.8	151.6	69.3	111.0	130.9	63.8	97.8	154.2	88.1	121.6	145.4	89.9	118.0
04	MASONRY	172.0	94.1	123.8	161.7	65.5	102.2	182.8	73.2	115.0	165.6	66.1	104.0	188.0	91.5	128.3	162.1	96.0	121.2
05	METALS	124.5	92.8	114.7	102.7	74.9	94.1	116.0	75.3	103.4	104.4	72.8	94.7	117.0	90.4	108.8	105.5	92.6	101.5
06	WOOD, PLASTICS & COMPOSITES	120.2	93.5	105.1	99.0	60.4	77.2	94.6	72.6	82.2	97.8	65.4	79.5	127.3	89.6	106.0	117.2	90.1	101.9
07	THERMAL & MOISTURE PROTECTION	118.4	96.5	109.5	105.6	65.5	89.3	110.4	71.4	94.6	108.9	63.6	90.5	115.7	87.4	104.2	113.1	89.2	103.4
08	OPENINGS	92.8	82.4	90.3	90.1	58.1	82.5	88.3	66.4	83.1	89.2	57.2	81.6	95.9	85.7	93.5	92.2	89.9	91.7
0920	Plaster & Gypsum Board	144.9	92.8	109.5	121.3	59.1	79.0	154.0	71.7	98.0	151.8	64.2	92.2	151.5	89.2	109.1	146.6	89.8	108.0
0950, 0980	Ceilings & Acoustic Treatment	115.8	92.8	100.5	97.4	59.1	71.9	102.5	71.7	82.0	100.9	64.2	76.5	102.5	89.2	93.6	100.8	89.8	93.5
0960	Flooring	138.7	101.4	127.5	120.8	64.4	104.0	118.0	68.5	103.2	117.8	73.4	104.5	136.3	99.2	125.2	127.7	106.8	121.5
0970, 0990	Wall Finishes & Painting/Coating	112.9	105.7	108.6	109.0	68.2	84.5	113.0	65.5	84.4	108.4	52.7	74.9	113.0	93.1	101.1	107.9	106.6	107.1
09	FINISHES	123.1	96.7	108.4	110.3	62.7	83.8	116.0	71.8	91.4	113.4	65.9	86.9	121.6	91.8	105.0	115.7	95.7	104.5
COVERS	DIVS. 10 - 14, 25, 28, 41, 43, 44, 46	131.1	88.6	122.6	131.1	66.1	118.1	131.1	68.2	118.5	131.1	66.9	118.3	131.1	74.5	119.8	131.1	99.7	124.8
21, 22, 23	FIRE SUPPRESSION, PLUMBING & HVAC	101.3	93.8	98.3	102.9	77.8	92.8	102.5	86.3	96.0	102.5	71.3	90.0	102.5	100.3	101.6	100.5	102.9	101.5
26, 27, 3370	ELECTRICAL, COMMUNICATIONS & UTIL.	113.8	86.5	99.6	115.4	64.3	88.7	117.7	65.9	90.7	113.4	63.3	87.3	117.7	93.0	104.8	115.6	92.6	103.6
MF2010	WEIGHTED AVERAGE	119.2	93.4	107.9	111.2	70.8	93.6	117.7	76.1	99.6	112.1	69.3	93.5	120.5	92.9	108.5	114.6	96.1	106.5

CANADA

DIVISION		OTTAWA, ONTARIO			OWEN SOUND, ONTARIO			PETERBOROUGH, ONTARIO			PORTAGE LA PRAIRIE, MANITOBA			PRINCE ALBERT, SASKATCHEWAN			PRINCE GEORGE, BRITISH COLUMBIA		
		MAT.	INST.	TOTAL	MAT.	INST.	TOTAL	MAT.	INST.	TOTAL	MAT.	INST.	TOTAL	MAT.	INST.	TOTAL	MAT.	INST.	TOTAL
015433	CONTRACTOR EQUIPMENT		103.8	103.8		101.6	101.6		101.6	101.6		103.9	103.9		100.2	100.2		105.6	105.6
0241, 31 - 34	SITE & INFRASTRUCTURE, DEMOLITION	110.8	105.0	106.7	115.3	100.9	105.2	114.9	100.8	105.0	113.7	99.9	104.0	107.8	96.7	100.0	121.8	104.0	109.3
0310	Concrete Forming & Accessories	126.4	90.9	95.8	125.1	87.5	92.6	124.9	90.1	94.9	124.3	73.6	80.6	107.3	61.7	67.9	113.0	87.5	91.0
0320	Concrete Reinforcing	144.0	93.0	118.5	174.2	86.5	130.4	167.9	84.9	126.5	155.3	54.9	105.2	114.6	63.8	89.2	112.0	79.2	95.6
0330	Cast-in-Place Concrete	161.6	99.0	135.6	166.0	84.3	132.1	162.0	90.1	132.2	149.7	77.1	119.6	131.6	72.0	106.9	145.8	103.1	128.0
03	CONCRETE	143.6	94.1	119.2	151.0	86.4	119.2	147.1	89.3	118.6	132.7	72.0	102.7	116.7	66.3	91.9	150.0	91.5	121.2
04	MASONRY	175.1	97.0	126.8	165.2	97.3	123.2	167.9	98.4	124.9	166.3	67.7	105.3	160.8	65.5	101.8	172.9	96.6	125.7
05	METALS	123.6	91.9	113.8	108.2	91.2	103.0	105.5	90.9	101.0	106.4	77.7	97.5	102.8	74.7	94.1	106.2	88.8	100.8
06	WOOD, PLASTICS & COMPOSITES	108.8	90.5	98.5	120.6	86.2	101.2	119.8	88.2	102.0	117.1	75.1	93.4	99.0	60.4	77.2	101.3	83.9	91.5
07	THERMAL & MOISTURE PROTECTION	127.5	92.7	113.4	112.4	88.2	102.6	112.6	93.1	104.7	106.1	72.8	92.6	105.5	64.4	88.8	115.3	87.9	104.1
08	OPENINGS	93.8	89.2	92.7	92.5	85.0	90.7	93.6	87.5	92.2	94.7	66.3	88.0	88.9	58.1	81.6	91.0	83.7	89.3
0920	Plaster & Gypsum Board	192.8	90.2	123.0	151.2	85.7	106.7	142.7	87.8	105.4	123.6	73.9	89.8	121.3	59.1	79.0	124.3	83.0	96.2
0950, 0980	Ceilings & Acoustic Treatment	134.7	90.2	105.1	95.2	85.7	88.9	97.4	87.8	91.0	97.4	59.1	71.9	97.4	63.0	81.8	97.4	83.0	87.8
0960	Flooring	126.9	99.2	118.6	130.7	100.7	121.7	130.5	99.2	121.1	130.5	71.5	112.8	120.8	64.4	104.0	126.1	77.6	111.6
0970, 0990	Wall Finishes & Painting/Coating	112.4	94.1	101.4	111.9	91.8	99.8	109.0	95.5	100.9	109.1	60.0	79.6	109.0	58.2	78.5	109.0	85.8	95.0
09	FINISHES	130.7	92.9	109.7	117.3	90.6	102.4	115.6	92.5	102.7	113.4	72.3	90.5	110.3	61.6	83.2	113.9	85.1	97.9
COVERS	DIVS. 10 - 14, 25, 28, 41, 43, 44, 46	131.1	97.1	124.3	139.2	75.0	126.4	131.1	76.0	120.1	131.1	68.7	118.6	131.1	66.1	118.1	131.1	97.1	124.3
21, 22, 23	FIRE SUPPRESSION, PLUMBING & HVAC	101.5	83.2	94.2	102.2	100.2	101.4	102.5	103.9	103.1	102.5	85.4	95.6	102.9	70.1	89.7	102.5	96.5	100.1
26, 27, 3370	ELECTRICAL, COMMUNICATIONS & UTIL.	109.0	94.4	101.4	117.9	91.7	104.2	112.9	92.6	102.3	115.1	61.5	87.1	115.4	64.3	88.7	113.1	83.9	97.8
MF2010	WEIGHTED AVERAGE	119.2	92.5	107.5	117.4	93.0	106.8	115.5	94.9	106.5	113.9	75.6	97.2	110.3	69.0	92.3	115.9	91.7	105.3

City Cost Indexes

CANADA

DIVISION		QUEBEC, QUEBEC			RED DEER, ALBERTA			REGINA, SASKATCHEWAN			RIMOUSKI, QUEBEC			ROUYN-NORANDA, QUEBEC			SAINT HYACINTHE, QUEBEC		
		MAT.	INST.	TOTAL	MAT.	INST.	TOTAL	MAT.	INST.	TOTAL	MAT.	INST.	TOTAL	MAT.	INST.	TOTAL	MAT.	INST.	TOTAL
015433	CONTRACTOR EQUIPMENT		104.4	104.4		103.7	103.7		100.2	100.2		102.5	102.5		102.5	102.5		102.5	102.5
0241, 31 - 34	SITE & INFRASTRUCTURE, DEMOLITION	103.7	100.0	101.1	111.6	101.9	104.8	123.5	98.4	105.9	96.1	99.7	98.6	95.7	100.5	99.1	96.1	100.5	99.2
0310	Concrete Forming & Accessories	131.8	93.4	98.6	139.3	84.3	91.9	114.1	90.4	93.7	130.3	92.8	98.0	130.3	89.8	95.4	130.3	89.8	95.4
0320	Concrete Reinforcing	136.6	96.4	116.5	155.3	74.0	114.7	132.1	83.4	107.8	107.7	96.4	102.1	151.1	81.3	116.3	151.1	81.3	116.3
0330	Cast-in-Place Concrete	142.8	100.9	125.4	149.7	101.0	129.5	173.6	94.9	140.9	135.7	99.2	120.5	129.0	99.3	116.7	133.3	99.3	119.2
03	CONCRETE	133.9	96.4	115.4	140.3	88.6	114.8	150.3	90.6	120.9	125.3	95.6	110.7	129.2	91.7	110.7	131.2	91.7	111.7
04	MASONRY	172.0	94.1	123.8	163.2	82.9	113.5	184.7	94.4	128.8	163.4	94.1	120.5	163.6	88.6	117.2	163.8	88.6	117.3
05	METALS	124.4	93.0	114.7	106.4	88.8	100.9	125.6	85.7	113.3	103.4	92.3	100.0	103.8	88.7	99.2	103.8	88.7	99.2
06	WOOD, PLASTICS & COMPOSITES	119.9	93.5	105.0	117.3	83.8	98.4	98.3	91.3	94.4	131.6	93.2	110.0	131.6	89.9	108.1	131.6	89.9	108.1
07	THERMAL & MOISTURE PROTECTION	118.9	96.6	109.9	130.7	88.0	113.4	117.9	85.6	104.8	106.2	95.9	102.0	106.2	91.8	100.3	106.6	91.8	100.6
08	OPENINGS	94.0	90.3	93.1	94.7	79.2	91.0	90.8	80.2	88.3	94.2	80.5	91.0	94.7	78.2	90.8	94.7	78.2	90.8
0920	Plaster & Gypsum Board	164.8	92.8	115.8	129.2	83.0	97.7	147.8	91.0	109.2	172.1	92.8	118.1	140.3	89.4	105.7	142.7	89.4	106.5
0950, 0980	Ceilings & Acoustic Treatment	122.8	92.8	102.8	97.4	83.0	87.8	123.9	91.0	102.0	96.5	92.8	94.0	96.5	89.4	91.8	96.5	89.4	91.8
0960	Flooring	134.3	101.4	124.5	133.2	94.3	121.6	132.5	66.7	112.8	130.5	101.4	121.8	130.5	101.4	121.8	130.5	101.4	121.8
0970, 0990	Wall Finishes & Painting/Coating	115.4	105.7	109.5	109.0	83.0	93.4	112.9	90.9	99.7	109.0	105.7	107.0	109.0	93.7	99.8	109.0	93.7	99.8
09	FINISHES	126.3	96.7	109.8	115.0	85.7	98.7	125.5	86.8	103.9	118.7	96.5	106.3	114.3	92.8	102.3	114.6	92.8	102.5
COVERS	DIVS. 10 - 14, 25, 28, 41, 43, 44, 46	131.1	88.7	122.6	131.1	94.1	123.7	131.1	73.5	119.6	131.1	88.0	122.5	131.1	87.3	122.4	131.1	87.3	122.4
21, 22, 23	FIRE SUPPRESSION, PLUMBING & HVAC	101.2	93.8	98.3	102.5	94.2	99.2	100.7	84.2	94.1	102.5	93.7	99.0	102.5	93.0	98.7	98.9	93.0	96.5
26, 27, 3370	ELECTRICAL, COMMUNICATIONS & UTIL.	110.9	86.5	98.2	106.1	87.5	96.4	115.4	91.4	102.9	113.2	86.5	99.3	113.2	74.3	92.9	114.0	74.3	93.3
MF2010	WEIGHTED AVERAGE	117.5	93.8	107.2	114.5	89.6	103.6	120.3	88.4	106.4	112.3	93.1	103.9	112.5	89.2	102.3	111.9	89.2	102.0

CANADA

DIVISION		SAINT JOHN, NEW BRUNSWICK			SARNIA, ONTARIO			SASKATOON, SASKATCHEWAN			SAULT STE MARIE, ONTARIO			SHERBROOKE, QUEBEC			SOREL, QUEBEC		
		MAT.	INST.	TOTAL	MAT.	INST.	TOTAL	MAT.	INST.	TOTAL	MAT.	INST.	TOTAL	MAT.	INST.	TOTAL	MAT.	INST.	TOTAL
015433	CONTRACTOR EQUIPMENT		101.9	101.9		101.6	101.6		100.2	100.2		101.6	101.6		102.5	102.5		102.5	102.5
0241, 31 - 34	SITE & INFRASTRUCTURE, DEMOLITION	98.5	98.5	98.5	113.4	100.9	104.6	108.7	98.4	101.5	103.7	100.4	101.4	96.2	100.5	99.2	96.3	100.6	99.3
0310	Concrete Forming & Accessories	123.2	66.3	74.1	123.7	97.4	101.0	107.4	90.4	92.7	113.0	87.0	90.6	130.3	89.8	95.4	130.3	90.0	95.5
0320	Concrete Reinforcing	139.2	61.9	100.6	119.4	86.3	102.9	115.5	83.4	99.5	107.8	84.6	96.2	151.1	81.3	116.3	142.9	81.4	112.2
0330	Cast-in-Place Concrete	134.1	80.7	111.9	149.5	103.7	130.5	141.4	94.9	122.1	134.3	88.5	115.3	133.3	99.3	119.2	134.4	99.3	119.8
03	CONCRETE	131.4	71.4	101.8	133.3	97.4	115.6	122.5	90.6	106.7	117.3	87.3	102.5	131.2	91.7	111.7	130.4	91.7	111.3
04	MASONRY	185.8	72.2	115.5	179.3	101.2	131.0	167.2	94.4	122.2	164.3	94.5	121.1	163.9	88.6	117.3	164.0	88.6	117.3
05	METALS	104.3	82.6	97.6	105.5	91.3	101.1	101.9	85.7	96.9	104.7	90.8	100.4	103.8	88.7	99.2	103.8	88.8	99.2
06	WOOD, PLASTICS & COMPOSITES	121.4	64.5	89.3	118.9	96.8	106.4	98.0	91.3	94.3	106.0	86.5	95.0	131.6	89.9	108.1	131.6	89.9	108.1
07	THERMAL & MOISTURE PROTECTION	109.2	69.6	93.1	112.6	96.7	106.2	106.8	85.6	98.2	111.5	89.4	102.5	106.2	91.8	100.3	106.2	91.8	100.3
08	OPENINGS	89.1	60.8	82.4	95.7	91.5	94.7	90.1	80.2	87.7	87.2	85.3	86.8	94.7	78.2	90.8	94.7	83.0	91.9
0920	Plaster & Gypsum Board	169.4	63.3	97.2	164.7	96.7	118.4	129.7	91.0	103.4	131.7	86.0	100.6	142.7	89.4	106.5	172.1	89.4	115.9
0950, 0980	Ceilings & Acoustic Treatment	107.2	63.3	78.0	102.7	96.7	98.7	119.7	91.0	100.6	97.4	86.0	89.8	96.5	89.4	91.8	96.5	89.4	91.8
0960	Flooring	128.4	73.4	112.0	130.5	108.5	123.9	120.7	66.7	104.6	123.6	102.7	117.3	130.5	101.4	121.8	130.5	101.4	121.8
0970, 0990	Wall Finishes & Painting/Coating	108.4	78.8	90.6	109.0	108.0	108.4	109.9	90.9	98.5	109.0	100.0	103.6	109.0	93.7	99.8	109.0	93.7	99.8
09	FINISHES	120.2	68.5	91.4	119.8	101.1	109.4	116.5	86.8	99.9	111.4	91.3	100.2	114.6	92.8	102.5	118.7	92.8	104.3
COVERS	DIVS. 10 - 14, 25, 28, 41, 43, 44, 46	131.1	67.3	118.3	131.1	77.5	120.4	131.1	73.5	119.6	131.1	98.4	124.6	131.1	87.3	122.4	131.1	87.3	122.4
21, 22, 23	FIRE SUPPRESSION, PLUMBING & HVAC	102.5	81.0	93.9	102.5	110.2	105.6	100.6	84.2	94.0	102.5	96.8	100.2	102.9	93.0	98.9	102.5	93.0	98.7
26, 27, 3370	ELECTRICAL, COMMUNICATIONS & UTIL.	121.9	90.3	105.4	115.7	94.9	104.8	117.1	91.4	103.7	114.4	93.0	103.2	113.2	74.3	92.9	113.2	74.3	92.9
MF2010	WEIGHTED AVERAGE	114.6	78.3	98.8	115.3	99.5	108.4	111.3	88.4	101.3	110.7	93.1	103.0	112.8	89.2	102.5	113.0	89.4	102.7

CANADA

DIVISION		ST CATHARINES, ONTARIO			ST JEROME, QUEBEC			ST JOHNS, NEWFOUNDLAND			SUDBURY, ONTARIO			SUMMERSIDE, PRINCE EDWARD ISLAND			SYDNEY, NOVA SCOTIA		
		MAT.	INST.	TOTAL	MAT.	INST.	TOTAL	MAT.	INST.	TOTAL	MAT.	INST.	TOTAL	MAT.	INST.	TOTAL	MAT.	INST.	TOTAL
015433	CONTRACTOR EQUIPMENT		101.6	101.6		102.5	102.5		103.1	103.1		101.6	101.6		101.5	101.5		101.5	101.5
0241, 31 - 34	SITE & INFRASTRUCTURE, DEMOLITION	101.8	101.8	101.8	95.7	100.5	99.1	116.4	99.9	104.8	101.9	101.3	101.5	112.2	96.4	101.1	100.1	98.9	99.2
0310	Concrete Forming & Accessories	115.9	93.9	96.9	130.3	89.8	95.4	113.8	82.4	86.7	111.9	87.4	90.7	99.1	59.6	65.0	98.8	73.2	76.8
0320	Concrete Reinforcing	104.3	93.0	98.7	151.1	81.3	116.3	161.9	74.4	118.2	105.1	93.0	99.0	171.5	48.3	110.0	173.8	48.2	111.1
0330	Cast-in-Place Concrete	134.4	100.1	120.1	129.0	99.3	116.7	166.6	87.6	133.8	135.5	96.2	119.2	155.8	62.0	116.9	126.6	73.2	104.5
03	CONCRETE	115.8	95.8	105.9	129.2	91.7	110.7	159.3	83.0	121.6	116.2	91.6	104.0	156.5	59.1	108.5	133.8	69.3	102.0
04	MASONRY	158.6	99.7	122.1	163.6	88.6	117.2	187.1	89.8	126.9	158.6	94.9	119.2	181.9	63.7	108.8	180.0	73.2	113.9
05	METALS	104.6	92.4	100.8	103.8	88.7	99.2	126.2	83.4	113.0	104.6	91.9	100.7	116.0	68.6	101.4	116.0	75.3	103.4
06	WOOD, PLASTICS & COMPOSITES	107.4	93.6	99.6	131.6	89.9	108.1	101.4	80.9	89.8	103.5	86.5	93.9	94.9	59.1	74.7	94.6	72.6	82.2
07	THERMAL & MOISTURE PROTECTION	112.1	95.9	105.5	106.2	91.8	100.3	118.6	84.9	105.0	111.5	90.3	102.9	109.9	62.5	90.7	110.4	71.4	94.6
08	OPENINGS	85.2	90.7	86.5	94.7	78.2	90.8	99.3	73.4	93.2	86.0	85.3	85.8	99.1	51.1	87.8	88.3	66.4	83.1
0920	Plaster & Gypsum Board	129.0	93.4	104.8	140.3	89.4	105.7	154.5	80.1	103.9	133.6	86.0	101.2	154.4	57.7	88.6	154.0	71.7	98.0
0950, 0980	Ceilings & Acoustic Treatment	100.8	93.4	95.9	96.5	89.4	91.8	116.8	80.1	92.4	95.5	86.0	89.2	102.5	57.7	72.7	102.5	71.7	82.0
0960	Flooring	123.0	100.7	116.3	130.5	101.4	121.8	122.3	60.3	103.8	120.9	102.7	115.4	118.1	64.0	101.9	118.0	68.5	103.2
0970, 0990	Wall Finishes & Painting/Coating	107.9	102.5	104.7	109.0	93.7	99.8	112.8	75.8	90.6	107.9	93.5	99.3	113.0	43.9	71.4	113.0	65.5	84.4
09	FINISHES	111.2	96.4	103.0	114.3	92.8	102.3	121.1	78.0	97.1	110.1	90.7	99.3	117.0	59.1	84.7	116.0	71.8	91.4
COVERS	DIVS. 10 - 14, 25, 28, 41, 43, 44, 46	131.1	76.3	120.1	131.1	87.3	122.4	131.1	73.2	119.5	131.1	98.5	124.6	131.1	66.4	118.2	131.1	68.2	118.5
21, 22, 23	FIRE SUPPRESSION, PLUMBING & HVAC	100.5	84.4	94.1	102.5	93.0	98.7	100.8	73.9	90.0	102.5	85.8	95.8	102.5	65.2	87.6	102.5	86.3	96.0
26, 27, 3370	ELECTRICAL, COMMUNICATIONS & UTIL.	116.0	94.5	104.7	113.8	74.3	93.2	115.9	72.4	93.2	113.1	94.8	103.6	116.7	53.6	83.7	117.7	65.9	90.7
MF2010	WEIGHTED AVERAGE	109.7	93.0	102.4	112.5	89.2	102.3	122.0	80.4	103.9	109.9	91.7	101.9	119.5	64.1	95.4	115.6	76.1	98.4

		\multicolumn{18}{c}{CANADA}															

DIVISION		THUNDER BAY, ONTARIO			TIMMINS, ONTARIO			TORONTO, ONTARIO			TROIS RIVIERES, QUEBEC			TRURO, NOVA SCOTIA			VANCOUVER, BRITISH COLUMBIA		
		MAT.	INST.	TOTAL	MAT.	INST.	TOTAL	MAT.	INST.	TOTAL	MAT.	INST.	TOTAL	MAT.	INST.	TOTAL	MAT.	INST.	TOTAL
015433	CONTRACTOR EQUIPMENT		101.6	101.6		101.6	101.6		103.9	103.9		102.5	102.5		101.5	101.5		112.1	112.1
0241, 31 - 34	SITE & INFRASTRUCTURE, DEMOLITION	106.8	101.7	103.2	114.9	100.4	104.7	137.0	105.8	115.1	100.9	100.6	100.7	100.1	98.9	99.2	120.3	108.2	111.8
0310	Concrete Forming & Accessories	123.7	90.8	95.3	124.9	89.0	93.9	128.5	99.5	103.4	133.2	90.0	95.9	97.1	73.3	76.5	122.0	91.6	95.8
0320	Concrete Reinforcing	93.2	91.9	92.5	167.9	84.4	126.2	151.3	93.2	122.3	173.8	81.4	127.6	142.9	48.2	95.6	142.2	81.7	112.0
0330	Cast-in-Place Concrete	147.9	98.5	127.4	162.0	88.4	131.5	150.7	108.2	133.1	131.1	99.3	117.9	168.3	73.3	128.8	157.7	99.2	133.4
03	CONCRETE	124.2	93.7	109.2	147.1	88.1	118.0	139.8	101.1	120.7	136.8	91.7	114.6	145.5	69.3	107.9	148.1	92.7	120.8
04	MASONRY	159.3	99.0	122.0	167.9	91.5	120.6	194.2	104.1	138.4	183.8	88.6	124.9	163.4	73.2	107.6	182.3	93.0	127.0
05	METALS	104.4	91.3	100.4	105.5	90.4	100.9	125.1	93.1	115.2	115.1	88.8	107.0	104.6	75.3	95.6	141.4	91.4	126.0
06	WOOD, PLASTICS & COMPOSITES	117.2	89.5	101.6	119.8	89.6	102.7	111.5	98.7	104.3	139.8	89.9	111.7	89.2	72.6	79.9	104.1	91.7	97.1
07	THERMAL & MOISTURE PROTECTION	112.4	93.0	104.5	112.6	87.4	102.4	124.3	99.6	114.2	109.3	91.8	102.2	107.5	71.4	92.9	130.0	88.5	113.2
08	OPENINGS	84.3	87.8	85.1	93.6	85.7	91.7	91.8	96.8	93.0	97.0	83.0	93.7	86.6	66.4	81.8	94.5	87.9	93.0
0920	Plaster & Gypsum Board	162.4	89.2	112.6	142.7	89.2	106.3	155.3	98.6	116.7	186.1	89.4	120.3	141.8	71.7	94.1	140.0	90.9	106.6
0950, 0980	Ceilings & Acoustic Treatment	95.5	89.2	91.3	97.4	89.2	91.9	125.4	98.6	107.6	101.6	89.4	93.5	97.4	71.7	80.3	150.0	90.9	110.7
0960	Flooring	127.7	107.8	121.7	130.5	99.2	121.1	132.8	110.4	126.1	136.3	101.4	125.8	113.1	68.5	99.8	129.2	97.7	119.8
0970, 0990	Wall Finishes & Painting/Coating	107.9	92.1	98.4	109.0	93.1	99.5	114.8	106.6	109.9	113.0	93.7	101.4	109.0	65.5	82.9	114.0	97.7	104.2
09	FINISHES	116.4	94.2	104.0	115.6	91.8	102.4	127.3	102.5	113.5	125.3	92.8	107.2	109.9	71.8	88.7	129.1	93.6	109.3
COVERS	DIVS. 10 - 14, 25, 28, 41, 43, 44, 46	131.1	76.0	120.1	131.1	74.5	119.8	131.1	102.1	125.3	131.1	87.3	122.4	131.1	68.2	118.5	131.1	96.9	124.3
21, 22, 23	FIRE SUPPRESSION, PLUMBING & HVAC	100.5	84.3	94.0	102.5	100.3	101.6	101.2	94.6	98.6	102.5	93.0	98.7	102.5	86.3	96.0	101.3	82.3	93.7
26, 27, 3370	ELECTRICAL, COMMUNICATIONS & UTIL.	114.4	93.0	103.2	114.4	93.0	103.2	111.5	96.6	103.7	118.0	74.3	95.2	112.9	65.9	88.4	111.3	81.3	95.6
MF2010	WEIGHTED AVERAGE	110.9	91.7	102.6	115.6	92.9	105.7	120.2	99.0	111.0	117.9	89.4	105.5	113.0	76.1	96.9	123.3	90.0	108.8

		\multicolumn{18}{c}{CANADA}															

DIVISION		VICTORIA, BRITISH COLUMBIA			WHITEHORSE, YUKON			WINDSOR, ONTARIO			WINNIPEG, MANITOBA			YARMOUTH, NOVA SCOTIA			YELLOWKNIFE, NWT		
		MAT.	INST.	TOTAL	MAT.	INST.	TOTAL	MAT.	INST.	TOTAL	MAT.	INST.	TOTAL	MAT.	INST.	TOTAL	MAT.	INST.	TOTAL
015433	CONTRACTOR EQUIPMENT		108.7	108.7		101.7	101.7		101.6	101.6		105.9	105.9		101.5	101.5		101.6	101.6
0241, 31 - 34	SITE & INFRASTRUCTURE, DEMOLITION	121.5	107.7	111.8	123.0	97.4	105.0	97.3	101.6	100.3	119.5	101.3	106.7	104.2	98.9	100.5	138.7	101.7	112.8
0310	Concrete Forming & Accessories	112.9	91.0	94.0	117.7	60.9	68.7	123.7	90.0	94.6	131.5	68.1	76.8	98.8	73.2	76.8	116.1	82.0	86.7
0320	Concrete Reinforcing	113.0	81.6	97.3	153.9	61.9	107.9	102.2	93.0	97.6	141.0	57.4	99.3	173.8	48.2	111.1	154.5	64.7	109.6
0330	Cast-in-Place Concrete	146.4	98.6	126.6	191.7	71.3	141.7	137.7	100.3	122.1	181.5	73.8	136.8	162.4	73.2	125.4	231.5	89.4	172.5
03	CONCRETE	152.5	91.9	122.6	158.2	65.4	112.4	117.6	94.1	106.0	153.4	68.7	111.6	150.8	69.3	110.6	176.9	81.7	129.9
04	MASONRY	170.1	93.0	122.4	205.6	63.9	117.9	158.7	97.8	121.0	179.6	69.3	111.4	182.6	73.2	114.9	209.1	76.2	126.8
05	METALS	104.3	87.4	99.1	129.3	74.6	112.5	104.5	91.8	100.6	143.3	74.6	122.1	116.0	75.3	103.4	129.3	79.8	114.0
06	WOOD, PLASTICS & COMPOSITES	101.6	91.6	96.0	100.8	59.6	77.5	117.2	89.1	101.4	111.1	69.0	87.3	94.6	72.6	82.2	100.3	83.7	90.9
07	THERMAL & MOISTURE PROTECTION	115.8	88.3	104.7	125.8	63.4	100.5	112.2	92.2	104.0	118.7	70.7	99.2	110.4	71.4	94.6	122.1	80.0	105.0
08	OPENINGS	91.8	84.1	90.0	95.4	56.8	86.3	84.1	88.1	85.0	93.1	61.6	85.6	88.3	66.4	83.1	94.3	70.5	88.7
0920	Plaster & Gypsum Board	121.7	90.9	100.7	146.1	58.2	86.3	147.7	88.8	107.6	139.7	67.5	90.6	154.0	71.7	98.0	176.0	83.1	112.8
0950, 0980	Ceilings & Acoustic Treatment	99.3	90.9	93.7	127.7	58.2	81.4	95.5	88.8	91.0	143.6	67.5	93.0	102.5	71.7	82.0	133.5	83.1	99.9
0960	Flooring	125.9	77.6	111.5	140.3	62.5	117.0	127.7	105.0	120.9	131.3	74.0	114.1	118.0	68.5	103.2	134.7	94.2	122.6
0970, 0990	Wall Finishes & Painting/Coating	109.9	97.7	102.6	124.2	57.2	83.9	107.9	95.4	100.4	112.9	55.4	78.4	113.0	65.5	84.4	121.6	82.0	97.8
09	FINISHES	114.1	90.1	100.7	130.9	60.6	91.7	114.0	93.5	102.6	127.8	68.3	94.6	116.0	71.8	91.4	134.9	83.7	106.4
COVERS	DIVS. 10 - 14, 25, 28, 41, 43, 44, 46	131.1	72.9	119.5	131.1	65.4	118.0	131.1	75.4	120.0	131.1	67.9	118.5	131.1	68.2	118.5	131.1	68.7	118.6
21, 22, 23	FIRE SUPPRESSION, PLUMBING & HVAC	102.5	82.2	94.4	102.9	76.4	92.3	100.5	83.9	93.9	101.2	65.7	87.0	102.5	86.3	96.0	103.0	96.0	100.2
26, 27, 3370	ELECTRICAL, COMMUNICATIONS & UTIL.	113.8	81.3	96.8	134.3	63.7	97.5	119.3	94.5	106.4	112.9	66.9	88.9	117.7	65.9	90.7	132.1	87.2	108.7
MF2010	WEIGHTED AVERAGE	115.9	88.1	103.8	126.3	69.8	101.7	110.2	91.7	102.2	123.6	70.8	100.6	117.6	76.1	99.6	128.8	85.8	110.1

Location Factors

Costs shown in RSMeans cost data publications are based on national averages for materials and installation. To adjust these costs to a specific location, simply multiply the base cost by the factor and divide by 100 for that city. The data is arranged alphabetically by state and postal zip code numbers. For a city not listed, use the factor for a nearby city with similar economic characteristics.

STATE/ZIP	CITY	MAT.	INST.	TOTAL
ALABAMA				
350-352	Birmingham	97.5	74.1	87.3
354	Tuscaloosa	96.7	62.0	81.6
355	Jasper	97.0	56.7	79.4
356	Decatur	96.6	59.1	80.3
357-358	Huntsville	96.7	70.8	85.4
359	Gadsden	96.7	60.8	81.1
360-361	Montgomery	97.9	58.1	80.5
362	Anniston	96.5	63.0	81.9
363	Dothan	96.9	50.5	76.7
364	Evergreen	96.5	51.9	77.1
365-366	Mobile	97.8	65.9	83.9
367	Selma	96.7	50.8	76.7
368	Phenix City	97.3	55.7	79.2
369	Butler	96.8	50.5	76.6
ALASKA				
995-996	Anchorage	121.3	114.3	118.3
997	Fairbanks	120.7	115.0	118.2
998	Juneau	120.6	113.5	117.5
999	Ketchikan	132.0	113.5	124.0
ARIZONA				
850,853	Phoenix	99.4	74.8	88.7
851,852	Mesa/Tempe	98.8	73.8	87.9
855	Globe	99.0	71.2	86.9
856-857	Tucson	97.6	73.6	87.1
859	Show Low	99.1	73.5	88.0
860	Flagstaff	101.0	72.9	88.8
863	Prescott	98.9	69.7	86.2
864	Kingman	97.3	73.2	86.8
865	Chambers	97.3	71.9	86.3
ARKANSAS				
716	Pine Bluff	96.9	62.9	82.1
717	Camden	94.5	50.4	75.3
718	Texarkana	95.7	49.9	75.7
719	Hot Springs	93.8	52.3	75.8
720-722	Little Rock	98.0	63.8	83.1
723	West Memphis	95.6	62.7	81.2
724	Jonesboro	96.3	59.4	80.2
725	Batesville	94.2	55.1	77.2
726	Harrison	95.5	51.1	76.1
727	Fayetteville	93.0	52.9	75.6
728	Russellville	94.1	54.4	76.8
729	Fort Smith	96.8	61.7	81.5
CALIFORNIA				
900-902	Los Angeles	100.1	116.5	107.3
903-905	Inglewood	95.7	113.4	103.4
906-908	Long Beach	97.1	113.4	104.2
910-912	Pasadena	96.3	113.9	104.0
913-916	Van Nuys	99.1	113.9	105.5
917-918	Alhambra	98.2	113.9	105.0
919-921	San Diego	100.2	109.1	104.1
922	Palm Springs	96.7	111.6	103.2
923-924	San Bernardino	94.6	111.2	101.9
925	Riverside	98.9	114.0	105.5
926-927	Santa Ana	96.5	111.6	103.1
928	Anaheim	98.9	114.1	105.6
930	Oxnard	99.6	114.2	105.9
931	Santa Barbara	98.9	114.6	105.7
932-933	Bakersfield	100.8	110.6	105.1
934	San Luis Obispo	99.3	111.6	104.7
935	Mojave	96.8	108.6	101.9
936-938	Fresno	100.0	114.5	106.3
939	Salinas	100.2	120.3	108.9
940-941	San Francisco	106.1	143.8	122.5
942,956-958	Sacramento	101.2	118.6	108.8
943	Palo Alto	99.5	133.1	114.2
944	San Mateo	101.7	132.7	115.2
945	Vallejo	100.6	127.2	112.2
946	Oakland	103.9	133.1	116.6
947	Berkeley	103.4	133.1	116.4
948	Richmond	102.5	131.2	115.0
949	San Rafael	104.2	133.6	117.0
950	Santa Cruz	104.1	120.4	111.2

STATE/ZIP	CITY	MAT.	INST.	TOTAL
CALIFORNIA (CONT'D)				
951	San Jose	102.2	134.6	116.3
952	Stockton	100.5	115.5	107.0
953	Modesto	100.4	114.9	106.7
954	Santa Rosa	100.3	133.4	114.7
955	Eureka	101.6	111.6	106.0
959	Marysville	100.8	115.1	107.0
960	Redding	103.6	114.6	108.4
961	Susanville	102.6	114.6	107.8
COLORADO				
800-802	Denver	101.6	82.6	93.3
803	Boulder	98.6	82.0	91.3
804	Golden	100.6	79.6	91.4
805	Fort Collins	102.1	79.6	92.3
806	Greeley	99.7	76.1	89.4
807	Fort Morgan	99.1	79.6	90.6
808-809	Colorado Springs	101.3	79.3	91.7
810	Pueblo	100.9	78.6	91.2
811	Alamosa	102.0	74.6	90.1
812	Salida	101.7	76.2	90.6
813	Durango	102.4	76.8	91.2
814	Montrose	101.2	76.5	90.5
815	Grand Junction	104.4	76.7	92.4
816	Glenwood Springs	102.2	79.3	92.2
CONNECTICUT				
060	New Britain	100.8	120.4	109.3
061	Hartford	102.0	120.6	110.1
062	Willimantic	101.4	120.6	109.7
063	New London	97.7	120.5	107.6
064	Meriden	99.6	120.5	108.7
065	New Haven	102.2	120.5	110.2
066	Bridgeport	101.8	120.2	109.8
067	Waterbury	101.4	120.1	109.6
068	Norwalk	101.4	126.9	112.5
069	Stamford	101.5	127.9	113.0
D.C.				
200-205	Washington	101.6	91.6	97.2
DELAWARE				
197	Newark	98.8	108.1	102.8
198	Wilmington	99.4	108.1	103.2
199	Dover	99.7	108.1	103.3
FLORIDA				
320,322	Jacksonville	97.2	67.7	84.3
321	Daytona Beach	97.4	74.4	87.3
323	Tallahassee	98.7	58.0	81.0
324	Panama City	98.5	59.1	81.3
325	Pensacola	100.8	62.4	84.1
326,344	Gainesville	98.7	68.0	85.3
327-328,347	Orlando	99.3	71.4	87.1
329	Melbourne	99.9	77.9	90.3
330-332,340	Miami	97.7	74.6	87.6
333	Fort Lauderdale	96.4	73.9	86.6
334,349	West Palm Beach	95.3	72.3	85.3
335-336,346	Tampa	97.9	79.8	90.0
337	St. Petersburg	100.2	64.7	84.7
338	Lakeland	97.1	79.3	89.4
339,341	Fort Myers	96.4	74.1	86.7
342	Sarasota	99.3	75.3	88.8
GEORGIA				
300-303,399	Atlanta	96.8	75.3	87.5
304	Statesboro	96.5	56.9	79.2
305	Gainesville	95.2	62.8	81.1
306	Athens	94.7	65.1	81.8
307	Dalton	96.4	60.1	80.6
308-309	Augusta	95.6	64.8	82.2
310-312	Macon	96.2	65.9	83.0
313-314	Savannah	98.4	62.0	82.6
315	Waycross	97.3	60.7	81.3
316	Valdosta	97.5	62.6	82.3
317,398	Albany	97.5	62.5	82.3
318-319	Columbus	97.4	66.6	84.0

535

STATE/ZIP	CITY	MAT.	INST.	TOTAL
HAWAII				
967	Hilo	110.4	121.0	115.0
968	Honolulu	114.9	121.0	117.6
STATES & POSS.				
969	Guam	131.2	58.9	99.7
IDAHO				
832	Pocatello	100.8	78.4	91.1
833	Twin Falls	101.9	68.5	87.4
834	Idaho Falls	99.6	75.5	89.1
835	Lewiston	108.3	85.5	98.4
836-837	Boise	99.8	79.4	90.9
838	Coeur d'Alene	107.6	83.3	97.0
ILLINOIS				
600-603	North Suburban	98.5	135.9	114.8
604	Joliet	98.4	140.7	116.8
605	South Suburban	98.5	135.9	114.8
606-608	Chicago	98.9	141.9	117.6
609	Kankakee	94.8	130.8	110.5
610-611	Rockford	98.9	123.5	109.6
612	Rock Island	96.6	99.8	98.0
613	La Salle	97.7	125.1	109.6
614	Galesburg	97.6	107.6	102.0
615-616	Peoria	99.9	110.4	104.5
617	Bloomington	96.9	109.9	102.6
618-619	Champaign	100.2	108.0	103.6
620-622	East St. Louis	95.5	109.3	101.5
623	Quincy	96.9	101.6	99.0
624	Effingham	96.3	106.3	100.6
625	Decatur	98.2	105.9	101.6
626-627	Springfield	99.0	107.2	102.5
628	Centralia	94.1	108.4	100.4
629	Carbondale	93.9	106.2	99.2
INDIANA				
460	Anderson	95.5	83.1	90.1
461-462	Indianapolis	98.3	85.1	92.6
463-464	Gary	96.8	110.1	102.6
465-466	South Bend	95.7	84.2	90.7
467-468	Fort Wayne	96.0	79.2	88.7
469	Kokomo	93.3	82.0	88.4
470	Lawrenceburg	93.3	79.3	87.2
471	New Albany	94.4	77.0	86.8
472	Columbus	96.6	80.9	89.7
473	Muncie	97.3	81.8	90.6
474	Bloomington	98.7	81.5	91.2
475	Washington	95.0	83.3	89.9
476-477	Evansville	96.5	86.6	92.1
478	Terre Haute	97.2	85.8	92.2
479	Lafayette	96.3	82.9	90.5
IOWA				
500-503,509	Des Moines	99.4	83.9	92.7
504	Mason City	96.5	66.6	83.5
505	Fort Dodge	96.6	58.8	80.2
506-507	Waterloo	98.3	70.9	86.4
508	Creston	96.7	77.9	88.5
510-511	Sioux City	99.1	73.6	88.0
512	Sibley	97.7	57.5	80.2
513	Spencer	99.2	57.5	81.0
514	Carroll	96.4	72.2	85.9
515	Council Bluffs	100.2	78.8	90.9
516	Shenandoah	97.1	72.1	86.2
520	Dubuque	98.5	80.4	90.6
521	Decorah	97.4	65.6	83.5
522-524	Cedar Rapids	99.4	83.4	92.4
525	Ottumwa	97.3	73.0	86.7
526	Burlington	96.6	76.9	88.0
527-528	Davenport	98.4	93.4	96.2
KANSAS				
660-662	Kansas City	99.4	95.7	97.7
664-666	Topeka	99.4	66.5	85.1
667	Fort Scott	97.7	74.3	87.5
668	Emporia	97.8	70.0	85.7
669	Belleville	99.4	65.6	84.7
670-672	Wichita	99.9	65.1	84.8
673	Independence	99.9	70.3	87.0
674	Salina	100.4	67.0	85.9
675	Hutchinson	95.8	65.4	82.5
676	Hays	99.6	65.6	84.8
677	Colby	100.2	65.6	85.1

STATE/ZIP	CITY	MAT.	INST.	TOTAL
KANSAS (CONT'D)				
678	Dodge City	101.7	66.5	86.4
679	Liberal	99.5	66.4	85.1
KENTUCKY				
400-402	Louisville	97.8	84.2	91.9
403-405	Lexington	97.0	82.4	90.7
406	Frankfort	97.9	82.4	91.1
407-409	Corbin	94.4	73.3	85.2
410	Covington	95.2	93.2	94.3
411-412	Ashland	94.0	100.1	96.6
413-414	Campton	95.1	80.2	88.6
415-416	Pikeville	96.2	90.0	93.5
417-418	Hazard	94.6	79.9	88.2
420	Paducah	93.2	86.2	90.2
421-422	Bowling Green	95.5	84.8	90.8
423	Owensboro	95.6	88.9	92.7
424	Henderson	93.0	87.9	90.8
425-426	Somerset	92.7	81.5	87.8
427	Elizabethtown	92.2	83.7	88.5
LOUISIANA				
700-701	New Orleans	101.5	69.8	87.7
703	Thibodaux	98.7	64.9	84.0
704	Hammond	96.3	54.9	78.3
705	Lafayette	98.5	62.7	82.9
706	Lake Charles	98.6	65.0	84.0
707-708	Baton Rouge	100.4	63.6	84.4
710-711	Shreveport	98.4	59.6	81.5
712	Monroe	96.7	56.4	79.2
713-714	Alexandria	96.8	57.8	79.8
MAINE				
039	Kittery	95.1	85.7	91.0
040-041	Portland	100.8	91.2	96.6
042	Lewiston	98.3	91.2	95.2
043	Augusta	100.3	80.8	91.8
044	Bangor	97.6	90.8	94.7
045	Bath	96.1	83.0	90.4
046	Machias	95.7	86.5	91.7
047	Houlton	95.9	86.6	91.8
048	Rockland	95.0	86.5	91.3
049	Waterville	96.2	80.8	89.5
MARYLAND				
206	Waldorf	97.5	82.2	90.9
207-208	College Park	97.6	83.9	91.7
209	Silver Spring	96.9	83.8	91.2
210-212	Baltimore	100.5	82.8	92.8
214	Annapolis	100.3	81.8	92.3
215	Cumberland	96.0	81.9	89.9
216	Easton	97.4	69.3	85.2
217	Hagerstown	96.9	82.8	90.7
218	Salisbury	97.8	63.3	82.8
219	Elkton	95.1	81.9	89.4
MASSACHUSETTS				
010-011	Springfield	100.0	109.8	104.3
012	Pittsfield	99.6	109.6	103.9
013	Greenfield	97.5	110.2	103.0
014	Fitchburg	96.3	124.6	108.6
015-016	Worcester	100.0	124.6	110.7
017	Framingham	95.5	131.0	111.0
018	Lowell	99.4	130.1	112.8
019	Lawrence	100.1	131.3	113.7
020-022, 024	Boston	101.8	138.1	117.6
023	Brockton	100.4	126.3	111.7
025	Buzzards Bay	94.9	126.7	108.8
026	Hyannis	97.6	126.7	110.3
027	New Bedford	99.6	126.8	111.4
MICHIGAN				
480,483	Royal Oak	94.2	107.5	100.0
481	Ann Arbor	96.4	107.9	101.4
482	Detroit	97.9	109.3	102.8
484-485	Flint	96.1	96.8	96.4
486	Saginaw	95.7	92.4	94.2
487	Bay City	95.8	92.5	94.4
488-489	Lansing	97.7	94.1	96.1
490	Battle Creek	96.8	87.4	92.7
491	Kalamazoo	97.1	85.9	92.2
492	Jackson	95.0	95.9	95.4
493,495	Grand Rapids	98.1	84.5	92.2
494	Muskegon	95.6	85.5	91.2

Location Factors

STATE/ZIP	CITY	MAT.	INST.	TOTAL
MICHIGAN (CONT'D)				
496	Traverse City	94.6	81.0	88.6
497	Gaylord	95.5	84.4	90.7
498-499	Iron mountain	97.5	87.4	93.1
MINNESOTA				
550-551	Saint Paul	100.2	118.8	108.3
553-555	Minneapolis	100.6	119.9	109.0
556-558	Duluth	99.0	108.6	103.2
559	Rochester	98.8	104.9	101.5
560	Mankato	96.6	102.4	99.1
561	Windom	95.3	93.3	94.4
562	Willmar	94.9	103.0	98.4
563	St. Cloud	96.3	117.8	105.7
564	Brainerd	96.3	102.3	98.9
565	Detroit Lakes	98.3	96.7	97.6
566	Bemidji	97.7	99.5	98.5
567	Thief River Falls	97.3	93.8	95.7
MISSISSIPPI				
386	Clarksdale	95.8	53.8	77.5
387	Greenville	99.4	63.1	83.6
388	Tupelo	97.3	56.1	79.4
389	Greenwood	97.0	53.3	78.0
390-392	Jackson	99.1	63.5	83.6
393	Meridian	96.2	65.3	82.7
394	Laurel	97.2	57.1	79.7
395	Biloxi	97.7	59.1	80.9
396	Mccomb	95.7	52.9	77.1
397	Columbus	97.1	55.1	78.8
MISSOURI				
630-631	St. Louis	99.6	107.6	103.1
633	Bowling Green	98.1	94.7	96.6
634	Hannibal	97.1	92.6	95.1
635	Kirksville	100.3	90.0	95.8
636	Flat River	99.0	95.8	97.6
637	Cape Girardeau	99.1	90.6	95.4
638	Sikeston	97.5	90.6	94.5
639	Poplar Bluff	97.0	90.3	94.1
640-641	Kansas City	101.6	106.7	103.8
644-645	St. Joseph	100.7	93.9	97.8
646	Chillicothe	97.7	96.2	97.1
647	Harrisonville	97.3	102.2	99.5
648	Joplin	99.5	81.0	91.5
650-651	Jefferson City	98.4	93.4	96.2
652	Columbia	98.6	93.5	96.4
653	Sedalia	98.1	92.6	95.7
654-655	Rolla	96.3	95.6	96.0
656-658	Springfield	99.9	82.1	92.1
MONTANA				
590-591	Billings	102.2	77.9	91.6
592	Wolf Point	101.7	74.9	90.0
593	Miles City	99.5	74.7	88.7
594	Great Falls	103.4	77.1	92.0
595	Havre	100.6	74.2	89.1
596	Helena	102.0	75.1	90.2
597	Butte	102.0	77.2	91.2
598	Missoula	99.4	76.4	89.4
599	Kalispell	98.4	76.3	88.8
NEBRASKA				
680-681	Omaha	99.6	79.3	90.7
683-685	Lincoln	99.5	75.3	89.0
686	Columbus	97.5	77.9	88.9
687	Norfolk	99.2	77.5	89.8
688	Grand Island	99.1	77.8	89.9
689	Hastings	98.5	81.2	90.9
690	Mccook	98.1	72.1	86.8
691	North Platte	98.5	79.6	90.2
692	Valentine	100.6	73.6	88.9
693	Alliance	100.4	71.6	87.9
NEVADA				
889-891	Las Vegas	100.2	109.4	104.2
893	Ely	98.7	103.4	100.8
894-895	Reno	99.0	93.8	96.7
897	Carson City	99.1	93.8	96.8
898	Elko	97.6	86.1	92.6
NEW HAMPSHIRE				
030	Nashua	100.6	93.6	97.6
031	Manchester	102.1	93.6	98.4

STATE/ZIP	CITY	MAT.	INST.	TOTAL
NEW HAMPSHIRE (CONT'D)				
032-033	Concord	100.8	92.3	97.1
034	Keene	97.4	74.0	87.2
035	Littleton	97.1	77.2	88.4
036	Charleston	96.9	70.1	85.2
037	Claremont	96.0	70.1	84.7
038	Portsmouth	98.2	94.2	96.4
NEW JERSEY				
070-071	Newark	101.9	125.6	112.2
072	Elizabeth	99.5	125.3	110.7
073	Jersey City	98.5	125.5	110.2
074-075	Paterson	100.2	125.2	111.1
076	Hackensack	98.4	125.5	110.2
077	Long Branch	97.8	124.1	109.2
078	Dover	98.7	125.4	110.3
079	Summit	98.7	125.3	110.3
080,083	Vineland	97.5	123.4	108.8
081	Camden	99.4	123.6	109.9
082,084	Atlantic City	98.1	123.5	109.2
085-086	Trenton	100.8	124.0	110.9
087	Point Pleasant	99.2	121.4	108.8
088-089	New Brunswick	99.7	125.2	110.8
NEW MEXICO				
870-872	Albuquerque	99.2	73.6	88.0
873	Gallup	99.3	73.6	88.1
874	Farmington	99.7	73.6	88.3
875	Santa Fe	100.5	73.6	88.8
877	Las Vegas	97.7	73.6	87.2
878	Socorro	97.3	73.6	87.0
879	Truth/Consequences	97.0	70.4	85.4
880	Las Cruces	95.8	70.4	84.7
881	Clovis	98.1	73.5	87.4
882	Roswell	99.8	73.6	88.4
883	Carrizozo	100.1	73.6	88.5
884	Tucumcari	98.8	73.5	87.8
NEW YORK				
100-102	New York	103.8	166.4	131.1
103	Staten Island	99.7	163.9	127.6
104	Bronx	98.0	163.9	126.7
105	Mount Vernon	97.9	133.3	113.3
106	White Plains	98.0	133.3	113.4
107	Yonkers	102.5	133.4	116.0
108	New Rochelle	98.2	133.3	113.5
109	Suffern	98.0	126.2	110.3
110	Queens	100.8	162.9	127.8
111	Long Island City	102.2	162.9	128.6
112	Brooklyn	102.6	162.9	128.9
113	Flushing	102.3	162.9	128.7
114	Jamaica	100.6	162.9	127.7
115,117,118	Hicksville	100.7	146.6	120.7
116	Far Rockaway	102.4	162.9	128.7
119	Riverhead	101.4	146.6	121.1
120-122	Albany	98.6	100.2	99.3
123	Schenectady	97.8	100.2	98.8
124	Kingston	100.7	118.3	108.4
125-126	Poughkeepsie	100.0	127.5	112.0
127	Monticello	99.4	116.5	106.8
128	Glens Falls	92.6	95.0	93.6
129	Plattsburgh	97.3	88.4	93.4
130-132	Syracuse	98.5	97.0	97.9
133-135	Utica	96.7	93.2	95.1
136	Watertown	98.1	91.9	95.4
137-139	Binghamton	98.1	99.4	98.6
140-142	Buffalo	99.7	105.8	102.4
143	Niagara Falls	96.9	103.9	100.0
144-146	Rochester	100.3	95.6	98.3
147	Jamestown	96.0	87.8	92.4
148-149	Elmira	95.9	93.1	94.7
NORTH CAROLINA				
270,272-274	Greensboro	98.8	56.3	80.3
271	Winston-Salem	98.5	57.1	80.5
275-276	Raleigh	98.2	55.1	79.4
277	Durham	100.2	54.4	80.2
278	Rocky Mount	96.0	53.2	77.4
279	Elizabeth City	96.9	52.2	77.4
280	Gastonia	98.4	56.1	80.0
281-282	Charlotte	98.7	57.6	80.8
283	Fayetteville	100.9	58.5	82.5
284	Wilmington	97.2	56.6	79.5
285	Kinston	95.3	53.7	77.2

Location Factors

STATE/ZIP	CITY	MAT.	INST.	TOTAL
NORTH CAROLINA (CONT'D)				
286	Hickory	95.6	54.9	77.9
287-288	Asheville	97.6	55.9	79.4
289	Murphy	96.4	49.2	75.8
NORTH DAKOTA				
580-581	Fargo	101.7	67.6	86.9
582	Grand Forks	101.8	51.6	79.9
583	Devils Lake	101.0	58.6	82.5
584	Jamestown	101.0	45.1	76.7
585	Bismarck	102.0	64.0	85.5
586	Dickinson	101.9	57.9	82.7
587	Minot	101.8	71.0	88.4
588	Williston	100.3	58.0	81.9
OHIO				
430-432	Columbus	98.6	89.7	94.7
433	Marion	95.0	83.7	90.1
434-436	Toledo	98.7	96.4	97.7
437-438	Zanesville	95.5	85.6	91.2
439	Steubenville	96.8	94.5	95.8
440	Lorain	98.8	90.4	95.2
441	Cleveland	98.9	100.0	99.4
442-443	Akron	99.8	93.4	97.0
444-445	Youngstown	99.1	88.3	94.4
446-447	Canton	99.2	85.0	93.0
448-449	Mansfield	96.6	85.6	91.8
450	Hamilton	97.4	84.5	91.8
451-452	Cincinnati	97.8	84.8	92.1
453-454	Dayton	97.5	83.2	91.3
455	Springfield	97.5	84.2	91.7
456	Chillicothe	96.4	92.2	94.5
457	Athens	99.1	80.6	91.1
458	Lima	99.6	84.4	93.0
OKLAHOMA				
730-731	Oklahoma City	97.3	65.4	83.4
734	Ardmore	95.2	62.3	80.9
735	Lawton	97.5	63.3	82.6
736	Clinton	96.5	63.6	82.2
737	Enid	97.2	60.9	81.4
738	Woodward	95.3	63.8	81.6
739	Guymon	96.3	60.5	80.7
740-741	Tulsa	97.4	62.2	82.1
743	Miami	94.0	68.3	82.8
744	Muskogee	96.6	56.3	79.0
745	Mcalester	93.7	61.1	79.5
746	Ponca City	94.3	62.5	80.5
747	Durant	94.3	64.0	81.1
748	Shawnee	95.7	62.2	81.1
749	Poteau	93.5	62.1	79.8
OREGON				
970-972	Portland	97.2	101.4	99.1
973	Salem	99.0	100.5	99.7
974	Eugene	97.1	100.4	98.5
975	Medford	98.4	98.2	98.3
976	Klamath Falls	98.2	98.2	98.2
977	Bend	97.3	100.5	98.7
978	Pendleton	93.1	102.5	97.2
979	Vale	91.3	89.8	90.7
PENNSYLVANIA				
150-152	Pittsburgh	99.7	105.8	102.3
153	Washington	96.7	103.2	99.5
154	Uniontown	97.1	102.2	99.4
155	Bedford	98.0	92.6	95.7
156	Greensburg	98.0	101.6	99.5
157	Indiana	96.9	100.1	98.3
158	Dubois	98.4	97.1	97.8
159	Johnstown	97.9	97.3	97.7
160	Butler	92.1	102.2	96.5
161	New Castle	92.1	97.2	94.4
162	Kittanning	92.5	103.4	97.3
163	Oil City	92.1	97.0	94.3
164-165	Erie	94.1	95.0	94.5
166	Altoona	94.3	92.7	93.6
167	Bradford	95.3	96.8	96.0
168	State College	94.9	93.0	94.1
169	Wellsboro	95.9	92.3	94.3
170-171	Harrisburg	99.6	95.9	98.0
172	Chambersburg	96.0	89.7	93.2
173-174	York	96.5	96.0	96.3
175-176	Lancaster	94.8	89.3	92.4

STATE/ZIP	CITY	MAT.	INST.	TOTAL
PENNSYLVANIA (CONT'D)				
177	Williamsport	93.6	79.3	87.4
178	Sunbury	95.5	96.3	95.8
179	Pottsville	94.7	98.0	96.1
180	Lehigh Valley	95.4	112.5	102.9
181	Allentown	97.6	108.8	102.4
182	Hazleton	95.0	98.8	96.7
183	Stroudsburg	94.9	102.0	98.0
184-185	Scranton	98.3	99.7	98.9
186-187	Wilkes-Barre	94.7	99.2	96.7
188	Montrose	94.4	97.1	95.6
189	Doylestown	94.6	120.8	106.0
190-191	Philadelphia	98.9	133.4	113.9
193	Westchester	95.8	121.7	107.1
194	Norristown	94.9	129.5	110.0
195-196	Reading	97.1	101.1	98.8
PUERTO RICO				
009	San Juan	122.1	24.6	79.7
RHODE ISLAND				
028	Newport	99.0	116.6	106.6
029	Providence	101.5	116.6	108.1
SOUTH CAROLINA				
290-292	Columbia	97.2	57.3	79.8
293	Spartanburg	96.4	57.4	79.5
294	Charleston	97.7	66.4	84.1
295	Florence	96.2	57.8	79.4
296	Greenville	96.2	56.5	78.9
297	Rock Hill	95.6	51.2	76.3
298	Aiken	96.4	70.4	85.1
299	Beaufort	97.1	45.7	74.7
SOUTH DAKOTA				
570-571	Sioux Falls	100.0	58.8	82.1
572	Watertown	98.5	50.2	77.5
573	Mitchell	97.5	49.5	76.6
574	Aberdeen	100.1	50.6	78.6
575	Pierre	100.8	59.2	82.7
576	Mobridge	98.1	49.8	77.1
577	Rapid City	100.0	60.3	82.7
TENNESSEE				
370-372	Nashville	97.9	74.1	87.5
373-374	Chattanooga	99.5	65.5	84.7
375,380-381	Memphis	98.5	71.3	86.7
376	Johnson City	99.2	56.7	80.7
377-379	Knoxville	96.2	60.5	80.6
382	Mckenzie	96.6	60.4	80.9
383	Jackson	98.3	62.5	82.7
384	Columbia	95.3	67.5	83.2
385	Cookeville	96.4	60.4	80.7
TEXAS				
750	Mckinney	99.3	48.9	77.3
751	Waxahackie	99.3	56.4	80.6
752-753	Dallas	100.1	65.8	85.2
754	Greenville	99.4	42.2	74.5
755	Texarkana	98.4	49.7	77.2
756	Longview	99.0	40.5	73.5
757	Tyler	98.9	54.6	79.6
758	Palestine	95.4	55.8	78.2
759	Lufkin	95.9	59.1	79.9
760-761	Fort Worth	99.6	62.5	83.5
762	Denton	98.2	46.6	75.7
763	Wichita Falls	97.8	55.0	79.2
764	Eastland	96.5	49.1	75.9
765	Temple	95.0	52.7	76.6
766-767	Waco	97.2	57.3	79.8
768	Brownwood	98.5	48.4	76.7
769	San Angelo	98.2	49.5	77.0
770-772	Houston	98.7	70.9	86.6
773	Huntsville	97.2	56.9	79.6
774	Wharton	98.1	53.8	78.8
775	Galveston	96.3	69.7	84.7
776-777	Beaumont	96.7	61.2	81.2
778	Bryan	94.1	63.0	80.5
779	Victoria	98.2	42.7	74.1
780	Laredo	97.1	53.7	78.2
781-782	San Antonio	97.5	64.5	83.1
783-784	Corpus Christi	99.7	53.7	79.6
785	Mc Allen	99.7	46.6	76.6
786-787	Austin	98.6	59.1	81.4

Location Factors

STATE/ZIP	CITY	MAT.	INST.	TOTAL
TEXAS (CONT'D)				
788	Del Rio	99.4	46.2	76.3
789	Giddings	96.4	55.6	78.6
790-791	Amarillo	98.2	60.0	81.6
792	Childress	97.9	56.9	80.0
793-794	Lubbock	100.0	56.3	81.0
795-796	Abilene	98.3	53.6	78.8
797	Midland	99.9	53.6	79.7
798-799,885	El Paso	97.2	50.6	76.9
UTAH				
840-841	Salt Lake City	102.2	66.8	86.8
842,844	Ogden	97.9	68.8	85.2
843	Logan	99.7	68.8	86.2
845	Price	100.1	65.3	84.9
846-847	Provo	100.1	66.7	85.6
VERMONT				
050	White River Jct.	97.9	79.7	90.0
051	Bellows Falls	96.6	94.7	95.8
052	Bennington	96.9	95.3	96.2
053	Brattleboro	97.2	94.5	96.0
054	Burlington	101.5	85.1	94.4
056	Montpelier	99.2	84.3	92.7
057	Rutland	99.2	85.1	93.0
058	St. Johnsbury	97.9	79.2	89.8
059	Guildhall	96.8	79.2	89.1
VIRGINIA				
220-221	Fairfax	99.3	84.4	92.8
222	Arlington	100.6	84.7	93.6
223	Alexandria	99.8	85.4	93.5
224-225	Fredericksburg	98.2	83.0	91.6
226	Winchester	98.7	82.1	91.5
227	Culpeper	98.6	80.8	90.9
228	Harrisonburg	98.8	68.1	85.5
229	Charlottesville	99.2	67.3	85.3
230-232	Richmond	100.4	68.9	86.7
233-235	Norfolk	100.0	68.8	86.4
236	Newport News	99.2	68.2	85.7
237	Portsmouth	98.6	66.1	84.5
238	Petersburg	98.6	68.9	85.6
239	Farmville	97.8	53.3	78.5
240-241	Roanoke	100.2	66.0	85.3
242	Bristol	98.1	55.6	79.6
243	Pulaski	97.7	58.3	80.6
244	Staunton	98.4	62.7	82.9
245	Lynchburg	98.6	66.1	84.5
246	Grundy	98.0	53.4	78.6
WASHINGTON				
980-981,987	Seattle	102.5	104.8	103.5
982	Everett	102.1	99.6	101.0
983-984	Tacoma	102.3	100.0	101.3
985	Olympia	101.0	100.0	100.6
986	Vancouver	103.1	93.2	98.8
988	Wenatchee	102.0	87.1	95.5
989	Yakima	102.5	94.4	99.0
990-992	Spokane	102.5	82.7	93.8
993	Richland	102.2	89.6	96.7
994	Clarkston	100.6	80.7	92.0
WEST VIRGINIA				
247-248	Bluefield	97.1	91.7	94.7
249	Lewisburg	98.4	91.4	95.4
250-253	Charleston	99.0	94.1	96.9
254	Martinsburg	98.3	86.1	93.0
255-257	Huntington	99.8	95.2	97.8
258-259	Beckley	96.9	92.0	94.8
260	Wheeling	99.8	93.8	97.2
261	Parkersburg	98.9	91.4	95.6
262	Buckhannon	98.2	93.6	96.2
263-264	Clarksburg	98.8	93.5	96.5
265	Morgantown	98.8	92.6	96.1
266	Gassaway	98.2	92.8	95.8
267	Romney	98.2	90.4	94.8
268	Petersburg	98.0	90.1	94.6
WISCONSIN				
530,532	Milwaukee	100.9	107.9	104.0
531	Kenosha	100.2	102.9	101.4
534	Racine	99.7	102.6	101.0
535	Beloit	99.6	101.9	100.6
537	Madison	99.5	99.8	99.6

STATE/ZIP	CITY	MAT.	INST.	TOTAL
WISCONSIN (CONT'D)				
538	Lancaster	97.4	93.5	95.7
539	Portage	96.1	97.9	96.9
540	New Richmond	96.2	95.9	96.1
541-543	Green Bay	100.5	94.1	97.7
544	Wausau	95.7	94.4	95.2
545	Rhinelander	98.7	92.4	95.9
546	La Crosse	97.2	95.0	96.3
547	Eau Claire	98.7	96.4	97.7
548	Superior	96.1	97.6	96.7
549	Oshkosh	96.3	92.4	94.6
WYOMING				
820	Cheyenne	100.1	65.7	85.2
821	Yellowstone Nat'l Park	97.7	65.2	83.5
822	Wheatland	98.7	64.6	83.9
823	Rawlins	100.2	67.8	86.1
824	Worland	98.3	64.1	83.4
825	Riverton	99.3	64.9	84.3
826	Casper	100.4	60.7	83.1
827	Newcastle	98.2	67.8	84.9
828	Sheridan	100.9	65.2	85.4
829-831	Rock Springs	101.8	65.4	85.9
CANADIAN FACTORS (reflect Canadian currency)				
ALBERTA				
	Calgary	124.7	95.4	111.9
	Edmonton	125.0	95.4	112.1
	Fort McMurray	124.1	94.2	111.1
	Lethbridge	118.7	93.7	107.8
	Lloydminster	113.9	90.3	103.6
	Medicine Hat	114.0	89.6	103.4
	Red Deer	114.5	89.6	103.6
BRITISH COLUMBIA				
	Kamloops	114.9	92.4	105.1
	Prince George	115.9	91.7	105.3
	Vancouver	123.3	90.0	108.8
	Victoria	115.9	88.1	103.8
MANITOBA				
	Brandon	118.7	77.3	100.7
	Portage la Prairie	113.9	75.6	97.2
	Winnipeg	123.6	70.8	100.6
NEW BRUNSWICK				
	Bathurst	112.1	68.6	93.1
	Dalhousie	112.1	68.8	93.2
	Fredericton	118.1	74.0	98.9
	Moncton	112.3	71.4	94.5
	Newcastle	112.1	69.3	93.5
	St. John	114.6	78.3	98.8
NEWFOUNDLAND				
	Corner Brook	122.2	69.0	99.0
	St Johns	122.0	80.4	103.9
NORTHWEST TERRITORIES				
	Yellowknife	128.8	85.8	110.1
NOVA SCOTIA				
	Bridgewater	113.4	76.1	97.2
	Dartmouth	119.7	76.1	100.7
	Halifax	119.9	80.6	102.8
	New Glasgow	117.7	76.1	99.6
	Sydney	115.6	76.1	98.4
	Truro	113.0	76.1	96.9
	Yarmouth	117.6	76.1	99.6
ONTARIO				
	Barrie	117.3	94.9	107.5
	Brantford	115.5	98.8	108.2
	Cornwall	115.5	95.1	106.7
	Hamilton	119.3	95.3	108.8
	Kingston	116.6	95.2	107.3
	Kitchener	111.6	92.0	103.1
	London	119.5	91.9	107.5
	North Bay	120.5	92.9	108.5
	Oshawa	114.6	96.1	106.5
	Ottawa	119.2	92.5	107.5
	Owen Sound	117.4	93.0	106.8
	Peterborough	115.5	94.9	106.5
	Sarnia	115.3	99.5	108.4

Location Factors

STATE/ZIP	CITY	MAT.	INST.	TOTAL
ONTARIO (CONT'D)				
	Sault Ste Marie	110.7	93.1	103.0
	St. Catharines	109.7	93.0	102.4
	Sudbury	109.9	91.7	101.9
	Thunder Bay	110.9	91.7	102.6
	Timmins	115.6	92.9	105.7
	Toronto	120.2	99.0	111.0
	Windsor	110.2	91.7	102.2
PRINCE EDWARD ISLAND				
	Charlottetown	120.5	64.0	95.9
	Summerside	119.5	64.1	95.4
QUEBEC				
	Cap-de-la-Madeleine	112.8	89.4	102.6
	Charlesbourg	112.8	89.4	102.6
	Chicoutimi	111.9	93.1	103.7
	Gatineau	112.5	89.2	102.3
	Granby	112.8	89.1	102.4
	Hull	112.6	89.2	102.4
	Joliette	113.0	89.4	102.7
	Laval	112.4	89.2	102.3
	Montreal	119.2	93.4	107.9
	Quebec	117.5	93.8	107.2
	Rimouski	112.3	93.1	103.9
	Rouyn-Noranda	112.5	89.2	102.3
	Saint Hyacinthe	111.9	89.2	102.0
	Sherbrooke	112.8	89.2	102.5
	Sorel	113.0	89.4	102.7
	St Jerome	112.5	89.2	102.3
	Trois Rivieres	117.9	89.4	105.5
SASKATCHEWAN				
	Moose Jaw	111.2	70.8	93.6
	Prince Albert	110.3	69.0	92.3
	Regina	120.3	88.4	106.4
	Saskatoon	111.3	88.4	101.3
YUKON				
	Whitehorse	126.3	69.8	101.7

R011105-05   Tips for Accurate Estimating

1. Use pre-printed or columnar forms for orderly sequence of dimensions and locations and for recording telephone quotations.

2. Use only the front side of each paper or form except for certain pre-printed summary forms.

3. Be consistent in listing dimensions: For example, length x width x height. This helps in rechecking to ensure that the total length of partitions is appropriate for the building area.

4. Use printed (rather than measured) dimensions where given.

5. Add up multiple printed dimensions for a single entry where possible.

6. Measure all other dimensions carefully.

7. Use each set of dimensions to calculate multiple related quantities.

8. Convert foot and inch measurements to decimal feet when listing. Memorize decimal equivalents to .01 parts of a foot (1/8″ equals approximately .01′).

9. Do not "round off" quantities until the final summary.

10. Mark drawings with different colors as items are taken off.

11. Keep similar items together, different items separate.

12. Identify location and drawing numbers to aid in future checking for completeness.

13. Measure or list everything on the drawings or mentioned in the specifications.

14. It may be necessary to list items not called for to make the job complete.

15. Be alert for: Notes on plans such as N.T.S. (not to scale); changes in scale throughout the drawings; reduced size drawings; discrepancies between the specifications and the drawings.

16. Develop a consistent pattern of performing an estimate. For example:
   a. Start the quantity takeoff at the lower floor and move to the next higher floor.
   b. Proceed from the main section of the building to the wings.
   c. Proceed from south to north or vice versa, clockwise or counterclockwise.
   d. Take off floor plan quantities first, elevations next, then detail drawings.

17. List all gross dimensions that can be either used again for different quantities, or used as a rough check of other quantities for verification (exterior perimeter, gross floor area, individual floor areas, etc.).

18. Utilize design symmetry or repetition (repetitive floors, repetitive wings, symmetrical design around a center line, similar room layouts, etc.). Note: Extreme caution is needed here so as not to omit or duplicate an area.

19. Do not convert units until the final total is obtained. For instance, when estimating concrete work, keep all units to the nearest cubic foot, then summarize and convert to cubic yards.

20. When figuring alternatives, it is best to total all items involved in the basic system, then total all items involved in the alternates. Therefore, you work with positive numbers in all cases. When adds and deducts are used, it is often confusing whether to add or subtract a portion of an item; especially on a complicated or involved alternate.

R011105-10   Unit Gross Area Requirements

The figures in the table below indicate typical ranges in square feet as a function of the "occupant" unit. This table is best used in the preliminary design stages to help determine the probable size requirement for the total project.

Building Type	Unit	Gross Area in S.F.		
		1/4	Median	3/4
Apartments	Unit	660	860	1,100
Auditorium & Play Theaters	Seat	18	25	38
Bowling Alleys	Lane		940	
Churches & Synagogues	Seat	20	28	39
Dormitories	Bed	200	230	275
Fraternity & Sorority Houses	Bed	220	315	370
Garages, Parking	Car	325	355	385
Hospitals	Bed	685	850	1,075
Hotels	Rental Unit	475	600	710
Housing for the elderly	Unit	515	635	755
Housing, Public	Unit	700	875	1,030
Ice Skating Rinks	Total	27,000	30,000	36,000
Motels	Rental Unit	360	465	620
Nursing Homes	Bed	290	350	450
Restaurants	Seat	23	29	39
Schools, Elementary	Pupil	65	77	90
Junior High & Middle		85	110	129
Senior High		102	130	145
Vocational		110	135	195
Shooting Ranges	Point		450	
Theaters & Movies	Seat		15	

R011105-50   Metric Conversion Factors

Description: This table is primarily for converting customary U.S. units in the left hand column to SI metric units in the right hand column. In addition, conversion factors for some commonly encountered Canadian and non-SI metric units are included.

If You Know		Multiply By		To Find
Length				
Inches	x	25.4[a]	=	Millimeters
Feet	x	0.3048[a]	=	Meters
Yards	x	0.9144[a]	=	Meters
Miles (statute)	x	1.609	=	Kilometers
Area				
Square inches	x	645.2	=	Square millimeters
Square feet	x	0.0929	=	Square meters
Square yards	x	0.8361	=	Square meters
Volume **(Capacity)**				
Cubic inches	x	16,387	=	Cubic millimeters
Cubic feet	x	0.02832	=	Cubic meters
Cubic yards	x	0.7646	=	Cubic meters
Gallons (U.S. liquids)[b]	x	0.003785	=	Cubic meters[c]
Gallons (Canadian liquid)[b]	x	0.004546	=	Cubic meters[c]
Ounces (U.S. liquid)[b]	x	29.57	=	Milliliters[c, d]
Quarts (U.S. liquid)[b]	x	0.9464	=	Liters[c, d]
Gallons (U.S. liquid)[b]	x	3.785	=	Liters[c, d]
Force				
Kilograms force[d]	x	9.807	=	Newtons
Pounds force	x	4.448	=	Newtons
Pounds force	x	0.4536	=	Kilograms force[d]
Kips	x	4448	=	Newtons
Kips	x	453.6	=	Kilograms force[d]
Pressure, Stress, Strength (Force per unit area)				
Kilograms force per square centimeter[d]	x	0.09807	=	Megapascals
Pounds force per square inch (psi)	x	0.006895	=	Megapascals
Kips per square inch	x	6.895	=	Megapascals
Pounds force per square inch (psi)	x	0.07031	=	Kilograms force per square centimeter[d]
Pounds force per square foot	x	47.88	=	Pascals
Pounds force per square foot	x	4.882	=	Kilograms force per square meter[d]
Bending Moment Or Torque				
Inch-pounds force	x	0.01152	=	Meter-kilograms force[d]
Inch-pounds force	x	0.1130	=	Newton-meters
Foot-pounds force	x	0.1383	=	Meter-kilograms force[d]
Foot-pounds force	x	1.356	=	Newton-meters
Meter-kilograms force[d]	x	9.807	=	Newton-meters
Mass				
Ounces (avoirdupois)	x	28.35	=	Grams
Pounds (avoirdupois)	x	0.4536	=	Kilograms
Tons (metric)	x	1000	=	Kilograms
Tons, short (2000 pounds)	x	907.2	=	Kilograms
Tons, short (2000 pounds)	x	0.9072	=	Megagrams[e]
Mass per Unit Volume				
Pounds mass per cubic foot	x	16.02	=	Kilograms per cubic meter
Pounds mass per cubic yard	x	0.5933	=	Kilograms per cubic meter
Pounds mass per gallon (U.S. liquid)[b]	x	119.8	=	Kilograms per cubic meter
Pounds mass per gallon (Canadian liquid)[b]	x	99.78	=	Kilograms per cubic meter
Temperature				
Degrees Fahrenheit	(F-32)/1.8		=	Degrees Celsius
Degrees Fahrenheit	(F+459.67)/1.8		=	Degrees Kelvin
Degrees Celsius	C+273.15		=	Degrees Kelvin

[a]The factor given is exact
[b]One U.S. gallon = 0.8327 Canadian gallon
[c]1 liter = 1000 milliliters = 1000 cubic centimeters
  1 cubic decimeter = 0.001 cubic meter
[d]Metric but not SI unit
[e]Called "tonne" in England and "metric ton" in other metric countries

R011105-60   Weights and Measures

Measures of Length
1 Mile = 1760 Yards = 5280 Feet
1 Yard = 3 Feet = 36 inches
1 Foot = 12 Inches
1 Mil = 0.001 Inch
1 Fathom = 2 Yards = 6 Feet
1 Rod = 5.5 Yards = 16.5 Feet
1 Hand = 4 Inches
1 Span = 9 Inches
1 Micro-inch = One Millionth Inch or 0.000001 Inch
1 Micron = One Millionth Meter + 0.00003937 Inch

Surveyor's Measure
1 Mile = 8 Furlongs = 80 Chains
1 Furlong = 10 Chains = 220 Yards
1 Chain = 4 Rods = 22 Yards = 66 Feet = 100 Links
1 Link = 7.92 Inches

Square Measure
1 Square Mile = 640 Acres = 6400 Square Chains
1 Acre = 10 Square Chains = 4840 Square Yards = 43,560 Sq. Ft.
1 Square Chain = 16 Square Rods = 484 Square Yards = 4356 Sq. Ft.
1 Square Rod = 30.25 Square Yards = 272.25 Square Feet = 625 Square Lines
1 Square Yard = 9 Square Feet
1 Square Foot = 144 Square Inches
An Acre equals a Square 208.7 Feet per Side

Cubic Measure
1 Cubic Yard = 27 Cubic Feet
1 Cubic Foot = 1728 Cubic Inches
1 Cord of Wood = 4 x 4 x 8 Feet = 128 Cubic Feet
1 Perch of Masonry = 16½ x 1½ x 1 Foot = 24.75 Cubic Feet

Avoirdupois or Commercial Weight
1 Gross or Long Ton = 2240 Pounds
1 Net or Short ton = 2000 Pounds
1 Pound = 16 Ounces = 7000 Grains
1 Ounce = 16 Drachms = 437.5 Grains
1 Stone = 14 Pounds

Shipping Measure
For Measuring Internal Capacity of a Vessel:
    1 Register Ton = 100 Cubic Feet

For Measurement of Cargo:
    Approximately 40 Cubic Feet of Merchandise is considered a Shipping Ton, unless that bulk would weigh more than 2000 Pounds, in which case Freight Charge may be based upon weight.

40 Cubic Feet = 32.143 U.S. Bushels = 31.16 Imp. Bushels

Liquid Measure
1 Imperial Gallon = 1.2009 U.S. Gallon = 277.42 Cu. In.
1 Cubic Foot = 7.48 U.S. Gallons

R011110-30 Engineering Fees

Typical **Structural Engineering Fees** based on type of construction and total project size. These fees are included in Architectural Fees.

Type of Construction	Total Project Size (in thousands of dollars)			
	$500	$500-$1,000	$1,000-$5,000	Over $5000
Industrial buildings, factories & warehouses	Technical payroll times 2.0 to 2.5	1.60%	1.25%	1.00%
Hotels, apartments, offices, dormitories, hospitals, public buildings, food stores		2.00%	1.70%	1.20%
Museums, banks, churches and cathedrals		2.00%	1.75%	1.25%
Thin shells, prestressed concrete, earthquake resistive		2.00%	1.75%	1.50%
Parking ramps, auditoriums, stadiums, convention halls, hangars & boiler houses		2.50%	2.00%	1.75%
Special buildings, major alterations, underpinning & future expansion		Add to above 0.5%	Add to above 0.5%	Add to above 0.5%

For complex reinforced concrete or unusually complicated structures, add 20% to 50%.

Typical **Mechanical and Electrical Engineering Fees** are based on the size of the subcontract. The fee structure for both are shown below. These fees are included in Architectural Fees.

Type of Construction	Subcontract Size							
	$25,000	$50,000	$100,000	$225,000	$350,000	$500,000	$750,000	$1,000,000
Simple structures	6.4%	5.7%	4.8%	4.5%	4.4%	4.3%	4.2%	4.1%
Intermediate structures	8.0	7.3	6.5	5.6	5.1	5.0	4.9	4.8
Complex structures	10.1	9.0	9.0	8.0	7.5	7.5	7.0	7.0

For renovations, add 15% to 25% to applicable fee.

R012153-10 Repair and Remodeling

Cost figures are based on new construction utilizing the most cost-effective combination of labor, equipment and material with the work scheduled in proper sequence to allow the various trades to accomplish their work in an efficient manner.

The costs for repair and remodeling work must be modified due to the following factors that may be present in any given repair and remodeling project.

1. Equipment usage curtailment due to the physical limitations of the project, with only hand-operated equipment being used.

2. Increased requirement for shoring and bracing to hold up the building while structural changes are being made and to allow for temporary storage of construction materials on above-grade floors.

3. Material handling becomes more costly due to having to move within the confines of an enclosed building. For multi-story construction, low capacity elevators and stairwells may be the only access to the upper floors.

4. Large amounts of cutting and patching and attempting to match the existing construction is required. It is often more economical to remove entire walls rather than create many new door and window openings. This sort of trade-off has to be carefully analyzed.

5. Cost of protection of completed work is increased since the usual sequence of construction usually cannot be accomplished.

6. Economies of scale usually associated with new construction may not be present. If small quantities of components must be custom fabricated due to job requirements, unit costs will naturally increase. Also, if only small work areas are available at a given time, job scheduling between trades becomes difficult and subcontractor quotations may reflect the excessive start-up and shut-down phases of the job.

7. Work may have to be done on other than normal shifts and may have to be done around an existing production facility which has to stay in production during the course of the repair and remodeling.

8. Dust and noise protection of adjoining non-construction areas can involve substantial special protection and alter usual construction methods.

9. Job may be delayed due to unexpected conditions discovered during demolition or removal. These delays ultimately increase construction costs.

10. Piping and ductwork runs may not be as simple as for new construction. Wiring may have to be snaked through walls and floors.

11. Matching "existing construction" may be impossible because materials may no longer be manufactured. Substitutions may be expensive.

12. Weather protection of existing structure requires additional temporary structures to protect building at openings.

13. On small projects, because of local conditions, it may be necessary to pay a tradesman for a minimum of four hours for a task that is completed in one hour.

All of the above areas can contribute to increased costs for a repair and remodeling project. Each of the above factors should be considered in the planning, bidding and construction stage in order to minimize the increased costs associated with repair and remodeling jobs.

R012909-80   Sales Tax by State

State sales tax on materials is tabulated below (5 states have no sales tax). Many states allow local jurisdictions, such as a county or city, to levy additional sales tax.

Some projects may be sales tax exempt, particularly those constructed with public funds.

State	Tax (%)	State	Tax (%)	State	Tax (%)	State	Tax (%)
Alabama	4	Illinois	6.25	Montana	0	Rhode Island	7
Alaska	0	Indiana	7	Nebraska	5.5	South Carolina	6
Arizona	5.6	Iowa	6	Nevada	6.85	South Dakota	4
Arkansas	6	Kansas	6.3	New Hampshire	0	Tennessee	7
California	7.25	Kentucky	6	New Jersey	7	Texas	6.25
Colorado	2.9	Louisiana	4	New Mexico	5.13	Utah	5.95
Connecticut	6.35	Maine	5	New York	4	Vermont	6
Delaware	0	Maryland	6	North Carolina	4.75	Virginia	5
District of Columbia	6	Massachusetts	6.25	North Dakota	5	Washington	6.5
Florida	6	Michigan	6	Ohio	5.5	West Virginia	6
Georgia	4	Minnesota	6.88	Oklahoma	4.5	Wisconsin	5
Hawaii	4	Mississippi	7	Oregon	0	Wyoming	4
Idaho	6	Missouri	4.23	Pennsylvania	6	Average	5.06 %

Sales Tax by Province (Canada)

GST - a value-added tax, which the government imposes on most goods and services provided in or imported into Canada. PST - a retail sales tax, which three of the provinces impose on the price of most goods and some

services. QST - a value-added tax, similar to the federal GST, which Quebec imposes. HST - Five provinces have combined their retail sales tax with the federal GST into one harmonized tax.

Province	PST (%)	QST (%)	GST(%)	HST(%)
Alberta	0	0	5	0
British Columbia	7	0	5	0
Manitoba	7	0	5	0
New Brunswick	0	0	0	13
Newfoundland	0	0	0	13
Northwest Territories	0	0	5	0
Nova Scotia	0	0	0	15
Ontario	0	0	0	13
Prince Edward Island	0	0	5	14
Quebec	0	9.975	5	0
Saskatchewan	5	0	5	0
Yukon	0	0	5	0

R012909-85　Unemployment Taxes and Social Security Taxes

State unemployment tax rates vary not only from state to state, but also with the experience rating of the contractor. The federal unemployment tax rate is 6.0% of the first $7,000 of wages. This is reduced by a credit of up to 5.4% for timely payment to the state. The minimum federal unemployment tax is 0.6% after all credits.

Social security (FICA) for 2014 is estimated at time of publication to be 7.65% of wages up to $115,800.

R012909-90　Overtime

One way to improve the completion date of a project or eliminate negative float from a schedule is to compress activity duration times. This can be achieved by increasing the crew size or working overtime with the proposed crew.

To determine the costs of working overtime to compress activity duration times, consider the following examples. Below is an overtime efficiency and

cost chart based on a five, six, or seven day week with an eight through twelve hour day. Payroll percentage increases for time and one half and double time are shown for the various working days.

Days per Week	Hours per Day	Production Efficiency					Payroll Cost Factors	
		1st Week	2nd Week	3rd Week	4th Week	Average 4 Weeks	@ 1-1/2 Times	@ 2 Times
5	8	100%	100%	100%	100%	100 %	100 %	100 %
	9	100	100	95	90	96.25	105.6	111.1
	10	100	95	90	85	91.25	110.0	120.0
	11	95	90	75	65	81.25	113.6	127.3
	12	90	85	70	60	76.25	116.7	133.3
6	8	100	100	95	90	96.25	108.3	116.7
	9	100	95	90	85	92.50	113.0	125.9
	10	95	90	85	80	87.50	116.7	133.3
	11	95	85	70	65	78.75	119.7	139.4
	12	90	80	65	60	73.75	122.2	144.4
7	8	100	95	85	75	88.75	114.3	128.6
	9	95	90	80	70	83.75	118.3	136.5
	10	90	85	75	65	78.75	121.4	142.9
	11	85	80	65	60	72.50	124.0	148.1
	12	85	75	60	55	68.75	126.2	152.4

R013113-40　Builder's Risk Insurance

Builder's Risk Insurance is insurance on a building during construction. Premiums are paid by the owner or the contractor. Blasting, collapse and underground insurance would raise total insurance costs above those listed. Floater policy for materials delivered to the job runs $.75 to $1.25 per $100 value. Contractor equipment insurance runs $.50 to $1.50 per $100 value. Insurance for miscellaneous tools to $1,500 value runs from $3.00 to $7.50 per $100 value.

Tabulated below are New England Builder's Risk insurance rates in dollars per $100 value for $1,000 deductible. For $25,000 deductible, rates can be reduced 13% to 34%. On contracts over $1,000,000, rates may be lower than those tabulated. Policies are written annually for the total completed value in place. For "all risk" insurance (excluding flood, earthquake and certain other perils) add $.025 to total rates below.

Coverage	Frame Construction (Class 1)			Brick Construction (Class 4)			Fire Resistive (Class 6)		
	Range		Average	Range		Average	Range		Average
Fire Insurance	$.350 to	$ .850	$.600	$.158 to	$.189	$.174	$.052 to	$.080	$.070
Extended Coverage	.115 to	.200	.158	.080 to	.105	.101	.081 to	.105	.100
Vandalism	.012 to	.016	.014	.008 to	.011	.011	.008 to	.011	.010
Total Annual Rate	$.477 to	$1.066	$.772	$.246 to	$.305	$.286	$.141 to	$.196	$.180

R013113-50   General Contractor's Overhead

There are two distinct types of overhead on a construction project: Project overhead and main office overhead. Project overhead includes those costs at a construction site not directly associated with the installation of construction materials. Examples of project overhead costs include the following:

1. Superintendent
2. Construction office and storage trailers
3. Temporary sanitary facilities
4. Temporary utilities
5. Security fencing
6. Photographs
7. Cleanup
8. Performance and payment bonds

The above project overhead items are also referred to as general requirements and therefore are estimated in Division 1. Division 1 is the first division listed in the CSI MasterFormat but it is usually the last division estimated. The sum of the costs in Divisions 1 through 49 is referred to as the sum of the direct costs.

All construction projects also include indirect costs. The primary components of indirect costs are the contractor's main office overhead and profit. The amount of the main office overhead expense varies depending on the following:

1. Owner's compensation
2. Project managers' and estimators' wages
3. Clerical support wages
4. Office rent and utilities
5. Corporate legal and accounting costs
6. Advertising
7. Automobile expenses
8. Association dues
9. Travel and entertainment expenses

These costs are usually calculated as a percentage of annual sales volume. This percentage can range from 35% for a small contractor doing less than $500,000 to 5% for a large contractor with sales in excess of $100 million.

R013113-60   Workers' Compensation Insurance Rates by Trade

The table below tabulates the national averages for workers' compensation insurance rates by trade and type of building. The average "Insurance Rate" is multiplied by the "% of Building Cost" for each trade. This produces the "Workers' Compensation" cost by % of total labor cost, to be added for each trade by building type to determine the weighted average workers' compensation rate for the building types analyzed.

Trade	Insurance Rate (% Labor Cost)			% of Building Cost			Workers' Compensation		
	Range		Average	Office Bldgs.	Schools & Apts.	Mfg.	Office Bldgs.	Schools & Apts.	Mfg.
Excavation, Grading, etc.	3.9 % to	18.7%	9.7%	4.8%	4.9%	4.5%	0.47%	0.48%	0.44%
Piles & Foundations	5.8 to	37	14.7	7.1	5.2	8.7	1.04	0.76	1.28
Concrete	4.6 to	38.6	13.1	5.0	14.8	3.7	0.66	1.94	0.48
Masonry	5.1 to	35	13.5	6.9	7.5	1.9	0.93	1.01	0.26
Structural Steel	5.8 to	108.2	34.6	10.7	3.9	17.6	3.70	1.35	6.09
Miscellaneous & Ornamental Metals	5.3 to	31.2	12.0	2.8	4.0	3.6	0.34	0.48	0.43
Carpentry & Millwork	5.5 to	34.9	15.4	3.7	4.0	0.5	0.57	0.62	0.08
Metal or Composition Siding	5.8 to	68.3	18.2	2.3	0.3	4.3	0.42	0.05	0.78
Roofing	5.8 to	93.3	31.3	2.3	2.6	3.1	0.72	0.81	0.97
Doors & Hardware	4.2 to	34.8	11.1	0.9	1.4	0.4	0.10	0.16	0.04
Sash & Glazing	5.3 to	36.8	13.1	3.5	4.0	1.0	0.46	0.52	0.13
Lath & Plaster	3.7 to	35	11.6	3.3	6.9	0.8	0.38	0.80	0.09
Tile, Marble & Floors	3.3 to	27.1	8.5	2.6	3.0	0.5	0.22	0.26	0.04
Acoustical Ceilings	3.4 to	36.7	8.1	2.4	0.2	0.3	0.19	0.02	0.02
Painting	5 to	33	11.5	1.5	1.6	1.6	0.17	0.18	0.18
Interior Partitions	5.5 to	34.9	15.4	3.9	4.3	4.4	0.60	0.66	0.68
Miscellaneous Items	2.5 to	150	14.4	5.2	3.7	9.7	0.75	0.53	1.40
Elevators	1.6 to	14.4	5.7	2.1	1.1	2.2	0.12	0.06	0.13
Sprinklers	3.2 to	18.3	7.2	0.5	—	2.0	0.04	—	0.14
Plumbing	2.6 to	17.5	7.1	4.9	7.2	5.2	0.35	0.51	0.37
Heat., Vent., Air Conditioning	4 to	18.8	8.8	13.5	11.0	12.9	1.19	0.97	1.14
Electrical	2.8 to	13.2	5.8	10.1	8.4	11.1	0.59	0.49	0.64
Total	1.6 % to	150%	—	100.0%	100.0%	100.0%	14.01%	12.66%	15.81%
				Overall Weighted Average	14.16%				

Workers' Compensation Insurance Rates by States

The table below lists the weighted average workers' compensation base rate for each state with a factor comparing this with the national average of 14%.

State	Weighted Average	Factor	State	Weighted Average	Factor	State	Weighted Average	Factor
Alabama	20.3%	145	Kentucky	14.9%	106	North Dakota	9.6%	69
Alaska	14.3	102	Louisiana	19.3	138	Ohio	8.6	61
Arizona	13.1	94	Maine	12.7	91	Oklahoma	12.9	92
Arkansas	8.6	61	Maryland	14.6	104	Oregon	12.3	88
California	26.8	191	Massachusetts	12.0	86	Pennsylvania	18.1	129
Colorado	7.5	54	Michigan	16.9	121	Rhode Island	12.6	90
Connecticut	24.6	176	Minnesota	23.5	168	South Carolina	19.6	140
Delaware	12.4	89	Mississippi	13.8	99	South Dakota	15.5	111
District of Columbia	11.5	82	Missouri	15.4	110	Tennessee	12.5	89
Florida	11.1	79	Montana	8.9	64	Texas	9.5	68
Georgia	29.2	209	Nebraska	18.3	131	Utah	9.1	65
Hawaii	8.4	60	Nevada	9.0	64	Vermont	13.6	97
Idaho	10.7	76	New Hampshire	20.8	149	Virginia	9.9	71
Illinois	28.1	201	New Jersey	15.0	107	Washington	10.9	78
Indiana	5.7	41	New Mexico	15.0	107	West Virginia	8.3	59
Iowa	15.1	108	New York	17.9	128	Wisconsin	13.6	97
Kansas	8.7	62	North Carolina	17.4	124	Wyoming	6.3	45
			Weighted Average for U.S. is	14.2% of payroll = 100%				

Rates in the following table are the base or manual costs per $100 of payroll for workers' compensation in each state. Rates are usually applied to straight time wages only and not to premium time wages and bonuses.

The weighted average skilled worker rate for 35 trades is 14%. For bidding purposes, apply the full value of workers' compensation directly to total labor costs, or if labor is 38%, materials 42% and overhead and profit 20% of total cost, carry 38/80 × 14% =6.6% of cost (before overhead and profit) into overhead. Rates vary not only from state to state but also with the experience rating of the contractor.

Rates are the most current available at the time of publication.

R013113-60 Workers' Compensation Insurance Rates by Trade and State (cont.)

State	Carpentry — 3 stories or less 5651	Carpentry — interior cab. work 5437	Carpentry — general 5403	Concrete Work — NOC 5213	Concrete Work — flat (flr, sdwk.) 5221	Electrical Wiring — inside 5190	Excavation — earth NOC 6217	Excavation — rock 6217	Glaziers 5462	Insulation Work 5479	Lathing 5443	Masonry 5022	Painting & Decorating 5474	Pile Driving 6003	Plastering 5480	Plumbing 5183
AL	34.03	17.01	24.59	13.79	10.37	8.87	16.42	16.42	19.75	19.84	9.24	23.79	15.91	18.57	15.88	8.73
AK	17.14	9.44	11.58	10.41	9.44	5.63	8.23	8.23	32.03	14.36	11.73	10.86	10.79	21.68	11.31	6.20
AZ	21.40	9.27	20.28	10.31	6.93	6.08	7.07	7.07	10.80	14.99	5.47	13.01	12.78	10.80	8.64	6.84
AR	12.80	4.77	9.18	6.63	5.16	4.03	6.82	6.82	8.14	7.67	3.99	8.30	7.08	8.10	13.00	4.55
CA	34.87	34.83	34.86	16.55	16.55	13.24	18.65	18.65	24.16	16.48	17.83	25.13	23.59	23.18	35.04	17.48
CO	8.95	4.93	6.07	6.68	4.75	2.92	5.49	5.49	6.00	7.27	3.95	8.29	6.56	6.51	5.69	4.16
CT	30.95	18.48	31.07	26.49	14.43	7.89	16.02	16.02	20.24	20.60	9.92	31.53	18.60	20.17	18.75	11.28
DE	12.46	12.46	9.93	10.07	8.80	4.53	7.70	7.70	10.90	9.93	6.60	11.53	13.39	15.03	10.90	6.60
DC	8.92	7.25	6.94	12.65	7.38	4.69	8.02	8.02	9.61	6.63	9.83	9.32	5.82	17.36	8.00	7.80
FL	11.32	8.62	11.92	12.51	6.20	6.17	7.30	7.30	10.56	10.45	5.53	10.92	10.23	17.84	9.68	5.63
GA	68.34	20.32	28.40	21.36	15.09	12.05	17.32	17.32	19.39	18.80	11.94	35.11	33.04	26.44	19.65	12.76
HI	8.51	6.92	12.69	7.05	4.02	4.47	5.40	5.40	8.22	5.20	6.26	7.01	6.19	7.32	8.33	5.31
ID	14.63	6.84	12.43	11.48	5.28	4.43	7.25	7.25	8.81	9.06	4.46	10.55	9.38	7.65	8.43	5.10
IL	33.34	21.75	28.57	38.59	16.83	10.03	14.53	14.53	25.55	21.41	14.32	28.94	18.57	24.53	32.84	12.51
IN	8.36	4.21	6.98	4.56	4.15	2.83	3.92	3.92	5.29	6.15	3.85	5.14	4.97	7.23	3.71	2.56
IA	15.46	11.35	19.21	14.85	11.16	5.20	11.01	11.01	15.76	9.89	6.48	14.30	10.13	11.38	21.73	7.80
KS	14.92	7.39	9.74	8.07	5.70	4.46	5.55	5.55	7.54	8.17	4.09	8.38	7.75	7.56	6.20	5.23
KY	18.75	13.00	20.00	8.50	6.50	6.00	14.00	14.00	22.12	11.15	7.60	7.50	9.00	16.03	12.93	5.00
LA	31.05	16.86	25.02	13.35	12.37	7.85	14.11	14.11	19.19	15.06	7.99	17.31	17.64	18.85	14.84	7.10
ME	12.24	8.87	16.18	16.43	8.70	4.74	12.34	12.34	14.82	12.79	5.09	10.88	12.44	9.00	9.99	5.90
MD	11.40	15.89	15.00	19.22	6.00	5.23	8.11	8.11	13.77	12.42	6.15	10.11	6.40	15.19	9.17	6.65
MA	8.68	5.23	9.61	18.85	6.24	2.84	4.35	4.35	9.58	7.78	5.27	10.55	5.09	12.92	4.68	3.50
MI	18.41	10.45	18.41	18.27	12.58	5.82	13.28	13.28	11.67	10.92	10.45	18.82	13.49	37.00	10.88	7.18
MN	20.27	20.82	34.13	11.58	13.15	5.83	11.00	11.00	36.85	18.37	11.55	15.59	17.55	27.58	11.58	8.09
MS	22.76	9.99	13.34	9.60	6.67	6.18	9.49	9.49	14.94	10.15	5.94	13.12	11.01	13.14	10.43	8.17
MO	19.66	11.10	12.81	12.38	10.34	6.15	9.96	9.96	9.47	11.55	7.44	15.13	11.42	13.38	12.15	9.00
MT	9.67	6.59	10.80	6.22	5.67	3.71	7.41	7.41	7.50	8.67	4.15	7.62	8.22	7.95	6.80	4.79
NE	21.80	13.15	21.48	18.23	11.20	8.03	17.38	17.38	12.05	21.53	8.35	18.73	13.00	13.93	13.68	8.73
NV	10.41	6.82	12.39	10.83	5.26	4.33	8.75	8.75	6.55	7.55	3.41	6.63	7.10	8.43	6.21	6.76
NH	24.24	11.06	19.63	28.40	16.47	5.63	12.61	12.61	14.12	12.63	7.30	16.19	20.93	18.53	11.20	9.49
NJ	18.69	13.18	18.69	15.51	11.69	5.13	10.83	10.83	11.21	12.63	15.13	18.71	11.28	15.79	15.13	6.96
NM	25.01	9.56	18.92	11.73	9.43	6.05	7.22	7.22	10.95	10.56	6.11	15.74	12.91	12.66	14.20	6.70
NY	12.54	10.74	19.18	23.71	17.15	7.85	12.16	12.16	18.72	10.88	12.86	21.24	14.99	20.13	10.02	10.83
NC	30.12	9.54	15.76	16.11	9.31	10.56	15.41	15.41	14.28	13.03	8.11	13.05	14.08	16.61	14.12	10.73
ND	9.54	4.31	9.54	5.43	5.43	3.75	4.13	4.13	10.50	10.50	9.93	6.52	6.00	10.59	9.93	6.11
OH	7.73	4.31	5.52	4.67	4.92	3.43	4.79	4.79	8.83	11.38	36.65	6.93	8.15	6.99	7.80	3.52
OK	18.82	10.32	12.49	13.50	7.66	6.82	9.48	9.48	11.79	11.31	5.13	14.63	10.66	11.24	14.01	6.82
OR	22.30	8.52	11.92	10.51	10.42	5.35	9.08	9.08	12.64	10.88	6.09	14.37	11.81	11.44	10.81	6.35
PA	19.22	19.22	15.60	19.54	13.58	7.39	11.11	11.11	14.34	15.60	14.34	16.04	17.13	19.44	14.34	9.19
RI	9.84	11.55	13.75	12.56	9.68	4.58	8.16	8.16	10.61	12.48	5.17	10.07	8.82	19.30	8.69	6.75
SC	29.32	13.80	17.10	12.84	8.21	9.82	12.01	12.01	12.92	14.13	9.57	15.39	16.10	15.67	14.47	9.40
SD	20.86	17.96	20.35	16.10	10.49	5.08	9.75	9.75	10.84	14.71	6.95	14.65	11.22	12.63	10.35	10.02
TN	22.71	10.14	11.54	10.75	7.08	5.49	10.05	10.05	11.07	10.34	6.60	11.69	8.27	12.08	11.04	5.17
TX	10.12	7.33	10.12	8.08	5.99	5.98	7.60	7.60	8.25	9.27	4.15	8.69	7.28	10.26	7.28	5.27
UT	14.72	6.59	7.50	7.34	6.57	3.56	6.79	6.79	7.45	7.15	6.41	8.96	7.73	6.72	5.56	4.17
VT	15.69	10.21	13.77	13.60	11.57	4.37	11.83	11.83	12.40	15.34	5.79	11.92	8.78	10.43	9.35	7.38
VA	12.38	8.14	8.89	10.51	5.76	4.50	7.40	7.40	8.02	7.27	5.45	7.98	8.93	8.31	6.69	5.07
WA	9.18	9.18	9.18	8.55	8.55	3.18	6.68	6.68	13.99	6.79	9.18	13.26	12.26	21.68	10.43	4.33
WV	14.85	5.54	8.34	7.68	4.51	3.53	6.46	6.46	6.74	6.29	4.33	8.39	6.03	8.09	6.54	3.52
WI	12.52	13.14	17.06	11.98	7.59	5.08	7.84	7.84	13.88	11.22	4.00	13.69	12.43	28.24	10.16	5.95
WY	5.82	5.82	5.82	5.82	5.82	5.82	5.82	5.82	5.82	5.82	5.82	5.82	5.82	5.82	5.82	5.82
AVG.	18.19	11.07	15.38	13.15	8.92	5.83	9.69	9.69	13.15	11.67	8.12	13.49	11.54	14.65	11.55	7.08

R013113-60 Workers' Compensation Insurance Rates by Trade and State (cont.)

State	Roofing	Sheet Metal Work (HVAC)	Steel Erection — door & sash	Steel Erection — inter., ornam.	Steel Erection — structure	Steel Erection — NOC	Tile Work — (interior ceramic)	Waterproofing	Wrecking	Boiler- making	Mill- wrights	Structural Painters	Elevator Constructor	Truck Drivers	Sprinkler Installer
	5551	5538	5102	5102	5040	5057	5348	9014	5701	3726	3724	5037	5160	7219	5188
AL	54.71	10.34	12.52	12.52	40.75	21.56	12.35	9.27	40.75	11.93	12.83	55.64	10.09	23.31	8.40
AK	25.49	8.05	7.82	7.82	29.28	26.40	5.33	5.77	29.28	8.56	8.24	31.37	4.06	11.75	5.13
AZ	29.22	10.31	13.61	13.61	26.91	12.89	4.50	6.21	26.91	9.07	7.41	59.76	5.09	8.01	5.48
AR	16.03	4.74	8.34	8.34	15.40	11.28	6.69	2.58	15.40	4.94	5.76	25.69	2.73	10.96	5.72
CA	71.13	18.78	18.41	18.41	35.28	19.69	14.59	23.59	19.69	8.98	13.58	35.28	5.07	27.13	18.32
CO	15.60	4.16	5.36	5.36	23.64	10.59	4.10	3.23	18.40	4.37	3.87	22.88	2.78	7.22	3.85
CT	58.74	12.93	21.24	21.24	65.13	29.83	15.63	6.43	65.13	19.44	14.12	66.07	11.07	22.67	10.49
DE	26.71	5.55	13.16	13.16	23.27	13.16	8.00	11.53	23.27	6.17	6.19	23.27	6.19	11.13	6.60
DC	17.05	6.45	23.33	23.33	26.02	7.91	6.08	4.38	26.02	15.31	7.62	35.16	9.65	14.34	4.00
FL	18.17	7.69	9.91	9.91	23.04	12.59	5.63	5.26	23.04	9.07	6.05	39.44	3.54	9.24	6.41
GA	93.33	16.61	22.17	22.17	60.89	25.88	13.50	11.64	60.89	19.54	14.79	72.60	7.53	19.39	15.17
HI	17.24	4.93	7.60	7.60	21.77	6.88	6.52	5.40	21.77	4.82	7.71	24.83	2.28	10.54	3.49
ID	24.73	7.46	12.31	12.31	25.42	10.44	6.15	4.84	25.42	5.30	5.61	29.98	3.26	9.46	4.65
IL	47.55	15.21	31.16	31.16	84.68	19.36	27.09	7.88	84.68	23.23	15.10	101.23	14.42	19.73	16.65
IN	11.57	4.32	5.32	5.32	11.59	4.80	3.26	2.77	11.59	3.85	3.70	13.20	1.58	5.75	3.20
IA	28.01	8.53	9.15	9.15	52.79	14.30	9.38	6.15	25.61	14.19	8.59	87.69	6.14	13.60	7.75
KS	16.26	5.64	6.71	6.71	18.97	11.93	5.23	4.63	11.96	6.01	7.60	26.96	3.22	7.88	5.30
KY	34.00	12.00	11.72	11.72	36.00	11.02	16.73	3.50	36.00	12.15	8.35	150.00	9.56	10.00	10.32
LA	43.53	14.84	20.14	20.14	44.26	12.79	9.99	7.91	44.20	15.48	10.94	55.57	6.11	23.25	11.22
ME	22.72	6.77	8.92	8.92	37.18	12.21	5.76	4.64	37.18	8.08	6.58	54.29	4.39	8.15	6.70
MD	29.05	11.90	11.92	11.92	59.02	17.92	6.49	5.84	32.91	14.58	9.95	43.60	4.46	16.02	11.88
MA	30.99	5.72	6.89	6.89	54.08	33.00	5.81	2.48	23.75	15.42	5.98	23.51	4.21	8.28	4.13
MI	41.18	9.83	12.28	12.28	37.00	11.27	11.73	6.46	37.00	9.41	8.67	11.27	5.13	9.09	8.21
MN	67.71	17.30	9.82	9.82	108.16	7.70	14.69	7.95	7.70	10.92	12.24	36.60	6.41	11.39	7.62
MS	31.61	7.46	15.08	15.08	38.45	7.53	8.45	5.85	38.45	7.77	9.97	38.74	4.78	52.03	7.53
MO	36.73	8.64	12.33	12.33	47.55	24.29	10.41	5.92	47.55	14.45	9.66	60.85	6.20	12.44	8.43
MT	29.42	4.80	6.98	6.98	18.72	5.89	5.47	4.64	5.89	5.03	5.91	28.63	3.01	6.06	3.67
NE	35.08	12.88	14.80	14.80	59.03	14.13	8.68	6.73	67.83	16.33	11.35	66.18	7.13	15.98	10.30
NV	16.59	7.34	7.06	7.06	19.32	12.20	4.42	4.23	12.20	4.79	6.15	28.67	4.30	10.30	5.47
NH	36.82	8.93	14.28	14.28	95.03	28.57	12.35	7.17	95.03	8.91	9.04	54.18	6.29	36.82	10.68
NJ	40.10	7.22	12.87	12.87	17.25	13.36	9.55	6.63	16.86	4.18	9.20	23.37	7.00	15.75	5.89
NM	35.07	9.63	11.79	11.79	45.52	16.76	7.73	7.53	45.52	8.00	8.00	42.38	8.41	14.60	7.75
NY	35.96	13.56	19.25	19.25	35.26	19.08	10.35	7.49	18.10	21.41	10.22	42.40	12.20	14.37	6.45
NC	35.39	11.87	14.43	14.43	45.05	22.54	7.62	6.10	45.05	18.41	9.71	78.32	8.90	17.84	10.60
ND	17.63	6.11	10.59	10.59	10.90	10.59	10.50	17.63	8.18	3.33	10.50	10.59	3.33	8.79	6.11
OH	14.89	4.00	6.58	6.58	11.33	5.84	4.85	4.58	5.84	2.76	4.52	24.79	2.13	10.02	3.68
OK	22.44	9.62	10.98	10.98	29.21	15.15	6.69	6.41	32.94	9.25	6.21	52.38	3.88	11.26	6.40
OR	27.22	6.85	7.08	7.08	22.04	13.44	9.10	6.73	22.04	9.10	9.28	11.81	3.70	14.41	6.02
PA	38.35	8.92	19.10	19.10	30.01	19.10	11.69	16.04	30.01	7.95	8.85	30.01	8.85	16.69	9.19
RI	23.59	8.48	7.37	7.37	35.11	24.46	5.33	4.98	22.15	11.33	6.32	39.60	4.85	10.13	5.85
SC	46.45	10.88	15.24	15.24	31.33	24.73	7.98	5.42	31.33	11.24	10.59	66.07	8.64	15.07	10.12
SD	34.39	7.29	14.74	14.74	33.52	23.05	6.19	5.30	33.52	22.67	11.38	66.74	6.69	10.62	8.14
TN	29.78	8.13	13.42	13.42	16.07	18.88	7.46	4.09	16.07	8.10	7.46	40.34	4.92	10.23	6.10
TX	16.22	11.69	7.60	7.60	22.99	9.48	4.68	5.57	7.13	4.30	5.95	7.34	3.34	12.75	5.27
UT	23.93	5.78	5.71	5.71	24.61	11.81	5.75	4.31	20.14	5.91	4.77	21.36	3.71	7.05	4.16
VT	27.01	7.30	12.01	12.01	34.39	18.99	7.01	7.90	34.39	8.67	7.28	48.29	6.07	13.40	6.27
VA	25.97	5.47	7.46	7.46	29.39	10.80	5.76	3.30	29.39	4.92	7.38	27.90	5.39	9.03	5.30
WA	21.14	4.47	8.09	8.09	8.09	8.09	7.86	21.14	8.09	2.77	6.15	12.26	2.46	8.52	4.70
WV	19.62	5.62	7.13	7.13	19.39	9.55	4.86	3.11	20.70	5.62	6.64	35.86	4.23	8.93	3.76
WI	28.04	9.43	12.08	12.08	20.72	16.12	13.78	5.07	20.72	9.68	8.34	24.08	4.11	10.15	4.64
WY	5.82	5.82	5.82	5.82	5.82	5.82	5.82	5.82	5.82	5.82	5.82	5.82	5.82	5.82	5.82
AVG.	31.29	8.79	11.95	11.95	34.64	15.21	8.54	6.86	29.21	9.87	8.39	41.47	5.67	13.67	7.24

R013113-60 Workers' Compensation (cont.) (Canada in Canadian dollars)

Province		Alberta	British Columbia	Manitoba	Ontario	New Brunswick	Newfndld. & Labrador	Northwest Territories	Nova Scotia	Prince Edward Island	Quebec	Saskat- chewan	Yukon
Carpentry—3 stories or less	Rate	5.71	3.27	3.98	4.55	3.17	7.81	4.99	6.85	6.42	12.03	3.43	6.63
	Code	42143	721028	40102	723	238130	4226	4-41	4226	403	80110	B1226	202
Carpentry—interior cab. work	Rate	1.89	2.80	3.98	4.55	3.17	4.07	4.99	5.87	3.94	12.03	3.43	6.63
	Code	42133	721021	40102	723	238350	4279	4-41	4274	402	80110	B1127	202
CARPENTRY—general	Rate	5.71	3.27	3.98	4.55	3.17	4.07	4.99	6.85	6.42	12.03	3.43	6.63
	Code	42143	721028	40102	723	238130	4299	4-41	4226	403	80110	B1202	202
CONCRETE WORK—NOC	Rate	3.88	3.57	6.50	18.31	3.17	7.81	4.99	5.75	3.94	1.92	2.99	6.63
	Code	42104	721010	40110	748	238110	4224	4-41	4224	402	80100	B1314	203
CONCRETE WORK—flat (flr. sidewalk)	Rate	3.88	3.57	6.50	18.31	3.17	7.81	4.99	5.75	3.94	11.92	2.99	6.63
	Code	42104	721010	40110	748	238110	4224	4-41	4224	402	80100	B1314	203
ELECTRICAL Wiring—inside	Rate	1.56	1.18	1.65	3.69	1.28	2.29	2.44	1.97	3.94	5.04	1.83	3.89
	Code	42124	721019	40203	704	238210	4261	4-46	4261	402	80170	B1105	206
EXCAVATION—earth NOC	Rate	1.86	3.79	4.00	5.29	3.17	2.93	3.88	3.89	3.94	6.40	2.19	3.89
	Code	40604	721031	40706	711	238190	4214	4-43	4214	402	80030	R1106	207
EXCAVATION—rock	Rate	1.86	3.79	4.00	5.29	3.17	2.93	3.88	3.89	3.94	6.40	2.19	3.89
	Code	40604	721031	40706	711	238190	4214	4-43	4214	402	80030	R1106	207
GLAZIERS	Rate	2.23	3.79	3.98	10.25	3.17	5.10	4.99	5.22	3.94	15.02	2.99	3.89
	Code	42121	715020	40109	751	238150	4233	4-41	4233	402	80150	B1304	212
INSULATION WORK	Rate	1.62	9.87	3.98	10.25	3.17	5.10	4.99	5.22	3.94	12.03	3.43	6.63
	Code	42184	721029	40102	751	238310	4234	4-41	4234	402	80110	B1207	202
LATHING	Rate	4.20	7.84	3.98	4.55	3.17	4.07	4.99	5.87	3.94	12.03	2.99	6.63
	Code	42135	721042	40102	723	238390	4279	4-41	4271	402	80110	B1316	202
MASONRY	Rate	3.88	3.30	3.98	12.70	3.17	5.10	4.99	5.22	3.94	11.92	2.99	6.63
	Code	42102	721037	40102	741	238140	4231	4-41	4231	402	80100	B1318	202
PAINTING & DECORATING	Rate	3.16	3.62	3.11	7.51	3.17	4.07	4.99	5.87	3.94	12.03	3.43	6.63
	Code	42111	721041	40105	719	238320	4275	4-41	4275	402	80110	B1201	202
PILE DRIVING	Rate	3.88	3.42	4.00	7.03	3.17	2.93	3.88	5.75	6.42	6.40	2.99	6.63
	Code	42159	722004	40706	732	238190	4129	4-43	4221	403	80030	B1310	202
PLASTERING	Rate	4.20	7.84	4.45	7.51	3.17	4.07	4.99	5.87	3.94	12.03	3.43	6.63
	Code	42135	721042	40108	719	238390	4271	4-41	4271	402	80110	B1221	202
PLUMBING	Rate	1.56	2.55	2.78	4.16	1.28	2.47	2.44	2.81	3.94	5.89	1.83	3.89
	Code	42122	721043	40204	707	238220	4241	4-46	4241	401	80160	B1101	214
ROOFING	Rate	4.71	5.42	7.22	14.80	3.17	7.81	4.99	9.43	3.94	18.52	2.99	6.63
	Code	42118	721036	40403	728	238160	4236	4-41	4236	402	80130	B1320	202
SHEET METAL WORK (HVAC)	Rate	1.56	2.55	7.22	4.16	1.28	2.47	2.44	2.81	3.94	5.89	1.83	3.89
	Code	42117	721043	40402	707	238220	4244	4-46	4244	401	80160	B1107	208
STEEL ERECTION—door & sash	Rate	1.62	10.62	6.82	18.31	3.17	7.81	4.99	5.22	3.94	16.70	2.99	6.63
	Code	42106	722005	40502	748	238120	4227	-37	4227	402	80080	B1322	202
STEEL ERECTION—inter., ornam.	Rate	1.62	10.62	6.82	18.31	3.17	7.81	4.99	5.22	3.94	16.70	2.99	6.63
	Code	42106	722005	40502	748	238120	4227	4-41	4227	402	80080	B1322	202
STEEL ERECTION—structure	Rate	1.62	10.62	6.82	18.31	3.17	7.81	4.99	5.22	3.94	16.70	2.99	6.63
	Code	42106	722005	40502	748	238120	4227	4-41	4227	402	80080	B1322	202
STEEL ERECTION—NOC	Rate	1.62	10.62	6.82	18.31	3.17	7.81	4.99	5.22	3.94	16.70	2.99	6.63
	Code	42106	722005	40502	748	238120	4227	4-41	4227	402	80080	B1322	202
TILE WORK—inter. (ceramic)	Rate	3.02	3.65	1.86	7.51	3.17	4.07	4.99	5.87	3.94	12.03	2.99	6.63
	Code	42113	721054	40103	719	238340	4276	4-41	4276	402	80110	B1301	202
WATERPROOFING	Rate	3.16	3.62	3.98	4.55	3.17	4.07	4.99	5.22	3.94	18.52	3.13	6.63
	Code	42139	721016	40102	723	238190	4299	4-41	4239	402	80130	B1217	202
WRECKING	Rate	1.86	5.64	6.90	18.31	3.17	2.93	3.88	3.89	3.94	12.03	2.19	6.63
	Code	40604	721005	40106	748	238190	4211	4-43	4211	402	80110	R1125	202

R013113-80   Performance Bond

This table shows the cost of a Performance Bond for a construction job scheduled to be completed in 12 months. Add 1% of the premium cost per month for jobs requiring more than 12 months to complete. The rates are "standard" rates offered to contractors that the bonding company considers financially sound and capable of doing the work. Preferred rates are offered by some bonding companies based upon financial strength of the contractor. Actual rates vary from contractor to contractor and from bonding company to bonding company. Contractors should prequalify through a bonding agency before submitting a bid on a contract that requires a bond.

Contract Amount	Building Construction Class B Projects		Highways & Bridges			
			Class A New Construction		Class A-1 Highway Resurfacing	
First $ 100,000 bid	$25.00 per M		$15.00 per M		$9.40 per M	
Next 400,000 bid	$ 2,500 plus	$15.00 per M	$ 1,500 plus	$10.00 per M	$ 940 plus	$7.20 per M
Next 2,000,000 bid	8,500 plus	10.00 per M	5,500 plus	7.00 per M	3,820 plus	5.00 per M
Next 2,500,000 bid	28,500 plus	7.50 per M	19,500 plus	5.50 per M	15,820 plus	4.50 per M
Next 2,500,000 bid	47,250 plus	7.00 per M	33,250 plus	5.00 per M	28,320 plus	4.50 per M
Over 7,500,000 bid	64,750 plus	6.00 per M	45,750 plus	4.50 per M	39,570 plus	4.00 per M

R015113-65   Temporary Power Equipment

Cost data for the temporary equipment was developed utilizing the following information.

1) Re-usable material-services, transformers, equipment and cords are based on new purchase and prorated to three projects.
2) PVC feeder includes trench and backfill.
3) Connections include disconnects and fuses.
4) Labor units include an allowance for removal.
5) No utility company charges or fees are included.
6) Concrete pads or vaults are not included.
7) Utility company conduits not included.

R015423-10   Steel Tubular Scaffolding

On new construction, tubular scaffolding is efficient up to 60' high or five stories. Above this it is usually better to use a hung scaffolding if construction permits. Swing scaffolding operations may interfere with tenants. In this case, the tubular is more practical at all heights.

In repairing or cleaning the front of an existing building the cost of tubular scaffolding per S.F. of building front increases as the height increases above the first tier. The first tier cost is relatively high due to leveling and alignment.

The minimum efficient crew for erecting and dismantling is three workers. They can set up and remove 18 frame sections per day up to 5 stories high. For 6 to 12 stories high, a crew of four is most efficient. Use two or more on top and two on the bottom for handing up or hoisting. They can

also set up and remove 18 frame sections per day. At 7' horizontal spacing, this will run about 800 S.F. per day of erecting and dismantling. Time for placing and removing planks must be added to the above. A crew of three can place and remove 72 planks per day up to 5 stories. For over 5 stories, a crew of four can place and remove 80 planks per day.

The table below shows the number of pieces required to erect tubular steel scaffolding for 1000 S.F. of building frontage. This area is made up of a scaffolding system that is 12 frames (11 bays) long by 2 frames high.

For jobs under twenty-five frames, add 50% to rental cost. Rental rates will be lower for jobs over three months duration. Large quantities for long periods can reduce rental rates by 20%.

Description of Component	Number of Pieces for 1000 S.F. of Building Front	Unit
5' Wide Standard Frame, 6'-4" High	24	Ea.
Leveling Jack & Plate	24	
Cross Brace	44	
Side Arm Bracket, 21"	12	
Guardrail Post	12	
Guardrail, 7' section	22	
Stairway Section	2	
Stairway Starter Bar	1	
Stairway Inside Handrail	2	
Stairway Outside Handrail	2	
Walk-Thru Frame Guardrail	2	

Scaffolding is often used as falsework over 15' high during construction of cast-in-place concrete beams and slabs. Two foot wide scaffolding is generally used for heavy beam construction. The span between frames depends upon the load to be carried with a maximum span of 5'.

Heavy duty shoring frames with a capacity of 10,000#/leg can be spaced up to 10' O.C. depending upon form support design and loading.

Scaffolding used as horizontal shoring requires less than half the material required with conventional shoring.

On new construction, erection is done by carpenters.

Rolling towers supporting horizontal shores can reduce labor and speed the job. For maintenance work, catwalks with spans up to 70' can be supported by the rolling towers.

R015433-10   Contractor Equipment

Rental Rates shown elsewhere in the book pertain to late model high quality machines in excellent working condition, rented from equipment dealers. Rental rates from contractors may be substantially lower than the rental rates from equipment dealers depending upon economic conditions; for older, less productive machines, reduce rates by a maximum of 15%. Any overtime must be added to the base rates. For shift work, rates are lower. Usual rule of thumb is 150% of one shift rate for two shifts; 200% for three shifts.

For periods of less than one week, operated equipment is usually more economical to rent than renting bare equipment and hiring an operator.

Costs to move equipment to a job site (mobilization) or from a job site (demobilization) are not included in rental rates, nor in any Equipment costs on any Unit Price line items or crew listings. These costs can be found elsewhere. If a piece of equipment is already at a job site, it is not appropriate to utilize mob/demob costs in an estimate again.

Rental rates vary throughout the country with larger cities generally having lower rates. Lease plans for new equipment are available for periods in excess of six months with a percentage of payments applying toward purchase.

Rental rates can also be treated as reimbursement costs for contractor-owned equipment. Owned equipment costs include depreciation, loan payments, interest, taxes, insurance, storage, and major repairs.

Monthly rental rates vary from 2% to 5% of the cost of the equipment depending on the anticipated life of the equipment and its wearing parts. Weekly rates are about 1/3 the monthly rates and daily rental rates about 1/3 the weekly rate.

The hourly operating costs for each piece of equipment include costs to the user such as fuel, oil, lubrication, normal expendables for the equipment, and a percentage of mechanic's wages chargeable to maintenance. The hourly operating costs listed do not include the operator's wages.

The daily cost for equipment used in the standard crews is figured by dividing the weekly rate by five, then adding eight times the hourly operating cost to give the total daily equipment cost, not including the operator. This figure is in the right hand column of the Equipment listings under Equipment Cost/Day.

Pile Driving rates shown for pile hammer and extractor do not include leads, crane, boiler or compressor. Vibratory pile driving requires an added field specialist during set-up and pile driving operation for the electric model. The hydraulic model requires a field specialist for set-up only. Up to 125 reuses of sheet piling are possible using vibratory drivers. For normal conditions, crane capacity for hammer type and size are as follows.

Crane Capacity	Hammer Type and Size		
	Air or Steam	Diesel	Vibratory
25 ton	to 8,750 ft.-lb.		70 H.P.
40 ton	15,000 ft.-lb.	to 32,000 ft.-lb.	170 H.P.
60 ton	25,000 ft.-lb.		300 H.P.
100 ton		112,000 ft.-lb.	

Cranes should be specified for the job by size, building and site characteristics, availability, performance characteristics, and duration of time required.

Backhoes & Shovels rent for about the same as equivalent size cranes but maintenance and operating expense is higher. Crane operators rate must be adjusted for high boom heights. Average adjustments: for 150' boom add 2% per hour; over 185', add 4% per hour; over 210', add 6% per hour; over 250', add 8% per hour and over 295', add 12% per hour.

Tower Cranes of the climbing or static type have jibs from 50' to 200' and capacities at maximum reach range from 4,000 to 14,000 pounds. Lifting capacities increase up to maximum load as the hook radius decreases.

Typical rental rates, based on purchase price are about 2% to 3% per month.

Erection and dismantling runs between 500 and 2000 labor hours. Climbing operation takes 10 labor hours per 20' climb. Crane dead time is about 5 hours per 40' climb. If crane is bolted to side of the building add cost of ties and extra mast sections. Climbing cranes have from 80' to 180' of mast while static cranes have 80' to 800' of mast.

Truck Cranes can be converted to tower cranes by using tower attachments. Mast heights over 400' have been used.

A single 100' high material **Hoist and Tower** can be erected and dismantled in about 400 labor hours; a double 100' high hoist and tower in about 600 labor hours. Erection times for additional heights are 3 and 4 labor hours per vertical foot respectively up to 150', and 4 to 5 labor hours per vertical foot over 150' high. A 40' high portable Buck hoist takes about 160 labor hours to erect and dismantle. Additional heights take 2 labor hours per vertical foot to 80' and 3 labor hours per vertical foot for the next 100'. Most material hoists do not meet local code requirements for carrying personnel.

A 150' high **Personnel Hoist** requires about 500 to 800 labor hours to erect and dismantle. Budget erection time at 5 labor hours per vertical foot for all trades. Local code requirements or labor scarcity requiring overtime can add up to 50% to any of the above erection costs.

Earthmoving Equipment: The selection of earthmoving equipment depends upon the type and quantity of material, moisture content, haul distance, haul road, time available, and equipment available. Short haul cut and fill operations may require dozers only, while another operation may require excavators, a fleet of trucks, and spreading and compaction equipment. Stockpiled material and granular material are easily excavated with front end loaders. Scrapers are most economically used with hauls between 300' and 1-1/2 miles if adequate haul roads can be maintained. Shovels are often used for blasted rock and any material where a vertical face of 8' or more can be excavated. Special conditions may dictate the use of draglines, clamshells, or backhoes. Spreading and compaction equipment must be matched to the soil characteristics, the compaction required and the rate the fill is being supplied.

R024119-10   Demolition Defined

Whole Building Demolition - Demolition of the whole building with no concern for any particular building element, component, or material type being demolished. This type of demolition is accomplished with large pieces of construction equipment that break up the structure, load it into trucks and haul it to a disposal site, but disposal or dump fees are not included. Demolition of below-grade foundation elements, such as footings, foundation walls, grade beams, slabs on grade, etc., is not included. Certain mechanical equipment containing flammable liquids or ozone-depleting refrigerants, electric lighting elements, communication equipment components, and other building elements may contain hazardous waste, and must be removed, either selectively or carefully, as hazardous waste before the building can be demolished.

Foundation Demolition - Demolition of below-grade foundation footings, foundation walls, grade beams, and slabs on grade. This type of demolition is accomplished by hand or pneumatic hand tools, and does not include saw cutting, or handling, loading, hauling, or disposal of the debris.

Gutting - Removal of building interior finishes and electrical/mechanical systems down to the load-bearing and sub-floor elements of the rough building frame, with no concern for any particular building element, component, or material type being demolished. This type of demolition is accomplished by hand or pneumatic hand tools, and includes loading into trucks, but not hauling, disposal or dump fees, scaffolding, or shoring. Certain mechanical equipment containing flammable liquids or ozone-depleting refrigerants, electric lighting elements, communication equipment components, and other building elements may contain hazardous waste, and must be removed, either selectively or carefully, as hazardous waste, before the building is gutted.

Selective Demolition - Demolition of a selected building element, component, or finish, with some concern for surrounding or adjacent elements, components, or finishes (see the first Subdivision (s) at the beginning of appropriate Divisions). This type of demolition is accomplished by hand or pneumatic hand tools, and does not include handling, loading, storing, hauling, or disposal of the debris, scaffolding, or shoring. "Gutting" methods may be used in order to save time, but damage that is caused to surrounding or adjacent elements, components, or finishes may have to be repaired at a later time.

Careful Removal - Removal of a piece of service equipment, building element or component, or material type, with great concern for both the removed item and surrounding or adjacent elements, components or finishes. The purpose of careful removal may be to protect the removed item for later re-use, preserve a higher salvage value of the removed item, or replace an item while taking care to protect surrounding or adjacent elements, components, connections, or finishes from cosmetic and/or structural damage. An approximation of the time required to perform this type of removal is 1/3 to 1/2 the time it would take to install a new item of like kind (see Reference Number R260105-30). This type of removal is accomplished by hand or pneumatic hand tools, and does not include loading, hauling, or storing the removed item, scaffolding, shoring, or lifting equipment.

Cutout Demolition - Demolition of a small quantity of floor, wall, roof, or other assembly, with concern for the appearance and structural integrity of the surrounding materials. This type of demolition is accomplished by hand or pneumatic hand tools, and does not include saw cutting, handling, loading, hauling, or disposal of debris, scaffolding, or shoring.

Rubbish Handling - Work activities that involve handling, loading or hauling of debris. Generally, the cost of rubbish handling must be added to the cost of all types of demolition, with the exception of whole building demolition.

Minor Site Demolition - Demolition of site elements outside the footprint of a building. This type of demolition is accomplished by hand or pneumatic hand tools, or with larger pieces of construction equipment, and may include loading a removed item onto a truck (check the Crew for equipment used). It does not include saw cutting, hauling or disposal of debris, and, sometimes, handling or loading.

R024119-20   Dumpsters

Dumpster rental costs on construction sites are presented in two ways.

The cost per week rental includes the delivery of the dumpster; its pulling or emptying once per week, and its final removal. The assumption is made that the dumpster contractor could choose to empty a dumpster by simply bringing in an empty unit and removing the full one. These costs also include the disposal of the materials in the dumpster.

The Alternate Pricing can be used when actual planned conditions are not approximated by the weekly numbers. For example, these lines can be used when a dumpster is needed for 4 weeks and will need to be emptied 2 or 3 times per week. Conversely the Alternate Pricing lines can be used when a dumpster will be rented for several weeks or months but needs to be emptied only a few times over this period.

R053100-10   Decking Descriptions

General - All Deck Products

Steel deck is made by cold forming structural grade sheet steel into a repeating pattern of parallel ribs. The strength and stiffness of the panels are the result of the ribs and the material properties of the steel. Deck lengths can be varied to suit job conditions, but because of shipping considerations, are usually less than 40 feet. Standard deck width varies with the product used but full sheets are usually 12″, 18″, 24″, 30″, or 36″. Deck is typically furnished in a standard width with the ends cut square. Any cutting for width, such as at openings or for angular fit, is done at the job site.

Deck is typically attached to the building frame with arc puddle welds, self-drilling screws, or powder or pneumatically driven pins. Sheet to sheet fastening is done with screws, button punching (crimping), or welds.

Composite Floor Deck

After installation and adequate fastening, floor deck serves several purposes. It (a) acts as a working platform, (b) stabilizes the frame, (c) serves as a concrete form for the slab, and (d) reinforces the slab to carry the design loads applied during the life of the building. Composite decks are distinguished by the presence of shear connector devices as part of the deck. These devices are designed to mechanically lock the concrete and deck together so that the concrete and the deck work together to carry subsequent floor loads. These shear connector devices can be rolled-in embossments, lugs, holes, or wires welded to the panels. The deck profile can also be used to interlock concrete and steel.

Composite deck finishes are either galvanized (zinc coated) or phosphatized/painted. Galvanized deck has a zinc coating on both the top and bottom surfaces. The phosphatized/painted deck has a bare (phosphatized) top surface that will come into contact with the concrete. This bare top surface can be expected to develop rust before the concrete is placed. The bottom side of the deck has a primer coat of paint.

Composite floor deck is normally installed so the panel ends do not overlap on the supporting beams. Shear lugs or panel profile shape often prevent a tight metal to metal fit if the panel ends overlap; the air gap caused by overlapping will prevent proper fusion with the structural steel supports when the panel end laps are shear stud welded.

Adequate end bearing of the deck must be obtained as shown on the drawings. If bearing is actually less in the field than shown on the drawings, further investigation is required.

Roof Deck

Roof deck is not designed to act compositely with other materials. Roof deck acts alone in transferring horizontal and vertical loads into the building frame. Roof deck rib openings are usually narrower than floor deck rib openings. This provides adequate support of rigid thermal insulation board.

Roof deck is typically installed to endlap approximately 2″ over supports. However, it can be butted (or lapped more than 2″) to solve field fit problems. Since designers frequently use the installed deck system as part of the horizontal bracing system (the deck as a diaphragm), any fastening substitution or change should be approved by the designer. Continuous perimeter support of the deck is necessary to limit edge deflection in the finished roof and may be required for diaphragm shear transfer.

Standard roof deck finishes are galvanized or primer painted. The standard factory applied paint for roof deck is a primer paint and is not intended to weather for extended periods of time. Field painting or touching up of abrasions and deterioration of the primer coat or other protective finishes is the responsibility of the contractor.

Cellular Deck

Cellular deck is made by attaching a bottom steel sheet to a roof deck or composite floor deck panel. Cellular deck can be used in the same manner as floor deck. Electrical, telephone, and data wires are easily run through the chase created between the deck panel and the bottom sheet.

When used as part of the electrical distribution system, the cellular deck must be installed so that the ribs line up and create a smooth cell transition at abutting ends. The joint that occurs at butting cell ends must be taped or otherwise sealed to prevent wet concrete from seeping into the cell. Cell interiors must be free of welding burrs, or other sharp intrusions, to prevent damage to wires.

When used as a roof deck, the bottom flat plate is usually left exposed to view. Care must be maintained during erection to keep good alignment and prevent damage.

Cellular deck is sometimes used with the flat plate on the top side to provide a flat working surface. Installation of the deck for this purpose requires special methods for attachment to the frame because the flat plate, now on the top, can prevent direct access to the deck material that is bearing on the structural steel. It may be advisable to treat the flat top surface to prevent slipping.

Cellular deck is always furnished galvanized or painted over galvanized.

Form Deck

Form deck can be any floor or roof deck product used as a concrete form. Connections to the frame are by the same methods used to anchor floor and roof deck. Welding washers are recommended when welding deck that is less than 20 gauge thickness.

Form deck is furnished galvanized, prime painted, or uncoated. Galvanized deck must be used for those roof deck systems where form deck is used to carry a lightweight insulating concrete fill.

Thermal & Moist. Protec.   R0784 Firestopping

R078413-30   Firestopping

Firestopping is the sealing of structural, mechanical, electrical and other penetrations through fire-rated assemblies. The basic components of firestop systems are safing insulation and firestop sealant on both sides of wall penetrations and the top side of floor penetrations.

Pipe penetrations are assumed to be through concrete, grout, or joint compound and can be sleeved or unsleeved. Costs for the penetrations and sleeves are not included. An annular space of 1″ is assumed. Escutcheons are not included.

Metallic pipe is assumed to be copper, aluminum, cast iron or similar metallic material. Insulated metallic pipe is assumed to be covered with a thermal insulating jacket of varying thickness and materials.

Non-metallic pipe is assumed to be PVC, CPVC, FR Polypropylene or similar plastic piping material. Intumescent firestop sealant or wrap strips are included. Collars on both sides of wall penetrations and a sheet metal plate on the underside of floor penetrations are included.

Ductwork is assumed to be sheet metal, stainless steel or similar metallic material. Duct penetrations are assumed to be through concrete, grout or joint compound. Costs for penetrations and sleeves are not included. An annular space of 1/2″ is assumed.

Multi-trade openings include costs for sheet metal forms, firestop mortar, wrap strips, collars and sealants as necessary.

Structural penetrations joints are assumed to be 1/2″ or less. CMU walls are assumed to be within 1-1/2″ of metal deck. Drywall walls are assumed to be tight to the underside of metal decking.

Metal panel, glass or curtain wall systems include a spandrel area of 5′ filled with mineral wool foil-faced insulation. Fasteners and stiffeners are included.

R220102-20   Labor Adjustment Factors

Labor Adjustment Factors are provided for Divisions 21, 22, and 23 to assist the mechanical estimator account for the various complexities and special conditions of any particular project. While a single percentage has been entered on each line of Division 22 01 02.20, it should be understood that these are just suggested midpoints of ranges of values commonly used by mechanical estimators. They may be increased or decreased depending on the severity of the special conditions.

The group for "existing occupied buildings", has been the subject of requests for explanation. Actually there are two stages to this group: buildings that are existing and "finished" but unoccupied, and those that also are occupied. Buildings that are "finished" may result in higher labor costs due to the workers having to be more careful not to damage finished walls, ceilings, floors, etc. and may necessitate special protective coverings and barriers. Also corridor bends and doorways may not accommodate long pieces of pipe or larger pieces of equipment. Work above an already hung ceiling can be very time consuming. The addition of occupants may force the work to be done on premium time (nights and/or weekends), eliminate the possible use of some preferred tools such as pneumatic drivers, powder charged drivers etc. The estimator should evaluate the access to the work area and just how the work is going to be accomplished to arrive at an increase in labor costs over "normal" new construction productivity.

R224000-10 Hot Water Consumption Rates

Type of Building	Size Factor	Maximum Hourly Demand	Average Day Demand
Apartment Dwellings	No. of Apartments: Up to 20 21 to 50 51 to 75 76 to 100 101 to 200 201 up	12.0 Gal. per apt. 10.0 Gal. per apt. 8.5 Gal. per apt. 7.0 Gal. per apt. 6.0 Gal. per apt. 5.0 Gal. per apt.	42.0 Gal. per apt. 40.0 Gal. per apt. 38.0 Gal. per apt. 37.0 Gal. per apt. 36.0 Gal. per apt. 35.0 Gal. per apt.
Dormitories	Men Women	3.8 Gal. per man 5.0 Gal. per woman	13.1 Gal. per man 12.3 Gal. per woman
Hospitals	Per bed	23.0 Gal. per patient	90.0 Gal. per patient
Hotels	Single room with bath Double room with bath	17.0 Gal. per unit 27.0 Gal. per unit	50.0 Gal. per unit 80.0 Gal. per unit
Motels	No. of units: Up to 20 21 to 100 101 Up	6.0 Gal. per unit 5.0 Gal. per unit 4.0 Gal. per unit	20.0 Gal. per unit 14.0 Gal. per unit 10.0 Gal. per unit
Nursing Homes		4.5 Gal. per bed	18.4 Gal. per bed
Office buildings		0.4 Gal. per person	1.0 Gal. per person
Restaurants	Full meal type Drive-in snack type	1.5 Gal./max. meals/hr. 0.7 Gal./max. meals/hr.	2.4 Gal. per meal 0.7 Gal. per meal
Schools	Elementary Secondary & High	0.6 Gal. per student 1.0 Gal. per student	0.6 Gal. per student 1.8 Gal. per student

For evaluation purposes, recovery rate and storage capacity are inversely proportional. Water heaters should be sized so that the maximum hourly demand anticipated can be met in addition to allowance for the heat loss from the pipes and storage tank.

R224000-20 Fixture Demands in Gallons Per Fixture Per Hour

Table below is based on 140° F final temperature except for dishwashers in public places (*) where 180° F water is mandatory.

Fixture	Apartment House	Club	Gym	Hospital	Hotel	Indust. Plant	Office	Private Home	School
Bathtubs	20	20	30	20	20			20	
Dishwashers, automatic	15	50-150*		50-150*	50-200*	20-100*		15	20-100*
Kitchen sink	10	20		20	30	20	20	10	20
Laundry, stationary tubs	20	28		28	28			20	
Laundry, automatic wash	75	75		100	150			75	
Private lavatory	2	2	2	2	2	2	2	2	2
Public lavatory	4	6	8	6	8	12	6		15
Showers	30	150	225	75	75	225	30	30	225
Service sink	20	20		20	30	20	20	15	20
Demand factor	0.30	0.30	0.40	0.25	0.25	0.40	0.30	0.30	0.40
Storage capacity factor	1.25	0.90	1.00	0.60	0.80	1.00	2.00	0.70	1.00

To obtain the probable maximum demand multiply the total demands for the fixtures (gal./fixture/hour) by the demand factor. The heater should have a heating capacity in gallons per hour equal to this maximum. The storage tank should have a capacity in gallons equal to the probable maximum demand multiplied by the storage capacity factor.

Fig. R238313-11

R238313-10 Heat Trace Systems

Before you can determine the cost of a HEAT TRACE installation, the method of attachment must be established. There are (4) common methods:

1. Cable is simply attached to the pipe with polyester tape every 12'.
2. Cable is attached with a continuous cover of 2" wide aluminum tape.
3. Cable is attached with factory extruded heat transfer cement and covered with metallic raceway with clips every 10'.
4. Cable is attached between layers of pipe insulation using either clips or polyester tape.

Example: Components for method 3 must include:

A. Heat trace cable by voltage and watts per linear foot.
B. Heat transfer cement, 1 gallon per 60 linear feet of cover.
C. Metallic raceway by size and type.
D. Raceway clips by size of pipe.

When taking off linear foot lengths of cable add the following for each valve in the system. (E)

In all of the above methods each component of the system must be priced individually.

SCREWED OR WELDED VALVE:			FLANGED VALVE:			BUTTERFLY VALVES:		
1/2"	=	6"	1/2"	=	1' -0"	1/2"	=	0'
3/4"	=	9"	3/4"	=	1' -6"	3/4"	=	0'
1"	=	1' -0"	1"	=	2' -0"	1"	=	1' -0"
1-1/2"	=	1' -6"	1-1/2"	=	2' -6"	1-1/2"	=	1' -6"
2"	=	2'	2"	=	2' -6"	2"	=	2' -0"
2-1/2"	=	2' -6"	2-1/2"	=	3' -0"	2-1/2"	=	2' -6"
3"	=	2' -6"	3"	=	3' -6"	3"	=	2' -6"
4"	=	4' -0"	4"	=	4' -0"	4"	=	3' -0"
6"	=	7' -0"	6"	=	8' -0"	6"	=	3' -6"
8"	=	9' -6"	8"	=	11' -0"	8"	=	4' -0"
10"	=	12' -6"	10"	=	14' -0"	10"	=	4' -0"
12"	=	15' -0"	12"	=	16' -6"	12"	=	5' -0"
14"	=	18' -0"	14"	=	19' -6"	14"	=	5' -6"
16"	=	21' -6"	16"	=	23' -0"	16"	=	6' -0"
18"	=	25' -6"	18"	=	27' -0"	18"	=	6' -6"
20"	=	28' -6"	20"	=	30' -0"	20"	=	7' -0"
24"	=	34' -0"	24"	=	36' -0"	24"	=	8' -0"
30"	=	40' -0"	30"	=	42' -0"	30"	=	10' -0"

R238313-10   Heat Trace Systems (cont.)

Add the following quantities of heat transfer cement to linear foot totals for each valve:

Nominal Valve Size	Gallons of Cement per Valve
1/2"	0.14
3/4"	0.21
1"	0.29
1-1/2"	0.36
2"	0.43
2-1/2"	0.70
3"	0.71
4"	1.00
6"	1.43
8"	1.48
10"	1.50
12"	1.60
14"	1.75
16"	2.00
18"	2.25
20"	2.50
24"	3.00
30"	3.75

The following must be added to the list of components to accurately price HEAT TRACE systems:

1. Expediter fitting and clamp fasteners (F)
2. Junction box and nipple connected to expediter fitting (G)
3. Field installed terminal blocks within junction box
4. Ground lugs
5. Piping from power source to expediter fitting
6. Controls
7. Thermostats
8. Branch wiring
9. Cable splices
10. End of cable terminations
11. Branch piping fittings and boxes

Deduct the following percentages from labor if cable lengths in the same area exceed:

150' to 250'	10%	351' to 500'	20%
251' to 350'	15%	Over 500'	25%

Add the following percentages to labor for elevated installations:

15' to 20' high	10%	31' to 35' high	40%
21' to 25' high	20%	36' to 40' high	50%
26' to 30' high	30%	Over 40' high	60%

R238313-20   Spiral-Wrapped Heat Trace Cable (Pitch Table)

In order to increase the amount of heat, occasionally heat trace cable is wrapped in a spiral fashion around a pipe; increasing the number of feet of heater cable per linear foot of pipe.

Engineers first determine the heat loss per foot of pipe (based on the insulating material, its thickness, and the temperature differential across it). A ratio is then calculated by the formula:

$$\text{Feet of Heat Trace per Foot of Pipe} = \frac{\text{Watts/Foot of Heat Loss}}{\text{Watts/Foot of the Cable}}$$

The linear distance between wraps (pitch) is then taken from a chart or table. Generally, the pitch is listed on a drawing leaving the estimator to calculate the total length of heat tape required. An approximation may be taken from this table.

Feet of Heat Trace Per Foot of Pipe

Pitch In Inches	1	1¼	1½	2	2½	3	4	6	8	10	12	14	16	18	20	24
3.5	1.80															
4	1.65															
5	1.46	1.60	1.80													
6	1.34	1.45	1.55	1.75												
7	1.25	1.35	1.43	1.57	1.75											
8	1.20	1.28	1.34	1.45	1.60	1.80										
9	1.16	1.23	1.28	1.37	1.51	1.68										
10	1.13	1.19	1.24	1.32	1.44	1.57	1.82									
15	1.06	1.08	1.10	1.15	1.21	1.29	1.42	1.78								
20	1.04	1.05	1.06	1.08	1.13	1.17	1.25	1.49	1.73							
25		1.04	1.04	1.06	1.08	1.11	1.17	1.33	1.51	1.72						
30				1.04	1.05	1.07	1.12	1.24	1.37	1.54	1.70	1.80				
35						1.06	1.09	1.17	1.28	1.42	1.54	1.64	1.78			
40						1.05	1.07	1.14	1.22	1.33	1.44	1.52	1.64	1.75		
50							1.05	1.09	1.15	1.22	1.29	1.35	1.44	1.53	1.64	1.83
60								1.06	1.11	1.16	1.21	1.25	1.31	1.39	1.46	1.62
70								1.05	1.08	1.12	1.17	1.19	1.24	1.30	1.35	1.47
80									1.06	1.09	1.13	1.15	1.19	1.24	1.28	1.38
90									1.04	1.06	1.10	1.13	1.16	1.19	1.23	1.32
100										1.05	1.08	1.10	1.13	1.15	1.19	1.23

Note: Common practice would normally limit the lower end of the table to 5% of additional heat and above 80% an engineer would likely opt for two (2) parallel cables.

R260105-30   Electrical Demolition (Removal for Replacement)

The purpose of this reference number is to provide a guide to users for electrical "removal for replacement" by applying the rule of thumb: 1/3 of new installation time (typical range from 20% to 50%) for removal. Remember to use reasonable judgment when applying the suggested percentage factor. For example:

Contractors have been requested to remove an existing fluorescent lighting fixture and replace with a new fixture utilizing energy saver lamps and electronic ballast:

In order to fully understand the extent of the project, contractors should visit the job site and estimate the time to perform the renovation work in accordance with applicable national, state and local regulation and codes.

The contractor may need to add extra labor hours to his estimate if he discovers unknown concealed conditions such as: contaminated asbestos ceiling, broken acoustical ceiling tile and need to repair, patch and touch-up paint all the damaged or disturbed areas, tasks normally assigned to general contractors. In addition, the owner could request that the contractors salvage the materials removed and turn over the materials to the owner or dispose of the materials to a reclamation station. The normal removal item is 0.5 labor hour for a lighting fixture and 1.5 labor-hours for new installation time. Revise the estimate times from 2 labor-hours work up to a minimum 4 labor-hours work for just fluorescent lighting fixture.

For removal of large concentrations of lighting fixtures in the same area, apply an "economy of scale" to reduce estimating labor hours.

Figure R260110-10   Typical Overhead Service Entrance

Utility Pole

Service Entrance Cap 3"

Building Wall

Galvanized Conduit 3"

Two Hole Pipe Clip 3"

Coupling

Grade Line

3" Locknut

Pullbox

3" Locknut & Bushing Insulated In Pullbox

Elbow

3" Galv. Conduit

3" PVC Conduit

Adapter

3" PVC To Galv. Adapter

4 - 350 kcmil **XHHW** In Above Conduit

3" Galv. Conduit Used For Safety On Pole And Through Foundation

For 600 AMP Service, Wiring 2 Parallel Runs of 3" Galv. Conduit to 600 AMP. Circuit Breaker Below

2 Locknuts And 1 Insulating Bushing Required Where #4 Or Larger Wire Used

Figure R260110-11   Typical Commercial Electric System

Panel First Floor 100 AMP.

Floor Box In Slab

Plate for Concrete Ring   Concrete Ring

Meter

600 AMP. Circuit Breaker

To Pull Box

CT Cabinet

200A CB

Cutaway Box Showing Wire Room

Ground Rod

Ground Clamp

Panel Basement 600 AMP.

NEMA-3R 200 AMP.

100 AMP.

A = 1" Conduit w/1-#1/0 Wire XHHW
B = 2-3" Conduits w/4-350kcmil XHHW in Each
C = 1¼" Conduit w/4 #3 THHN
D = 1¼" Conduit w/3 #1 XHHW
E = 1" Conduit w/4 #6 THHN
F = ½" Conduit w/2 #12 THHN

Condenser Unit

Elevator Controller

562

Figure R260110-12 Preliminary Procedure

1. Determine building size and use.
2. Develop total load in watts.
   a. Lighting
   b. Receptacle
   c. Air Conditioning
   d. Elevator
   e. Other power requirements
3. Determine best voltage available from utility company.
4. Determine cost from tables for loads (a) thru (e) above.
5. Determine size of service from formulas (R260110-17).
6. Determine costs for service, panels, and feeders from tables.

Figure R260110-13 Office Building 90' x 210', 3 story, w/garage

Garage Area = 18,900 S.F.
Office Area = 56,700 S.F.
Elevator = 2 @ 125 FPM

Tables	Power Required	Watts
R260110-14	Garage Lighting .5 Watt/S.F.	9,450
	Office Lighting 3 Watts/S.F.	170,100
R260110-14	Office Receptacles 2 Watts/S.F.	113,400
R260110-14, R262213-27	Low Rise Office A.C. 4.3 Watts/S.F.	243,810
R260110-15, 16	Elevators - 2 @ 20 HP = 2 @ 17,404 Watts/Ea.	34,808
R260110-14	Misc. Motors + Power 1.2 Watts/S.F.	68,040
	Total	639,608 Watts

Voltage Available

277/480V, 3 Phase, 4 Wire

Formula

R260110-17

$$\text{Amperes} = \frac{\text{Watts}}{\text{Volts x Power Factor x 1.73}} = \frac{639,608}{480V \text{ x .8 x 1.73}} = 963 \text{ Amps}$$

Use 1200 Amp Service

System	Description	Unit Cost	Unit	Total
D5020 210 0200	Garage Lighting (Interpolated)	$1.30	S.F.	$ 24,570
D5020 210 0540	Office Lighting (Using 32 W lamps)	7.86	S.F.	445,662
D5020 115 0880	Receptacle-Undercarpet	3.78	S.F.	214,326
D5020 140 0280	Air Conditioning	0.61	S.F.	34,587
D5020 135 0320	Misc. Pwr.	0.32	S.F.	18,144
D5020 145 2120	Elevators - 2 @ 20HP	2,825.00	Ea.	5,650
D5010 120 0480	Service-1200 Amp (add 25% for 277/480V)	62,900.00	Ea.	62,900
D5010 230 0480	Feeder - Assume 200 Ft.	332.00	L.F.	66,400
D5010 240 0320	Panels - 1200 Amp (add 20% for 277/480V)	32,350.00	Ea.	32,350
D5030 910 0400	Fire Detection	33,700.00	Ea.	33,700
	Total			$938,289
	or			$938,300

Table R260110-14 Nominal Watts Per S.F. for Electric Systems for Various Building Types

Type Construction	1. Lighting	2. Devices	3. HVAC	4. Misc.	5. Elevator	Total Watts
Apartment, luxury high rise	2	2.2	3	1		
Apartment, low rise	2	2	3	1		
Auditorium	2.5	1	3.3	.8		
Bank, branch office	3	2.1	5.7	1.4		
Bank, main office	2.5	1.5	5.7	1.4		
Church	1.8	.8	3.3	.8		
College, science building	3	3	5.3	1.3		
College, library	2.5	.8	5.7	1.4		
College, physical education center	2	1	4.5	1.1		
Department store	2.5	.9	4	1		
Dormitory, college	1.5	1.2	4	1		
Drive-in donut shop	3	4	6.8	1.7		
Garage, commercial	.5	.5	0	.5		
Hospital, general	2	4.5	5	1.3		
Hospital, pediatric	3	3.8	5	1.3		
Hotel, airport	2	1	5	1.3		
Housing for the elderly	2	1.2	4	1		
Manufacturing, food processing	3	1	4.5	1.1		
Manufacturing, apparel	2	1	4.5	1.1		
Manufacturing, tools	4	1	4.5	1.1		
Medical clinic	2.5	1.5	3.2	1		
Nursing home	2	1.6	4	1		
Office building, hi rise	3	2	4.7	1.2		
Office building, low rise	3	2	4.3	1.2		
Radio-TV studio	3.8	2.2	7.6	1.9		
Restaurant	2.5	2	6.8	1.7		
Retail store	2.5	.9	5.5	1.4		
School, elementary	3	1.9	5.3	1.3		
School, junior high	3	1.5	5.3	1.3		
School, senior high	2.3	1.7	5.3	1.3		
Supermarket	3	1	4	1		
Telephone exchange	1	.6	4.5	1.1		
Theater	2.5	1	3.3	.8		
Town Hall	2	1.9	5.3	1.3		
U.S. Post Office	3	2	5	1.3		
Warehouse, grocery	1	.6	0	.5		

Rule of Thumb: 1 KVA = 1 HP (Single Phase)

Three Phase:

Watts = 1.73 x Volts x Current x Power Factor x Efficiency

$$\text{Horsepower} = \frac{\text{Volts x Current x 1.73 x Power Factor}}{746 \text{ Watts}}$$

See R260120 for an additional reference

Table R260110-15 Horsepower Requirements for Elevators with 3 Phase Motors

Type	Maximum Travel Height in Ft.	Travel Speeds in FPM	Capacity of Cars in Lbs.		
			1200	1500	1800
Hydraulic	70	70	10	15	15
		85	15	15	15
		100	15	15	20
		110	20	20	20
		125	20	20	20
		150	25	25	25
		175	25	30	30
		200	30	30	40
Geared Traction	300	200			
		350			
			2000	2500	3000
Hydraulic	70	70	15	20	20
		85	20	20	25
		100	20	25	30
		110	20	25	30
		125	25	30	40
		150	30	40	50
		175	40	50	50
		200	40	50	60
Geared Traction	300	200	10	10	15
		350	15	15	23
			3500	4000	4500
Hydraulic	70	70	20	25	30
		85	25	30	30
		100	30	40	40
		110	40	40	50
		125	40	50	50
		150	50	50	60
		175	60		
		200	60		
Geared Traction	300	200	15		23
		350	23		35

The power factor of electric motors varies from 80% to 90% in larger size motors. The efficiency likewise varies from 80% on a small motor to 90% on a large motor.

Table R260110-16 Watts per Motor

90% Power Factor & Efficiency @ 200 or 460V			
HP	Watts	HP	Watts
10	9024	30	25784
15	13537	40	33519
20	17404	50	41899
25	21916	60	49634

Table R260110-17 Electrical Formulas

Ohm's Law

Ohm's Law is a method of explaining the relation existing between voltage, current, and resistance in an electrical circuit. It is practically the basis of all electrical calculations. The term "electromotive force" is often used to designate pressure in volts. This formula can be expressed in various forms.

To find the current in amperes:

$$\text{Current} = \frac{\text{Voltage}}{\text{Resistance}} \quad \text{or} \quad \text{Amperes} = \frac{\text{Volts}}{\text{Ohms}} \quad \text{or} \quad I = \frac{E}{R}$$

The flow of current in amperes through any circuit is equal to the voltage or electromotive force divided by the resistance of that circuit.

To find the pressure or voltage:

$$\text{Voltage} = \text{Current} \times \text{Resistance} \quad \text{or} \quad \text{Volts} = \text{Amperes} \times \text{Ohms}$$
$$\text{or} \quad E = I \times R$$

The voltage required to force a current through a circuit is equal to the resistance of the circuit multiplied by the current.

To find the resistance:

$$\text{Resistance} = \frac{\text{Voltage}}{\text{Current}} \quad \text{or} \quad \text{Ohms} = \frac{\text{Volts}}{\text{Amperes}} \quad \text{or} \quad R = \frac{E}{I}$$

The resistance of a circuit is equal to the voltage divided by the current flowing through that circuit.

Power Formulas

One horsepower = 746 watts   One kilowatt = 1000 watts

The power factor of electric motors varies from 80% to 90% in the larger size motors.

Single-Phase Alternating Current Circuits

Power in Watts = Volts x Amperes x Power Factor

To find current in amperes:

$$\text{Current} = \frac{\text{Watts}}{\text{Volts} \times \text{Power Factor}} \quad \text{or}$$

$$\text{Amperes} = \frac{\text{Watts}}{\text{Volts} \times \text{Power Factor}} \quad \text{or} \quad I = \frac{W}{E \times PF}$$

To find current of a motor, single phase:

$$\text{Current} = \frac{\text{Horsepower} \times 746}{\text{Volts} \times \text{Power Factor} \times \text{Efficiency}} \quad \text{or}$$

$$I = \frac{HP \times 746}{E \times PF \times \text{Eff.}}$$

To find horsepower of a motor, single phase:

$$\text{Horsepower} = \frac{\text{Volts} \times \text{Current} \times \text{Power Factor} \times \text{Efficiency}}{746 \text{ Watts}}$$

$$HP = \frac{E \times I \times PF \times \text{Eff.}}{746}$$

To find power in watts of a motor, single phase:

Watts = Volts x Current x Power Factor x Efficiency   or

Watts = E x I x PF x Eff.

To find single phase kVA:

$$\text{1 Phase kVA} = \frac{\text{Volts} \times \text{Amps}}{1000}$$

Three-Phase Alternating Current Circuits

Power in Watts = Volts x Amperes x Power Factor x 1.73

To find current in amperes in each wire:

$$\text{Current} = \frac{\text{Watts}}{\text{Voltage} \times \text{Power Factor} \times 1.73} \quad \text{or}$$

$$\text{Amperes} = \frac{\text{Watts}}{\text{Volts} \times \text{Power Factor} \times 1.73} \quad \text{or} \quad I = \frac{W}{E \times PF \times 1.73}$$

To find current of a motor, 3 phase:

$$\text{Current} = \frac{\text{Horsepower} \times 746}{\text{Volts} \times \text{Power Factor} \times \text{Efficiency} \times 1.73} \quad \text{or}$$

$$I = \frac{HP \times 746}{E \times PF \times \text{Eff.} \times 1.73}$$

To find horsepower of a motor, 3 phase:

$$\text{Horsepower} = \frac{\text{Volts} \times \text{Current} \times 1.73 \times \text{Power Factor}}{746 \text{ Watts}}$$

$$HP = \frac{E \times I \times 1.73 \times PF}{746}$$

To find power in watts of a motor, 3 phase:

Watts = Volts x Current x 1.73 x Power Factor x Efficiency   or

Watts = E x I x 1.73 x PF x Eff.

To find 3 phase kVA:

$$\text{3 phase kVA} = \frac{\text{Volts} \times \text{Amps} \times 1.73}{1000} \quad \text{or}$$

$$kVA = \frac{V \times A \times 1.73}{1000}$$

Power Factor (PF) is the percentage ratio of the measured watts (effective power) to the volt-amperes (apparent watts)

$$\text{Power Factor} = \frac{\text{Watts}}{\text{Volts} \times \text{Amperes}} \times 100\%$$

A Conceptual Estimate of the costs for a building when final drawings are not available can be quickly figured by using **Table R260110-18 Cost Per S.F. for Electrical Systems for Various Building Types.** The following definitions apply to this table.

1. **Service and Distribution:** This system includes the incoming primary feeder from the power company, main building transformer, metering arrangement, switchboards, distribution panel boards, stepdown transformers, and power and lighting panels. Items marked (*) include the cost of the primary feeder and transformer. In all other projects the cost of the primary feeder and transformer is paid for by the local power company.

2. **Lighting:** Includes all interior fixtures for decor, illumination, exit and emergency lighting. Fixtures for exterior building lighting are included, but parking area lighting is not included unless mentioned. See also Section D5020 for detailed analysis of lighting requirements and costs.

3. **Devices:** Includes all outlet boxes, receptacles, switches for lighting control, dimmers and cover plates.

4. **Equipment Connections:** Includes all materials and equipment for making connections for heating, ventilating and air conditioning, food service and other motorized items requiring connections.

5. **Basic Materials:** This category includes all disconnect power switches not part of service equipment, raceways for wires, pull boxes, junction boxes, supports, fittings, grounding materials, wireways, busways, and wire and cable systems.

6. **Special Systems:** Includes installed equipment only for the particular system such as fire detection and alarm, sound, emergency generator and others as listed in the table.

Table R260110-18 Cost per S.F. for Electric Systems for Various Building Types

Type Construction	1. Service & Distrib.	2. Lighting	3. Devices	4. Equipment Connections	5. Basic Materials	6. Special Systems Fire Alarm & Detection	Telephone	Television Distribution
Apartment, Low Rise	$0.80	$0.87	$0.22	$0.29	$ 2.17	$ 0.51	0.8	$0.22
Apartment, High Rise	2.39	1.33	0.40	0.53	4.12	0.80	1.33	0.27
Assisted Living	4.10	1.27	0.28	0.57	2.69	1.13	0.71	0.71
Auditorium	2.40	4.14	0.44	0.65	6.97	1.31	1.09	
Bank	4.10	2.93	0.29	0.88	6.44	3.51	2.34	
Church	1.06	3.71	0.35	0.88	6.35	1.77	0.71	
College, Classroom	3.66	3.23	0.43	0.65	5.81	0.86	0.86	0.86
College, Laboratory	6.53	6.97	0.87	0.87	13.06	2.61	2.18	
Community Center	1.74	2.70	0.39	0.77	5.79	2.51	0.77	0.39
Department Store	1.07	2.97	0.12	0.47	4.03	0.71	0.47	
Dormitory, Low Rise	0.91	1.83	0.61	0.61	5.02	1.22	1.37	0.30
Factory	1.33	5.47	0.17	0.50	5.64	1.33	0.5	0.17
Funeral Home	1.55	1.76	0.21	0.52	3.31	1.24	1.35	
Garage, Commercial	3.34	0.91	0.15	0.30	2.73	5.77	$0.76	
Garage, Parking	0.20	0.57	0.12	0.04	1.75	0.08	0.0	
Gymnasium	1.02	2.76	0.29	0.58	5.80	1.60	0.44	
Hospital, Low Rise	2.58	3.87	1.72	1.72	14.61	3.01	2.15	0.86
Hospital, High Rise	4.56	3.73	1.66	1.24	16.57	2.07	1.66	0.41
Housing, Elderly	4.10	1.27	0.28	0.57	2.69	1.13	0.71	0.71
Library	1.65	5.20	0.47	1.42	9.69	2.13	0.71	
Medical Office	2.32	1.35	0.58	0.39	4.45	1.55	1.55	0.39
Motel	1.45	1.01	0.56	0.11	3.69	1.23	1.12	0.22
Nursing Home	2.29	2.50	0.42	0.62	5.41	1.25	1.46	1.46
Office, Low Rise	3.03	2.19	0.34	0.67	5.22	1.85	1.18	0.17
Office, High Rise	0.71	2.27	0.28	0.43	5.26	1.42	1.14	0.14
Post Office	2.92	1.55	0.17	0.52	5.85	3.61	1.55	
Restaurant	8.53	3.10	0.26	1.03	6.46	4.14	0.52	
Restaurant, Fast Food	6.09	3.60	0.28	1.11	8.58	5.54	0.55	
Retail Store	1.80	2.28	0.12	0.48	4.07	1.08	0.36	
School, Elementary	1.61	2.87	0.36	0.54	5.92	2.15	0.9	0.36
School, Junior High	1.45	2.89	0.54	0.72	5.78	0.90	1.08	0.36
School, High	1.35	3.38	0.45	0.68	6.99	1.13	1.35	1.35
Supermarket	3.08	2.66	0.14	0.42	4.47	1.40	0.28	
Telephone Exchange	6.65	5.23	0.95	2.38	14.26	10.93	5.7	
Theater (Movie)	2.73	4.97	0.25	0.75	5.72	3.48	0.99	
Town Hall	1.59	3.37	0.6	1.19	7.54	1.59	1.59	
Warehouse	0.33	1.25	0.25	0.17	3	2	0.42	

*Includes cost of primary feeder and transformer. Cont'd on next page.

COST ASSUMPTIONS:

Each of the projects analyzed in Table R260110-18 was bid within the last 10 years in the northeastern part of the United States. Bid prices have been adjusted to Jan. 1 levels. The list of projects is by no means all-inclusive, yet by carefully examining the various systems for a particular building type, certain cost relationships will emerge. The use of Square Foot/Cubic Costs, in the Reference Section, along with the Project Size Modifier, should produce a budget S.F. cost for the electrical portion of a job that is consistent with the amount of design information normally available at the conceptual estimate stage.

Table R260110-18 Cost per S.F. for Electric Systems for Various Building Types (cont.)

Type Construction	6. Special Systems, (cont.)						
	Intercom Systems	Sound Systems	Video Monitoring	Internet	Emergency Generator	Access Control	Master Clock Sys.
Apartment, Low Rise	0.65			0.29	$0.14	0.29	
Apartment, High Rise	0.93			0.53	0.13	$0.53	
Assisted Living	1.70	0.14	0.42		0.14	0.28	
Auditorium		2.40	0.87		1.31	0.22	
Bank		2.93		2.64	0.59	2.64	
Church		2.29			0.18	0.35	
College, Classroom	0.86	1.08		1.72	0.43	0.43	0.65
College, Laboratory	2.18			4.35	0.44	1.31	2.18
Community Center		1.93		0.97	0.39	0.97	
Department Store			0.83	0.36		0.83	
Dormitory, Low Rise	1.37			1.52		0.46	
Factory			0.50		0.17	0.5	0.33
Funeral Home				0.1	0.10	0.21	
Garage, Commercial					0.46	0.76	
Garage, Parking			0.41		0.04	0.81	
Gymnasium		1.31			0.15	0.58	
Hospital, Low Rise	2.58	0.43	0.86	0.43	6.44	1.29	0.43
Hospital, High Rise	1.24	0.41	0.41	$0.41	5.39	1.24	0.41
Housing, Elderly	1.7	0.14	0.42		0.14	0.28	
Library		0.47		0.95	0.47	0.47	
Medical Office	2.32		0.97	1.74	0.77	0.97	
Motel				1.23	0.11	0.45	
Nursing Home	3.54	0.21	0.83		0.42	0.42	
Office, Low Rise			0.34	1.35	0.17	0.34	
Office, High Rise			0.43	1.28	0.43	0.43	
Post Office				0.52	0.17	0.34	
Restaurant				0.52	0.52	0.78	
Restaurant, Fast Food				0.55	0.55	0.83	
Retail Store		0.12	0.72	0.12	0.12	0.72	
School, Elementary	1.08	0.90		0.36		0.54	0.36
School, Junior High	1.26	0.9		0.54	0.36	0.54	0.72
School, High	1.35	1.13	0.9	0.45	0.68	0.68	0.68
Supermarket		0.14	0.56	0.14		0.7	
Telephone Exchange					0.48	0.95	
Theater (Movie)		4.48		0.25	0.25	0.99	
Town Hall				1.79	0.2	0.4	
Warehouse			0.42	0.08		0.42	

*Includes cost of primary feeder and transformer. Cont'd on next page.

General: Variations in the following square foot costs are due to the type of structural systems of the buildings, geographical location, local electrical codes, designer's preference for specific materials and equipment, and the owner's particular requirements.

Table R260110-18 Cost per S.F. for Total Electric Systems for Various Building Types (cont.)

Type Construction	Additional Information	Total Floor Area in Square Feet	Total Cost per Square Foot for Total Electric Systems
Apartment, Low Rise	Stories = 3, Height = 10', Perimeter = 715'	40,200	7.23
Apartment, High Rise	Stories = 15, Height = 10', Perimeter = 350'	115,000	13.29
Assisted Living	Stories = 1, Height = 10', Perimeter = 400'	10,000	14.14
Auditorium	Stories = 1, Height = 24', Perimeter = 747'	28,000	21.78
Bank	Stories = 1, Height = 14', Perimeter = 168'	2,700	29.29
Church	Stories = 1, Height = 24', Perimeter = 562'	17,700	17.65
College, Classroom	Stories = 2, Height = 12', Perimeter = 630'	50,000	21.50
College, Laboratory	Stories = 1, Height = 12', Perimeter = 550'	27,500	43.54
Community Center	Stories = 1, Height = 12', Perimeter = 453'	10,000	19.31
Department Store	Stories = 1, Height = 14', Perimeter = 1039'	85,800	11.87
Dormitory, Low Rise	Stories = 3, Height = 12', Perimeter = 1008'	63,000	15.23
Factory	Stories = 1, Height = 20', Perimeter = 730'	30,000	16.58
Funeral Home	Stories = 1, Height = 12', Perimeter = 424'	10,000	10.35
Garage, Commercial	Stories = 1, Height = 12', Perimeter = 153'	1,400	15.18
Garage, Parking	Stories = 5, Height = 10', Perimeter = 723'	145,000	4.06
Gymnasium	Stories = 1, Height = 25', Perimeter = 660'	22,000	14.50
Hospital, Low Rise	Stories = 3, Height = 12', Perimeter = 566'	55,000	42.96
Hospital, High Rise	Stories = 6, Height = 12', Perimeter = 866'	200,000	41.43
Housing, Elderly	Stories = 1, Height = 10', Perimeter = 400'	10,000	14.14
Library	Stories = 2, Height = 14', Perimeter = 435'	22,000	23.63
Medical Office	Stories = 2, Height = 10', Perimeter = 240'	7,000	19.36
Motel	Stories = 3, Height = 9', Perimeter = 606'	49,000	11.17
Nursing Home	Stories = 2, Height = 10', Perimeter = 384'	21,000	20.81
Office, Low Rise	Stories = 3, Height = 12', Perimeter = 360'	20,000	16.85
Office, High Rise	Stories = 15, Height = 10', Perimeter = 634'	311,200	14.21
Post Office	Stories = 1, Height = 14', Perimeter = 486'	13,000	17.20
Restaurant	Stories = 1, Height = 12', Perimeter = 174'	2,900	25.85
Restaurant, Fast Food	Stories = 1, Height = 10', Perimeter = 260'	4,000	27.69
Retail Store	Stories = 1, Height = 14', Perimeter = 360'	8,000	11.98
School, Elementary	Stories = 1, Height = 15', Perimeter = 1325'	39,500	17.94
School, Junior High	Stories = 2, Height = 15', Perimeter = 1700'	110,000	18.07
School, High	Stories = 2, Height = 15', Perimeter = 2411'	158300	22.56
Supermarket	Stories = 1, Height = 18', Perimeter = 584'	30600	13.97875
Telephone Exchange	Stories = 1, Height = 12', Perimeter = 286'	5000	47.52
Theater (Movie)	Stories = 1, Height = 20', Perimeter = 513'	14000	24.8625
Town Hall	Stories = 3, Height = 12', Perimeter = 356'	20000	19.845
Warehouse	Stories = 1, Height = 24', Perimeter = 700'	30000	8.33

*Includes cost of primary feeder and transformer.

Table R260120-10 ASHRAE Climate Zone Map

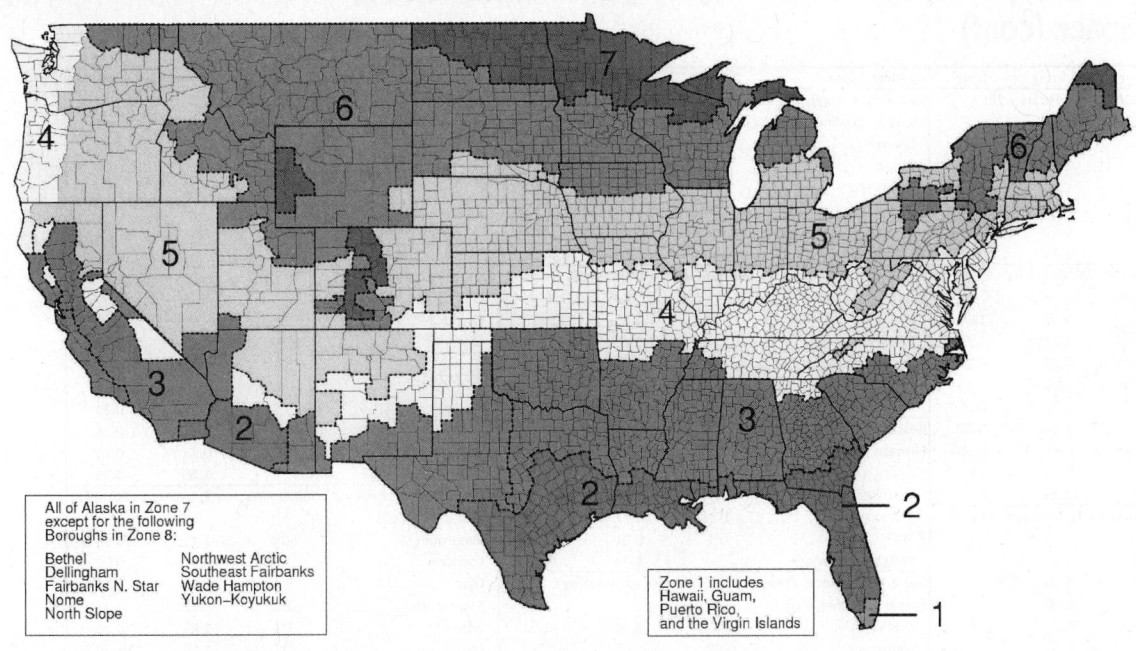

All of Alaska in Zone 7 except for the following Boroughs in Zone 8:

Bethel	Northwest Arctic
Dellingham	Southeast Fairbanks
Fairbanks N. Star	Wade Hampton
Nome	Yukon–Koyukuk
North Slope	

Zone 1 includes Hawaii, Guam, Puerto Rico, and the Virgin Islands

Table R260120-20 Watts per Square Foot for Lighting and Misc. Power by Building Type & Space

Building / Space	Percent	Lighting	Misc
Apartment, Low: Area = 20,000 SF; Floors = 2; Heating = DX Coils; Cooling = DX Coils			
Room	85%	0.50	0.10
Storage	8%	1.19	0.00
Laundry	7%	1.28	0.15
Weighted Average		0.61	0.10
Apartment, Mid: Area = 40,000 SF; Floors = 5; Heating = DX Coils; Cooling = DX Coils			
Room	71%	0.50	0.30
Corridor	16%	0.57	0.00
Storage	7%	1.19	0.00
Laundry	6%	1.28	0.15
Average		0.61	0.22
Apartment, High: Area = 72,000 SF; Floors = 8; Heating = DX Coils; Cooling = DX Coils			
Room	71%	0.50	0.30
Corridor	16%	0.57	0.00
Storage	7%	1.19	0.00
Laundry	6%	1.28	0.15
Weighted Average		0.61	0.22
Assisted Living: Area = 10,000 SF; Floors = 1; Heating = Furnace; Cooling = DX Coils			
Room	70%	1.46	0.25
Dining	15%	1.85	0.10
Food Prep	10%	1.19	1.00
Care	5%	1.34	1.00
Weighted Average		1.49	0.34

Building / Space	Percent	Lighting	Misc
Auditorum: Area = 70,000 SF; Floors = 1; Heating = Hot Water Coils; Cooling = Chilled Water Coils			
Theater	70%	0.90	
Lobby	15%	1.50	
Office	5%	1.30	
Corridor	5%	0.60	
MEP	5%	0.70	
Weighted Average		0.99	
Bank: Area = 25,000 SF; Floors = 2; Heating = Furnace; Cooling = DX Coils			
Bank	60%	2.33	0.45
Corridor	10%	0.57	0.00
Office	10%	1.49	0.75
Lobby	5%	1.52	0.25
Restrooms	5%	0.77	0.10
Conference	4%	0.92	0.10
MEP	4%	0.81	0.10
Production	2%	1.64	0.70
Weighted Average		1.82	0.38
Community Center: Area = 35,000 SF; Floors = 1; Heating = Furnace; Cooling = DX Coils			
Convention	50%	1.27	
Dining	15%	1.85	
Food Prep	15%	1.19	
Storage	10%	1.19	
Corridor	5%	0.57	
Restrooms	5%	0.77	
Weighted Average		1.28	
Convention Center: Area = 150,000 SF; Floors = 2; Heating = Hot Water Coils; Cooling = Chilled Water Coils			
Convention	65%	1.27	0.25
Conference	20%	0.92	0.10
Dining	5%	1.85	0.10
Corridor	4%	0.57	0.00
Food Prep	3%	1.19	1.00
Restrooms	3%	0.77	0.10
Weighted Average		1.18	0.22

Building / Space	Percent	Lighting	Misc
Day Care: Area = 2,500 SF; Floors = 1; Heating = Furnace; Cooling = DX Coils			
Classroom	45%	1.45	
Corridor	15%	0.57	
Library	15%	1.67	
Food Prep	10%	1.19	
Restrooms	7%	0.77	
Office	5%	1.49	
Storage	3%	1.19	
Weighted Average		1.27	
Fitness Center: Area = 20,000 SF; Floors = 1; Heating = Furnace; Cooling = DX Coils			
Exercise	80%	1.13	0.10
Corridor	5%	0.57	0.00
MEP	5%	0.81	0.10
Office	5%	1.24	0.75
Restrooms	7%	0.77	0.10
Weighted Average		1.07	0.13
Hospital: Area = 250,000 SF; Floors = 4; Heating = Hot Water Coils; Cooling = Chilled Water Coils			
Medical	60%	1.34	1.00
Laboratory	15%	3.19	1.00
Corridor	10%	0.57	0.00
Laundry	5%	1.28	0.15
MEP	5%	0.81	0.10
Restrooms	5%	0.77	0.10
Weighted Average		1.48	0.77
Hotel: Area = 180,000 SF; Floors = 10; Heating = Furnace; Cooling = DX Coils			
Room	75%	1.46	
Meeting Center	15%	1.27	
Corridor	8%	0.57	
MEP	2%	0.81	
Weighted Average		1.35	

Steps:
1. Determine climate zone using R260120-10
2. Use R260120-20 to determine lighting and miscellaneous power for appropriate building type
3. Use corresponding building type and climate zone in R260120-30 to determine other systems

*Values for R260120-20, & -30 were derived by performing an energy analysis using the eQUEST® software from James J. Hirsch & Associates. eQUEST is a trademark of James J. Hirsch

**Additional systems may be necessary

Table R260120-20 Watts per Square Foot for Lighting and Misc. Power by Building Type & Space (cont)

Building / Space	Percent	Lighting	Misc	Building / Space	Percent	Lighting	Misc	Building / Space	Percent	Lighting	Misc
Manufacturing, General: Area = 100,000 SF; Floors = 10; Heating = Furnace; Cooling = DX Coils				**Restaurant, Bar/Lounge: Area = 4,000 SF; Floors = 1; Heating = Furnace; Cooling = DX Coils**				**Store, Warehouse: Area = 150,000 SF; Floors = 1; Heating = Furnace; Cooling = DX Coils**			
Work	60%	2.66	1.00	Lounge	35%	1.85	0.10	Retail	75%	2.38	0.25
Storage	25%	1.19	0.00	Dining	30%	1.85	0.10	Storage	20%	1.19	0.00
Office	10%	1.24	0.75	Food Prep	25%	1.19	1.00	Office	4%	1.24	0.75
Restrooms	3%	0.77	0.10	MEP	5%	0.81	0.10	Restrooms	1%	0.77	0.10
MEP	2%	0.81	0.10	Restrooms	5%	0.77	0.10	**Weighted Average**		**2.08**	**0.22**
Weighted Average		**2.06**	**0.68**	**Weighted Average**		**1.58**	**0.33**	**School, Elementary: Area = 35,000 SF; Floors = 1; Heating = Furnace; Cooling = DX Coils**			
Manufacturing, High Tech: Area = 200,000 SF; Floors = 1; Heating = Furnace; Cooling = DX Coils				**Restaurant, Full: Area = 5,000 SF; Floors = 1; Heating = Furnace; Cooling = DX Coils**				Classroom	50%	1.45	0.50
Work	45%	3.19	1.00	Dining	60%	1.85	0.10	Corridor	15%	0.57	0.00
Office	27%	1.24	0.75	Food Prep	30%	1.19	1.00	Gym	10%	1.13	0.10
Corridor	18%	0.57	0.00	Restrooms	5%	0.77	0.10	Dining	5%	1.85	0.10
Dining	3%	1.85	0.10	MEP	5%	0.81	0.10	Food Prep	5%	1.19	1.00
Conference	2%	0.92	0.10	**Weighted Average**		**1.55**	**0.37**	Library	5%	1.78	0.25
Server	2%	1.24	5.00	**Restaurant, Quick: Area = 2,500 SF; Floors = 1; Heating = Furnace; Cooling = DX Coils**				Auditorum	5%	1.62	0.25
MEP	2%	0.81	0.10	Dining	50%	1.85	0.10	Restrooms	5%	0.77	0.10
Food Prep	1%	1.19	1.00	Food Prep	35%	1.19	1.00	**Weighted Average**		**1.28**	**0.35**
Weighted Average		**2.00**	**0.77**	Restrooms	10%	0.77	0.10	**School, Middle: Area = 75,000 SF; Floors = 1; Heating = Furnace; Cooling = DX Coils**			
Medical Clinc: Area = 20,000 SF; Floors = 1; Heating = Furnace; Cooling = DX Coils				MEP	5%	0.81	0.10	Classroom	50%	1.45	0.50
Clinic	50%	1.40		**Weighted Average**		**1.46**	**0.42**	Corridor	14%	0.57	0.00
General Office	15%	1.30		**Store, Conveniece: Area = 3,000 SF; Floors = 1; Heating = Furnace; Cooling = DX Coils**				Gym	9%	1.13	0.10
Private Office	10%	1.30		Retail	80%	2.38	0.25	Auditorium	7%	1.62	0.25
Laboratory	10%	1.30		Office	10%	1.49	0.75	Dining	6%	1.85	0.10
Storage	5%	0.60		Restrooms	10%	0.77	0.10	Food Prep	5%	1.19	1.00
Restrooms	5%	0.60		**Weighted Average**		**2.13**	**0.29**	Library	5%	1.78	0.25
MEP	5%	0.70		**Store, Department: Area = 120,000 SF; Floors = 2; Heating = Hot Water Coils; Cooling = Chilled Water Coils**				Restrooms	4%	0.77	0.10
Weighted Average		**1.25**		Retail	55%	2.38	0.25	**Weighted Average**		**1.31**	**0.35**
Motel: Area = 10,000 SF; Floors = 1; Heating = DX Coils; Cooling = DX Coils				Exhibit Area	15%	1.03	0.25	**School, High: Area = 150,000 SF; Floors = 2; Heating = Furnace; Cooling = DX Coils**			
Room	95%	1.46	0.25	Storage	16%	1.19	0.00	Classroom	45%	1.45	0.50
Office	3%	1.24	0.75	Office	5%	1.24	0.75	Corridor	10%	0.57	0.00
MEP	2%	0.81	0.10	Smoking Lounge	2%	1.85	0.10	Gym	10%	1.13	0.10
Weighted Average		**1.44**	**0.26**	Restrooms	2%	0.77	0.10	Dining	10%	1.85	0.10
Museum: Area = 20,000 SF; Floors = 1; Heating = Hot Water Coils; Cooling = Chilled Water Coils				MEP	5%	0.81	0.10	Food Prep	5%	1.19	1.00
Exhibit Area	65%	1.03	0.25	**Weighted Average**		**1.81**	**0.22**	Library	5%	1.78	0.25
Theater	15%	1.02	0.25	**Store, Large: Area = 70,000 SF; Floors = 1; Heating = Furnace; Cooling = DX Coils**				Auditorum	10%	1.62	0.25
Lobby	5%	1.77	0.25	Retail	65%	2.38	0.25	Restrooms	5%	0.77	0.10
MEP	5%	0.81	0.10	Storage	20%	1.19	0.00	**Weighted Average**		**1.36**	**0.34**
Office	5%	1.49	0.75	Office	5%	1.24	0.75	**School, College: Area = 250,000 SF; Floors = 4; Heating = Furnace; Cooling = DX Coils**			
Restrooms	5%	0.77	0.10	Dining	3%	1.85	0.10	Classroom	50%	1.45	0.50
Weighted Average		**1.06**	**0.26**	Food Prep	1%	1.19	1.00	Office	15%	1.49	0.75
Office, Low: Area = 25,000 SF; Floors = 2; Heating = Furnace; Cooling = DX Coils				Lounge	2%	1.85	0.10	Corridor	12%	0.57	0.00
Office	70%	1.30	0.75	Restrooms	2%	0.77	0.10	Gym	10%	1.13	0.10
Corridor	10%	0.60	0.00	MEP	2%	0.81	0.10	Library	5%	1.78	0.25
Lobby	5%	1.10	0.25	**Weighted Average**		**1.98**	**0.22**	Dining	3%	1.85	0.10
Restrooms	5%	0.60	0.10	**Store, Small: Area = 5,000 SF; Floors = 1; Heating = Furnace; Cooling = DX Coils**				Restrooms	3%	0.77	0.10
Conference	4%	1.60	0.10	Retail	58%	2.38	0.25	Food Prep	2%	1.19	1.00
MEP	4%	0.70	0.10	Exhibit Area	15%	1.03	0.25	**Weighted Average**		**1.32**	**0.41**
Production	2%	1.50	0.70	Storage	15%	1.19	0.00	**Service Station: Area = 2,000 SF; Floors = 1; Heating = Furnace; Cooling = DX Coils**			
Weighted Average		**1.18**	**0.56**	Office	5%	1.24	0.75	Workshop	70%	1.31	0.50
Office, Mid: Area = 125,000 SF; Floors = 4; Heating = Hot Water Coils; Cooling = Chilled Water Coils				Restrooms	5%	0.77	0.10	Office	10%	1.49	0.75
Office Open	40%	1.24	0.75	MEP	2%	0.81	0.10	Restrooms	10%	0.77	0.10
Office Private	30%	1.49	0.75	**Weighted Average**		**1.83**	**0.23**	Retail	10%	2.38	0.25
Corridor	10%	0.57	0.00	**Store, Strip Mall: Area = 15,000 SF; Floors = 1; Heating = Furnace; Cooling = DX Coils**				**Weighted Average**		**1.38**	**0.46**
Lobby	5%	1.52	0.25	Retail	38%	2.38	0.25	**Storage, High Bay: Area = 200,000 SF; Floors = 1; Heating = Furnace; Cooling = DX Coils**			
Restrooms	5%	0.77	0.10	Atrium	20%	1.55	0.13	Storage	85%	1.19	0.00
Conference	4%	0.92	0.10	Exhibit Area	15%	1.03	0.25	Office	10%	1.24	0.75
MEP	4%	0.81	0.10	Storage	15%	1.19	0.00	Restrooms	3%	0.77	0.10
Production	2%	1.64	0.70	Office	5%	1.24	0.75	MEP	2%	0.81	0.10
Weighted Average		**1.22**	**0.56**	Restrooms	5%	0.77	0.10	**Weighted Average**		**1.17**	**0.08**
Office, High: Area = 250,000 SF; Floors = 8; Heating = Hot Water Coils; Cooling = Chilled Water Coils				MEP	2%	0.81	0.10	**Storage, Low Bay: Area = 70,000 SF; Floors = 1; Heating = Furnace; Cooling = DX Coils**			
Office Open	45%	1.24	0.75	**Weighted Average**		**1.66**	**0.20**	Storage	85%	1.19	0.00
Office Private	25%	1.49	0.75					Office	10%	1.24	0.75
Corridor	10%	0.57	0.00					Restrooms	3%	0.77	0.10
Lobby	5%	1.52	0.25					MEP	2%	0.81	0.10
Conference	4%	0.92	0.10					**Weighted Average**		**1.17**	**0.08**
Production	2%	1.64	0.70								
Restrooms	5%	0.77	0.10								
MEP	4%	0.81	0.10								
Weighted Average		**1.20**	**0.56**								

Table R260120-30 Watts per S.F. for Various Systems by Building Type & Climate Zone

Building / Zone	Heating	Cooling	Heat rejection	Auxiliary (pumps)	Vent fan	Supp heat pump	Building / Zone	Heating	Cooling	Heat rejection	Auxiliary (pumps)	Vent fan	Supp heat pump
Apartment, Low							**Community Center**						
Zone 1	0.42	3.65	0.00	0.00	0.12	1.47	Zone 1	0.00	2.70	0.00	0.00	0.44	0.00
Zone 2	0.63	3.53	0.00	0.00	0.12	4.01	Zone 2	0.00	2.66	0.00	0.00	0.44	0.00
Zone 3	0.64	3.45	0.00	0.00	0.13	4.63	Zone 3	0.00	2.66	0.00	0.00	0.45	0.00
Zone 4	0.65	3.09	0.00	0.00	0.13	7.41	Zone 4	0.00	2.44	0.00	0.00	0.45	0.00
Zone 5	0.67	2.88	0.00	0.00	0.12	10.30	Zone 5	0.00	2.42	0.00	0.00	0.44	0.00
Zone 6	0.69	2.66	0.00	0.00	0.13	10.68	Zone 6	0.00	2.16	0.00	0.00	0.44	0.00
Zone 7	0.85	2.11	0.00	0.00	0.12	11.76	Zone 7	0.00	1.91	0.00	0.00	0.44	0.00
Average	**0.65**	**3.05**	**0.00**	**0.00**	**0.13**	**7.18**	**Average**	**0.00**	**2.42**	**0.00**	**0.00**	**0.45**	**0.00**
Apartment, Mid							**Convention Center**						
Zone 1	0.00	1.90	0.00	0.00	0.11	0.00	Zone 1	0.00	1.46	0.25	0.27	0.31	0.00
Zone 2	0.43	1.82	0.00	0.00	0.11	2.12	Zone 2	0.00	1.44	0.25	0.27	0.29	0.00
Zone 3	0.48	1.69	0.00	0.00	0.11	2.32	Zone 3	0.00	1.43	0.24	0.26	0.30	0.00
Zone 4	0.67	1.67	0.00	0.00	0.12	4.25	Zone 4	0.00	1.42	0.23	0.26	0.28	0.00
Zone 5	0.70	1.54	0.00	0.00	0.12	4.79	Zone 5	0.00	1.30	0.29	0.31	0.22	0.00
Zone 6	0.72	1.45	0.00	0.00	0.12	5.07	Zone 6	0.00	1.17	0.12	0.29	0.16	0.00
Zone 7	0.93	1.39	0.00	0.00	0.12	5.11	Zone 7	0.00	0.90	0.20	0.31	0.16	0.00
Average	**0.56**	**1.63**	**0.00**	**0.00**	**0.12**	**3.38**	**Average**	**0.00**	**1.30**	**0.22**	**0.28**	**0.25**	**0.00**
Apartment, High							**Day Care**						
Zone 1	0.00	2.11	0.00	0.00	0.14	0.00	Zone 1	0.00	3.47	0.00	0.02	0.29	0.00
Zone 2	0.00	2.03	0.00	0.00	0.13	0.00	Zone 2	0.00	3.42	0.00	0.02	0.28	0.00
Zone 3	0.48	1.88	0.00	0.00	0.14	1.75	Zone 3	0.00	3.40	0.00	0.02	0.29	0.00
Zone 4	0.57	1.77	0.00	0.00	0.14	4.22	Zone 4	0.00	3.29	0.00	0.02	0.29	0.00
Zone 5	0.66	1.64	0.00	0.00	0.14	5.40	Zone 5	0.00	3.08	0.00	0.02	0.29	0.00
Zone 6	0.71	1.63	0.00	0.00	0.15	5.64	Zone 6	0.00	2.29	0.00	0.02	0.29	0.00
Zone 7	0.91	1.18	0.00	0.00	0.14	5.75	Zone 7	0.00	2.12	0.00	0.02	0.24	0.00
Average	**0.48**	**1.75**	**0.00**	**0.00**	**0.14**	**3.25**	**Average**	**0.05**	**3.01**	**0.05**	**0.07**	**0.29**	**0.05**
Assisted Living							**Fitness Center**						
Zone 1	0.00	2.65	0.00	0.04	0.31	0.00	Zone 1	0.00	4.97	0.00	0.00	0.59	0.00
Zone 2	0.00	2.63	0.00	0.04	0.31	0.00	Zone 2	0.00	4.94	0.00	0.00	0.59	0.00
Zone 3	0.00	2.57	0.00	0.04	0.31	0.00	Zone 3	0.00	4.80	0.00	0.00	0.60	0.00
Zone 4	0.00	2.50	0.00	0.04	0.31	0.00	Zone 4	0.00	4.76	0.00	0.00	0.60	0.00
Zone 5	0.00	2.42	0.00	0.04	0.30	0.00	Zone 5	0.00	4.75	0.00	0.00	0.59	0.00
Zone 6	0.00	1.99	0.00	0.04	0.32	0.00	Zone 6	0.00	4.13	0.00	0.00	0.60	0.00
Zone 7	0.00	1.90	0.00	0.04	0.30	0.00	Zone 7	0.00	3.27	0.00	0.00	0.60	0.00
Average	**0.00**	**2.38**	**0.00**	**0.04**	**0.31**	**0.00**	**Average**	**0.00**	**4.52**	**0.00**	**0.00**	**0.59**	**0.00**
Auditorium							**Hospital**						
Zone 1	0.00	2.32	0.34	0.37	1.03	0.00	Zone 1	0.00	1.54	0.23	0.26	0.80	0.00
Zone 2	0.00	2.32	0.34	0.38	1.03	0.00	Zone 2	0.00	1.47	0.22	0.25	0.80	0.00
Zone 3	0.00	2.29	0.33	0.38	1.04	0.00	Zone 3	0.00	1.44	0.22	0.25	0.81	0.00
Zone 4	0.00	2.26	0.29	0.35	1.05	0.00	Zone 4	0.00	1.42	0.22	0.25	0.81	0.00
Zone 5	0.00	2.20	0.34	0.39	1.04	0.00	Zone 5	0.00	1.37	0.23	0.27	0.81	0.00
Zone 6	0.00	1.71	0.14	0.34	1.05	0.00	Zone 6	0.00	1.26	0.15	0.25	0.82	0.00
Zone 7	0.00	1.38	0.23	0.36	1.04	0.00	Zone 7	0.00	1.07	0.19	0.26	0.82	0.00
Average	**0.00**	**2.07**	**0.29**	**0.37**	**1.04**	**0.00**	**Average**	**0.00**	**1.36**	**0.21**	**0.26**	**0.81**	**0.00**
Bank							**Hotel**						
Zone 1	0.00	3.32	0.00	0.00	0.60	0.00	Zone 1	0.00	1.77	0.00	0.01	0.26	0.00
Zone 2	0.00	3.28	0.00	0.00	0.61	0.00	Zone 2	0.00	1.64	0.00	0.01	0.26	0.00
Zone 3	0.00	3.28	0.00	0.00	0.61	0.00	Zone 3	0.00	1.62	0.00	0.01	0.27	0.00
Zone 4	0.00	3.21	0.00	0.00	0.60	0.00	Zone 4	0.00	1.62	0.00	0.01	0.27	0.00
Zone 5	0.00	3.06	0.00	0.00	0.59	0.00	Zone 5	0.00	1.51	0.00	0.01	0.27	0.00
Zone 6	0.00	2.82	0.00	0.00	0.61	0.00	Zone 6	0.00	1.25	0.00	0.01	0.28	0.00
Zone 7	0.00	2.66	0.00	0.00	0.60	0.00	Zone 7	0.00	1.23	0.00	0.01	0.28	0.00
Average	**0.00**	**3.09**	**0.00**	**0.00**	**0.60**	**0.00**	**Average**	**0.00**	**1.52**	**0.00**	**0.01**	**0.27**	**0.00**

Table R260120-30 Watts per S.F. for Various Systems by Building Type & Climate Zone (cont.)

Building / Zone	Heating	Cooling	Heat rejection	Auxiliary (pumps)	Vent fan	Supp heat pump	Building / Zone	Heating	Cooling	Heat rejection	Auxiliary (pumps)	Vent fan	Supp heat pump
Manufacturing, General							**Office, Mid**						
Zone 1	0.00	1.46	0.00	0.00	0.13	0.00	Zone 1	0.00	1.61	0.31	0.33	0.45	0.00
Zone 2	0.00	1.45	0.00	0.00	0.13	0.00	Zone 2	0.00	1.60	0.29	0.32	0.46	0.00
Zone 3	0.00	1.43	0.00	0.00	0.13	0.00	Zone 3	0.00	1.59	0.28	0.34	0.42	0.00
Zone 4	0.00	1.39	0.00	0.00	0.13	0.00	Zone 4	0.00	1.56	0.27	0.33	0.43	0.00
Zone 5	0.00	1.39	0.00	0.00	0.13	0.00	Zone 5	0.00	1.55	0.34	0.36	0.37	0.00
Zone 6	0.00	1.21	0.00	0.00	0.13	0.00	Zone 6	0.00	1.29	0.16	0.34	0.31	0.00
Zone 7	0.00	1.00	0.00	0.00	0.13	0.00	Zone 7	0.00	1.10	0.22	0.35	0.32	0.00
Average	**0.00**	**1.33**	**0.00**	**0.00**	**0.13**	**0.00**	**Average**	**0.00**	**1.47**	**0.27**	**0.34**	**0.40**	**0.00**
Manufacturing, High Tech							**Office, High**						
Zone 1	0.00	2.15	0.00	0.00	0.17	0.00	Zone 1	0.00	1.70	0.30	0.33	0.44	0.00
Zone 2	0.00	2.12	0.00	0.00	0.17	0.00	Zone 2	0.00	1.59	0.29	0.32	0.41	0.00
Zone 3	0.00	2.04	0.00	0.00	0.17	0.00	Zone 3	0.00	1.59	0.32	0.34	0.42	0.00
Zone 4	0.00	2.01	0.00	0.00	0.17	0.00	Zone 4	0.00	1.58	0.28	0.35	0.42	0.00
Zone 5	0.00	1.88	0.00	0.00	0.17	0.00	Zone 5	0.00	1.58	0.32	0.36	0.41	0.00
Zone 6	0.00	1.63	0.00	0.00	0.17	0.00	Zone 6	0.00	1.11	0.16	0.29	0.33	0.00
Zone 7	0.00	1.50	0.00	0.00	0.17	0.00	Zone 7	0.00	0.86	0.23	0.31	0.37	0.00
Average	**0.00**	**1.90**	**0.00**	**0.00**	**0.17**	**0.00**	**Average**	**0.00**	**1.43**	**0.27**	**0.33**	**0.40**	**0.00**
Medial Clinic							**Restaurant, Bar/Lounge**						
Zone 1	0.00	3.15	0.00	0.00	0.54	0.00	Zone 1	0.00	3.07	0.00	0.06	0.24	0.00
Zone 2	0.00	3.09	0.00	0.00	0.53	0.00	Zone 2	0.00	2.87	0.00	0.06	0.25	0.00
Zone 3	0.00	3.07	0.00	0.00	0.54	0.00	Zone 3	0.00	2.85	0.00	0.06	0.24	0.00
Zone 4	0.00	3.02	0.00	0.00	0.54	0.00	Zone 4	0.00	2.83	0.00	0.06	0.24	0.00
Zone 5	0.00	2.99	0.00	0.00	0.54	0.00	Zone 5	0.00	2.80	0.00	0.06	0.24	0.00
Zone 6	0.00	2.58	0.00	0.00	0.55	0.00	Zone 6	0.00	1.88	0.00	0.06	0.24	0.00
Zone 7	0.00	2.45	0.00	0.00	0.52	0.00	Zone 7	0.00	1.87	0.00	0.06	0.23	0.00
Average	**0.00**	**2.91**	**0.00**	**0.00**	**0.54**	**0.00**	**Average**	**0.00**	**2.59**	**0.00**	**0.06**	**0.24**	**0.00**
Motel							**Restaurant, Full**						
Zone 1	0.97	2.68	0.00	0.00	0.19	2.15	Zone 1	0.00	3.37	0.00	0.05	0.38	0.00
Zone 2	1.19	2.52	0.00	0.00	0.18	4.43	Zone 2	0.00	3.30	0.00	0.05	0.39	0.00
Zone 3	1.19	2.52	0.00	0.00	0.19	4.81	Zone 3	0.00	3.25	0.00	0.05	0.39	0.00
Zone 4	1.19	2.42	0.00	0.00	0.21	7.39	Zone 4	0.00	3.20	0.00	0.05	0.39	0.00
Zone 5	1.28	2.26	0.00	0.00	0.21	7.82	Zone 5	0.00	3.03	0.00	0.05	0.38	0.00
Zone 6	1.29	2.07	0.00	0.00	0.23	8.89	Zone 6	0.00	2.35	0.00	0.05	0.39	0.00
Zone 7	1.39	1.97	0.00	0.00	0.23	9.51	Zone 7	0.00	2.33	0.00	0.05	0.38	0.00
Average	**1.21**	**2.35**	**0.00**	**0.00**	**0.21**	**6.43**	**Average**	**0.00**	**2.98**	**0.00**	**0.05**	**0.38**	**0.00**
Museum							**Restaurant, Quick**						
Zone 1	0.00	3.77	0.00	0.11	0.67	0.00	Zone 1	0.00	3.78	0.00	0.08	0.48	0.00
Zone 2	0.00	3.71	0.00	0.12	0.77	0.00	Zone 2	0.00	3.72	0.00	0.08	0.50	0.00
Zone 3	0.00	3.59	0.00	0.12	0.72	0.00	Zone 3	0.00	3.64	0.00	0.08	0.50	0.00
Zone 4	0.00	3.47	0.00	0.14	0.74	0.00	Zone 4	0.00	3.60	0.00	0.08	0.50	0.00
Zone 5	0.00	3.42	0.00	0.16	0.51	0.00	Zone 5	0.00	3.42	0.00	0.08	0.49	0.00
Zone 6	0.00	2.39	0.00	0.15	0.23	0.00	Zone 6	0.00	2.74	0.00	0.08	0.51	0.00
Zone 7	0.00	1.69	0.00	0.17	0.24	0.00	Zone 7	0.00	2.69	0.00	0.08	0.49	0.00
Average	**0.00**	**3.15**	**0.00**	**0.14**	**0.55**	**0.00**	**Average**	**0.00**	**3.37**	**0.00**	**0.08**	**0.50**	**0.00**
Office, Low							**Store, Convenience**						
Zone 1	0.00	2.90	0.00	0.02	0.29	0.00	Zone 1	0.00	4.08	0.00	0.02	0.29	0.00
Zone 2	0.00	2.90	0.00	0.02	0.30	0.00	Zone 2	0.00	3.66	0.00	0.02	0.30	0.00
Zone 3	0.00	2.81	0.00	0.02	0.30	0.00	Zone 3	0.00	3.60	0.00	0.02	0.30	0.00
Zone 4	0.00	2.70	0.00	0.02	0.29	0.00	Zone 4	0.00	3.46	0.00	0.02	0.29	0.00
Zone 5	0.00	2.66	0.00	0.02	0.29	0.00	Zone 5	0.00	3.28	0.00	0.02	0.29	0.00
Zone 6	0.00	2.04	0.00	0.02	0.30	0.00	Zone 6	0.00	2.66	0.00	0.02	0.27	0.00
Zone 7	0.00	1.82	0.00	0.02	0.29	0.00	Zone 7	0.00	2.50	0.00	0.02	0.24	0.00
Average	**0.00**	**2.55**	**0.00**	**0.02**	**0.29**	**0.00**	**Average**	**0.00**	**3.32**	**0.00**	**0.02**	**0.28**	**0.00**

Table R260120-30 Watts per S.F. for Various Systems by Building Type & Climate Zone (cont.)

Building / Zone	Heating	Cooling	Heat rejection	Auxiliary (pumps)	Vent fan	Supp heat pump	Building / Zone	Heating	Cooling	Heat rejection	Auxiliary (pumps)	Vent fan	Supp heat pump
Store, Department							**School, Middle**						
Zone 1	0.00	1.65	0.20	0.23	0.63	0.00	Zone 1	0.00	2.48	0.00	0.00	0.17	0.00
Zone 2	0.00	1.61	0.20	0.23	0.63	0.00	Zone 2	0.00	2.45	0.00	0.00	0.17	0.00
Zone 3	0.00	1.59	0.19	0.22	0.64	0.00	Zone 3	0.00	2.41	0.00	0.00	0.18	0.00
Zone 4	0.00	1.52	0.19	0.23	0.64	0.00	Zone 4	0.00	2.35	0.00	0.00	0.17	0.00
Zone 5	0.00	1.51	0.21	0.24	0.63	0.00	Zone 5	0.00	2.26	0.00	0.00	0.17	0.00
Zone 6	0.00	1.32	0.13	0.21	0.63	0.00	Zone 6	0.00	1.61	0.00	0.00	0.17	0.00
Zone 7	0.00	0.98	0.17	0.23	0.63	0.00	Zone 7	0.00	1.49	0.00	0.00	0.17	0.00
Average	**0.00**	**1.45**	**0.18**	**0.23**	**0.63**	**0.00**	**Average**	**0.00**	**2.15**	**0.00**	**0.00**	**0.17**	**0.00**
Store, Large							**School, High**						
Zone 1	0.00	2.68	0.00	0.00	0.16	0.00	Zone 1	0.00	3.24	0.00	0.01	0.22	0.00
Zone 2	0.00	2.61	0.00	0.00	0.17	0.00	Zone 2	0.00	3.08	0.00	0.01	0.22	0.00
Zone 3	0.00	2.61	0.00	0.00	0.17	0.00	Zone 3	0.00	3.05	0.00	0.01	0.23	0.00
Zone 4	0.00	2.56	0.00	0.00	0.16	0.00	Zone 4	0.00	3.00	0.00	0.01	0.23	0.00
Zone 5	0.00	2.53	0.00	0.00	0.16	0.00	Zone 5	0.00	2.94	0.00	0.01	0.23	0.00
Zone 6	0.00	1.84	0.00	0.00	0.16	0.00	Zone 6	0.00	1.98	0.00	0.01	0.23	0.00
Zone 7	0.00	1.50	0.00	0.00	0.16	0.00	Zone 7	0.00	1.81	0.00	0.01	0.23	0.00
Average	**0.00**	**2.33**	**0.00**	**0.00**	**0.16**	**0.00**	**Average**	**0.00**	**2.73**	**0.00**	**0.01**	**0.23**	**0.00**
Store, Small							**School, College**						
Zone 1	0.00	3.60	0.00	0.01	0.25	0.00	Zone 1	0.00	2.92	0.00	0.00	0.20	0.00
Zone 2	0.00	3.51	0.00	0.01	0.26	0.00	Zone 2	0.00	2.91	0.00	0.00	0.20	0.00
Zone 3	0.00	3.38	0.00	0.01	0.25	0.00	Zone 3	0.00	2.78	0.00	0.00	0.20	0.00
Zone 4	0.00	3.32	0.00	0.01	0.25	0.00	Zone 4	0.00	2.72	0.00	0.00	0.20	0.00
Zone 5	0.00	3.23	0.00	0.01	0.24	0.00	Zone 5	0.00	2.43	0.00	0.00	0.20	0.00
Zone 6	0.00	2.49	0.00	0.01	0.23	0.00	Zone 6	0.00	1.77	0.00	0.00	0.20	0.00
Zone 7	0.00	2.17	0.00	0.01	0.21	0.00	Zone 7	0.00	1.51	0.00	0.00	0.19	0.00
Average	**0.00**	**3.10**	**0.00**	**0.01**	**0.24**	**0.00**	**Average**	**0.00**	**2.43**	**0.00**	**0.00**	**0.20**	**0.00**
Store, Strip Mall							**Service Station**						
Zone 1	0.00	3.36	0.00	0.02	0.25	0.00	Zone 1	0.00	10.42	0.00	0.03	0.47	0.00
Zone 2	0.00	3.30	0.00	0.02	0.25	0.00	Zone 2	0.00	9.01	0.00	0.03	0.46	0.00
Zone 3	0.00	3.14	0.00	0.02	0.25	0.00	Zone 3	0.00	7.82	0.00	0.03	0.49	0.00
Zone 4	0.00	3.14	0.00	0.02	0.25	0.00	Zone 4	0.00	7.02	0.00	0.03	0.47	0.00
Zone 5	0.00	2.24	0.00	0.02	0.24	0.00	Zone 5	0.00	6.13	0.00	0.03	0.46	0.00
Zone 6	0.00	1.96	0.00	0.02	0.25	0.00	Zone 6	0.00	4.32	0.00	0.03	0.45	0.00
Zone 7	0.00	1.64	0.00	0.02	0.24	0.00	Zone 7	0.00	4.18	0.00	0.03	0.40	0.00
Average	**0.00**	**2.52**	**0.00**	**0.02**	**0.25**	**0.00**	**Average**	**0.00**	**6.99**	**0.00**	**0.03**	**0.46**	**0.00**
Store, Warehouse							**Storage, High Bay**						
Zone 1	0.00	2.99	0.00	0.00	0.21	0.00	Zone 1	0.00	1.98	0.00	0.00	0.17	0.00
Zone 2	0.00	2.97	0.00	0.00	0.21	0.00	Zone 2	0.00	1.97	0.00	0.00	0.18	0.00
Zone 3	0.00	2.94	0.00	0.00	0.21	0.00	Zone 3	0.00	1.90	0.00	0.00	0.18	0.00
Zone 4	0.00	2.83	0.00	0.00	0.21	0.00	Zone 4	0.00	1.89	0.00	0.00	0.18	0.00
Zone 5	0.00	2.70	0.00	0.00	0.21	0.00	Zone 5	0.00	1.85	0.00	0.00	0.17	0.00
Zone 6	0.00	2.10	0.00	0.00	0.21	0.00	Zone 6	0.00	1.29	0.00	0.00	0.17	0.00
Zone 7	0.00	1.78	0.00	0.00	0.21	0.00	Zone 7	0.00	1.10	0.00	0.00	0.17	0.00
Average	**0.00**	**2.61**	**0.00**	**0.00**	**0.21**	**0.00**	**Average**	**0.00**	**1.71**	**0.00**	**0.00**	**0.17**	**0.00**
School, Elementary							**Storage, Low Bay**						
Zone 1	0.00	2.73	0.00	0.01	0.21	0.00	Zone 1	0.00	1.87	0.00	0.00	0.16	0.00
Zone 2	0.00	2.69	0.00	0.01	0.21	0.00	Zone 2	0.00	1.85	0.00	0.00	0.16	0.00
Zone 3	0.00	2.63	0.00	0.01	0.21	0.00	Zone 3	0.00	1.82	0.00	0.00	0.16	0.00
Zone 4	0.00	2.61	0.00	0.01	0.20	0.00	Zone 4	0.00	1.77	0.00	0.00	0.16	0.00
Zone 5	0.00	2.51	0.00	0.01	0.20	0.00	Zone 5	0.00	1.76	0.00	0.00	0.16	0.00
Zone 6	0.00	1.71	0.00	0.01	0.20	0.00	Zone 6	0.00	1.20	0.00	0.00	0.16	0.00
Zone 7	0.00	1.64	0.00	0.01	0.20	0.00	Zone 7	0.00	1.00	0.00	0.00	0.16	0.00
Average	**0.00**	**2.36**	**0.00**	**0.01**	**0.20**	**0.00**	**Average**	**0.00**	**1.61**	**0.00**	**0.00**	**0.16**	**0.00**

R260519-20 Armored Cable

Armored Cable – Quantities are taken off in the same manner as wire.

Bx Type Cable – Productivities are based on an average run of 50' before terminating at a box fixture, etc. Each 50' section includes field preparation of (2) ends with hacksaw, identification and tagging of wire. Set up is open coil type without reels attaching cable to snake and pulling across a suspended ceiling or open face wood or steel studding, price does not include drilling of studs.

Cable in Tray – Productivities are based on an average run of 100 L.F. with set up of pulling equipment for (2) 90° bends, attaching cable to pull-in means, identification and tagging of wires, set up of reels. Wire termination to breakers equipment, etc., are not included.

Job Conditions – Productivities are based on new construction to a height of 15' using rolling staging in an unobstructed area. Material staging is assumed to be within 100' of work being performed.

R260519-80 Undercarpet Systems

Takeoff Procedure for Power Systems: List components for each fitting type, tap, splice, and bend on your quantity takeoff sheet. Each component must be priced separately. Start at the power supply transition fittings and survey each circuit for the components needed. List the quantities of each component under a specific circuit number. Use the floor plan layout scale to get cable footage.

Reading across the list, combine the totals of each component in each circuit and list the total quantity in the last column. Calculate approximately 5% for scrap for items such as cable, top shield, tape, and spray adhesive. Also provide for final variations that may occur on-site.

Suggested guidelines are:
1. Equal amounts of cable and top shield should be priced.
2. For each roll of cable, price a set of cable splices.
3. For every 1 ft. of cable, price 2-1/2 ft. of hold-down tape.
4. For every 3 rolls of hold-down tape, price 1 can of spray adhesive.

Adjust final figures wherever possible to accommodate standard packaging of the product. This information is available from the distributor.

Each transition fitting requires:
1. 1 base
2. 1 cover
3. 1 transition block

Each floor fitting requires:
1. 1 frame/base kit
2. 1 transition block
3. 2 covers (duplex/blank)

Each tap requires:
1. 1 tap connector for each conductor
2. 1 pair insulating patches
3. 2 top shield connectors

Each splice requires:
1. 1 splice connector for each conductor
2. 1 pair insulating patches
3. 3 top shield connectors

Each cable bend requires:
1. 2 top shield connectors

Each cable dead end (outside of transition block) requires:
1. 1 pair insulating patches

Labor does not include:
1. Patching or leveling uneven floors.
2. Filling in holes or removing projections from concrete slabs.
3. Sealing porous floors.
4. Sweeping and vacuuming floors.
5. Removal of existing carpeting.
6. Carpet square cut-outs.
7. Installation of carpet squares.

Takeoff Procedures for Telephone Systems: After reviewing floor plans identify each transition. Number or letter each cable run from that fitting.

Start at the transition fitting and survey each circuit for the components needed. List the cable type, terminations, cable length, and floor fitting type under the specific circuit number. Use the floor plan layout scale to get the cable footage. Add some extra length (next higher increment of 5 feet) to preconnectorized cable.

Transition fittings require:
1. 1 base plate
2. 1 cover
3. 1 transition block

Floor fittings require:
1. 1 frame/base kit
2. 2 covers
3. Modular jacks

Reading across the list, combine the list of components in each circuit and list the total quantity in the last column. Calculate the necessary scrap factors for such items as tape, bottom shield and spray adhesive. Also provide for final variations that may occur on-site.

Adjust final figures whenever possible to accommodate standard packaging. Check that items such as transition fittings, floor boxes, and floor fittings that are to utilize both power and telephone have been priced as combination fittings, so as to avoid duplication.

Make sure to include marking of floors and drilling of fasteners if fittings specified are not the adhesive type.

Labor does not include:
1. Conduit or raceways before transition of floor boxes
2. Telephone cable before transition boxes
3. Terminations before transition boxes
4. Floor preparation as described in power section

Be sure to include all cable folds when pricing labor.

Takeoff Procedure for Data Systems: Start at the transition fittings and take off quantities in the same manner as the telephone system, keeping in mind that data cable does not require top or bottom shields.

The data cable is simply cross-taped on the cable run to the floor fitting.

Data cable can be purchased in either bulk form in which case coaxial connector material and labor must be priced, or in preconnectorized cut lengths.

Data cable cannot be folded and must be notched at 1 inch intervals. A count of all turns must be added to the labor portion of the estimate. (Note: Some manufacturers have prenotched cable.)

Notching required:
1. 90 degree turn requires 8 notches per side.
2. 180 degree turn requires 16 notches per side.

Floor boxes, transition boxes, and fittings are the same as described in the power and telephone procedures.

Since undercarpet systems require special hand tools, be sure to include this cost in proportion to number of crews involved in the installation.

Job Conditions: Productivity is based on new construction in an unobstructed area. Staging area is assumed to be within 200' of work being performed.

R260519-90 Wire

Wire quantities are taken off by either measuring each cable run or by extending the conduit and raceway quantities times the number of conductors in the raceway. Ten percent should be added for waste and tie-ins. Keep in mind that the unit of measure of wire is C.L.F. not L.F. as in raceways so the formula would read:

$$\frac{(\text{L.F. Raceway x No. of Conductors}) \times 1.10}{100} = \text{C.L.F.}$$

Price per C.L.F. of wire includes:
1. Set up wire coils or spools on racks
2. Attaching wire to pull in means
3. Measuring and cutting wire
4. Pulling wire into a raceway
5. Identifying and tagging

Price does not include:
1. Connections to breakers, panelboards or equipment
2. Splices

Job Conditions: Productivity is based on new construction to a height of 15′ using rolling staging in an unobstructed area. Material staging is assumed to be within 100′ of work being performed.

Economy of Scale: If more than three wires at a time are being pulled, deduct the following percentages from the labor of that grouping:

4-5 wires	25%
6-10 wires	30%
11-15 wires	35%
over 15	40%

If a wire pull is less than 100′ in length and is interrupted several times by boxes, lighting outlets, etc., it may be necessary to add the following lengths to each wire being pulled:

Junction box to junction box	2 L.F.
Lighting panel to junction box	6 L.F.
Distribution panel to sub panel	8 L.F.
Switchboard to distribution panel	12 L.F.
Switchboard to motor control center	20 L.F.
Switchboard to cable tray	40 L.F.

Measure of Drops and Riser: It is important when taking off wire quantities to include the wire for drops to electrical equipment. If heights of electrical equipment are not clearly stated, use the following guide:

	Bottom A.F.F.	Top A.F.F.	Inside Cabinet
Safety switch to 100A	5′	6′	2′
Safety switch 400 to 600A	4′	6′	3′
100A panel 12 to 30 circuit	4′	6′	3′
42 circuit panel	3′	6′	4′
Switch box	3′	3′6″	1′
Switchgear	0′	8′	8′
Motor control centers	0′	8′	8′
Transformers - wall mount	4′	8′	2′
Transformers - floor mount	0′	12′	4′

R260519-91   Maximum Circuit Length (approximate) for Various Power Requirements Assuming THW, Copper Wire @ 75°C, Based Upon a 4% Voltage Drop

Maximum Circuit Length: Table R260519-91 indicates typical maximum installed length a circuit can have and still maintain an adequate voltage level at the point of use. The circuit length is similar to the conduit length.

If the circuit length for an ampere load and a copper wire size exceeds the length obtained from Table R260519-91, use the next largest wire size to compensate voltage drop.

Example: A 130 ampere load at 480 volts, 3 phase, 3 wire with No. 1 wire can be run a maximum of 555 L.F. and provide satisfactory operation. If the same load is to be wired at the end of a 625 L.F. circuit, then a larger wire must be used.

| Amperes | Wire Size | Maximum Circuit Length in Feet | | | | |
| | | 2 Wire, 1 Phase | | 3 Wire, 3 Phase | | |
		120V	240V	240V	480V	600V
15	14*	50	105	120	240	300
	14	50	100	120	235	295
20	12*	60	125	145	290	360
	12	60	120	140	280	350
30	10*	65	130	155	305	380
	10	65	130	150	300	375
50	8	60	125	145	285	355
65	6	75	150	175	345	435
85	4	90	185	210	425	530
115	2	110	215	250	500	620
130	1	120	240	275	555	690
150	1/0	130	260	305	605	760
175	2/0	140	285	330	655	820
200	3/0	155	315	360	725	904
230	4/0	170	345	395	795	990
255	250	185	365	420	845	1055
285	300	195	395	455	910	1140
310	350	210	420	485	975	1220
380	500	245	490	565	1130	1415

*Solid Conductor

Note: The circuit length is the one-way distance between the origin and the load.

R260519-92 Minimum Copper and Aluminum Wire Size Allowed for Various Types of Insulation

Minimum Wire Sizes

Amperes	Copper THW THWN or XHHW	Copper THHN XHHW *	Aluminum THW XHHW	Aluminum THHN XHHW *	Amperes	Copper THW THWN or XHHW	Copper THHN XHHW *	Aluminum THW XHHW	Aluminum THHN XHHW *
15A	#14	#14	#12	#12	195	3/0	2/0	250kcmil	4/0
20	#12	#12	#10	#10	200	3/0	3/0	250kcmil	4/0
25	#10	#10	#10	#10	205	4/0	3/0	250kcmil	4/0
30	#10	#10	#8	#8	225	4/0	3/0	300kcmil	250kcmil
40	#8	#8	#8	#8	230	4/0	4/0	300kcmil	250kcmil
45	#8	#8	#6	#8	250	250kcmil	4/0	350kcmil	300kcmil
50	#8	#8	#6	#6	255	250kcmil	4/0	400kcmil	300kcmil
55	#6	#8	#4	#6	260	300kcmil	4/0	400kcmil	350kcmil
60	#6	#6	#4	#6	270	300kcmil	250kcmil	400kcmil	350kcmil
65	#6	#6	#4	#4	280	300kcmil	250kcmil	500kcmil	350kcmil
75	#4	#6	#3	#4	285	300kcmil	250kcmil	500kcmil	400kcmil
85	#4	#4	#2	#3	290	350kcmil	250kcmil	500kcmil	400kcmil
90	#3	#4	#2	#2	305	350kcmil	300kcmil	500kcmil	400kcmil
95	#3	#4	#1	#2	310	350kcmil	300kcmil	500kcmil	500kcmil
100	#3	#3	#1	#2	320	400kcmil	300kcmil	600kcmil	500kcmil
110	#2	#3	1/0	#1	335	400kcmil	350kcmil	600kcmil	500kcmil
115	#2	#2	1/0	#1	340	500kcmil	350kcmil	600kcmil	500kcmil
120	#1	#2	1/0	1/0	350	500kcmil	350kcmil	700kcmil	500kcmil
130	#1	#2	2/0	1/0	375	500kcmil	400kcmil	700kcmil	600kcmil
135	1/0	#1	2/0	1/0	380	500kcmil	400kcmil	750kcmil	600kcmil
150	1/0	#1	3/0	2/0	385	600kcmil	500kcmil	750kcmil	600kcmil
155	2/0	1/0	3/0	3/0	420	600kcmil	500kcmil		700kcmil
170	2/0	1/0	4/0	3/0	430		500kcmil		750kcmil
175	2/0	2/0	4/0	3/0	435		600kcmil		750kcmil
180	3/0	2/0	4/0	4/0	475		600kcmil		

*Dry Locations Only

Notes:

1. Size #14 to 4/0 is in AWG units (American Wire Gauge).
2. Size 250 to 750 is in kcmil units (Thousand Circular Mils).
3. Use next higher ampere value if exact value is not listed in table.
4. For loads that operate continuously increase ampere value by 25% to obtain proper wire size.
5. Refer to Table R260519-91 for the maximum circuit length for the various size wires.
6. Table R260519-92 has been written for estimating purpose only, based on ambient temperature of 30°C (86° F); for ambient temperature other than 30°C (86° F), ampacity correction factors will be applied.

R260519-93   Metric Equivalent, Wire

United States		European	
Size AWG or kcmil	**Area** Cir. Mils.(cmil) mm²	**Size** mm²	**Area** Cir. Mils.
18	1620/.82	.75	1480
16	2580/1.30	1.0	1974
14	4110/2.08	1.5	2961
12	6530/3.30	2.5	4935
10	10,380/5.25	4	7896
8	16,510/8.36	6	11,844
6	26,240/13.29	10	19,740
4	41,740/21.14	16	31,584
3	52,620/26.65	25	49,350
2	66,360/33.61	–	–
1	83,690/42.39	35	69,090
1/0	105,600/53.49	50	98,700
2/0	133,100/67.42	–	–
3/0	167,800/85.00	70	138,180
4/0	211,600/107.19	95	187,530
250	250,000/126.64	120	236,880
300	300,000/151.97	150	296,100
350	350,000/177.30	–	–
400	400,000/202.63	185	365,190
500	500,000/253.29	240	473,760
600	600,000/303.95	300	592,200
700	700,000/354.60	–	–
750	750,000/379.93	–	–

R260519-94   Size Required and Weight (Lbs./1000 L.F.) of Aluminum and Copper THW Wire by Ampere Load

Amperes	Copper Size	Aluminum Size	Copper Weight	Aluminum Weight
15	14	12	24	11
20	12	10	33	17
30	10	8	48	39
45	8	6	77	52
65	6	4	112	72
85	4	2	167	101
100	3	1	205	136
115	2	1/0	252	162
130	1	2/0	324	194
150	1/0	3/0	397	233
175	2/0	4/0	491	282
200	3/0	250	608	347
230	4/0	300	753	403
255	250	400	899	512
285	300	500	1068	620
310	350	500	1233	620
335	400	600	1396	772
380	500	750	1732	951

R260526-80 Grounding

Grounding When taking off grounding systems, identify separately the type and size of wire.

Example:

> Bare copper & size
>
> Bare aluminum & size
>
> Insulated copper & size
>
> Insulated aluminum & size

Count the number of ground rods and their size.

Example:

> 1. 8' grounding rod – 5/8" dia. 20 Ea.
> 2. 10' grounding rod – 5/8" dia. 12 Ea.
> 3. 15' grounding rod – 3/4" dia. 4 Ea.

Count the number of connections; the size of the largest wire will determine the productivity.

Example:

> Braze a #2 wire to a #4/0 cable
>
> The 4/0 cable will determine the L.H. and cost to be used.

Include individual connections to:
1. Ground rods
2. Building steel
3. Equipment
4. Raceways

Price does not include:
1. Excavation
2. Backfill
3. Sleeves or raceways used to protect grounding wires
4. Wall penetrations
5. Floor cutting
6. Core drilling

Job Conditions: Productivity is based on a ground floor area, using cable reels in an unobstructed area. Material staging area assumed to be within 100' of work being performed.

R260533-20 Conduit To 15' High

List conduit by quantity, size, and type. Do not deduct for lengths occupied by fittings, since this will be allowance for scrap. Example:
- A. Aluminum — size
- B. Rigid galvanized — size
- C. Steel intermediate (IMC) — size
- D. Rigid steel, plastic-coated 20 Mil. — size
- E. Rigid steel, plastic-coated 40 Mil. — size
- F. Electric metallic tubing (EMT) — size
- G. PVC Schedule 40 — size

Types (A) thru (E) listed above contain the following per 100 L.F.:
1. (11) Threaded couplings
2. (11) Beam-type hangers
3. (2) Factory sweeps
4. (2) Fiber bushings
5. (4) Locknuts
6. (2) Field threaded pipe terminations
7. (2) Removal of concentric knockouts

Type (F) contains per 100 L.F.:
1. (11) Set screw couplings
2. (11) Beam clamps
3. (2) Field bends on 1/2" and 3/4" diameter
4. (2) Factory sweeps for 1" and above
5. (2) Set screw steel connectors
6. (2) Removal of concentric knockouts

Type (G) contains per 100 L.F.:
1. (11) Field cemented couplings
2. (34) Beam clamps

3. (2) Factory sweeps
4. (2) Adapters
5. (2) Locknuts
6. (2) Removal of concentric knockouts

Labor-hours for all conduit to 15' include:
1. Unloading by hand
2. Hauling by hand to an area up to 200' from loading dock
3. Setup of rolling staging
4. Installation of conduit and fittings as described in Conduit models (A) thru (G)

Not included in the material and labor are:
1. Staging rental or purchase
2. Structural modifications
3. Wire
4. Junction boxes
5. Fittings in excess of those described in conduit models (A) thru (G)
6. Painting of conduit

Fittings

Only those fittings listed above are included in the linear foot totals, although they should be listed separately from conduit lengths, without prices, to ensure proper quantities for material procurement.

If the fittings required exceed the quantities included in the model conduit runs, then material and labor costs must be added to the difference. If actual needs per 100 L.F. of conduit are: (2) sweeps, (4) LBs and (1) field bend, then, (4) LBs and (1) field bend must be priced additionally.

R260533-21 Hangers

It is sometimes desirable to substitute an alternate style of hanger if the support being used is not the type described in the conduit models.

One approach is the substitution method:

1. Find the cost of the type hanger described in the conduit model.
2. Calculate the cost of the desired type hanger (it may be necessary to calculate individual components such as drilling, expansion shields, etc.).
3. Calculate the cost difference (delta) between the two types of hangers.
4. Multiply the cost delta by the number of hangers in the model.
5. Divide the total delta cost for hangers in the model by the length of the model to find the delta cost per L.F. for that model.
6. Modify the given unit costs per L.F. for the model by the delta cost per L.F.

Another approach to hanger configurations would be to start with the conduit only and add all the supports and any other items as separate lines. This procedure is most useful if the project involves racking many runs of conduit on a single hanger, for instance, a trapeze type hanger.

Example: Five (5) 2″ RGS conduits, 50 L.F. each, are to be run on trapeze hangers from one pull box to another. The run includes one 90° bend.

1. List the hangers' components to create an assembly cost for each 2′-wide trapeze.
2. List the components for the 50′ conduit run, noting that 6 trapeze supports will be required.

Job Conditions: Productivities are based on new construction to 15′ high, using scaffolding in an unobstructed area. Material storage is assumed to be within 100′ of work being performed.

Add to labor for elevated installations:

15′ to 20′ High–10%	30′ to 35′ High–30%
20′ to 25′ High–20%	35′ to 40′ High–35%
25′ to 30′ High–25%	Over 40′ High–40%

Add these percentages to the L.F. labor cost, but not to fittings. Add these percentages only to quantities exceeding the different height levels, rather than the total conduit quantities.

Linear foot price for labor does not include penetrations in walls or floors and must be added to the estimate

R260533-22   Conductors in Conduit

Table below lists maximum number of conductors for various sized conduit using THW, TW or THWN insulations.

Copper Wire Size	1/2″ TW	1/2″ THW	1/2″ THWN	3/4″ TW	3/4″ THW	3/4″ THWN	1″ TW	1″ THW	1″ THWN	1-1/4″ TW	1-1/4″ THW	1-1/4″ THWN	1-1/2″ TW	1-1/2″ THW	1-1/2″ THWN	2″ TW	2″ THW	2″ THWN	2-1/2″ TW	2-1/2″ THW	2-1/2″ THWN	3″ THW	3″ THWN	3-1/2″ THW	3-1/2″ THWN	4″ THW	4″ THWN
#14	9	6	13	15	10	24	25	16	39	44	29	69	60	40	94	99	65	154	142	93		143		192			
#12	7	4	10	12	8	18	19	13	29	35	24	51	47	32	70	78	53	114	111	76	164	117		157			
#10	5	4	6	9	6	11	15	11	18	26	19	32	36	26	44	60	43	73	85	61	104	95	160	127		163	
#8	2	1	3	4	3	5	7	5	9	12	10	16	17	13	22	28	22	36	40	32	51	49	79	66	106	85	136
#6		1	1		2	4		4	6		7	11		10	15		16	26		23	37	36	57	48	76	62	98
#4		1	1		1	2		3	4		5	7		7	9		12	16		17	22	27	35	36	47	47	60
#3		1	1		1	1		2	3		4	6		6	8		10	13		15	19	23	29	31	39	40	51
#2		1	1		1	1		2	3		4	5		5	7		9	11		13	16	20	25	27	33	34	43
#1					1	1		1	1		3	3		4	5		6	8		9	12	14	18	19	25	25	32
1/0					1	1		1	1		2	3		3	4		5	7		8	10	12	15	16	21	21	27
2/0					1	1		1	1		1	2		3	3		5	6		7	8	10	13	14	17	18	22
3/0					1	1		1	1		1	1		2	3		4	5		6	7	9	11	12	14	15	18
4/0						1		1	1		1	1		1	2		3	4		5	6	7	9	10	12	13	15
250 kcmil								1	1		1	1		1	1		2	3		4	4	6	7	8	10	10	12
300								1	1		1	1		1	1		2	3		3	4	5	6	7	8	9	11
350									1		1	1		1	1		1	2		3	3	4	5	6	7	8	9
400											1	1		1	1		1	1		2	3	4	5	5	6	7	8
500											1	1		1	1		1	1		1	2	3	4	4	5	6	7
600												1		1	1		1	1		1	1	3	3	4	4	5	5
700														1	1		1	1		1	1	2	3	3	4	4	5
750														1	1		1	1		1	1	2	2	3	3	4	4

R260533-23 Metric Equivalent, Conduit

United States		European	
U.S. vs. European Conduit – Approximate Equivalents			
Trade Size	Inside Diameter Inch/mm	Trade Size	Inside Diameter mm
½	.622/15.8	11	16.4
¾	.824/20.9	16	19.9
1	1.049/26.6	21	25.5
1¼	1.380/35.0	29	34.2
1½	1.610/40.9	36	44.0
2	2.067/52.5	42	51.0
2½	2.469/62.7		
3	3.068/77.9		
3½	3.548/90.12		
4	4.026/102.3		
5	5.047/128.2		
6	6.065/154.1		

R260533-24   Conduit Weight Comparisons (Lbs. per 100 ft.) Empty

Type	1/2″	3/4″	1″	1-1/4″	1-1/2″	2″	2-1/2″	3″	3-1/2″	4″	5″	6″
Rigid Aluminum	28	37	55	72	89	119	188	246	296	350	479	630
Rigid Steel	79	105	153	201	249	332	527	683	831	972	1314	1745
Intermediate Steel (IMC)	60	82	116	150	182	242	401	493	573	638		
Electrical Metallic Tubing (EMT)	29	45	65	96	111	141	215	260	365	390		
Polyvinyl Chloride, Schedule 40	16	22	32	43	52	69	109	142	170	202	271	350
Polyvinyl Chloride Encased Burial						38		67	88	105	149	202
Fibre Duct Encased Burial						127		164	180	206	400	511
Fibre Duct Direct Burial						150		251	300	354		
Transite Encased Burial						160		240	290	330	450	550
Transite Direct Burial						220		310		400	540	640

R260533-25   Conduit Weight Comparisons (Lbs. per 100 ft.) with Maximum Cable Fill*

Type	1/2″	3/4″	1″	1-1/4″	1-1/2″	2″	2-1/2″	3″	3-1/2″	4″	5″	6″
Rigid Galvanized Steel (RGS)	104	140	235	358	455	721	1022	1451	1749	2148	3083	4343
Intermediate Steel (IMC)	84	113	186	293	379	611	883	1263	1501	1830		
Electrical Metallic Tubing (EMT)	54	116	183	296	368	445	641	930	1215	1540		

*Conduit & Heaviest Conductor Combination

R260533-30   Labor for Couplings and Fittings

The labor included in the unit price lines for couplings is for their installation separate from a fitting. However, the labor included in the unit price lines for fittings covers their complete installation, which may include the labor installation of the coupling(s). If coupling(s) are required to complete a fitting installation, only the material for the required number of couplings needs to be added to the appropriate unit price lines.

R260533-60 Wireway

When "taking off" Wireway, list by size and type.

Example:

1. Screw cover, unflanged + size
2. Screw cover, flanged + size
3. Hinged cover, flanged + size
4. Hinged cover, unflanged + size

Each 10′ length on Wireway contains:

1. 10′ of cover either screw or hinged type
2. (1) Coupling or flange gasket
3. (1) Wall type mount

All fittings must be priced separately.

Substitution of hanger types is done the same as described in R260533-21, "HANGERS," keeping in mind that the wireway model is based on 10′ sections instead of a 100′ conduit run.

Labor-hours for wireway include:

1. Unloading by hand
2. Hauling by hand up to 100′ from loading dock
3. Measuring and marking
4. Mounting wall bracket using (2) anchor type lead fasteners
5. Installing wireway on brackets, to 15′ high (For higher elevations use factors in R260533-20)

Job Conditions: Productivity is based on new construction, to a height of 15′ using rolling staging in an unobstructed area.

Material staging area is assumed to be within 100′ of work being performed.

R260533-65 Outlet Boxes

Outlet boxes should be included on the same takeoff sheet as branch piping or devices to better explain what is included in each circuit.

Each unit price in this section is a stand alone item and contains no other component unless specified. For example, to estimate a duplex outlet, components that must be added are:

1. 4″ square box
2. 4″ plaster ring
3. Duplex receptacle
4. Device cover

The method of mounting outlet boxes is (2) plastic shield fasteners.

Outlet boxes plastic, labor-hours include:

1. Marking box location on wood studding
2. Mounting box

Economy of Scale – For large concentrations of plastic boxes in the same area deduct the following percentages from labor-hour totals:

1	to	10	0%
11	to	25	20%
26	to	50	25%
51	to	100	30%
	over	100	35%

Note: It is important to understand that these percentages are not used on the total job quantities, but only areas where concentrations exceed the levels specified.

R260533-70 Pull Boxes and Cabinets

List cabinets and pull boxes by NEMA type and size.

Example:	TYPE	SIZE
	NEMA 1	6″W x 6″H x 4″D
	NEMA 3R	6″W x 6″H x 4″D

Labor-hours for wall mount (indoor or outdoor) installations include:

1. Unloading and uncrating
2. Handling of enclosures up to 200′ from loading dock using a dolly or pipe rollers
3. Measuring and marking
4. Drilling (4) anchor type lead fasteners using a hammer drill
5. Mounting and leveling boxes

Note: A plywood backboard is not included.

Labor-hours for ceiling mounting include:

1. Unloading and uncrating
2. Handling boxes up to 100′ from loading dock

3. Measuring and marking
4. Drilling (4) anchor type lead fasteners using a hammer drill
5. Installing and leveling boxes to a height of 15′ using rolling staging

Labor-hours for free standing cabinets include:

1. Unloading and uncrating
2. Handling of cabinets up to 200′ from loading dock using a dolly or pipe rollers
3. Marking of floor
4. Drilling (4) anchor type lead fasteners using a hammer drill
5. Leveling and shimming

Labor-hours for telephone cabinets include:

1. Unloading and uncrating
2. Handling cabinets up to 200′ using a dolly or pipe rollers
3. Measuring and marking
4. Mounting and leveling, using (4) lead anchor type fasteners

R260533-75 Weight Comparisons of Common Size Cast Boxes in Lbs.

Size NEMA 4 or 9	Cast Iron	Cast Aluminum	Size NEMA 7	Cast Iron	Cast Aluminum
6″ x 6″ x 6″	17	7	6″ x 6″ x 6″	40	15
8″ x 6″ x 6″	21	8	8″ x 6″ x 6″	50	19
10″ x 6″ x 6″	23	9	10″ x 6″ x 6″	55	21
12″ x 12″ x 6″	52	20	12″ x 6″ x 6″	100	37
16″ x 16″ x 6″	97	36	16″ x 16″ x 6″	140	52
20″ x 20″ x 6″	133	50	20″ x 20″ x 6″	180	67
24″ x 18″ x 8″	149	56	24″ x 18″ x 8″	250	93
24″ x 24″ x 10″	238	88	24″ x 24″ x 10″	358	133
30″ x 24″ x 12″	324	120	30″ x 24″ x 10″	475	176
36″ x 36″ x 12″	500	185	30″ x 24″ x 12″	510	189

Fig. R260536-11

R260536-10   Cable Tray

Cable Tray - When taking off cable tray it is important to identify separately the different types and sizes involved in the system being estimated. (Fig. R260536-11)

- A. – Ladder Type, galvanized or aluminum
- B. – Trough Type, galvanized or aluminum
- C. – Solid Bottom, galvanized or aluminum

The unit of measure is calculated in linear feet; do not deduct from this footage any length occupied by fittings, this will be the only allowance for scrap. Be sure to include all vertical drops to panels, switch gear, etc.

Hangers – Included in the linear footage of cable tray is

- D. – 1 – Pair of connector plates per 12 L.F.
- E. – 1 – Pair clamp type hangers and 4′ of 3/8″ threaded rod per 12 L.F.

Not included are structural supports, which must be priced in addition to the hangers.

Fittings – Identify separately the different types of fittings

1.) Ladder Type, galvanized or aluminum
2.) Trough Type, galvanized or aluminum
3.) Solid Bottom Type, galvanized or aluminum

The configuration, radius and rung spacing must also be listed. The unit of measure is "Ea."

- F. – Elbow, vertical Ea.
- G. – Elbow, horizontal Ea.
- H. – Tee, vertical Ea.
- I. – Cross, horizontal Ea.
- J. – Wye, horizontal Ea.
- K. – Tee, horizontal Ea.
- L. – Reducing fitting Ea.

Depending on the use of the system other examples of units which must be included are:

- M. – Divider strip L.F.
- N. – Drop-outs Ea.

- O. – End caps Ea.
- P. – Panel connectors Ea.

Wire and cable are not included and should be taken off separately, see Unit Price sections.

Job Conditions – Unit prices are based on a new installation to a work plane of 15′ using rolling staging.

Add to labor for elevated installations:

15′ to 20′ High	10%
20′ to 25′ High	20%
25′ to 30′ High	25%
30′ to 35′ High	30%
35′ to 40′ High	35%
Over 40′ High	40%

Add these percentages for L.F. totals but not to fittings. Add percentages to only those quantities that fall in the different elevations, in other words, if the total quantity of cable tray is 200′ but only 75′ is above 15′ then the 10% is added to the 75′ only.

Linear foot costs do not include penetrations through walls and floors which must be added to the estimate.

Cable Tray Covers

Covers – Cable tray covers are taken off in the same manner as the tray itself, making distinctions as to the type of cover. (Fig. R260536-11)

- Q. – Vented, galvanized or aluminum
- R. – Solid, galvanized or aluminum

Cover configurations are taken off separately noting type, specific radius and widths.

Note: Care should be taken to identify from plans and specifications exactly what is being covered. In many systems only vertical fittings are covered to retain wire and cable.

Reference Tables

R260539-30   Conduit In Concrete Slab

List conduit by quantity, size and type.

Example:

  A. Rigid galvanized steel + size

  B. P.V.C. + size

Rigid galvanized steel (A) contains per 100 L.F.:

1. (20) Ties to slab reinforcing
2. (11) Threaded steel couplings
3. (2) Factory sweeps
4. (2) Field threaded conduit terminations
5. (2) Fiber bushings + locknuts
6. (2) Removal of concentric knockouts

P.V.C. (B) contains per 100 L.F.:

1. (20) Ties to slab reinforcing
2. (11) Field cemented couplings
3. (2) Factory sweeps
4. (2) Adapters
5. (2) Removal of concentric knockouts

R260539-40   Conduit In Trench

Conduit in trench is galvanized steel and contains per 100 L.F.:

1. (11) Threaded couplings
2. (2) Factory sweeps
3. (2) Fiber bushings + (4) locknuts
4. (2) Field threaded conduit terminations
5. (2) Removal of concentric knockouts

Note:

Conduit in Unit Price sections do not include:

1. Floor cutting
2. Excavation or backfill
3. Grouting or patching

Conduit fittings in excess of those listed in the above Conduit model must be added. (Refer to R260533-20 for Procedure example.)

Fig. R260543-51

R260543-50 Underfloor Duct

When pricing Underfloor Duct it is important to identify and list each component, since costs vary significantly from one type of fitting to another. Do not deduct boxes or fittings from linear foot totals; this will be your allowance for scrap.

The first step is to identify the system as either:

    FIG. R260543-51 Single Level

    FIG. R260543-52 Dual Level

Single Level System

Include on your "takeoff sheet" the following unit price items, making sure to distinquish between Standard and Super duct:

  A. Feeder duct (blank) in L.F.
  B. Distribution duct (Inserts 2′ on center) in L.F.
  C. Elbows (Vertical) Ea.
  D. Elbows (Horizontal) Ea.
  E. Cabinet connector Ea.
  F. Single duct junction box Ea.
  G. Double duct junction box Ea.
  H. Triple duct junction box Ea.
  I. Support, single cell Ea.
  J. Support, double cell Ea.
  K. Support, triple cell Ea.
  L. Carpet pan Ea.
  M. Terrazzo pan Ea.
  N. Insert to conduit adapter Ea.
  O. Conduit adapter Ea.
  P. Low tension outlet Ea.
  Q. High tension outlet Ea.
  R. Galvanized nipple Ea.
  S. Wire per C.L.F.
  T. Offset (Duct type) Ea.

**Dual Level System + Labor
see next page**

Fig. R260543-52

R260543-50   Underfloor Duct (cont.)

Dual Level

Include the following when "taking off" Dual Level systems:

Distinguish between Standard and Super duct.

- A. Feeder duct (blank) in L.F.
- B. Distribution duct (Inserts 2' on center) in L.F.
- C. Elbows (Vertical) Ea.
- D. Elbows (Horizontal) Ea.
- E. Cabinet connector Ea.
- F. Single duct, 2 level, junction box Ea.
- G. Double duct, 2 level, junction box Ea.
- H. Support, single cell Ea.
- I. Support, double cell Ea.
- J. Support, triple cell Ea.
- K. Carpet pan Ea.
- L. Terrazzo pan Ea.
- M. Insert to conduit adapter Ea.
- N. Conduit adapter Ea.
- O. Low tension outlet Ea.
- P. High tension outlet Ea.
- Q. Wire per C.L.F.

Note: Make sure to include risers in linear foot totals. High tension outlets include box, receptacle, covers and related mounting hardware.

Labor-hours for both Single and Dual Level systems include:

1. Unloading and uncrating
2. Hauling up to 200' from loading dock

3. Measuring and marking
4. Setting raceway and fittings in slab or on grade
5. Leveling raceway and fittings

Labor-hours do not include:

1. Floor cutting
2. Excavation or backfill
3. Concrete pour
4. Grouting or patching
5. Wire or wire pulls
6. Additional outlets after concrete is poured
7. Piping to or from Underfloor Duct

Note: Installation is based on installing up to 150' of duct. If quantities exceed this, deduct the following percentages:

1. 150' to 250' - 10%
2. 250' to 350' - 15%
3. 350' to 500' - 20%
4. over 500' - 25%

Deduct these percentages from labor only.

Deduct these percentages from straight sections only.

Do not deduct from fittings or junction boxes.

Job Conditions: Productivity is based on new construction.

Underfloor duct to be installed on first three floors.

Material staging area within 100' of work being performed.

Area unobstructed and duct not subject to physical damage.

R260580-75 Motor Connections

Motor connections should be listed by size and type of motor. Included in the material and labor cost is:

1. (2) Flex connectors
2. 18″ of flexible metallic wireway
3. Wire identification and termination
4. Test for rotation
5. (2) or (3) Conductors

Price does not include:

1. Mounting of motor
2. Disconnect Switch
3. Motor Starter
4. Controls
5. Conduit or wire ahead of flex

Note: When "Taking off" Motor connections, it is advisable to list connections on the same quantity sheet as Motors, Motor Starters and controls.

R260913-80 Switchboard Instruments

Switchboard instruments are added to the price of switchboards according to job specifications. This equipment is usually included when ordering "Gear" from the manufacturer and will arrive factory installed.

Included in the labor cost is:

1. Internal wiring connections
2. Wire identification
3. Wire tagging

Transition sections include:

1. Uncrating
2. Hauling sections up to 100′ from loading dock
3. Positioning sections
4. Leveling sections

5. Bolting enclosures
6. Bolting vertical bus bars

Price does not include:

1. Equipment pads
2. Steel channels embedded or grouted in concrete
3. Special knockouts
4. Rigging

Job Conditions: Productivity is based on new construction, equipment to be installed on the first floor within 200′ of the loading dock.

R262213-10   Electric Circuit Voltages

General: The following method provides the user with a simple non-technical means of obtaining comparative costs of wiring circuits. The circuits considered serve the electrical loads of motors, electric heating, lighting and transformers, for example, that require low voltage 60 Hertz alternating current.

The method used here is suitable only for obtaining estimated costs. It is **not** intended to be used as a substitute for electrical engineering design applications.

Conduit and wire circuits can represent from twenty to thirty percent of the total building electrical cost. By following the described steps and using the tables the user can translate the various types of electric circuits into estimated costs.

Wire Size: Wire size is a function of the electric load which is usually listed in one of the following units:

1. Amperes (A)
2. Watts (W)
3. Kilowatts (kW)
4. Volt amperes (VA)
5. Kilovolt amperes (kVA)
6. Horsepower (HP)

These units of electric load must be converted to amperes in order to obtain the size of wire necessary to carry the load. To convert electric load units to amperes one must have an understanding of the voltage classification of the power source and the voltage characteristics of the electrical equipment or load to be energized. The seven A.C. circuits commonly used are illustrated in Figures R262213-11 thru R262213-17 showing the tranformer load voltage and the point of use voltage at the point on the circuit where the load is connected. The difference between the source and point of use voltages is attributed to the circuit voltage drop and is considered to be approximately 4%.

Motor Voltages: Motor voltages are listed by their point of use voltage and not the power source voltage.

For example: 460 volts instead of 480 volts
200 instead of 208 volts
115 volts instead of 120 volts

Lighting and Heating Voltages: Lighting and heating equipment voltages are listed by the power source voltage and not the point of wire voltage.

For example: 480, 277, 120 volt lighting
480 volt heating or air conditioning unit
208 volt heating unit

Transformer Voltages: Transformer primary (input) and secondary (output) voltages are listed by the power source voltage.

For example: Single phase 10 kVA
Primary 240/480 volts
Secondary 120/240 volts

In this case, the primary voltage may be 240 volts with a 120 volts secondary or may be 480 volts with either a 120V or a 240V secondary.

For example: Three phase 10 kVA
Primary 480 volts
Secondary 208Y/120 volts

In this case the transformer is suitable for connection to a circuit with a 3 phase 3 wire or 3 phase 4 wire circuit with a 480 voltage. This application will provide a secondary circuit of 3 phase 4 wire with 208 volts between phase wires and 120 volts between any phase wire and the neutral (white) wire.

R262213-11

3 Wire, 1 Phase, 120/240 Volt System

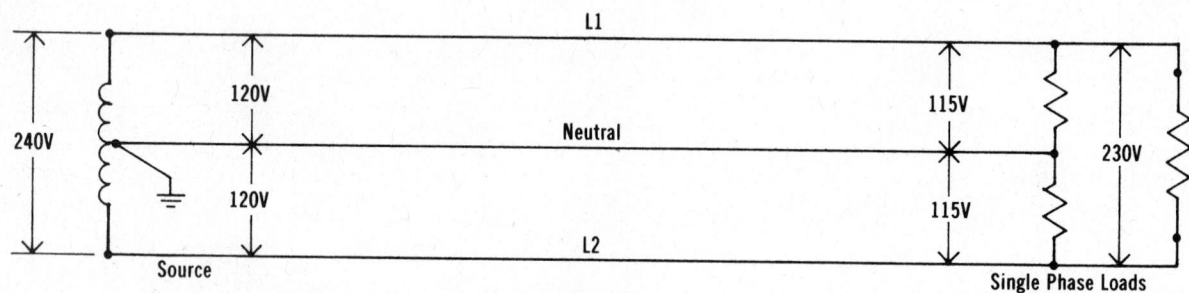

R262213-12

4 wire, 3 Phase, 208Y/120 Volt System

R262213-13 Electric Circuit Voltages (cont.)

3 Wire, 3 Phase 240 Volt System

R262213-14

4 Wire, 3 Phase, 240/120 Volt System

R262213-15

3 Wire, 3 Phase 480 Volt System

R262213-16

4 Wire, 3 Phase, 480Y/277 Volt System

R262213-17　Electric Circuit Voltages (cont.)

3 Wire, 3 Phase, 600 Volt System

R262213-20　kW Value/Cost Determination

General: Lighting and electric heating loads are expressed in watts and kilowatts.

Cost Determination:

The proper ampere values can be obtained as follows:

1. Convert watts to kilowatts
(watts 1000 ÷ kilowatts)
2. Determine voltage rating of equipment.
3. Determine whether equipment is single phase or three phase.
4. Refer to Table R262213-21 to find ampere value from kW, Ton and Btu/hr. values.
5. Determine type of wire insulation – TW, THW, THWN.
6. Determine if wire is copper or aluminum.
7. Refer to Table R260519-92 to obtain copper or aluminum wire size from ampere values.
8. Next refer to Table R260533-22 for the proper conduit size to accommodate the number and size of wires in each particular case.
9. Next refer to unit cost data for the per linear foot cost of the conduit.
10. Next refer to unit cost data for the per linear foot cost of the wire. Multiply cost of wire per L.F. x number of wires in the circuits to obtain total wire cost per L.F.

11. Add values obtained in Step 9 and 10 for total cost per linear foot for conduit and wire x length of circuit = Total Cost.

Notes:

1. 1 Phase refers to single phase, 2 wire circuits.
2. 3 Phase refers to three phase, 3 wire circuits.
3. For circuits which operate continuously for 3 hours or more, multiply the ampere values by 1.25 for a given kw requirement.
4. For kW ratings not listed, add ampere values.

For example: Find the ampere value of 9 kW at 208 volt, single phase.

$$\begin{array}{r} 4\text{ kW} = 19.2\text{A} \\ 5\text{ kW} = 24.0\text{A} \\ \hline 9\text{ kW} = 43.2\text{A} \end{array}$$

5. "Length of Circuit" refers to the one way distance of the run, not to the total sum of wire lengths.

R262213-21 Ampere Values as Determined by kW Requirements, BTU/HR or Ton, Voltage and Phase Values

			Ampere Values						
			120V	208V		240V		277V	480V
kW	Ton	BTU/HR	1 Phase	1 Phase	3 Phase	1 Phase	3 Phase	1 Phase	3 Phase
0.5	.1422	1,707	4.2A	2.4A	1.4A	2.1A	1.2A	1.8A	0.6A
0.75	.2133	2,560	6.2	3.6	2.1	3.1	1.9	2.7	.9
1.0	.2844	3,413	8.3	4.9	2.8	4.2	2.4	3.6	1.2
1.25	.3555	4,266	10.4	6.0	3.5	5.2	3.0	4.5	1.5
1.5	.4266	5,120	12.5	7.2	4.2	6.3	3.1	5.4	1.8
2.0	.5688	6,826	16.6	9.7	5.6	8.3	4.8	7.2	2.4
2.5	.7110	8,533	20.8	12.0	7.0	10.4	6.1	9.1	3.1
3.0	.8532	10,239	25.0	14.4	8.4	12.5	7.2	10.8	3.6
4.0	1.1376	13,652	33.4	19.2	11.1	16.7	9.6	14.4	4.8
5.0	1.4220	17,065	41.6	24.0	13.9	20.8	12.1	18.1	6.1
7.5	2.1331	25,598	62.4	36.0	20.8	31.2	18.8	27.0	9.0
10.0	2.8441	34,130	83.2	48.0	27.7	41.6	24.0	36.5	12.0
12.5	3.5552	42,663	104.2	60.1	35.0	52.1	30.0	45.1	15.0
15.0	4.2662	51,195	124.8	72.0	41.6	62.4	37.6	54.0	18.0
20.0	5.6883	68,260	166.4	96.0	55.4	83.2	48.0	73.0	24.0
25.0	7.1104	85,325	208.4	120.2	70.0	104.2	60.0	90.2	30.0
30.0	8.5325	102,390		144.0	83.2	124.8	75.2	108.0	36.0
35.0	9.9545	119,455		168.0	97.1	145.6	87.3	126.0	42.1
40.0	11.3766	136,520		192.0	110.8	166.4	96.0	146.0	48.0
45.0	12.7987	153,585			124.8	187.5	112.8	162.0	54.0
50.0	14.2208	170,650			140.0	208.4	120.0	180.4	60.0
60.0	17.0650	204,780			166.4		150.4	216.0	72.0
70.0	19.9091	238,910			194.2		174.6		84.2
80.0	22.7533	273,040			221.6		192.0		96.0
90.0	25.5975	307,170					225.6		108.0
100.0	28.4416	341,300							120.0

R262213-25   kVA Value/Cost Determination

General: Control transformers are listed in VA. Step-down and power transformers are listed in kVA.

Cost Determination:
1. Convert VA to kVA. Volt amperes (VA) ÷ 1000 = Kilovolt amperes (kVA).
2. Determine voltage rating of equipment.
3. Determine whether equipment is single phase or three phase.
4. Refer to Table R262213-26 to find ampere value from kVA value.
5. Determine type of wire insulation – TW, THW, THWN.
6. Determine if wire is copper or aluminum.
7. Refer to Table R260519-92 to obtain copper or aluminum wire size from ampere values.

Example: A transformer rated 10 kVA 480 volts primary, 240 volts secondary, 3 phase has the capacity to furnish the following:
1. Primary amperes = 10 kVA x 1.20 = 12 amperes (from Table R262213-26)
2. Secondary amperes = 10 kVA x 2.40 = 24 amperes (from Table R262213-26)

Note: Transformers can deliver generally 125% of their rated kVA. For instance, a 10 kVA rated transformer can safely deliver 12.5 kVA.

8. Next refer to Table R260533-22 for the proper conduit size to accommodate the number and size of wires in each particular case.
9. Next refer to unit price data for the per linear foot cost of the conduit.
10. Next refer to unit price data for the per linear foot cost of the wire. Multiply cost of wire per L.F. x number of wires in the circuits to obtain total wire cost.
11. Add values obtained in Step 9 and 10 for total cost per linear foot for conduit and wire x length of circuit = Total Cost.

R262213-26   Multiplier Values for kVA to Amperes Determined by Voltage and Phase Values

Volts	2 Wire, 1 Phase	3 Wire, 3 Phase
115	8.70	
120	8.30	
230	4.30	2.51
240	4.16	2.40
200	5.00	2.89
208	4.80	2.77
265	3.77	2.18
277	3.60	2.08
460	2.17	1.26
480	2.08	1.20
575	1.74	1.00
600	1.66	0.96

R262213-27 Central Air Conditioning Watts per S.F., BTUs per Hour per S.F. of Floor Area and S.F. per Ton of Air Conditioning

Type Building	Watts per S.F.	BTUH per S.F.	S.F. per Ton	Type Building	Watts per S.F.	BTUH per S.F.	S.F. per Ton	Type Building	Watts per S.F.	BTUH per S.F.	S.F. per Ton
Apartments, Individual	3	26	450	Dormitory, Rooms	4.5	40	300	Libraries	5.7	50	240
Corridors	2.5	22	550	Corridors	3.4	30	400	Low Rise Office, Ext.	4.3	38	320
Auditoriums & Theaters	3.3	40	300/18*	Dress Shops	4.9	43	280	Interior	3.8	33	360
Banks	5.7	50	240	Drug Stores	9	80	150	Medical Centers	3.2	28	425
Barber Shops	5.5	48	250	Factories	4.5	40	300	Motels	3.2	28	425
Bars & Taverns	15	133	90	High Rise Off.-Ext. Rms.	5.2	46	263	Office (small suite)	4.9	43	280
Beauty Parlors	7.6	66	180	Interior Rooms	4.2	37	325	Post Office, Int. Office	4.9	42	285
Bowling Alleys	7.8	68	175	Hospitals, Core	4.9	43	280	Central Area	5.3	46	260
Churches	3.3	36	330/20*	Perimeter	5.3	46	260	Residences	2.3	20	600
Cocktail Lounges	7.8	68	175	Hotels, Guest Rooms	5	44	275	Restaurants	6.8	60	200
Computer Rooms	16	141	85	Public Spaces	6.2	55	220	Schools & Colleges	5.3	46	260
Dental Offices	6	52	230	Corridors	3.4	30	400	Shoe Stores	6.2	55	220
Dept. Stores, Basement	4	34	350	Industrial Plants, Offices	4.3	38	320	Shop'g. Ctrs., Sup. Mkts.	4	34	350
Main Floor	4.5	40	300	General Offices	4	34	350	Retail Stores	5.5	48	250
Upper Floor	3.4	30	400	Plant Areas	4.5	40	300	Specialty Shops	6.8	60	200

*Persons per ton

12,000 BTUH = 1 ton of air conditioning

R262213-60 Oil Filled Transformers

Transformers in this section include:
1. Rigging (as required)
2. Rental of crane and operator
3. Setting of oil filled transformer
4. (4) Anchor bolts, nuts and washers in concrete pad

Price does not include:
1. Primary and secondary terminations
2. Transformer pad
3. Equipment grounding
4. Cable
5. Conduit locknuts or bushings

Transformers in Unit Price sections for dry type, back-boost and isolating transformers include:
1. Unloading and uncrating
2. Hauling transformer to within 200' of loading dock

3. Setting in place
4. Wall mounting hardware
5. Testing

Price does not include:
1. Structural supports
2. Suspension systems
3. Welding or fabrication
4. Primary & secondary terminations

Add the following percentages to the labor for ceiling mounted transformers:

10' to 15'	= + 15%
15' to 25'	= + 30%
Over 25'	= + 35%

Job Conditions: Productivities are based on new construction. Installation is assumed to be on the first floor, in an obstructed area to a height of 10'. Material staging area is within 100' of final transformer location.

R262213-65   Transformer Weight (Lbs.) by kVA

Oil Filled 3 Phase 5/15 KV To 480/277			
kVA	Lbs.	kVA	Lbs.
150	1800	1000	6200
300	2900	1500	8400
500	4700	2000	9700
750	5300	3000	15000

Dry 240/480 To 120/240 Volt			
1 Phase		3 Phase	
kVA	Lbs.	kVA	Lbs.
1	23	3	90
2	36	6	135
3	59	9	170
5	73	15	220
7.5	131	30	310
10	149	45	400
15	205	75	600
25	255	112.5	950
37.5	295	150	1140
50	340	225	1575
75	550	300	1870
100	670	500	2850
167	900	750	4300

Electrical   R2624 Switchboards & Panelboards

R262416-50   Load Centers and Panelboards

When pricing Load Centers list panels by size and type. List Breakers in a separate column of the "Quantity Sheet," and define by phase and ampere rating.

Material and Labor prices include breakers; for example: a 100A, 3-Wire, 102/240V, 18 circuit panel w/ main breaker, as described in the unit cost section, contains 18 single pole 20A breakers.

If you do not choose to include a full panel of single pole breakers, use the following method to adjust material and labor costs.

Example: In an 18 circuit panel only 16 single pole breakers are desired, requiring that the cost of 2 breakers be subtracted from the panel cost.

1. Go to the appropriate unit cost section of the book to find the unit prices of the given circuit breaker type.
2. Modify those costs as follows: Bare material price x 0.50   Bare labor cost x 0.60.
3. Multiply those modified bare costs by 2 (breakers in this example).
4. Subtract the modified costs for the 2 breakers from the given cost of the panel.

Labor-hours for Load Center installation includes:

1. Unloading, uncrating, and handling enclosures 200' from unloading area
2. Measuring and marking

3. Drilling (4) lead anchor type fasteners using a hammer drill
4. Mounting and leveling panel to a height of 6'
5. Preparation and termination of feeder cable to lugs or main breaker
6. Branch circuit identification
7. Lacing using tie wraps
8. Testing and load balancing
9. Marking panel directory

Not included in the material and labor are:

1. Modifications to enclosure
2. Structural supports
3. Additional lugs
4. Plywood backboards
5. Painting or lettering

Note: Knockouts are included in the price of terminating pipe runs and need not be added to the Load Center costs.

Job Conditions: Productivity is based on new construction to a height of 6', in an unobstructed area. Material staging area is assumed to be within 100' of work being performed.

Fig. R262419-61

R262419-60 Motor Control Centers

When taking off Motor Control Centers, list the size, type and height of structures.

Example:
1. 600A, 22,000 RMS, 72″ high
2. 600A, back to back, 72″ high

Next take off individual starters; the number of structures can also be determined by adding the height in inches of starters divided by the height of the structure, and list on the same quantity sheet as the structures. Identify starters by Type, Horsepower rating, Size and Height in inches.

Example:
A. Class 1, Type B, FVNR starter, 25 H.P., 18″ high.
   Add to the list with Starters, factory installed controls.

Example:

B. Pilot lights
C. Push buttons
D. Auxiliary contacts

Identify starters and structures as either copper or aluminum and by the NEMA type of enclosure.

When pricing starters and structures, be sure to add or deduct adjustments, using lines in the unit price section.

Included in the cost of Motor Control Structures are:
1. Uncrating
2. Hauling to location within 100′ of loading dock
3. Setting structures
4. Leveling
5. Aligning
6. Bolting together structure frames
7. Bolting horizontal bus bars

Labor-hours do not include:
1. Equipment pad
2. Steel channels embedded or grouted in concrete
3. Pull boxes
4. Special knockouts
5. Main switchboard section
6. Transition section
7. Instrumentation
8. External control wiring
9. Conduit or wire

Material for Starters includes:
1. Circuit breaker or fused disconnect
2. Magnetic motor starter
3. Control transformer
4. Control fuse and fuse block

Labor-hours for Starters include:
1. Handling
2. Installing starter within structure
3. Internal wiring connections
4. Lacing within enclosure
5. Testing
6. Phasing

Job Conditions: Productivity is based on new construction. Motor control location assumed to be on first floor in an unobstructed area. Material staging area within 100′ of final location.

Note: Additional labor-hours must be added if M.C.C. is to be installed on other than the first floor, or if rigging is required.

R262419-65   Motor Starters and Controls

Motor starters should be listed on the same "Quantity Sheet" as Motors and Motor Connections. Identify each starter by:

1. Size
2. Voltage
3. Type

Example:

  A. FVNR, 480V, 2HP, Size 00
  B. FVNR, 480V, 5HP, Size 0, Combination type
  C. FVR, 480V, Size 2

The NEMA type of enclosure should also be identified.

Included in the labor-hours are:

1. Unloading, uncrating and handling of starters up to 200' from loading area
2. Measuring and marking
3. Drilling (4) anchor type lead fasteners, using a hammer drill
4. Mounting and leveling starter
5. Connecting wire or cable to line and load sides of starter (when already lugged)
6. Installation of (3) thermal type heaters
7. Testing

The following is not included unless specified in the unit price description.

1. Control transformer
2. Controls, either factory or field installed
3. Conduit and wire to or from starter
4. Plywood backboard
5. Cable terminations

The following material and labor has been included for Combination type starters:

1. Unloading, uncrating and handling of starter up to 200' of loading dock
2. Measuring and marking
3. Drilling (4) anchor type lead fasteners using a hammer drill
4. Mounting and leveling
5. Connecting prepared cable conductors
6. Installation of (3) dual element cartridge type fuses
7. Installation of (3) thermal type heaters
8. Test for rotation

MOTOR CONTROLS

When pricing motor controls make sure you consider the type of control system being utilized. If the controls are factory installed and located in the enclosure itself, then you would add to your starter price the items 5200 thru 5800 in the Unit Price section. If control voltage is different from line voltage, add the material and labor cost of a control transformer.

For external control of starters include the following items:

1. Raceways
2. Wire & terminations
3. Control enclosures
4. Fittings
5. Push button stations
6. Indicators

Job Conditions: Productivity is based on new construction to a height of 10'.

Material staging area is assumed to be within 100' of work being performed.

Fig. R262419-81

R262419-80 Distribution Section

After "Taking off" the Switchboard section, include on the same "Quantity sheet" the Distribution section; identify by:
1. Voltage
2. Ampere rating
3. Type

Example:

(Fig. R262419-81) (B) Distribution Section

Included in the labor costs of the Distribution section is:
1. Uncrating
2. Hauling to 200′ of loading dock
3. Setting of distribution panel
4. Leveling & shimming

5. Anchoring of equipment to pad or floor
6. Bolting of horizontal bus bars between
7. Testing of equipment

Not included in the Distribution section is:
1. Breakers (C)
2. Equipment pads
3. Steel channels embedded or grouted in concrete
4. Pull boxes
5. Special knockouts
6. Transition section
7. Conduit or wire

R262419-82 Feeder Section

List quantities on the same sheet as the Distribution section. Identify breakers by: (Fig. R262419-81)
1. Frame type
2. Number of poles
3. Ampere rating

Installation includes:
1. Handling

2. Placing breakers in distribution panel
3. Preparing wire or cable
4. Lacing wire
5. Marking each phase with colored tape
6. Marking panel legend
7. Testing
8. Balancing

R262419-84   Switchgear

It is recommended that "Switchgear" or those items contained in the following sections be quoted from equipment manufacturers as a package price.

    Switchboard instruments

    Switchboard distribution sections

    Switchboard feeder sections

    Switchboard Service Disc.

    Switchboard In-plant Dist.

Included in these sections are the most common types and sizes of factory assembled equipment.

The recommended procedure for low voltage switchgear would be to price (Fig. R262419-81) (A) Main Switchboard

Identify by:

1. Voltage
2. Amperage
3. Type

Example:

1. 120/208V, 4-wire, 600A, nonfused
2. 277/480V, 4-wire, 600A, nonfused
3. 120/208V, 4-wire, 400A w/fused switch & CT compartment
4. 277/480V, 4-wire, 400A, w/fused switch & CT compartment
5. 120/208V, 4-wire, 800A, w/pressure switch & CT compartment
6. 277/480V, 4-wire, 800A, w/molded CB & CT compartment

Included in the labor costs for Switchboards are:

1. Uncrating
2. Hauling to 200' of loading dock
3. Setting equipment
4. Leveling and shimming
5. Anchoring
6. Cable identification
7. Testing of equipment

Not included in the Switchboard price is:

1. Rigging
2. Equipment pads
3. Steel channels embedded or grouted in concrete
4. Special knockouts
5. Transition or Auxiliary sections
6. Instrumentation
7. External control
8. Conduit and wire
9. Conductor terminations

Fig. R16450-101

R262513-10 Aluminum Bus Duct

When taking off bus duct identify the system as either:
1. Aluminum
2. Copper

List straight lengths by type and size
  A. Feeder — 800 A
  B. Plug in — 800 A

Do not measure thru fittings as you would on conduit, since there is no allowance for scrap in bus duct systems.

If upon taking off linear foot quantities of bus duct you find your quantities are not divisible evenly by 10 ft., then the remainder must be priced as a special item and quoted from the manufactuer. Do not use the bare material cost per L.F. for these special items. You can, however, safely use the bare labor cost per L.F. for the entire length.

Identify fittings by type and ampere rating.

Example:

C. Switchboard stub 800 A

D. Elbows 800 A

E. End box 800 A

F. Cable tap box 800 A

G. Tee Fittings 800 A

H. Hangers

Plug-in Units – List separately plug-in units and identify by type and ampere rating

I. Plug-in switches 600 Volt 3 phase 60 A

J. Plug-in molded case C.B. 60 A

K. Combination starter FVNR NEMA 1

L. Combination contactor & fused switch NEMA 1

M. Combination fusible switch & lighting control 60 A

Labor-hours for feeder and plug-in sections include:
1. Unloading and uncrating
2. Hauling up to 200 ft. from loading dock
3. Measuring and marking
4. Setup of rolling staging
5. Installing hangers
6. Hanging and bolting sections
7. Aligning and leveling
8. Testing

Labor-hours do not include:
1. Modifications to existing structure for hanger supports
2. Threaded rod in excess of 2 ft.
3. Welding
4. Penetrations thru walls
5. Staging rental

Deduct the following percentages from labor only:

       150 ft. to 250 ft. — 10%
       251 ft. to 350 ft. — 15%
       351 ft. to 500 ft. — 20%
          Over 500 ft. — 25%

Deduct percentage only if runs are contained in the same area.

Example: If the job entails running 100 ft. in 5 different locations do not deduct 20%, but if the duct is being run in 1 area and the quantity is 500 ft. then you would deduct 20%.

Deduct only from straight lengths, not fittings or plug-in units.

R262513-10   Aluminum Bus Duct (cont.)

Add to labor for elevated installations:

15 ft. to 20 ft. high	10%
21 ft. to 25 ft. high	20%
26 ft. to 30 ft. high	30%
31 ft. to 35 ft. high	40%
36 ft. to 40 ft. high	50%
Over 40 ft. high	60%

Bus Duct Fittings:

Labor-hours for fittings include:
1. Unloading and uncrating
2. Hauling up to 200 ft. from loading dock
3. Installing, fitting, and bolting all ends to in-place sections

Plug-in units include:
1. Unloading and uncrating
2. Hauling up to 200 ft. from loading dock
3. Installing plug-in into in-place duct
4. Setup of rolling staging
5. Connecting load wire to lugs
6. Marking wire
7. Checking phase rotation

Labor-hours for plug-ins do not include:
1. Conduit runs from plug-in
2. Wire from plug-in
3. Conduit termination

Economy of Scale – For large concentrations of plug-in units in the same area deduct the following percentages:

11	to	25	15%
26	to	50	20%
51	to	75	25%
76	to	100	30%
100	and	over	35%

Job Conditions: Productivities are based on new construction in an unobstructed first floor area to a height of 15 ft. using rolling staging.

Material staging area is within 100 ft. of work being performed.
Add to the duct fittings and hangers:
   Plug-in switches (fused)
   Plug-in breakers
   Combination starters
   Combination contactors
   Combination fusible switch and lighting control

Labor-hours for duct and fittings include:
1. Unloading and uncrating
2. Hauling up to 200 ft. from loading dock
3. Measuring and marking
4. Installing duct runs
5. Leveling
6. Sound testing

Labor-hours for plug-ins include:
1. Unloading and uncrating
2. Hauling up to 200 ft. from loading dock
3. Installing plug-ins
4. Preparing wire
5. Wire connections and marking

R262513-15   Weight (Lbs./L.F.) of 4 Pole Aluminum and Copper Bus Duct by Ampere Load

Amperes	Aluminum Feeder	Copper Feeder	Aluminum Plug-In	Copper Plug-In
225			7	7
400			8	13
600	10	10	11	14
800	10	19	13	18
1000	11	19	16	22
1350	14	24	20	30
1600	17	26	25	39
2000	19	30	29	46
2500	27	43	36	56
3000	30	48	42	73
4000	39	67		
5000		78		

R262716-40   Standard Electrical Enclosure Types

NEMA Enclosures

Electrical enclosures serve two basic purposes; they protect people from accidental contact with enclosed electrical devices and connections, and they protect the enclosed devices and connections from specified external conditions. The National Electrical Manufacturers Association (NEMA) has established the following standards. Because these descriptions are not intended to be complete representations of NEMA listings, consultation of NEMA literature is advised for detailed information.

The following definitions and descriptions pertain to NONHAZARDOUS locations.

NEMA   Type 1: General purpose enclosures intended for use indoors, primarily to prevent accidental contact of personnel with the enclosed equipment in areas that do not involve unusual conditions.

NEMA   Type 2: Dripproof indoor enclosures intended to protect the enclosed equipment against dripping noncorrosive liquids and falling dirt.

NEMA   Type 3: Dustproof, raintight and sleet-resistant (ice-resistant) enclosures intended for use outdoors to protect the enclosed equipment against wind-blown dust, rain, sleet, and external ice formation.

NEMA   Type 3R: Rainproof and sleet-resistant (ice-resistant) enclosures which are intended for use outdoors to protect the enclosed equipment against rain. These enclosures are constructed so that the accumulation and melting of sleet (ice) will not damage the enclosure and its internal mechanisms.

NEMA   Type 3S: Enclosures intended for outdoor use to provide limited protection against wind-blown dust, rain, and sleet (ice) and to allow operation of external mechanisms when ice-laden.

NEMA   Type 4: Watertight and dust-tight enclosures intended for use indoors and out – to protect the enclosed equipment against splashing water, see page of water, falling or hose-directed water, and severe external condensation.

NEMA   Type 4X: Watertight, dust-tight, and corrosion-resistant indoor and outdoor enclosures featuring the same provisions as Type 4 enclosures, plus corrosion resistance.

NEMA   Type 5: Indoor enclosures intended primarily to provide limited protection against dust and falling dirt.

NEMA   Type 6: Enclosures intended for indoor and outdoor use – primarily to provide limited protection against the entry of water during occasional temporary submersion at a limited depth.

NEMA   Type 6R: Enclosures intended for indoor and outdoor use – primarily to provide limited protection against the entry of water during prolonged submersion at a limited depth.

NEMA   Type 11: Enclosures intended for indoor use – primarily to provide, by means of oil immersion, limited protection to enclosed equipment against the corrosive effects of liquids and gases.

NEMA   Type 12: Dust-tight and driptight indoor enclosures intended for use indoors in industrial locations to protect the enclosed equipment against fibers, flyings, lint, dust, and dirt, as well as light splashing, see page, dripping, and external condensation of noncorrosive liquids.

NEMA   Type 13: Oil-tight and dust-tight indoor enclosures intended primarily to house pilot devices, such as limit switches, foot switches, push buttons, selector switches, and pilot lights, and to protect these devices against lint and dust, see page, external condensation, and sprayed water, oil, and noncorrosive coolant.

The following definitions and descriptions pertain to HAZARDOUS, or CLASSIFIED, locations:

NEMA   Type 7: Enclosures intended to use in indoor locations classified as Class 1, Groups A, B, C, or D, as defined in the National Electrical Code.

NEMA   Type 9: Enclosures intended for use in indoor locations classified as Class 2, Groups E, F, or G, as defined in the National Electrical Code.

R262726-90   Wiring Devices

Wiring devices should be priced on a separate takeoff form which includes boxes, covers, conduit and wire.

Labor-hours for devices include:
1. Stripping of wire
2. Attaching wire to device using terminators on the device itself, lugs, set screws etc.
3. Mounting of device in box

Labor-hours do not include:
1. Conduit
2. Wire
3. Boxes
4. Plates

Economy of Scale – for large concentrations of devices in the same area deduct the following percentages from labor-hours:

1	to	10	0%
11	to	25	20%
26	to	50	25%
51	to	100	30%
	over	100	35%

R262726-90   Wiring Devices (cont.)

NEMA No.	15 R	20 R	30 R	50 R	60 R
1 125V 2 Pole, 2 Wire					
2 250V 2 Pole, 2 Wire					
5 125V 2 Pole, 3 Wire					
6 250V 2 Pole, 3 Wire					
7 277V, AC 2 Pole, 3 Wire					
10 125/250V 3 Pole, 3 Wire					
11 3 Phase 250V 3 Pole, 3 Wire					
14 125/250V 3 Pole, 4 Wire					
15 3 Phase 250V 3 Pole, 4 Wire					
18 3 Phase 208Y/120V 4 Pole, 4 Wire					

R262726-90 Wiring Devices (cont.)

NEMA No.	15 R	20 R	30 R	NEMA No.	15 R	20 R	30 R
L 1 125V 2 Pole, 2 Wire				**L 13** 3 Phase 600V 3 Pole, 3 Wire			
L 2 250V 2 Pole, 2 Wire				**L 14** 125/250V 3 Pole, 4 Wire			
L 5 125 V 2 Pole, 3 Wire				**L 15** 3 Phase 250 V 3 Pole, 4 Wire			
L 6 250 V 2 Pole, 3 Wire				**L 16** 3 Phase 480V 3 Pole, 4 Wire			
L 7 227 V, AC 2 Pole, 3 Wire				**L 17** 3 Phase 600 V 3 Pole, 4 Wire			
L 8 480 V 2 Pole, 3 Wire				**L 18** 3 Phase 208Y/120V 4 Pole, 4 Wire			
L 9 600 V 2 Pole, 3 Wire				**L 19** 3 Phase 480Y/277V 4 Pole, 4 Wire			
L 10 125 /250V 3 Pole, 3 Wire				**L 20** 3 Phase 600Y/347V 4 Pole, 4 Wire			
L 11 3 Phase 250 V 3 Pole, 3 Wire				**L 21** 3 Phase 208Y/120V 4 Pole, 5 Wire			
L 12 3 Phase 480 V 3 Pole, 3 Wire				**L 22** 3 Phase 480Y/277V 4 Pole, 5 Wire			
				L 23 3 Phase 600Y/347V 4 Pole, 5 Wire			

R262816-80   Safety Switches

List each Safety Switch by type, ampere rating, voltage, single or three phase, fused or nonfused.

Example:

A. General duty, 240V, 3-pole, fused
B. Heavy Duty, 600V, 3-pole, nonfused
C. Heavy Duty, 240V, 2-pole, fused
D. Heavy Duty, 600V, 3-pole, fused

Also include NEMA enclosure type and identify as:

1. Indoor
2. Weatherproof
3. Explosionproof

Installation of Safety Switches includes:

1. Unloading and uncrating
2. Handling disconnects up to 200′ from loading dock
3. Measuring and marking location
4. Drilling (4) anchor type lead fasteners, using a hammer drill
5. Mounting and leveling Safety Switch
6. Installing (3) fuses
7. Phasing and tagging line and load wires

Price does not include:

1. Modifications to enclosure
2. Plywood backboard
3. Conduit or wire
4. Fuses
5. Termination of wires

Job Conditions: Productivities are based on new construction to an installed height of 6′ above finished floor.

Material staging area is assumed to be within 100′ of work in progress.

R263213-45   Generator Weight (Lbs.) by kW

3 Phase 4 Wire /480 Volt			
Gas		Diesel	
kW	Lbs.	kW	Lbs.
7.5	600	30	1800
10	630	50	2230
15	960	75	2250
30	1500	100	3840
65	2350	125	4030
85	2570	150	5500
115	4310	175	5650
170	6530	200	5930
		250	6320
		300	7840
		350	8220
		400	10750
		500	11900

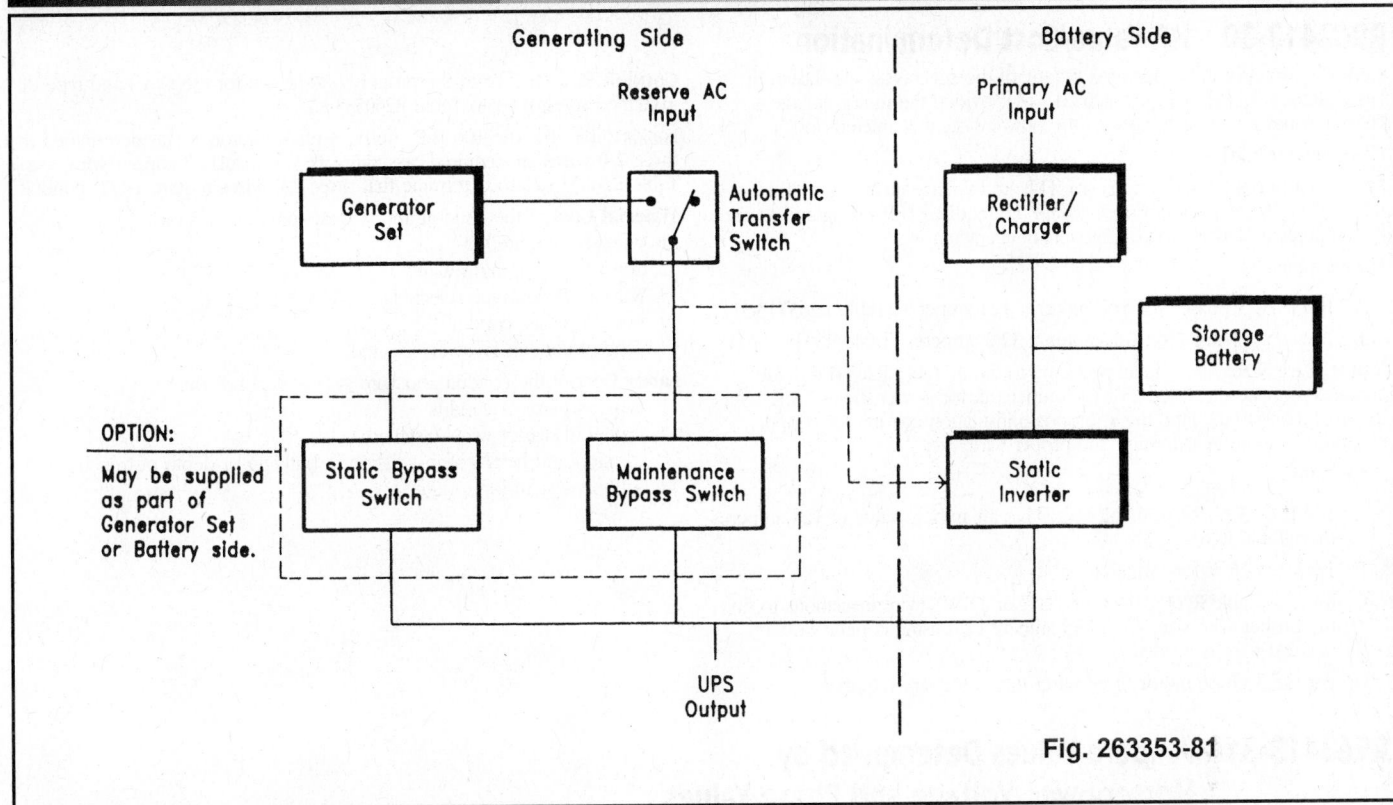

Fig. 263353-81

R263353-80   Uninterruptible Power Supply Systems

General: Uninterruptible Power Supply (UPS) Systems are used to provide power for legally required standby systems. They are also designed to protect computers and provide optional additional coverage for any or all loads in a facility. Figure R263353-81 shows a typical configuration for a UPS System with generator set.

Cost Determination:

It is recommended that UPS System material costs be obtained from equipment manufacturers as a package. Installation costs should be obtained from a vendor or contractor.

The recommended procedure for pricing UPS Systems would be to identify:
1. Frequency – by Hertz (Hz.)
   For example: 50, 60, and 400/415 Hz.
2. Apparent power and real power
   For example: 200 kVA/170 kW
3. Input/Output voltage   For example: 120V, 208V or 240V

4. Phase – either single or three-phase
5. Options and accessories – For extended run time more batteries would be required. Other accessories include battery cabinets, battery racks, remote control panel, power distribution unit (PDU), and warranty enhancement plans.

Note:
1. Larger systems can be configured by paralleling two or more standard small size single modules.
   For example: two 15 kVA modules, combined together and configured to 30 kVA.
2. Maximum input current during battery recharge is typically 15% higher than normal current.
3. UPS Systems weights vary depending on options purchased and power rating in kVA.

R263413-30   HP Value/Cost Determination

General: Motors can be powered by any of the seven systems shown in Figure R262213-11 thru Figure R262213-17 provided the motor voltage characteristics are compatible with the power system characteristics.

Cost Determination:

Motor Amperes for the various size H.P. and voltage are listed in Table R263413-31. To find the amperes, locate the required H.P. rating and locate the amperes under the appropriate circuit characteristics.

For example:

 A. 100 H.P., 3 phase, 460 volt motor = 124 amperes (Table R263413-31)

 B. 10 H.P., 3 phase, 200 volt motor = 32.2 amperes (Table R263413-31)

Motor Wire Size: After the amperes are found in Table R263413-31 the amperes must be increased 25% to compensate for power losses. Next refer to Table R260519-92. Find the appropriate insulation column for copper or aluminum wire to determine the proper wire size.

For example:

 A. 100 H.P., 3 phase, 460 volt motor has an ampere value of 124 amperes from Table R263413-31

 B. 124A x 1.25 = 155 amperes

 C. Refer to Table R260519-92 for THW or THWN wire insulations to find the proper wire size. For a 155 ampere load using copper wire a size 2/0 wire is needed.

 D. For the 3 phase motor three wires of 2/0 size are required.

Conduit Size: To obtain the proper conduit size for the wires and type of insulation used, refer to Table R260533-22.

For example: For the 100 H.P., 460V, 3 phase motor, it was determined that three 2/0 wires are required. Assuming THWN insulated copper wire, use Table R260533-22 to determine that three 2/0 wires require 1-1/2″ conduit.

Material Cost of the conduit and wire system depends on:

 1. Wire size required
 2. Copper or aluminum wire
 3. Wire insulation type selected
 4. Steel or plastic conduit
 5. Type of conduit raceway selected.

Labor Cost of the conduit and wire system depends on:

 1. Type and size of conduit
 2. Type and size of wires installed
 3. Location and height of installation in building or depth of trench
 4. Support system for conduit.

R263413-31   Ampere Values Determined by Horsepower, Voltage and Phase Values

H.P.	Amperes							
	Single Phase			**Three Phase**				
	115V	208V	230V	200V	208V	230V	460V	575V
1/6	4.4A	2.4A	2.2A					
1/4	5.8	3.2	2.9					
1/3	7.2	4.0	3.6					
1/2	9.8	5.4	4.9	2.5A	2.4A	2.2A	1.1A	0.9A
3/4	13.8	7.6	6.9	3.7	3.5	3.2	1.6	1.3
1	16	8.8	8	4.8	4.6	4.2	2.1	1.7
1-1/2	20	11	10	6.9	6.6	6.0	3.0	2.4
2	24	13.2	12	7.8	7.5	6.8	3.4	2.7
3	34	18.7	17	11.0	10.6	9.6	4.8	3.9
5	56	30.8	28	17.5	16.7	15.2	7.6	6.1
7-1/2	80	44	40	25.3	24.2	22	11	9
10	100	55	50	32.2	30.8	28	14	11
15				48.3	46.2	42	21	17
20				62.1	59.4	54	27	22
25				78.2	74.8	68	34	27
30				92	88	80	40	32
40				120	114	104	52	41
50				150	143	130	65	52
60				177	169	154	77	62
75				221	211	192	96	77
100				285	273	248	124	99
125				359	343	312	156	125
150				414	396	360	180	144
200				552	528	480	240	192
250							302	242
300							361	289
350							414	336
400							477	382

Reprinted with permission from NFPA 70-2002, the National Electrical Code,® Copyright © 2001, National Fire Protection Association.
This reprinted material is not the complete and official position of the National Fire Protection Association, on the referenced subject, which is represented only by the standard in its entirety.

National Electrical Code® and NEC® are registered trademarks of the National Fire Protection Association Inc., Quincy, MA 02269.

R263413-30   Cost Determination (cont.)

Magnetic starters, switches, and motor connection:

To complete the cost picture from H.P. to Costs additional items must be added to the cost of the conduit and wire system to arrive at a total cost.

1. Assembly Table D5020-160 Magnetic Starters Installed Cost lists the various size starters for single phase and three phase motors.
2. Assembly Table D5020-165 Heavy Duty Safety Switches Installed Cost lists safety switches required at the beginning of a motor circuit and also one required in the vicinity of the motor location.
3. Assembly Table D5020-170 Motor Connection lists the various costs for single and three phase motors.

Worksheet to obtain total motor wiring costs:

It is assumed that the motors or motor driven equipment are furnished and installed under other sections for this estimate and the following work is done under this section:

1. Conduit
2. Wire (add 10% for additional wire beyond conduit ends for connections to switches, boxes, starters, etc.)
3. Starters
4. Safety switches
5. Motor connections

Figure   R263413-32

Worksheet for Motor Circuits				Cost	
Item	Type	Size	Quantity	Unit	Total
Wire					
Conduit					
Switch					
Starter					
Switch					
Motor Connection					
Other					
Total Cost					

R263413-33   Maximum Horsepower for Starter Size by Voltage

Starter Size	Maximum HP (3φ)			
	208V	240V	480V	600V
00	1½	1½	2	2
0	3	3	5	5
1	7½	7½	10	10
2	10	15	25	25
3	25	30	50	50
4	40	50	100	100
5		100	200	200
6		200	300	300
7		300	600	600
8		450	900	900
8L		700	1500	1500

Electrical | R2636 Transfer Switches

R263623-60   Automatic Transfer Switches

When taking off Automatic transfer switches identify by voltage, amperage and number of poles.

Example: Automatic transfer switch, 480V, 3 phase, 30A, NEMA 1 enclosure

Labor-hours for transfer switches include:

1. Unloading, uncrating and handling switches up to 200' from loading dock
2. Measuring and marking
3. Drilling (4) lead type anchors, using a hammer drill
4. Mounting and leveling
5. Circuit identification
6. Testing and load balancing

Labor-hours do not include:

1. Modifications in enclosure
2. Structural supports
3. Additional lugs
4. Plywood backboard
5. Painting or lettering
6. Conduit runs to or from transfer switch
7. Wire
8. Termination of wires

Electrical | R2651 Interior Lighting

R265113-40   Interior Lighting Fixtures

When taking off interior lighting fixtures, it is advisable to set up your quantity worksheet to conform to the lighting schedule as it appears on the print. Include the alpha-numeric code plus the symbol on your work sheet.

Take off a particular section or floor of the building and count each type of fixture before going on to another type. It would also be advantageous to include on the same worksheet the pipe, wire, fittings and circuit number associated with each type of lighting fixture. This will help you identify the costs associated with any particular lighting system and in turn make material purchases more specific as to when and how much to order under the classification of lighting.

By taking off lighting first you can get a complete "WALK THRU" of the job. This will become helpful when doing other phases of the project.

Materials for a recessed fixture include:

1. Fixture
2. Lamps
3. 6' of jack chain
4. (2) S hooks
5. (2) Wire nuts

Labor for interior recessed fixtures include:

1. Unloading by hand
2. Hauling by hand to an area up to 200' from loading dock
3. Uncrating
4. Layout
5. Installing fixture
6. Attaching jack chain & S hooks
7. Connecting circuit power
8. Reassembling fixture
9. Installing lamps
10. Testing

Material for surface mounted fixtures includes:

1. Fixture
2. Lamps
3. Either (4) lead type anchors, (4) toggle bolts, or (4) ceiling grid clips
4. (2) Wire nuts

Material for pendent mounted fixtures includes:

1. Fixture
2. Lamps
3. (2) Wire nuts

4. Rigid pendents as required by type of fixtures
5. Canopies as required by type of fixture

Labor hours include the following for both surface and pendent fixtures:

1. Unloading by hand
2. Hauling by hand to an area up to 200' from loading dock
3. Uncrating
4. Layout and marking
5. Drilling (4) holes for either lead anchors or toggle bolts using a hammer drill
6. Installing fixture
7. Leveling fixture
8. Connecting circuit power
9. Installing lamps
10. Testing

Labor for surface or pendent fixtures does not include:

1. Conduit
2. Boxes or covers
3. Connectors
4. Fixture whips
5. Special support
6. Switching
7. Wire

Economy of Scale: For large concentrations of lighting fixtures in the same area deduct the following percentages from labor:

25	to	50	fixtures	15%
51	to	75	fixtures	20%
76	to	100	fixtures	25%
101 and over				30%

Job Conditions: Productivity is based on new construction in a unobstructed first floor location, using rolling staging to 15' high.

Material staging is assumed to be within 100' of work being performed.

Add the following percentages to labor for elevated installations:

15'	to	20'	high	10%
21'	to	25'	high	20%
26'	to	30'	high	30%
31'	to	35'	high	40%
36'	to	40'	high	50%
41' and over				60%

R265723-05   Comparison - Operation of High Intensity Discharge Lamps

Lamp Type	Wattage	Life (Hours)	Circuit Wattage [1]	Average Initial Lumens	L.L.D. [2]	Mean Lumens [3]
M.V.	100 DX	24,000	125	4,000	61%	2,440
L.P.S.	SOX-35	18,000	65	4,800	100%	4,800
H.P.S.	LU-70	12,000	84	5,800	90%	5,220
M.H.	No Equivalent					
M.V.	175 DX	24,000	210	8,500	66%	5,676
M.V.	250 DX	24,000	295	13,000	66%	7,986
L.P.S.	SOX-55	18,000	82	8,000	100%	8,000
H.P.S.	LU-100	12,000	120	9,500	90%	8,550
M.H.	No Equivalent					
M.V.	400 DX	24,000	465	24,000	64%	14,400
L.P.S.	SOX-90	18,000	141	13,500	100%	13,500
H.P.S.	LU-150	16,000	188	16,000	90%	14,400
M.H.	MH-175	7,500	210	14,000	73%	10,200
M.V.	No Equivalent					
L.P.S.	SOX-135	18,000	147	22,500	100%	22,500
H.P.S.	LU-250	20,000	310	25,500	92%	23,205
M.H.	MH-250	7,500	295	20,500	78%	16,000
M.V.	1000 DX	24,000	1,085	63,000	61%	37,820
L.P.S.	SOX-180	18,000	248	33,000	100%	33,000
H.P.S.	LU-400	20,000	480	50,000	90%	45,000
M.H.	MH-400	15,000	465	34,000	72%	24,600
H.P.S.	LU-1000	15,000	1,100	140,000	91%	127,400

1. Includes ballast losses and average lamp watts
2. Lamp lumen depreciation (% of initial light output at 70% rated life)
3. Lamp lumen output at 70% rated life (L.L.D. x initial)
4. Cost of yearly operation = Cicuit Wattage/1000 x annual operating hours x average cost per KWH

M.V. = Mercury Vapor
L.P.S. = Low pressure sodium
H.P.S. = High pressure sodium
L.H. = Metal halide

R265723-10   For Other than Regular Cool White (CW) Lamps

	Multiply Material Costs as Follows:				
Regular Lamps	Cool white deluxe (CWX)	x 1.35	Energy Saving Lamps	Cool white (CW/ES)	x 1.35
	Warm white deluxe (WWX)	x 1.35		Cool white deluxe (CWX/ES)	x 1.65
	Warm white (WW)	x 1.30		Warm white (WW/ES)	x 1.55
	Natural (N)	x 2.05		Warm white deluxe (WWX/ES)	x 1.65

R265723-20   Lamp Comparison Chart with Enclosed Floodlight, Ballast, & Lamp for Pole Mounting

Type	Watts	Initial Lumens	Lumens per Watt	Lumens @ 40% Life	Life (Hours)
Incandescent	150	2,880	19	85	750
	300	6,360	21	84	750
	500	10,850	22	80	1,000
	1,000	23,740	24	80	1,000
	1,500	34,400	23	80	1,000
Tungsten	500	10,950	22	97	2,000
Halogen	1,500	35,800	24	97	2,000
Fluorescent	40	3,150	79	88	20,000
Cool	110	9,200	84	87	12,000
White	215	16,000	74	81	12,000
Deluxe	250	12,100	48	86	24,000
Mercury	400	22,500	56	85	24,000
	1,000	63,000	63	75	24,000
Metal	175	14,000	80	77	7,500
Halide	400	34,000	85	75	15,000
	1,000	100,000	100	83	10,000
	1,500	155,000	103	92	1,500
High	70	5,800	83	90	20,000
Pressure	100	9,500	95	90	20,000
Sodium	150	16,000	107	90	24,000
	400	50,000	125	90	24,000
	1,000	140,000	140	90	24,000
Low	55	4,600	131	98	18,000
Pressure	90	12,750	142	98	18,000
Sodium	180	33,000	183	98	18,000

Color: High Pressure Sodium — Slightly Yellow
       Low Pressure Sodium — Yellow
       Mercury Vapor — Green-Blue
       Metal Halide — Blue White

Note: Pole not included.

R265723-25   Energy Efficiency Rating for Luminaires

The energy efficiency program for luminaires recommends the use of a metric called the Luminaire Efficacy Rating (LER). The LER value expresses the total luminaires generated by a lamp, to the watts consumed by the lamp.

Lamp Type	Lumens per Watt
Incandescent	17
Tungsten Halogen	14 - 20
Fluorescent	50 - 104
Metal Halide	64 - 96
High Pressure Sodium	76 - 116

R271323-40   Fiber Optics

Fiber optic systems use optical fiber such as plastic, glass, or fused silica, a transparent material, to transmit radiant power (i.e., light) for control, communication, and signaling applications. The types of fiber optic cables can be nonconductive, conductive, or composite. The composite cables contain fiber optics and current-carrying electrical conductors. The configuration for one of the fiber optic systems is as follows:

The transceiver module acts as transmitting and receiving equipment in a common house, which converts electrical energy to light energy or vice versa.

Pricing the fiber optic system is not an easy task. The performance of the whole system will affect the cost significantly. New specialized tools and techniques decrease the installing cost tremendously. In the fiber optic section of Means Electrical Cost Data, a benchmark for labor-hours and material costs is set up so that users can adjust their costs according to unique project conditions.

Units for Measure: Fiber optic cable is measured in hundred linear feet (C.L.F.) or industry units of measure - meter (m) or kilometer (km). The connectors are counted as units (EA.)

Material Units: Generally, the material costs include only the cable. All the accessories shall be priced separately.

Labor Units: The following procedures are generally included for the installation of fiber optic cables:

- Receiving
- Material handling
- Setting up pulling equipment
- Measuring and cutting cable
- Pulling cable

These additional items are listed and extended: Terminations

Takeoff Procedure: Cable should be taken off by type, size, number of fibers, and number of terminations required. List the lengths of each type of cable on the takeoff sheets. Total and add 10% for waste. Transfer the figures to a cost analysis sheet and extend.

R271513-75   High Performance Cable

There are several categories used to describe high performance cable. The following information includes a description of categories CAT 3, 5, 5e, 6, and 7, and details classifications of frequency and specific standards. The category standards have evolved under the sponsorship of organizations such as the Telecommunication Industry Association (TIA), the Electronic Industries Alliance (EIA), the American National Standards Institute (ANSI), the International Organization for Standardization (ISO), and the International Electrotechnical Commission (IEC), all of which have catered to the increasing complexities of modern network technology. For network cabling, users must comply with national or international standards. A breakdown of these categories is as follows:

Category 3: Designed to handle frequencies up to 16 MHz.

Category 5: (TIA/EIA 568A) Designed to handle frequencies up to 100 MHz.

Category 5e: Additional transmission performance to exceed Category 5.

Category 6: Development by TIA and other international groups to handle frequencies of 250 MHz.

Category 7: Development to handle a frequency range up to 600 MHz.

R337116-60 Average Transmission Line Material Requirements (Per Mile)

Item	Unit	Flat 69kV Wood	Flat 161kV Wood	Flat 161kV Steel	Flat 500kV Steel	Rolling 69kV Wood	Rolling 161kV Wood	Rolling 161kV Steel	Rolling 500kV Steel	Mountain 69kV Wood	Mountain 161kV Wood	Mountain 161kV Steel	Mountain 500kV Steel
Conductor: 397,500 – Cir. Mil., 26/7–ACSR													
Structures	Ea.	12[1]				9[2]				7[2]			
Poles	Ea.	12				18				14			
Crossarms	Ea.	24[3]				9[4]				7[4]			
Conductor	Ft.	15,990				15,990				15,990			
Insulators[5]	Ea.	180				135				105			
Ground Wire	Ft.	5,330				10,660				10,660			
Conductor: 636,000 – Cir. Mil., 26/7–ACSR													
Structures	Ea.	13[1]	11[6]			10[2]	9[2]	6[9]		8[2]	8[2]	6[9]	
Excavation	C.Y.	–	–			–	–	120		–	–	120	
Concrete	C.Y.	–	–			–	–	10		–	–	10	
Steel Towers	Tons	–	–			–	–	32		–	–	32	
Poles	Ea.	13	11			20	18	–		16	16	–	
Crossarms	Ea.	26[3]	33[7]			10[4]	9[8]			8[4]	8[8]		
Conductor	Ft.	15,990	15,990			15,990	15,990	15,990		15,990	15,990	15,990	
Insulators[5]	Ea.	195	165			150	297	297		120	264	297	
Ground Wire	Ft.	5330	5330			10,660	10,660	10,660		10,660	10,660	10,660	
Conductor: 954,000 – Cir. Mil., 45/7–ACSR													
Structures	Ea.	14[1]	12[6]		4[11]	10[2]	9[2]	6[9]	4[11]	8[2]	8[2]	6[9]	4[13]
Excavation	C.Y.	–	–		200	–	–	125	214	–	–	125	233
Concrete	C.Y.	–	–		20	–	–	10	21	–	–	10	21
Steel Towers	Tons	–	–		57	–	–	33	57	–	–	33	63
Poles	Ea.	14	12		–	20	18	–	–	16	16	–	–
Crossarms	Ea.	28[10]	36[7]		–	10[4]	9[8]	–	–	8[4]	8[8]	–	–
Conductor	Ft.	15,990	15,990		47,970[12]	15,990	15,990	15,990	47,970[12]	15,990	15,990	15,990	47,970[12]
Insulators[5]	Ea.	210	180		288	150	297	297	288	120	264	297	576
Ground Wire	Ft.	5330	5330		10,660	10,660	10,660	10,660	10,660	10,660	10,660	10,660	10,660
Conductor: 1,351,500 – Cir. Mil., 45/7–ACSR													
Structures	Ea.	8[14]				8[14]							
Excavation	C.Y.	220				220							
Concrete	C.Y.	28				28							
Steel Towers	Ton	46				46							
Conductor	Ft.	31,680[15]				31,680[15]							
Insulators	Ea.	528				528							
Ground Wire	Ft.	10,660				10,660							

1. Single pole two-arm suspension type construction
2. Two-pole wood H-frame construction
3. 4¾" x 5¾" x 8' and 4¾" x 5¾" x 10' wood crossarm
4. 6" x 8" x 26'-0" wood crossarm
5. 5¾" x 10" disc insulator
6. Single pole construction with 3 fiberglass crossarms (5 fog- type insulators per phase)
7. 7'-0" fiberglass crossarms
8. 6" x 10" x 35'-0" wood crossarm
9. Laced steel tower, single circuit construction
10. 5" x 7" x 8' and 5" x 7" x 10' wood crossarm
11. Laced steel tower, single circuit 500-kV construction
12. Bundled conductor (3 sub-conductors per phase)
13. Laced steel tower, single circuit restrained phases (500-kV)
14. Laced steel tower, double circuit construction
15. Both sides of double circuit strung

Note: To allow for sagging, a mile (5280 Ft.) of transmission line uses 5330 Ft. of conductor per wire (called a wire mile).

R337119-30 Concrete for Conduit Encasement

Table below lists C.Y. of concrete for 100 L.F. of trench. Conduits separation center to center should meet 7.5″ (N.E.C.).

Number of Conduits	1	2	3	4	6	8	9	Number of Conduits
Trench Dimension	11.5″ x 11.5″	11.5″ x 19″	11.5″ x 27″	19″ x 19″	19″ x 27″	19″ x 38″	27″ x 27″	Trench Dimension
Conduit Diameter 2.0″	3.29	5.39	7.64	8.83	12.51	17.66	17.72	Conduit Diameter 2.0″
2.5″	3.23	5.29	7.49	8.62	12.19	17.23	17.25	2.5″
3.0″	3.15	5.13	7.24	8.29	11.71	16.59	16.52	3.0″
3.5″	3.08	4.97	7.02	7.99	11.26	15.98	15.84	3.5″
4.0″	2.99	4.80	6.76	7.65	10.74	15.30	15.07	4.0″
5.0″	2.78	4.37	6.11	6.78	9.44	13.57	13.12	5.0″
6.0″	2.52	3.84	5.33	5.74	7.87	11.48	10.77	6.0″

Change Orders

Change Order Considerations

A change order is a written document, usually prepared by the design professional, and signed by the owner, the architect/engineer, and the contractor. A change order states the agreement of the parties to: an addition, deletion, or revision in the work; an adjustment in the contract sum, if any; or an adjustment in the contract time, if any. Change orders, or "extras" in the construction process occur after execution of the construction contract and impact architects/ engineers, contractors, and owners.

Change orders that are properly recognized and managed can ensure orderly, professional, and profitable progress for all who are involved in the project. There are many causes for change orders and change order requests. In all cases, change orders or change order requests should be addressed promptly and in a precise and prescribed manner. The following paragraphs include information regarding change order pricing and procedures.

The Causes of Change Orders

Reasons for issuing change orders include:

- Unforeseen field conditions that require a change in the work
- Correction of design discrepancies, errors, or omissions in the contract documents
- Owner-requested changes, either by design criteria, scope of work, or project objectives
- Completion date changes for reasons unrelated to the construction process
- Changes in building code interpretations, or other public authority requirements that require a change in the work
- Changes in availability of existing or new materials and products

Procedures

Properly written contract documents must include the correct change order procedures for all parties—owners, design professionals and contractors—to follow in order to avoid costly delays and litigation.

Being "in the right" is not always a sufficient or acceptable defense. The contract provisions requiring notification and documentation must be adhered to within a defined or reasonable time frame.

The appropriate method of handling change orders is by a written proposal and acceptance by all parties involved. Prior to starting work on a project, all parties should identify their authorized agents who may sign and accept change orders, as well as any limits placed on their authority.

Time may be a critical factor when the need for a change arises. For such cases, the contractor might be directed to proceed on a "time and materials" basis, rather than wait for all paperwork to be processed—a delay that could impede progress. In this situation, the contractor must still follow the prescribed change order procedures including, but not limited to, notification and documentation.

All forms used for change orders should be dated and signed by the proper authority. Lack of documentation can be very costly, especially if legal judgments are to be made, and if certain field personnel are no longer available. For time and material change orders, the contractor should keep accurate daily records of all labor and material allocated to the change.

Owners or awarding authorities who do considerable and continual building construction (such as the federal government) realize the inevitability of change orders for numerous reasons, both predictable and unpredictable. As a result, the federal government, the American Institute of Architects (AIA), the Engineers Joint Contract Documents Committee (EJCDC) and other contractor, legal, and technical organizations have developed standards and procedures to be followed by all parties to achieve contract continuance and timely completion, while being financially fair to all concerned.

In addition to the change order standards put forth by industry associations, there are also many books available on the subject.

Pricing Change Orders

When pricing change orders, regardless of their cause, the most significant factor is when the change occurs. The need for a change may be perceived in the field or requested by the architect/engineer *before* any of the actual installation has begun, or may evolve or appear *during* construction when the item of work in question is partially installed. In the latter cases, the original sequence of construction is disrupted, along with all contiguous and supporting systems. Change orders cause the greatest impact when they occur *after* the installation has been completed and must be uncovered, or even replaced. Post-completion changes may be caused by necessary design changes, product failure, or changes in the owner's requirements that are not discovered until the building or the systems begin to function.

Specified procedures of notification and record keeping must be adhered to and enforced regardless of the stage of construction: *before, during,* or *after* installation. Some bidding documents anticipate change orders by requiring that unit prices including overhead and profit percentages—for additional as well as deductible changes—be listed. Generally these unit prices do not fully take into account the ripple effect, or impact on other trades, and should be used for general guidance only.

When pricing change orders, it is important to classify the time frame in which the change occurs. There are two basic time frames for change orders: *pre-installation change orders,* which occur before the start of construction, and *post-installation change orders,* which involve reworking after the original installation. Change orders that occur between these stages may be priced according to the extent of work completed using a combination of techniques developed for pricing *pre-* and *post-installation* changes.

The following factors are the basis for a check list to use when preparing a change order estimate.

Factors To Consider When Pricing Change Orders

As an estimator begins to prepare a change order, the following questions should be reviewed to determine their impact on the final price.

General

- *Is the change order work* pre-installation *or* post-installation?

Change order work costs vary according to how much of the installation has been completed. Once workers have the project scoped in their minds, even though they have not started, it can be difficult to refocus.

Consequently they may spend more than the normal amount of time understanding the change. Also, modifications to work in place, such as trimming or refitting, usually take more time than was initially estimated. The greater the amount of work in place, the more reluctant workers are to change it. Psychologically they may resent the change and as a result the rework takes longer than normal. Post-installation change order estimates must include demolition of existing work as required to accomplish the change. If the work is performed at a later time, additional obstacles, such as building finishes, may be present which must be protected. Regardless of whether the change

occurs pre-installation or post-installation, attempt to isolate the identifiable factors and price them separately. For example, add shipping costs that may be required pre-installation or any demolition required post-installation. Then analyze the potential impact on productivity of psychological and/or learning curve factors and adjust the output rates accordingly. One approach is to break down the typical workday into segments and quantify the impact on each segment. The following chart may be useful as a guide:

	Activities (Productivity) Expressed as Percentages of a Workday		
Task	Means Mechanical Cost Data (for New Construction)	Pre-Installation Change Orders	Post-Installation Change Orders
1. Study plans	3%	6%	6%
2. Material procurement	3%	3%	3%
3. Receiving and storing	3%	3%	3%
4. Mobilization	5%	5%	5%
5. Site movement	5%	5%	8%
6. Layout and marking	8%	10%	12%
7. Actual installation	64%	59%	54%
8. Clean-up	3%	3%	3%
9. Breaks—non-productive	6%	6%	6%
Total	100%	100%	100%

Change Order Installation Efficiency

The labor-hours expressed (for new construction) are based on average installation time, using an efficiency level of approximately 60%–65%. For change order situations, adjustments to this efficiency level should reflect the daily labor-hour allocation for that particular occurrence.

If any of the specific percentages expressed in the above chart do not apply to a particular project situation, then those percentage points should be reallocated to the appropriate task(s). Example: Using data for new construction, assume there is no new material being utilized. The percentages for Tasks 2 and 3 would therefore be reallocated to other tasks. If the time required for Tasks 2 and 3 can now be applied to installation, we can add the time allocated for *Material Procurement* and *Receiving and Storing* to the *Actual Installation* time for new construction, thereby increasing the Actual Installation percentage.

This chart shows that, due to reduced productivity, labor costs will be higher than those for new construction by 5%–15% for pre-installation change orders and by 15%–25% for post-installation change orders. Each job and change order is unique and must be examined individually. Many factors, covered elsewhere in this section, can each have a significant impact on productivity and change order costs. All such factors should be considered in every case.

- *Will the change substantially delay the original completion date?*

A significant change in the project may cause the original completion date to be extended. The extended schedule may subject the contractor to new wage rates dictated by relevant labor contracts. Project supervision and other project overhead must also be extended beyond the original completion date. The schedule extension may also put installation into a new weather season. For example, underground piping scheduled for October installation was delayed until January. As a result, frost penetrated the trench area, thereby changing the degree of difficulty of the task. Changes and delays may have a ripple effect throughout the project. This effect must be analyzed and negotiated with the owner.

- *What is the net effect of a deduct change order?*

In most cases, change orders resulting in a deduction or credit reflect only bare costs. The contractor may retain the overhead and profit based on the original bid.

Materials

- *Will you have to pay more or less for the new material, required by the change order, than you paid for the original purchase?*

The same material prices or discounts will usually apply to materials purchased for change orders as new construction. In some instances, however, the contractor may forfeit the advantages of competitive pricing for change orders. Consider the following example:

A contractor purchased over $20,000 worth of fan coil units for an installation, and obtained the maximum discount. Some time later it was determined the project required an additional matching unit. The contractor has to purchase this unit from the original supplier to ensure a match. The supplier at this time may not discount the unit because of the small quantity, and the fact that he is no longer in a competitive situation. The impact of quantity on purchase can add between 0% and 25% to material prices and/or subcontractor quotes.

- *If materials have been ordered or delivered to the job site, will they be subject to a cancellation charge or restocking fee?*

Check with the supplier to determine if ordered materials are subject to a cancellation charge. Delivered materials not used as result of a change order may be subject to a restocking fee if returned to the supplier. Common restocking charges run between 20% and 40%. Also, delivery charges to return the goods to the supplier must be added.

Labor

- *How efficient is the existing crew at the actual installation?*

Is the same crew that performed the initial work going to do the change order? Possibly the change consists of the installation of a unit identical to one already installed; therefore, the change should take less time. Be sure to consider this potential productivity increase and modify the productivity rates accordingly.

- *If the crew size is increased, what impact will that have on supervision requirements?*

Under most bargaining agreements or management practices, there is a point at which a working foreman is replaced by a nonworking foreman. This replacement increases project overhead by adding a nonproductive worker. If additional workers are added to accelerate the project or to perform changes while maintaining the schedule, be sure to add additional supervision time if warranted. Calculate the hours involved and the additional cost directly if possible.

- *What are the other impacts of increased crew size?*

The larger the crew, the greater the potential for productivity to decrease. Some of the factors that cause this productivity loss are: overcrowding (producing restrictive conditions in the working space), and possibly a shortage of any special tools and equipment required. Such factors affect not only the crew working on the elements directly involved in the change order, but other crews whose movement may also be hampered.

As the crew increases, check its basic composition for changes by the addition or deletion of apprentices or nonworking foreman, and quantify the potential effects of equipment shortages or other logistical factors.

- *As new crews, unfamiliar with the project, are brought onto the site, how long will it take them to become oriented to the project requirements?*

The orientation time for a new crew to become 100% effective varies with the site and type of project. Orientation is easiest at a new construction site, and most difficult at existing, very restrictive renovation sites. The type of work also affects orientation time. When all elements of the work are exposed, such as concrete or masonry work, orientation is decreased. When the work is concealed or less visible, such as existing electrical systems, orientation takes longer. Usually orientation can be accomplished in one day or less. Costs for added orientation should be itemized and added to the total estimated cost.

- *How much actual production can be gained by working overtime?*

Short term overtime can be used effectively to accomplish more work in a day. However, as overtime is scheduled to run beyond several weeks, studies have shown marked decreases in output. The following chart shows the effect of long term overtime on worker efficiency. If the anticipated change requires extended overtime to keep the job on schedule, these factors can be used as a guide to predict the impact on time and cost. Add project overhead, particularly supervision, that may also be incurred.

Days per Week	Hours per Day	Production Efficiency					Payroll Cost Factors	
		1 Week	2 Weeks	3 Weeks	4 Weeks	Average 4 Weeks	@ 1-1/2 Times	@ 2 Times
5	8	100%	100%	100%	100%	100	100%	100%
	9	100	100	95	90	96.25	105.6	111.1
	10	100	95	90	85	91.25	110.0	120.0
	11	95	90	75	65	81.25	113.6	127.3
	12	90	85	70	60	76.25	116.7	133.3
6	8	100	100	95	90	96.25	108.3	116.7
	9	100	95	90	85	92.50	113.0	125.9
	10	95	90	85	80	87.50	116.7	133.3
	11	95	85	70	65	78.75	119.7	139.4
	12	90	80	65	60	73.75	122.2	144.4
7	8	100	95	85	75	88.75	114.3	128.6
	9	95	90	80	70	83.75	118.3	136.5
	10	90	85	75	65	78.75	121.4	142.9
	11	85	80	65	60	72.50	124.0	148.1
	12	85	75	60	55	68.75	126.2	152.4

Effects of Overtime

Caution: Under many labor agreements, Sundays and holidays are paid at a higher premium than the normal overtime rate.

The use of long-term overtime is counterproductive on almost any construction job; that is, the longer the period of overtime, the lower the actual production rate. Numerous studies have been conducted, and while they have resulted in slightly different numbers, all reach the same conclusion. The figure above tabulates the effects of overtime work on efficiency.

As illustrated, there can be a difference between the *actual* payroll cost per hour and the *effective* cost per hour for overtime work. This is due to the reduced production efficiency with the increase in weekly hours beyond 40. This difference between actual and effective cost results from overtime work over a prolonged period. Short-term overtime work does not result in as great a reduction in efficiency and, in such cases, effective cost may not vary significantly from the actual payroll cost. As the total hours per week are increased on a regular basis, more time is lost due to fatigue, lowered morale, and an increased accident rate.

As an example, assume a project where workers are working 6 days a week, 10 hours per day. From the figure above (based on productivity studies), the average effective productive hours over a 4-week period are:

$$0.875 \times 60 = 52.5$$

Depending upon the locale and day of week, overtime hours may be paid at time and a half or double time. For time and a half, the overall (average) *actual* payroll cost (including regular and overtime hours) is determined as follows:

$$\frac{40 \text{ reg. hrs.} + (20 \text{ overtime hrs.} \times 1.5)}{60 \text{ hrs.}} = 1.167$$

Based on 60 hours, the payroll cost per hour will be 116.7% of the normal rate at 40 hours per week. However, because the effective production (efficiency) for 60 hours is reduced to the equivalent of 52.5 hours, the effective cost of overtime is calculated as follows:

For time and a half:

$$\frac{40 \text{ reg. hrs.} + (20 \text{ overtime hrs.} \times 1.5)}{52.5 \text{ hrs.}} = 1.33$$

Installed cost will be 133% of the normal rate (for labor).

Thus, when figuring overtime, the actual cost per unit of work will be higher than the apparent overtime payroll dollar increase, due to the reduced productivity of the longer workweek. These efficiency calculations are true only for those cost factors determined by hours worked. Costs that are applied weekly or monthly, such as equipment rentals, will not be similarly affected.

Equipment

- *What equipment is required to complete the change order?*

Change orders may require extending the rental period of equipment already on the job site, or the addition of special equipment brought in to accomplish the change work. In either case, the additional rental charges and operator labor charges must be added.

Summary

The preceding considerations and others you deem appropriate should be analyzed and applied to a change order estimate. The impact of each should be quantified and listed on the estimate to form an audit trail.

Change orders that are properly identified, documented, and managed help to ensure the orderly, professional and profitable progress of the work. They also minimize potential claims or disputes at the end of the project.

Estimating Tips

- The cost figures in this Square Foot Cost section were derived from approximately 11,000 projects contained in the RSMeans database of completed construction projects. They include the contractor's overhead and profit, but do not generally include architectural fees or land costs. The figures have been adjusted to January of the current year. New projects are added to our files each year, and outdated projects are discarded. For this reason, certain costs may not show a uniform annual progression. In no case are all subdivisions of a project listed.

- These projects were located throughout the U.S. and reflect a tremendous variation in square foot (S.F.) and cubic foot (C.F.) costs. This is due to differences, not only in labor and material costs, but also in individual owners' requirements. For instance, a bank in a large city would have different features than one in a rural area. This is true of all the different types of buildings analyzed. Therefore, caution should be exercised when using these square foot costs. For example, for courthouses, costs in the database are local courthouse costs and will not apply to the larger, more elaborate federal courthouses. As a general rule, the projects in the 1/4 column do not include any site work or equipment, while the projects in the 3/4 column may include both equipment and site work. The median figures do not generally include site work.

- None of the figures "go with" any others. All individual cost items were computed and tabulated separately. Thus, the sum of the median figures for plumbing, HVAC and electrical will not normally total up to the total mechanical and electrical costs arrived at by separate analysis and tabulation of the projects.

- Each building was analyzed as to total and component costs and percentages. The figures were arranged in ascending order with the results tabulated as shown. The 1/4 column shows that 25% of the projects had lower costs and 75% had higher. The 3/4 column shows that 75% of the projects had lower costs and 25% had higher. The median column shows that 50% of the projects had lower costs and 50% had higher.

- There are two times when square foot costs are useful. The first is in the conceptual stage when no details are available. Then square foot costs make a useful starting point. The second is after the bids are in and the costs can be worked back into their appropriate categories for information purposes. As soon as details become available in the project design, the square foot approach should be discontinued and the project priced as to its particular components. When more precision is required, or for estimating the replacement cost of specific buildings, the current edition of *RSMeans Square Foot Costs* should be used.

- In using the figures in this section, it is recommended that the median column be used for preliminary figures if no additional information is available. The median figures, when multiplied by the total city construction cost index figures (see city cost indexes) and then multiplied by the project size modifier at the end of this section, should present a fairly accurate base figure, which would then have to be adjusted in view of the estimator's experience, local economic conditions, code requirements, and the owner's particular requirements. There is no need to factor the percentage figures, as these should remain constant from city to city. All tabulations mentioning air conditioning had at least partial air conditioning.

- The editors of this book would greatly appreciate receiving cost figures on one or more of your recent projects, which would then be included in the averages for next year. All cost figures received will be kept confidential, except that they will be averaged with other similar projects to arrive at square foot cost figures for next year's book. See the last page of the book for details and the discount available for submitting one or more of your projects.

50 17 00 | S.F. Costs

			UNIT	UNIT COSTS			% OF TOTAL			
				1/4	MEDIAN	3/4	1/4	MEDIAN	3/4	
01	0010	**APARTMENTS Low Rise (1 to 3 story)**	S.F.	73	92.50	123				01
	0020	Total project cost	C.F.	6.55	8.70	10.75				
	0100	Site work	S.F.	5.35	8.55	15	6.05%	10.55%	13.95%	
	0500	Masonry		1.44	3.55	5.80	1.54%	3.92%	6.50%	
	1500	Finishes		7.75	10.65	13.15	9.05%	10.75%	12.85%	
	1800	Equipment		2.40	3.63	5.40	2.71%	3.99%	5.95%	
	2720	Plumbing		5.70	7.30	9.30	6.65%	8.95%	10.05%	
	2770	Heating, ventilating, air conditioning		3.63	4.47	6.55	4.20%	5.60%	7.60%	
	2900	Electrical		4.25	5.65	7.65	5.20%	6.65%	8.35%	
	3100	Total: Mechanical & Electrical	↓	15.10	19.60	24	16.05%	18.20%	23%	
	9000	Per apartment unit, total cost	Apt.	68,000	104,000	153,500				
	9500	Total: Mechanical & Electrical	"	12,900	20,300	26,500				
02	0010	**APARTMENTS Mid Rise (4 to 7 story)**	S.F.	97	117	144				02
	0020	Total project costs	C.F.	7.55	10.45	14.30				
	0100	Site work	S.F.	3.88	7.85	15.15	5.25%	6.70%	9.20%	
	0500	Masonry		4.05	8.90	12.15	5.10%	7.25%	10.50%	
	1500	Finishes		12.70	17	20	10.70%	13.50%	17.70%	
	1800	Equipment		2.71	4.22	5.55	2.54%	3.47%	4.31%	
	2500	Conveying equipment		2.20	2.70	3.26	2.05%	2.27%	2.69%	
	2720	Plumbing		5.70	9.10	9.65	5.70%	7.20%	8.95%	
	2900	Electrical		6.40	8.70	10.55	6.35%	7.20%	8.95%	
	3100	Total: Mechanical & Electrical	↓	20.50	25.50	31	18.25%	21%	23%	
	9000	Per apartment unit, total cost	Apt.	109,500	129,500	214,000				
	9500	Total: Mechanical & Electrical	"	20,700	24,000	25,100				
03	0010	**APARTMENTS High Rise (8 to 24 story)**	S.F.	110	127	152				03
	0020	Total project costs	C.F.	10.70	12.45	14.65				
	0100	Site work	S.F.	3.99	6.45	9	2.58%	4.84%	6.15%	
	0500	Masonry		6.35	11.55	14.40	4.74%	9.65%	11.05%	
	1500	Finishes		12.20	15.25	18	9.75%	11.80%	13.70%	
	1800	Equipment		3.54	4.35	5.75	2.78%	3.49%	4.35%	
	2500	Conveying equipment		2.50	3.79	5.15	2.23%	2.78%	3.37%	
	2720	Plumbing		7.05	9.55	11.70	6.80%	7.20%	10.45%	
	2900	Electrical		7.55	9.55	12.90	6.45%	7.65%	8.80%	
	3100	Total: Mechanical & Electrical	↓	22.50	29	35	17.95%	22.50%	24.50%	
	9000	Per apartment unit, total cost	Apt.	114,000	126,000	174,500				
	9500	Total: Mechanical & Electrical	"	22,700	28,200	29,900				
04	0010	**AUDITORIUMS**	S.F.	115	158	230				04
	0020	Total project costs	C.F.	7.15	9.95	14.30				
	2720	Plumbing	S.F.	6.80	10	11.95	5.85%	7.20%	8.70%	
	2900	Electrical		9.20	13.30	22	6.85%	9.50%	11.40%	
	3100	Total: Mechanical & Electrical	↓	61	81.50	99	24.50%	27.50%	31%	
05	0010	**AUTOMOTIVE SALES**	S.F.	84.50	115	143				05
	0020	Total project costs	C.F.	5.55	6.70	8.65				
	2720	Plumbing	S.F.	3.86	6.70	7.30	2.89%	6.05%	6.50%	
	2770	Heating, ventilating, air conditioning		5.95	9.10	9.80	4.61%	10%	10.35%	
	2900	Electrical		6.80	10.70	15.70	7.25%	8.80%	12.15%	
	3100	Total: Mechanical & Electrical	↓	19	30.50	36.50	17.30%	20.50%	22%	
06	0010	**BANKS**	S.F.	166	207	263				06
	0020	Total project costs	C.F.	11.85	16.10	21				
	0100	Site work	S.F.	19	30	42.50	7.90%	12.95%	17%	
	0500	Masonry		7.55	16.15	29	3.36%	6.90%	10.05%	
	1500	Finishes		14.85	22.50	28	5.85%	8.65%	11.70%	
	1800	Equipment		6.25	13.75	28	1.34%	5.55%	10.50%	
	2720	Plumbing		5.20	7.45	10.85	2.82%	3.90%	4.93%	
	2770	Heating, ventilating, air conditioning		9.85	13.20	17.60	4.86%	7.15%	8.50%	
	2900	Electrical		15.75	21	27.50	8.25%	10.20%	12.20%	
	3100	Total: Mechanical & Electrical	↓	38	50.50	61	16.55%	19.45%	24%	
	3500	See also division 11020 & 11030 (MF2004 11 16 00 & 11 17 00)								

	50 17 00	S.F. Costs		UNIT COSTS			% OF TOTAL			
			UNIT	1/4	MEDIAN	3/4	1/4	MEDIAN	3/4	
13	0010	**CHURCHES**	S.F.	112	143	185				13
	0020	Total project costs	C.F.	6.90	8.75	11.50				
	1800	Equipment	S.F.	1.21	3.17	6.80	.93%	2.11%	4.31%	
	2720	Plumbing		4.35	6.10	8.95	3.51%	4.96%	6.25%	
	2770	Heating, ventilating, air conditioning		10.15	13.25	18.75	7.50%	10%	12%	
	2900	Electrical		9.55	12.95	17.65	7.30%	8.80%	10.90%	
	3100	Total: Mechanical & Electrical	↓	29.50	39	51.50	18.30%	22%	24.50%	
	3500	See also division 11040 (MF2004 11 91 00)								
15	0010	**CLUBS, COUNTRY**	S.F.	120	144	182				15
	0020	Total project costs	C.F.	9.65	11.75	16.25				
	2720	Plumbing	S.F.	7.25	10.75	24.50	5.60%	7.90%	10%	
	2900	Electrical		9.40	12.90	16.80	7%	8.95%	11%	
	3100	Total: Mechanical & Electrical	↓	50	62.50	66	19%	26.50%	29.50%	
17	0010	**CLUBS, SOCIAL Fraternal**	S.F.	101	138	184				17
	0020	Total project costs	C.F.	5.95	9.05	10.80				
	2720	Plumbing	S.F.	6	7.50	11.30	5.60%	6.90%	8.55%	
	2770	Heating, ventilating, air conditioning		7.05	10.50	13.45	8.20%	9.25%	14.40%	
	2900	Electrical		6.85	11.80	13.55	5.95%	9.30%	10.55%	
	3100	Total: Mechanical & Electrical	↓	37	40	51	21%	23%	23.50%	
18	0010	**CLUBS, Y.M.C.A.**	S.F.	132	164	247				18
	0020	Total project costs	C.F.	5.55	9.25	13.80				
	2720	Plumbing	S.F.	7.60	15.10	16.95	5.65%	7.60%	10.85%	
	2900	Electrical		9.85	12.75	22	6.45%	8%	9.20%	
	3100	Total: Mechanical & Electrical	↓	40	45.50	74.50	20.50%	21.50%	28.50%	
19	0010	**COLLEGES Classrooms & Administration**	S.F.	116	164	212				19
	0020	Total project costs	C.F.	6.95	12.35	19.85				
	0500	Masonry	S.F.	9.15	16.85	20.50	5.10%	8.05%	10.50%	
	2720	Plumbing		5.20	12.65	23.50	5.10%	6.60%	8.95%	
	2900	Electrical		9.85	15.55	21.50	7.70%	9.85%	12%	
	3100	Total: Mechanical & Electrical	↓	41.50	57	76	23%	28%	31.50%	
21	0010	**COLLEGES Science, Engineering, Laboratories**	S.F.	225	264	305				21
	0020	Total project costs	C.F.	12.95	18.85	21.50				
	1800	Equipment	S.F.	6.10	28.50	31	2%	6.45%	12.65%	
	2900	Electrical		18.60	26.50	40.50	7.10%	9.40%	12.10%	
	3100	Total: Mechanical & Electrical	↓	69	82	127	28.50%	31.50%	41%	
	3500	See also division 11600 (MF2004 11 53 00)								
23	0010	**COLLEGES Student Unions**	S.F.	144	195	236				23
	0020	Total project costs	C.F.	8	10.50	13				
	3100	Total: Mechanical & Electrical	S.F.	54	58.50	69	23.50%	26%	29%	
25	0010	**COMMUNITY CENTERS**	S.F.	120	147	199				25
	0020	Total project costs	C.F.	7.75	11.05	14.35				
	1800	Equipment	S.F.	2.40	4.70	7.75	1.47%	3.01%	5.30%	
	2720	Plumbing		5.60	9.85	13.45	4.85%	7%	8.95%	
	2770	Heating, ventilating, air conditioning		8.45	13.10	18.80	6.80%	10.35%	12.90%	
	2900	Electrical		10.05	13.25	19.50	7.15%	8.90%	10.40%	
	3100	Total: Mechanical & Electrical	↓	33	42	60	18.80%	23%	30%	
28	0010	**COURT HOUSES**	S.F.	171	202	278				28
	0020	Total project costs	C.F.	13.15	15.70	19.80				
	2720	Plumbing	S.F.	8.15	11.35	12.90	5.95%	7.45%	8.20%	
	2900	Electrical		18.20	20.50	30	9.05%	10.65%	12.15%	
	3100	Total: Mechanical & Electrical	↓	51	66.50	73	22.50%	26.50%	30%	
30	0010	**DEPARTMENT STORES**	S.F.	63.50	85.50	108				30
	0020	Total project costs	C.F.	3.39	4.40	5.95				
	2720	Plumbing	S.F.	1.97	2.49	3.78	1.82%	4.21%	5.90%	
	2770	Heating, ventilating, air conditioning	↓	4.79	8.90	13.35	8.20%	9.10%	14.80%	

		50 17 00 | S.F. Costs	UNIT	UNIT COSTS			% OF TOTAL			
				1/4	MEDIAN	3/4	1/4	MEDIAN	3/4	
30	2900	Electrical	S.F.	7.25	9.95	11.75	9.05%	12.15%	14.95%	30
	3100	Total: Mechanical & Electrical	↓	12.80	16.35	28.50	13.20%	21.50%	50%	
31	0010	**DORMITORIES Low Rise (1 to 3 story)**	S.F.	120	166	208				31
	0020	Total project costs	C.F.	6.80	10.95	16.45				
	2720	Plumbing	S.F.	7.20	9.65	12.20	8.05%	9%	9.65%	
	2770	Heating, ventilating, air conditioning		7.65	9.15	12.20	4.61%	8.05%	10%	
	2900	Electrical		7.95	12.10	16.55	6.40%	8.65%	9.50%	
	3100	Total: Mechanical & Electrical	↓	41.50	44.50	69	22%	25%	27%	
	9000	Per bed, total cost	Bed	51,000	56,500	121,000				
32	0010	**DORMITORIES Mid Rise (4 to 8 story)**	S.F.	147	192	237				32
	0020	Total project costs	C.F.	16.25	17.85	21.50				
	2900	Electrical	S.F.	15.65	17.80	24	8.20%	10.20%	11.95%	
	3100	Total: Mechanical & Electrical	"	44	87.50	88.50	25%	30.50%	35.50%	
	9000	Per bed, total cost	Bed	21,000	47,900	277,000				
34	0010	**FACTORIES**	S.F.	55.50	83	128				34
	0020	Total project costs	C.F.	3.58	5.35	8.85				
	0100	Site work	S.F.	6.35	11.60	18.40	6.95%	11.45%	17.95%	
	2720	Plumbing		3.01	5.60	9.25	3.73%	6.05%	8.10%	
	2770	Heating, ventilating, air conditioning		5.85	8.40	11.35	5.25%	8.45%	11.35%	
	2900	Electrical		6.95	10.95	16.75	8.10%	10.50%	14.20%	
	3100	Total: Mechanical & Electrical	↓	19.80	30.50	40	21%	28.50%	35.50%	
36	0010	**FIRE STATIONS**	S.F.	111	153	209				36
	0020	Total project costs	C.F.	6.45	8.90	11.85				
	0500	Masonry	S.F.	15.10	29	37.50	7.50%	10.95%	15.55%	
	1140	Roofing		3.61	9.75	11.10	1.90%	4.94%	5.05%	
	1580	Painting		2.80	4.18	4.28	1.37%	1.57%	2.07%	
	1800	Equipment		1.38	2.64	4.87	.62%	1.63%	3.42%	
	2720	Plumbing		6.20	9.85	13.95	5.85%	7.35%	9.45%	
	2770	Heating, ventilating, air conditioning		6.15	9.95	15.35	5.15%	7.40%	9.40%	
	2900	Electrical		7.95	14.10	19	6.90%	8.60%	10.60%	
	3100	Total: Mechanical & Electrical	↓	40.50	52	59	18.40%	23%	27%	
37	0010	**FRATERNITY HOUSES & Sorority Houses**	S.F.	111	142	195				37
	0020	Total project costs	C.F.	11	11.50	13.85				
	2720	Plumbing	S.F.	8.35	9.60	17.55	6.80%	8%	10.85%	
	2900	Electrical	↓	7.30	15.80	19.35	6.60%	9.90%	10.65%	
38	0010	**FUNERAL HOMES**	S.F.	117	159	289				38
	0020	Total project costs	C.F.	11.90	13.25	25.50				
	2900	Electrical	S.F.	5.15	9.45	10.35	3.58%	4.44%	5.95%	
39	0010	**GARAGES, COMMERCIAL (Service)**	S.F.	66	102	141				39
	0020	Total project costs	C.F.	4.35	6.40	9.35				
	1800	Equipment	S.F.	3.73	8.40	13.05	2.21%	4.62%	6.80%	
	2720	Plumbing		4.58	7.05	12.85	5.45%	7.85%	10.65%	
	2730	Heating & ventilating		4.51	7.80	9.95	5.25%	6.85%	8.20%	
	2900	Electrical		6.30	9.60	13.80	7.15%	9.25%	10.85%	
	3100	Total: Mechanical & Electrical	↓	11.85	26.50	39	12.35%	17.40%	26%	
40	0010	**GARAGES, MUNICIPAL (Repair)**	S.F.	97	129	182				40
	0020	Total project costs	C.F.	6.05	7.70	13.25				
	0500	Masonry	S.F.	5.10	17.85	27.50	4.03%	9.15%	12.50%	
	2720	Plumbing		4.36	8.35	15.75	3.59%	6.70%	7.95%	
	2730	Heating & ventilating		7.45	10.80	21	6.15%	7.45%	13.50%	
	2900	Electrical		7.15	11.30	16.40	6.65%	8.15%	11.15%	
	3100	Total: Mechanical & Electrical	↓	33	46	66.50	21.50%	23%	28.50%	
41	0010	**GARAGES, PARKING**	S.F.	38.50	55	95				41
	0020	Total project costs	C.F.	3.55	4.82	7				

50 17 00	S.F. Costs	UNIT	UNIT COSTS			% OF TOTAL		
			1/4	MEDIAN	3/4	1/4	MEDIAN	3/4
41 2720	Plumbing	S.F.	.65	1.66	2.56	1.72%	2.70%	3.85%	**41**
2900	Electrical		2.08	2.74	4	4.52%	5.40%	6.35%
3100	Total: Mechanical & Electrical	↓	4.03	5.90	7.35	7%	8.90%	11.05%
3200								
9000	Per car, total cost	Car	16,000	20,000	25,500			
43 0010	**GYMNASIUMS**	S.F.	105	140	193				**43**
0020	Total project costs	C.F.	5.25	7.10	8.75			
1800	Equipment	S.F.	2.50	4.68	8.45	1.76%	3.26%	6.70%
2720	Plumbing		5.75	7.75	10.20	4.65%	6.40%	7.75%
2770	Heating, ventilating, air conditioning		6.25	10.90	22	5.15%	9.05%	11.10%
2900	Electrical		8.35	11	14.65	6.75%	8.50%	10.30%
3100	Total: Mechanical & Electrical	↓	28.50	40	48.50	19.75%	23.50%	29%
3500	See also division 11480 (MF2004 11 67 00)							
46 0010	**HOSPITALS**	S.F.	202	253	350				**46**
0020	Total project costs	C.F.	15.30	19.05	27			
1800	Equipment	S.F.	5.10	9.85	17	.80%	2.53%	4.80%
2720	Plumbing		17.40	24.50	31.50	7.60%	9.10%	10.85%
2770	Heating, ventilating, air conditioning		25.50	32.50	45.50	7.80%	12.95%	16.65%
2900	Electrical		22.50	30	44.50	10%	11.75%	14.10%
3100	Total: Mechanical & Electrical	↓	65	90.50	134	28%	33.50%	37%
9000	Per bed or person, total cost	Bed	233,500	322,000	371,000			
9900	See also division 11700 (MF2004 11 71 00)							
48 0010	**HOUSING For the Elderly**	S.F.	99	126	154				**48**
0020	Total project costs	C.F.	7.10	9.85	12.55			
0100	Site work	S.F.	6.90	10.75	15.75	5.05%	7.90%	12.10%
0500	Masonry		1.87	11.30	16.50	1.30%	6.05%	11%
1800	Equipment		2.40	3.30	5.25	1.88%	3.23%	4.43%
2510	Conveying systems		2.24	3.24	4.40	1.78%	2.20%	2.81%
2720	Plumbing		7.40	9.40	11.85	8.15%	9.55%	10.50%
2730	Heating, ventilating, air conditioning		3.78	5.35	8	3.30%	5.60%	7.25%
2900	Electrical		7.40	10.05	12.85	7.30%	8.50%	10.25%
3100	Total: Mechanical & Electrical	↓	25.50	31.50	40.50	18.10%	22.50%	29%
9000	Per rental unit, total cost	Unit	92,500	108,000	120,000			
9500	Total: Mechanical & Electrical	"	20,600	23,600	27,600			
50 0010	**HOUSING Public (Low Rise)**	S.F.	83.50	116	151				**50**
0020	Total project costs	C.F.	6.60	9.25	11.50			
0100	Site work	S.F.	10.60	15.30	25	8.35%	11.75%	16.50%
1800	Equipment		2.27	3.70	5.65	2.26%	3.03%	4.24%
2720	Plumbing		6	7.95	10.05	7.15%	9.05%	11.60%
2730	Heating, ventilating, air conditioning		3.02	5.85	6.40	4.26%	6.05%	6.45%
2900	Electrical		5.05	7.50	10.45	5.10%	6.55%	8.25%
3100	Total: Mechanical & Electrical	↓	24	31	34.50	14.50%	17.55%	26.50%
9000	Per apartment, total cost	Apt.	91,500	104,000	131,000			
9500	Total: Mechanical & Electrical	"	19,500	24,100	26,700			
51 0010	**ICE SKATING RINKS**	S.F.	71.50	167	183				**51**
0020	Total project costs	C.F.	5.25	5.35	6.20			
2720	Plumbing	S.F.	2.66	5	5.10	3.12%	3.23%	5.65%
2900	Electrical	↓	7.65	11.70	12.40	6.30%	10.15%	15.05%
52 0010	**JAILS**	S.F.	185	280	360				**52**
0020	Total project costs	C.F.	19.60	27.50	32			
1800	Equipment	S.F.	8.45	25	42.50	2.80%	6.95%	9.80%
2720	Plumbing		21	28	37	7%	8.90%	13.35%
2770	Heating, ventilating, air conditioning		19.60	26	50.50	7.50%	9.45%	17.75%
2900	Electrical		23	31	39	9.40%	11.70%	15.25%
3100	Total: Mechanical & Electrical	↓	61	108	128	28%	30%	34%
53 0010	**LIBRARIES**	S.F.	140	186	243				**53**
0020	Total project costs	C.F.	9.40	11.75	15			

		50 17 00 | S.F. Costs	UNIT	UNIT COSTS			% OF TOTAL			
				1/4	MEDIAN	3/4	1/4	MEDIAN	3/4	
53	0500	Masonry	S.F.	9.30	19	31.50	5.60%	7.15%	10.95%	53
	1800	Equipment		1.88	5.05	7.60	.28%	1.39%	4.07%	
	2720	Plumbing		5.05	7.20	10	3.38%	4.60%	5.70%	
	2770	Heating, ventilating, air conditioning		11.20	19	23	7.80%	10.95%	12.80%	
	2900	Electrical		14.10	18.65	25	8.40%	10.55%	12%	
	3100	Total: Mechanical & Electrical		46	56	65.50	21%	24%	28%	
54	0010	**LIVING, ASSISTED**	S.F.	127	151	176				54
	0020	Total project costs	C.F.	10.70	12.50	14.20				
	0500	Masonry	S.F.	3.68	4.44	5.45	2.36%	3.16%	3.86%	
	1800	Equipment		2.81	3.35	4.30	2.12%	2.45%	2.87%	
	2720	Plumbing		10.65	14.25	14.75	6.05%	8.15%	10.60%	
	2770	Heating, ventilating, air conditioning		12.65	13.20	14.45	7.95%	9.35%	9.70%	
	2900	Electrical		12.45	14	16.15	9%	9.95%	10.65%	
	3100	Total: Mechanical & Electrical		34	40.50	47	24%	28.50%	31.50%	
55	0010	**MEDICAL CLINICS**	S.F.	129	159	203				55
	0020	Total project costs	C.F.	9.45	12.25	16.30				
	1800	Equipment	S.F.	1.65	7.35	11.40	1.05%	2.94%	6.35%	
	2720	Plumbing		8.60	12.10	16.15	6.15%	8.40%	10.10%	
	2770	Heating, ventilating, air conditioning		10.25	13.40	19.70	6.65%	8.85%	11.35%	
	2900	Electrical		11.10	15.75	20.50	8.10%	10%	12.20%	
	3100	Total: Mechanical & Electrical		35.50	48	66	22%	27%	33.50%	
	3500	See also division 11700 (MF2004 11 71 00)								
57	0010	**MEDICAL OFFICES**	S.F.	121	151	185				57
	0020	Total project costs	C.F.	9.05	12.30	16.60				
	1800	Equipment	S.F.	4.01	7.95	11.15	.66%	4.87%	6.50%	
	2720	Plumbing		6.70	10.35	13.95	5.60%	6.80%	8.50%	
	2770	Heating, ventilating, air conditioning		8.10	11.70	15.45	6.10%	8%	9.70%	
	2900	Electrical		9.70	14.20	19.75	7.50%	9.80%	11.70%	
	3100	Total: Mechanical & Electrical		26.50	38.50	55	19.30%	23%	28.50%	
59	0010	**MOTELS**	S.F.	76.50	111	145				59
	0020	Total project costs	C.F.	6.80	9.15	14.95				
	2720	Plumbing	S.F.	7.75	9.90	11.80	9.45%	10.60%	12.55%	
	2770	Heating, ventilating, air conditioning		4.74	7.05	12.65	5.60%	5.60%	10%	
	2900	Electrical		7.25	9.15	11.40	7.45%	9.05%	10.45%	
	3100	Total: Mechanical & Electrical		21.50	31	53	18.50%	24%	25.50%	
	5000									
	9000	Per rental unit, total cost	Unit	39,000	74,000	80,000				
	9500	Total: Mechanical & Electrical	"	7,600	11,500	13,400				
60	0010	**NURSING HOMES**	S.F.	120	155	193				60
	0020	Total project costs	C.F.	9.45	11.80	16.15				
	1800	Equipment	S.F.	3.28	5	8.40	2%	3.62%	4.99%	
	2720	Plumbing		10.30	15.60	18.80	8.75%	10.10%	12.70%	
	2770	Heating, ventilating, air conditioning		10.85	16.45	22	9.70%	11.45%	11.80%	
	2900	Electrical		11.90	14.95	20.50	9.40%	10.60%	12.50%	
	3100	Total: Mechanical & Electrical		28.50	39.50	66.50	26%	28%	30%	
	9000	Per bed or person, total cost	Bed	53,500	67,000	86,000				
61	0010	**OFFICES Low Rise (1 to 4 story)**	S.F.	101	132	172				61
	0020	Total project costs	C.F.	7.25	9.95	13.10				
	0100	Site work	S.F.	8.15	14.20	21	5.95%	9.70%	13.55%	
	0500	Masonry		3.92	7.90	14.40	2.61%	5.40%	8.45%	
	1800	Equipment		.91	2.11	5.75	.57%	1.50%	3.42%	
	2720	Plumbing		3.62	5.60	8.20	3.66%	4.50%	6.10%	
	2770	Heating, ventilating, air conditioning		8	11.15	16.35	7.20%	10.30%	11.70%	
	2900	Electrical		8.30	11.85	16.85	7.45%	9.65%	11.40%	
	3100	Total: Mechanical & Electrical		23	31.50	48	18.20%	22%	27%	
62	0010	**OFFICES Mid Rise (5 to 10 story)**	S.F.	107	130	176				62
	0020	Total project costs	C.F.	7.60	9.70	13.75				

		**50 17 00	S.F. Costs**		UNIT COSTS			% OF TOTAL			
			UNIT	1/4	MEDIAN	3/4	1/4	MEDIAN	3/4	
62	2720	Plumbing	S.F.	3.24	5	7.20	2.83%	3.74%	4.50%	62
	2770	Heating, ventilating, air conditioning		8.15	11.65	18.60	7.65%	9.40%	11%	
	2900	Electrical		7.95	10.20	15.40	6.35%	7.80%	10%	
	3100	Total: Mechanical & Electrical	↓	20.50	26.50	50.50	18.95%	21%	27.50%	
63	0010	OFFICES High Rise (11 to 20 story)	S.F.	132	166	204				63
	0020	Total project costs	C.F.	9.20	11.50	16.50				
	2900	Electrical	S.F.	8	9.75	14.50	5.80%	7.85%	10.50%	
	3100	Total: Mechanical & Electrical	↓	26	34.50	58.50	16.90%	23.50%	34%	
64	0010	POLICE STATIONS	S.F.	158	208	265				64
	0020	Total project costs	C.F.	12.60	15.40	21				
	0500	Masonry	S.F.	15.80	26	33	7.80%	9.10%	11.35%	
	1800	Equipment		2.27	11.05	17.50	.98%	3.35%	6.70%	
	2720	Plumbing		8.85	17.60	22	5.65%	6.90%	10.75%	
	2770	Heating, ventilating, air conditioning		12.95	18.35	26	5.85%	10.55%	11.70%	
	2900	Electrical		17.25	25	32.50	9.80%	11.70%	14.50%	
	3100	Total: Mechanical & Electrical	↓	65.50	69	92	28.50%	31.50%	32.50%	
65	0010	POST OFFICES	S.F.	125	154	196				65
	0020	Total project costs	C.F.	7.50	9.50	10.80				
	2720	Plumbing	S.F.	5.05	6.95	8.80	4.24%	5.30%	5.60%	
	2770	Heating, ventilating, air conditioning		7.85	10.85	12.05	6.65%	7.15%	9.35%	
	2900	Electrical		10.30	14.50	17.20	7.25%	9%	11%	
	3100	Total: Mechanical & Electrical	↓	29.50	39	44	16.25%	18.80%	22%	
66	0010	POWER PLANTS	S.F.	865	1,150	2,100				66
	0020	Total project costs	C.F.	24	52	111				
	2900	Electrical	S.F.	61	130	193	9.30%	12.75%	21.50%	
	8100	Total: Mechanical & Electrical	↓	152	495	1,100	32.50%	32.50%	52.50%	
67	0010	RELIGIOUS EDUCATION	S.F.	101	131	161				67
	0020	Total project costs	C.F.	5.60	8	10				
	2720	Plumbing	S.F.	4.20	5.95	8.40	4.40%	5.30%	7.10%	
	2770	Heating, ventilating, air conditioning		10.60	12	16.95	10.05%	11.45%	12.35%	
	2900	Electrical		7.40	11.25	14.85	7.60%	9.05%	10.35%	
	3100	Total: Mechanical & Electrical	↓	32.50	42.50	52	22%	23%	27%	
69	0010	RESEARCH Laboratories & Facilities	S.F.	154	217	315				69
	0020	Total project costs	C.F.	11.20	22	26.50				
	1800	Equipment	S.F.	6.70	13.15	32	.94%	4.58%	9.10%	
	2720	Plumbing		10.95	19.30	31	6.15%	8.30%	10.80%	
	2770	Heating, ventilating, air conditioning		13.55	45.50	54	7.25%	16.50%	17.50%	
	2900	Electrical		18.10	29	47.50	9.55%	11.15%	14.60%	
	3100	Total: Mechanical & Electrical	↓	57.50	102	144	29.50%	37%	41%	
70	0010	RESTAURANTS	S.F.	146	188	245				70
	0020	Total project costs	C.F.	12.25	16.10	21				
	1800	Equipment	S.F.	8.65	23	35	6.10%	13%	15.65%	
	2720	Plumbing		11.55	14	18.40	6.10%	8.15%	9%	
	2770	Heating, ventilating, air conditioning		14.70	20.50	24.50	9.20%	12%	12.40%	
	2900	Electrical		15.40	19	24.50	8.35%	10.55%	11.55%	
	3100	Total: Mechanical & Electrical	↓	47.50	51	66	21%	25%	29.50%	
	9000	Per seat unit, total cost	Seat	5,350	7,125	8,450				
	9500	Total: Mechanical & Electrical	"	1,350	1,775	2,100				
72	0010	RETAIL STORES	S.F.	68	91.50	121				72
	0020	Total project costs	C.F.	4.62	6.60	9.15				
	2720	Plumbing	S.F.	2.47	4.12	7.05	3.26%	4.60%	6.80%	
	2770	Heating, ventilating, air conditioning		5.35	7.30	10.95	6.75%	8.75%	10.15%	
	2900	Electrical		6.15	8.40	12.10	7.25%	9.90%	11.60%	
	3100	Total: Mechanical & Electrical	↓	16.30	21	28	17.05%	21%	23.50%	
74	0010	SCHOOLS Elementary	S.F.	111	138	169				74
	0020	Total project costs	C.F.	7.25	9.30	12.05				
	0500	Masonry	S.F.	7.85	16.95	25	4.89%	10.50%	14%	
	1800	Equipment	↓	2.58	4.93	9.30	1.83%	3.13%	4.61%	

		50 17 00 | S.F. Costs	UNIT	UNIT COSTS			% OF TOTAL			
				1/4	MEDIAN	3/4	1/4	MEDIAN	3/4	
74	2720	Plumbing	S.F.	6.40	9.05	12.05	5.70%	7.15%	9.35%	74
	2730	Heating, ventilating, air conditioning		9.60	15.25	21.50	8.15%	10.80%	14.90%	
	2900	Electrical		10.50	14	17.85	8.45%	10.05%	11.85%	
	3100	Total: Mechanical & Electrical	↓	38	46.50	59	25%	27.50%	30%	
	9000	Per pupil, total cost	Ea.	12,800	19,000	42,500				
	9500	Total: Mechanical & Electrical	"	3,600	4,550	11,500				
76	0010	**SCHOOLS Junior High & Middle**	S.F.	115	142	173				76
	0020	Total project costs	C.F.	7.30	9.40	10.55				
	0500	Masonry	S.F.	12.30	18.40	22.50	7.55%	11.10%	14.30%	
	1800	Equipment		3.14	5.90	8.65	1.80%	3.03%	4.80%	
	2720	Plumbing		6.70	8.25	10.20	5.30%	6.80%	7.25%	
	2770	Heating, ventilating, air conditioning		13.35	16.20	28.50	8.90%	11.55%	14.20%	
	2900	Electrical		11.70	14.45	18.25	8.05%	9.55%	10.60%	
	3100	Total: Mechanical & Electrical	↓	36.50	47	58.50	23%	25.50%	29.50%	
	9000	Per pupil, total cost	Ea.	14,500	19,100	25,600				
78	0010	**SCHOOLS Senior High**	S.F.	120	147	184				78
	0020	Total project costs	C.F.	7.20	10.10	17.05				
	1800	Equipment	S.F.	3.17	7.40	10.25	1.88%	2.67%	4.30%	
	2720	Plumbing		6.70	10.05	18.35	5.60%	6.90%	8.30%	
	2770	Heating, ventilating, air conditioning		11.50	15.65	24	8.95%	11.60%	15%	
	2900	Electrical		12.05	16.05	23.50	8.70%	10.35%	12.50%	
	3100	Total: Mechanical & Electrical	↓	39.50	47	76	24%	26.50%	28.50%	
	9000	Per pupil, total cost	Ea.	11,200	22,900	28,600				
80	0010	**SCHOOLS Vocational**	S.F.	96.50	140	173				80
	0020	Total project costs	C.F.	6	8.60	11.90				
	0500	Masonry	S.F.	4.24	14	21.50	3.20%	4.61%	10.95%	
	1800	Equipment		3.02	7.50	10.35	1.24%	3.10%	4.26%	
	2720	Plumbing		6.15	9.20	13.55	5.40%	6.90%	8.55%	
	2770	Heating, ventilating, air conditioning		8.65	16.05	27	8.60%	11.90%	14.65%	
	2900	Electrical		10.05	13.75	18.80	8.45%	11%	13.20%	
	3100	Total: Mechanical & Electrical	↓	38	66.50	85	27.50%	29.50%	31%	
	9000	Per pupil, total cost	Ea.	13,400	36,000	53,500				
83	0010	**SPORTS ARENAS**	S.F.	84.50	113	174				83
	0020	Total project costs	C.F.	4.58	8.20	10.60				
	2720	Plumbing	S.F.	4.22	7.40	15.65	4.35%	6.35%	9.40%	
	2770	Heating, ventilating, air conditioning		10.55	12.45	17.30	8.80%	10.20%	13.55%	
	2900	Electrical	↓	7.80	11.90	15.40	7.65%	9.75%	11.90%	
85	0010	**SUPERMARKETS**	S.F.	78	90	106				85
	0020	Total project costs	C.F.	4.34	5.25	7.95				
	2720	Plumbing	S.F.	4.35	5.50	6.40	5.40%	6%	7.45%	
	2770	Heating, ventilating, air conditioning		6.40	8.50	10.35	8.60%	8.65%	9.60%	
	2900	Electrical		9.15	11.15	13.25	10.40%	12.45%	13.60%	
	3100	Total: Mechanical & Electrical	↓	25	26.50	35.50	23.50%	27.50%	28.50%	
86	0010	**SWIMMING POOLS**	S.F.	126	295	450				86
	0020	Total project costs	C.F.	10.10	12.60	13.75				
	2720	Plumbing	S.F.	11.65	13.35	17.90	4.80%	9.70%	20.50%	
	2900	Electrical		9.60	15.35	34	6.05%	6.95%	7.60%	
	3100	Total: Mechanical & Electrical	↓	60	79	104	11.15%	14.10%	23.50%	
87	0010	**TELEPHONE EXCHANGES**	S.F.	157	238	310				87
	0020	Total project costs	C.F.	10.40	16.75	23				
	2720	Plumbing	S.F.	7.05	10.90	16	4.52%	5.80%	6.90%	
	2770	Heating, ventilating, air conditioning		16.40	33	41	11.80%	16.05%	18.40%	
	2900	Electrical		17.05	27	48	10.90%	14%	17.85%	
	3100	Total: Mechanical & Electrical	↓	50.50	95.50	135	29.50%	33.50%	44.50%	
91	0010	**THEATERS**	S.F.	99.50	135	199				91
	0020	Total project costs	C.F.	4.87	7.20	10.55				

50 17 00	S.F. Costs		UNIT	UNIT COSTS			% OF TOTAL			
				1/4	MEDIAN	3/4	1/4	MEDIAN	3/4	
91	2720	Plumbing	S.F.	3.28	3.80	15.55	2.92%	4.70%	6.80%	**91**
	2770	Heating, ventilating, air conditioning		10.25	12.40	15.35	8%	12.25%	13.40%	
	2900	Electrical		9.20	12.45	25.50	8.05%	9.35%	12.25%	
	3100	Total: Mechanical & Electrical		23.50	32	37.50	21.50%	25.50%	27.50%	
94	0010	**TOWN HALLS City Halls & Municipal Buildings**	S.F.	110	149	195				**94**
	0020	Total project costs	C.F.	7.95	12.40	18.10				
	2720	Plumbing	S.F.	4.91	9.20	16.10	4.31%	5.95%	7.95%	
	2770	Heating, ventilating, air conditioning		8.85	17.60	26	7.05%	9.05%	13.45%	
	2900	Electrical		11.15	15.40	21	8.05%	9.45%	11.65%	
	3100	Total: Mechanical & Electrical		38.50	49	75	22%	26.50%	31%	
97	0010	**WAREHOUSES & Storage Buildings**	S.F.	44	63	93				**97**
	0020	Total project costs	C.F.	2.30	3.60	5.95				
	0100	Site work	S.F.	4.53	9	13.55	6.05%	12.95%	19.55%	
	0500	Masonry		2.49	6.25	13.45	3.60%	7.35%	12%	
	1800	Equipment		.67	1.52	8.55	.72%	1.69%	5.55%	
	2720	Plumbing		1.46	2.62	4.91	2.90%	4.80%	6.55%	
	2730	Heating, ventilating, air conditioning		1.67	4.71	6.30	2.41%	5%	8.90%	
	2900	Electrical		2.60	4.89	8	5.15%	7.20%	10.05%	
	3100	Total: Mechanical & Electrical		7.25	11.10	22	13.30%	18.90%	26%	
99	0010	**WAREHOUSE & OFFICES Combination**	S.F.	54	72	99				**99**
	0020	Total project costs	C.F.	2.76	4.01	5.90				
	1800	Equipment	S.F.	.93	1.81	2.69	.42%	1.19%	2.40%	
	2720	Plumbing		2.08	3.63	5.40	3.74%	4.76%	6.30%	
	2770	Heating, ventilating, air conditioning		3.29	5.15	7.20	5%	5.65%	10.05%	
	2900	Electrical		3.68	5.45	8.50	5.85%	8%	10%	
	3100	Total: Mechanical & Electrical		10.25	15.55	24.50	14.55%	19.95%	24.50%	

Square Foot Project Size Modifier

One factor that affects the S.F. cost of a particular building is the size. In general, for buildings built to the same specifications in the same locality, the larger building will have the lower S.F. cost. This is due mainly to the decreasing contribution of the exterior walls, plus the economy of scale usually achievable in larger buildings. The area conversion scale shown below will give a factor to convert costs for the typical size building to an adjusted cost for the particular project.

The square foot base size table lists the median costs, most typical project size in our accumulated data, and the range in size of the projects.

The size factor for your project is determined by dividing your project area in S.F. by the typical project size for the particular building type. With this factor, enter the area conversion scale at the appropriate size factor and determine the appropriate cost multiplier for your building size.

Example: Determine the cost per S.F. for a 100,000 S.F. Mid-rise apartment building.

$$\frac{\text{Proposed building area} = 100,000 \text{ S.F.}}{\text{Typical size from below} = 50,000 \text{ S.F.}} = 2.00$$

Enter area conversion scale at 2.0, intersect curve, read horizontally the appropriate cost multiplier of .94. Size adjusted cost becomes .94 × $117.00 = $110.00 based on national average costs.

Note: For size factors less than .50, the cost multiplier is 1.1
For size factors greater than 3.5, the cost multiplier is .90

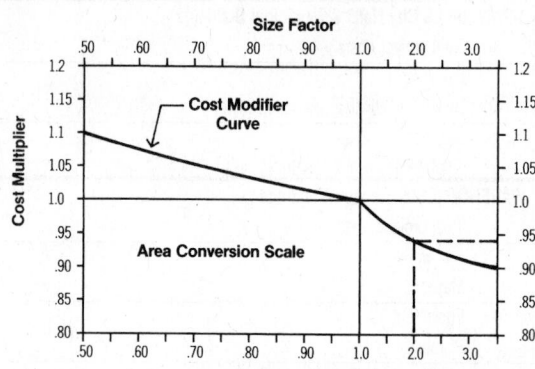

Square Foot Base Size							
Building Type	Median Cost per S.F.	Typical Size Gross S.F.	Typical Range Gross S.F.	Building Type	Median Cost per S.F.	Typical Size Gross S.F.	Typical Range Gross S.F.
Apartments, Low Rise	$ 92.50	21,000	9,700 - 37,200	Jails	$ 280.00	40,000	5,500 - 145,000
Apartments, Mid Rise	117.00	50,000	32,000 - 100,000	Libraries	186.00	12,000	7,000 - 31,000
Apartments, High Rise	127.00	145,000	95,000 - 600,000	Living, Assisted	151.00	32,300	23,500 - 50,300
Auditoriums	158.00	25,000	7,600 - 39,000	Medical Clinics	159.00	7,200	4,200 - 15,700
Auto Sales	115.00	20,000	10,800 - 28,600	Medical Offices	151.00	6,000	4,000 - 15,000
Banks	207.00	4,200	2,500 - 7,500	Motels	111.00	40,000	15,800 - 120,000
Churches	143.00	17,000	2,000 - 42,000	Nursing Homes	155.00	23,000	15,000 - 37,000
Clubs, Country	144.00	6,500	4,500 - 15,000	Offices, Low Rise	132.00	20,000	5,000 - 80,000
Clubs, Social	138.00	10,000	6,000 - 13,500	Offices, Mid Rise	130.00	120,000	20,000 - 300,000
Clubs, YMCA	164.00	28,300	12,800 - 39,400	Offices, High Rise	166.00	260,000	120,000 - 800,000
Colleges (Class)	164.00	50,000	15,000 - 150,000	Police Stations	208.00	10,500	4,000 - 19,000
Colleges (Science Lab)	264.00	45,600	16,600 - 80,000	Post Offices	154.00	12,400	6,800 - 30,000
College (Student Union)	195.00	33,400	16,000 - 85,000	Power Plants	1150.00	7,500	1,000 - 20,000
Community Center	147.00	9,400	5,300 - 16,700	Religious Education	131.00	9,000	6,000 - 12,000
Court Houses	202.00	32,400	17,800 - 106,000	Research	217.00	19,000	6,300 - 45,000
Dept. Stores	85.50	90,000	44,000 - 122,000	Restaurants	188.00	4,400	2,800 - 6,000
Dormitories, Low Rise	166.00	25,000	10,000 - 95,000	Retail Stores	91.50	7,200	4,000 - 17,600
Dormitories, Mid Rise	192.00	85,000	20,000 - 200,000	Schools, Elementary	138.00	41,000	24,500 - 55,000
Factories	83.00	26,400	12,900 - 50,000	Schools, Jr. High	142.00	92,000	52,000 - 119,000
Fire Stations	153.00	5,800	4,000 - 8,700	Schools, Sr. High	147.00	101,000	50,500 - 175,000
Fraternity Houses	142.00	12,500	8,200 - 14,800	Schools, Vocational	140.00	37,000	20,500 - 82,000
Funeral Homes	159.00	10,000	4,000 - 20,000	Sports Arenas	113.00	15,000	5,000 - 40,000
Garages, Commercial	102.00	9,300	5,000 - 13,600	Supermarkets	90.00	44,000	12,000 - 60,000
Garages, Municipal	129.00	8,300	4,500 - 12,600	Swimming Pools	295.00	20,000	10,000 - 32,000
Garages, Parking	55.00	163,000	76,400 - 225,300	Telephone Exchange	238.00	4,500	1,200 - 10,600
Gymnasiums	140.00	19,200	11,600 - 41,000	Theaters	135.00	10,500	8,800 - 17,500
Hospitals	253.00	55,000	27,200 - 125,000	Town Halls	149.00	10,800	4,800 - 23,400
House (Elderly)	126.00	37,000	21,000 - 66,000	Warehouses	63.00	25,000	8,000 - 72,000
Housing (Public)	116.00	36,000	14,400 - 74,400	Warehouse & Office	72.00	25,000	8,000 - 72,000
Ice Rinks	167.00	29,000	27,200 - 33,600				

Abbreviations

A	Area Square Feet; Ampere	Brk., brk	Brick	Csc	Cosecant
AAFES	Army and Air Force Exchange Service	brkt	Bracket	C.S.F.	Hundred Square Feet
ABS	Acrylonitrile Butadiene Stryrene; Asbestos Bonded Steel	Brng.	Bearing	CSI	Construction Specifications Institute
		Brs.	Brass		
A.C., AC	Alternating Current; Air-Conditioning; Asbestos Cement; Plywood Grade A & C	Brz.	Bronze	CT	Current Transformer
		Bsn.	Basin	CTS	Copper Tube Size
		Btr.	Better	Cu	Copper, Cubic
		Btu	British Thermal Unit	Cu. Ft.	Cubic Foot
ACI	American Concrete Institute	BTUH	BTU per Hour	cw	Continuous Wave
ACR	Air Conditioning Refrigeration	Bu.	bushels	C.W.	Cool White; Cold Water
ADA	Americans with Disabilities Act	BUR	Built-up Roofing	Cwt.	100 Pounds
AD	Plywood, Grade A & D	BX	Interlocked Armored Cable	C.W.X.	Cool White Deluxe
Addit.	Additional	°C	degree centegrade	C.Y.	Cubic Yard (27 cubic feet)
Adj.	Adjustable	c	Conductivity, Copper Sweat	C.Y./Hr.	Cubic Yard per Hour
af	Audio-frequency	C	Hundred; Centigrade	Cyl.	Cylinder
AFUE	Annual Fuel Utilization Efficiency	C/C	Center to Center, Cedar on Cedar	d	Penny (nail size)
AGA	American Gas Association	C-C	Center to Center	D	Deep; Depth; Discharge
Agg.	Aggregate	Cab	Cabinet	Dis., Disch.	Discharge
A.H., Ah	Ampere Hours	Cair.	Air Tool Laborer	Db	Decibel
A hr.	Ampere-hour	Cal.	caliper	Dbl.	Double
A.H.U., AHU	Air Handling Unit	Calc	Calculated	DC	Direct Current
A.I.A.	American Institute of Architects	Cap.	Capacity	DDC	Direct Digital Control
AIC	Ampere Interrupting Capacity	Carp.	Carpenter	Demob.	Demobilization
Allow.	Allowance	C.B.	Circuit Breaker	d.f.t.	Dry Film Thickness
alt., alt	Alternate	C.C.A.	Chromate Copper Arsenate	d.f.u.	Drainage Fixture Units
Alum.	Aluminum	C.C.F.	Hundred Cubic Feet	D.H.	Double Hung
a.m.	Ante Meridiem	cd	Candela	DHW	Domestic Hot Water
Amp.	Ampere	cd/sf	Candela per Square Foot	DI	Ductile Iron
Anod.	Anodized	CD	Grade of Plywood Face & Back	Diag.	Diagonal
ANSI	American National Standards Institute	CDX	Plywood, Grade C & D, exterior glue	Diam., Dia	Diameter
				Distrib.	Distribution
APA	American Plywood Association	Cefi.	Cement Finisher	Div.	Division
Approx.	Approximate	Cem.	Cement	Dk.	Deck
Apt.	Apartment	CF	Hundred Feet	D.L.	Dead Load; Diesel
Asb.	Asbestos	C.F.	Cubic Feet	DLH	Deep Long Span Bar Joist
A.S.B.C.	American Standard Building Code	CFM	Cubic Feet per Minute	dlx	Deluxe
Asbe.	Asbestos Worker	CFRP	Carbon Fiber Reinforced Plastic	Do.	Ditto
ASCE.	American Society of Civil Engineers	c.g.	Center of Gravity	DOP	Dioctyl Phthalate Penetration Test (Air Filters)
A.S.H.R.A.E.	American Society of Heating, Refrig. & AC Engineers	CHW	Chilled Water; Commercial Hot Water	Dp., dp	Depth
ASME	American Society of Mechanical Engineers	C.I., CI	Cast Iron	D.P.S.T.	Double Pole, Single Throw
		C.I.P., CIP	Cast in Place	Dr.	Drive
ASTM	American Society for Testing and Materials	Circ.	Circuit	DR	Dimension Ratio
		C.L.	Carload Lot	Drink.	Drinking
Attchmt.	Attachment	CL	Chain Link	D.S.	Double Strength
Avg., Ave.	Average	Clab.	Common Laborer	D.S.A.	Double Strength A Grade
AWG	American Wire Gauge	Clam	Common maintenance laborer	D.S.B.	Double Strength B Grade
AWWA	American Water Works Assoc.	C.L.F.	Hundred Linear Feet	Dty.	Duty
Bbl.	Barrel	CLF	Current Limiting Fuse	DWV	Drain Waste Vent
B&B, BB	Grade B and Better; Balled & Burlapped	CLP	Cross Linked Polyethylene	DX	Deluxe White, Direct Expansion
		cm	Centimeter	dyn	Dyne
B&S	Bell and Spigot	CMP	Corr. Metal Pipe	e	Eccentricity
B.&W.	Black and White	CMU	Concrete Masonry Unit	E	Equipment Only; East; emissivity
b.c.c.	Body-centered Cubic	CN	Change Notice	Ea.	Each
B.C.Y.	Bank Cubic Yards	Col.	Column	EB	Encased Burial
BE	Bevel End	CO₂	Carbon Dioxide	Econ.	Economy
B.F.	Board Feet	Comb.	Combination	E.C.Y	Embankment Cubic Yards
Bg. cem.	Bag of Cement	comm.	Commercial, Communication	EDP	Electronic Data Processing
BHP	Boiler Horsepower; Brake Horsepower	Compr.	Compressor	EIFS	Exterior Insulation Finish System
		Conc.	Concrete	E.D.R.	Equiv. Direct Radiation
B.I.	Black Iron	Cont., cont	Continuous; Continued, Container	Eq.	Equation
bidir.	bidirectional	Corr.	Corrugated	EL	elevation
Bit., Bitum.	Bituminous	Cos	Cosine	Elec.	Electrician; Electrical
Bit., Conc.	Bituminous Concrete	Cot	Cotangent	Elev.	Elevator; Elevating
Bk.	Backed	Cov.	Cover	EMT	Electrical Metallic Conduit; Thin Wall Conduit
Bkrs.	Breakers	C/P	Cedar on Paneling		
Bldg., bldg	Building	CPA	Control Point Adjustment	Eng.	Engine, Engineered
Blk.	Block	Cplg.	Coupling	EPDM	Ethylene Propylene Diene Monomer
Bm.	Beam	CPM	Critical Path Method		
Boil.	Boilermaker	CPVC	Chlorinated Polyvinyl Chloride	EPS	Expanded Polystyrene
bpm	Blows per Minute	C.Pr.	Hundred Pair	Eqhv.	Equip. Oper., Heavy
BR	Bedroom	CRC	Cold Rolled Channel	Eqlt.	Equip. Oper., Light
Brg.	Bearing	Creos.	Creosote	Eqmd.	Equip. Oper., Medium
Brhe.	Bricklayer Helper	Crpt.	Carpet & Linoleum Layer	Eqmm.	Equip. Oper., Master Mechanic
Bric.	Bricklayer	CRT	Cathode-ray Tube	Eqol.	Equip. Oper., Oilers
		CS	Carbon Steel, Constant Shear Bar Joist	Equip.	Equipment
				ERW	Electric Resistance Welded

629

E.S.	Energy Saver	H	High Henry	Lath.	Lather
Est.	Estimated	HC	High Capacity	Lav.	Lavatory
esu	Electrostatic Units	H.D., HD	Heavy Duty; High Density	lb.; #	Pound
E.W.	Each Way	H.D.O.	High Density Overlaid	L.B., LB	Load Bearing; L Conduit Body
EWT	Entering Water Temperature	HDPE	High density polyethelene plastic	L. & E.	Labor & Equipment
Excav.	Excavation	Hdr.	Header	lb./hr.	Pounds per Hour
excl	Excluding	Hdwe.	Hardware	lb./L.F.	Pounds per Linear Foot
Exp., exp	Expansion, Exposure	H.I.D., HID	High Intensity Discharge	lbf/sq.in.	Pound-force per Square Inch
Ext., ext	Exterior; Extension	Help.	Helper Average	L.C.L.	Less than Carload Lot
Extru.	Extrusion	HEPA	High Efficiency Particulate Air	L.C.Y.	Loose Cubic Yard
f.	Fiber stress		Filter	Ld.	Load
F	Fahrenheit; Female; Fill	Hg	Mercury	LE	Lead Equivalent
Fab., fab	Fabricated; fabric	HIC	High Interrupting Capacity	LED	Light Emitting Diode
FBGS	Fiberglass	HM	Hollow Metal	L.F.	Linear Foot
F.C.	Footcandles	HMWPE	high molecular weight	L.F. Nose	Linear Foot of Stair Nosing
f.c.c.	Face-centered Cubic		polyethylene	L.F. Rsr	Linear Foot of Stair Riser
f'c.	Compressive Stress in Concrete;	HO	High Output	Lg.	Long; Length; Large
	Extreme Compressive Stress	Horiz.	Horizontal	L & H	Light and Heat
F.E.	Front End	H.P., HP	Horsepower; High Pressure	LH	Long Span Bar Joist
FEP	Fluorinated Ethylene Propylene	H.P.F.	High Power Factor	L.H.	Labor Hours
	(Teflon)	Hr.	Hour	L.L., LL	Live Load
F.G.	Flat Grain	Hrs./Day	Hours per Day	L.L.D.	Lamp Lumen Depreciation
F.H.A.	Federal Housing Administration	HSC	High Short Circuit	lm	Lumen
Fig.	Figure	Ht.	Height	lm/sf	Lumen per Square Foot
Fin.	Finished	Htg.	Heating	lm/W	Lumen per Watt
FIPS	Female Iron Pipe Size	Htrs.	Heaters	LOA	Length Over All
Fixt.	Fixture	HVAC	Heating, Ventilation & Air-	log	Logarithm
FJP	Finger jointed and primed		Conditioning	L-O-L	Lateralolet
Fl. Oz.	Fluid Ounces	Hvy.	Heavy	long.	longitude
Flr.	Floor	HW	Hot Water	L.P., LP	Liquefied Petroleum; Low Pressure
FM	Frequency Modulation;	Hyd.; Hydr.	Hydraulic	L.P.F.	Low Power Factor
	Factory Mutual	Hz	Hertz (cycles)	LR	Long Radius
Fmg.	Framing	I.	Moment of Inertia	L.S.	Lump Sum
FM/UL	Factory Mutual/Underwriters Labs	IBC	International Building Code	Lt.	Light
Fdn.	Foundation	I.C.	Interrupting Capacity	Lt. Ga.	Light Gauge
FNPT	Female National Pipe Thread	ID	Inside Diameter	L.T.L.	Less than Truckload Lot
Fori.	Foreman, Inside	I.D.	Inside Dimension; Identification	Lt. Wt.	Lightweight
Foro.	Foreman, Outside	I.F.	Inside Frosted	L.V.	Low Voltage
Fount.	Fountain	I.M.C.	Intermediate Metal Conduit	M	Thousand; Material; Male;
fpm	Feet per Minute	In.	Inch		Light Wall Copper Tubing
FPT	Female Pipe Thread	Incan.	Incandescent	M²CA	Meters Squared Contact Area
Fr	Frame	Incl.	Included; Including	m/hr.; M.H.	Man-hour
F.R.	Fire Rating	Int.	Interior	mA	Milliampere
FRK	Foil Reinforced Kraft	Inst.	Installation	Mach.	Machine
FSK	Foil/scrim/kraft	Insul., insul	Insulation/Insulated	Mag. Str.	Magnetic Starter
FRP	Fiberglass Reinforced Plastic	I.P.	Iron Pipe	Maint.	Maintenance
FS	Forged Steel	I.P.S., IPS	Iron Pipe Size	Marb.	Marble Setter
FSC	Cast Body; Cast Switch Box	IPT	Iron Pipe Threaded	Mat; Mat'l.	Material
Ft., ft	Foot; Feet	I.W.	Indirect Waste	Max.	Maximum
Ftng.	Fitting	J	Joule	MBF	Thousand Board Feet
Ftg.	Footing	J.I.C.	Joint Industrial Council	MBH	Thousand BTU's per hr.
Ft lb.	Foot Pound	K	Thousand; Thousand Pounds;	MC	Metal Clad Cable
Furn.	Furniture		Heavy Wall Copper Tubing, Kelvin	MCC	Motor Control Center
FVNR	Full Voltage Non-Reversing	K.A.H.	Thousand Amp. Hours	M.C.F.	Thousand Cubic Feet
FVR	Full Voltage Reversing	kcmil	Thousand Circular Mils	MCFM	Thousand Cubic Feet per Minute
FXM	Female by Male	KD	Knock Down	M.C.M.	Thousand Circular Mils
Fy.	Minimum Yield Stress of Steel	K.D.A.T.	Kiln Dried After Treatment	MCP	Motor Circuit Protector
g	Gram	kg	Kilogram	MD	Medium Duty
G	Gauss	kG	Kilogauss	MDF	Medium-density fibreboard
Ga.	Gauge	kgf	Kilogram Force	M.D.O.	Medium Density Overlaid
Gal., gal.	Gallon	kHz	Kilohertz	Med.	Medium
gpm, GPM	Gallon per Minute	Kip	1000 Pounds	MF	Thousand Feet
Galv., galv	Galvanized	KJ	Kiljoule	M.F.B.M.	Thousand Feet Board Measure
GC/MS	Gas Chromatograph/Mass	K.L.	Effective Length Factor	Mfg.	Manufacturing
	Spectrometer	K.L.F.	Kips per Linear Foot	Mfrs.	Manufacturers
Gen.	General	Km	Kilometer	mg	Milligram
GFI	Ground Fault Interrupter	KO	Knock Out	MGD	Million Gallons per Day
GFRC	Glass Fiber Reinforced Concrete	K.S.F.	Kips per Square Foot	MGPH	Thousand Gallons per Hour
Glaz.	Glazier	K.S.I.	Kips per Square Inch	MH, M.H.	Manhole; Metal Halide; Man-Hour
GPD	Gallons per Day	kV	Kilovolt	MHz	Megahertz
gpf	Gallon per flush	kVA	Kilovolt Ampere	Mi.	Mile
GPH	Gallons per Hour	kVAR	Kilovar (Reactance)	MI	Malleable Iron; Mineral Insulated
GPM	Gallons per Minute	KW	Kilowatt	MIPS	Male Iron Pipe Size
GR	Grade	KWh	Kilowatt-hour	mj	Mechanical Joint
Gran.	Granular	L	Labor Only; Length; Long;	m	Meter
Grnd.	Ground		Medium Wall Copper Tubing	mm	Millimeter
GVW	Gross Vehicle Weight	Lab.	Labor	Mill.	Millwright
GWB	Gypsum wall board	lat	Latitude	Min., min.	Minimum, minute

Abbreviations

Misc.	Miscellaneous	PDCA	Painting and Decorating Contractors of America	SC	Screw Cover
ml	Milliliter, Mainline			SCFM	Standard Cubic Feet per Minute
M.L.F.	Thousand Linear Feet	P.E., PE	Professional Engineer; Porcelain Enamel; Polyethylene; Plain End	Scaf.	Scaffold
Mo.	Month			Sch., Sched.	Schedule
Mobil.	Mobilization			S.C.R.	Modular Brick
Mog.	Mogul Base	P.E.C.I.	Porcelain Enamel on Cast Iron	S.D.	Sound Deadening
MPH	Miles per Hour	Perf.	Perforated	SDR	Standard Dimension Ratio
MPT	Male Pipe Thread	PEX	Cross linked polyethylene	S.E.	Surfaced Edge
MRGWB	Moisture Resistant Gypsum Wallboard	Ph.	Phase	Sel.	Select
		P.I.	Pressure Injected	SER, SEU	Service Entrance Cable
MRT	Mile Round Trip	Pile.	Pile Driver	S.F.	Square Foot
ms	Millisecond	Pkg.	Package	S.F.C.A.	Square Foot Contact Area
M.S.F.	Thousand Square Feet	Pl.	Plate	S.F. Flr.	Square Foot of Floor
Mstz.	Mosaic & Terrazzo Worker	Plah.	Plasterer Helper	S.F.G.	Square Foot of Ground
M.S.Y.	Thousand Square Yards	Plas.	Plasterer	S.F. Hor.	Square Foot Horizontal
Mtd., mtd., mtd	Mounted	plf	Pounds Per Linear Foot	SFR	Square Feet of Radiation
Mthe.	Mosaic & Terrazzo Helper	Pluh.	Plumbers Helper	S.F. Shlf.	Square Foot of Shelf
Mtng.	Mounting	Plum.	Plumber	S4S	Surface 4 Sides
Mult.	Multi; Multiply	Ply.	Plywood	Shee.	Sheet Metal Worker
M.V.A.	Million Volt Amperes	p.m.	Post Meridiem	Sin.	Sine
M.V.A.R.	Million Volt Amperes Reactance	Pntd.	Painted	Skwk.	Skilled Worker
MV	Megavolt	Pord.	Painter, Ordinary	SL	Saran Lined
MW	Megawatt	pp	Pages	S.L.	Slimline
MXM	Male by Male	PP, PPL	Polypropylene	Sldr.	Solder
MYD	Thousand Yards	P.P.M.	Parts per Million	SLH	Super Long Span Bar Joist
N	Natural; North	Pr.	Pair	S.N.	Solid Neutral
nA	Nanoampere	P.E.S.B.	Pre-engineered Steel Building	SO	Stranded with oil resistant inside insulation
NA	Not Available; Not Applicable	Prefab.	Prefabricated		
N.B.C.	National Building Code	Prefin.	Prefinished	S-O-L	Socketolet
NC	Normally Closed	Prop.	Propelled	sp	Standpipe
NEMA	National Electrical Manufacturers Assoc.	PSF, psf	Pounds per Square Foot	S.P.	Static Pressure; Single Pole; Self-Propelled
		PSI, psi	Pounds per Square Inch		
NEHB	Bolted Circuit Breaker to 600V.	PSIG	Pounds per Square Inch Gauge	Spri.	Sprinkler Installer
NFPA	National Fire Protection Association	PSP	Plastic Sewer Pipe	spwg	Static Pressure Water Gauge
NLB	Non-Load-Bearing	Pspr.	Painter, Spray	S.P.D.T.	Single Pole, Double Throw
NM	Non-Metallic Cable	Psst.	Painter, Structural Steel	SPF	Spruce Pine Fir; Sprayed Polyurethane Foam
nm	Nanometer	P.T.	Potential Transformer		
No.	Number	P. & T.	Pressure & Temperature	S.P.S.T.	Single Pole, Single Throw
NO	Normally Open	Ptd.	Painted	SPT	Standard Pipe Thread
N.O.C.	Not Otherwise Classified	Ptns.	Partitions	Sq.	Square; 100 Square Feet
Nose.	Nosing	Pu	Ultimate Load	Sq. Hd.	Square Head
NPT	National Pipe Thread	PVC	Polyvinyl Chloride	Sq. In.	Square Inch
NQOD	Combination Plug-on/Bolt on Circuit Breaker to 240V.	Pvmt.	Pavement	S.S.	Single Strength; Stainless Steel
		PRV	Pressure Relief Valve	S.S.B.	Single Strength B Grade
N.R.C., NRC	Noise Reduction Coefficient/ Nuclear Regulator Commission	Pwr.	Power	sst, ss	Stainless Steel
		Q	Quantity Heat Flow	Sswk.	Structural Steel Worker
N.R.S.	Non Rising Stem	Qt.	Quart	Sswl.	Structural Steel Welder
ns	Nanosecond	Quan., Qty.	Quantity	St.; Stl.	Steel
nW	Nanowatt	Q.C.	Quick Coupling	STC	Sound Transmission Coefficient
OB	Opposing Blade	r	Radius of Gyration	Std.	Standard
OC	On Center	R	Resistance	Stg.	Staging
OD	Outside Diameter	R.C.P.	Reinforced Concrete Pipe	STK	Select Tight Knot
O.D.	Outside Dimension	Rect.	Rectangle	STP	Standard Temperature & Pressure
ODS	Overhead Distribution System	recpt.	receptacle	Stpi.	Steamfitter, Pipefitter
O.G.	Ogee	Reg.	Regular	Str.	Strength; Starter; Straight
O.H.	Overhead	Reinf.	Reinforced	Strd.	Stranded
O&P	Overhead and Profit	Req'd.	Required	Struct.	Structural
Oper.	Operator	Res.	Resistant	Sty.	Story
Opng.	Opening	Resi.	Residential	Subj.	Subject
Orna.	Ornamental	RF	Radio Frequency	Subs.	Subcontractors
OSB	Oriented Strand Board	RFID	Radio-frequency identification	Surf.	Surface
OS&Y	Outside Screw and Yoke	Rgh.	Rough	Sw.	Switch
OSHA	Occupational Safety and Health Act	RGS	Rigid Galvanized Steel	Swbd.	Switchboard
		RHW	Rubber, Heat & Water Resistant; Residential Hot Water	S.Y.	Square Yard
Ovhd.	Overhead			Syn.	Synthetic
OWG	Oil, Water or Gas	rms	Root Mean Square	S.Y.P.	Southern Yellow Pine
Oz.	Ounce	Rnd.	Round	Sys.	System
P.	Pole; Applied Load; Projection	Rodm.	Rodman	t.	Thickness
p.	Page	Rofc.	Roofer, Composition	T	Temperature; Ton
Pape.	Paperhanger	Rofp.	Roofer, Precast	Tan	Tangent
P.A.P.R.	Powered Air Purifying Respirator	Rohe.	Roofer Helpers (Composition)	T.C.	Terra Cotta
PAR	Parabolic Reflector	Rots.	Roofer, Tile & Slate	T & C	Threaded and Coupled
P.B., PB	Push Button	R.O.W.	Right of Way	T.D.	Temperature Difference
Pc., Pcs.	Piece, Pieces	RPM	Revolutions per Minute	Tdd	Telecommunications Device for the Deaf
P.C.	Portland Cement; Power Connector	R.S.	Rapid Start		
P.C.F.	Pounds per Cubic Foot	Rsr	Riser	T.E.M.	Transmission Electron Microscopy
PCM	Phase Contrast Microscopy	RT	Round Trip	temp	Temperature, Tempered, Temporary
		S.	Suction; Single Entrance; South	TFFN	Nylon Jacketed Wire
		SBS	Styrene Butadiere Styrene		

631

Abbreviations

TFE	Tetrafluoroethylene (Teflon)	U.L., UL	Underwriters Laboratory	w/	With
T. & G.	Tongue & Groove;	Uld.	unloading	W.C., WC	Water Column; Water Closet
	Tar & Gravel	Unfin.	Unfinished	W.F.	Wide Flange
Th., Thk.	Thick	UPS	Uninterruptible Power Supply	W.G.	Water Gauge
Thn.	Thin	URD	Underground Residential	Wldg.	Welding
Thrded	Threaded		Distribution	W. Mile	Wire Mile
Tilf.	Tile Layer, Floor	US	United States	W-O-L	Weldolet
Tilh.	Tile Layer, Helper	USGBC	U.S. Green Building Council	W.R.	Water Resistant
THHN	Nylon Jacketed Wire	USP	United States Primed	Wrck.	Wrecker
THW.	Insulated Strand Wire	UTMCD	Uniform Traffic Manual For Control	W.S.P.	Water, Steam, Petroleum
THWN	Nylon Jacketed Wire		Devices	WT., Wt.	Weight
T.L., TL	Truckload	UTP	Unshielded Twisted Pair	WWF	Welded Wire Fabric
T.M.	Track Mounted	V	Volt	XFER	Transfer
Tot.	Total	VA	Volt Amperes	XFMR	Transformer
T-O-L	Threadolet	V.C.T.	Vinyl Composition Tile	XHD	Extra Heavy Duty
tmpd	Tempered	VAV	Variable Air Volume	XHHW,	Cross-Linked Polyethylene Wire
TPO	Thermoplastic Polyolefin	VC	Veneer Core	XLPE	Insulation
T.S.	Trigger Start	VDC	Volts Direct Current	XLP	Cross-linked Polyethylene
Tr.	Trade	Vent.	Ventilation	Xport	Transport
Transf.	Transformer	Vert.	Vertical	Y	Wye
Trhv.	Truck Driver, Heavy	V.F.	Vinyl Faced	yd	Yard
Trlr	Trailer	V.G.	Vertical Grain	yr	Year
Trlt.	Truck Driver, Light	VHF	Very High Frequency	Δ	Delta
TTY	Teletypewriter	VHO	Very High Output	%	Percent
TV	Television	Vib.	Vibrating	~	Approximately
T.W.	Thermoplastic Water Resistant	VLF	Vertical Linear Foot	Ø	Phase; diameter
	Wire	VOC	Volitile Organic Compound	@	At
UCI	Uniform Construction Index	Vol.	Volume	#	Pound; Number
UF	Underground Feeder	VRP	Vinyl Reinforced Polyester	<	Less Than
UGND	Underground Feeder	W	Wire; Watt; Wide; West	>	Greater Than
UHF	Ultra High Frequency			Z	zone
U.I.	United Inch				

Index

A

Abandon catch basin 22
Accent light 293
  light connector 293
  light transformer 293
Access card key 48
  control card type 48
  control equipment 48
  control fingerprint scanner 49
  control security 49
  control video camera 318
  doors and panels 48
  floor 52
  flooring 52
  keypad 48
  road and parking area 16
Accessories bathroom 54
Acoustical sealant 45
Adapter compression
equipment 117
  conduit 137
  connector 271
  EMT 137
  Greenfield 137
Adjustable frequency drives 278
Adjustment factor 8
  factor labor 74
Aerial lift 447
  lift truck 452
Aggregate spreader 444, 451
Air circuit breaker 350
  compressor 447
  compressor control system . . . 82
  compressor portable 447
  conditioner general 80
  conditioner receptacle 200
  conditioner wiring 203
  conditioning fan 83
  conditioning
ventilating 81, 83, 84, 86
  handling fan 83
  handling troffer 290
  hose 448
  quality 58
  spade 448
  supply pneumatic control 82
  terminal 285
Air conditioning power
central 385
Airfield wind cone 354
Airfoil fan centrifugal 84
Airport lighting 354
Alarm exit control 317
  fire 317, 319
  panels and devices 319
  panels and devices fire 319
  residential 202
  sprinkler 319
  standpipe 319
Aluminum bus duct 599, 600
  bus duct 234
  cable 112
  cable connector 114
  cable tray 170, 178, 182
  conduit 133
  handrail 52
  light pole 440
  low lamp bus duct 240
  service entrance cable 119
  shielded cable 109
  substation bus assemblies 349
  wire compression adapter 117
Ammeter 203
Anchor bolt 27
  chemical 31, 34

epoxy 31, 34
expansion 34
hollow wall 34
lead screw 34
machinery 27
metal nailing 34
nylon nailing 34
plastic screw 35
pole 342
screw 34
toggle bolt 34
tower 340
wedge 35
Angle plug 269
Antenna TV 311
  wire TV 308
Apartment call system 312
  intercom door 312
Appliance 57
  plumbing 77
  repair 19
  residential 57, 202
Approach ramp 52
Arc fault circuit breaker 217
Armored cable 111, 574
  cable connector 115
Arrester lightning 206, 285, 350
Asbestos disposal 24
Asphalt distributor 448
  paver 449
  plant portable 451
Attachment ripper 451
Attachments skidsteer 447
Attic ventilation fan 86
Auger earth 444
Automatic circuit closing 257
  fire-suppression 72
  timed thermostat 82
  transfer switches 284, 608
  voltage regulator 282
  wall switch 204
Autopsy equipment 62
Autoscrub machine 448
Axial flow fan 83

B

Backer rod 43
Backfill 324
  trench 322
Backhoe 445
  bucket 446
  extension 451
  trenching 428, 430, 432
Baler 62
Ball wrecking 451
Ballast emergency 292
  fixture 292
  high intensity discharge 294
  replacement 294
Bank capacitor 346
Bare floor conduit 293
Barge construction 454
Bar hanger outlet box 148
Barrels flasher 448
  reflectorized 448
Barricade 17
  flasher 448
  tape 17
Barrier and enclosure 17
  dust 17
  slipform paver 451
  X-ray 68
Base stabilizer 452
  transformer 297

Baseball scoreboard 61
Baseboard heat electric 90
  heater electric 90
Baseplate scaffold 14
  shoring 14
Basic meter device 256
Basketball scoreboard 61
Bath steam 67
Bathroom accessories 54
  exhaust fan 85
  heater & fan 86
Battery inverter 280
  substation 349
Beam clamp channel 130
  clamp conduit 132
  flange clip 76
Bedding pipe 324
Bell system 271
  transformer door 271
Bend EMT field 133, 134, 136
Bender duct 346
Biometric identity access 48
Bituminous paver 449
Blaster shot 450
Blind structural bolt 36
Block manhole 340
Blower HVAC fan 83
  insulation 449
Board control 61
Boat work 454
Boiler 87
  control system 81
  electric 87, 367
  electric steam 87
  gas-fired 87
  general 74
  hot water 87
  hot water electric 367
  steam electric 366
Bollard light 299
Bolt & hex nut steel 35
  anchor 27
  blind structural 36
Bolted switch pressure 212
Bolt-on circuit breaker 217
Bond performance 11, 552
Bonding plates 286
Boom truck 452
Booster fan 84
Border light 296
Borer horizontal 451
Boring horizontal 338
  machine horizontal 444
  service 338
Box & wiring device . . 128-132, 148,
         149, 259, 266, 267, 307
  conduit 159
  connector EMT 136
  electrical 151
  electrical pull 149
  explosionproof 151
  junction 261
  low voltage relay 265
  outlet 148
  outlet cover 148
  outlet steel 147
  pull 149
  storage 13
  terminal wire 261
  termination 312
  trench 450
  utility 338
Bracing shoring 14
Bracket pipe 75
  scaffold 14
Branch circuit fusible switch 214

circuit breaker 215
circuit breaker HIC 214
meter device 257
Brazed connection 127
Break glass station 319
Breaker circuit 272
  low voltage switchgear 206
  motor operated switchboard . . 213
  pavement 447
Breather conduit
explosionproof 160
Brick cart 448
  catch basin 340
  manhole 340
Broiler 59
Broom finish concrete 30
  sidewalk 450
Brush chipper 445
  cutter 445, 446
Buck-boost transformer 210
Bucket backhoe 446
  clamshell 445
  concrete 444
  dragline 445
  excavator 451
  pavement 451
Buggy concrete 444
Builder's risk insurance 546
Building demolition selective 22
  directory 312
  permit 11
  portable 13
  temporary 13
Bulb incandescent 302
Bulldozer 324, 446
Bullfloat concrete 444
Buncher feller 445
Bunk house trailer 13
Burglar alarm indicating panel . . . 317
Burial cable direct 119
Buried conduit PVC 343
Bus aluminum duct 599, 600
  bar aluminum 213
  duct 600
  substation 349
Bus duct 234, 235
  aluminum 234
  aluminum low lamp 240
  cable tap box 235, 240, 249
  circuit breaker 240, 250
  circuit breaker low amp 241
  copper 242, 245, 247-250
  copper lighting 241
  copper low amp 241
  end closure 252
  feed low amp 241
  feeder 234, 238, 240, 242, 243
  fitting 23-238, 240, 242,
         244, 245, 248, 251-254
  ground bus 251
  ground neutralizer 248
  hanger 253
  lighting 241
  plug 246
  plug & capacitor 247
  plug & contactor 247
  plug & starter 246
  plug circuit breaker 247
  plug in 234, 238, 241, 242, 248
  plug in low amp 240
  plug low amp 241
  reducer 236, 243
  roof flange 251
  service head 252
  switch 249
  switch & control 250

switch low amp 241
switchboard stub 235
tap box 240, 245, 249
tap box circuit breaker 241
transformer 250
wall flange 235, 251
weather stop 251
Bushing conduit 154
Busway 234
cleaning tool 254
connection switchboard 212
feedrail 254
trolley 254
Butt fusion machine 448
Button enclosure oiltight push .. 259
enclosure sloping front push .. 259
Butyl rubber 44
Buzzer system 312
BX cable 111

C

Cabinets & enclosures 260
convector heater 92
current transformer 258
electric 258
electrical floormount 258
electrical hinged 150
heater electric 92
hotel 54
medicine 54
oil tight 260
pedestal 259
strip 312
telephone wiring 307
transformer 258
unit heater 93
Cable aluminum 112
aluminum shielded 109
armored 111
BX 111
channel aluminum 186
channel fitting 186
channel hanger 186
channel vented 186
clamp 175
coaxial 310
connector 286
connector armored 115
connector nonmetallic 114
connector split bolt 116
control 124
copper 111, 124
copper shielded 109
crimp termination 116
electric 111, 113
electrical 121, 124
fire alarm 316
flat 110
grip 125
hangers 76
high voltage in conduit 369
in conduit support 157
jack 453
lug switchgear 205
MC 113
medium-voltage 109
mineral insulated 118
multipair 308
pulling 450
PVC jacket 113, 124
reel w/receptacle 267
sheathed nonmetallic 119
sheathed Romex 119

splicing high-voltage 115
tap box bus duct 235, 240, 249
tensioning 450
termination high-voltage 109
termination indoor 108
termination medium-voltage ... 108
termination mineral insulated .. 118
termination outdoor 108
termination padmount 108
terminations 116
tie 125
trailer 450
tray 124, 167, 583
tray alum cover 188
tray aluminum 170, 178, 182
tray cable 112, 113
tray divider 191
tray fitting 167-174, 176-185
tray galvanized cover 186, 187
tray galvanized steel .. 167, 176, 180
tray ladder type 167
tray raceway 167
tray solid bottom 176
tray trough 180
tray trough type 180
tray vented 180
tray wall bracket 175
trays utility substations 349
undercarpet system 110
underground feeder 119
underground URD 119
UTP 308
wire THWN 124
Cables & fittings
communication 308
Call system apartment 312
system emergency 312
system nurse 313
Camera TV closed circuit 318
Cap service entrance 120
Capacitor bank 346
indoor 281
static 347
station 346
synchronous 346
transmission 346
Car tram 450
Carbon dioxide extinguisher ... 72
Card control entrance system ... 417
key access 48
system entrance 48
Care facility cable health 114
Carpet computer room 52
lighting 293
Cart brick 448
concrete 444
Cast boxes 582
in place concrete 28, 29
iron manhole cover 339
iron pull box 151
Caster scaffold 15
Catchbasins 339
basin brick 340
basin cleaner 451
basin precast 340
basin removal 22
basin vacuum 451
Category 3 jack 308
5 connector 308
5 jack 308
5e jack 308
6 jack 308
Cathodic protection 286
Catwalk scaffold 15
Caulking 44

polyurethane 44
sealant 44
Ceiling fan 83
heater 58
luminous 301
Cell fuel 356
Cellular decking 37
Cement duct 346
PVC conduit 140
underlayment self-leveling 30
Center meter 255
Central AC power system 385
Centrifugal airfoil fan 84
pump 449
type HVAC fan 84
Chain fall hoist 453
link fence 17
saw 450
trencher 322, 447
Chandelier 293
Chan-L-Wire system 264
Channel aluminum cable 186
beam clamp 130
concrete insert 130
conduit 131
fitting 130
hanger cable 186
junction box 130
plastic coated 132
steel 129
strap conduit 130
strap EMT 130
support 130
vented cable 186
w/conduit clip 131
Charge disposal 24
powder 36
Chemical anchor 31, 34
spreader 451
toilet 450
Chime door 271
Chipper brush 445
log 446
stump 446
Chipping hammer 448
Chloride accelerator concrete 29
Cinema equipment 60
Circline fixture 294
Circuit closing automatic 257
closing manual 257
length 576
Circuit breaker 272
air 350
air high-voltage 350
bolt-on 217
bus duct 240, 250
bus duct plug 247
bus duct tap box 241
CLF 250
control center 226
disconnect 273
explosionproof 273
feeder section 214
gas 350
HIC 217
HIC branch 214
light contactor 225
load center 218
low amp bus duct 241
motor operated 218
motor starter 221-223, 225
oil 350
panel 218
panelboard 220
plug-in 217

power 350
shunt trip 217
switch 218
vacuum 350
Circular saw 450
Clamp cable 175
ground rod 126
jay 131
pipe 75
riser 75
water pipe ground 127
Clamshell bucket 445
Clean control joint 27
room 66
Cleaner catch basin 451
crack 451
steam 450
Cleaning & disposal
equipment 59
tool busway 254
up 18
Clearance pole 348
CLF circuit breaker 250
Climbing crane 452
hydraulic jack 453
Clip channel w/conduit 131
Clock equipment 313
timer 202
wall 64
Closed circuit television
system 318
Coaxial cable 310
cable & fittings 309
connector 309
Coffee urn 59
CO_2 fire extinguishing system ... 72
Cold mix paver 451
Color wheel 296
Combination device 199
Commercial dishwasher 59
electric radiation 426
electric service 370
heater electric 90
water heater 77, 365
water heater electric 77
Commissioning 18
Communication cables
& fittings 308
component 19
optical fiber 307
outlet 310
system 312
Compaction 334
soil 324
Compactor earth 445
landfill 446
plate 334
Compensation workers' 10
Component control 81, 82
Component/DDC system
control 80
Compression equipment
adapter 117
fitting 138
seals 43
Compressor air 447
tank mounted 82
Computer connector 271
floor 52
grade voltage regulator 282
isolation transformer 282
plug 271
power cond transformer 282
receptacle 270
regulator transformer 282

room carpet 52
transformer 282
Concrete broom finish 30
bucket 444
buggy 444
bullfloat 444
cart 444
cast-in-place 28, 29
catch basin 340
chloride accelerator 29
conveyer 444
core drilling 31
demo 22
direct chute 29
drill 164
drilling 32
equipment rental 444
fiber reinforcing 29
float finish 30
floor edger 444
floor grinder 444
floor patching 26
hand hole 345
hand trowel finish 30
hole cutting 166
hole drilling 165
hot weather 29
impact drilling 32
insert 130
insert channel 130
insert hanger 75
lance 449
machine trowel finish 30
manhole 340, 345
mixer 444
non-chloride accelerator 29
paver 449
placement footing 29
placing 29
pump 444
ready mix 29
reinforcing 27
removal 22
retarder 29
saw 444
saw blade 31
short load 29
slab conduit 193
slab cutting 31
spreader 449
trowel 444
truck 444
truck holding 29
utility vault 338
vibrator 444
wall patching 26
water reducer 29
winter 29
Conditioner general air 80
Conductor 109, 307, 309
& grounding . . 108, 109, 112, 114,
116, 118, 119, 121, 126, 308
high voltage shielded 369
line 347
overhead 347
sagging 347
substation 349
underbuilt 348
Conductors 580
Conduit 579-581, 584
& fitting flexible 197
aluminum 133
bare floor 293
beam clamp 132
body 160

box 159
bushing 154
channel 131
concrete slab 193
connector 164
coupling explosionproof 158
electrical 133, 141
encasement 613
fitting 133, 154, 162
fitting fire stop 157
flexible coupling 158
flexible metallic 162
Greenfield 162, 163
hanger 128, 161
high installation 99, 140
hub 161
in slab PVC 193
in trench electrical 194
in trench steel 194
intermediate fitting 135
intermediate steel 134
nipple 141
plastic coated 135
PVC 139
PVC buried 343
PVC fitting 139
raceway 133
rigid expansion coupling 156
rigid in slab 194
rigid steel 133, 141
riser clamp 129
sealing fitting 161
sealtite 163
steel fitting 135
substation 349
support 128, 138
w/coupling 141
weight 581
Connection brazed 127
motor 197
temporary 12
wire weld 127
Connector accent light 293
adapter 271
cable 286
category 5 308
coaxial 309
computer 271
conduit 164
locking 270
nonmetallic cable 114
screw-on 116
sealtite 164
split bolt cable 116
wire 116
wiring 270
Constant voltage transformer 282
Construction barge 454
cost index 9
management fee 8
photography 11
temporary 17
Contactor 276
& fusible switch 250
control switchboard 213
lighting 276
Contingency 8
Contractor overhead 18
Control & motor starter 228
basic section motor 227
board 61
bus duct lighting 250
cable 124
center motor 221-225, 227
component 81, 82

component/DDC systems 80
DDC 80
DDC system 81
enclosure 260
hand/off/automatic 276
install 18
joint 27
joint clean 27
joint sawn 27
joint sealant 27
pilot light motor 227
replace 18
station 276
station stop/start 276
switch 277
system 80
system air compressor 82
system boiler 81
system cooling 81
system pneumatic 81
system split system 81
system VAV box 82
system ventilator 81
wire low-voltage 266
wire weatherproof 266
Control center auxiliary contact . . 227
circuit breaker 226
component 221-225
fusible switch 226
pilot light 227
push button 227
Controller time system 313
Convector cabinet heater 92
heater electric 92
Converter stations 349
Cooking range 57
Cooler beverage 58
Cooling control system 81
Copper bus duct . . 242, 245, 247-250
cable 111, 124
ground wire 126
lighting bus duct 241
low amp bus duct 241
shielded cable 109
wire 121, 122
Cord portable 120
replace 19
Core drill 444
drilling concrete 31
Corrosion resistant enclosure 260
Corrosion resistant receptacle . . . 268
Cost mark-up 10
Coupling and fitting 581
EMT 136
expansion rigid conduit 156
Cover cable tray alum 188
cable tray galvanized 186, 187
manhole 339
outlet box 148
plate receptacle 268
stair tread 17
Crack cleaner 451
filler 451
Crane climbing 452
crawler 452
crew daily 14
hydraulic 452
tower 14, 452
truck mounted 452
Crawler crane 452
drill rotary 448
shovel 446
Crematory 62
Crew daily crane 14
forklift 14

Crimp termination cable 116
Critical care isolating panel 209
CT & PT compartment
switchboard 213
substation 347
Curb builder 444
extruder 445
slipform paver 451
Current limiting fuse 272
transformer 203, 347
transformer cabinet 258
Cutout demolition 22
slab 22
Cutter brush 445, 446
Cutting and drilling 164
concrete floor saw 31
concrete slab 31
torch 450

D

Daily crane crews 14
Data communication 419
outlets 310
Daylight control 204
dimming control system 415
on/off control system 416
Daylighting sensor 204
DDC controls 80
Deck metal floor 37
roof 40
Decking 556
cellular 37
floor 37
Decorator device 199
switch 199
Deep therapy room 68
Dehumidifier 448
Delivery charge 16
Demo concrete 22
transmission &
distribution 104
Demolish remove pavement
and curb 22
Demolition 555
communications 306
electrical 96
electrical power 102
grounding 99
hammer 445
lighting 107
low-voltage distribution 104
pavement 22
safety and security 316
selective building 22
site 22
sound & video 306
torch cutting 23
wiring methods 99
Derrick crane guyed 453
crane stiffleg 453
Detection intrusion 317
system 317
tape 339
Detector explosive 317
fire & heat 319
infrared 317
metal 316
motion 317
smoke 319
temperature rise 319
ultrasonic motion 317
Detention lighting fixture 296
Detour sign 17

Device & box wiring .. 128-132, 148,
 151, 259, 267, 307
 branch meter 257
 combination 199
 decorator 199
 GFI 200
 load management 275
 motor control center 227
 receptacle 200
 residential 198
 wiring 150, 265, 266
Dewater 323
Dewatering 324
 equipment 453
Diaphragm pump 449
Diesel generator 279
 generator set 279
 hammer 445
 tugboat 454
Diesel-engine-driven
generator 280
Digital movie equipment 60
 video recorder 318
Dimmer switch 198, 266
Dimming control modular 204
Direct burial cable 119
 burial PVC 343
 chute concrete 29
Directional drill horizontal 451
 drilling 338
Directory building 312
Disc harrow 445
Discharge hose 449
Disconnect & motor-starter 232
 circuit breaker 273
 safety switch 275
 switch 351
Dishwasher commercial 59
 residential 57
Disk insulator 347
Dispenser hot water 77
 ticket 56
Dispensing equipment food 59
Dispersion nozzle 72
Disposal asbestos 24
 charges 24
 garbage 58
Distiller water 61
Distribution pole galvanized 341
 section 597
 switchboard 213
Distributor asphalt 448
Ditching 324
Divider cable tray 191
Diving stand 66
Dog house switchboard 213
Domestic hot water 364
Door and panels access 48
 apartment intercom 312
 bell residential 201
 buzzer system 312
 chime 271
 fire 319
 frame lead lined 68
 opener remote 312
 release 312
 release speaker 312
 special 48
 switch burglar alarm 317
Doorbell system 271
Dozer 446
Dragline bucket 445
Drain conduit explosionproof . . . 160
Drill concrete 164
 core 444
 quarry 448

steel 448
track 447
Drilling concrete 32
 concrete impact 32
 directional 338
 horizontal 338
 plaster 52
 rig 444
 steel 35
Dripproof motor 303
 motor and starter 392-394
Driver post 451
 sheeting 448
Driveway removal 22
Dry cleaning equipment 56
 wall leaded 68
Drycleaner 56
Dryer receptacle 201
Dry-type transformer 207
Drywall cutout 22
Duct aluminum bus 599, 600
 and manhole underground . . . 343
 bender 346
 cement 346
 electric 194, 248
 electrical underfloor 196
 fitting trench 195
 fitting underfloor 196
 grille 52
 heater electric 88
 plastic wiring 262
 steel trench 195
 trench 194
 underfloor 196
 underground 345
 utility 345
 weatherproof
 feeder 237, 240, 242
Dump charges 23
 truck 447
Dumpsters 22, 555
Dumptruck off highway 447
Duplex receptacle 267
Dust barrier 17

E

Earth auger 444
 compactor 445
 scraper 446
 vibrator 324
Earthwork equipment 324
 equipment rental 444
Edger concrete floor 444
Educational TV studio 318
Electric ballast 292
 baseboard heater 90
 baseboard radiation 425
 board system 425
 boiler 87, 366
 boiler large hydronic 367
 boiler small hydronic 366
 cabinet 258
 cabinet heater 92
 cable 111, 113
 capacitor 281
 circuit voltages 588-590
 commercial water heater 77
 convector heater 92
 duct 194, 248
 duct heater 88
 feeder 248
 fixture 289, 294, 296, 299, 302
 fixture exterior 298
 fixture hanger 287

furnace 88
generator 448
heater 58
heating 90
hot air heater 92
infrared heater 90
lamp 301
metallic tubing 136, 141
meter 203
motors 303, 304
panelboard 219, 594
pool heater 88
receptacle 267
relay 277
service 272
service 3 phase/4 wire . . . 370, 371
service installation 370
switch 265, 273, 275
systems 566-570
unit heater 91
utility 345
vehicle charging 280
water heater 77
water heater residential . . . 77, 364
water heater commercial 365
Electrical & telephone
underground 345
 box 151
 cable 121, 124
 conduit 133, 141, 435
 conduit Greenfield 162, 163
 demolition 96, 561, 571-573
 duct underfloor 197
 fee 8
 field bend 133, 134, 136
 floormount cabinet 258
 installation drilling 164
 knockout 166
 laboratory 60
 maintenance 18
 metering smart 204
 nonmetallic tubing 162
 nonmetallic tubing (ENT) 162
 outlet box 147
 pole 343
 tubing 133
 undercarpet 110
 wire 121
Electricity temporary 11
Electronic rack enclosure 259
Electroytic grounding 128
Elevated conduit 99, 140
 installation add 74
 pipe add 74
Elevator fee 8
Emergency ballast 292
 call system 312
 equipment laboratory 60
 light test 19
 lighting 295
 lighting unit 295
Employer liability 10
EMT 136
 adapter 137
 box connector 136
 coupling 136
 fitting 136
 spacer 138
 support 128, 138
Emulsion sprayer 448
Enclosed bus assemblies 234
 motor 303
Enclosures & cabinets 260
 corrosion resistant 260
 electronic rack 259
 fiberglass 260

oil tight 260
panel 259
polyester 261
pushbutton 261
wireway 262
Energy monitoring system 424
 saving lighting devices 204
Engineering fees 8, 544
Entrance cable service 120
 card system 48
 weather cap 157
Epoxy anchor 31, 34
 welded wire 28
Equipment 444
 access control 48
 cinema 60
 cleaning & disposal 59
 clock 313
 dry cleaning 56
 earthwork 324
 fire-extinguishing 72
 food preparation 59
 foundation formwork 26
 insurance 10
 laboratory emergency 60
 laundry 56
 medical 61, 62
 medical sterilizing 61
 mobilization 16
 pad 28
 parking 56
 parking collection 56
 parking control 56
 parking gate 56
 parking ticket 56
 refrigerated storage 58
 rental 444, 554
 rental concrete 444
 rental earthwork 444
 rental general 447
 rental highway 451
 rental lifting 452
 rental marine 454
 rental wellpoint 453
 theater and stage 61
 voice 306
 waste handling 62
Ericson conduit coupling 156
Estimate electrical heating 90
Estimating 541
Excavating trench 322
 utility trench 322
 utility trench plow 323
Excavation 322
 hand 322-324
 machine 324
 structural 323
 tractor 322
 trench 322, 428-430, 432
Excavator bucket 451
 hydraulic 444
Exhaust granular fan 84
 hood 58
Exhauster industrial 84
 roof 84
Exit control alarm 317
 light 295
 lighting 295
Exothermic wire to grounding . . 127
Expansion anchor 34
 coupling rigid conduit 156
 fitting 137
 shield 34
Explosionproof box 151
 breather conduit 160
 circuit breaker 273

conduit box 159
conduit coupling 158
conduit fitting 159
drain conduit 160
fixture 295
fixture incandescent 296
lighting 296
plug 266
receptacle 266
reducer conduit 160
safety switch 274
sealing conduit 160
sealing hub conduit 160
toggle switch 266
Explosive detection equipment . . 317
detector 317
Extension backhoe 451
ladder 449
Exterior fixture LED 298
fixture lighting 298
floodlamp 300
LED fixtures 298
lighting 438
lighting fixture 298
Extinguisher fire 72
FM200 fire 72
Extra work 10
Extruder curb 445

F

Facial scanner unit 49
Factor 8
Fan air conditioning 83
air handling 83
attic ventilation 86
axial flow 83
bathroom exhaust 85
blower HVAC 83
booster 84
ceiling 83
centrifugal type HVAC 84
exhaust granular 84
house ventilation 86
HVAC axial flow 83
in-line 84
kitchen exhaust 86
low sound 83
paddle 202
propeller exhaust 85
residential 202
roof 84
utility set 83, 84
vaneaxial 83
ventilation 202
ventilator 83
wall 85
wall exhaust 85
wiring 202
Fee engineering 8
Feeder bus duct 234, 238,
240, 242, 243
cable underground 119
duct weatherproof . . . 237, 240, 242
electric 248
in conduit w/neutral 374
installation system 374
section branch circuit 214
section circuit breaker 214
section frame 214
section fusible switch 215
section switchboard 214
switchboard 215
Feedrail busway 254
cleaning tool 255

Feller buncher 445
Fence and gate 336
chain link 17
chain link gate & post 336
plywood 17
temporary 17
wire 17
Fiber optic cable 307
Fiber optics 611
Fiber reinforcing concrete 29
Fiberglass enclosure 260
light pole 297
panel 17
wireway 262
Field bend electrical . . . 133, 134, 136
bend EMT 133, 134, 136
office 13
office expense 13
personnel 10
Fill 324
by borrow & utility bedding . . 324
Filler crack 451
joint 44
Filtration equipment 66
Fingerprint scanner unit 48
Finishing floor concrete 30
Fire & heat detectors 319
alarm 317, 319
alarm cable 316
alarm maintenance 316
alarm panel and devices 319
call pullbox 319
detection system 418
door 319
extinguisher 72
extinguishing 368
horn 319
panel 319
signal bell 319
stop fitting 157
suppression 368
Fire-extinguishing equipment . . . 72
Firestopping 42, 556
Fire-suppression automatic 72
Fitting bus duct . . . 235-238, 240, 242,
244, 245, 248, 251-254
cable channel 186
cable tray 167-174, 176-185
compression 138
conduit 133, 154, 162
conduit intermediate 135
conduit PVC 139
conduit steel 135
EMT 136
expansion 137
Greenfield 162
poke-thru 149
PVC duct 346
sealtite 164
special 125
underfloor duct 196
underground duct 346
wireway 152
Fixture ballast 292
clean 19
electric 289
explosionproof 295
fluorescent 289, 294
fluorescent installed 398
hanger 287
high intensity discharge 407
incandescent 291
incandescent vaportight 292
interior light 287, 289, 296
landscape 299
lantern 294

LED exterior 298
lighting 298
metal halide 291
metal halide explosionproof . . 295
mirror light 292
residential 201, 294
sodium high pressure . . . 295, 300
type/S.F. HID 403, 405, 407
type/S.F. incandescent 400
vandalproof 292
vaporproof 288
watt/S.F. fluorescent 402
watt/S.F. HID 404, 406, 408
watt/S.F. incandescent 401
wet location 290
whip 300
wire 124
Flange clip beam 76
Flasher barrels 448
barricade 448
Flatbed truck 447, 450
truck crane 452
Flexible conduit & fitting 197
coupling conduit 158
metallic conduit 162
Float finish concrete 30
Floater equipment 10
Floodlamp exterior 300
Floodlights LED 300
pole mounted 300
trailer 448
tripod 448
Floor access 52
cleaning 18
concrete finishing 30
decking 37
pedestal 52
sander 450
stand kit 260
Fluorescent fixture . . . 289, 294, 398
fixture high bay 402
fixture system 398
light by wattage . . . 399, 415, 416
Fluoroscopy room 68
FM200 fire extinguisher 72
fire suppression 368
system 368
Foam spray rig 448
Folder laundry 56
Food dispensing equipment 59
mixer 59
preparation equipment 59
warmer 59
Football scoreboard 61
Footing concrete placement 29
formwork continuous 26
formwork spread 26
reinforcing 27
spread 28
tower 340
Forklift 448
crew 14
Formwork continuous footing . . . 26
equipment foundation 26
slab edge 26
slab haunch 29
slab on grade 26
slab thickened edge 29
slab trench 26
slab turndown 29
sleeve 26
spread footing 26
Foundation mat 28
mat concrete placement 30
pole 342
tower 340

Frame scaffold 14
Framing pipe support 37
Freestanding enclosure 260
Frequency drive adjustable 278
drive variable 278
Fuel cell 356
Full spectrum fixture 301
Furnace cutoff switch 267
electric 88
hot air 88
Fuse 271, 350
current limiting 272
duel element 271
high capacity 272
motor control center 225
plug 272
substation 350
transmission 350
Fusible switch branch circuit . . . 214
switch control center 226
switch feeder section 215
Fusion machine butt 448

G

Galvanized distribution pole 341
welded wire 28
Gang operated switch 351
Garbage disposal 58
Gas circuit breaker 350
fired boiler 87
generator 280
generator set 280
water heater residential 78
Gasoline generator 280
Gate parking 56
General air conditioner 80
boiler 74
contractor's overhead 547
equipment rental 447
maintenance 18
Generator 420
diesel 279, 420
diesel-engine-driven 280
electric 448
emergency 280
gas 280
gasoline 280
gas-operated 420
set 280
steam 61
system 420
test 19
weight 604
GFI receptacle 200
Glass break alarm 317
lead 68
lined water heater 58
Glassware sterilizer 60
washer 60
Gradall 445
Grader motorized 445
Grading 330
Greenfield adapter 137
conduit 162, 163
fitting 162
Grille duct 52
Grinder concrete floor 444
Grip cable 125
strain relief 125
Ground bus 237
bus to bus duct 251
clamp water pipe 127
fault indicating receptacle 267
fault protection 203

insulated wire 127
neutralizer bus duct 248
rod 125
rod clamp 126
rod wire 128
wire 126
wire overhead 348
Ground-fault protector
switchboard 212
Grounding 125
& conductor . . 108, 109, 112, 114,
116, 118, 119, 121, 126, 308
electroytic 128
substation 349
switch 351
system 351
transformer 349
wire bar and rod 351
wire brazed 127
Groung rods 579
Grout pump 450
topping 30
Guard lamp 302
Guardrail scaffold 14
temporary 17
Guyed derrick crane 453
tower 351
Guying power pole 342
Gypsum board leaded 68

H

Hammer chipping 448
demolition 445
diesel 445
drill rotary 448
hydraulic 445, 449
pile 445
vibratory 446
Hammermill 451
Hand excavation 322-324
hole concrete 345
scanner unit 48
trowel finish concrete 30
Handling extra labor 302
transformer 209
waste 62
Hand/off/automatic control 276
Handrail aluminum 52
Hanger & support pipe 75
bus duct 253
cable 76
conduit 128, 161
EMT 128
fixture 287
pipe 75
plastic coated 132
rod pipe 75
rod threaded 129
steel 128
trough 185
U-bolt pipe 76
Harrow disc 445
Hauling 324
cycle 324
Haunch slab 29
Hazardous waste handling 62
Header pipe wellpoint 453
Health care facility cable 114
Heat electric baseboard 90
pump residential 203
temporary 13
trace system 91
Heater & fan bathroom 86

cabinet convector 92
ceiling mount unit 93
electric 58
electric commercial 90
electric hot air 92
gas residential water 78
infrared 90
infrared quartz 91
quartz 91
sauna 67
space 449
swimming pool 88
vertical discharge 92
wall mounted 93
water 58, 203
water residential electric 77
Heating 87, 88
electric 90
electric baseboard 426
estimate electrical 90
hot air 88
hydronic 87
industrial 87
kettle 452
panel radiant 91
Heavy duty shoring 14
duty switch 273
Helicopter 453
Hex bolt steel 35
HID fixture high bay 403
fixture interior 287
fixture low bay 411
High installation conduit 99, 140
intensity discharge ballast 294
intensity discharge lamp 301
intensity discharge lamps 609
intensity discharge
lighting 287, 295
intensity fixture 406
output lamp 287
High performance cable 611
High pressure fixture
sodium 287, 291
pressure sodium
lighting 287, 291
voltage shielded conductor . . . 369
High voltage circuit breaker air . . 350
transmission 341
Highway equipment rental 451
Historical Cost Index 9
Hockey scoreboard 61
Hoist chain fall 453
personnel 453
tower 453
Hole cutting electrical 166
drilling electrical 165
Hollow wall anchor 34
Home theater 311
Hood exhaust 58
range 58
Hook-up electric service . . . 370, 424
mechanical service 424
Horizontal borer 451
boring 338
boring machine 444
directional drill 451
drilling 338
Horn fire 319
Hose air 448
discharge 449
suction 449
water 449
Hospital kitchen equipment 59
power 209
Hot air furnace 88

air heater electric 92
air heating 88
water boiler 87
water dispenser 77
water heating 87
weather concrete 29
Hotel cabinet 54
Hot water commercial 365
system 364, 365
House telephone 312
ventilation fan 86
Housing instrument 261
HP value/cost
determination 606, 607
Hub conduit 161
Humidifier 58
HV cable system 369
HVAC axial flow fan 83
power circulator &
ventilator 85
Hydraulic crane 452
excavator 444
hammer 445, 449
jack 453
jack climbing 453
Hydromulcher 451
Hydronic heating 87

I

Impact wrench 448
Incandescent bulb 302
explosionproof fixture 296
exterior lamp 302
fixture 291, 294
fixture system 400
interior lamp 302
Index construction cost 9
Indicating panel burglar alarm . . . 317
Indicator light 277
Induction lighting 298
Industrial address system 312
exhauster 84
heating 87
lighting 290
Infrared broiler 59
detector 317
heater 90
heater electric 90
quartz heater 91
In-line fan 84
Insert hanger concrete 75
Installation add elevated 74
Instrument enclosure 260
housing 261
switchboard 203
Insulation blower 449
Insulator disk 347
line 347
post 346
substation 349
transmission 347
Insurance 10, 546
builder risk 10
equipment 10
public liability 10
Intercom 312
system 418
Interior HID fixture 287
LED fixture 293
light fixture 287, 289, 296
lighting 288
Intermediate metallic
conduit 134, 141

Interrupter switch load 205
Interval timer 199
Intrusion detection 317
system 317
In-use cover 149
Invert manhole 339
Inverter battery 280
Ironer laundry 56
Isolating panel 209
transformer 209
Isolation transformer 282

J

Jack cable 453
category 3 308
category 5 308
category 5e 308
category 6 308
coaxial 310
hydraulic 453
Jackhammer 447
Jacking 338
Jay clamp 131
Job condition 9
Joint control 27
filler 44
sealer 43, 44
Jumper transmission 347
Junction box 261
box channel 130
Junction box underfloor duct 196

K

Kettle 59
heating 452
tar 450, 452
Keypad access 48
Kitchen equipment 59
equipment fee 8
exhaust fan 86
unit commercial 57
Knockout electrical 166
kVA value/cost determination . . . 592
kW value/cost
determination 590, 592

L

Lab grade voltage regulator 283
Labor adjustment factors 74
index 9
modifiers 74
restricted situation 209, 302
Laboratory safety equipment 60
Ladder extension 449
rolling 15
type cable tray 167
Lag screw 36
screw shield 34
Laminated lead 68
Lamp 587, 610
electric 301
guard 302
high intensity discharge 301
high output 287
incandescent exterior 302
incandescent interior 302
LED 302
mercury vapor 301
metal halide 301
post 297

quartz line 302
replace 19
slimline 301
sodium high pressure 301
sodium low pressure 301
tungsten halogen 302
Lampholder 267
Lamphouse 60
Lamps 609
high intensity discharge 609
Lance concrete 449
Landfill compactor 446
Landscape fixture 299
light 299
Lantern fixture 294
Laser level 449
Latex caulking 44
Laundry equipment 56
folder 56
ironer 56
Lawn mower 449
LB conduit body 160
Lead glass 68
gypsum board 68
lined door frame 68
plastic 68
screw anchor 34
sheet 68
shielding 68
Leads pile 445
LED fixture exterior 298
fixtures interior 293
floodlights 300
lamps 302
lighting parking 298
luminaire roadway 298
luminaire walkway 300
Level laser 449
Leveling jack shoring 14
Liability employer 10
insurance 547
Lift aerial 447
scissor 447
truck aerial 452
Lifting equipment rental 452
Light accent 293
bollard 299
border 296
exit 295
fixture interior 287, 289, 296
fixture troffer 290
indicator 277
landscape 299
nurse call 313
pilot 267
pole 343, 440
pole aluminum 297
pole fiberglass 297
pole steel 297
pole wood 297
post 294
relamp 19
runway centerline 354
safety 293
strobe 296
switch 383
temporary 13
tower 449
track 296
underwater 66
Light-contactor circuit breaker . . . 225
Lighting 287, 290, 291, 294, 301
airport 354
bus duct 241
carpet 293

contactor 276
control bus duct 250
emergency 295
equipment maintenance 19
exit 295
explosionproof 296
exterior fixture 298
fixture 298
fixture detention 296
fixture exterior 298
fixture replacement 96
high intensity discharge . . 287, 295
high pressure sodium 291
incandescent 291
induction 298
industrial 290
interior 288
metal halide 288, 291
outlet 201
picture 292
pole 297
residential 201
stage 296
stair 293
strip 290
surgical 61
temporary 13
theatrical 296
track 296
unit emergency 295
Lightning arrester 206, 285, 350
arrester switchgear 206
protection 285
suppressor 198
Line conductor 347
insulator 347
pole 342
pole transmission 342
remover traffic 452
right of way transmission 341
spacer 347
tower & fixtures 340
tower special 341
transmission 341
Liquid-filled transformer 205
Load center circuit breaker 218
center indoor 216
center plug-in breaker 216
center rainproof 216
center residential 198
centers 594
interrupter switch 205
management switch 275
Loader skidsteer 447
tractor 446
wheeled 446
windrow 452
Locking receptacle 201
receptacle electric 267
Locomotive tunnel 452
Log chipper 446
skidder 446
Low sound fan 83
Lowbed trailer 452
Low-voltage distribution
demolition 104
receptacle 265
silicon rectifier 265
switching 265
switchplate 265
transformer 265
wiring 266
Lug switchboard 211
terminal 116
Luminaire replacement 96

roadway 298
walkway 299, 300
Luminous ceiling 301

M

Machine auto-scrub 448
excavation 324
screw 36
trowel finish concrete 30
welding 451
X-ray 317
Machinery anchor 27
Magnetic motor starter 228
starter 389
starter & motor 392-394
Mail box call system 312
Maintenance detection
and alarm 316
electrical 18
fire alarm 316
general 18
Management fee construction 8
Manhole 340
brick 340
concrete 340, 345
cover 339
electric service 345
frame and cover 339
invert 339
raise 339
removal 22
step 340
substation 349
Manual circuit closing 257
starter & motor 393, 394
transfer switch 284
Marine equipment rental 454
Mark-up cost 10
Mason scaffold 14
Masonry manhole 340
saw 450
Mass notification system 320
Master clock system 418
T.V. antenna system 419
Mat concrete placement
foundation 30
foundation 28
Material index 9
MC cable 113
MCC test 19
Mechanical 557
fee 8
Medical equipment 61, 62
sterilizing equipment 61
Medicine cabinet 54
Medium-voltage cable 109
Mercury vapor lamp 301
Metal detector 316
floor deck 37
halide explosionproof
fixture 295
halide fixture 291
halide lamp 301
halide lighting 288, 291
nailing anchor 34
support assemblies 52
Metallic conduit
intermediate 134, 141
Meter center 255
center rainproof 256
device basic 256
device branch 257
electric 203

socket 255
socket 5 jaw 257
Metric conversion factors 542
Microphone 312
wire 308
Microtunneling 338
Microwave detector 317
oven 57
Mineral insulated cable 118
Minor site demolition 22
Mirror light fixture 292
Misc power system 384
Miscellaneous power 384
Mixer concrete 444
food 59
mortar 444, 449
plaster 449
road 451
Mobilization equipment 16
or demobilization 16
Modification to cost 9
Modifiers labor 74
switchboard 213
Modular dimming control 204
flexible wiring system 121
Moil point 448
Monitor 318
voltage 204
Mortar mixer 444, 449
Mortuary equipment 62
Motion detector 317
Motor 303
actuated valve 82
added labor 302
and starter 392
bearing oil 19
connections 197, 391, 587
control basic section 227
control center 221-225, 227
control center device 227
control center fuse 225
control centers 595
control incoming line 226
control pilot light 227
controlled valve 82
dripproof 303
electric 303, 304
enclosed 303
feeder 388
feeder system 388
installation 386
operated circuit breaker 218
starter 221, 227-229
starter & control 228
starter & disconnect 232
starter enclosed & heated 228
starter magnetic 228
starter safety switch 275
starter w/circuit protector 228
starter w/fused switch 228
starter w/pushbutton 228
starters and controls 596
test 19
Motor-control-center structure . . . 227
Motorized grader 445
Motors 593
Motor starter
circuit breaker 221-223, 225
Mounting board plywood 40
Movie equipment 60
equipment digital 60
Mower lawn 449
Muck car tunnel 452
Mud pump 444
trailer 451

Mulcher power 446
Multipair cable 308
Multi-channel rack enclosure ... 307

N

Nail lead head 68
Nailer pneumatic 448, 449
Nailing anchor 34
NEMA enclosures 601
Network hub 310
  switch 310
Nipple conduit 141
  plastic coated conduit 146
Non-automatic transfer switch ... 284
Non-chloride accelerator
concrete 29
Nozzle dispersion 72
Nurse call light 313
  call systems 313
  speaker station 313
Nylon nailing anchor 34

O

Occupancy sensors 204
Off highway dumptruck 447
Office & storage space 13
  field 13
  floor 52
  overhead 10
  trailer 13
Oil circuit breaker 350
  tight cabinet 260
  tight enclosure 260
Oil-filled transformer 205
Oil-fired water heater
residential 78
Omitted work 10
Operated switch gang 351
Operating room equipment 61
  room isolating panel 209
  room power & lighting 209
Operation maintenance
electrical 96
  maintenance lighting 96
Outlet box 148
  box bar hanger 148
  box cover 148
  box plastic 149
  box steel 147
  boxes 582
  communication 310
  lighting 201
Oven microwave 57
Overhaul 23
Overhead & profit 10
  conductor 347
  contractor 18, 547
  distribution system 264
  ground wire 348
  high-voltage wiring 347
  office 10
Overtime 9, 10, 546
Oxygen lance cutting 23

P

Packaging waste 24
Pad equipment 28
Paddle blade air circulator 85
  fan 202

Padmount cable termination 108
Paint sprayer 449
  striper 451, 452
Panels and devices alarm 319
  critical care isolating 209
  enclosure 259
  fiberglass 17
  fire 319
  isolating 209
  operating room isolating 209
  radiant heat 91
Panelboard 219, 376, 377
  electric 219, 594
  installation system 376
  spacer 218
  w/circuit breaker 217, 219
  w/lug only 220
Paneling cutout 22
Parking collection equipment 56
  control equipment 56
  gate 56
  gate equipment 56
  LED lighting 298
  ticket equipment 56
Patching concrete floor 26
  concrete wall 26
Patient nurse call 313
Pavement breaker 447
  bucket 451
  demolition 22
  planer 452
  profiler 452
  widener 451
Paver asphalt 449
  bituminous 449
  cold mix 451
  concrete 449
  shoulder 451
Pedestal cabinet 259
  floor 52
Performance bond 11, 552
Permit building 11
Personnel field 10
  hoist 453
Photoelectric control 204
Photography 11
  construction 11
  time lapse 11
Photovoltaic 278
  power 421, 423
  rack system 279
  system grid connected 423
Pickup truck 450
Pick-up vacuum 444
Picture lighting 292
Pigtail 300
Pile driving 554
  hammer 445
  leads 445
Pilot light 267
  light control center 227
  light enclosure 261
Pin powder 36
Pipe & fitting 75
  add elevated 74
  bedding 324
  bedding trench 323
  bracket 75
  clamp 75
  hanger 75
  hanger & support 75
  hanger rod 75
  hanger U-bolt 76
  sleeve plastic 26
  sleeve sheet metal 26

sleeve steel 26
strap 161
support 75
support framing 37
Piping excavation 428, 430, 432
Pit excavation 323
  sump 324
Placing concrete 29
Planer pavement 452
Plank scaffolding 15
Plant screening 446
Plaster cutout 22
  drilling 52
  mixer 449
Plastic coated conduit 135
  coated conduit nipple 146
  coating touch-up 161
  lead 68
  outlet box 149
  screw anchor 35
Plate bonding 286
  compactor 334
  roadway 452
  vibrating 334
  wall switch 267
Platform trailer 450
Plow vibrator 447
Plug & capacitor bus duct 247
  & contactor bus duct 247
  & starter bus duct 246
  angle 269
  bus duct 246
  coaxial 309
  computer 271
  explosionproof 266
  fuse 272
  in bus duct 234, 238,
               241, 242, 248
  in circuit breaker 217, 250
  in low amp bus duct 240
  in switch 249
  in tandem circuit breaker 217
  locking 269
  low amp bus duct 241
  wiring device 268
Plugmold raceway 264
Plumbing appliance 77
  laboratory 60
Plywood fence 17
  mounting board 40
  sheathing roof & walls 40
  sidewalk 18
Pneumatic control systems 81
  nailer 448, 449
Point moil 448
Poke-thru fitting 149
Pole aluminum light 297
  anchor 342
  clearance 348
  cross arm 343
  electrical 343
  fiberglass light 297
  foundation 342
  light 343
  lighting 297
  steel electrical utility 341
  steel light 297
  telephone 343
  tele-power 264
  utility 297, 343
  wood 342, 343
  wood electrical utility 342
  wood light 297
Police connect panel 317
Polyester enclosure 261

Polyurethane caulking 44
Pool heater electric 88
  heater swimming 88
  swimming 66
Portable air compressor 447
  asphalt plant 451
  building 13
  cord 120
Post driver 451
  insulator 346
  lamp 297
  light 294
Potential transformer 347
  transformer fused 213
Pothead switchgear 205
Powder actuated tool 36
  charge 36
  pin 36
Power circuit breaker 350
  conditioner transformer .. 280, 282
  equipment 552
  hospital 209
  mulcher 446
  pole guying 342
  supply systems 605
  supply uninterruptible 280
  system & capacitor 281, 284
  system undercarpet 110
  temporary 11, 13
  transformer 349
  transmission tower 340
  trowel 444
  wiring 272
Precast catch basin 340
Pressure sodium lighting
high 287
  switch switchboard 211, 212
  washer 449
Primary transformer 349
Prison work inside 8
Process air handling fan 83
Profiler pavement 452
Project overhead 10
  sign 18
Propeller exhaust fan 85
Protection lightning 285
  winter 17
Protective device 348
PT substation 347
Public address system 312
Pull box 149
  box electrical 149
  box explosionproof 151
  rope 346
  section switchboard 213
Pump centrifugal 449
  concrete 444
  contractor 324
  diaphragm 449
  grout 450
  mud 444
  shotcrete 444
  submersible 449
  trash 450
  water 449
  wellpoint 453
Pumping 324
Push button control center 227
  button enclosure oiltight 259
  button enclosure sloping
  front 259
Pushbutton enclosure 261
Push button switch 277
Putlog scaffold 15
PVC conduit 139

conduit cement 140
conduit fitting 139
conduit in slab 193
direct burial 343
duct fitting 346
underground duct 345

Q

Quarry drill 448
Quartz heater 91
line lamp 302

R

Raceway 110, 133, 134, 136-140,
142-147, 152-159, 161-173, 176-186,
188, 191-193, 195, 197, 264
cable tray 167
conduit 133
plugmold 264
surface 263
wiremold 263
wireway 153
Raceway/box utility
substations 349
Rack enclosure electronic 259
system photovoltaic 279
Radiant heating panel 91
Radiation electric 425
electric baseboard 426
Radiator thermostat control
system 81
Radio frequency shielding 69
tower 351
Rainproof meter center 256
Raise manhole 339
manhole frame 339
Raised floor 52
Rake tractor 446
Rammer/tamper 445
Ramp approach 52
temporary 16
Range cooking 57
hood 58
receptacle 201, 267
Reactors 350
substation 350
Ready mix concrete 29, 544
Receptacle 378
air conditioner 200
computer 270
corrosion resistant 268
cover plate 268
device 200
dryer 201
duplex 267
electric 267
electric locking 267
explosionproof 266
floor 379
GFI 200
ground fault indicating 267
isolated ground 267
locking 201
low-voltage 265
range 201, 267
surge suppresser 267
switch 380
telephone 201
television 201
weatherproof 200
Recorder videotape 318

Rectifier low-voltage silicon 265
Reducer bus duct 236, 243
conduit explosionproof 160
Reflectorized barrels 448
Refrigerated storage equipment . . . 58
Regulator automatic voltage 282
voltage 282, 283, 347
Reinforcing concrete 27
footing 27
Relamp light 19
Relay box low-voltage 265
electric 277
frame low-voltage 265
Release door 312
Remote power pack 204
Removal catch basin 22
concrete 22
driveway 22
sidewalk 22
Rental equipment 444
Replacement luminaire 96
Residential alarm 202
appliance 57, 202
applications. 198
device 198
dishwasher 57
door bell 201
electric radiation 425
fan 202
fixture 201, 294
gas water heater 78
heat pump 203
lighting 201
load center 198
oil-fired water heater 78
roof jack 86
service 198
smoke detector 202
switch 198
transition 86
wall cap 86
washer 58
water heater 58, 78, 203
water heater electric 77
water-heater 364
wiring 198, 202
Resistor substation 350
Restricted situation labor . . . 209, 302
Retarder concrete 29
Rigid conduit in trench 194
in slab conduit 194
steel conduit 135, 141
Ripper attachment 451
Riser clamp 75
clamp conduit 129
pipe wellpoint 454
Road mixer 451
sweeper 451
temporary 16
Roadway LED luminaire 298
plate 452
Rock trencher 447
Rocker switch 266
Rod backer 43
ground 125
pipe hanger 75
threaded 75
threaded hanger 129
Roller compaction 324
sheepsfoot 324, 446
tandem 446
vibratory 446
Rolling ladder 15
tower scaffold 15
Romex copper 119

Roof exhauster 84
fan 84
jack residential 86
sheathing 40
Room clean 66
Rotary crawler drill 448
hammer drill 448
Rototiller 446
Rubbish handling 22
Runway centerline light 354

S

Safety equipment laboratory 60
light 293
switch 273, 390
switch disconnect 275
switch explosionproof 274
switch motor starter 275
switches 604
Sagging conductor 347
Salamander 450
Sales tax 9, 545
Sandblasting equipment 450
Sander floor 450
Sauna 67
Saw blade concrete 31
chain 450
circular 450
concrete 444
cutting concrete floor 31
cutting slab 31
masonry 450
Sawn control joint 27
Scaffold baseplate 14
bracket 14
caster 15
catwalk 15
frame 14
guardrail 14
mason 14
putlog 15
rolling tower 15
specialties 15
stairway 15
wood plank 15
Scaffolding 553
plank 15
tubular 14
Scanner unit facial 49
unit fingerprint 48
School TV 318
Scissor lift 447
Scoreboard baseball 61
basketball 61
football 61
hockey 61
Scraper earth 446
Screed, gas engine, 8HP
vibrating 444
Screening plant 446
Screw anchor 34
lag 36
machine 36
Screw-on connector 116
Seal compression 43
thru-wall 157
Sealant acoustical 45
caulking 44
control joint 27
Sealer joint 43, 44
Sealing conduit explosionproof . . . 160
fitting conduit 161
hub conduit explosionproof . . 160

Sealtite conduit 163
connector 164
fitting 164
Sectionalizing switch 348
Security access control 49
system 417
X-ray equipment 317
Self-supporting tower 351
Service boring 338
distribution pole steel 341
electric 272
entrance cable aluminum 119
entrance cap 120, 157
entrance electric 370
head bus duct 252
residential 198
Sheathed nonmetallic cable 119
Romex cable 119
Sheathing 40
roof 40
roof & walls plywood 40
wall 40
Sheepsfoot roller 324, 446
Sheet lead 68
Sheeting driver 448
Shield expansion 34
lag screw 34
undercarpet 111
Shielded aluminum cable 109
copper cable 109
sound wire 308
Shielding lead 68
radio frequency 69
Shift work 9
Shoring baseplate 14
bracing 14
heavy duty 14
leveling jack 14
Short load concrete 29
Shot blaster 450
Shotcrete pump 444
Shoulder paver 451
Shovel crawler 446
Shredder compactor 62
Shunt trip circuit breaker 217
trip switchboard 212
Sidewalk broom 450
removal 22
temporary 18
Sign 18
detour 17
project 18
Signal bell fire 319
Silicone 44
Single pole disconnect switch . . . 351
Siren 317
Site demolition 22
lighting 438, 439
Skidder log 446
Skidsteer attachments 447
loader 447
Slab cutout 22
edge formwork 26
haunch 29
haunch formwork 29
on grade 29
on grade formwork 26
saw cutting 31
thickened edge 29
thickened edge formwork 29
trench formwork 26
turndown 29
turndown formwork 29
Sleeve formwork 26
plastic pipe 26

sheet metal pipe 26
steel pipe 26
Sliding mirror 54
Slipform paver barrier 451
    paver curb 451
Small tools 16
Smart electrical metering 204
Smoke detector 319
Socket meter 255
    meter 5 jaw 257
Sodium high pressure fixture .. 287, 291, 295, 300
    high pressure lamp 301
    low pressure fixture 298
    low pressure lamp 301
Soil compaction 324
    tamping 324
Solar energy- photovoltaics ... 278
Sound and video cable & fitting 308
    movie 60
    system 312, 418
    system speaker 312
    wire shielded 308
Space heater 449
    office & storage 13
Spacer EMT 138
    line 347
    panelboard 218
    transmission 347
Spade air 448
    tree 447
Speaker door release 312
    movie 60
    sound system 312
    station nurse 313
Special door 48
    nurse call systems 313
    wire & fittings 124
Specialty scaffold 15
Splicing high-voltage cable ... 115
Split bus switchboard 213
    system control system 81
Spotlight 296
Spotter 330
Spray rig foam 448
Sprayer emulsion 448
    paint 449
Spread footing 28
    footing formwork 26
Spreader aggregate 444, 451
    chemical 451
    concrete 449
Sprinkler alarm 319
Stabilizer base 452
Stage lighting 296
Staging swing 16
Stair lighting 293
    temporary protection 17
Stairway scaffold 15
Stand kit floor 260
Standpipe alarm 319
Starter and motor 392-394
    board & switch 206, 211-213, 215-217, 220, 221, 229, 272-274, 277, 303
    motor 221, 227-229
    replace 19
    w/pushbutton motor 228
Static capacitor 347
Station capacitor 346
    control 276
Steam bath 67
    boiler electric 87
    cleaner 450
    jacketed kettle 59

Steamer 59
Steel bolt & hex nut 35
    channel 129
    conduit in slab 194
    conduit in trench 194
    conduit intermediate 134
    conduit rigid 133
    drill 448
    drilling 35
    hanger 128
    hex bolt 35
    light pole 440
    outlet box 147
    service distribution pole . 341
    tower 351
    underground duct 345
Step manhole 340
Sterilizer glassware 60
Stiffleg derrick crane 453
Stop-start control station 276
Storage box 13
Storm drainage manholes
frames 340
Strain relief grip 125
Stranded wire 121
Strap channel w/conduit 130
    channel w/EMT 130
    support rigid conduit 128
    tie 125
Streetlight 298
Strip cabinet 312
    footing 28
    lighting 290
Striper paint 451, 452
Strobe light 296
Structural excavation 323
Structure motor-control-center 227
Stub switchboard 249
Stump chipper 446
Subcontractor O&P 10
Submersible pump 449
Substation backup batteries ... 349
    battery 349
    bus 349
    conductor 349
    conduit 349
    converter stations 349
    CT 347
    fuse 350
    grounding 349
    insulator 349
    manhole 349
    PT 347
    reactor 350
    resistor 350
    transformer 349
    transmission 349
Suction hose 449
Sump pit 324
Support cable in conduit 157
    channel 130
    conduit 128, 138
    EMT 128, 138
    framing pipe 37
    pipe 75
    rigid conduit strap 128
Suppressor lightning 198
    transformer 282
    transient 283
Surface raceway 263
Surge arresters 350
Surgical lighting 61
Surveillance system TV 318
Sweeper road 451
Swimming pool 66
    pool heater 88

Swing staging 16
Switch 351
    box 148
    box plastic 149
    bus duct 249
    circuit breaker 218
    control 277
    decorator 199
    dimmer 198, 266
    disconnect 351
    electric 265, 273, 275
    furnace cutoff 267
    fusible 226
    fusible & starter 250
    gear 597
    general duty 273
    grounding 351
    heavy duty 273
    load management 275
    low amp bus duct 241
    plate wall 267
    pressure bolted 212
    push-button 277
    residential 198
    rocker 266
    safety 273
    sectionalizing 348
    single pole disconnect 351
    switchboard 212
    time 275
    toggle 266
    transfer accessory 285
    transmission 348, 351
Switchboard 211
    breaker motor operated 213
    busway connection 212
    contactor control 213
    CT & PT compartment 213
    distribution 213
    dog house 213
    electric 276
    feeder 215
    feeder section 214
    ground-fault protector 212
    incoming 210
    instrument 203
    lug 211
    main circuit breaker 211
    modifier 213
    pressure switch 211, 212
    pull section 213
    shunt trip 212
    split bus 213
    stub 249
    stub bus duct 235
    switch fusible 212
    transition section 213
    w/CT compartment 210
Switchgear 205, 375
    breaker low-voltage 206
    cable lug 205
    low-voltage components 206
    pothead 205
    temperature alarm 206
    transformer 206
Switching and routing
equipment 310
    low-voltage 265
Switchplate low-voltage 265
Swithgear installation system . 375
Synchronous capacitor 346
System baseboard electric 425
    baseboard heat commercial . 426
    boiler hot water 367
    boiler steam 366
    buzzer 312

cable in conduit w/neutral ... 369
Chan-L-Wire 264
communication and
alarm 418, 419
control 80, 81
electric feeder 374
electric service 370
fire detection 418
fluorescent fixture 398
fluorescent watt/S.F. 399
FM200 fire suppression 368
heat trace 91
high bay HID fixture 403
high intensity fixture ... 402-414
incandescent fixture 401
incandescent fixture/S.F. 400
intercom 418
light pole 440
master clock 418
overhead distribution 264
receptacle 379, 380, 382
receptacle watt/S.F. 378
sitework trenching .. 428, 430, 432
switch wall 380, 382
telephone underfloor 417
toggleswitch 380, 382
T.V. 311
T.V. surveillance 318
UHF 311
undercarpet data
processing 111
undercarpet power 110
undercarpet telephone 110
underfloor duct 379
VAV box control 82
VHF 311
wall switch 383

T

Tamper 448
Tamping soil 324
Tandem roller 446
Tank water 451
Tap box bus duct ... 240, 245, 249
    box electrical 235
Tape barricade 17
    detection 339
    underground 339
Tar kettle 450, 452
Tarpaulin 17
Taxes 9, 546
    sales 9
    social security 9
    unemployment 9
Telephone house 312
    manhole 345
    pole 343
    receptacle 201
    systems 417
    TTY 311
    undercarpet 110
    wire 308
    wiring cabinet 307
Tele-power pole 264
Television equipment 311
    receptacle 201
    system 311
Temperature rise detector 319
Temporary barricade 17
    building 13
    connection 12
    construction 13, 17
    electricity 11
    facility 17

fence 17
guardrail 17
heat 13
light 13
lighting 13
power 11, 13
ramp 16
road 16
toilet 450
utility 13
Terminal air 285
lug 116
wire box 261
Termination block and
patch panel 307
box 312
cable 116
high-voltage cable 109
medium-voltage cable 108
mineral insulated cable 118
Theater and stage equipment 61
home 311
Theatrical lighting 296
Thermostat 82
automatic timed 82
contactor combination 91
control system radiator 81
integral 91
wire 124, 203
Thickened edge slab 29
Threaded rod 75
Thru-wall seal 157
THW wire 578
Ticket dispenser 56
Tie cable 125
strap 125
Time lapse photography 11
switch 275
system controller 313
Timer clock 202
interval 199
switch ventilator 86
Toaster 59
Toggle switch 266
switch explosionproof 266
Toggle bolt anchor 34
Toilet chemical 450
temporary 450
Tool powder actuated 36
small 16
Topping grout 30
Torch cutting 450
cutting demolition 23
Tower & fixtures line 340
anchor 340
crane 14, 452, 554
footing 340
foundation 340
hoist 453
light 449
radio 351
transmission 340
Track drill 447
light 296
lighting 296
Tractor 324
loader 446
rake 446
truck 450
Traffic line remover 452
Trailer bunk house 13
floodlight 448
lowbed 452
mud 451
office 13
platform 450

truck 450
water 450
Tram car 450
Transceiver 307
Transfer switch accessory 285
switch automatic 284
switch manual 284
switch non-automatic 284
Transformer 205, 210
& bus duct 207, 209, 235, 241,
242, 250, 251, 254
accent light 293
added labor 209
base 297
buck-boost 210
bus duct 250
cabinet 258
computer 282
constant voltage 282
current 203, 347
door bell 271
dry-type 207, 208
fused potential 203
grounding 349
handling 209
high-voltage 208
isolating 209
isolation 282
K rated 207
liquid-filled 205
low-voltage 265
maintenance 19
oil-filled 205
potential 347
power 349
power conditioner 282
primary 349
substation 349
suppressor 282
switchgear 206
transmission 349
UPS 280
weight 594
Transient suppressor 283
voltage suppressor 282
Transition residential 86
section switchboard 213
Transmission capacitor 346
fuse 350
high-voltage 341
insulator 347
jumper 347
line 341, 612
line pole 342
line right of way 341
spacer 347
substation 349
switch 348, 351
tower 340
transformer 349
Trash pump 450
Tray cable 167
Tree spade 447
Trench backfill 322
box 450
duct 194
duct fitting 195
duct steel 195
excavating 322
excavation 322
utility 322
Trencher chain 322, 447
rock 447
wheel 447
Trenching 324, 429
common earth 428, 429

loam & sandy clay 430, 431
sand & gravel 432, 433
Tripod floodlight 448
Troffer air handling 290
direct/indirect 290
light fixture 290
Trolley busway 254
Trough hanger 185
Trowel concrete 444
power 444
Truck boom 452
concrete 444
crane flatbed 452
dump 447
flatbed 447, 450
holding concrete 29
mounted crane 452
pickup 450
tractor 450
trailer 450
vacuum 451
winch 451
TTY telephone 311
telephone equipment 311
Tubing electric metallic . . . 136, 141
electrical 133
Tubular scaffolding 14
Tugboat diesel 454
Tungsten halogen lamp 302
Tunnel locomotive 452
muck car 452
ventilator 452
Turbine wind 356
Turndown slab 29
TV antenna 311
antenna wire 308
closed circuit camera 318
systems 311

U

U-bolt pipe hanger 76
U.F. feeder cable underground . . 119
underground cable 119
UHF system 311
Ultrasonic motion detector 317
Underbuilt circuit 348
conductor 348
Undercarpet electrical 110
power system 110
shield 111
telephone system 110
Underdrain 340
Underfloor duct 196, 585, 586
duct electrical 196
duct fitting 196
duct junction-box 196
duct telephone system 417
Underground cable UF 119
duct 345
duct fitting 346
ducts and manholes 343
electric service 372 373
electrical & telephone 345
electrical conduit 435
power feed 434
tape 339
UF feeder cable 119
Underlayment self-leveling
cement 30
Underwater light 66
Unemployment tax 9
Uninterruptible power supply . . . 280
Unit commercial kitchen 57
gross area requirements 541

heater cabinet 93
heater electric 91
URD cable underground 119
Utility accessories 339
boxes 338
duct 345
electric 345
excavation 428, 430, 432
pole 297, 343
pole steel electrical 341
pole wood electrical 342
set fan 83, 84
sitework 345
temporary 13
trench 322
trench excavating 322
trench plow excavating 323
UTP cable 308

V

Vacuum catch basin 451
circuit breaker 350
pick-up 444
truck 451
wet/dry 451
Valve motor actuated 82
motor controlled 82
Vandalproof fixture 292
Vaneaxial fan 83
Vaportight fixture
incandescent 292
Variable frequency drives 278
Vehicle charging electric 280
Ventilating air
conditioning 81, 83, 84, 86
Ventilation fan 202
Ventilator control system 81
fan 83
timer switch 86
tunnel 452
Vertical discharge heater 92
VHF system 311
Vibrating plate 334
screed, gas engine, 8HP 444
Vibrator concrete 444
earth 324
plate 324
plow 447
Vibratory hammer 446
roller 446
Video surveillance 318
Videotape recorder 318
Voice data outlets 310
data plate 310
equipment 306
Voltage monitor 204
regulator 282, 283, 347
regulator computer grade 282
regulator lab grade 283
regulator std. grade 283
Voltmeter 203

W

Walkway LED luminaire 300
luminaire 299, 300
Wall cap residential 86
clock 64
exhaust fan 85
fan 85
flange bus duct 235, 251
heater 58
sheathing 40

switch installed/ea. 382
switch installed/S.F. 380
switch plate 267
Washer and extractor 57
  commercial 57
  pressure 449
  residential 58
Waste handling 62
  handling equipment 62
  packaging 24
Water dispenser hot 77
  distiller 61
  heater 58, 203
  heater commercial 77
  heater electric 77
  heater gas residential 78
  heater residential 58, 78, 203
  heater residential electric 77
  heating hot 87
  hose 449
  pipe ground clamp 127
  pump 449
  pumping 324
  reducer concrete 29
  tank 451
  trailer 450
Water-heater electric 364, 365
Water-tight enclosure 260
Watt meter 203
Weather cap entrance 157

Weatherproof bus duct 237, 242
  control wire 266
  receptacle 200
Wedge anchor 35
Weights and measures 543
Welded wire epoxy 28
  wire fabric 28
  wire galvanized 28
Welding machine 451
Wellpoint discharge pipe 453
  equipment rental 453
  header pipe 453
  pump 453
  riser pipe 454
Wet/dry vacuum 451
Wheel color 296
  trencher 447
Wheelbarrow 451
Wheeled loader 446
Whip fixture 300
Widener pavement 451
Winch truck 451
Wind cone 354
  turbine 356
Windrow loader 452
Winter concrete 29
  protection 17
Wire 575, 577, 578
  connector 116
  copper 121, 122

copper ground 126
electrical 121
fence 17
fixture 124
ground 126
ground insulated 127
ground rod 128
low-voltage control 266
stranded 121
telephone 308
thermostat 124, 203
THW 121
THWN-THHN 122
to grounding exothermic 127
trough 260
T.V. antenna 308
XHHW copper 122
XLPE-use aluminum 123
XLPE-use copper 123
Wireless time system 313
Wiremold raceway 263
  raceway non-metallic 265
Wireway 151, 582
  enclosure 262
  fiberglass 262
  fitting 152
  raceway 153
Wiring air conditioner 203
  connector 270
  device 150, 265, 266

device & boxes 128-132, 148,
                 151, 259, 267, 307
device plug 268
devices 601-603
duct plastic 262
fan 202
low-voltage 266
methods 207
power 272
residential 198, 202
Wood electrical utility poles 342
  plank scaffold 15
  pole 342, 343
Work boat 454
  extra 10
  inside prisons 8
Workers' compensation . 10, 548-551
Wrecking ball 451
Wrench impact 448

X

X-ray barrier 68
  equipment 317
  equipment security 317
  machine 317
  protection 68
  system 209

Table of Contents
Annual Cost Guides
Seminars
New Titles
Reference Books
Order Form

Reed Construction Data/RSMeans— a tradition of excellence in construction cost information and services since 1942

For more information visit the RSMeans website at **www.rsmeans.com**

Unit prices according to the latest MasterFormat!

Book Selection Guide

The following table provides definitive information on the content of each cost data publication. The number of lines of data provided in each unit price or assemblies division, as well as the number of reference tables and crews, is listed for each book. The presence of other elements such as equipment rental costs, historical cost indexes, city cost indexes, square foot models, or cross-referenced indexes is also indicated. You can use the table to help select the RSMeans book that has the quantity and type of information you most need in your work.

Unit Cost Divisions	Building Construction	Mechanical	Electrical	Commercial Renovation	Square Foot	Site Work Landsc.	Green Building	Interior	Concrete Masonry	Open Shop	Heavy Construction	Light Commercial	Facilities Construction	Plumbing	Residential
1	582	395	411	508		521	196	319	463	581	511	244	1054	411	180
2	789	279	85	741		1005	180	399	222	788	746	481	1233	289	257
3	1689	347	231	1087		1477	995	329	2040	1688	1685	479	1787	317	385
4	963	21	0	932		729	179	620	1159	931	621	521	1176	0	435
5	1873	158	155	1068		811	1779	1073	716	1873	1034	954	1874	204	723
6	2404	18	18	2062		110	560	1523	278	2400	126	2085	2076	22	2612
7	1531	215	128	1569		584	747	516	509	1528	26	1282	1633	227	1018
8	2107	81	45	2530		268	1140	1768	112	2093	0	2109	2842	0	1511
9	1929	72	26	1726		313	449	2014	387	1870	15	1644	2179	54	1430
10	1006	17	10	623		215	27	818	158	1006	29	511	1098	235	224
11	1026	208	165	489		126	54	878	28	1010	0	230	1046	169	110
12	557	0	2	327		248	151	1669	12	545	19	364	1718	23	317
13	729	117	114	246		347	128	253	66	728	253	82	756	71	80
14	277	36	0	224		0	0	261	0	277	0	12	297	16	6
21	90	0	16	37		0	0	250	0	87	0	68	427	432	220
22	1159	7329	156	1170		1578	1099	849	20	1123	1687	861	7171	9082	712
23	1182	6901	586	886		160	898	741	38	1106	111	759	5013	1830	434
26	1379	499	10134	1039		816	661	1142	55	1351	562	1305	10083	444	621
27	72	0	298	34		13	0	60	0	72	39	52	280	0	4
28	97	58	137	72		0	21	79	0	100	0	41	152	44	26
31	1498	735	610	803		3206	289	7	1208	1443	3285	600	1560	660	611
32	782	54	8	859		4396	352	362	286	753	1824	407	1660	166	455
33	509	1020	542	209		2036	41	0	231	508	2024	120	1585	1211	118
34	109	0	47	4		192	0	0	31	64	195	0	130	0	0
35	18	0	0	0		327	0	0	0	18	442	0	83	0	0
41	60	0	0	32		0	0	22	0	60	30	0	67	14	0
44	75	83	0	0		0	0	0	0	0	0	0	75	75	0
46	23	16	0	0		274	151	0	0	23	264	0	33	33	0
48	12	0	27	0		0	27	0	0	12	19	12	12	0	12
Totals	24527	18659	13951	19277		19752	10124	15952	8019	24038	15547	15223	49100	16029	12501

Assem Div	Building Construction	Mechanical	Electrical	Commercial Renovation	Square Foot	Site Work Landscape	Assemblies	Green Building	Interior	Concrete Masonry	Heavy Construction	Light Commercial	Facilities Construction	Plumbing	Asm Div	Residential
A		15	0	188	150	577	598	0	0	536	571	154	24	0	1	368
B		0	0	848	2498	0	5658	56	329	1975	368	2089	174	0	2	211
C		0	0	647	872	0	1251	0	1576	146	0	762	249	0	3	588
D		1067	943	712	1856	72	2532	330	822	0	0	1342	1106	1081	4	851
E		0	0	86	260	0	300	0	5	0	0	257	5	0	5	392
F		0	0	0	114	0	114	0	0	0	0	114	3	0	6	357
G		522	442	318	262	3740	663	0	0	534	1558	199	288	758	7	307
															8	760
															9	80
															10	0
															11	0
															12	0
Totals		1604	1385	2799	6012	4389	11116	386	2732	3191	2497	4917	1849	1839		3914

Reference Section	Building Construction Costs	Mechanical	Electrical	Commercial Renovation	Square Foot	Site Work Landscape	Assem.	Green Building	Interior	Concrete Masonry	Open Shop	Heavy Construction	Light Commercial	Facilities Construction	Plumbing	Resi.
Reference Tables	yes	yes	yes	yes	no	yes	yes	yes	yes	yes	yes	yes	yes	yes	yes	yes
Models					111			25					50			28
Crews	561	561	561	540		561		558	561	561	538	561	538	540	561	538
Equipment Rental Costs	yes	yes	yes	yes		yes		yes	yes	yes	yes	yes	yes	yes	yes	yes
Historical Cost Indexes	yes	yes	yes	yes	yes	yes	yes	yes	yes	yes	yes	yes	yes	yes	yes	no
City Cost Indexes	yes	yes	yes	yes	yes	yes	yes	yes	yes	yes	yes	yes	yes	yes	yes	yes

RSMeans Building Construction Cost Data 2014

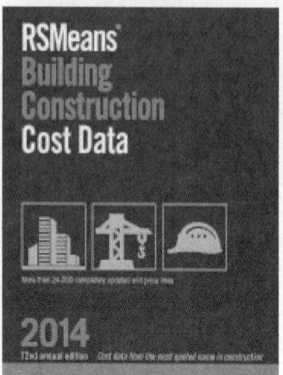

Offers you unchallenged unit price reliability in an easy-to-use format. Whether used for verifying complete, finished estimates or for periodic checks, it supplies more cost facts better and faster than any comparable source. More than 24,000 unit prices have been updated for 2014. The City Cost Indexes and Location Factors cover more than 930 areas, for indexing to any project location in North America. Order and get *RSMeans Quarterly Update Service* FREE.

$194.95 | Available Sept. 2013 | Catalog no. 60014

RSMeans Green Building Cost Data 2014

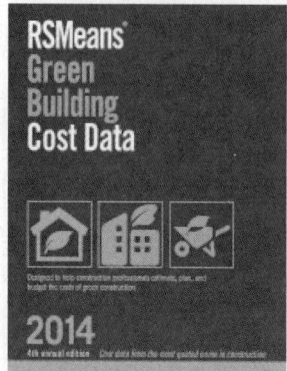

Estimate, plan, and budget the costs of green building for both new commercial construction and renovation work with this fourth edition of *RSMeans Green Building Cost Data*. More than 10,000 unit costs for a wide array of green building products plus assemblies costs. Easily identified cross references to LEED and Green Globes building rating systems criteria.

$160.95 | Available Nov. 2013 | Catalog no. 60554

RSMeans Mechanical Cost Data 2014

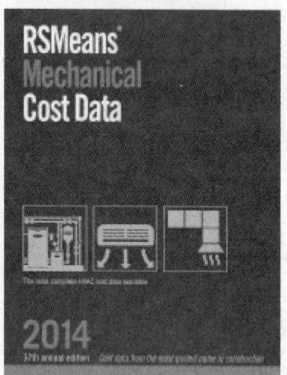

Total unit and systems price guidance for mechanical construction . . . materials, parts, fittings, and complete labor cost information. Includes prices for piping, heating, air conditioning, ventilation, and all related construction.

Plus new 2014 unit costs for:

- Thousands of installed HVAC/controls assemblies components
- "On-site" Location Factors for more than 930 cities and towns in the U.S. and Canada
- Crews, labor, and equipment

$191.95 | Available Oct. 2013 | Catalog no. 60024

RSMeans Facilities Construction Cost Data 2014

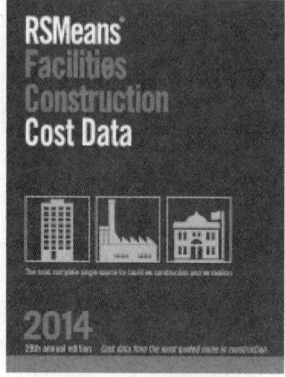

For the maintenance and construction of commercial, industrial, municipal, and institutional properties. Costs are shown for new and remodeling construction and are broken down into materials, labor, equipment, and overhead and profit. Special emphasis is given to sections on mechanical, electrical, furnishings, site work, building maintenance, finish work, and demolition.

More than 49,000 unit costs, plus assemblies costs and a comprehensive Reference Section are included.

$496.95 | Available Nov. 2013 | Catalog no. 60204

RSMeans Square Foot Costs 2014
Accurate and Easy-to-Use

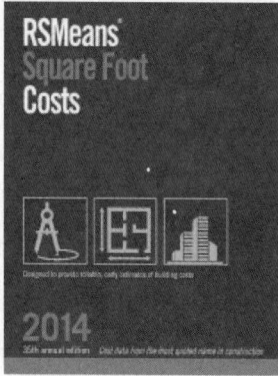

- **Updated price information** based on nationwide figures from suppliers, estimators, labor experts, and contractors
- New green building models
- Realistic graphics, offering true-to-life illustrations of building projects
- Extensive information on using square foot cost data, including sample estimates and alternate pricing methods

$205.95 | Available Oct. 2013 | Catalog no. 60054

RSMeans Commercial Renovation Cost Data 2014
Commercial/Multifamily Residential

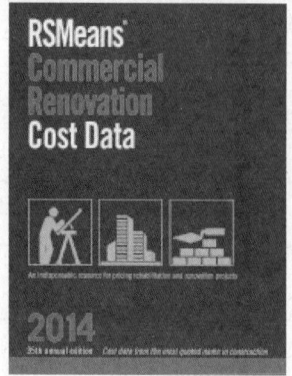

Use this valuable tool to estimate commercial and multifamily residential renovation and remodeling.

Includes: New costs for hundreds of unique methods, materials, and conditions that only come up in repair and remodeling, PLUS:

- Unit costs for more than 19,000 construction components
- Installed costs for more than 2,000 assemblies
- More than 930 "on-site" localization factors for the U.S. and Canada

$156.95 | Available Oct. 2013 | Catalog no. 60044

RSMeans Electrical Cost Data 2014

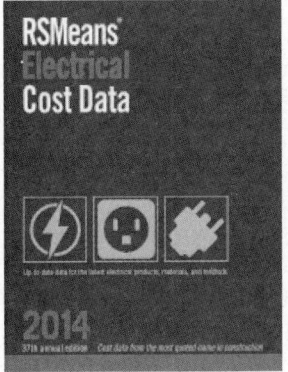

Pricing information for every part of electrical cost planning. More than 13,000 unit and systems costs with design tables; clear specifications and drawings; engineering guides; illustrated estimating procedures; complete labor-hour and materials costs for better scheduling and procurement; and the latest electrical products and construction methods.

- A variety of special electrical systems, including cathodic protection
- Costs for maintenance, demolition, HVAC/mechanical, specialties, equipment, and more

$196.95 | Available Oct. 2013 | Catalog no. 60034

RSMeans Assemblies Cost Data 2014

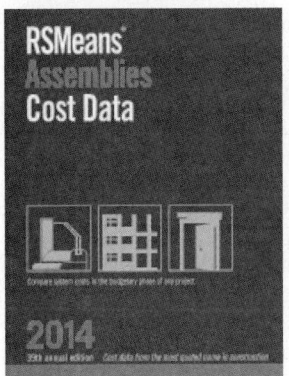

RSMeans Assemblies Cost Data takes the guesswork out of preliminary or conceptual estimates. Now you don't have to try to calculate the assembled cost by working up individual component costs. We've done all the work for you.

Presents detailed illustrations, descriptions, specifications, and costs for every conceivable building assembly—over 350 types in all—arranged in the easy-to-use UNIFORMAT II system. Each illustrated "assembled" cost includes a complete grouping of materials and associated installation costs, including the installing contractor's overhead and profit.

$315.95 | Available Sept. 2013 | Catalog no. 60064

RSMeans Residential Cost Data 2014

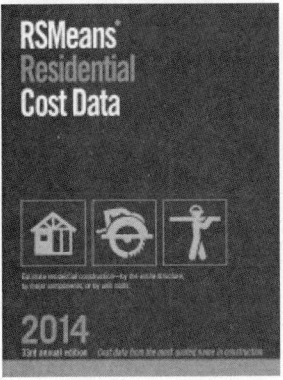

Contains square foot costs for 28 basic home models with the look of today, plus hundreds of custom additions and modifications you can quote right off the page. Includes costs for more than 700 residential systems. Complete with blank estimating forms, sample estimates, and step-by-step instructions.

Contains line items for cultured stone and brick, PVC trim lumber, and TPO roofing.

$139.95 | Available Oct. 2013 | Catalog no. 60174

RSMeans Electrical Change Order Cost Data 2014

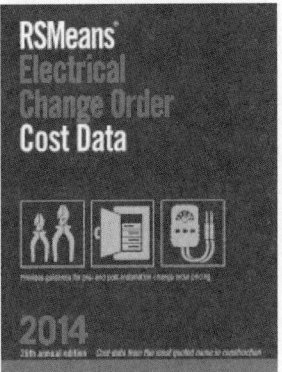

RSMeans Electrical Change Order Cost Data provides you with electrical unit prices exclusively for pricing change orders—based on the recent, direct experience of contractors and suppliers. Analyze and check your own change order estimates against the experience others have had doing the same work. It also covers productivity analysis and change order cost justifications. With useful information for calculating the effects of change orders and dealing with their administration.

$187.95 | Available Dec. 2013 | Catalog no. 60234

RSMeans Open Shop Building Construction Cost Data 2014

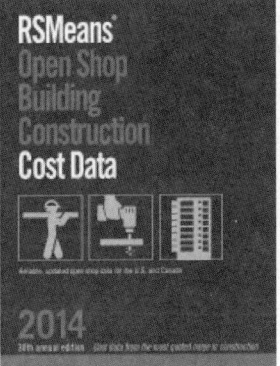

The latest costs for accurate budgeting and estimating of new commercial and residential construction . . . renovation work . . . change orders . . . cost engineering.

RSMeans Open Shop "BCCD" will assist you to:

- Develop benchmark prices for change orders
- Plug gaps in preliminary estimates and budgets
- Estimate complex projects
- Substantiate invoices on contracts
- Price ADA-related renovations

$165.95 | Available Dec. 2013 | Catalog no. 60154

RSMeans Site Work & Landscape Cost Data 2014

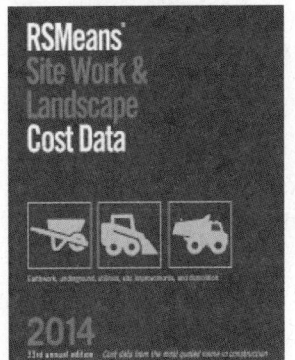

Includes unit and assemblies costs for earthwork, sewerage, piped utilities, site improvements, drainage, paving, trees and shrubs, street openings/repairs, underground tanks, and more. Contains 78 types of assemblies costs for accurate conceptual estimates.

Includes:

- Estimating for infrastructure improvements
- Environmentally-oriented construction
- ADA-mandated handicapped access
- Hazardous waste line items

$187.95 | Available Dec. 2013 | Catalog no. 60284

RSMeans Facilities Maintenance & Repair Cost Data 2014

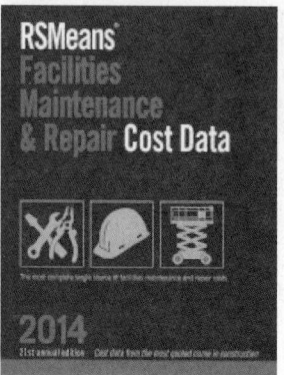

RSMeans Facilities Maintenance & Repair Cost Data gives you a complete system to manage and plan your facility repair and maintenance costs and budget efficiently. Guidelines for auditing a facility and developing an annual maintenance plan. Budgeting is included, along with reference tables on cost and management, and information on frequency and productivity of maintenance operations.

The only nationally recognized source of maintenance and repair costs. Developed in cooperation with the Civil Engineering Research Laboratory (CERL) of the Army Corps of Engineers.

$434.95 | Available Nov. 2013 | Catalog no. 60304

RSMeans Concrete & Masonry Cost Data 2014

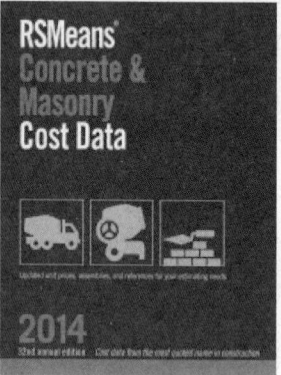

Provides you with cost facts for virtually all concrete/masonry estimating needs, from complicated formwork to various sizes and face finishes of brick and block—all in great detail. The comprehensive Unit Price Section contains more than 8,000 selected entries. Also contains an Assemblies [Cost] Section, and a detailed Reference Section that supplements the cost data.

$177.95 | Available Dec. 2013 | Catalog no. 60114

RSMeans Construction Cost Indexes 2014

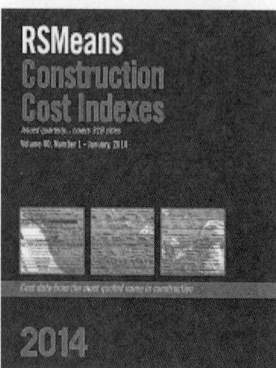

What materials and labor costs will change unexpectedly this year? By how much?

- Breakdowns for 318 major cities
- National averages for 30 key cities
- Expanded five major city indexes
- Historical construction cost indexes

$362.00 per year (subscription) | Catalog no. 50144
$93.50 individual quarters | Catalog no. 60144 A,B,C,D

RSMeans Light Commercial Cost Data 2014

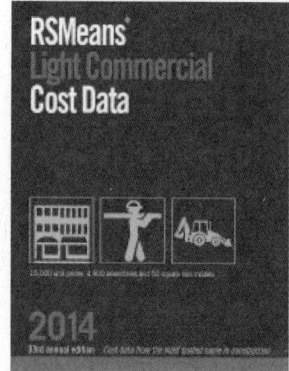

Specifically addresses the light commercial market, which is a specialized niche in the construction industry. Aids you, the owner/designer/contractor, in preparing all types of estimates—from budgets to detailed bids. Includes new advances in methods and materials.

Assemblies Section allows you to evaluate alternatives in the early stages of design/planning.

More than 15,000 unit costs ensure that you have the prices you need . . . when you need them.

$144.95 | Available Nov. 2013 | Catalog no. 60184

RSMeans Labor Rates for the Construction Industry 2014

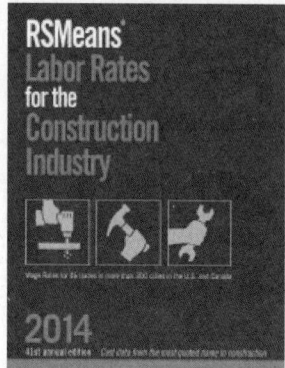

Complete information for estimating labor costs, making comparisons, and negotiating wage rates by trade for more than 300 U.S. and Canadian cities. With 46 construction trades listed by local union number in each city, and historical wage rates included for comparison. Each city chart lists the county and is alphabetically arranged with handy visual flip tabs for quick reference.

$424.95 | Available Dec. 2013 | Catalog no. 60124

RSMeans Interior Cost Data 2014

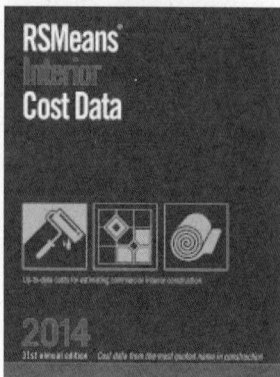

Provides you with prices and guidance needed to make accurate interior work estimates. Contains costs on materials, equipment, hardware, custom installations, furnishings, and labor costs . . . for new and remodel commercial and industrial interior construction, including updated information on office furnishings, and reference information.

$196.95 | Available Nov. 2013 | Catalog no. 60094

Annual Cost Guides

For more information visit the RSMeans website at **www.rsmeans.com**

Unit prices according to the latest MasterFormat!

RSMeans Heavy Construction Cost Data 2014

A comprehensive guide to heavy construction costs. Includes costs for highly specialized projects such as tunnels, dams, highways, airports, and waterways. Information on labor rates, equipment, and materials costs is included. Features unit price costs, systems costs, and numerous reference tables for costs and design.

$192.95 | Available Dec. 2013 | Catalog no. 60164

RSMeans Plumbing Cost Data 2014

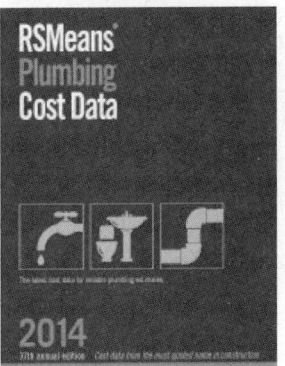

Comprehensive unit prices and assemblies for plumbing, irrigation systems, commercial and residential fire protection, point-of-use water heaters, and the latest approved materials. This publication and its companion, *RSMeans Mechanical Cost Data*, provide full-range cost estimating coverage for all the mechanical trades.

Now contains updated costs for potable water and radiant heat systems and high efficiency fixtures.

$194.95 | Available Oct. 2013 | Catalog no. 60214

2014 RSMeans Seminar Schedule

Note: call for exact dates and details.

Location	Dates	Location	Dates
Seattle, WA	January and August	Jacksonville FL	September
Dallas/Ft. Worth, TX	January	Dallas, TX	September
Austin, TX	February	Charleston, SC	October
Anchorage, AK	March and September	Houston, TX	October
Las Vegas, NV	March and October	Atlanta, GA	November
Washington, DC	April and September	Baltimore, MD	November
Phoenix, AZ	April	Orlando, FL	November
Kansas City, MO	April	San Diego, CA	December
Toronto	May	San Antonio, TX	December
Denver, CO	May	Raleigh, NC	December
San Francisco, CA	June		
Bethesda, MD	June		
Columbus, GA	June		
El Segundo, CA	August		

1-800-334-3509, Press 1

Professional Development

For more information visit the RSMeans website at **www.rsmeans.com**

eLearning Training Sessions

Learn how to use *RSMeans Online*™ or *RSMeans CostWorks*® CD from the convenience of your home or office. Our eLearning training sessions let you join a training conference call and share the instructors' desktops, so you can view the presentation and step-by-step instructions on your own computer screen. The live webinars are held from 9 a.m. to 4 p.m. eastern standard time, with a one-hour break for lunch.

For these sessions, you must have a computer with high speed Internet access and a compatible Web browser. Learn more at www.rsmeansonline.com or call for a schedule: 1-800-334-3509 and press 1. Webinars are generally held on selected Wednesdays each month.

RSMeans Online™
$299 per person

RSMeans CostWorks® CD
$299 per person

For more information visit the RSMeans website at **www.rsmeans.com**

RSMeans Online™ Training

Construction estimating is vital to the decision-making process at each state of every project. RSMeansOnline works the way you do. It's systematic, flexible and intuitive. In this one day class you will see how you can estimate any phase of any project faster and better.

Some of what you'll learn:
- Customizing RSMeansOnline
- Making the most of RSMeans "Circle Reference" numbers
- How to integrate your cost data
- Generate reports, exporting estimates to MS Excel, sharing, collaborating and more

Also available: RSMeans Online™ training webinar

Facilities Construction Estimating

In this *two-day* course, professionals working in facilities management can get help with their daily challenges to establish budgets for all phases of a project.

Some of what you'll learn:
- Determining the full scope of a project
- Identifying the scope of risks and opportunities
- Creative solutions to estimating issues
- Organizing estimates for presentation and discussion
- Special techniques for repair/remodel and maintenance projects
- Negotiating project change orders

Who should attend: facility managers, engineers, contractors, facility tradespeople, planners, and project managers.

Scheduling with Microsoft Project

Two days of hands-on training gives you a look at the basics to putting it all together with MS Project to create a complete project schedule. Learn to work better and faster, manage changes and updates, enhance tracking, generate actionable reports, and boost control of your time and budget.

Some of what you'll learn:
- Defining task/activity relationships: link tasks, establish predecessor/successor relationships and create logs
- Baseline scheduling: the essential what/when
- Using MS Project to determine the critical path
- Integrate RSMeans data with MS Project
- The dollar-loaded schedule: enhance the reliability of this "must-know" technique for scheduling public projects.

Who should attend: contractors' project-management teams, architects and engineers, project owners and their representatives, and those interested in improving their project planning prod scheduling and management skills.

Maintenance & Repair Estimating for Facilities

This *two-day* course teaches attendees how to plan, budget, and estimate the cost of ongoing and preventive maintenance and repair for existing buildings and grounds.

Some of what you'll learn:
- The most financially favorable maintenance, repair, and replacement scheduling and estimating
- Auditing and value engineering facilities
- Preventive planning and facilities upgrading
- Determining both in-house and contract-out service costs
- Annual, asset-protecting M&R plan

Who should attend: facility managers, maintenance supervisors, buildings and grounds superintendents, plant managers, planners, estimators, and others involved in facilities planning and budgeting.

Practical Project Management for Construction Professionals

In this *two-day* course, acquire the essential knowledge and develop the skills to effectively and efficiently execute the day-to-day responsibilities of the construction project manager.

Covers:
- General conditions of the construction contract
- Contract modifications: change orders and construction change directives
- Negotiations with subcontractors and vendors
- Effective writing: notification and communications
- Dispute resolution: claims and liens

Who should attend: architects, engineers, owner's representatives, project managers.

Mechanical & Electrical Estimating

This *two-day* course teaches attendees how to prepare more accurate and complete mechanical/electrical estimates, avoiding the pitfalls of omission and double-counting, while understanding the composition and rationale within the RSMeans mechanical/electrical database.

Some of what you'll learn:
- The unique way mechanical and electrical systems are interrelated
- M&E estimates—conceptual, planning, budgeting, and bidding stages
- Order of magnitude, square foot, assemblies, and unit price estimating
- Comparative cost analysis of equipment and design alternatives

Who should attend: architects, engineers, facilities managers, mechanical and electrical contractors, and others who need a highly reliable method for developing, understanding, and evaluating mechanical and electrical contracts.

Unit Price Estimating

This interactive *two-day* seminar teaches attendees how to interpret project information and process it into final, detailed estimates with the greatest accuracy level.

The most important credential an estimator can take to the job is the ability to visualize construction and estimate accurately.

Some of what you'll learn:

- Interpreting the design in terms of cost
- The most detailed, time-tested methodology for accurate pricing
- Key cost drivers—material, labor, equipment, staging, and subcontracts
- Understanding direct and indirect costs for accurate job cost accounting and change order management

Who should attend: corporate and government estimators and purchasers, architects, engineers, and others who need to produce accurate project estimates.

RSMeans CostWorks® CD Training

This one-day course helps users become more familiar with the functionality of *RSMeans CostWork*s program. Each menu, icon, screen, and function found in the program is explained in depth. Time is devoted to hands-on estimating exercises.

Some of what you'll learn:

- Searching the database using all navigation methods
- Exporting RSMeans data to your preferred spreadsheet format
- Viewing crews, assembly components, and much more
- Automatically regionalizing the database

This training session requires you to bring a laptop computer to class.

When you register for this course you will receive an outline for your laptop requirements.

Also offering web training for CostWorks CD!

Facilities Est. Using RSMeans CostWorks® CD

Combines hands-on skill building with best estimating practices and real-life problems. Brings you up-to-date with key concepts, and provides tips, pointers, and guidelines to save time and avoid cost oversights and errors.

Some of what you'll learn:

- Estimating process concepts
- Customizing and adapting RSMeans cost data
- Establishing scope of work to account for all known variables
- Budget estimating: when, why, and how
- Site visits: what to look for—what you can't afford to overlook
- How to estimate repair and remodeling variables

This training session requires you to bring a laptop computer to class.

Who should attend: facility managers, architects, engineers, contractors, facility tradespeople, planners, project managers and anyone involved with JOC, SABRE, or IDIQ.

Conceptual Estimating Using RSMeans CostWorks® CD

This *two day* class uses the leading industry data and a powerful software package to develop highly accurate conceptual estimates for your construction projects. All attendees must bring a laptop computer loaded with the current year *Square Foot Model Costs* and the *Assemblies Cost Data* CostWorks titles.

Some of what you'll learn:

- Introduction to conceptual estimating
- Types of conceptual estimates
- Helpful hints
- Order of magnitude estimating
- Square foot estimating
- Assemblies estimating

Who should attend: architects, engineers, contractors, construction estimators, owner's representatives, and anyone looking for an electronic method for performing square foot estimating.

Assessing Scope of Work for Facility Construction Estimating

This *two-day* practical training program addresses the vital importance of understanding the SCOPE of projects in order to produce accurate cost estimates for facility repair and remodeling.

Some of what you'll learn:

- Discussions of site visits, plans/specs, record drawings of facilities, and site-specific lists
- Review of CSI divisions, including means, methods, materials, and the challenges of scoping each topic
- Exercises in SCOPE identification and SCOPE writing for accurate estimating of projects
- Hands-on exercises that require SCOPE, take-off, and pricing

Who should attend: corporate and government estimators, planners, facility managers, and others who need to produce accurate project estimates.

Unit Price Est. Using RSMeans CostWorks® CD

Step-by-step instruction and practice problems to identify and track key cost drivers—material, labor, equipment, staging, and subcontractors—for each specific task. Learn the most detailed, time-tested methodology for accurately "pricing" these variables, their impact on each other and on total cost.

Some of what you'll learn:

- Unit price cost estimating
- Order of magnitude, square foot, and assemblies estimating
- Quantity takeoff
- Direct and indirect construction costs
- Development of contractor's bill rates
- How to use *RSMeans Building Construction Cost Data*

This training session requires you to bring a laptop computer to class.

Who should attend: architects, engineers, corporate and government estimators, facility managers, and government procurement staff.

For more information visit the RSMeans website at **www.rsmeans.com**

Registration Information

Register early . . . Save up to $100! Register 30 days before the start date of a seminar and save $100 off your total fee. *Note: This discount can be applied only once per order. It cannot be applied to team discount registrations or any other special offer.*

How to register Register by phone today! The RSMeans toll-free number for making reservations is **1-800-334-3509, Press 1.**

Two day seminar registration fee - $935. One-day *RSMeans* CostWorks® training registration fee - $375. To register by mail, complete the registration form and return, with your full fee, to: RSMeans Seminars, 700 Longwater Drive, Norwell, MA 02061.

Government pricing All federal government employees save off the regular seminar price. Other promotional discounts cannot be combined with the government discount.

Team discount program for two to four seminar registrations. Call for pricing: 1-800-334-3509, Press 1

Multiple course discounts When signing up for two or more courses, call for pricing.

Refund policy Cancellations will be accepted up to ten business days prior to the seminar start. There are no refunds for cancellations received later than ten working days prior to the first day of the seminar. A $150 processing fee will be applied for all cancellations. Written notice of cancellation is required. Substitutions can be made at any time before the session starts. **No-shows are subject to the full seminar fee.**

AACE approved courses Many seminars described and offered here have been approved for 14 hours (1.4 recertification credits) of credit by the AACE International Certification Board toward meeting the continuing education requirements for recertification as a Certified Cost Engineer/Certified Cost Consultant.

AIA Continuing Education We are registered with the AIA Continuing Education System (AIA/CES) and are committed to developing quality learning activities in accordance with the CES criteria. Many seminars meet the AIA/CES criteria for Quality Level 2. AIA members may receive (14) learning units (LUs) for each two-day RSMeans course.

Daily course schedule The first day of each seminar session begins at 8:30 a.m. and ends at 4:30 p.m. The second day begins at 8:00 a.m. and ends at 4:00 p.m. Participants are urged to bring a hand-held calculator, since many actual problems will be worked out in each session.

Continental breakfast Your registration includes the cost of a continental breakfast, and a morning and afternoon refreshment break. These informal segments allow you to discuss topics of mutual interest with other seminar attendees. (You are free to make your own lunch and dinner arrangements.)

Hotel/transportation arrangements RSMeans arranges to hold a block of rooms at most host hotels. To take advantage of special group rates when making your reservation, be sure to mention that you are attending the RSMeans seminar. You are, of course, free to stay at the lodging place of your choice. **(Hotel reservations and transportation arrangements should be made directly by seminar attendees.)**

Important Class sizes are limited, so please register as soon as possible.

Note: Pricing subject to change.

Registration Form

ADDS-1000

Call 1-800-334-3509, Press 1 to register or FAX this form 1-800-632-6732. Visit our website: www.rsmeans.com

Please register the following people for the RSMeans construction seminars as shown here. We understand that we must make our own hotel reservations if overnight stays are necessary.

☐ Full payment of $_____ enclosed.

☐ Bill me.

Please print name of registrant(s).

(To appear on certificate of completion)

_____

_____

_____

_____

P.O. #: _____
GOVERNMENT AGENCIES MUST SUPPLY PURCHASE ORDER NUMBER OR
TRAINING FORM.

Firm name_____

Address_____

City/State/Zip_____

Telephone no._____ Fax no._____

E-mail address_____

Charge registration(s) to: ☐ MasterCard ☐ VISA ☐ American Express

Account no._____ Exp. date_____

Cardholder's signature_____

Seminar name_____

Seminar City _____

Please mail check to: RSMeans Seminars, 700 Longwater Drive, Norwell, MA 02061 USA

For more information visit the RSMeans website at **www.rsmeans.com**

Solar Energy: Technologies & Project Delivery for Buildings

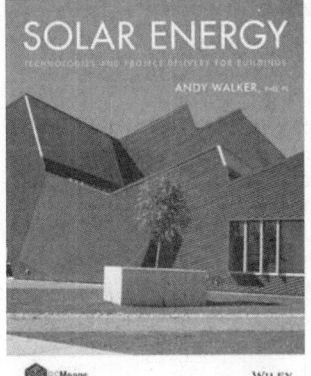

by Andy Walker

An authoritative reference on the design of solar energy systems in building projects, with applications, operating principles, and simple tools for the construction, engineering, and design professionals, the book simplifies the solar design and engineering process and provides sample documentation for the complete design of a solar energy system for buildings.

$85.00 | 320 pages, hardover | Catalog no. 67365

Project Scheduling & Management for Construction, 4th Edition

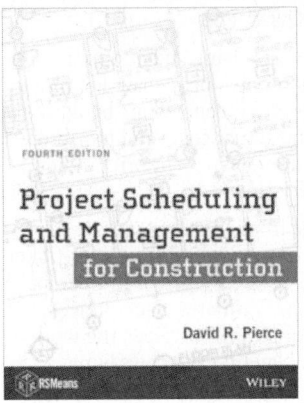

by David R. Pierce

First published in 1988 by RSMeans, the new edition of *Project Scheduling and Management for Construction* has been substantially revised for both professionals and students enrolled in construction management and civil engineering programs. While retaining its emphasis on developing practical, professional-level scheduling skills, the new edition is a relatable, real world case study.

$95.00 | 272 pages, softcover | Catalog no. 67367

Estimating Building Costs, 2nd Edition
For the Residential & Light Commercial Construction Professional

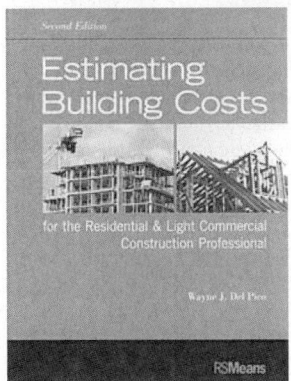

by Wayne J. Del Pico

This book guides readers through the entire estimating process, explaining in detail how to put together a reliable estimate that can be used not only for budgeting, but also for developing a schedule, managing a project, dealing with contingencies, and ultimately making a profit.

Completely revised and updated to reflect the CSI MasterFormat 2010 system, this practical guide describes estimating techniques for each building system and how to apply them according to the latest industry standards.

$75.00 | 398 pages, softcover | Catalog no. 67343A

Project Control: Integrating Cost & Schedule in Construction

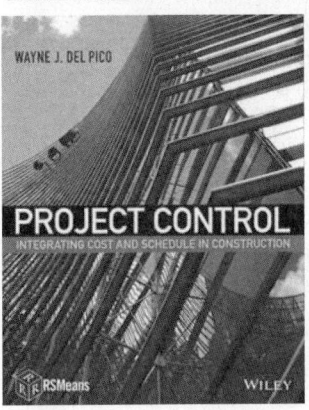

by Wayne J. Del Pico

Written by a seasoned professional in the field, *Project Control: Integrating Cost and Schedule in Construction* fills a void in the area of project control as applied in the construction industry today. It demonstrates how productivity models for an individual project are created, monitored, and controlled, and how corrective actions are implemented as deviations from the baseline occur.

$95.00 | 240 pages, softcover | Catalog no. 67366

How to Estimate with Means Data & CostWorks, 4th Edition

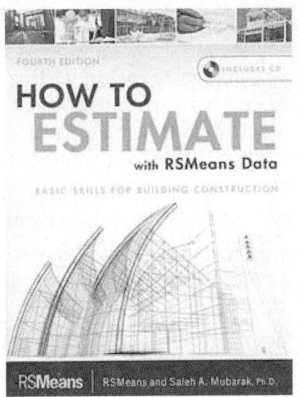

by RSMeans and Saleh A. Mubarak, Ph.D.

This step-by-step guide takes you through all the major construction items with extensive coverage of site work, concrete and masonry, wood and metal framing, doors and windows, and other divisions. The only construction cost estimating handbook that uses the most popular source of construction data, RSMeans, this indispensible guide features access to the instructional version of CostWorks in electronic form, enabling you to practice techniques to solve real-world estimating problems.

$75.00 | 292 pages, softcover | Includes CostWorks CD
Catalog no. 67324C

RSMeans Cost Data, Student Edition

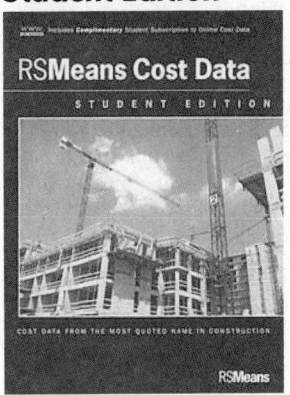

This book provides a thorough introduction to cost estimating in a self-contained print and online package. With clear explanations and a hands-on, example-driven approach, it is the ideal reference for students and new professionals.

Features include:

- Commercial and residential construction cost data in print and online formats
- Complete how-to guidance on the essentials of cost estimating
- A supplemental website with plans, problem sets, and a full sample estimate

$99.00 | 512 pages, softcover | Catalog no. 67363

For more information visit the RSMeans website at **www.rsmeans.com**

Value Engineering: Practical Applications

For Design, Construction, Maintenance & Operations

by Alphonse Dell'Isola, PE

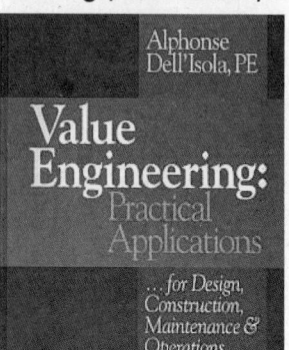

A tool for immediate application—for engineers, architects, facility managers, owners, and contractors. Includes making the case for VE—the management briefing; integrating VE into planning, budgeting, and design; conducting life cycle costing; using VE methodology in design review and consultant selection; case studies; VE workbook; and a life cycle costing program on disk.

$85.00 | Over 450 pages, illustrated, softcover | Catalog no. 67319A

Risk Management for Design & Construction

Introduces risk as a central pillar of project management and shows how a project manager can prepare to deal with uncertainty. Experts in the field use clear, straightforward terminology to demystify the concepts of project uncertainty and risk. Includes:

- Integrated cost and schedule risk analysis
- An introduction to a ready-to-use system of analyzing a project's risks and tools to proactively manage risks
- A methodology that was developed and used by the Washington State Department of Transportation
- Case studies and examples on the proper application of principles
- Information about combining value analysis with risk analysis

$125.00 | Over 250 pages, softcover | Catalog no. 67359

How Your House Works, 2nd Edition

by Charlie Wing

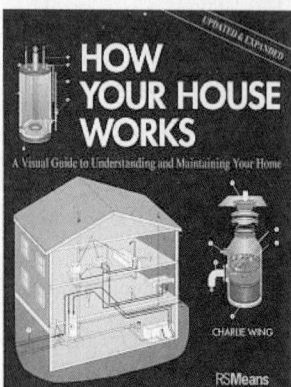

Knowledge of your home's systems helps you control repair and construction costs and makes sure the correct elements are being installed or replaced. This book uncovers the mysteries behind just about every major appliance and building element in your house. See-through, cross-section drawings in full color show you exactly how these things should be put together and how they function, including what to check if they don't work. It just might save you having to call in a professional.

$22.95 | 192 pages, softcover | Catalog no. 67351A

Complete Book of Framing, 2nd Edition

by Scot Simpson

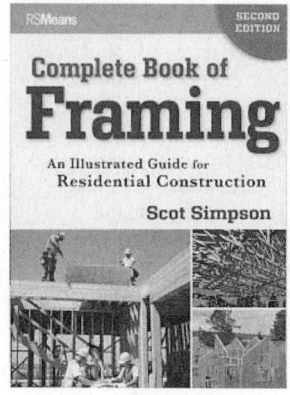

This updated, easy-to-learn guide to rough carpentry and framing is written by an expert with more than thirty years of framing experience. Starting with the basics, this book begins with types of lumber, nails, and what tools are needed, followed by detailed, fully illustrated steps for framing each building element. Framer-Friendly Tips throughout the book show how to get a task done right—and more easily.

$29.95 | 352 pages, softcover | Catalog no. 67353A

Green Home Improvement

by Daniel D. Chiras, PhD

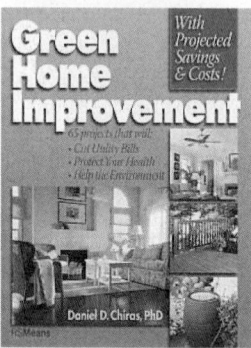

With energy costs rising and environmental awareness increasing, people are looking to make their homes greener. This book, with 65 projects and actual costs and projected savings help homeowners prioritize their green improvements.

Projects range from simple water savers that cost only a few dollars, to bigger-ticket items such as HVAC systems. With color photos and cost estimates, each project compares options and describes the work involved, the benefits, and the savings.

$34.95 | 320 pages, illustrated, softcover | Catalog no. 67355

Life Cycle Costing for Facilities

by Alphonse Dell'Isola and Dr. Steven Kirk

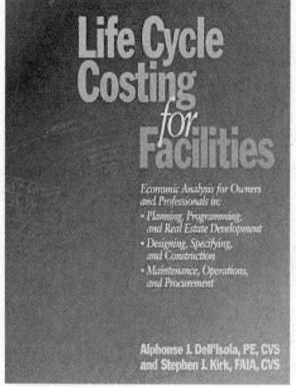

Guidance for achieving higher quality design and construction projects at lower costs! Cost-cutting efforts often sacrifice quality to yield the cheapest product. Life cycle costing enables building designers and owners to achieve both. The authors of this book show how LCC can work for a variety of projects — from roads to HVAC upgrades to different types of buildings.

$99.95 | 396 pages, hardcover | Catalog no. 67341

Reference Books

Builder's Essentials: Plan Reading & Material Takeoff
For Residential and Light Commercial Construction
by Wayne J. DelPico

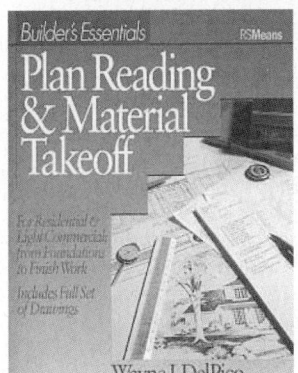

A valuable tool for understanding plans and specs, and accurately calculating material quantities. Step-by-step instructions and takeoff procedures based on a full set of working drawings.

$45.00 | Over 420 pages, softcover | Catalog no. 67307

Concrete Repair & Maintenance Illustrated
by Peter Emmons

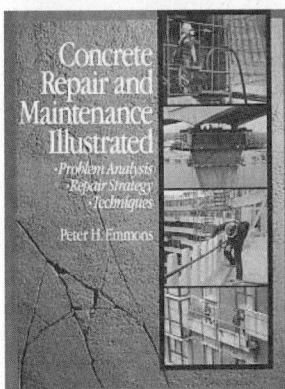

Hundreds of illustrations show users how to analyze, repair, clean, and maintain concrete structures for optimal performance and cost effectiveness. From parking garages to roads and bridges to structural concrete, this comprehensive book describes the causes, effects, and remedies for concrete wear and failure. Invaluable for planning jobs, selecting materials, and training employees, this book is a must-have for concrete specialists, general contractors, facility managers, civil and structural engineers, and architects.

$75.00 | 300 pages, illustrated, softcover | Catalog no. 67146

Electrical Estimating Methods, 3rd Edition

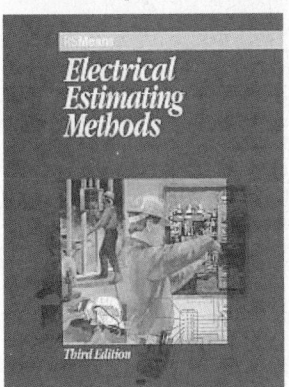

Expanded edition includes sample estimates and cost information in keeping with the latest version of the CSI MasterFormat and UNIFORMAT II. Complete coverage of fiber optic and uninterruptible power supply electrical systems, broken down by components, and explained in detail. Includes a new chapter on computerized estimating methods. A practical companion to *RSMeans Electrical Cost Data*.

$75.00 | Over 325 pages, hardcover | Catalog no. 67230B

Means Unit Price Estimating Methods, 4th Edition

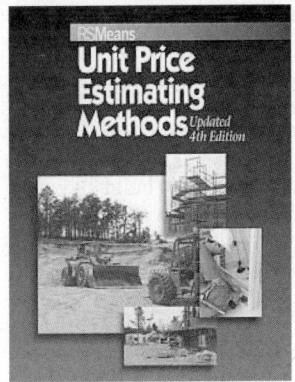

Includes cost data and estimating examples, updated to reflect changes to the CSI numbering system and new features of RSMeans cost data. It describes the most productive, universally accepted ways to estimate, and uses checklists and forms to illustrate shortcuts and timesavers. A model estimate demonstrates procedures. A new chapter explores computer estimating alternatives.

$75.00 | Over 350 pages, illustrated, softcover | Catalog no. 67303B

Square Foot & UNIFORMAT Assemblies Estimating Methods, 3rd Edition

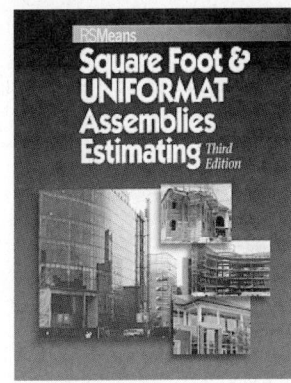

Develop realistic square foot and assemblies costs for budgeting and construction funding. The new edition features updated guidance on square foot and assemblies estimating using UNIFORMAT II. An essential reference for anyone who performs conceptual estimates.

$75.00 | Over 300 pages, illustrated, softcover | Catalog no. 67145B

Residential & Light Commercial Construction Standards, 3rd Edition
Updated by RSMeans and contributing authors

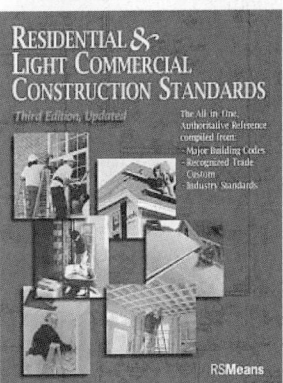

This book provides authoritative requirements and recommendations compiled from leading professional associations, industry publications, and building code organizations. This all-in-one reference helps establish a standard for workmanship, quickly resolve disputes, and avoid defect claims. Updated third edition includes new coverage of green building, seismic, hurricane, and mold-resistant construction.

$65.00 | Over 550 pages, illustrated, softcover | Catalog no. 67322B

Reference Books

For more information visit the RSMeans website at **www.rsmeans.com**

Construction Business Management

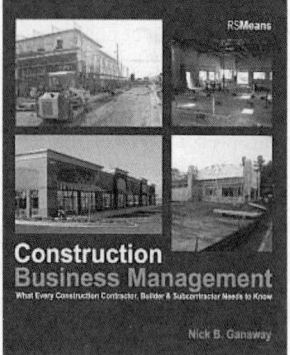

by Nick Ganaway

Only 43% of construction firms stay in business after four years. Make sure your company thrives with valuable guidance from a pro with 25 years of success as a commercial contractor. Find out what it takes to build all aspects of a business that is profitable, enjoyable, and enduring. With a bonus chapter on retail construction.

$55.00 | 200 pages, softcover | Catalog no. 67352

The Practice of Cost Segregation Analysis

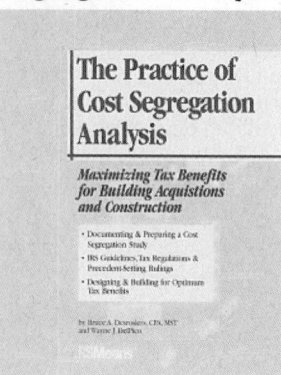

by Bruce A. Desrosiers and Wayne J. DelPico

This expert guide walks you through the practice of cost segregation analysis, which enables property owners to defer taxes and benefit from "accelerated cost recovery" through depreciation deductions on assets that are properly identified and classified.

With a glossary of terms, sample cost segregation estimates for various building types, key information resources, and updates via a dedicated website, this book is a critical resource for anyone involved in cost segregation analysis.

$105.00 | Over 225 pages | Catalog no. 67345

Green Building: Project Planning & Cost Estimating, 3rd Edition

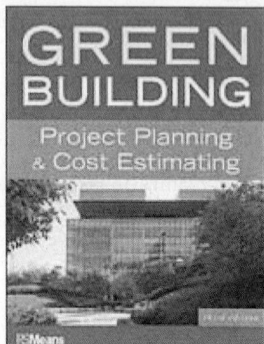

Since the widely read first edition of this book, green building has gone from a growing trend to a major force in design and construction.

This new edition has been updated with the latest in green building technologies, design concepts, standards, and costs. Full-color with all new case studies—plus a new chapter on commercial real estate.

$110.00 | Over 450 pages, softcover | Catalog no. 67338B

Project Scheduling & Management for Construction, 3rd Edition

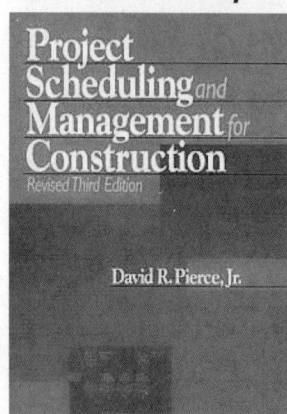

by David R. Pierce, Jr.

A comprehensive yet easy-to-follow guide to construction project scheduling and control—from vital project management principles through the latest scheduling, tracking, and controlling techniques. The author is a leading authority on scheduling, with years of field and teaching experience at leading academic institutions. Spend a few hours with this book and come away with a solid understanding of this essential management topic.

$64.95 | Over 300 pages, illustrated, hardcover | Catalog no. 67247B

Preventive Maintenance for Multi-Family Housing

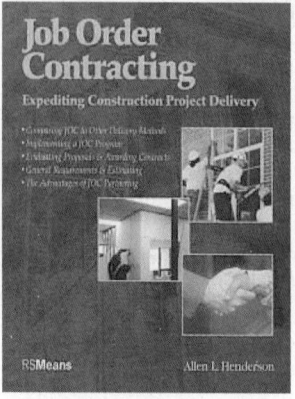

by John C. Maciha

Prepared by one of the nation's leading experts on multi-family housing.

This complete PM system for apartment and condominium communities features expert guidance, checklists for buildings and grounds maintenance tasks and their frequencies, a reusable wall chart to track maintenance, and a dedicated website featuring customizable electronic forms. A must-have for anyone involved with multi-family housing maintenance and upkeep.

$95.00 | 225 pages | Catalog no. 67346

Job Order Contracting: Expediting Construction Project Delivery

by Allen L. Henderson

Expert guidance to help you implement JOC—fast becoming the preferred project delivery method for repair and renovation, minor new construction, and maintenance projects in the public sector and in many states and municipalities. The author, a leading JOC expert and practitioner, shows how to:

- Establish a JOC program
- Evaluate proposals & award contracts
- Handle general requirements and estimating
- Partner for maximum benefits

$89.95 | 192 pages, illustrated, hardcover | Catalog no. 67348

Reference Books

For more information visit the RSMeans website at **www.rsmeans.com**

Construction Supervision

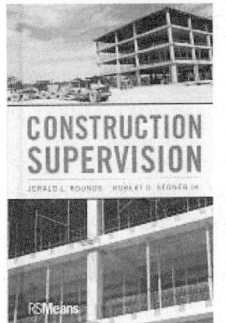

This brand new title inspires supervisory excellence with proven tactics and techniques applied by thousands of construction supervisors over the past decade. Recognizing the unique and critical role the supervisor plays in project success, the book's leadership guidelines carve out a practical blueprint for motivating work performance and increasing productivity through effective communication.

Features:

- A unique focus on field supervision and crew management
- Coverage of supervision from the foreman to the superintendent level
- An overview of technical skills whose mastery will build confidence and success for the supervisor
- A detailed view of "soft" management and communication skills

$95.00 | Over 450 pages, softcover | Catalog no. 67358

The Homeowner's Guide to Mold

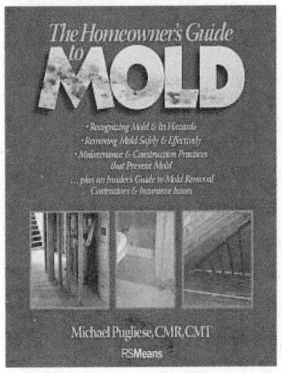

By Michael Pugliese

Expert guidance to protect your health and your home.

Mold, whether caused by leaks, humidity or flooding, is a real health and financial issue—for homeowners and contractors.

This full-color book explains:

- Construction and maintenance practices to prevent mold
- How to inspect for and remove mold
- Mold remediation procedures and costs
- What to do after a flood
- How to deal with insurance companies

$21.95 | 144 pages, softcover | Catalog no. 67344

Landscape Estimating Methods, 5th Edition

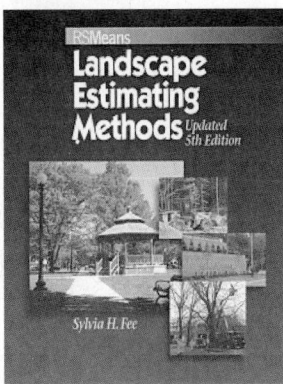

Answers questions about preparing competitive landscape construction estimates, with up-to-date cost estimates and the new MasterFormat classification system. Expanded and revised to address the latest materials and methods, including new coverage on approaches to green building. Includes:

- Step-by-step explanation of the estimating process
- Sample forms and worksheets that save time and prevent errors

$75.00 | Over 350 pages, softcover | Catalog no. 67295C

Building & Renovating Schools

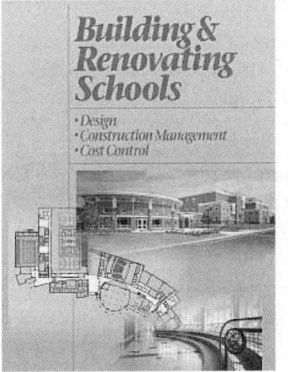

This all-inclusive guide covers every step of the school construction process—from initial planning, needs assessment, and design, right through moving into the new facility. A must-have resource for anyone concerned with new school construction or renovation. With square foot cost models for elementary, middle, and high school facilities, and real-life case studies of recently completed school projects.

The contributors to this book—architects, construction project managers, contractors, and estimators who specialize in school construction—provide start-to-finish, expert guidance on the process.

$99.95 | Over 425 pages, hardcover | Catalog no. 67342

Universal Design Ideas for Style, Comfort & Safety

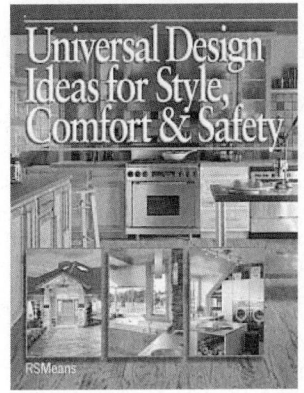

by RSMeans and Lexicon Consulting, Inc.

Incorporating universal design when building or remodeling helps people of any age and physical ability more fully and safely enjoy their living spaces. This book shows how universal design can be artfully blended into the most attractive homes. It discusses specialized products like adjustable countertops and chair lifts, as well as simple ways to enhance a home's safety and comfort. With color photos and expert guidance, every area of the home is covered. Includes budget estimates that give an idea how much projects will cost.

$21.95 | 160 pages, illustrated, softcover | Catalog no. 67354

Plumbing Estimating Methods, 3rd Edition

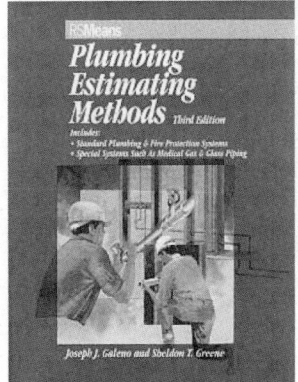

by Joseph J. Galeno and Sheldon T. Greene

Updated and revised! This practical guide walks you through a plumbing estimate, from basic materials and installation methods through change order analysis. *Plumbing Estimating Methods* covers residential, commercial, industrial, and medical systems, and features sample takeoff and estimate forms and detailed illustrations of systems and components.

$65.00 | 330+ pages, softcover | Catalog no. 67283B

Understanding & Negotiating Construction Contracts

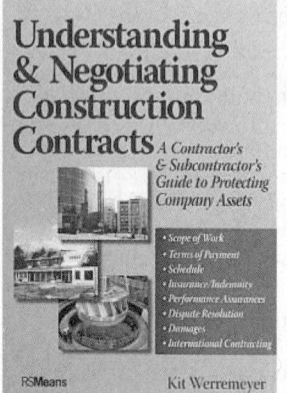

by Kit Werremeyer

Take advantage of the author's 30 years' experience in small-to-large (including international) construction projects. Learn how to identify, understand, and evaluate high risk terms and conditions typically found in all construction contracts—then negotiate to lower or eliminate the risk, improve terms of payment, and reduce exposure to claims and disputes. The author avoids "legalese" and gives real-life examples from actual projects.

$80.00 | 300 pages, softcover | Catalog no. 67350

The Gypsum Construction Handbook

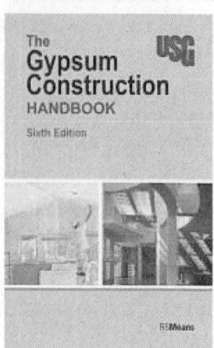

by USG

An invaluable reference of construction procedures for gypsum drywall, cement board, veneer plaster, and conventional plaster. This new edition includes the newest product developments, installation methods, fire- and sound-rated construction information, illustrated framing-to-finish application instructions, estimating and planning information, and more. Great for architects and engineers; contractors, builders, and dealers; apprentices and training programs; building inspectors and code officials; and anyone interested in gypsum construction. Features information on tools and safety practices, a glossary of construction terms, and a list of important agencies and associations. Also available in Spanish.

$40.00 | Over 575 pages, illustrated, softcover | Catalog no. 67357A

Mechanical Estimating Methods, 4th Edition

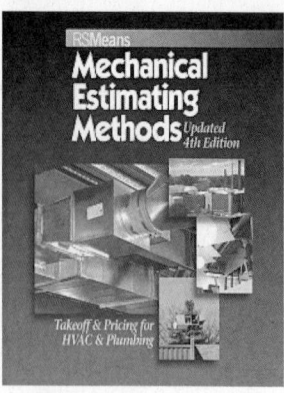

Completely updated, this guide assists you in making a review of plans, specs, and bid packages, with suggestions for takeoff procedures, listings, substitutions, and pre-bid scheduling for all components of HVAC. Includes suggestions for budgeting labor and equipment usage. Compares materials and construction methods to allow you to select the best option.

$70.00 per copy | Over 350 pages, illustrated, softcover | Catalog no. 67294B

Means Illustrated Construction Dictionary

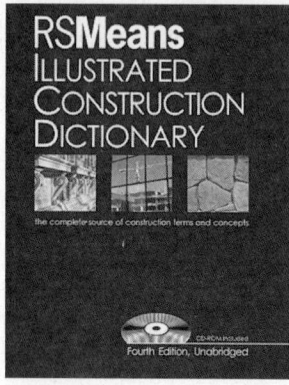

Unabridged 4th Edition, with CD-ROM

Long regarded as the industry's finest, *Means Illustrated Construction Dictionary* is now even better. With nearly 20,000 terms and more than 1,400 illustrations and photos, it is the clear choice for the most comprehensive and current information. The companion CD-ROM that comes with this new edition adds extra features such as: larger graphics and expanded definitions.

$99.95 | Over 790 pages, illust., hardcover | Catalog no. 67292B

RSMeans Estimating Handbook, 3rd Edition

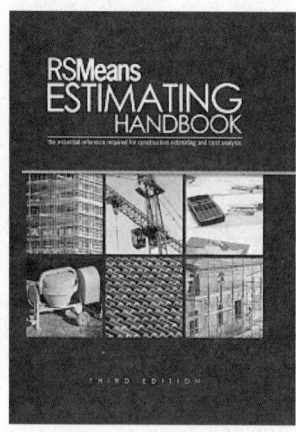

Widely used in the industry for tasks ranging from routine estimates to special cost analysis projects, this handbook has been completely updated and reorganized with new and expanded technical information.

RSMeans Estimating Handbook will help construction professionals:

- Evaluate architectural plans and specifications
- Prepare accurate quantity takeoffs
- Compare design alternatives and costs
- Perform value engineering
- Double-check estimates and quotes
- Estimate change orders

$110.00 | Over 900 pages, hardcover | Catalog No. 67276B

2014 Order Form

For more information visit the RSMeans website at **www.rsmeans.com**

Qty.	Book no.	COST ESTIMATING BOOKS	Unit Price	Total
	60064	Assemblies Cost Data 2014	$315.95	
	60014	Building Construction Cost Data 2014	194.95	
	60044	Commercial Renovaion Cost Data 2014	156.95	
	60114	Concrete & Masonry Cost Data 2014	177.95	
	50144	Construction Cost Indexes 2014 (subscription)	362.00	
	60144A	Construction Cost Index–January 2014	93.50	
	60144B	Construction Cost Index–April 2014	93.50	
	60144C	Construction Cost Index–July 2014	93.50	
	60144D	Construction Cost Index–October 2014	93.50	
	60344	Contr. Pricing Guide: Resid. R & R Costs 2014	39.95	
	60234	Electrical Change Order Cost Data 2014	187.95	
	60034	Electrical Cost Data 2014	196.95	
	60204	Facilities Construction Cost Data 2014	496.95	
	60304	Facilities Maintenance & Repair Cost Data 2014	434.95	
	60554	Green Building Cost Data 2014	160.95	
	60164	Heavy Construction Cost Data 2014	192.95	
	60094	Interior Cost Data 2014	196.95	
	60124	Labor Rates for the Const. Industry 2014	424.95	
	60184	Light Commercial Cost Data 2014	144.95	
	60024	Mechanical Cost Data 2014	191.95	
	60154	Open Shop Building Const. Cost Data 2014	165.95	
	60214	Plumbing Cost Data 2014	194.95	
	60174	Residential Cost Data 2014	139.95	
	60284	Site Work & Landscape Cost Data 2014	187.95	
	60054	Square Foot Costs 2014	205.95	
	62013	Yardsticks for Costing (2013)	185.95	
	62014	Yardsticks for Costing (2014)	189.95	
		REFERENCE BOOKS		
	67329	Bldrs Essentials: Best Bus. Practices for Bldrs	40.00	
	67307	Bldrs Essentials: Plan Reading & Takeoff	35.95	
	67342	Building & Renovating Schools	99.95	
	67261A	The Building Prof. Guide to Contract Documents	75.00	
	67339	Building Security: Strategies & Costs	95.00	
	67353A	Complete Book of Framing, 2nd Ed.	29.95	
	67146	Concrete Repair & Maintenance Illustrated	75.00	
	67352	Construction Business Management	55.00	
	67364	Construction Law	110.00	
	67358	Construction Supervision	95.00	
	67363	Cost Data Student Edition	99.00	
	67314	Cost Planning & Est. for Facil. Maint.	95.00	
	67230B	Electrical Estimating Methods, 3rd Ed.	75.00	
	67343A	Est. Bldg. Costs for Resi. & Lt. Comm., 2nd Ed.	75.00	
	67276B	Estimating Handbook, 3rd Ed.	110.00	
	67318	Facilities Operations & Engineering Reference	115.00	
	67301	Facilities Planning & Relocation	89.95	
	67338B	Green Building: Proj. Planning & Cost Est., 3rd Ed.	110.00	
	67355	Green Home Improvement	34.95	
	67357A	The Gypsum Construction Handbook	40.00	
	67148	Heavy Construction Handbook	110.00	
	67308E	Home Improvement Costs–Int. Projects, 9th Ed.	24.95	
	67309E	Home Improvement Costs–Ext. Projects, 9th Ed.	24.95	

Qty.	Book no.	REFERENCE BOOKS (Cont.)	Unit Price	Total
	67344	Homeowner's Guide to Mold	21.95	
	67324C	How to Est.w/Means Data & CostWorks, 4th Ed.	75.00	
	67351A	How Your House Works, 2nd Ed.	22.95	
	67282A	Illustrated Const. Dictionary, Condensed, 2nd Ed.	65.00	
	67292B	Illustrated Const. Dictionary, w/CD-ROM, 4th Ed.	99.95	
	67362	Illustrated Const. Dictionary, Student Ed.	55.00	
	67348	Job Order Contracting	89.95	
	67295C	Landscape Estimating Methods, 5th Ed.	75.00	
	67341	Life Cycle Costing for Facilities	99.95	
	67294B	Mechanical Estimating Methods, 4th Ed.	70.00	
	67283B	Plumbing Estimating Methods, 3rd Ed.	65.00	
	67345	Practice of Cost Segregation Analysis	105.00	
	67366	Project Control: Integrating Cost & Sched.	95.00	
	67346	Preventive Maint. for Multi-Family Housing	95.00	
	67367	Project Scheduling & Manag. for Constr., 4th Ed.	95.00	
	67322B	Resi. & Light Commercial Const. Stds., 3rd Ed.	65.00	
	67359	Risk Management for Design & Construction	125.00	
	67365	Solar Energy: Technology & Project Delivery	85.00	
	67145B	Sq. Ft. & Assem. Estimating Methods, 3rd Ed.	75.00	
	67321	Total Productive Facilities Management	85.00	
	67350	Understanding and Negotiating Const. Contracts	80.00	
	67303B	Unit Price Estimating Methods, 4th Ed.	75.00	
	67354	Universal Design	21.95	
	67319A	Value Engineering: Practical Applications	85.00	

MA residents add 6.25% state sales tax ____

Shipping & Handling** ____

Total (U.S. Funds)* ____

Prices are subject to change and are for U.S. delivery only. *Canadian customers may call for current prices. **Shipping & handling charges: Add 7% of total order for check and credit card payments. Add 9% of total order for invoiced orders.

Send order to: ADDV-1000

Name (please print) _____

Company _____

☐ Company

☐ Home Address _____

_____

City/State/Zip _____

Phone # _____ P.O. # _____

(Must accompany all orders being billed)

Mail to: RSMeans, 700 Longwater Drive, Norwell, MA 02061

RSMeans Project Cost Report

By filling out this report, your project data will contribute to the database that supports the RSMeans® Project Cost Square Foot Data. When you fill out this form, RSMeans will provide a $50 discount off one of the RSMeans products advertised in the preceding pages. Please complete the form including all items where you have cost data, and all the items marked (☑).

$50.00 Discount per product for each report you submit

Project Description (NEW construction only, please)

☑ Building Use (Office School . . .) _____

☑ Address (City, State)_____

_____

☑ Total Building Area (SF) _____

☑ Ground Floor (SF) _____

☑ Frame (Wood, Steel . . .) _____

☑ Exterior Wall (Brick, Tilt-up . . .) _____

☑ Basement: (checkone)   ☐ Full   ☐ Partial   ☐ None

☑ Number of Stories _____

☑ Floor-to-Floor Height_____

☑ Volume (C.F.)_____

  % Air Conditioned _____ Tons_____

  Total Project Cost $ _____

Owner _____

Architect_____

General Contractor _____

☑ Bid Date _____

☑ Typical Bay Size _____

☑ Occupant Capacity_____

☑ Labor Force: _____ % Union _____ % Non-Union

☑ Project Description (Circle one number in each line.)

  1. Economy 2. Average 3. Custom 4. Luxury

  1. Square 2. Rectangular 3. Irregular 4. Very Irregular

Comments _____

_____

A	☑	General Conditions	$	
B	☑	Site Work	$	
C	☑	Concrete	$	
D	☑	Masonry	$	
E	☑	Metals	$	
F	☑	Wood & Plastics	$	
G	☑	Thermal &Moisture Protection	$	
GR		Roofing & Flashing	$	
H	☑	Doors and Windows	$	
J	☑	Finishes	$	
JP		Painting & Wall Covering	$	

K	☑	Specialties	$	
L	☑	Equipment	$	
M	☑	Furnishings	$	
N	☑	SpecialConstruction	$	
P	☑	ConveyingSystems	$	
Q	☑	Mechanical	$	
QP		Plumbing	$	
QB		HVAC	$	
R	☑	Electrical	$	
S	☑	Mech./Elec. Combined	$	

Please specify the RSMeans product you wish to receive. Complete the address information.

Product Name _____

Product Number _____

Your Name _____

Title _____

Company_____
  ☐ Company
  ☐ Home   Street Address_____

City, State, Zip _____

Email Address _____

Return by mail or fax 888-492-6770.

Method of Payment:

Credit Card # _____

Expiration Date _____

Check_____

Purchase Order _____

Phone Number _____
(if you would prefer a sales call)

RSMeans
a division of Reed Construction Data
Square Foot Costs Department
700 Longwater Drive
Norwell, MA 02061
www.RSMeans.com